W9-AEY-598

ALGEBRA'S COMMON GRAPHS

Identity Function

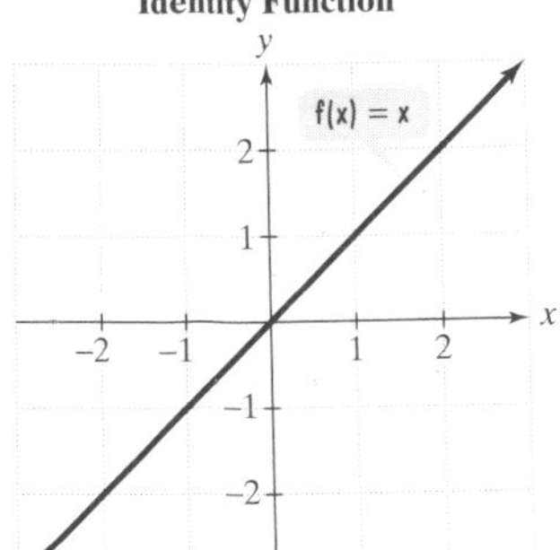

Standard Quadratic Function

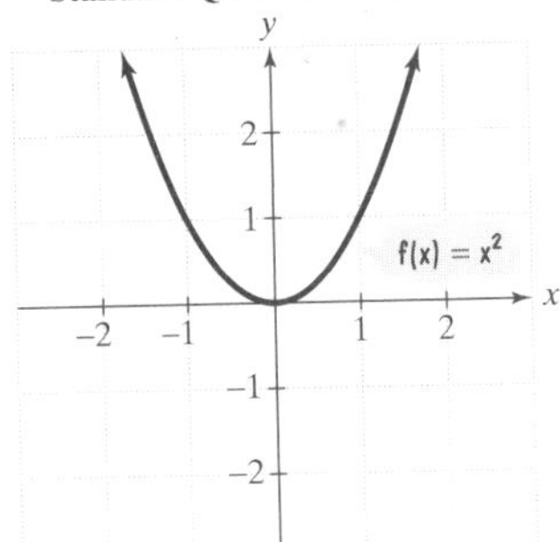

Standard Cubic Function

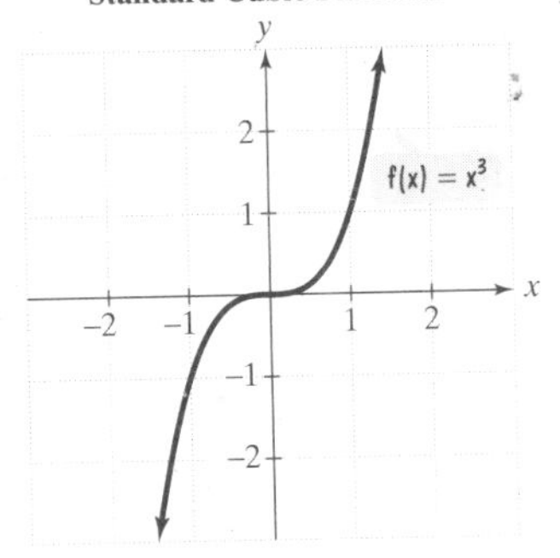

Absolute Value Function

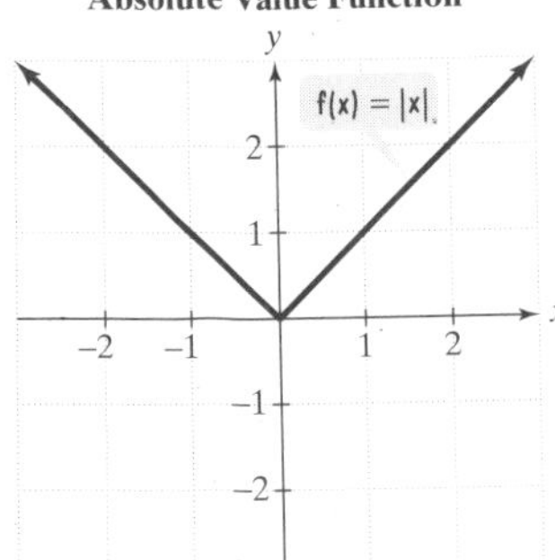

Square Root Function

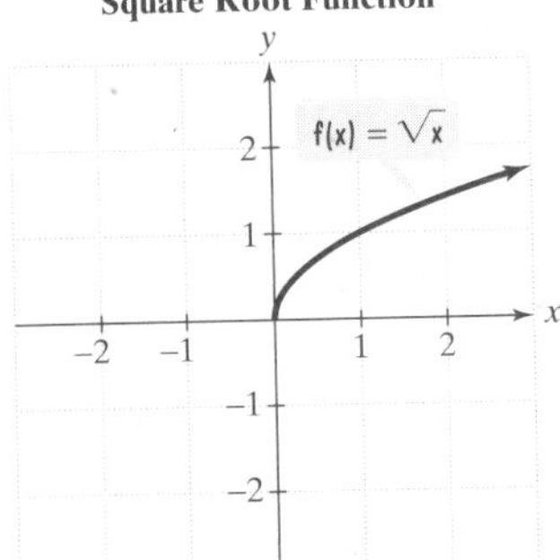

Greatest Integer Function

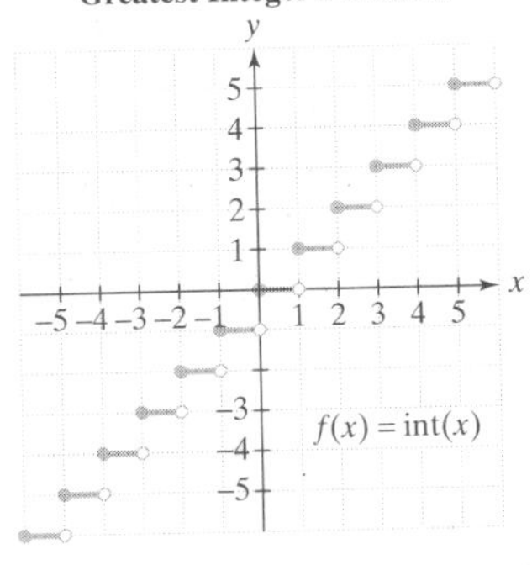

TRANSFORMATIONS

In each case, c represents a positive real number.

Function		Draw the graph of f and:
Vertical translations	$y = f(x) + c$ $y = f(x) - c$	Shift f upward c units. Shift f downward c units.
Horizontal translations	$y = f(x - c)$ $y = f(x + c)$	Shift f to the right c units. Shift f to the left c units.
Reflections	$y = -f(x)$ $y = f(-x)$	Reflect f about the x-axis. Reflect f about the y-axis.
Stretching or Shrinking	$y = cf(x); c > 1$ $y = cf(x); 0 < x < 1$	Stretch f, multiplying each of its y-values by c. Shrink f, multiplying each of its y-values by c.

DISTANCE AND MIDPOINT FORMULAS

1. The distance from (x_1, y_1) to (x_2, y_2) is

$$\sqrt{(x_2 - x_1)^2 + (y_2 - y_1)^2}.$$

2. The midpoint of the line segment with endpoints (x_1, y_1) and (x_2, y_2) is

$$\left(\frac{x_1 + x_2}{2}, \frac{y_1 + y_2}{2}\right).$$

QUADRATIC FORMULA

The solutions to $ax^2 + bx + c = 0$ with $a \neq 0$ are

$$x = \frac{-b \pm \sqrt{b^2 - 4ac}}{2a}.$$

FUNCTIONS

1. Linear Function: $f(x) = mx + b$
   Graph is a line with slope m and y-intercept b.

(continued on inside back cover)

SECOND EDITION

College Algebra

Robert Blitzer
Miami-Dade Community College

PRENTICE HALL
Upper Saddle River, NJ 07458

Blitzer, Robert.
College algebra / Robert Blitzer.-- 2nd ed.
p. cm.
Includes index.
ISBN 0-13-087828-6
1. Algebra. I. Title.
QA152.2.B584 2001
512.9--dc21 00-062368

Editor-in-Chief and Acquisitions Editor: ***Sally Yagan***
Senior Development Editor: ***Shana Ederer***
Editor-in-Chief, Development: ***Carol Trueheart***
Editorial/Production Supervision: ***Bayani Mendoza de Leon***
Assistant Vice President of Production and Manufacturing: ***David W. Riccardi***
Senior Managing Editor: ***Linda Mihatov Behrens***
Executive Managing Editor: ***Kathleen Schiaparelli***
Manufacturing Buyer: ***Alan Fischer***
Manufacturing Manager: ***Trudy Pisciotti***
Director of Marketing: ***John Tweeddale***
Senior Marketing Manager: ***Patrice Lumumba Jones***
Marketing Assistant: ***Vince Jansen***
Associate Editor, Mathematics/Statistics Media: ***Audra J. Walsh***
Director of Creative Services: ***Paul Belfanti***
Art Director: ***Maureen Eide***
Assistant to the Art Director: ***John Christiana***
Art Editor: ***Grace Hazeldine***
Photo Researcher: ***Melinda Alexander***
Photo Editor: ***Beth Boyd***
Cover Designer: ***Maureen Eide***
Interior Design and Layout: ***Lorraine Castellano***
Editorial Assistant: ***Meisha Welch***
Cover image: *False clown anemonefish (Amphiprion ocellaris), Australia / Stuart Westmorland / Tony Stone Images*

©2001, 1998 by Prentice-Hall, Inc.
Upper Saddle River, New Jersey 07458

All rights reserved. No part of this book may be reproduced, in any form or by any other means, without permission in writing from the publisher.

Printed in the United States of America

10 9 8 7 6 5 4 3

ISBN 0-13-087828-6

Prentice-Hall International (UK) Limited, *London*
Prentice-Hall of Australia Pty. Limited, *Sydney*
Prentice-Hall Canada Inc., *Toronto*
Prentice-Hall Hispanoamericana, S.A., *Mexico*
Prentice-Hall of India Private Limited, *New Delhi*
Prentice-Hall of Japan, Inc., *Tokyo*
Pearson Education Asia Pte. Ltd.
Editora Prentice-Hall do Brasil, Ltda., *Rio de Janeiro*

Contents

Chapter 2

Functions and Graphs 167

Chapter 3

Polynomial and Rational Functions 258

Chapter 4

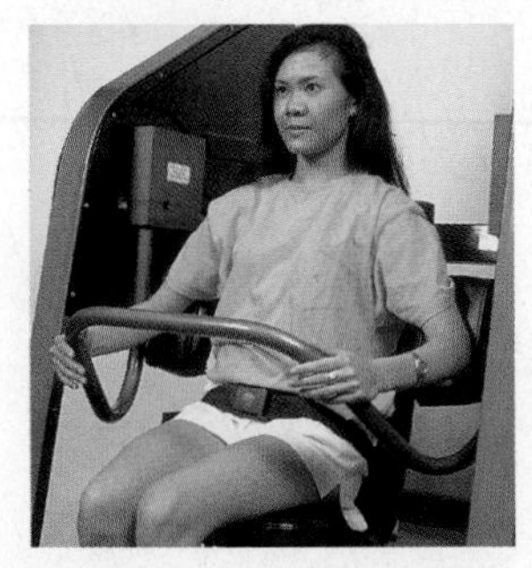

Exponential and Logarithmic Functions 350

Chapter 5

Systems of Equations and Inequalities 415

Chapter 6

Matrices and Determinants 481

Chapter 7

Conic Sections 552

Chapter 8

Sequences, Induction, and Probability 599

Appendix

Where Did That Come From? Selected Proofs A1

Preface

Today's college algebra students are a diverse group. Some are going on into precalculus, calculus, or other math sequences, whereas others will complete their math requirements with this course. This text is designed and written to help both sets of students succeed. The book has three fundamental goals: First, to help students acquire a solid foundation in college algebra, preparing them for other college courses such as calculus, business calculus, finite mathematics, and computer science; second, to show students how algebra can model and solve authentic real-world problems; and third, to enable students to develop problem-solving skills, fostering critical thinking within a varied and interesting setting.

Writing The Second Edition

A source of frustration for me and my colleagues is that very few students read their textbook. When I ask students why they do not take full advantage of the text, their responses generally fall into two categories:

- "I cannot follow the explanations."
- "The applications are not interesting."

I thought about both of these objections in writing every page of the Second Edition.

"I can't follow the explanations." For many of my students, textbook explanations are too compressed. The chapters in the Second Edition have been extensively rewritten to make them more accessible. I have paid close attention to ensuring that the amount of detail and depth of coverage is appropriate for a liberal arts college algebra course. Every section has been rewritten to contain a better range of simple, intermediate, and challenging examples. Voice balloons allow for more specific annotations in examples, further clarifying procedures and concepts. A more open format with a softer color palette gives the book a less crowded look than the First Edition.

"The applications are not interesting." One of the things I enjoy most about teaching in a large urban community college is the diversity of who my students are and what interests them. Real-world data that celebrate this variety are used to bring relevance to examples, discussions, and applications. All data from the previous edition have been replaced to include data that extend as far up to the present as possible. I selected all updated real-world data to be interesting and intriguing to students. By connecting algebra to the whole spectrum of their interests, it is my intent to show students that their world is profoundly mathematical and, indeed, π is in the sky.

Student Supplements

Student Solutions Manual (0-13-089410-9); (8941K-1) Includes fully worked out solutions to most of the odd-numbered exercises in the text as well as all exercises in chapter tests and all review exercises.

MathPak Integrated Learning Environment (0-13-088267-4); (8826G-0) Contains the College Algebra MathPro 4.0 along with a passcode-protected Website specifically designed to accompany this text. This product combines the series' key supplements into a comprehensive, easy-to-navigate package. Materials on the Website include but are not limited to: Section-by-section reading quizzes, Section-by-Section Powerpoint downloads, additional chapter projects, Chapter Quizzes and Tests, Student Solutions Manuals presented by chapter (exactly what is in the print version), Chapter Destinations and to interesting math Websites, and Graphing Calculator Manuals for the full line of TI's, Sharp, HP, and Casio Calculators.

MathPro4 Network Version The best algorithmic tutorial software on the market – MathPro steps the student objective by objective, section by section, throughout the entire College Algebra text, including the Appendix review materials. Students can benefit from over 100 QuickTime instructional segments video.

MathPro4 Network Version (0-13-088269-0); (8826J-3)

Review Videos (0-13-088264-X); (8826D-7) Section-by-section videos written by and highlighting Jacquelyn White of St. Leo College. Each segment covers approximately 20 minutes of the key concepts and examples for each section. Each set of videos comes with a permissions letter allowing the school to duplicate for specific campus needs.

Precalculus Investigations/Simundza, et al. (013-010954-1); (1095D-6) A three year NSF-funded project integrates an applied approach to the topics in the Precalculus curriculum via applied projects. The investigations reflect the AMATYC and NCTM Standards in both curriculum content and pedagogy.

Companion Website `www.prenhall.com/blitzer` This CW address will lead to the bridge page for all of the Blitzer titles. On the CW sites (which are different than the MathPak sites) are the following: Chapter Quizzes, Chapter Tests, Projects, Graphing Calculator Manual, Destinations, and PowerPoints.

WebCT/Blackboard Contains all the materials from the MathPak website (i.e., no MathPro) plus testing materials. Can be made available in Blackboard on adoption.

Instructor Supplements

Instructor's Edition (0-13-089417-6); (8941G-0) Includes full student text as well as the full set of answers at the back of the text for both odds and even exercises.

Instructor's Resource Manual (0-13-089418-4); (8941H-8) The College Algebra IRM contains the full solutions to the even-numbered exercises in the text. The Precalculus IRM contains solutions to both the odd- and the even-numbered exercises in the text.

TestGen-EQ WIN/MAC CD (0-13-088260-7); (8826K-1) New to Prentice Hall Mathematics is the use of TestGen EQ for our mathematics testing. TestGen-EQ is a fully algorithmic, easy-to-use software program written and based on the section objectives in the text.

Test Item File (0-13-089421-4); (8942A-2) A hard-copy version of materials derived from the TestGen-EQ program.

Acknowledgments

I wish to express my appreciation to all the reviewers for their helpful criticisms and suggestions, frequently transmitted with wit, humor, and intelligence. In particular, I would like to thank:

Reviewers for the First Edition

Howard Anderson — *Skagit Valley College*
John Anderson — *Illinois Valley Community College*
Michael H. Andreoli — *Miami Dade Community College-North Campus*
Warren J. Burch — *Brevard Community College*
Alice Burstein — *Middlesex Community College*
Sandra Pryor Clarkson — *Hunter College*
Sally Copeland — *Johnson County Community College*
Robert A. Davies — *Cuyahoga Community College*
Ben Divers, Jr. — *Ferrum College*
Irene Doo — *Austin Community College*
Charles C. Edgar — *Onondaga Community College*
Susan Forman — *Bronx Community College*
Gary Glaze — *Eastern Washington University*
Jay Graening — *University of Arkansas*
Robert B. Hafer — *Brevard Community College*
Mary Lou Hammond — *Spokane Community College*
Donald Herrick — *Northern Illinois University*
Beth Hooper — *Golden West College*
Tracy Hoy — *College of Lake County*
Gary Knippenberg — *Lansing Community College*
Mary Koehler — *Cuyahoga Community College*
Hank Martel — *Broward Community College*
John Robert Martin — *Tarrant County Junior College*
Irwin Metviner — *State University of New York at Oild Westbury*
Allen R. Newhart — *Parkersburg Community College*
Peg Pankowski — *Community College of Allegheny County—South Campus*
Nancy Ressler — *Oakton Community College*
Gayle Smith — *Lane Community College*
Dick Spangler — *Tacoma Community College*
Janette Summers — *University of Arkansas*
Robert Thornton — *Loyola University*
Lucy C. Thrower — *Francis Marion College*
Andrew Walker — *North Seattle Community College*

Reviewers for the Second Edition

Christopher N. Hay-Jahans	*University of South Dakota*
Cynthia Glickman	*Community College of Southern Nevada*
Dan Van Peursem	*University of South Dakota*
David White	*The Victoria College*
Debra A. Pharo	*Northwestern Michigan College*
Diana Colt	*University of Minnesota-Duluth*
Donald Gordon	*Manatee Community College*
Joel K. Haack	*University of Northern Iowa*
Kayoko Yates Barnhill	*Clark College*
Lloyd Best	*Pacific Union College*
Nancy Raye Johnson	*Manatee Community College*
Richard E. Van Lommel	*California State University-Sacramento*
Sudhir Kumar Goel	*Valdosta State University*
Yvelyne Germain-McCarthy	*University of New Orleans*

Additional acknowledgments are extended to Jacquelyn White for creating the dynamic video tape series covering every section of the book; the team at Laurel Technical Services for the Herculean task of solving all the book's exercises, preparing the answer section and solutions manuals, as well as serving as accuracy checker; Melinda Alexander, photo researcher, for obtaining the book's photographs; the team of graphic artists and mathematicians at Scientific Illustrators, whose superb illustrations and graphs provide visual support to the verbal portions of the text; Prepare Inc., the book's compositor, for inputting hundreds of pages with hardly an error; Professor Elizabeth Farber, who took time from her teaching and writing schedule to help accuracy check the final pages; and Bayani Mendoza de Leon, whose talents as production editor contributed to the book's wonderful look.

Most of all, I wish to thank Sally Yagan and Shana Ederer. Shana, my development editor, contributed invaluable edits and suggestions that resulted in a finished product that is both accessible and up-to-date. Her influence on this book is extraordinary, guiding and coordinating every detail of this project. Sally, editor-in-chief of mathematics at Prentice Hall, is the key person in making this book a reality, and I am grateful to have had an editor with her experience, insight, and professionalism.

Sally Yagan and Shana Ederer are members of the terrific team at Prentice Hall who made this book possible, including ESM Co-President Tim Bozik and President Paul Corey. Thank you Patrice Lumumba Jones, Senior Marketing Manager, for your innovative marketing efforts as well as the Prentice Hall sales force for your confidence in and enthusiasm for the book.

To the Student

I've written this book so that you can learn about the power of algebra and how it relates directly to your life outside the classroom. All concepts are carefully explained, important definitions and procedures are set off in boxes, and worked-out examples that present solutions in a step-by-step manner appear in every section. Each Example is followed by a similar matched problem, called a CheckPoint, for you to try so that you can actively participate in the learning process as you read the book. (Answers to all CheckPoints appear in the back of the book.) Study Tips offer hints and suggestions and often point out common errors to avoid. A great deal of attention has been given to applying algebra to your life to make your learning experience both interesting and relevant.

As you begin your studies, I would like to offer some specific suggestions for using this book and for being successful in this course:

1. **Attend all lectures.** No book is intended to be a substitute for valuable insights and interactions that occur in the classroom. In addition to arriving for lecture on time and being prepared, you will find it useful to read the section before it is covered in lecture. This will give you a clear idea of the new material that will be discussed.
2. **Read the book.** Read each section with pen (or pencil) in hand. Move through the illustrative examples with great care. These worked-out examples provide a model for doing exercises in the exercise sets. As you proceed through the reading, do not give up if you do not understand every single word. Things will become clearer as you read on and see how various procedures are applied to specific worked-out examples.
3. **Work problems every day and check your answers.** The way to learn mathematics is by doing mathematics, which means working the CheckPoints and assigned exercises in the exercise sets. The more exercises you work, the better you will understand the material.
4. **Prepare for chapter exams.** After completing a chapter, study the summary, work the exercises in the Chapter Review, and work the exercises in the Chapter Test. Answers to all these exercises are given in the back of the book.
5. **Use the supplements available with this book.** A solutions manual containing worked-out solutions to the book's odd-numbered exercises and all review exercises, a dynamic web page, and video tapes created for every section of the book are among the supplements created to help you tap into the power of mathematics. Ask you instructor or bookstore what supplements are available and where you can find them.

It is my hope that that you will enjoy the pages of this book as you empower yourself with the algebra needed to succeed in college, your career, and in your life.

Regards,

Bob

Robert Blitzer

A Guide to Using This Text

Relevant Chapter Openers
Every chapter highlights a scenario from everyday life and how the algebra relates to it. These scenarios are revisited later in the chapter.

Functions and Graphs

Chapter 2

'Tis the season and you've waited until the last minute to mail your holiday gifts. Your only option is overnight express mail. You realize that the cost of mailing a gift depends on its weight, but the mailing costs seem somewhat odd. Your packages that weigh 1.1 pounds, 1.5 pounds, and 2 pounds cost $15.75 each to send overnight. Packages that weigh 2.01 pounds and 3 pounds cost you $18.50 each. Finally, your heaviest gift is barely over 3 pounds and its mailing cost is $21.25. What sort of system is this in which costs increase by $2.75, stepping from $15.75 to $18.50 and from $18.50 to $21.25?

The cost of mailing a package depends on its weight. The probability that you and another person in a room share the same birthday depends on the number of people in the room. In both these situations, the relationship between variables can be illustrated with the notion of a *function*. Understanding this concept will give you a new perspective on many ordinary situations.

167

Page 167

Section Objectives

The learning objectives focus the students' study habits and also are the foundation for the algorithms found in MathPro (tutorial software) and in the Test-Gen EQ (test generator software). Objectives reappear in the margin at their point of use.

Section Openers

Each and every section opens with a unique application of algebra in students' lives outside the classroom. These scenarios are revisited later in the section.

Current Real-World Data

Relevant current data are used to illustrate the power of the algebra to real issues and contemporary information. Data are used throughout the examples, exercises, and discussions. The data were selected to be interesting and intriguing to students.

SECTION 2.1 Lines and Slope

Objectives

1. Compute a line's slope.
2. Write the point-slope equation of a line.
3. Write and graph the slope-intercept equation of a line.
4. Recognize equations of horizontal and vertical lines.
5. Recognize and use the general form of a line's equation.
6. Model data with linear equations.

Good news: Projections indicate that in the next decades we'll live longer and move somewhere warmer where we'll shop online and chat on our tiny video cell phones. Figure 2.1 shows projected online shopping per U.S. online household through 2004. The graph is composed of two line segments. The segment on the right is steeper than the one on the left. This shows that online shopping is expected to increase more per year in 2001–2004 than in 1999–2001.

Data often fall on or near a line. In this section we will use equations to model such data and make predictions. We begin with a discussion of a line's steepness.

Online Spending: Yearly Spending per Online Household

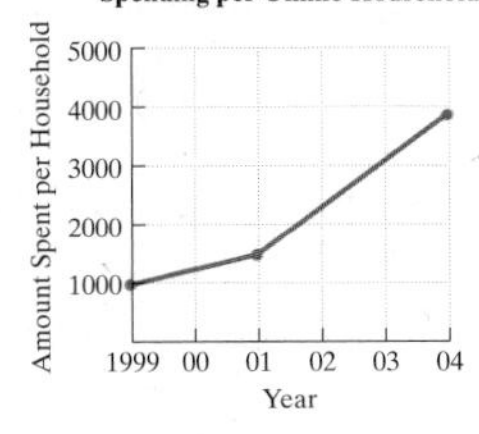

Source: Forrester Research

Figure 2.1

The Slope of a Line

Mathematicians have developed a useful measure of the steepness of a line, called the **slope** of the line. Slope compares the vertical change (the **rise**) to the horizontal change (the **run**) when moving from one fixed point to another along the line. To calculate the slope of a line, mathematicians use a ratio comparing the change in y (the rise) to the change in x (the run).

Definition of Slope

The **slope** of the line through the distinct points (x_1, y_1) and (x_2, y_2) is

$$\frac{\text{Change in } y}{\text{Change in } x} = \frac{\text{Rise}}{\text{Run}} = \frac{y_2 - y_1}{x_2 - x_1}$$

where $x_2 - x_1 \neq 0$.

It is common notation to let the letter m represent the slope of a line. The letter m is used because it is the first letter of the French verb *monter*, meaning to rise, or to ascend.

Page 168

Discovery

These exercises, found throughout the text, encourage students to explore problems in order to better understand them and their solutions. These are a great way to stimulate class-time exploration of concepts.

Discovery

The study of how changing a function's equation can affect its graph can be explored with a graphing utility. Use your graphing utility to verify the hand-drawn graphs as you read this section.

Page 221

Explanatory Art

Much of the art program has text "balloons" that highlight what the equations or graphs "mean" in English. This helps students not lose sight of what the problem is asking for.

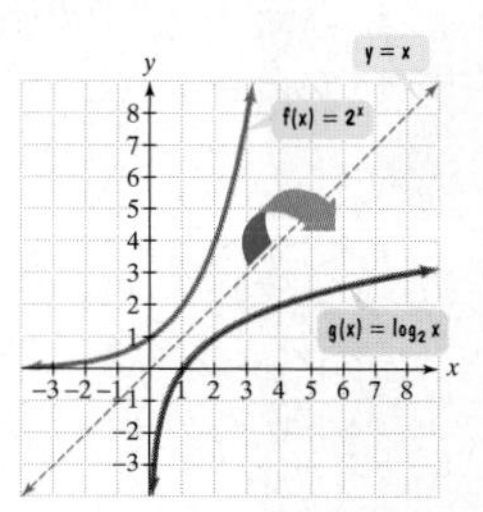

Figure 4.6 The graphs of $f(x) = 2^x$ and its inverse function

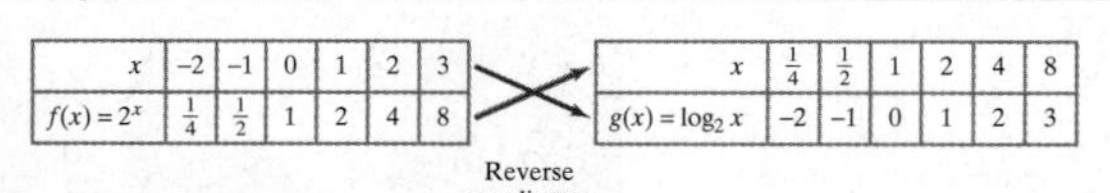

x	-2	-1	0	1	2	3
$f(x) = 2^x$	$\frac{1}{4}$	$\frac{1}{2}$	1	2	4	8

x	$\frac{1}{4}$	$\frac{1}{2}$	1	2	4	8
$g(x) = \log_2 x$	-2	-1	0	1	2	3

Reverse coordinates.

We now plot the ordered pairs in both tables, connecting them with smooth curves. Figure 4.6 shows the graphs of $f(x) = 2^x$ and its inverse function $g(x) = \log_2 x$. The graph of the inverse can also be drawn by reflecting the graph of $f(x) = 2^x$ about the line $y = x$.

Graph $f(x) = 3^x$ and $g(x) = \log_3 x$ in the same rectangular coordinate system.

Figure 4.7 illustrates the relationship between the graph of the exponential function, shown in blue and its inverse, the logarithmic function, shown in red, for bases greater than 1 and for bases between 0 and 1.

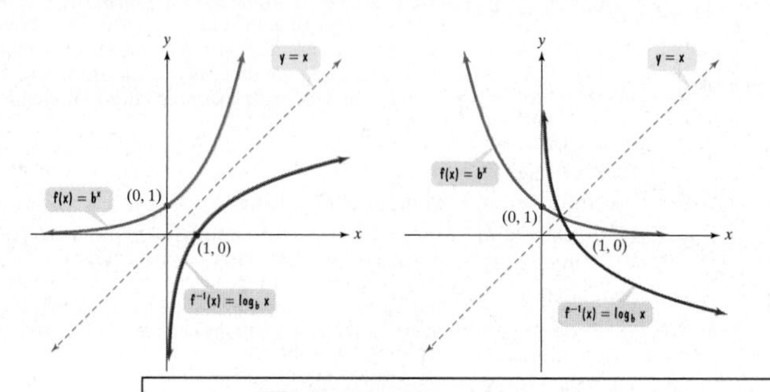

Figure 4.7 Gra

Page 366

Examples

Each example is titled—making clear the purpose of the example.

Examples are clearly written and provide students with detailed step-by-step solutions. No steps are omitted and each step is clearly explained.

Essays

Enrichment Essays provide historical, interdisciplinary, and otherwise interesting connections throughout the text.

Carbon Dating and Artistic Development

The artistic community was electrified by the discovery in 1995 of spectacular cave paintings in a limestone cavern in France. Carbon dating of the charcoal from the site showed that the images, created by artists of remarkable talent, were 30,000 years old, making them the oldest cave paintings ever found. The artists seemed to have used the cavern's natural contours to heighten a sense of perspective. The quality of the painting suggests that the art of early humans did not mature steadily from primitive to sophisticated in any simple linear fashion.

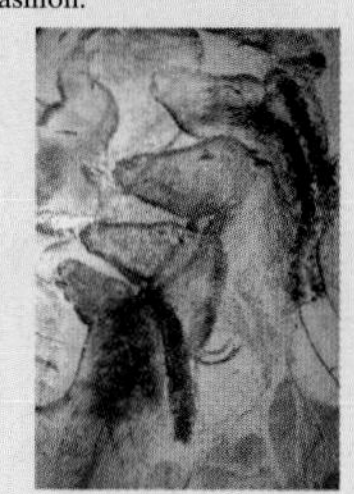

Our next example involves exponential decay and its use in determining the age of fossils and artifacts. The method is based on considering the percentage of carbon-14 remaining in the fossil or artifact. Carbon-14 decays exponentially with a *half-life* of approximately 5715 years. The **half-life** of a substance is the time required for half of a given sample to disintegrate. Thus, after 5715 years a given amount of carbon-14 will have decayed to half the original amount. Carbon dating is useful for artifacts or fossils up to 80,000 years old. Older objects do not have enough carbon-14 left to date age accurately.

EXAMPLE 2 Carbon-14 Dating: The Dead Sea Scrolls

a. Use the fact that after 5715 years a given amount of carbon-14 will have decayed to half the original amount to find the exponential decay model for carbon-14.

b. In 1947, earthenware jars containing what are known as the Dead Sea Scrolls were found by an Arab Bedouin herdsman. Analysis indicated that the scroll wrappings contained 76% of their original carbon-14. Estimate the age of the Dead Sea Scrolls.

Solution We begin with the exponential decay model $A = A_0e^{kt}$. We know that $k < 0$ because the problem involves the decay of carbon-14. After 5715 years ($t = 5715$), the amount of carbon-14 present, A, is half the original amount A_0. Thus we can substitute $\frac{A_0}{2}$ for A in the exponential decay model. This will enable us to find k, the decay rate.

a. $\frac{A_0}{2} = A_0e^{k5715}$ — After 5715 years ($t = 5715$), $A = \frac{A_0}{2}$ (because the amount present, A, is half the original amount, A_0).

$\frac{1}{2} = e^{5715k}$ — Divide both sides of the equation by A_0.

$\ln\frac{1}{2} = \ln e^{5715k}$ — Take the natural logarithm of both sides.

$\ln\frac{1}{2} = 5715k$ — $\ln e^x = x$

$k = \frac{\ln\frac{1}{2}}{5715} \approx -0.000121$ — Solve for k.

Substituting for k in the decay model, the model for carbon-14 is $A = A_0e^{-0.000121t}$.

Page 398

CheckPoints

CheckPoints offer students the opportunity to test their understanding of the example by working a similar exercise while they are reading the material. The answers to all the CheckPoints are given in the answer section

> **Check Point 1** Find the domain and the range of the relation $\{(20, 157.4), (30, 231.8), (100, 752.6), (200, 1496.6)\}$.
>
> As you worked Checkpoint 1, did you wonder if the numbers in each ordered pair represented anything? Think snakes! The first number in each ordered pair is a snake's tail length, in millimeters, and the second number is its body length, also in millimeters. Consider, for example, the ordered pair $(30, 231.8)$.
>
> $(30, 231.8)$
>
> 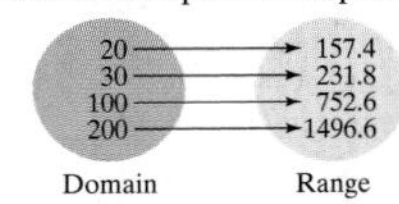
>
>
> The relation in the snake example can be pictured as follows:
>
> | Domain | | Range |
> |---|---|---|
> | 20 | → | 157.4 |
> | 30 | → | 231.8 |
> | 100 | → | 752.6 |
> | 200 | → | 1496.6 |
>
> Body Length (millimeters): 0, 1000, 2000; Tail Length (millimeters): 0, 100, 200; points (200, 1496.6), (100, 752.6), (30, 231.8), (20, 157.4)
>
> **Figure 2.19** The graph of a relation showing a correspondence between a snake's tail length and its body length
>
> A scatter plot, like the one shown in Figure 2.19, is a way to represent the relation.

Page 191

Study Tips

Study Tip boxes offer suggestions for problem solving, point out common student errors, and provide informal tips and suggestions. These invaluable hints appear in abundance throughout the book.

> **Study Tip**
>
> The word *range* can mean many things, from a chain of mountains to a cooking stove. For functions, it means the set of images of the domain. For graphing utilities, it means the setting used for the viewing rectangle. Try not to confuse these meanings.

Page 193

> **Study Tip**
>
> The notation $f(x)$ does *not* mean "f times x." The notation describes the value of the function at x.

Page 195

Thorough, yet Optional Technology

Although the use of graphing utilities is optional, they are utilized in technology boxes to enable students to visualize, discover, and explore algebraic concepts. The use of graphing utilities is also reinforced in the technology exercises appearing in the exercise sets for those who want this option. With the book's early introduction to graphing, students can look at the calculator screens in the technology boxes and gain an increased understanding of an example's solution even if they are not using a graphing utility in the course.

> **2** Solve logarithmic equations.
>
> **Technology**
>
> The graphs of
>
> $y_1 = \log_4(x + 3)$ and $y_2 = 2$
>
> have an intersection point whose x-coordinate is 13. This verifies that $\{13\}$ is the solution set for $\log_4(x + 3) = 2$.
>
> *Note*: Because
>
> $$\log_b x = \frac{\ln x}{\ln b}$$
>
> (change-of-base property), we entered y_1 using
>
> $$y_1 = \log_4(x + 3) = \frac{\ln(x + 3)}{\ln 4}$$
>
> 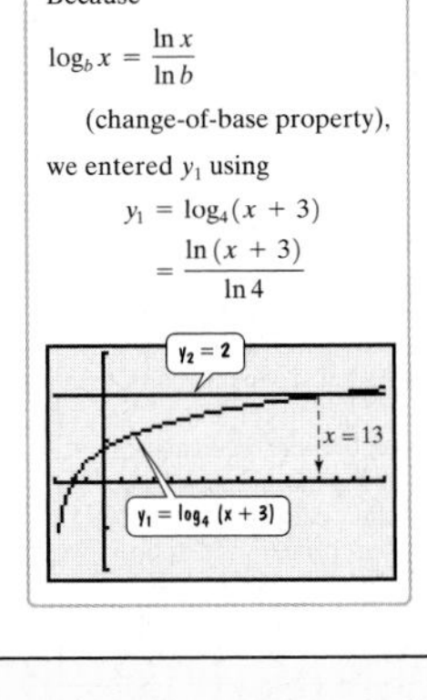
>
>
> **Logarithmic Equations**
>
> A **logarithmic equation** is an equation containing a variable in a logarithmic expression. Examples of logarithmic equations include
>
> $$\log_4(x + 3) = 2 \quad \text{and} \quad \ln 2x = 3.$$
>
> If a logarithmic equation is in the form $\log_b x = c$, we can solve the equation by rewriting it in its equivalent exponential form $b^c = x$. Example 5 illustrates how this is done.
>
> **EXAMPLE 5 Solving a Logarithmic Equation**
>
> Solve: $\log_4(x + 3) = 2$.
>
> **Solution** We first rewrite the equation as an equivalent equation in exponential form using the fact that $\log_b x = c$ means $b^c = x$.
>
> $$\log_4(x + 3) = 2 \quad \text{means} \quad 4^2 = x + 3$$
>
> Logarithms are exponents.
>
> Now we solve the equivalent equation for x.
>
> $4^2 = x + 3$ This is the equivalent equation.
>
> $16 = x + 3$ Square 16.
>
> $13 = x$ Subtract 3 from both sides.
>
> **Check**
>
> $\log_4(x + 3) = 2$ This is the given logarithmic equation.
>
> $\log_4(13 + 3) \stackrel{?}{=} 2$ Substitute 13 for x.
>
> $\log_4 16 \stackrel{?}{=} 2$
>
> $2 = 2$ ✓ $\log_4 16 = 2$ because $4^2 = 16$.
>
> This true statement indicates that the solution set is $\{13\}$.

Page 387

Exercise Sets

An extensive collection of exercises is included in all end-of-section and end-of-chapter materials. Within each category type, the exercises are organized by level. The category types found are: Practice Exercises, Application Exercises, Writing in Mathematics, Technology Exercises, Critical Thinking Exercises, and Group Exercises.

EXERCISE SET 4.1

Practice Exercises

In Exercises 1–10, approximate each number using a calculator. Round your answer to three decimal places.

1. $2^{3.4}$ **2.** $3^{2.4}$ **3.** $3^{\sqrt{5}}$ **4.** $5^{\sqrt{3}}$ **5.** $4^{-1.5}$
6. $6^{-1.2}$ **7.** $e^{2.3}$ **8.** $e^{3.4}$ **9.** $e^{-0.95}$ **10.** $e^{-0.75}$

In Exercises 11–18, graph each function by making a table of coordinates. If applicable, use a graphing utility to confirm your hand-drawn graph.

21.

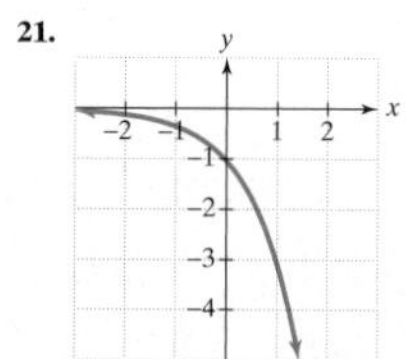

Pages 359-361

Application Exercises

Use a calculator with an $\boxed{x^y}$ key or a $\boxed{\wedge}$ key to solve Exercises 39–46.

39. The exponential function $f(x) = 67.38(1.026)^x$ describe the population of Mexico, $f(x)$, in millions, x years afte 1980.
a. Substitute 0 for x and, without using a calculator, fin Mexico's population in 1980.
b. Substitute 27 for x and use your calculator to fin Mexico's population in the year 2007 as predicted b this function.
c. Find Mexico's population in the year 2034 as predict ed by this function.
d. Find Mexico's population in the year 2061 as predict ed by this function.
e. What appears to be happening to Mexico's populatio every 27 years?

40. The 1986 explosion at the Chernobyl nuclear power plan in the former Soviet Union sent about 1000 kilograms c radioactive cesium-137 into the atmosphere. The func $f(x) = 1000(0.5)^{x/30}$ describes the amount, $f(x)$, i rams, of cesium-137 remaining in Chernobyl x years 1986. If even 100 kilograms of cesium-137 remain in nobyl's atmosphere, the area is considered unsafe uman habitation. Find $f(80)$ and determine if Cher- l will be safe for human habitation by 2066.

Writing in Mathematics

52. What is an exponential function?
53. What is the natural exponential function?
54. Use a calculator to evaluate $\left(1 + \frac{1}{x}\right)^x$ for $x = 10, 100, 1000, 10{,}000, 100{,}000$, and $1{,}000{,}000$. Describe what happens to the expression as x increases.
55. Write an example similar to Example 6 on page 358 in which continuous compounding at a slightly lower yearly interest rate is a better investment than compounding n times per year.
56. Describe how you could use the graph of $f(x) = 2^x$ to obtain a decimal approximation for $\sqrt{2}$.
57. The exponential function $y = 2^x$ is one-to-one and has an inverse function. Try finding the inverse function by

Technology Exercises

59. Graph $y = 13.49(0.967)^x - 1$, the function for the O-rings expected to fail at x°F, in a $[0, 90, 10]$ by $[0, 20, 5]$ viewing rectangle. If NASA engineers had used this function and its graph, is it likely they would have allowed the *Challenger* to be launched when the temperature was 31°F? Explain.

60. The student–teacher ratio in U.S. elementary and secondary schools can be modeled by $y = 25.34(0.987)^x$, where x re resents the number of years since 1959 and y represents t student–teacher ratio. Graph the function in a $[1, 40, 1]$ $[0, 26, 1]$ viewing rectangle. When did the student–teach ratio become less than 21 students per teacher?

61. You have \$10,000 to invest. One bank pays 5% interc compounded quarterly and the other pays 4.5% interc compounded monthly.
a. Use the formula for compound interest to write a fur tion for the balance in each account at any time t.
b. Use a graphing utility to graph both functions in appropriate viewing rectangle. Based on the grapl which bank offers the better return on your money

62. a. Graph $y = e^x$ and $y = 1 + x + \frac{x^2}{2}$ in the same vie ing rectangle.
b. Graph $y = e^x$ and $y = 1 + x + \frac{x^2}{2} + \frac{x^3}{6}$ in the same viewing rectangle.
c. Graph $y = e^x$ and $y = 1 + x + \frac{x^2}{2} + \frac{x^3}{6} + \frac{x^4}{24}$ in the same viewing rectangle.
d. Describe what you observe in parts (a)–(c). Try generalizing this observation.

Critical Thinking Exercises

63. Which one of the following is true?
a. As the number of compounding periods increases a fixed investment, the amount of money in the count over a fixed interval of time will increase w out bound.
b. The functions $f(x) = 3^{-x}$ and $g(x) = -3^x$ have same graph.
c. $e = 2.718$.
d. The functions $f(x) = \left(\frac{1}{3}\right)^x$ and $g(x) = 3^{-x}$ have same graph.

Group Exercises

56. This activity is intended for three or four people who would like to take up weightlifting. Each person in the group should record the maximum number of pounds that he or she can lift at the end of each week for the first 10 consecutive weeks. Use the Logarithmic REGression option of a graphing utility to obtain a model showing the amount of weight that group members can lift from week 1 to week 10. Graph each of the models in the same viewing rectangle to observe similarities and differences among weight-growth patterns of each member. Use the functions to predict the amount of weight that group members will be able to lift in the future. If the group continues to work out together, check the accuracy of these predictions.

57. Each group member should consult an almanac, newspaper, magazine, or the Internet to find data that can be modeled by exponential or logarithmic functions. Group members should select the two sets of data that are most interesting and relevant. For each data set selected, find a model that best fits the data. Each group member

Page 408

End-of-Chapter Materials

Chapter Summaries and Review Exercises

Each section has its own focused summary and review exercises. These provide students with a good review for a chapter test. Beginning with Chapter 2, each chapter concludes with a collection of cumulative review exercises, providing students with the opportunity for continuous review.

CHAPTER SUMMARY, REVIEW, AND TEST

Summary

4.1 Exponential Functions

a. The exponential function with base b is defined by $f(x) = b^x$, where $b > 0$ and $b \neq 1$.

b. Characteristics of exponential functions and graphs for $0 < b < 1$ and $b > 1$ are shown in the box on page 354.

c. Transformations involving exponential functions are summarized in Table 4.1 on page 354.

d. The natural exponential function: $f(x) = e^x$. The irrational number e is called the natural base, where $e \approx 2.7183$.

e. Formulas for compound interest: After t years, the balance A in an account with principal P and annual interest rate r (in decimal form) is given by one of the following formulas:

1. For n compoundings per year: $A = P\left(1 + \frac{r}{n}\right)^{nt}$
2. For continuous compounding: $A = Pe^{rt}$

4.2 Logarithmic Functions

a. Definition of the logarithmic function: For $x > 0$ and $b > 0$, $b \neq 1$, $y = \log_b x$ is equivalent to $b^y = x$. The function $f(x) = \log_b x$ is the logarithmic function with base b. This function is the inverse function of the exponential function with base b.

b. Graphs of logarithmic functions for $b > 1$ and $0 < b < 1$ are shown in Figure 4.7 on page 366. Characteristics of the graphs are summarized in the box that follows the figure.

c. Transformations involving logarithmic functions are summarized in Table 4.3 on page 367.

d. The domain of a logarithmic function is the set of all positive real numbers. The domain of $f(x) = \log_b(x + c)$ consists of all x for which $x + c > 0$.

e. Common and natural logarithms: $f(x) = \log x$ means $f(x) = \log_{10} x$ and is the common logarithmic function. $f(x) = \ln x$ means $f(x) = \log_e x$ and is the natural logarithmic function.

f. Basic Logarithmic Properties

Base b ($b > 0, b \neq 1$)	Base 10 (Common Logarithms)	Base e (Natural Logarithms)
$\log_b 1 = 0$	$\log 1 = 0$	$\ln 1 = 0$
$\log_b b = 1$	$\log 10 = 1$	$\ln e = 1$
$\log_b b^x = x$	$\log 10^x = x$	$\ln e^x = x$
$b^{\log_b x} = x$	$10^{\log x} = x$	$e^{\ln x} = x$

Chapter 4 Test

1. Graph $f(x) = 2^x$ and $g(x) = 2^{x+1}$ in the same rectangular coordinate system.
2. Graph $f(x) = \log_2 x$ and $g(x) = \log_2(x - 1)$ in the same rectangular coordinate system.
3. Write in exponential form: $\log_5 125 = 3$.
4. Write in logarithmic form: $\sqrt{36} = 6$.
5. Find the domain of $f(x) = \ln(3 - x)$.

In Exercises 6–7, use properties of logarithms to expand each logarithmic expression as much as possible. Where possible, evaluate logarithmic expressions without using a calculator.

6. $\log_4(64x^5)$
7. $\log_3 \frac{\sqrt[3]{x}}{81}$

In Exercises 8–9, write each expression as a single logarithm.

8. $6 \log x + 2 \log y$
9. $\ln 7 - 3 \ln x$
10. Use a calculator to evaluate $\log_{15} 71$ to four decimal places.

In Exercises 11–16, solve each equation.

11. $5^x = 1.4$
12. $400e^{0.005x} = 1600$
13. $e^{2x} - 6e^x + 5 = 0$
14. $\log_6(4x - 1) = 3$
15. $\log x + \log(x + 15) = 2$
16. $2 \ln 3x = 8$
17. Suppose you have $3000 to invest. Which investment yields the greater return over 10 years: 6.5% compounded semianually or 6% compounded continuously? How much more (to the nearest dollar) is yielded by the better investment?
18. On the decibel scale, the loudness of a sound, in decibels, is given by $D = 10 \log \frac{I}{I_0}$, where I is the intensity of the sound, in watts per meter², and I_0 is the intensity of a sound barely audible to the human ear. If the intensity of a sound is $10^{12}I_0$, what is its loudness in decibels? (Such a sound is potentially damaging to the ear.)
19. The percentage of married men in the United States who are employed is modeled by $P = 89.18e^{-0.004t}$. The model indicates that P% of married men were employed t years after 1959.
    a. What percentage of married men were employed in 1959?
    b. Is the percentage of married men who are employed increasing or decreasing? Explain.
    c. In what year were 77% of U.S. married men employed?

Cumulative Review Exercises (Chapters 1–4)

Solve each equation in Exercises 1–5.

1. $|3x - 4| = 2$
2. $\sqrt{2x - 5} - \sqrt{x - 3} = 1$
3. $x^4 + x^3 - 3x^2 - x + 2 = 0$
4. $e^{5x} - 32 = 96$
5. $\log_2(x + 5) + \log_2(x - 1) = 4$

Solve each inequality in Exercises 6–7. Express the answer in interval notation.

6. $14 - 5x \geq -6$
7. $|2x - 4| \leq 2$
8. Write the point-slope form and the slope-intercept form of the line passing through $(1, 3)$ and $(3, -3)$.
9. If $f(x) = x^2$ and $g(x) = x + 2$, find $(f \circ g)(x)$ and $(g \circ f)(x)$.
10. If $f(x) = 2x - 7$, find $f^{-1}(x)$.
11. Divide $x^3 + 5x^2 + 3x - 10$ by $x + 2$.
12. Use the Rational Zero Theorem to list all possible rational zeros for $f(x) = 4x^3 - 7x - 3$.
13. The value of y varies directly as the square of x. If $x = 3$ when $y = 12$, find y when $x = 15$.
14. Solve $x^3 - 4x^2 + 6x - 4 = 0$ given that $1 + i$ is a root.

In Exercises 15–18, graph each equation.

15. $(x - 3)^2 + (y + 2)^2 = 4$
16. $f(x) = (x - 2)^2 - 1$
17. $f(x) = \frac{x^2 - 1}{x^2 - 4}$
18. $f(x) = (x - 2)^2(x + 1)$
19. You are paid time-and-a-half for each hour worked over 40 hours a week. Last week you worked 50 hours and earned $660. What is your normal hourly salary?
20. The formula $F = 1 - k \ln(t + 1)$ models the fraction of people, F, who remember all the words in a list of nonsense words t hours after memorizing the list. After 3 hours only half the people could remember all the words. Determine the value of k and then predict the fraction of people in the group who will remember all the words after 6 hours.

Pages 408-414

Applications Index

D

E

F

G

H

I

J

T

U

V

W

Y

Prerequisites: Fundamental Concepts of Algebra

Chapter P

This chapter reviews fundamental concepts of algebra that are prerequisites for the study of college algebra. Algebra, like all of mathematics, provides the tools to help you recognize, classify, and explore the hidden patterns of your world, revealing its underling structure. Throughout the new millennium, literacy in algebra will be a prerequisite for functioning in a meaningful way personally, professionally, and as a citizen.

Listening to the radio on the way to work, you hear candidates in the upcoming election discussing the problem of the country's 5.5 trillion dollar deficit. It seems like this is a real problem, but then you realize that you don't really know what that number means. How can you look at this deficit in the proper perspective? If the national debt were evenly divided among all citizens of the country, how much would each citizen have to pay? Does the deficit seem like such a significant problem now?

SECTION P.1 Real Numbers and Algebraic Expressions

Objectives

1. Recognize subsets of the real numbers.
2. Use inequality symbols.
3. Evaluate absolute value.
4. Use absolute value to express distance.
5. Evaluate algebraic expressions.
6. Identity properties of the real numbers.
7. Simplify algebraic expressions.

The U.N. building is designed with three golden rectangles.

The United Nations Building in New York was designed to represent its mission of promoting world harmony. Viewed from the front, the building looks like three rectangles stacked upon each other. In each rectangle, the ratio of the width to height is $\sqrt{5} + 1$ to 2, approximately 1.618 to 1. The ancient Greeks believed that such a rectangle, called a **golden rectangle**, was the most visually pleasing of all rectangles.

The ratio 1.618 to 1 is approximate because $\sqrt{5}$ is an irrational number, a special kind of real number. Irrational? Real? Let's make sense of all this by describing the kinds of numbers you will encounter in this course.

1 Recognize subsets of the real numbers.

The Set of Real Numbers

Before we describe the set of real numbers, let's be sure you are familiar with some basic ideas about sets. A **set** is a collection of objects whose contents can be clearly determined. The objects in a set are called the **elements** of the set. For example, the set of numbers used for counting can be represented by

"well defined"

Roster Form →

$$\{1, 2, 3, 4, 5, \ldots\}.$$

The braces, { }, indicate that we are representing a set. This form of representing a set uses commas to separate the elements of the set. The set of numbers used for counting is called the set of **natural numbers**. The three dots after the 5 indicate that there is no final element and that the listing goes on forever. (in the same way)

The sets that make up the real numbers are summarized in Table P.1. We refer to these sets as **subsets** of the real numbers, meaning that all elements in each subset are also elements in the set of real numbers.

Notice the use of the symbol ≈ in the examples of irrational numbers. The symbol means "is approximately equal to." Thus,

$$\sqrt{2} \approx 1.414214.$$

We can verify that this is only an approximation by multiplying 1.414214 by itself The product is very close to but not exactly 2:

$$1.414214 \times 1.414214 = 2.0000012378.$$

Technology

A calculator with a square root key gives a decimal approximation for $\sqrt{2}$, not the exact value.

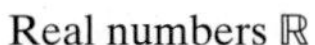

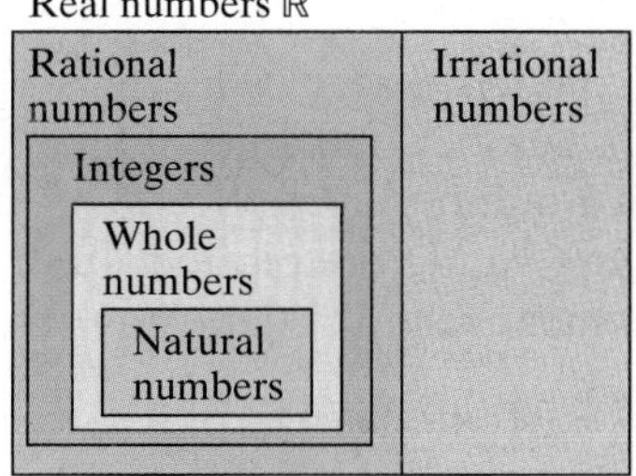

This diagram shows that every real number is rational or irrational.

Table P.1 Important Subsets of the Real Numbers

Name	Description	Examples
Natural numbers ℕ	$\{1, 2, 3, 4, 5, \ldots\}$ These numbers are used for counting.	$2, 3, 5, 17$
Whole numbers 𝕎	$\{0, 1, 2, 3, 4, 5, \ldots\}$ The whole numbers add 0 to the set of natural numbers.	$0, 2, 3, 5, 17$
Integers ℤ	$\{\ldots, -5, -4, -3, -2, -1, 0, 1, 2, 3, 4, 5, \ldots\}$ The integers add the negatives of the natural numbers to the set of whole numbers.	$-17, -5, -3, -2, 0, 2, 3, 5, 17$
Rational numbers ℚ	These numbers can he expressed as an integer divided by a nonzero integer: $\frac{a}{b}$: a and b are integers: $b \neq 0$. Rational numbers can be expressed as terminating or repeating decimals.	$-17 = \frac{-17}{1}, -5 = \frac{-5}{1}, -3, -2,$ $0, 2, 3, 5, 17,$ $\frac{2}{5} = 0.4, \frac{-2}{3} = -0.6666\cdots = -0.\overline{6}$
Irrational numbers 𝕀	This is the set of numbers whose decimal representations are neither terminating nor repeating. Irrational numbers cannot be expressed as a quotient of integers.	$\sqrt{2} \approx 1.414214$ $-\sqrt{3} \approx -1.73205$ $\pi \approx 3.142$ $-\frac{\pi}{2} \approx -1.571$

The set of **real numbers** is formed by combining the rational numbers and the irrational numbers. Thus, every real number is either rational or irrational.

The Real Number Line

The **real number line** is a graph used to represent the set of real numbers. An arbitrary point, called the **origin**, is labeled 0; units to the right of the origin are **positive** and units to the left of the origin are **negative**. The real number line is shown in Figure P.1.

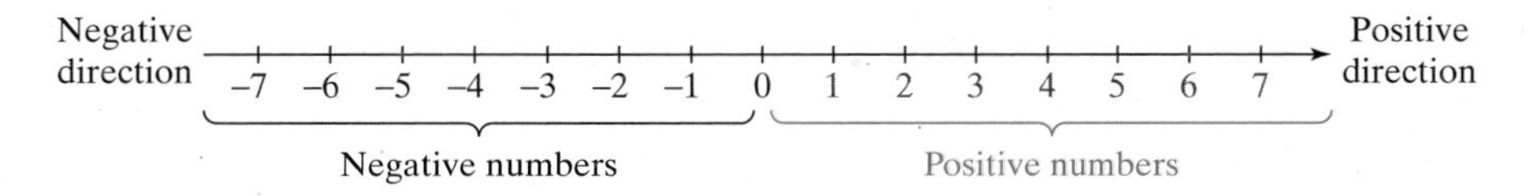

Figure P.1 The real number line

Real numbers are **graphed** on a number line by placing a dot at the correct location for each number. The integers are easiest to locate. In Figure P.2 we've graphed the integers −3, 0, and 4.

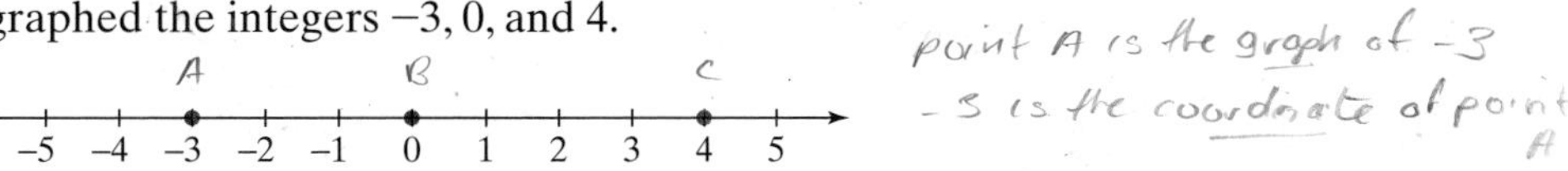

Figure P.2 Graphing −3, 0, and 4 on number line

Every real number corresponds to a point on the number line and every point on the number line corresponds to a real number. We say there is a **one-to-one correspondence** between all the real numbers and all points on a real number line. If you draw a point on the real number line corresponding to a real number, you are **plotting** the real number. In Figure P.2, we are plotting the real numbers −3, 0, and 4.

2 Use inequality symbols.

Ordering the Real Numbers

On the real number line, the real numbers increase from left to right. The lesser of two real numbers is the one farther to the left on a number line. The greater of two real numbers is the one farther to the right on a number line.

Look at the number line in Figure P.3. The integers 2 and 5 are plotted. Observe that 2 is to the left of 5 on the number line. This means that 2 is less than 5:

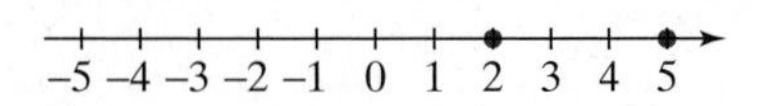

Figure P.3

$2 < 5$: 2 is less than 5 because 2 is to the *left* of 5 on the number line.

In Figure P.3, we can also observe that 5 is to the right of 2 on the number line. This means that 5 is greater than 2.

$5 > 2$: 5 is greater than 2 because 5 is to the right of 2 on the number line.

The symbols $<$ and $>$ are called **inequality symbols**. They may be combined with an equal sign, as shown in the following table.

Study Tip

The symbols $<$ and $>$ always point to the lesser of the two real numbers when the inequality is true.

$2 < 5$ The symbol points to 2, the lesser number.

$5 > 2$ The symbol points to 2, the lesser number.

Symbols	Meaning	Example	Explanation
$a \le b$	a is less than or equal to b.	$3 \le 7$	Because $3 < 7$
		$7 \le 7$	Because $7 = 7$
$b \ge a$	b is greater than or equal to a.	$7 \ge 3$	Because $7 > 3$
		$-5 \ge -5$	Because $-5 = -5$

3 Evaluate absolute value.

Absolute Value

Absolute value describes the distance from 0 on a real number line. If a represents a real number, the symbol $|a|$ represents its absolute value, read "the absolute value of a." For example, the real number line in Figure P.4 shows that

$$|-3| = 3 \quad \text{and} \quad |5| = 5.$$

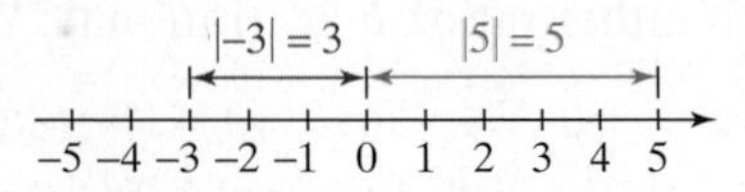

Figure P.4 Absolute value as the distance from 0

The absolute value of -3 is 3 because -3 is 3 units from 0 on the number line. The absolute value of 5 is 5 because 5 is 5 units from 0 on the number line. The absolute value of a positive real number or 0 is the number itself. The absolute value of a negative real number, such as -3, is positive.

We can define the absolute value of the real number x without referring to a number line. The algebraic definition of the absolute value of x is given as follows:

Definition of Absolute Value

$$|x| = \begin{cases} x & \text{if } x \ge 0 \\ -x & \text{if } x < 0 \end{cases}$$

If x is nonnegative (that is $x \ge 0$), the absolute value of x is the number itself. For example:

$$|5| = 5 \qquad |\pi| = \pi \qquad \left|\frac{1}{3}\right| = \frac{1}{3} \qquad |0| = 0$$

Zero is the only number whose absolute value is 0.

If x is a negative number (that is, $x < 0$), the absolute value of x is the opposite of x. This makes the absolute value positive. For example,

$$|-3| = -(-3) = 3 \qquad |-\pi| = -(-\pi) = \pi \qquad \left|-\frac{1}{3}\right| = -\left(-\frac{1}{3}\right) = \frac{1}{3}.$$

This middle step is usually omitted.

EXAMPLE 1 Evaluating Absolute Value

Rewrite each expression without absolute value bars.

a. $|\sqrt{3} - 1|$ b. $|2 - \pi|$ c. $\dfrac{|x|}{x}$ if $x < 0$

Solution

a. Because $\sqrt{3} \approx 1.7$, the expression inside the absolute value bars is positive. The absolute value of a positive number is the number itself. Thus,

$$|\sqrt{3} - 1| = \sqrt{3} - 1.$$

b. Because $\pi \approx 3.14$, the number inside the absolute value bars is negative. The absolute value of x when $x < 0$ is $-x$. Thus,

$$|2 - \pi| = -(2 - \pi) = \pi - 2.$$

c. If $x < 0$, then $|x| = -x$. Thus,

$$\frac{|x|}{x} = \frac{-x}{x} = -1.$$

Check Point 1 Rewrite each expression without absolute value bars.

a. $|1 - \sqrt{2}|$ **b.** $|\pi - 3|$ **c.** $\dfrac{|x|}{x}$ if $x > 0$

Discovery

Verify the triangle inequality if $a = 4$ and $b = 5$. Verify the triangle inequality if $a = 4$ and $b = -5$.

When does equality occur in the triangle inequality and when does inequality occur? Verify your observation with additional number pairs.

Next, we list several basic properties of absolute value. Each of these properties can be derived from the definition of absolute value.

Properties of Absolute Value

For all real numbers a and b,

1. $|a| \geq 0$ **2.** $|-a| = |a|$ **3.** $a \leq |a|$

4. $|ab| = |a||b|$ **5.** $\left|\dfrac{a}{b}\right| = \dfrac{|a|}{|b|}$, $b \neq 0$

6. $|a + b| \leq |a| + |b|$ (called the triangle inequality)

4 Use absolute value to express distance.

Distance Between Points on a Real Number Line

Absolute value is used to find the distance between two points on a real number line. If a and b are any real numbers, the **distance between a and b** is the absolute value of their difference. For example, the distance between 4 and 10 is 6. Using absolute value, we find this distance in one of two ways:

$$|10 - 4| = |6| = 6 \quad \text{or} \quad |4 - 10| = |-6| = 6.$$

The distance between 4 and 10 on the real number line is 6.

Notice that we obtain the same distance regardless of the order in which we subtract.

* Consider also: "Directed Distance"

Distance Between Two Points on the Real Number Line

If a and b are any two points on a real number line, then the distance between a and b is given by

$$|a - b| \quad \text{or} \quad |b - a|.$$

EXAMPLE 2 Distance Between Two Points on a Number Line

Find the distance between −5 and 3 on the real number line.

Solution Because the distance between a and b is given by $|a - b|$, the distance between −5 and 3 is

$$|-5 - 3| = |-8| = 8.$$

$a = -5$ $b = 3$

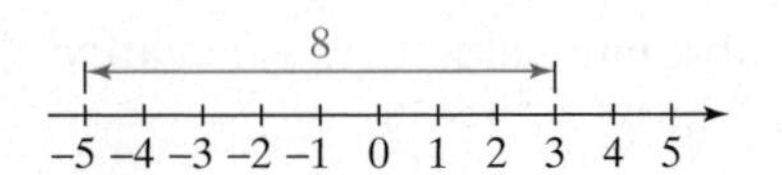

Figure P.5 The distance between −5 and 3 is 8.

Figure P.5 verifies that there are 8 units between −5 and 3 on the real number line. We obtain the same distance if we reverse the order of the subtraction:

$$|3 - (-5)| = |8| = 8.$$

Check Point 2 Find the distance between −4 and 5 on the real number line.

Algebraic Expressions

Algebra uses letters, such as x and y, to represent real numbers. Such letters are called **variables**. For example, imagine that you are basking in the sun on the beach. We can let x represent the number of minutes that you can stay in the sun without burning with no sunscreen. With a number 6 sunscreen, exposure time without burning is six times as long, or 6 times x. This can be written $6 \cdot x$, but it is usually expressed as $6x$. Placing a number and a letter next to one another indicates multiplication.

Notice that $6x$ combines the number 6 and the variable x using the operation of multiplication. A combination of variables and numbers using the operations of addition, subtraction, multiplication, or division, as well as powers or roots, is called an **algebraic expression**. Here are some examples of algebraic expressions:

$$x + 6, \quad x - 6, \quad 6x, \quad \frac{x}{6}, \quad 3x + 5, \quad \sqrt{x} + 7.$$

5 Evaluate algebraic expressions.

Evaluating Algebraic Expressions

Evaluating an algebraic expression means to find the value of the expression for a given value of the variable. For example, we can evaluate $6x$ (from the sunscreen example) when $x = 15$. We substitute 15 for x. We obtain $6 \cdot 15$, or 90. This means if you can stay in the sun for 15 minutes without burning when you don't put on any lotion, then with a number 6 lotion, you can "cook" for 90 minutes without burning.

Many algebraic expressions involve more than one operation. Evaluating an algebraic expression without a calculator involves carefully applying the following order of operations agreement.

The Order of Operations Agreement

1. Perform operations within the innermost parentheses and work outward. If the algebraic expression involves division, treat the numerator and the denominator as if they were each enclosed in parentheses.
2. Perform multiplication or division as they occur, working from left to right.
3. Perform addition or subtraction as they occur, working from left to right.

EXAMPLE 3 Evaluating an Algebraic Expression

The algebraic expression $2.35x + 179.5$ describes the population of the United States, in millions, x years after 1960. Evaluate the expression when $x = 40$. Describe what the answer means in practical terms.

Solution We begin by substituting 40 for x. Because $x = 40$, we will be finding the U.S. population 40 years after 1960, in the year 2000.

$$2.35x + 179.5$$

Replace x by 40.

$$= 2.35(40) + 179.5$$

$$= 94 + 179.5 \quad \text{Perform the multiplication: } 2.35(40) = 94.$$

$$= 273.5 \quad \text{Perform the addition.}$$

Thus, in 2000 the population of the United States was 273.5 million.

Check Point 3 Evaluate: $2.35x + 179.5$ when $x = 20$. Describe what your answer means in practical terms.

6 Identify properties of the real numbers.

Properties of Real Numbers and Algebraic Expressions

When you use your calculator to add two real numbers, you can enter them in any order. The fact that two real numbers can be added in any order is called the **commutative property of addition**. You probably use this property, as well as other properties of real numbers listed in Table P.2 on the next page, without giving it much thought. The properties of the real numbers are especially useful when working with algebraic expressions. For each property listed in Table P.2, a, b, and c represent real numbers, variables, or algebraic expressions.

The Associative Property and the English Language

In the English language, phrases can take on different meanings depending on the way the words are associated. For example, (man eating) tiger does not mean the same thing as man (eating tiger).

Table P.2 Properties of the Real Numbers

Name	Meaning	Examples
Commutative Property of Addition	Two real numbers can be added in any order. $a + b = b + a$	• $13 + 7 = 7 + 13$ • $13x + 7 = 7 + 13x$
Commutative Property of Multiplication	Two real numbers can be multiplied in any order. $ab = ba$	• $\sqrt{2} \cdot \sqrt{5} = \sqrt{5} \cdot \sqrt{2}$ • $x \cdot 6 = 6x$
Associative Property of Addition	If three real numbers are added, it makes no difference which two are added first. $(a + b) + c = a + (b + c)$	• $3 + (8 + x) = (3 + 8) + x = 11 + x$
Associative Property of Multiplication	If three real numbers are multiplied, it makes no difference which two are multiplied first. $(a \cdot b) \cdot c = a \cdot (b \cdot c)$	• $-2(3x) = (-2 \cdot 3)x = -6x$
Distributive Property of Multiplication over Addition	Multiplication distributes over addition. $a \cdot (b + c) = a \cdot b + a \cdot c$	• $7(4 + \sqrt{3}) = 7 \cdot 4 + 7 \cdot \sqrt{3} = 28 + 7\sqrt{3}$ • $5(3x + 7) = 5 \cdot 3x + 5 \cdot 7 = 15x + 35$
Identity Property of Addition	Zero can be deleted from a sum. $a + 0 = a$ $0 + a = a$	• $\sqrt{3} + 0 = \sqrt{3}$ • $0 + 6x = 6x$
Identity Property of Multiplication	One can be deleted from a product. $a \cdot 1 = a$ $1 \cdot a = a$	• $1 \cdot \pi = \pi$ • $13x \cdot 1 = 13x$
Inverse Property of Addition	The sum of a real number and its additive inverse gives 0, the additive identity. $a + (-a) = 0$ $(-a) + a = 0$	• $\sqrt{5} + (-\sqrt{5}) = 0$ • $6x + (-6x) = 0$ • $(-4y) + 4y = 0$
Inverse Property of Multiplication	The product of a nonzero real number and its multiplicative inverse gives 1, the multiplicative identity. $a \cdot \frac{1}{a} = 1, \quad a \neq 0$ $\frac{1}{a} \cdot a = 1, \quad a \neq 0$	• $7 \cdot \frac{1}{7} = 1$ • $\left(\frac{1}{x - 3}\right)(x - 3) = 1, \; x \neq 3$

Commutative Words and Sentences

The commutative property states that a change in order produces no change in the answer. The words and sentences listed here are commutative; they read the same from left to right and from right to left!

dad	Draw, o coward!	Revolting is error. Resign it, lover.
repaper	Dennis sinned.	Naomi, did I moan?
never odd or even	Ma is a nun, as I am.	Al lets Della call Ed Stella.

The properties in Table P.2 apply to the operations of addition and multiplication. Subtraction and division are defined in terms of addition and multiplication.

Definitions of Subtraction and Division

Let a and b represent real numbers.

Subtraction: $a - b = a + (-b)$ negative

We call $-b$ the **additive inverse** or **opposite** of b.

Division: $a \div b = a \cdot \frac{1}{b}$, where $b \neq 0$

We call $\frac{1}{b}$ the **multiplicative inverse** or **reciprocal** of b. The quotient of a and b, $a \div b$, can be written in the form $\frac{a}{b}$, where a is the **numerator** and b the **denominator** of the fraction.

Because subtraction is defined in terms of adding an inverse, the distributive property can be applied to subtraction:

$$a(b - c) = ab - ac$$

$$(b - c)a = ba - ca.$$

For example,

$$4(2x - 5) = 4 \cdot 2x - 4 \cdot 5 = 8x - 20.$$

7 Simplify algebraic expressions.

Simplifying Algebraic Expressions

The **terms** of an algebraic expression are those parts that are separated by addition. For example, consider the algebraic expression

$$7x - 9y - 3,$$

which can be expressed as

$$7x + (-9y) + (-3).$$

This expression contains three terms, namely $7x$, $-9y$, and -3.

The numerical part of a term is called its **numerical coefficient**. In the term $7x$, the 7 is the numerical coefficient. In the term $-9y$, the -9 is the numerical coefficient.

A term that consists of just a number is called a **constant term**. The constant term of $7x - 9y - 3$ is -3.

A term indicates a product. The expressions that are multiplied to form the term are called its **factors**. **Like terms** have the same variable factors with the same exponents on the variables. For example, $7x$ and $3x$ are like terms because they have the same variable factor, x. The distributive property (in reverse) can be used to add these terms:

$$7x + 3x = (7 + 3)x = 10x.$$

Study Tip

To add like terms, add their numerical coefficients. Use this result as the numerical coefficient of the terms' common variable(s).

An algebraic expression is **simplified** when parentheses have been removed and like terms have been combined.

EXAMPLE 4 Simplifying an Algebraic Expression

Simplify: $6(2x - 4y) + 10(4x + 3y)$.

Solution

$$
\begin{aligned}
&6(2x - 4y) + 10(4x + 3y) && \\
&= 6 \cdot 2x - 6 \cdot 4y + 10 \cdot 4x + 10 \cdot 3y && \text{Use the distributive property to remove the parentheses.} \\
&= 12x - 24y + 40x + 30y && \text{Multiply.} \\
&= (12x + 40x) + (30y - 24y) && \text{Group like terms.} \\
&= 52x + 6y && \text{Combine like terms.}
\end{aligned}
$$

Check Point 4 Simplify: $7(4x - 3y) + 2(5x + y)$.

Properties of Negatives

The distributive property can be extended to cover more than two terms within parentheses. For example,

This sign represents subtraction.

This sign tells us that −3 is negative.

$$
\begin{aligned}
-3(4x - 2y + 6) &= -3 \cdot 4x - (-3) \cdot 2y - 3 \cdot 6 \\
&= -12x - (-6y) - 18 \\
&= -12x + 6y - 18
\end{aligned}
$$

The voice balloons illustrate that negative signs can appear side by side. They can represent the operation of subtraction or the fact that a real number is negative. Here is a list of properties of negatives and how they are applied to algebraic expressions.

Properties of Negatives

Let a and b represent real numbers, variables, or algebraic expressions.

Property	Examples
1. $(-1)a = -a$	$(-1)4xy = -4xy$
2. $-(-a) = a$	$-(-6y) = 6y$
3. $(-a)b = -ab$	$(-7)4xy = -7 \cdot 4xy = -28xy$
4. $a(-b) = -ab$	$5x(-3y) = -5x \cdot 3y = -15xy$
5. $-(a + b) = -a - b$	$-(7x + 6y) = -7x - 6y$
6. $-(a - b) = -a + b = b - a$	$-(3x - 7y) = -3x + 7y = 7y - 3x$

Do you notice that properties 5 and 6 in the box are related? In general, expressions within parentheses that are preceded by a negative can be simplified by dropping the parentheses and changing the sign of every term inside the parentheses.

For example,

$$-(3x - 2y + 5z - 6) = -3x + 2y - 5z + 6.$$

EXERCISE SET P.1

Practice Exercises

In Exercises 1–4, list all numbers from the given set that are a. natural numbers, b. whole numbers, c. integers, d. rational numbers, e. irrational numbers.

1. $\{-9, -\frac{4}{5}, 0, 0.25, \sqrt{3}, 9.2, \sqrt{100}\}$
2. $\{-7, -0.\overline{6}, 0, \sqrt{49}, \sqrt{50}\}$
3. $\{-11, -\frac{5}{6}, 0, 0.75, \sqrt{5}, \pi, \sqrt{64}\}$
4. $\{-5, -0.\overline{3}, 0, \sqrt{2}, \sqrt{4}\}$
5. Give an example of a whole number that is not a natural number.
6. Give an example of a rational number that is not an integer.
7. Give an example of a number that is an integer, a whole number, and a natural number.
8. Give an example of a number that is a rational number, an integer, and a real number.

Determine whether each statement in Exercises 9–14 is true or false.

9. $-13 \le -2$
10. $-6 > 2$
11. $4 \ge -7$
12. $-13 < -5$
13. $-\pi \ge -\pi$
14. $-3 > -13$

In Exercises 15–22, rewrite each expression without absolute value bars.

15. $|300|$
16. $|-203|$
17. $|12 - \pi|$
18. $|7 - \pi|$
19. $|\sqrt{2} - 5|$
20. $|\sqrt{5} - 13|$
21. $\frac{-3}{|-3|}$
22. $\frac{-7}{|-7|}$

In Exercises 23–30, express the distance between the given numbers using absolute value. Then find the distance by evaluating the absolute value expression.

23. 2 and 17
24. 4 and 15
25. −2 and 5
26. −6 and 8
27. −19 and −4
28. −26 and −3
29. −3.6 and −1.4
30. −5.4 and −1.2

In Exercises 31–38, evaluate each algebraic expression for the given value of the variable.

31. $5x + 7;\quad x = 4$
32. $9x + 6;\quad x = 5$
33. $4(x + 3) - 11;\quad x = -5$
34. $6(x + 5) - 13;\quad x = -7$
35. $\frac{5}{9}(F - 32);\quad F = 77$
36. $\frac{5}{9}(F - 32);\quad F = 50$
37. $\frac{5(x + 2)}{2x - 14};\quad x = 10$
38. $\frac{7(x - 3)}{2x - 16};\quad x = 9$

In Exercises 39–46, state the name of the property illustrated.

39. $6 + (-4) = (-4) + 6$
40. $11 \cdot (7 + 4) = 11 \cdot 7 + 11 \cdot 4$
41. $6 + (2 + 7) = (6 + 2) + 7$
42. $6 \cdot (2 \cdot 3) = 6 \cdot (3 \cdot 2)$
43. $(2 + 3) + (4 + 5) = (4 + 5) + (2 + 3)$
44. $7 \cdot (11 \cdot 8) = (11 \cdot 8) \cdot 7$
45. $2(-8 + 6) = -16 + 12$
46. $-8(3 + 11) = -24 + (-88)$

In Exercises 47–52, simplify each algebraic expression.

47. $5(3x + 4) - 4$
48. $2(5x + 4) - 3$
49. $5(3x - 2) + 12x$
50. $2(5x - 1) + 14x$
51. $7(3y - 5) + 2(4y + 3)$
52. $4(2y - 6) + 3(5y + 10)$

In Exercises 53–58, write each algebraic expression without parentheses.

53. $-(-14x)$
54. $-(-17y)$
55. $-(2x - 3y - 6)$
56. $-(5x - 13y - 1)$
57. $\frac{1}{3}(3x) + [(4y) + (-4y)]$
58. $\frac{1}{2}(2y) + [(-7x) + 7x]$

Application Exercises

59. Are first putting on your left shoe and then putting on your right shoe commutative?
60. Are first getting undressed and then taking a shower commutative?
61. Give an example of two things that you do that are not commutative.
62. Give an example of two things that you do that are commutative.
63. The algebraic expression $962x + 18{,}667$ describes average yearly earnings in United States x years after 1990. Evaluate the algebraic expression when $x = 7$. Describe what the answer means in practical terms.

64. The algebraic expression $1527x + 31{,}290$ describes average yearly earnings for elementary and secondary teachers in the United States x years after 1990. Evaluate the algebraic expression when $x = 10$. Describe what the answer means in practical terms.

65. The optimum heart rate is the rate that a person should achieve during exercise for the exercise to be most beneficial. The algebraic expression

$$0.6(220 - a)$$

describes a person's optimum heart rate in beats per minute, where a represents the age of the person.

a. Use the distributive property to rewrite the algebraic expression without parentheses.

b. Use each form of the algebraic expression to determine the optimum heart rate for a 20-year-old runner.

Writing in Mathematics

Writing about mathematics will help you to learn mathematics. For all writing exercises in this book, use complete sentences to respond to the question. Some writing exercises can be answered in a sentence; others require a paragraph or two. You can decide how much you need to write as long as your writing clearly and directly answers the question in the exercise. Standard references such as a dictionary and a thesaurus should be helpful.

66. How do the whole numbers differ from the natural numbers?

67. Can a real number be both rational and irrational? Explain your answer.

68. If you are given two real numbers, explain how to determine which one is the lesser.

69. How can $\frac{|x|}{x}$ be equal to 1 or -1?

70. What is an algebraic expression? Give an example with your explanation.

71. Why is $3(x + 7) - 4x$ not simplified? What must be done to simplify the expression?

72. You can transpose the letters in the word "conversation" to form the phrase "voices rant on." From "total abstainers" we can form "sit not at ale bars." What two algebraic properties do each of these transpositions (called anagrams) remind you of? Explain your answer.

Critical Thinking Exercises

73. Which one of the following statements is true?

a. Every rational number is an integer.

b. Some whole numbers are not integers.

c. Some rational numbers are not positive.

d. Irrational numbers cannot be negative.

74. Which of the following is true?

a. The term x has no numerical coefficient.

b. $5 + 3(x - 4) = 8(x - 4) = 8x - 32$

c. $-x - x = -x + (-x) = 0$

d. $x - 0.02(x + 200) = 0.98x - 4$

In Exercises 75–77, insert either $<$ or $>$ in the box between the numbers to make the statement true.

75. $\sqrt{2} \,\square\, 1.5$

76. $-\pi \,\square\, -3.5$

77. $-\frac{3.14}{2} \,\square\, -\frac{\pi}{2}$

78. A business that manufactures small alarm clocks has weekly fixed costs of \$5000. The average cost per clock for the business to manufacture x clocks is described by

$$\frac{0.5x + 5000}{x}.$$

a. Find the average cost when $x = 100, 1000,$ and $10{,}000$.

b. Like all other businesses, the alarm clock manufacturer must make a profit. To do this, each clock must be sold for at least 50¢ more than what it costs to manufacture. Due to competition from a larger company, the clocks can be sold for \$1.50 each and no more. Our small manufacturer can only produce 2000 clocks weekly. Does this business have much of a future? Explain.

SECTION P.2 Exponents and Scientific Notation

Objectives

1. Understand and use integer exponents.
2. Use properties of exponents.
3. Simplify exponential expressions.
4. Use scientific notation.

1 Understand and use integer exponents.

Although people do a great deal of talking, the total output since the beginning of gabble to the present day, including all baby talk, love songs, and congressional debates, only amounts to about 10 million billion words. This can be expressed as 16 factors of 10, or 10^{16} words.

Exponents such as 2, 3, 4, and so on are used to indicate repeated multiplication. For example,

$$10^2 = 10 \cdot 10 = 100,$$

$$10^3 = 10 \cdot 10 \cdot 10 = 1000, \quad 10^4 = 10 \cdot 10 \cdot 10 \cdot 10 = 10{,}000.$$

The 10 that is repeated when multiplying is called the **base**. The small numbers above and to the right of the base are called **exponents**. The exponent tells the number of times the base is to be used when multiplying. In 10^3, the base is 10 and the exponent is 3.

Any number with an exponent of 1 is the number itself. Thus, $10^1 = 10$.

Multiplications that are expressed in exponential notation are read as follows:

10^1: "ten to the first power"

10^2: "ten to the second power" or "ten squared"

10^3: "ten to the third power" or "ten cubed"

10^4: "ten to the fourth power"

10^5: "ten to the fifth power"

Any real number can be used as the base. Thus,

$$7^2 = 7 \cdot 7 = 49 \text{ and } (-3)^4 = (-3)(-3)(-3)(-3) = 81.$$

The bases are 7 and −3, respectively. Do not confuse $(-3)^4$ and -3^4.

$$-3^4 = -3 \cdot 3 \cdot 3 \cdot 3 = -81$$

The negative is not taken to the power because it is not inside parentheses.

Powers Of Ten

$10 = 10^1$
$100 = 10^2$
$1000 = 10^3$
$10{,}000 = 10^4$
$100{,}000 = 10^5$
$1{,}000{,}000 = 10^6$ million
$10{,}000{,}000 = 10^7$
$100{,}000{,}000 = 10^8$
$1{,}000{,}000{,}000 = 10^9$ billion

Technology

You can use a calculator to evaluate exponential expressions. For example, to evaluate 5^3, press the following keys:

Scientific Calculator

5 [x^y] 3 [=]

Graphing Calculator

5 [∧] 3 [ENTER]

Although calculators have special keys to evaluate powers of ten and squaring bases, you can always use one of the sequences shown here.

EXAMPLE 1 Evaluating an Exponential Expression

Evaluate: $(-2)^3 \cdot 3^2$.

Solution

$$(-2)^3 \cdot 3^2 = (-2)(-2)(-2) \cdot 3 \cdot 3 = -8 \cdot 9 = -72$$

This is $(-2)^3$, read "−2 cubed."

This is 3^2, read "3 squared."

Check Point 1 Evaluate: $(-4)^3 \cdot 2^2$.

The formal algebraic definition of a natural number exponent summarizes our discussion.

Definition of a Natural Number Exponent

If b is a real number and n is a natural number,

$$b^n = \underbrace{b \cdot b \cdot b \cdot \cdots \cdot b}_{b \text{ appears as a factor } n \text{ times.}}.$$

Exponent: n. Base: b.

b^n is read "the nth power of b" or "b to the nth power." Thus, the nth power of b is defined as the product of n factors of b.

Furthermore, $b^1 = b$.

Negative Integers as Exponents

A nonzero base can be raised to a negative power using the following definition.

The Negative Exponent Rule

If b is any real number other than 0 and n is a natural number, then

$$b^{-n} = \frac{1}{b^n}.$$

EXAMPLE 2 Evaluating Expressions Containing Negative Exponents

Evaluate: a. 5^{-3} b. $\frac{1}{4^{-2}}$

Solution

a. $5^{-3} = \frac{1}{5^3} = \frac{1}{5 \cdot 5 \cdot 5} = \frac{1}{125}$

b. $\frac{1}{4^{-2}} = \frac{1}{\frac{1}{4^2}} = 4^2 = 4 \cdot 4 = 16$

Study Tip

When a negative integer appears as an exponent, switch the position of the base (from numerator to denominator or from denominator to numerator) and make the exponent positive.

Check Point 2 Evaluate: **a.** 2^{-3} **b.** $\frac{1}{6^{-2}}$

Zero as an Exponent

A nonzero base can be raised to the 0 power using the following definition.

The Zero Exponent Rule

If b is any real number other than 0,

$$b^0 = 1.$$

Here are three examples involving simplification using the zero exponent rule:

$$7^0 = 1 \qquad (-5)^0 = 1 \qquad -5^0 = -1.$$

Only 5 is raised to the zero power.

2 Use properties of exponents.

The Product Rule

Consider the multiplication of two exponential expressions such as $2^4 \cdot 2^3$. We are multiplying 4 factors of 2 and 3 factors of 2. We have a total of 7 factors of 2. Thus,

$$2^4 \cdot 2^3 = 2^7.$$

We can quickly find the exponent on the product, 7, by adding 4 and 3, the original exponents. This suggests the following rule.

The Product Rule

$$b^m \cdot b^n = b^{m+n}$$

Add exponents when multiplying with the same base.

A Number with 369 Million Digits

The largest number that can be expressed with only three digits is

$9^{(9^9)}$ or $9^{387,420,489}$

This number begins with 428124773 . . . , has 369 million digits, and would take around 70 years to read.

EXAMPLE 3 Using the Product Rule

Use the product rule to simplify each expression.

a. $2^2 \cdot 2^3$ b. $4^2 \cdot 4^{-5}$ c. $x^{-3} \cdot x^7$

Solution

a. $2^2 \cdot 2^3 = 2^{2+3} = 2^5 = 32$

b. $4^2 \cdot 4^{-5} = 4^{2+(-5)} = 4^{-3} = \frac{1}{4^3} = \frac{1}{64}$

c. $x^{-3} \cdot x^7 = x^{-3+7} = x^4$

Check Point 3 Use the product rule to simplify each expression.

a. $3^3 \cdot 3^2$ **b.** $2^4 \cdot 2^{-7}$ **c.** $x^{-5} \cdot x^{11}$

The Power Rule

The next property of exponents applies when an expression containing a power is itself raised to a power.

The Power Rule (Powers to Powers)

$$(b^m)^n = b^{mn}$$

Multiply exponents when an exponential expression is raised to a power.

EXAMPLE 4 Using the Power Rule

Use the power rule to simplify each expression.

a. $(2^2)^3$ b. $(y^5)^6$ c. $(x^{-3})^4$

Solution

a. $(2^2)^3 = 2^{2 \cdot 3} = 2^6 = 64$

b. $(y^5)^6 = y^{5 \cdot 6} = y^{30}$

c. $(x^{-3})^4 = x^{-3 \cdot 4} = x^{-12} = \frac{1}{x^{12}}$

Check Point 4 Use the power rule to simplify each expression.

a. $(3^3)^2$ **b.** $(y^7)^4$ **c.** $(x^{-4})^2$

The Quotient Rule

The next property of exponents applies when we are dividing exponential expressions with the same base.

The Quotient Rule

$$\frac{b^m}{b^n} = b^{m-n}$$

Subtract exponents when dividing with the same base.

EXAMPLE 5 Using the Quotient Rule

Use the quotient rule to simplify each expression.

a. $\frac{2^8}{2^4}$ b. $\frac{x^3}{x^7}$ c. $\frac{y^9}{y^{-5}}$

Solution

a. $\frac{2^8}{2^4} = 2^{8-4} = 2^4 = 16$

b. $\frac{x^3}{x^7} = x^{3-7} = x^{-4} = \frac{1}{x^4}$

c. $\frac{y^9}{y^{-5}} = y^{9-(-5)} = y^{9+5} = y^{14}$

Study Tip

$\frac{4^3}{4^5}$ and $\frac{4^5}{4^3}$ represent different numbers:

$$\frac{4^3}{4^5} = 4^{3-5} = 4^{-2} = \frac{1}{4^2} = \frac{1}{16}$$

$$\frac{4^5}{4^3} = 4^{5-3} = 4^2 = 16.$$

Check Point 5

Use the quotient rule to simplify each expression.

a. $\frac{3^6}{3^4}$ **b.** $\frac{x^5}{x^{12}}$ **c.** $\frac{y^2}{y^{-7}}$

Products Raised to Powers

The next property of exponents applies when we are raising a product to a power.

Products to Powers

$$(ab)^n = a^n b^n$$

When a product is raised to a power, raise each factor to the power.

EXAMPLE 6 Raising a Product to a Power

Simplify: $(-2y)^4$.

Solution $(-2y)^4 = (-2)^4 y^4 = 16y^4$

Check Point 6

Simplify: $(-4x)^3$.

The rule for products of powers can be extended to cover three or more factors. For example,

$$(-2xy)^3 = (-2)^3 x^3 y^3 = -8x^3 y^3.$$

Quotients Raised to Powers

Our final exponential property applies when we are raising a quotient to a power.

Quotients to Powers

$$\left(\frac{a}{b}\right)^n = \frac{a^n}{b^n}$$

When a quotient is raised to a power, raise the numerator and the denominator to the power.

EXAMPLE 7 Raising Quotients to Powers

Simplify by raising each quotient to the given power.

a. $\left(\frac{2}{5}\right)^4$ b. $\left(-\frac{3}{x}\right)^3$

Solution

a. $\left(\frac{2}{5}\right)^4 = \frac{2^4}{5^4} = \frac{16}{625}$

b. $\left(-\frac{3}{x}\right)^3 = \frac{(-3)^3}{x^3} = \frac{-27}{x^3}$

Check Point 7 Simplify: **a.** $\left(\frac{3}{4}\right)^3$ b. $\left(-\frac{2}{y}\right)^5$

3 Simplify exponential expressions.

Simplifying Exponential Expressions

Properties of exponents are used to simplify exponential expressions. Here is a summary of the properties we have discussed.

Properties of Exponents

1. $b^{-n} = \frac{1}{b^n}$ **2.** $b^0 = 1$ **3.** $b^m \cdot b^n = b^{m+n}$ **4.** $(b^m)^n = b^{mn}$

5. $\frac{b^m}{b^n} = b^{m-n}$ **6.** $(ab)^n = a^n b^n$ **7.** $\left(\frac{a}{b}\right)^n = \frac{a^n}{b^n}$

An exponential expression is **simplified** when

- No parentheses appear.
- No powers are raised to powers.
- Each base occurs only once.
- No negative exponents appear.

EXAMPLE 8 Simplifying Exponential Expressions

Simplify:

a. $(-3x^4y^5)^3$ b. $(-7xy^4)(-2x^5y^6)$ c. $\frac{-35x^2y^4}{5x^6y^{-8}}$ d. $\left(\frac{4x^2}{y}\right)^{-3}$

Solution

a. $(-3x^4y^5)^3 = (-3)^3(x^4)^3(y^5)^3$ Raise each factor in the product to the power.

$= (-3)^3x^{4\cdot3}y^{5\cdot3}$ Multiply powers to powers.

$= -27x^{12}y^{15}$ $(-3)^3 = (-3)(-3)(-3) = -27$

b. $(-7xy^4)(-2x^5y^6) = (-7)(-2)xx^5y^4y^6$ Group factors with the same base.

$= 14x^{1+5}y^{4+6}$ When multiplying expressions with the same base, add the exponents.

$= 14x^6y^{10}$ Simplify.

c. $\dfrac{-35x^2y^4}{5x^6y^{-8}} = \left(\dfrac{-35}{5}\right)\left(\dfrac{x^2}{x^6}\right)\left(\dfrac{y^4}{y^{-8}}\right)$ Group factors with the same base.

$= -7x^{2-6}y^{4-(-8)}$ When dividing expression with the same base, subtract the exponents.

$= -7x^{-4}y^{12}$ Simplify. Notice that $4 - (-8) = 4 + 8 = 12$.

$= \dfrac{-7y^{12}}{x^4}$ Move x^{-4}, the factor with the negative exponent, from the numerator to the denominator.

d. $\left(\dfrac{4x^2}{y}\right)^{-3} = \dfrac{4^{-3}(x^2)^{-3}}{y^{-3}}$ Raise each factor inside the parentheses to the -3 power.

$= \dfrac{4^{-3}x^{-6}}{y^{-3}}$ Multiply powers to powers.

$= \dfrac{y^3}{4^3x^6}$ Move factors with negative exponents from the numerator to the denominator (or vice versa) by changing the sign of the exponent.

$= \dfrac{y^3}{64x^6}$ $4^3 = 4 \cdot 4 \cdot 4 = 64$

Visualizing Powers of 3

The triangles contain $3, 3^2, 3^3$, and 3^4 circles.

Check Point 8

Simplify:

a. $(2x^3y^6)^4$ **b.** $(-6x^2y^5)(3xy^3)$ **c.** $\dfrac{100x^{12}y^2}{20x^{16}y^{-4}}$ **d.** $\left(\dfrac{5x}{y^4}\right)^{-2}$

Study Tip

Try to avoid the following common errors that can occur when simplifying exponential expressions.

Incorrect	Description of Error	Correct
$b^3b^4 = b^{12}$	Exponents should be added, not multiplied.	$b^3b^4 = b^7$
$3^n \cdot 3^m = 9^{n+m}$	The common base should be retained, not multiplied.	$3^n \cdot 3^m = 3^{n+m}$
$\dfrac{5^{16}}{5^4} = 5^4$	Exponents should be subtracted, not divided.	$\dfrac{5^{16}}{5^4} = 5^{12}$
$(4a)^3 = 4a^3$	Both factors should be cubed.	$(4a)^3 = 64a^3$
$b^{-n} = -\dfrac{1}{b^n}$	Only the exponent should change sign.	$b^{-n} = \dfrac{1}{b^n}$
$(a + b)^{-1} = \dfrac{1}{a} + \dfrac{1}{b}$	The exponent applies to the entire expression $a + b$.	$(a + b)^{-1} = \dfrac{1}{a + b}$

4 Use scientific notation.

Scientific Notation

The national debt of the United States is about \$5.5 trillion. A stack of \$1 bills equaling the national debt would rise to twice the distance from the Earth to the moon. Because a trillion is 10^{12}, the national debt can be expressed as

$$5.5 \times 10^{12}.$$

The number 5.5×10^{12} is written in a form called **scientific notation.** A number in scientific notation is expressed as a number greater than or equal to 1 and less than 10 multiplied by some power of 10. It is customary to use the multiplication symbol, ×, rather than a dot in scientific notation.

Here are two examples of numbers in scientific notation:

Each day, 2.6×10^7 pounds of dust from the atmosphere settle on Earth.

The diameter of a hydrogen atom is 1.016×10^{-8} centimeter.

We can use the exponent on the 10 to change a number in scientific notation to decimal notation. If the exponent is *positive*, move the decimal point in the number to the *right* the same number of places as the exponent. If the exponent is *negative*, move the decimal point in the number to the *left* the same number of places as the exponent.

EXAMPLE 9 Converting from Scientific to Decimal Notation

Write each number in decimal notation.

a. 2.6×10^7 b. 1.016×10^{-8}

Solution

a. We express 2.6×10^7 in decimal notation by moving the decimal point in 2.6 seven places to the right. We need to add six zeros.

$$2.6 \times 10^7 = 26{,}000{,}000$$

b. We express 1.016×10^{-8} in decimal notation by moving the decimal point in 1.016 eight places to the left. We need to add seven zeros to the right of the decimal point.

$$1.016 \times 10^{-8} = 0.00000001016$$

Check Point 9 Write each number in decimal notation.

a. 7.4×10^9 **b.** 3.017×10^{-6}

To convert from decimal notation to scientific notation, we reverse the procedure of Example 9.

- Move the decimal point in the given number to obtain a number greater than or equal to 1 and less than 10.
- The number of places the decimal point moves gives the exponent on 10; the exponent is positive if the given number is greater than 10 and negative if the given number is between 0 and 1.

EXAMPLE 10 Converting from Decimal Notation to Scientific Notation

Write each number in scientific notation.

a. 4,600,000 b. 0.00023

Solution

a. $4{,}600{,}000 = 4.6 \times 10^?$ $\xrightarrow{\text{Decimal point moves 6 places.}}$ 4.6×10^6

b. $0.00023 = 2.3 \times 10^{-?}$ $\xrightarrow{\text{Decimal point moves 4 places.}}$ 2.3×10^{-4}

Check Point 10 Write each number in scientific notation.

a. 7,410,000,000 **b.** 0.000000092

Computations with Scientific Notation

The product and quotient rules for exponents can be used to multiply or divide numbers that are expressed in scientific notation. For example, here's how to find the product of 3.4×10^9 and 2×10^{-5}.

$$\begin{aligned}(3.4 \times 10^9)(2 \times 10^{-5}) &= (3.4 \times 2) \times (10^9 \times 10^{-5}) \\ &= 6.8 \times 10^{9+(-5)} \\ &= 6.8 \times 10^4 \quad \text{or} \quad 68,000\end{aligned}$$

Technology

On a graphing calculator, you can use the EE (enter exponent) key to find the product of 3.4×10^9 and 2×10^{-5}:

3.4 [EE] 9 [×] 2 [EE] [(−)] 5 [ENTER]

If your calculator is in the scientific notation mode, it will display 6.8 E4; in the normal mode it will display 68000.

In our next example, we use the quotient of two numbers in scientific notation to help put a number into perspective. The number is our national debt. The United States began accumulating large deficits in the 1980s. To finance the deficit, the government had borrowed \$5.5 trillion as of the end of 1998. The graph in Figure P.6 shows the national debt increasing over time.

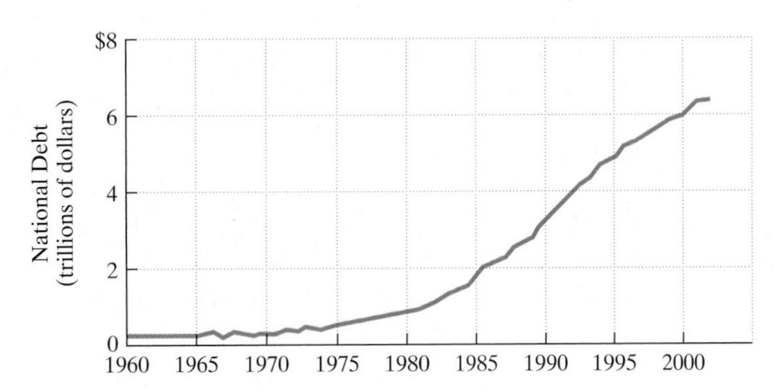

Figure P.6 The national debt

Source: Office of Management and Budget

EXAMPLE 11 The National Debt

As of the end of 1998, the national debt was \$5.5 trillion, or 5.5×10^{12} dollars. At that time, the U.S. population was approximately 270,000,000 (270 million), or 2.7×10^8. If the national debt were evenly divided among every individual in the United States, how much would each citizen have to pay?

Solution The amount each citizen must pay is the total debt, 5.5×10^{12} dollars, divided by the number of citizens, 2.7×10^8.

$$\begin{aligned}\frac{5.5 \times 10^{12}}{2.7 \times 10^8} &= \left(\frac{5.5}{2.7}\right) \times \left(\frac{10^{12}}{10^8}\right) \\ &\approx 2.04 \times 10^{12-8} \\ &= 2.04 \times 10^4 \\ &= 20,400\end{aligned}$$

Technology

Here is the keystroke sequence for solving Example 11 using a graphing calculator:

5.5 [EE] 12 [÷] 2.7 [EE] 8 [ENTER]

Every U.S. citizen would have to pay about $20,400 to the federal government to pay off the national debt. A family of three would owe $61,200!

Check Point 11 Approximately 2×10^4 people run in the New York City Marathon each year. Each runner runs a distance of 26 miles. Write the total distance covered by all the runners (assuming that each person completes the marathon) in scientific notation.

EXERCISE SET P.2

Practice Exercises

Evaluate each exponential expression in Exercises 1–22.

1. $5^2 \cdot 2$
2. $6^2 \cdot 2$
3. $(-2)^6$
4. $(-2)^4$
5. -2^6
6. -2^4
7. $(-3)^0$
8. $(-9)^0$
9. -3^0
10. -9^0
11. 4^{-3}
12. 2^{-6}
13. $2^2 \cdot 2^3$
14. $3^3 \cdot 3^2$
15. $(2^2)^3$
16. $(3^3)^2$
17. $\dfrac{2^8}{2^4}$
18. $\dfrac{3^8}{3^4}$
19. $3^{-3} \cdot 3$
20. $2^{-3} \cdot 2$
21. $\dfrac{2^3}{2^7}$
22. $\dfrac{3^4}{3^7}$

Simplify each exponential expression in Exercises 23–60.

23. $x^{-2}y$
24. xy^{-3}
25. x^0y^5
26. x^7y^0
27. $x^3 \cdot x^7$
28. $x^{11} \cdot x^5$
29. $x^{-5} \cdot x^{10}$
30. $x^{-6} \cdot x^{12}$
31. $(x^3)^7$
32. $(x^{11})^5$
33. $(x^{-5})^3$
34. $(x^{-6})^4$
35. $\dfrac{x^{14}}{x^7}$
36. $\dfrac{x^{30}}{x^{10}}$
37. $\dfrac{x^{14}}{x^{-7}}$
38. $\dfrac{x^{30}}{x^{-10}}$
39. $(8x^3)^2$
40. $(6x^4)^2$
41. $\left(-\dfrac{4}{x}\right)^3$
42. $\left(-\dfrac{6}{y}\right)^3$
43. $(-3x^2y^5)^2$
44. $(-3x^4y^6)^3$
45. $(3x^4)(2x^7)$
46. $(11x^5)(9x^{12})$
47. $(-9x^3y)(-2x^6y^4)$
48. $(-5x^4y)(-6x^7y^{11})$
49. $\dfrac{8x^{20}}{2x^4}$
50. $\dfrac{20x^{24}}{10x^6}$
51. $\dfrac{25a^{13}b^4}{-5a^2b^3}$
52. $\dfrac{35a^{14}b^6}{-7a^7b^3}$
53. $\dfrac{14b^7}{7b^{14}}$
54. $\dfrac{20b^{10}}{10b^{20}}$
55. $(4x^3)^{-2}$
56. $(10x^2)^{-3}$
57. $\dfrac{24x^3y^5}{32x^7y^{-9}}$
58. $\dfrac{10x^4y^9}{30x^{12}y^{-3}}$
59. $\left(\dfrac{5x^3}{y}\right)^{-2}$
60. $\left(\dfrac{3x^4}{y}\right)^{-3}$

In Exercises 61–68, write each number in decimal notation.

61. 4.7×10^3
62. 9.12×10^5
63. 4×10^6
64. 7×10^6
65. 7.86×10^{-4}
66. 4.63×10^{-5}
67. 3.18×10^{-6}
68. 5.84×10^{-7}

In Exercises 69–76, write each number in scientific notation.

69. 3600
70. 2700
71. 220,000,000
72. 370,000,000,000
73. 0.027
74. 0.014
75. 0.000763
76. 0.000972

In Exercises 77–84, perform the indicated operation and express the answer in decimal notation.

77. $(2 \times 10^3)(3 \times 10^2)$
78. $(5 \times 10^2)(4 \times 10^4)$
79. $(4.1 \times 10^2)(3 \times 10^{-4})$
80. $(1.2 \times 10^3)(2 \times 10^{-5})$
81. $\dfrac{12 \times 10^6}{4 \times 10^2}$
82. $\dfrac{20 \times 10^{26}}{10 \times 10^{15}}$
83. $\dfrac{6.3 \times 10^3}{3 \times 10^5}$
84. $\dfrac{9.6 \times 10^2}{3 \times 10^{-3}}$

Application Exercises

In Exercises 85–88, use 10^6 for one million and 10^9 for one billion to rewrite the number in each statement in scientific notation.

85. In 1999, the U.S. government collected $1,694,300 million.
86. In 1999, the U.S. government spent $1,751,800 million.

87. The federal government is expected to provide nearly \$60 billion in student aid in 2002.
88. In 1998, U.S. consumers spent \$5,493.7 billion.
89. If the population of the United States is 2.7×10^8 and each person spends about \$120 per year on ice cream, express the total annual spending on ice cream in scientific notation.

Writing in Mathematics

90. Describe what it means to raise a number to a power. In your description, include a discussion of the difference between -5^2 and $(-5)^2$.
91. Explain the product rule for exponents. Use $2^3 \cdot 2^5$ in your explanation.
92. Explain the power rule for exponents. Use $(3^2)^4$ in your explanation.
93. Explain the quotient rule for exponents. Use $\frac{5^8}{5^2}$ in your explanation.
94. Why is $(-3x^2)(2x^{-5})$ not simplified? What must be done to simplify the expression?
95. How do you know if a number is written in scientific notation?
96. Explain how to convert from scientific to decimal notation and give an example.
97. Explain how to convert from decimal to scientific notation and give an example.

Critical Thinking Exercises

98. Which one of the following is true?
    a. $4^{-2} < 4^{-3}$
    b. $5^{-2} > 2^{-5}$
    c. $(-2)^4 = 2^{-4}$
    d. $5^2 \cdot 5^{-2} > 2^5 \cdot 2^{-5}$
99. The mad Dr. Frankenstein has gathered enough bits and pieces (so to speak) for $2^{-1} + 2^{-2}$ of his creature-to-be. Write a fraction that represents the amount of his creature that must still be obtained.
100. If $b^A = MN$, $b^C = M$, and $b^D = N$, what is the relationship among A, C, and D?

Group Exercise

101. **Putting Numbers into Perspective.** A large number can be put into perspective by comparing it with another number. For example, we put the \$5.5 trillion national debt into perspective by comparing it to the number of U.S. citizens. The total distance covered by all the runners in the New York City Marathon (checkpoint Example 11 on page 22) can be put into perspective by comparing this distance with, say, the distance from New York to San Francisco.

     For this project, each group member should consult an almanac, a newspaper, or the World Wide Web to find a number greater than one million. Explain to other members of the group the context in which the large number is used. Express the number in scientific notation. Then put the number into perspective by comparing it with another number.

SECTION P.3 *Radicals and Rational Exponents*

Objectives

1. Evaluate square roots.
2. Use the product rule to simplify square roots.
3. Use the quotient rule to simplify square roots.
4. Add and subtract square roots.
5. Rationalize denominators.
6. Evaluate and perform operations with higher roots.
7. Understand and use rational exponents.

What is the maximum speed at which a racing cyclist can turn a corner without tipping over? The answer, in miles per hour, is given by the algebraic expression $4\sqrt{x}$, where x is the radius of the corner, in feet. Algebraic expressions containing roots describe phenomena as diverse as a wild animal's territorial area, evaporation on a lake's surface, and Albert Einstein's bizarre concept of how an astronaut moving close to the speed of light would barely age relative to friends watching from Earth. No description of your world can be complete without roots and radicals. In this section, we review the basics of radical expressions and the use of rational exponents to indicate radicals.

1 Evaluate square roots.

Square Roots

The **principal square root** of a nonnegative real number b, written $\sqrt{b}$, is that number whose square equals b. For example,

$$\sqrt{100} = 10 \text{ because } 10^2 = 100 \quad \text{and} \quad \sqrt{0} = 0 \text{ because } 0^2 = 0.$$

Observe that the principal square root of a positive number is positive and the principal square root of 0 is 0.

The symbol $\sqrt{}$ that we use to denote the principal square root is called a **radical sign**. The number under the radical sign is called the **radicand**. Together we refer to the radical sign and its radicand as a **radical**.

The following definition summarizes our discussion.

Definition of the Principal Square Root

If a is a nonnegative real number, the nonnegative number b such that $b^2 = a$, denoted by $b = \sqrt{a}$, is the **principal square root** of a.

In the real number system, negative numbers do not have square roots. For example, $\sqrt{-9}$ is not a real number because there is no real number whose square is -9.

If a number is nonnegative $(a \geq 0)$, then $(\sqrt{a})^2 = a$. For example,

$$(\sqrt{2})^2 = 2, \quad (\sqrt{3})^2 = 3, \quad (\sqrt{4})^2 = 4, \quad \text{and } (\sqrt{5})^2 = 5.$$

A number that is the square of a rational number is called a **perfect square**. For example,

64 is a perfect square because $64 = 8^2$.

$\frac{1}{9}$ is a perfect square because $\frac{1}{9} = \left(\frac{1}{3}\right)^2$.

The following rule can be used to find square roots of perfect squares.

Square Roots of Perfect Squares

$$\sqrt{a^2} = |a|$$

For example, $\sqrt{6^2} = 6$ and $\sqrt{(-6)^2} = |-6| = 6$.

2 Use the product rule to simplify square roots.

The Product Rule for Square Roots

A square root is **simplified** when its radicand has no factors other than 1 that are perfect squares. For example, $\sqrt{500}$ is not simplified because it can be expressed as $\sqrt{100 \cdot 5}$ and $\sqrt{100}$ is a perfect square. The **product rule for square roots** can be used to simplify $\sqrt{500}$.

The Product Rule for Square Roots

If a and b represent nonnegative real numbers, then

$$\sqrt{ab} = \sqrt{a}\sqrt{b} \text{ and } \sqrt{a}\sqrt{b} = \sqrt{ab}.$$

The square root of a product is the product of the square roots.

Example 1 shows how the product rule is used to remove from the square root any perfect squares that occur as factors.

EXAMPLE 1 Using the Product Rule to Simplify Square Roots

Simplify: a. $\sqrt{500}$ b. $\sqrt{6x} \cdot \sqrt{3x}$

Solution

a. $\sqrt{500} = \sqrt{100 \cdot 5}$ 100 is the largest perfect square factor of 500.

$= \sqrt{100}\sqrt{5}$ $\sqrt{ab} = \sqrt{a}\sqrt{b}$

$= 10\sqrt{5}$ $\sqrt{100} = 10$

b. $\sqrt{6x} \cdot \sqrt{3x} = \sqrt{6x \cdot 3x}$ $\sqrt{a}\sqrt{b} = \sqrt{ab}$

$= \sqrt{18x^2}$ Multiply.

$= \sqrt{9x^2 \cdot 2}$ 9 is the largest perfect square factor of 18.

$= \sqrt{9x^2}\sqrt{2}$ $\sqrt{ab} = \sqrt{a}\sqrt{b}$

$= \sqrt{9}\sqrt{x^2}\sqrt{2}$ Split $\sqrt{9x^2}$ into two square roots.

$= 3|x|\sqrt{2}$ $\sqrt{9} = 3$ (because $3^2 = 9$) and $\sqrt{x^2} = |x|$.

Check Point 1 Simplify: **a.** $\sqrt{3^2}$ **b.** $\sqrt{5x} \cdot \sqrt{10x}$

3 Use the quotient rule to simplify square roots.

The Quotient Rule for Square Roots

Another property for square roots involves division.

The Quotient Rule for Square Roots

If a and b represent nonnegative real numbers and $b \neq 0$, then

$$\frac{\sqrt{a}}{\sqrt{b}} = \sqrt{\frac{a}{b}} \text{ and } \sqrt{\frac{a}{b}} = \frac{\sqrt{a}}{\sqrt{b}}.$$

The square root of a quotient is the quotient of the square roots.

EXAMPLE 2 Using the Quotient Rule to Simplify Square Roots

Simplify: a. $\sqrt{\frac{100}{9}}$ b. $\frac{\sqrt{48x^3}}{\sqrt{6x}}$

Solution

a. $\sqrt{\frac{100}{9}} = \frac{\sqrt{100}}{\sqrt{9}} = \frac{10}{3}$

b. $\frac{\sqrt{48x^3}}{\sqrt{6x}} = \sqrt{\frac{48x^3}{6x}} = \sqrt{8x^2} = \sqrt{4x^2}\sqrt{2} = \sqrt{4}\sqrt{x^2}\sqrt{2} = 2|x|\sqrt{2}$

Check Point 2 Simplify: **a.** $\sqrt{\frac{25}{16}}$ **b.** $\frac{\sqrt{150x^3}}{\sqrt{2x}}$

4 Add and subtract square roots.

Adding and Subtracting Square Roots

Two or more square roots can be combined provided that they have the same radicand. Such radicals are called **like radicals**. For example,

$$7\sqrt{11} + 6\sqrt{11} = (7 + 6)\sqrt{11} = 13\sqrt{11}.$$

A Radical Idea: Time Is Relative

What does travel in space have to do with radicals? Imagine that in the future we will be able to travel at velocities approaching the speed of light (approximately 186,000 miles per second). According to Einstein's theory of relativity, time would pass more quickly on Earth than it would in the moving spaceship. The expression

$$R_f\sqrt{1 - \left(\frac{v}{c}\right)^2}$$

gives the aging rate of an astronaut relative to the aging rate of a friend on Earth, R_f. In the expression, v is the astronaut's speed and c is the speed of light. As the astronaut's speed approaches the speed of light, we can substitute c for v:

$$R_f\sqrt{1 - \left(\frac{v}{c}\right)^2} \quad \text{Let } v = c.$$

$$= R_f\sqrt{1 - 1^2}$$

$$= R_f\sqrt{0} = 0$$

Close to the speed of light, the astronaut's aging rate relative to a friend on earth is nearly 0. What does this mean? As we age here on Earth, the space traveler would barely get older. The space traveler would return to a futuristic world in which friends and loved ones would be long dead.

EXAMPLE 3 Adding and Subtracting Like Radicals

Add or subtract as indicated:

a. $7\sqrt{2} + 5\sqrt{2}$ b. $\sqrt{5x} - 7\sqrt{5x}$

Solution

a. $7\sqrt{2} + 5\sqrt{2} = (7 + 5)\sqrt{2}$ Apply the distributive property.

$= 12\sqrt{2}$ Simplify.

b. $\sqrt{5x} - 7\sqrt{5x} = 1\sqrt{5x} - 7\sqrt{5x}$ Write $\sqrt{5x}$ as $1\sqrt{5x}$.

$= (1 - 7)\sqrt{5x}$ Apply the distributive property.

$= -6\sqrt{5x}$ Simplify.

Check Point 3 Add or subtract as indicated:

a. $8\sqrt{13} + 9\sqrt{13}$ **b.** $\sqrt{17x} - 20\sqrt{17x}$

In some cases, radicals can be combined once they have been simplified. For example, to add $\sqrt{2}$ and $\sqrt{8}$, we can write $\sqrt{8}$ as $\sqrt{4 \cdot 2}$ because 4 is a perfect square factor of 8.

$$\sqrt{2} + \sqrt{8} = \sqrt{2} + \sqrt{4 \cdot 2} = 1\sqrt{2} + 2\sqrt{2} = (1 + 2)\sqrt{2} = 3\sqrt{2}$$

EXAMPLE 4 Combining Radicals That First Require Simplification

Add or subtract as indicated:

a. $7\sqrt{3} + \sqrt{12}$ b. $4\sqrt{50x} - 6\sqrt{32x}$

Solution

a. $7\sqrt{3} + \sqrt{12}$

$= 7\sqrt{3} + \sqrt{4 \cdot 3}$ Split 12 into two factors such that one is a perfect square.

$= 7\sqrt{3} + 2\sqrt{3}$ $\sqrt{4 \cdot 3} = \sqrt{4}\sqrt{3} = 2\sqrt{3}$

$= (7 + 2)\sqrt{3}$ Apply the distributive property. You will find that this step is usually done mentally.

$= 9\sqrt{3}$ Simplify.

b. $4\sqrt{50x} - 6\sqrt{32x}$

$= 4\sqrt{25 \cdot 2x} - 6\sqrt{16 \cdot 2x}$ — 25 is the largest perfect square factor of 50 and 16 is the largest perfect square factor of 32.

$= 4 \cdot 5\sqrt{2x} - 6 \cdot 4\sqrt{2x}$ — $\sqrt{25 \cdot 2} = \sqrt{25}\sqrt{2} = 5\sqrt{2}$ and $\sqrt{16 \cdot 2} = \sqrt{16}\sqrt{2} = 4\sqrt{2}$

$= 20\sqrt{2x} - 24\sqrt{2x}$ — Multiply.

$= (20 - 24)\sqrt{2x}$ — Apply the distributive property.

$= -4\sqrt{2x}$ — Simplify.

Check Point 4 Add or subtract as indicated:

a. $5\sqrt{27} + \sqrt{12}$ **b.** $6\sqrt{18x} - 4\sqrt{8x}$

5 Rationalize denominators.

Rationalizing Denominators

You can use a calculator to compare the approximate values for $\frac{1}{\sqrt{3}}$ and $\frac{\sqrt{3}}{3}$. The two approximations are the same. This is not a coincidence:

$$\frac{1}{\sqrt{3}} = \frac{1}{\sqrt{3}} \cdot \boxed{\frac{\sqrt{3}}{\sqrt{3}}} = \frac{\sqrt{3}}{\sqrt{9}} = \frac{\sqrt{3}}{3}.$$

Any number divided by itself is 1. Multiplication by 1 does not change the value of $\frac{1}{\sqrt{3}}$.

This process involves rewriting a radical to remove the square root from the denominator without changing the value of the radical. The process is called **rationalizing the denominator**. If the denominator contains the square root of a natural number that is not a perfect square, multiply the numerator and denominator by the smallest number that produces the square root of a perfect square in the denominator.

EXAMPLE 5 Rationalizing Denominators

Rationalize the denominator: a. $\frac{15}{\sqrt{6}}$ b. $\frac{12}{\sqrt{8}}$

Solution

a. If we multiply numerator and denominator by $\sqrt{6}$, the denominator becomes $\sqrt{6} \cdot \sqrt{6} = \sqrt{36} = 6$. Therefore, we multiply by 1, choosing $\frac{\sqrt{6}}{\sqrt{6}}$ for 1:

$$\frac{15}{\sqrt{6}} = \frac{15}{\sqrt{6}} \cdot \frac{\sqrt{6}}{\sqrt{6}} = \frac{15\sqrt{6}}{\sqrt{36}} = \frac{15\sqrt{6}}{6} = \frac{5\sqrt{6}}{2}$$

Multiply by 1.

Simplify: $\frac{15}{6} = \frac{15 \div 3}{6 \div 3} = \frac{5}{2}$.

b. The *smallest* number that will produce a perfect square in the denominator of $\frac{12}{\sqrt{8}}$ is $\sqrt{2}$, because $\sqrt{8} \cdot \sqrt{2} = \sqrt{16} = 4$. We multiply by 1, choosing $\frac{\sqrt{2}}{\sqrt{2}}$ for 1.

$$\frac{12}{\sqrt{8}} = \frac{12}{\sqrt{8}} \cdot \frac{\sqrt{2}}{\sqrt{2}} = \frac{12\sqrt{2}}{\sqrt{16}} = \frac{12\sqrt{2}}{4} = 3\sqrt{2}$$

Check Point 5 Rationalize the denominator: **a.** $\frac{5}{\sqrt{3}}$ **b.** $\frac{6}{\sqrt{12}}$

How can we rationalize a denominator if the denominator contains two terms? In general,

$$(\sqrt{a} + \sqrt{b})(\sqrt{a} - \sqrt{b}) = (\sqrt{a})^2 - (\sqrt{b})^2 = a - b.$$

Notice that the product does not contain a radical. Here are some specific examples.

The Denominator Contains:	Multiply by:	The New Denominator Contains:
$7 + \sqrt{5}$	$7 - \sqrt{5}$	$7^2 - (\sqrt{5})^2 = 49 - 5 = 44$
$\sqrt{3} - 6$	$\sqrt{3} + 6$	$(\sqrt{3})^2 - 6^2 = 3 - 36 = -33$
$\sqrt{7} + \sqrt{3}$	$\sqrt{7} - \sqrt{3}$	$(\sqrt{7})^2 - (\sqrt{3})^2 = 7 - 3 = 4$

EXAMPLE 6 Rationalizing a Denominator Containing Two Terms

Rationalize the denominator: $\frac{7}{5 + \sqrt{3}}$.

Solution If we multiply the numerator and denominator by $5 - \sqrt{3}$, the denominator will not contain a radical. Therefore, we multiply by 1, choosing $\frac{5 - \sqrt{3}}{5 - \sqrt{3}}$ for 1:

$$\frac{7}{5 + \sqrt{3}} = \frac{7}{5 + \sqrt{3}} \cdot \frac{5 - \sqrt{3}}{5 - \sqrt{3}} = \frac{7(5 - \sqrt{3})}{5^2 - (\sqrt{3})^2} = \frac{7(5 - \sqrt{3})}{25 - 3}$$

Multiply by 1.

$$= \frac{7(5 - \sqrt{3})}{22} \text{ or } \frac{35 - 7\sqrt{3}}{22}.$$

In either form of the answer, there is no radical in the denominator.

Check Point 6 Rationalize the denominator: $\frac{8}{4 + \sqrt{5}}$.

6 Evaluate and perform operations with higher roots.

Other Kinds of Roots

We define the **principal *n*th root** of a real number a, symbolized by $\sqrt[n]{a}$, as follows:

Definition of the Principal *n*th Root of a Real Number

$$\sqrt[n]{a} = b \text{ means that } b^n = a.$$

If n, the **index**, is even, then a is nonnegative $(a \geq 0)$ and b is also nonnegative $(b \geq 0)$. If n is odd, a and b can be any real numbers.

For example,

$$\sqrt[3]{64} = 4 \text{ because } 4^3 = 64 \quad \text{and} \quad \sqrt[5]{-32} = -2 \text{ because } (-2)^5 = -32.$$

The same vocabulary that we learned for square roots applies to nth roots. The symbol $\sqrt[n]{a}$ is called a **radical** and a is called the **radicand**.

A number that is the nth power of a rational number is called a **perfect *n*th power**. For example, 8 is a perfect third power, or perfect cube, because $8 = 2^3$. In general, one of the following rules can be used to find nth roots of perfect nth powers.

Finding *n*th Roots of Perfect *n*th Powers

If n is odd, $\sqrt[n]{a^n} = a$.

If n is even, $\sqrt[n]{a^n} = |a|$.

For example,

$$\sqrt[3]{(-2)^3} = -2 \quad \text{and} \quad \sqrt[4]{(-2)^4} = |-2| = 2.$$

Absolute value is not needed with odd roots, but is necessary with even roots.

The Product and Quotient Rules for Other Roots

The product and quotient rules apply to cube roots, fourth roots, and all higher roots.

The Product and Quotient Rules for *n*th Roots

For all real numbers, where the indicated roots represent real numbers,

$$\sqrt[n]{a} \cdot \sqrt[n]{b} = \sqrt[n]{ab} \quad \text{and} \quad \frac{\sqrt[n]{a}}{\sqrt[n]{b}} = \sqrt[n]{\frac{a}{b}}, \quad b \neq 0.$$

EXAMPLE 7 Simplifying, Multiplying, and Dividing Higher Roots

Simplify: a. $\sqrt[3]{24}$ b. $\sqrt[4]{8} \cdot \sqrt[4]{4}$ c. $\sqrt[4]{\frac{81}{16}}$

Solution

a. $\sqrt[3]{24} = \sqrt[3]{8 \cdot 3}$ Find the largest *perfect cube* that is a factor of 24. $\sqrt[3]{8} = 2$, so 8 is a perfect cube and is the largest perfect cube factor of 24.

$= \sqrt[3]{8} \cdot \sqrt[3]{3}$ $\sqrt[n]{ab} = \sqrt[n]{a}\sqrt[n]{b}$

$= 2\sqrt[3]{3}$

b. $\sqrt[4]{8} \cdot \sqrt[4]{4} = \sqrt[4]{8 \cdot 4}$ $\sqrt[n]{a} \cdot \sqrt[n]{b} = \sqrt[n]{ab}$

$= \sqrt[4]{32}$ Find the largest *perfect fourth power* that is a factor of 32.

$= \sqrt[4]{16 \cdot 2}$ $\sqrt[4]{16} = 2$, so 16 is a perfect fourth power and is the largest perfect fourth power that is a factor of 32.

$= \sqrt[4]{16} \cdot \sqrt[4]{2}$ $\sqrt[n]{ab} = \sqrt[n]{a} \cdot \sqrt[n]{b}$

$= 2\sqrt[4]{2}$

c. $\sqrt[4]{\frac{81}{16}} = \frac{\sqrt[4]{81}}{\sqrt[4]{16}}$ $\sqrt[n]{\frac{a}{b}} = \frac{\sqrt[n]{a}}{\sqrt[n]{b}}$

$= \frac{3}{2}$ $\sqrt[4]{81} = 3$ because $3^4 = 81$ and $\sqrt[4]{16} = 2$ because $2^4 = 16$.

Check Point 7 Simplify: **a.** $\sqrt[3]{40}$ **b.** $\sqrt[5]{8} \cdot \sqrt[5]{8}$ **c.** $\sqrt[3]{\frac{125}{27}}$

We have seen that adding and subtracting square roots often involves simplifying terms. The same idea applies to adding and subtracting *n*th roots.

EXAMPLE 8 Combining Cube Roots

Subtract: $5\sqrt[3]{16} - 11\sqrt[3]{2}$.

Solution

$5\sqrt[3]{16} - 11\sqrt[3]{2}$

$= 5\sqrt[3]{8 \cdot 2} - 11\sqrt[3]{2}$ Because $\sqrt[3]{8} = 2$, 8 is the largest perfect cube that is a factor of 16.

$= 5 \cdot 2\sqrt[3]{2} - 11\sqrt[3]{2}$ $\sqrt[3]{8 \cdot 2} = \sqrt[3]{8}\sqrt[3]{2} = 2\sqrt[3]{2}$

$= 10\sqrt[3]{2} - 11\sqrt[3]{2}$ Multiply.

$= (10 - 11)\sqrt[3]{2}$ Apply the distributive property.

$= -1\sqrt[3]{2}$ or $-\sqrt[3]{2}$ Simplify.

Check Point 8 Subtract: $3\sqrt[3]{81} - 4\sqrt[3]{3}$.

7 Understand and use rational exponents.

Rational Exponents

Animals in the wild have regions to which they confine their movement, called their territorial area. Territorial area, in square miles, is related to an animal's body weight. If an animal weighs W pounds, its territorial area is

$$W^{141/100}$$

square miles.

W to the *what* power?! How can we interpret the information given by this algebraic expression?

In the last part of this section, we turn our attention to rational exponents such as $\frac{141}{100}$ and their relationship to roots of real numbers.

Definition of Rational Exponents

If $\sqrt[n]{a}$ represents a real number and $n \geq 2$ is an integer, then

$$a^{1/n} = \sqrt[n]{a}.$$

Furthermore,

$$a^{-1/n} = \frac{1}{a^{1/n}} = \frac{1}{\sqrt[n]{a}}, \; a \neq 0.$$

EXAMPLE 9 Using the Definition of $a^{1/n}$

Simplify: a. $64^{1/2}$ b. $8^{1/3}$ c. $64^{-1/3}$

Solution

a. $64^{1/2} = \sqrt{64} = 8$

b. $8^{1/3} = \sqrt[3]{8} = 2$

c. $64^{-1/3} = \dfrac{1}{64^{1/3}} = \dfrac{1}{\sqrt[3]{64}} = \dfrac{1}{4}$

Check Point 9 Simplify: **a.** $81^{1/2}$ **b.** $27^{1/3}$ **c.** $32^{-1/5}$

Note that every rational exponent in Example 9 has a numerator of 1 or −1. We now define rational exponents with any integer in the numerator.

Definition of Rational Exponents

If $\sqrt[n]{a}$ represents a real number, $\dfrac{m}{n}$ is a rational number reduced to lowest terms, and $n \geq 2$ is an integer, then

$$a^{m/n} = (\sqrt[n]{a})^m = \sqrt[n]{a^m}.$$

The exponent m/n consists of two parts: the denominator n is the root and the numerator m is the exponent. Furthermore,

$$a^{-m/n} = \frac{1}{a^{m/n}}.$$

EXAMPLE 10 Using the Definition of $a^{m/n}$

Simplify: a. $27^{2/3}$ b. $9^{3/2}$ c. $16^{-3/4}$

Solution

a. $27^{2/3} = (\sqrt[3]{27})^2 = 3^2 = 9$

The denominator of $\frac{2}{3}$ is the root and the numerator is the exponent.

b. $9^{3/2} = (\sqrt{9})^3 = 3^3 = 27$

c. $16^{-3/4} = \dfrac{1}{16^{3/4}} = \dfrac{1}{(\sqrt[4]{16})^3} = \dfrac{1}{2^3} = \dfrac{1}{8}$

Check Point 10 Simplify: **a.** $4^{3/2}$ **b.** $32^{-2/5}$

Properties of exponents can be applied to expressions containing rational exponents.

EXAMPLE 11 Simplifying Expressions with Rational Exponents

Simplify using properties of exponents:

a. $(5x^{1/2})(7x^{3/4})$ b. $\dfrac{32x^{5/3}}{16x^{3/4}}$

Solution

a. $(5x^{1/2})(7x^{3/4}) = 5 \cdot 7x^{1/2} \cdot x^{3/4}$ *Group factors with the same base.*

$= 35x^{(1/2)+(3/4)}$ *When multiplying expressions with the same base, add the exponents.*

$= 35x^{5/4}$ $\frac{1}{2} + \frac{3}{4} = \frac{2}{4} + \frac{3}{4} = \frac{5}{4}$

b. $\dfrac{32x^{5/3}}{16x^{3/4}} = \left(\dfrac{32}{16}\right)\left(\dfrac{x^{5/3}}{x^{3/4}}\right)$ *Group factors with the same base.*

$= 2x^{(5/3)-(3/4)}$ *When dividing expressions with the same base, subtract the exponents.*

$= 2x^{11/12}$ $\frac{5}{3} - \frac{3}{4} = \frac{20}{12} - \frac{9}{12} = \frac{11}{12}$

Check Point 11 Simplify: **a.** $(2x^{4/3})(5x^{8/3})$ **b.** $\dfrac{20x^4}{5x^{3/2}}$

Rational exponents are sometimes useful for simplifying radicals by reducing their index.

EXAMPLE 12 **Reducing the Index of a Radical**

Simplify: $\sqrt[9]{x^3}$.

Solution $\sqrt[9]{x^3} = x^{3/9} = x^{1/3} = \sqrt[3]{x}$

Check Point 12 Simplify: $\sqrt[6]{x^3}$.

EXERCISE SET P.3

Practice Exercises

Evaluate each expression in Exercises 1–7 or indicate that the root is not a real number.

1. $\sqrt{36}$
2. $\sqrt{25}$
3. $\sqrt{-36}$
4. $\sqrt{-25}$
5. $\sqrt{(-13)^2}$
6. $\sqrt{(-17)^2}$

Use the product rule to simplify the expressions in Exercises 7–16.

7. $\sqrt{50}$
8. $\sqrt{27}$
9. $\sqrt{45x^2}$
10. $\sqrt{125x^2}$
11. $\sqrt{2x} \cdot \sqrt{6x}$
12. $\sqrt{10x} \cdot \sqrt{8x}$
13. $\sqrt{x^3}$
14. $\sqrt{y^3}$
15. $\sqrt{2x^2} \cdot \sqrt{6x}$
16. $\sqrt{6x} \cdot \sqrt{3x^2}$

Use the quotient rule to simplify the expressions in Exercises 17–24.

17. $\sqrt{\frac{1}{81}}$
18. $\sqrt{\frac{1}{49}}$
19. $\sqrt{\frac{49}{16}}$
20. $\sqrt{\frac{121}{9}}$
21. $\frac{\sqrt{48x^3}}{\sqrt{3x}}$
22. $\frac{\sqrt{72x^3}}{\sqrt{8x}}$
23. $\frac{\sqrt{150x^4}}{\sqrt{3x}}$
24. $\frac{\sqrt{24x^4}}{\sqrt{3x}}$

In Exercises 25–34, add or subtract terms whenever possible.

25. $7\sqrt{3} + 6\sqrt{3}$
26. $8\sqrt{5} + 11\sqrt{5}$
27. $6\sqrt{17x} - 8\sqrt{17x}$
28. $4\sqrt{13x} - 6\sqrt{13x}$
29. $\sqrt{8} + 3\sqrt{2}$
30. $\sqrt{20} + 6\sqrt{5}$
31. $\sqrt{50x} - \sqrt{8x}$
32. $\sqrt{63x} - \sqrt{28x}$
33. $3\sqrt{18} + 5\sqrt{50}$
34. $4\sqrt{12} - 2\sqrt{75}$

In Exercises 35–44, rationalize the denominator.

35. $\frac{1}{\sqrt{7}}$
36. $\frac{2}{\sqrt{10}}$
37. $\frac{\sqrt{2}}{\sqrt{5}}$
38. $\frac{\sqrt{7}}{\sqrt{3}}$
39. $\frac{13}{3 + \sqrt{11}}$
40. $\frac{3}{3 + \sqrt{7}}$
41. $\frac{7}{\sqrt{5} - 2}$
42. $\frac{5}{\sqrt{3} - 1}$
43. $\frac{6}{\sqrt{5} + \sqrt{3}}$
44. $\frac{11}{\sqrt{7} - \sqrt{3}}$

Evaluate each expression in Exercises 45–54 or indicate that the root is not a real number.

45. $\sqrt[3]{125}$
46. $\sqrt[3]{8}$
47. $\sqrt[3]{-8}$
48. $\sqrt[3]{-125}$
49. $\sqrt[4]{-16}$
50. $\sqrt[4]{-81}$
51. $\sqrt[4]{(-3)^4}$
52. $\sqrt[4]{(-2)^4}$
53. $\sqrt[5]{(-3)^5}$
54. $\sqrt[5]{(-2)^5}$

Simplify the radical expressions in Exercises 55–62.

55. $\sqrt[3]{32}$
56. $\sqrt[3]{150}$
57. $\sqrt[3]{x^4}$
58. $\sqrt[3]{x^5}$
59. $\sqrt[3]{9} \cdot \sqrt[3]{6}$
60. $\sqrt[3]{12} \cdot \sqrt[3]{4}$
61. $\frac{\sqrt[5]{64x^6}}{\sqrt[5]{2x}}$
62. $\frac{\sqrt[4]{162x^5}}{\sqrt[4]{2x}}$

In Exercises 63–70, evaluate each expression without using a calculator.

63. $36^{1/2}$
64. $121^{1/2}$
65. $8^{1/3}$
66. $27^{1/3}$
67. $125^{2/3}$
68. $8^{2/3}$
69. $32^{-4/5}$
70. $16^{-5/2}$

In Exercises 71–78, simplify using properties of exponents.

71. $(7x^{1/3})(2x^{1/4})$
72. $(3x^{2/3})(4x^{3/4})$
73. $\frac{20x^{1/2}}{5x^{1/4}}$
74. $\frac{72x^{3/4}}{9x^{1/3}}$
75. $(x^{2/3})^3$
76. $(x^{4/5})^5$
77. $(25x^4y^6)^{1/2}$
78. $(125x^9y^6)^{1/3}$

In Exercises 79–84, simplify by reducing the index of the radical.

79. $\sqrt[4]{5^2}$

80. $\sqrt[4]{7^2}$

81. $\sqrt[3]{x^6}$

82. $\sqrt[4]{x^{12}}$

83. $\sqrt[6]{x^4}$

84. $\sqrt[9]{x^6}$

Application Exercises

85. The algebraic expression $2\sqrt{5L}$ is used to estimate the speed of a car prior to an accident, in miles per hour, based on the length of its skid marks L, in feet. Find the speed of a car that left skid marks 40 feet long, and write the answer in simplified radical form.

86. The time, in seconds, that it takes an object to fall a distance d, in feet, is given by the algebraic expression $\sqrt{\frac{d}{16}}$. Find how long it will take a ball dropped from the top of a building 320 feet tall to hit the ground. Write the answer in simplified radical form.

87. The early Greeks believed that the most pleasing of all rectangles were golden rectangles whose ratio of width to height is

$$\frac{w}{h} = \frac{2}{\sqrt{5} - 1}.$$

Rationalize the denominator for this ratio and then use a calculator to approximate the answer correct to the nearest hundredth.

88. The amount of evaporation, in inches per day, of a large body of water can be described by the algebraic expression

$$\frac{w}{20\sqrt{a}}$$

where

a = surface area of the water in square miles

w = average wind speed of the air over the water, in miles per hour.

Determine the evaporation on a lake whose surface area is 9 square miles on a day when the wind speed over the water is 10 miles per hour.

89. In the Peanuts cartoon shown above, Woodstock appears to be working steps mentally. Fill in the missing steps that show how to go from $\frac{7\sqrt{2 \cdot 2 \cdot 3}}{6}$ to $\frac{7}{3}\sqrt{3}$.

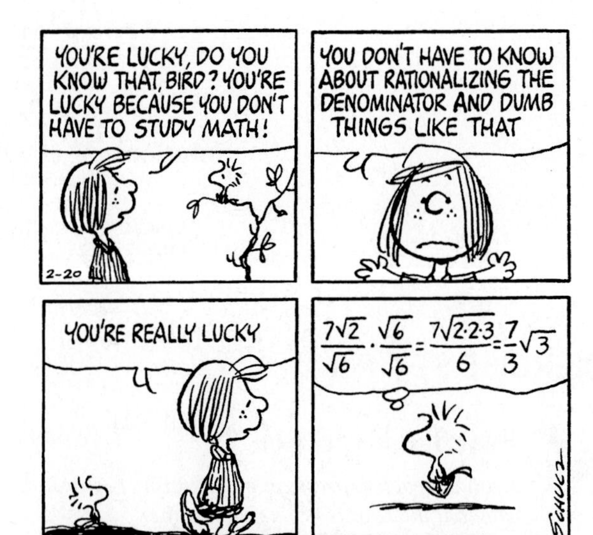

PEANUTS reprinted by permission of United Feature Syndicate, Inc.

90. The algebraic expression $63.25x^{1/4}$ describes the average sale price, in thousands of dollars, of single-family homes in the U.S. Midwest x years after 1981. Evaluate the algebraic expression when $x = 16$. Describe what the answer means in practical terms.

91. The algebraic expression $0.07d^{3/2}$ describes the duration of a storm, in hours, whose diameter is d miles. Evaluate the algebraic expression when $d = 9$. Describe what the answer means in practical terms.

Writing in Mathematics

92. Explain how to simplify $\sqrt{10} \cdot \sqrt{5}$.

93. Explain how to add $\sqrt{3} + \sqrt{12}$.

94. Describe what it means to rationalize a denominator. Use both $\frac{1}{\sqrt{5}}$ and $\frac{1}{5 + \sqrt{5}}$ in your explanation.

95. What difference is there in simplifying $\sqrt[3]{(-5)^3}$ and $\sqrt[4]{(-5)^4}$?

96. What does $a^{m/n}$ mean?

97. Describe the kinds of numbers that have rational fifth roots.

98. Why must a and b represent nonnegative numbers when we write $\sqrt{a} \cdot \sqrt{b} = \sqrt{ab}$? Is it necessary to use this restriction in the case of $\sqrt[3]{a} \cdot \sqrt[3]{b} = \sqrt[3]{ab}$? Explain.

Technology Exercises

99. The algebraic expression $60.19x^{0.025}$ describes the expected lifespan of African American men x years after 1969. Use a calculator to find the expected lifespan from 1970 through 2000. During what year did the expected lifespan of African American men first exceed 65 years?

100. The territorial area of an animal in the wild is defined to be the area of the region to which the animal confines its movements. The algebraic expression $W^{1.41}$ describes the territorial area, in square miles, of an animal that weighs W pounds. Use a calculator to find the territorial area of animals weighing 25, 50, 150, 200, 250, and 300 pounds. What do the values that you obtain with your calculator indicate about the relationship between body weight and territorial area?

Critical Thinking Exercises

101. Which one of the following is true?

a. Neither $(-8)^{1/2}$ nor $(-8)^{1/3}$ represent real numbers.

b. $\sqrt{x^2 + y^2} = x + y$

c. $8^{-1/3} = -2$

d. $2^{1/2} \cdot 2^{1/2} = 2$

In Exercises 102–103, fill in each box to make the statement true.

102. $(5 + \sqrt{\square})(5 - \sqrt{\square}) = 22$

103. $\sqrt{\square x^{\square}} = 5x^7$

104. Find exact value of $\sqrt{13 + \sqrt{2} + \frac{7}{3 + \sqrt{2}}}$ without the use of a calculator.

105. Place the correct symbol, $>$ or $<$, in the box between each of the given numbers. *Do not use a calculator.* Then check your result with a calculator.

a. $3^{1/2} \square 3^{1/3}$ **b.** $\sqrt{7} + \sqrt{18} \square \sqrt{7 + 18}$

SECTION P.4 Polynomials

Objectives

1. Understand the vocabulary of polynomials.
2. Add and subtract polynomials.
3. Multiply polynomials.
4. Use FOIL in polynomial multiplication.
5. Use special products in polynomial multiplication.
6. Perform operations with polynomials in several variables.

Runny nose? Sneezing? You are probably familiar with the unpleasant onset of a cold. We "catch cold" when the cold virus enters our bodies, where it multiplies. Fortunately, at a certain point the virus begins to die. The algebraic expression $-0.75x^4 + 3x^3 + 5$ describes the billions of viral particles in our bodies after x days of invasion. The expression enables mathematicians to determine the day on which there is a maximum number of viral particles and, consequently, the day we feel sickest.

The algebraic expression $-0.75x^4 + 3x^3 + 5$ is an example of a polynomial. A **polynomial** is a single term or the sum of two or more terms containing variables with whole number exponents. This particular polynomial contains three terms. Equations containing polynomials are used in such diverse areas as science, business, medicine, psychology, and sociology. In this section, we review basic ideas about polynomials and their operations.

1 Understand the vocabulary of polynomials.

The Vocabulary of Polynomials

Consider the polynomial

$$7x^3 - 9x^2 + 13x - 6.$$

We can express this polynomial as

$$7x^3 + (-9x^2) + 13x + (-6).$$

The polynomial contains four terms. It is customary to write the terms in the order of descending powers of the variables. This is the **standard form** of a polynomial.

We begin this section by limiting our discussion to polynomials containing only one variable. Each term of a polynomial in x is of the form ax^n. The **degree** of ax^n is n. For example, the degree of the term $7x^3$ is 3.

Study Tip

We can express 0 in many ways, including $0x$, $0x^2$, and $0x^3$. It is impossible to assign a single exponent on the variable. This is why 0 has no defined degree.

The Degree of ax^n

If $a \neq 0$, the degree of ax^n is n. The degree of a nonzero constant is 0. The constant 0 has no defined degree.

Here is an example of a polynomial and the degree of each of its four terms.

$$6x^4 - 3x^3 + 2x - 5$$

degree 4 | degree 3 | degree 1 | degree of nonzero constant: 0

Notice that the exponent on x for the term $2x$ is understood to be 1: $2x^1$. For this reason, the degree of $2x$ is 1. You can think of -5 as $-5x^0$; thus, its degree is 0.

A polynomial with exactly one term is called a **monomial**. A **binomial** is a polynomial that has two terms, each with a different exponent. A **trinomial** is a polynomial with three terms, each with a different exponent. Polynomials with four or more terms have no special names.

The **degree of a polynomial** is the highest degree of all the terms of the polynomial. For example, $4x^2 + 3x$ is a binomial of degree 2 because the degree of the first term is 2, and the degree of the other term is less than 2. Also, $7x^5 - 2x^2 + 4$ is a trinomial of degree 5 because the degree of the first term is 5, and the degrees of the other terms are less than 5.

Up to now, we have used x to represent the variable in a polynomial. However, any letter can be used. For example,

$7x^5 - 3x^3 + 8$ is a polynomial (in x) of degree 5.
$6y^3 + 4y^2 - y + 3$ is a polynomial (in y) of degree 3.
$z^7 + \sqrt{2}$ is a polynomial (in z) of degree 7.

Not every algebraic expression is a polynomial. Algebraic expressions whose variables do not contain whole number exponents such as

$$3x^{-2} + 7 \quad \text{and} \quad 5x^{3/2} + 9x^{1/2} + 2$$

are not polynomials. Furthermore, a quotient of polynomials such as

$$\frac{x^2 + 2x + 5}{x^3 - 7x^2 + 9x - 3}$$

is not a polynomial because the form of a polynomial involves only addition and subtraction of terms, not division.

We can tie together the threads of our discussion with the formal definition of a polynomial in one variable. In this definition, the coefficients of the terms are represented by a_n (read "a sub n"), a_{n-1} (read "a sub n minus 1"), a_{n-2}, and so on. The small letters to the lower right of each a are called **subscripts** and are *not exponents*. Subscripts are used to distinguish one constant from another when a large and undetermined number of such constants are needed.

Definition of a Polynomial in *x*

A **polynomial in *x*** is an algebraic expression of the form

$$a_n x^n + a_{n-1}x^{n-1} + a_{n-2}x^{n-2} + \cdots + a_1 x + a_0,$$

where $a_n, a_{n-1}, a_{n-2}, \ldots, a_1$ and a_0 are real numbers, $a_n \neq 0$, and n is a nonnegative integer. The polynomial is of **degree *n***, a_n is the **leading coefficient**, and a_0 is the **constant term**.

2 Add and subtract polynomials.

Adding and Subtracting Polynomials

Polynomials are added and subtracted by combining like terms. For example, we can combine the monomials $-9x^3$ and $13x^3$ using addition as follows:

$$-9x^3 + 13x^3 = (-9 + 13)x^3 = 4x^3.$$

EXAMPLE 1 Adding and Subtracting Polynomials

Perform the indicated operations and simplify:

a. $(-9x^3 + 7x^2 - 5x + 3) + (13x^3 + 2x^2 - 8x - 6)$
b. $(7x^3 - 8x^2 + 9x - 6) - (2x^3 - 6x^2 - 3x + 9)$

Solution

a. $(-9x^3 + 7x^2 - 5x + 3) + (13x^3 + 2x^2 - 8x - 6)$

$= (-9x^3 + 13x^3) + (7x^2 + 2x^2) + (-5x - 8x) + (3 - 6)$ Group like terms.

$= 4x^3 + 9x^2 + (-13x) + (-3)$ Combine like terms.

$= 4x^3 + 9x^2 - 13x - 3$

b. $(7x^3 - 8x^2 + 9x - 6) - (2x^3 - 6x^2 - 3x + 9)$

$= (7x^3 - 8x^2 + 9x - 6) + (-2x^3 + 6x^2 + 3x - 9)$ Rewrite subtraction as addition of the additive inverse. Be sure to change the sign of each term inside parentheses preceded by the negative sign.

$= (7x^3 - 2x^3) + (-8x^2 + 6x^2) + (9x + 3x) + (-6 - 9)$ Group like terms.

$= 5x^3 + (-2x^2) + 12x + (-15)$ Combine like terms.

$= 5x^3 - 2x^2 + 12x - 15$

Study Tip

You can also arrange like terms in columns and combine vertically:

$$\begin{array}{r} 7x^3 - 8x^2 + 9x - 6 \\ -2x^3 + 6x^2 + 3x - 9 \\ \hline 5x^3 - 2x^2 + 12x - 15 \end{array}$$

The like terms can be combined by adding their coefficients.

Check Point 1

Perform the indicated operations and simplify:

a. $(-17x^3 + 4x^2 - 11x - 5) + (16x^3 - 3x^2 + 3x - 15)$

b. $(13x^3 - 9x^2 - 7x + 1) - (-7x^3 + 2x^2 - 5x + 9)$

3 Multiply polynomials.

Multiplying Polynomials

The product of two monomials is obtained by using properties of exponents. For example,

$$(-8x^6)(5x^3) = -8 \cdot 5x^{6+3} = -40x^9$$

Multiply coefficients and add exponents.

Furthermore, we can use the distributive property to multiply a monomial and a polynomial that is not a monomial. For example,

$$3x^4(2x^3 - 7x + 3) = 3x^4 \cdot 2x^3 - 3x^4 \cdot 7x + 3x^4 \cdot 3 = 6x^7 - 21x^5 + 9x^4.$$

monomial trinomial

How do we multiply two polynomials if neither is a monomial? For example, consider

$$(2x + 3)(x^2 + 4x + 5).$$

binomial trinomial

One way to perform this multiplication is to distribute $2x$ throughout the trinomial

$$2x(x^2 + 4x + 5)$$

and 3 throughout the trinomial

$$3(x^2 + 4x + 5).$$

Then combine the like terms that result. In general, the product of two polynomials is the polynomial obtained by multiplying each term of one polynomial by each term of the other polynomial and then combining like terms.

EXAMPLE 2 Multiplying a Binomial and a Trinomial

Multiply: $(2x + 3)(x^2 + 4x + 5)$.

Solution

$(2x + 3)(x^2 + 4x + 5)$

$= 2x(x^2 + 4x + 5) + 3(x^2 + 4x + 5)$ — Use the distributive property to multiply the trinomial by each term of the binomial.

$= 2x \cdot x^2 + 2x \cdot 4x + 2x \cdot 5 + 3x^2 + 3 \cdot 4x + 3 \cdot 5$ — Use the distributive property.

$= 2x^3 + 8x^2 + 10x + 3x^2 + 12x + 15$ — Multiply the monomials.

$= 2x^3 + 11x^2 + 22x + 15$ — Combine like terms.

Another method for solving Example 2 is to use a vertical format similar to that used for multiplying whole numbers.

$$\begin{array}{rrrr} & x^2 + & 4x + & 5 \\ & & 2x + & 3 \\ \hline & 3x^2 + & 12x + & 15 \\ 2x^3 + & 8x^2 + & 10x & \\ \hline 2x^3 + & 11x^2 + & 22x + & 15 \end{array}$$

Write like terms in the same column.

$3(x^2 + 4x + 5)$

$2x(x^2 + 4x + 5)$

Combine like terms.

Check Point 2 Multiply: $(5x - 2)(3x^2 - 5x + 4)$.

4 Use FOIL in polynomial multiplication.

The Product of Two Binomials: FOIL

Frequently we need to find the product of two binomials. We can use a method called FOIL, which is based on the distributive property, to do so. For example, we can find the product of the binomials $3x + 2$ and $4x + 5$ as follows:

$$(3x + 2)(4x + 5) = 3x(4x + 5) + 2(4x + 5)$$

First, distribute $3x$ over $4x + 5$. Then distribute 2.

$$= 3x(4x) + 3x(5) + 2(4x) + 2(5)$$
$$= 12x^2 + 15x + 8x + 10.$$

Two binomials can be quickly multiplied by using the FOIL method, in which F represents the product of the **first** terms in each binomial, O represents the product of the **outside** terms, I represents the product of the two **inside** terms, and L represents the product of the **last**, or second, terms in each binomial.

first, last, inside, outside

F O I L

$$(3x + 2)(4x + 5) = 12x^2 + 15x + 8x + 10$$
$$= 12x^2 + 23x + 10$$

Combine like terms.

In general, here's how to use the FOIL method to find the product of $ax + b$ and $cx + d$:

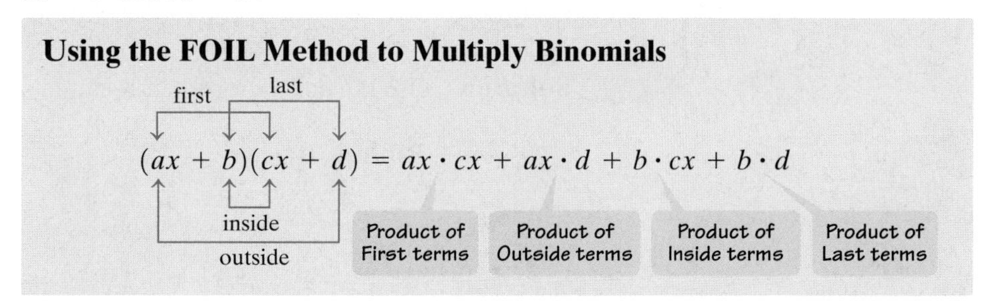
Using the FOIL Method to Multiply Binomials

$$(ax + b)(cx + d) = ax \cdot cx + ax \cdot d + b \cdot cx + b \cdot d$$

EXAMPLE 3 Using the FOIL Method

Multiply: $(3x + 4)(5x - 3)$.

Solution

first, last, inside, outside

F O I L

$$(3x + 4)(5x - 3) = 3x \cdot 5x + 3x(-3) + 4 \cdot 5x + 4(-3)$$
$$= 15x^2 - 9x + 20x - 12$$
$$= 15x^2 + 11x - 12$$

Combine like terms.

Check Point 3 Multiply: $(7x - 5)(4x - 3)$.

5 Use special products in polynomial multiplication.

Multiplying the Sum and Difference of Two Terms

We can use the FOIL method to multiply $A + B$ and $A - B$ as follows:

$$(A + B)(A - B) = \overset{F}{A^2} - \overset{O}{AB} + \overset{I}{AB} - \overset{L}{B^2} = A^2 - B^2.$$

Notice that the outside and inside products have a sum of 0 and the terms cancel. The FOIL multiplication provides us with a quick rule for multiplying the sum and difference of two terms, referred to as a special-product formula.

The Product of the Sum and Difference of Two Terms

$$(A + B)(A - B) = A^2 - B^2$$

The product of the sum and the difference of the same two terms. is The square of the first term minus the square of the second term.

EXAMPLE 4 Finding the Product of the Sum and Difference of Two Terms

Find each product by using the preceding rule.

a. $(4y + 3)(4y - 3)$ b. $(5a^4 + 6)(5a^4 - 6)$

Solution Use the special-product formula shown.

$(A + B)(A - B)$	$=$	A^2	$-$	B^2		
		First term squared	$-$	Second term squared	$=$	Product
a. $(4y + 3)(4y - 3)$	$=$	$(4y)^2$	$-$	3^2	$=$	$16y^2 - 9$
b. $(5a^4 + 6)(5a^4 - 6)$	$=$	$(5a^4)^2$	$-$	6^2	$=$	$25a^8 - 36$

Check Point 4 Find each product:

a. $(7x + 8)(7x - 8)$ **b.** $(2y^3 - 5)(2y^3 + 5)$

The Square of a Binomial

Let us find $(A + B)^2$, the square of a binomial sum. To do so, we begin with the FOIL method and look for a general rule.

$$(A + B)^2 = (A + B)(A + B) = \underset{F}{A \cdot A} + \underset{O}{A \cdot B} + \underset{I}{A \cdot B} + \underset{L}{B \cdot B}$$
$$= A^2 + 2AB + B^2$$

This result implies the following rule, which is another example of a special-product formula.

Study Tip

Caution! The square of a sum is *not* the sum of the squares.

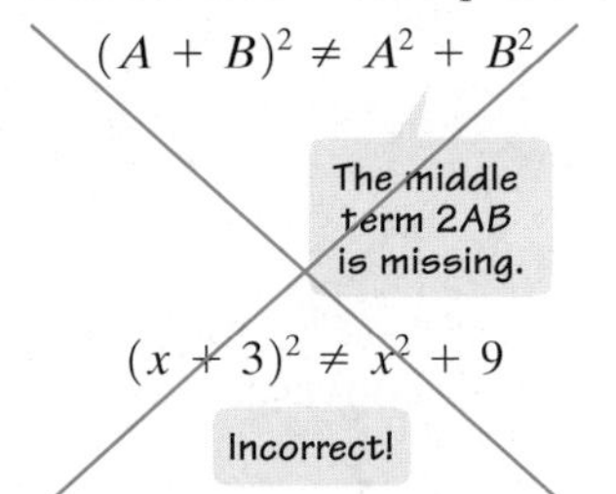

Show that $(x + 3)^2$ and $x^2 + 9$ are not equal by substituting 5 for x in each expression and simplifying.

The Square of a Binomial Sum

$$(A + B)^2 = A^2 + 2AB + B^2$$

The square of a binomial sum is first term squared plus 2 times the product of the terms plus last term squared.

EXAMPLE 5 Finding the Square of a Binomial Sum

Square each binomial using the preceding rule.

a. $(x + 3)^2$ b. $(3x + 7)^2$

Solution Use the special-product formula shown.

$$(A + B)^2 = A^2 + 2AB + B^2$$

	(First Term)^2	+	2 · Product of the Terms	+	(Last Term)^2	= Product
a. $(x + 3)^2 =$	x^2	+	$2 \cdot x \cdot 3$	+	3^2	$= x^2 + 6x + 9$
b. $(3x + 7)^2 =$	$(3x)^2$	+	$2(3x)(7)$	+	7^2	$= 9x^2 + 42x + 49$

Check Point 5 Square each binomial:

a. $(x + 10)^2$ **b.** $(5x + 4)^2$

Using the FOIL method on $(A - B)^2$, the square of a binomial difference, we obtain the following rule.

The Square of a Binomial Difference

$$(A - B)^2 = A^2 - 2AB + B^2$$

The square of a binomial difference is first term squared minus 2 times the product of the terms plus last term squared.

EXAMPLE 6 Finding the Square of a Binomial Difference

Square each binomial using the preceding rule.

a. $(x - 4)^2$ b. $(5y - 6)^2$

Solution Use the special-product formula shown.

$$(A - B)^2 = A^2 - 2AB + B^2$$

	(First Term)^2	−	2 · Product of the Terms	+	(Last Term)^2	= Product
a. $(x - 4)^2 =$	x^2	−	$2 \cdot x \cdot 4$	+	4^2	$= x^2 - 8x + 16$
b. $(5y - 6)^2 =$	$(5y)^2$	−	$2(5y)(6)$	+	6^2	$= 25y^2 - 60y + 36$

Check Point 6 Square each binomial:

a. $(x - 9)^2$ **b.** $(7x - 3)^2$

Special Products

There are several products that occur so frequently that it's convenient to memorize the form or pattern of these formulas.

Special Products

Let A and B represent real numbers, variables, or algebraic expressions.

Special Product	**Example**
Sum and Difference of Two Terms	
$(A + B)(A - B) = A^2 - B^2$	$(2x + 3)(2x - 3) = (2x)^2 - 3^2 = 4x^2 - 9$
Squaring a Binomial	
$(A + B)^2 = A^2 + 2AB + B^2$	$(y + 5)^2 = y^2 + 2 \cdot y \cdot 5 + 5^2 = y^2 + 10y + 25$
$(A - B)^2 = A^2 - 2AB + B^2$	$(3x - 4)^2 = (3x)^2 - 2 \cdot 3x \cdot 4 + 4^2 = 9x^2 - 24x + 16$
Cubing a Binomial	
$(A + B)^3 = A^3 + 3A^2B + 3AB^2 + B^3$	$(x + 4)^3 = x^3 + 3x^2(4) + 3x(4)^2 + 4^3 = x^3 + 12x^2 + 48x + 64$
$(A - B)^3 = A^3 - 3A^2B + 3AB^2 - B^3$	$(x - 2)^3 = x^3 - 3x^2(2) + 3x(2)^2 - 2^3 = x^3 - 6x^2 + 12x - 8$

Study Tip

Although it's convenient to memorize these forms, the FOIL method can be used on all five examples in the box. To cube $x + 4$, you can first square $x + 4$ using FOIL and then multiply this result by $x + 4$. In short, you do not necessarily have to utilize these special formulas. What is the advantage of knowing and using these forms?

6 Perform operations with polynomials in several variables.

Polynomials in Several Variables

The next time you visit the lumber yard and go rummaging through piles of wood, think *polynomials*, although polynomials a bit different from those we have encountered so far. The construction industry uses a polynomial in two variables to determine the number of board feet that can be manufactured from a tree with a diameter of x inches and a length of y feet. This polynomial is

$$\tfrac{1}{4}x^2y - 2xy + 4y.$$

In general, a **polynomial in two variables**, x and y, contains the sum of one or more monomials in the form ax^ny^m. The constant a is the **coefficient**. The exponents n and m represent whole numbers. The **degree** of the monomial ax^ny^m is $n + m$. We'll use the polynomial from the construction industry to illustrate these ideas.

The coefficients are $\frac{1}{4}$, -2, and 4.

$$\tfrac{1}{4}x^2y \qquad -2xy \qquad +4y$$

Degree of monomial: $2 + 1 = 3$	Degree of monomial: $1 + 1 = 2$	Degree of monomial: $0 + 1 = 1$

The degree of a polynomial in two variables is the highest degree of all its terms. For the preceding polynomial, the degree is 3.

Polynomials containing two or more variables can be added, subtracted, and multiplied just like polynomials that contain only one variable.

EXAMPLE 7 Subtracting Polynomials in Two Variables

Subtract as indicated:

$$(5x^3 - 9x^2y + 3xy^2 - 4) - (3x^3 - 6x^2y - 2xy^2 + 3)$$

Solution

$(5x^3 - 9x^2y + 3xy^2 - 4) - (3x^3 - 6x^2y - 2xy^2 + 3)$

$= (5x^3 - 9x^2y + 3xy^2 - 4) + (-3x^3 + 6x^2y + 2xy^2 - 3)$

Change the sign of each term in the second polynomial and add the two polynomials.

$= (5x^3 - 3x^3) + (-9x^2y + 6x^2y) + (3xy^2 + 2xy^2) + (-4 - 3)$

Group like terms.

$= 2x^3 - 3x^2y + 5xy^2 - 7$ Combine like terms by combining coefficients and keeping the same variable factors.

Check Point 7 Subtract: $(x^3 - 4x^2y + 5xy^2 - y^3) - (x^3 - 6x^2y + y^3)$.

EXAMPLE 8 Multiplying Polynomials in Two Variables

Multiply: a. $(x + 4y)(3x - 5y)$ b. $(5x + 3y)^2$

Solution We will perform the multiplication in part (a) using the FOIL method. We will multiply in part (b) using the formula for the square of a binomial, $(A + B)^2$.

a. $(x + 4y)(3x - 5y)$ Multiply these binomials using the FOIL method.

F O I L

$$= (x)(3x) + (x)(-5y) + (4y)(3x) + (4y)(-5y)$$
$$= 3x^2 - 5xy + 12xy - 20y^2$$
$$= 3x^2 + 7xy - 20y^2 \quad \text{Combine like terms.}$$

$(A + B)^2 = A^2 + 2 \cdot A \cdot B + B^2$

b. $$(5x + 3y)^2 = (5x)^2 + 2(5x)(3y) + (3y)^2$$
$$= 25x^2 + 30xy + 9y^2$$

Check Point 8 Multiply:

a. $(7x - 6y)(3x - y)$ **b.** $(x^2 + 5y)^2$

EXERCISE SET P.4

Practice Exercises

In Exercises 1–4, is the algebraic expression a polynomial? If it is, write the polynomial in standard form.

1. $2x + 3x^2 - 5$ **2.** $2x + 3x^{-1} - 5$

3. $\dfrac{2x + 3}{x}$ **4.** $x^2 - x^3 + x^4 - 5$

In Exercises 5–8, find the degree of the polynomial.

5. $3x^2 - 5x + 4$ **6.** $-4x^3 + 7x^2 - 11$

7. $x^2 - 4x^3 + 9x - 12x^4 + 63$

8. $x^2 - 8x^3 + 15x^4 + 91$

In Exercises 9–14, perform the indicated operations. Write the resulting polynomial in standard form and indicate its degree.

9. $(-6x^3 + 5x^2 - 8x + 9) + (17x^3 + 2x^2 - 4x - 13)$

10. $(-7x^3 + 6x^2 - 11x + 13) + (19x^3 - 11x^2 + 7x - 17)$

11. $(17x^3 - 5x^2 + 4x - 3) - (5x^3 - 9x^2 - 8x + 11)$

12. $(18x^4 - 2x^3 - 7x + 8) - (9x^4 - 6x^3 - 5x + 7)$

13. $(5x^2 - 7x - 8) + (2x^2 - 3x + 7) - (x^2 - 4x - 3)$

14. $(8x^2 + 7x - 5) - (3x^2 - 4x) - (-6x^3 - 5x^2 + 3)$

In Exercises 15–54, find each product.

15. $(x + 1)(x^2 - x + 1)$ **16.** $(x + 5)(x^2 - 5x + 25)$

17. $(2x - 3)(x^2 - 3x + 5)$ **18.** $(2x - 1)(x^2 - 4x + 3)$

19. $(x + 7)(x + 3)$ **20.** $(x + 8)(x + 5)$

21. $(x - 5)(x + 3)$ **22.** $(x - 1)(x + 2)$

23. $(3x + 5)(2x + 1)$ **24.** $(7x + 4)(3x + 1)$

25. $(2x - 3)(5x + 3)$ **26.** $(2x - 5)(7x + 2)$

27. $(5x^2 - 4)(3x^2 - 7)$ **28.** $(7x^2 - 2)(3x^2 - 5)$

29. $(x + 3)(x - 3)$ **30.** $(x + 5)(x - 5)$

31. $(3x + 2)(3x - 2)$ **32.** $(2x + 5)(2x - 5)$

33. $(5 - 7x)(5 + 7x)$ **34.** $(4 - 3x)(4 + 3x)$

35. $(4x^2 + 5x)(4x^2 - 5x)$ **36.** $(3x^2 + 4x)(3x^2 - 4x)$

37. $(x + 2)^2$ **38.** $(x + 5)^2$

39. $(2x + 3)^2$ **40.** $(3x + 2)^2$

41. $(x - 3)^2$ **42.** $(x - 4)^2$

43. $(4x^2 - 1)^2$ **44.** $(5x^2 - 3)^2$

45. $(7 - 2x)^2$ **46.** $(9 - 5x)^2$

47. $(x + 1)^3$ **48.** $(x + 2)^3$

49. $(2x + 3)^3$ **50.** $(3x + 4)^3$

51. $(x - 3)^3$ **52.** $(x - 1)^3$

53. $(3x - 4)^3$ **54.** $(2x - 3)^3$

In Exercises 55–62, perform the indicated operations. Indicate the degree of the resulting polynomial.

55. $(5x^2y - 3xy) + (2x^2y - xy)$

56. $(-2x^2y + xy) + (4x^2y + 7xy)$

57. $(4x^2y + 8xy + 11) + (-2x^2y + 5xy + 2)$

58. $(7x^4y^2 - 5x^2y^2 + 3xy) + (-18x^4y^2 - 6x^2y^2 - xy)$

59. $(x^3 + 7xy - 5y^2) - (6x^3 - xy + 4y^2)$

60. $(x^4 - 7xy - 5y^3) - (6x^4 - 3xy + 4y^3)$

61. $(3x^4y^2 + 5x^3y - 3y) - (2x^4y^2 - 3x^3y - 4y + 6x)$

62. $(5x^4y^2 + 6x^3y - 7y) - (3x^4y^2 - 5x^3y - 6y + 8x)$

In Exercises 63–76, find each product.

63. $(x + 5y)(7x + 3y)$ **64.** $(x + 9y)(6x + 7y)$

65. $(x - 3y)(2x + 7y)$ **66.** $(3x - y)(2x + 5y)$

67. $(3xy - 1)(5xy + 2)$

68. $(7x^2y + 1)(2x^2y - 3)$

69. $(7x + 5y)^2$

70. $(9x + 7y)^2$

71. $(x^2y^2 - 3)^2$

72. $(x^2y^2 - 5)^2$

73. $(x - y)(x^2 + xy + y^2)$

74. $(x + y)(x^2 - xy + y^2)$

75. $(3x + 5y)(3x - 5y)$

76. $(7x + 3y)(7x - 3y)$

Application Exercises

77. The polynomial $0.018x^2 - 0.757x + 9.047$ describes the amount, in thousands of dollars, that a person earning x thousand dollars a year feels underpaid. Evaluate the polynomial when $x = 40$. Describe what the answer means in practical terms.

78. The polynomial $104.5x^2 - 1501.5x + 6016$ describes the death rate per year per 100,000 men for men averaging x hours of sleep each night. Evaluate the polynomial when $x = 10$. Describe what the answer means in practical terms.

79. The polynomial $-0.02A^2 + 2A + 22$ is used by coaches to get athletes fired up so that they can perform well. The polynomial represents the performance level related to various levels of enthusiasm, from $A = 1$ (almost no enthusiasm) to $A = 100$ (maximum level of enthusiasm). Evaluate the polynomial when $A = 20$, $A = 50$, and $A = 80$. Describe what happens to performance as we get more and more fired up.

80. The polynomial

$$0.0001x^3 - 0.0043x^2 + 0.089x + 2.66$$

describes the number of pounds of waste produced each day by every American x years after 1960. (The bar graph illustrates daily waste production for eight years.) Evaluate the polynomial when $x = 10$. Describe what the answer means in practical terms.

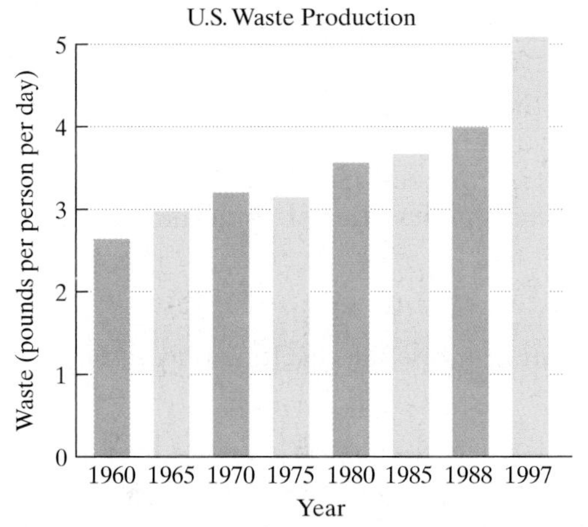

Source: U.S. Environmental Protection Agency.

81. The number of people who catch a cold t weeks after January 1 is $5t - 3t^2 + t^3$. The number of people who recover t weeks after January 1 is $t - t^2 + \frac{1}{3}t^3$. Write a polynomial in standard form for the number of people who are still ill with a cold t weeks after January 1.

82. The weekly cost, in thousands of dollars, of producing x stereo headphones is $30x + 50$. The weekly revenue, in thousands of dollars, of selling x stereo headphones is $90x^2 - x$. Write a polynomial in standard form for the weekly profit, in thousands of dollars, for producing and selling x stereo headphones.

In Exercises 83–84, write a polynomial in standard form that represents the area of the shaded region of each figure.

83.

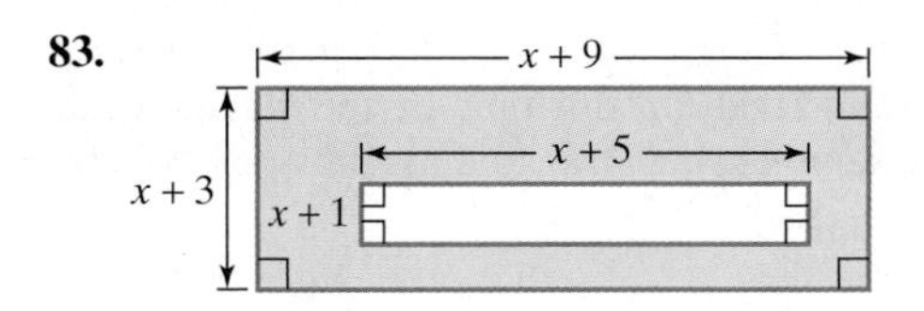

84.

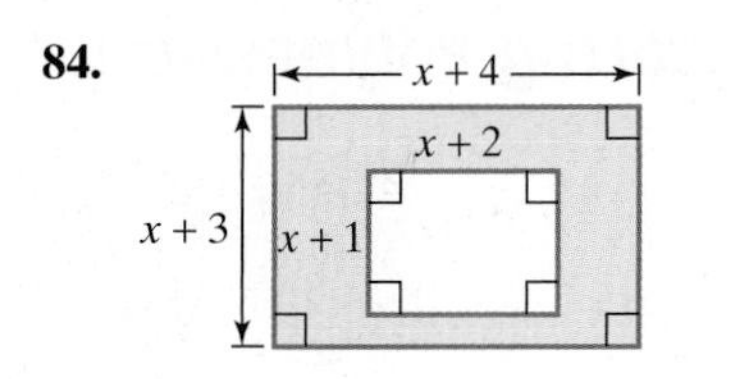

Writing in Mathematics

85. What is a polynomial in x?

86. Explain how to subtract polynomials.

87. Explain how to multiply two binomials using the FOIL method. Give an example with your explanation.

88. Explain how to find the product of the sum and difference of two terms. Give an example with your explanation.

89. Explain how to square a binomial difference. Give an example with your explanation.

90. Explain how to find the degree of a polynomial in two variables.

91. For Exercise 79, explain why performance levels do what they do as we get more and more fired up. If possible, describe an example of a time when you were too enthused and thus did poorly at something you were hoping to do well.

Technology Exercises

92. The common cold is caused by a rhinovirus. The polynomial

$$-0.75x^4 + 3x^3 + 5$$

describes the billions of viral particles in our bodies after x days of invasion. Use a calculator to find the number of viral particles after 0 days (the time of the cold's onset), 1 day, 2 days, 3 days, and 4 days. After how many days is the number of viral particles at a maximum and consequently the day we feel the sickest? By when should we feel completely better?

93. The polynomial $-3.08x^2 + 40.35x + 305.89$ describes the annual number of aggravated assaults in the United States per 100,000 people x years after 1986. Use a calculator to find the number of aggravated assaults per 100,000 people from 1986 to 2000. For this time period, during what year was the number of aggravated assaults per 100,000 people the greatest?

Critical Thinking Exercises

In Exercises 94–97, perform the indicated operations.

94. $(x - y)^2 - (x + y)^2$

95. $[(7x + 5) + 4y][(7x + 5) - 4y]$

96. $[(3x + y) + 1]^2$

97. $(x + y)(x - y)(x^2 + y^2)$

98. Express the area of the plane figure shown as a polynomial in standard form.

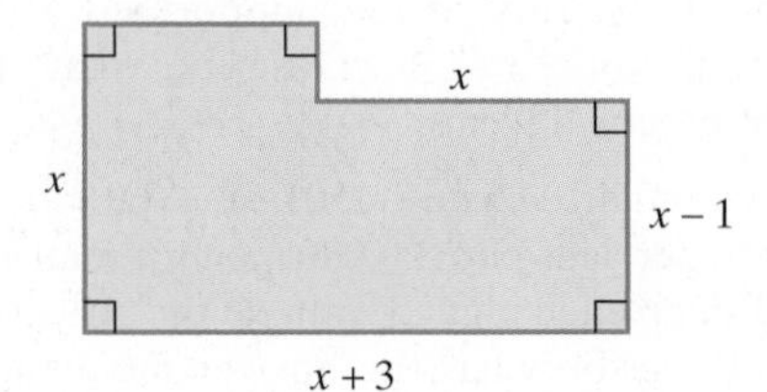

SECTION P.5 Factoring Polynomials

Objectives

1. Factor out the greatest common factor of a polynomial.
2. Factor by grouping.
3. Factor trinomials.
4. Factor the difference of squares.
5. Factor perfect square trinomials.
6. Factor the sum and difference of cubes.
7. Factor completely.

A two-year-old boy is asked, "Do you have a brother?" He answers, "Yes." "What is your brother's name?" "Tom." Asked if Tom has a brother, the two-year-old replies, "No." The child can go in the direction from self to brother, but he cannot reverse this direction and move from brother back to self.

As our intellects develop, we learn to reverse the direction of our thinking. Reversibility of thought is found throughout algebra. For example, we can multiply polynomials and show that

$$(2x + 1)(3x - 2) = 6x^2 - x - 2.$$

We can also reverse this process and express the resulting polynomial as

$$6x^2 - x - 2 = (2x + 1)(3x - 2).$$

Factoring is the process of writing a polynomial as the product of two or more polynomials. The factors of $6x^2 - x - 2$ are $2x + 1$ and $3x - 2$.

In this section, we will be **factoring over the set of integers**, meaning that the coefficients in the factors are integers. Polynomials that cannot be factored using integer coefficients are called **irreducible over the integers**, or **prime**.

The goal in factoring a polynomial is to use one or more factoring techniques until each of the polynomial's factors is prime or irreducible. In this situation, the polynomial is said to be **factored completely**.

We will now discuss basic techniques for factoring polynomials.

1 Factor out the greatest common factor of a polynomial.

Common Factors

In any factoring problem, the first step is to look for the **greatest common factor**. The greatest common factor is an expression of the highest degree that divides each term of the polynomial. The distributive property in the reverse direction

$$ab + ac = a(b + c)$$

can be used to factor out the greatest common factor.

EXAMPLE 1 Factoring out the Greatest Common Factor

Factor: a. $18x^3 + 27x^2$ b. $x^2(x + 3) + 5(x + 3)$

Solution

a. We begin by determining the greatest common factor. 9 is the greatest integer that divides 18 and 27. Furthermore, x^2 is the greatest expression that divides x^3 and x^2. Thus, the greatest common factor of the two terms in the polynomial is $9x^2$.

$18x^3 + 27x^2$

$= 9x^2(2x) + 9x^2(3)$ Express each term with the greatest common factor as a factor.

$= 9x^2(2x + 3)$ Factor out the greatest common factor.

b. In this situation, the greatest common factor is the common binomial factor $(x + 3)$. We factor out this common factor as follows.

$x^2(x + 3) + 5(x + 3) = (x + 3)(x^2 + 5)$ Factor out the common binomial factor.

Check Point 1 Factor:

a. $10x^3 - 4x^2$ **b.** $2x(x - 7) + 3(x - 7)$

2 Factor by grouping.

Factoring by Grouping

Some polynomials have only a greatest common factor of 1. However, by a suitable rearrangement of the terms, it still may be possible to factor. This process, called **factoring by grouping**, is illustrated in Example 2.

EXAMPLE 2 Factoring by Grouping

Factor: $x^3 + 4x^2 + 3x + 12$.

Solution Group terms that have a common factor:

$$\boxed{x^3 + 4x^2} + \boxed{3x + 12}.$$

Common factor is x^2. Common factor is 3.

Discovery

In Example 2, group the terms as follows:

$$(x^3 + 3x) + (4x^2 + 12).$$

Factor out the greatest common factor from each group and complete the factoring process. Describe what happens. What can you conclude?

We now factor the given polynomial as follows.

$$x^3 + 4x^2 + 3x + 12$$

$= (x^3 + 4x^2) + (3x + 12)$ — Group terms with common factors.

$= x^2(x + 4) + 3(x + 4)$ — Factor out the greatest common factor from the grouped terms. The remaining two terms have $x + 4$ as a common binomial factor.

$= (x + 4)(x^2 + 3)$ — Factor $(x + 4)$ out of both terms.

Thus, $x^3 + 4x^2 + 3x + 12 = (x + 4)(x^2 + 3)$. Check the factorization by multiplying the right side of the equation using the FOIL method. If the factorization is correct, you will obtain the original polynomial.

Check Point 2 Factor: $x^3 + 5x^2 - 2x - 10$.

3 Factor trinomials.

Factoring Trinomials

To factor a trinomial of the form $ax^2 + bx + c$, a little trial and error may be necessary.

A Strategy for Factoring $ax^2 + bx + c$

(Assume, for the moment, that there is no greatest common factor.)

1. Find two **F**irst terms whose product is ax^2:

$$(\square x + \quad)(\square x + \quad) = ax^2 + bx + c$$

2. Find two **L**ast terms whose product is c:

$$(x + \square)(x + \square) = ax^2 + bx + c$$

3. By trial and error, perform steps 1 and 2 until the sum of the **O**utside product and **I**nside product is bx:

$$(\square x + \square)(\square x + \square) = ax^2 + bx + c$$

I — O — (sum of O + I)

If no such combinations exist, the polynomial is prime.

EXAMPLE 3 Factoring Trinomials Whose Leading Coefficients Are 1

Factor: a. $x^2 + 6x + 8$ b. $x^2 + 3x - 18$

Solution

a. The factors of the first term are x and x:

$$(x \quad)(x \quad)$$

To find the second term of each factor, we must find two numbers whose product is 8 and whose sum is 6.

Factors of 8	8, 1	4, 2	−8, −1	−4, −2
Sum of Factors	9	6	−9	−6

This is the desired sum.

From the table in the margin, we see that 4 and 2 are the required integers. Thus,

$$x^2 + 6x + 8 = (x + 4)(x + 2) \quad \text{or} \quad (x + 2)(x + 4).$$

b. We begin with

$$x^2 + 3x - 18 = (x \qquad)(x \qquad).$$

Factors of -18	$18, -1$	$-18, 1$	$9, -2$	$-9, 2$	$6, -3$	$-6, 3$
Sum of factors	17	-17	7	-7	3	-3

This is the desired sum.

To find the second term of each factor, we must find two numbers whose product is -18 and whose sum is 3. From the table in the margin, we see that 6 and -3 are the required integers. Thus,

$$x^2 + 3x - 18 = (x + 6)(x - 3)$$
$$\text{or} \quad (x - 3)(x + 6).$$

Check Point 3

Factor:

a. $x^2 + 13x + 40$ **b.** $x^2 - 5x - 14$

EXAMPLE 4 Factoring a Trinomial Whose Leading Coefficient Is Not 1

Factor: $8x^2 - 10x - 3$.

Solution

Step 1 **Find two First terms whose product is $8x^2$.**

$$8x^2 - 10x - 3 \stackrel{?}{=} (8x \qquad)(x \qquad)$$
$$8x^2 - 10x - 3 \stackrel{?}{=} (4x \qquad)(2x \qquad)$$

Step 2 **Find two Last terms whose product is -3.** The possible factors are $1(-3)$ and $-1(3)$.

Step 3 **Try various combinations of these factors.** The correct factorization of $8x^2 - 10x - 3$ is the one in which the sum of the *O*utside and *I*nside products is equal to $-10x$. Here is a list of the possible factors.

Possible Factors of $8x^2 - 10x - 3$	Sum of Outside and Inside Products (Should Equal $-10x$)
$(8x + 1)(x - 3)$	$-24x + x = -23x$
$(8x - 3)(x + 1)$	$8x - 3x = 5x$
$(8x - 1)(x + 3)$	$24x - x = 23x$
$(8x + 3)(x - 1)$	$-8x + 3x = -5x$
$(4x + 1)(2x - 3)$	$-12x + 2x = -10x$
$(4x - 3)(2x + 1)$	$4x - 6x = -2x$
$(4x - 1)(2x + 3)$	$12x - 2x = 10x$
$(4x + 3)(2x - 1)$	$-4x + 6x = 2x$

This is the required middle term.

Thus,

$$8x^2 - 10x - 3 = (4x + 1)(2x - 3) \quad \text{or} \quad (2x - 3)(4x + 1).$$

Show that this factorization is correct by multiplying the factors using the FOIL method. You should obtain the original trinomial.

Check Point 4 Factor: $6x^2 + 19x - 7$.

4 Factor the difference of squares.

Factoring the Difference of Two Squares

A method for factoring the difference of two squares is obtained by reversing the special product for the sum and difference of two terms.

The Difference of Two Squares

If A and B are real numbers, variables, or algebraic expressions, then

$$A^2 - B^2 = (A + B)(A - B).$$

In words: The difference of the squares of two terms factors as the product of a sum and the difference of those terms.

EXAMPLE 5 Factoring the Difference of Two Squares

Factor: a. $x^2 - 4$ b. $81x^2 - 49$

Solution We must express each term as the square of some monomial. Then we use the formula for factoring $A^2 - B^2$.

a. $x^2 - 4^2 = x^2 - 2^2 = (x + 2)(x - 2)$

$A^2 - B^2 = (A + B)(A - B)$

b. $81x^2 - 49 = (9x)^2 - 7^2 = (9x + 7)(9x - 7)$

Check Point 5 Factor:

a. $x^2 - 81$ **b.** $36x^2 - 25$

We have seen that a polynomial is factored completely when it is written as the product of prime polynomials. To be sure that you have factored completely, check to see whether the factors can be factored.

EXAMPLE 6 A Repeated Factorization

Factor completely: $x^4 - 81$.

Solution

$x^4 - 81 = (x^2)^2 - 9^2$ Express as the difference of two squares.

$= (x^2 + 9)(x^2 - 9)$ The factors are the sum and difference of the squared terms.

$= (x^2 + 9)(x^2 - 3^2)$ The factor $x^2 - 9$ is the difference of two squares and can be factored.

$= (x^2 + 9)(x + 3)(x - 3)$ The factors of $x^2 - 9$ are the sum and difference of the squared terms.

Study Tip

Factoring $x^4 - 81$ as

$$(x^2 + 9)(x^2 - 9)$$

is not a complete factorization. The second factor $x^2 - 9$ is itself a difference of two squares and can be factored.

Check Point 6 Factor completely: $81x^4 - 16$.

5 Factor perfect square trinomials.

Factoring Perfect Square Trinomials

Our next factoring technique is obtained by reversing the special products for squaring binomials. The trinomials that are factored using this technique are called **perfect square trinomials**.

Factoring Perfect Square Trinomials

Let A and B be real numbers, variables, or algebraic expressions.

1. $A^2 + 2AB + B^2 = (A + B)^2$

Same sign

2. $A^2 - 2AB + B^2 = (A - B)^2$

Same sign

The two items in the box show that perfect square trinomials come in two forms: one in which the middle term is positive and one in which the middle term is negative. Here's how to recognize a perfect square trinomial:

1. The first and last terms are positive perfect squares.

2. The middle term is twice the product of the square roots of the first and last terms.

EXAMPLE 7 Factoring Perfect Square Trinomials

Factor: a. $x^2 + 6x + 9$ b. $25x^2 - 60x + 36$

Solution

a. $x^2 + 6x + 9 = x^2 + 2 \cdot x \cdot 3 + 3^2 = (x + 3)^2$ The middle term has a positive sign.

$A^2 + 2AB + B^2 = (A + B)^2$

b. We suspect that $25x^2 - 60x + 36$ is a perfect square trinomial because $25x^2 = (5x)^2$ and $36 = 6^2$. The middle term can be expressed as twice the product of $5x$ and 6.

$$25x^2 - 60x + 36 = (5x)^2 - 2 \cdot 5x \cdot 6 + 6^2 = (5x - 6)^2$$

$A^2 - 2AB + B^2 = (A - B)^2$

Check Point 7 Factor:

a. $x^2 + 14x + 49$ **b.** $16x^2 - 56x + 49$

6 Factor the sum and difference of cubes.

Factoring the Sum and Difference of Two Cubes

We can use the following formulas to factor the sum or the difference of two cubes.

Factoring the Sum and Difference of Two Cubes

1. Factoring the Sum of Two Cubes

$$A^3 + B^3 = (A + B)(A^2 - AB + B^2)$$

2. Factoring the Difference of Two Cubes

$$A^3 - B^3 = (A - B)(A^2 + AB + B^2)$$

EXAMPLE 8 Factoring Sums and Differences of Two Cubes

Factor: a. $x^3 + 8$ b. $64x^3 - 125$

Solution

a. $x^3 + 8 = x^3 + 2^3 = (x + 2)(x^2 - x \cdot 2 + 2^2) = (x + 2)(x^2 - 2x + 4)$

$A^3 + B^3 = (A + B)(A^2 - AB + B^2)$

b. $64x^3 - 125 = (4x)^3 - 5^3 = (4x - 5)[(4x)^2 + (4x)(5) + 5^2]$

$A^3 - B^3 = (A - B)(A^2 + AB + B^2)$

$= (4x - 5)(16x^2 + 20x + 25)$

Check Point 8 Factor:

a. $x^3 + 1$ **b.** $125x^3 - 8$

7 Factor completely.

Factoring Completely

Some polynomials can be factored using more than one technique. Always begin by trying to factor out the greatest common factor.

EXAMPLE 9 Factoring Completely

Factor: a. $2a^3 + 8a^2 + 8a$ b. $x^3 - 5x^2 - 4x + 20$

Solution

a. $2a^3 + 8a^2 + 8a$

$= 2a(a^2 + 4a + 4)$ Factor out the greatest common factor.

$= 2a(a + 2)^2$ Factor the perfect square trinomial.

b. $x^3 - 5x^2 - 4x + 20$

$= (x^3 - 5x^2) + (-4x + 20)$ Group the terms with common factors.

$= x^2(x - 5) - 4(x - 5)$ Factor from each group.

$= (x - 5)(x^2 - 4)$ Factor out the common binomial factor, $(x - 5)$.

$= (x - 5)(x + 2)(x - 2)$ Factor completely by factoring $x^2 - 4$ as the difference of two squares.

Check Point 9 Factor:

a. $2x^3 - 24x^2 + 72x$ **b.** $x^3 - 4x^2 - 9x + 36$

EXERCISE SET P.5

Practice Exercises

In Exercises 1–10, factor out the greatest common factor.

1. $18x + 27$
2. $16x - 24$
3. $3x^2 + 6x$
4. $4x^2 - 8x$
5. $9x^4 - 18x^3 + 27x^2$
6. $6x^4 - 18x^3 + 12x^2$
7. $x(x + 5) + 3(x + 5)$
8. $x(2x + 1) + 4(2x + 1)$
9. $x^2(x - 3) + 12(x - 3)$
10. $x^2(2x + 5) + 17(2x + 5)$

In Exercises 11–16, factor by grouping.

11. $x^3 - 2x^2 + 5x - 10$
12. $x^3 - 3x^2 + 4x - 12$
13. $x^3 - x^2 + 2x - 2$
14. $x^3 + 6x^2 - 2x - 12$
15. $3x^3 - 2x^2 - 6x + 4$
16. $x^3 - x^2 - 5x + 5$

In Exercises 17–30, factor each trinomial, or state that the trinomial is prime.

17. $x^2 + 5x + 6$
18. $x^2 + 8x + 15$
19. $x^2 - 2x - 15$
20. $x^2 - 4x - 5$
21. $x^2 - 8x + 15$
22. $x^2 - 14x + 45$
23. $3x^2 - x - 2$
24. $2x^2 + 5x - 3$
25. $3x^2 - 25x - 28$
26. $3x^2 - 2x - 5$
27. $6x^2 - 11x + 4$
28. $6x^2 - 17x + 12$
29. $4x^2 + 16x + 15$
30. $8x^2 + 33x + 4$

In Exercises 31–40, factor the difference of two squares.

31. $x^2 - 100$
32. $x^2 - 144$
33. $36x^2 - 49$
34. $64x^2 - 81$
35. $9x^2 - 25y^2$
36. $36x^2 - 49y^2$
37. $x^4 - 16$
38. $x^4 - 1$
39. $16x^4 - 81$
40. $81x^4 - 1$

In Exercises 41–48, factor any perfect square trinomials, or state that the polynomial is prime.

41. $x^2 + 2x + 1$
42. $x^2 + 4x + 4$
43. $x^2 - 14x + 49$
44. $x^2 - 10x + 25$
45. $4x^2 + 4x + 1$
46. $25x^2 + 10x + 1$
47. $9x^2 - 6x + 1$
48. $64x^2 - 16x + 1$

In Exercises 49–56, factor using the formula for the sum or difference of two cubes.

49. $x^3 + 27$
50. $x^3 + 64$
51. $x^3 - 64$
52. $x^3 - 27$
53. $8x^3 - 1$
54. $27x^3 - 1$
55. $64x^3 + 27$
56. $8x^3 + 125$

In Exercises 57–76, factor completely, or state that the polynomial is prime.

57. $3x^3 - 3x$
58. $5x^3 - 45x$
59. $4x^2 - 4x - 24$
60. $6x^2 - 18x - 60$
61. $2x^4 - 162$
62. $7x^4 - 7$
63. $x^3 + 2x^2 - 9x - 18$
64. $x^3 + 3x^2 - 25x - 75$
65. $2x^2 - 2x - 112$
66. $6x^2 - 6x - 12$
67. $x^3 - 4x$
68. $9x^3 - 9x$
69. $x^2 + 64$
70. $x^2 + 36$
71. $x^3 + 2x^2 - 4x - 8$
72. $x^3 + 2x^2 - x - 2$
73. $y^5 - 81y$
74. $y^5 - 16y$
75. $20y^4 - 45y^2$
76. $48y^4 - 3y^2$

Application Exercises

77. You dive directly upward from a board that is 32 feet high. After t seconds, your height above the water is described by the polynomial $-16t^2 + 16t + 32$. Factor the polynomial completely.

78. If x represents a positive integer, factor $x^3 + 3x^2 + 2x$ to show that the trinomial represents the product of three consecutive integers.

In Exercises 79–80, find the formula for the area of the shaded region and express it in factored form.

79.

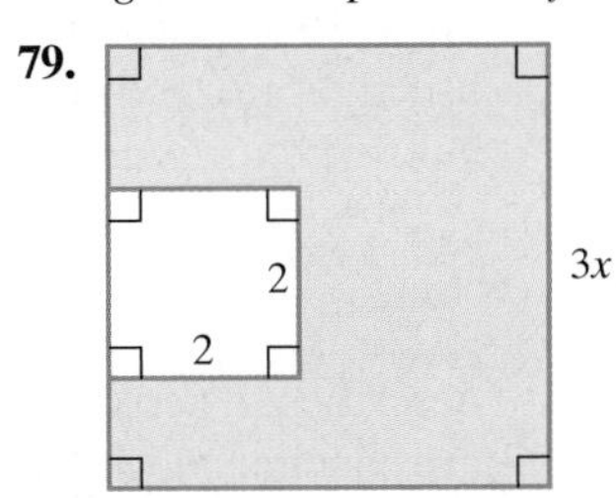

80.

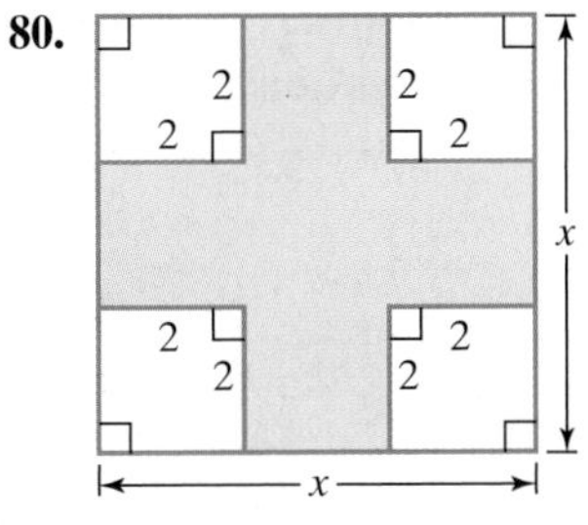

Writing in Mathematics

81. Use an example and explain how to factor out the greatest common factor of a polynomial.

82. Suppose that a polynomial contains four terms. Explain how to use factoring by grouping to factor the polynomial.

83. Explain how to factor $3x^2 + 10x + 8$.

84. Explain how to factor the difference of two squares. Provide an example with your explanation.

85. What is a perfect square trinomial and how is it factored?

86. Explain how to factor $x^3 + 1$.

87. What does it mean to factor completely?

88. For Exercise 77, explain how to use your factored polynomial to figure out how many seconds it will take for you to hit the water.

Critical Thinking Exercises

89. Which one of the following is true?

a. Because $x^2 + 1$ is irreducible over the integers, it follows that $x^3 + 1$ is also irreducible.

b. One correct factored form for $x^2 - 4x + 3$ is $x(x - 4) + 3$.

c. $x^3 - 64 = (x - 4)^3$

d. None of the above is true.

In Exercises 90–93, factor completely.

90. $x^{2n} + 6x^n + 8$

91. $-x^2 - 4x + 5$

92. $x^2 + 8x + 16 - 25y^2$

93. $x^4 - y^4 - 2x^3y + 2xy^3$

In Exercises 94–95, find all integers b so that the trinomial can be factored.

94. $x^2 + bx + 15$

95. $x^2 + 4x + b$

Group Exercise

96. Without looking at any factoring problems in the book, create five factoring problems. Make sure that some of your problems require at least two factoring strategies. Next, exchange problems with another person in your group. Work to factor your partner's problems. Evaluate the problems as you work: Are they too easy? Too difficult? Can the polynomials really be factored? Share your response with the person who wrote the problems. Finally, grade each other's work in factoring the polynomials. Each factoring problem is worth 20 points. You may award partial credit. If you take off points, explain why points are deducted and how you decided to take off a particular number of points for the error(s) that you found.

SECTION P.6 Rational Expressions

Objectives

1. Specify numbers that must be excluded from the domain of rational expressions.
2. Simplify rational expressions.
3. Multiply rational expressions.
4. Divide rational expressions.
5. Add and subtract rational expressions.
6. Simplify complex rational expressions.

How do we describe the costs of reducing environmental pollution? We often use algebraic expressions involving quotients of polynomials. For example, the algebraic expression

$$\frac{250x}{100 - x}$$

describes the cost, in millions of dollars, to remove x percent of the pollutants that are discharged into a river. Removing a modest percentage of pollutants, say

Discovery

What happens if you try substituting 100 for x in

$$\frac{250x}{100 - x}?$$

What does this tell you about the cost of cleaning up all of the river's pollutants?

40%, is far less costly than removing a substantially greater percentage, such as 95%. We see this by evaluating the algebraic expression for $x = 40$ and $x = 95$.

Evaluating $\frac{250x}{100 - x}$ for

$x = 40$: Cost is $\frac{250(40)}{100 - 40} \approx 167$

$x = 95$: Cost is $\frac{250(95)}{100 - 95} = 4750$

The cost increases from approximately $167 million to a possibly prohibitive $4750 million, or $4.75 billion. Costs spiral upward as the percentage of removed pollutants increases.

Many algebraic expressions that describe costs of environmental projects are examples of rational expressions. First we will define rational expressions. Then we will review how to perform operations with such expressions.

1 Specify numbers that must be excluded from the domain of rational expressions.

Rational Expressions

A **rational expression** is the quotient of two polynomials. Some examples are

$$\frac{x - 2}{4}, \quad \frac{4}{x - 2}, \quad \frac{x}{x^2 - 1}, \quad \text{and} \quad \frac{x^2 + 1}{x^2 + 2x - 3}.$$

The set of real numbers for which an algebraic expression is defined is the **domain** of the expression. Because rational expressions indicate division and division by zero is undefined, we must exclude numbers from a rational expression's domain that make the denominator zero.

EXAMPLE 1 Excluding Numbers from the Domain

Find all the numbers that must be excluded from the domain of each rational expression.

a. $\frac{4}{x - 2}$ b. $\frac{x}{x^2 - 1}$

Solution To determine the numbers that must be excluded from each domain, examine the denominators.

a. $\frac{4}{x - 2}$ b. $\frac{x}{x^2 - 1} = \frac{x}{(x + 1)(x - 1)}$

This denominator would equal zero if $x = 2$.

This factor would equal zero if $x = -1$.

This factor would equal zero if $x = 1$.

For the rational expression in part (a), we must exclude 2 from the domain. For the rational expression in part (b), we must exclude both -1 and 1 from the domain. These excluded numbers are often written to the right of a rational expression.

$$\frac{4}{x - 2}, x \neq 2 \qquad \frac{x}{x^2 - 1}, x \neq -1, x \neq 1$$

Check Point 1 Find all the numbers that must be excluded from each rational expression's domain.

a. $\frac{7}{x+5}$ **b.** $\frac{x}{x^2-36}$

2 Simplify rational expressions.

Simplifying Rational Expressions

A rational expression is **simplified** if its numerator and denominator have no common factors other than 1 or −1. The following procedure can be used to simplify rational expressions.

Simplifying Rational Expressions

1. Factor the numerator and denominator completely.
2. Divide both the numerator and denominator by the common factors.

EXAMPLE 2 Simplifying Rational Expressions

Simplify: a. $\frac{x^3+x^2}{x+1}$ b. $\frac{x^2+6x+5}{x^2-25}$

Solution

a. $\frac{x^3+x^2}{x+1} = \frac{x^2(x+1)}{x+1}$ Factor the numerator. Because the denominator is $x+1$, $x \neq -1$.

$= \frac{x^2\cancel{(x+1)}^{1}}{\cancel{x+1}_{1}}$ Divide out the common factor of $x+1$.

$= x^2, x \neq -1$ Denominators of 1 need not be written because $\frac{a}{1} = a$.

b. $\frac{x^2+6x+5}{x^2-25} = \frac{(x+5)(x+1)}{(x+5)(x-5)}$ Factor the numerator and denominator. Because the denominator is $(x+5)(x-5)$, $x \neq -5$ and $x \neq 5$.

$= \frac{\cancel{(x+5)}^{1}(x+1)}{\cancel{(x+5)}_{1}(x-5)}$ Divide out the common factor of $x+5$.

$= \frac{x+1}{x-5}, \quad x \neq -5 \text{ and } x \neq 5$

Check Point 2 Simplify:

a. $\frac{x^3+3x^2}{x+3}$ **b.** $\frac{x^2-1}{x^2+2x+1}$

3 Multiply rational expressions.

Multiplying Rational Expressions

The product of two rational expressions is the product of their numerators over the product of their denominators. Here is a step-by-step procedure for multiplying rational expressions.

Multiplying Rational Expressions

1. Factor all numerators and denominators completely.
2. Divide both the numerator and denominator by common factors.
3. Multiply the remaining factors in the numerator and multiply the remaining factors in the denominator.

EXAMPLE 3 Multiplying Rational Expressions

Multiply and simplify:

$$\frac{x-7}{x-1} \cdot \frac{x^2-1}{3x-21}.$$

Solution

$$\frac{x-7}{x-1} \cdot \frac{x^2-1}{3x-21}$$

$$= \frac{x-7}{x-1} \cdot \frac{(x+1)(x-1)}{3(x-7)}$$

Factor all numerators and denominators. Because the denominator has factors of $x - 1$ and $x - 7$, $x \neq 1$ and $x \neq 7$.

$$= \frac{\overset{1}{\cancel{x-7}}}{\underset{1}{\cancel{x-1}}} \cdot \frac{(x+1)\overset{1}{\cancel{(x-1)}}}{3\underset{1}{\cancel{(x-7)}}}$$

Divide both the numerator and the denominator by common factors.

$$= \frac{x+1}{3}, x \neq 1, x \neq 7$$

Multiply the remaining factors in the numerator and denominator.

These excluded numbers from the domain must also be excluded from the simplified expression's domain.

Check Point 3 Multiply and simplify:

$$\frac{x+3}{x^2-4} \cdot \frac{x^2-x-6}{x^2+6x+9}.$$

4 Divide rational expressions.

Dividing Rational Expressions

We find the quotient of two rational expressions by inverting the divisor and multiplying.

EXAMPLE 4 Dividing Rational Expressions

Divide and simplify:

$$\frac{x^2-2x-8}{x^2-9} \div \frac{x-4}{x+3}.$$

Solution

$$\frac{x^2 - 2x - 8}{x^2 - 9} \div \frac{x - 4}{x + 3}$$

$$= \frac{x^2 - 2x - 8}{x^2 - 9} \cdot \frac{x + 3}{x - 4}$$ Invert the divisor and multiply.

$$= \frac{(x - 4)(x + 2)}{(x + 3)(x - 3)} \cdot \frac{x + 3}{x - 4}$$ Factor throughout. For nonzero denominators, $x \neq -3$, $x \neq 3$, and $x \neq 4$.

$$= \frac{\overset{1}{\cancel{(x - 4)}}(x + 2)}{\underset{1}{\cancel{(x + 3)}}(x - 3)} \cdot \frac{\overset{1}{\cancel{(x + 3)}}}{\underset{1}{\cancel{(x - 4)}}}$$ Divide both the numerator and denominator by common factors.

$$= \frac{x + 2}{x - 3}, x \neq -3, x \neq 3, x \neq 4$$ Multiply the remaining factors in the numerator and the denominator.

Check Point 4 Divide and simplify:

$$\frac{x^2 - 2x + 1}{x^3 + x} \div \frac{x^2 + x - 2}{3x^2 + 3}.$$

5 Add and subtract rational expressions.

Adding and Subtracting Rational Expressions with the Same Denominator

We add or subtract rational expressions with the same denominator by (1) adding or subtracting the numerators, (2) placing this result over the common denominator, and (3) simplifying, if possible.

EXAMPLE 5 Subtracting Rational Expressions with the Same Denominator

Subtract: $\frac{5x + 1}{x^2 - 9} - \frac{4x - 2}{x^2 - 9}$.

Study Tip

Example 5 shows that when a numerator is being subtracted, we must subtract every term in that expression.

Solution

$$\frac{5x + 1}{x^2 - 9} - \frac{4x - 2}{x^2 - 9} = \frac{5x + 1 - (4x - 2)}{x^2 - 9}$$ Subtract numerators and include parentheses to indicate that both terms are subtracted. Place this difference over the common denominator.

$$= \frac{5x + 1 - 4x + 2}{x^2 - 9}$$ Remove parentheses and then change the sign of each term.

$$= \frac{x + 3}{x^2 - 9}$$ Combine like terms.

$$= \frac{\overset{1}{\cancel{x + 3}}}{\underset{1}{\cancel{(x + 3)}}(x - 3)}$$ Factor and simplify ($x \neq -3$ and $x \neq 3$).

$$= \frac{1}{x - 3}, x \neq -3, x \neq 3$$

Check Point 5 Subtract: $\frac{x}{x + 1} - \frac{3x + 2}{x + 1}$.

Adding and Subtracting Rational Expressions with Different Denominators

Rational expressions that have no common factors in their denominators can be added or subtracted using one of the following properties:

$$\frac{a}{b}+\frac{c}{d}=\frac{ad+bc}{bd} \qquad \frac{a}{b}-\frac{c}{d}=\frac{ad-bc}{bd}, \; b \neq 0, d \neq 0.$$

The least common denominator, bd, is the product of the distinct factors in the two denominators.

EXAMPLE 6 Subtracting Rational Expressions Having No Common Factors in Their Denominators

Subtract: $\dfrac{x+2}{2x-3}-\dfrac{4}{x+3}$.

Solution We need to find the least common denominator. This is the product of the distinct factors in each denominator, namely $(2x-3)(x+3)$. We can therefore use the subtraction property given above as follows:

$$\frac{a}{b}-\frac{c}{d}=\frac{ad-bc}{bd}$$

$$\frac{x+2}{2x-3}-\frac{4}{x+3}=\frac{(x+2)(x+3)-(2x-3)4}{(2x-3)(x+3)}$$

Observe that $a = x+2$, $b = 2x-3$, $c = 4$, and $d = x+3$.

$$=\frac{x^2+5x+6-(8x-12)}{(2x-3)(x+3)}$$

Multiply.

$$=\frac{x^2+5x+6-8x+12}{(2x-3)(x+3)}$$

Remove parentheses and then change the sign of each term.

$$=\frac{x^2-3x+18}{(2x-3)(x+3)}, \; x \neq \frac{3}{2}, \; x \neq -3$$

Combine like terms in the numerator.

Check Point 6 Add: $\dfrac{3}{x+1}+\dfrac{5}{x-1}$.

When adding and subtracting rational expressions that have different denominators with one or more common factors in the denominators, it is efficient to find the least common denominator first.

Finding the Least Common Denominator

1. Factor each denominator completely.
2. List the factors of the first denominator.
3. Add to the list in step 2 any factors of the second denominator that do not appear in the list.
4. Form the product of each different factor from the list in step 3. This product is the least common denominator.

EXAMPLE 7 Finding the Least Common Denominator

Find the least common denominator that is needed to add or subtract the rational expressions

$$\frac{7}{5x^2 + 15x} \quad \text{and} \quad \frac{9}{x^2 + 6x + 9}.$$

Solution

Step 1 Factor each denominator completely.

$$5x^2 + 15x = 5x(x + 3)$$

$$x^2 + 6x + 9 = (x + 3)^2$$

Step 2 List the factors of the first denominator.

$$5, x, (x + 3)$$

Step 3 Add any unlisted factors from the second denominator. The second denominator is $(x + 3)^2$ or $(x + 3)(x + 3)$. One factor of $x + 3$ is already in our list, but the other factor is not. We add $x + 3$ to the list. We have

$$5, x, (x + 3), (x + 3).$$

Step 4 The least common denominator is the product of all factors in the final list. Thus,

$$5x(x + 3)(x + 3)$$

is the least common denominator.

Check Point 7 What is the least common denominator for denominators of $x^2 - 6x + 9$ and $x^2 - 9$?

Finding the least common denominator for two (or more) rational expressions is the first step needed to add or subtract the expressions.

Adding and Subtracting Rational Expressions That Have Different Denominators With Shared Factors

1. Find the least common denominator.
2. Write all rational expressions in terms of the least common denominator. To do so, multiply both the numerator and the denominator of each rational expression by any factor(s) needed to convert the denominator into the least common denominator.
3. Add or subtract the numerators, placing the resulting expression over the least common denominator.
4. If necessary, simplify the resulting rational expression.

EXAMPLE 8 Adding Rational Expressions with Different Denominators

Add: $\dfrac{x+3}{x^2+x-2}+\dfrac{2}{x^2-1}$.

Solution

Step 1 Find the least common denominator. Start by factoring the denominators.

$$x^2+x-2=(x+2)(x-1)$$
$$x^2-1=(x+1)(x-1)$$

The factors of the first denominator are $x+2$ and $x-1$. The only factor from the second denominator that is unlisted is $x+1$. Thus, the least common denominator is

$$(x+2)(x-1)(x+1).$$

Step 2 Write all rational expressions in terms of the least common denominator. We do so by multiplying both the numerator and the denominator by any factor(s) needed to convert the denominator into the least common denominator.

$$\frac{x+3}{x^2+x-2}+\frac{2}{x^2-1}$$

$$=\frac{x+3}{(x+2)(x-1)}+\frac{2}{(x+1)(x-1)}$$

The least common denominator is $(x+2)(x-1)(x+1)$.

$$=\frac{(x+3)(x+1)}{(x+2)(x-1)(x+1)}+\frac{2(x+2)}{(x+2)(x-1)(x+1)}$$

Rewrite each rational expression with the least common denominator. Multiply the numerator and the denominator by whatever extra factors are required to form $(x+2)(x-1)(x+1)$.

Step 3 Add numerators, putting this sum over the least common denominator.

$$=\frac{(x+3)(x+1)+2(x+2)}{(x+2)(x-1)(x+1)}$$

$$=\frac{x^2+4x+3+2x+4}{(x+2)(x-1)(x+1)}$$

Multiply in the numerator.

$$=\frac{x^2+6x+7}{(x+2)(x-1)(x+1)},\ x\neq-2,\ x\neq1,\ x\neq-1$$

Combine like terms in the numerator.

Step 4 If necessary, simplify. Because the numerator is prime, no further simplification is possible.

Check Point 8 Subtract: $\dfrac{x}{x^2-10x+25}-\dfrac{x-4}{2x-10}$.

6 Simplify complex rational expressions.

Complex Rational Expressions

Complex rational expressions have numerators or denominators containing one or more rational expressions. Here are two examples of such expressions:

$$\frac{1+\frac{1}{x}}{1-\frac{1}{x}} \qquad \frac{\frac{1}{x+h}-\frac{1}{x}}{h}$$

Separate rational expressions occur in the numerator and denominator.

Separate rational expressions occur in the numerator.

One method for simplifying a complex rational expression is to combine its numerator into a single expression and combine its denominator into a single expression. Then perform the division by inverting the denominator and multiplying.

EXAMPLE 9 Simplifying a Complex Rational Expression

Simplify: $\dfrac{1+\frac{1}{x}}{1-\frac{1}{x}}$.

Solution

$$\frac{1+\frac{1}{x}}{1-\frac{1}{x}} = \frac{\frac{x}{x}+\frac{1}{x}}{\frac{x}{x}-\frac{1}{x}}, \; x \neq 0$$

The terms in the numerator and in the denominator are each combined by performing the addition and subtraction. The least common denominator is x.

$$= \frac{\frac{x+1}{x}}{\frac{x-1}{x}}$$

Perform the addition in the numerator and the subtraction in the denominator.

$$= \frac{x+1}{x} \div \frac{x-1}{x}$$

Rewrite the main fraction bar as ÷.

$$= \frac{x+1}{x} \cdot \frac{x}{x-1}$$

Invert the divisor and multiply ($x \neq 0$ and $x \neq 1$).

$$= \frac{x+1}{\cancel{x}_1} \cdot \frac{\cancel{x}^1}{x-1}$$

Divide both the numerator and denominator by the common factor, x.

$$= \frac{x+1}{x-1}, \; x \neq 0, x \neq 1$$

Multiply the remaining factors in the numerator and in the denominator.

Check Point 9 Simplify: $\dfrac{\frac{1}{x}-\frac{3}{2}}{\frac{1}{x}+\frac{3}{4}}$.

A second method for simplifying a complex rational expression is to find the least common denominator of all the rational expressions in its numerator and denominator. Then multiply each term in its numerator and denominator by this least common denominator. Here we use this method to simplify the complex rational expression in Example 9.

$$\frac{1+\frac{1}{x}}{1-\frac{1}{x}} = \frac{\left(1+\frac{1}{x}\right)}{\left(1-\frac{1}{x}\right)} \cdot \frac{x}{x}$$

The least common denominator of all the rational expressions is x. Multiply the numerator and denominator by x. Because $\frac{x}{x} = 1$, we are not changing the complex fraction ($x \neq 0$).

$$= \frac{1 \cdot x + \frac{1}{x} \cdot x}{1 \cdot x - \frac{1}{x} \cdot x}$$

Use the distributive property. Be sure to distribute x to every term.

$$= \frac{x+1}{x-1},\ x \neq 0,\ x \neq 1$$

Multiply. The complex rational expression is now simplified.

EXERCISE SET P.6

Practice Exercises

In Exercises 1–6, find all numbers that must be excluded from the domain of each rational expression.

1. $\frac{7}{x-3}$
2. $\frac{13}{x+9}$
3. $\frac{x+5}{x^2-25}$
4. $\frac{x+7}{x^2-49}$
5. $\frac{x-1}{x^2+11x+10}$
6. $\frac{x-3}{x^2+4x-45}$

In Exercises 7–14, simplify each rational expression. Find all numbers that must be excluded from the domain of the simplified rational expression.

7. $\frac{3x-9}{x^2-6x+9}$
8. $\frac{4x-8}{x^2-4x+4}$
9. $\frac{x^2-12x+36}{4x-24}$
10. $\frac{x^2-8x+16}{3x-12}$
11. $\frac{y^2+7y-18}{y^2-3y+2}$
12. $\frac{y^2-4y-5}{y^2+5y+4}$
13. $\frac{x^2+12x+36}{x^2-36}$
14. $\frac{x^2-14x+49}{x^2-49}$

In Exercises 15–30, multiply or divide as indicated.

15. $\frac{x-2}{3x+9} \cdot \frac{2x+6}{2x-4}$
16. $\frac{6x+9}{3x-15} \cdot \frac{x-5}{4x+6}$
17. $\frac{x^2-9}{x^2} \cdot \frac{x^2-3x}{x^2+x-12}$
18. $\frac{x^2-4}{x^2-4x+4} \cdot \frac{2x-4}{x+2}$
19. $\frac{x^2-5x+6}{x^2-2x-3} \cdot \frac{x^2-1}{x^2-4}$
20. $\frac{x^2+5x+6}{x^2+x-6} \cdot \frac{x^2-9}{x^2-x-6}$
21. $\frac{x^3-8}{x^2-4} \cdot \frac{x+2}{3x}$
22. $\frac{x^2+6x+9}{x^3+27} \cdot \frac{1}{x+3}$
23. $\frac{x+1}{3} \div \frac{3x+3}{7}$
24. $\frac{x+5}{7} \div \frac{4x+20}{9}$
25. $\frac{x^2-4}{x} \div \frac{x+2}{x-2}$
26. $\frac{x^2-4}{x-2} \div \frac{x+2}{4x-8}$
27. $\frac{4x^2+10}{x-3} \div \frac{6x^2+15}{x^2-9}$
28. $\frac{x^2+x}{x^2-4} \div \frac{x^2-1}{x^2+5x+6}$
29. $\frac{x^2-25}{2x-2} \div \frac{x^2+10x+25}{x^2+4x-5}$
30. $\frac{x^2-4}{x^2+3x-10} \div \frac{x^2+5x+6}{x^2+8x+15}$

In Exercises 31–50, add or subtract as indicated.

31. $\frac{4x+1}{6x+5} + \frac{8x+9}{6x+5}$
32. $\frac{3x+2}{3x+4} + \frac{3x+6}{3x+4}$
33. $\frac{x^2-2x}{x^2+3x} + \frac{x^2+x}{x^2+3x}$
34. $\frac{x^2-4x}{x^2-x-6} + \frac{4x-4}{x^2-x-6}$
35. $\frac{4x-10}{x-2} - \frac{x-4}{x-2}$
36. $\frac{2x+3}{3x-6} - \frac{3-x}{3x-6}$
37. $\frac{x^2+3x}{x^2+x-12} - \frac{x^2-12}{x^2+x-12}$
38. $\frac{x^2-4x}{x^2-x-6} - \frac{x-6}{x^2-x-6}$
39. $\frac{3}{x+4} + \frac{6}{x+5}$
40. $\frac{8}{x-2} + \frac{2}{x-3}$
41. $\frac{3}{x+1} - \frac{3}{x}$
42. $\frac{4}{x} - \frac{3}{x+3}$

43. $\dfrac{2x}{x+2} + \dfrac{x+2}{x-2}$

44. $\dfrac{3x}{x-3} - \dfrac{x+4}{x+2}$

45. $\dfrac{x+5}{x-5} + \dfrac{x-5}{x+5}$

46. $\dfrac{x+3}{x-3} + \dfrac{x-3}{x+3}$

47. $\dfrac{4}{x^2+6x+9} + \dfrac{4}{x+3}$

48. $\dfrac{3}{5x+2} + \dfrac{5x}{25x^2-4}$

49. $\dfrac{3x}{x^2+3x-10} - \dfrac{2x}{x^2+x-6}$

50. $\dfrac{x}{x^2-2x-24} - \dfrac{x}{x^2-7x+6}$

In Exercise 51–60, simplify each complex rational expression.

51. $\dfrac{\frac{x}{3}-1}{x-3}$

52. $\dfrac{\frac{x}{4}-1}{x-4}$

53. $\dfrac{1+\frac{1}{x}}{3-\frac{1}{x}}$

54. $\dfrac{8+\frac{1}{x}}{4-\frac{1}{x}}$

55. $\dfrac{\frac{1}{x}+\frac{1}{y}}{x+y}$

56. $\dfrac{1-\frac{1}{x}}{xy}$

57. $\dfrac{x-\frac{x}{x+3}}{x+2}$

58. $\dfrac{x-3}{x-\frac{3}{x-2}}$

59. $\dfrac{\frac{3}{x-2}-\frac{4}{x+2}}{\frac{7}{x^2-4}}$

60. $\dfrac{\frac{x}{x-2}+1}{\frac{3}{x^2-4}+1}$

Application Exercises

61. The polynomial $-0.14t^2 + 0.51t + 31.6$ describes the U.S. population (in millions) age 65 and older t years after 1990. The polynomial $0.54t^2 + 12.64t + 107.1$ describes the total yearly cost of Medicare (in billions of dollars) t years after 1990. Write a rational expression that describes the average cost of Medicare per person age 65 or older t years after 1990.

62. The polynomial

$$6t^4 - 207t^3 + 2128t^2 - 6622t + 15{,}220$$

describes the annual number of drug convictions in the United States t years after 1984. The polynomial

$$28t^4 - 711t^3 + 5963t^2 - 1695t + 27{,}424$$

describes the annual number of drug arrests in the United States t years after 1984. Write a rational expression that describes the conviction rate for drug arrests in the United States t years after 1984.

63. The rational expression

$$\frac{130x}{100-x}$$

describes the cost, in millions of dollars, to inoculate x percent of the population against a particular strain of flu.

a. Evaluate the expression for $x = 40$, $x = 80$, and $x = 90$. Describe the meaning of each evaluation in terms of percentage inoculated and cost.

b. For what value of x is the function undefined?

c. What happens to the cost as x approaches 100%? How can you interpret this observation?

64. Doctors use the rational expression

$$\frac{DA}{A+12}$$

to determine the dosage of a drug prescribed for children. In this expression, A = child's age, and D = adult dosage. What is the difference in the child's dosage for a 7-year-old child and a 3-year-old child? Express the answer as a single rational expression in terms of D. Then describe what your answer means in terms of the variables in the rational expression.

65. The average speed on a round-trip commute having a one-way distance d is given by the complex rational expression

$$\frac{2d}{\frac{d}{r_1}+\frac{d}{r_2}}$$

in which r_1 and r_2 are the speeds on the outgoing and return trips, respectively. Simplify the expression. Then find the average speed for a person who drives from home to work at 30 miles per hour and returns on the same route averaging 20 miles per hour. Explain why the answer is not 25 miles per hour.

Writing in Mathematics

66. What is a rational expression?

67. Explain how to determine what numbers must be excluded from the domain of a rational expression.

68. Explain how to simplify a rational expression.

69. Explain how to multiply rational expressions.

70. Explain how to divide rational expressions.

71. Explain how to add or subtract rational expressions with the same denominators.

72. Explain how to add rational expressions having no common factors in their denominators. Use $\dfrac{3}{x+5} + \dfrac{7}{x+2}$ in your explanation.

73. Explain how to find the least common denominator for denominators of $x^2 - 100$ and $x^2 - 20x + 100$.

74. Describe two ways to simplify $\dfrac{\dfrac{3}{x}+\dfrac{2}{x^2}}{\dfrac{1}{x^2}+\dfrac{2}{x}}$.

Explain the error in Exercises 75–77. Then rewrite the right side of the equation to correct the error that now exists.

75. $\dfrac{1}{a}+\dfrac{1}{b}=\dfrac{1}{a+b}$

76. $\dfrac{1}{x}+7=\dfrac{1}{x+7}$

77. $\dfrac{a}{x}+\dfrac{a}{b}=\dfrac{a}{x+b}$

78. A politician claims that each year the conviction rate for drug arrests in the United States is increasing. Explain how to use the polynomials in Exercise 62 to verify this claim.

Technology Exercise

79. The polynomial

$$413.48t^2 + 185.72t + 24{,}031.95$$

describes the amount of Medicaid payments, in millions of dollars, t years after 1980. The polynomial

$$0.004t^2 + 0.02t^3 + 0.01t^2 - 0.24t + 21.66$$

describes the annual number of Medicaid recipients in the United States t years after 1980. Use a calculator to find the amount paid per recipient of Medicaid each year from 1988 to 2000. In what year did the amount paid per recipient fall below $900?

Critical Thinking Exercises

80. Which one of the following is true?

a. $\dfrac{x^2-25}{x-5}=x-5$

b. $\dfrac{x}{y}\div\dfrac{y}{x}=1$, if $x \neq 0$ and $y \neq 0$.

c. The least common denominator needed to find $\dfrac{1}{x}+\dfrac{1}{x+3}$ is $x+3$.

d. The rational expression

$$\frac{x^2-16}{x-4}$$

is not defined for $x = 4$. However, as x gets closer and closer to 4, the value of the expression approaches 8.

In Exercises 81–82, find the missing expression.

81. $\dfrac{3x}{x-5}+\dfrac{\square}{5-x}=\dfrac{7x+1}{x-5}$

82. $\dfrac{4}{x-2}-\boxed{\dfrac{\quad}{\quad}}=\dfrac{2x+8}{(x-2)(x+1)}$

83. In one short sentence, five words or less, explain what

$$\frac{\dfrac{1}{x}+\dfrac{1}{x^2}+\dfrac{1}{x^3}}{\dfrac{1}{x^4}+\dfrac{1}{x^5}+\dfrac{1}{x^6}}$$

does to each number x.

SECTION P.7 Complex Numbers

Objectives

1. Add and subtract complex numbers.
2. Multiply complex numbers.
3. Divide complex numbers.
4. Perform operations with square roots of negative numbers.

© 2000 Roz Chast from Cartoonbank.com. All rights reserved.

Who is this kid warning us about our eyeballs turning black if we attempt to find the square root of −9? Don't believe what you hear on the street. Although square roots of negative numbers are not real numbers, they do play a significant role in algebra. In this section, we move beyond the real numbers and discuss square roots with negative radicands.

The Imaginary Unit *i*

In Chapter 1, we'll be studying equations whose solutions involve the square roots of negative numbers. Because the square of a real number is never negative, there is no real number x such that $x^2 = -1$. To provide a setting in which such equations have solutions, mathematicians invented an expanded system of numbers, the complex numbers. The imaginary number i, defined to be a solution to the equation $x^2 = -1$, is the basis of this new set.

The Imaginary Unit *i*

The imaginary unit i is defined as

$$i = \sqrt{-1}, \quad \text{where} \quad i^2 = -1.$$

Using the imaginary unit i, we can express the square root of any negative number as a real multiple of i. For example,

$$\sqrt{-25} = i\sqrt{25} = 5i.$$

We can check this result by squaring $5i$ and obtaining −25.

$$(5i)^2 = 5^2i^2 = 25(-1) = -25$$

A new system of numbers, called **complex numbers**, is based on adding multiples of i, such as $5i$, to the real numbers.

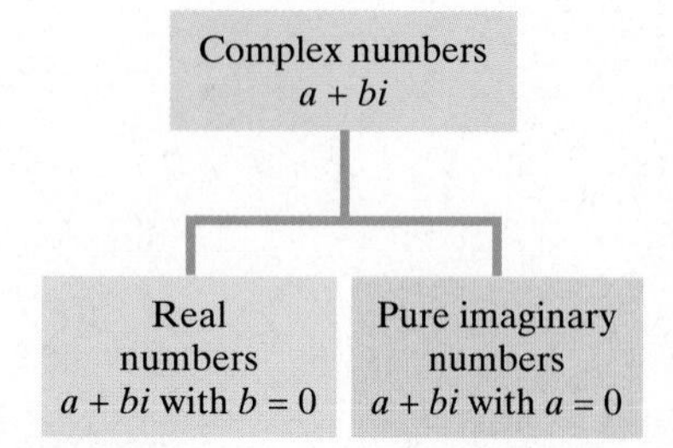

Figure P.7 The complex number system

Complex Numbers

The set of all numbers in the form

$$a + bi$$

with real numbers a and b, and i, the imaginary unit, is called the set of **complex numbers**. The real number a is called the **real part**, and the real number b is called the **imaginary part**, of the complex number $a + bi$. If $a = 0$ and $b \neq 0$, then the complex number bi is called a **pure imaginary number** (Figure P.7).

A complex number is said to be **simplified** if it is expressed in the **standard form** $a + bi$. If b is a radical, we usually write i before b. For example, we write $7 + i\sqrt{5}$ rather than $7 + \sqrt{5}\,i$, which could easily be confused with $7 + \sqrt{5i}$.

Expressed in standard form, two complex numbers are equal if and only if their real parts are equal and their imaginary parts are equal.

Equality of Complex Numbers

$a + bi = c + di$ if and only if $a = c$ and $b = d$.

1 Add and subtract complex numbers.

Operations with Complex Numbers

The form of a complex number $a + bi$ is like the binomial $a + bx$. Consequently, we can add, subtract, and multiply complex numbers using the same methods we used for binomials, remembering that $i^2 = -1$.

Adding and Subtracting Complex Numbers

1. $(a + bi) + (c + di) = (a + c) + (b + d)i$
   In words, this says that you add complex numbers by adding their real parts, adding their imaginary parts, and expressing the sum as a complex number.
2. $(a + bi) - (c + di) = (a - c) + (b - d)i$
   In words, this says that you subtract complex numbers by subtracting their real parts, subtracting their imaginary parts, and expressing the difference as a complex number.

EXAMPLE 1 Adding and Subtracting Complex Numbers

Perform the indicated operations, writing the result in standard form.

a. $(5 - 11i) + (7 + 4i)$ b. $(-5 + 7i) - (-11 - 6i)$

Study Tip

The following examples, using the same integers as in Example 1, show how operations with complex numbers are just like operations with polynomials.

a. $(5 - 11x) + (7 + 4x)$
$= 12 - 7x$

b. $(-5 + 7x) - (-11 - 6x)$
$= -5 + 7x + 11 + 6x$
$= 6 + 13x$

Solution

a. $(5 - 11i) + (7 + 4i)$

$= 5 - 11i + 7 + 4i$ Remove the parentheses.

$= 5 + 7 - 11i + 4i$ Group real and imaginary terms.

$= (5 + 7) + (-11 + 4)i$

$= 12 - 7i$ Add real parts and add imaginary parts.

b. $(-5 + 7i) - (-11 - 6i)$

$= -5 + 7i + 11 + 6i$ Remove the parentheses.

$= -5 + 11 + 7i + 6i$ Group real and imaginary terms.

$= (-5 + 11) + (7 + 6)i$

$= 6 + 13i$

Check Point 1 Add or subtract as indicated.

a. $(5 - 2i) + (3 + 3i)$ **b.** $(2 + 6i) - (12 - 4i)$

2 Multiply complex numbers.

Multiplication of complex numbers is performed the same way as multiplication of polynomials, using the distributive property and the FOIL method. After completing the multiplication, we replace i^2 with -1. This idea is illustrated in the next example.

EXAMPLE 2 Multiplying Complex Numbers

Find the products: a. $4i(3 - 5i)$ b. $(7 - 3i)(-2 - 5i)$

Solution

a. $4i(3 - 5i) = 4i(3) - 4i(5i)$ Distribute $4i$ throughout the parentheses.
$= 12i - 20i^2$ Multiply.
$= 12i - 20(-1)$ Replace i^2 with -1.
$= 20 + 12i$ Simplify to $12i + 20$ and write in standard form.

b. $(7 - 3i)(-2 - 5i)$

F O I L

$= -14 - 35i + 6i + 15i^2$ Use the FOIL method.
$= -14 - 35i + 6i + 15(-1)$ $i^2 = -1$
$= -14 - 15 - 35i + 6i$ Group real and imaginary terms.
$= -29 - 29i$ Combine real and imaginary terms.

Check Point 2 Find the products:

a. $7i(2 - 9i)$ **b.** $(5 + 4i)(6 - 7i)$

3 Divide complex numbers.

Complex Conjugates and Division

It is possible to multiply complex numbers and obtain a real number. This occurs when we multiply $a + bi$ and $a - bi$.

F O I L

$$(a + bi)(a - bi) = a^2 - abi + abi - b^2i^2$$ Use the FOIL method.
$$= a^2 - b^2(-1)$$ $i^2 = -1$
$$= a^2 + b^2$$ Notice that this product eliminates i.

For the complex number $a + bi$, we define its **complex conjugate** to be $a - bi$. The multiplication of complex conjugates results in a real number.

Conjugate of a Complex Number

The **complex conjugate** of the number $a + bi$ is $a - bi$, and the complex conjugate of $a - bi$ is $a + bi$. The multiplication of complex conjugates gives a real number.

$$(a + bi)(a - bi) = a^2 + b^2$$
$$(a - bi)(a + bi) = a^2 + b^2$$

Complex conjugates are used to divide complex numbers. By multiplying the numerator and the denominator of the division by the complex conjugate of the denominator, you will obtain a real number in the denominator.

Complex Numbers on a Postage Stamp

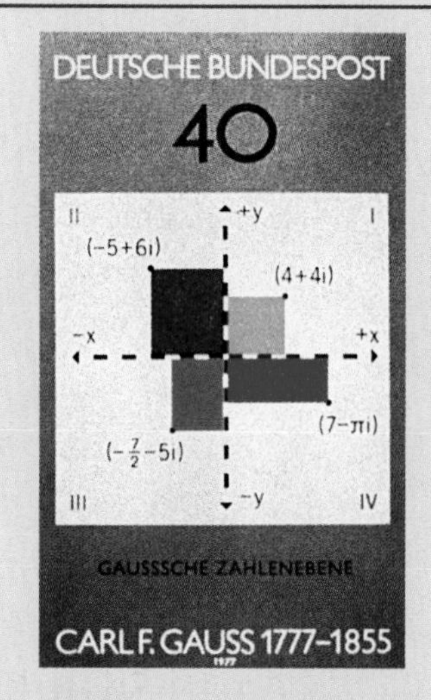

This stamp honors the work done by the German mathematician Carl Friedrich Gauss (1777–1855) with complex numbers. Gauss represented complex numbers as points in the plane.

EXAMPLE 3 Using Complex Conjugates to Divide Complex Numbers

Divide: $7 + 4i$ by $2 - 5i$.

Solution We first write the problem as $\frac{7 + 4i}{2 - 5i}$. The complex conjugate of the denominator, $2 - 5i$, is $2 + 5i$, so we multiply the numerator and the denominator by $2 + 5i$.

$$\frac{7 + 4i}{2 - 5i} = \frac{(7 + 4i)}{(2 - 5i)} \cdot \frac{(2 + 5i)}{(2 + 5i)}$$

Multiply the numerator and the denominator by the complex conjugate of the denominator.

F O I L

$$= \frac{14 + 35i + 8i + 20i^2}{2^2 + 5^2}$$

Use the FOIL method in the numerator and $(a - bi)(a + bi) = a^2 + b^2$ in the denominator.

$$= \frac{14 + 43i + 20(-1)}{29}$$

Replace i^2 by -1.

$$= \frac{-6 + 43i}{29}$$

Combine real terms in the numerator.

$$= -\frac{6}{29} + \frac{43}{29}i$$

Observe that the quotient is expressed in the standard form $a + bi$, with $a = -\frac{6}{29}$ and $b = \frac{43}{29}$.

Check Point 3 Divide: $\frac{5 + 4i}{4 - 2i}$.

4 Perform operations with square roots of negative numbers.

Roots of Negative Numbers

The square of $4i$ and the square of $-4i$ both result in -16.

$$(4i)^2 = 16i^2 = 16(-1) = -16 \qquad (-4i)^2 = 16i^2 = -16$$

Consequently, in the complex number system -16 has two square roots, namely, $4i$ and $-4i$. We call $4i$ the **principal square root** of -16.

Principal Square Root of a Negative Number

For any positive number real number b, the **principal square root** of the negative number $-b$ is defined by

$$\sqrt{-b} = i\sqrt{b}.$$

EXAMPLE 4 Operations Involving Square Roots of Negative Numbers

Perform the indicated operations and write the result in standard form:

a. $\sqrt{-18} - \sqrt{-8}$ b. $(-1 + \sqrt{-5})^2$ c. $\frac{-25 + \sqrt{-50}}{15}$

Study Tip

Do not apply the properties

$$\sqrt{b}\sqrt{c} = \sqrt{bc}$$

and

$$\frac{\sqrt{b}}{\sqrt{c}} = \sqrt{\frac{b}{c}}$$

to the pure imaginary numbers because these properties can only be used when b and c are positive.

Correct:

$$\begin{aligned}\sqrt{-25}\sqrt{-4} &= i\sqrt{25}\,i\sqrt{4}\\ &= (5i)(2i)\\ &= 10i^2\\ &= 10(-1)\\ &= -10\end{aligned}$$

Incorrect:

$$\begin{aligned}\sqrt{-25}\sqrt{-4} &= \sqrt{(-25)(-4)}\\ &= \sqrt{100}\\ &= 10\end{aligned}$$

One way to avoid confusion is to represent square roots of negative numbers in terms of i before performing any operations.

Solution Begin by expressing all square roots of negative numbers in terms of i.

a. $$\begin{aligned}\sqrt{-18} - \sqrt{-8} &= i\sqrt{18} - i\sqrt{8} = i\sqrt{9 \cdot 2} - i\sqrt{4 \cdot 2}\\ &= 3i\sqrt{2} - 2i\sqrt{2} = i\sqrt{2}\end{aligned}$$

$(A + B)^2 = A^2 + 2AB + B^2$

b. $$\begin{aligned}(-1 + \sqrt{-5})^2 = (-1 + i\sqrt{5})^2 &= (-1)^2 + 2(-1)(i\sqrt{5}) + (i\sqrt{5})^2\\ &= 1 - 2i\sqrt{5} + 5i^2\\ &= 1 - 2i\sqrt{5} + 5(-1)\\ &= -4 - 2i\sqrt{5}\end{aligned}$$

c. $$\frac{-25 + \sqrt{-50}}{15}$$

$= \frac{-25 + i\sqrt{50}}{15}$ $\quad \sqrt{-b} = i\sqrt{b}$

$= \frac{-25 + 5i\sqrt{2}}{15}$ $\quad \sqrt{50} = \sqrt{25 \cdot 2} = 5\sqrt{2}$

$= \frac{-25}{15} + \frac{5i\sqrt{2}}{15}$ $\quad$ Write the complex number in standard form.

$= -\frac{5}{3} + i\frac{\sqrt{2}}{3}$ $\quad$ Simplify.

Check Point 4 Perform the indicated operations and write the result in standard form.

a. $\sqrt{-27} + \sqrt{-48}$ **b.** $(-2 + \sqrt{-3})^2$ **c.** $\frac{-14 + \sqrt{-12}}{2}$

EXERCISE SET P.7

Practice Exercises

In Exercises 1–8, add or subtract as indicated and write the result in standard form.

1. $(7 + 2i) + (1 - 4i)$ **2.** $(-2 + 6i) + (4 - i)$
3. $(3 + 2i) - (5 - 7i)$ **4.** $(-7 + 5i) - (-9 - 11i)$
5. $6 - (-5 + 4i) - (-13 - 11i)$
6. $7 - (-9 + 2i) - (-17 - 6i)$
7. $8i - (14 - 9i)$ **8.** $15i - (12 - 11i)$

In Exercises 9–20, find each product and write the result in standard form.

9. $-3i(7i - 5)$ **10.** $-8i(2i - 7)$
11. $(-5 + 4i)(3 + 7i)$ **12.** $(-4 - 8i)(3 + 9i)$
13. $(7 - 5i)(-2 - 3i)$ **14.** $(8 - 4i)(-3 + 9i)$
15. $(3 + 5i)(3 - 5i)$ **16.** $(2 + 7i)(2 - 7i)$
17. $(-5 + 3i)(-5 - 3i)$ **18.** $(-7 - 4i)(-7 + 4i)$
19. $(2 + 3i)^2$ **20.** $(5 - 2i)^2$

In Exercises 21–28, divide and express the result in standard form.

21. $\frac{2}{3 - i}$ **22.** $\frac{3}{4 + i}$
23. $\frac{2i}{1 + i}$ **24.** $\frac{5i}{2 - i}$
25. $\frac{8i}{4 - 3i}$ **26.** $\frac{-6i}{3 + 2i}$
27. $\frac{2 + 3i}{2 + i}$ **28.** $\frac{3 - 4i}{4 + 3i}$

In Exercises 29–44, perform the indicated operations and write the result in standard form.

29. $\sqrt{-64} - \sqrt{-25}$ **30.** $\sqrt{-81} - \sqrt{-144}$
31. $5\sqrt{-16} + 3\sqrt{-81}$ **32.** $5\sqrt{-8} + 3\sqrt{-18}$
33. $(-2 + \sqrt{-4})^2$ **34.** $(-5 - \sqrt{-9})^2$
35. $(-3 - \sqrt{-7})^2$ **36.** $(-2 + \sqrt{-11})^2$

37. $\dfrac{-8+\sqrt{-32}}{24}$

38. $\dfrac{-12+\sqrt{-28}}{32}$

39. $\dfrac{-6-\sqrt{-12}}{48}$

40. $\dfrac{-15-\sqrt{-18}}{33}$

41. $\sqrt{-8}(\sqrt{-3}-\sqrt{5})$

42. $\sqrt{-12}(\sqrt{-4}-\sqrt{2})$

43. $(3\sqrt{-5})(-4\sqrt{-12})$

44. $(3\sqrt{-7})(2\sqrt{-8})$

Writing in Mathematics

45. What is i?
46. Explain how to add complex numbers. Provide an example with your explanation.
47. Explain how to multiply complex numbers and give an example.
48. What is the complex conjugate of $2+3i$? What happens when you multiply this complex number by its complex conjugate?
49. Explain how to divide complex numbers. Provide an example with your explanation.
50. A stand-up comedian uses algebra in some jokes, including one about a telephone recording that announces "You have just reached an imaginary number. Please multiply by i and dial again." Explain the joke.

Explain the error in Exercises 51–52.

51. $\sqrt{-9}+\sqrt{-16}=\sqrt{-25}=i\sqrt{25}=5i$
52. $(\sqrt{-9})^2=\sqrt{-9}\cdot\sqrt{-9}=\sqrt{81}=9$

Critical Thinking Exercises

53. Which one of the following is true?
    a. Some irrational numbers are not complex numbers.
    b. $(3+7i)(3-7i)$ is an imaginary number.
    c. $\dfrac{7+3i}{5+3i}=\dfrac{7}{5}$
    d. In the complex number system, x^2+y^2 (the sum of two squares) can be factored as $(x+yi)(x-yi)$.

In Exercises 54–56, perform the indicated operations and write the result in standard form.

54. $(8+9i)(2-i)-(1-i)(1+i)$

55. $\dfrac{4}{(2+i)(3-i)}$

56. $\dfrac{1+i}{1+2i}+\dfrac{1-i}{1-2i}$

57. Evaluate x^2-2x+2 for $x=1+i$.

SECTION P.8 *Graphs and Graphing Utilities*

Objectives

1. Plot points in the rectangular coordinate system.
2. Graph equations in the rectangular coordinate system.
3. Interpret information about a graphing utility's viewing rectangle.
4. Use a graph to determine intercepts.
5. Find the distance between two points.
6. Find the midpoint of a line segment.
7. Interpret information given by graphs.

The beginning of the seventeenth century was a time of innovative ideas and enormous intellectual progress in Europe. English theatergoers enjoyed a succession of exciting new plays by Shakespeare. William Harvey proposed the radical notion that the heart was a pump for blood rather than the center of emotion. Galileo, with his new-fangled invention called the telescope, supported the theory of Polish astronomer Copernicus that the sun, not the Earth, was the center of the solar system. Monteverdi was writing the world's first grand operas. French mathematicians Pascal and Fermat invented a new field of mathematics called probability theory.

Into this arena of intellectual electricity stepped French aristocrat René Descartes (1596–1650). Descartes, propelled by the creativity surrounding him, developed a new branch of mathematics that brought together algebra and geometry in a unified way—a way that visualized numbers as points on a graph, equations as geometric figures, and geometric figures as equations. This new branch of mathematics, called *analytic geometry*, established Descartes as one of the founders of modern thought and among the most original mathematicians and philosophers of any age. We begin this section by looking at Descartes's deceptively simple idea, called the **rectangular coordinate system** or (in his honor) the **Cartesian coordinate system**.

1 Plot points in the rectangular coordinate system.

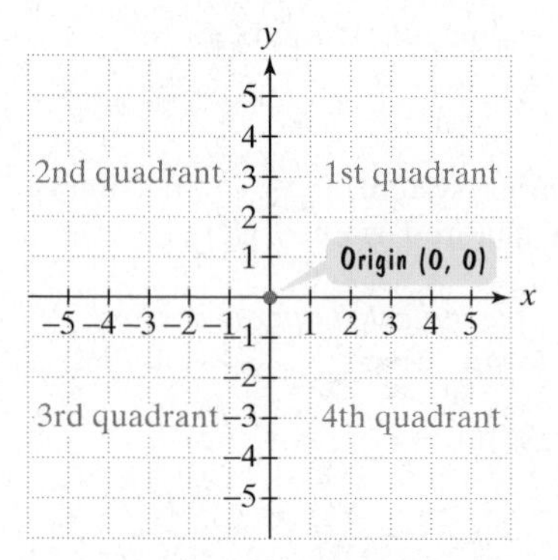

Figure P.8 The rectangular coordinate system

Points and Ordered Pairs

Descartes used two number lines that intersect at right angles at their zero points, as shown in Figure P.8. The horizontal number line is the ***x*-axis**. The vertical number line is the ***y*-axis**. The point of intersection of these axes is the **origin**. Positive numbers are shown to the right and above the origin. Negative numbers are shown to the left and below the origin. The axes divide the plane into four quarters, called **quadrants**. The points located on the axes are not in any quadrant.

Each point in the rectangular coordinate system corresponds to an **ordered pair** of real numbers, (x, y). Examples of such pairs are $(4, 2)$ and $(-5, -3)$. The first number in each pair, called the ***x*-coordinate**, denotes the distance and direction from the origin along the x-axis. The second number, called the ***y*-coordinate**, denotes vertical distance and direction along a line parallel to the y-axis or along the y-axis itself.

Figure P.9 shows how we **plot**, or locate, the points corresponding to the ordered pairs $(4, 2)$ and $(-5, -3)$. We plot $(4, 2)$ by going 4 units from 0 to the right along the x-axis. Then we go 2 units up parallel to the y-axis. We plot $(-5, -3)$ by going 5 units from 0 to the left along the x-axis and 3 units down parallel to the y-axis. The phrase "the point corresponding to the ordered pair $(-5, -3)$" is often abbreviated as "the point $(-5, -3)$."

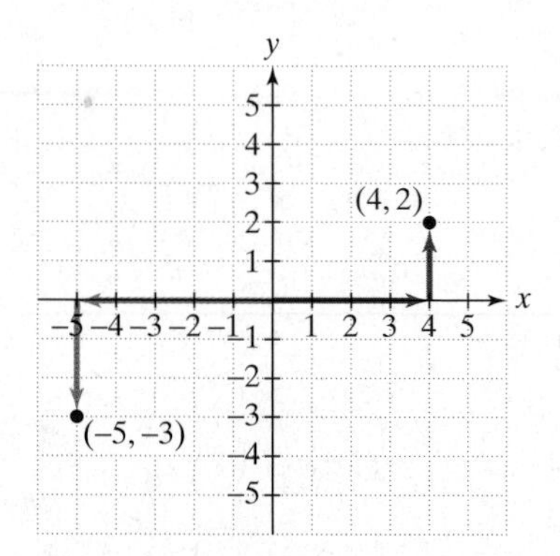

Figure P.9 Plotting $(4, 2)$ and $(-5, -3)$

EXAMPLE 1 Plotting Points in the Rectangular Coordinate System

Plot the points $A(-3, 5)$, $B(2, -4)$, $C(5, 0)$, $D(-5, -3)$, $E(0, 4)$, and $F(0, 0)$.

Solution See Figure P.10. We plot the points in the following way:

$A(-3, 5)$:	3 units left, 5 units up
$B(2, -4)$:	2 units right, 4 units down
$C(5, 0)$:	5 units right, 0 units up or down
$D(-5, -3)$:	5 units left, 3 units down
$E(0, 4)$:	0 units right or left, 4 units up
$F(0, 0)$:	0 units right or left, 0 units up or down

The origin is represented by $(0, 0)$.

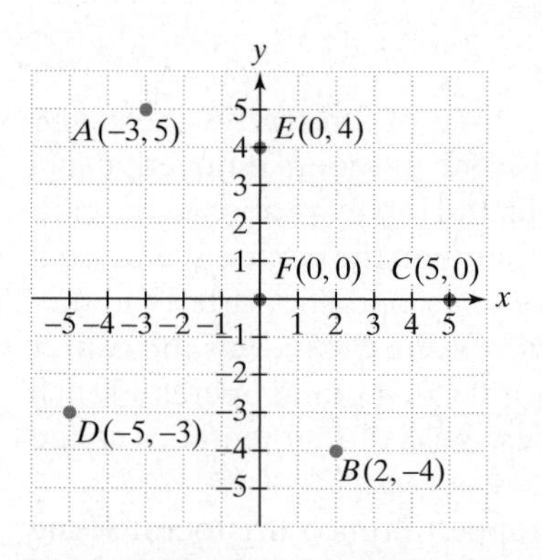

Figure P.10 Plotting points

Check Point 1 Plot the points

$A(-2, 4)$, $B(4, -2)$, $C(-3, 0)$, and $D(0, -3)$.

2 Graph equations in the rectangular coordinate system.

Graphs of Equations

A relationship between two quantities can be expressed as an **equation in two variables**, such as

$$y = x^2 - 4.$$

A **solution** to this equation is an ordered pair of real numbers with the following property: When the x-coordinate is substituted for x and the y-coordinate is substituted for y in the equation, we obtain a true statement. For example, if we let $x = 3$, then $y = 3^2 - 4 = 9 - 4 = 5$. The ordered pair $(3, 5)$ is a solution to the equation $y = x^2 - 4$. We also say that $(3, 5)$ **satisfies** the equation.

We can generate as many ordered-pair solutions as desired to $y = x^2 - 4$ by substituting numbers for x and then finding the values for y. The **graph of the equation** is the set of all points whose coordinates satisfy the equation.

One method for graphing an equation such as $y = x^2 - 4$ is the **point-plotting method**. First, we find several ordered pairs that are solutions to the equation. Next, we plot these ordered pairs as points in the rectangular coordinate system. Finally, we connect the points with a smooth curve or line. This often gives us a picture of all ordered pairs that satisfy the equation.

EXAMPLE 2 Graphing an Equation Using the Point-Plotting Method

Graph $y = x^2 - 4$. Select integers for x, starting with -3 and ending with 3.

Solution For each value of x we find the corresponding value for y.

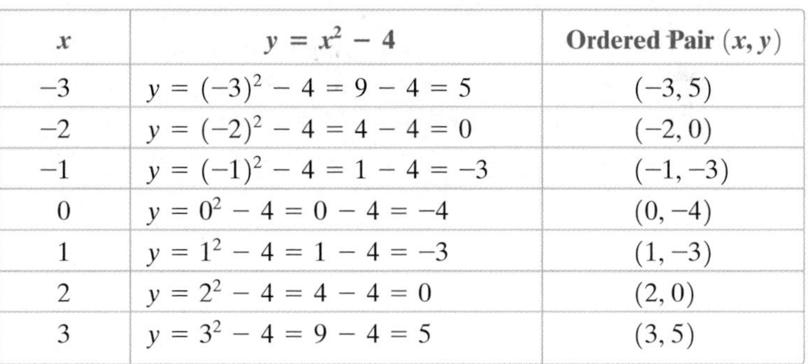
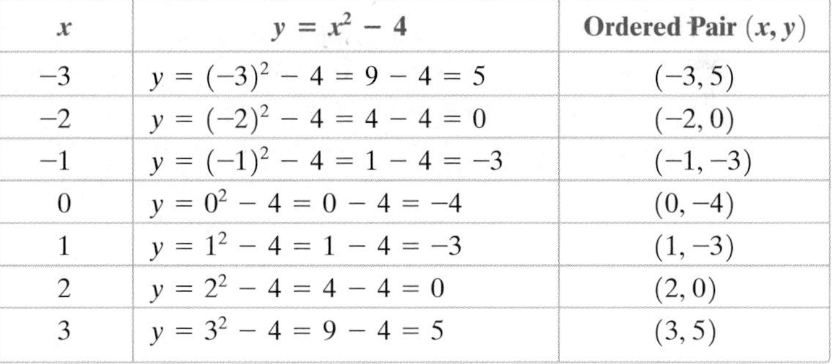

x	$y = x^2 - 4$	Ordered Pair (x, y)
-3	$y = (-3)^2 - 4 = 9 - 4 = 5$	$(-3, 5)$
-2	$y = (-2)^2 - 4 = 4 - 4 = 0$	$(-2, 0)$
-1	$y = (-1)^2 - 4 = 1 - 4 = -3$	$(-1, -3)$
0	$y = 0^2 - 4 = 0 - 4 = -4$	$(0, -4)$
1	$y = 1^2 - 4 = 1 - 4 = -3$	$(1, -3)$
2	$y = 2^2 - 4 = 4 - 4 = 0$	$(2, 0)$
3	$y = 3^2 - 4 = 9 - 4 = 5$	$(3, 5)$

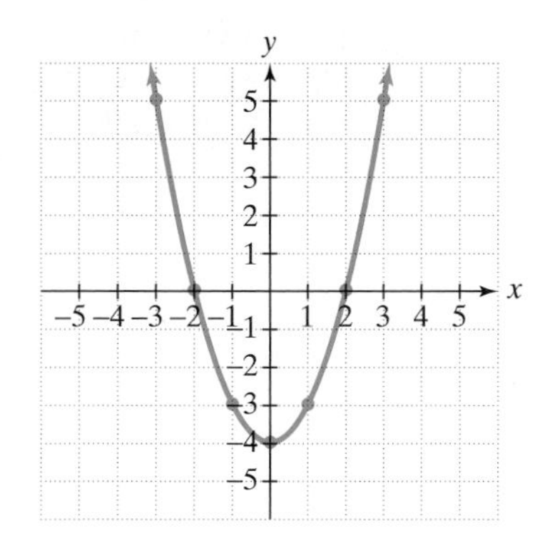

Figure P.11 The graph of $y = x^2 - 4$

Now we plot the seven points and join them with a smooth curve, as shown in Figure P.11. The graph of $y = x^2 - 4$ is a curve where the part of the graph to the right of the y-axis is a reflection of the part to the left of it and vice versa. The arrows on the left and the right of the curve indicate that it extends indefinitely in both directions.

Check Point 2 Graph $y = 2x - 4$. Select integers for x, starting with -1 and ending with 3.

3 Interpret information about a graphing utility's viewing rectangle.

Graphing calculators or graphing software packages for computers are referred to as **graphing utilities** or graphers. A graphing utility is a powerful tool that quickly generates the graph of an equation in two variables. Figure P.12 on page 74 shows two such graphs for the equations in Example 2 and the checkpoint example.

What differences do you notice between these graphs and the graphs that we (and you) drew by hand? They do seem a bit "jittery." Arrows do not appear

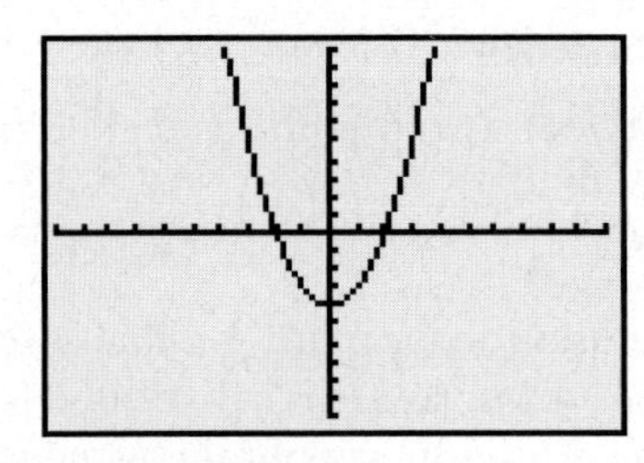

Figure P.12(a)
The graph of $y = x^2 - 4$

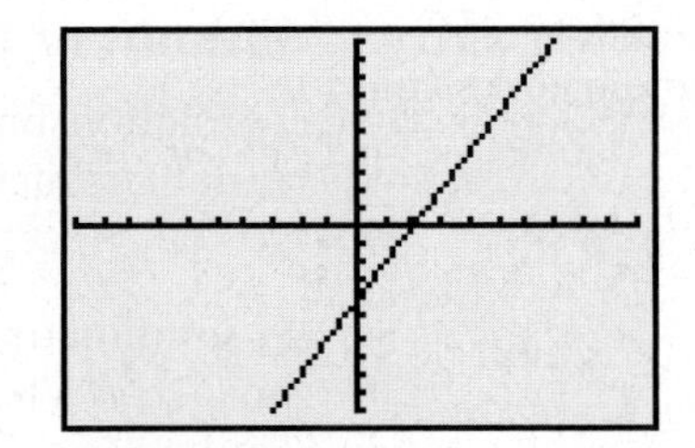

Figure P.12(b)
The graph of $y = 2x - 4$

on the left and right ends of the graphs. Furthermore, numbers are not given along the axes. For both graphs in Figure P.12, the x-axis extends from -10 to 10 and the y-axis also extends from -10 to 10. The distance represented by each consecutive tick mark is one unit. We say that the **viewing rectangle** is $[-10, 10, 1]$ by $[-10, 10, 1]$.

$[-10,$ $10,$ $1]$ by $[-10,$ $10,$ $1].$

$[-10,$	$10,$	$1]$	$[-10,$	$10,$	$1].$
The minimum x-value along the x-axis is −10.	The maximum x-value along the x-axis is 10.	Distance between consecutive tick marks on the x-axis is one unit.	The minimum y-value along the y-axis is −10.	The maximum y-value along the y-axis is 10.	Distance between consecutive tick marks on the y-axis is one unit.

To graph an equation in x and y using a graphing utility, enter the equation and specify the size of the viewing rectangle. The size of the viewing rectangle sets minimum and maximum values for both the x- and y-axes. Enter these values, as well as the values between consecutive tick marks, on the respective axes. The $[-10, 10, 1]$ by $[-10, 10, 1]$ viewing rectangle used in Figure P.12 is called the **standard viewing rectangle**.

EXAMPLE 3 Understanding the Viewing Rectangle

What is the meaning of a $[-2, 3, 0.5]$ by $[-10, 20, 5]$ viewing rectangle?

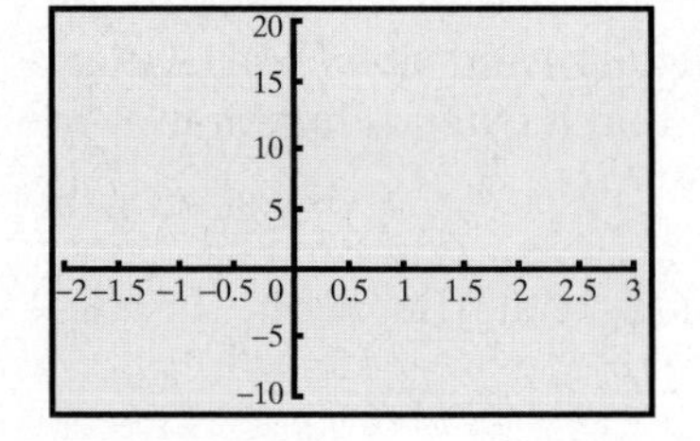

Figure P.13 A $[-2, 3, 0.5]$ by $[-10, 20, 5]$ viewing rectangle

Solution We begin with $[-2, 3, 0.5]$, which describes the x-axis. The minimum x-value is -2 and the maximum x-value is 3. The distance between consecutive tick marks is 0.5.

Next, consider $[-10, 20, 5]$, which describes the y-axis. The minimum y-value is -10 and the maximum y-value is 20. The distance between consecutive tick marks is 5.

Figure P.13 illustrates a $[-2, 3, 0.5]$ by $[-10, 20, 5]$ viewing rectangle. To make things clearer, we've placed numbers by each tick mark. These numbers do not appear on the axes when you use a graphing utility to graph an equation.

Check Point 3 What is the meaning of a $[-100, 100, 50]$ by $[-100, 100, 10]$ viewing rectangle? Create a figure like the one in Figure P.13 that illustrates this viewing rectangle.

On most graphing utilities, the display screen is two-thirds as high as it is wide. By using a square setting, you can make the x and y tick marks be equally spaced. (This does not occur in the standard viewing rectangle.) Graphing utilities can also *zoom in* and *zoom out*. When you zoom in, you see a smaller portion of the graph, but you do so in greater detail. When you zoom out, you see a larger portion of the graph. Thus, zooming out may help you to develop a better understanding of the overall character of the graph. With practice, you will become more comfortable with graphing equations in two variables using your graphing utility. You will also develop a better sense of the size of the viewing rectangle that will reveal needed information about a particular graph.

4 Use a graph to determine intercepts.

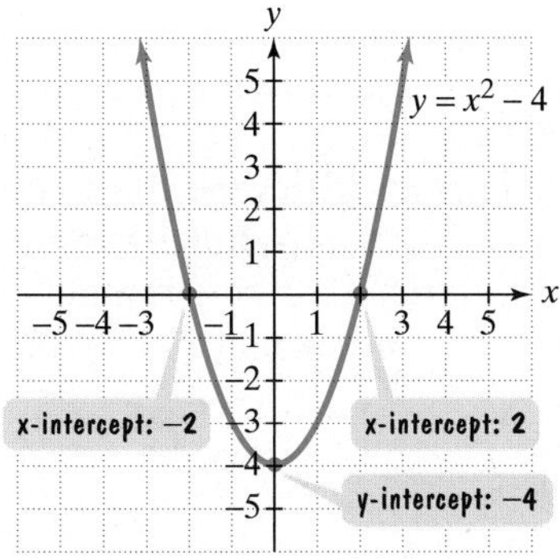

Figure P.14 Intercepts of $y = x^2 - 4$

Intercepts

An ***x*-intercept** of a graph is an x-coordinate of the point where the graph intersects the x-axis. For example, look at the graph of $y = x^2 - 4$ in Figure P.14. The graph crosses the x-axis at $(-2, 0)$ and $(2, 0)$. Thus, the x-intercepts are -2 and 2. **The *y*-coordinate corresponding to a graph's *x*-intercept is always zero.**

A ***y*-intercept** of a graph is a y-coordinate of the point where the graph intersects the y-axis. The graph of $y = x^2 - 4$ in Figure P.14 shows that the graph crosses the y-axis at $(0, -4)$. Thus, the y-intercept is -4. **The *x*-coordinate corresponding to a graph's *y*-intercept is always zero.**

Figure P.15 illustrates that a graph may have no intercepts or several intercepts.

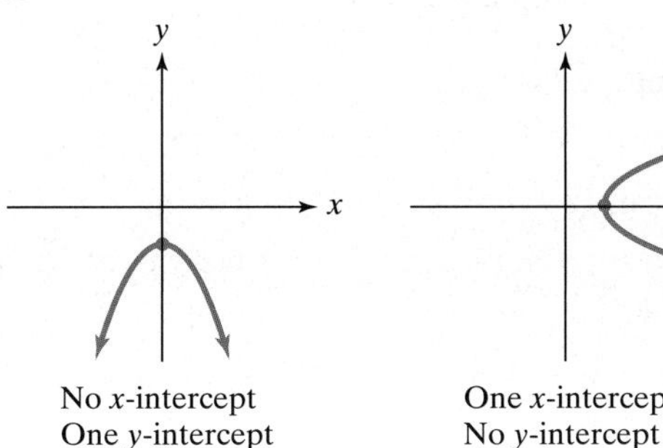

No x-intercept
One y-intercept

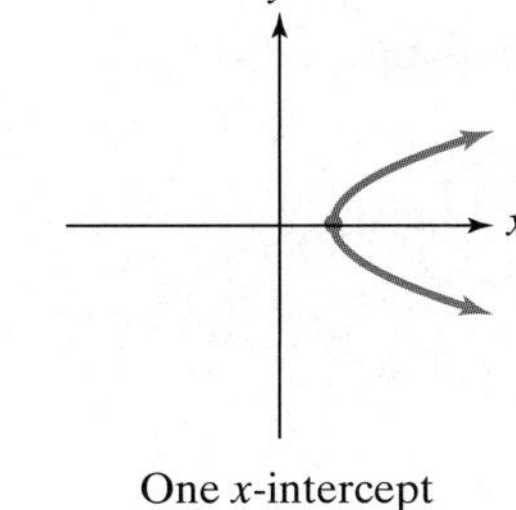

One x-intercept
No y-intercept

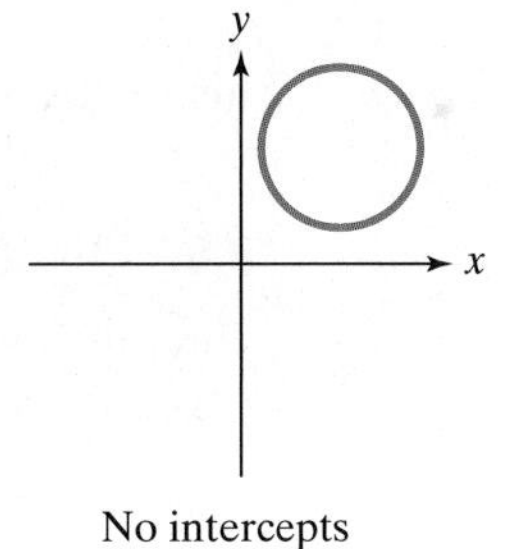

No intercepts

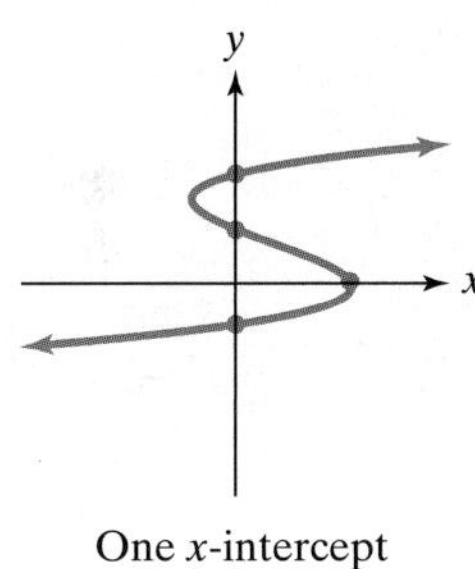

One x-intercept
Three y-intercepts

Figure P.15

5 Find the distance between two points.

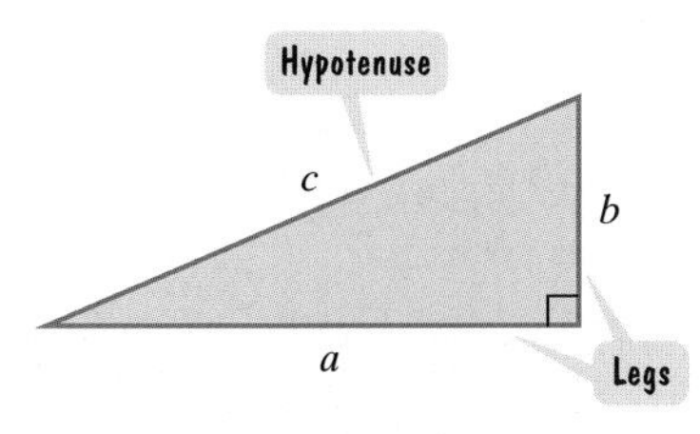

Figure P.16 The Pythagorean Theorem: $a^2 + b^2 = c^2$

The Distance Formula

We can use the **Pythagorean Theorem** to develop a formula that gives the distance between any two points in the rectangular coordinate system. Recall that the Pythagorean Theorem involves a triangle with a right (90°) angle, called a **right triangle**. The side opposite the right angle is called the **hypotenuse**. The other two sides are called the **legs**. In Figure P.16, a and b represent the lengths of the two legs of the triangle. Likewise, c represents the length of the hypotenuse. Based on this figure, we express the Pythagorean Theorem as

$$a^2 + b^2 = c^2.$$

In any right triangle, the sum of the squares of the lengths of the legs is the square of the length of the hypotenuse.

Now we are ready to find the distance between the two points $P_1(x_1, y_1)$ and $P_2(x_2, y_2)$ in the rectangular coordinate system. The two points are illustrated in Figure P.17 on the next page.

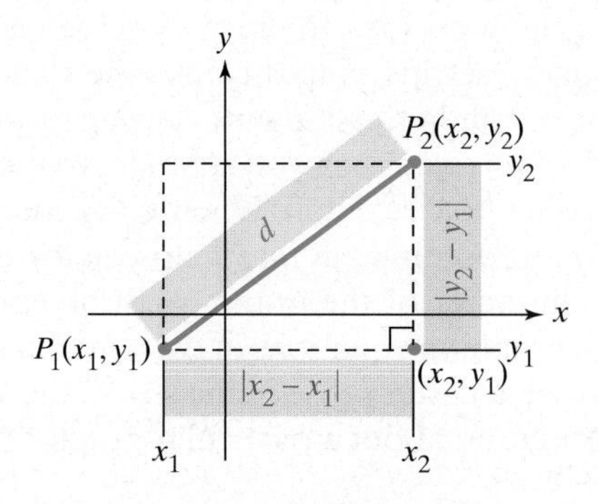

Figure P.17

The distance that we need to find is represented by d and shown in blue. Notice that the distance between two points on the dashed horizontal line is the absolute value of the difference between the x-coordinates of the two points. This distance, $|x_2 - x_1|$, is shown in pink. Similarly, the distance between two points on the dashed vertical line is the absolute value of the difference between the y-coordinates of the two points. This distance, $|y_2 - y_1|$, is also shown in pink.

Because the dashed lines are horizontal and vertical, a right triangle is formed. Thus, we can use the Pythagorean Theorem to find distance d. By the Pythagorean Theorem,

$$d^2 = |x_2 - x_1|^2 + |y_2 - y_1|^2$$
$$d = \sqrt{|x_2 - x_1|^2 + |y_2 - y_1|^2}$$
$$d = \sqrt{(x_2 - x_1)^2 + (y_2 - y_1)^2}.$$

This result is the **distance formula**.

The Distance Formula

The distance d between the points (x_1, y_1) and (x_2, y_2) in the rectangular coordinate system is

$$d = \sqrt{(x_2 - x_1)^2 + (y_2 - y_1)^2}.$$

When using the distance formula, it does not matter which point you call (x_1, y_1) and which you call (x_2, y_2).

EXAMPLE 4 Using the Distance Formula

Find the distance between $(-1, -3)$ and $(2, 3)$.

Solution Letting $(x_1, y_1) = (-1, -3)$ and $(x_2, y_2) = (2, 3)$, we obtain

$$d = \sqrt{(x_2 - x_1)^2 + (y_2 - y_1)^2}$$ Use the distance formula.

$$= \sqrt{[2 - (-1)]^2 + [3 - (-3)]^2}$$ Substitute the given values.

$$= \sqrt{(2 + 1)^2 + (3 + 3)^2}$$ Perform subtractions within the grouping symbols.

$$= \sqrt{3^2 + 6^2}$$

$$= \sqrt{9 + 36}$$ Square 3 and 6.

$$= \sqrt{45}$$ Add.

$$= 3\sqrt{5} \approx 6.7$$ $\sqrt{45} = \sqrt{9 \cdot 5} = \sqrt{9}\sqrt{5} = 3\sqrt{5}$

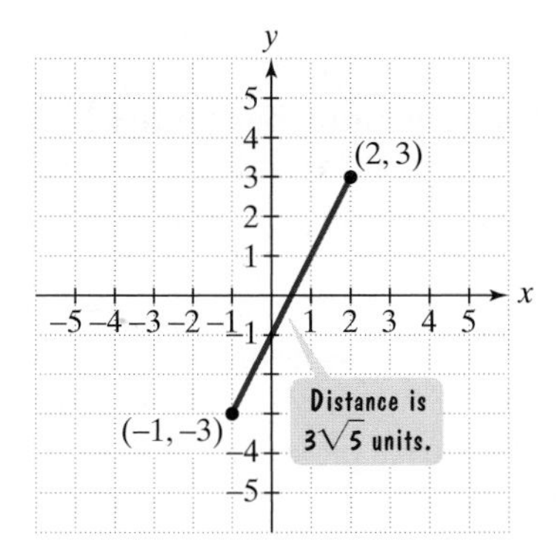

Figure P.18 Finding the distance between two points

The distance between the given points is $3\sqrt{5}$ units, or approximately 6.7 units. The situation is illustrated in Figure P.18.

Check Point 4 Find the distance between $(2, -2)$ and $(5, 2)$.

6 Find the midpoint of a line segment.

The Midpoint Formula

The distance formula can be used to prove a formula for finding the midpoint of a line segment between two given points. The formula is as follows.

> **The Midpoint Formula**
>
> Consider a line segment whose endpoints are (x_1, y_1) and (x_2, y_2). The coordinates of the segment's midpoint are
>
> $$\left(\frac{x_1 + x_2}{2}, \frac{y_1 + y_2}{2}\right).$$
>
> To find the midpoint, take the average of the two x-coordinates and of the two y-coordinates.

EXAMPLE 5 Using the Midpoint Formula

Find the midpoint of the line segment with endpoints $(1, -6)$ and $(-8, -4)$.

Solution To find the coordinates of the midpoint, we average the coordinates of the endpoints.

$$\text{Midpoint} = \left(\frac{1 + (-8)}{2}, \frac{-6 + (-4)}{2}\right) = \left(\frac{-7}{2}, \frac{-10}{2}\right) = \left(-\frac{7}{2}, -5\right)$$

Average the x-coordinates. Average the y-coordinates.

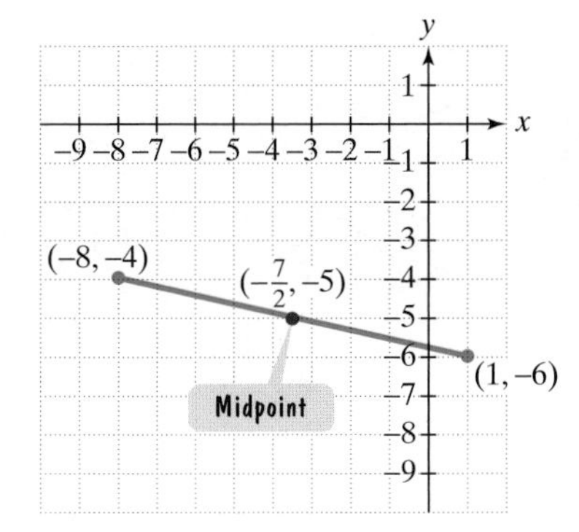

Figure P.19 Finding a line segment's midpoint

Figure P.19 illustrates that the point $\left(-\frac{7}{2}, -5\right)$ is midway between the points $(1, -6)$ and $(-8, -4)$.

Check Point 5 Find the midpoint of the line segment with endpoints $(1, 2)$ and $(7, -3)$.

7 Interpret information given by graphs.

Interpreting Information Given by Graphs

Singers Ricky Martin and Jennifer Lopez, television journalist Soledad O'Brien, boxer Oscar De La Hoya, Internet entrepreneur Carlos Cardona, actor John Leguizamo, and political organizer Luigi Crespo are members of a generation of young Hispanics that is changing the way America looks, feels, thinks, eats, dances, and votes. There are 31 million Hispanics in the United States, pumping \$300 billion a year into the economy. By 2050, the Hispanic population is expected to reach 96 million—an increase of more than 200 percent. Hispanics are also younger than the rest of the nation: A third are under 18. The graph in Figure P.20 shows

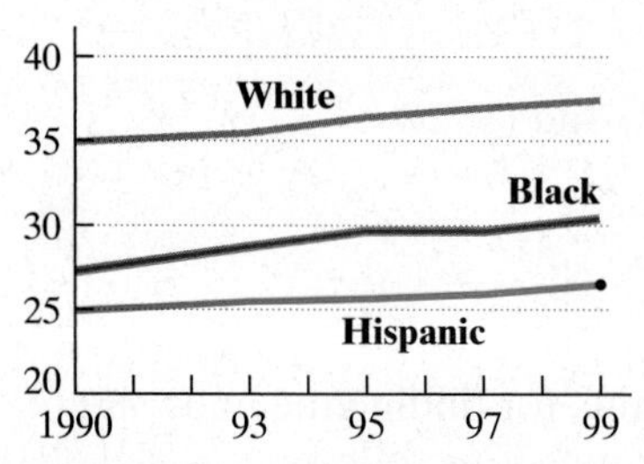

Figure P.20 *Source:* U.S. Census Bureau

the average age of whites, African Americans, and Americans of Hispanic origin.

Magazines and newspapers often display information using **line graphs** like those in Figure P.20. Line graphs are often used to illustrate trends over time. Some measure of time, such as months or years, frequently appears on the horizontal axis. Amounts are generally listed on the vertical axis.

A line graph displays information in the first quadrant of a rectangular coordinate system. By identifying points on line graphs and their coordinates, you can interpret specific information given by the graph. For example, we've shown a point on the far right of the line graph in Figure P.20 for Hispanics. The point is directly above 1999 on the horizontal axis, so its x-coordinate is 1999. The point lies between 25 and 30, at approximately 27, in relation to the horizontal axis. Thus, its y-coordinate is 27. The coordinates (1999, 27) tell us that in 1999 the average age of U.S. Hispanics was approximately 27.

EXAMPLE 6 Interpreting Information Given by Graphs

The graph in Figure P.21(a) shows the ticket sales from concert tours in North America from 1990 through 1997. For the period shown, in which year did sales reach a maximum? Estimate the sales for that year.

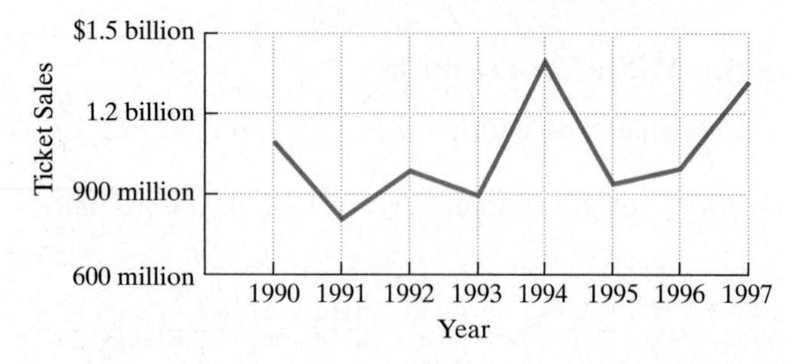

Source: Pollstar

Figure P.21(a)

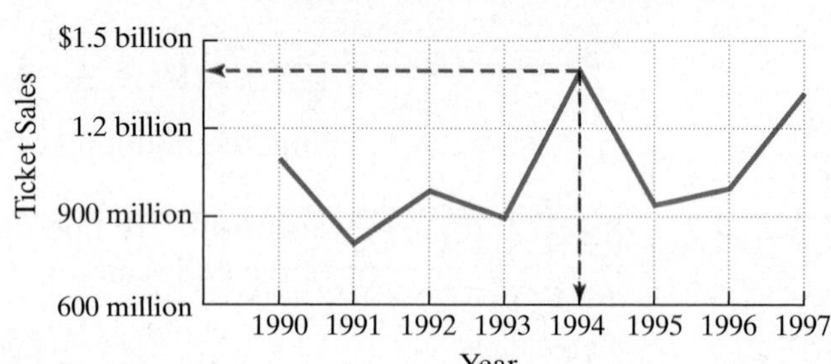

Source: Pollstar

Figure P.21(b)

Solution Maximum sales correspond to the highest point on the graph. This point is identified in Figure P.21(b). The coordinates of this point are approximately (1994, 1.4). This means that in 1994 ticket sales reached a maximum. The sales for that year were approximately $1.4 billion.

Check Point 6 Use the graph in Figure P.21(a) to determine in which year sales reached a minimum. Estimate the sales for that year.

EXERCISE SET P.8

Practice Exercises

In Exercises 1–10, plot the given point in a rectangular coordinate system.

1. $(1, 4)$

2. $(2, 5)$

3. $(-2, 3)$

4. $(-1, 4)$

5. $(-3, -5)$

6. $(-4, -2)$

7. $(4, -1)$

8. $(3, -2)$

9. $(-4, 0)$

10. $(0, -3)$

Graph each equation in Exercises 11–22. Let $x = -3, -2, -1, 0, 1, 2$, and 3.

11. $y = x^2 - 2$ **12.** $y = x^2 + 2$

13. $y = x - 2$ **14.** $y = x + 2$

15. $y = 2x + 1$ **16.** $y = 2x - 4$

17. $y = -\frac{1}{2}x$ **18.** $y = -\frac{1}{2}x + 2$

19. $y = x^3$ **20.** $y = x^3 - 1$

21. $y = |x|$ **22.** $y = |x| + 2$

In Exercises 23–26, match the viewing rectangle with the correct figure. Then label the tick marks in the figure to illustrate this viewing rectangle.

23. $[-5, 5, 1]$ by $[-5, 5, 1]$
24. $[-10, 10, 2]$ by $[-4, 4, 2]$
25. $[-20, 80, 10]$ by $[-30, 70, 10]$
26. $[-40, 40, 20]$ by $[-1000, 1000, 100]$

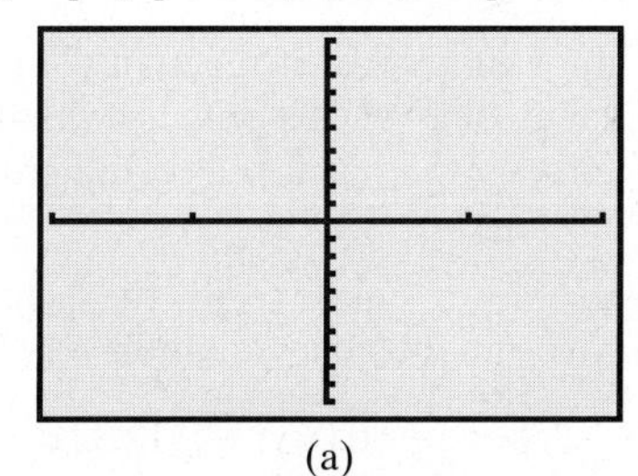

(a)

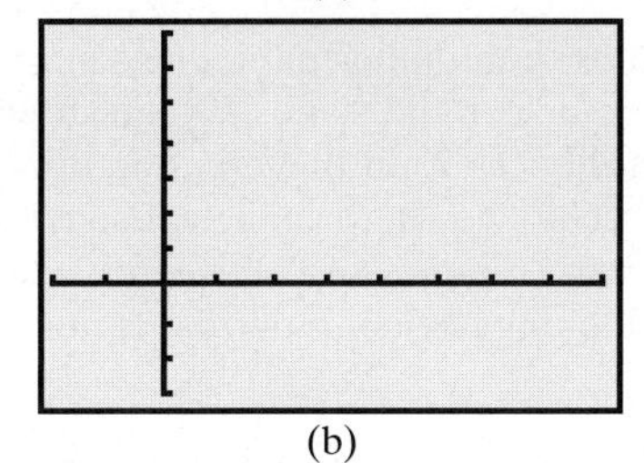

(b)

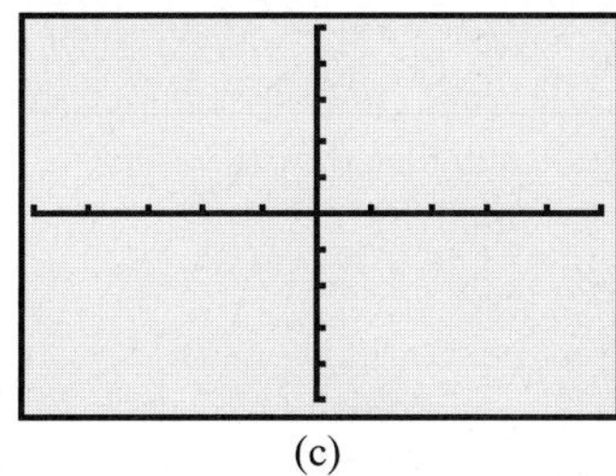

(c)

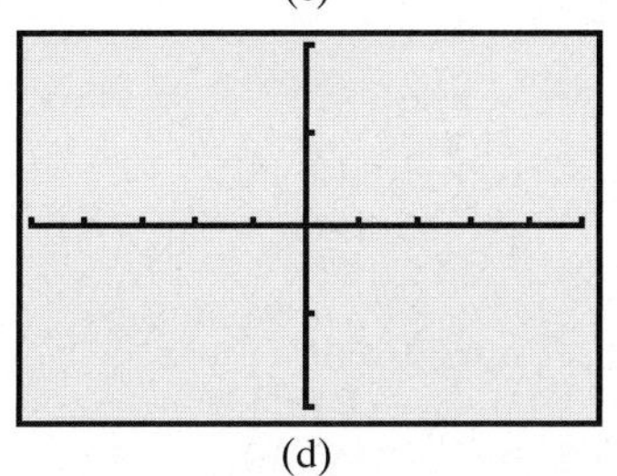

(d)

In Exercises 27–32, use the graph and a. determine the x-intercepts, if any. b. Determine the y-intercepts, if any. For each graph, tick marks along the axes represent one unit each.

27.

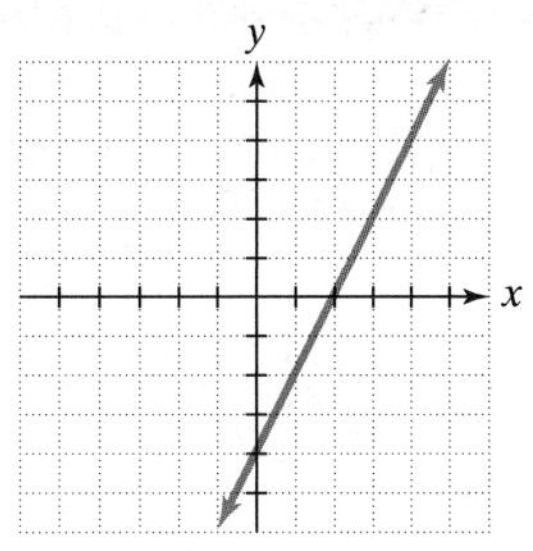

28.

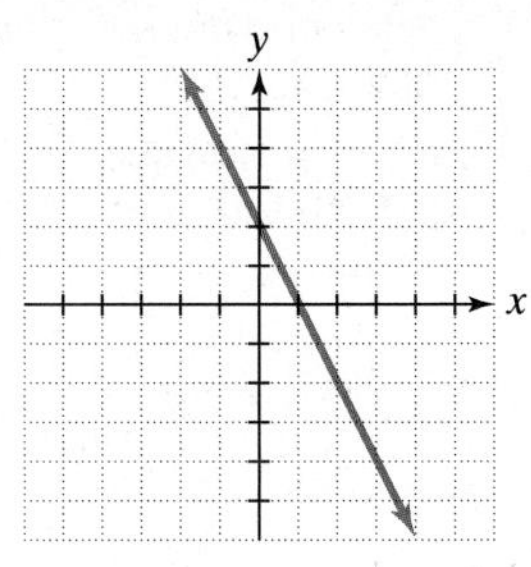

29.

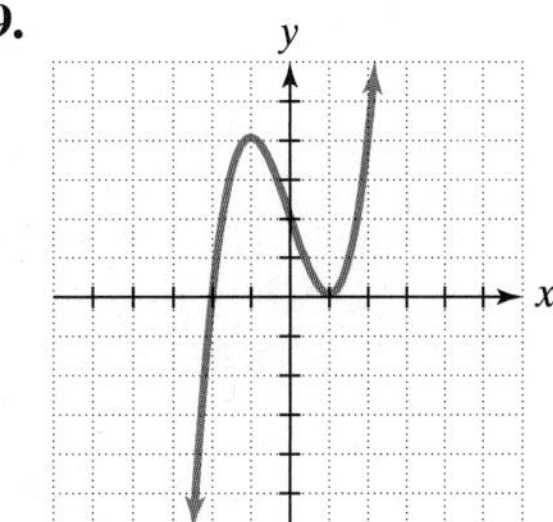

30.

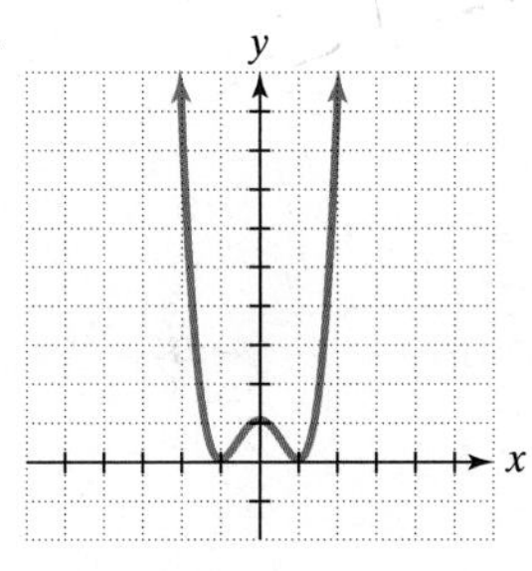

31.

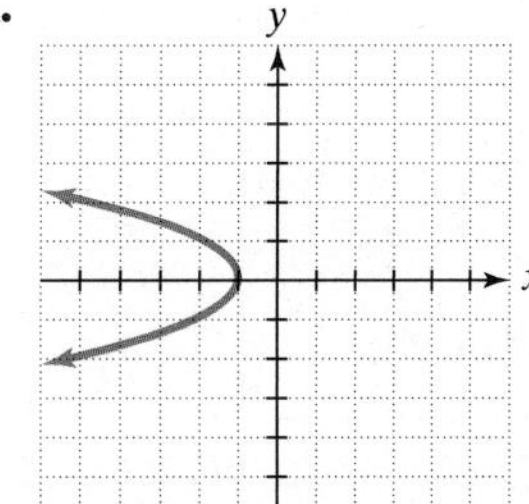

32.

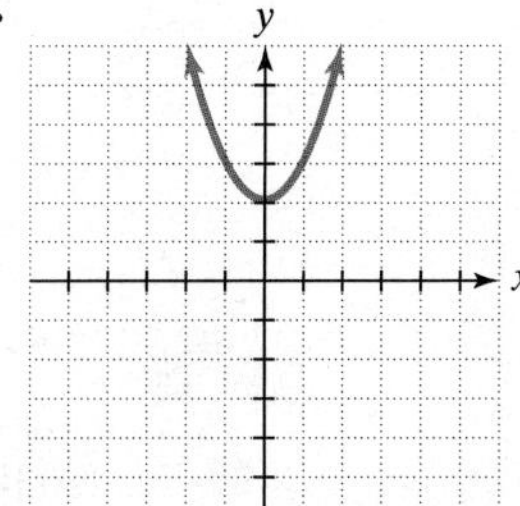

In Exercises 33–38, a. Plot the points. b. Find the distance between the points. c. Find the midpoint of the line segment joining the points.

33. $(4, -3), (-6, 2)$ **34.** $(6, -3), (-4, -5)$
35. $(3, 2), (6, 7)$ **36.** $(-3, 6), (3, 4)$
37. $(1, -2), (-3, 6)$ **38.** $(5, 7), (2, 3)$

Application Exercises

The line graph shows the U.S. unemployment rate from 1960 through 1997. Use the graph to solve Exercises 39–42.

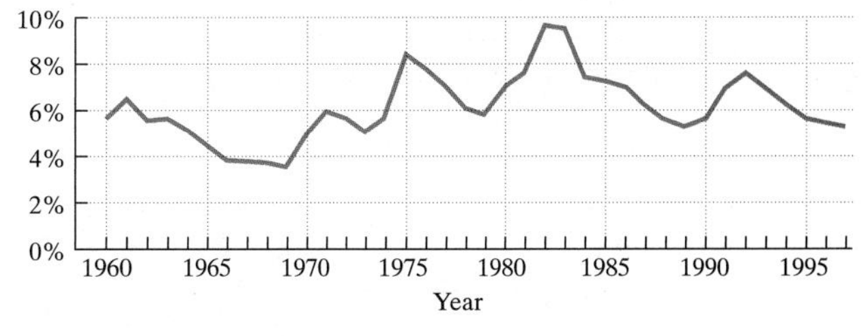

Source: U.S. Bureau of Labor Statistics

39. Find an estimate for the unemployment rate in 1970.
40. Find an estimate for the unemployment rate in 1980.
41. For the period shown, when did the unemployment rate reach a maximum? What is a reasonable estimate for the rate during that year?
42. For the period shown, when did the unemployment rate reach a minimum? What is a reasonable estimate for the rate during that year?

The line graph shows the population of the United States from 1970 to 2000 for people under 16. In Exercises 43–46, find (or estimate) the coordinates of the given point. Then interpret the coordinates in terms of the information given by the graph.

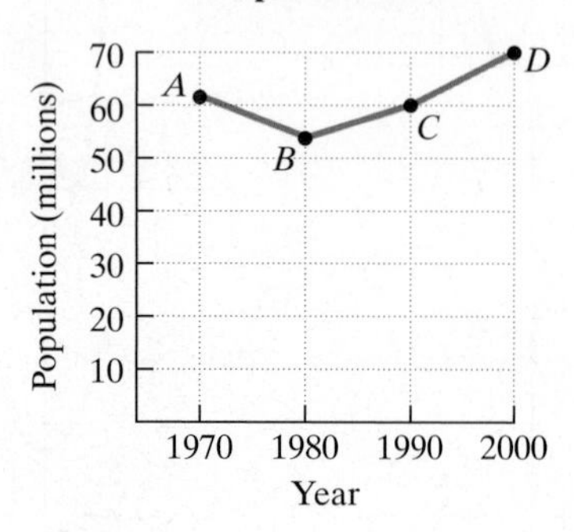

Source: U.S. Census Bureau

43. *A* **44.** *B* **45.** *C* **46.** *D*

47. The graph shows the height (y, in feet) of a ball dropped from the top of the Empire State Building at different times.
   a. What is a reasonable estimate of the height of the Empire State Building?
   b. What is a reasonable estimate, to the nearest tenth of a second, of how long it takes for the ball to hit the ground?

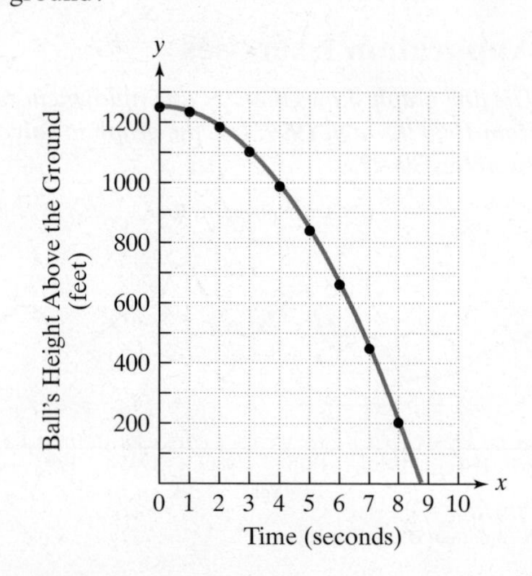

Writing in Mathematics

48. What is the rectangular coordinate system?
49. Explain how to plot a point in the rectangular coordinate system. Give an example with your explanation.
50. Explain why $(5, -2)$ and $(-2, 5)$ do not represent the same point.
51. Explain how to graph an equation in the rectangular coordinate system.
52. What does a $[-20, 2, 1]$ by $[-4, 5, 0.5]$ viewing rectangle mean?
53. How is the distance formula used to find the distance between two given points?
54. How do you find the midpoint of a line segment if its endpoints are known?

Technology Exercises

55. Use a graphing utility to verify each of your hand-drawn graphs in Exercises 11–22. Experiment with the range setting to make the graph displayed by the graphing utility resemble your hand-drawn graph as much as possible.
56. How many people are employed by the executive branch of the U.S. government? The equation

$$y = 0.09x^4 - 3.78x^3 + 50.06x^2 - 212.32x + 3101.16$$

approximates the number of civilians employed by the executive branch, in thousands, x years after 1979. Use a graphing utility to graph the equation in a $[0, 20, 1]$ by $[2500, 3500, 500]$ viewing rectangle. Then describe something about the relationship between x and y that is revealed by looking at the graph that is not obvious from the equation.

A graph of an equation is a complete graph *if it shows all of the important features of the graph. Use a graphing utility to graph the equations in Exercises 57–60 in each of the given viewing rectangles. Then choose which viewing rectangle gives a complete graph.*

57. $y = x^2 + 10$
   a. $[-5, 5, 1]$ by $[-5, 5, 1]$
   b. $[-10, 10, 1]$ by $[-10, 10, 1]$
   c. $[-10, 10, 1]$ by $[-50, 50, 1]$
58. $y = 0.1x^4 - x^3 + 2x^2$
   a. $[-5, 5, 1]$ by $[-8, 2, 1]$
   b. $[-10, 10, 1]$ by $[-10, 10, 1]$
   c. $[-8, 16, 1]$ by $[-16, 8, 1]$
59. $y = \sqrt{x + 18}$
   a. $[-10, 10, 1]$ by $[-10, 10, 1]$
   b. $[-50, 50, 10]$ by $[-10, 10, 1]$
   c. $[-10, 10, 1]$ by $[-50, 50, 10]$

60. $y = x^3 - 30x + 20$
a. $[-10, 10, 1]$ by $[-10, 10, 1]$
b. $[-10, 10, 1]$ by $[-50, 50, 10]$
c. $[-10, 10, 1]$ by $[-50, 100, 10]$

Critical Thinking Exercises

61. Which one of the following is true?
a. If the coordinates of a point satisfy the inequality $xy > 0$, then (x, y) must be in quadrant I.
b. The ordered pair $(2, 5)$ satisfies $3y - 2x = -4$.
c. If a point is on the x-axis, it is neither up nor down, so $x = 0$.
d. None of the above is true.

62. Show that the points $A(1, 1 + d)$, $B(3, 3 + d)$, and $C(6, 6 + d)$ are collinear (lie along a straight line) by showing that the distance from A to B plus the distance from B to C equals the distance from A to C.

63. Prove the midpoint formula by using the following procedure.
a. Show that the distance between (x_1, y_1) and $\left(\frac{x_1 + x_2}{2}, \frac{y_1 + y_2}{2}\right)$ is equal to the distance between (x_2, y_2) and $\left(\frac{x_1 + x_2}{2}, \frac{y_1 + y_2}{2}\right)$.
b. Use the procedure from Exercise 62 and the distances from part (a) to show that the points (x_1, y_1), $\left(\frac{x_1 + x_2}{2}, \frac{y_1 + y_2}{2}\right)$, and (x_2, y_2) are collinear.

CHAPTER SUMMARY, REVIEW, AND TEST

Summary: Basic Formulas

Definition of Absolute Value

$$|x| = \begin{cases} x & \text{if } x \geq 0 \\ -x & \text{if } x < 0 \end{cases}$$

Distance Between Points a and b on a Number Line

$$|a - b| \quad \text{or} \quad |b - a|$$

Properties of Algebra

Commutative	$a + b = b + a, \quad ab = ba$
Associative	$(a + b) + c = a + (b + c)$ $(ab)c = a(bc)$
Distributive	$a(b + c) = ab + ac$
Identity	$a + 0 = a \quad a \cdot 1 = a$
Inverse	$a + (-a) = 0 \quad a \cdot \frac{1}{a} = 1, a \neq 0$

Properties of Exponents

$$b^{-n} = \frac{1}{b^n}, \quad b^0 = 1, \quad b^m \cdot b^n = b^{m+n},$$

$$(b^m)^n = b^{mn}, \quad \frac{b^m}{b^n} = b^{m-n}, \quad (ab)^n = a^n b^n, \quad \left(\frac{a}{b}\right)^n = \frac{a^n}{b^n}$$

Product and Quotient Rules for nth Roots

$$\sqrt[n]{a} \cdot \sqrt[n]{b} = \sqrt[n]{ab} \qquad \frac{\sqrt[n]{a}}{\sqrt[n]{b}} = \sqrt[n]{\frac{a}{b}}$$

Rational Exponents

$$a^{1/n} = \sqrt[n]{a}, \quad a^{-1/n} = \frac{1}{a^{1/n}} = \frac{1}{\sqrt[n]{a}},$$

$$a^{m/n} = (\sqrt[n]{a})^m = \sqrt[n]{a^m}, \quad a^{-m/n} = \frac{1}{a^{m/n}}$$

Special Products

$$(A + B)(A - B) = A^2 - B^2$$
$$(A + B)^2 = A^2 + 2AB + B^2$$
$$(A - B)^2 = A^2 - 2AB + B^2$$
$$(A + B)^3 = A^3 + 3A^2B + 3AB^2 + B^3$$
$$(A - B)^3 = A^3 - 3A^2B + 3AB^2 - B^3$$

Factoring Formulas

$$A^2 - B^2 = (A + B)(A - B)$$
$$A^2 + 2AB + B^2 = (A + B)^2$$
$$A^2 - 2AB + B^2 = (A - B)^2$$
$$A^3 + B^3 = (A + B)(A^2 - AB + B^2)$$
$$A^3 - B^3 = (A - B)(A^2 + AB + B^2)$$

Complex Numbers

$$i = \sqrt{-1} \text{ and } i^2 = -1, \quad (a + bi)(a - bi) = a^2 + b^2$$

Distance and Midpoint Formulas

$$d = \sqrt{(x_2 - x_1)^2 + (y_2 - y_1)^2}$$

$$\text{midpoint} = \left(\frac{x_1 + x_2}{2}, \frac{y_1 + y_2}{2}\right)$$

Review Exercises

You can use these review exercises, like the review exercises at the end of each chapter, to test your understanding of the chapter's topics. However, you can also use these exercises as a prerequisite test to check your mastery of the fundamental algebra skills needed in this book.

P.1

1. Consider the set

$$\{-17, -\tfrac{9}{13}, 0, 0.75, \sqrt{2}, \pi, \sqrt{81}\}$$

List all numbers from the set that are a. natural numbers, b. whole numbers, c. integers, d. rational numbers, e. irrational numbers.

In Exercises 2–4, rewrite each expressions without absolute value bars.

2. $|-103|$ **3.** $|\sqrt{2} - 1|$

4. $|3 - \sqrt{17}|$

5. Express the distance between the numbers -17 and 4 using absolute value. Then evaluate the absolute value.

In Exercises 6–7, evaluate each algebraic expression for the given value of the variable.

6. $\frac{5}{9}(F - 32); F = 68$ **7.** $\frac{8(x + 5)}{3x + 8}, x = 2$

In Exercises 8–13, state the name of the property illustrated.

8. $3 + 17 = 17 + 3$ **9.** $(6 \cdot 3) \cdot 9 = 6 \cdot (3 \cdot 9)$

10. $\sqrt{3}(\sqrt{5} + \sqrt{3}) = \sqrt{15} + 3$

11. $(6 \cdot 9) \cdot 2 = 2 \cdot (6 \cdot 9)$

12. $\sqrt{3}(\sqrt{5} + \sqrt{3}) = (\sqrt{5} + \sqrt{3})\sqrt{3}$

13. $(3 \cdot 7) + (4 \cdot 7) = (4 \cdot 7) + (3 \cdot 7)$

In Exercises 14–15, simplify each algebraic expression.

14. $3(7x - 5y) - 2(4y - x + 1)$

15. $\frac{1}{5}(5x) + [(3y) + (-3y)] - (-x)$

P.2

Evaluate each exponential expression in Exercises 16–19.

16. $(-3)^3(-2)^2$ **17.** $2^{-4} + 4^{-1}$

18. $5^{-3} \cdot 5$ **19.** $\frac{3^3}{3^6}$

Simplify each exponential expression in Exercises 20–23.

20. $(-2x^4y^3)^3$ **21.** $(-5x^3y^2)(-2x^{-11}y^{-2})$

22. $(2x^3)^{-4}$ **23.** $\frac{7x^5y^6}{28x^{15}y^{-2}}$

In Exercises 24–25, write each number in decimal notation.

24. 3.74×10^4 **25.** 7.45×10^{-5}

In Exercises 26–27, write each number in scientific notation.

26. 3,590,000 **27.** 0.00725

In Exercises 28–29, perform the indicated operation and write the answer in decimal notation.

28. $(3 \times 10^3)(1.3 \times 10^2)$ **29.** $\frac{6.9 \times 10^3}{3 \times 10^5}$

30. If you earned \$1 million per year ($\$10^6$), how long would it take to accumulate \$1 billion ($\$10^9$)?

31. If the population of the United States is 2.7×10^8 and each person spends about \$150 per year going to the movies (or renting movies), express the total annual spending on movies in scientific notation.

P.3

Use the product rule to simplify the expressions in Exercises 32–35.

32. $\sqrt{300}$ **33.** $\sqrt{12x^2}$

34. $\sqrt{10x} \cdot \sqrt{2x}$ **35.** $\sqrt{r^3}$

Use the quotient rule to simplify the expressions in Exercises 36–37.

36. $\sqrt{\frac{121}{4}}$ **37.** $\frac{\sqrt{96x^3}}{\sqrt{2x}}$

In Exercises 38–40, add or subtract terms whenever possible.

38. $7\sqrt{5} + 13\sqrt{5}$ **39.** $2\sqrt{50} + 3\sqrt{8}$

40. $4\sqrt{72} - 2\sqrt{48}$

In Exercises 41–44, rationalize the denominator.

41. $\frac{30}{\sqrt{5}}$ **42.** $\frac{\sqrt{2}}{\sqrt{3}}$

43. $\frac{5}{6 + \sqrt{3}}$ **44.** $\frac{14}{\sqrt{7} - \sqrt{5}}$

Evaluate each expression in Exercises 45–48 or indicate that the root is not a real number.

45. $\sqrt[3]{125}$ **46.** $\sqrt[5]{-32}$

47. $\sqrt[4]{-125}$ **48.** $\sqrt[4]{(-5)^4}$

Simplify the radical expressions in Exercises 49–53.

49. $\sqrt[3]{81}$ **50.** $\sqrt[3]{y^5}$

51. $\sqrt[4]{8} \cdot \sqrt[4]{10}$ **52.** $4\sqrt[3]{16} + 5\sqrt[3]{2}$

53. $\frac{\sqrt[4]{32x^5}}{\sqrt[4]{16x}}$

In Exercises 54–59, evaluate each expression.

54. $16^{1/2}$ **55.** $25^{-1/2}$

56. $125^{1/3}$ **57.** $27^{-1/3}$

58. $64^{2/3}$ **59.** $27^{-4/3}$

In Exercises 60–62, simplify using properties of exponents.

60. $(5x^{2/3})(4x^{1/4})$

61. $\dfrac{15x^{3/4}}{5x^{1/2}}$

62. $(125x^6)^{2/3}$

63. Simplify by reducing the index of the radical: $\sqrt[6]{y^3}$.

P.4

In Exercises 64–65, perform the indicated operations. Write the resulting polynomial in standard form and indicate its degree.

64. $(-6x^3 + 7x^2 - 9x + 3) + (14x^3 + 3x^2 - 11x - 7)$

65. $(13x^4 - 8x^3 + 2x^2) - (5x^4 - 3x^3 + 2x^2 - 6)$

In Exercises 66–72, find each product.

66. $(3x - 2)(4x^2 + 3x - 5)$

67. $(3x - 5)(2x + 1)$

68. $(4x + 5)(4x - 5)$

69. $(2x + 5)^2$

70. $(3x - 4)^2$

71. $(2x + 1)^3$

72. $(5x - 2)^3$

In Exercises 73–74, perform the indicated operations. Indicate the degree of the resulting polynomial.

73. $(7x^2 - 8xy + y^2) + (-8x^2 - 9xy - 4y^2)$

74. $(13x^3y^2 - 5x^2y - 9x^2) - (-11x^3y^2 - 6x^2y + 3x^2 - 4)$

In Exercises 75–79, find each product.

75. $(x + 7y)(3x - 5y)$

76. $(3x - 5y)^2$

77. $(3x^2 + 2y)^2$

78. $(7x + 4y)(7x - 4y)$

79. $(a - b)(a^2 + ab + b^2)$

P.5

In Exercises 80–95, factor completely, or state that the polynomial is prime.

80. $15x^3 + 3x^2$

81. $x^2 - 11x + 28$

82. $15x^2 - x - 2$

83. $64 - x^2$

84. $x^2 + 16$

85. $3x^4 - 9x^3 - 30x^2$

86. $20x^7 - 36x^3$

87. $x^3 - 3x^2 - 9x + 27$

88. $16x^2 - 40x + 25$

89. $x^4 - 16$

90. $y^3 - 8$

91. $x^3 + 64$

92. $3x^4 - 12x^2$

93. $27x^3 - 125$

94. $x^5 - x$

95. $x^3 + 5x^2 - 2x - 10$

P.6

In Exercises 96–98, simplify each rational expression. Also, list all numbers that must be excluded from the domain.

96. $\dfrac{x^3 + 2x^2}{x + 2}$

97. $\dfrac{x^2 + 3x - 18}{x^2 - 36}$

98. $\dfrac{x^2 + 2x}{x^2 + 4x + 4}$

In Exercises 99–101, multiply or divide as indicated.

99. $\dfrac{x^2 + 6x + 9}{x^2 - 4} \cdot \dfrac{x + 3}{x - 2}$

100. $\dfrac{6x + 2}{x^2 - 1} \div \dfrac{3x^2 + x}{x - 1}$

101. $\dfrac{x^2 - 5x - 24}{x^2 - x - 12} \div \dfrac{x^2 - 10x + 16}{x^2 + x - 6}$

In Exercises 102–105, add or subtract as indicated.

102. $\dfrac{2x - 7}{x^2 - 9} - \dfrac{x - 10}{x^2 - 9}$

103. $\dfrac{3x}{x + 2} + \dfrac{x}{x - 2}$

104. $\dfrac{x}{x^2 - 9} + \dfrac{x - 1}{x^2 - 5x + 6}$

105. $\dfrac{4x - 1}{2x^2 + 5x - 3} - \dfrac{x + 3}{6x^2 + x - 2}$

In Exercises 106–108, simplify each complex rational expression.

106. $\dfrac{\frac{1}{x} - \frac{1}{2}}{\frac{1}{3} - \frac{x}{6}}$

107. $\dfrac{3 + \frac{12}{x}}{1 - \frac{16}{x^2}}$

108. $\dfrac{3 - \frac{1}{x + 3}}{3 + \frac{1}{x + 3}}$

P.7

In Exercises 109–118, perform the indicated operations and write the result in standard form.

109. $(8 - 3i) - (17 - 7i)$

110. $4i(3i - 2)$

111. $(7 - 5i)(2 + 3i)$

112. $(3 - 4i)^2$

113. $(7 + 8i)(7 - 8i)$

114. $\dfrac{6}{5 + i}$

115. $\dfrac{3 + 4i}{4 - 2i}$

116. $\sqrt{-32} - \sqrt{-18}$

117. $(-2 + \sqrt{-100})^2$

118. $\dfrac{4 + \sqrt{-8}}{2}$

P.8

Graph each equation in Exercises 119–121. Let $x = -3, -2, -1, 0, 1, 2,$ and 3.

119. $y = 2x - 2$

120. $y = x^2 - 3$

121. $y = x$

122. What does a $[-20, 40, 10]$ by $[-5, 5, 1]$ viewing rectangle mean? Draw axes with tick marks and label the tick marks to illustrate this viewing rectangle.

In Exercises 123–125, use the graph and determine the x-intercepts, if any, and the y-intercepts, if any. For each graph, tick marks along the axes represent one unit each.

123.

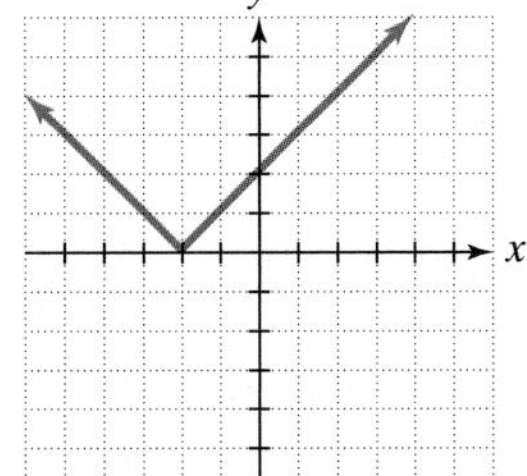

124.

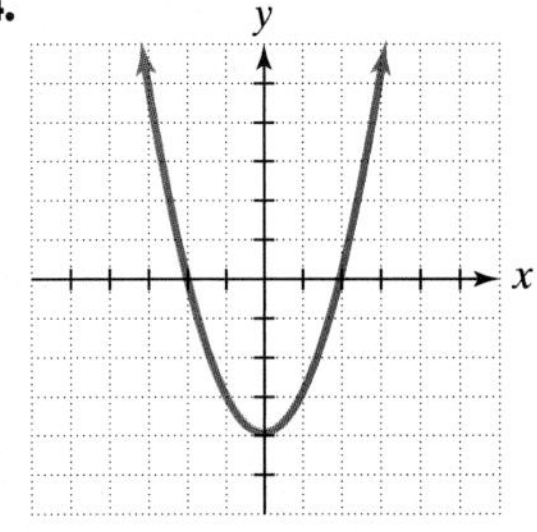

125.

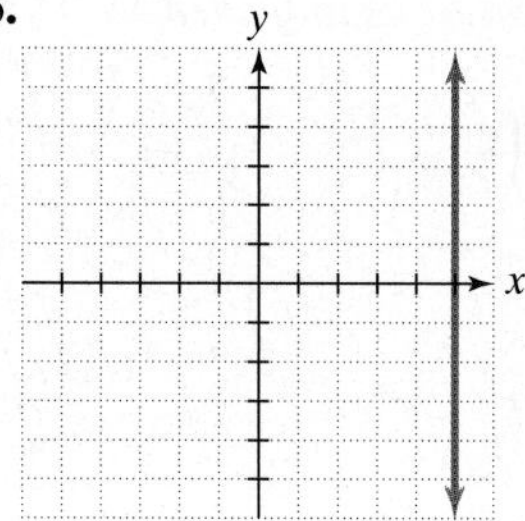

In Exercises 126–127, a. Plot the points. b. Find the distance between the points. c. Find the midpoint of the line segment joining the points.

126. $(-1, 0), (2, 4)$ **127.** $(2, -3), (4, 2)$

128. The line graphs show the number of applicants to U.S. law schools and medical schools.

a. For the period shown, when did the number of law school applicants reach a maximum? What is a reasonable estimate for the number of applicants during that year?

b. Find an estimate for the number of medical school applicants in 1990.

c. Estimate the coordinates of point A. Then interpret the coordinates in terms of the information given by the graph.

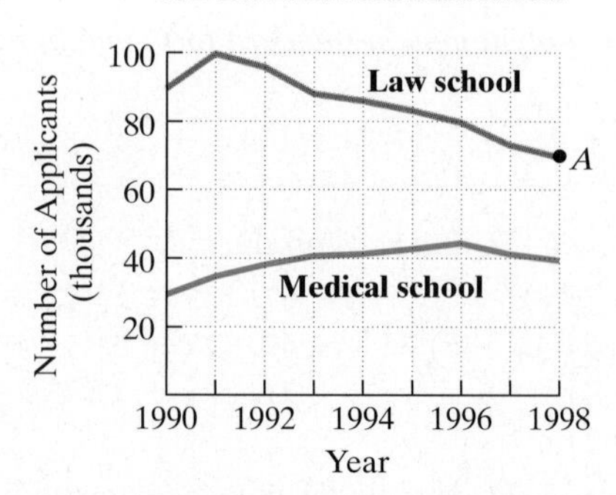

Source: U.S. Education Department

Chapter P Test

1. List all the rational numbers in this set.

$$\{-7, -\tfrac{4}{5}, 0, 0.25, \sqrt{3}, \sqrt{4}, \tfrac{22}{7}, \pi\}$$

In Exercises 2–3, state the name of the property illustrated.

2. $3(2 + 5) = 3(5 + 2)$ **3.** $6(7 + 4) = 6 \cdot 7 + 6 \cdot 4$

4. Express in scientific notation: 0.00076.

Simplify each expression in Exercises 5–11.

5. $9(10x - 2y) - 5(x - 4y + 3)$

6. $\dfrac{30x^3y^4}{6x^9y^{-4}}$ **7.** $\sqrt{6r}\sqrt{3r}$

8. $4\sqrt{50} - 3\sqrt{18}$ **9.** $\dfrac{3}{5 + \sqrt{2}}$

10. $\sqrt[3]{16x^4}$ **11.** $\dfrac{x^2 + 2x - 3}{x^2 - 3x + 2}$

12. Evaluate: $27^{-5/3}$

In Exercises 13–14, find each product.

13. $(2x - 5)(x^2 - 4x + 3)$ **14.** $(5x + 3y)^2$

In Exercises 15–19, factor completely, or state that the polynomial is prime.

15. $x^2 - 9x + 18$ **16.** $x^3 + 2x^2 + 3x + 6$

17. $25x^2 - 9$ **18.** $36x^2 - 84x + 49$

19. $y^3 - 125$

In Exercises 20–23, perform the operations and simplify, if possible.

20. $\dfrac{2x + 8}{x - 3} \div \dfrac{x^2 + 5x + 4}{x^2 - 9}$ **21.** $\dfrac{x}{x + 3} + \dfrac{5}{x - 3}$

22. $\dfrac{2x + 3}{x^2 - 7x + 12} - \dfrac{2}{x - 3}$ **23.** $\dfrac{\frac{1}{x} - \frac{1}{3}}{\frac{1}{x}}$

In Exercises 24–26, perform the indicated operations and write the result in standard form.

24. $(6 - 7i)(2 + 5i)$ **25.** $\dfrac{5}{2 - i}$

26. $2\sqrt{-49} + 3\sqrt{-64}$

27. Graph $y = x^2 - 4$ by letting x equal integers from -3 through 3.

28. Find the distance between $(2, 9)$ and $(6, 3)$ in simplified radical form.

Equations, Inequalities, and Mathematical Models

Chapter 1

Formulas like those that describe the height a child will attain as an adult are frequently obtained from actual data. Formulas can be used to explain what is happening in the present and to make predictions about what might occur in the future. Knowing how to create and use formulas will help you recognize patterns, logic, and order in a world that can appear chaotic to the untrained eye. In many ways, algebra will provide you with a new way of looking at your world.

Sitting in the biology department office, you overhear two of the professors discussing the possible adult heights of their respective children. Looking at the blackboard that they've been writing on, you see that there are formulas that can estimate the height a child will attain as an adult. If the child is x years old and h inches tall, that child's adult height, H, in inches, is approximated by one of the following formulas.

$$\text{Girls:} \quad H = \frac{h}{0.00028x^3 - 0.0071x^2 + 0.0926x + 0.3524}$$

$$\text{Boys:} \quad H = \frac{h}{0.00011x^3 - 0.0032x^2 + 0.0604x + 0.3796}$$

SECTION 1.1 *Linear Equations*

Objectives

1. Solve linear equations in one variable.
2. Solve equations with constants in denominators.
3. Solve equations with variables in denominators.
4. Recognize identities, conditional equations, and inconsistent equations.

Unfortunately, many of us have been fined for driving over the speed limit. The amount of the fine depends on how fast we are speeding. Suppose that a highway has a speed limit of 60 miles per hour. The amount that speeders are fined, F, is described by the statement of equality

$$F = 10x - 600,$$

where x is the speed in miles per hour. We can use this statement to determine the fine, F, for a speeder traveling at, say, 70 miles per hour. We substitute 70 for x in the given statement and then find the corresponding value for F.

$$F = 10(70) - 600 = 700 - 600 = 100$$

Thus, a person caught driving 70 miles per hour gets a $100 fine.

A friend, whom we shall call Leadfoot, borrows your car and returns a few hours later with a $400 speeding fine. Leadfoot is furious, protesting that the car was barely driven over the speed limit. Should you believe Leadfoot?

In order to decide if Leadfoot is telling the truth, use $F = 10x - 600$. Leadfoot was fined $400, so substitute 400 for F:

$$400 = 10x - 600.$$

In Example 1, we will find the value for x. This variable represents Leadfoot's speed, which resulted in the $400 fine.

An **equation** consists of two algebraic expressions joined by an equal sign. Thus, $400 = 10x - 600$ is an example of an equation. The equal sign divides the equation into two parts, the left side and the right side:

$$\underbrace{\boxed{400}}_{\text{Left side}} = \underbrace{\boxed{10x - 600}}_{\text{Right side}}$$

The two sides of an equation can be reversed. So, we can also express this equation as

$$10x - 600 = 400.$$

The form of this equation is $ax + b = c$, with $a = 10$, $b = -600$, and $c = 400$. Any equation in this form is called a **linear equation in one variable**. The exponent on the variable in such an equation is 1. In this section, we will study how to solve linear equations.

1 Solve linear equations in one variable.

Solving Linear Equations in One Variable

We begin by restating the definition of a linear equation in one variable.

Definition of a Linear Equation

A **linear equation in one variable x** is an equation that can be written in the form

$$ax + b = 0$$

where a and b are real numbers and $a \neq 0$.

An example of a linear equation in one variable is $4x + 12 = 0$. **Solving an equation** in x involves determining all values of x that result in a true statement when substituted into the equation. Such values are **solutions** or **roots** of the equation. For example, substitute -3 into $4x + 12 = 0$. We obtain $4(-3) + 12 = 0$, or $-12 + 12 = 0$. This simplifies to the true statement $0 = 0$. Thus, -3 is a solution of the equation $4x + 12 = 0$. We also say that -3 **satisfies** the equation $4x + 12 = 0$, because when we substitute -3 for x, a true statement results. The set of all such solutions is called the equation's **solution set**. For example, the solution set of the equation $4x + 12 = 0$ is $\{-3\}$.

Equations that have the same solution set are called **equivalent equations**. For example, the equations $4x + 12 = 0$, $4x = -12$, and $x = -3$ are equivalent equations because the solution set for each is $\{-3\}$. To solve a linear equation in x, we transform the equation into an equivalent equation one or more times. Our final equivalent equation should be in the form $x = d$, where d is a real number. By inspection, we can see that the solution set for this equation is $\{d\}$.

To generate equivalent equations, we will use the following principles.

Study Tip

We can solve equations such as $3(x - 6) = 5x$ for a variable. However, we cannot solve for a variable in an algebraic expression such as $3(x - 6)$. We *simplify* algebraic expressions.

Correct

Simplify: $3(x - 6)$.

$3(x - 6) = 3x - 18$

Incorrect

Simplify: $3(x - 6)$.

$3(x - 6) = 0$

$3x - 18 = 0$

$3x - 18 + 18 = 0 + 18$

$3x = 18$

$\frac{3x}{3} = \frac{18}{3}$

$x = 6$

Generating Equivalent Equations

An equation can be transformed into an **equivalent equation** by one or more of the following operations.

	Example
1. Simplify an expression by removing grouping symbols and combining like terms.	$3(x - 6) = 6x - x$ $3x - 18 = 5x$
2. Add (or subtract) the same real number or variable expression on *both* sides of the equation.	$3x - 18 = 5x$ $3x - 18 - 3x = 5x - 3x$ $-18 = 2x$ Subtract 3x from both sides of the equation.
3. Multiply (or divide) on *both* sides of the equation by the same *nonzero* quantity.	$-18 = 2x$ $\frac{-18}{2} = \frac{2x}{2}$ $-9 = x$ Divide both sides of the equation by 2.
4. Interchange the two sides of the equation.	$-9 = x$ $x = -9$

If you look closely at the equations in the box, you will notice that we have solved the equation $3(x - 6) = 6x - x$. The final equation, with x isolated by itself on the left side, shows that $\{-9\}$ is the solution set. The idea in solving a linear equation is to get the variable by itself on one side of the equal sign and a number by itself on the other side.

EXAMPLE 1 Solving a Linear Equation (Is Leadfoot Telling the Truth?)

Solve the equation: $10x - 600 = 400$.

Study Tip

If your proposed solution is incorrect, you will get a false statement when you check your answer. For example, 65 is not a solution of $10x - 600 = 400$. Look what happens when we substitute 65 for x:

$$10x - 600 = 400$$
$$10(65) - 600 \stackrel{?}{=} 400$$
$$650 - 600 \stackrel{?}{=} 400$$
$$50 = 400 \quad \text{False.}$$

Solution Remember that x represents Leadfoot's speed that resulted in the \$400 fine. Our goal is to get x by itself on the left side. We do this by adding 600 to both sides to get $10x$ by itself. Then we isolate x from $10x$ by dividing both sides of the equation by 10.

$$10x - 600 = 400 \quad \text{This is the given equation.}$$
$$10x - 600 + 600 = 400 + 600 \quad \text{Add 600 to both sides.}$$
$$10x = 1000 \quad \text{Combine like terms.}$$
$$\frac{10x}{10} = \frac{1000}{10} \quad \text{Divide both sides by 10.}$$
$$x = 100$$

Can this possibly be correct? Was Leadfoot doing 100 miles per hour in the car he borrowed from you? To find out, check the proposed solution, 100, in the original equation. In other words, evaluate when $x = 100$.

Check

$$10x - 600 = 400 \quad \text{This is the original equation.}$$
$$10(100) - 600 \stackrel{?}{=} 400 \quad \text{Substitute 100 for x.}$$
$$1000 - 600 \stackrel{?}{=} 400 \quad \text{Multiply: } 10(100) = 1000.$$
$$400 = 400 \quad \text{Subtract: } 1000 - 600 = 400.$$

The true statement $400 = 400$ indicates that 100 is the solution. This verifies that the solution set is $\{100\}$. Leadfoot was doing an outrageous 100 miles per hour, and lied by claiming that your car was barely driven over the speed limit.

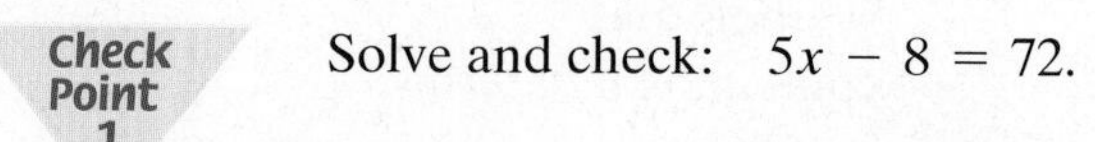

Check Point 1 Solve and check: $5x - 8 = 72$.

The compact, symbolic notation of algebra enables us to use a clear step-by-step method for solving equations, designed to avoid the confusion shown in the painting. *Source*: Squeak Carnwath *Equations* 1981, oil on cotton canvas 96 in. h x 72 in. w.

We now present a step-by-step procedure for solving a linear equation in one variable. Not all of these steps are necessary to solve every equation.

Solving a Linear Equation

1. Simplify the algebraic expression on each side.
2. Collect all the variable terms on one side and all the constant terms on the other side.
3. Isolate the variable and solve.
4. Check the proposed solution in the original equation.

EXAMPLE 2 Solving a Linear Equation

Solve the equation: $2(x - 3) - 17 = 13 - 3(x + 2)$.

Solution

Step 1 Simplify the algebraic expression on each side.

$$2(x - 3) - 17 = 13 - 3(x + 2) \quad \text{This is the given equation.}$$
$$2x - 6 - 17 = 13 - 3x - 6 \quad \text{Use the distributive property.}$$
$$2x - 23 = -3x + 7 \quad \text{Combine like terms.}$$

Discovery

Solve the equation in Example 2 by collecting terms with the variable on the right and numerical terms on the left. What do you observe?

Step 2 Collect variable terms on one side and constant terms on the other side. We will collect variable terms on the left by adding $3x$ to both sides. We will collect the numbers on the right by adding 23 to both sides.

$$2x - 23 + 3x = -3x + 7 + 3x \quad \text{Add } 3x \text{ to both sides.}$$
$$5x - 23 = 7 \quad \text{Simplify.}$$
$$5x - 23 + 23 = 7 + 23 \quad \text{Add 23 to both sides.}$$
$$5x = 30 \quad \text{Simplify.}$$

Step 3 Isolate the variable and solve. We isolate the variable x by dividing both sides by 5.

$$\frac{5x}{5} = \frac{30}{5} \quad \text{Divide both sides by 5.}$$
$$x = 6 \quad \text{Simplify.}$$

Step 4 Check the proposed solution in the original equation. Substitute 6 for x in the original equation.

$$2(x - 3) - 17 = 13 - 3(x + 2) \quad \text{This is the original equation.}$$
$$2(6 - 3) - 17 \stackrel{?}{=} 13 - 3(6 + 2) \quad \text{Substitute 6 for } x.$$
$$2(3) - 17 \stackrel{?}{=} 13 - 3(8) \quad \text{Simplify inside parentheses.}$$
$$6 - 17 \stackrel{?}{=} 13 - 24 \quad \text{Multiply.}$$
$$-11 = -11 \checkmark \quad \text{This true statement indicates that 6 is the solution.}$$

The solution set is $\{6\}$.

Check Point 2 Solve and check: $4(2x + 1) - 29 = 3(2x - 5)$.

2 Solve equations with constants in denominators.

Linear Equations with Fractions

Equations are easier to solve when they do not contain fractions. How do we solve equations involving fractions? We begin by multiplying both sides of the equation by the least common denominator. The least common denominator is the smallest number that all the denominators will divide into. Example 3 shows how this is done.

Technology

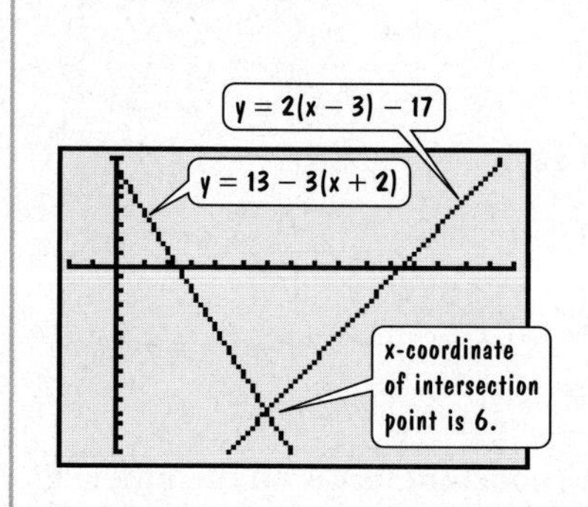

You can use a graphing utility to check the solution to a linear equation in one variable. **Graph the left side and graph the right side. The solution is the x-coordinate of the point where the graphs intersect.** For example, to verify that 6 is the solution of

$$2(x-3)-17=13-3(x+2),$$

graph these two equations in the same viewing rectangle:

$$y=2(x-3)-17$$

$$\text{and}\quad y=13-3(x+2)$$

Choose a large enough range setting so that you can see where the graphs intersect. The viewing rectangle on the left shows that the x-coordinate of the intersection point is 6, verifying that $\{6\}$ is the solution set for the given equation.

EXAMPLE 3 Solving a Linear Equation Involving Fractions

Solve the equation: $\frac{3x}{2}=\frac{x}{5}-\frac{39}{5}$.

Solution The denominators are 2, 5, and 5. The smallest number that is divisible by 2, 5, and 5 is 10. We begin by multiplying both sides of the equation by 10, the least common denominator.

$$\frac{3x}{2}=\frac{x}{5}-\frac{39}{5}$$ This is the given equation.

$$10\cdot\frac{3x}{2}=10\left(\frac{x}{5}-\frac{39}{5}\right)$$ Multiply both sides by 10.

$$10\cdot\frac{3x}{2}=10\cdot\frac{x}{5}-10\cdot\frac{39}{5}$$ Use the distributive property. Be sure to multiply all terms by 10.

$$\overset{5}{\cancel{10}}\cdot\frac{3x}{\underset{1}{\cancel{2}}}=\overset{2}{\cancel{10}}\cdot\frac{x}{\underset{1}{\cancel{5}}}-\overset{2}{\cancel{10}}\cdot\frac{39}{\underset{1}{\cancel{5}}}$$ Divide out common factors in the multiplication.

$$15x=2x-78$$ Complete the multiplication. The fractions are now cleared.

At this point, we have an equation similar to those we previously solved. Collect the variable terms on one side and the constant terms on the other side.

$$15x-2x=2x-2x-78$$ Subtract $2x$ to get the x-terms on the left.

$$13x=-78$$ Simplify.

Isolate x by dividing both sides by 13.

$$\frac{13x}{13}=\frac{-78}{13}$$ Divide both sides by 13.

$$x=-6$$ Simplify.

Check the proposed solution in the original equation. Substitute -6 for x in the original equation. You should obtain $-9=-9$. This true statement verifies the solution set is $\{-6\}$.

Check Point 3 Solve and check: $\frac{x}{4} = \frac{2x}{3} + \frac{5}{6}$.

3 Solve equations with variables in denominators.

Equations Involving Rational Expressions

In Example 3 we solved a linear equation with constants in denominators. Now, let's consider an equation such as

$$\frac{1}{x} = \frac{1}{5} + \frac{3}{2x}.$$

Can you see how this equation differs from the fractional equation that we solved earlier? The variable, x, appears in two of the denominators. The procedure for solving this equation still involves multiplying each side by the least common denominator. However, we must avoid any values of the variable that make a denominator zero. For example, examine the denominators in the equation

$$\frac{1}{x} = \frac{1}{5} + \frac{3}{2x}.$$

This denominator would equal zero if $x = 0$.

This denominator would equal zero if $x = 0$.

We see that x cannot equal zero. With this in mind, let's solve the equation.

EXAMPLE 4 Solving an Equation Involving Rational Expressions

Solve: $\frac{1}{x} = \frac{1}{5} + \frac{3}{2x}$.

Solution The denominators are x, 5, and $2x$. The least common denominator is $10x$. We begin by multiplying both sides of the equation by $10x$. We will also write the restriction that x cannot equal zero to the right of the equation.

$$\frac{1}{x} = \frac{1}{5} + \frac{3}{2x}, \quad x \neq 0$$ This is the given equation.

$$10x \cdot \frac{1}{x} = 10x\left(\frac{1}{5} + \frac{3}{2x}\right)$$ Multiply both sides by 10x.

$$10x \cdot \frac{1}{x} = 10x \cdot \frac{1}{5} + 10x \cdot \frac{3}{2x}$$ Use the distributive property. Be sure to multiply all terms by 10x.

$$10\cancel{x} \cdot \frac{1}{\cancel{x}} = \overset{2}{\cancel{10}}x \cdot \frac{1}{\underset{1}{\cancel{5}}} + \overset{5}{\cancel{10}}\cancel{x} \cdot \frac{3}{\underset{1}{\cancel{2x}}}$$ Divide out common factors in the multiplication.

$$10 = 2x + 15$$ Simplify.

Observe that the resulting equation,

$$10 = 2x + 15,$$

is now cleared of fractions. With the variable term, $2x$, already on the right, we will collect constant terms on the left by subtracting 15 from both sides.

$$10 - 15 = 2x + 15 - 15 \quad \text{Subtract 15 from both sides.}$$

$$-5 = 2x \quad \text{Simplify.}$$

Finally, we isolate the variable, x, by dividing both sides by 2.

$$\frac{-5}{2} = \frac{2x}{2} \quad \text{Divide both sides by 2.}$$

$$-\frac{5}{2} = x \quad \text{Simplify.}$$

We check our solution by substituting $-\frac{5}{2}$ into the original equation or by using a calculator. With a calculator, evaluate each side of the equation for $x = -\frac{5}{2}$, or for $x = -2.5$. Note that the original restriction that $x \neq 0$ is met. The solution set is $\{-\frac{5}{2}\}$.

Check Point 4 Solve: $\frac{5}{2x} = \frac{17}{18} - \frac{1}{3x}$.

EXAMPLE 5 Solving an Equation Involving Rational Expressions

Solve: $\frac{x}{x-3} = \frac{3}{x-3} + 9$.

Solution We must avoid any values of the variable x that make a denominator zero.

$$\frac{x}{x-3} = \frac{3}{x-3} + 9$$

These denominators are zero if $x - 3 = 0$, or in other words, if $x = 3$.

We see that x cannot equal 3. With denominators of $x - 3$, $x - 3$, and 1, the least common denominator is $x - 3$. We multiply both sides of the equation by $x - 3$. We also write the restriction that x cannot equal 3 to the right of the equation.

$$\frac{x}{x-3} = \frac{3}{x-3} + 9, \quad x \neq 3 \quad \text{This is the given equation.}$$

$$(x-3) \cdot \frac{x}{x-3} = (x-3)\left[\frac{3}{x-3} + 9\right] \quad \text{Multiply both sides by } x - 3.$$

$$(x-3) \cdot \frac{x}{x-3} = (x-3) \cdot \frac{3}{x-3} + (x-3) \cdot 9 \quad \text{Use the distributive property.}$$

$$\cancel{(x-3)} \cdot \frac{x}{\cancel{x-3}} = \cancel{(x-3)} \cdot \frac{3}{\cancel{x-3}} + (x-3) \cdot 9 \quad \text{Divide out common factors in the multiplications.}$$

$$x = 3 + (x-3) \cdot 9 \quad \text{Simplify.}$$

The resulting equation, which can be expressed as

$$x = 3 + 9(x - 3),$$

is cleared of fractions. We now solve for x.

$$x = 3 + 9x - 27 \quad \text{Use the distributive property.}$$
$$x = 9x - 24 \quad \text{Combine numerical terms.}$$
$$x - 9x = 9x - 24 - 9x \quad \text{Subtract 9x from both sides.}$$
$$-8x = -24 \quad \text{Simplify.}$$
$$\frac{-8x}{-8} = \frac{-24}{-8} \quad \text{Solve for x, dividing both sides by } -8.$$
$$x = 3 \quad \text{Simplify.}$$

The proposed solution, 3, is *not* a solution because of the restriction that $x \neq 3$. There is *no solution to this equation*. The solution set for this equation contains no elements and is called the empty set, written $\varnothing$.

Check Point 5 Solve: $\frac{x}{x-2} = \frac{2}{x-2} - \frac{2}{3}$.

4 Recognize identities, conditional equations, and inconsistent equations.

Types of Equations

We tend to place things in categories, allowing us to order and structure the world. For example, you can categorize yourself by your age group, your ethnicity, your academic major, or your gender. Equations can be placed into categories that depend on their solution sets.

An equation that is true for all real numbers for which both sides are defined is called an **identity**. An example of an identity is

$$x + 3 = x + 2 + 1.$$

Every number plus 3 is equal to that number plus 2 plus 1. Therefore, the solution set to this equation is the set of all real numbers. Another example of an identity is

$$\frac{2x}{x} = 2.$$

Because division by 0 is undefined, this equation is true for all real number values of x except 0. The solution set is the set of nonzero real numbers.

An equation that is not an identity but that is true for at least one real number is called a **conditional equation**. The equation $10x - 600 = 400$ is an example of a conditional equation. The equation is not an identity and is true only if $x = 100$.

An **inconsistent equation** is an equation that is not true for even one real number. An example of an inconsistent equation is

$$x = x + 7.$$

There is no number that is equal to itself plus 7. Some inconsistent equations are less obvious than this. Consider the equation in Example 5,

$$\frac{x}{x-3} = \frac{2}{x-3} + 9.$$

This equation is not true for any real number and has no solution. Thus, it is inconsistent.

EXAMPLE 6 Categorizing an Equation

Determine whether the equation

$$2(x + 1) = 2x + 3$$

is an identity, a conditional equation, or an inconsistent equation.

Solution Let's see what happens if we try solving the equation. Applying the distributive property on the left side, we obtain

$$2x + 2 = 2x + 3.$$

Does something look strange? Can doubling a number and increasing the product by 2 give the same result as doubling the same number and increasing the product by 3? No. Let's continue solving the equation by subtracting $2x$ from both sides.

$$2x + 2 - 2x = 2x + 3 - 2x$$
$$2 = 3$$

The false statement $2 = 3$ verifies that the given equation is inconsistent.

Check Point 6 Determine whether the equation

$$2(x + 1) = 2x + 2$$

is an identity, a conditional equation, or an inconsistent equation.

EXERCISE SET 1.1

Practice Exercises

In Exercises 1–16, solve and check each linear equation.

1. $5x - 8 = 72$

2. $6x - 3 = 63$

3. $11x - (6x - 5) = 40$

4. $5x - (2x - 10) = 35$

5. $2x - 7 = 6 + x$

6. $3x + 5 = 2x + 13$

7. $7x + 4 = x + 16$

8. $13x + 14 = 12x - 5$

9. $3(x - 2) + 7 = 2(x + 5)$

10. $2(x - 1) + 3 = x - 3(x + 1)$

11. $3(x - 4) - 4(x - 3) = x + 3 - (x - 2)$

12. $2 - (7x + 5) = 13 - 3x$

13. $16 = 3(x - 1) - (x - 7)$

14. $5x - (2x + 2) = x + (3x - 5)$

15. $25 - [2 + 5y - 3(y + 2)] = -3(2y - 5) - [5(y - 1) - 3y + 3]$

16. $45 - [4 - 2y - 4(y + 7)] = -4(1 + 3y) - [4 - 3(y + 2) - 2(2y - 5)]$

Exercises 17–30 contain equations with constants in denominators. Solve each equation by multiplying both sides by the least common denominator, thereby clearing fractions.

17. $\frac{x}{3} = \frac{x}{2} - 2$

18. $\frac{x}{5} = \frac{x}{6} + 1$

19. $20 - \frac{x}{3} = \frac{x}{2}$

20. $\frac{x}{5} - \frac{1}{2} = \frac{x}{6}$

21. $\frac{3x}{5} = \frac{2x}{3} + 1$

22. $\frac{x}{2} = \frac{3x}{4} + 5$

23. $\frac{3x}{5} - x = \frac{x}{10} - \frac{5}{2}$

24. $2x - \frac{2x}{7} = \frac{x}{2} + \frac{17}{2}$

25. $\frac{x + 3}{6} = \frac{3}{8} + \frac{x - 5}{4}$

26. $\frac{x + 1}{4} = \frac{1}{6} + \frac{2 - x}{3}$

27. $\frac{x}{4} = 2 + \frac{x - 3}{3}$

28. $5 + \frac{x - 2}{3} = \frac{x + 3}{8}$

29. $\frac{x + 1}{3} = 5 - \frac{x + 2}{7}$

30. $\frac{3x}{5} - \frac{x - 3}{2} = \frac{x + 2}{3}$

Exercises 31–50 contain equations with variables in denominators. For each equation, **a.** *Write the value or values of the variable that make a denominator zero. These are the restrictions on the variable.* **b.** *Keeping the restrictions in mind, solve the equation by multiplying both sides by the least common denominator.*

31. $\frac{4}{x} = \frac{5}{2x} + 3$

32. $\frac{5}{x} = \frac{10}{3x} + 4$

33. $\frac{2}{x} + 3 = \frac{5}{2x} + \frac{13}{4}$

34. $\frac{7}{2x} - \frac{5}{3x} = \frac{22}{3}$

35. $\frac{2}{3x} + \frac{1}{4} = \frac{11}{6x} - \frac{1}{3}$

36. $\frac{5}{2x} - \frac{8}{9} = \frac{1}{18} - \frac{1}{3x}$

37. $\frac{x-2}{2x} + 1 = \frac{x+1}{x}$

38. $\frac{4}{x} = \frac{9}{5} - \frac{7x-4}{5x}$

39. $\frac{1}{x-1} + 5 = \frac{11}{x-1}$

40. $\frac{3}{x+4} - 7 = \frac{-4}{x+4}$

41. $\frac{8x}{x+1} = 4 - \frac{8}{x+1}$

42. $\frac{2}{x-2} = \frac{x}{x-2} - 2$

43. $\frac{3}{2x-2} + \frac{1}{2} = \frac{2}{x-1}$

44. $\frac{3}{x+3} = \frac{5}{2x+6} + \frac{1}{x-2}$

45. $\frac{3}{x+2} + \frac{2}{x-2} = \frac{8}{(x+2)(x-2)}$

46. $\frac{5}{x+2} + \frac{3}{x-2} = \frac{12}{(x+2)(x-2)}$

47. $\frac{2}{x+1} - \frac{1}{x-1} = \frac{2x}{x^2-1}$

48. $\frac{4}{x+5} + \frac{2}{x-5} = \frac{32}{x^2-25}$

49. $\frac{1}{x-4} - \frac{5}{x+2} = \frac{6}{x^2-2x-8}$

50. $\frac{6}{x+3} - \frac{5}{x-2} = \frac{-20}{x^2+x-6}$

In Exercises 51–58, determine whether each equation is an identity, a conditional equation, or an inconsistent equation.

51. $4(x-7) = 4x - 28$

52. $4(x-7) = 4x + 28$

53. $2x + 3 = 2x - 3$

54. $\frac{7x}{x} = 7$

55. $4x + 5x = 8x$

56. $8x + 2x = 9x$

57. $\frac{2x}{x-3} = \frac{6}{x-3} + 4$

58. $\frac{3}{x-3} = \frac{x}{x-3} + 3$

The equations in Exercises 59–68 combine the types of equations we have discussed in this section. Solve each equation or state that it is true for all real numbers or no real numbers.

59. $\frac{x+5}{2} - 4 = \frac{2x-1}{3}$

60. $\frac{x+2}{7} = 5 - \frac{x+1}{3}$

61. $\frac{2}{x-2} = 3 + \frac{x}{x-2}$

62. $\frac{6}{x+3} + 2 = \frac{-2x}{x+3}$

63. $8x - (3x+2) + 10 = 3x$

64. $2(x+2) + 2x = 4(x+1)$

65. $\frac{2}{x} + \frac{1}{2} = \frac{3}{4}$

66. $\frac{3}{x} - \frac{1}{6} = \frac{1}{3}$

67. $\frac{4}{x-2} + \frac{3}{x+5} = \frac{7}{(x+5)(x-2)}$

68. $\frac{1}{x-1} = \frac{1}{(2x+3)(x-1)} + \frac{4}{2x+3}$

Application Exercises

69. The equation $d = 5000c - 525{,}000$ describes the relationship between the annual number of deaths (d) in the United States from heart disease and the average cholesterol level (c) of blood. (Cholesterol level, c, is expressed in milligrams per deciliter of blood.)
    a. In 1990, 500,000 Americans died from heart disease. Substitute 500,000 for d in the given equation and then solve for c to determine the average cholesterol level in 1990.
    b. Suppose that the average cholesterol level for people in the United States could be reduced to 180. Substitute 180 for c in the given formula and then compute the value for d to determine the number of annual deaths from heart disease with this reduced cholesterol level. Compared to the number of deaths in 1990, how many lives would be saved by this cholesterol reduction?

70. There is a relationship between the vocabulary of a child and its age. The equation $60A - V = 900$ describes this relationship, where A is the age of the child in months and V is the number of words that the child uses. Suppose that a child uses 1500 words. Substitute 1500 for V in the equation to determine the child's age in months.

71. The equation

$$p = 15 + \frac{15d}{33}$$

describes the pressure of sea water (p, in pounds per square foot) at a depth of d feet below the surface. The record depth for breath-held diving, by Francisco Ferreras (Cuba) off Grand-Bahama Island, on November 14,1993, involved pressure of 201 pounds per square foot. To what depth did Ferreras descend on this ill-advised venture? (He was underwater for 2 minutes and 9 seconds!)

72. The equation $P = -0.5d + 100$ describes the percentage (P) of lost hikers found in search and rescue missions when members of the search team walk parallel to one another separated by a distance of d yards. If a search and rescue team finds 70% of lost hikers, substitute 70 for P in the equation and find the parallel distance of separation between members of the search party.

Writing in Mathematics

73. What is a linear equation in one variable? Give an example of this type of equation.
74. What does it mean to solve an equation?
75. What is the solution set of an equation?
76. What are equivalent equations? Give an example.
77. What is the difference between solving an equation such as $2(x-4) + 5x = 34$ and simplifying an algebraic expression such as $2(x-4) + 5x$? If there is a difference, which topic should be taught first? Why?

78. Suppose that you solve $\frac{x}{5} - \frac{x}{2} = 1$ by multiplying both sides by 20, rather than the least common denominator of 5 and 2 (namely, 10). Describe what happens. If you get the correct solution, why do you think we clear the equation of fractions by multiplying by the *least* common denominator?

79. Suppose you are an algebra teacher grading the following solution on an examination:

$$-3(x - 6) = 2 - x$$
$$-3x - 18 = 2 - x$$
$$-2x - 18 = 2$$
$$-2x = -16$$
$$x = 8$$

You should note that 8 checks, and the solution set is $\{8\}$. The student who worked the problem therefore wants full credit. Can you find any errors in the solution? If full credit is 10 points, how many points should you give the student? Justify your position.

80. Explain how to determine the restrictions on the variable for the equation

$$\frac{3}{x + 5} + \frac{4}{x - 2} = \frac{7}{(x + 5)(x - 2)}.$$

81. What is an identity? Give an example.

82. What is a conditional equation? Give an example.

83. What is an inconsistent equation? Give an example.

Technology Exercises

For Exercises 84–87, use your graphing utility to graph each side of the equations in the same viewing rectangle. Based on the resulting graph, label each equation as conditional, inconsistent, or an identity. If the equation is conditional, use the x-coordinate of the intersection point to find the solution set. Verify this value by direct substitution into the equation.

84. $2(x - 6) + 3x = x + 6$

85. $9x + 3 - 3x = 2(3x + 1)$

86. $2(x + \frac{1}{2}) = 5x + 1 - 3x$

87. $\frac{2x - 1}{3} - \frac{x - 5}{6} = \frac{x - 3}{4}$

Critical Thinking Exercises

88. Which one of the following is true?

a. The equation $-7x = x$ has no solution.

b. The equations $\frac{x}{x - 4} = \frac{4}{x - 4}$ and $x = 4$ are equivalent.

c. The equations $3y - 1 = 11$ and $3y - 7 = 5$ are equivalent.

d. If a and b are any real numbers, then $ax + b = 0$ always has only one number in its solution set.

89. Solve for x: $ax + b = c$.

90. Write three equations that are equivalent to $x = 5$.

91. If x represents a number, write an English sentence about the number that results in an inconsistent equation.

92. Find b such that $\frac{7x + 4}{b} + 13 = x$ will have a solution set given by $\{-6\}$.

93. Find b such that $\frac{4x - b}{x - 5} = 3$ will have a solution set given by $\varnothing$.

Group Exercise

94. In your group, describe the best procedure for solving an equation like

$$0.47x + \frac{19}{4} = -0.2 + \frac{2}{5}x.$$

Use this procedure to actually solve the equation. Then compare procedures with other groups working on this problem. Which group devised the most streamlined method?

SECTION 1.2 Formulas and Applications

Objectives

1. Solve problems using formulas.
2. Use linear equations to solve problems.
3. Solve for a variable in a formula.

Could you live to be 125? The number of Americans ages 100 or older could approach 850,000 by 2050. Some scientists predict that by 2100, our descendants could live to be 200 years of age. In this section, we will see how equations can be used to make these kinds of predictions as we turn to applications of linear equations.

1 Solve problems using formulas.

Formulas and Modeling Data

The graph in Figure 1.1 shows life expectancy in the United States by year of birth. For example, we can use the graph to find life expectancy for women born in 1980. Find the two bars for 1980 and then look at the bar on the right, representing females. The number printed on this bar is 77.4. Thus, the life expectancy for women born in 1980 is 77.4 years.

Life Expectancy by Year of Birth

Year	Males	Females
1950	65.6	71.1
1960	66.6	73.1
1970	67.1	74.7
1980	70.0	77.4
1990	71.8	78.8
2000	73.0	80.0

Source: U.S. Bureau of the Census

Figure 1.1 Life expectancy by year of birth

The data for U.S. women in Figure 1.1 can be approximated by the equation

$$E = 0.215t + 71.05,$$

where the variable E represents life expectancy for women born t years after 1950. This equation is an example of a *formula*. A **formula** is an equation that uses letters to express relationships between two or more variables. The given formula expresses the relationship between the number of years born after 1950, t, and life expectancy for U.S. women, E.

EXAMPLE 1 Using a Formula

Use the formula

$$E = 0.215t + 71.05$$

to determine the year of birth for which U.S. women can expect to live 77.5 years.

Solution We are given that the life expectancy for women is 77.5 years, so substitute 77.5 for E in the formula and solve for t.

$$E = 0.215t + 71.05 \quad \text{This is the given formula.}$$

$$77.5 = 0.215t + 71.05 \quad \text{Replace E by 77.5 and solve for t.}$$

$$77.5 - 71.05 = 0.215t + 71.05 - 71.05 \quad \text{Isolate the term containing t by subtracting 71.05 from both sides.}$$

$$6.45 = 0.215t \quad \text{Simplify.}$$

$$\frac{6.45}{0.215} = \frac{0.215t}{0.215} \quad \text{Divide both sides by 0.215.}$$

$$30 = t \quad \text{Simplify.}$$

The formula indicates that U.S. women born 30 years after 1950, or in 1980, can expect to live 77.5 years.

Table 1.1 Life Expectancy for U.S. Women

Birth Year	Actual Value	Value Predicted by $E = 0.215t + 71.05$
1950	71.1	71.05
1960	73.1	73.2
1970	74.7	75.35
1980	77.4	77.5
1990	78.8	79.65
2000	80.0	81.8

Check Point 1 The formula $D = 0.2F - 1$ describes death rate from breast cancer per 100,000 women, D, and daily fat intake, F, in grams. The death rate of American women from breast cancer is 19 women per 100,000. What is the daily fat intake for women in America?

If you look back at the actual data in Figure 1.1, you will see that women born in 1980 can expect to live 77.4 years, not 77.5 as predicted by the formula. In developing formulas that describe, or *model*, data, mathematicians strive for both accuracy and simplicity. The formula $E = 0.215t + 71.05$ is relatively simple to use, but as we can see from Table 1.1, it is not an entirely accurate description of the data. Furthermore, we may not want to project the formula past the year 2000 because unforseen progress in conquering breast cancer and other diseases could have an impact on actual life expectancy.

2 Use linear equations to solve problems.

Problem Solving with Linear Equations

Americans love their pets. The number of cats in the United States exceeds the number of dogs by 7.5 million. The number of cats and dogs combined is 114.7 million. So, how many dogs and cats are there in the United States?

Before answering the question, let's see if we can write a critical sentence that describes, or *models*, the problem's conditions. The **verbal model** is

The number of dogs in the U.S.	plus	The number of cats in the U.S.	equals	114.7 million.
?	+	?	=	114.7 (million).

The question marks under the voice balloons indicate that we need algebraic expressions for these unknowns. Once we obtain these expressions, we will have an equation that models the verbal conditions. We call this equation a *mathematical model*. A **mathematical model** is a formula or algebraic equation that can be formed from a verbal model. Earlier, we saw that a mathematical model can be formed using actual data.

Here is a step-by-step strategy for solving problems using mathematical models that are created from verbal models.

> **Strategy for Problem Solving**
>
> **Step 1** Read the problem carefully. Attempt to state the problem in your own words and state what the problem is looking for. Let x (or any variable) represent one of the quantities in the problem.
>
> **Step 2** If necessary, write expressions for any other unknown quantities in the problem in terms of x.
>
> **Step 3** Form a verbal model of the problem's conditions and then write an equation in x that translates the verbal model.
>
> **Step 4** Solve the equation written in step 3 and answer the question in the problem.
>
> **Step 5** Check the proposed solution in the original wording of the problem.

EXAMPLE 2 Pet Population

The number of cats in the United States exceeds the number of dogs by 7.5 million. The number of cats and dogs combined is 114.7 million. Determine the number of dogs and cats in the United States.

U.S. Pet Population

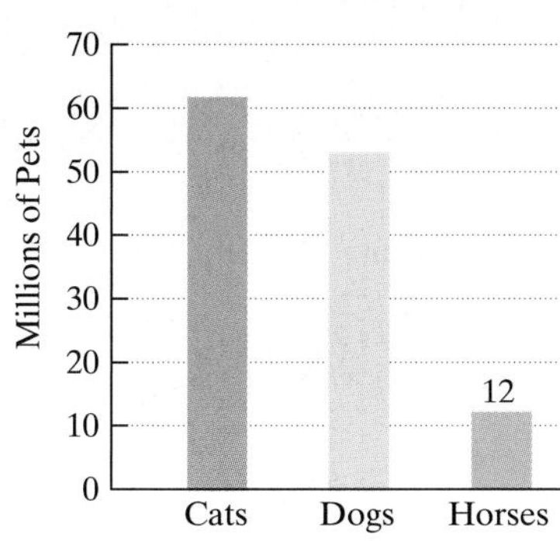

Source: American Veterinary Medical Association

Americans spend more than \$21 billion a year on their pets. 31.4% of households have cats and 34.3% have dogs.

Solution

Step 1 Let x represent one of the quantities. We know something about the number of cats; the cat population exceeds the dog population by 7.5 million. This means that there are 7.5 million more cats than dogs. We will let

$x =$ the number (in millions) of dogs in the United States.

Step 2 Represent other quantities in terms of x. The other unknown quantity is the number of cats. Because there are 7.5 million more cats than dogs, let

$x + 7.5 =$ the number (in millions) of cats in the United States.

Step 3 Write an equation in x that describes the conditions. The number of cats and dogs combined is 114.7 million.

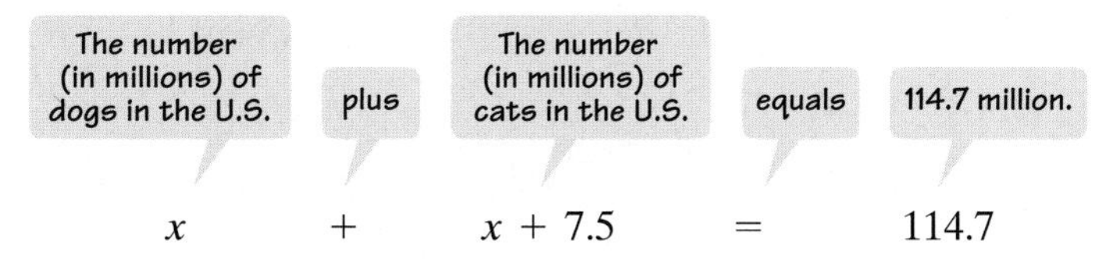

Step 4 Solve the equation and answer the question.

$$x + x + 7.5 = 114.7$$ This is the equation that models the verbal conditions.

$$2x + 7.5 = 114.7$$ Combine like terms on the left side.

$$2x + 7.5 - 7.5 = 114.7 - 7.5 \quad \text{Subtract 7.5 from both sides.}$$

$$2x = 107.2 \quad \text{Simplify.}$$

$$\frac{2x}{2} = \frac{107.2}{2} \quad \text{Divide both sides by 2.}$$

$$x = 53.6 \quad \text{Simplify.}$$

Because x represents the number (in millions) of dogs, there are 53.6 million dogs in the United States. Because $x + 7.5$ represents the number (in millions) of cats, there are $53.6 + 7.5$, or 61.1 million cats in the United States.

Step 5 Check the proposed solution in the original wording of the problem. The problem states that the number of cats and dogs combined is 114.7 million. By adding 53.6 million, the dog population, and 61.1 million, the cat population, we do, indeed, obtain a sum of 114.7 million.

Two of the top-selling music albums of all time are *Jagged Little Pill* (Alanis Morissette) and *Saturday Night Fever* (Bee Gees). The Morissette album sold 5 million more copies than that of the Bee Gees. Combined, the two albums sold 27 million copies. Determine the number of sales for each of the albums.

EXAMPLE 3 Renting a Car

Rent-a-Heap Agency charges \$125 per week plus \$0.20 per mile to rent a small car. How many miles can you travel for \$335?

Solution

Step 1 Let x represent one of the quantities. Because we are asked to find the number of miles we can travel for \$335, let

$$x = \text{the number of miles.}$$

Step 2 Represent other quantities in terms of x. There are no other unknown quantities to find, so we can skip this step.

Step 3 Write an equation in x that describes the conditions.

The weekly charge of \$125	plus	the charge of \$0.20 per mile for x miles	equals	the total \$335 rental charge.
125	$+$	$0.20x$	$=$	335

Step 4 Solve the equation and answer the question.

$$125 + 0.20x = 335 \quad \text{This is the equation that models the verbal conditions.}$$

$$125 + 0.20x - 125 = 335 - 125 \quad \text{Subtract 125 from both sides.}$$

$$0.20x = 210 \quad \text{Simplify.}$$

$$\frac{0.20x}{0.20} = \frac{210}{0.20} \quad \text{Divide both sides by 0.20.}$$

$$x = 1050 \quad \text{Simplify.}$$

You can travel 1050 miles for \$335.

Step 5 Check the proposed solution in the original wording of the problem. Traveling 1050 miles should result in a total rental charge of \$335. The mileage charge of \$0.20 per mile is

$$\$0.20(1050) = \$210.$$

Adding this to the \$125 weekly charge gives a total rental charge of

$$\$125 + \$210 = \$335.$$

Because this results in the given rental charge of \$335, this verifies that you can travel 1050 miles.

Check Point 3 Healthy Bodies, a fitness club, charges \$440 per year, plus \$1.75 per hour used. If you purchase a \$540 annual membership, how many hours of use does this allow?

Our next example involves simple interest. The annual simple interest that an investment earns is given by the formula

$$I = Pr$$

where I is the simple interest, P is the principal, and r is the simple interest rate, expressed in decimal form. Suppose, for example, that you deposit \$2000 ($P = 2000$) in a savings account that has a simple interest rate of 6% ($r = 0.06$). The annual simple interest is computed as follows:

$$I = Pr = (2000)(0.06) = 120.$$

The annual interest is \$120.

EXAMPLE 4 Solving a Simple Interest Problem

You inherit \$16,000 with the stipulation that for the first year the money must be invested in two stocks paying 6% and 8% annual interest, respectively. How much should be invested at each rate if the total interest earned for the year is to be \$1180?

Solution

Step 1 Let x represent one of the quantities.

Let x = the amount invested at 6%.

Step 2 Represent other quantities in terms of x. The other quantity that we seek is the amount to be invested at 8%. Because the total amount to be invested is \$16,000, and we already used up x,

$16{,}000 - x$ = the amount invested at 8%.

Step 3 Write an equation in x that describes the conditions. The interest for the two investments combined must be \$1180. Interest is Pr or rP for each investment.

Interest from 6% investment	plus	interest from 8% investment	is	\$1180.
$0.06x$	$+$	$0.08(16{,}000 - x)$	$=$	1180
rate times principal		rate times principal		

Study Tip

Look at the expression in step 2. Notice that when you add x and $16{,}000 - x$, you get 16,000, the total investment. In many word problems, a total amount is divided into two parts.

Step 4 Solve the equation and answer the question.

$0.06x + 0.08(16{,}000 - x) = 1180$	This is the equation that models the verbal conditions.
$0.06x + 1280 - 0.08x = 1180$	Use the distributive property.
$-0.02x + 1280 = 1180$	Combine like terms.
$-0.02x + 1280 - 1280 = 1180 - 1280$	Subtract 1280 from both sides.
$-0.02 = -100$	Simplify.
$\frac{-0.02x}{-0.02} = \frac{-100}{-0.02}$	Divide both sides by −0.02.
$x = 5000$	Simplify.

Because x represents the amount invested at 6%, $5000 should be invested at 6%. Because $16{,}000 - x$ represents the amount invested at 8%, $16,000 − $5000, or $11,000, should be invested at 8%.

Step 5 Check the proposed solution in the original wording of the problem. The problem states that the total interest should be $1180. The interest earned on $5000 at 6% is ($5000)(0.06), or $300. The interest earned on $11,000 at 8% is ($11,000)(0.08), or $880. The total interest is $300 + $880, or $1180, exactly as it should be.

Suppose that you invest $25,000, part at 9% simple interest and the remainder at 12%. If the total yearly interest from these investments was $2550, find the amount invested at each rate.

Solving geometry problems usually requires a knowledge of basic geometric ideas and formulas. Formulas for area, perimeter, and volume are given in Table 1.2.

Table 1.2 Common Formulas for Area, Perimeter, and Volume

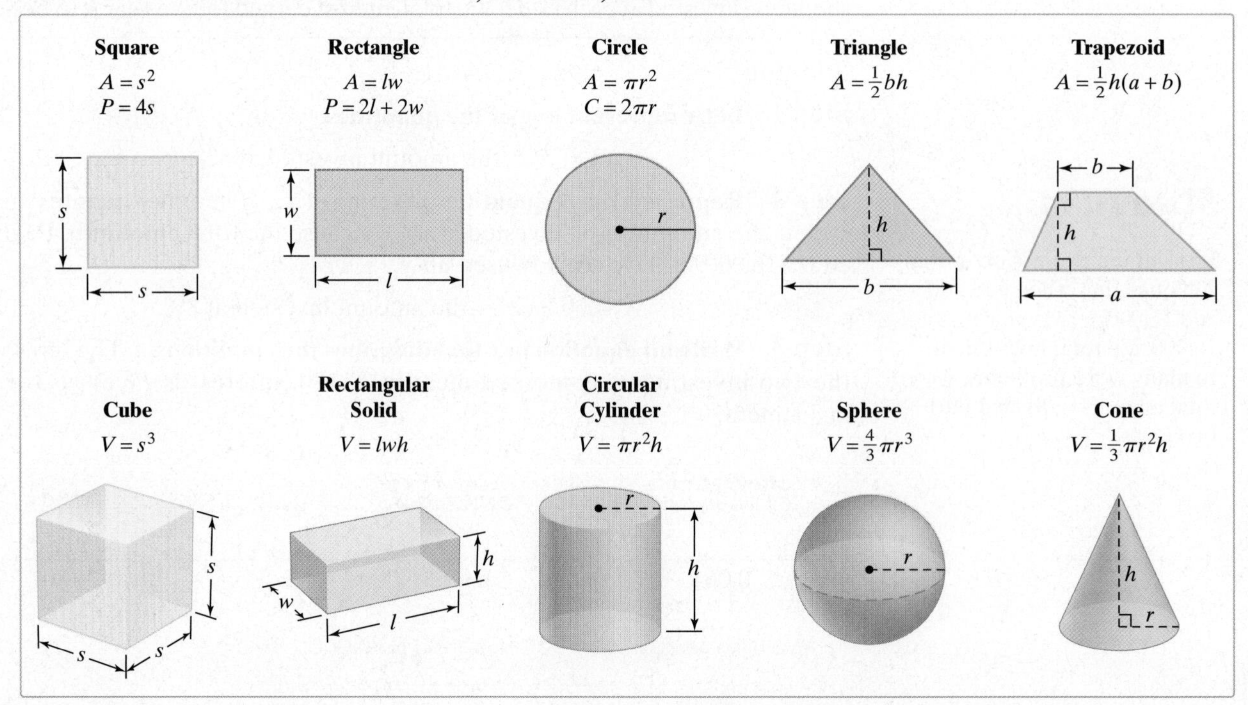

We will be using the formula for the perimeter of a rectangle, $P = 2l + 2w$, in our next example. A helpful verbal model for this formula is 2 times length plus 2 times width is a rectangle's perimeter.

EXAMPLE 5 Finding the Dimensions of a Soccer Field

A rectangular soccer field is twice as long as it is wide. If the perimeter of a soccer field is 300 yards, what are the field's dimensions?

Solution

Step 1 Let x represent one of the quantities. We know something about the length; the field is twice as long as it is wide. We will let

$$x = \text{the width.}$$

Step 2 Represent other quantities in terms of x. Because the field is twice as long as it is wide, let

$$2x = \text{the length.}$$

Figure 1.2 illustrates the soccer field and its dimensions.

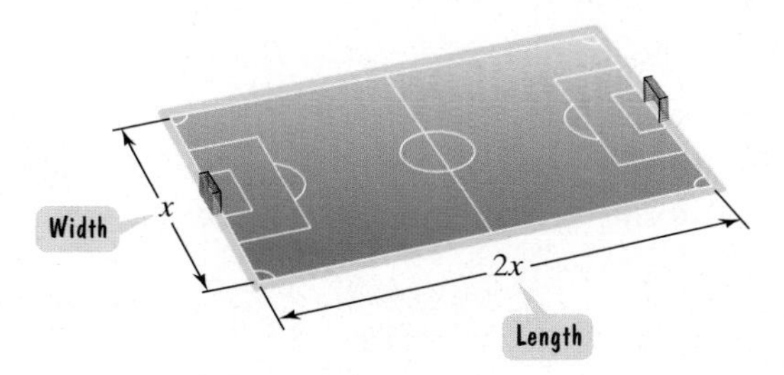

Figure 1.2

Step 3 Write an equation in x that describes the conditions. Because the perimeter of a soccer field is 300 yards,

Twice the length	plus	twice the width	is	the perimeter.
$2 \cdot 2x$	$+$	$2 \cdot x$	$=$	$300.$

Step 4 Solve the equation and answer the question.

$$2 \cdot 2x + 2 \cdot x = 300 \quad \text{This is the equation that models the verbal conditions.}$$

$$4x + 2x = 300 \quad \text{Multiply.}$$

$$6x = 300 \quad \text{Combine like terms.}$$

$$\frac{6x}{6} = \frac{300}{6} \quad \text{Divide both sides by 6.}$$

$$x = 50 \quad \text{Simplify.}$$

Thus,

$$\text{Width} = x = 50$$

$$\text{Length} = 2x = 2(50) = 100.$$

The dimensions of a soccer field are 50 yards by 100 yards.

Step 5 Check the proposed solution in the original wording of the problem. The perimeter of the soccer field using the dimensions that we found is 2(50 feet) + 2(100 feet) = 100 feet + 200 feet, or 300 feet. Because the problem's wording tells us that the perimeter is 300 feet, our dimensions are correct.

Check Point 5 A rectangular swimming pool is three times as long as it is wide. If the perimeter of the pool is 320 feet, what are the pool's dimensions?

3 Solve for a variable in a formula.

Solving for a Variable in a Formula

When working with formulas, such as the geometric formulas shown in Table 1.2, it is often necessary to solve for a specified variable. This is done by isolating the specified variable on one side of the equation. Begin by isolating all terms with the specified variable on one side of the equation and all terms without the specified variable on the other side. The next example shows how to do this.

EXAMPLE 6 Solving for a Variable in a Formula

Solve the formula $2l + 2w = P$ for w.

Solution First, isolate $2w$ on the left by subtracting $2l$ from both sides. Then solve for w by dividing both sides by 2.

We need to isolate w

$$2l + 2w = P$$ This is the given formula.

$$2l - 2l + 2w = P - 2l$$ Isolate $2w$ by subtracting $2l$ from both sides.

$$2w = P - 2l$$ Simplify.

$$\frac{2w}{2} = \frac{P - 2l}{2}$$ Isolate w by dividing both sides by 2.

$$w = \frac{P - 2l}{2}$$ Simplify.

Check Point 6 Solve $y = mx + b$ for m.

EXERCISE SET 1.2

Practice Exercises

In Exercises 1–14, let x represent the number. Write each English phrase as an algebraic expression.

1. The sum of a number and 9
2. A number increased by 13
3. A number subtracted from 20
4. 13 less than a number
5. 8 decreased by 5 times a number
6. 14 less than the product of 6 and a number
7. The quotient of 15 and a number
8. The quotient of a number and 15
9. The sum of twice a number and 20
10. Twice the sum of a number and 20
11. 30 subtracted from 7 times a number
12. The quotient of 12 and a number, decreased by 3 times the number
13. Four times the sum of a number and 12
14. Five times the difference of a number and 6

In Exercises 15–20, let x represent the number. Use the given conditions to write an equation. Solve the equation and find the number.

15. A number increased by 40 is equal to 450. Find the number.
16. The sum of a number and 29 is 54. Find the number.
17. Seven subtracted from five times a number is 123. Find the number.
18. Eight subtracted from six times a number is 184. Find the number.
19. Nine times a number is 30 more than three times that number. Find the number.
20. Five more than four times a number is that number increased by 35. Find the number.

Application Exercises

Medical researchers have found that the desirable heart rate R, in beats per minute, for beneficial exercise is approximated by the formulas

$$R = 143 - 0.65A \quad \text{for women}$$

$$R = 165 - 0.75A \quad \text{for men}$$

where A is the person's age. Use these formulas to solve Exercises 21–22.

21. If the desirable heart rate for a woman is 117 beats per minute, how old is she? How is the solution shown on the line graph at the top of the next column?

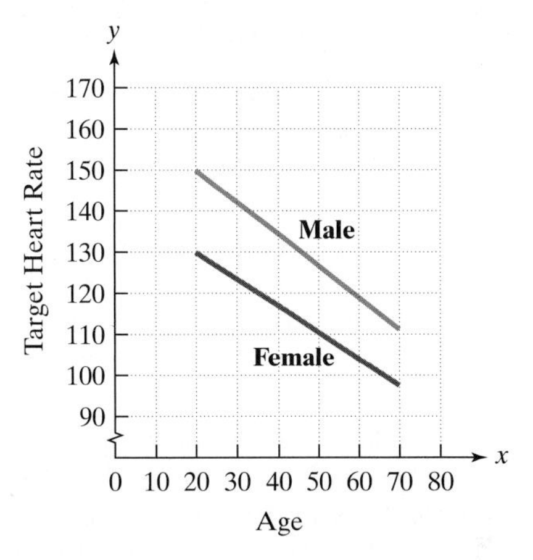

22. If the desirable heart rate for a man is 147 beats per minute, how old is he? How is the solution shown by the line graph?
23. The Indianapolis 500 is a car race in which specially built cars compete by racing 500 miles around the 2.5-mile track. Using the actual speeds of the winners from 1980 through 1992, mathematicians obtained the formula $y = 2.5x + 198.73$, in which x represents the number of years after 1980 and y represents the winning speed. How many years after 1980 is the winning speed predicted to be 273.73 miles per hour? What year will that be?
24. The International Panel on Climate Change is a U.N.-sponsored body made up of more than 1500 leading experts from 60 nations. According to their recent findings, increased levels of atmospheric carbon dioxide are affecting our climate. Global warming is under way and the effects could be catastrophic. The formula $C = 1.44t + 280$ describes carbon dioxide concentration C, in parts per million, t years after 1939. The preindustrial carbon dioxide concentration of 280 parts per million remained fairly constant until World War II, increasing after that due primarily to the burning of fossil fuels related to energy consumption. When will the concentration be double the preindustrial level?
25. A woman's height (h) is related to the length of the femur (f) (the bone from the knee to the hip socket) by the formula $f = 0.432h - 10.44$. Both h and f are measured in inches. A partial skeleton is found of a woman in which the femur is 16 inches long. Police find the skeleton in an area where a woman slightly over 5 feet tall has been missing for over a year. Can the partial skeleton be that of the missing woman? Explain.
26. The formula

$$\frac{W}{2} - 3H = 53$$

describes the recommended weight W, in pounds, for a male, where H represents the man's height in inches over 5 feet. What is the recommended weight for a man who is 6 feet, 3 inches tall?

In Exercises 27–56, use the five-step strategy given in the box on page 99 to solve each problem.

27. During the 1998 baseball season, Mark McGwire hit four more home runs than Sammy Sosa. Combined, the two athletes hit 136 home runs. Determine the number of home runs hit by McGwire and Sosa.

28. In 1999, the most populous countries in the world were China and India. In that year, China's population exceeded India's by 269 million. Combined, the two countries had a population of 2265 million. Determine the 1999 population for China and India.

29. The first Super Bowl was played between the Green Bay Packers and the Kansas City Chiefs in 1967. Only once, in 1991, were the winning and losing scores in the Super Bowl consecutive integers. If the sum of the scores was 39, what were the scores?

30. The longest-lived U.S. presidents are John Adams (age 90), Herbert Hoover (also 90), and Harry Truman (88). Behind them are James Madison, Thomas Jefferson, and Richard Nixon. The latter three men lived a total of 249 years, and their ages at the time of death form consecutive odd integers. For how long did Nixon, Jefferson, and Madison live?

31. The graph reflects fear of crime in ten selected countries. The percentage of people feeling unsafe in their neighborhoods after dark in the United States exceeds twice that of Sweden by 14%. In the two countries combined, 54.5% of the public feel unsafe walking in their neighborhood after dark. Find the percentage of people feeling unsafe after dark for Sweden and the United States.

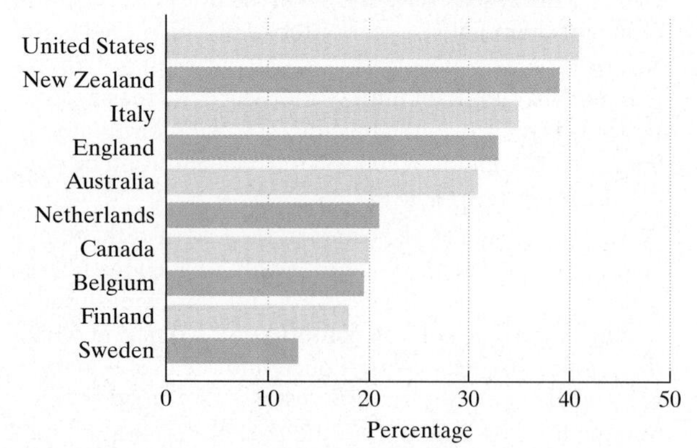

Source: Ministry of Justice, The Netherlands

32. In 1999, Americans in 68 urban areas wasted almost 7 billion gallons of fuel sitting in traffic. The graph at the top of the next column shows the number of hours in traffic per year for the average motorist in ten cities. The average motorist in Los Angeles spends 32 hours less than twice that of the average motorist in Miami stuck in traffic each year. In the two cities combined, 139 hours are spent by the average motorist per year in traffic. How many hours are wasted in traffic by the average motorist in Los Angeles and Miami?

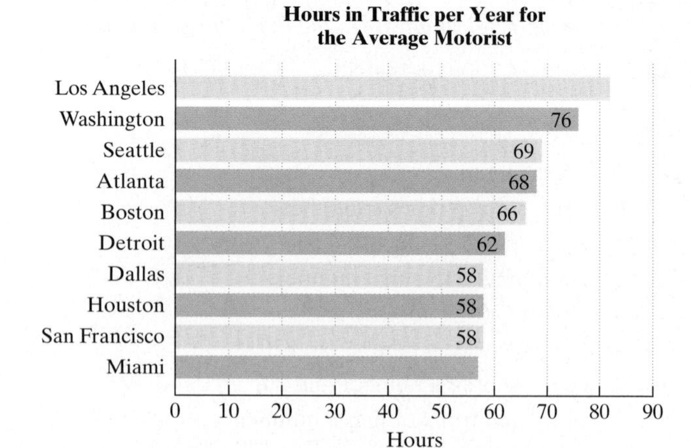

Source: Texas Transportation Institute

33. A car rental agency charges $200 per week plus $0.15 per mile to rent a car. How many miles can you travel in one week for $320?

34. A car rental agency charges $180 per week plus $0.25 per mile to rent a car. How many miles can you travel in one week for $395?

35. The average weight for female infants at birth is 7 pounds, with a monthly weight gain of 1.5 pounds. After how many months does a baby girl weigh 16 pounds?

36. In 1995, the average yearly salary for teachers in the United States was $38,556. If the salary increases by $1496 per year, in which year will the salary reach $56,508?

37. Answer the question in the following *Peanuts* cartoon strip. (*Note:* You may not use the answer given in the cartoon!)

PEANUTS reprinted by permission of United Features Syndicate, Inc.

38. Every year, approximately 1760 Americans suffer spinal cord injuries due to falls. This represents 22% of the total number of Americans who suffer spinal cord injuries yearly. Determine the number of Americans who suffer spinal cord injuries each year.

Causes of U.S. Spinal Cord Injuries

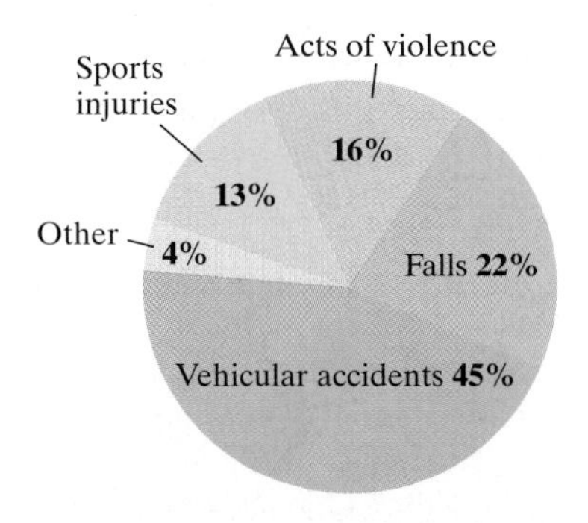

Source: *U.S. News and World Report*

39. The bus fare in a city is \$1.25. People who use the bus have the option of purchasing a monthly coupon book for \$21.00. With the coupon book, the fare is reduced to \$0.50.

a. Let x represent the number of times in a month the bus is used. Write algebraic expressions for the total monthly costs of using the bus x times both with and without the coupon book.

b. Determine the number of times in a month the bus must be used so that the total monthly cost without the coupon book is the same as the total monthly cost with the coupon book.

40. A coupon book for a bridge costs \$21 per month. The toll for the bridge is normally \$2.50, but it is reduced to \$1 for people who have purchased the coupon book.

a. Let x represent the number of times in a month the bridge is used. Write algebraic expressions for the total monthly costs of using the bridge x times both with and without the coupon book.

b. Determine the number of times in a month the bridge must be crossed so that the total monthly cost without the coupon book is the same as the total monthly cost with the coupon book.

41. You inherit \$25,000 with the stipulation that for the first year the money must be invested in two stocks paying 9% and 12% annual interest, respectively. How much should be invested at each rate if the total interest earned for the year is to be \$2250?

42. You inherit \$18,750 with the stipulation that for the first year the money must be invested in two stocks paying 10% and 12% annual interest, respectively. How much should be invested at each rate if the total interest earned for the year is to be \$2117?

43. The length of the rectangular tennis court at Wimbledon is 6 feet longer than twice the width. If the court's perimeter is 228 feet, what are the court's dimensions?

44. The length of a rectangular basketball court is 6 feet less than twice the width. If the court's perimeter is 288 feet, what are the court's dimensions?

45. A bookcase is to be constructed as shown in the figure. The length is to be 3 times the height. If 60 feet of lumber is available for the entire unit, find the length and height of the bookcase.

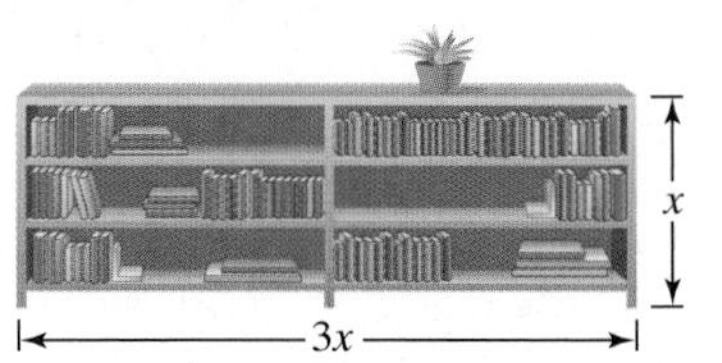

46. The height of the bookcase in the figure is 3 feet longer than the length of a shelf. If 18 feet of lumber is available for the entire unit, find the length and height of the unit.

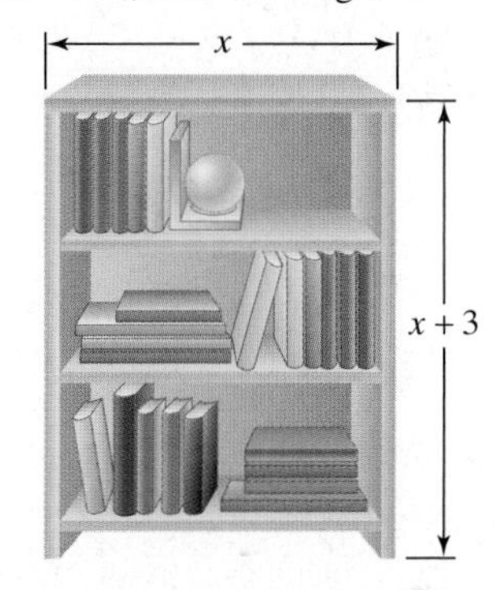

47. An automobile repair shop charged a customer \$448, listing \$63 for parts and the remainder for labor. If the cost of labor is \$35 per hour, how many hours of labor did it take to repair the car?

48. A repair bill on a yacht came to \$1603, including \$532 for parts and the remainder for labor. If the cost of labor is \$63 per hour, how many hours of labor did it take to repair the yacht?

49. After a 35% price reduction, a graphing calculator sold for \$81.90. What was the calculator's price before the reduction?

50. After a 12% price reduction, a car sold for \$17,600. What was the car's price before the reduction?

51. Inclusive of a 6.5% sales tax, a television sold for \$788.10. Find the price of the television before the tax was added.

52. Inclusive of a 6.5% sales tax, a car sold for \$17,466. Find the price of the car before the tax was added.

53. Markup is the amount added to the dealer's cost of an item to arrive at the selling price of that item. The selling price of a refrigerator is \$584. If the markup is 25% of the dealer's cost, what is the dealer's cost of the refrigerator?

54. A calculator costs a dealer \$80. Determine the selling price if the markup is 20% of the selling price.

55. An HMO pamphlet contains the following recommended weight for women: "Give yourself 100 pounds for the first 5 feet plus 5 pounds for every inch over 5 feet tall."

Using this description, what height corresponds to an ideal weight of 135 pounds?

56. A job pays an annual salary of $33,150, which includes a holiday bonus of $750. If paychecks are issued twice a month, what is the gross amount for each paycheck?

In Exercises 57–72, solve each formula for the specified variable. Do you recognize the formula? If so, what does it describe?

57. $A = lw$ for w

58. $D = RT$ for R

59. $A = \frac{1}{2}bh$ for b

60. $V = \frac{1}{3}Bh$ for B

61. $I = Prt$ for P

62. $C = 2\pi r$ for r

63. $E = mc^2$ for m

64. $V = \pi r^2 h$ for h

65. $T = D + pm$ for p

66. $P = C + MC$ for M

67. $A = \frac{1}{2}h(a + b)$ for a

68. $A = \frac{1}{2}h(a + b)$ for b

69. $S = P + Prt$ for r

70. $S = P + Prt$ for t

71. $B = \dfrac{F}{S - V}$ for S

72 $S = \dfrac{C}{1 - r}$ for r

Writing in Mathematics

73. What is a formula?

74. We discussed formulas in this section after we considered procedures for solving linear equations. Doesn't working with a formula simply mean substituting given numbers into the formula and using the order of operations? Is it necessary to know how to solve equations to work with formulas? Explain.

75. In your own words, describe a step-by-step approach for solving algebraic word problems.

76. Did you have some difficulties solving some of the problems that were assigned in this exercise set? Discuss what you did if this happened to you. Did your course of action enhance your ability to solve algebraic word problems?

Technology Exercises

77. The average hourly rate (y) for public school cafeteria workers in the United States in 1980 was $3.82 per hour. This rate has increased steadily by $0.30 per hour each year since 1980.

a. Write a mathematical model for the hourly rate x years after 1980.

b. Use a graphing utility to graph the model in a $[0, 12, 1]$ by $[0, 8, 1]$ viewing rectangle.

c. Use the trace feature to trace along the curve to determine in what year the hourly wage was $6.22.

d. Verify your observation in part (c) algebraically by setting the model equal to 6.22 and solving for x.

78. A tennis club offers two payment options. Members can pay a monthly fee of $30 plus $5 per hour for court rental time. The second option has no monthly fee, but court time costs $7.50 per hour.

a. Write a mathematical model representing total monthly costs for each option for x hours of court rental time.

b. Use a graphing utility to graph the two models in a $[0, 15, 1]$ by $[0, 120, 6]$ viewing rectangle.

c. Use your utility's trace or intersection feature to determine where the two graphs intersect. Describe what the coordinates of this intersection point represent in practical terms.

d. Verify part (c) using an algebraic approach by setting the two models equal to one another and determining how many hours one has to rent the court so that the two plans result in identical monthly costs.

Critical Thinking Exercises

79. A school board plans to merge two schools into one school of 1000 students in which 42% of the students will be African American. One of the schools has a 10% African American student body and the other has a 90% African American student body. What is the student population in each of the two schools?

80. The price of a dress is reduced by 40%. When the dress still does not sell, it is reduced by 40% of the reduced price. If the price of the dress after both reductions is $72, what was the original price?

81. In a film, the actor Charles Coburn plays an elderly "uncle" character criticized for marrying a woman when he is 3 times her age. He wittily replies, "Ah, but in 20 years time I shall only be twice her age." How old is the "uncle" and the woman?

82. Suppose that we agree to pay you 8¢ for every problem in this chapter that you solve correctly and fine you 5¢ for every problem done incorrectly. If at the end of 26 problems we do not owe each other any money, how many problems did you solve correctly?

83. It was wartime when the Ricardos found out Mrs. Ricardo was pregnant. Ricky Ricardo was drafted and made out a will, deciding that $14,000 in a savings account was to be divided between his wife and his child-to-be. Rather strangely, and certainly with gender bias, Ricky stipulated that if the child were a boy, he would get twice the amount of the mother's portion. If it were a girl, the mother would get twice the amount the girl was to receive. We'll never know what Ricky was thinking of, for (as fate would have it) he did not return from war. Mrs. Ricardo gave birth to twins—a boy and a girl. How was the money divided?

84. Solve for P: $A = P + Prt$.

Group Exercise

85. One of the best ways to learn how to *solve* a word problem in algebra is to *design* word problems of your own. Creating a word problem makes you very aware of precisely how much information is needed to solve the problem. You must also focus on the best way to present information to a reader and on how much information to give. As you write your problem, you gain skills that will help you solve problems created by others.

The group should design five different word problems that can be solved using an algebraic equation. All of the problems should be on different topics. For example, the group should not have more than one problem on simple interest. The group should turn in both the problems and their algebraic solutions.

SECTION 1.3 *Quadratic Equations*

Objectives

1. Solve quadratic equations by factoring.
2. Solve quadratic equations by the square root method.
3. Solve quadratic equations by completing the square.
4. Solve quadratic equations using the quadratic formula.
5. Use the discriminant to determine the kinds of solutions.
6. Determine the most efficient method to use when solving a quadratic equation.
7. Solve problems modeled by quadratic equations.

The crocodile, an endangered species, was the subject of a protection program at Florida's Everglades National Park. Park rangers used the formula

$$P = -10x^2 + 475x + 3500$$

to estimate the crocodile population, P, after x years of the protection program. Their goal was to bring the population up to 7250. To find out how long the program had to be continued for this to occur, we need to substitute 7250 for P in the formula and solve for x:

$$7250 = -10x^2 + 475x + 3500.$$

Do you see how this equation differs from a linear equation? The exponent on x is 2. Solving such an equation involves finding the set of numbers that will make the equation a true statement. In this section, we study a number of methods for solving equations in the form $ax^2 + bx + c = 0$. We also look at applications of these equations.

The Standard Form of a Quadratic Equation

We begin by defining a quadratic equation.

Definition of a Quadratic Equation

A **quadratic equation** in x is an equation that can be written in the **standard form**

$$ax^2 + bx + c = 0$$

where a, b, and c are real numbers with $a \neq 0$. A quadratic equation in x is also called a **second-degree polynomial equation** in x.

An example of a quadratic equation in standard form is $x^2 - 7x + 10 = 0$. The coefficient of x^2 is $1(a = 1)$, the coefficient of x is $-7(b = -7)$, and the constant term is $10(c = 10)$.

1 Solve quadratic equations by factoring.

Solving Quadratic Equations by Factoring

We can factor the left side of the quadratic equation $x^2 - 7x + 10 = 0$. We obtain $(x - 5)(x - 2) = 0$. If a quadratic equation has zero on one side and a factored expression on the other side, it can be solved using the **zero-product principle**.

The Zero-Product Principle

If the product of two algebraic expressions is zero, then at least one of the factors is equal to zero.

$$\text{If } AB = 0, \quad \text{then } A = 0 \text{ or } B = 0.$$

For example, consider the equation $(x - 5)(x - 2) = 0$. According to the zero-product principle, this product can be zero only if at least one of the factors is zero. We set each individual factor equal to zero and solve each resulting equation for x.

$$(x - 5)(x - 2) = 0$$

$$x - 5 = 0 \quad \text{or} \quad x - 2 = 0$$

$$x = 5 \qquad\qquad x = 2$$

We can check each of these proposed solutions in the original quadratic equation, $x^2 - 7x + 10 = 0$.

Check 5:

$$5^2 - 7 \cdot 5 + 10 \stackrel{?}{=} 0$$

$$25 - 35 + 10 \stackrel{?}{=} 0$$

$$0 = 0 \checkmark$$

Check 2:

$$2^2 - 7 \cdot 2 + 10 \stackrel{?}{=} 0$$

$$4 - 14 + 10 \stackrel{?}{=} 0$$

$$0 = 0 \checkmark$$

The resulting true statements indicate that the solutions are 5 and 2. The solution set is $\{5, 2\}$. Note that with a quadratic equation, we can have two solutions, compared to the linear equation that had one.

Solving a Quadratic Equation by Factoring

1. If necessary, rewrite the equation in the form $ax^2 + bx + c = 0$, moving all terms to one side, thereby obtaining zero on the other side.
2. Factor.

3. Apply the zero-product principle, setting each factor equal to zero.
4. Solve the equations in step 3.
5. Check the solutions in the original equation.

EXAMPLE 1 Solving Quadratic Equations by Factoring

Solve by factoring and then using the zero-product principle.

a. $4x^2 - 2x = 0$ b. $2x^2 + 7x = 4$

Solution

a. We begin with $4x^2 - 2x = 0$.

Step 1 Move all terms to one side and obtain zero on the other side. All terms are already on the left and zero is on the other side, so we can skip this step.

Step 2 Factor. We factor out $2x$ from the two terms on the left side.

$$4x^2 - 2x = 0 \quad \text{This is the given equation.}$$
$$2x(2x - 1) = 0 \quad \text{Factor.}$$

Steps 3 and 4 Set each factor equal to zero and solve the resulting equations.

$$2x = 0 \quad \text{or} \quad 2x - 1 = 0$$
$$x = 0 \qquad\qquad 2x = 1$$
$$x = \tfrac{1}{2}$$

Step 5 Check the solutions in the original equation.

Check 0:

$$4x^2 - 2x = 0$$
$$4 \cdot 0^2 - 2 \cdot 0 \stackrel{?}{=} 0$$
$$0 - 0 \stackrel{?}{=} 0$$
$$0 = 0 \checkmark$$

Check $\frac{1}{2}$:

$$4x^2 - 2x = 0$$
$$4\left(\tfrac{1}{2}\right)^2 - 2\left(\tfrac{1}{2}\right) \stackrel{?}{=} 0$$
$$4\left(\tfrac{1}{4}\right) - 2\left(\tfrac{1}{2}\right) \stackrel{?}{=} 0$$
$$1 - 1 \stackrel{?}{=} 0$$
$$0 = 0 \checkmark$$

The solution set is $\{0, \frac{1}{2}\}$.

b. Next, we solve $2x^2 + 7x = 4$.

Step 1 Move all terms to one side and obtain zero on the other side. Subtract 4 from both sides and write the equation in standard form.

$$2x^2 + 7x - 4 = 4 - 4$$
$$2x^2 + 7x - 4 = 0$$

Step 2 Factor.

$$2x^2 + 7x - 4 = 0$$
$$(2x - 1)(x + 4) = 0$$

Steps 3 and 4 Set each factor equal to zero and solve each resulting equation.

$$2x - 1 = 0 \quad \text{or} \quad x + 4 = 0$$
$$2x = 1 \qquad\qquad x = -4$$
$$x = \tfrac{1}{2}$$

Step 5 Check the solutions in the original equation.

Check $\frac{1}{2}$:

$$2x^2 + 7x = 4$$
$$2\left(\tfrac{1}{2}\right)^2 + 7\left(\tfrac{1}{2}\right) \stackrel{?}{=} 4$$
$$\tfrac{1}{2} + \tfrac{7}{2} \stackrel{?}{=} 4$$
$$4 = 4 \checkmark$$

Check -4:

$$2x^2 + 7x = 4$$
$$2(-4)^2 + 7(-4) \stackrel{?}{=} 4$$
$$32 + (-28) \stackrel{?}{=} 4$$
$$4 = 4 \checkmark$$

The solution set is $\{-4, \frac{1}{2}\}$.

Check Point 1 Solve by factoring and then using the zero-product principle.

a. $3x^2 - 9x = 0$ **b.** $2x^2 + x = 1$

Technology

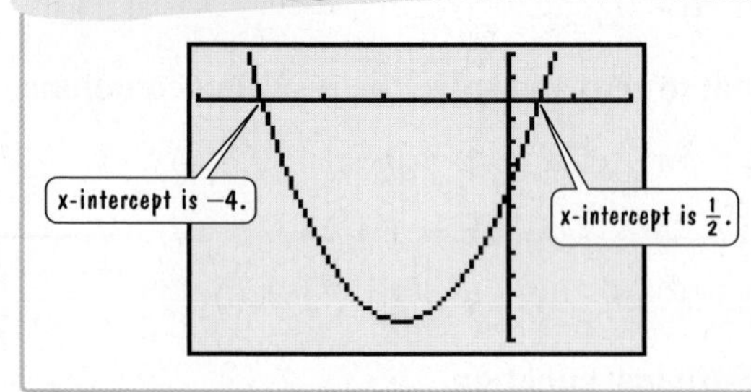

You can use a graphing utility to check the real solutions to a quadratic equation. **The solutions to $ax^2 + bx + c = 0$ correspond to the x-intercepts for the graph of $y = ax^2 + bx + c$.** For example, to check the solutions of $2x^2 + 7x = 4$, or $2x^2 + 7x - 4 = 0$, graph $y = 2x^2 + 7x - 4 = 0$, as shown on the left. Note that it is important to have all nonzero terms on one side of the quadratic equation before entering it into the graphing utility. The x-intercepts are -4 and $\frac{1}{2}$, verifying $\{-4, \frac{1}{2}\}$ as the solution set.

2 Solve quadratic equations by the square root method.

Solving Quadratic Equations by the Square Root Method

Quadratic equations of the form $u^2 = d$, where $d > 0$ and u is an algebraic expression, can be solved by the **square root method**. First, isolate the squared expression u^2 on one side of the equation and the number d on the other side. Then take the square root of both sides. Remember, there are two numbers whose square is d. One number is positive and one is negative.

We can use factoring to verify that $u^2 = d$ has two solutions.

$$u^2 = d \qquad \text{This is the given equation.}$$
$$u^2 - d = 0 \qquad \text{Move all terms to one side and obtain zero on the other side.}$$
$$(u + \sqrt{d})(u - \sqrt{d}) = 0 \qquad \text{Factor.}$$
$$u + \sqrt{d} = 0 \quad \text{or} \quad u - \sqrt{d} = 0 \qquad \text{Set each factor equal to zero.}$$
$$u = -\sqrt{d} \qquad u = \sqrt{d} \qquad \text{Solve the resulting equations.}$$

Because the solutions differ only in sign, we can write them in abbreviated notation as $u = \pm\sqrt{d}$. We read this as "u equals positive or negative the square root of d."

Now that we have verified these solutions, we can solve $u^2 = d$ directly by taking square roots. This process is called **the square root method**.

The Square Root Method

If u is an algebraic expression and d is a positive real number, then $u^2 = d$ has exactly two solutions:

$$\text{If } u^2 = d, \quad \text{then } u = \sqrt{d} \text{ or } u = -\sqrt{d}.$$

Equivalently,

$$\text{If } u^2 = d, \quad \text{then } u = \pm\sqrt{d}.$$

EXAMPLE 2 Solving Quadratic Equations by the Square Root Method

Solve by the square root method:

a. $4x^2 = 20$ b. $(x - 2)^2 = 6$

Solution

a. In order to apply the square root method, we need a squared expression by itself on one side of the equation.

$$4x^2 = 20$$

We want x^2 by itself.

We can get x^2 by itself if we divide both sides by 4.

$$\frac{4x^2}{4} = \frac{20}{4}$$

$$x^2 = 5$$

Now, we can apply the square root method.

$$x = \pm\sqrt{5}$$

By checking both values in the original equation, we can confirm that the solution set is $\{-\sqrt{5}, \sqrt{5}\}$.

b. $(x - 2)^2 = 6$

The squared expression is by itself.

With the squared expression by itself, we can apply the square root method.

$$x - 2 = \pm\sqrt{6}$$

We solve for x by adding 2 to both sides.

$$x = 2 \pm \sqrt{6}$$

By checking both values in the original equation, we can confirm that the solution set is $\{2 + \sqrt{6}, 2 - \sqrt{6}\}$.

Check Point 2 Solve by the square root method:

a. $3x^2 = 21$ **b.** $(x + 5)^2 = 11$

3 Solve quadratic equations by completing the square.

Completing the Square

How do we solve an equation in the form $ax^2 + bx + c = 0$ if the equation cannot be factored? We cannot use the zero-product principle in such a case. However, we can convert the equation into an equivalent equation that can be solved using the square root method. This is accomplished by **completing the square**.

Completing the Square

If $x^2 + bx$ is a binomial, then by adding $\left(\frac{b}{2}\right)^2$, which is the square of half the coefficient of x, a perfect square trinomial will result. That is,

$$x^2 + bx + \left(\frac{b}{2}\right)^2 = \left(x + \frac{b}{2}\right)^2.$$

EXAMPLE 3 Completing the Square

What term should be added to the binomial $x^2 + 8x$ so that it becomes a perfect square trinomial?

Solution The term that should be added is the square of half the coefficient of x. The coefficient of x is 8. Thus, we will add $\left(\frac{8}{2}\right)^2 = 4^2$. A perfect square trinomial is the result.

$$x^2 + 8x + 4^2 = x^2 + 8x + 16 = (x + 4)^2$$

(half)^2

Check Point 3 What term should be added to the binomial $x^2 - 14x$ so that it becomes a perfect square trinomial? Factor the trinomial.

We can solve any quadratic equation by completing the square. If the coefficient of the x^2 term is one, we add the square of half the coefficient of x to both sides of the equation. **When you add a constant term to one side of the equation to complete the square, be certain to add the same constant to the other side of the equation.** These ideas are illustrated in Example 4.

EXAMPLE 4 Solving a Quadratic Equation by Completing the Square

Solve by completing the square: $x^2 - 6x + 2 = 0$.

Solution We begin the procedure by isolating the binomial, $x^2 - 6x$, so that we can complete the square. Thus, we subtract 2 from both sides of the equation.

$$x^2 - 6x + 2 = 0$$

$$x^2 - 6x + 2 - 2 = 0 - 2$$

$$x^2 - 6x = -2$$

We need to add a constant to this binomial that will make it a perfect square trinomial.

What constant should we add? Add the square of half the coefficient of x.

$$x^2 - 6x = -2$$

-6 is the coefficient of x.
$\left(\frac{-6}{2}\right)^2 = (-3)^2 = 9$

Thus, we need to add 9 to $x^2 - 6x$. In order to keep the equation balanced, we must add 9 to both sides.

$$x^2 - 6x = -2$$ This is the quadratic equation with the binomial isolated.

$$x^2 - 6x + 9 = -2 + 9$$ Complete the square, adding 9 to both sides.

$$(x - 3)^2 = 7$$ Factor the perfect square trinomial.

$$x - 3 = \pm\sqrt{7}$$ Apply the square root method.

$$x = 3 \pm \sqrt{7}$$ Add 3 to both sides.

In this step we have converted our equation into one that can be solved by the square root method.

The solution set is $\{3 + \sqrt{7}, 3 - \sqrt{7}\}$.

Check Point 4 Solve by completing the square: $x^2 - 2x - 2 = 0$.

If the coefficient of the x^2 term in a quadratic equation is not one, you must divide each side of the equation by this coefficient before completing the square. For example, to solve $3x^2 - 2x - 4 = 0$ by completing the square, first divide every term by 3:

$$\frac{3x^2}{3} - \frac{2x}{3} - \frac{4}{3} = \frac{0}{3}$$

$$x^2 - \frac{2}{3}x - \frac{4}{3} = 0.$$

Now that the coefficient of x^2 is one, we can solve by completing the square using the method of Example 4.

4 Solve quadratic equations using the quadratic formula.

Solving Quadratic Equations Using the Quadratic Formula

We can use the method of completing the square to derive a formula that can be used to solve all quadratic equations. The derivation given here also shows a particular quadratic equation, $3x^2 - 2x - 4 = 0$, to specifically illustrate each of the steps.

Deriving the Quadratic Formula

Standard Form of a Quadratic Equation	Comment	A Specific Example
$ax^2 + bx + c = 0, \quad a \neq 0$	This is the given equation.	$3x^2 - 2x - 4 = 0$
$x^2 + \frac{b}{a}x + \frac{c}{a} = 0$	Divide both sides by the coefficient of x^2.	$x^2 - \frac{2}{3}x - \frac{4}{3} = 0$
$x^2 + \frac{b}{a}x = -\frac{c}{a}$	Isolate the binomial by adding $-\frac{c}{a}$ on both sides.	$x^2 - \frac{2}{3}x = \frac{4}{3}$
$x^2 + \frac{b}{a}x + \left(\frac{b}{2a}\right)^2 = -\frac{c}{a} + \left(\frac{b}{2a}\right)^2$ (half)^2 $x^2 + \frac{b}{a}x + \frac{b^2}{4a^2} = -\frac{c}{a} + \frac{b^2}{4a^2}$	Complete the square. Add the square of half the coefficient of x to both sides.	$x^2 - \frac{2}{3}x + \left(\frac{1}{3}\right)^2 = \frac{4}{3} + \left(\frac{1}{3}\right)^2$ (half)^2 $x^2 - \frac{2}{3}x + \frac{1}{9} = \frac{4}{3} + \frac{1}{9}$
$\left(x + \frac{b}{2a}\right)^2 = -\frac{c}{a} \cdot \frac{4a}{4a} + \frac{b^2}{4a^2}$	Factor on the left and obtain a common denominator on the right.	$\left(x - \frac{1}{3}\right)^2 = \frac{4}{3} \cdot \frac{3}{3} + \frac{1}{9}$
$\left(x + \frac{b}{2a}\right)^2 = \frac{-4ac + b^2}{4a^2}$ $\left(x + \frac{b}{2a}\right)^2 = \frac{b^2 - 4ac}{4a^2}$	Add fractions on the right.	$\left(x - \frac{1}{3}\right)^2 = \frac{12 + 1}{9}$ $\left(x - \frac{1}{3}\right)^2 = \frac{13}{9}$
$x + \frac{b}{2a} = \pm\sqrt{\frac{b^2 - 4ac}{4a^2}}$	Apply the square root method.	$x - \frac{1}{3} = \pm\sqrt{\frac{13}{9}}$
$x + \frac{b}{2a} = \pm\frac{\sqrt{b^2 - 4ac}}{2\|a\|}$	Take the square root of the quotient, simplifying the denominator.	$x - \frac{1}{3} = \pm\frac{\sqrt{13}}{3}$
$x = \frac{-b}{2a} \pm \frac{\sqrt{b^2 - 4ac}}{2\|a\|}$	Solve for x by subtracting $\frac{b}{2a}$ from both sides.	$x = \frac{1}{3} \pm \frac{\sqrt{13}}{3}$
$x = \frac{-b \pm \sqrt{b^2 - 4ac}}{2\|a\|}$	Combine fractions on the right.	$x = \frac{1 \pm \sqrt{13}}{3}$

Because the same real numbers are represented by $\pm 2|a|$ and $\pm 2a$, we can omit the absolute value sign in the last step. The resulting formula is called the **quadratic formula**.

To Die at Twenty

Can the equations

$$7x^5 + 12x^3 - 9x + 4 = 0$$

and

$$8x^6 - 7x^5 + 4x^3 - 19 = 0$$

be solved using a formula similar to the quadratic formula? The first equation has five solutions and the second has six solutions, but they cannot be found using a formula. How do we know? In 1832, a 20-year-old Frenchman, Evariste Galois, wrote down a proof showing that there is no general formula to solve equations when the exponent on the variable is 5 or greater. Galois was jailed as a political activist several times while still a teenager. The day after his brilliant proof he fought a duel over a woman. The duel was a political setup. As he lay dying, Galois told his brother, Alfred, of the manuscript that contained his proof: "Mathematical manuscripts are in my room. On the table. Take care of my work. Make it known. Important. Don't cry, Alfred. I need all my courage—to die at twenty." (Our source is Leopold Infeld's biography of Galois, *Whom the Gods Love*. Some historians, however, dispute the story of Galois's ironic death the very day after his algebraic proof. Mathematical truths seem more reliable than historical ones!)

The Quadratic Formula

The solutions of a quadratic equation in standard form $ax^2 + bx + c = 0$, with $a \neq 0$, are given by the **quadratic formula**

$$x = \frac{-b \pm \sqrt{b^2 - 4ac}}{2a}.$$

x equals negative b, plus or minus the square root of $b^2 - 4ac$, all divided by 2a.

To use the quadratic formula, write the quadratic equation in standard form if necessary. Then determine the numerical values for a (the coefficient of the squared term), b (the coefficient of the x term), and c (the constant term). Substitute the values of a, b, and c in the quadratic formula and evaluate the expression. The $\pm$ sign indicates that there are two solutions of the equation.

EXAMPLE 5 Solving a Quadratic Equation Using the Quadratic Formula

Solve using the quadratic formula: $2x^2 - 6x + 1 = 0$.

Solution The given equation is in standard form. Begin by identifying the values for a, b, and c.

$$2x^2 - 6x + 1 = 0$$

$a = 2$ $b = -6$ $c = 1$

$$x = \frac{-b \pm \sqrt{b^2 - 4ac}}{2a}$$ Use the quadratic formula: $a = 2$, $b = -6$, and $c = 1$.

$$= \frac{-(-6) \pm \sqrt{(-6)^2 - 4(2)(1)}}{2 \cdot 2}$$ Substitute the values for a, b, and c.

$$= \frac{6 \pm \sqrt{36 - 8}}{4}$$ $-(-6) = 6$ and $(-6)^2 = (-6)(-6) = 36$.

$$= \frac{6 \pm \sqrt{28}}{2}$$ Complete the subtraction under the radical.

$$= \frac{6 \pm 2\sqrt{7}}{4}$$ $\sqrt{28} = \sqrt{4 \cdot 7} = \sqrt{4}\sqrt{7} = 2\sqrt{7}$.

$$= \frac{2(3 \pm \sqrt{7})}{4}$$ Factor out 2 from the numerator.

$$= \frac{3 \pm \sqrt{7}}{2}$$ Divide the numerator and denominator by 2.

The solution set is $\left\{\frac{3 + \sqrt{7}}{2}, \frac{3 - \sqrt{7}}{2}\right\}$.

Check Point 5 Solve using the quadratic formula:

$$2x^2 + 2x - 1 = 0.$$

We have seen that a graphing utility can be used to check the solutions to the quadratic equation $ax^2 + bx + c = 0$. The x-intercepts of the graph of

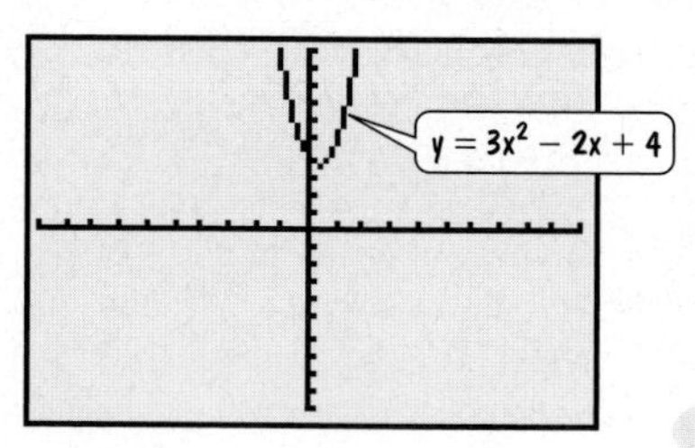

Figure 1.3 This graph has no x-intercepts.

$y = ax^2 + bx + c$ are the solutions. However, take a look at the graph of $y = 3x^2 - 2x + 4$, shown in Figure 1.3. Notice that the graph has no x-intercepts. Can you guess what this means about the solutions of the quadratic equation $3x^2 - 2x + 4 = 0$? If you're not sure, we'll answer this question in the next example.

EXAMPLE 6 Solving a Quadratic Equation Using the Quadratic Formula

Solve using the quadratic-formula: $3x^2 - 2x + 4 = 0$.

Study Tip

See Section P.7, pages 65–70, to review complex numbers.

Solution The given equation is in standard form. Begin by identifying the values for a, b, and c.

$$3x^2 - 2x + 4 = 0$$

$a = 3 \quad b = -2 \quad c = 4$

$$x = \frac{-b \pm \sqrt{b^2 - 4ac}}{2a}$$ Use the quadratic formula $a = 3$, $b = -2$, and $c = 4$.

$$= \frac{-(-2) \pm \sqrt{(-2)^2 - 4(3)(4)}}{2(3)}$$ Substitute the values for a, b, and c.

$$= \frac{2 \pm \sqrt{4 - 48}}{6}$$ $-(-2) = 2$ and $(-2)^2 = (-2)(-2) = 4$.

$$= \frac{2 \pm \sqrt{-44}}{6}$$ Because the number under the radical sign is negative, the solutions will not be real numbers.

$$= \frac{2 \pm 2i\sqrt{11}}{6}$$ $\sqrt{-44} = \sqrt{4(11)(-1)} = 2i\sqrt{11}$

$$= \frac{2(1 \pm i\sqrt{11})}{6}$$ Factor 2 from the numerator.

$$= \frac{1 \pm i\sqrt{11}}{3}$$ Divide numerator and denominator by 2.

You can check that these solutions are correct using operations with complex numbers. The solutions are complex conjugates and the solution set is

$$\left\{\frac{1 + i\sqrt{11}}{3}, \frac{1 - i\sqrt{11}}{3}\right\}.$$

Hence, **complex imaginary solutions mean that the graph will not have any x-intercepts.**

Check Point 6 Solve using the quadratic formula:

$$x^2 - 2x + 2 = 0.$$

5 Use the discriminant to determine the kinds of solutions.

The Discriminant

The quantity $b^2 - 4ac$, which appears under the radical sign in the quadratic formula, is called the **discriminant**. In Example 5 the discriminant was 28, a positive number that is not a perfect square. The equation had two solutions that were irrational numbers. In Example 6 the discriminant was −44, a negative number. The equation had solutions involving the imaginary number i. In this case our graph had no x-intercepts.

These observations are generalized in Table 1.3.

Table 1.3 The Discriminant and the Nature of the Solutions to $ax^2 + bx + c = 0$

Discriminant $b^2 - 4ac$	Kinds of Solutions to $ax^2 + bx + c = 0$	Graph of $y = ax^2 + bx + c$
$b^2 - 4ac > 0$	two unequal real solutions	Two x-intercepts
$b^2 - 4ac = 0$	one real solution (a repeated solution)	One x-intercept
$b^2 - 4ac < 0$	No real solution; two complex imaginary solutions	No x-intercepts

EXAMPLE 7 Using the Discriminant

Compute the discriminant of $4x^2 - 8x + 1 = 0$. What does the discriminant indicate about the kinds of solutions?

Solution Begin by identifying the values for a, b, and c.

$$4x^2 - 8x + 1 = 0$$

$a = 4$ $b = -8$ $c = 1$

Now, compute $b^2 - 4ac$, the discriminant.

$$b^2 - 4ac = (-8)^2 - 4 \cdot 4 \cdot 1 = 64 - 16 = 48$$

The discriminant is 48. Because the discriminant is positive, the equation $4x^2 - 8x + 1 = 0$ has two unequal real solutions.

Check Point 7 Compute the discriminant of $3x^2 - 2x + 5 = 0$. What does the discriminant indicate about the kinds of solutions?

6 Determine the most efficient method to use when solving a quadratic equation.

Determining Which Method to Use

All quadratic equations can be solved by the quadratic formula. However, if an equation is in the form $u^2 = d$, such as $x^2 = 5$ or $(2x + 3)^2 = 8$, it is faster to use the square root method, taking the square root of both sides. If the equation is not in the form $u^2 = d$, write the quadratic equation in standard form $\left(ax^2 + bx + c = 0\right)$. Try to solve the equation by the factoring method. If $ax^2 + bx + c$ cannot be factored, then solve the quadratic equation by the quadratic formula.

Because we used the method of completing the square to derive the quadratic formula, we no longer need it for solving quadratic equations. However, we will use completing the square later in the book to help graph certain kinds of equations.

Table 1.4 summarizes our observations about which technique to use when solving a quadratic equation.

Table 1.4 Determining the Most Efficient Technique to Use When Solving a Quadratic Equation

Description and Form of the Quadratic Equation	Most Efficient Solution Method	Example
$ax^2 + bx + c = 0$ and $ax^2 + bx + c$ can be factored easily.	Factor and use the zero-product principle.	$3x^2 + 5x - 2 = 0$ $(3x - 1)(x + 2) = 0$ $3x - 1 = 0$ or $x + 2 = 0$ $x = \frac{1}{3}$ $\quad x = -2$
$ax^2 + c = 0$ The quadratic equation has no linear (x) term. $(b = 0)$	Solve for x^2 and apply the square root method.	$4x^2 - 7 = 0$ $4x^2 = 7$ $x^2 = \frac{7}{4}$ $x = \pm\frac{\sqrt{7}}{2}$
$u^2 = d$; u is a first-degree polynomial.	Use the square root method.	$(x + 4)^2 = 5$ $x + 4 = \pm\sqrt{5}$ $x = -4 \pm \sqrt{5}$
$ax^2 + bx + c = 0$ and $ax^2 + bx + c$ cannot be factored or the factoring is too difficult.	Use the quadratic formula: $x = \frac{-b \pm \sqrt{b^2 - 4ac}}{2a}$	$x^2 - 2x - 6 = 0$ $x = \frac{2 \pm \sqrt{4 - 4(1)(-6)}}{2(1)}$ $= \frac{2 \pm \sqrt{28}}{2} = \frac{2 \pm \sqrt{4}\sqrt{7}}{2}$ $= \frac{2 \pm 2\sqrt{7}}{2} = \frac{2(1 \pm \sqrt{7})}{2}$ $= 1 \pm \sqrt{7}$

7 Solve problems modeled by quadratic equations.

Applications

It's been one of those days! Traffic is really backed up on the highway. Finally, you see the source of the traffic jam—a minor fender-bender. Still stuck in traffic, you notice that the driver appears to be quite young. This might seem like a strange observation. After all, what does a driver's age have to do with his or her chance of getting into an accident?

Oddly enough a driver's age does have something to do with his or her chance of getting into a car accident. The formula $N = 0.4x^2 - 36x + 1000$ approximates the number of accidents, N, per 50 million miles driven, for a driver who is x years old. The formula models data for drivers ages 16 to 74 years, inclusively. Notice that this formula contains an expression in the form $ax^2 + bx + c$ on the right. If a formula contains such an expression, we can write and solve a quadratic equation to answer questions about the variable x. Our next example shows how this is done.

EXAMPLE 8 Using a Quadratic Equation to Answer a Question About the Variable in a Formula

Use the formula $N = 0.4x^2 - 36x + 1000$ to answer this question: What is the age of a driver predicted to have 312 accidents per 50 million miles driven?

Solution We must find x, a driver's age, with $N = 312$ accidents per 50 million miles driven. Use the formula and substitute 312 for N.

$$N = 0.4x^2 - 36x + 1000 \qquad 312 = 0.4x^2 - 36x + 1000$$

Substitute 312 for N.

Let's write the quadratic equation on the right in standard form. Subtract 312 from both sides.

$$312 - 312 = 0.4x^2 - 36x + 1000 - 312$$
$$0 = 0.4x^2 - 36x + 688$$

Equivalently, $0.4x^2 - 36x + 1000 = 0$. The most efficient technique for solving this equation is the quadratic formula. Identify the values for a, b, and c.

$$0.4x^2 - 36x + 688 = 0$$

$a = 0.4$ $b = -36$ $c = 688$

Now, substitute these values into the quadratic formula.

$$x = \frac{-b \pm \sqrt{b^2 - 4ac}}{2a} = \frac{-(-36) \pm \sqrt{(-36)^2 - 4(0.4)(688)}}{2(0.4)}$$

$$= \frac{36 \pm \sqrt{195.2}}{0.8}$$

Thus,

$$x = \frac{36 + \sqrt{195.2}}{0.8} \quad \text{or} \quad x = \frac{36 - \sqrt{195.2}}{0.8}$$

$$\approx 62 \qquad\qquad \approx 28$$

Use a calculator to obtain an approximation to the nearest whole number.

Drivers who are about 28 and 62 years old are predicted to have 312 accidents per 50 million miles driven.

Check Point 8 As we mentioned in the introduction to this section, rangers at a national park used the formula

$$P = -10x^2 + 475x + 3500$$

to estimate the crocodile population, P, after x years of a protection program. How many years will it take to bring the population up to 7250?

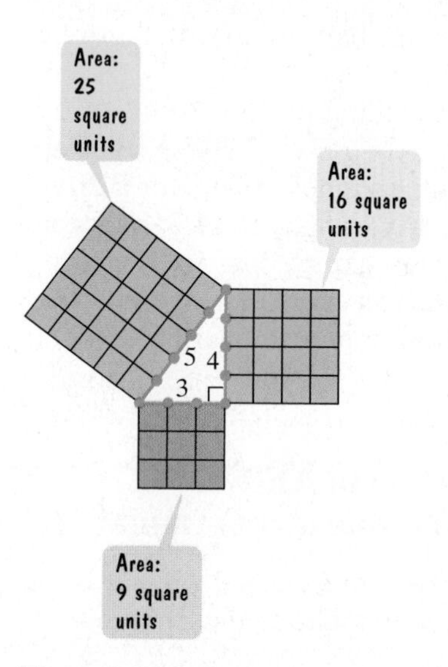

Figure 1.4 The area of the large square equals the sum of the areas of the smaller squares.

In our next example, we will be using the *Pythagorean Theorem* to obtain a verbal model. The ancient Greek philosopher and mathematician Pythagoras (approximately 582–500 B.C.) founded a school whose motto was "All is number." Pythagoras is best remembered for his work with the **right triangle**, a triangle with one angle measuring 90°. The side opposite the 90° angle is called the **hypotenuse**. The other sides are called **legs**. Pythagoras found that if he constructed squares on each of the legs, as well as a larger square on the hypotenuse, the sum of the areas of the smaller squares is equal to the area of the larger square. This is illustrated in Figure 1.4.

This relationship is usually stated in terms of the lengths of the three sides of a right triangle and is called the **Pythagorean Theorem**.

The Pythagorean Theorem

The sum of the squares of the lengths of the legs of a right triangle equals the square of the length of the hypotenuse.

If the legs have lengths a and b, and the hypotenuse has length c, then

$$a^2 + b^2 = c^2.$$

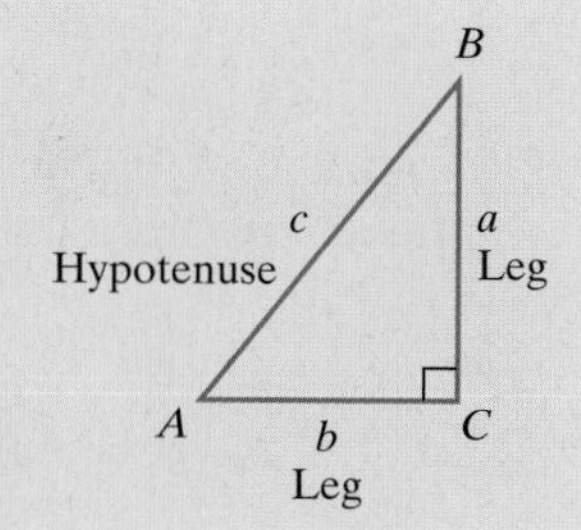

EXAMPLE 9 Using the Pythagorean Theorem

In a 25-inch television set, the length of the screen's diagonal is 25 inches. If the screen's height is 15 inches, what is its width?

Solution Figure 1.5 shows a right triangle that is formed by the height, width, and diagonal. We can find w, the screen's width, using the Pythagorean Theorem.

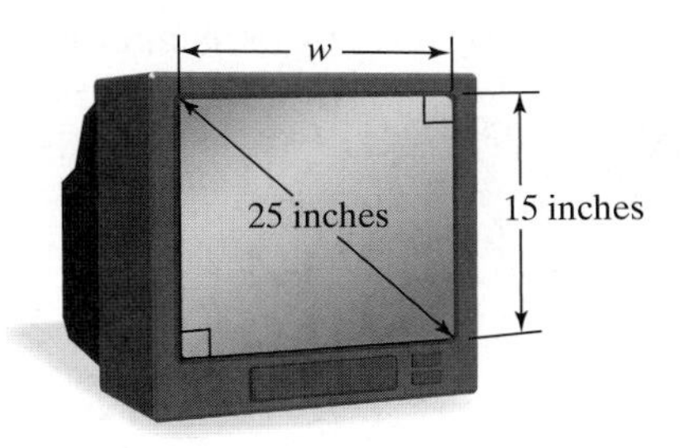

Figure 1.5 A right triangle is formed by the television's height, width, and diagonal.

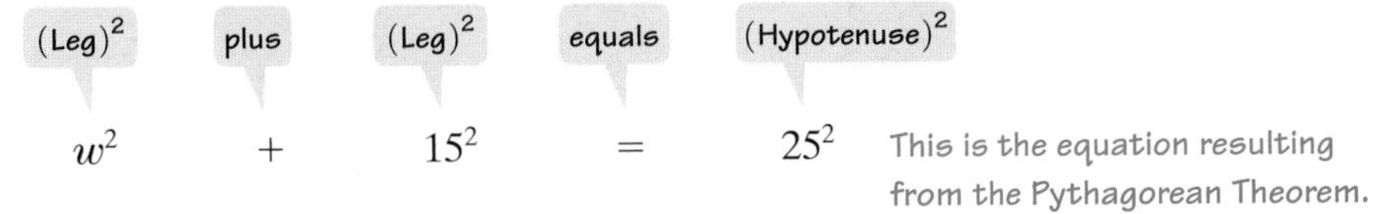

$$w^2 \quad + \quad 15^2 \quad = \quad 25^2$$ This is the equation resulting from the Pythagorean Theorem.

The equation $w^2 + 15^2 = 25^2$ can be solved most efficiently by the square root method.

$w^2 + 15^2 = 25^2$ — This is the equation that models the verbal conditions.

$w^2 + 225 = 625$ — Square 15 and 25.

$w^2 + 225 - 225 = 625 - 225$ — Isolate w^2 by subtracting 225 from both sides.

$w^2 = 400$ — Simplify.

$w = \pm\sqrt{400}$ — Apply the square root method.

$w = \pm 20$

Because w represents the width of the television's screen, this dimension must be positive. We reject -20. Thus, the width of the television is 20 inches.

Check Point 9 What is the width in a 15-inch television set whose height is 9 inches?

EXERCISE SET 1.3

Practice Exercises

Solve each equation in Exercises 1–14 by factoring and then using the zero-product principle.

1. $x^2 - 3x - 10 = 0$
2. $x^2 - 13x + 36 = 0$
3. $x^2 = 8x - 15$
4. $x^2 = -11x - 10$
5. $6x^2 + 11x - 10 = 0$
6. $9x^2 + 9x + 2 = 0$
7. $3x^2 - 2x = 8$
8. $4x^2 - 13x = -3$
9. $3x^2 + 12x = 0$
10. $5x^2 - 20x = 0$
11. $2x(x - 3) = 5x^2 - 7x$
12. $16x(x - 2) = 8x - 25$
13. $7 - 7x = (3x + 2)(x - 1)$
14. $10x - 1 = (2x + 1)^2$

Solve each equation in Exercises 15–26 by the square root method.

15. $3x^2 = 27$
16. $5x^2 = 45$
17. $5x^2 + 1 = 51$
18. $3x^2 - 1 = 47$
19. $(x + 2)^2 = 25$
20. $(x - 3)^2 = 36$
21. $(3x + 2)^2 = 9$
22. $(4x - 1)^2 = 16$
23. $(5x - 1)^2 = 7$
24. $(8x - 3)^2 = 5$
25. $(3x - 4)^2 = 8$
26. $(2x + 8)^2 = 27$

In Exercises 27–38, determine the constant that should be added to the binomial so that it becomes a perfect square trinomial. Then factor the trinomial.

27. $x^2 + 12x$
28. $x^2 + 16x$
29. $x^2 - 10x$
30. $x^2 - 14x$
31. $x^2 + 3x$
32. $x^2 + 5x$
33. $x^2 - 7x$
34. $x^2 - 9x$
35. $x^2 - \frac{2}{3}x$
36. $x^2 + \frac{4}{5}x$

37. $x^2 - \frac{1}{3}x$

38. $x^2 - \frac{1}{4}x$

Solve each equation in Exercises 39–54 by completing the square.

39. $x^2 + 6x = 7$

40. $x^2 + 6x = -8$

41. $x^2 - 2x = 2$

42. $x^2 + 4x = 12$

43. $x^2 - 6x - 11 = 0$

44. $x^2 - 2x - 5 = 0$

45. $x^2 + 4x + 1 = 0$

46. $x^2 + 6x - 5 = 0$

47. $x^2 + 3x - 1 = 0$

48. $x^2 - 3x - 5 = 0$

49. $2x^2 - 7x + 3 = 0$

50. $2x^2 + 5x - 3 = 0$

51. $4x^2 - 4x - 1 = 0$

52. $2x^2 - 4x - 1 = 0$

53. $3x^2 - 2x - 2 = 0$

54. $3x^2 - 5x - 10 = 0$

Solve each equation in Exercises 55–64 using the quadratic formula.

55. $x^2 + 8x + 15 = 0$

56. $x^2 + 8x + 12 = 0$

57. $x^2 + 5x + 3 = 0$

58. $x^2 + 5x + 2 = 0$

59. $3x^2 - 3x - 4 = 0$

60. $5x^2 + x - 2 = 0$

61. $4x^2 = 2x + 7$

62. $3x^2 = 6x - 1$

63. $x^2 - 6x + 10 = 0$

64. $x^2 - 2x + 17 = 0$

Compute the discriminant of each equation in Exercises 65–72. What does the discriminant indicate about the kinds of solutions?

65. $x^2 - 4x - 5 = 0$

66. $4x^2 - 2x + 3 = 0$

67. $2x^2 - 11x + 3 = 0$

68. $2x^2 + 11x - 6 = 0$

69. $x^2 - 2x + 1 = 0$

70. $3x^2 = 2x - 1$

71. $x^2 - 3x - 7 = 0$

72. $3x^2 + 4x - 2 = 0$

Solve each equation in Exercises 73–94 by the method of your choice.

73. $2x^2 - x = 1$

74. $3x^2 - 4x = 4$

75. $5x^2 + 2 = 11x$

76. $5x^2 = 6 - 13x$

77. $3x^2 = 60$

78. $2x^2 = 250$

79. $x^2 - 2x = 1$

80. $2x^2 + 3x = 1$

81. $(2x + 3)(x + 4) = 1$

82. $(2x - 5)(x + 1) = 2$

83. $(3x - 4)^2 = 16$

84. $(2x + 7)^2 = 25$

85. $3x^2 - 12x + 12 = 0$

86. $9 - 6x + x^2 = 0$

87. $4x^2 - 16 = 0$

88. $3x^2 - 27 = 0$

89. $x^2 - 6x + 13 = 0$

90. $x^2 - 4x + 29 = 0$

91. $x^2 = 4x - 7$

92. $5x^2 = 2x - 3$

93. $2x^2 - 7x = 0$

94. $2x^2 + 5x = 3$

Application Exercises

95. The formula $M = 0.0075x^2 - 0.2676x + 14.8$ models the fuel efficiency of passenger cars, M, in miles per gallon, x years after 1940. Environmentalists pressured automobile manufacturers for a fuel efficiency of 45 miles per gallon by the year 2000. In which year will fuel efficiency reach 45 miles per gallon according to the formula?

96. The formula $N = 0.036x^2 - 2.8x + 58.14$ models the number of deaths per year, N, per thousand people, for people who are x years old, where $40 \le x \le 60$. Find, to the nearest whole number, the age at which 12 people per 1000 die annually.

The Internet is the world's largest communications network. Although millions of computer owners access the Internet using phone lines, the process of downloading files can be slow and tedious. Cable-TV modems dramatically speed up this process. By contrast to phone modems, which transmit 56,000 bits per second, cable modems are capable of transmitting 10 million bits per second. The graph shows the millions of Internet users in the United States with this new technology. The data can be modeled by the formula $N = 0.4x^2 + 0.5$, where N represents the millions of people in the United States using cable modems x years after 1996. Use this formula to solve Exercises 97–98.

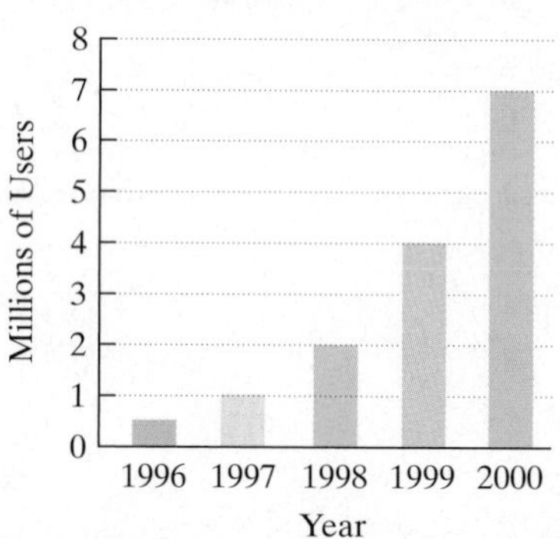

Source: *The New York Times*

97. According to the formula, in which year will 4.1 million Americans use cable TV modems? How well does the formula describe the actual number of users for that year?

98. According to the formula, in which year will 20.1 million Americans use cable TV modems?

99. The formula $N = 29{,}035t^2 + 429{,}200$ describes the leading golf winnings in the United States t years after 1983. The leading golf winner for one of the years modeled by the formula was Greg Norman, who won \$690,515. In what year did this occur?

100. The weight of a human fetus is given by the formula $W = 3t^2$, where W is the weight in grams and t is the time in weeks, $t \ge 0$ and $t \le 39$. After how many weeks does the fetus weigh 300 grams?

The data and the accompanying graph show the number of inmates in U.S. state and federal prisons from 1980 through 1998. The data can be modeled by the formula $N = 2x^2 + 22x + 320$, *in which N represents the number of inmates, in thousands, in U.S. state and federal prisons x years after 1980. Use this formula to solve Exercises 101–102.*

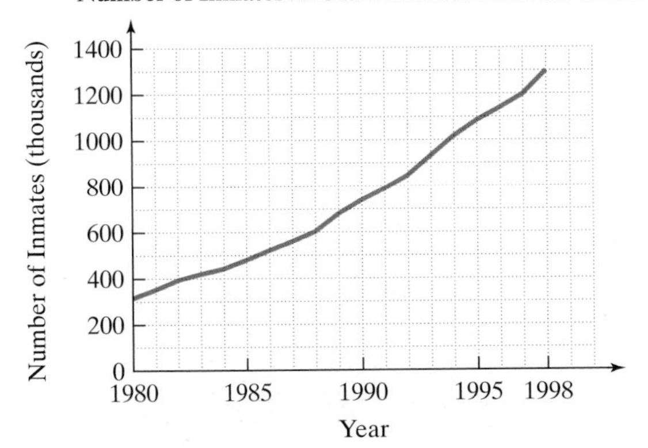

Source: U.S. Justice Department

Year	Number of Inmates
1980	315,974
1981	353,167
1982	394,374
1983	419,820
1984	443,398
1985	480,568
1986	522,084
1987	560,812
1988	603,732
1989	680,907
1990	739,980
1991	789,610
1992	846,277
1993	932,074
1994	1,016,691
1995	1,085,022
1996	1,138,984
1997	1,197,590
1998	1,302,019

101. According to the formula, in which year were there 740 thousand inmates in U.S. state and federal prisons? What was the actual number for that year? How well does the formula describe the actual number of inmates for that year?

102. According to the formula, in which year were there 1100 thousand inmates in U.S. state and federal prisons? What was the actual number for that year? How well does the formula describe the actual number of inmates for that year?

The data and the accompanying graph show the cumulative number of deaths from AIDS in the United States from 1990 through 1998. The data can be modeled by the formula $N = -1.65x^2 + 51.8x + 111.44$, *in which N represents the cumulative number of U.S. AIDS deaths, in thousands, x years after 1990. Use this formula to solve Exercises 103–104.*

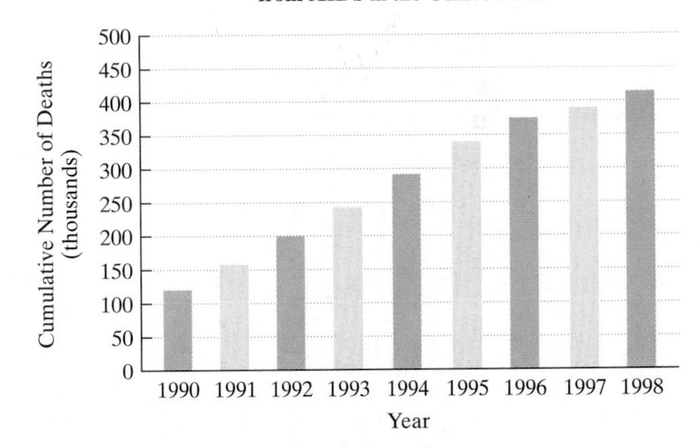

Source: Centers for Disease Control

Year	Cumulative Number of Deaths from AIDS, in Thousands
1990	121
1991	158
1992	199
1993	243
1994	292
1995	340
1996	375
1997	390
1998	414

103. According to the formula, in which year did the cumulative number of deaths from AIDS in the United States reach 330 thousand? What was the actual number for that year? How well does the formula describe the situation for that year?

104. According to the formula, in which year will the total number of U.S. AIDS deaths reach 500 thousand?

105. A baseball diamond is actually a square with 90-foot sides. What is the distance from home plate to second base?

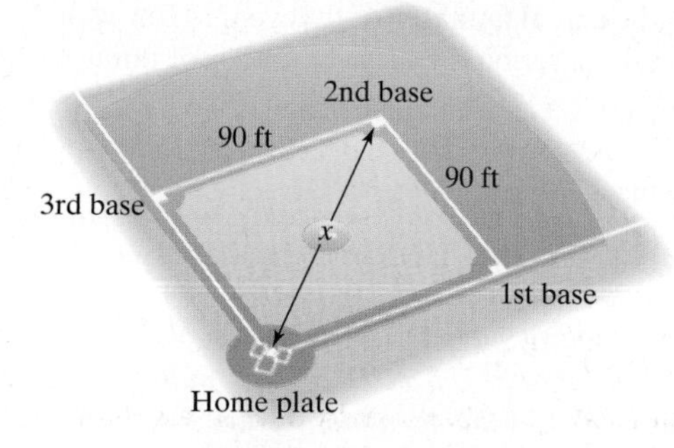

106. A 20-foot ladder is 15 feet from the house. How far up the house does the ladder reach?

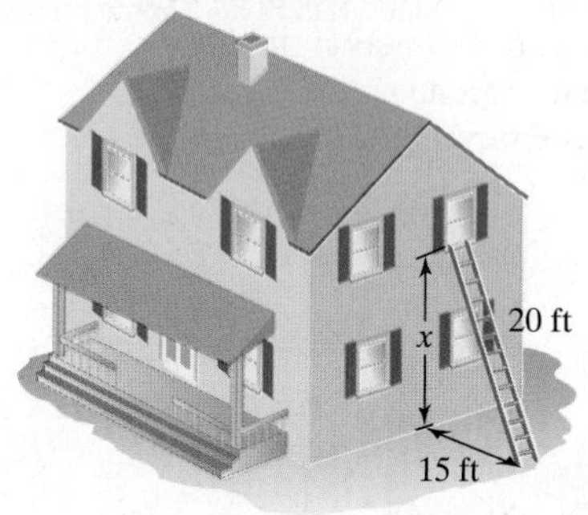

107. An 8-foot tree is supported by two wires that extend from the top of the tree to a point on the ground located 15 feet from the base of the tree. Find the total length of the two support wires.

108. A vertical pole is supported by three wires. Each wire is 13 yards long and is anchored 5 yards from the base of the pole. How far up the pole will the wires be attached?

109. The length of a rectangular garden is 5 feet greater than the width. The area of the garden is 300 square feet. Find the length and the width.

110. A rectangular parking lot has a length that is 3 yards greater than the width. The area of the rectangular lot is 180 square yards. Find the length and the width.

111. A machine produces open boxes using square sheets of metal. The figure illustrates that the machine cuts equal-sized squares measuring 2 inches on a side from the corners and then shapes the metal into an open box by turning up the sides. If each box must have a volume of 200 cubic inches, find the size of the length and width of the open box.

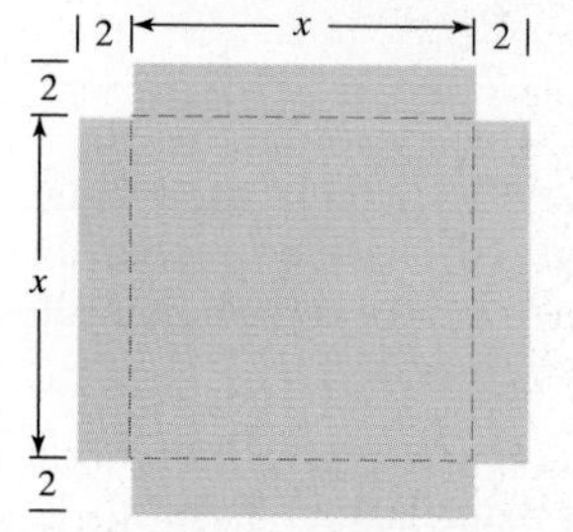

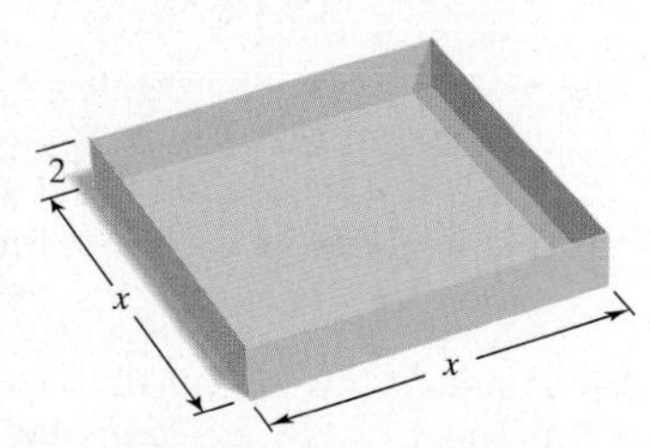

112. A machine produces open boxes using square sheets of metal. The machine cuts equal-sized squares measuring 3 inches on a side from the corners and then shapes the metal into an open box by turning up the sides. If each box must have a volume of 75 cubic inches, find the size of the length and width of the open box.

Writing in Mathematics

113. What is a quadratic equation?

114. Explain how to solve $x^2 + 6x + 8 = 0$ using factoring and the zero-product principle.

115. Explain how to solve $x^2 + 6x + 8 = 0$ by completing the square.

116. Explain how to solve $x^2 + 6x + 8 = 0$ using the quadratic formula.

117. How is the quadratic formula derived?

118. What is the discriminant and what information does it provide about a quadratic equation?

119. If you are given a quadratic equation, how do you determine which method to use to solve it?

120. If $(x + 2)(x - 4) = 0$ indicates that $x + 2 = 0$ or $x - 4 = 0$, explain why $(x + 2)(x - 4) = 6$ does not mean $x + 2 = 6$ or $x - 4 = 6$. Could we solve the equation using $x + 2 = 3$ and $x - 4 = 2$ because $3 \cdot 2 = 6$?

Technology Exercises

121. If you have access to a calculator that solves quadratic equations, consult the owner's manual to determine how to use this feature. Then use your calculator to solve any five of the equations in Exercises 55–64.

122. Graph the formula in Exercise 95,

$$y = 0.0075x^2 - 0.2676x + 14.8$$

using a $[0, 80, 20]$ by $[0, 50, 10]$ viewing rectangle. Move the cursor along the graph, using the TRACE and ZOOM features to estimate the year in which automobile fuel efficiency was the poorest. What was the gas mileage in that year? Use the Internet or your library to find pictures of the most popular cars for that year and write a sentence relating the most popular cars of the time with the fuel efficiency indicated by the formula's graph.

Critical Thinking Exercises

123. Which one of the following is true?

a. The equation $(2x - 3)^2 = 25$ is equivalent to $2x - 3 = 5$.

b. Every quadratic equation has two distinct numbers in its solution set.

c. A quadratic equation whose coefficients are real numbers can never have a solution set containing one real number and one complex nonreal number.

d. The equation $ax^2 + c = 0$ cannot be solved by the quadratic formula.

124. Solve the equation: $x^2 + 2\sqrt{3}x - 9 = 0$.

125. Write a quadratic equation in standard form whose solution set is $\{-3, 5\}$.

126. A person throws a rock upward from the edge of an 80-foot cliff. The height, h, in feet, of the rock above the water at the bottom of the cliff after t seconds is described by the formula

$$h = -16t^2 + 64t + 80.$$

How long will it take for the rock to reach the water?

127. The personnel manager of a roller skate company knows that the company's weekly revenue is a function of the price of each pair of skates, modeled by $R = -2x^2 + 36x$, where x represents the dollar price of a pair of skates and R represents weekly revenue in tens of thousands of dollars. A job applicant promises the personnel manager an advertising campaign guaranteed to generate \$190,000 in weekly revenue. Substitute 19 for R in the given model, compute the discriminant, and then explain why the applicant will or will not be hired in the advertising department.

Group Exercise

128. Each group member should find an algebraic formula that contains an expression in the form $ax^2 + bx + c$ on one side that he or she finds intriguing. Consult college algebra books or liberal arts mathematics books to do so. Group members should select four of the formulas. For each formula selected, write and solve a problem similar to Exercises 95 and 96 in this exercise set.

SECTION 1.4 Other Types of Equations

Objectives

1. Solve polynomial equations by factoring.
2. Solve radical equations.
3. Solve equations with rational exponents.
4. Solve equations that are quadratic in form.
5. Solve equations involving absolute value.

The Galápagos Islands are a volcanic chain of islands lying 600 miles west of Ecuador. They are famed for their extraordinary wildlife, which includes a rare flightless cormorant, marine iguanas, and giant tortoises weighing more than 600 pounds. It was here that naturalist Charles Darwin began to formulate his theory of evolution. Darwin made an enormous collection of the islands' plant species. The formula

$$S = 28.5\sqrt[3]{x}$$

describes the number of plant species, S, on the various islands of the Galápagos chain in terms of the area, x, in square miles, of a particular island.

How can we find the area of a Galápagos island with 57 species of plants? Substitute 57 for S in the formula and solve for x:

$$57 = 28.5\sqrt[3]{x}.$$

The resulting equation contains a variable in the radicand and is called a *radical equation*. In this section, in addition to radical equations, we will show you how to solve certain kinds of polynomial equations, equations involving rational exponents, and equations involving absolute value.

1 Solve polynomial equations by factoring.

Polynomial Equations

The linear and quadratic equations that we studied in the first three sections of this chapter can be thought of as polynomial equations of degrees 1 and 2, respectively. By contrast, consider the following polynomial equations of degree greater than 2.

$$3x^4 = 27x^2 \qquad\qquad x^3 + x^2 = 4x + 4$$

This equation is of degree 4 because 4 is the largest exponent.

This equation is of degree 3 because 3 is the largest exponent.

We can solve these equations by moving all terms to one side, thereby obtaining zero on the other side. We then use factoring and the zero-product principle.

EXAMPLE 1 Solving a Polynomial Equation by Factoring

Solve by factoring: $3x^4 = 27x^2$.

Solution

Step 1 Move all terms to one side and obtain zero on the other side. Subtract $27x^2$ from both sides.

$$3x^4 - 27x^2 = 27x^2 - 27x^2$$
$$3x^4 - 27x^2 = 0$$

Step 2 Factor. We can factor $3x^2$ from each term.

$$3x^4 - 27x^2 = 0$$
$$3x^2(x^2 - 9) = 0$$

Steps 3 and 4 Set each factor equal to zero and solve the resulting equations.

$$3x^2 = 0 \quad \text{or} \quad x^2 - 9 = 0$$
$$x^2 = 0 \qquad\qquad x^2 = 9$$
$$x = \pm\sqrt{0} \qquad\qquad x = \pm\sqrt{9}$$
$$x = 0 \qquad\qquad x = \pm 3$$

Step 5 Check the solutions in the original equation. Check the three solutions, 0, −3, and 3, by substituting them into the original equation. Can you verify that the solution set is $\{-3, 0, 3\}$?

Study Tip

In solving $3x^4 = 27x^2$, be careful not to divide both sides by x^2. If you do, you'll lose 0 as a solution. In general, do not divide both sides of an equation by a variable because that variable might take on the value 0 and you cannot divide by 0.

Check Point 1 Solve by factoring: $4x^4 = 12x^2$.

EXAMPLE 2 Solving a Polynomial Equation by Factoring

Solve by factoring: $x^3 + x^2 = 4x + 4$.

Solution

Step 1 Move all terms to one side and obtain zero on the other side. Subtract $4x + 4$ from both sides.

$$x^3 + x^2 - 4x - 4 = 4x + 4 - 4x - 4$$
$$x^3 + x^2 - 4x - 4 = 0$$

Step 2 Factor. Use factoring by grouping. Group terms that have a common factor.

$$\boxed{x^3 + x^2} + \boxed{-4x - 4} = 0$$

Common factor is x^2.

Common factor is -4.

$$x^2(x + 1) - 4(x + 1) = 0$$ Factor x^2 from the first two terms and -4 from the last two terms.

$$(x + 1)(x^2 - 4) = 0$$ Factor out the common binomial, $x + 1$, from each term.

Steps 3 and 4 Set each factor equal to zero and solve the resulting equations.

$$x + 1 = 0 \quad \text{or} \quad x^2 - 4 = 0$$
$$x = -1 \qquad\qquad x^2 = 4$$
$$x = \pm\sqrt{4} = \pm 2$$

Step 5 Check the solutions in the original equation. Check the three solutions, $-1, -2$, and 2, by substituting them into the original equation. Can you verify that the solution set is $\{-2, -1, 2\}$?

Technology

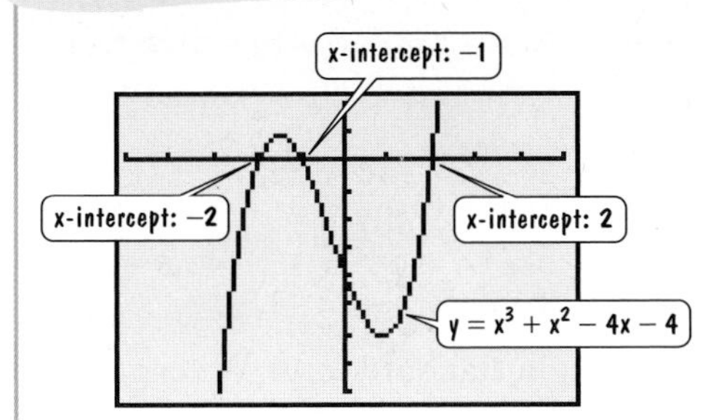

You can use a graphing utility to check the solutions to $x^3 + x^2 - 4x - 4 = 0$. Graph $y = x^3 + x^2 - 4x - 4$, as shown on the left. The x-intercepts are $-2, -1$, and 2, corresponding to the equation's solutions.

Check Point 2 Solve by factoring: $2x^3 + 3x^2 = 8x + 12$.

2 Solve radical equations.

Equations Involving Radicals

A **radical equation** is an equation in which the variable occurs in a square root, cube root, or any higher root. An example of a radical equation is

$$28.5\sqrt[3]{x} = 57.$$

The variable occurs in a cube root.

This equation can be used to find the area, x, of a Galápagos island with 57 species of plants. First, we isolate the radical by dividing both sides of the equation by 28.5.

$$\frac{28.5\sqrt[3]{x}}{28.5} = \frac{57}{28.5}$$
$$\sqrt[3]{x} = 2$$

Next we eliminate the radical by raising each side of the equation to a power equal to the index of the radical. Because the index is 3, we cube both sides of the equation.

$$\left(\sqrt[3]{x}\right)^3 = 2^3$$
$$x = 8$$

Thus, a Galápagos island with 57 species of plants has an area of 8 square miles.

The Galápagos equation shows that solving equations involving radicals involves raising both sides of the equation to a power equal to the radical's index. All solutions of the original equation are also solutions of the resulting equation. However, the resulting equation may have some extra solutions that do not satisfy the original equation. Because the resulting equation may not be equivalent to the original equation, we must check each proposed solution by substituting it into the original equation. Let's see exactly how this works.

EXAMPLE 3 Solving an Equation Involving a Radical

Solve: $x + \sqrt{26 - 11x} = 4$.

Solution To solve this equation, we isolate the radical expression $\sqrt{26 - 11x}$ on one side of the equation. By squaring both sides of the equation, we can then eliminate the square root.

$$x + \sqrt{26 - 11x} = 4 \quad \text{This is the given equation.}$$
$$x + \sqrt{26 - 11x} - x = 4 - x \quad \text{Isolate the radical by subtracting } x \text{ from both sides.}$$
$$\sqrt{26 - 11x} = 4 - x \quad \text{Simplify.}$$
$$\left(\sqrt{26 - 11x}\right)^2 = (4 - x)^2 \quad \text{Square both sides.}$$
$$26 - 11x = 16 - 8x + x^2 \quad \text{Use } (A - B)^2 = A^2 - 2AB + B^2 \text{ to square } 4 - x.$$

Next, we need to write this quadratic equation in standard form. We can obtain zero on the left side by subtracting 26 and adding $11x$ on both sides.

$$26 - 26 - 11x + 11x = 16 - 26 - 8x + 11x + x^2$$
$$0 = x^2 + 3x - 10 \quad \text{Simplify.}$$
$$0 = (x + 5)(x - 2) \quad \text{Factor.}$$
$$x + 5 = 0 \quad \text{or} \quad x - 2 = 0 \quad \text{Set each factor equal to zero.}$$
$$x = -5 \qquad\qquad x = 2 \quad \text{Solve for } x.$$

We have not completed the solution process. Although -5 and 2 satisfy the squared equation, there is no guarantee that they satisfy the original equation. Thus, we must check the proposed solutions. We can do this using a graphing utility (see the technology box in the margin) or by substituting both proposed solutions into the given equation

$$x + \sqrt{26 - 11x} = 4.$$

Study Tip

Be sure to square *both sides* of an equation. Do *not* square each term.

Correct:

$$\left(\sqrt{26 - 11}\right)^2 = (4 - x)^2$$

Incorrect:

$$\left(\sqrt{26 - 11}\right)^2 = 4^2 - x^2$$

Technology

The graph of

$$y = x + \sqrt{26 - 11x} - 4$$

is shown in a $[-10, 3, 1]$ by $[-4, 3, 1]$ viewing rectangle. The x-intercepts are -5 and 2, verifying $\{-5, 2\}$ as the solution set.

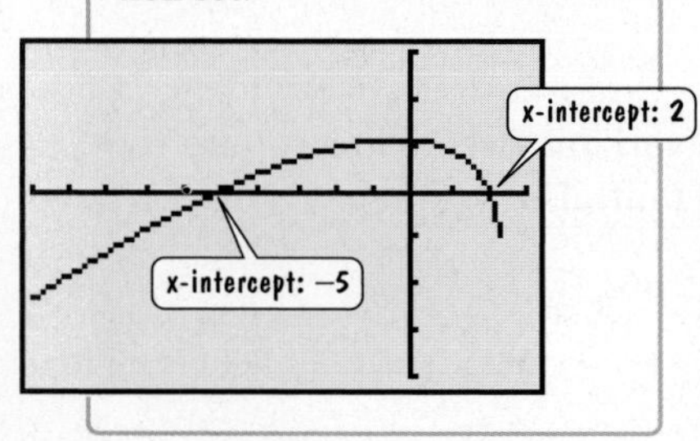

Check −5:

$-5 + \sqrt{26 - 11(-5)} \stackrel{?}{=} 4$

$-5 + \sqrt{81} \stackrel{?}{=} 4$

$-5 + 9 \stackrel{?}{=} 4$

$4 = 4$ ✓

Check 2:

$2 + \sqrt{26 - 11(2)} \stackrel{?}{=} 4$ — Substitute −5 and 2, respectively.

$2 + \sqrt{4} \stackrel{?}{=} 4$ — Simplify.

$2 + 2 \stackrel{?}{=} 4$

$4 = 4$ ✓ — Both −5 and 2 are solutions.

The solution set is $\{-5, 2\}$.

Check Point 3 Solve and check: $\sqrt{6x + 7} - x = 2$.

Study Tip

Don't forget to check for extraneous solutions when solving equations by raising both sides to an even power. Here's a simple example:

$x = 4$

$x^2 = 16$ — Square both sides.

$x = \pm\sqrt{16}$ — Use the square root method.

$x = \pm 4$

However, −4 does not check in $x = 4$. Thus, −4 is an extraneous solution.

When solving a radical equation, extra solutions may be introduced when you raise both sides of the equation to an even power. Such solutions are called **extraneous solutions**.

The solution of radical equations with two or more square root expressions involves isolating a radical, squaring both sides, and then repeating this process. Let's consider an equation containing two square root expressions.

EXAMPLE 4 Solving an Equation Involving Two Radicals

Solve: $\sqrt{3x + 1} - \sqrt{x + 4} = 1$.

Solution

$\sqrt{3x + 1} - \sqrt{x + 4} = 1$ — This is the given equation.

$\sqrt{3x + 1} = \sqrt{x + 4} + 1$ — Isolate one of the radicals by adding $\sqrt{x + 4}$ to both sides.

$\left(\sqrt{3x + 1}\right)^2 = \left(\sqrt{x + 4} + 1\right)^2$ — Square both sides.

Squaring the expression on the right side of the equation can be a bit tricky. We need to use the formula

$$(A + B)^2 = A^2 + 2AB + B^2.$$

Focusing on just the right side, here is how the squaring is done.

$$(A + B)^2 = A^2 + 2\ A\ B + B^2$$

$$\left(\sqrt{x + 4} + 1\right)^2 = \left(\sqrt{x + 4}\right)^2 + 2 \cdot \sqrt{x + 4} \cdot 1 + 1^2$$

This simplifies to $x + 4 + 2\sqrt{x + 4} + 1$. Thus, our equation can be written as follows.

$3x + 1 = x + 4 + 2\sqrt{x + 4} + 1$

$3x + 1 = x + 5 + 2\sqrt{x + 4}$ — Combine numerical terms on the right.

$2x - 4 = 2\sqrt{x + 4}$ — Isolate $2\sqrt{x + 4}$, the radical term, by subtracting $x + 5$ from both sides.

$x - 2 = \sqrt{x + 4}$ — Divide both sides by 2.

$(x - 2)^2 = \left(\sqrt{x + 4}\right)^2$ — Square both sides.

$x^2 - 4x + 4 = x + 4$ Multiply.

$x^2 - 5x = 0$ Write the quadratic equation in standard form by subtracting $x + 4$ from both sides.

$x(x - 5) = 0$ Factor.

$x = 0$ or $x - 5 = 0$ Set each factor equal to zero.

$x = 0$ $\quad x = 5$ Solve for x.

Complete the solution process by checking both proposed solutions. We can do this using a graphing utility (see the technology box in the margin) or by substituting both proposed solutions in the given equation.

Technology

The graph of

$y = \sqrt{3x+1} - \sqrt{x+4} - 1$

has only one x-intercept at 5. This verifies that the solution set for the given equation is $\{5\}$.

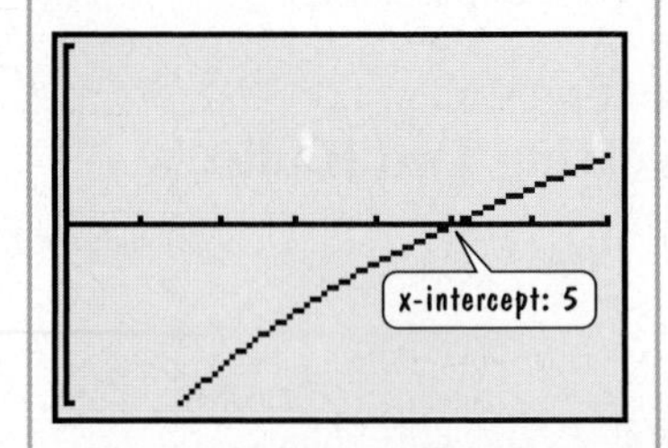

Check 0:

$$\sqrt{3x+1} - \sqrt{x+4} = 1$$
$$\sqrt{3 \cdot 0 + 1} - \sqrt{0+4} \stackrel{?}{=} 1$$
$$\sqrt{1} - \sqrt{4} \stackrel{?}{=} 1$$
$$1 - 2 \stackrel{?}{=} 1$$
$$-1 = 1 \quad \text{False}$$

Check 5:

$$\sqrt{3x+1} - \sqrt{x+4} = 1$$
$$\sqrt{3 \cdot 5 + 1} - \sqrt{5+4} \stackrel{?}{=} 1$$
$$\sqrt{16} - \sqrt{9} \stackrel{?}{=} 1$$
$$4 - 3 \stackrel{?}{=} 1$$
$$1 = 1 \checkmark$$

The check indicates that 0 is not a solution. It is an extraneous solution brought about by squaring each side of the equation. The only solution is 5, and the solution set is $\{5\}$.

Check Point 4 Solve and check: $\sqrt{x+5} - \sqrt{x-3} = 2$.

Radicals and Windchill

The way that we perceive the temperature on a cold day depends on both air temperature and wind speed. The windchill temperature is what the air temperature would have to be with no wind to achieve the same chilling effect on the skin. The formula that describes windchill temperature, W, in terms of the velocity of the wind, v, in miles per hour, and the actual air temperature, t, in degrees Fahrenheit, is

$$W = 91.4 - \frac{(10.5 + 6.7\sqrt{v} - 0.45v)(457 - 5t)}{110}.$$

Use your calculator to describe how cold the air temperature feels (that is, the windchill temperature) when the temperature is 15° Fahrenheit and the wind is 5 miles per hour. Contrast this with a temperature of 40° Fahrenheit and a wind blowing at 50 miles per hour.

3 Solve equations with rational exponents.

Because $\sqrt[n]{b}$ can be expressed as $b^{1/n}$, radical equations can be written using rational exponents. For example, the Galápagos equation

$$28.5\sqrt[3]{x} = 57$$

can be written

$$28.5x^{1/3} = 57.$$

We solve this equation exactly as we did when it was expressed in radical form. First, isolate $x^{1/3}$.

$$\frac{28.5x^{1/3}}{28.5} = \frac{57}{28.5}$$

$$x^{1/3} = 2$$

Complete the solution process by raising both sides to the third power.

$$\left(x^{1/3}\right)^3 = 2^3$$

$$x = 8$$

In general, a radical equation with rational exponents can be solved by: (1) isolating the expression with the rational exponent, and (2) raising both sides of the equation to a power that is the reciprocal of the rational exponent. Be sure to *complete the solution process* by *checking all proposed solutions in the original equation* to find out if they are actual solutions or extraneous solutions.

EXAMPLE 5 Solving an Equation Involving a Rational Exponent

Solve: $3x^{3/4} - 6 = 0.$

Solution Our goal is to isolate $x^{3/4}$. Then we can raise both sides of the equation to the $\frac{4}{3}$ power because $\frac{4}{3}$ is the reciprocal of $\frac{3}{4}$.

$3x^{3/4} - 6 = 0$ This is the given equation; we will isolate $x^{3/4}$.

$3x^{3/4} = 6$ Add 6 to both sides.

$\frac{3x^{3/4}}{3} = \frac{6}{3}$ Divide both sides by 3.

$x^{3/4} = 2$ Simplify.

$\left(x^{3/4}\right)^{4/3} = 2^{4/3}$ Raise both sides to the $\frac{4}{3}$ power.

$x = 2^{4/3}$ Simplify the left side: $\left(x^{3/4}\right)^{4/3} = x^{\frac{3\cdot 4}{4\cdot 3}} = x^{\frac{12}{12}} = x^1 = x.$

The proposed solution is $2^{4/3}$. Complete the solution process by checking this value in the given equation.

$3x^{3/4} - 6 = 0$ This is the original equation.

$3\left(2^{4/3}\right)^{3/4} - 6 \stackrel{?}{=} 0$ Substitute the proposed solution.

$3 \cdot 2 - 6 \stackrel{?}{=} 0$ $\left(2^{4/3}\right)^{3/4} = 2^{\frac{4\cdot 3}{3\cdot 4}} = 2^{\frac{12}{12}} = 2^1 = 2.$

$0 = 0$ ✓ Thus, $2^{4/3}$ is a solution.

The solution is $2^{4/3} = \sqrt[3]{2^4} \approx 2.52$. The solution set is $\{2^{4/3}\}$.

Check Point 5 Solve and check: $5x^{3/2} - 25 = 0.$

4 Solve equations that are quadratic in form.

Equations That Are Quadratic in Form

Some equations that are not quadratic can be written as quadratic equations using an appropriate substitution. Here are some examples.

Given Equation	Substitution	New Equation
$x^4 - 8x^2 - 9 = 0$ or $(x^2)^2 - 8x^2 - 9 = 0$	$t = x^2$	$t^2 - 8t - 9 = 0$
$5x^{2/3} + 11x^{1/3} + 2 = 0$ or $5(x^{1/3})^2 + 11x^{1/3} + 2 = 0$	$t = x^{1/3}$	$5t^2 + 11t + 2 = 0$

An equation that is **quadratic in form** is one that can be expressed as a quadratic equation using an appropriate substitution. Both of the preceding given equations are quadratic in form.

For equations that are quadratic in form, the exponent in one of the terms is half that of the other term. By letting t equal the variable to the half power, a quadratic equation in t will result. Now it's easy. Solve this quadratic equation for t. Finally, use your substitution to find the values for the variable in the given equation. Example 6 shows how this is done.

EXAMPLE 6 Solving an Equation That Is Quadratic in Form

Solve: $x^4 - 8x^2 - 9 = 0$.

Solution Notice that the exponent on x^2 is half that of the exponent on x^4. We let t equal the variable to the power that is half of 4. Thus,

$$\text{let } t = x^2.$$

Now we write the given equation as a quadratic equation in t and solve for t.

$$x^4 - 8x^2 - 9 = 0 \quad \text{This is the given equation.}$$

$$(x^2)^2 - 8x^2 - 9 = 0 \quad \text{The given equation contains } x^2 \text{ and } x^2 \text{ squared.}$$

$$t^2 - 8t - 9 = 0 \quad \text{Replace } x^2 \text{ by } t.$$

$$(t - 9)(t + 1) = 0 \quad \text{Factor.}$$

$$t - 9 = 0 \quad \text{or} \quad t + 1 = 0 \quad \text{Apply the zero-product principle.}$$

$$t = 9 \qquad t = -1 \quad \text{Solve for } t.$$

We're not done! Why not? We were asked to solve for x and we have values for t. We use the original substitution, $t = x^2$, to solve for x. Replace t by x^2 in each equation shown.

$$x^2 = 9 \qquad x^2 = -1$$

$$x = \pm\sqrt{9} \qquad x = \pm\sqrt{-1}$$

$$x = \pm 3 \qquad x = \pm i$$

The solution set is $\{-3, 3, -i, i\}$.

Check Point 6 Solve: $x^4 - 5x^2 + 6 = 0$.

EXAMPLE 7 Solving an Equation That Is Quadratic in Form

Solve: $5x^{2/3} + 11x^{1/3} + 2 = 0$.

Solution Notice that the exponent on $x^{1/3}$ is half that of the exponent on $x^{2/3}$. We let t equal the variable to the power that is half of 2/3. Thus,

$$\text{let } t = x^{1/3}.$$

Now we write the given equation as a quadratic equation in t and solve for t.

$$5x^{2/3} + 11x^{1/3} + 2 = 0 \quad \text{This is the given equation.}$$

$$5(x^{1/3})^2 + 11(x^{1/3}) + 2 = 0 \quad \text{The given equation contains } x^{1/3} \text{ and } x^{1/3} \text{ squared.}$$

$$5t^2 + 11t + 2 = 0 \quad \text{Replace } x^{1/3} \text{ by } t.$$

$$(5t + 1)(t + 2) = 0 \quad \text{Factor.}$$

$$5t + 1 = 0 \quad \text{or} \quad t + 2 = 0 \quad \text{Set each factor equal to 0.}$$

$$5t = -1 \qquad t = -2 \quad \text{Solve for } t.$$

$$t = -\tfrac{1}{5}$$

Use the original substitution, $t = x^{1/3}$, to solve for x. Replace t by $x^{1/3}$ in each of the preceding equations.

$$x^{1/3} = -\frac{1}{5} \qquad x^{1/3} = -2 \quad \text{Replace } t \text{ with } x^{1/3}.$$

$$(x^{1/3})^3 = \left(-\frac{1}{5}\right)^3 \qquad (x^{1/3})^3 = (-2)^3 \quad \text{Solve for } x \text{ by cubing both sides of each equation.}$$

$$x = -\frac{1}{125} \qquad x = -8$$

Check these values to verify that the solution set is $\{-\frac{1}{125}, -8\}$.

Check Point 7 Solve: $3x^{2/3} - 11x^{1/3} - 4 = 0$.

5 Solve equations involving absolute value.

Equations Involving Absolute Value

We solve equations containing absolute value using the fact that the expression inside the absolute value bars can be either positive or negative. For example, the equation

$$|2x - 3| = 11$$

is satisfied if $2x - 3$ is either 11 or -11, resulting in the two equations

$$2x - 3 = 11 \quad \text{or} \quad 2x - 3 = -11.$$

Rewriting an Absolute Value Equation without Absolute Value Bars

If c is a positive real number and X represents any algebraic expression, then $|X| = c$ is equivalent to $X = c$ or $X = -c$.

EXAMPLE 8 Solving an Equation Involving Absolute Value

Solve: $|2x - 3| = 11$.

Solution

$|2x - 3| = 11$ — This is the given equation.

$2x - 3 = 11$ or $2x - 3 = -11$ — Rewrite the equation without absolute value bars.

$2x = 14$ $\quad$ $2x = -8$ — Add 3 to both sides of each equation.

$x = 7$ $\quad$ $x = -4$ — Divide both sides of each equation by 2.

Discovery

Graph $y = |2x - 3|$ and $y = 11$ in a $[-10, 10, 1]$ by $[-1, 15, 1]$ viewing rectangle. How is the solution set of $|2x - 3| = 11$, namely $\{-4, 7\}$, shown by the graphs?

Check

$|2x - 3| = 11$ — This is the original equation.

$|2(7) - 3| \stackrel{?}{=} 11$ $\quad$ $|2(-4) - 3| \stackrel{?}{=} 11$ — Substitute the proposed solutions.

$|14 - 3| \stackrel{?}{=} 11$ $\quad$ $|-8 - 3| \stackrel{?}{=} 11$ — Perform operations inside the absolute value bars.

$|11| \stackrel{?}{=} 11$ $\quad$ $|-11| \stackrel{?}{=} 11$

$11 = 11$ ✓ $\quad$ $11 = 11$ ✓ — These true statements indicate that 7 and -4 are solutions.

The solution set is $\{-4, 7\}$.

Check Point 8 Solve: $|2x - 1| = 5$.

EXERCISE SET 1.4

Practice Exercises

Solve each polynomial equation in Exercises 1–10 by factoring and then using the zero-product principle.

1. $3x^4 - 48x^2 = 0$
2. $5x^4 - 20x^2 = 0$
3. $2x^4 = 16x$
4. $3x^4 = 81x$
5. $3x^3 + 2x^2 = 12x + 8$
6. $4x^3 - 12x^2 = 9x - 27$
7. $2x - 3 = 8x^3 - 12x^2$
8. $x + 1 = 9x^3 + 9x^2$
9. $4y^3 - 2 = y - 8y^2$
10. $9y^3 + 8 = 4y + 18y^2$

Solve each radical equation in Exercises 11–28. Check all proposed solutions.

11. $\sqrt{3x + 18} = x$
12. $\sqrt{20 - 8x} = x$
13. $\sqrt{x + 3} = x - 3$
14. $\sqrt{x + 10} = x - 2$
15. $\sqrt{2x + 13} = x + 7$
16. $\sqrt{6x + 1} = x - 1$
17. $x - \sqrt{2x + 5} = 5$
18. $x - \sqrt{x + 11} = 1$
19. $\sqrt{3x + 10} = x + 4$
20. $\sqrt{x} - 3 = x - 9$
21. $\sqrt{x + 8} - \sqrt{x - 4} = 2$
22. $\sqrt{x + 5} - \sqrt{x - 3} = 2$
23. $\sqrt{x - 5} - \sqrt{x - 8} = 3$
24. $\sqrt{2x - 3} - \sqrt{x - 2} = 1$
25. $\sqrt{2x + 3} + \sqrt{x - 2} = 2$
26. $\sqrt{x + 2} + \sqrt{3x + 7} = 1$
27. $\sqrt{3\sqrt{x + 1}} = \sqrt{3x - 5}$
28. $\sqrt{1 + 4\sqrt{x}} = 1 + \sqrt{x}$

Solve and check each equation with rational exponents in Exercises 29–36.

29. $x^{3/2} = 8$
30. $x^{3/2} = 27$
31. $(x - 4)^{3/2} = 27$
32. $(x + 5)^{3/2} = 8$
33. $6x^{5/2} - 12 = 0$
34. $8x^{5/3} - 24 = 0$
35. $(x^2 - x - 4)^{3/4} - 2 = 6$
36. $(x^2 - 3x + 3)^{3/2} - 1 = 0$

Solve each equation in Exercises 37–56 by making an appropriate substitution.

37. $x^4 - 5x^2 + 4 = 0$
38. $x^4 - 13x^2 + 36 = 0$
39. $9x^4 = 25x^2 - 16$
40. $4x^4 = 13x^2 - 9$
41. $x^6 + 8x^3 + 15 = 0$
42. $x^6 + 5x^3 + 6 = 0$
43. $5x^6 + x^3 = 18$
44. $3x^6 - 4x^3 = 15$

45. $x^{2/3} - x^{1/3} - 6 = 0$
46. $2x^{2/3} + 7x^{1/3} - 15 = 0$
47. $x^{3/2} - 2x^{3/4} + 1 = 0$
48. $x^{2/5} + x^{1/5} - 6 = 0$
49. $2x - 3x^{1/2} + 1 = 0$
50. $x + 3x^{1/2} - 4 = 0$
51. $(x - 5)^2 - 4(x - 5) - 21 = 0$
52. $(x + 3)^2 + 7(x + 3) - 18 = 0$
53. $(x^2 - x)^2 - 14(x^2 - x) + 24 = 0$
54. $(x^2 - 2x)^2 - 11(x^2 - 2x) + 24 = 0$
55. $\left(y - \frac{8}{y}\right)^2 + 5\left(y - \frac{8}{y}\right) - 14 = 0$
56. $\left(y - \frac{10}{y}\right)^2 + 6\left(y - \frac{10}{y}\right) - 27 = 0$

Solve each equation in Exercises 57–62 by first rewriting the equation as two equations without absolute value bars.

57. $|x| = 8$
58. $|x| = 6$
59. $|x - 2| = 7$
60. $|x + 1| = 5$
61. $|2x - 1| = 5$
62. $|2x - 3| = 11$

Solve each equation in Exercises 63–72 by the method of your choice.

63. $x + 2\sqrt{x} - 3 = 0$
64. $x^3 + 3x^2 - 4x - 12 = 0$
65. $(x + 4)^{3/2} = 8$
66. $(x^2 - 1)^2 - 2(x^2 - 1) = 3$
67. $\sqrt{4x + 15} - 2x = 0$
68. $x^{2/5} - 1 = 0$
69. $|x^2 + 2x - 36| = 12$
70. $\sqrt{3x + 1} - \sqrt{x - 1} = 2$
71. $x^3 - 2x^2 = x - 2$
72. $|x^2 + 6x + 1| = 8$

Application Exercises

73. For a group of 50,000 births, the number of people, N, surviving to age x is modeled by the formula $N = 5000\sqrt{100 - x}$. To what age will 40,000 people in the group survive?

74. Psychologists use the formula $N = 2\sqrt{Q} - 9$ to determine the number of nonsense syllables, N, that a subject with an IQ of Q can repeat. If a subject can repeat 14 nonsense syllables, what is that person's IQ? Round to the nearest whole number.

75. The formula $N = 1220\sqrt[3]{x + 42} + 4900$ models the number of congressional aides in the House of Representatives x years after 1930. In what year were approximately 9780 aides assigned to the House of Representatives?

76. Police use the model $v = \sqrt{24L}$ to estimate the speed of a car (v, in miles per hour) prior to an accident based on the length of its skid marks (L, in feet) on dry pavement. If a car is traveling at 60 miles per hour, what is the length of its skid marks?

77. In Albert Einstein's special theory of relativity, time slows down from the point of view of an observer watching an object moving at a velocity close to the speed of light. The formula

$$T = \frac{T_0}{\sqrt{1 - \frac{v^2}{c^2}}}$$

relates the passage of time T on a futuristic starship moving at v miles per second to the passage of time for a stationary observer on Earth, T_0. (c, the speed of light, is approximately 186,000 miles per second.) Suppose that 4 hours pass on the starship, but for you, on Earth, 2 hours have passed. How fast is the starship traveling?

78. The graph shows the remaining life expectancies for males and females in the United States in 1998. The formula $E = \sqrt{0.66A^2 - 110.55A + 4680.24}$ is an approximate model for some of the data in the graph, where A represents current age and E stands for remaining life expectancy (in years).

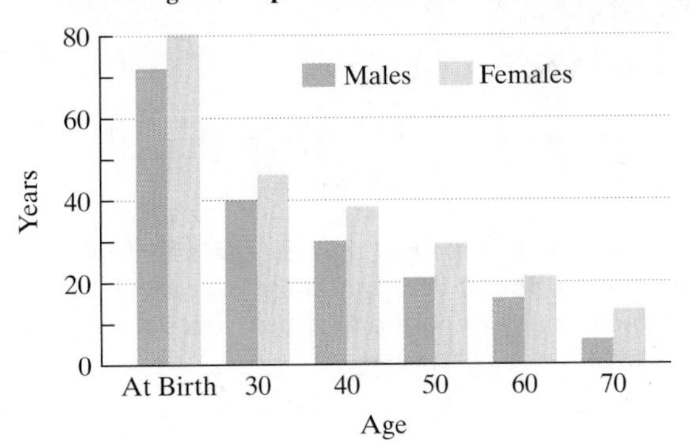

Source: Department of Health and Human Services

a. Is the formula a better model for males or for females?
b. If $E = 60$, find A and describe what your result means in terms of the variables modeled by the formula.

79. Laser Records marketing research department determines that weekly demand for a boxed set of CDs by the Jumping Artichokes depends on the price per set, modeled by the formula $p = 30 - \sqrt{0.01x + 1}$, where p represents the price of the set and x represents the number of sets sold each week at price p.
a. At what price will there be no demand for the CD sets?
b. Approximately how many CD sets will sell weekly at a price of \$27.76?

Use the Pythagorean Theorem to solve Exercises 80–81.

80. Two vertical poles of lengths 6 feet and 8 feet stand 10 feet apart (see the figure on page 138). A cable reaches from the top of one pole to some point on the ground

between the poles and then to the top of the other pole. Where should this point be located to use 18 feet of cable?

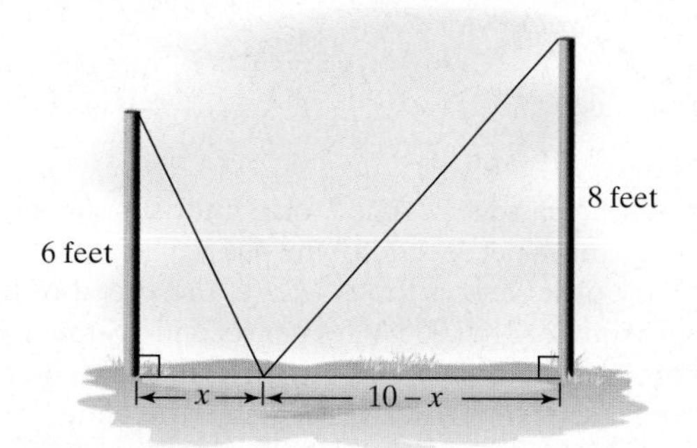

81. Twelve miles separate towns A and B, located 6 miles and 3 miles, respectively, from a major expressway. Two new roads are to be built from A to the expressway and then to B. (See the figure.)

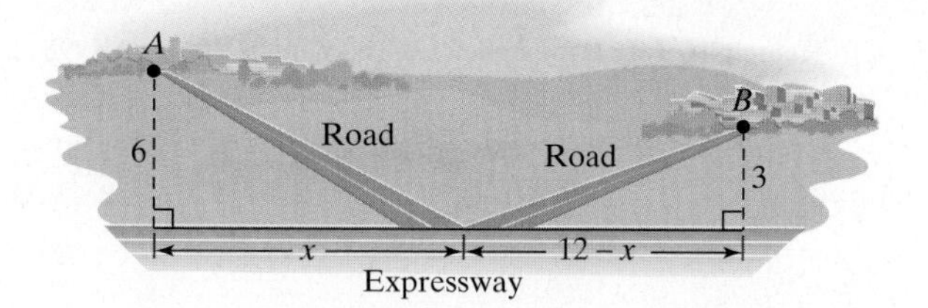

a. Find x if the length of the new roads is 15 miles.

b. Write a verbal description for the road crew telling them where to position the new roads based on your answer to part (a).

Writing in Mathematics

82. Without actually solving the equation, give a general description of how to solve $x^3 - 5x^2 - x + 5 = 0$

83. In solving $\sqrt{3x + 4} - \sqrt{2x + 4} = 2$, why is it a good idea to isolate a radical term? What if we don't do this and simply square each side? Describe what happens.

84. What is an extraneous solution to a radical equation?

85. Explain how to recognize an equation that is quadratic in form. Provide two original examples with your explanation.

86. Describe two methods for solving this equation: $x - 5\sqrt{x} + 4 = 0$.

87. Explain how to solve an equation involving absolute value.

88. Explain why the procedure that you explained in Exercise 87 does not apply to the equation $|x - 2| = -3$. What is the solution set for this equation?

89. Reread Exercise 79. Suppose you are writing a report on the relationship between the price of the CDs and the units that will sell. Assume that you are the only one in the company who can understand the formula $p = 30 - \sqrt{0.01x + 1}$. Consequently, you must describe the relationship between p and x strictly using words. Write a description, minimizing the use of mathematical terminology.

Technology Exercises

Use a graphing utility to solve the equations in Exercises 90–94. Check by direct substitution.

90. $x^3 + 3x^2 - x - 3 = 0$

91. $-x^4 + 4x^3 - 4x^2 = 0$

92. $x^4 - 2x^2 + 1 = 0$

93. $\sqrt{2x + 13} - x - 5 = 0$

94. $\sqrt{4 - x} - \sqrt{x + 6} - 2 = 0$

95. Use a graphing utility to graph the formula in Exercise 73. In particular, graph $y = 5000\sqrt{100 - x}$ in a $[0, 100, 10]$ by $[0, 50{,}000, 5000]$ viewing rectangle. Then use the TRACE feature to trace along the curve until you reach the point that visually shows the solution to Exercise 73.

96. Use a graphing utility to graph the formula in Exercise 74. In particular, graph $y = 2\sqrt{x} - 9$ in a $[0, 180, 10]$ by $[0, 20, 1]$ viewing rectangle. Then use the TRACE feature to trace along the curve until you reach the point that visually shows the solution to Exercise 74.

Critical Thinking Exercises

97. Which one of the following is true?

a. Squaring both sides of $\sqrt{y + 4} + \sqrt{y - 1} = 5$ leads to $y + 4 + y - 1 = 25$, an equation with no radicals.

b. The equation $(x^2 - 2x)^9 - 5(x^2 - 2x)^3 + 6 = 0$ is quadratic in form and should be solved by letting
$$t = (x^2 - 2x)^3.$$

c. If a radical equation has two proposed solutions and one of these values is not a solution, the other value is also not a solution.

d. None of these statements is true.

98. Solve: $\sqrt{6x - 2} = \sqrt{2x + 3} - \sqrt{4x - 1}$.

99. Solve *without* squaring both sides:
$$5 - \frac{2}{x} = \sqrt{5 - \frac{2}{x}}.$$

100. Solve for x: $\sqrt[3]{x\sqrt{x}} = 9$.

101. Solve for x: $x^{5/6} + x^{2/3} - 2x^{1/2} = 0$.

SECTION 1.5 Linear Inequalities

Objectives

1. Graph an inequality's solution set.
2. Use set-builder and interval notations.
3. Use properties of inequalities to solve inequalities.
4. Solve compound inequalities.
5. Solve inequalities involving absolute value.

Rent-a-Heap, a car rental company, charges \$125 per week plus \$0.20 per mile to rent one of their cars. Suppose you are limited by how much money you can spend for the week: You can spend at most \$335. If we let x represent the number of miles you drive the heap in a week, we can write an inequality that models the given conditions.

The weekly charge of \$125	plus	the charge of \$0.20 per mile for x miles	must be less than or equal to	\$335.
125	$+$	$0.20x$	$\leq$	335

Using the commutative property of addition, we can express this inequality as $0.20x + 125 \leq 335$. The form of this inequality is $ax + b \leq c$, with $a = 0.20$, $b = 125$, and $c = 335$. Any inequality in this form is called a **linear inequality in one variable**. The greatest exponent on the variable in such an equality is 1. The symbol between $ax + b$ and c can be $\leq$ (is less than or equal to), $<$ (is less than), $\geq$ (is greater than or equal to), or $>$ (is greater than).

In this section, we will study how to solve linear inequalities such as $0.20x + 125 \leq 335$. **Solving an inequality** is the process of finding the set of numbers that make the inequality a true statement. These numbers are called the **solutions** of the inequality, and we say that they **satisfy** the inequality. The set of all solutions is called the **solution set** of the inequality. We begin by discussing how to graph and how to represent these solution sets.

1 Graph an inequality's solution set.

Graphs of Inequalities; Interval Notation

There are infinitely many solutions to the inequality $x > -4$, namely all real numbers that are greater than -4. Although we cannot list all the solutions, we can make a drawing on a number line that represents these solutions. Such a drawing is called the **graph of the inequality**.

Graphs of solutions to linear inequalities are shown on a number line by shading all points representing numbers that are solutions. Parentheses indicate endpoints that are not solutions. Square brackets indicate endpoints that are solutions.

EXAMPLE 1 Graphing Inequalities

Graph the solutions of:

a. $x < 3$ b. $x \geq -1$ c. $-1 < x \leq 3$.

> **Study Tip**
>
> Because an inequality symbol points to the smaller number, $x < 3$ (x is less than 3) may be expressed as $3 > x$ (3 is greater than x).

Solution

a. The solutions of $x < 3$ are all real numbers that are less than 3. They are graphed on a number line by shading all points to the left of 3. The parenthesis at 3 indicates that 3 is not a solution, but numbers such as 2.9999 and 2.6 are. The arrow shows that the graph extends indefinitely to the left.

−4 −3 −2 −1 0 1 2 3 4 x

b. The solutions of $x \geq -1$ are all real numbers that are greater than or equal to -1. We shade all points to the right of -1 and the point for -1 itself. The bracket at -1 shows that -1 is a solution for the given inequality. The arrow shows that the graph extends indefinitely to the right.

−4 −3 −2 −1 0 1 2 3 4 x

c. The inequality $-1 < x \leq 3$ is read "-1 is less than x *and* x is less than or equal to 3," or "x is greater than -1 *and* less than or equal to 3." The solutions of $-1 < x \leq 3$ are all real numbers between -1 and 3, not including -1 but including 3. The parenthesis at -1 indicates that -1 is not a solution. By contrast, the bracket at 3 shows that 3 is a solution. Shading indicates the other solutions.

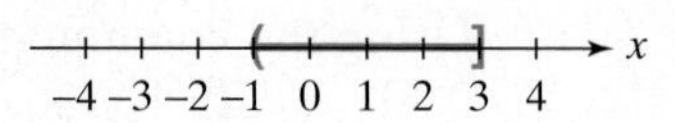

Check Point 1 Graph the solutions of:

a. $x \leq 2$ **b.** $x > -4$ **c.** $2 \leq x < 6$

2 Use set-builder and interval notations.

Now that we know how to graph the solution set for an inequality such as $x > -4$, let's see how to represent the solution set. One method is with **set-builder notation**. Using this method, the solution set for $x > -4$ can be expressed as

$$\{x \mid x > -4\}.$$

The set of all x — such that

We read this as "the set of all real numbers x such that x is greater than -4."

Another method used to represent solution sets for inequalities is **interval notation**. Using this notation, the solution set for $x > -4$ is expressed as $(-4, \infty)$. The parenthesis at -4 indicates that -4 is not included in the interval. The infinity symbol, ∞, does not represent a real number. It indicates that the interval extends indefinitely to the right.

Table 1.5 lists nine possible types of intervals used to describe subsets of real numbers.

Table 1.5 Intervals on the Real Number Line

Let a and b be real numbers such that $a < b$.		
Interval Notation	**Set-Builder Notation**	**Graph**
(a, b)	$\{x \mid a < x < b\}$	a b x
$[a, b]$	$\{x \mid a \leq x \leq b\}$	a b x
$[a, b)$	$\{x \mid a \leq x < b\}$	a b x
$(a, b]$	$\{x \mid a < x \leq b\}$	a b x
(a, ∞)	$\{x \mid x > a\}$	a x
$[a, \infty)$	$\{x \mid x \geq a\}$	a x
$(-\infty, b)$	$\{x \mid x < b\}$	b x
$(-\infty, b]$	$\{x \mid x \leq b\}$	b x
$(-\infty, \infty)$	$\mathbb{R}$ (set of all real numbers)	x

EXAMPLE 2 Intervals and Inequalities

Express the intervals in terms of inequalities and graph:

a. $(-1, 4]$ b. $[2.5, 4]$ c. $(-4, \infty)$

Solution

a. $(-1, 4] = \{x \mid -1 < x \leq 4\}$

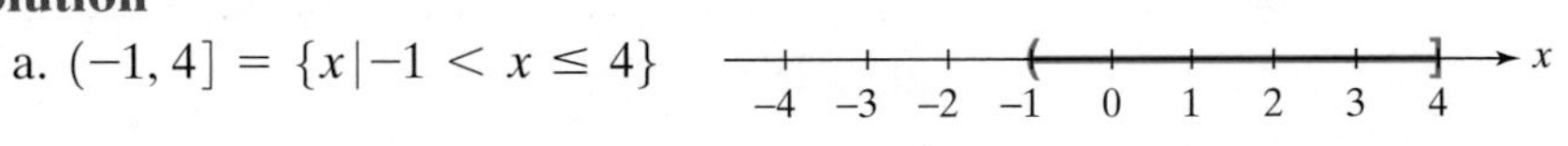

b. $[2.5, 4] = \{x \mid 2.5 \leq x \leq 4\}$

−4 −3 −2 −1 0 1 2 3 4 x

c. $(-4, \infty) = \{x \mid x > -4\}$

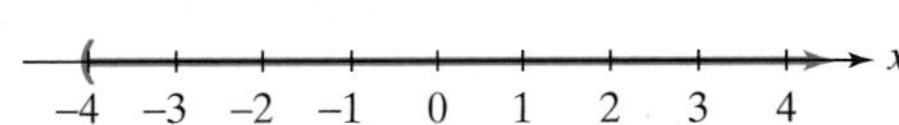

Check Point 2 Express the intervals in terms of inequalities and graph:

a. $[-2, 5)$ **b.** $[1, 3.5]$ **c.** $(-\infty, -1)$

3 Use properties of inequalities to solve inequalities.

Solving Linear Inequalities

Back to our question: How many miles can you drive on your Rent-a-Heap car if you can spend at most \$335 per week? We answer the question by solving

$$0.20x + 125 \leq 335$$

for x. The solution procedure is nearly identical to that for solving

$$0.20x + 125 = 335.$$

Our goal is to get x by itself on the left side. We do this by isolating $0.20x$, subtracting 125 from both sides:

$$0.20x + 125 \leq 335$$
$$0.20x + 125 - 125 \leq 335 - 125$$
$$0.20x \leq 210$$

Finally, we isolate x from $0.20x$ by dividing both sides of the inequality by 0.20:

$$\frac{0.20x}{0.20} \leq \frac{210}{0.20}$$

$$x \leq 1050$$

With at most \$335 per week to spend, you can travel at most 1050 miles.

We started with the inequality $0.20x + 125 \leq 335$ and obtained the inequality $x \leq 1050$ in the final step. Both of these inequalities have the same solution set, namely $\{x | x \leq 1050\}$. Inequalities such as these, with the same solution set, are said to be **equivalent**.

We isolated x from $0.20x$ by dividing both sides of $0.20x \leq 210$ by 0.20, a positive number. Let's see what happens if we divide both sides of an inequality by a negative number. Consider the inequality $10 < 14$. Divide 10 and 14 by -2:

$$\frac{10}{-2} = -5 \quad \text{and} \quad \frac{14}{-2} = -7.$$

Because -5 lies to the right of -7 on the number line, -5 is greater than -7:

$$-5 > -7.$$

Notice that the direction of the inequality symbol is reversed:

$$10 < 14$$

$$-5 > -7$$

In general, **when we multiply or divide both sides of an inequality by a negative number, the direction of the inequality symbol is reversed.** When we reverse the direction of the inequality symbol, we say that we change the *sense* of the inequality.

We can isolate a variable in a linear inequality the same way we can isolate a variable in a linear equation. The following properties are used to create equivalent inequalities.

Properties of Inequalities

Property	The Property in Words	Example
Addition and Subtraction Properties If $a < b$, then $a + c < b + c$. If $a < b$, then $a - c < b - c$.	If the same quantity is added to or subtracted from both sides of an inequality, the resulting inequality is equivalent to the original one.	$2x + 3 < 7$ Subtract 3: $2x + 3 - 3 < 7 - 3$ Simplify: $2x < 4$
Positive Multiplication and Division Properties If $a < b$ and c is positive, then $ac < bc$. If $a < b$ and c is positive, then $\frac{a}{c} < \frac{b}{c}$.	If we multiply or divide both sides of an inequality by the same positive quantity, the resulting inequality is equivalent to the original one.	$2x < 4$ Divide by 2: $\frac{2x}{2} < \frac{4}{2}$ Simplify: $x < 2$
Negative Multiplication and Division Properties If $a < b$ and c is negative, then $ac > bc$. If $a < b$ and c is negative, then $\frac{a}{c} > \frac{b}{c}$.	If we multiply or divide both sides of an inequality by the same negative quantity and reverse the direction of the inequality symbol, the result is an equivalent inequality.	$-4x < 20$ Divide by -4 and reverse the sense of the inequality: $\frac{-4x}{-4} > \frac{20}{-4}$ Simplify: $x > -5$

EXAMPLE 3 Solving a Linear Inequality

Solve and graph the solution set on a number line:

$$3 - 2x < 11.$$

Solution

$3 - 2x < 11$ This is the given inequality.

$3 - 2x - 3 < 11 - 3$ Subtract 3 from both sides.

$-2x < 8$ Simplify.

$\frac{-2x}{-2} > \frac{8}{-2}$ Divide both sides by -2 and reverse the sense of the inequality.

$x > -4$ Simplify.

The solution set consists of all real numbers that are greater than -4, expressed as $\{x \mid x > -4\}$ in set-builder notation. The interval notation for this solution set is $(-4, \infty)$. The graph of the solution set is shown as follows:

Check Point 3 Solve and graph the solution set on a number line:

$2 - 3x \leq 5.$

EXAMPLE 4 Solving a Linear Inequality

Solve and graph the solution set: $7x + 15 \geq 13x + 51.$

Solution We will collect variable terms on the left and constant terms on the right.

$7x + 15 \geq 13x + 51$ This is the given inequality.

$7x + 15 - 13x \geq 13x + 51 - 13x$ Subtract 13x from both sides.

$-6x + 15 \geq 51$ Simplify.

$-6x + 15 - 15 \geq 51 - 15$ Subtract 15 from both sides.

$-6x \geq 36$ Simplify.

$\frac{-6x}{-6} \leq \frac{36}{-6}$ Divide both sides by -6 and reverse the sense of the inequality.

$x \leq -6$ Simplify.

The solution set consists of all real numbers that are less than or equal to -6, expressed as $\{x \mid x \leq -6\}$. The interval notation for this solution set is $(-\infty, -6]$. The graph of the solution set is shown as follows:

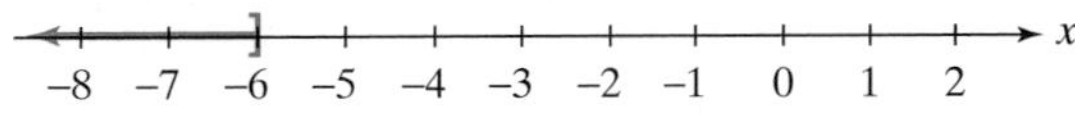

Check Point 4 Solve and graph the solution set: $6 - 3x \leq 5x - 2.$

Study Tip

You can solve

$$7x + 15 \geq 13x + 51$$

by isolating x on the right side. Subtract $7x$ from both sides:

$$7x + 15 - 7x \geq 13x + 51 - 7x$$

$$15 \geq 6x + 51$$

Now subtract 51 from both sides.

$$15 - 51 \geq 6x + 51 - 51$$

$$-36 \geq 6x$$

Finally, divide both sides by 6.

$$\frac{-36}{6} \geq \frac{6x}{6}$$

$$-6 \geq x$$

This last inequality means the same thing as

$$x \leq -6.$$

Technology

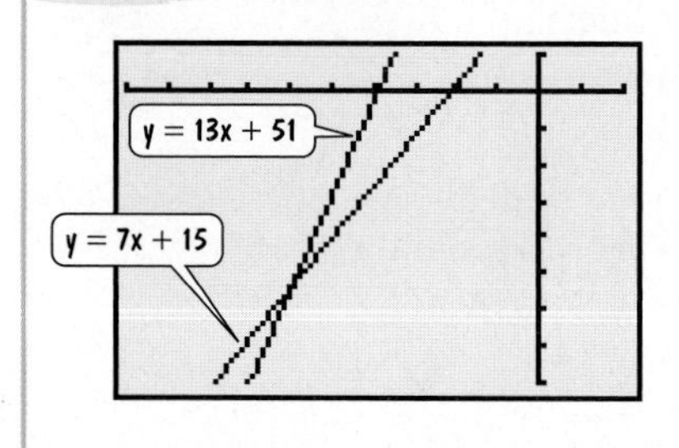

You can use a graphing utility to verify that $(-\infty, -6]$ is the solution set for

$$7x + 15 \geq 13x + 51.$$

For what values of x does the graph of y = 7x + 15 | lie above or on | the graph of y = 13x + 51?

The graphs are shown on the left in a $[-10, 2, 1]$ by $[-40, 5, 5]$ viewing rectangle. Notice that the graph of $y = 7x + 15$ lies above or on the graph of $y = 13x + 51$ when $x \leq -6$.

4 Solve compound inequalities.

Solving Compound Inequalities

We now consider two inequalities such as

$$-3 < 2x + 1 \text{ and } 2x + 1 \leq 3$$

expressed as a **compound inequality**

$$-3 < 2x + 1 \leq 3.$$

This double inequality form enables us to solve both inequalities at once. With three parts to a compound inequality, our goal is to **isolate *x* in the middle**.

EXAMPLE 5 Solving a Compound Inequality

Solve and graph the solution set:

$$-3 < 2x + 1 \leq 3.$$

Solution We would like to isolate x in the middle. We can do this by first subtracting 1 from all three parts of the compound inequality. Then we isolate x from $2x$ by dividing all three parts of the inequality by 2.

$$-3 < 2x + 1 \leq 3 \quad \text{This is the given inequality.}$$

$$-3 - 1 < 2x + 1 - 1 \leq 3 - 1 \quad \text{Subtract 1 from all three parts.}$$

$$-4 < 2x \leq 2 \quad \text{Simplify.}$$

$$\frac{-4}{2} < \frac{2x}{2} \leq \frac{2}{2} \quad \text{Divide each part by 2.}$$

$$-2 < x \leq 1 \quad \text{Simplify.}$$

The solution set consists of all real numbers greater than -2 and less than or equal to 1, represented by $\{x \mid -2 < x \leq 1\}$ in set-builder notation and $(-2, 1]$ in interval notation. The graph is shown as follows:

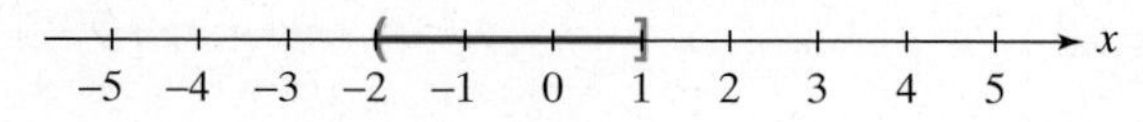

Check Point 5 Solve and graph the solution set: $1 \leq 2x + 3 < 11$.

5 Solve inequalities involving absolute value.

Solving Inequalities with Absolute Value

We have seen that $|x|$ describes the distance of x from zero on a real number line. We can use this geometric interpretation to solve an inequality such as

$$|x| < 2.$$

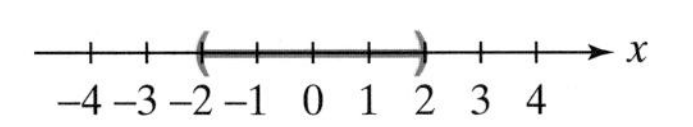

Figure 1.6 $|x| < 2$, so $-2 < x < 2$.

Figure 1.7 $|x| > 2$, so $x < -2$ or $x > 2$.

This means that the distance of x from 0 is *less than* 2, as shown in Figure 1.6. The interval shows values of x that lie less than 2 units from 0. Thus, x can lie between -2 and 2. That is, x is greater than -2 and less than 2. We write $(-2, 2)$ or $\{x|-2 < x < 2\}$.

Some absolute value inequalities use the "greater than" symbol. For example, $|x| > 2$ means that the distance of x from 0 is *greater than* 2, as shown in Figure 1.7. Thus, x can be less than -2 *or* greater than 2. We write $x < -2$ or $x > 2$.

These observations suggest the following principles for solving inequalities with absolute value.

Study Tip

In the $|X| < c$ case, we have one compound inequality to solve. In the $|X| > c$ case, we have two separate inequalities to solve.

Solving an Absolute Value Inequality

If X is an algebraic expression and c is a positive number,

1. The solutions of $|X| < c$ are the numbers that satisfy $-c < X < c$.
2. The solutions of $|X| > c$ are the numbers that satisfy $X < -c$ or $X > c$.

These rules are valid if $<$ is replaced by $\leq$ and $>$ is replaced by $\geq$.

EXAMPLE 6 Solving an Absolute Value Inequality with <

Solve and graph: $|x - 4| < 3$.

Solution

$|X| < c$ *means* $-c < X < c$.

$$|x - 4| < 3 \quad \text{means} \quad -3 < x - 4 < 3$$

We solve the compound inequality by adding 4 to all three parts.

$$-3 < x - 4 < 3$$
$$-3 + 4 < x - 4 + 4 < 3 + 4$$
$$1 < x < 7$$

The solution set is all real numbers greater than 1 and less than 7, denoted by $\{x|1 < x < 7\}$ or $(1, 7)$. The graph of the solution set is shown as follows:

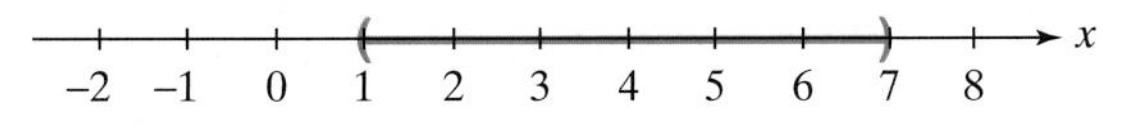

Check Point 6 Solve and graph: $|x - 2| < 5$.

EXAMPLE 7 Solving an Absolute Value Inequality with $\geq$

Solve and graph: $|2x + 3| \geq 5$.

Solution

$|X| \geq c$ means $X \leq -c$ or $X \geq c$.

$$|2x + 3| \geq 5 \quad \text{means} \quad 2x + 3 \leq -5 \quad \text{or} \quad 2x + 3 \geq 5$$

We solve each of these inequalities separately.

$2x + 3 \leq -5$	or	$2x + 3 \geq 5$	These are the inequalities without absolute value bars.
$2x + 3 - 3 \leq -5 - 3$	or	$2x + 3 - 3 \geq 5 - 3$	Subtract 3 from both sides.
$2x \leq -8$	or	$2x \geq 2$	Simplify.
$\frac{2x}{2} \leq \frac{-8}{2}$	or	$\frac{2x}{2} \geq \frac{2}{2}$	Divide both sides by 2.
$x \leq -4$	or	$x \geq 1$	Simplify.

Study Tip

The graph of the solution set for $|X| > c$ will be divided into two intervals. The graph of the solution set for $|X| < c$ will be a single interval.

The solution set is $\{x | x \leq -4 \text{ or } x \geq 1\}$, that is, all x in $(-\infty, -4]$ or $[1, \infty)$. The graph of the solution set is shown as follows:

Check Point 7 Solve and graph: $|2x - 5| \geq 3$.

Applications

Our next example shows how to use an inequality to select the better deal between two pricing options. We will use our five-step strategy for solving problems using mathematical models.

EXAMPLE 8 Creating and Comparing Mathematical Models

Acme Car rental agency charges \$4 a day plus \$0.15 a mile, whereas Interstate rental agency charges \$20 a day and \$0.05 a mile. How many miles must be driven to make the daily cost of an Acme rental a better deal than an Interstate rental?

Solution

Step 1 Let x represent one of the quantities. We are looking for the number of miles that must be driven in a day to make Acme the better deal. Thus,

let x = the number of miles driven in a day.

Step 2 Represent other quantities in terms of x. We are not asked to find another quantity, so we can skip this step.

Step 3 **Write an inequality in x that describes the conditions.**

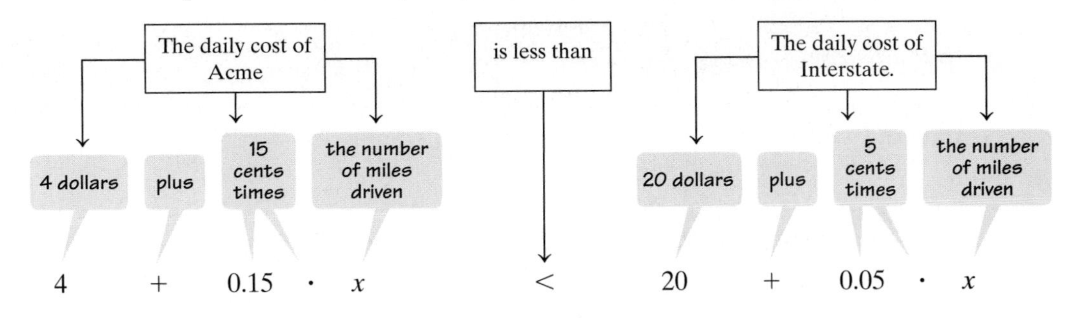

Step 4 **Solve the inequality and answer the question.**

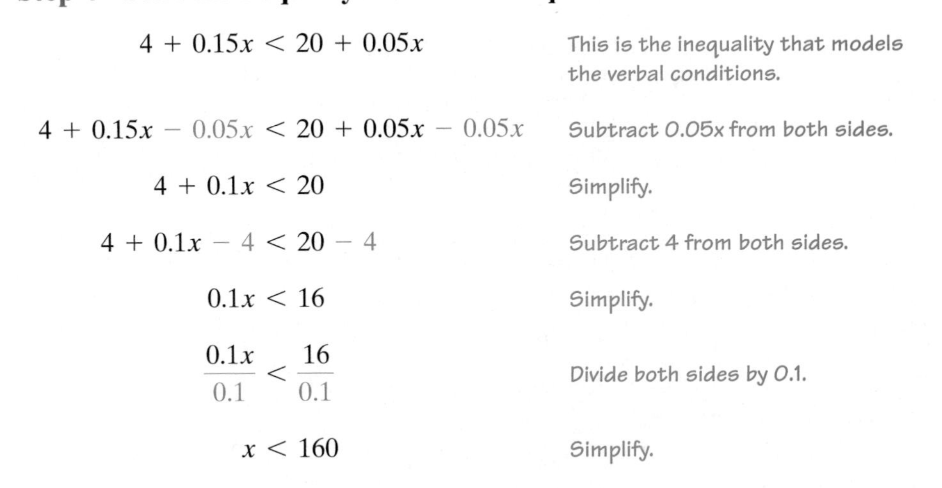

Thus, driving fewer than 160 miles per day makes Acme the better deal.

Step 5 **Check the proposed solution in the original wording of the problem.** One way to do this is to take a mileage less than 160 miles per day to see if Acme is the better deal. Suppose that 150 miles are driven in a day.

$$\text{Cost for Acme} = 4 + 0.15(150) = 26.50$$

$$\text{Cost for Interstate} = 20 + 0.05(150) = 27.50$$

and Acme has a lower daily cost, making Acme the better deal.

Check Point 8 A car can be rented from Basic Rental for \$260 per week with no extra charge for mileage. Continental charges \$80 per week plus 25 cents for each mile driven to rent the same car. How many miles should be driven in a week to make the rental cost for Basic Rental a better deal than Continental's?

EXERCISE SET 1.5

Practice Exercises

In Exercises 1–12, graph the solutions of each inequality on a number line.

1. $x > 6$
2. $x > -2$
3. $x < -4$
4. $x < 0$
5. $x \geq -3$
6. $x \geq -5$
7. $x \leq 4$
8. $x \leq 7$
9. $-2 < x \leq 5$
10. $-3 \leq x < 7$
11. $-1 < x < 4$
12. $-7 \leq x \leq 0$

In Exercises 13–26, express each interval in terms of an inequality and graph the interval on a number line.

13. $(1, 6]$
14. $(-2, 4]$
15. $[-5, 2)$
16. $[-4, 3)$
17. $[-3, 1]$
18. $[-2, 5]$
19. $(2, \infty)$
20. $(3, \infty)$
21. $[-3, \infty)$
22. $[-5, \infty)$
23. $(-\infty, 3)$
24. $(-\infty, 2)$
25. $(-\infty, 5.5)$
26. $(-\infty, 3.5]$

Solve each linear inequality in Exercises 27–48 and graph the solution set on a number line.

27. $5x + 11 < 26$
28. $2x + 5 < 17$
29. $3x - 7 \geq 13$
30. $8x - 2 \geq 14$
31. $-9x \geq 36$
32. $-5x \leq 30$
33. $8x - 11 \leq 3x - 13$
34. $18x + 45 \leq 12x - 8$
35. $4(x + 1) + 2 \geq 3x + 6$
36. $8x + 3 > 3(2x + 1) + x + 5$
37. $2x - 11 < -3(x + 2)$
38. $-4(x + 2) > 3x + 20$
39. $1 - (x + 3) \geq 4 - 2x$
40. $5(3 - x) \leq 3x - 1$
41. $\frac{x}{4} - \frac{3}{5} \leq \frac{x}{2} + 1$
42. $\frac{3x}{10} + 1 \geq \frac{1}{5} - \frac{x}{10}$
43. $1 - \frac{x}{2} > 4$
44. $7 - \frac{4}{5}x < \frac{3}{5}$
45. $\frac{x - 4}{6} \geq \frac{x - 2}{9} + \frac{5}{18}$
46. $\frac{4x - 3}{6} + 2 \geq \frac{2x - 1}{12}$
47. $4(3x - 2) - 3x < 3(1 + 3x) - 7$
48. $3(x - 8) - 2(10 - x) > 5(x - 1)$

Solve each inequality in Exercises 49–56 by isolating the variable by itself in the middle. Graph the solution set on a number line.

49. $6 < x + 3 < 8$
50. $7 < x + 5 < 11$
51. $-3 \leq x - 2 < 1$
52. $-6 < x - 4 \leq 1$
53. $-11 < 2x - 1 \leq -5$
54. $3 \leq 4x - 3 < 19$
55. $-3 \leq \frac{2}{3}x - 5 < -1$
56. $-6 \leq \frac{1}{2}x - 4 < -3$

Solve each inequality in Exercises 57–84 by first rewriting each one as an equivalent inequality without absolute value bars. Graph the solution set on a number line.

57. $|x| < 3$
58. $|x| < 5$
59. $|x - 1| \leq 2$
60. $|x + 3| \leq 4$
61. $|2x - 6| < 8$
62. $|3x + 5| < 17$
63. $|2(x - 1) + 4| \leq 8$
64. $|3(x - 1) + 2| \leq 20$
65. $\left|\frac{2y + 6}{3}\right| < 2$
66. $\left|\frac{3(x - 1)}{4}\right| < 6$
67. $|x| > 3$
68. $|x| > 5$
69. $|x - 1| \geq 2$
70. $|x + 3| \geq 4$
71. $|3x - 8| > 7$
72. $|5x - 2| > 13$
73. $\left|\frac{2x + 2}{4}\right| \geq 2$
74. $\left|\frac{3x - 3}{9}\right| \geq 1$
75. $\left|3 - \frac{2}{3}x\right| > 5$
76. $\left|3 - \frac{3}{4}x\right| > 9$
77. $3|x - 1| + 2 \geq 8$
78. $-2|4 - x| \geq -4$
79. $3 < |2x - 1|$
80. $5 \geq |4 - x|$
81. $12 < \left|-2x + \frac{6}{7}\right| + \frac{3}{7}$
82. $1 < \left|x - \frac{11}{3}\right| + \frac{7}{3}$
83. $4 + \left|3 - \frac{x}{3}\right| \geq 9$
84. $\left|2 - \frac{x}{2}\right| - 1 \leq 1$

Application Exercises

The list on the next page shown ranks the ten best-educated cities in the United States measured by the percentage of the population with 16 or more years of education. Let x represent the percentage of the population with 16 or more years of education. In Exercises 85–90 write the name or names of the city or cities described by the given inequality or interval.

85. $x \geq 34.4\%$
86. $x > 35.0\%$
87. $x < 30.0\%$
88. $x \leq 30.3\%$

89. $[30.0\%, 34.4\%]$ **90.** $(30.6\%, 35.0\%]$

Most Educated

City	% with 16 + Years of Education
1. Raleigh, NC	40.6%
2. Seattle, WA	37.9
3. San Francisco, CA	35.0
4. Austin, TX	34.4
5. Washington, DC	33.3
6. Lexington-Fayette, KY	30.6
7. Minneapolis, MN	30.3
8. Boston, MA	30.0
9. Arlington, TX	30.0
10. San Diego, CA	29.8

Source: U.S. Census Bureau

91. The bar graph shows the number of people in the United States with various disorders of mental illness. If x represents millions of people, which disorders are described by $11 \le 3x - 4 \le 56$ cases?

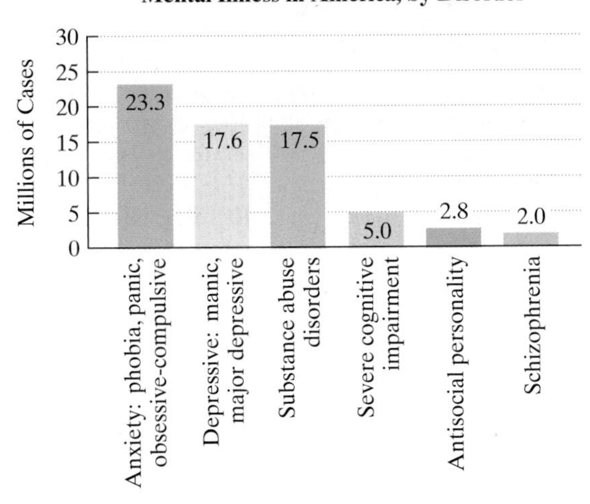

Source: National Institute of Mental Health.

92. The bar graph at the top of the next column shows revenues, in billions of dollars, for the pet pharmaceutical industry for three selected years. Let x represent revenues in billions of dollars per year. Which year or years shown in the graph is/are described by the inequality

$$5 < 4x - 1 < 11?$$

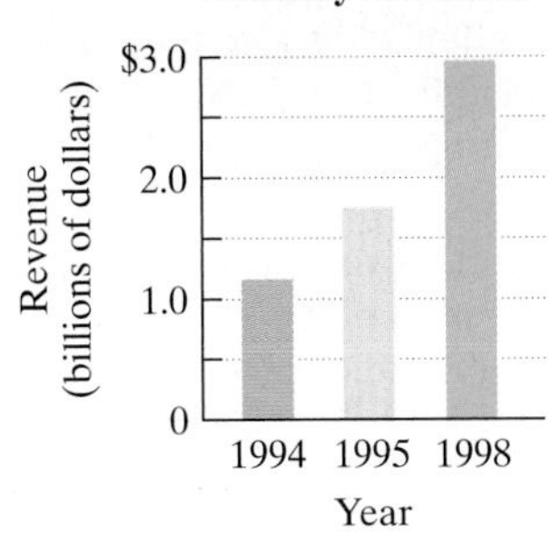

Source: American Veterinary Medical Association

93. Using data from 1996–1998, the number of liposuctions in the United States can be modeled by the formula $y = 30x + 113$, where y is the number of liposuctions, in thousands, x years after 1996. According to this formula, when will the number of liposuctions exceed 623 thousand?

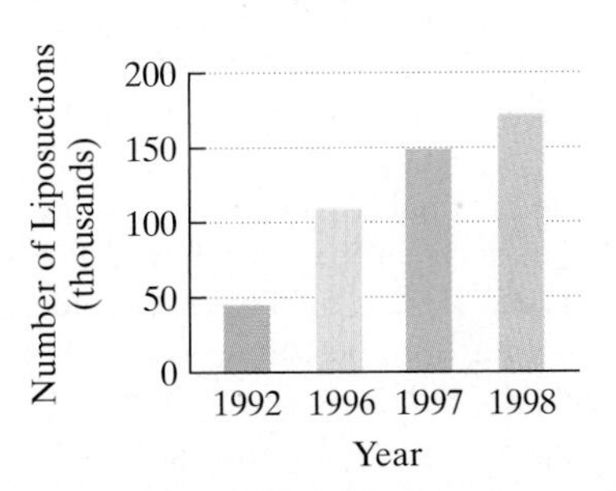

Source: U.S. Department of Health and Human Services

94. Lower interest rates have fueled larger mortgage loans. The formula $y = 3.5x + 58$ models the data shown in the graph, where y is the size of the loan, in thousands of dollars, x years after 1980. According to this formula, when will the average mortgage loan exceed $142 thousand?

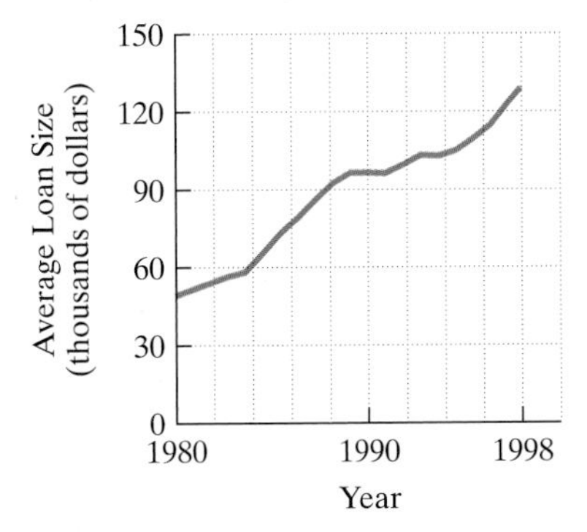

Source: Mortgage Bankers Association of America

95. Using current trends, future costs of Medicare can be modeled by $C = 18x + 250$, where x represents the number of years after 2000 and C represents the cost of Medicare, in billions of dollars. Use a compound inequality to determine the years when Medicare costs will range from 322 to 412 billion dollars.

96. The formula for converting Celsius temperature (C) to Fahrenheit (F) is $F = \frac{9}{5}C + 32$. If Fahrenheit temperature exceeds 77°, what does this mean in terms of Celsius temperature?

97. A local bank charges \$8 per month plus 5¢ per check. The credit union charges \$2 per month plus 8¢ per check. How many checks should be written each month to make the credit union a better deal?

98. A city commission has proposed two tax bills. The first bill requires that a homeowner pay \$1800 plus 3% of the assessed home value in taxes, The second bill requires taxes of \$200 plus 8% of the assessed home value. What price range of home assessment would make the first bill a better deal?

99. On two examinations, you have grades of 86 and 88. There is an optional final examination, which counts as one grade. You decide to take the final in order to get a course grade of A, meaning a final average of at least 90.

a. What must you get on the final to earn an A in the course?

b. By taking the final, if you do poorly, you might risk the B that you have in the course based on the first two exam grades. If your final average is less than 80, you will lose your B in the course. Describe the grades on the final that will cause this to happen.

100. A company that manufactures small clocks has fixed costs of \$75,000 per month. It costs the company \$3 to manufacture each clock. If the clocks sell for \$18 each, how many should be manufactured and sold monthly to make a profit?

101. A company that manufactures running shoes has fixed costs of \$65,000 per month. It costs the company \$20 to manufacture each pair of running shoes. If the shoes sell for \$85 a pair, how many should be manufactured and sold monthly to make a profit?

102. If a coin is tossed 100 times, we would expect approximately 50 of the outcomes to be heads. It can be demonstrated that a coin is unfair if h, the number of outcomes that result in heads, satisfies $\left|\frac{h - 50}{5}\right| \geq 1.645$. Describe the number of outcomes that determine an unfair coin that is tossed 100 times.

Writing in Mathematics

103. When graphing the solutions of an inequality, what does a parenthesis signify? What does a bracket signify?

104. When solving an inequality, when is it necessary to change the sense of the inequality? Give an example.

105. Describe ways in which solving a linear inequality is similar to solving a linear equation.

106. Describe ways in which solving a linear inequality is different than solving a linear equation.

107. What is a compound inequality and how is it solved?

108. Describe how to solve an absolute value inequality involving the symbol $<$. Give an example.

109. Describe how to solve an absolute value inequality involving the symbol $>$. Give an example.

110. Explain why $|x| < -4$ has no solution.

111. Describe the solution set of $|x| > -4$.

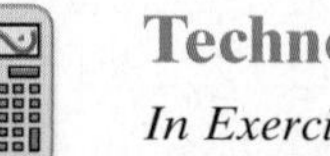

Technology Exercises

In Exercises 112–113, solve each inequality using a graphing utility. Graph each side separately. Then determine the values of x for which the graph on the left side lies above the graph on the right side.

112. $-3(x - 6) > 2x - 2$ **113.** $-2(x + 4) > 6x + 16$

Use the same technique employed in Exercises 112–113 to solve each inequality in Exercises 114–115. In each case, what conclusion can you draw? What happens if you try solving the inequalities algebraically?

114. $12x - 10 > 2(x - 4) + 10x$

115. $2x + 3 > 3(2x - 4) - 4x$

116. A bank offers two checking account plans. Plan A has a base service charge of \$4.00 per month plus 10¢ per check. Plan B charges a base service charge of \$2.00 per month plus 15¢ per check.

a. Write models for the total monthly costs for each plan if x checks are written.

b. Use a graphing utility to graph the models in the same viewing rectangle. Use a [0, 50, 1] by [0, 10, 1] viewing rectangle.

c. Use the graphs (and the TRACE or intersection feature) to determine for what number of checks per month plan A will be better than plan B.

d. Verify the result of part (c) algebraically by solving an inequality.

Critical Thinking Exercises

117. Which one of the following is true?

a. The first step in solving $|2x - 3| > -7$ is to rewrite the inequality as $2x - 3 > -7$ or $2x - 3 < 7$.

b. The smallest real number in the solution set of $2x > 6$ is 4.

c. All irrational numbers satisfy $|x - 4| > 0$.

d. None of these statements is true.

118. What's wrong with this argument? Suppose x and y represent two real numbers, where $x > y$:

$2 > 1$	This is a true statement.
$2(y - x) > 1(y - x)$	Multiply both sides by $y - x$.
$2y - 2x > y - x$	Use the distributive property.
$y > x$	Subtract y from both sides. Add $2x$ to both sides.

The final inequality, $y > x$, is impossible because we were initially given $x > y$.

119. The graphs of $y = 6$, $y = 3(-x - 5) - 9$, and $y = 0$ are shown in the figure. The graph was obtained using a graphing utility with x ranging from -12 to 1 and y ranging from -2 to 8. Use the graph to write the solution set for the compound inequality

$$0 < 3(-x - 5) - 9 < 6.$$

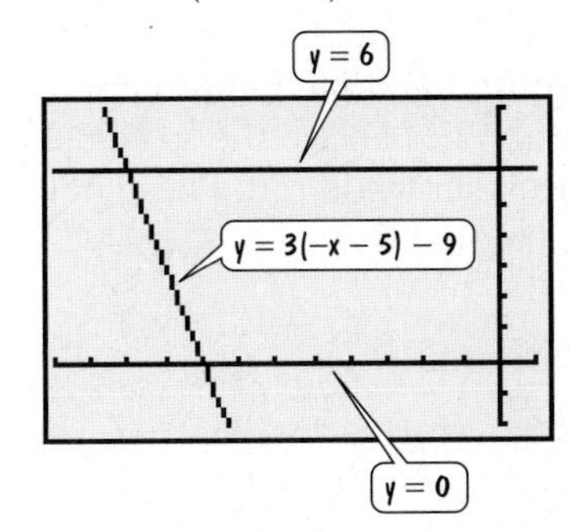

120. The percentage, p, of defective products manufactured by a company is given as $|p - 0.3\%| \leq 0.2\%$. If 100,000 products are manufactured and the company offers a \$5 refund for each defective product, describe the company's cost for refunds.

Group Exercise

121. Each group member should research one situation that provides two different pricing options. These can involve areas such as public transportation options (with or without coupon books) or long-distance telephone plans or anything of interest. Be sure to bring in all the details for each option. At a second group meeting, select the two pricing situations that are most interesting and relevant. Using each situation, write a word problem about selecting the better of the two options. The word problem should be one that can be solved using a linear inequality. The group should turn in the two problems and their solutions.

SECTION 1.6 Quadratic and Rational Inequalities

Objectives

1. Solve quadratic inequalities.
2. Solve rational inequalities.
3. Solve problems modeled by nonlinear inequalities.

Not afraid of heights and cutting-edge excitement? How about sky diving? Behind your exhilarating experience is the world of algebra. After you jump from the airplane, your height above the ground at every instant of your fall can be described by a formula involving a variable that is squared. At some point, you'll need to open your parachute. How can you determine when you must do so? Let x represent the number of seconds you are falling. You can compute when to open the parachute by solving an inequality that takes on the form $ax^2 + bx + c < 0$. Such an inequality is called a **quadratic inequality**.

Definition of a Quadratic Inequality

A **quadratic inequality** is any inequality that can be put in one of the forms

$$ax^2 + bx + c < 0 \qquad ax^2 + bx + c > 0$$
$$ax^2 + bx + c \le 0 \qquad ax^2 + bx + c \ge 0$$

where a, b, and c are real numbers and $a \neq 0$.

In this section we establish the basic techniques for solving quadratic inequalities. We will use these techniques to solve inequalities containing quotients, called **rational inequalities**. Finally, we will consider a formula that models the position of any free-falling object. As a sky diver, you could be that free-falling object!

1 Solve quadratic inequalities.

Solving Quadratic Inequalities

Graphs can help us to estimate the solutions of quadratic inequalities. The graph of $y = x^2 - 7x + 10$ is shown in Figure 1.8. The x-intercepts, 2 and 5, are **boundary points** between where the graph lies above the x-axis, shown in blue, and where the graph lies below the x-axis, shown in red. These boundary points play a critical role in solving quadratic inequalities.

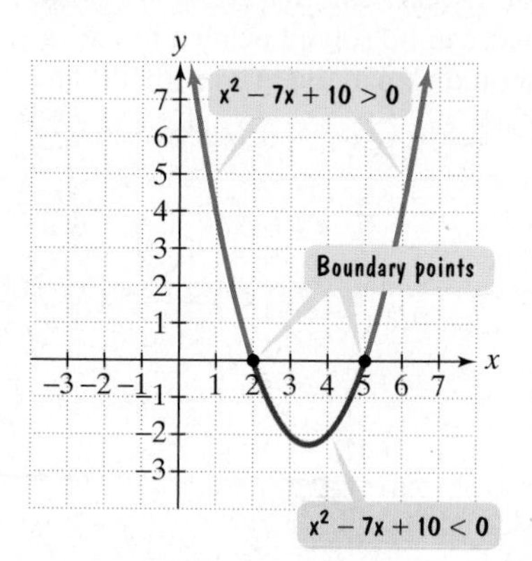

Figure 1.8 The graph of $y = x^2 - 7x + 10$. The blue parts of the graph lie above the x-axis: $x^2 - 7x + 10 > 0$. By contrast, the red part lies below the x-axis: $x^2 - 7x + 10 < 0$.

Procedure for Solving Quadratic Inequalities

1. Express the inequality in the standard form
$$ax^2 + bx + c > 0 \quad \text{or} \quad ax^2 + bx + c < 0.$$
2. Solve the equation $ax^2 + bx + c = 0$. The real solutions are the **boundary points**.
3. Locate these boundary points on a number line, thereby dividing the number line into **test intervals**.
4. Choose one representative number within each test interval. If substituting that value into the original inequality produces a true statement, then all real numbers in the test interval belong to the solution set. If substituting that value into the original inequality produces a false statement, then no real numbers in the test interval belong to the solution set.
5. Write the solution set, selecting the interval(s) that produced a true statement. The graph of the solution set on a number line usually appears as

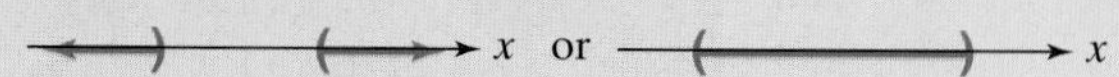

This procedure is valid if $<$ is replaced by $\le$ and $>$ is replaced by $\ge$.

EXAMPLE 1 Solving a Quadratic Inequality

Solve and graph the solution set on a real number line: $x^2 - 7x + 10 < 0$.

Solution

Step 1 Write the inequality in standard form. The inequality is given in this form, so this step has been done for us.

Step 2 Solve the related quadratic equation. This equation is obtained by replacing the inequality sign by an equal sign. Thus, we will solve $x^2 - 7x + 10 = 0$.

$$x^2 - 7x + 10 = 0 \quad \text{This is the related quadratic equation.}$$
$$(x - 2)(x - 5) = 0 \quad \text{Factor.}$$
$$x - 2 = 0 \quad \text{or} \quad x - 5 = 0 \quad \text{Set each factor equal to 0.}$$
$$x = 2 \quad \text{or} \quad x = 5 \quad \text{Solve for } x.$$

The boundary points are 2 and 5.

Step 3 Locate the boundary points on a number line. The number line with the boundary points is shown as follows:

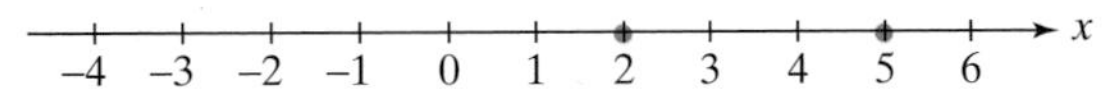

The boundary points divide the number line into three test intervals, namely $(-\infty, 2)$, $(2, 5)$, and $(5, \infty)$.

Step 4 Take one representative number within each test interval and substitute that number into the original inequality.

Test Interval	Representative Number	Substitute into $x^2 - 7x + 10 < 0$	Conclusion
$(-\infty, 2)$	0	$0^2 - 7 \cdot 0 + 10 \overset{?}{<} 0$ $10 < 0$, False	$(-\infty, 2)$ does not belong to the solution set.
$(2, 5)$	3	$3^2 - 7 \cdot 3 + 10 \overset{?}{<} 0$ $9 - 21 + 10 \overset{?}{<} 0$ $-2 < 0$, True	$(2, 5)$ belongs to the solution set.
$(5, \infty)$	6	$6^2 - 7 \cdot 6 + 10 \overset{?}{<} 0$ $36 - 42 + 10 \overset{?}{<} 0$ $4 < 0$, False	$(5, \infty)$ does not belong to the solution set.

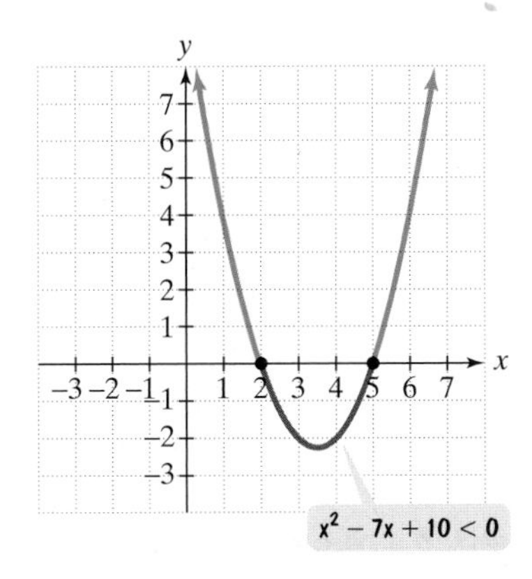

Figure 1.9 The graph lies below the x-axis between the boundary points 2 and 5, in the interval $(2, 5)$

Step 5 The solution set is the interval that produced a true statement. Our analysis shows that the solution set is the interval $(2, 5)$. The graph in Figure 1.9 confirms that $x^2 - 7x + 10 < 0$ (lies below the x-axis) in this interval. The graph of the solution set on a number line is shown as follows:

−4 −3 −2 −1 0 1 2 3 4 5 6 x

Check Point 1 Solve and graph the solution set:

$x^2 + 2x - 3 < 0$.

EXAMPLE 2 Solving a Quadratic Inequality

Solve and graph the solution set on a real number line: $2x^2 + x \geq 15$.

Solution

Step 1 Write the inequality in standard form. We can write $2x^2 + x \geq 15$ by subtracting 15 from both sides. This will give us zero on the right.

$$2x^2 + x - 15 \geq 15 - 15$$

$$2x^2 + x - 15 \geq 0$$

Step 2 Solve the related quadratic equation. This equation is obtained by replacing the inequality sign by an equal sign. Thus, we will solve $2x^2 + x - 15 = 0$.

$$2x^2 + x - 15 = 0 \quad \text{This is the related quadratic equation.}$$

$$(2x - 5)(x + 3) = 0 \quad \text{Factor.}$$

$$2x - 5 = 0 \quad \text{or} \quad x + 3 = 0 \quad \text{Set each factor equal to 0.}$$

$$x = \tfrac{5}{2} \quad \text{or} \quad x = -3 \quad \text{Solve for } x.$$

The boundary points are -3 and $\frac{5}{2}$.

Step 3 Locate the boundary points on a number line. The number line with the boundary points is shown as follows:

$-5 \quad -4 \quad -3 \quad -2 \quad -1 \quad 0 \quad 1 \quad 2 \quad 3 \quad 4 \quad 5 \quad x$ (points marked at -3 and $\frac{5}{2}$)

The boundary points divide the number line into three test intervals. Including the boundary points (because of the given greater than or equal to sign), the intervals are $(-\infty, -3]$, $[-3, \frac{5}{2}]$, and $[\frac{5}{2}, \infty)$.

Step 4 Take one representative number within each test interval and substitute that number into the original inequality.

Test Interval	Representative Number	Substitute into $2x^2 + x \geq 15$	Conclusion
$(-\infty, -3]$	-4	$2(-4)^2 + (-4) \stackrel{?}{\geq} 15$ $28 \geq 15$, True	$(-\infty, -3]$ belongs to the solution set.
$\left[-3, \frac{5}{2}\right]$	0	$2 \cdot 0^2 + 0 \stackrel{?}{\geq} 15$ $0 \geq 15$, False	$\left[-3, \frac{5}{2}\right]$ does not belong to the solution set.
$\left[\frac{5}{2}, \infty\right)$	3	$2 \cdot 3^2 + 3 \stackrel{?}{\geq} 15$ $21 \geq 15$, True	$\left[\frac{5}{2}, \infty\right)$ belongs to the solution set.

Step 5 The solution set are the intervals that produced a true statement. Our analysis shows that the solution set is

$$(-\infty, -3] \text{ or } \left[\frac{5}{2}, \infty\right).$$

Technology

The solution set for

$$2x^2 + x \geq 15$$

or, equivalently,

$$2x^2 + x - 15 \geq 0$$

can be verified with a graphing utility. The graph of $y = 2x^2 + x - 15$ was obtained using a $[-10, 10, 1]$ by $[-16, 6, 1]$ viewing rectangle. The graph lies above or on the x-axis, representing $\geq$, for all x in $(-\infty, -3]$ or $[\frac{5}{2}, \infty)$.

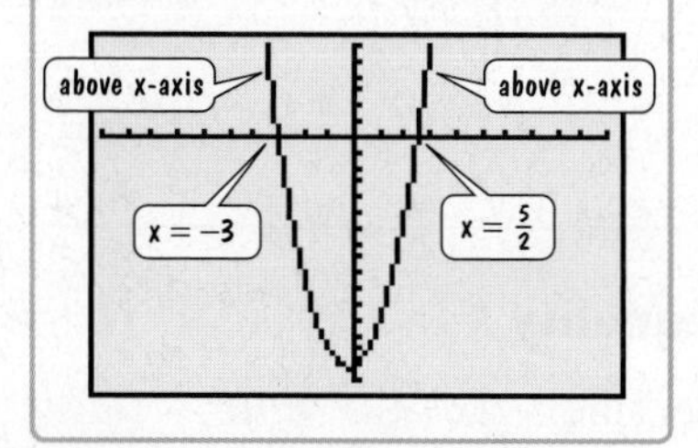

The graph of the solution set on a number line is shown as follows:

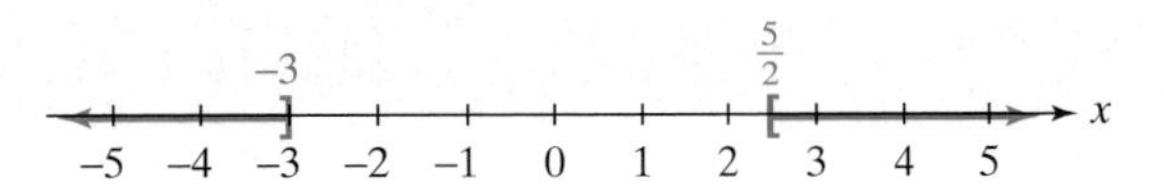

Check Point 2 Solve and graph the solution set: $x^2 - x \geq 20$.

2 Solve rational inequalities.

Solving Rational Inequalities

Inequalities that involve quotients can be solved in the same manner as quadratic inequalities. For example, the inequalities

$$(x + 3)(x - 7) > 0 \quad \text{and} \quad \frac{x + 3}{x - 7} > 0$$

are similar in that both are positive under the same conditions. To be positive, each of these inequalities must have two positive linear expressions

$$x + 3 > 0 \quad \text{and} \quad x - 7 > 0$$

or two negative linear expressions

$$x + 3 < 0 \quad \text{and} \quad x - 7 < 0.$$

Consequently, we use boundary points to divide the number line into test intervals. Then we select one representative number in each interval to determine whether that interval belongs to the solution set. Example 3 illustrates how this is done.

Study Tip

Many students want to solve

$$\frac{x + 3}{x - 7} > 0$$

by first multiplying both sides by $x - 7$ to clear fractions. This is incorrect. The problem is that $x - 7$ contains a variable and can be positive or negative, depending on the value of x. Thus, we do not know whether or not to reverse the sense of the inequality.

EXAMPLE 3 Using Test Numbers to Solve a Rational Inequality

Solve and graph the solution set: $\frac{x + 3}{x - 7} > 0$.

Solution We begin by finding values of x that make the numerator and denominator 0.

$x + 3 = 0 \qquad x - 7 = 0$ — Set the numerator and denominator equal to 0.

$x = -3 \qquad x = 7$ — Solve.

The boundary points are −3 and 7. We locate these numbers on a number line as follows:

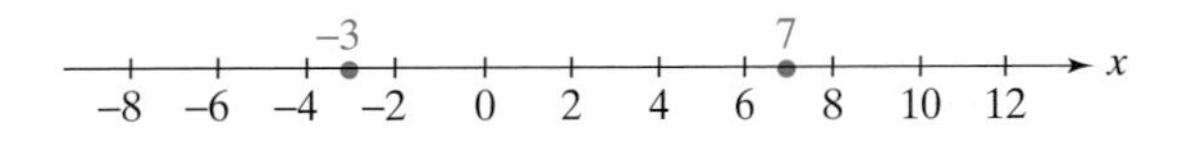

These boundary points divide the number line into three test intervals, namely $(-\infty, -3)$, $(-3, 7)$, and $(7, \infty)$. Now, we take one representative number from each test interval and substitute that number into the original inequality.

Test Interval	Representative Number	Substitute into $\frac{x+3}{x-7} > 0$	Conclusion
$(-\infty, -3)$	-4	$\frac{-4+3}{-4-7} \overset{?}{>} 0$ $\frac{-1}{-11} \overset{?}{>} 0$ $\frac{1}{11} > 0$, True	$(-\infty, -3)$ belongs to the solution set.
$(-3, 7)$	0	$\frac{0+3}{0-7} \overset{?}{>} 0$ $-\frac{3}{7} > 0$, False	$(-3, 7)$ does not belong to the solution set.
$(7, \infty)$	8	$\frac{8+3}{8-7} \overset{?}{>} 0$ $11 > 0$, True	$(7, \infty)$ belongs to the solution set.

Our analysis shows that the solution set is

$$(-\infty, -3) \quad \text{or} \quad (7, \infty).$$

The graph of the solution set on a number line is shown as follows:

−3 7
−8 −6 −4 −2 0 2 4 6 8 10 12 x

Check Point 3 Solve and graph the solution set: $\frac{x-5}{x+2} > 0$.

The first step in solving a rational inequality is to bring all terms to one side, obtaining zero on the other side. Then express the nonzero side as a single quotient. At this point, we follow the same procedure as in Example 3.

EXAMPLE 4 Solving a Rational Inequality

Solve and graph the solution set: $\frac{x+1}{x+3} \leq 2$.

Solution

Step 1 Express the inequality so that one side is zero and the other side is a single quotient. We subtract 2 from both sides to obtain zero on the right.

$$\frac{x+1}{x+3} \leq 2 \qquad \text{This is the given inequality.}$$

$$\frac{x+1}{x+3} - 2 \leq 0 \qquad \text{Subtract 2 from both sides, obtaining 0 on the right.}$$

Study Tip

Do not begin solving

$$\frac{x+1}{x+3} \leq 2$$

by multiplying both sides by $x + 3$. We do not know if $x + 3$ is positive or negative. Thus, we do not know whether or not to reverse the sense of the inequality.

$$\frac{x+1}{x+3} - \frac{2(x+3)}{x+3} \le 0$$ The least common denominator is $x + 3$. Express 2 in terms of this denominator.

$$\frac{x+1-2(x+3)}{x+3} \le 0$$ Subtract rational expressions.

$$\frac{x+1-2x-6}{x+3} \le 0$$ Apply the distributive property.

$$\frac{-x-5}{x+3} \le 0$$ Simplify.

Step 2 Find boundary points by setting the numerator and the denominator equal to zero.

$$-x - 5 = 0 \qquad x + 3 = 0$$ Set the numerator and denominator equal to 0. These are the values that make the previous quotient zero or undefined.

$$x = -5 \qquad x = -3$$ Solve for x.

The boundary points are −5 and −3. Because equality is included in the given less-than-or-equal-to symbol, we include the value of x that causes the quotient $\frac{-x-5}{x+3}$ to be zero. Thus, −5 is included in the solution set. By contrast, we do not include −3 in the solution set because −3 makes the denominator zero.

Step 3 Locate boundary points on a number line. The number line, with the boundary points, is shown as follows:

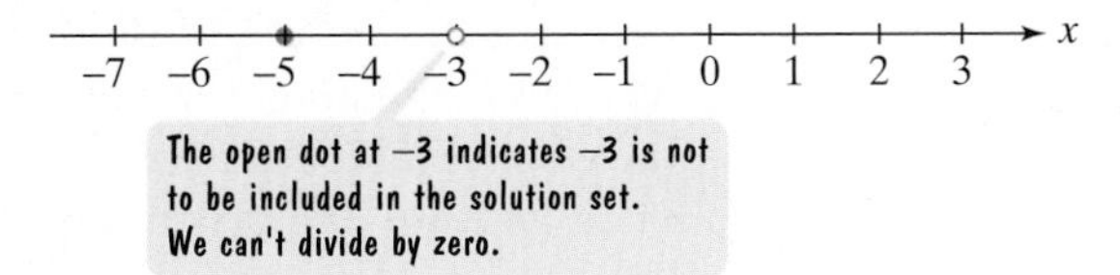

The boundary points divide the number line into three test intervals, namely $(-\infty, -5]$, $[-5, -3)$, and $(-3, \infty)$.

Step 4 Take one representative number within each test interval and substitute that number into the original inequality.

Test Interval	Representative Number	Substitute into $\frac{x+1}{x+3} \le 2$	Conclusion
$(-\infty, -5]$	−6	$\frac{-6+1}{-6+3} \overset{?}{\le} 2$ $\frac{5}{3} \le 2$, True	$(-\infty, -5]$ belongs to the solution set.
$[-5, -3)$	−4	$\frac{-4+1}{-4+3} \overset{?}{\le} 2$ $3 \le 2$, False	$[-5, -3)$ does not belong to the solution set.
$(-3, \infty)$	0	$\frac{0+1}{0+3} \overset{?}{\le} 2$ $\frac{1}{3} \le 2$, True	$(-3, \infty)$ belongs to the solution set.

Discovery

Because $(x + 3)^2$ is positive, it is possible so solve

$$\frac{x+1}{x+3} \le 2$$

by first multiplying both sides by $(x + 3)^2$ (where $x \ne -3$). This will not reverse the sense of the inequality and will clear the fraction. Try using this solution method and compare it to the one on pages 156–158.

Step 5 The solution set consists of the intervals that produced a true statement. Our analysis shows that the solution set is

$$(-\infty, -5] \quad \text{or} \quad (-3, \infty).$$

The graph of the solution set on a number line is shown as follows:

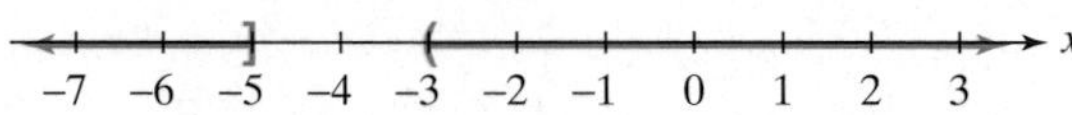

Check Point 4 Solve and graph the solution set: $\frac{2x}{x+1} \le 1.$

Applications

3 Solve problems modeled by nonlinear inequalities.

Intriguing signs point out that the world is profoundly mathematical. For example, did you know that every time you throw an object vertically upward, its changing height above the ground can be described by a mathematical formula? The same formula can be used to describe objects that are falling, such as the sky divers shown in the opening to this section.

The Position Formula for a Free-Falling Object Near Earth's Surface

An object that is falling or vertically projected into the air has its height in feet above the ground given by

$$s = -16t^2 + v_0 t + s_0$$

where s is the height in feet, v_0 is the original velocity (initial velocity) of the object in feet per second, t is the time that the object is in motion in seconds, and s_0 is the original height (initial height) of the object in feet.

In Example 5, we solve a quadratic inequality in a problem about the position of a free-falling object.

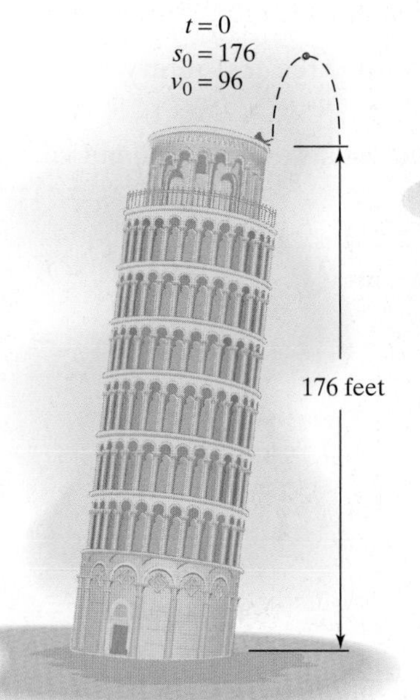

Figure 1.10 Throwing a ball from 176 feet with a velocity of 96 feet per second

EXAMPLE 5 Using the Position Model

A ball is thrown vertically upward from the top of the Leaning Tower of Pisa (176 feet high) with an initial velocity of 96 feet per second (Figure 1.10). During which time period will the ball's height exceed that of the tower?

Solution

$s = -16t^2 + v_0 t + s_0$ This is the position formula for a free-falling object.

$s = -16t^2 + 96t + 176$ Because v_0 (initial velocity) $= 96$ and s_0 (initial position) $= 176$, substitute these values into the formula.

When will the ball's height | exceed that | of the tower?

$$-16t^2 + 96t + 176 \quad > \quad 176$$

$-16t^2 + 96t + 176 > 176$ This is the inequality implied by the problem's question. We must find t.

$-16t^2 + 96t > 0$ Subtract 176 from both sides.

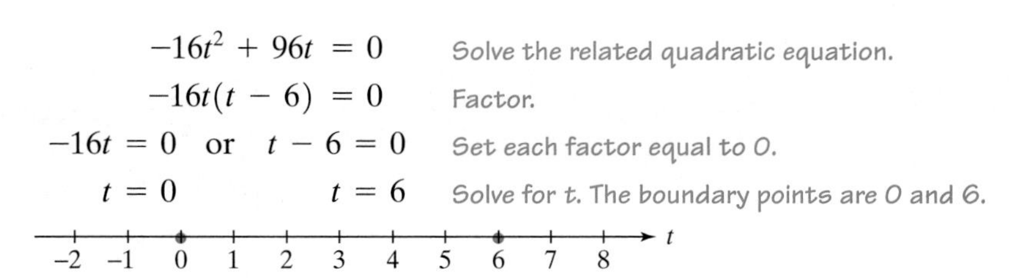

Locate these values on a number line, with $t \geq 0$.

The intervals are $(0, 6)$ and $(6, \infty)$, although the time interval should not extend to infinity but rather to the value of t when the ball hits the ground. (By setting $-16t^2 + 96t + 176$ equal to zero, we find $t \approx 7.47$; the ball hits the ground after approximately 7.47 seconds.)

We use $(0, 6)$ and $(6, 7.47)$ for our test intervals.

Test Interval	Representative Number	Substitute into $-16t^2 + 96t > 0$	Conclusion
$(0, 6)$	1	$-16 \cdot 1^2 + 96 \cdot 1 \overset{?}{>} 0$ $80 > 0$, True	$(0, 6)$ belongs to the solution set.
$(6, 7.47)$	7	$-16 \cdot 7^2 + 96 \cdot 7 \overset{?}{>} 0$ $-112 > 0$, False	$(6, 7.47)$ does not belong to the solution set.

The ball's height exceeds that of the tower between 0 and 6 seconds, excluding $t = 0$ and $t = 6$.

An object is propelled straight up from ground level with an initial velocity of 80 feet per second. Its height at time t is described by

$$s = -16t^2 + 80t,$$

where the height, s, is measured in feet and the time, t, is measured in seconds. In which time interval will the object be more than 64 feet above the ground?

EXERCISE SET 1.6

Practice Exercises

Solve each quadratic inequality in Exercises 1–26, and graph the solution set on a real number line. Express each solution set in interval notation.

1. $(x - 4)(x + 2) > 0$
2. $(x + 3)(x - 5) > 0$
3. $(x - 7)(x + 3) \leq 0$
4. $(x + 1)(x - 7) \leq 0$
5. $x^2 - 5x + 4 > 0$
6. $x^2 - 4x + 3 < 0$
7. $x^2 + 5x + 4 > 0$
8. $x^2 + x - 6 > 0$
9. $x^2 - 6x + 9 < 0$
10. $x^2 - 2x + 1 > 0$
11. $x^2 - 6x + 8 \leq 0$
12. $x^2 - 2x - 3 \geq 0$
13. $3x^2 + 10x - 8 \leq 0$
14. $9x^2 + 3x - 2 \geq 0$
15. $2x^2 + x < 15$
16. $6x^2 + x > 1$
17. $4x^2 + 7x < -3$
18. $3x^2 + 16x < -5$
19. $5x \leq 2 - 3x^2$
20. $4x^2 + 1 \geq 4x$
21. $x^2 - 4x \geq 0$
22. $x^2 + 2x < 0$
23. $2x^2 + 3x > 0$
24. $3x^2 - 5x \leq 0$
25. $-x^2 + x \geq 0$
26. $-x^2 + 2x \geq 0$

Solve each rational inequality in Exercises 27–42, and graph the solution set on a real number line. Express each solution set in interval notation.

27. $\dfrac{x - 4}{x + 3} > 0$
28. $\dfrac{x + 5}{x - 2} > 0$

29. $\frac{x+3}{x+4} < 0$

30. $\frac{x+5}{x+2} < 0$

31. $\frac{-x+2}{x-4} \geq 0$

32. $\frac{-x-3}{x+2} \leq 0$

33. $\frac{4-2x}{3x+4} \leq 0$

34. $\frac{3x+5}{6-2x} \geq 0$

35. $\frac{x}{x-3} > 0$

36. $\frac{x+4}{x} > 0$

37. $\frac{x+1}{x+3} < 2$

38. $\frac{x}{x-1} > 2$

39. $\frac{x+4}{2x-1} \leq 3$

40. $\frac{1}{x-3} < 1$

41. $\frac{x-2}{x+2} \leq 2$

42. $\frac{x}{x+2} \geq 2$

Application Exercises

Use the position formula

$$s = -16t^2 + v_0t + s_0$$

$(v_0 = \text{initial velocity}, s_0 = \text{initial position}, t = \text{time})$

to answer Exercises 43–46.

43. A projectile is fired straight upward from ground level with an initial velocity of 80 feet per second. During which interval of time will the projectile's height exceed 96 feet?

44. A projectile is fired straight upward from ground level with an initial velocity of 128 feet per second. During which interval of time will the projectile's height exceed 128 feet?

45. A ball is thrown upward with a velocity of 64 feet per second from the top edge of a building 80 feet high. For how long is the ball higher than 96 feet?

46. A diver leaps into the air at 20 feet per second from a diving board that is 10 feet above the water. For how many seconds is the diver at least 12 feet above the water?

47. The formula

$$H = \frac{15}{8}x^2 - 30x + 200$$

describes heart rate, H, in beats per minute, x minutes after a strenuous workout.

a. What is the heart rate immediately following the workout?

b. Describe the interval of time after a strenuous workout in which heart rate exceeds 110 beats per minute.

48. The data in the bar graph at the top of the next column can be modeled by the formula

$$y = 0.6x^2 - 2.9x + 3.2$$

where y represents the number of U.S. cellular telephone subscribers, in hundreds of thousands, x years after 1985. When will the number of cellular subscribers exceed 185,200,000 or 185.2 hundred thousand?

U.S. Cellular Telephone Subscribers

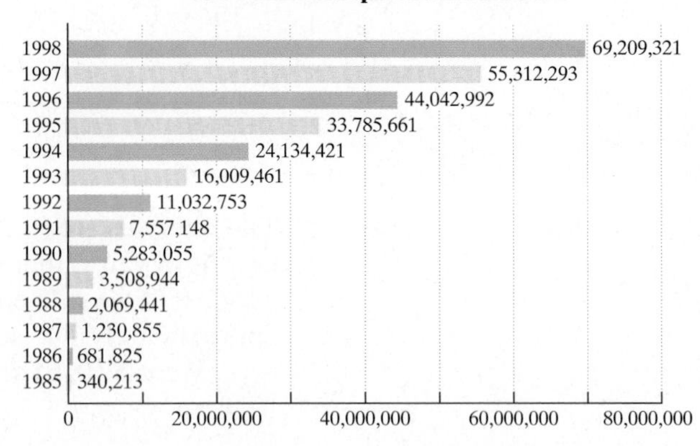

Source: Cellular Telephone Industry of America

49. The data in the bar graph shown below can be modeled by the formula

$$y = -0.22x^2 + 4.32x + 26$$

where y represents the number of international visitors, in millions, x years after 1986. According to this model, when will the number of international visitors to the United States be less than 41.3 million?

International Visitors to the United States

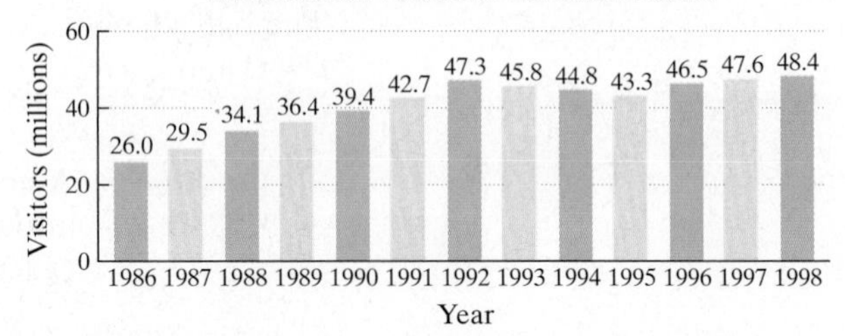

Source: U.S. Department of Commerce

50. The average cost per unit, $(\bar{C})$, of producing x units of a product is modeled by

$$\bar{C} = \frac{150{,}000 + 0.25x}{x}.$$

How many units must be produced so that the average cost of producing each unit does not exceed \$1.75?

51. The cost of removing $p\%$ of the bacteria from a river is given by the formula

$$C = \frac{4p}{100 - p}$$

where C is measured in hundreds of thousands of dollars. If less than \$600,000 is spent, what percentage of the bacteria can be removed?

Writing in Mathematics

52. What is a quadratic inequality?

53. What is a rational inequality?

54. Describe similarities and differences between the solutions of

$$(x - 2)(x + 5) \geq 0 \quad \text{and} \quad \frac{x - 2}{x + 5} \geq 0.$$

Technology Exercises

Solve each inequality in Exercises 55–60 using a graphing utility.

55. $x^2 + 3x - 10 > 0$

56. $2x^2 + 5x - 3 \leq 0$

57. $\dfrac{x - 4}{x - 1} \leq 0$

58. $\dfrac{x + 2}{x - 3} \leq 2$

59. $\dfrac{1}{x + 1} \leq \dfrac{2}{x + 4}$

60. $x^3 + 2x^2 - 5x - 6 > 0$

61. In a study of the winter moth in Nova Scotia, the number of eggs, N, in a female moth depended on her abdominal width, W, in millimeters, approximated by $N = 14W^3 - 17W^2 - 6W + 34$, where $1.5 \leq W \leq 3.5$. Graph the model on your graphing utility and use the TRACE and ZOOM features to describe the abdominal width of a moth with more than 46 eggs.

Critical Thinking Exercises

62. Which one of the following is true?

a. The solution set to $x^2 > 25$ is $(5, \infty)$.

b. The inequality $\dfrac{x - 2}{x + 3} < 2$ can be solved by multiplying both sides by $x + 3$, resulting in the equivalent inequality $x - 2 < 2(x + 3)$.

c. $(x + 3)(x - 1) \geq 0$ and $\dfrac{x + 3}{x - 1} \geq 0$ have the same solution set.

d. None of these statements is true.

63. Write a quadratic inequality whose solution set is $[-3, 5]$.

64. Write a rational inequality whose solution set is $(-\infty, -4)$ or $[3, \infty)$.

In Exercises 65–68, use inspection to describe each inequality's solution set. Do not solve any of the inequalities.

65. $(x - 2)^2 > 0$

66. $(x - 2)^2 \leq 0$

67. $(x - 2)^2 < -1$

68. $\dfrac{1}{(x - 2)^2} > 0$

69. The graphing calculator screen at the top of the next column shows the graph of $y = 4x^2 - 8x + 7$.

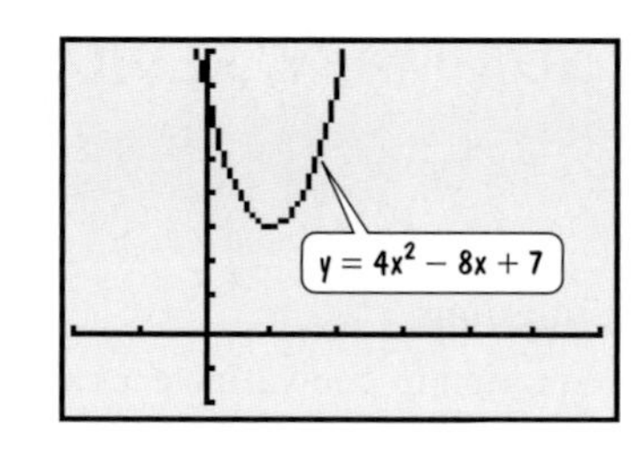

a. Use the graph to describe the solution set for $4x^2 - 8x + 7 > 0$.

b. Use the graph to describe the solution set for $4x^2 - 8x + 7 < 0$.

c. Use an algebraic approach to verify each of your descriptions in parts (a) and (b).

70. The graphing calculator screen shows the graph of $y = \sqrt{27 - 3x^2}$. Write and solve a quadratic inequality that explains why the graph only appears for $-3 \leq x \leq 3$.

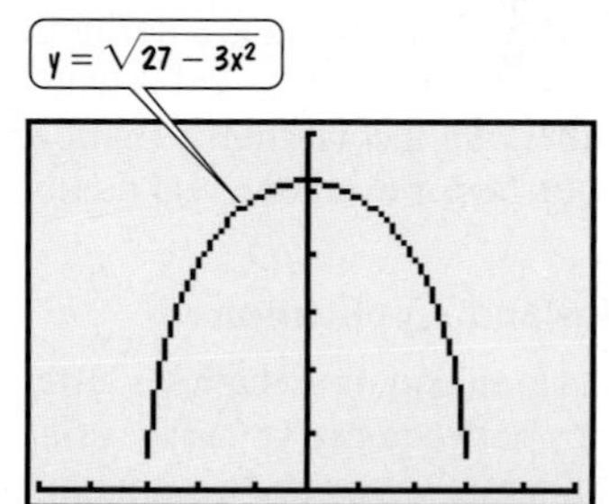

Group Exercise

71. This exercise is intended as a group learning experience and is appropriate for groups of three to five people. Before working on the various parts of the problem, reread the description of the position formula on page 158.

a. Drop a ball from a height of 3 feet, 6 feet, and 12 feet. Record the number of seconds it takes for the ball to hit the ground.

b. For each of the three initial positions, use the position formula to determine the time required for the ball to hit the ground.

c. What factors might result in differences between the times that you recorded and the times indicated by the formula?

d. What appears to be happening to the time required for a free-falling object to hit the ground as its initial height is doubled? Verify this observation algebraically and with a graphing utility.

e. Repeat part (a) using a sheet of paper rather than a ball. What differences do you observe? What factor seems to be ignored in the position formula?

f. What is meant by the acceleration of gravity and how does this number appear in the position formula for a free-falling object?

CHAPTER SUMMARY, REVIEW, AND TEST

Summary

1.1 Linear Equations

a. A linear equation in one variable x can be written in the form $ax + b = 0, a \neq 0$.

b. The procedure for solving a linear equation is given in the box on page 88.

c. If an equation contains fractions, begin by multiplying both sides by the least common denominator, thereby clearing fractions.

d. If an equation contains rational expressions with variable denominators, avoid in the solution set any values of the variable that make a denominator zero.

e. An identity is an equation that is true for all real numbers for which both sides are defined. A conditional equation is not an identity and is true for at least one real number. An inconsistent equation is an equation that is not true for even one real number.

1.2 Formulas and Applications

a. A formula is an equation that uses letters to express relationships between two or more variables.

b. A mathematical model is a formula or algebraic equation that can be formed from a verbal model or from actual data.

c. A five-step procedure for solving problems using mathematical models is given in the box on page 99.

1.3 Quadratic Equations

a. A quadratic equation in x can be written in the standard form $ax^2 + bx + c = 0, a \neq 0$.

b. The procedure for solving a quadratic equation by factoring and the zero-product principle is given in the box on pages 110–111.

c. The procedure for solving a quadratic equation by the square root method is given in the box on page 113.

d. All quadratic equations can be solved by completing the square. Isolate the binomial with the two variable terms on one side of the equation. If the coefficient of the x^2 term is not one, divide each side of the equation by this coefficient. Then add the square of half the coefficient of x to both sides.

e. All quadratic equations can be solved by the quadratic formula

$$x = \frac{-b \pm \sqrt{b^2 - 4ac}}{2a}.$$

The formula is derived by completing the square of the equation $ax^2 + bx + c = 0$.

f. The discriminant, $b^2 - 4ac$, indicates the kinds of solutions to the quadratic equation $ax^2 + bx + c = 0$, shown in Table 1.3 on page 119.

g. Table 1.4 on page 120 shows the most efficient technique to use when solving a quadratic equation.

1.4 Other Types of Equations

a. Some polynomial equations of degree 3 or greater can be solved by moving all terms to one side, obtaining zero on the other side, factoring, and using the zero-product principle. Factoring by grouping is often used.

b. A radical equation is an equation in which the variable occurs in a square root, cube root, and so on. A radical equation can be solved by isolating the radical and raising both sides of the equation to a power equal to the radical's index. When raising both sides to an even power, check all proposed solutions in the original equation. Eliminate extraneous solutions from the solution set.

c. A radical equation with rational exponents can be solved by isolating the expression with the rational exponent and raising both sides of the equation to a power that is the reciprocal of the rational exponent. Check for possible extraneous solutions.

d. An equation is quadratic in form if it can be written in the form $at^2 + bt + c = 0$, where t is an algebraic expression and $a \neq 0$. Solve for t and use the substitution that resulted in this equation to find the values for the variable in the given equation.

e. Absolute value equations in the form $|X| = c, c > 0$, can be solved by rewriting the equation without absolute value bars: $X = c$ or $X = -c$.

1.5 Linear Inequalities

a. A linear inequality in one variable x can be expressed as $ax + b \leq c$, $ax + b < c$, $ax + b \geq c$, or $ax + b > c$, $a \neq 0$.

b. Graphs of solutions to inequalities are shown on a number line by shading all points representing numbers that are solutions. Parentheses exclude endpoints and square brackets include endpoints.

c. Solution sets to inequalities can be expressed in set-builder or interval notation. Table 1.5 on page 141 compares the notations.

d. A linear inequality is solved using a procedure similar to solving a linear equation. However, when multiplying or dividing by a negative number, reverse the sense of the inequality.

e. A compound inequality with three parts can be solved by isolating x in the middle.

f. Inequalities involving absolute value can be solved by rewriting the inequalities without absolute value bars. The ways to do this are shown in the box on page 145.

1.6 Quadratic and Rational Inequalities

a. A quadratic inequality can be expressed as $ax^2 + bx + c < 0$, $ax^2 + bx + c > 0$, $ax^2 + bx + c \leq 0$, or $ax^2 + bx + c \geq 0$, $a \neq 0$.

b. A procedure for solving quadratic inequalities is given in the box on page 152.

c. Inequalities involving quotients are called rational inequalities. The procedure for solving such inequalities begins with expressing them so that one side is zero and the other side is a single quotient. Find boundary points by setting the numerator and denominator equal to zero. Then follow a procedure similar to that for solving quadratic inequalities.

Review Exercises

1.1

In Exercises 1–6, solve and check each linear equation.

1. $2x - 5 = 7$

2. $5x + 20 = 3x$

3. $7(x - 4) = x + 2$

4. $1 - 2(6 - x) = 3x + 2$

5. $2(x - 4) + 3(x + 5) = 2x - 2$

6. $2x - 4(5x + 1) = 3x + 17$

Exercises 7–11 contain equations with constants in denominators. Solve each equation and check by the method of your choice.

7. $\frac{2x}{3} = \frac{x}{6} + 1$

8. $\frac{x}{2} - \frac{1}{10} = \frac{x}{5} + \frac{1}{2}$

9. $\frac{2x}{3} = 6 - \frac{x}{4}$

10. $\frac{x}{4} = 2 + \frac{x - 3}{3}$

11. $\frac{3x + 1}{3} - \frac{13}{2} = \frac{1 - x}{4}$

Exercises 12–15 contain equations with variables in denominators. a. List the value or values representing restriction(s) on the variable. b. Solve the equation.

12. $\frac{9}{4} - \frac{1}{2x} = \frac{4}{x}$

13. $\frac{7}{x - 5} + 2 = \frac{x + 2}{x - 5}$

14. $\frac{1}{x - 1} - \frac{1}{x + 1} = \frac{2}{x^2 - 1}$

15. $\frac{4}{x + 2} + \frac{2}{x - 4} = \frac{30}{x^2 - 2x - 8}$

In Exercises 16–18, determine whether each equation is an identity, a conditional equation, or an inconsistent equation.

16. $\frac{1}{x + 5} = 0$

17. $7x + 13 = 4x - 10 + 3x + 23$

18. $7x + 13 = 3x - 10 + 2x + 23$

19. The formula $M = 420x + 720$ models the data for the amount of money lost to credit card fraud worldwide, M, expressed in millions of dollars, x years after 1989. In which year did losses amount to 4080 million dollars?

1.2

20. Suppose you were to list in order, from least to most, the family income for every U.S. family. The median income is the income in the middle of this list of ranked data. This income can be modeled by the formula

$$I = 1321.7(x - 1980) + 21{,}153.$$

In this formula, I represents median family income in the United States and x is the actual year, beginning in 1980. When was the median income \$47,587?

In Exercises 21–27, use the five-step strategy given in the box on page 99 to solve each problem.

21. The bus fare in a city is \$1.50. People who use the bus have the option of purchasing a monthly coupon book for \$25.00. With the coupon book, the fare is reduced to \$0.25. Determine the number of times in a month the bus must be used so that the total monthly cost without the coupon book is the same as the total monthly cost with the coupon book.

22. Los Angeles has more unhealthy air days per year than any other U.S. city. On average, the number of unhealthy air days per year in Los Angeles exceeds five times that of New York City by 29 days. If Los Angeles and New York combined have 185 unhealthy air days per year, determine the number of unhealthy days for the two cities.

23. You inherit \$10,000 with the stipulation that for the first year the money must be invested in two stocks paying 8% and 12% annual interest, respectively. How much should be invested at each rate if the total interest earned for the year is to be \$950?

24. The length of a rectangular football field is 14 meters more than twice the width. If the perimeter is 346 meters, find the field's dimensions.

25. A salesperson earns \$300 per week plus 5% commission of sales. How much must be sold to earn \$800 in a week?

26. After a 45% price reduction, a VCR sold for \$247.50. What was the price before the reduction?

27. A study entitled *Performing Arts—The Economic Dilemma* documents the relationship between the number of concerts given by a major orchestra and the attendance per concert. For each additional concert given per year, attendance per concert drops by approximately eight people. If 50 concerts are given, attendance per concert is 2987 people. How many concerts should be given to ensure an audience of 2627 people at each concert?

In Exercises 28–29, solve each formula for the specified variable.

28. $V = \frac{1}{3}Bh$ for h

29. $F = f(1 - M)$ for M

1.3

Solve each equation in Exercises 30–31 by factoring and then using the zero-product principle.

30. $2x^2 + 15x = 8$

31. $5x^2 + 20x = 0$

Solve each equation in Exercises 32–33 by the square root method.

32. $2x^2 - 3 = 125$

33. $(3x - 4)^2 = 18$

In Exercises 34–35, determine the constant that should be added to the binomial so that it becomes a perfect square trinomial. Then factor the trinomial.

34. $x^2 + 20x$

35. $x^2 - 3x$

Solve each equation in Exercises 36–37 by completing the square.

36. $x^2 - 12x + 27 = 0$

37. $3x^2 - 12x + 11 = 0$

Solve each equation in Exercises 38–40 using the quadratic formula.

38. $x^2 = 2x + 4$

39. $x^2 - 2x + 19 = 0$

40. $2x^2 = 3 - 4x$

Compute the discriminant of each equation in Exercises 41–42. What does the discriminant indicate about the kinds of solutions?

41. $x^2 - 4x + 13 = 0$

42. $9x^2 = 2 - 3x$

Solve each equation in Exercises 43–48 by the method of your choice.

43. $2x^2 - 11x + 5 = 0$

44. $(3x + 5)(x - 3) = 5$

45. $3x^2 - 7x + 1 = 0$

46. $x^2 - 9 = 0$

47. $(x - 3)^2 - 25 = 0$

48. $3x^2 - x + 2 = 0$

49. The weight of a human fetus is modeled by the formula $W = 3t^2$, where W is the weight in grams and t is the time in weeks, $0 \leq t \leq 39$. After how many weeks does the fetus weigh 1200 grams?

50. The formula $N = 0.337x^2 - 2.265x + 3.962$ models the number of mountain bike owners N, in millions, in the United States x years after 1980, where $3 \leq x \leq 20$. In which year (to the nearest year) did 10.9 million Americans own mountain bikes?

51. A billboard is 15 feet longer than it is high and has space for 324 square feet of advertising. What are the billboard's dimensions?

52. A building casts a shadow that is double the length of its height. If the distance from the end of the shadow to the top of the building is 300 meters, how high is the building?

1.4

Solve each polynomial equation in Exercises 53–54.

53. $2x^4 = 50x^2$

54. $2x^3 - x^2 - 18x + 9 = 0$

Solve each radical equation in Exercises 55–56.

55. $\sqrt{2x - 3} + x = 3$

56. $\sqrt{x - 4} + \sqrt{x + 1} = 5$

Solve the equations with rational exponents in Exercises 57–58.

57. $3x^{3/4} - 24 = 0$

58. $(x - 7)^{3/2} = 125$

Solve each equation in Exercises 59–60 by making an appropriate substitution.

59. $x^4 - 5x^2 + 4 = 0$

60. $x^{1/2} + 3x^{1/4} - 10 = 0$

Solve the equations containing absolute value in Exercises 61–62.

61. $|2x + 1| = 7$

62. $2|x - 3| - 7 = 10$

Solve each equation in Exercises 63–66 by the method of your choice.

63. $3x^{4/3} - 5x^{2/3} + 2 = 0$

64. $2\sqrt{x - 1} = x$

65. $|2x - 5| - 3 = 0$

66. $x^3 + 2x^2 = 9x + 18$

67. The distance to the horizon that you can see, measured in miles, on the top of a mountain H feet high is modeled by the formula $D = \sqrt{2H}$. You've hiked to the top of a mountain with views extending 50 miles to the horizon. How high is the mountain?

1.5

In Exercises 68–70, graph the solutions of each inequality on a number line.

68. $x > 5$

69. $x \leq 1$

70. $-3 \leq x < 0$

In Exercises 71–73, express each interval in terms of an inequality, and graph the interval on a number line.

71. $(-2, 3]$

72. $[-1.5, 2]$

73. $(-1, \infty)$

Solve each linear inequality in Exercises 74–79 and graph the solution set on a number line.

74. $-6x + 3 \leq 15$

75. $6x - 9 \geq -4x - 3$

76. $\frac{x}{3} - \frac{3}{4} - 1 > \frac{x}{2}$

77. $6x + 5 > -2(x - 3) - 25$

78. $3(2x - 1) - 2(x - 4) \geq 7 + 2(3 + 4x)$

79. $7 < 2x + 3 \leq 9$

Solve each inequality in Exercises 80–82 by first rewriting each one as an equivalent inequality without absolute value bars. Graph the solution set on a number line.

80. $|2x + 3| \leq 15$

81. $\left|\frac{2x + 6}{3}\right| > 2$

82. $|2x + 5| - 7 \geq -6$

The graph indicates that the United States has the world's highest incarceration rate. If x represents the incarceration rate per 100,000 population, list the country or countries that satisfy each inequality in Exercises 83–84.

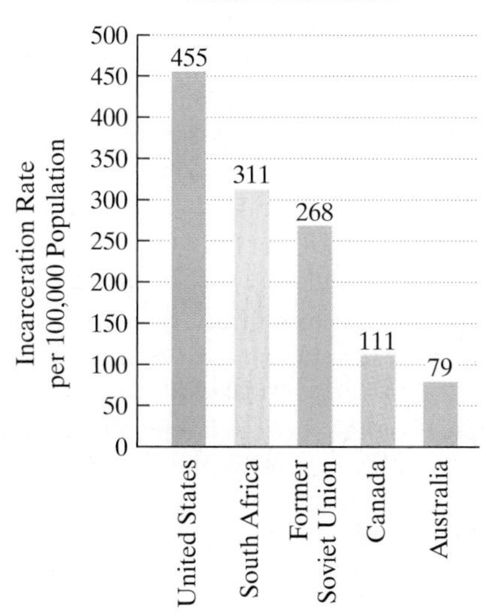

Source: FBI

83. $437 \le 4x - 7 \le 1229$ **84.** $|x - 320| > 80$

85. Approximately 90% of the population sleeps h hours daily, where h is modeled by the inequality $|h - 6.5| \le 1$. Write a sentence describing the range for the number of hours that most people sleep. Do *not* use the phrase "absolute value" in your description.

86. The formula for converting Fahrenheit temperature (F) to Celsius temperature (C) is $C = \frac{5}{9}(F - 32)$. If Celsius temperature ranges from 15° to 35°, inclusively, what is the range for the Fahrenheit temperature?

87. A person can choose between two charges on a checking account. The first method involves a fixed cost of $11 per month plus 6¢ for each check written. The second method involves a fixed cost of $4 per month plus 20¢ for each check written. How many checks should be written to make the first method a better deal?

88. In 1984, approximately 1644 thousand turntables were sold in the United States. This number has decreased by around 82 thousand each year since then. By contrast, annual sales of compact disc players has increased by 496 thousand each year, with approximately 284 thousand units sold in 1984. What was the first year in which the sales of compact disc players exceeded those of turntables?

1.6

Solve each quadratic inequality in Exercises 89–90, and graph the solution set on a real number line. Express each solution set in interval notation.

89. $2x^2 + 7x \le 4$ **90.** $2x^2 > 6x - 3$

Solve each rational inequality in Exercises 91–92, and graph the solution set on a real number line. Express each solution set in interval notation.

91. $\frac{x - 6}{x + 2} > 0$ **92.** $\frac{x + 3}{x - 4} \le 5$

93. Use the position formula

$$s = -16t^2 + v_0 t + s_0$$

initial velocity (v_0) initial height (s_0)

to solve this problem. A projectile is fired vertically upward from ground level with an initial velocity of 48 feet per second. During which time period will the projectile's height exceed 32 feet?

Chapter 1 Test

Find the solution set for each equation in Exercises 1–13.

1. $7(x - 2) = 4(x + 1) - 21$

2. $\frac{2x - 3}{4} = \frac{x - 4}{2} - \frac{x + 1}{4}$

3. $\frac{2}{x - 3} - \frac{4}{x + 3} = \frac{8}{x^2 - 9}$

4. $2x^2 - 3x - 2 = 0$ **5.** $(3x - 1)^2 = 75$

6. $x(x - 2) = 4$ **7.** $4x^2 = 8x - 5$

8. $x^3 - 4x^2 - x + 4 = 0$ **9.** $\sqrt{x - 3} + 5 = x$

10. $\sqrt{x + 4} + \sqrt{x - 1} = 5$ **11.** $5x^{3/2} - 10 = 0$

12. $x^{2/3} - 9x^{1/3} + 8 = 0$ **13.** $\left|\frac{2}{3}x - 6\right| = 2$

Solve each inequality in Exercises 14–19. Express the answer in interval notation and graph the solution set on a number line.

14. $3(x + 4) \ge 5x - 12$ **15.** $\frac{x}{6} + \frac{1}{8} \le \frac{x}{2} - \frac{3}{4}$

16. $-3 \le \frac{2x + 5}{3} < 6$ **17.** $|3x + 2| \ge 3$

18. $x^2 < x + 12$ **19.** $\frac{2x + 1}{x - 3} > 3$

20. The monthly benefit (B) of a retirement plan is given by

$$B = \frac{2}{5}w + \frac{1}{125}n$$

where

w = an employee's average monthly salary

n = the number of years an employee worked for the company

a. Solve the formula for n.

b. Use your answer from part (a) to find the number of years an employee whose average monthly salary is \$800 must work to receive a monthly benefit of \$512.

21. Approximate population and growth figures for two states are given as follows.

	1980 Population (in Thousands)	Yearly Growth (in Thousands)
Arizona	2795	89
South Carolina	3071	43

When did Arizona have the same population as South Carolina?

22. The formula $y = 420x + 720$ describes the amount of money, y, in millions of dollars, lost to credit card fraud worldwide x years after 1989. In which year did losses amount to 4080 million dollars?

23. With a 9% raise, a physical therapist will earn \$45,780 annually. What is the therapist's salary prior to this raise?

24. You invest \$6000, part at 9% and the remainder at 6%. If the total yearly interest from the two investments is \$480, find the amount invested at each rate.

25. A vertical pole is to be supported by a wire that is 26 feet long and anchored 24 feet from the base of the pole. How far up the pole should the wire be attached?

26. A student has grades on three examinations of 76, 80, and 72. What must the student earn on a fourth examination in order to have an average of at least 80?

27. A computer online service charges a flat monthly rate of \$20 or a monthly rate of \$5 plus 15 cents for every hour spent online. How many hours online each month will make the second option a better deal?

Functions and Graphs

The cost of mailing a package depends on its weight. The probability that you and another person in a room share the same birthday depends on the number of people in the room. In both these situations, the relationship between variables can be illustrated with the notion of a *function*. Understanding this concept will give you a new perspective on many ordinary situations.

'Tis the season and you've waited until the last minute to mail your holiday gifts. Your only option is overnight express mail. You realize that the cost of mailing a gift depends on its weight, but the mailing costs seem somewhat odd. Your packages that weigh 1.1 pounds, 1.5 pounds, and 2 pounds cost $15.75 each to send overnight. Packages that weigh 2.01 pounds and 3 pounds cost you $18.50 each. Finally, your heaviest gift is barely over 3 pounds and its mailing cost is $21.25. What sort of system is this in which costs increase by $2.75, stepping from $15.75 to $18.50 and from $18.50 to $21.25?

SECTION 2.1 *Lines and Slope*

Objectives

1. Compute a line's slope.
2. Write the point-slope equation of a line.
3. Write and graph the slope-intercept equation of a line.
4. Recognize equations of horizontal and vertical lines.
5. Recognize and use the general form of a line's equation.
6. Model data with linear equations.

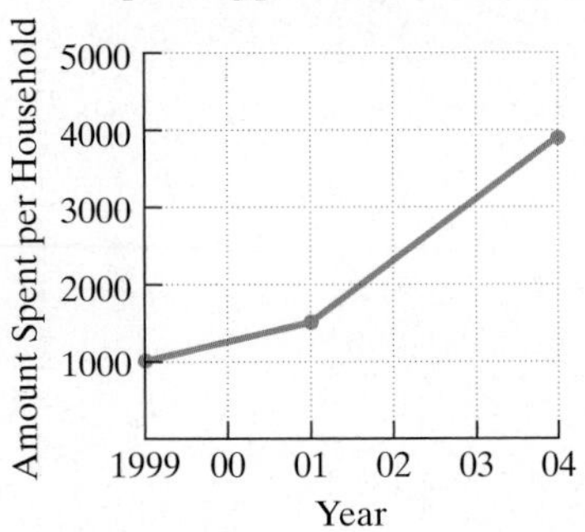

Source: Forrester Research

Figure 2.1

Good news: Projections indicate that in the next decades we'll live longer and move somewhere warmer where we'll shop online and chat on our tiny video cell phones. Figure 2.1 shows projected online shopping per U.S. online household through 2004. The graph is composed of two line segments. The segment on the right is steeper than the one on the left. This shows that online shopping is expected to increase more per year in 2001–2004 than in 1999–2001.

Data often fall on or near a line. In this section we will use equations to model such data and make predictions. We begin with a discussion of a line's steepness.

The Slope of a Line

Mathematicians have developed a useful measure of the steepness of a line, called the **slope** of the line. Slope compares the vertical change (the **rise**) to the horizontal change (the **run**) when moving from one fixed point to another along the line. To calculate the slope of a line, mathematicians use a ratio comparing the change in y (the rise) to the change in x (the run).

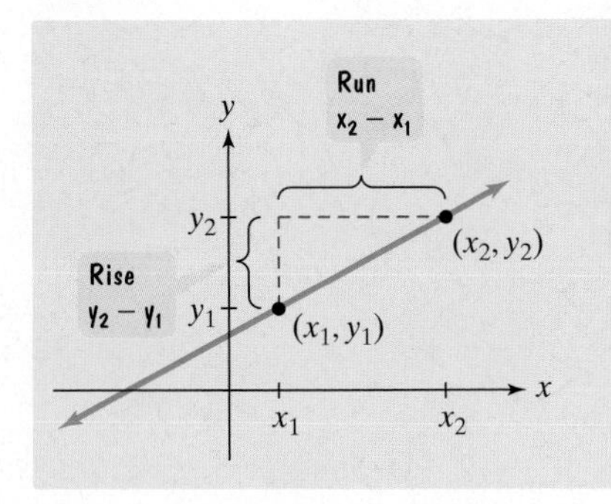

Definition of Slope

The **slope** of the line through the distinct points (x_1, y_1) and (x_2, y_2) is

$$\frac{\text{Change in } y}{\text{Change in } x} = \frac{\text{Rise}}{\text{Run}} = \frac{y_2 - y_1}{x_2 - x_1}$$

where $x_2 - x_1 \neq 0$.

It is common notation to let the letter m represent the slope of a line. The letter m is used because it is the first letter of the French verb *monter*, meaning to rise, or to ascend.

1 Compute a line's slope.

EXAMPLE 1 Using the Definition of Slope

Find the slope of the line passing through each pair of points.

a. $(-3, -1)$ and $(-2, 4)$ b. $(-3, 4)$ and $(2, -2)$

Solution

a. Let $(x_1, y_1) = (-3, -1)$ and $(x_2, y_2) = (-2, 4)$. We obtain a slope of

$$m = \frac{\text{Change in } y}{\text{Change in } x} = \frac{y_2 - y_1}{x_2 - x_1} = \frac{4 - (-1)}{-2 - (-3)} = \frac{5}{1} = 5.$$

The situation is illustrated in Figure 2.2(a). The slope of the line is 5, indicating that there is a vertical change, a rise, of 5 units for each horizontal change, a run, of 1 unit. The slope is positive, and the line rises from left to right.

Study Tip

When computing slope, it makes no difference which point you call (x_1, y_1) and which point you call (x_2, y_2). If we let $(x_1, y_1) = (-2, 4)$ and $(x_2, y_2) = (-3, -1)$, the slope is still 5:

$$m = \frac{y_2 - y_1}{x_2 - x_1} = \frac{-1 - 4}{-3 - (-2)} = \frac{-5}{-1} = 5.$$

However, you should not subtract in one order in the numerator $(y_2 - y_1)$ and then in a different order in the denominator $(x_1 - x_2)$. The slope is *not*

$$\frac{-1 - 4}{-2 - (-3)} = \frac{-5}{1} = -5. \quad \text{Incorrect.}$$

b. We can let $(x_1, y_1) = (-3, 4)$ and $(x_2, y_2) = (2, -2)$. The slope of the line shown in Figure 2.2(b) is computed as follows:

$$m = \frac{-2 - 4}{2 - (-3)} = \frac{-6}{5} = -\frac{6}{5}.$$

The slope of the line is $-\frac{6}{5}$. For every vertical change of -6 units (6 units down), there is a corresponding horizontal change of 5 units. The slope is negative and the line falls from left to right.

Slope and the Streets of San Francisco

San Francisco's Filbert Street has a slope of 0.613, meaning that for every horizontal distance of 100 feet, the street ascends 61.3 feet vertically. With its 31.5° angle of inclination, the street is too steep to pave and is only accessible by wooden stairs.

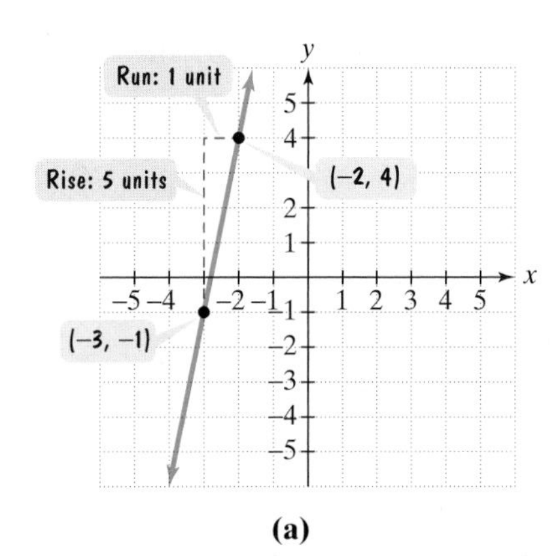

(a)

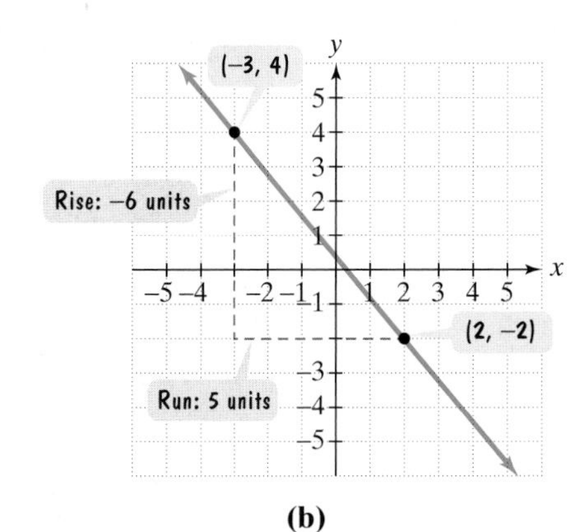

(b)

Figure 2.2 Visualizing Slope

Check Point 1 Find the slope of the line passing through each pair of points.

a. $(-3, 4)$ and $(-4, -2)$ **b.** $(4, -2)$ and $(-1, 5)$

Example 1 illustrates that a line with a positive slope is rising from left to right and a line with a negative slope is falling from left to right. By contrast, a horizontal line neither rises nor falls and has a slope of zero. A vertical line has no horizontal change, so $x_2 - x_1 = 0$ in the formula for slope. Because we cannot divide by zero, the slope of a vertical line is undefined. This discussion is summarized in Table 2.1.

Table 2.1 Possibilities for a Line's Slope

Positive Slope	Negative Slope	Zero Slope	Undefined Slope
$m > 0$	$m < 0$	$m = 0$	m is undefined.
Line rises from left to right.	Line falls from left to right.	Line is horizontal.	Line is vertical.

2 Write the point-slope equation of a line.

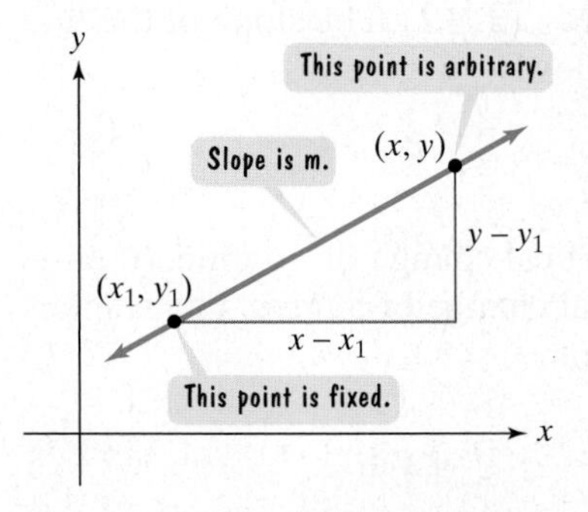

Figure 2.3 A line passing through (x_1, y_1) with slope m

The Point-Slope Form of the Equation of a Line

We can use the slope of a line to obtain various forms of the line's equation. For example, consider a nonvertical line that has a slope of m and contains the point (x_1, y_1). Now, let (x, y) represent any other point on the line, shown in Figure 2.3. Keep in mind that the point (x, y) is arbitrary and is not in one fixed position. By contrast, the point (x_1, y_1) is fixed. Regardless of where the point (x, y) is located, the shape of the triangle in Figure 2.3 remains the same. Thus, the ratio for slope stays a constant m. This means that for all points along the line,

$$m = \frac{y - y_1}{x - x_1}, \quad x \neq x_1.$$

We can clear the fraction by multiplying both sides by $x - x_1$, the least common denominator.

$$m(x - x_1) = \frac{y - y_1}{\cancel{x - x_1}} \cdot \cancel{x - x_1}$$

$$m(x - x_1) = y - y_1 \qquad \text{Simplify.}$$

Now, if we reverse the two sides, we obtain the **point-slope form** of the equation of a line.

Point-Slope Form of the Equation of a Line

The point-slope equation of a nonvertical line of slope m that passes through the point (x_1, y_1) is

$$y - y_1 = m(x - x_1).$$

For example, an equation of the line passing through $(1, 1)$ with a slope of 2 $(m = 2)$ is

$$y - 1 = 2(x - 1).$$

After we obtain the point-slope form of a line, it is customary to express the equation with y isolated on one side of the equal sign. Example 2 illustrates how this is done.

EXAMPLE 2 Writing the Point-Slope Equation of a Line

Write the point-slope form of the equation of the line passing through $(-1, 3)$ with a slope of 4. Then solve the equation for y.

Solution We use the point-slope equation of a line with $m = 4$, $x_1 = -1$, and $y_1 = 3$.

$$y - y_1 = m(x - x_1)$$ This is the point-slope form of the equation.

$$y - 3 = 4[x - (-1)]$$ Substitute the given values.

$$y - 3 = 4(x + 1)$$ We now have the point-slope form of the equation for the given line.

We can solve this equation for y by applying the distributive property on the right side.

$$y - 3 = 4x + 4$$

Finally, we add 3 to both sides.

$$y = 4x + 7$$

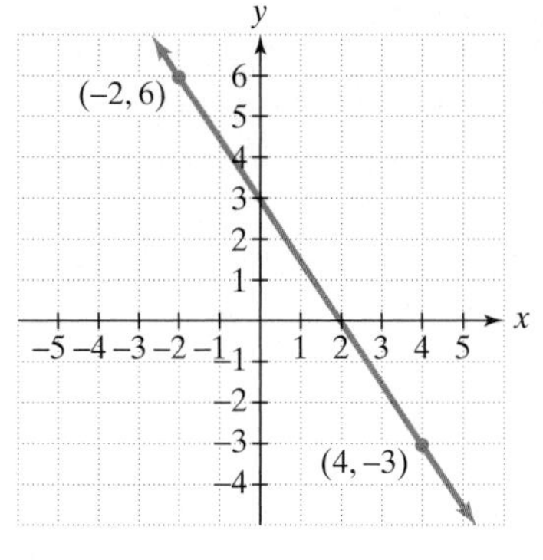

Figure 2.4 Write the point-slope equation of this line.

Discovery

You can use either point for (x_1, y_1) when you write a line's point-slope equation. Rework Example 3 using $(-2, 6)$ for (x_1, y_1). Once you solve for y, you should obtain the same equation as the one shown in the last line of the solution on page 172.

Check Point 2 Write the point-slope form of the equation of the line passing through $(2, -5)$ with a slope of 6. Then solve the equation for y.

EXAMPLE 3 Writing the Point-Slope Equation of a Line

Write the point-slope form of the equation of the line passing through the points $(4, -3)$ and $(-2, 6)$. (See Figure 2.4.) Then solve the equation for y.

Solution To use the point-slope form, we need to find the slope. The slope is the change in the y-coordinates divided by the corresponding change in the x-coordinates.

$$m = \frac{6 - (-3)}{-2 - 4} = \frac{9}{-6} = -\frac{3}{2}$$ This is the definition of slope using $(4, -3)$ and $(-2, 6)$.

We can take either point on the line to be (x_1, y_1). Let's use $(x_1, y_1) = (4, -3)$. Now, we are ready to write the point-slope equation.

$$y - y_1 = m(x - x_1)$$ This is the point-slope form of the equation.

$$y - (-3) = -\tfrac{3}{2}(x - 4)$$ Substitute: $(x_1, y_1) = (4, -3)$ and $m = -\frac{3}{2}$.

$$y + 3 = -\tfrac{3}{2}(x - 4)$$ Simplify.

We now have the point-slope form of the equation of the line shown in Figure 2.4. Now, we solve this equation for y.

$y + 3 = -\frac{3}{2}(x - 4)$ This is the point-slope form of the equation.

$y + 3 = -\frac{3}{2}x + 6$ Use the distributive property.

$y = -\frac{3}{2}x + 3$ Subtract 3 from both sides.

Write the point-slope form of the equation of the line passing through the points $(-2, -1)$ and $(-1, -6)$. Then solve the equation for y.

3 Write and graph the slope-intercept equation of a line.

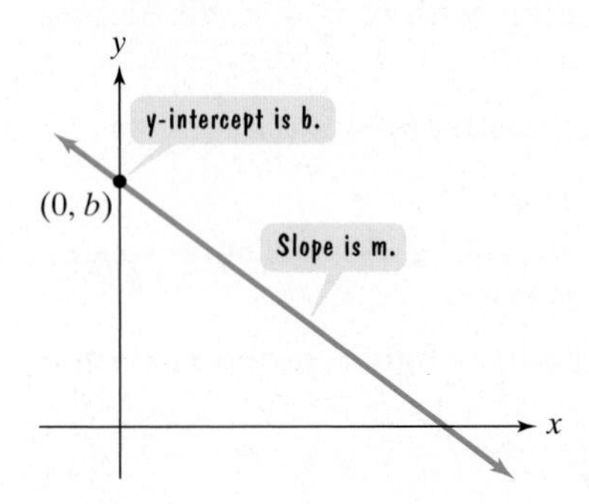

Figure 2.5 A line with slope m and y-intercept b

The Slope-Intercept Form of the Equation of a Line

Let's write the point-slope form of the equation of a line whose y-intercept is b with slope m. The line is shown in Figure 2.5. Because the y-intercept is b, the line intersects the y-axis at $(0, b)$. We use the point-slope form with $x_1 = 0$ and $y_1 = b$.

$$y - y_1 = m(x - x_1)$$

Let $y_1 = b$. Let $x_1 = 0$.

We obtain

$$y - b = m(x - 0).$$

Simplifying on the right side gives us

$$y - b = mx.$$

Finally, we solve for y by adding b to both sides.

$$y = mx + b$$

Thus, if a line's equation is written with y isolated on one side, the x-coefficient is the line's slope and the constant term is the y-intercept. This form of a line's equation is called the **slope-intercept form** of a line.

Slope-Intercept Form of the Equation of a Line

The **slope-intercept equation** of a nonvertical line with slope m and y-intercept b is

$$y = mx + b.$$

EXAMPLE 4 Graphing by Using the Slope and y-Intercept

Graph the line whose equation is $y = \frac{2}{3}x + 2$.

Solution The equation of the line is in the form $y = mx + b$. We can find the slope, m, by identifying the coefficient of x. We can find the y-intercept, b, by identifying the constant term.

$$y = \frac{2}{3}x + 2$$

The slope is $\frac{2}{3}$. The y-intercept is 2.

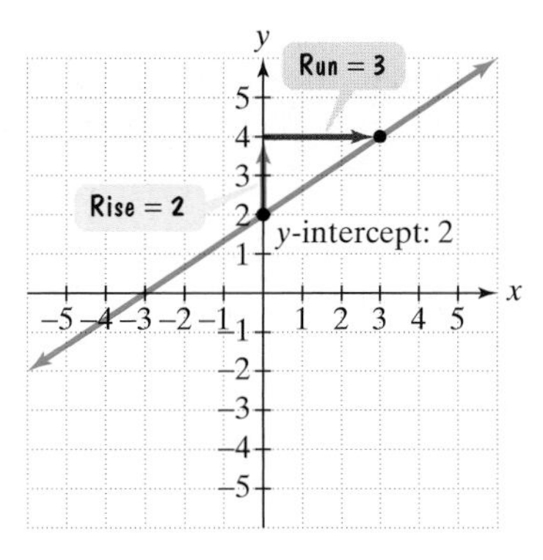

Figure 2.6 The graph of $y = \frac{2}{3}x + 2$

We need two points in order to graph the line. We can use the y-intercept, 2, to obtain the first point $(0, 2)$. Plot this point on the y-axis, shown in Figure 2.6.

We know the slope and one point on the line. We can use the slope, $\frac{2}{3}$, to determine a second point on the line. By definition,

$$m = \frac{2}{3} = \frac{\text{Rise}}{\text{Run}}.$$

We plot the second point on the line by starting at $(0, 2)$, the first point. Based on the slope, we move 2 units *up* (the rise) and 3 units to the *right* (the run). This puts us at a second point on the line, shown in Figure 2.6.

We use a straightedge to draw a line through the two points. The graph of $y = \frac{2}{3}x + 2$ is shown in Figure 2.6.

Graphing $y = mx + b$ by Using the Slope and y-Intercept

1. Plot the y-intercept on the y-axis. This is the point $(0, b)$.
2. Obtain a second point using the slope, m. Write m as a fraction, and use rise over run starting at the y-intercept to plot this point.
3. Use a straightedge to draw a line through the two points. Draw arrowheads at the ends of the line to show that the line continues indefinitely in both directions.

Check Point 4 Graph the line whose equation is $y = \frac{3}{5}x + 1$.

4 Recognize equations of horizontal and vertical lines.

Equations of Horizontal and Vertical Lines

Some things change very little. For example, from 1985 to the present, the number of Americans participating in downhill skiing has remained relatively constant, indicated by the graph shown in Figure 2.7. Shown in the figure is a horizontal line that passes through or near most of the data points.

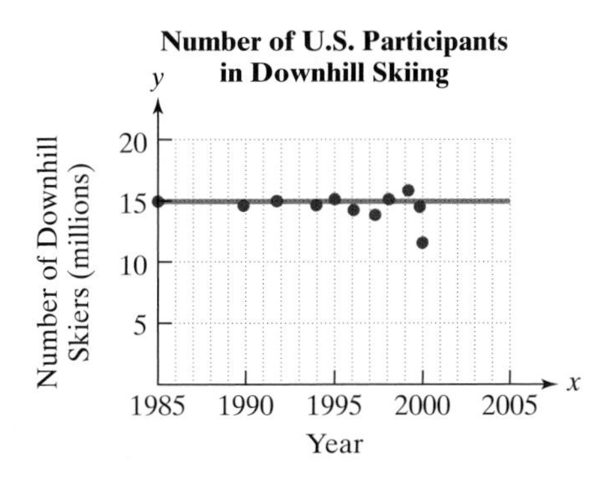

Source: National Ski Areas Association

Figure 2.7

We can use $y = mx + b$, the slope-intercept form of a line's equation, to write the equation of the horizontal line in Figure 2.7. We need the line's slope,

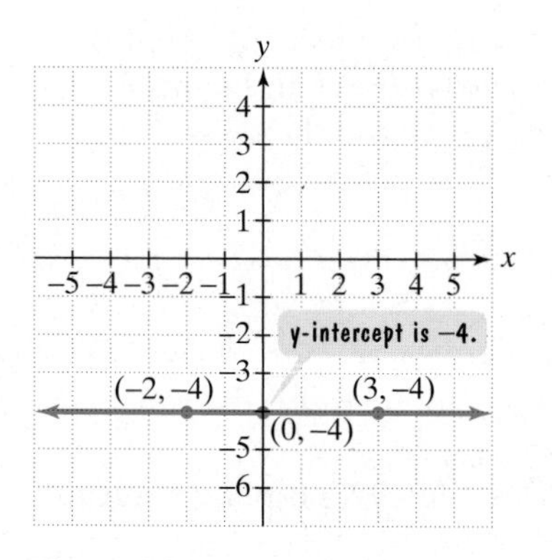

Figure 2.8 The graph of $y = -4$

m, and its y-intercept, b. Because the line is horizontal, $m = 0$. The line intersects the y-axis at 15, so $b = 15$. Thus, an equation that models the number of participants in downhill skiing for the period shown is

$$y = 0x + 15, \quad \text{or} \quad y = 15.$$

The popularity of downhill skiing remained relatively constant in the United States from 1985 to 2000 at approximately 15 million participants each year.

In general, if a line is horizontal, its slope is zero: $m = 0$. Thus, the equation $y = mx + b$ becomes $y = b$, where b is the y-intercept. For example, the graph of $y = -4$ is a horizontal line with a y-intercept of -4. The graph is shown in Figure 2.8. Three of the points along the line are shown and labeled. No matter what the x-coordinate is, the corresponding y-coordinate for every point on line is -4.

Equation of a Horizontal Line

A horizontal line is given by an equation of the form

$$y = b$$

where b is the y-intercept.

Next, let's see what we can discover about a vertical line by looking at an example.

EXAMPLE 5 Graphing a Vertical Line

Graph $x = 5$ in the rectangular coordinate system.

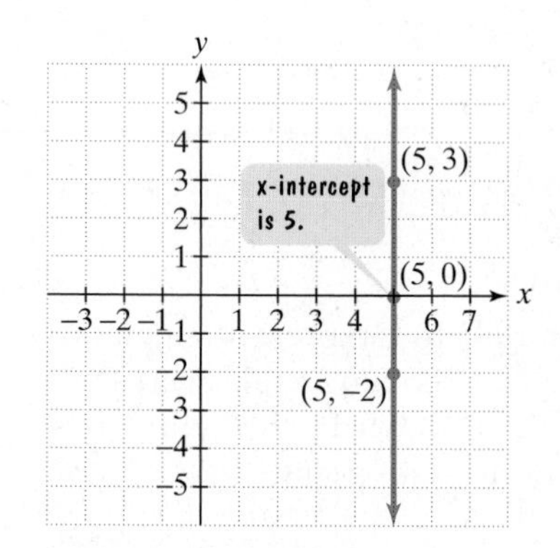

Figure 2.9 The graph $x = 5$

Solution All points on the graph of $x = 5$ have a value of x that is always 5. No matter what the y-coordinate is, the corresponding x-coordinate for every point on the line is 5. Let us select three of the possible values of y: -2, 0, and 3. So, three of the points on the graph of $x = 5$ are $(5, -2)$, $(5, 0)$, and $(5, 3)$. Plot each of these three points. Drawing a line that passes through the three points gives the vertical line shown in Figure 2.9.

Equation of a Vertical Line

A vertical line is given by an equation of the form

$$x = a$$

where a is the x-intercept.

Check Point 5 Graph $x = -1$ in the rectangular coordinate system.

5 Recognize and use the general form of a line's equation.

The General Form of the Equation of a Line

The vertical line whose equation is $x = 5$ cannot be written in slope-intercept form, $y = mx + b$, because its slope is undefined. However, every line has an equation that can be expressed in the form $Ax + By + C = 0$. For example,

$x = 5$ can be expressed as $1x + 0y - 5 = 0$, or $x - 5 = 0$. The equation $Ax + By + C = 0$ is called the **general form** of the equation of a line.

General Form of the Equation of a Line

Every line has an equation that can be written in the **general form**

$$Ax + By + C = 0$$

where A, B, and C are three real numbers, and A and B are not both zero.

If the equation of a line is given in general form, it is possible to find the slope, m, and the y-intercept, b, for the line. We solve the equation for y, transforming it into the slope-intercept form $y = mx + b$. In this form, the coefficient of x is the slope of the line, and the constant term is its y-intercept.

EXAMPLE 6 Finding the Slope and the y-Intercept

Find the slope and the y-intercept of the line whose equation is $2x - 3y + 6 = 0$.

Solution The equation is given in general form. We begin by rewriting it in the form $y = mx + b$. We need to solve for y.

$$2x - 3y + 6 = 0 \quad \text{This is the given equation.}$$

$$2x + 6 = 3y \quad \text{To isolate the } y\text{-term, add } 3y \text{ on both sides.}$$

$$3y = 2x + 6 \quad \text{Reverse the two sides. (This step is optional.)}$$

$$y = \frac{2}{3}x + 2 \quad \text{Divide both sides by 3.}$$

The coefficient of x, $\frac{2}{3}$, is the slope and the constant term, 2, is the y-intercept. This is the form of the equation that we graphed in Example 4 on page 173.

Check Point 6 Find the slope and the y-intercept of the line whose equation is $3x + 6y - 12 = 0$. Then use the y-intercept and the slope to graph the equation.

We've covered a lot of territory. Let's take a moment to summarize the various forms for equations of lines.

Equations of Lines

1. Point-slope form: $y - y_1 = m(x - x_1)$
2. Slope-intercept form: $y = mx + b$
3. Horizontal line: $y = b$
4. Vertical line: $x = a$
5. General form: $Ax + By + C = 0$

6 Model data with linear equations.

Applications

Linear equations are useful for modeling data that fall on or near a line. For example, Table 2.2 on page 176 gives the population of the United States, in millions, in the indicated year. The data are displayed as a set of five points in Figure 2.10.

Table 2.2

Year	x (Year after 1960)	y (U.S. Population) (in millions)
1960	0	179.3
1970	10	203.3
1980	20	226.5
1990	30	250.0
1998	38	268.9

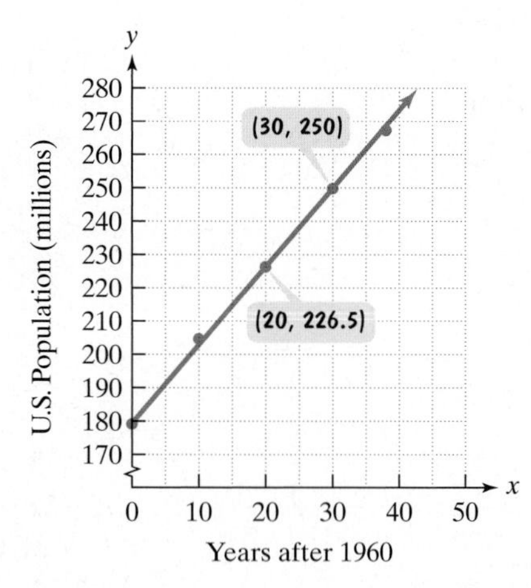

Figure 2.10

Data presented in a visual form as a set of points is called a **scatter plot**. Also shown in Figure 2.10 is a line that passes through or near the five points. By writing the equation of this line, we can obtain a model of the data and make predictions about the population of the United States in the future.

Technology

You can use a graphing utility to obtain a model for a scatter plot in which the data points fall on or near a straight line. The line that best fits the data is called the **regression line**. After entering the data in Table 2.2, a graphing utility displays a scatter plot of the data and the regression line.

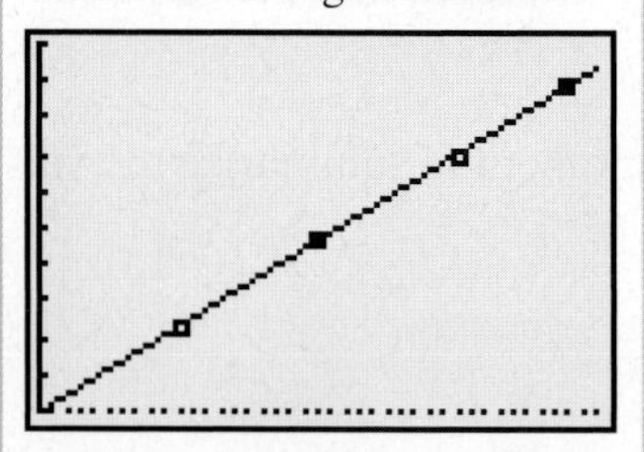

[0, 40, 1] by [180, 280, 10]

Also displayed is the regression line's equation.

EXAMPLE 7 Modeling U.S. Population

Write the equation of the line shown in Figure 2.10. Use the equation to predict U.S. population in 2010.

Solution The line in Figure 2.10 passes through (20, 226.5) and (30, 250). We start by finding the slope.

$$m = \frac{\text{change in } y}{\text{change in } x} = \frac{250 - 226.5}{30 - 20} = \frac{23.5}{10} = 2.35$$

Now, we write the line's equation.

$$y - y_1 = m(x - x_1)$$ Begin with the point-slope form.

$$y - 250 = 2.35(x - 30)$$ Either ordered pair can be (x_1, y_1). Let $(x_1, y_1) = (30, 250)$. From above, $m = 2.35$.

$$y - 250 = 2.35x - 70.5$$ Apply the distributive property on the right.

$$y = 2.35x + 179.5$$ Add 250 to both sides and solve for y.

The linear equation that models U.S. population, y, in millions, x years after 1960 is

$$y = 2.35x + 179.5.$$

Now, let's use this equation to predict U.S. population in 2010. Because 2010 is 50 years after 1960, substitute 50 for x and compute y.

$$y = 2.35(50) + 179.5 = 297$$

Our equation predicts that the population of the United States in the year 2010 will be 297 million. (The projected figure from the U.S. Census Bureau is 297.716 million.)

If an equation in slope-intercept form describes some real-world situation, the slope and the y-intercept can be interpreted in terms of that situation. For example, $y = 2.35x + 179.5$ models U.S. population, y, in millions, x years after 1960. The slope, 2.35, indicates that U.S. population is increasing by 2.35 million people per year. 179.5 is the y-intercept. At the beginning, in 1960, U.S. population was 179.5 million.

Check Point 7 Use the data points (10, 203.3) and (20, 226.5) from Table 2.2 to write an equation that models U.S. population x years after 1960. Use the equation to predict U.S. population in 2020.

EXERCISE SET 2.1

Practice Exercises

In Exercises 1–10, find the slope of the line passing through each pair of points or state that the slope is undefined. Then indicate whether the line through the points rises, falls, is horizontal, or is vertical.

1. $(4, 7)$ and $(8, 10)$
2. $(2, 1)$ and $(3, 4)$
3. $(-2, 1)$ and $(2, 2)$
4. $(-1, 3)$ and $(2, 4)$
5. $(4, -2)$ and $(3, -2)$
6. $(4, -1)$ and $(3, -1)$
7. $(-2, 4)$ and $(-1, -1)$
8. $(6, -4)$ and $(4, -2)$
9. $(5, 3)$ and $(5, -2)$
10. $(3, -4)$ and $(3, 5)$

In Exercises 11–38, use the given conditions to write an equation for each line in point-slope form and slope-intercept form.

11. Slope $= 2$, passing through $(3, 5)$
12. Slope $= 4$, passing through $(1, 3)$
13. Slope $= 6$, passing through $(-2, 5)$
14. Slope $= 8$, passing through $(4, -1)$
15. Slope $= -3$, passing through $(-2, -3)$
16. Slope $= -5$, passing through $(-4, -2)$
17. Slope $= -4$, passing through $(-4, 0)$
18. Slope $= -2$, passing through $(0, -3)$
19. Slope $= -1$, passing through $\left(-\frac{1}{2}, -2\right)$
20. Slope $= -1$, passing through $\left(-4, -\frac{1}{4}\right)$
21. Slope $= \frac{1}{2}$, passing through the origin
22. Slope $= \frac{1}{3}$, passing through the origin
23. Slope $= -\frac{2}{3}$, passing through $(6, -2)$
24. Slope $= -\frac{3}{5}$, passing through $(10, -4)$
25. Passing through $(1, 2)$ and $(5, 10)$
26. Passing through $(3, 5)$ and $(8, 15)$
27. Passing through $(-3, 0)$ and $(0, 3)$
28. Passing through $(-2, 0)$ and $(0, 2)$
29. Passing through $(-3, -1)$ and $(2, 4)$
30. Passing through $(-2, -4)$ and $(1, -1)$
31. Passing through $(-3, -2)$ and $(3, 6)$
32. Passing through $(-3, 6)$ and $(3, -2)$
33. Passing through $(-3, -1)$ and $(4, -1)$
34. Passing through $(-2, -5)$ and $(6, -5)$
35. Passing through $(2, 4)$ with x-intercept $= -2$
36. Passing through $(1, -3)$ with x-intereept $= -1$
37. x-intercept $= -\frac{1}{2}$ and y-intercept $= 4$
38. x-intercept $= 4$ and y-intercept $= -2$

In Exercises 39–46, give the slope and y-intercept of each line whose equation is given. Then graph the line.

39. $y = 2x + 1$
40. $y = 3x + 2$
41. $y = -2x + 1$
42. $y = -3x + 2$
43. $y = \dfrac{3}{4}x - 2$
44. $y = \dfrac{3}{4}x - 3$
45. $y = -\dfrac{3}{5}x + 7$
46. $y = -\dfrac{2}{5}x + 6$

In Exercises 47–52, graph each equation in the rectangular coordinate system.

47. $y = -2$
48. $y = 4$
49. $x = -3$
50. $x = 5$
51. $y = 0$
52. $x = 0$

In Exercises 53–60,

a. *Rewrite the given equation in slope-intercept form.*
b. *Give the slope and y-intercept.*
c. *Graph the equation.*

53. $3x + y - 5 = 0$
54. $4x + y - 6 = 0$
55. $2x + 3y - 18 = 0$
56. $4x + 6y + 12 = 0$
57. $8x - 4y - 12 = 0$
58. $6x - 5y - 20 = 0$
59. $3x - 9 = 0$
60. $4y + 28 = 0$

Application Exercises

61. As shown in the graph on page 178, the percentage of people in the United States satisfied with their lives remains relatively constant for all age groups. If x represents a person's age and y represents the percentage of people satisfied with their lives at that age, write an equation that reasonably models the data.

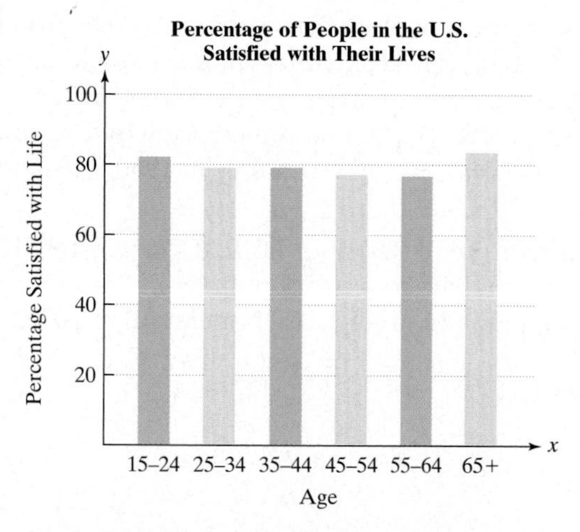

Source: *Culture Shift in Advanced Industrial Society*, Princeton University Press

62. The graph shows the life expectancy in years for U.S. women whose year of birth is indicated on the x-axis. Find the slope of the line passing through the points whose coordinates are shown on the graph. Describe what the slope represents.

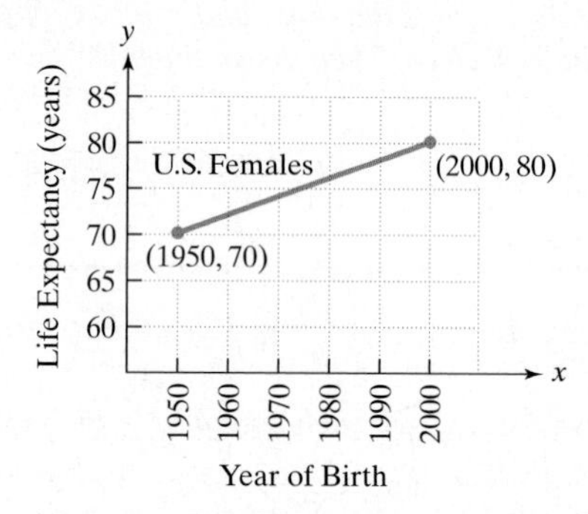

63. The graph shows U.S. population projections from 2000 through 2050. Use the equation $y = 2.35x + 179.5$, in which x is the number of years after 1960 and y is the U.S. population, in millions, to determine how well the equation models the projections for 2030, 2040, and 2050.

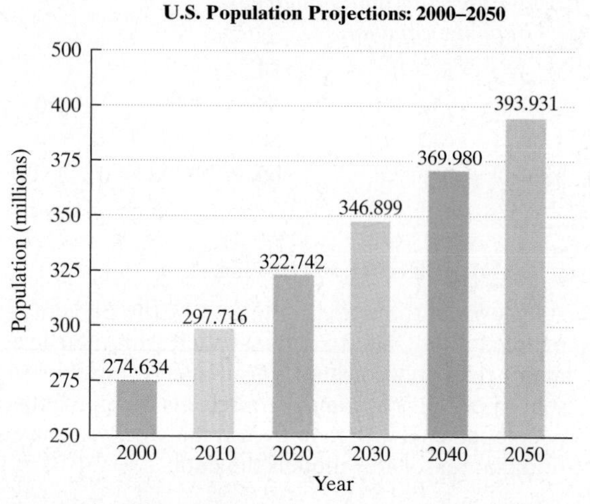

Source: U.S. Census Bureau

64. The figure shows projected online shopping per U.S. online household through 2004.

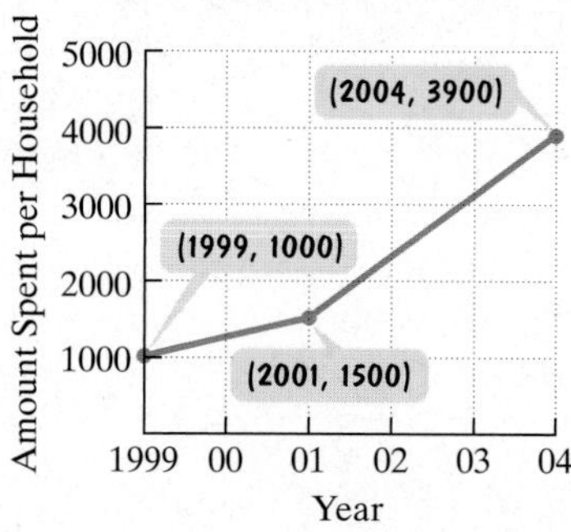

Source: Forrester Research

a. Find the slope of the line passing through (1999, 1000) and (2001, 1500). What does this represent in terms of the increase in online shopping per year?

b. Repeat part (a) for the data points (2001, 1500) and (2004, 3900).

c. Write the point-slope form of the equation of the line passing through (2001, 1500) and (2004, 3900).

d. Use the equation in part (c) to write the slope-intercept form of the equation.

e. Use the equation from part (c) to project the amount that will be spent shopping online per online household in 2010.

In Exercises 65–66, two measurements are given for variables having a linear relationship. For each exercise, write the point-slope form of the equation of the line on which these measurements fall. Then use the point-slope form of the equation to write the slope-intercept form of the equation. Finally, use this equation to answer the question.

65.

x (Number of Years after 1990)	y (Total Consumer Spending in the United States, in Billions of Dollars)
3	4459.2
7	5493.7

Source: U.S. Commerce Department

How much will consumers in the United States spend in the year 2020?

66.

x (Number of Years after 1985)	y (Total of All Health-Care Expenditures in the United States, in Billions of Dollars)
3	546
5	666

Source: U.S. Health Care Financing Administration

What will health-care expenditures in the United States be in the year 2010?

67. A business discovers a linear relationship between the number of shirts it can sell and the price per shirt. In particular, 20,000 shirts can be sold at \$19 each, and 2000 of the same shirts can be sold at \$55 each. Write the slope-intercept equation of the *demand line* through the ordered pairs (20,000 shirts, \$19) and (2000 shirts, \$55). Then determine the number of shirts that can be sold at \$50 each.

68. In 1965, radioactive wastes seeping into the Columbia River exposed citizens of eight Oregon counties and the city of Portland to radioactive contamination. In an article in the *Journal of Environmenial Health* (May–June, 1965), the authors formulated an index that measured the proximity of the residents to the contamination. The ordered pair for Columbia County (6.4, 178) indicates that its index is 6.4 and there are 178 cancer deaths per 100,000 residents. The corresponding ordered pair for Clatsop County is (8.3, 210). What is the predicted number of cancer deaths for Portland, with an index of 11.6?

69. Is there a relationship between education and prejudice? With increased education, does a person's level of prejudice tend to decrease? The scatter plot shows ten data points, each representing the number of years of school completed and the score on a test measuring prejudice for each subject. Higher scores on this 1-to-10 test indicate greater prejudice. Also shown is the regression line, the line that best fits the data. Use two points on this line to write both its point-slope and slope-intercept equations. Then use the slope-intercept equation to predict the score on the prejudice test for a person with seven years of education.

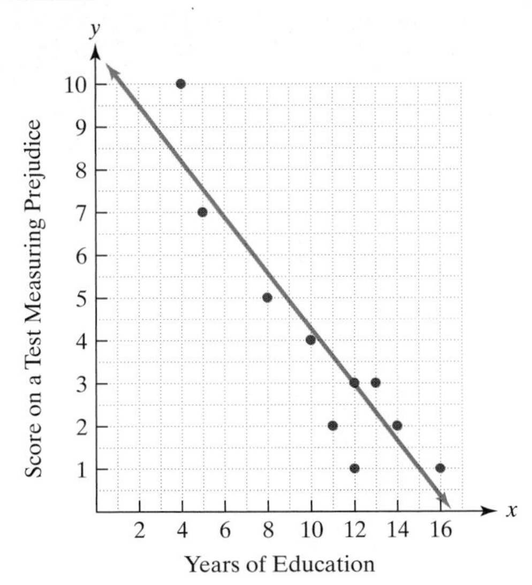

70. The scatter plot at the top of the next column shows the relationship between the percentage of married women of child-bearing age using contraceptives and the births per woman in selected countries. Also shown is the regression line. Use two points on this line to write both its point-slope and slope-intercept equations. Then find the number of births per woman if 90% of married women of child-bearing age use contraceptives.

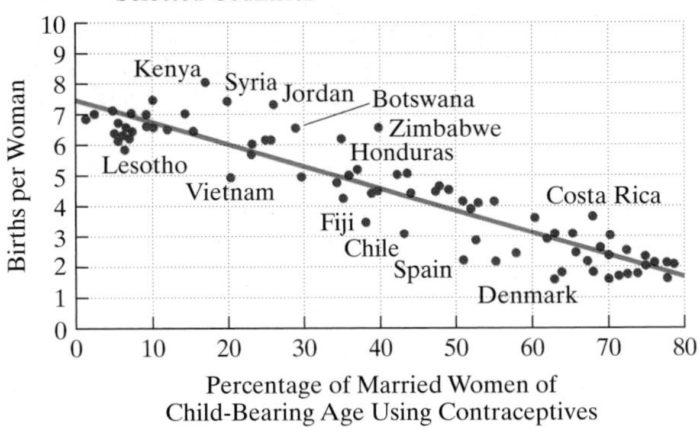

Source: Population Reference Bureau

Writing in Mathematics

71. What is the slope of a line and how is it found?

72. Describe how to write the equation of a line if two points along the line are known.

73. Explain how to derive the slope-intercept form of a line's equation, $y = mx + b$, from the point-slope form

$$y - y_1 = m(x - x_1).$$

74. Explain how to graph the equation $x = 2$. Can this equation be expressed in slope-intercept form? Explain.

75. Explain how to use the general form of a line's equation to find the line's slope and y-intercept.

76. Look back at Figure 2.1 on page 168. Do you think that the line through the points corresponding to 2001 and 2004 will describe online spending per online household in 2040? Explain your answer.

77. Take a second look at the scatter plot in Exercise 69. Although there is a relationship between education and prejudice, we cannot necessarily conclude that increased education causes a person's level of prejudice to decrease. Offer two or more possible explanations for the data in the scatter plot.

Technology Exercises

Use a graphing utility to graph each equation in Exercises 78–81. Then use the TRACE *feature to trace along the line and find the coordinates of two points. Use these points to compute the line's slope. Check your result by using the coefficient of x in the line's equation.*

78. $y = 2x + 4$

79. $y = -3x + 6$

80. $y = -\frac{1}{2}x - 5$

81. $y = \frac{3}{4}x - 2$

82. a. Use the statistical menu of your graphing utility to enter the ten data points shown in the scatter plot in Exercise 69.

b. Use the DRAW menu and the scatter plot capability to draw a scatter plot of the data points like the one shown in Exercise 69.

c. Select the linear regression option. Your utility should give you values for a and b for the equation of the regression line, $y = ax + b$. You may also be given a *correlation coefficient*, r. Values of r close to 1 indicate that the points can be described by a linear relationship and the regression line has a positive slope. Values of r close to -1 indicate that the points can be described by a linear relationship and the regression line has a negative slope. Values of r close to 0 indicate no linear relationship between the variables.

d. Use the appropriate sequence (consult your manual) to graph the regression equation on top of the points in the scatter plot.

Critical Thinking Exercises

83. Which one of the following is true?

a. A linear equation with nonnegative slope has a graph that rises from left to right.

b. The equations $y = 4x$ and $y = -4x$ have graphs that are perpendicular lines.

c. The line whose equation is $5x + 6y - 30 = 0$ passes through the point $(6, 0)$ and has slope $-\frac{5}{6}$.

d. The graph of $y = 7$ in the rectangular coordinate system is the single point $(7, 0)$.

84. Prove that the equation of a line passing through $(a, 0)$ and $(0, b)$ $(a \neq 0, b \neq 0)$ can be written in the form $\frac{x}{a} + \frac{y}{b} = 1$. Why is this called the *intercept form* of a line?

85. Use the figure at the top of the next column to make the following lists.

a. List the slopes m_1, m_2, m_3, and m_4 in order of decreasing size.

b. List the y-intercepts b_1, b_2, b_3, and b_4 in order of decreasing size.

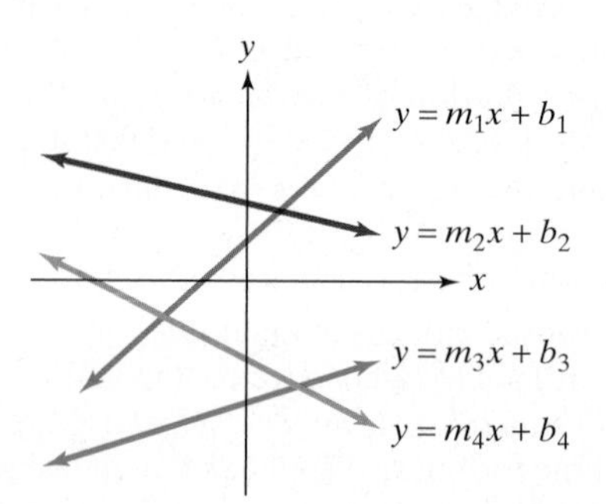

86. Excited about the success of celebrity stamps, post office officials were rumored to have put forth a plan to institute two new types of thermometers. On these new scales, °E represents degrees Elvis and °M represents degrees Madonna. If it is known that $40°E = 25°M$, $280°E = 125°M$, and degrees Elvis is linearly related to degrees Madonna, write an equation expressing E in terms of M.

Group Activity Exercise

87. Group members should consult an almanac, newspaper, magazine, or the Internet to find data that lie approximately on or near a straight line. Working by hand or using a graphing utility, construct a scatter plot for the data. If working by hand, draw a line that approximately fits the data and then write its equation. If using a graphing utility, obtain the equation of the regression line. Then use the equation of the line to make a prediction about what might happen in the future. Are there circumstances that might affect the accuracy of this prediction? List some of these circumstances.

SECTION 2.2 *Parallel and Perpendicular Lines and Circles*

Objectives

1. Find slopes and equations of parallel and perpendicular lines.
2. Write the standard form of a circle's equation.
3. Give the center and radius of a circle whose equation is in standard form.
4. Convert the general form of a circle's equation to standard form.

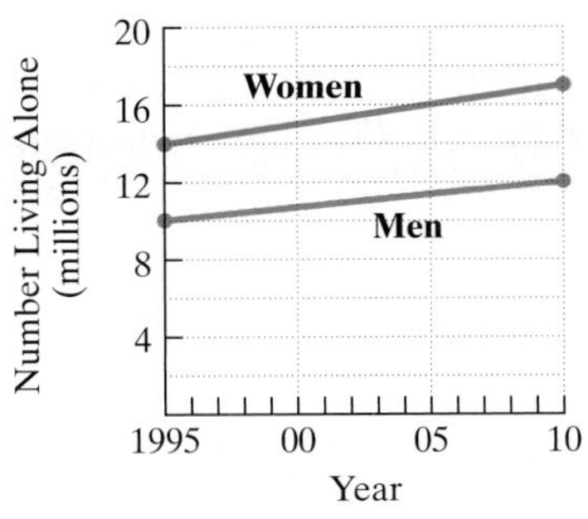

Source: Forrester Research

Figure 2.11

A best guess at the look of our nation in the next decades indicates that the number of men and women living alone will increase each year. Figure 2.11 shows that by 2010, approximately 12 million men and 17 million women will be living alone. Can you tell by the line graphs in the figure if the yearly increase for women is the same as the yearly increase for men? We begin this section by showing how we can use the slope of each line to answer this question.

Parallel and Perpendicular Lines

Two nonintersecting lines that lie in the same plane are parallel. If two lines do not intersect, the ratio of the vertical change to the horizontal change is the same for each line. Because two parallel lines have the same "steepness," they must have the same slope.

1 Find slopes and equations of parallel and perpendicular lines.

Slope and Parallel Lines

1. If two nonvertical lines are parallel, then they have the same slope.
2. If two distinct nonvertical lines have the same slope, then they are parallel.
3. Two distinct vertical lines, both with undefined slopes, are parallel.

EXAMPLE 1 Writing Equations of a Line Parallel to a Given Line

Write an equation of the line passing through $(-3, 2)$ and parallel to the line whose equation is $y = 2x + 1$. Express the equation in point-slope form and slope-intercept form.

Solution The situation is illustrated in Figure 2.12. We are looking for the equation of the line shown on the left. How do we obtain this equation? Notice that the line passes through the point $(-3, 2)$. Using the point-slope form of the line's equation, we have $x_1 = -3$ and $y_1 = 2$.

$$y - y_1 = m(x - x_1)$$

$y_1 = 2$ $\quad$ $x_1 = -3$

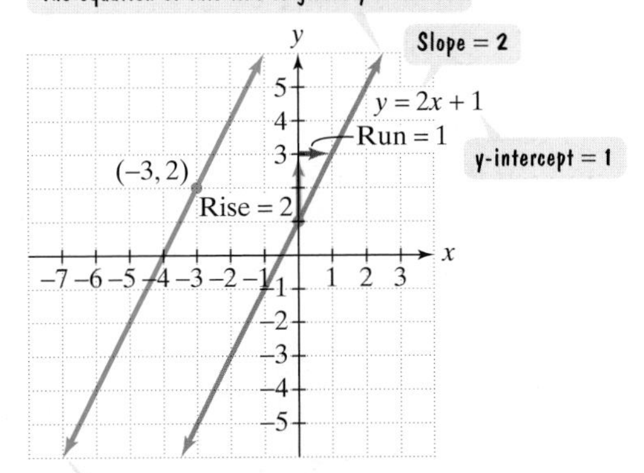

Figure 2.12 Writing equations of a line parallel to a given line

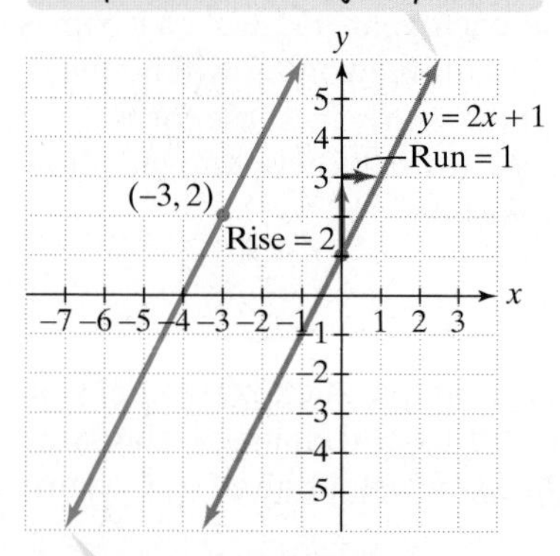

Figure 2.12 Shown again so that you do not have to turn back a page

Now, the only thing missing from the equation is m, the slope of the line on the left. Do we know anything about the slope of either line in Figure 2.12? The answer is yes; we know the slope of the line on the right, whose equation is given.

$$y = 2x + 1$$

The slope of the line on the right in Figure 2.12 is 2.

Parallel lines have the same slope. Because the slope of the line with the given equation is 2, $m = 2$ for the line whose equation we must write.

$$y - y_1 = m(x - x_1)$$

$y_1 = 2$ $m = 2$ $x_1 = -3$

The point-slope form of the line's equation is

$$y - 2 = 2[x - (-3)] \text{ or}$$
$$y - 2 = 2(x + 3).$$

Solving for y, we obtain the slope-intercept form of the equation.

$$y - 2 = 2x + 6$$ Apply the distributive property.

$$y = 2x + 8$$ Add 2 to both sides. This is the slope-intercept form, $y = mx + b$, of the equation.

Check Point 1 Write an equation of the line passing through $(-2, 5)$ and parallel to the line whose equation is $y = 3x + 1$. Express the equation in point-slope form and slope-intercept form.

Two lines that intersect at a right angle (90°) are said to be **perpendicular**, shown in Figure 2.13. There is a relationship between the slopes of perpendicular lines.

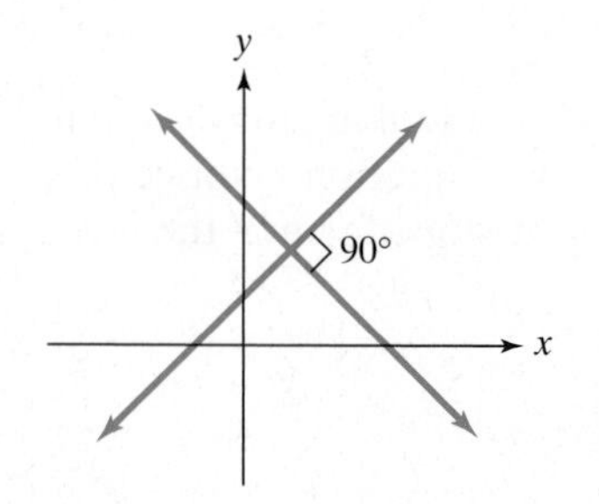

Figure 2.13 Perpendicular lines

Slope and Perpendicular Lines

1. If two nonvertical lines are perpendicular, then the product of their slopes is -1.
2. If the product of the slopes of two lines is -1, then the lines are perpendicular.
3. A horizontal line having zero slope is perpendicular to a vertical line having undefined slope.

An equivalent way of stating this relationship is to say that one line is perpendicular to another line if its slope is the *negative reciprocal* of the slope of the other. For example, if a line has slope 5, any line having slope $-\frac{1}{5}$ is perpendicular to it. Similarly, if a line has slope $-\frac{3}{4}$, any line having slope $\frac{4}{3}$ is perpendicular to it.

EXAMPLE 2 Finding the Slope of a Line Perpendicular to a Given Line

Find the slope of any line that is perpendicular to the line whose equation is $x + 4y - 8 = 0$.

Solution We begin by writing the equation of the given line in slope-intercept form. Solve for y.

$$x + 4y - 8 = 0 \quad \text{This is the given equation.}$$

$$4y = -x + 8 \quad \text{To isolate the y-term, subtract x and add 8 on both sides.}$$

$$y = -\tfrac{1}{4}x + 2 \quad \text{Divide both sides by 4.}$$

Slope is $-\frac{1}{4}$.

The given line has slope $-\frac{1}{4}$. Any line perpendicular to this line has a slope that is the negative reciprocal of $-\frac{1}{4}$. Thus, the slope of any perpendicular line is 4.

Check Point 2 Find the slope of any line that is perpendicular to the line whose equation is $x + 3y - 12 = 0$.

Circles

It's a good idea to know your way around a circle. Clocks, angles, maps, and compasses are based on circles. Circles occur everywhere in nature: in ripples on water, patterns on a butterfly's wings, and cross sections of trees. Some consider the circle to be the most pleasing of all shapes.

The rectangular coordinate system gives us a unique way of knowing a circle. It enables us to translate a circle's geometric definition into an algebraic equation. We begin with this geometric definition.

Definition of a Circle

A **circle** is the set of all points in a plane that are equidistant from a fixed point called the **center**. The fixed distance from the circle's center to any point on the circle is called the **radius**.

Figure 2.14 on page 184 is our starting point for obtaining a circle's equation. We've placed the circle into a rectangular coordinate system. The circle's center is (h, k) and its radius is r. We let (x, y) represent the coordinates of any point on the circle.

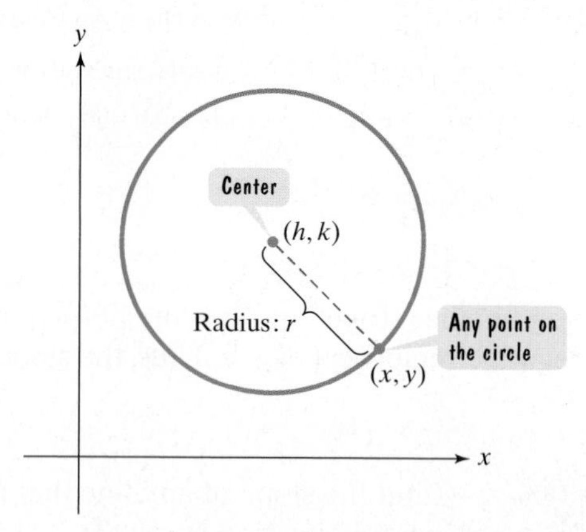

Figure 2.14 A circle centered at (h, k) with radius r

What does the geometric definition of a circle tell us about point (x, y) in Figure 2.14? The point is on the circle if and only if its distance from the center is r. We can use the distance formula to express this idea algebraically:

The distance between (x, y) and (h, k) — is always — r.

$$\sqrt{(x-h)^2+(y-k)^2} = r$$

Squaring both sides of this equation yields the **standard form of the equation of a circle**.

The Standard Form of the Equation of a Circle

The **standard form of the equation of a circle** with center (h, k) and radius r is

$$(x-h)^2+(y-k)^2=r^2.$$

2 Write the standard form of a circle's equation.

EXAMPLE 3 Finding the Standard Form of a Circle's Equation

Write the standard form of the equation of the circle with center $(0, 0)$ and radius 2. Graph the circle.

Solution The center is $(0, 0)$. Because the center is represented as (h, k) in the standard form of the equation, $h = 0$ and $k = 0$. The radius is 2, so we will let $r = 2$ in the equation.

$$(x-h)^2+(y-k)^2=r^2 \quad \text{This is the standard form of a circle's equation.}$$

$$(x-0)^2+(y-0)^2=2^2 \quad \text{Substitute 0 for } h\text{, 0 for } k\text{, and 2 for } r.$$

$$x^2+y^2=4 \quad \text{Simplify.}$$

The standard form of the equation of the circle is $x^2 + y^2 = 4$. Figure 2.15 shows the graph.

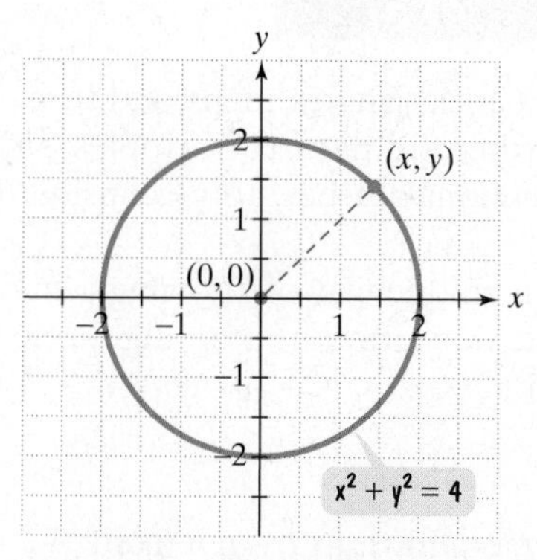

Figure 2.15 The graph of $x^2 + y^2 = 4$

Check Point 3 Write the standard form of the equation of the circle with center $(0, 0)$ and radius 4.

Technology

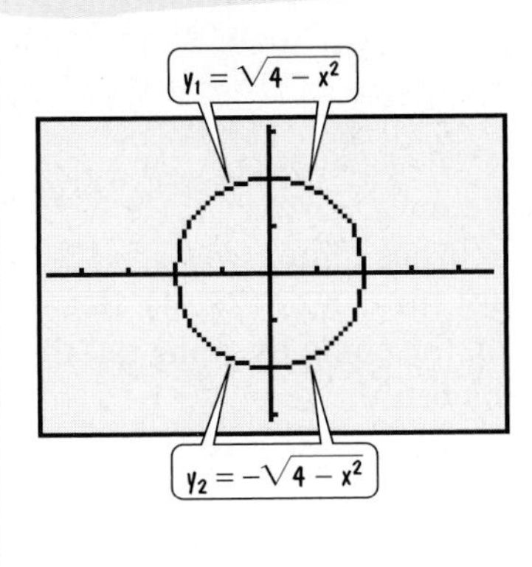

To graph a circle with a graphing utility, first solve the equation for y.

$$x^2 + y^2 = 4$$
$$y^2 = 4 - x^2$$
$$y = \pm\sqrt{4 - x^2}$$

Graph the two equations

$$y_1 = \sqrt{4 - x^2} \quad \text{and} \quad y_2 = -\sqrt{4 - x^2}$$

in the same viewing rectangle. The graph of $y_1 = \sqrt{4 - x^2}$ is the top semicircle because y is always positive. The graph of $y_2 = -\sqrt{4 - x^2}$ is the bottom semicircle because y is always negative. Use a ZOOM SQUARE setting so that the circle looks like a circle. (Many graphing utilities have problems connecting the two semicircles because the segments directly across horizontally from the center become nearly vertical.)

Example 3 and Checkpoint 3 involved circles centered at the origin. The standard form of the equation of all such circles is $x^2 + y^2 = r^2$, where r is the circle's radius. Now, let's consider a circle whose center is not at the origin.

EXAMPLE 4 Finding the Standard Form of a Circle's Equation

Write the standard form of the equation of the circle with center $(-2, 3)$ and radius 4.

Solution The center is $(-2, 3)$. Because the center is represented as (h, k) in the standard form of the equation, $h = -2$ and $k = 3$. The radius is 4, so we will let $r = 4$ in the equation.

$$(x - h)^2 + (y - k)^2 = r^2 \quad \text{This is the standard form of a circle's equation.}$$
$$[x - (-2)]^2 + (y - 3)^2 = 4^2 \quad \text{Substitute } -2 \text{ for } h, 3 \text{ for } k, \text{ and } 4 \text{ for } r.$$
$$(x + 2)^2 + (y - 3)^2 = 16 \quad \text{Simplify.}$$

The standard form of the equation of the circle is $(x + 2)^2 + (y - 3)^2 = 16$.

Check Point 4 Write the standard form of the equation of the circle with center $(5, -6)$ and radius 10.

3 Give the center and radius of a circle whose equation is in standard form.

EXAMPLE 5 Using the Standard Form of a Circle's Equation to Graph the Circle

Find the center and radius of the circle whose equation is

$$(x - 2)^2 + (y + 4)^2 = 9$$

and graph the equation.

Solution In order to graph the circle, we need to know its center, (h, k), and its radius r. We can find the values for h, k, and r by comparing the given equation to the standard form of the equation of a circle.

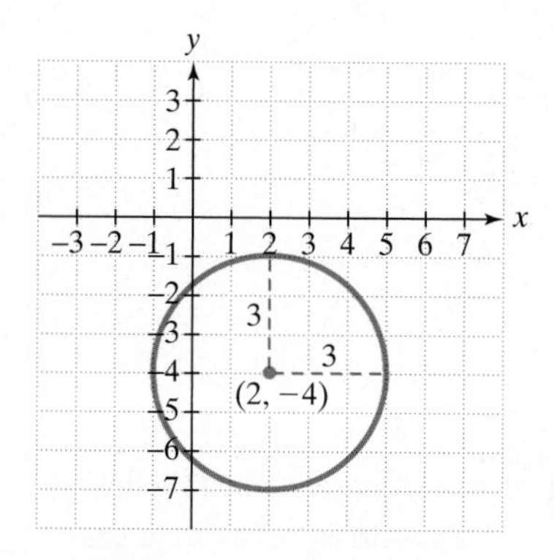

Figure 2.16 The graph of $(x - 2)^2 + (y + 4)^2 = 9$

$$(x - 2)^2 + (y + 4)^2 = 9$$
$$(x - 2)^2 + (y - (-4))^2 = 3^2$$

This is $(x - h)^2$, with $h = 2$. This is $(y - k)^2$, with $k = -4$. This is r^2, with $r = 3$.

We see that $h = 2$, $k = -4$, and $r = 3$. Thus, the circle has center $(h, k) = (2, -4)$ and a radius of 3 units. To graph this circle, first plot the center $(2, -4)$. Because the radius is 3, you can locate at least four points on the circle by going out three units to the right, to the left, up, and down from the center.

Two such points, to the right and to the left of $(2, -4)$, are $(5, -4)$ and $(-1, -4)$, respectively.

Using these points, we obtain the graph in Figure 2.16.

Check Point 5 Find the center and radius of the circle whose equation is

$$(x + 3)^2 + (y - 1)^2 = 4$$

and graph the equation.

If we square $x - 2$ and $y + 4$ in the standard form of the equation of Example 5, we obtain another form for the circle's equation.

$$(x - 2)^2 + (y + 4)^2 = 9 \quad \text{This is the standard form of the equation from Example 3.}$$
$$x^2 - 4x + 4 + y^2 + 8y + 16 = 9 \quad \text{Square } x - 2 \text{ and } y + 4.$$
$$x^2 + y^2 - 4x + 8y + 20 = 9 \quad \text{Combine numerical terms and rearrange terms.}$$
$$x^2 + y^2 - 4x + 8y + 11 = 0 \quad \text{Subtract 9 from both sides.}$$

This result suggests that an equation in the form $x^2 + y^2 + Dx + Ey + F = 0$ can represent a circle. This is called the **general form of the equation of a circle**.

The General Form of the Equation of a Circle

The **general form of the equation of a circle is**

$$x^2 + y^2 + Dx + Ey + F = 0.$$

4 Convert the general form of a circle's equation to standard form.

We can convert the general form of the equation of a circle to the standard form $(x - h)^2 + (y - k)^2 = r^2$. We do so by completing the square on x and y. Let's see how this is done.

EXAMPLE 6 Converting the General Form of a Circle's Equation to Standard Form and Graphing the Circle

Write in standard form and graph: $x^2 + y^2 + 4x - 6y - 23 = 0$.

Solution Because we plan to complete the square on both x and y, let's rearrange terms so that x-terms are arranged in descending order, y-terms are arranged in descending order, and the constant term appears on the right.

$x^2 + y^2 + 4x - 6y - 23 = 0$	This is the given equation.
$(x^2 + 4x \qquad) + (y^2 - 6y \qquad) = 23$	Rewrite in anticipation of completing the square.
$(x^2 + 4x + 4) + (y^2 - 6y + 9) = 23 + 4 + 9$	Complete the square on x: $\frac{1}{2} \cdot 4 = 2$ and $2^2 = 4$, so add 4 to both sides. Complete the square on y: $\frac{1}{2}(-6) = -3$ and $(-3)^2 = 9$, so add 9 to both sides.
$(x + 2)^2 + (y - 3)^2 = 36$	Factor on the left and add on the right.

Remember that numbers added on the left side must also be added on the right side.

This last equation is in standard form. We can identify the circle's center and radius by comparing this equation to the standard form of the equation of a circle, $(x - h)^2 + (y - k)^2 = r^2$.

$$(x + 2)^2 + (y - 3)^2 = 36$$

$$(x - (-2))^2 + (y - 3)^2 = 6^2$$

This is $(x - h)^2$, with $h = -2$. This is $(y - k)^2$, with $k = 3$. This is r^2, with $r = 6$.

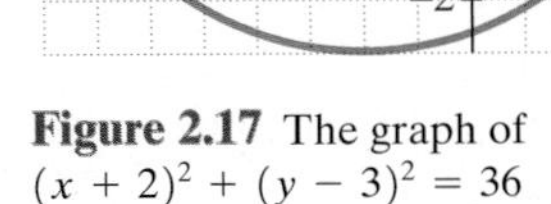

Figure 2.17 The graph of $(x + 2)^2 + (y - 3)^2 = 36$

We use the center, $(h, k) = (-2, 3)$, and the radius, $r = 6$, to graph the circle. The graph is shown in Figure 2.17.

Technology

To graph $x^2 + y^2 + 4x - 6y - 23 = 0$, rewrite the equation as a quadratic equation in y.

$$y^2 - 6y + (x^2 + 4x - 23) = 0$$

Now solve for y using the quadratic formula, with $a = 1$, $b = -6$, and $c = x^2 + 4x - 23$.

$$y = \frac{-b \pm \sqrt{b^2 - 4ac}}{2a} = \frac{-(-6) \pm \sqrt{(-6)^2 - 4 \cdot 1(x^2 + 4x - 23)}}{2 \cdot 1} = \frac{6 \pm \sqrt{36 - 4(x^2 + 4x - 23)}}{2}$$

Because we will enter these equations, there is no need to simplify. Enter

$$y_1 = \frac{6 + \sqrt{36 - 4(x^2 + 4x - 23)}}{2}$$

and

$$y_2 = \frac{6 - \sqrt{36 - 4(x^2 + 4x - 23)}}{2}.$$

Use a ZOOM SQUARE setting. The graph is shown on the right.

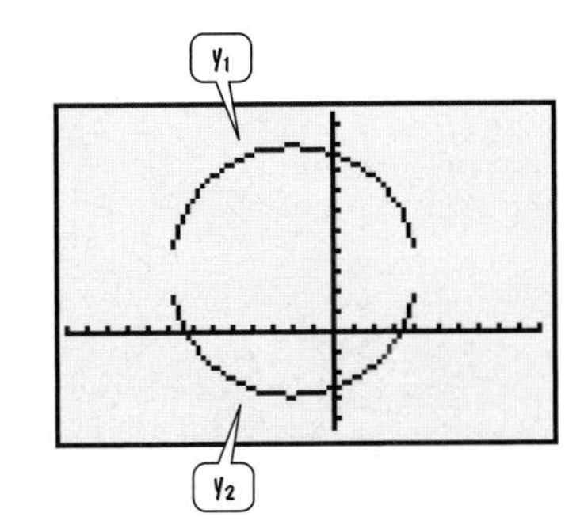

Check Point 6 Write in standard form and graph:

$$x^2 + y^2 + 4x - 4y - 1 = 0.$$

EXERCISE SET 2.2

Practice Exercises

In Exercises 1–16, the equation of a line is given. Find the slope of a line that is (a) parallel to the line with the given equation; and (b) perpendicular to the line with the given equation.

1. $y = 5x$
2. $y = 3x$
3. $y = -7x$
4. $y = -9x$
5. $y = \frac{1}{2}x + 3$
6. $y = \frac{1}{4}x - 5$
7. $y = -\frac{2}{5}x - 1$
8. $y = -\frac{3}{7}x - 2$
9. $4x + y = 7$
10. $8x + y = 11$
11. $2x + 4y - 8 = 0$
12. $3x + 2y - 6 = 0$
13. $2x - 3y - 5 = 0$
14. $3x - 4y + 7 = 0$
15. $x = 6$
16. $y = 9$

In Exercises 17–20, write an equation for line L in point-slope form and slope-intercept form.

17.

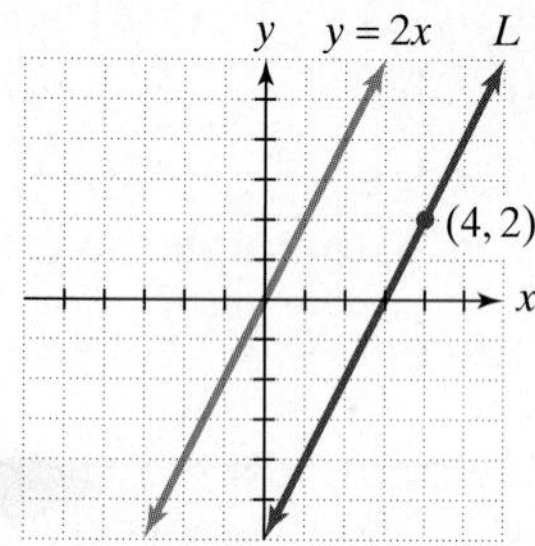

L is parallel to $y = 2x$.

18.

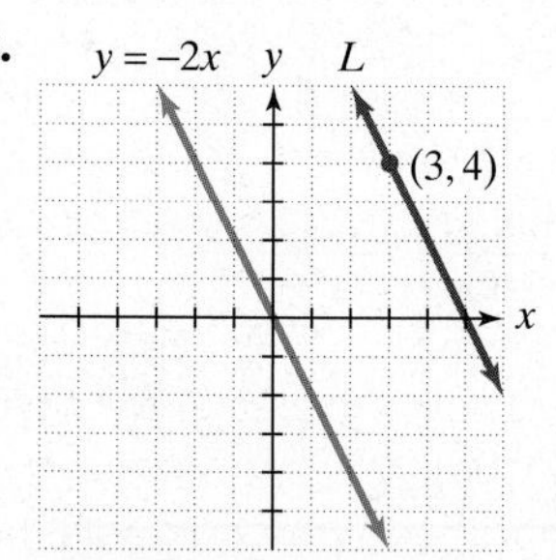

L is parallel to $y = -2x$.

19.

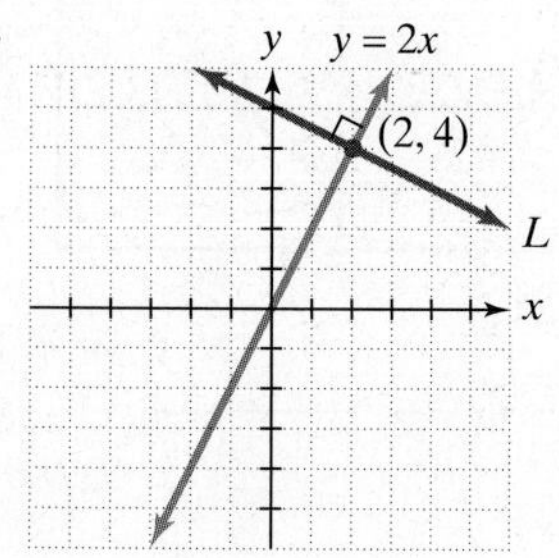

L is perpendicular to $y = 2x$.

20.

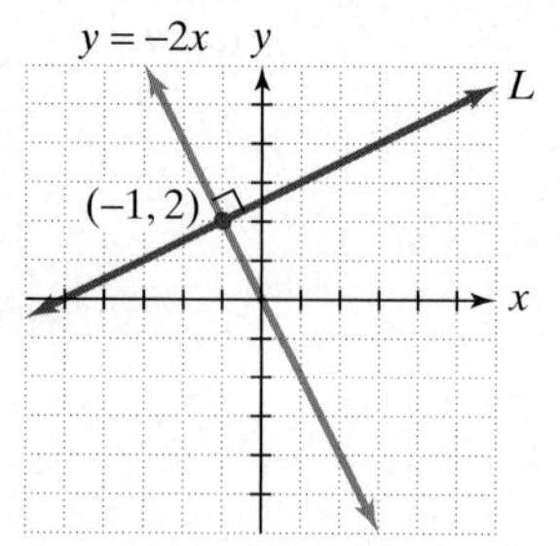

L is perpendicular to $y = -2x$.

In Exercises 21–28, use the given conditions to write an equation for each line in point-slope form and slope-intercept form.

21. Passing through $(-8, -10)$ and parallel to the line whose equation is $y = -4x + 3$

22. Passing through $(-2, -7)$ and parallel to the line whose equation is $y = -5x + 4$

23. Passing through $(2, -3)$ and perpendicular to the line whose equation is $y = \frac{1}{5}x + 6$

24. Passing through $(-4, 2)$ and perpendicular to the line whose equation is $y = \frac{1}{3}x + 7$

25. Passing through $(-2, 2)$ and parallel to the line whose equation is $2x - 3y - 7 = 0$

26. Passing through $(-1, 3)$ and parallel to the line whose equation is $3x - 2y - 5 = 0$

27. Passing through $(4, -7)$ and perpendicular to the line whose equation is $x - 2y - 3 = 0$

28. Passing through $(5, -9)$ and perpendicular to the line whose equation is $x + 7y - 12 = 0$

In Exercises 29–38, write the standard form of the equation of the circle with the given center and radius.

29. Center $(0, 0)$, $r = 7$
30. Center $(0, 0)$, $r = 8$
31. Center $(3, 2)$, $r = 5$
32. Center $(2, -1)$, $r = 4$
33. Center $(-1, 4)$, $r = 2$
34. Center $(-3, 5)$, $r = 3$
35. Center $(-3, -1)$, $r = \sqrt{3}$
36. Center $(-5, -3)$, $r = \sqrt{5}$
37. Center $(-4, 0)$, $r = 10$
38. Center $(-2, 0)$, $r = 6$

In Exercises 39–46, give the center and radius of the circle described by the equation and graph each equation.

39. $x^2 + y^2 = 16$
40. $x^2 + y^2 = 49$
41. $(x - 3)^2 + (y - 1)^2 = 36$
42. $(x - 2)^2 + (y - 3)^2 = 16$
43. $(x + 3)^2 + (y - 2)^2 = 4$
44. $(x + 1)^2 + (y - 4)^2 = 25$
45. $(x + 2)^2 + (y + 2)^2 = 4$
46. $(x + 4)^2 + (y + 5)^2 = 36$

In Exercises 47–54, complete the square and write the equation in standard form. Then give the center and radius of each circle and graph the equation.

47. $x^2 + y^2 + 6x + 2y + 6 = 0$
48. $x^2 + y^2 + 8x + 4y + 16 = 0$
49. $x^2 + y^2 - 10x - 6y - 30 = 0$
50. $x^2 + y^2 - 4x - 12y - 9 = 0$
51. $x^2 + y^2 + 8x - 2y - 8 = 0$
52. $x^2 + y^2 + 12x - 6y - 4 = 0$
53. $x^2 - 2x + y^2 - 15 = 0$
54. $x^2 + y^2 - 6y - 7 = 0$

Application Exercises

55. The line graph shows the number of people in the United States projected to live alone through 2010.

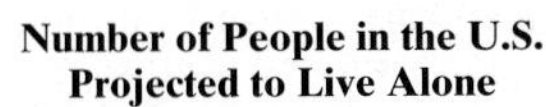

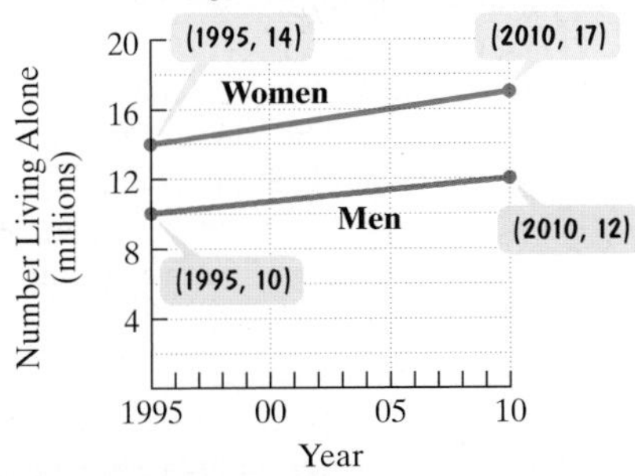

Source: Forrester Research

a. Find the slope of the line for U.S. women.
b. Find the slope of the line for U.S. men.
c. Are the lines parallel? What does this mean in terms of the yearly increase for women and the yearly increase for men?

56. The amount spent annually in college bookstores in the United States can be modeled by $y = 0.19x + 1.67$, where x represents the number of years since 1982 and y represents the amount spent in billions of dollars. If the graph of this equation is parallel to a line representing the amount spent annually in bookstores in the United States since 1982, what does this mean in terms of the yearly increase for spending on books?

57. We refer to the driveway in the figure shown at the top of the next column as being *circular*, meaning that it is bounded by two circles. The figure indicates that the radius of the larger circle is 52 feet and the radius of the smaller circle is 38 feet. All points on the circular driveway satisfy the following compound inequality:

all points on the smaller circle	$\le$	(x, y) on the driveway	$\le$	all points on the larger circle.

a. Rewrite the left portion of this inequality by writing the equation of the smaller circle.
b. Rewrite the right portion of this inequality by writing the equation of the larger circle.

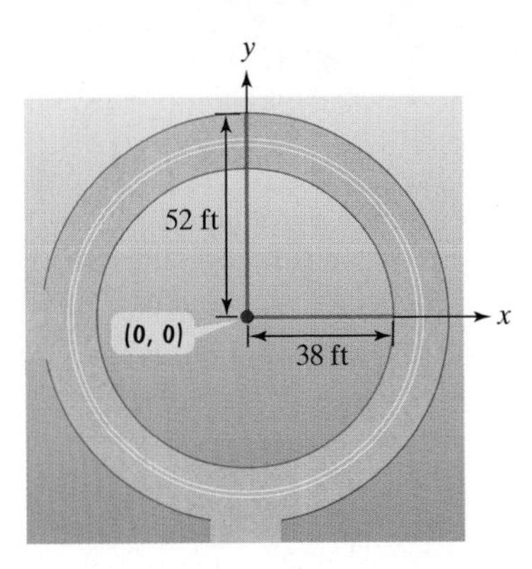

58. The circle formed by the middle lane of a circular running track can be described algebraically by $x^2 + y^2 = 4$, where all measurements are in miles. If you run around the track twice, approximately how many miles have you covered?

Writing in Mathematics

59. If two lines are parallel, describe the relationship between their slopes.

60. If two lines are perpendicular, describe the relationship between their slopes.

61. The number of multiple births in the United States (twins, triplets, etc.) per 1000 live births can be modeled by $y = 0.463x + 18.888$, where x represents the number of years since 1980 and y represents multiple births per 1000 live births. Explain why the equation of this line cannot be parallel to the line representing the number of births in the United States since 1980.

62. If you know a point on a line and you know the equation of a line perpendicular to this line, explain how to write the line's equation.

63. What is a circle? Without using variables, describe how your definition of a circle can be used to obtain a form of its equation.

64. Give an example of a circle's equation in standard form. Describe how to find the center and radius for this circle.

65. How is the standard form of a circle's equation obtained from its general form?

66. Does $(x - 3)^2 + (y - 5)^2 = 0$ represent the equation of a circle? If not, describe the graph of this equation.

67. Does $(x - 3)^2 + (y - 5)^2 = -25$ represent the equation of a circle? What sort of set is the graph of this equation?

Technology Exercises

68. The lines whose equations are $y = \frac{1}{3}x + 1$ and $y = -3x - 2$ are perpendicular because the product of their slopes, $\frac{1}{3}$ and -3, respectively, is -1.
   a. Use a graphing utility to graph the equations. Do the lines appear to be perpendicular?
   b. Now use the zoom square feature of your utility. Describe what happens to the graphs. Explain why this is so.

In Exercises 69–71, use a graphing utility to graph each circle whose equation is given.

69. $x^2 + y^2 = 25$

70. $(y + 1)^2 = 36 - (x - 3)^2$

71. $x^2 + 10x + y^2 - 4y - 20 = 0$

Critical Thinking Exercises

72. Which one of the following is true?
   a. The equation of the circle whose center is at the origin with radius 16 is $x^2 + y^2 = 16$.
   b. The graph of $(x - 3)^2 + (y + 5)^2 = 36$ is a circle with a radius 6 centered at $(-3, 5)$.
   c. The graph of $(x - 4) + (y + 6) = 25$ is a circle with a radius 5 centered at $(4, -6)$.
   d. None of the above is true.

In Exercises 73–74, write the point-slope form and the slope-intercept form of the equation for each line described.

73. Having an x-intercept of -3 and perpendicular to the line passing through $(0, 0)$ and $(6, -2)$

74. Perpendicular to $3x - 2y = 4$ with the same y-intercept

In Exercises 75–76, write the standard form and the general form of the equation of each circle.

75. Center at $(3, -5)$ and passing through the point $(-2, 1)$

76. Passing through $(-7, 2)$ and $(1, 2)$; these points lie on the line that passes through the circle's center.

77. Find the area of the region bounded by the graphs of $(x - 2)^2 + (y + 3)^2 = 25$ and $(x - 2)^2 + (y + 3)^2 = 36$.

78. A **tangent line** to a circle is a line that intersects the circle at exactly one point. The tangent line is perpendicular to the radius of the circle at this point of contact. Write the point-slope equation of a line tangent to the circle whose equation is $x^2 + y^2 = 25$ at the point $(3, -4)$.

SECTION 2.3 Introduction to Functions

Objectives

1. Find the domain and range of a relation.
2. Determine whether a relation is a function.
3. Determine whether an equation represents a function.
4. Evaluate a function.
5. Understand and use piecewise functions.
6. Find the domain of a function.
7. Interpret function values for functions that model data.

Enjoy talking on the phone? In 1999, nearly 80 million Americans were chatting up a storm on their mobile phones, an increase of 300% from 1994, when only 20 million Americans were using mobile phones. And who can blame them? The graph in Figure 2.18 shows the decrease in the monthly average U.S. mobile-phone bills from 1994 through 1998. With video mobile phones by 2020, there seems to be no limit to the ways in which we keep in touch.

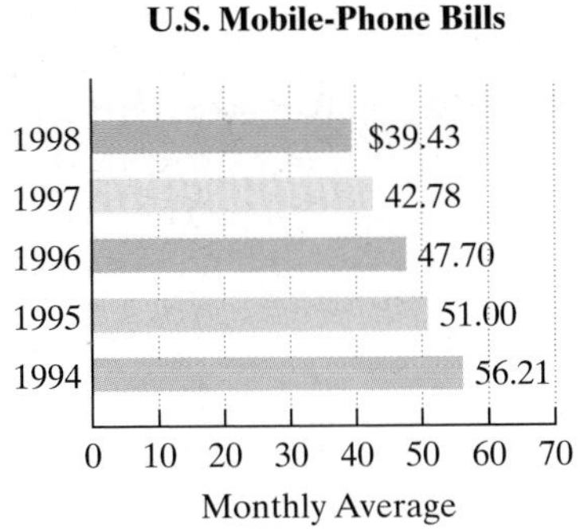

Figure 2.18

If we let x represent a year and y the monthly average mobile-phone bill, the graph in Figure 2.18 shows a correspondence between the two variables x and y. We can write this correspondence using a set of ordered pairs:

$$\{(1994, 56.21), (1995, 51.00), (1996, 47.70), (1997, 42.78), (1998, 39.43)\}$$

The mathematical term for a set of ordered pairs is a **relation**.

Definition of a Relation

A **relation** is any set of ordered pairs. The set of all first components of the ordered pairs is called the **domain** of the relation, and the set of all second components is called the **range** of the relation.

1 Find the domain and range of a relation.

EXAMPLE 1 Analyzing U.S. Mobile-Phone Bills as a Relation

Find the domain and range of the relation

$$\{(1994, 56.21), (1995, 51.00), (1996, 47.70), (1997, 42.78), (1998, 39.43)\}.$$

Solution The domain is the set of all first components. Thus, the domain is

$$\{1994, 1995, 1996, 1997, 1998\}.$$

The range is the set of all second components. Thus, the range is

$$\{56.21, 51.00, 47.70, 42.78, 39.43\}.$$

Check Point 1 Find the domain and the range of the relation $\{(20, 157.4), (30, 231.8), (100, 752.6), (200, 1496.6)\}$.

As you worked Checkpoint 1, did you wonder if the numbers in each ordered pair represented anything? Think snakes! The first number in each ordered pair is a snake's tail length, in millimeters, and the second number is its body length, also in millimeters. Consider, for example, the ordered pair (30, 231.8).

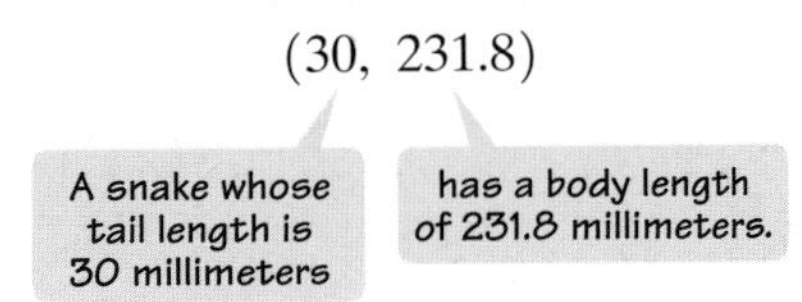

The relation in the snake example can be pictured as follows:

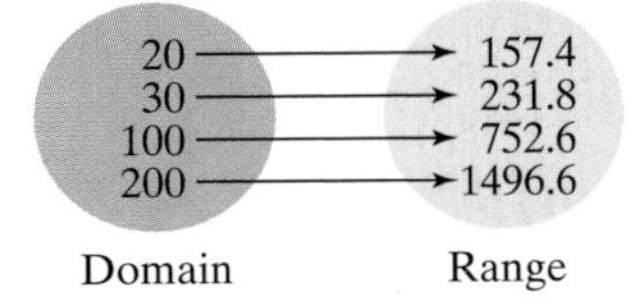

A scatter plot, like the one shown in Figure 2.19, is a way to represent the relation.

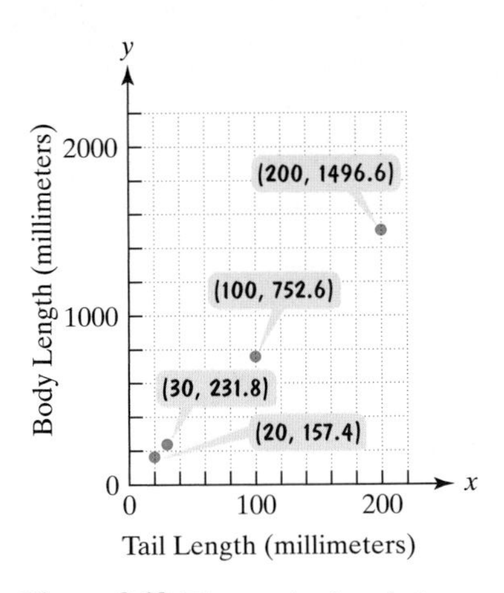

Figure 2.19 The graph of a relation showing a correspondence between a snake's tail length and its body length

Functions

The SAT is the test that everyone loves to hate. The scatter plot in Figure 2.20 on page 192 shows a relation indicating a correspondence between SAT scores and grade point averages for the first year in college for a group of randomly selected college students. The domain is the set of SAT scores for the students.

The range is the set of their grade point averages. Is it possible for two students with the same SAT score to have different grade point averages? Look for two or more data points that are aligned vertically. We see that there are two students who have the same SAT score, 700, but their grade point averages are different. One student has a grade point average of approximately 2.4 and the other a grade point average of approximately 3.7. These students are represented by the following ordered pairs:

$$(700,\ 2.4) \qquad (700,\ 3.7).$$

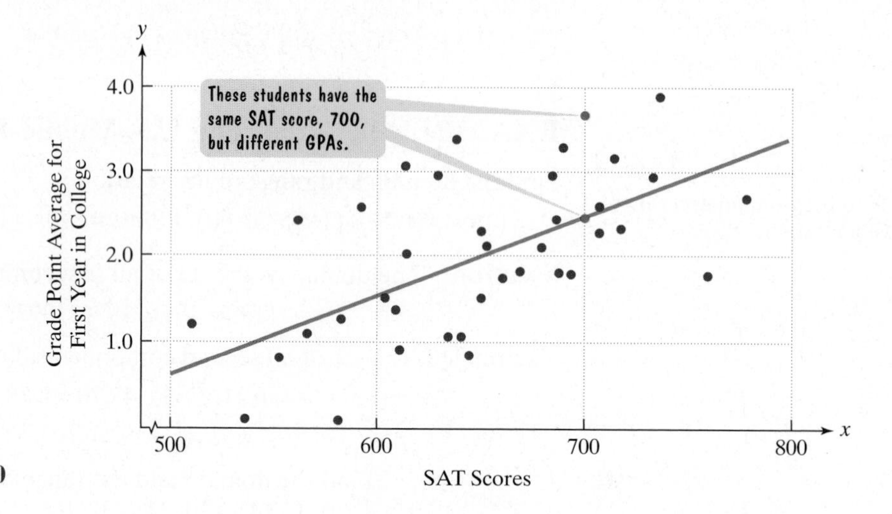

Figure 2.20

A relation in which each member of the domain corresponds to exactly one member of the range is a **function**. The relation in Figure 2.20, the SAT–grade point average scatter plot, is not a function because at least one member of the domain corresponds to two members of the range.

$$(700,\ 2.4) \qquad (700,\ 3.7)$$

The member of the domain, 700, corresponds to two members of the range, 2.4 and 3.7. Because a function is a relation in which **no two ordered pairs have the same first component and different second components**, the ordered pairs (700, 2.4) and (700, 3.7) are not ordered pairs of a function.

Same first components

$$(700,\ 2.4) \qquad (700,\ 3.7)$$

Different second components

Definition of a Function

A **function** is a correspondence between two sets X and Y that assigns to each element x of set X exactly one element y of set Y. For each element x in X, the corresponding element y in Y is called the **value** of the function at x. The set X is called the **domain** of the function, and the set of all function values, Y, is called the **range** of the function.

2 Determine whether a relation is a function.

EXAMPLE 2 Determining Whether a Relation is a Function

Determine whether each relation is a function.

a. $\{(1, 6), (2, 6), (3, 8), (4, 9)\}$ b. $\{(6, 1), (6, 2), (8, 3), (9, 4)\}$

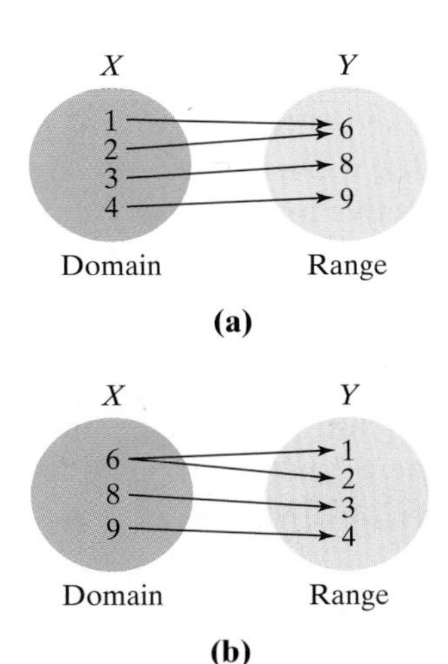

Figure 2.21

Solution We begin by making a figure for each relation that shows set X, the domain, and set Y, the range, shown in Figure 2.21.

a. Figure 2.21(a) shows that every element in the domain corresponds to exactly one element in the range. The element 1 in the domain corresponds to the element 6 in the range. Furthermore, 2 corresponds to 6, 3 corresponds to 8, and 4 corresponds to 9. No two ordered pairs in the given relation have the same first component and different second components. Thus, the relation is a function.

b. Figure 2.21(b) shows that 6 corresponds to both 1 and 2. If any element in the domain corresponds to more than one element in the range, the relation is not a function. This relation is not a function; two ordered pairs have the same first component and different second components.

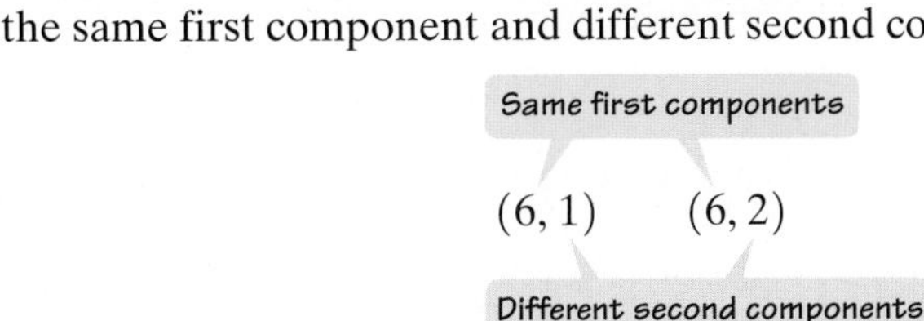

Study Tip

The word *range* can mean many things, from a chain of mountains to a cooking stove. For functions, it means the set of all function values. For graphing utilities, it means the setting used for the viewing rectangle. Try not to confuse these meanings.

Look at Figure 2.21 again. The fact that 1 and 2 in the domain have the same image, 6, in the range does not violate the definition of a function. A function can have two different first components with the same second component. By contrast, a relation is not a function when two different ordered pairs have the same first component and different second components. Thus, the relation in Example 2(b) is not a function.

Check Point 2 Determine whether each relation is a function.

a. $\{(1, 2), (3, 4), (5, 6), (5, 8)\}$

b. $\{(1, 2), (3, 4), (6, 5), (8, 5)\}$

Functions as Equations

Functions are usually given in terms of equations rather than as sets of ordered pairs. Earlier we noted that, for a particular snake, its total body length is a function of its tail length. The function is modeled by the equation

$$y = 7.44x + 8.6.$$

The variable x represents the snake's tail length, in millimeters. The variable y represents the snake's total body length, in millimeters. The variable y is a function of the variable x. For each value of x, there is one and only one value of y. The variable x is called the **independent variable** because it can be assigned any value from the domain. Thus, x can be assigned any positive number representing the snake's tail length. The variable y is called the **dependent variable** because its value depends on x. A snake's total body length depends on its tail length. The value of the dependent variable, y, is calculated after selecting a value for the independent variable, x.

3 Determine whether an equation represents a function.

We have seen that not every set of ordered pairs defines a function. Similarly, not all equations with the variables x and y define a function. If an equation is solved for y and more than one value of y can be obtained for a given x, then the equation does not define y as a function of x.

EXAMPLE 3 Determining Whether an Equation Represents a Function

Determine whether each equation defines y as a function of x.

a. $x^2 + y = 4$ b. $x^2 + y^2 = 4$

Solution Solve each equation for y in terms of x. If two or more values of y can be obtained for a given x, the equation is not a function.

a.

$x^2 + y = 4$	This is the given equation.
$x^2 + y - x^2 = 4 - x^2$	Solve for y by subtracting x^2 from both sides.
$y = 4 - x^2$	Simplify.

From this last equation we can see that for each value of x, there is one and only one value of y. For example, if $x = 1$, then $y = 4 - 1^2 = 3$. The equation defines y as a function of x.

b.

$x^2 + y^2 = 4$	This given equation describes a circle.
$x^2 + y^2 - x^2 = 4 - x^2$	Isolate y^2 by subtracting x^2 from both sides.
$y^2 = 4 - x^2$	Simplify.
$y = \pm\sqrt{4 - x^2}$	Apply the square root method.

The $\pm$ in this last equation shows that for certain values of x (all values between -2 and 2), there are two values of y. For example, if $x = 1$, then $y = \pm\sqrt{4 - 1^2} = \pm\sqrt{3}$. For this reason, the equation does not define y as a function of x.

Check Point 3 Solve each equation for y and then determine whether the equation defines y as a function of x.
a. $2x + y = 6$ **b.** $x^2 + y^2 = 1$

4 Evaluate a function.

Function Notation

When an equation represents a function, the function is often named by a letter such as f, g, h, F, G, or H. Any letter can be used to name a function. Suppose that f names a function. Think of the domain as the set of the function's inputs and the range as the set of the function's outputs. As shown in Figure 2.22, the input is represented by x and the output by $f(x)$. The special notation $f(x)$, read "f of x" or "f at x," represents the **value of the function at the number x**.

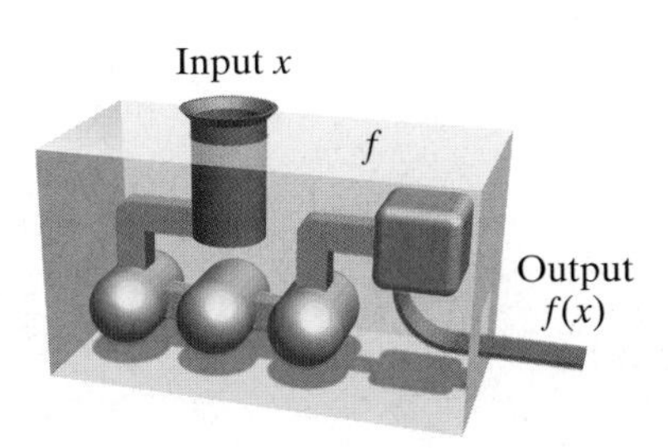

Figure 2.22 A function as a machine with inputs and outputs

Let's make this clearer by considering a specific example. We know that the equation $y = 4 - x^2$ defines y as a function of x. We'll name the function f. Now, we can apply our new function notation.

Input	Output	Equation	
x	$f(x)$	$f(x) = 4 - x^2$	We read this equation as "f of x equals $4 - x^2$."

Suppose that we are interested in finding $f(3)$, the function's output when the input is 3. To find the value of the function at 3, we substitute 3 for x. We are **evaluating the function** at 3.

$$f(x) = 4 - x^2 \quad \text{This is the given function.}$$
$$f(3) = 4 - 3^2 \quad \text{The input is 3.}$$
$$= 4 - 9$$
$$= -5$$

Study Tip

The notation $f(x)$ does *not* mean "f times x." The notation describes the value of the function at x.

The statement $f(3) = -5$ tells us that the value of the function at 3 is -5. When the function's input is 3, its output is -5. To find other function values, such as $f(-2), f(5)$, or $f(7)$, substitute the specified input values for x into the function's equation.

If a function is named f and x represents the independent variable, the notation $f(x)$ corresponds to the y-value for a given x. Thus,

$$f(x) = 4 - x^2 \quad \text{and} \quad y = 4 - x^2$$

define the same function. This function may be written as

$$y = f(x) = 4 - x^2.$$

EXAMPLE 4 Evaluating a Function

If $f(x) = x^2 + 3x + 5$, evaluate:

a. $f(2)$ b. $f(x + 3)$ c. $f(-x)$

Solution We substitute 2, $x + 3$, and $-x$ for x in the definition of f. When replacing x with a variable or an algebraic expression, you might find it helpful to think of the function's equation as

$$f(x) = x^2 + 3x + 5.$$

a. We find $f(2)$ by substituting 2 for x in the equation.

$$f(2) = 2^2 + 3 \cdot 2 + 5 = 4 + 6 + 5 = 15$$

Thus, $f(2) = 15$.

b. We find $f(x + 3)$ by substituting $x + 3$ for x in the equation.

$$f(x + 3) = (x + 3)^2 + 3(x + 3) + 5$$

Equivalently,

$$\begin{aligned} f(x + 3) &= (x + 3)^2 + 3(x + 3) + 5 \\ &= x^2 + 6x + 9 + 3x + 9 + 5 && \text{Square } x + 3 \text{ using } (A + B)^2 = A^2 + 2AB + B^2. \text{ Distribute 3 throughout the parentheses.} \\ &= x^2 + 9x + 23. && \text{Combine like terms.} \end{aligned}$$

c. We find $f(-x)$ by substituting $-x$ for x in the equation.

$$f(-x) = (-x)^2 + 3(-x) + 5$$

Equivalently,

$$\begin{aligned} f(-x) &= (-x)^2 + 3(-x) + 5 \\ &= x^2 - 3x + 5. \end{aligned}$$

Discovery

Using $f(x) = x^2 + 3x + 5$ and the answers in parts (b) and (c):

1. Is $f(x + 3)$ equal to $f(x) + f(3)$?
2. Is $f(-x)$ equal to $-f(x)$?

Check Point 4 If $f(x) = x^2 - 2x + 7$, evaluate:

a. $f(-5)$ **b.** $f(x + 4)$ **c.** $f(-x)$

5 Understand and use piecewise functions.

Piecewise Functions

The early part of the twentieth century was the golden age of immigration in America. More than 13 million people migrated to the United States between 1900 and 1914. By 1910, foreign-born residents accounted for 15% of the total U.S. population. The graph in Figure 2.23 shows the percentage of Americans who were foreign born throughout the twentieth century.

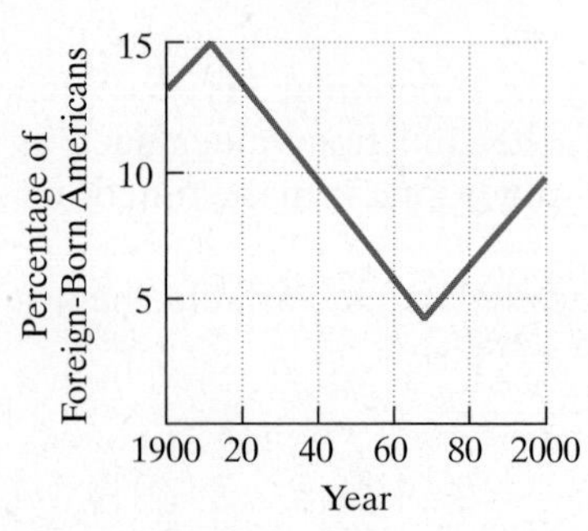

Source: U.S. Census Bureau

Figure 2.23

We can model the data from 1910 through 2000 with two functions, one from 1910 through 1970, years in which the percentage was decreasing, and one from 1970 through 2000, years in which the percentage was increasing. These two trends can be approximated by the function

$$P(t) = \begin{cases} -\dfrac{11}{60}t + 15 & \text{if } 0 \le t < 60 \\ \dfrac{1}{5}t - 8 & \text{if } 60 \le t \le 90 \end{cases}$$

in which t represents the number of years after 1910 and $P(t)$ is the percentage of foreign-born Americans. A function such as this that is defined by two (or more) equations over a specified domain is called a **piecewise function**.

EXAMPLE 5 Evaluating a Piecewise Function

Use the function $P(t)$, described previously, to find and interpret:

a. $P(30)$ b. $P(80)$

Solution

a. To find $P(30)$, we let $t = 30$. Because 30 is less than 60, we use the first line of the piecewise function.

$$P(t) = -\tfrac{11}{60}t + 15 \quad \text{This is the function's equation for } 0 \le t < 60.$$

$$P(30) = -\tfrac{11}{60} \cdot 30 + 15 \quad \text{Replace } t \text{ with } 60.$$

$$= 9.5$$

This means that 30 years after 1910, in 1940, 9.5% of Americans were foreign born.

b. To find $P(80)$, we let $t = 80$. Because 80 is between 60 and 90, we use the second line of the piecewise function.

$$P(t) = \frac{1}{5}t - 8 \quad \text{This is the function's equation for } 60 \le t \le 90.$$

$$P(80) = \tfrac{1}{5} \cdot 80 - 8 \quad \text{Replace } t \text{ with } 80.$$

$$= 8$$

This means that 80 years after 1910, in 1990, 8% of Americans were foreign born.

Check Point 5

If $f(x) = \begin{cases} x^2 + 3 & \text{if } x < 0 \\ 5x + 3 & \text{if } x \ge 0 \end{cases}$, find:

a. $f(-5)$ **b.** $f(6)$

6 Find the domain of a function.

The Domain of a Function

Let's reconsider the function that models the percentage of foreign-born Americans t years after 1910, up through and including 2000. The domain of this function is

$$\{0, 1, 2, 3, \ldots, 90\}.$$

0 years after 1910 is 1910.

3 years after 1910 is 1913.

90 years after 1910 brings the domain up to the year 2000.

Functions that model data often have their domains explicitly given along with the function's equation. However, for most functions, only an equation is given, and the domain is not specified. In cases like this, the domain of f is the largest set of real numbers for which the value of $f(x)$ is a real number. For example, consider the function

$$f(x) = \frac{1}{x - 3}.$$

Because division by 0 is undefined (and not a real number), the denominator $x - 3$ cannot be 0. Thus, x cannot equal 3. The domain of the function consists of all real numbers other than 3, represented by $\{x | x \neq 3\}$.

Just as the domain of a function must exclude real numbers that cause division by zero, it must also exclude real numbers that result in an even root of a negative number. For example, consider the function

$$g(x) = \sqrt{x}.$$

The equation tells us to take the square root of x. Because only nonnegative numbers have real square roots, the expression under the radical sign, x, must be greater than or equal to 0. The domain of g is $\{x | x \geq 0\}$ or the interval $[0, \infty)$.

Finding a Function's Domain

If a function f does not model data or verbal conditions, its domain is the largest set of real numbers for which the value of $f(x)$ is a real number. Exclude from a function's domain real numbers that cause division by zero and real numbers that result in an even root of a negative number.

EXAMPLE 6 Finding the Domain of a Function

Find the domain of each function:

a. $f(x) = x^2 - 7x$ b. $g(x) = \dfrac{6x}{x^2 - 9}$ c. $h(x) = \sqrt{3x + 12}$

Solution

a. The function $f(x) = x^2 - 7x$ contains neither division nor an even root. The domain of f is the set of all real numbers.

b. The function $g(x) = \dfrac{6x}{x^2 - 9}$ contains division. Because division by 0 is undefined, we must exclude from the domain values of x that cause $x^2 - 9$ to be 0. Thus, x cannot equal -3 or 3. The domain of function g is $\{x | x \neq -3, x \neq 3\}$.

c. The function $h(x) = \sqrt{3x + 12}$ contains an even root. Because only nonnegative numbers have real square roots, the quantity under the radical sign, $3x + 12$, must be greater than or equal to 0.

$$3x + 12 \geq 0$$
$$3x \geq -12$$
$$x \geq -4$$

The domain of h is $\{x | x \geq -4\}$ or the interval $[-4, \infty)$.

Technology

You can graph a function and visually determine its domain. For example, $h(x) = \sqrt{3x + 12}$, or $y = \sqrt{3x + 12}$, appears only for $x \geq -4$, verifying $[-4, \infty)$ as the domain.

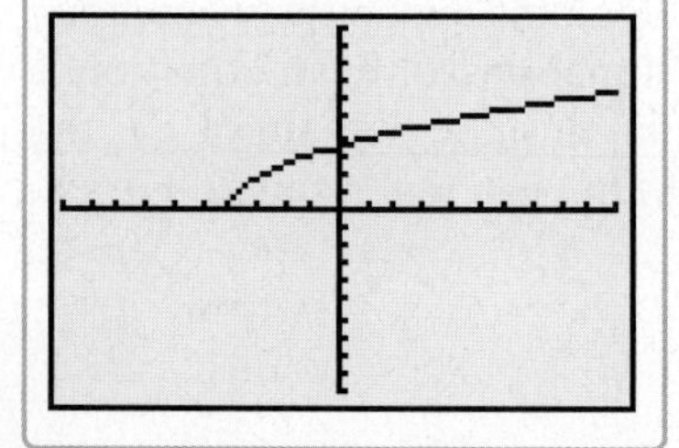

Check Point 6 Find the domain of each function:

a. $f(x) = x^2 + 3x - 17$ **b.** $g(x) = \dfrac{5x}{x^2 - 49}$

c. $h(x) = \sqrt{9x - 27}$

7 Interpret function values for functions that model data.

Applications

Like equations, functions can be obtained from verbal models or from actual data. We'll have lots of practice doing this throughout the book. For now, let's make sure that we can find and interpret function values for functions that were obtained from modeling data.

EXAMPLE 7 Evaluating a Function and Interpreting the Result

The function

$$f(x) = 0.0075x^2 - 0.2672x + 14.8$$

models the average number of miles per gallon of U.S. automobiles, y, x years after 1940. Find and interpret $f(18)$.

Solution

$f(x) = 0.0075x^2 - 0.2672x + 14.8$ This is the given function.

$f(18) = 0.0075(18)^2 - 0.2672(18) + 14.8$ Replace each occurrence of x by 18.

$= 12.4204$

We see that $f(18) = 12.4204$. Because 18 represents the number of years after 1940, this means that in 1958, U.S. automobiles averaged approximately 12.4 miles per gallon.

Check Point 7 Use the function in Example 7 to find and interpret $f(50)$.

EXERCISE SET 2.3

Practice Exercises

In Exercises 1–8, determine whether each relation is a function. Give the domain and range for each relation.

1. $\{(1,2), (3,4), (5,5)\}$
2. $\{(4,5), (6,7), (8,8)\}$
3. $\{(3,4), (3,5), (4,4), (4,5)\}$
4. $\{(5,6), (5,7)\ (6,6), (6,7)\}$
5. $\{(-3,-3), (-2,-2), (-1,-1), (0,0)\}$
6. $\{(-7,-7), (-5,-5), (-3,-3), (0,0)\}$
7. $\{(1,4), (1,5), (1,6)\}$
8. $\{(4,1), (5,1), (6,1)\}$

In Exercises 9–20, determine whether each equation defines y as a function of x.

9. $x + y = 16$
10. $x + y = 25$
11. $x^2 + y = 16$
12. $x^2 + y = 25$
13. $x^2 + y^2 = 16$
14. $x^2 + y^2 = 25$
15. $x = y^2$
16. $4x = y^2$
17. $y = \sqrt{x + 4}$
18. $y = -\sqrt{x + 4}$
19. $x + y^3 = 8$
20. $x + y^3 = 27$

In Exercises 21–32, evaluate each function at the given values of the independent variable and simplify.

21. $f(x) = 4x + 5$
    a. $f(6)$ **b.** $f(x + 1)$ **c.** $f(-x)$
22. $f(x) = 3x + 7$
    a. $f(4)$ **b.** $f(x + 1)$ **c.** $f(-x)$
23. $g(x) = x^2 + 2x + 3$
    a. $g(-1)$ **b.** $g(x + 5)$ **c.** $g(-x)$
24. $g(x) = x^2 - 10x - 3$
    a. $g(-1)$ **b.** $g(x + 2)$ **c.** $g(-x)$
25. $h(x) = x^4 - x^2 + 1$
    a. $h(2)$ **b.** $h(-1)$ **c.** $h(-x)$ **d.** $h(3a)$
26. $h(x) = x^3 - x + 1$
    a. $h(3)$ **b.** $h(-2)$ **c.** $h(-x)$ **d.** $h(3a)$
27. $f(r) = \sqrt{r + 6} + 3$
    a. $f(-6)$ **b.** $f(10)$ **c.** $f(x - 6)$
28. $f(r) = \sqrt{25 - r} - 6$
    a. $f(16)$ **b.** $f(-24)$ **c.** $f(25 - 2x)$

29. $f(x) = \dfrac{4x^2 - 1}{x^2}$
a. $f(2)$ **b.** $f(-2)$ **c.** $f(-x)$

30. $f(x) = \dfrac{4x^3 + 1}{x^3}$
a. $f(2)$ **b.** $f(-2)$ **c.** $f(-x)$

31. $f(x) = \dfrac{x}{|x|}$
a. $f(6)$ **b.** $f(-6)$ **c.** $f(r^2)$

32. $f(x) = \dfrac{|x + 3|}{x + 3}$
a. $f(5)$ **b.** $f(-5)$ **c.** $f(-9 - x)$

In Exercises 33–44, find and simplify:
a. $f(a)$
b. $f(a + h)$
c. $\dfrac{f(a + h) - f(a)}{h}, h \neq 0$
d. $f(a) + f(h)$

33. $f(x) = 4x$
34. $f(x) = 7x$
35. $f(x) = 3x + 7$
36. $f(x) = 6x + 1$
37. $f(x) = -5x - 3$
38. $f(x) = -8x - 9$
39. $f(x) = x^2$
40. $f(x) = 2x^2$
41. $f(x) = 6$
42. $f(x) = 7$
43. $f(x) = \dfrac{1}{x}$
44. $f(x) = \dfrac{1}{2x}$

In Exercises 45–50, evaluate each piecewise function at the given values of the independent variable.

45. $f(x) = \begin{cases} 3x + 5 & \text{if } x < 0 \\ 4x + 7 & \text{if } x \geq 0 \end{cases}$
a. $f(-2)$ **b.** $f(0)$ **c.** $f(3)$

46. $f(x) = \begin{cases} 6x - 1 & \text{if } x < 0 \\ 7x + 3 & \text{if } x \geq 0 \end{cases}$
a. $f(-3)$ **b.** $f(0)$ **c.** $f(4)$

47. $g(x) = \begin{cases} x + 3 & \text{if } x \geq -3 \\ -(x + 3) & \text{if } x < -3 \end{cases}$
a. $g(0)$ **b.** $g(-6)$ **c.** $g(-3)$

48. $g(x) = \begin{cases} x + 5 & \text{if } x \geq -5 \\ -(x + 5) & \text{if } x < -5 \end{cases}$
a. $g(0)$ **b.** $g(-6)$ **c.** $g(-5)$

49. $h(x) = \begin{cases} \dfrac{x^2 - 9}{x - 3} & \text{if } x \neq 3 \\ 6 & \text{if } x = 3 \end{cases}$
a. $h(5)$ **b.** $h(0)$ **c.** $h(3)$

50. $h(x) = \begin{cases} \dfrac{x^2 - 25}{x - 5} & \text{if } x \neq 5 \\ 10 & \text{if } x = 5 \end{cases}$
a. $h(7)$ **b.** $h(0)$ **c.** $h(5)$

In Exercises 51–72, find the domain of each function.

51. $f(x) = 4x^2 - 3x + 1$
52. $f(x) = 8x^2 - 5x + 2$
53. $g(x) = \dfrac{3}{x - 4}$
54. $g(x) = \dfrac{2}{x + 5}$
55. $h(x) = \dfrac{7x}{x^2 - 16}$
56. $h(x) = \dfrac{12x}{x^2 - 36}$
57. $f(x) = \dfrac{2}{(x + 3)(x - 7)}$
58. $f(x) = \dfrac{15}{(x + 8)(x - 3)}$
59. $H(r) = \dfrac{4}{r^2 + 11r + 24}$
60. $H(r) = \dfrac{5}{6r^2 + r - 2}$
61. $f(t) = \dfrac{3}{t^2 + 4}$
62. $f(t) = \dfrac{5}{t^2 + 9}$
63. $f(x) = \sqrt{x - 3}$
64. $f(x) = \sqrt{x + 2}$
65. $f(x) = \dfrac{1}{\sqrt{x - 3}}$
66. $f(x) = \dfrac{1}{\sqrt{x + 2}}$
67. $g(x) = \sqrt{5x + 35}$
68. $g(x) = \sqrt{7x - 70}$
69. $f(x) = \sqrt{24 - 2x}$
70. $f(x) = \sqrt{84 - 6x}$
71. $f(x) = \sqrt{x^2 - 5x - 14}$
72. $f(x) = \sqrt{x^2 - 5x - 24}$

Application Exercises

It seems that Phideau's medical bills are costing us an arm and a paw. The graph shows veterinary costs, in billions of dollars, for dogs and cats in five selected years. Use the graph to solve Exercises 73–74.

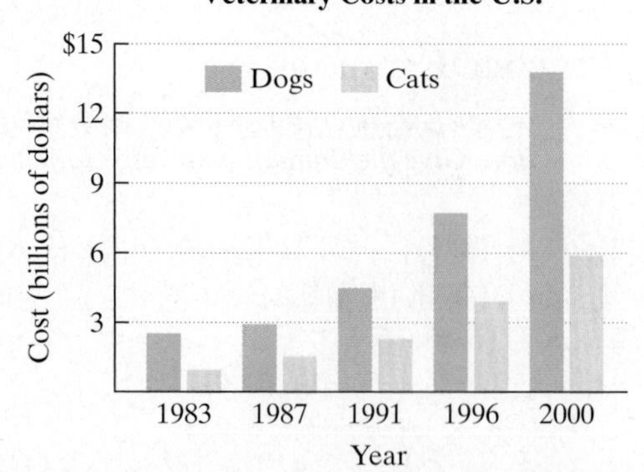

Source: American Veterinary Medical Association

73. Write five ordered pairs that approximate veterinary costs for dogs for the years shown. Find the domain and the range of the relation. Is this relation a function? Explain your answer.

74. Write five ordered pairs that approximate veterinary costs for cats for the years shown. Find the domain and the range of the relation. Is this relation a function? Explain your answer.

The number of women enrolled in U.S. colleges can be modeled by the function $f(x) = 0.07x + 4.1$, where x represents the number of years since 1984 and $f(x)$ represents enrollment in millions. The number of men enrolled in U.S. colleges can be modeled by the function $g(x) = 0.01x + 3.9$, where x represents the number of years since 1984 and $g(x)$ represents enrollment, in millions. In Exercises 75–78, use these functions to find and interpret:

75. $f(16)$ **76.** $g(16)$

77. $f(20) - g(20)$ **78.** $f(25) - g(25)$

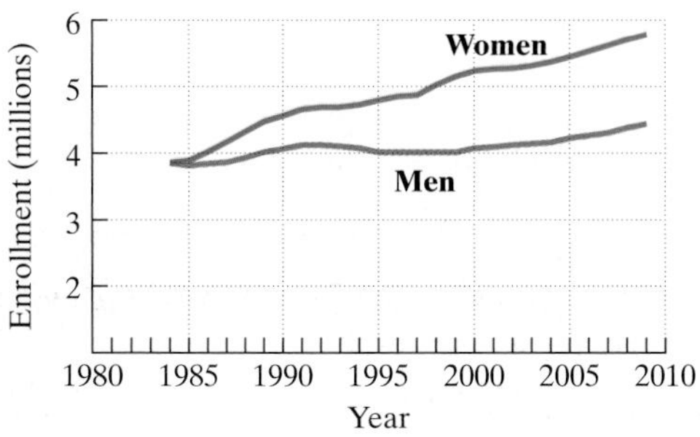

Source: Department of Education

The number of lawyers in the United States can be modeled by the function

$$f(x) = \begin{cases} 6.5x + 200 & \text{if } 0 \le x < 23 \\ 26.2x - 252 & \text{if } x \ge 23 \end{cases}$$

where x represents the number of years since 1951 and $f(x)$ represents the number of lawyers, in thousands. In Exercises 79–82, use this function to find and interpret:

79. $f(0)$ **80.** $f(10)$

81. $f(50)$ **82.** $f(60)$

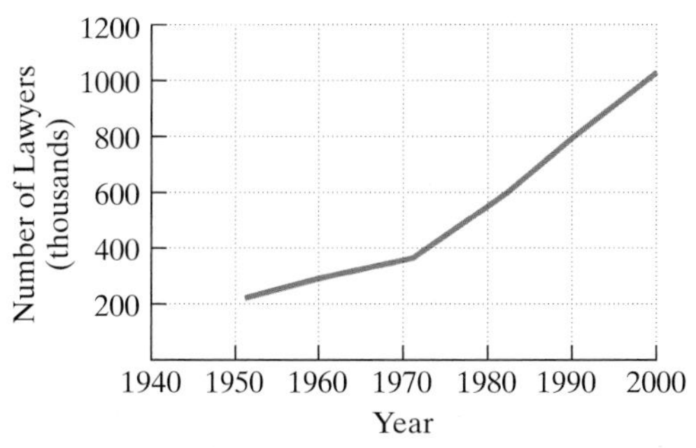

Source: Hudson Institute

83. On the average, infant girls weigh 7 pounds at birth and gain 1.5 pounds each month for the first six months. The function $f(x) = 1.5x + 7$ models this, where x represents the infant's age, in months, $x \le 6$, and $f(x)$ describes the baby's weight, in pounds. Use the function to find $f(0)$, $f(2)$, $f(4)$, and $f(6)$. Describe what these results mean. Identify each of your computations as an appropriate point on the graph at the top of the next column.

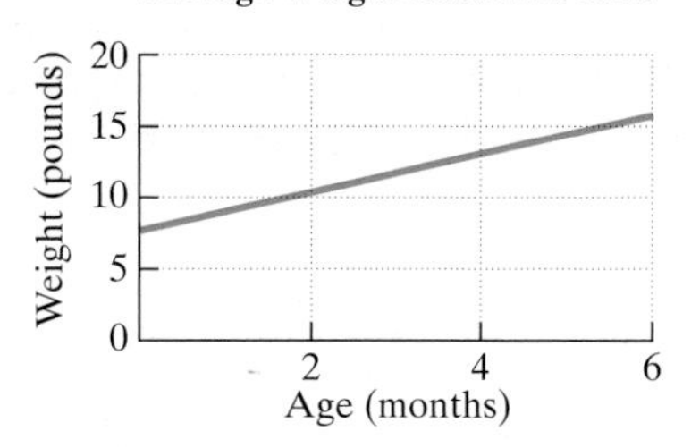

We're eating more and getting heavier. The number of calories consumed each day per person in the United States can be modeled by

$$f(x) = 0.89x^2 - 1.93x + 3306.27$$

where x is the number of years since 1974 and $f(x)$ represents the number of calories consumed each day per person. In Exercises 84–87, use this function to find and interpret:

84. $f(0)$ **85.** $f(10)$

86. $f(15) - f(0)$ **87.** $f(15) - f(10)$

Getting Heavier: Percentage of U.S. Men and Women Overweight

Men	**1999**	**2025**
Overweight	62%	73%
Obese	19%	37%
Women		
Overweight	47%	75%
Obese	19%	35%

Source: Beth Israel Medical Center

88. During a particular year, the taxes owed by a married person filing separately with an adjusted gross income of x dollars is given by the piecewise function

$$T(x) = \begin{cases} 0.15x & \text{if } 0 \le x < 17{,}900 \\ 0.28(x - 17{,}900) + 2685 & \text{if } 17{,}900 \le x < 43{,}250 \\ 0.31(x - 43{,}250) + 9783 & \text{if } x \ge 43{,}250 \end{cases}$$

Find and interpret $T(70{,}000) - T(40{,}000)$.

89. The function

$$f(x) = \begin{cases} 0.0005x^2 + 0.025x + 8.8 & \text{if } 0 \le x < 30 \\ 0.0202x^2 - 1.58x + 39.2 & \text{if } x \ge 30 \end{cases}$$

models the average number of miles (in thousands) driven per car in the United States, per year, x years after 1940. Find and interpret

a. $f(15)$ **b.** $f(50)$

90. A car was purchased for \$22,500. The value of the car decreases by \$3200 per year for the first seven years. Write a function V that describes the value of the car after x years, where $0 \le x \le 7$. Then find and interpret $V(3)$.

91. A car was purchased for \$17,900. The value of the car decreases by \$2100 per year for the first six years. Write a function V that describes the value of the car after x years, where $0 \le x \le 6$. Then find and interpret $V(4)$.

Writing in Mathematics

92. If a relation is represented by a set of ordered pairs, explain how to determine whether the relation is a function.

93. How do you determine if an equation in x and y defines y as a function of x?

94. A student in introductory algebra hears that functions are studied in subsequent algebra courses. The student asks you what a function is. Provide the student with a clear, relatively concise response.

95. Describe one advantage of using $f(x)$ rather than y in a function's equation.

96. What is a piecewise function?

97. How is the domain of a function determined?

98. For people filing a single return, federal income tax is a function of adjusted gross income because for each value of adjusted gross income there is a specific tax to be paid. On the other hand, the price of a house is not a function of the lot size on which the house sits because houses on same-sized lots can sell for many different prices.

a. Describe an everyday situation between variables that is a function.

b. Describe an everyday situation between variables that is not a function.

Technology Exercises

Use a graphing utility to find the domain of each function in Exercises 99–101. Then verify your observation algebraically.

99. $f(x) = \sqrt{x - 1}$

100. $g(x) = \sqrt{2x + 6}$

101. $h(x) = \sqrt{15 - 3x}$

102. Graph $y = 0.89x^2 - 1.93x + 3306.27$, the model for caloric consumption used in Exercises 84–87, in a $[0, 15, 1]$ by $[3200, 3500, 15]$ viewing rectangle. Describe one bit of information revealed by the graph of the function that is not obvious by looking at its equation.

Critical Thinking Exercises

103. Write a function defined by an equation in x whose domain is $\{x \mid x \ne -4, x \ne 11\}$.

104. Write a function defined by an equation in x whose domain is $[-6, \infty)$.

105. Give an example of an equation that does not define y as a function of x but that does define x as a function of y.

106. If $f(x) = ax^2 + bx + c$ and $r_1 = \frac{-b + \sqrt{b^2 - 4ac}}{2a}$, find $f(r_1)$ without doing any algebra and explain how you arrived at your result.

Group Exercise

107. Almanacs, newspapers, magazines, and the Internet contain bar graphs and line graphs that describe how things are changing over time. For example, the graph in Figure 2.18 on page 191 shows how mobile-phone bills are changing over time. Find a bar or line graph showing yearly changes that you find intriguing. Describe to the group what interests you about this data. The group should select their two favorite graphs. For each graph selected:

a. Rewrite the data so that they are presented as a relation in the form of a set of ordered pairs.

b. Determine whether the relation in part (a) is a function. Explain why the relation is a function, or why it is not.

SECTION 2.4 *Graphs of Functions*

Objectives

1. Graph functions by plotting points.
2. Obtain information about a function from its graph.
3. Use the vertical line test to identify functions.
4. Identify intervals on which a function increases, decreases, or is constant.
5. Identify even or odd functions and recognize their symmetries.
6. Recognize graphs of common functions.
7. Graph step functions.
8. Obtain information from graphs of functions in applied situations.

Have you ever seen a gas-guzzling car from the 1950s, with its huge fins and overstated design? The worst year for automobile fuel efficiency was 1958, when cars averaged a dismal 12.4 miles per gallon. We ended the last section with a function that modeled automobile fuel efficiency over time. If we graph the function's equation, we will get a much better idea of the relationship between time and fuel efficiency. In this section we will learn how to use the graph of a function to obtain useful information about the function.

Graphs of Functions

A graph enables us to visualize a function's behavior. The graph shows the relationship between the function's two variables more clearly than the function's equation does. The **graph of a function** is the graph of its ordered pairs. For example, the graph of $f(x) = \sqrt{x}$ is the set of points (x, y) in the rectangular coordinate system satisfying the equation $y = \sqrt{x}$. Thus, one way to graph a function is by plotting several of its ordered pairs and drawing a line or smooth curve through them. With the function's graph, we can picture its domain on the x-axis and its range on the y-axis. Our first example illustrates how this is done.

1 Graph functions by plotting points.

EXAMPLE 1 Graphing a Function by Plotting Points

Graph $f(x) = x^2 + 1$. To do so, use integer values of x from the set $\{-3, -2, -1, 0, 1, 2, 3\}$ to obtain seven ordered pairs. Plot each ordered pair and draw a smooth curve through the points. Use the graph to specify the function's domain and range.

Solution The graph of $f(x) = x^2 + 1$ is, by definition, the graph of $y = x^2 + 1$. We begin by setting up a partial table of coordinates.

x	$f(x) = x^2 + 1$	(x, y) or $(x, f(x))$
-3	$f(-3) = (-3)^2 + 1 = 10$	$(-3, 10)$
-2	$f(-2) = (-2)^2 + 1 = 5$	$(-2, 5)$
-1	$f(-1) = (-1)^2 + 1 = 2$	$(-1, 2)$
0	$f(0) = 0^2 + 1 = 1$	$(0, 1)$
1	$f(1) = 1^2 + 1 = 2$	$(1, 2)$
2	$f(2) = 2^2 + 1 = 5$	$(2, 5)$
3	$f(3) = 3^2 + 1 = 10$	$(3, 10)$

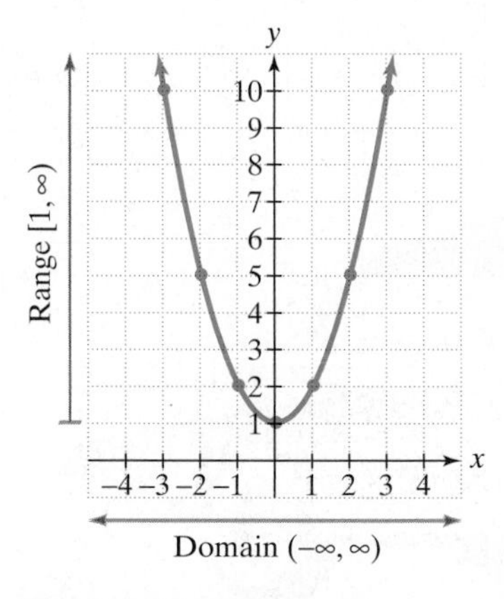

Figure 2.24 The graph of $f(x) = x^2 + 1$

Now, we plot the seven points and draw a smooth curve through them, as shown in Figure 2.24. The graph of f has a cuplike shape. The points on the graph of f have x-coordinates that extend indefinitely far to the left and to the right. Thus, the domain consists of all real numbers, represented by $(-\infty, \infty)$. By contrast, the points on the graph have y-coordinates that start at 1 and extend indefinitely upward. Thus, the range consists of all real numbers greater than or equal to 1, represented by $[1, \infty)$.

Check Point 1 Graph $f(x) = x^2 - 2$, using integers from -3 to 3 for x in the partial table of coordinates. Use the graph to specify the function's domain and range.

Technology

Does your graphing utility have a TABLE feature? If so, you can use it to create tables of coordinates for a function. You will need to enter the equation of the function and specify the starting value for x TblStart and the increment between successive x-values ΔTbl. For the table of coordinates in Example 1, we start the table at $x = -3$ and increment by 1. Using the up- or down-arrow keys, you can scroll through the table and determine as many ordered pairs of the graph as desired.

2 Obtain information about a function from its graph.

Obtaining Information from Graphs

You can obtain information about a function from its graph. At the right or left of a graph, you will find closed dots, open dots, or arrows.

- A closed dot indicates that the graph does not extend beyond this point and the point belongs to the graph.
- An open dot indicates that the graph does not extend beyond this point and the point does not belong to the graph.
- An arrow indicates that the graph extends indefinitely in the direction in which the arrow points.

EXAMPLE 2 Obtaining Information from a Function's Graph

Use the graph of the function f, shown in Figure 2.25, to answer the following questions.

a. What are the function values $f(-1)$ and $f(1)$?

b. What is the domain of f?

c. What is the range of f?

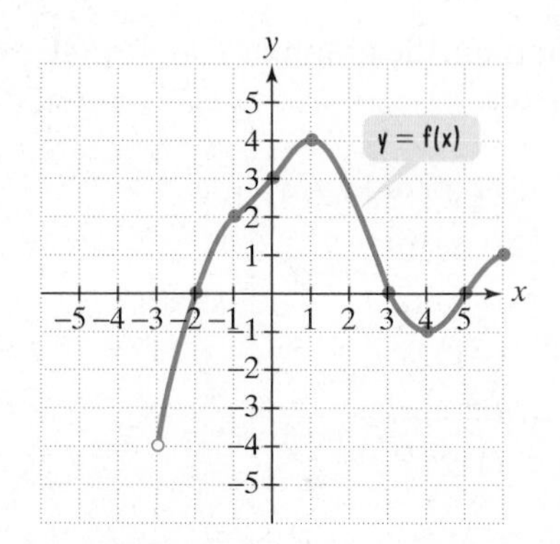

Figure 2.25

Solution

a. Because $(-1, 2)$ is a point on the graph of f, the y-coordinate, 2, is the value of the function at the x-coordinate, -1. Thus, $f(-1) = 2$. Similarly, because $(1, 4)$ is also a point on the graph of f, this indicates that $f(1) = 4$.

b. The open dot on the left shows that $x = -3$ is not in the domain of f. By contrast, the closed dot on the right shows that $x = 6$ is in the domain of f. We determine the domain of f by noticing that the points on the graph of f have x-coordinates between -3, excluding -3, and 6, including 6. For each number x between -3 and 6, there is a point $(x, f(x))$ on the graph. Thus, the domain of f is $\{x|-3 < x \leq 6\}$ or the interval $(-3, 6]$.

c. The points on the graph all have y-coordinates between -4, not including -4, and 4, including 4. The graph does not extend below $y = -4$ or above $y = 4$. Thus, the range of f is $\{y|-4 < y \leq 4\}$ or the interval $(-4, 4]$.

Check Point 2 Use the graph of function f, shown below, to find $f(4)$, the domain, and the range.

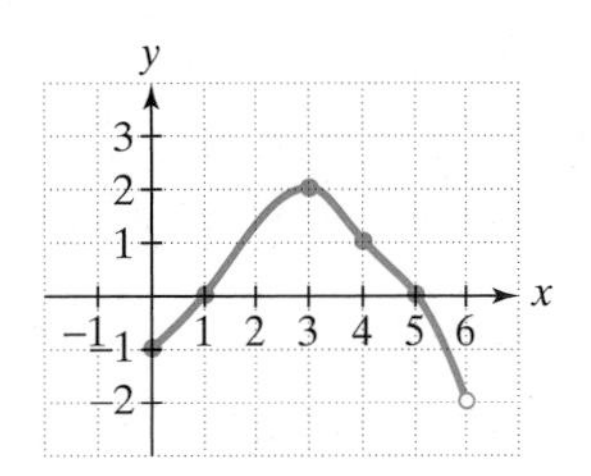

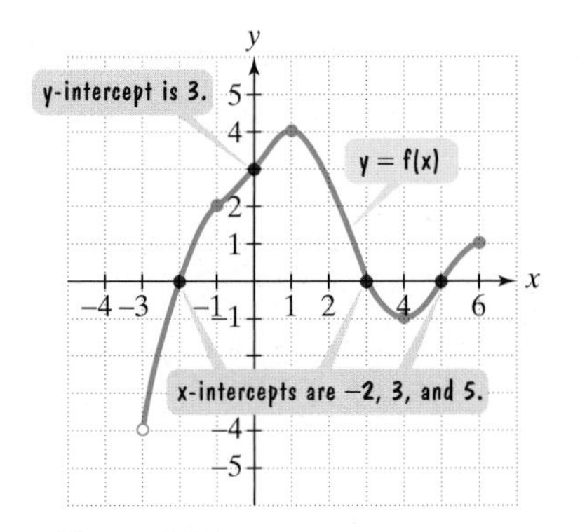

Figure 2.26 Identifying intercepts

Figure 2.26 illustrates how we can identify a graph's intercepts. To find the x-intercepts, look for the points at which the graph crosses the x-axis. There are three such points: $(-2, 0)$, $(3, 0)$, and $(5, 0)$. Thus, the x-intercepts are -2, 3, and 5. We express this in function notation by writing $f(-2) = 0$, $f(3) = 0$, and $f(5) = 0$.

To find the y-intercept, look for the point at which the graph crosses the y-axis. This occurs at $(0, 3)$. Thus, the y-intercept is 3. We express this in function notation by writing $f(0) = 3$.

By the definition of a function, for each value of x we can have at most one value for y. What does this mean in terms of intercepts? A function can have more than one x-intercept but at most one y-intercept.

3 Use the vertical line test to identify functions.

The Vertical Line Test

Not every graph in the rectangular coordinate system is the graph of a function. The definition of a function specifies that no value of x can be paired with two or more different values of y. Consequently, if a graph contains two or more different points with the same first coordinate, the graph cannot represent a function. This is illustrated in Figure 2.27. Observe that points sharing a common first coordinate are vertically above or below each other.

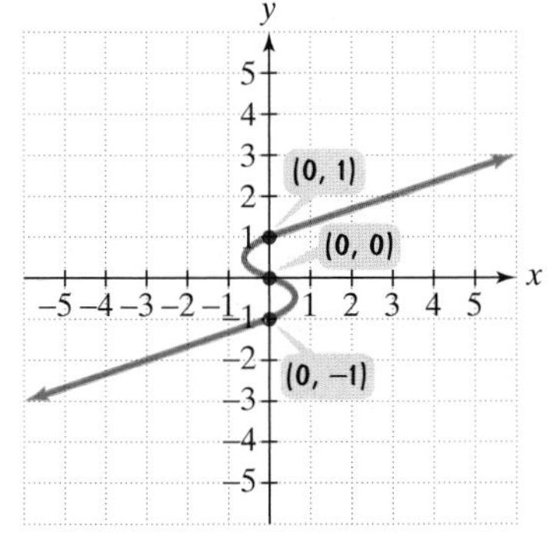

Figure 2.27 y is not a function of x because 0 is paired with three values of y, namely, 1, 0, and -1.

This observation is the basis of a useful test for determining whether a graph defines y as a function of x. The test is called the **vertical line test**.

The Vertical Line Test for Functions

If any vertical line intersects a graph in more than one point, the graph does not define y as a function of x.

EXAMPLE 3 Using the Vertical Line Test

Use the vertical line test to identify graphs in which y is a function of x.

a.
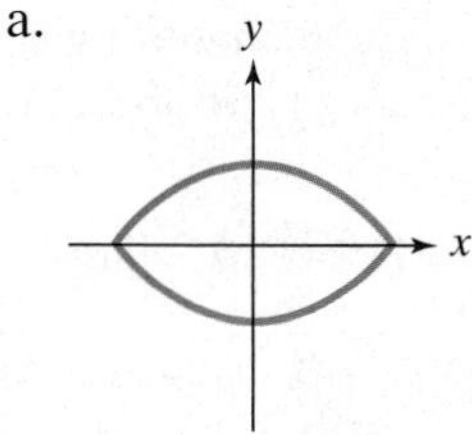

b.
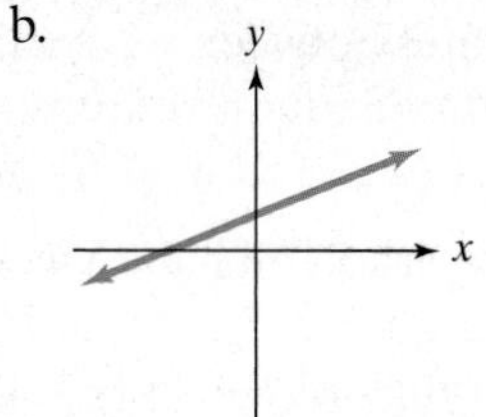

c.
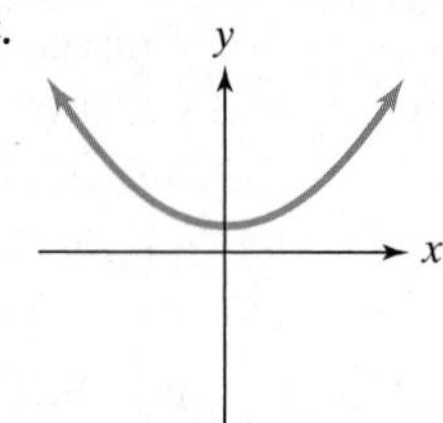

d.
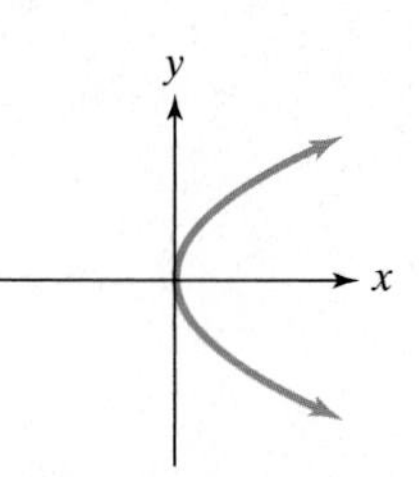

Solution y is a function of x for the graphs in (b) and (c).

a.
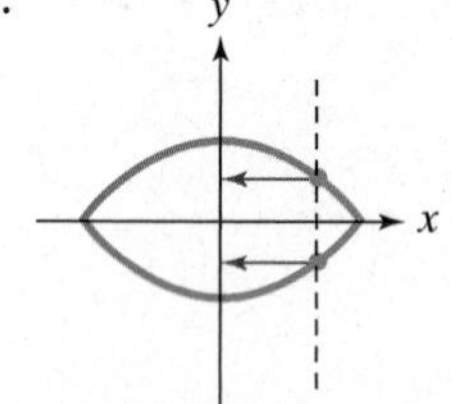

y **is not a function** of x. Two values of y correspond to an x-value.

b.
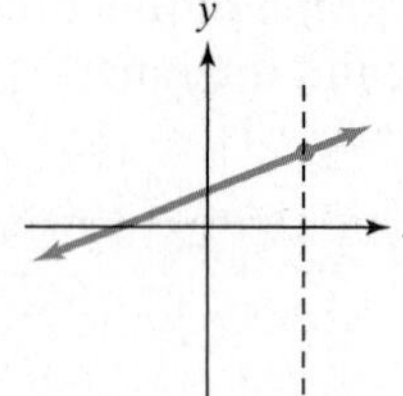

y **is a function** of x.

c.
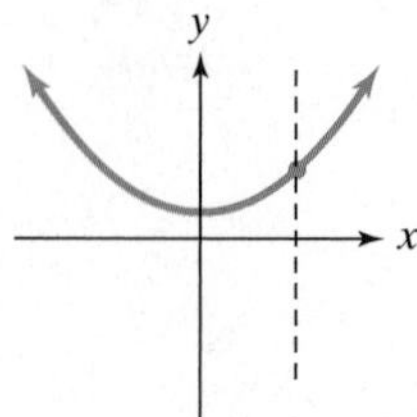

y **is a function** of x.

d.
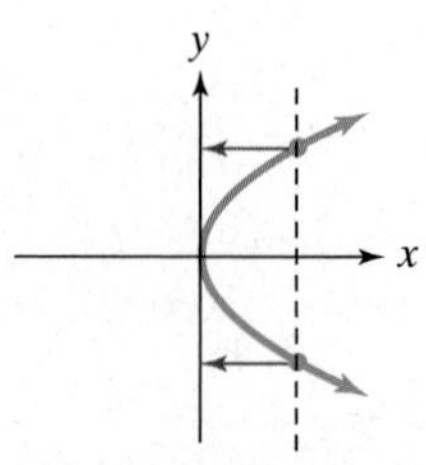

y **is not a function** of x. Two values of y correspond to an x-value.

Check Point 3 Use the vertical line test to identify graphs in which y is a function of x.

a.
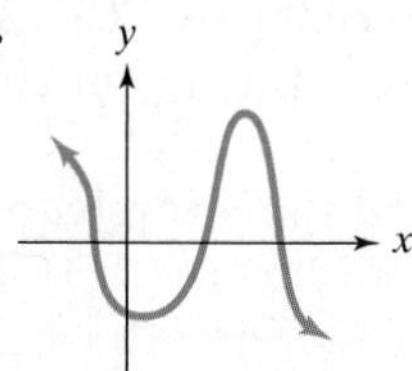

b.
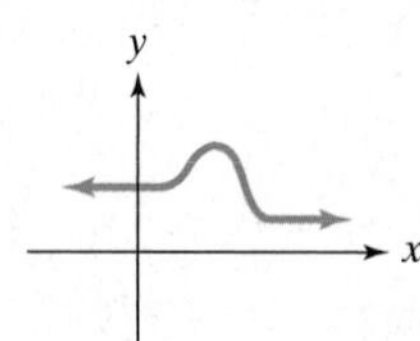

c.
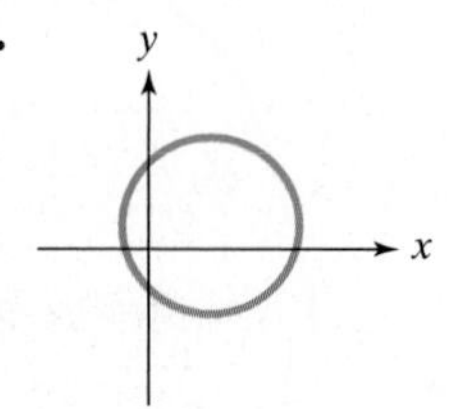

4 Identify intervals on which a function increases, decreases, or is constant.

Increasing and Decreasing Functions

Too late for that flu shot now! It's only 8 A.M. and you're feeling lousy. Your temperature is 101°F. Fascinated by the way that algebra models the world (your author is projecting a bit here), you decide to construct graphs showing your body temperature as a function of the time of day. You decide to let x represent the number of hours after 8 A.M. and $f(x)$ your temperature at time x.

At 8 A.M. your temperature is 101°F and you are not feeling well. However, your temperature starts to decrease. It reaches normal (98.6°F) by 11 A.M. Feeling energized, you construct the graph shown on the right, indicating decreasing temperature for $\{x \mid 0 < x < 3\}$ or on the interval $(0, 3)$.

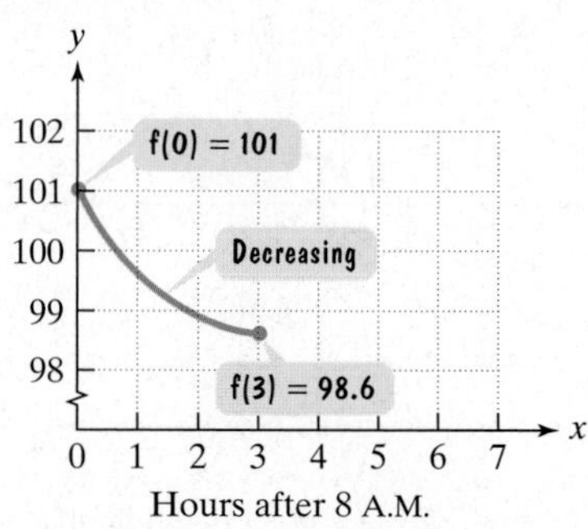

Temperature decreases on (0, 3), reaching 98.6° by 11 A.M.

Did creating that first graph drain you of your energy? Your temperature starts to rise after 11 A.M. By 1 P.M., 5 hours after 8 A.M., your temperature reaches 100°F. However, you keep plotting points on your graph. At right, we can see that your temperature increases for $\{x|3 < x < 5\}$ or on the interval (3, 5).

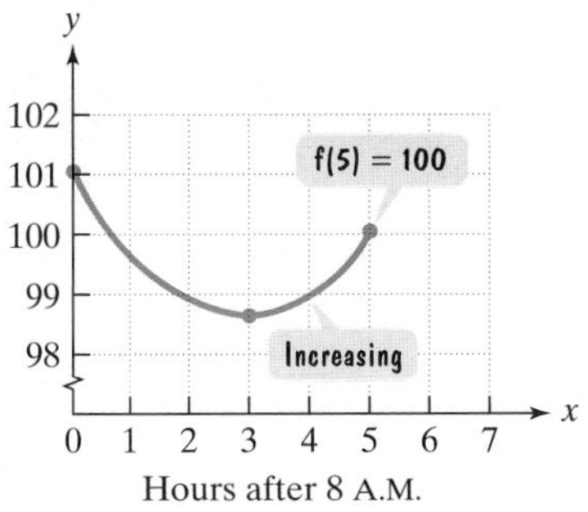

Temperature increases in (3, 5).

By 3 P.M., your temperature is no worse than it was at 1 P.M.: It is still 100°F. (Of course, it's no better, either.) Your temperature remained the same, or constant, for $\{x|5 < x < 7\}$ or on the interval (5, 7).

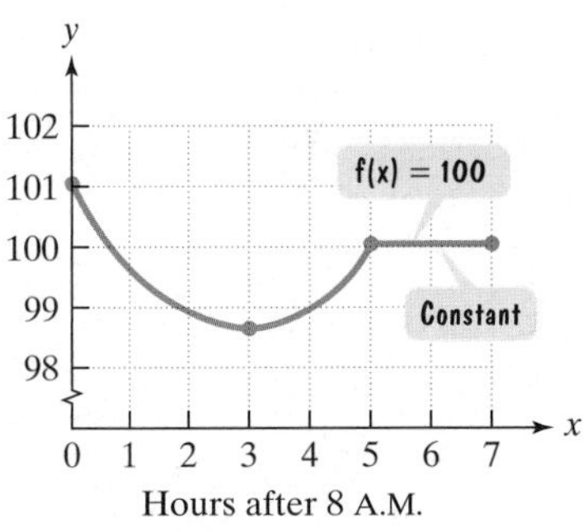

Temperature remains constant at 100° in (5, 7).

The time-temperature flu scenario illustrates that a function f is increasing when its graph rises, decreasing when its graph falls, and remains constant when it neither rises nor falls. Let's now provide a more algebraic description for these intuitive concepts.

Increasing, Decreasing, and Constant Functions

1. A function is **increasing** on an interval if for any x_1 and x_2 in the interval, where $x_1 < x_2$, then $f(x_1) < f(x_2)$.
2. A function is **decreasing** on an interval if for any x_1 and x_2 in the interval, where $x_1 < x_2$, then $f(x_1) > f(x_2)$.
3. A function is **constant** on an interval if for any x_1 and x_2 in the interval, where $x_1 < x_2$, then $f(x_1) = f(x_2)$.

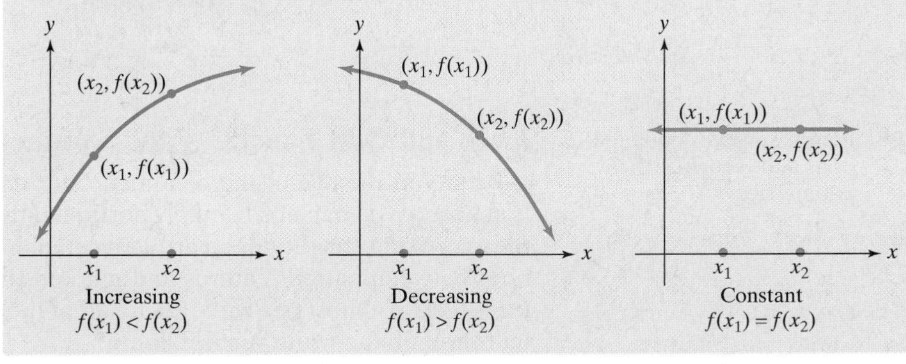

EXAMPLE 4 Intervals on Which a Function Increases, Decreases, or Is Constant

Describe the increasing, decreasing, or constant behavior of each function whose graph is shown.

a.

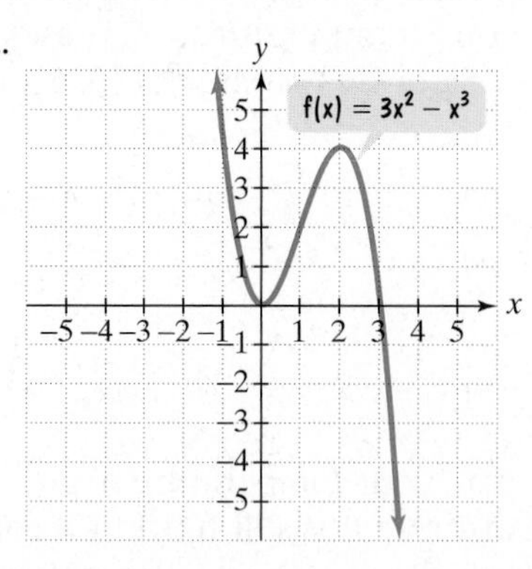

b.

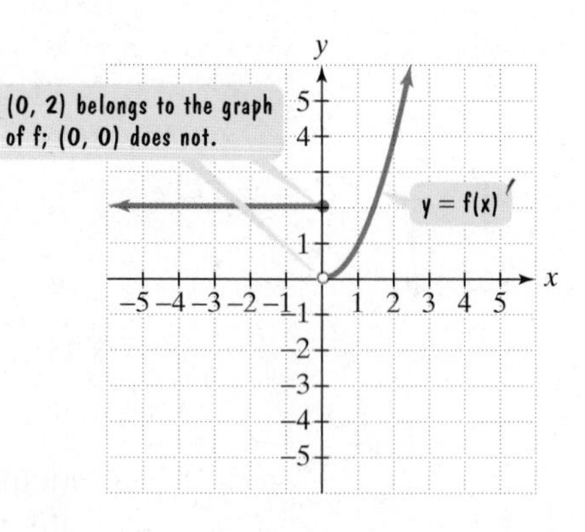

Solution

a. The function is decreasing on the interval $(-\infty, 0)$, increasing on the interval $(0, 2)$, and decreasing on the interval $(2, \infty)$.

b. Although the function's equations are not given, the graph indicates that the function is defined in two pieces. The part of the graph to the left of the y-axis shows that the function is constant on the interval $(-\infty, 0)$. The part to the right of the y-axis shows that the function is increasing on the interval $(0, \infty)$.

Check Point 4 Describe the increasing, decreasing, or constant behavior of the function whose graph is shown.

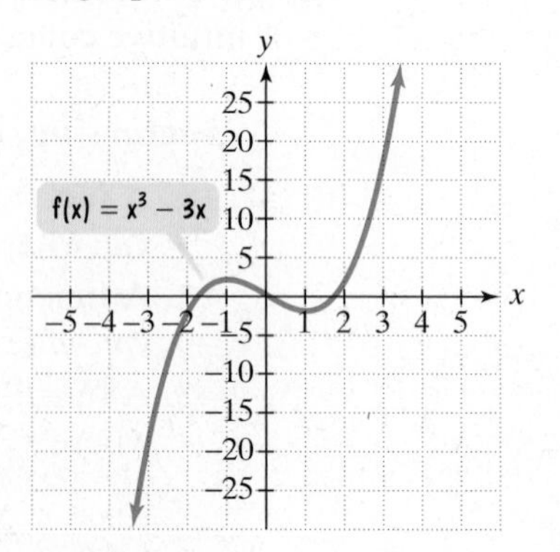

5 Identify even or odd functions and recognize their symmetries.

Even and Odd Functions and Symmetry

Is beauty in the eye of the beholder? Or are there certain objects (or people) that are so well balanced and proportioned that they are universally pleasing to the eye? What constitutes an attractive human face? In Figure 2.28, we've drawn lines between paired features and marked the midpoints. Notice how the features line up almost perfectly. Each half of the face is a mirror image of the other half through the white vertical line.

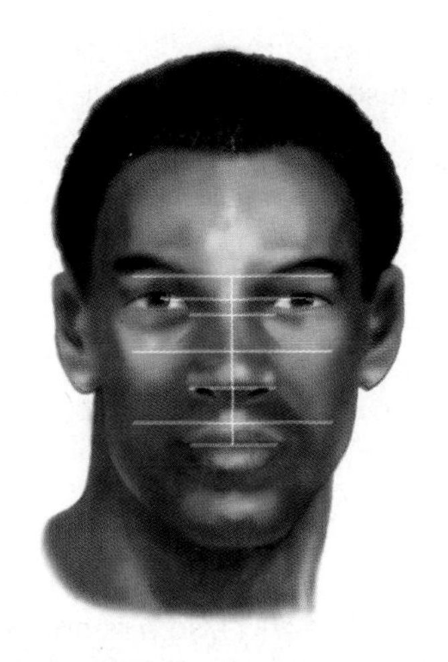

Figure 2.28 To most people, an attractive face is one in which each half is an almost perfect mirror image of the other half.

Did you know that graphs of some equations exhibit exactly the kind of symmetry shown by the attractive face in Figure 2.28? The word *symmetry* comes from the Greek *symmetria*, meaning "the same measure." We can identify graphs with symmetry by looking at a function's equation and determining if the function is *even* or *odd*.

Definition of Even and Odd Functions

The function f is an **even function** if

$$f(-x) = f(x) \quad \text{for all } x \text{ in the domain of } f.$$

The right side of the equation of an even function does not change if x is replaced with $-x$.

The function f is an **odd function** if

$$f(-x) = -f(x) \quad \text{for all } x \text{ in the domain of } f.$$

Every term in the right side of the equation of an odd function changes sign if x is replaced by $-x$.

EXAMPLE 5 Identifying Even or Odd Functions

Identify each of the following functions as even, odd, or neither.

a. $f(x) = x^3$ b. $g(x) = x^4 - 2x^2$ c. $h(x) = x^2 + 2x + 1$

Solution In each case, replace x with $-x$ and simplify. If the right side of the equation stays the same, the function is even. If every term on the right changes sign, the function is odd.

a. We use the given function's equation, $f(x) = x^3$, to find $f(-x)$.

Use $f(x) = x^3$.

Replace x with −x. Replace x with −x.

$$f(-x) = (-x)^3 = (-x)(-x)(-x) = -x^3$$

There is only one term in the equation $f(x) = x^3$, and the term changed signs when we replaced x with $-x$. Because $f(-x) = -f(x)$, f is an odd function.

b. We use the given function's equation, $g(x) = x^4 - 2x^2$, to find $g(-x)$.

Use $g(x) = x^4 - 2x^2$.

Replace x with −x.

$$g(-x) = (-x)^4 - 2(-x)^2 = (-x)(-x)(-x)(-x) - 2(-x)(-x)$$
$$= x^4 - 2x^2.$$

The right side of the equation of the given function, $g(x) = x^4 - 2x^2$, did not change when we replaced x with $-x$. Because $g(-x) = g(x)$, g is an even function.

c. We use the given function's equation, $h(x) = x^2 + 2x + 1$, to find $h(-x)$.

Use $h(x) = x^2 + 2x + 1$.

Replace x with −x.

$$h(-x) = (-x)^2 + 2(-x) + 1 = x^2 - 2x + 1$$

The right side of the equation of the given function, $h(x) = x^2 + 2x + 1$, changed when we replaced x with $-x$. Thus, $h(-x) \neq h(x)$, so h is not an even function. The sign of *each* of the three terms in the equation for $h(x)$ did not change when we replaced x with $-x$. Only the second term changed signs. Thus, $h(-x) \neq -h(x)$, so h is not an odd function. We conclude that h is neither an even nor an odd function.

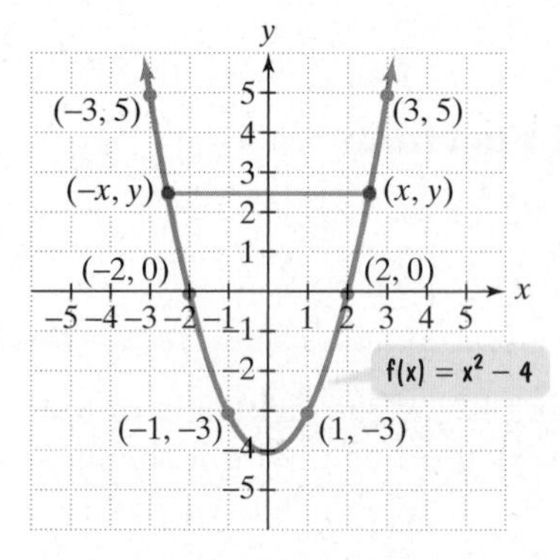

Figure 2.29 y-axis symmetry with $f(-x) = f(x)$

Check Point 5 Determine whether each of the following functions is even, odd, or neither.
a. $f(x) = x^2 + 6$ **b.** $g(x) = 7x^3 - x$ **c.** $h(x) = x^5 + 1$

Now, let's see what even and odd functions tell us about a function's graph. Begin with the even function $f(x) = x^2 - 4$, shown in Figure 2.29. The function is even because

$$f(-x) = (-x)^2 - 4 = x^2 - 4 = f(x).$$

Examine the pairs of points shown, such as (3, 5) and (−3, 5). Notice that we obtain the same y-coordinate whenever we evaluate the function at a value of x and the value of x equal to its opposite. Like the attractive face, each half of the graph is a mirror image of the other half through the y-axis. If we were to fold the paper along the y-axis, the two halves of the graph would coincide. This causes the graph to be *symmetric with respect to the y-axis*. A graph is **symmetric with respect to the y-axis** if, for every point (x, y) on the graph, the point $(-x, y)$ is also on the graph. All even functions have graphs with this kind of symmetry.

Even Functions and y-Axis Symmetry

The graph of an even function in which $f(-x) = f(x)$ is symmetric with respect to the y-axis.

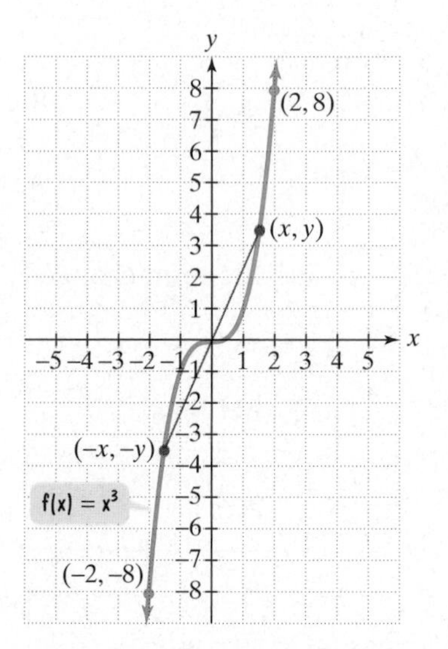

Figure 2.30 Origin symmetry with $f(-x) = -f(x)$

Now, consider the graph of the function $f(x) = x^3$. In Example 5, we saw that $f(-x) = -f(x)$, so this is an odd function. Although the graph in Figure 2.30 is not symmetric with respect to the y-axis, it is symmetric in another way. Look at the pairs of points, such as (2, 8) and (−2, −8). For each point (x, y) on the graph, the point $(-x, -y)$ is also on the graph. The points (2, 8) and (−2, −8) are reflections of one another in the origin. This means that

- the points are the same distance from the origin, and
- the points lie on a line through the origin.

A graph is **symmetric with respect to the origin** if, for every point (x, y) on the graph, the point $(-x, -y)$ is also on the graph. Observe that the first- and third-quadrant portions of $f(x) = x^3$ are reflections of one another in the origin. Notice that $f(x)$ and $f(-x)$ have opposite signs, so that $f(-x) = -f(x)$. All odd functions have graphs with origin symmetry.

Odd Functions and Origin Symmetry

The graph of an odd function in which $f(-x) = -f(x)$ is symmetric with respect to the origin.

Symmetry and Your World

Sixfold rotational symmetry

Fivefold rotational symmetry

Flowers with rotational symmetries

Origin symmetry and y-axis symmetry are just two examples of a powerful concept whose workings appear in art, nature, the sciences, poetry, and architecture. The two halves of a bridge span, the wings of a bird or an aircraft, the blades of a propeller—all have symmetry. Many flowers have rotational symmetry. The flower on the far left has fivefold rotational symmetry; after five equal turns, the flower is restored to its original position. By contrast, the flower on the immediate left has sixfold rotational symmetry.

6 Recognize graphs of common functions.

Graphs of Common Functions

Table 2.3 on page 212 gives names to six frequently encountered functions in algebra. The table shows each function's graph and lists characteristics of the function. Study the shape of each graph and take a few minutes to verify the function's characteristics from its graph. Knowing these graphs is essential for understanding later graphing techniques.

Discovery

Use a graphing utility to verify the six graphs shown in Table 2.3.

Table 2.3 Algebra's Common Graphs

Constant Function

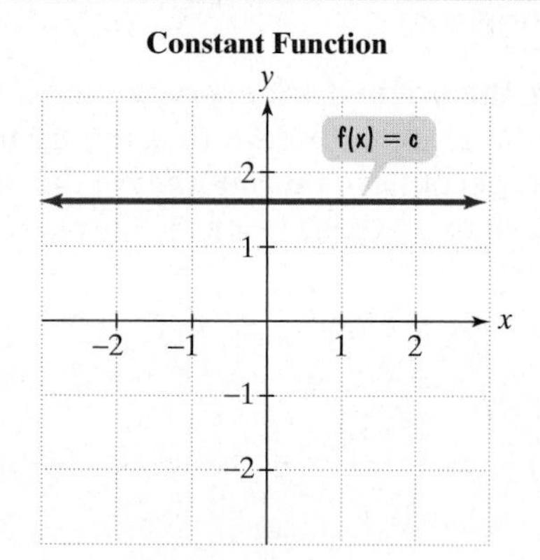

- Domain: $(-\infty, \infty)$
- Range: the single number c
- Constant on $(-\infty, \infty)$
- Even function

Identity Function

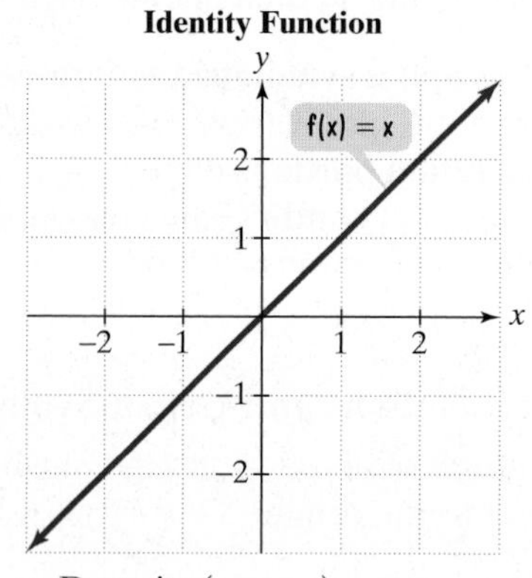

- Domain: $(-\infty, \infty)$
- Range: $(-\infty, \infty)$
- Increasing on $(-\infty, \infty)$
- Odd function

Standard Quadratic Function

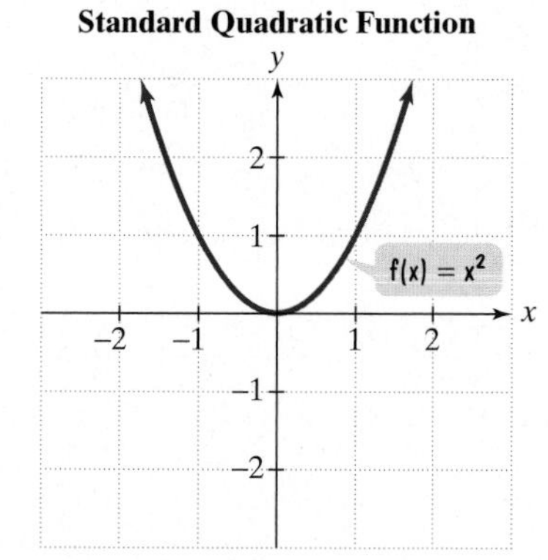

- Domain: $(-\infty, \infty)$
- Range: $[0, \infty)$
- Decreasing on $(-\infty, 0)$ and increasing on $(0, \infty)$
- Even function

Standard Cubic Function

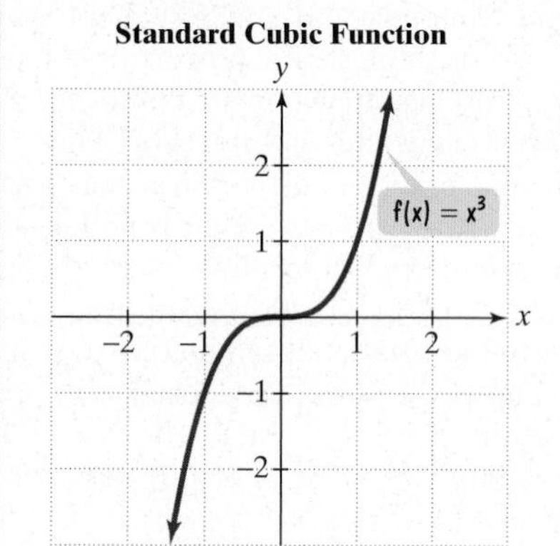

- Domain: $(-\infty, \infty)$
- Range: $(-\infty, \infty)$
- Increasing on $(-\infty, \infty)$
- Odd function

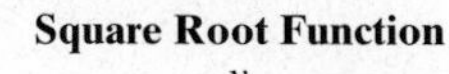

Square Root Function

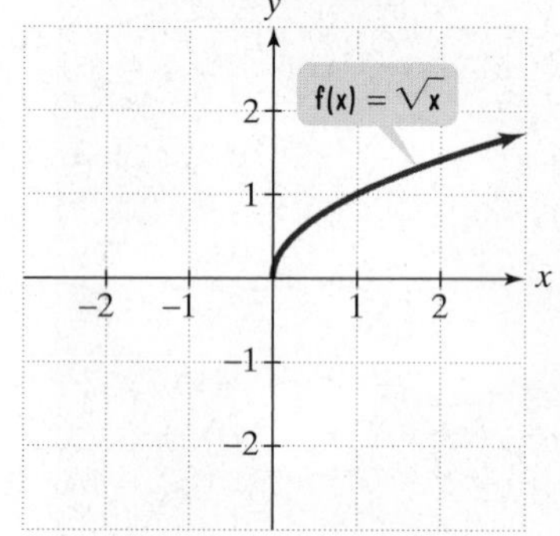

- Domain: $[0, \infty)$
- Range: $[0, \infty)$
- Increasing on $(0, \infty)$
- Neither even nor odd

Absolute Value Function

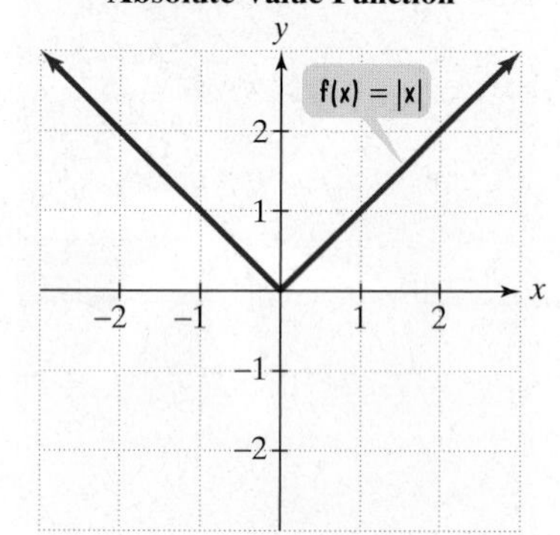

- Domain: $(-\infty, \infty)$
- Range: $[0, \infty)$
- Decreasing on $(-\infty, 0)$ and increasing on $(0, \infty)$
- Even function

7 Graph step functions.

Step Functions

Have you ever mailed a letter that seemed heavier than usual? Perhaps you worried that the letter would not have enough postage. Costs for mailing a letter weighing up to 5 ounces are given in Table 2.4. If your letter weighs an ounce or less, the cost is \$0.33. If your letter weighs 1.05 ounces, 1.50 ounces, 1.90 ounces, or 2.00 ounces, the cost "steps" to \$0.55. The cost does not take on any value between \$0.33 and \$0.55. If your letter weighs 2.05 ounces, 2.50 ounces, 2.90 ounces, or 3 ounces, the cost "steps" to \$0.77. Cost increases are \$0.22 per step.

Table 2.4 Cost of First-Class Mail (Effective January 10, 1999)

Weight Not Over	Cost
1 ounce	\$0.33
2 ounces	0.55
3 ounces	0.77
4 ounces	0.99
5 ounces	1.21

Source: U.S. Postal Service

Now, let's see what the graph of the function that models this situation looks like. Let

x = the weight of the letter in ounces, and

$y = f(x)$ = the cost of mailing a letter weighing x ounces.

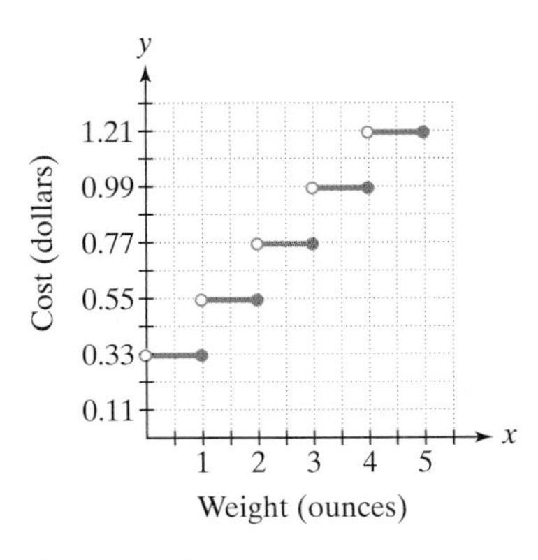

Figure 2.31

The graph is shown in Figure 2.31. Notice how it consists of a series of steps that jump vertically 22 units at each integer. The graph is constant between each pair of consecutive integers.

Mathematicians have a function that describes situations where function values graphically form discontinuous steps. The function is called the **greatest integer function**, symbolized by $\text{int}(x)$ or $[\![x]\!]$. And what is $\text{int}(x)$?

$$\text{int}(x) = \text{the greatest integer that is less than or equal to } x.$$

For example,

$$\text{int}(1) = 1, \quad \text{int}(1.3) = 1, \quad \text{int}(1.5) = 1, \quad \text{int}(1.9) = 1.$$

1 is the greatest integer that is less than or equal to 1, 1.3, 1.5, and 1.9.

Here are some additional examples:

$$\text{int}(2) = 2, \quad \text{int}(2.3) = 2, \quad \text{int}(2.5) = 2, \quad \text{int}(2.9) = 2.$$

2 is the greatest integer that is less than or equal to 2, 2.3, 2.5, and 2.9.

Notice how we jumped from 1 to 2 in the function values for $\text{int}(x)$. In particular,

$$\text{If } 1 \le x < 2, \quad \text{then} \quad \text{int}(x) = 1.$$

$$\text{If } 2 \le x < 3, \quad \text{then} \quad \text{int}(x) = 2.$$

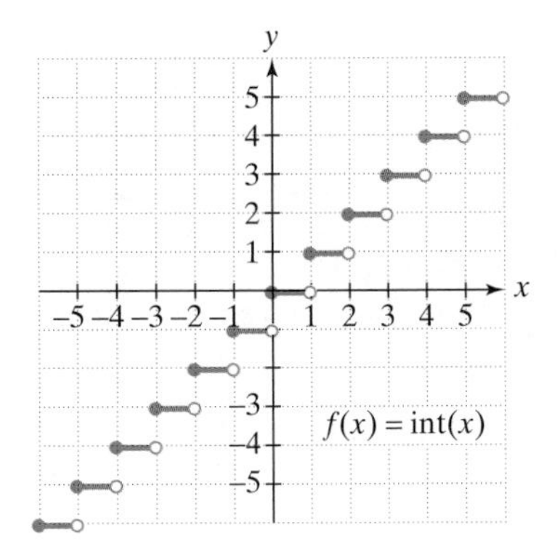

Figure 2.32 The graph of the greatest integer function

The graph of $f(x) = \text{int}(x)$ is shown in Figure 2.32. The graph of the greatest integer function jumps vertically one unit at each integer. However, the graph is constant between each pair of consecutive integers. The rightmost of the horizontal steps shown in the graph illustrates that

$$\text{If } 5 \le x < 6, \quad \text{then} \quad \text{int}(x) = 5.$$

In general,

$$\text{If } n \le x < n + 1, \text{where } n \text{ is an integer}, \quad \text{then} \quad \text{int}(x) = n.$$

By contrast to the graph for the cost of first-class mail, the graph of the greatest integer function includes the point on the left of each horizontal step, but does not include the point on the right. The domain of $f(x) = \text{int}(x)$ is the set of all real numbers, $(-\infty, \infty)$. The range is the set of all integers.

Technology

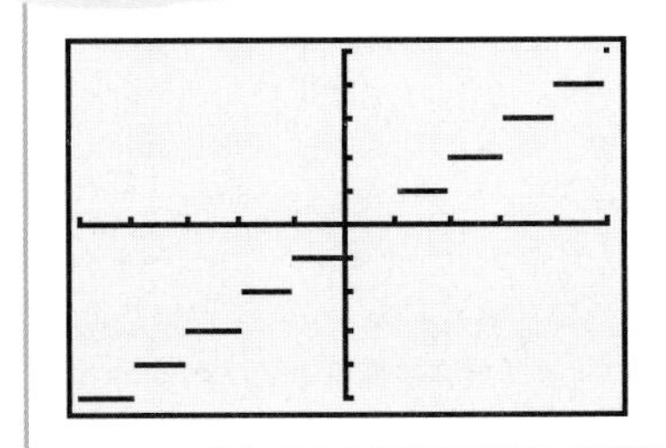

The graph of $f(x) = \text{int}(x)$, shown on the left, was obtained with a graphing utility. By graphing in "dot" mode, we can see the discontinuities at the integers. By looking at the graph, it is impossible to tell that, for each step, the point on the left is included and the point on the right is not. We must trace along the graph to obtain such information.

8 Obtain information from graphs of functions in applied situations.

Applications

We return to the function that models the fuel efficiency of U.S. automobiles over time. In the next example, we'll see what we can learn about this function from its graph.

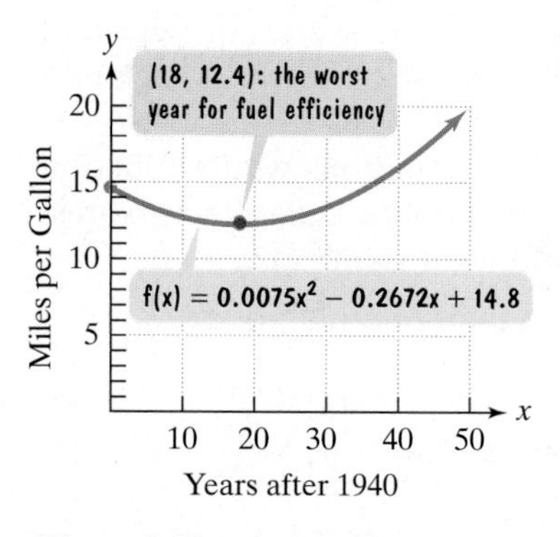

Figure 2.33 Fuel efficiency of U.S. automobiles over time

EXAMPLE 6 A Graph for the Fuel Efficiency Function

We have seen the function

$$f(x) = 0.0075x^2 - 0.2672x + 14.8$$

that models the average number of miles per gallon of U.S. automobiles, $f(x)$, x years after 1940. The graph of this function is shown as a continuous curve in Figure 2.33. (It can also be shown as a series of points, each point representing a year and miles per gallon for that year.)

a. On which interval is f decreasing, and what does this mean?
b. On which interval is f increasing, and what does this mean?

Solution Note the voice balloon pointing to (18, 12.4). It tells us that 18 years after 1940, in 1958, fuel efficiency was at its lowest point ever—a dismal 12.4 miles per gallon. This information, and the shape of the graph, enables us to find where f is decreasing and where it is increasing.

a. Function f is decreasing on the interval (0, 18).
This means that fuel efficiency was decreasing from 1940 through 1958.
b. Function f is increasing on the interval (18, 50).
This means that fuel efficiency was increasing from 1958 through 1990.

Check Point 6

When a person receives a drug injected into a muscle, the concentration of the drug in the body, measured in milligrams per 100 milliliters, is a function of the time elapsed since the injection, measured in hours. Figure 2.34 shows the graph of such a function, where x = hours since the injection and $f(x)$ = drug concentration at time x.

a. On which interval is f increasing, and what does this mean?
b. On which interval is f decreasing, and what does this mean?
c. What is the drug's maximum concentration and when does this occur?
d. What happens by the end of 13 hours?

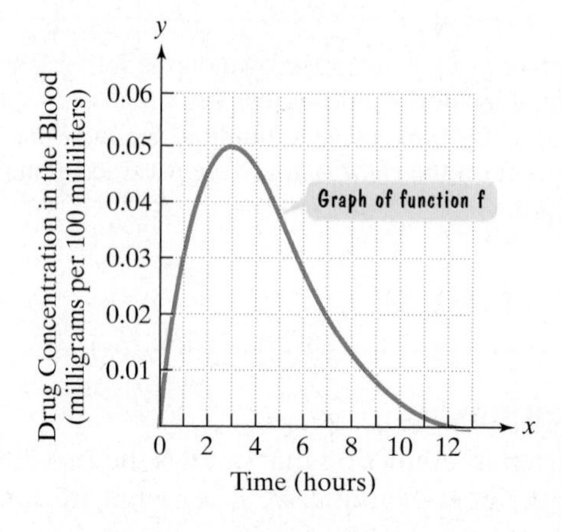

Figure 2.34 Concentration of a drug as a function of time

EXERCISE SET 2.4

Practice Exercises

Graph the function in Exercises 1–14. Use the integer values of x given to the right of the function to obtain ordered pairs. Use the graph to specify the function's domain and range.

1. $f(x) = x^2 + 2$ $\quad x = -3, -2, -1, 0, 1, 2, 3$
2. $f(x) = x^2 - 1$ $\quad x = -3, -2, -1, 0, 1, 2, 3$
3. $g(x) = \sqrt{x} - 1$ $\quad x = 0, 1, 4, 9$
4. $g(x) = \sqrt{x} + 2$ $\quad x = 0, 1, 4, 9$
5. $h(x) = \sqrt{x - 1}$ $\quad x = 1, 2, 5, 10$
6. $h(x) = \sqrt{x + 2}$ $\quad x = -2, -1, 2, 7$
7. $f(x) = |x| - 1$ $\quad x = -3, -2, -1, 0, 1, 2, 3$
8. $f(x) = |x| + 1$ $\quad x = -3, -2, -1, 0, 1, 2, 3$
9. $g(x) = |x - 1|$ $\quad x = -3, -2, -1, 0, 1, 2, 3$
10. $g(x) = |x + 1|$ $\quad x = -3, -2, -1, 0, 1, 2, 3$
11. $f(x) = 5$ $\quad x = -3, -2, -1, 0, 1, 2, 3$
12. $f(x) = 3$ $\quad x = -3, -2, -1, 0, 1, 2, 3$
13. $f(x) = x^3 - 2$ $\quad x = -2, -1, 0, 1, 2$
14. $f(x) = x^3 + 2$ $\quad x = -2, -1, 0, 1, 2$

In Exercises 15–30, use the graph to determine **a.** *the function's domain;* **b.** *the function's range;* **c.** *the x-intercepts, if any;* **d.** *the y-intercept, if any; and* **e.** *the function values indicated below some of the graphs.*

15.

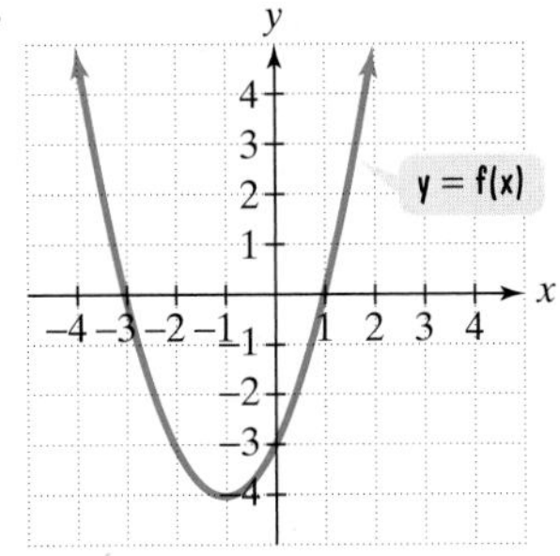

16.

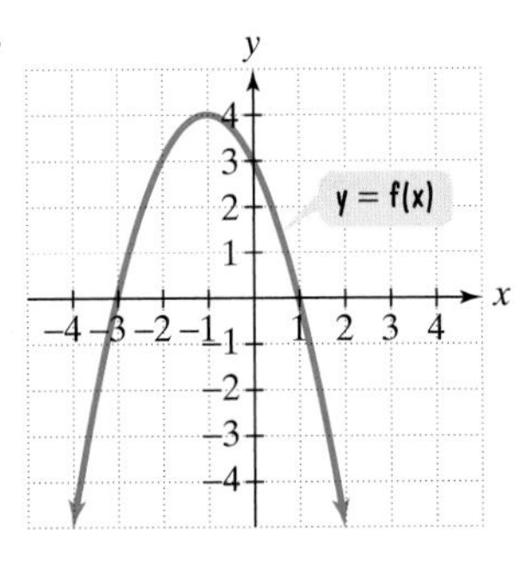

17.

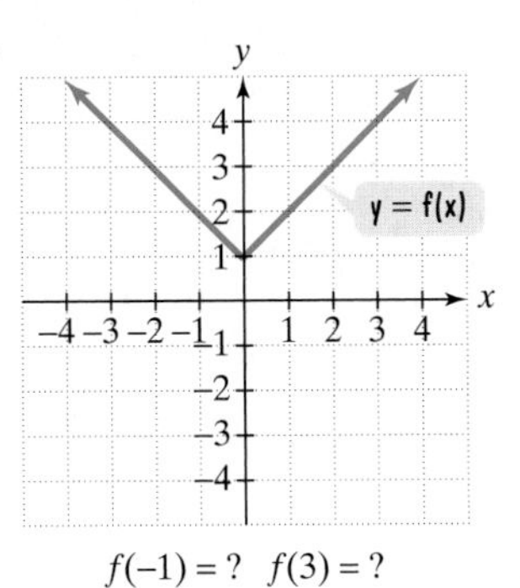

$f(-1) = ?\quad f(3) = ?$

18.

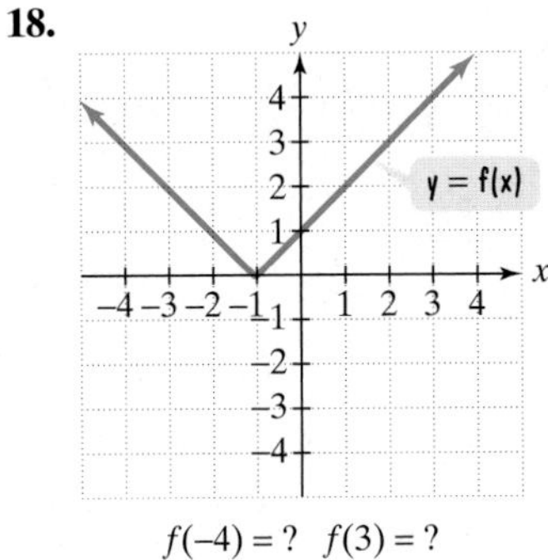

$f(-4) = ?\quad f(3) = ?$

19.

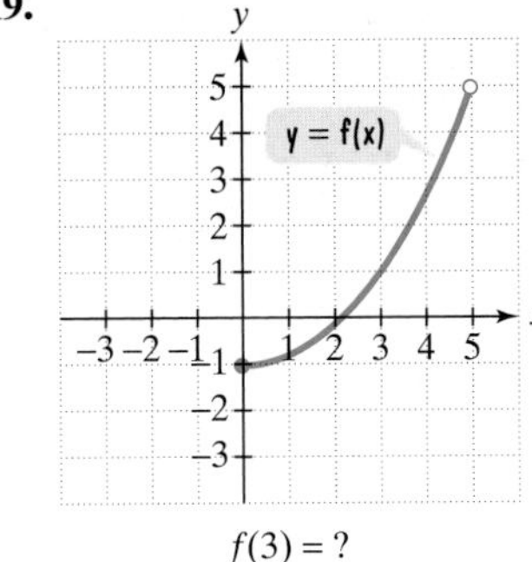

$f(3) = ?$

20.

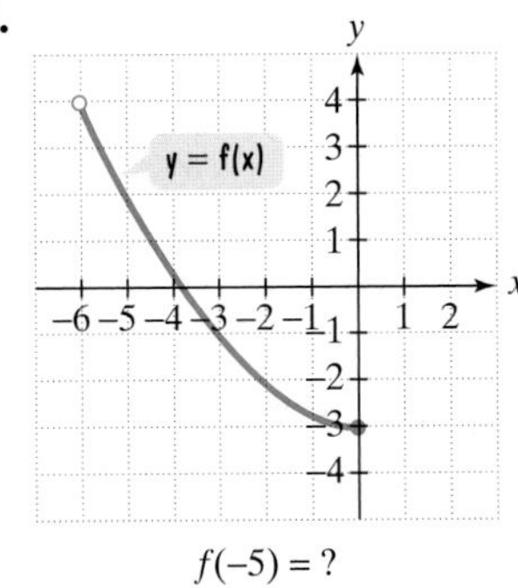

$f(-5) = ?$

21.

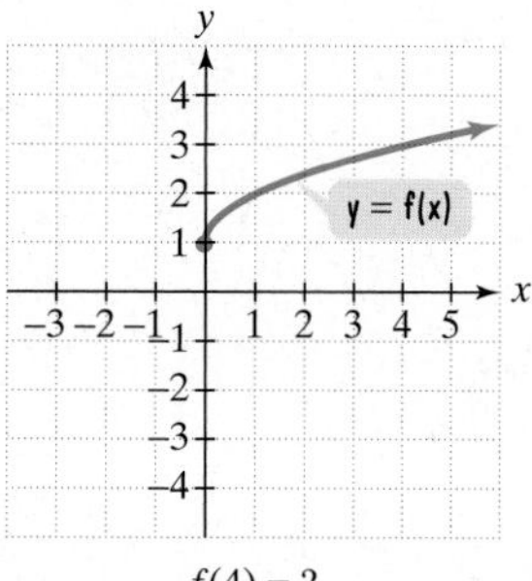

$f(4) = ?$

22.

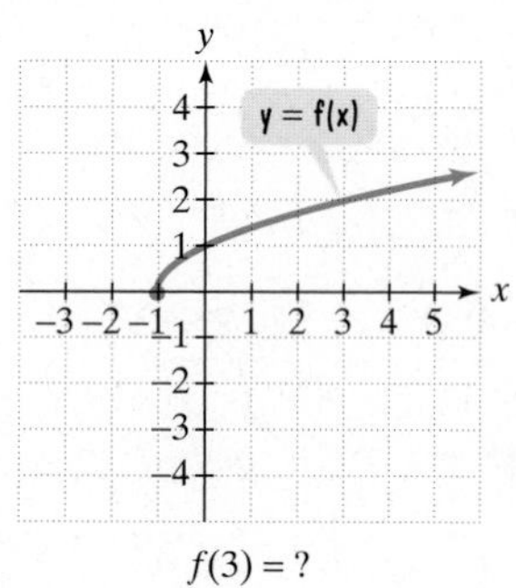

$f(3) = ?$

23.

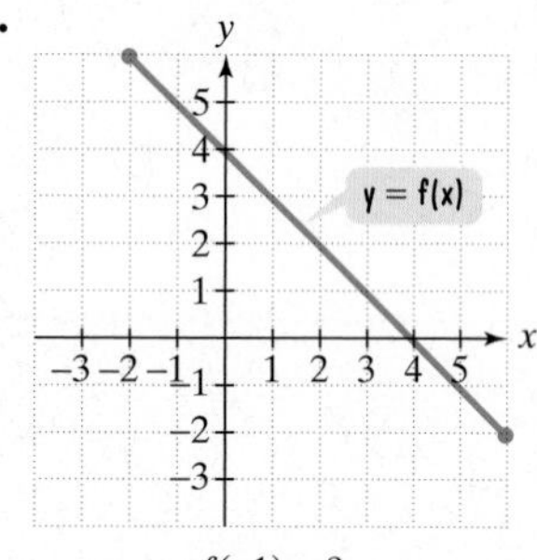

$f(-1) = ?$

24.

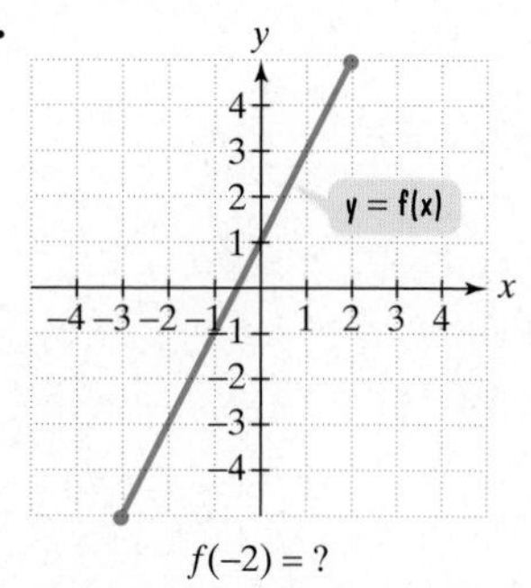

$f(-2) = ?$

25.

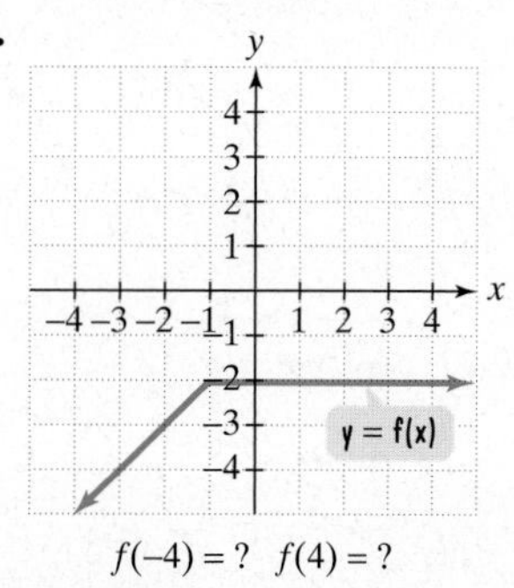

$f(-4) = ?$ $f(4) = ?$

26.

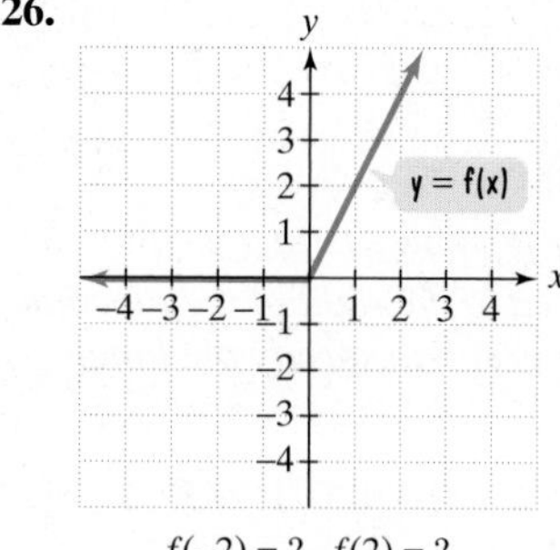

$f(-2) = ?$ $f(2) = ?$

27.

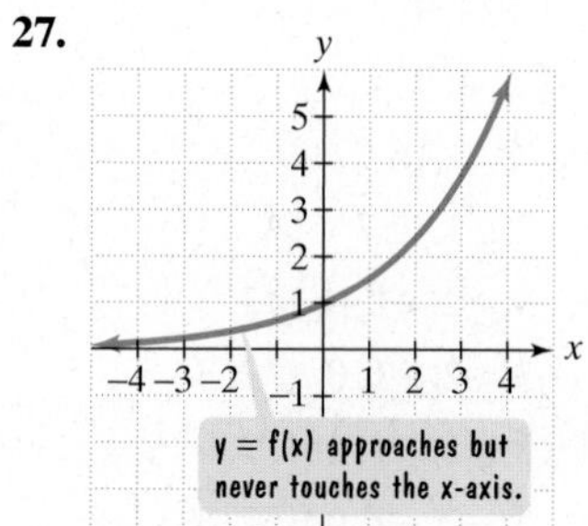

28.

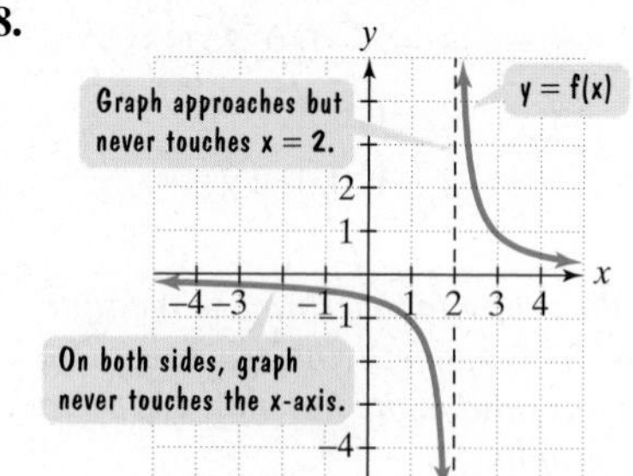

29.

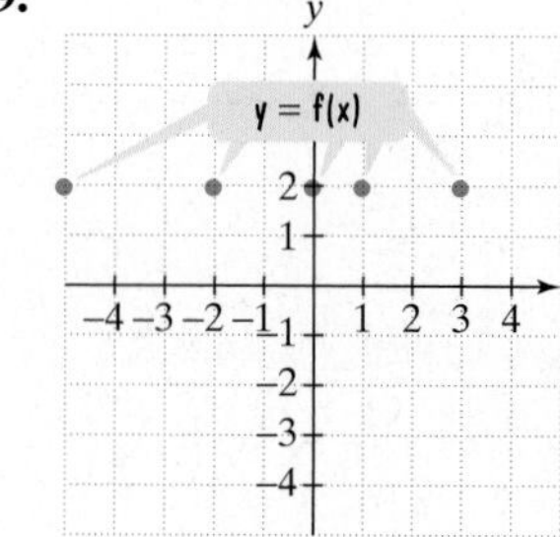

30.

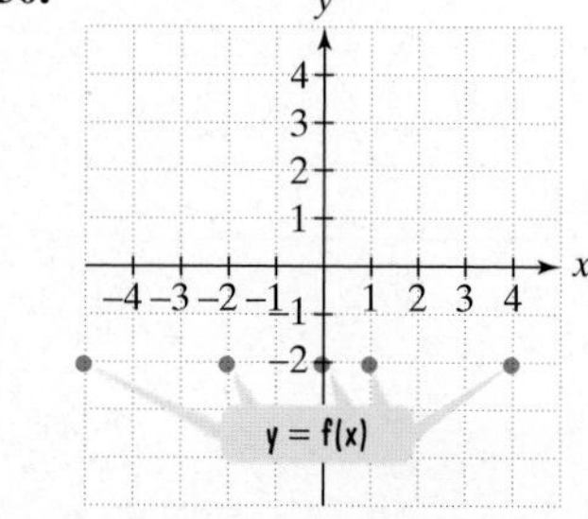

In Exercises 31–38, use the vertical line test to identify graphs in which y is a function of x.

31.

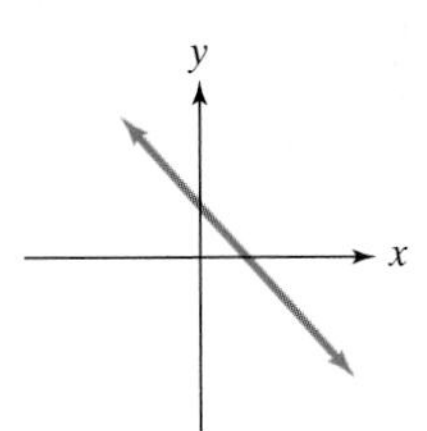

32.

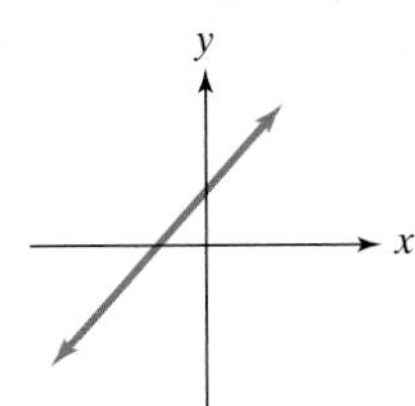

33.

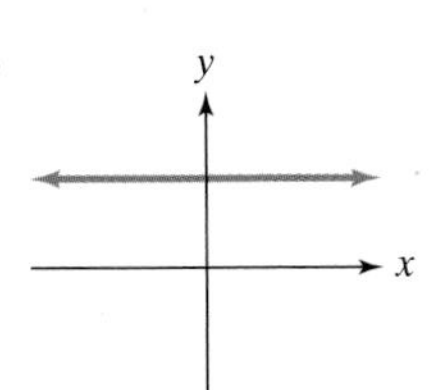

34.

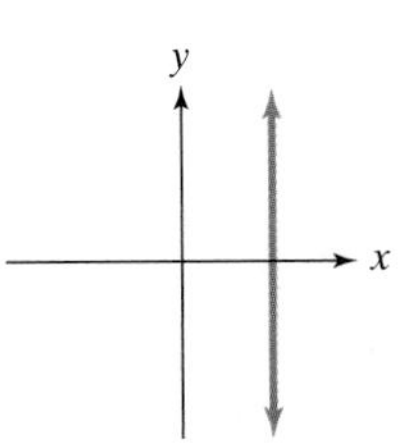

35.

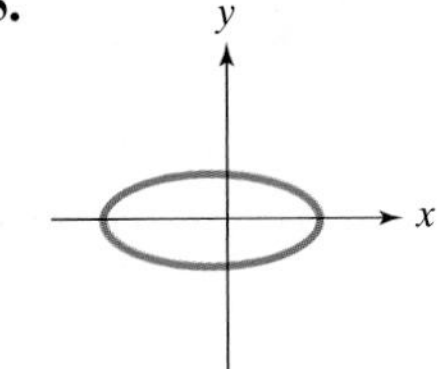

36.

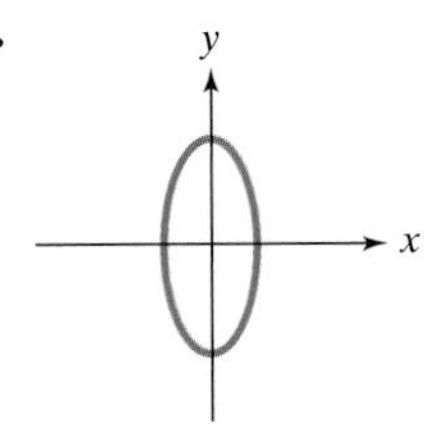

37.

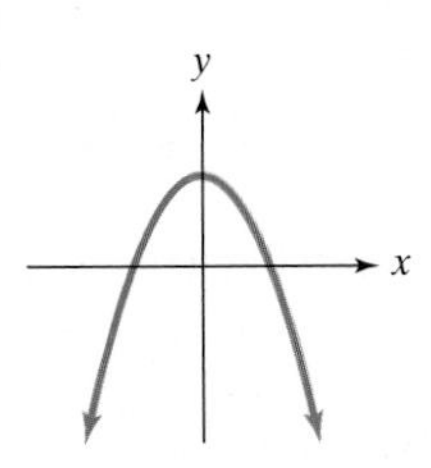

38.

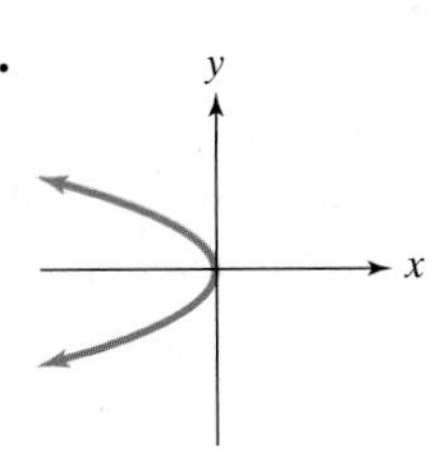

In Exercises 39–50, use the graph to determine:

a. *intervals on which the function is increasing, if any.*
b. *intervals on which the function is decreasing, if any.*
c. *intervals on which the function is constant, if any.*

39. Use the graph in Exercise 15.
40. Use the graph in Exercise 16.
41. Use the graph in Exercise 21.
42. Use the graph in Exercise 22.
43. Use the graph in Exercise 23.
44. Use the graph in Exercise 24.
45. Use the graph in Exercise 25.
46. Use the graph in Exercise 26.

47.

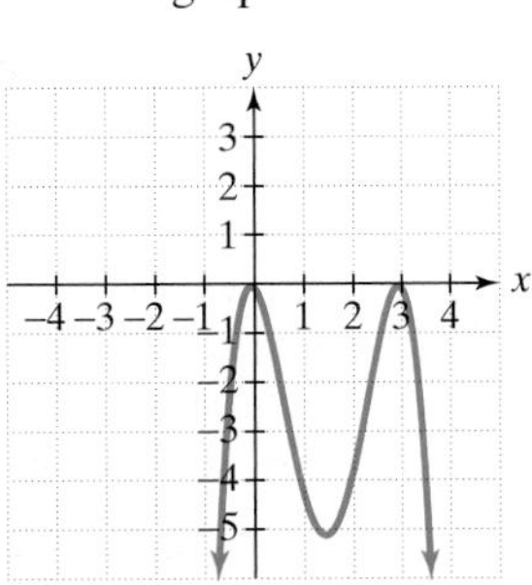

48.

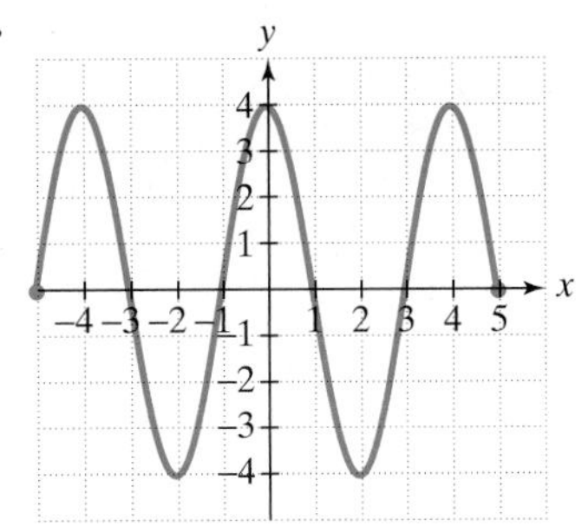

49.

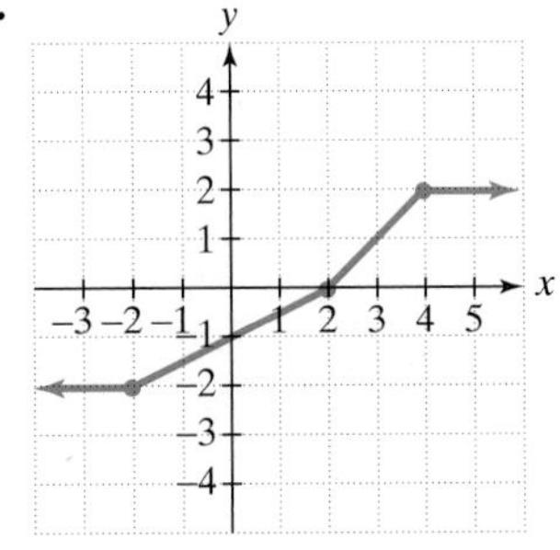

50.

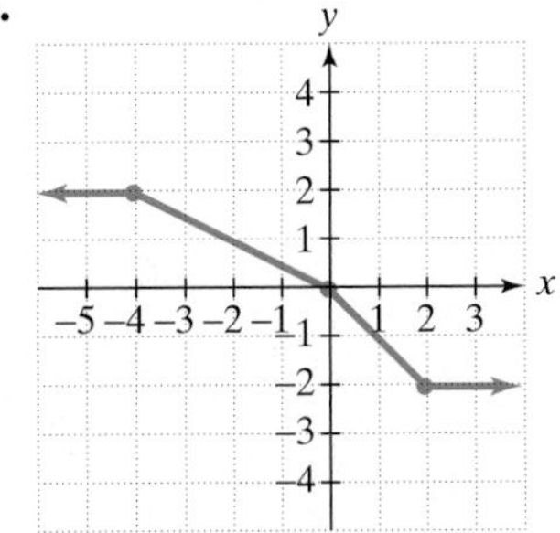

In Exercises 51–62, determine whether each function is even, odd, or neither.

51. $f(x) = x^3 + x$
52. $f(x) = x^3 - x$
53. $g(x) = x^2 + x$
54. $g(x) = x^2 - x$
55. $h(x) = x^2 - x^4$
56. $h(x) = 2x^2 + x^4$
57. $f(x) = x^2 - x^4 + 1$
58. $f(x) = 2x^2 + x^4 + 1$
59. $f(x) = \frac{1}{5}x^6 - 3x^2$
60. $f(x) = 2x^3 - 6x^5$
61. $f(x) = x\sqrt{1 - x^2}$
62. $f(x) = x^2\sqrt{1 - x^2}$

In Exercises 63–66, use possible symmetry to determine whether each graph is the graph of an even function, an odd function, or a function that is neither even nor odd.

63.

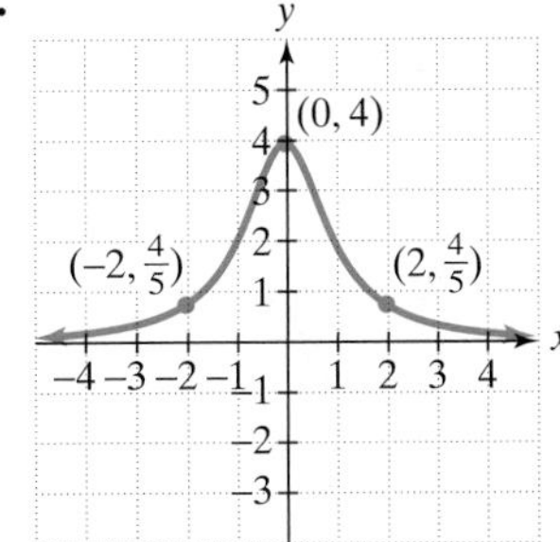

64.

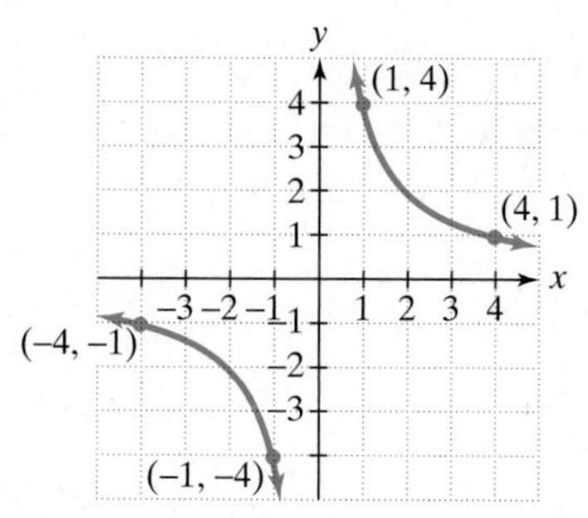

65.

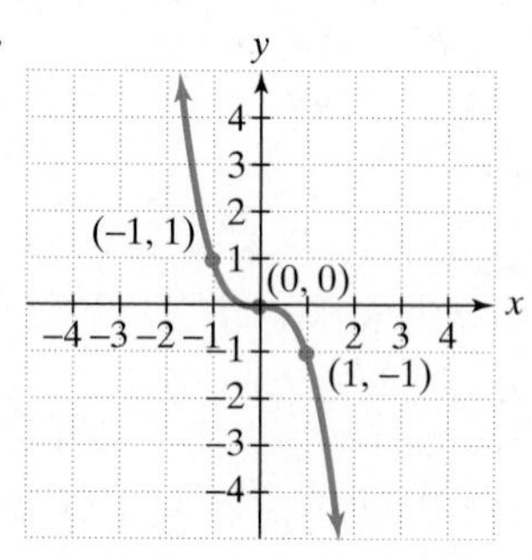

66.

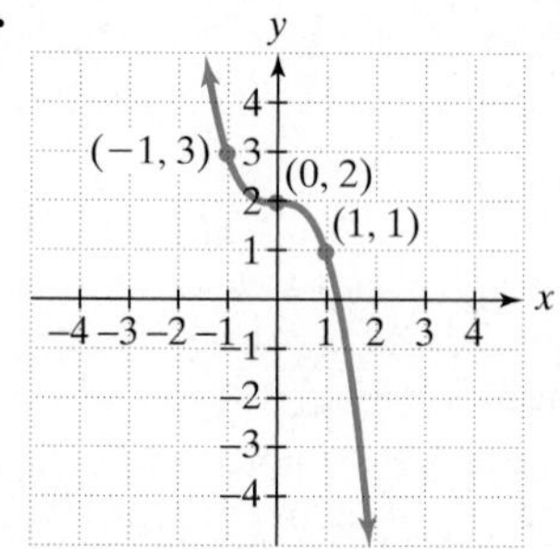

In Exercises 67–72, use the shape of each graph to name the function. Select from the following names: constant function, identity function, standard quadratic function, standard cubic function, square root function, absolute value function, greatest integer function.

67.

68.

69.

70.

71.

72.

In Exercises 73–78, if $f(x) = int(x)$, find each function value.

73. $f(1.06)$

74. $f(2.99)$

75. $f(\frac{1}{3})$

76. $f(-1.5)$

77. $f(-2.3)$

78. $f(-99.001)$

Application Exercises

The graph shows the function $y = f(x)$, where $f(x)$ is defense spending in year x for $1988 \le x \le 1998$. Use the graph to solve Exercises 79–82.

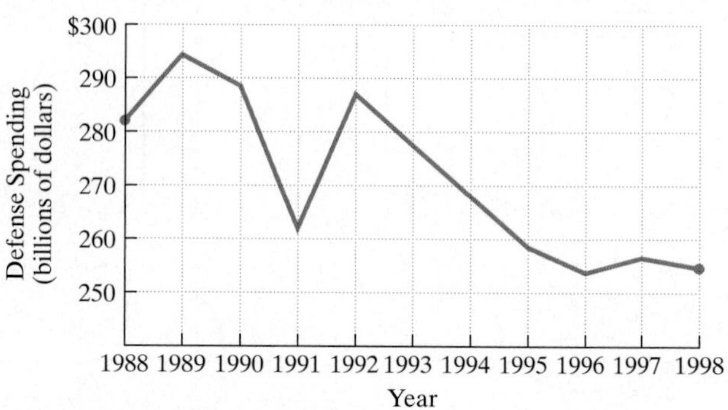

Source: Office of Management and Budget

79. Estimate the function value $f(1989)$. What is significant about this function value?

80. Estimate the function value $f(1996)$. What is significant about this function value?

81. In which time intervals is defense spending increasing?

82. In which time intervals is defense spending decreasing?

83. The function $f(x) = 0.4x^2 - 36x + 1000$ models the number of accidents per 50 million miles driven as a function of age, x, in years, where $16 \le x \le 74$. The graph of f is shown.

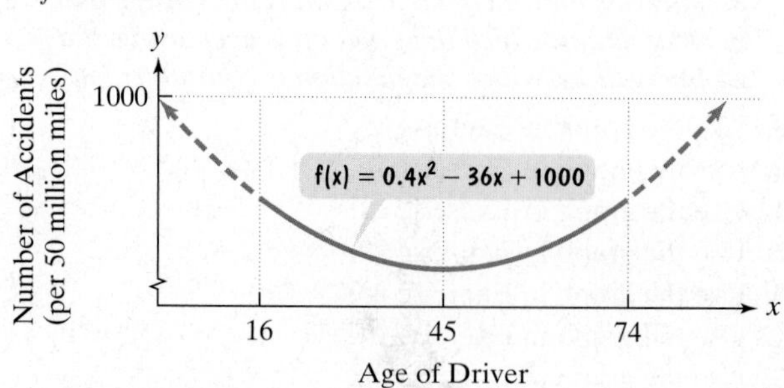

a. State the intervals on which the function is increasing and decreasing and describe what this means in terms of the variables modeled by the function.

b. For what value of x does the graph reach its lowest point? What is the minimum value of y? Describe the practical significance of this minimum value.

c. The domain of f is $[16, 74]$. Use the function's equation to determine the range of f. What is the practical significance of this range in terms of the meaning of $f(x)$ in the given function?

84. Based on a study by Vance Tucker (*Scientific American*, May 1969) the power expenditure of migratory birds in flight is a function of their flying speed, x, in miles per hour, modeled by $f(x) = 0.67x^2 - 27.74x + 387$. Power expenditure, $f(x)$, is measured in calories, and migratory birds generally fly between 12 and 30 miles per hour. The graph of f is shown in the figure, with a domain of $[12, 30]$.

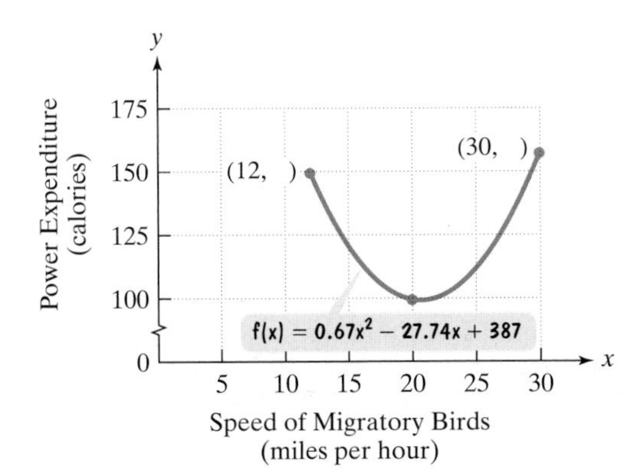

a. State the intervals on which the function is increasing and decreasing and describe what this means in terms of the variables modeled by the function.

b. For what approximate value of x does the graph reach its lowest point? What is the minimum value of y? Describe the practical significance of this minimum value.

c. The domain of f is $[12, 30]$. Use the function's equation to find the range of f. What is the practical significance of this range in terms of the meaning of $f(x)$ in the given function?

85. The cost of a telephone call between two cities is \$0.10 for the first minute and \$0.05 for each additional minute or portion of a minute. Draw a graph of the cost, C, in dollars, of the phone call as a function of time, t, in minutes, on the interval $(0, 5]$.

86. A cargo service charges a flat fee of \$4 plus \$1 for each pound or fraction of a pound to mail a package. Let $C(x)$ represent the cost to mail a package that weighs x pounds. Graph the cost function on the interval $(0, 5]$.

87. Researchers at Yale University have suggested that levels of passion and commitment in human relations are functions of time. Based on the shapes of the graphs shown, which do you think depicts passion and which represents commitment? Explain how you arrived at your answer.

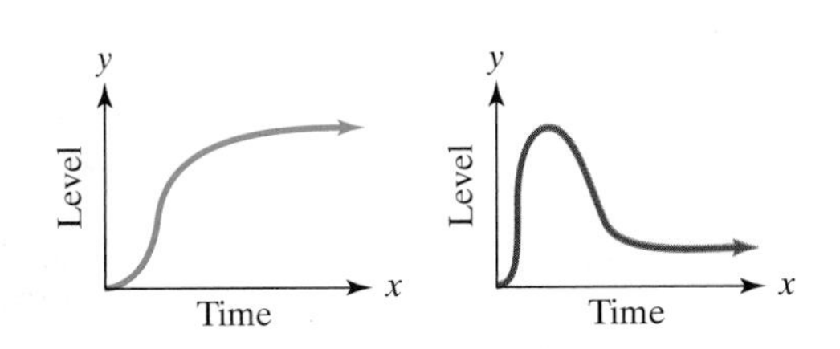

88. Although the level of air pollution varies from day to day and from hour to hour, during the summer the level of air pollution is a function of the time of day. The function

$$f(x) = 0.1x^2 - 0.4x + 0.6$$

describes the level of air pollution (in parts per million [ppm]), where x corresponds to the number of hours after 9 A.M.

a. Construct a table of values using integers from 0 to 5 for x, and graph the function from 0 to 5.

b. Researchers have determined that a level of 0.3 ppm of pollutants in the air can be hazardous to your health. Based on the graph, at what time of day between 9 A.M. and 2 P.M. should runners exercise to avoid unsafe air?

Writing in Mathematics

89. Discuss one disadvantage to using point plotting as a method for graphing functions.

90. Explain how to use a function's graph to find the function's domain and range.

91. Explain how the vertical line test is used to determine whether a graph is a function.

92. What does it mean if function f is increasing on an interval?

93. If you are given a function's equation, how do you determine if the function is even, odd, or neither?

94. If you are given a function's graph, how do you determine if the function is even, odd, or neither?

95. What is a step function? Give an example of an everyday situation that can be modeled using such a function. Do not use the cost-of-mail example.

96. Explain how to find $\text{int}(-3.000004)$.

Technology Exercises

97. The function $f(x) = -0.00002x^3 + 0.008x^2 - 0.3x + 6.95$ models the number of annual physician visits by a person of age x.

a. Use a graphing utility to graph the function in a $[0, 100, 5]$ by $[0, 60, 3]$ viewing rectangle.

b. What does the shape of the graph indicate about the relationship between one's age and the number of annual physician visits?

c. Use the TRACE or minimum function capability to find the coordinates of the lowest point on the graph of the function. What does this mean?

98. The function

$$C(x) = \begin{cases} -0.35x + 220 & \text{for } 0 \le x \le 20 \\ -0.80x + 229 & \text{for } x > 20 \end{cases}$$

describes the number of milligrams of cholesterol per deciliter of blood for American adults x years after 1960.

a. Graph the function using a graphing utility in a $[0, 100, 20]$ by $[0, 250, 50]$ viewing rectangle.
b. What does the graph indicate about cholesterol level from 1960 to the present?
c. Was the goal of lowering cholesterol to a level under 200 reached before the year 2000?

In Exercises 99–104, use a graphing utility to graph each function. Use a $[-5, 5, 1]$ by $[-5, 5, 1]$ viewing rectangle. Then find the intervals on which the function is increasing, decreasing, or constant.

99. $f(x) = x^3 - 6x^2 + 9x + 1$

100. $g(x) = |4 - x^2|$

101. $h(x) = |x - 2| + |x + 2|$

102. $f(x) = x^{1/3}(x - 4)$

103. $g(x) = x^{2/3}$

104. $h(x) = 2 - x^{2/5}$

105. a. Graph the functions $f(x) = x^n$ for $n = 2, 4,$ and 6 in a $[-2, 2, 1]$ by $[-1, 3, 1]$ viewing rectangle.
b. Graph the functions $f(x) = x^n$ for $n = 1, 3,$ and 5 in a $[-2, 2, 1]$ by $[-2, 2, 1]$ viewing rectangle.
c. If n is even, where is the graph of $f(x) = x^n$ increasing and where is it decreasing?
d. If n is odd, what can you conclude about the graph of $f(x) = x^n$ in terms of increasing or decreasing behavior?
e. Graph all six functions in a $[-1, 3, 1]$ by $[-1, 3, 1]$ viewing rectangle. What do you observe about the graphs in terms of how flat or how steep they are?

Critical Thinking Exercises

106. Which one of the following is true based on the graph of f in the figure?

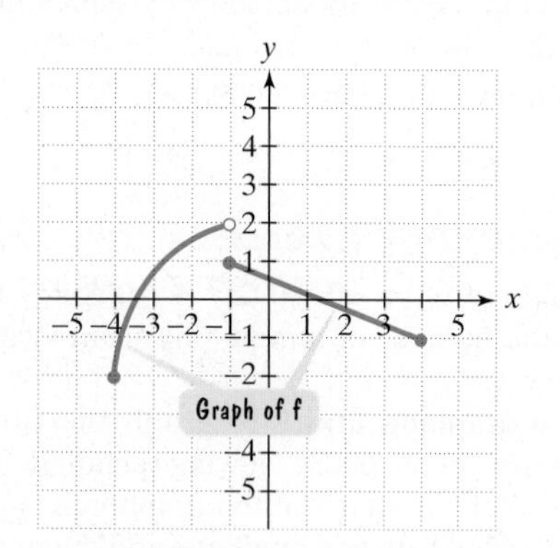

a. The domain of f is $[-4, 1)$ or $(1, 4]$.
b. The range of f is $[-2, 2]$.
c. $f(-1) - f(4) = 2$
d. $f(0) = 2.1$

107. Describe a situation that can be modeled by the graph shown.

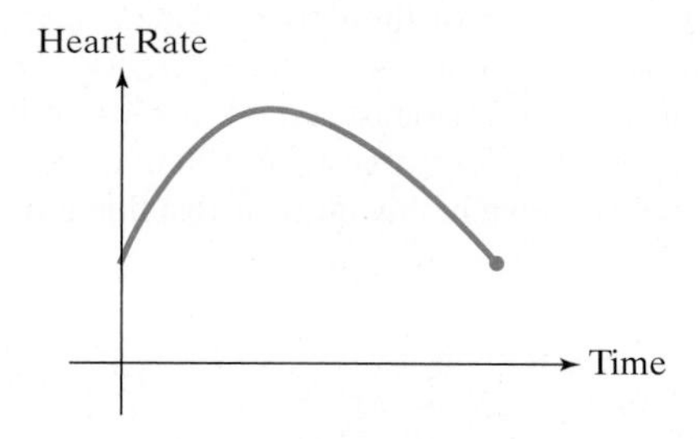

108. Sketch the graph of f using the following properties. (More than one correct graph is possible.) f is a piecewise function that is decreasing on $(-\infty, 2)$, $f(2) = 0$, f is increasing on $(2, \infty)$, and the range of f is $[0, \infty)$.

109. Define a piecewise function on the intervals $(-\infty, 2]$, $(2, 5)$, and $[5, \infty)$ that does not "jump" at 2 or 5 such that one piece is a constant function, another piece is an increasing function, and the third piece is a decreasing function.

110. Suppose that $h(x) = \dfrac{f(x)}{g(x)}$. The function f can be even, odd, or neither. The same is true for the function g.
a. Under what conditions is h definitely an even function?
b. Under what conditions is h definitely an odd function?

111. Take another look at the cost of first-class mail and its graph (Table 2.4 and Figure 2.31 on pages 212–213). Change the description of the heading in the left column of Table 2.4 so that the graph includes the point on the left of each horizontal step, but does not include the point on the right.

Group Exercise

112. In Exercise 87, passion and commitment are graphed over time. For this activity, you will be creating a graph of a particular experience that involved your feelings of love, anger, sadness. or any other emotion you choose. The horizontal axis should be labeled time and the vertical axis the emotion you are graphing. You will not be using your algebra skills to create your graph; however, you should try to make the graph as precise as possible. You may use negative numbers on the vertical axis, if appropriate. After each group member has created a graph, pool together all of the graphs and study them to see if there are any similarities in the graphs for a particular emotion or for all emotions.

SECTION 2.5 Transformations and Combinations of Functions

Objectives

1. Use vertical shifts to graph functions.
2. Use horizontal shifts to graph functions.
3. Use reflections to graph functions.
4. Use vertical stretching and shrinking to graph functions.
5. Graph functions involving a sequence of transformations.
6. Combine functions arithmetically, specifying domains.

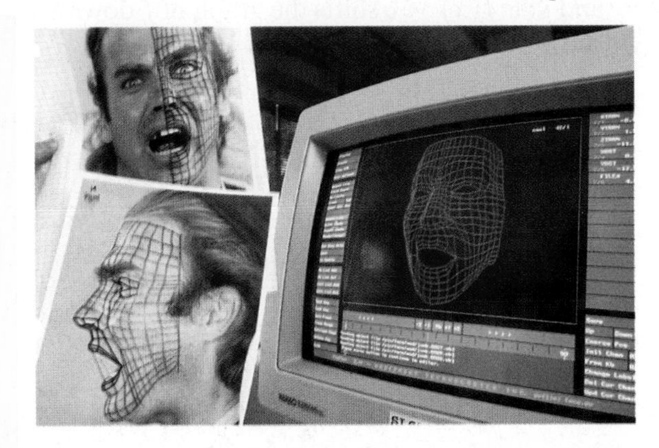

Have you seen *Terminator 2*, *The Mask*, or *The Matrix*? These were among the first films to use spectacular effects in which a character or object having one shape was transformed in a fluid fashion into a quite different shape. The name for such a transformation is **morphing**. The effect allows a real actor to be seamlessly transformed into a computer-generated animation. The animation can be made to perform impossible feats before it is morphed back to the conventionally filmed image.

Like transformed movie images, the graph of one function can be turned into the graph of a different function. To do this, we need to rely on a function's equation. Knowing that a graph is a transformation of a familiar graph makes graphing easier.

Discovery

The study of how changing a function's equation can affect its graph can be explored with a graphing utility. Use your graphing utility to verify the hand-drawn graphs as you read this section.

1 Use vertical shifts to graph functions.

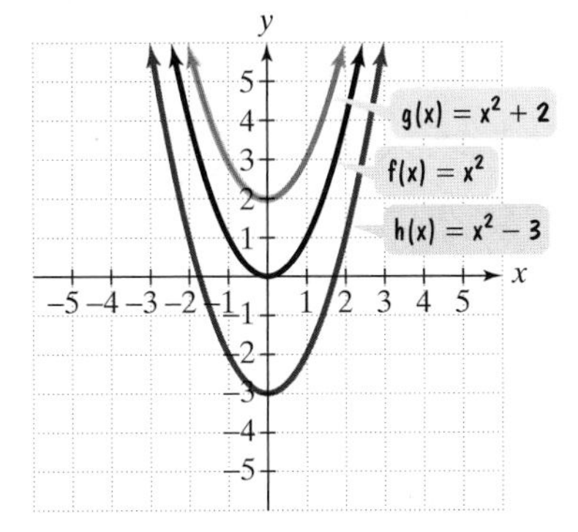

Figure 2.35 Vertical shifts

The graph of $f(x) = x^2$ can be gradually morphed into the graph of $g(x) = x^2 + 2$ by using animation to graph $f(x) = x^2 + c$ for $0 \le c \le 2$. By selecting many values for c, we can create an animated sequence in which change appears to occur continuously.

Vertical Shifts

Let's begin by looking at three graphs whose shapes are the same. Figure 2.35 shows the graphs. The black graph in the middle is the standard quadratic function $f(x) = x^2$. Now, look at the blue graph on the top. The equation of this graph, $g(x) = x^2 + 2$, adds 2 to the right side of $f(x) = x^2$. What effect does this have on the graph of f? It shifts the graph vertically up by 2 units.

$$g(x) = x^2 + 2 = f(x) + 2$$

The graph of g shifts the graph of f up 2 units.

Finally, look at the red graph on the bottom of Figure 2.35. The equation of this graph, $h(x) = x^2 - 3$, subtracts 3 from the right side of $f(x) = x^2$. What effect does this have on the graph of f? It shifts the graph vertically down by 3 units.

$$h(x) = x^2 - 3 = f(x) - 3$$

The graph of h shifts the graph of f down 3 units.

In general, if c is positive, $y = f(x) + c$ shifts the graph of f upward c units and $y = f(x) - c$ shifts the graph of f downward c units. These are called **vertical shifts** of the graph of f.

Vertical Shifts

Let f be a function and c a positive real number.

- The graph of $y = f(x) + c$ is the graph of $y = f(x)$ shifted c units vertically upward.
- The graph of $y = f(x) - c$ is the graph of $y = f(x)$ shifted c units vertically downward.

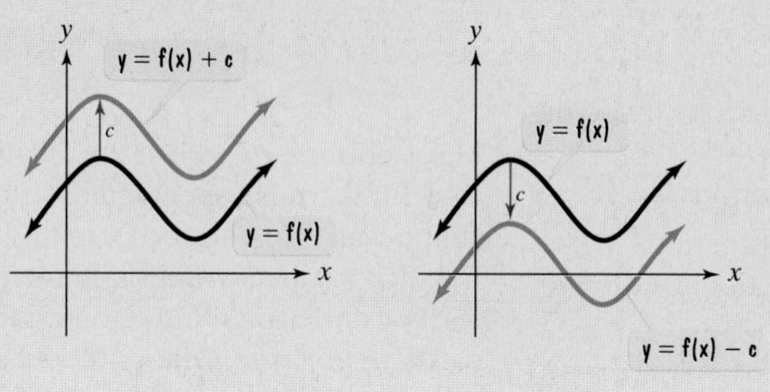

EXAMPLE 1 Vertical Shift Down

Use the graph of $f(x) = |x|$ to obtain the graph of $g(x) = |x| - 4$.

Solution The graph of $g(x) = |x| - 4$ has the same shape as the graph of $f(x) = |x|$. However, it is shifted down vertically 4 units. We have constructed a table showing some of the coordinates for f and g. The graphs of f and g are shown in Figure 2.36.

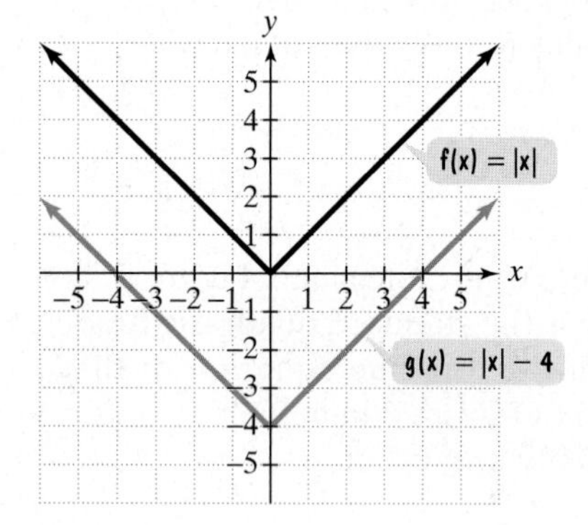

Figure 2.36

x	$y = f(x) = \|x\|$	$y = g(x)$ $= \|x\| - 4 = f(x) - 4$
-2	$\|-2\| = 2$	$\|-2\| - 4 = -2$
-1	$\|-1\| = 1$	$\|-1\| - 4 = -3$
0	$\|0\| = 0$	$\|0\| - 4 = -4$
1	$\|1\| = 1$	$\|1\| - 4 = -3$
2	$\|2\| = 2$	$\|2\| - 4 = -2$

Check Point 1 Use the graph of $f(x) = |x|$ to obtain the graph of $g(x) = |x| + 3$.

2 Use horizontal shifts to graph functions.

Horizontal Shifts

We return to the graph of $f(x) = x^2$, the standard quadratic function. In Figure 2.37, the graph of function f is in the middle of the three graphs. Note that there are graphs to the right and left of f. By contrast to the vertical shift situation,

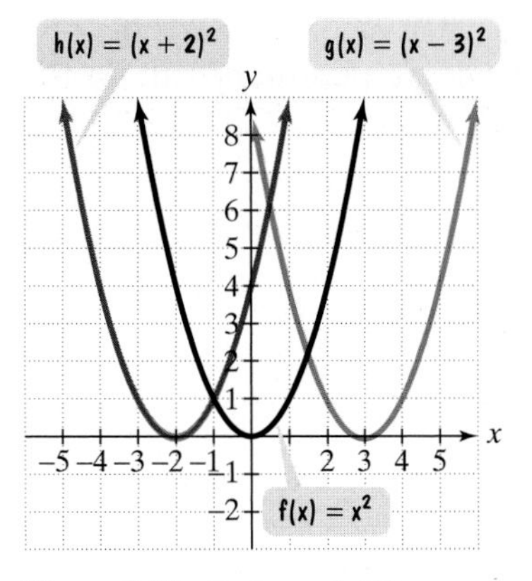

Figure 2.37 Horizontal shifts

this time there are graphs to the left and to the right of the graph of f. Look at the blue graph on the right. The equation of this graph, $g(x) = (x - 3)^2$, subtracts 3 from each value of x in the domain of $f(x) = x^2$. What effect does this have on the graph of f? It shifts the graph horizontally to the right by 3 units.

$$g(x) = (x - 3)^2 = f(x - 3)$$

The graph of g | shifts the graph of f 3 units to the right.

Now, look at the red graph on the left in Figure 2.37. The equation of this graph, $h(x) = (x + 2)^2$, adds 2 to each value of x in the domain of $f(x) = x^2$. What effect does this have on the graph of f? It shifts the graph horizontally to the left by 2 units.

$$h(x) = (x + 2)^2 = f(x + 2)$$

The graph of h | shifts the graph of f 2 units to the left.

In general, if c is positive, $y = f(x + c)$ shifts the graph of f to the left c units and $y = f(x - c)$ shifts the graph of f to the right c units. These are called **horizontal shifts** of the graph of f.

Study Tip

We know that positive numbers are to the right of zero on a number line and negative numbers are to the left of zero. This positive-negative orientation does not apply to horizontal shifts. A *positive* number causes a shift to the *left* and a *negative* number causes a shift to the *right*.

Horizontal Shifts

Let f be a function and c a positive real number.

- The graph of $y = f(x + c)$ is the graph of $y = f(x)$ shifted to the left c units.
- The graph of $y = f(x - c)$ is the graph of $y = f(x)$ shifted to the right c units.

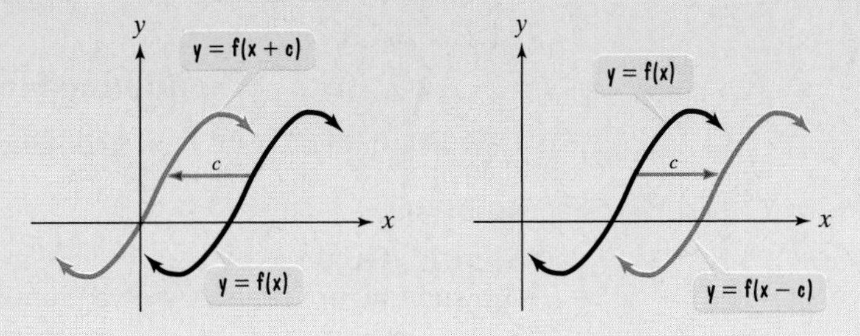

EXAMPLE 2 Horizontal Shift to the Left

Use the graph of $f(x) = \sqrt{x}$ to obtain the graph of $g(x) = \sqrt{x + 5}$.

Solution Compare the equations for $f(x) = \sqrt{x}$ and $g(x) = \sqrt{x + 5}$. The equation for g adds 5 to each value of x in the domain of f.

$$y = g(x) = \sqrt{x + 5} = f(x + 5)$$

The graph of g | shifts the graph of f 5 units to the left.

The graph of $g(x) = \sqrt{x + 5}$ has the same shape as the graph of $f(x) = \sqrt{x}$. However, it is shifted horizontally to the left 5 units. We have created tables

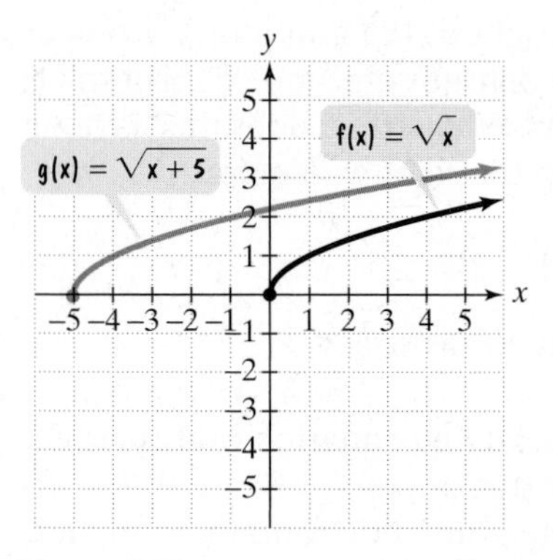

Figure 2.38 Shifting $f(x) = \sqrt{x}$ five units left

showing some of the coordinates for f and g. As shown in Figure 2.38, every point in the graph of g is exactly 5 units to the left of a corresponding point on the graph of f.

x	$y = f(x) = \sqrt{x}$
0	$\sqrt{0} = 0$
1	$\sqrt{1} = 1$
4	$\sqrt{4} = 2$

x	$y = g(x) = \sqrt{x+5}$
−5	$\sqrt{-5+5} = \sqrt{0} = 0$
−4	$\sqrt{-4+5} = \sqrt{1} = 1$
−1	$\sqrt{-1+5} = \sqrt{4} = 2$

Check Point 2 Use the graph of $f(x) = \sqrt{x}$ to obtain the graph of $g(x) = \sqrt{x-4}$.

Some functions can be graphed by combining horizontal and vertical shifts. The function should be a variation of a function whose equation you know how to graph, such as the standard quadratic function, the standard cubic function, the square root function, or the absolute value function.

In our next example, we will use the graph of the standard quadratic function $f(x) = x^2$ to obtain the graph of $h(x) = (x+1)^2 - 3$. We will graph three functions:

$$f(x) = x^2 \qquad g(x) = (x+1)^2 \qquad h(x) = (x+1)^2 - 3$$

Start by graphing the standard quadratic function.

Shift the graph of f horizontally one unit to the left.

Shift the graph of g vertically down 3 units.

EXAMPLE 3 Combining Horizontal and Vertical Shifts

Use the graph of $f(x) = x^2$ to obtain the graph of $h(x) = (x+1)^2 - 3$.

Solution

Step 1 Graph $f(x) = x^2$. The graph of the standard quadratic function is shown in Figure 2.39(a). We've identified three points on the graph.

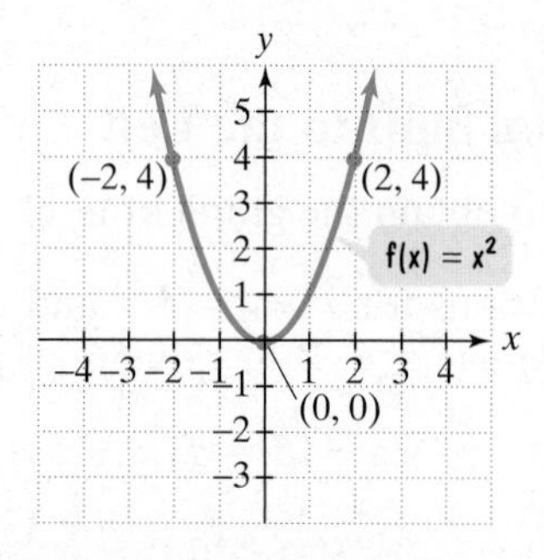

(a) The graph of $f(x) = x^2$

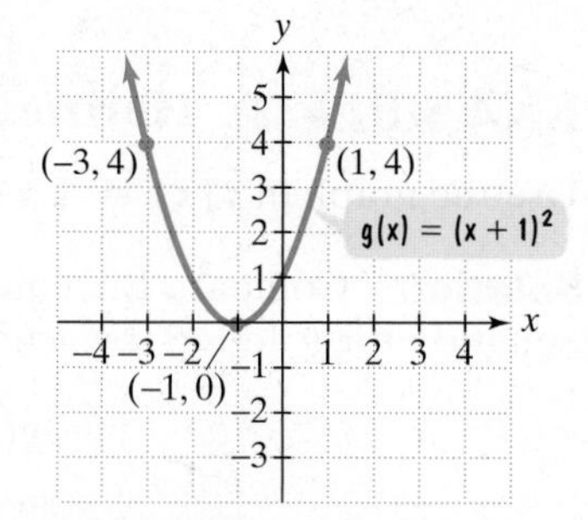

(b) The graph of $g(x) = (x+1)^2$

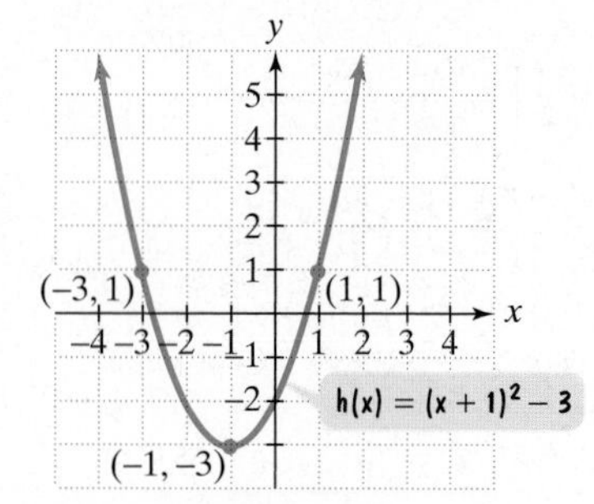

(c) The graph of $h(x) = (x+1)^2 - 3$

Figure 2.39

Discovery

Work Example 3 by first shifting the graph of $f(x) = x^2$ three units down, graphing $g(x) = x^2 - 3$. Now, shift this graph one unit left to graph $h(x) = (x + 1)^2 - 3$. Did you obtain the graph in Figure 2.39(c)? What can you conclude?

Step 2 Graph $g(x) = (x + 1)^2$. Because we add 1 to each value of x in the domain of the standard quadratic function $f(x) = x^2$, we shift the graph of f horizontally one unit to the left. This is shown in Figure 2.39(b). Notice that every point in the graph in Figure 2.39(b) has an x-coordinate that is one less than the x-coordinate for the corresponding point in the graph in Figure 2.39(a).

Step 3 Graph $h(x) = (x + 1)^2 - 3$. Because we subtract 3, we shift the graph in Figure 2.39(b) vertically down 3 units. The graph is shown in Figure 2.39(c). Notice that every point in the graph in Figure 2.39(c) has a y-coordinate that is three less than the y-coordinate of the corresponding point in the graph in Figure 2.39(b).

Check Point 3 Use the graph of $f(x) = \sqrt{x}$ to obtain the graph of $h(x) = \sqrt{x - 1} - 2$.

3 Use reflections to graph functions.

Reflections of Graphs

This photograph shows a reflection of an old bridge in a Maryland river. This perfect reflection occurs because the surface of the water is absolutely still. A mild breeze rippling the water's surface would distort the reflection.

Is it possible for graphs to have mirror-like qualities? Yes. Figure 2.40 shows the graphs of $f(x) = x^2$ and $g(x) = -x^2$. The graph of g is a **reflection about the x-axis** of the graph of f. In general, the graph of $y = -f(x)$ reflects the graph of f about the x-axis. Thus, the graph of g is a reflection of the graph of f about the x-axis because

$$g(x) = -x^2 = -f(x).$$

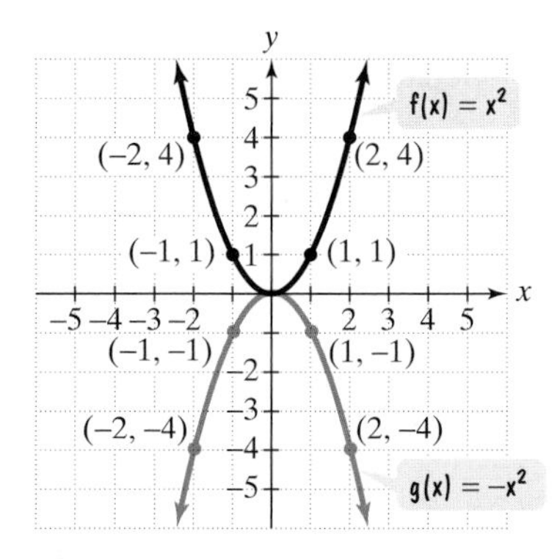

Figure 2.40 Reflections about the x-axis

Reflection About the x-Axis

The graph of $y = -f(x)$ is the graph of $y = f(x)$ reflected about the x-axis.

EXAMPLE 4 Reflection about the x-Axis

Use the graph of $f(x) = \sqrt{x}$ to obtain the graph of $g(x) = -\sqrt{x}$.

Solution Compare the equations for $f(x) = \sqrt{x}$ and $g(x) = -\sqrt{x}$. The graph of g is a reflection about the x-axis of the graph of f because

$$g(x) = -\sqrt{x} = -f(x).$$

We have created a table showing some of the coordinates for f and g. The graphs of f and g are shown in Figure 2.41.

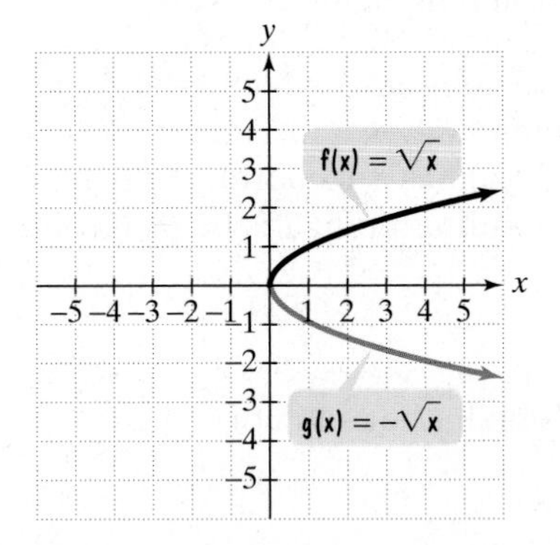

Figure 2.41 Reflecting $f(x) = \sqrt{x}$ about the x-axis

x	$f(x) = \sqrt{x}$	$g(x) = -\sqrt{x}$
0	$\sqrt{0} = 0$	$-\sqrt{0} = 0$
1	$\sqrt{1} = 1$	$-\sqrt{1} = -1$
4	$\sqrt{4} = 2$	$-\sqrt{4} = -2$

Check Point 4

Use the graph of $f(x) = |x|$ to obtain the graph of $g(x) = -|x|$.

It is also possible to reflect graphs about the y-axis.

Reflection about the y-Axis

The graph of $y = f(-x)$ is the graph of $y = f(x)$ reflected about the y-axis.

EXAMPLE 5 Reflection about the y-Axis

Use the graph of $f(x) = \sqrt{x}$ to obtain the graph of $h(x) = \sqrt{-x}$.

Solution Compare the equations for $f(x) = \sqrt{x}$ and $h(x) = \sqrt{-x}$. The graph of h is a reflection about the y-axis of the graph of f because

$$h(x) = \sqrt{-x} = f(-x).$$

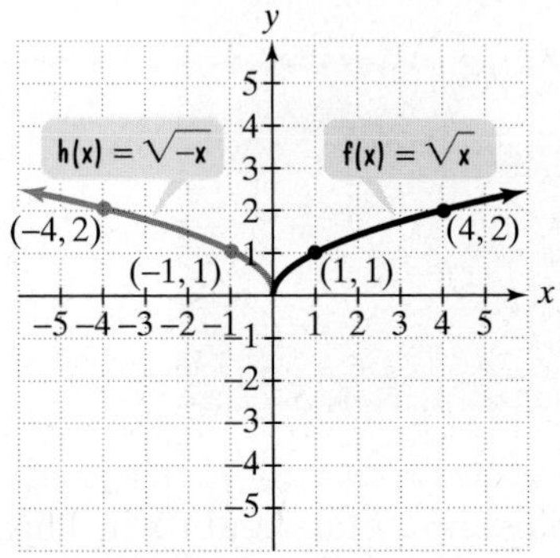

We have created tables showing some of the coordinates for f and h. The graphs of f and h are shown in Figure 2.42.

Figure 2.42 Reflecting $f(x) = \sqrt{x}$ about the y-axis

x	$f(x) = \sqrt{x}$
0	$\sqrt{0} = 0$
1	$\sqrt{1} = 1$
4	$\sqrt{4} = 2$

x	$h(x) = \sqrt{-x}$
0	$\sqrt{-0} = \sqrt{0} = 0$
−1	$\sqrt{-(-1)} = \sqrt{1} = 1$
−4	$\sqrt{-(-4)} = \sqrt{4} = 2$

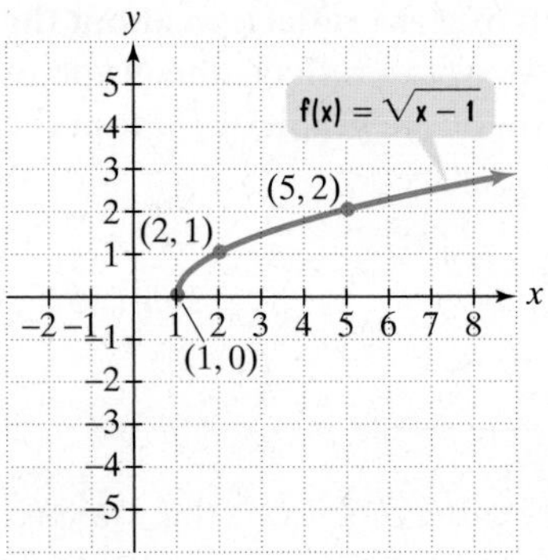

Figure 2.43

Check Point 5

Use the graph of $f(x) = \sqrt{x - 1}$ in Figure 2.43 to obtain the graph of $h(x) = \sqrt{-x - 1}$.

4 Use vertical stretching and shrinking to graph functions.

Vertical Stretching and Shrinking

Morphing does much more than move an image horizontally, vertically, or about an axis. An object having one shape is transformed into a different shape. Horizontal shifts, vertical shifts, and reflections do not change the basic shape of a graph. How can we shrink and stretch graphs, thereby altering their basic shapes?

Look at the three graphs in Figure 2.44. The black graph in the middle is the graph of the standard quadratic function, $f(x) = x^2$. Now, look at the blue graph on the top. The equation of this graph is $g(x) = 2x^2$. Thus, for each x, the y-coordinate of g is 2 times as large as the corresponding y-coordinate on the graph of f. The result is a narrower graph. We say that the graph of g is obtained by vertically *stretching* the graph of f. Now, look at the red graph on the bottom. The equation of this graph is $h(x) = \frac{1}{2}x^2$, or $h(x) = \frac{1}{2}f(x)$. Thus, for each x, the y-coordinate of h is one-half as large as the corresponding y-coordinate on the graph of f. The result is a wider graph. We say that the graph of h is obtained by vertically *shrinking* the graph of f.

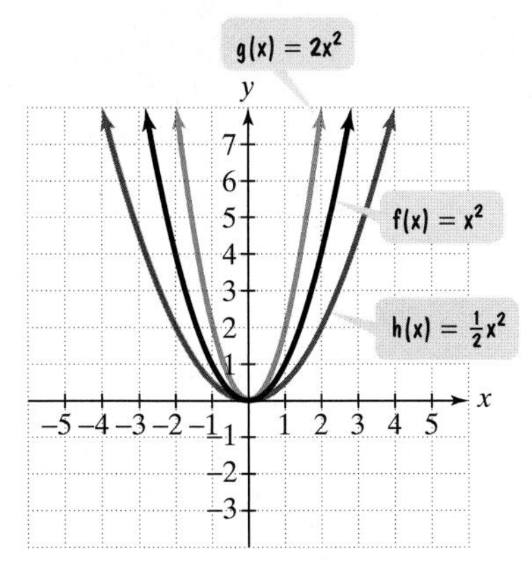

Figure 2.44 Stretching and shrinking $f(x) = x^2$

These observations can be summarized as follows.

Stretching and Shrinking Graphs

Let f be a function and c a positive real number.

- If $c > 1$, the graph of $y = cf(x)$ is the graph of $y = f(x)$ vertically stretched by multiplying each of its y-coordinates by c.
- If $0 < c < 1$, the graph of $y = cf(x)$ is the graph of $y = f(x)$ vertically shrunk by multiplying each of its y-coordinates by c.

EXAMPLE 6 Vertically Stretching a Graph

Use the graph of $f(x) = |x|$ to obtain the graph of $g(x) = 2|x|$.

Solution The graph of $g(x) = 2|x|$ is obtained by vertically stretching the graph of $f(x) = |x|$. We have constructed a table showing some of the coordinates for f and g. Observe that the y-coordinate on the graph of g is twice as large as the corresponding y-coordinate on the graph of f. The graphs of f and g are shown in Figure 2.45.

x	$f(x) = \|x\|$	$g(x) = 2\|x\| = 2f(x)$
−2	$\|-2\| = 2$	$2\|-2\| = 4$
−1	$\|-1\| = 1$	$2\|-1\| = 2$
0	$\|0\| = 0$	$2\|0\| = 0$
1	$\|1\| = 1$	$2\|1\| = 2$
2	$\|2\| = 2$	$2\|2\| = 4$

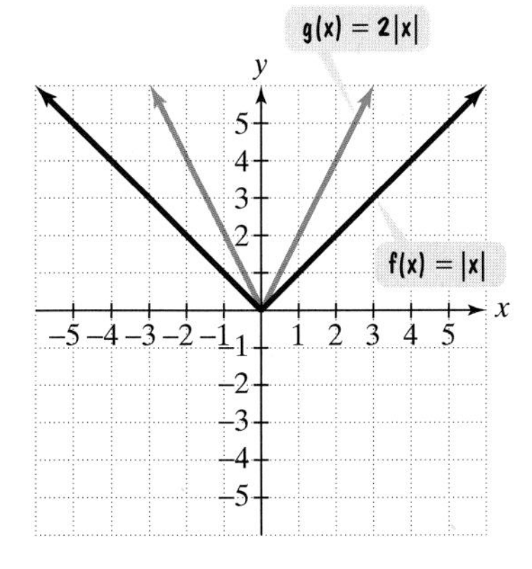

Figure 2.45 Stretching $f(x) = |x|$

Check Point 6 Use the graph of $f(x) = |x|$ to obtain the graph of $g(x) = 3|x|$.

EXAMPLE 7 Vertically Shrinking a Graph

Use the graph of $f(x) = |x|$ to obtain the graph of $h(x) = \frac{1}{2}|x|$.

Solution The graph of $h(x) = \frac{1}{2}|x|$ is obtained by vertically shrinking the graph of $f(x) = |x|$. We have constructed a table showing some of the coordinates for f and h. Observe that the y-coordinate on the graph of h is one-half the corresponding y-coordinate on the graph of f. The graphs of f and h are shown in Figure 2.46.

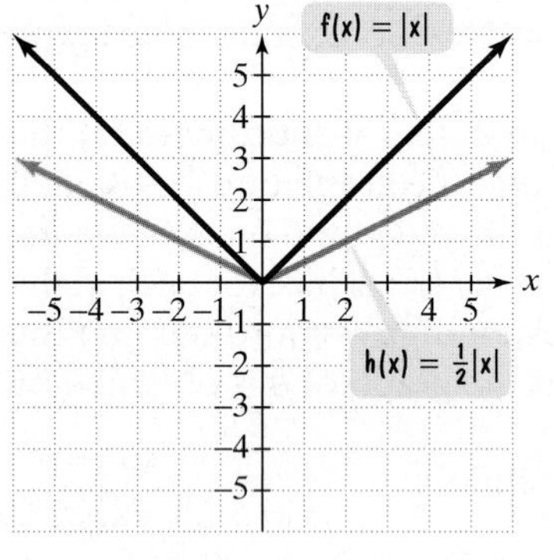

Figure 2.46 Shrinking $f(x) = |x|$

x	$f(x) = \|x\|$	$h(x) = \frac{1}{2}\|x\| = \frac{1}{2}f(x)$
-2	$\|-2\| = 2$	$\frac{1}{2}\|-2\| = 1$
-1	$\|-1\| = 1$	$\frac{1}{2}\|-1\| = \frac{1}{2}$
0	$\|0\| = 0$	$\frac{1}{2}\|0\| = 0$
1	$\|1\| = 1$	$\frac{1}{2}\|1\| = \frac{1}{2}$
2	$\|2\| = 2$	$\frac{1}{2}\|2\| = 1$

Check Point 7 Use the graph of $f(x) = |x|$ to obtain the graph of $h(x) = \frac{1}{4}|x|$.

5 Graph functions involving a sequence of transformations.

Sequences of Transformations

Table 2.5 summarizes the procedures for transforming the graph of $y = f(x)$.

Table 2.5 Summary of Transformations
In each case, c represents a positive real number.

To Graph:	Draw the Graph of f and:	Changes in the Equation of $y = f(x)$
Vertical shifts		
$y = f(x) + c$	Raise the graph of f by c units.	c is added to $f(x)$.
$y = f(x) - c$	Lower the graph of f by c units.	c is subtracted from $f(x)$.
Horizontal shifts		
$y = f(x + c)$	Shift the graph of f to the left c units.	x is replaced by $x + c$.
$y = f(x - c)$	Shift the graph of f to the right c units.	x is replaced by $x - c$.
Reflection about the x-axis $y = -f(x)$	Reflect the graph of f about the x-axis.	$f(x)$ is multiplied by -1.
Reflection about the y-axis $y = f(-x)$	Reflect the graph of f about the y-axis.	x is replaced by $-x$.
Vertical stretching or shrinking		
$y = cf(x), c > 1$	Multiply each y-coordinate of $y = f(x)$ by c, vertically stretching the graph of f.	$f(x)$ is multiplied by c, $c > 1$.
$y = cf(x), 0 < c < 1$	Multiply each y-coordinate of $y = f(x)$ by c, vertically shrinking the graph of f.	$f(x)$ is multiplied by c, $0 < c < 1$.

A function involving more than one transformation can be graphed by performing transformations in the following order.

1. Horizontal shifting
2. Vertical stretching or shrinking
3. Reflecting
4. Vertical shifting

EXAMPLE 8 Graphing Using a Sequence of Transformations

Use the graph of $f(x) = \sqrt{x}$ to graph $g(x) = \sqrt{1 - x} + 3$.

Solution The following sequence of steps is illustrated in Figure 2.47. We begin with the graph of $f(x) = \sqrt{x}$.

Step 1 Horizontal Shifting Graph $y = \sqrt{x + 1}$. Because x is replaced by $x + 1$, the graph of $f(x) = \sqrt{x}$ is shifted 1 unit to the left.

Step 2 Vertical Stretching or Shrinking Because the equation $y = \sqrt{x + 1}$ is not multiplied by a constant in $g(x) = \sqrt{1 - x} + 3$, no stretching or shrinking is involved.

Step 3 Reflecting We are interested in graphing $y = \sqrt{1 - x} + 3$, or $y = \sqrt{-x + 1} + 3$. We have now graphed $y = \sqrt{x + 1}$. We can graph $y = \sqrt{-x + 1}$ by noting that x is replaced by $-x$. Thus, we graph $y = \sqrt{-x + 1}$ by reflecting the graph of $y = \sqrt{x + 1}$ about the y-axis.

Step 4 Vertical Shifting We can use the graph of $y = \sqrt{1 - x}$ to get the graph of $g(x) = \sqrt{1 - x} + 3$. Because 3 is added, shift the graph of $y = \sqrt{1 - x}$ up by 3 units.

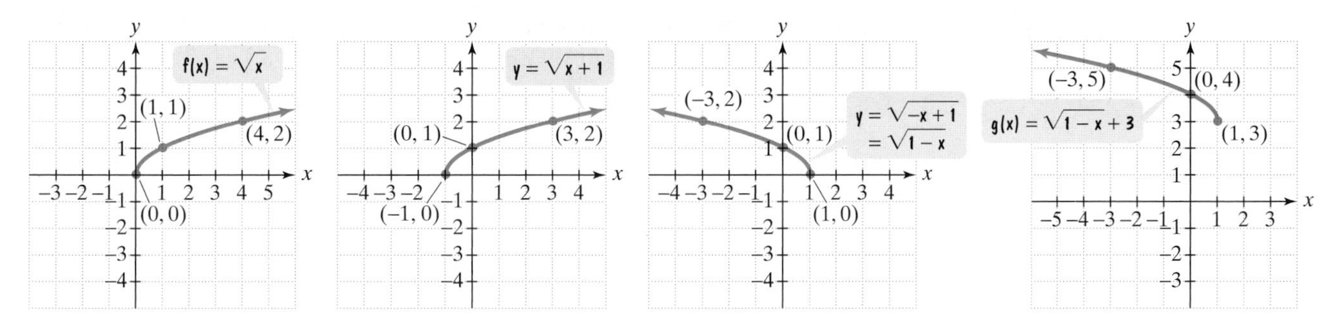

Figure 2.47 Using $f(x) = \sqrt{x}$ to graph $g(x) = \sqrt{1 - x} + 3$.

Check Point 8 Use the graph of $f(x) = x^2$ to graph $g(x) = -(x - 2)^2 + 3$.

6 Combine functions arithmetically, specifying domains.

Combinations of Functions

The graph in Figure 2.48 on page 230 shows the number of wars by region from 1990 through 1997. Have you seen data displayed in this style? The numbers on the bars do not represent the total number of wars for each year. Rather, they

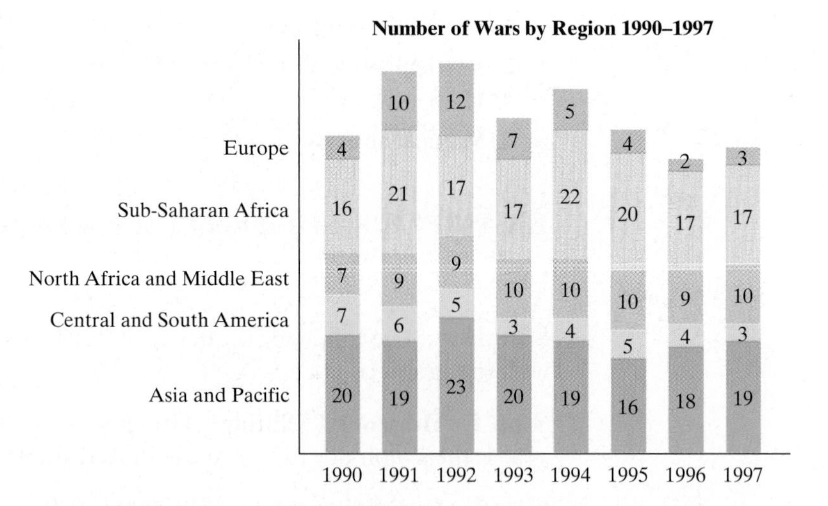

Figure 2.48
Source: International Peace Research Institute, Oslo

represent the number of wars by region for the year. Each bar's height represents the *total number* of wars per year. For example, in 1990 there were

$$20 + 7 + 7 + 16 + 4$$

or 54 wars. We can think of this addition as the addition of function values. We do this by introducing the following functions.

Let $f(x)$ = the number of wars in Asia and the Pacific in year x.

Let $g(x)$ = the number of wars in Central and South America in year x.

Using Figure 2.48, we see that

Look at the green portion of each bar.

$$f(1990) = 20 \quad f(1991) = 19 \quad f(1992) = 23 \quad f(1993) = 20, \text{ and so on}$$

and

Look at the light blue portion of each bar.

$$g(1990) = 7 \quad g(1991) = 6 \quad g(1992) = 5 \quad g(1993) = 3, \text{ and so on.}$$

We can add these function values by introducing a new function, $f + g$, defined by the addition of $f(x)$ and $g(x)$. Thus,

$(f + g)(x) = f(x) + g(x) =$ the number of wars in Asia and the Pacific in year x plus the number of wars in Central and South America in year x.

For example,

$$(f + g)(1990) = f(1990) + g(1990) = 20 + 7 = 27.$$

In 1990, there were a total of 27 wars in Asia, the Pacific, Central America, and South America combined.

We can also subtract these function values by introducing a new function, $f - g$, defined by the subtraction of $f(x)$ and $g(x)$. Thus,

$(f - g)(x) = f(x) - g(x) =$ the number of wars in Asia and the Pacific in year x minus the number of wars in Central and South America in year x.

For example,

$$(f - g)(1990) = f(1990) - g(1990) = 20 - 7 = 13.$$

In 1990, there were 13 more wars in Asia and the Pacific than in Central and South America.

Suppose that we have data for Asia and the Pacific for the years 1990–1993 only. The domain of f under these conditions is $\{1990, 1991, 1992, 1993\}$. Further suppose that we have data for Central and South America for the years 1992–1995 only. The domain of g under these conditions is $\{1992, 1993, 1994, 1995\}$. We can add or subtract f and g only for the values of x that are in the domain of f *and* in the domain of g. These common values of x are 1992 and 1993. Functions can be combined only if there are numbers that are common to the domains of both functions.

We see that functions, like numbers, can be added and subtracted. Functions can also be multiplied or divided. Because functions are usually given as equations, we perform these operations by performing operations with the algebraic expressions that appear on the right side of the equations. For example, we can combine the following two functions using addition:

$$f(x) = 2x + 1 \quad \text{and} \quad g(x) = x^2 - 4.$$

To do so, we add the terms to the right of the equal sign for $f(x)$ to the terms to the right of the equal sign for $g(x)$. Here is how it's done:

$$\begin{aligned}(f + g)(x) &= f(x) + g(x) \\ &= (2x + 1) + (x^2 - 4) && \text{Add terms for } f(x) \text{ and } g(x). \\ &= 2x - 3 + x^2 && \text{Combine like terms.} \\ &= x^2 + 2x - 3 && \text{Arrange terms in descending powers of } x.\end{aligned}$$

The name of this new function is $f + g$. Thus, the sum $f + g$ is the function defined by $(f + g)(x) = x^2 + 2x - 3$. The domain of $f + g$ consists of the numbers x that are in the domain of f and in the domain of g. Because neither f nor g contains division or even roots, the domain of each function is the set of all real numbers. Thus, the domain of $f + g$ is also the set of all real numbers.

EXAMPLE 9 Finding the Sum of Two Functions

Let $f(x) = x^2 - 3$ and $g(x) = 4x + 5$. Find

a. $(f + g)(x)$ b. $(f + g)(3)$

Solution

a. $(f + g)(x) = f(x) + g(x) = (x^2 - 3) + (4x + 5) = x^2 + 4x + 2$. Thus, $(f + g)(x) = x^2 + 4x + 2$.

b. We find $(f + g)(3)$ by substituting 3 for x in the equation for $f + g$.

$$(f + g)(x) = x^2 + 4x + 2 \quad \text{This is the equation for } f + g.$$

Substitute 3 for x.

$$(f + g)(3) = 3^2 + 4 \cdot 3 + 2 = 9 + 12 + 2 = 23$$

Check Point 9

Let $f(x) = 3x^2 + 4x - 1$ and $g(x) = 2x + 7$. Find

a. $(f + g)(x)$ **b.** $(f + g)(4)$

Here is a general definition for function addition.

The Sum of Functions

Let f and g be two functions. The **sum** $\boldsymbol{f + g}$ is the function defined by

$$(f + g)(x) = f(x) + g(x).$$

The domain of $f + g$ is the set of all real numbers that are common to the domain of f and the domain of g.

EXAMPLE 10 Adding Functions and Determining the Domain

Let $f(x) = \sqrt{x + 3}$ and $g(x) = \sqrt{x - 2}$. Find:

a. $(f + g)(x)$ b. the domain of $f + g$

Solution

a. $(f + g)(x) = f(x) + g(x) = \sqrt{x + 3} + \sqrt{x - 2}$.

b. The domain of $f + g$ is the set of all real numbers that are common to the domain of f and the domain of g. Thus, we must find the domains of f and g. We will do so for f first.

Note that $f(x) = \sqrt{x + 3}$ is a function involving the square root of $x + 3$. Because the square root of a negative quantity is not a real number, the value of $x + 3$ must be nonnegative. Thus, the domain of f is all x such that $x + 3 \geq 0$. Equivalently, the the domain is $\{x | x \geq -3\}$, or $[-3, \infty)$.

Likewise, $g(x) = \sqrt{x - 2}$ is also a square root function. Because the square root of a negative quantity is not a real number, the value of $x - 2$ must be nonnegative. Thus, the domain of g is all x such that $x - 2 \geq 0$. Equivalently, the domain is $\{x | x \geq 2\}$, or $[2, \infty)$.

Now, we can use a number line to determine the the domain of $f + g$. Figure 2.49 shows the domain of f in blue and the domain of g in red. Can you see that all real numbers greater than or equal to 2 are common to both domains? This is shown in purple on the number line. Thus, the domain of $f + g$ is $[2, \infty)$.

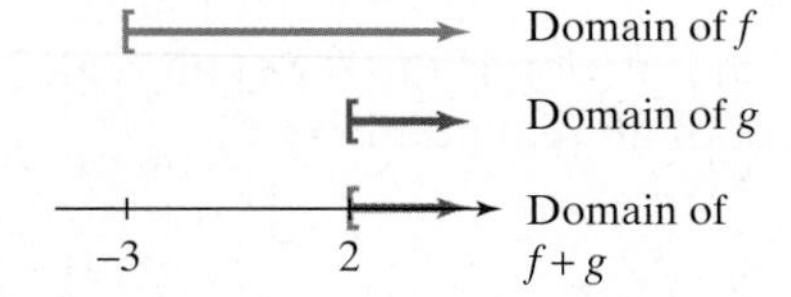

Figure 2.49 Finding the domain of the sum $f + g$

Technology

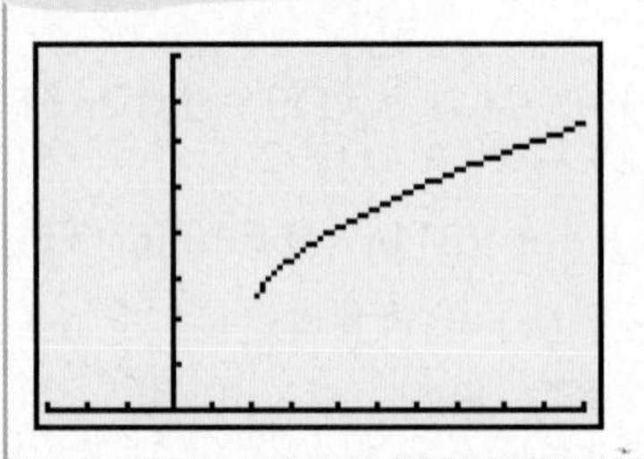

The graph on the left is the graph of

$$y = \sqrt{x + 3} + \sqrt{x - 2}$$

in a $[-3, 10, 1]$ by $[0, 8, 1]$ viewing rectangle. The graph reveals what we discovered algebraically in Example 10(b). The domain of this function is $[2, \infty)$.

Check Point 10 Let $f(x) = \sqrt{x - 3}$ and $g(x) = \sqrt{x + 1}$. Find:

a. $(f + g)(x)$ **b.** the domain of $f + g$

We can also combine functions using subtraction, multiplication, and division by performing operations with the algebraic expressions that appear on the right side of the equations. For example, the functions $f(x) = x + 3$ and

$g(x) = x - 1$ can be combined to form the difference, product, and quotient of f and g. Here's how it's done.

Difference: $f - g$

$$(f - g)(x) = f(x) - g(x)$$
$$= x + 3 - (x - 1) = x + 3 - x + 1 = 4$$

Product: fg

$$(fg)(x) = f(x) \cdot g(x)$$
$$= (x + 3)(x - 1) = x^2 + 2x - 3$$

Quotient: $\frac{f}{g}$

$$\left(\frac{f}{g}\right)(x) = \frac{f(x)}{g(x)} = \frac{x + 3}{x - 1}, \quad x \neq 1$$

Just like the domain for $f + g$, the domain for each of these functions consists of all real numbers that are common to the domains of f and g. In the case of the quotient function $\frac{f(x)}{g(x)}$, we must remember not to divide by 0, so we add the further restriction that $g(x) \neq 0$.

The following definitions summarize our discussion.

Definitions: Sum, Difference, Product, and Quotient of Functions

Let f and g be two functions. The **sum** $f + g$, the **difference** $f - g$, the **product** fg, and the **quotient** $\frac{f}{g}$ are functions whose domains are the set of all real numbers common to the domains of f and g, defined as follows:

1. Sum: $(f + g)(x) = f(x) + g(x)$

2. Difference: $(f - g)(x) = f(x) - g(x)$

3. Product: $(fg)(x) = f(x) \cdot g(x)$

4. Quotient: $\left(\frac{f}{g}\right)(x) = \frac{f(x)}{g(x)}$, provided $g(x) \neq 0$

EXAMPLE 11 Combining Functions

If $f(x) = 2x - 1$ and $g(x) = x^2 + x - 2$, find:

a. $(f - g)(x)$ b. $(fg)(x)$ c. $\left(\frac{f}{g}\right)(x)$

Determine the domain for each function.

Solution

a. $(f - g)(x) = f(x) - g(x)$ — This is the definition of the difference $f - g$.

$= (2x - 1) - (x^2 + x - 2)$ — Subtract $g(x)$ from $f(x)$.

$= 2x - 1 - x^2 - x + 2$ — Perform the subtraction.

$= -x^2 + x + 1$ — Combine like terms and arrange terms in descending powers of x.

b. $(fg)(x) = (2x - 1)(x^2 + x - 2)$ — This is the definition of the product fg.

$= 2x(x^2 + x - 2) - 1(x^2 + x - 2)$ — Multiply each term in the second factor by $2x$ and -1, respectively.

$= 2x^3 + 2x^2 - 4x - x^2 - x + 2$ — Use the distributive property.

$= 2x^3 + (2x^2 - x^2) + (-4x - x) + 2$ — Rearrange terms so that like terms are adjacent.

$= 2x^3 + x^2 - 5x + 2$ — Combine like terms.

c. $\left(\frac{f}{g}\right)(x) = \frac{f(x)}{g(x)}$ — This is the definition of the quotient $\frac{f}{g}$.

$= \frac{2x - 1}{x^2 + x - 2}$ — Divide the algebraic expressions for $f(x)$ and $g(x)$.

Because the equations for f and g do not involve division or contain even roots, the domain of both f and g is the set of all real numbers. Thus, the domain of $f - g$ and fg is the set of all real numbers. However, for $\frac{f}{g}$, the denominator cannot equal zero. We can factor the denominator as follows:

$$\left(\frac{f}{g}\right)(x) = \frac{2x - 1}{x^2 + x - 2} = \frac{2x - 1}{(x + 2)(x - 1)}$$

Because $x + 2 \neq 0$, $x \neq -2$.

Because $x - 1 \neq 0$, $x \neq 1$.

We see that the domain for $\frac{f}{g}$ is the set of all real numbers except -2 and 1: $\{x \mid x \neq -2 \text{ and } x \neq 1\}$.

Check Point 11 If $f(x) = x - 5$ and $g(x) = x^2 - 1$, find:

a. $(f - g)(x)$ **b.** $(fg)(x)$ **c.** $\left(\frac{f}{g}\right)(x)$

Determine the domain for each function.

EXERCISE SET 2.5

Practice Exercises

In Exercises 1–10, begin by graphing the standard quadratic function, $f(x) = x^2$. Then use transformations of this graph to graph the given function.

1. $g(x) = x^2 - 2$
2. $g(x) = x^2 - 1$
3. $g(x) = (x - 2)^2$
4. $g(x) = (x - 1)^2$
5. $h(x) = -(x - 2)^2$
6. $h(x) = -(x - 1)^2$
7. $h(x) = (x - 2)^2 + 1$
8. $h(x) = (x - 1)^2 + 2$
9. $g(x) = 2(x - 2)^2$
10. $g(x) = \frac{1}{2}(x - 1)^2$

In Exercises 11–22, begin by graphing the square root function, $f(x) = \sqrt{x}$. Then use transformations of this graph to graph the given function.

11. $g(x) = \sqrt{x} + 2$
12. $g(x) = \sqrt{x} + 1$
13. $g(x) = \sqrt{x + 2}$
14. $g(x) = \sqrt{x + 1}$

15. $h(x) = -\sqrt{x+2}$
16. $h(x) = -\sqrt{x+1}$
17. $h(x) = \sqrt{-x+2}$
18. $h(x) = \sqrt{-x+1}$
19. $g(x) = \frac{1}{2}\sqrt{x+2}$
20. $g(x) = 2\sqrt{x+1}$
21. $h(x) = \sqrt{x+2} - 2$
22. $h(x) = \sqrt{x+1} - 1$

In Exercises 23–34, begin by graphing the absolute value function, $f(x) = |x|$. Then use transformations of this graph to graph the given function.

23. $g(x) = |x| + 4$
24. $g(x) = |x| + 3$
25. $g(x) = |x+4|$
26. $g(x) = |x+3|$
27. $h(x) = |x+4| - 2$
28. $h(x) = |x+3| - 2$
29. $h(x) = -|x+4|$
30. $h(x) = -|x+3|$
31. $g(x) = -|x+4| + 1$
32. $g(x) = -|x+4| + 2$
33. $h(x) = 2|x+4|$
34. $h(x) = 2|x+3|$

In Exercises 35–44, begin by graphing the standard cubic function, $f(x) = x^3$. Then use transformations of this graph to graph the given function.

35. $g(x) = x^3 - 3$
36. $g(x) = x^3 - 2$
37. $g(x) = (x-3)^3$
38. $g(x) = (x-2)^3$
39. $h(x) = -x^3$
40. $h(x) = -(x-2)^3$
41. $h(x) = \frac{1}{2}x^3$
42. $h(x) = \frac{1}{4}x^3$
43. $r(x) = (x-3)^3 + 2$
44. $r(x) = (x-2)^3 + 1$

In Exercises 45–52, use the graph of the function f to sketch the graph of the given function g.

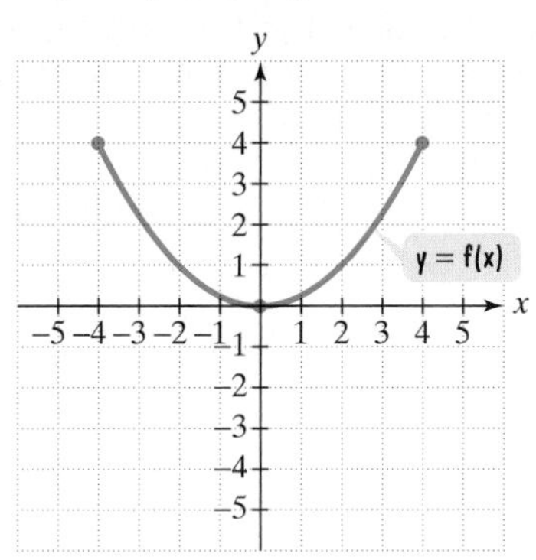

45. $g(x) = f(x) + 1$
46. $g(x) = f(x) + 2$
47. $g(x) = f(x+1)$
48. $g(x) = f(x+2)$
49. $g(x) = -f(x)$
50. $g(x) = \frac{1}{2}f(x)$
51. $g(x) = \frac{1}{2}f(x+1)$
52. $g(x) = -f(x+2)$

53. If $f(x) = 2x^2 - 5$ and $g(x) = 3x + 7$, find:
 a. $(f+g)(x)$ **b.** $(f+g)(4)$.

54. If $f(x) = 3x^2 - 2x + 1$ and $g(x) = 4x - 1$, find:
 a. $(f+g)(x)$ **b.** $(f+g)(5)$.

55. Let $f(x) = \sqrt{x-6}$ and $g(x) = \sqrt{x+2}$. Find:
 a. $(f+g)(x)$ **b.** the domain of $f+g$.

56. Let $f(x) = \sqrt{x-8}$ and $g(x) = \sqrt{x+5}$. Find:
 a. $(f+g)(x)$ **b.** the domain of $f+g$.

In Exercises 57–68, find $f+g$, $f-g$, fg, and $\frac{f}{g}$. Determine the domain for each function.

57. $f(x) = 2x + 3,\quad g(x) = x - 1$
58. $f(x) = 3x - 4,\quad g(x) = x + 2$
59. $f(x) = x - 5,\quad g(x) = 3x^2$
60. $f(x) = x - 6,\quad g(x) = 5x^2$
61. $f(x) = 2x^2 - x - 3,\quad g(x) = x + 1$
62. $f(x) = 6x^2 - x - 1,\quad g(x) = x - 1$
63. $f(x) = \sqrt{x},\quad g(x) = x - 4$
64. $f(x) = \dfrac{1}{x},\quad g(x) = x - 5$
65. $f(x) = 2 + \dfrac{1}{x},\quad g(x) = \dfrac{1}{x}$
66. $f(x) = 6 - \dfrac{1}{x},\quad g(x) = \dfrac{1}{x}$
67. $f(x) = \sqrt{x+4},\quad g(x) = \sqrt{x-1}$
68. $f(x) = \sqrt{x+6},\quad g(x) = \sqrt{x-3}$

Application Exercises

Consider the following functions:

$f(x)$ = population of the world's more-developed regions in year x

$g(x)$ = the population of the world's less-developed regions in year x

$h(x)$ = total world population in year x.

Use these functions and the graph shown to answer Exercises 69–72.

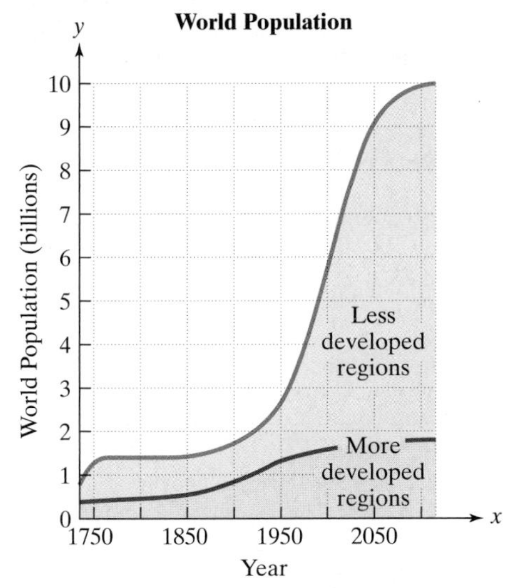

Source: Population Reference Bureau

69. What does the function $f + g$ represent?
70. What does the function $h - g$ represent?
71. Use the graph to estimate $(f+g)(2000)$.
72. Use the graph to estimate $(h-g)(2000)$.

73. A company that sells radios has yearly fixed costs of \$600,000. It costs the company \$45 to produce each radio. Each radio will sell for \$65. The company's costs and revenue are modeled by the functions

$C(x) = 600{,}000 + 45x$ This function models the company's costs.

$R(x) = 65x$ This function models the company's revenue.

Find and interpret $(R - C)(20{,}000)$, $(R - C)(30{,}000)$ and $(R - C)(40{,}000)$.

74. The function $f(t) = -0.14t^2 + 0.51t + 31.6$ models the U.S. population, $f(t)$, in millions, ages 65 and older t years after 1990. The function $g(t) = 0.54t^2 + 12.64t + 107.1$ models the total yearly cost of Medicare, $g(t)$, in billions of dollars, t years after 1990.

a. What does the function $\frac{g}{f}$ represent?

b. Find and interpret $\frac{g}{f}(10)$.

75. Consider two functions M and F that represent the number of male and female members of the House of Representatives for the years 1977, 1981, 1991, 1994, and 1999. Sketch the graphs of M and F in the same rectangular coordinate system, using the data in the bar graph. Each graph should consist of five points whose first coordinates are the years and second coordinates are the numbers of representatives, male or female. Now, add to the graphs in your coordinate system the graph of $M + F$. What constant function do you obtain? What is the significance of this constant?

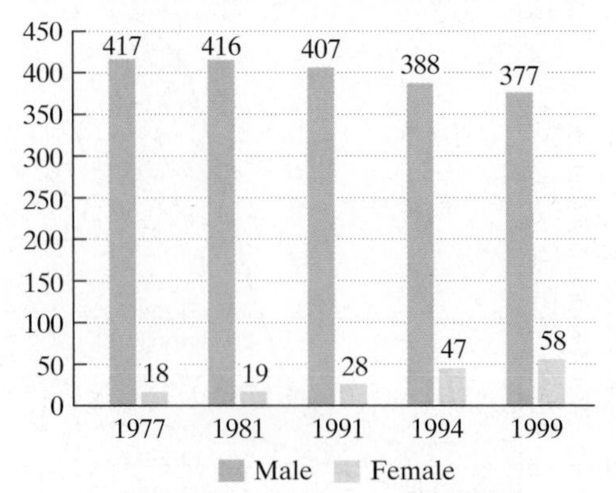

76. A department store has two locations in a city. From 1998 through 2002, the profits for each of the store's two branches are modeled by the functions $f(x) = -0.44x + 13.62$ and $g(x) = 0.51x + 11.14$. In each model, x represents the number of years after 1998 and f and g represent the profit in millions of dollars.

a. What is the slope for f? Describe what this means.

b. What is the slope for g? Describe what this means.

c. Find $f + g$. What is the slope for this function? What does this mean?

Writing in Mathematics

77. What must be done to a function's equation so that its graph is shifted vertically upward?

78. What must be done to a function's equation so that its graph is shifted horizontally to the right?

79. What must be done to a function's equation so that its graph is reflected about the x-axis?

80. What must be done to a function's equation so that its graph is reflected about the y-axis?

81. What must be done to a function's equation so that its graph is stretched?

82. If the equations of two functions are given, explain how to obtain the quotient function and its domain.

83. A company's profit is given by the function $y = P(x)$, where x represents the amount spent on advertising and P represents weekly profits, both expressed in hundreds of dollars.

a. Describe a situation that might occur in the company that would result in the graph of its profit function undergoing a vertical shift.

b. Now, consider the function $y = D(x)$, where x represents the amount spent on advertising and D represents weekly profits, both expressed in dollars rather than hundreds of dollars. If D and P are both graphed on the same axes, describe the relationship between the two graphs.

Technology Exercises

84. a. Use a graphing utility to graph $f(x) = x^2 + 1$.

b. Graph $f(x) = x^2 + 1$, $g(x) = f(2x)$, $h(x) = f(3x)$, and $k(x) = f(4x)$ on the same viewing rectangle.

c. Describe the relationship among the graphs of f, g, h, and k with emphasis on different values of x for points on all four graphs that give the same y-coordinate.

d. Generalize by describing the relationship between the graph of f and the graph of g, where $g(x) = f(cx)$ for $c > 1$.

e. Try out your generalization by sketching the graphs of $f(cx)$ for $c = 1$, $c = 2$, $c = 3$, and $c = 4$ for a function of your choice.

85. a. Use a graphing utility to graph $f(x) = x^2 + 1$.

b. Graph $f(x) = x^2 + 1$, and $g(x) = f(\frac{1}{2}x)$, and $h(x) = f(\frac{1}{4}x)$ on the same viewing rectangle.

c. Describe the relationship among the graphs of f, g, and h with emphasis on different values of x for points on all three graphs that give the same y-coordinate.

d. Generalize by describing the relationship between the graph of f and the graph of g, where $g(x) = f(cx)$ for $0 < c < 1$.

e. Try out your generalization by sketching the graphs of $f(cx)$ for $c = 1$, and $c = \frac{1}{2}$, and $c = \frac{1}{4}$ for a function of your choice.

Critical Thinking Exercises

86. Which one of the following is true?

a. If $f(x) = |x|$ and $g(x) = |x + 3| + 3$, then the graph of g is a translation of three units to the right and three units upward of the graph of f.

b. If $f(x) = -\sqrt{x}$ and $g(x) = \sqrt{-x}$, then f and g have identical graphs.

c. If $f(x) = x^2$ and $g(x) = 5(x^2 - 2)$, then the graph of g can be obtained from the graph of f by stretching f five units followed by a downward shift of two units.

d. If $f(x) = x^3$ and $g(x) = -(x - 3)^3 - 4$, then the graph of g can be obtained from the graph of f by moving f three units to the right, reflecting in the x-axis, and then moving the resulting graph down four units.

In Exercises 87–90, functions f and g are graphed in the same rectangular coordinate system. If g is obtained from f through a sequence of transformations, find an equation for g.

87.

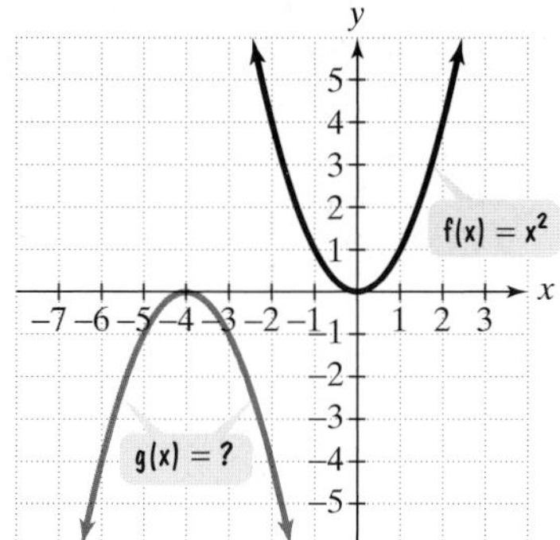

88.

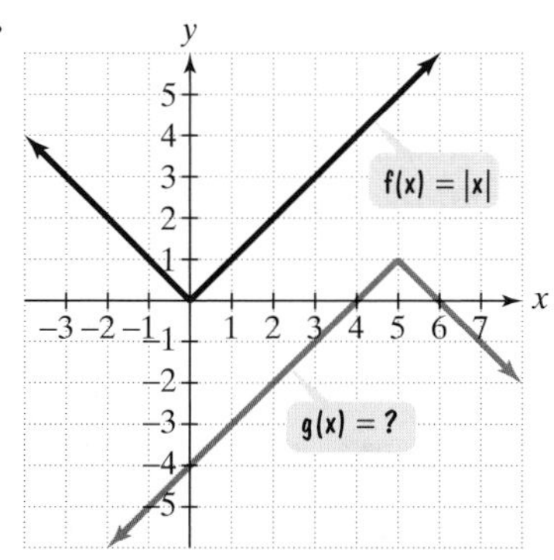

89.

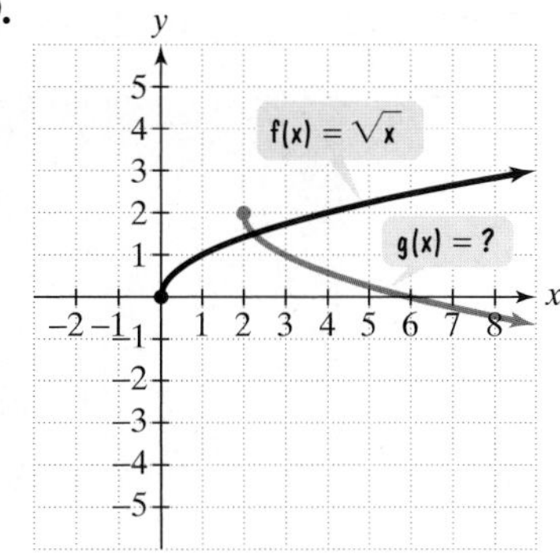

90.

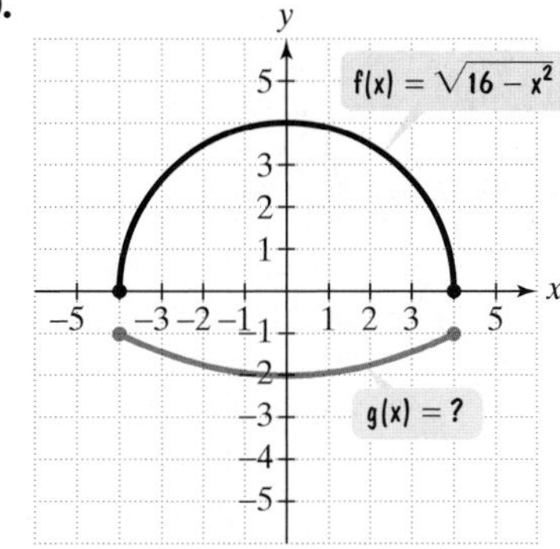

For Exercises 91–94, assume that (a, b) is a point on the graph of f. What is the corresponding point on the graph of each of the following functions?

91. $y = f(-x)$

92. $y = 2f(x)$

93. $y = f(x - 3)$

94. $y = f(x) - 3$

Group Exercises

95. Consult an almanac, newspaper, magazine, or the Internet to find data displayed in a bar graph in the style of Figure 2.48 on page 230. Using the two bar graphs that group members find most interesting, introduce two or more functions that are related to the graphs. Then write and solve a problem involving function addition and function subtraction for each selected graph. If you are not sure where to begin, reread page 230 or look at Exercises 69–72 in this exercise set.

96. This activity is a group research project on morphing and should result in a presentation made by group members to the entire class. Be sure to include morphing images that will intrigue class members. You should have no problem finding an array of fascinating images online. Also include a discussion of films using spectacular morphing effects. Rent videos of these films and show appropriate excerpts.

SECTION 2.6 Composite and Inverse Functions

Objectives

1. Form composite functions.
2. Verify inverse functions.
3. Find the inverse of a function.
4. Use the horizontal line test to determine if a function has an inverse function.
5. Use the graph of a one-to-one function to graph its inverse function.

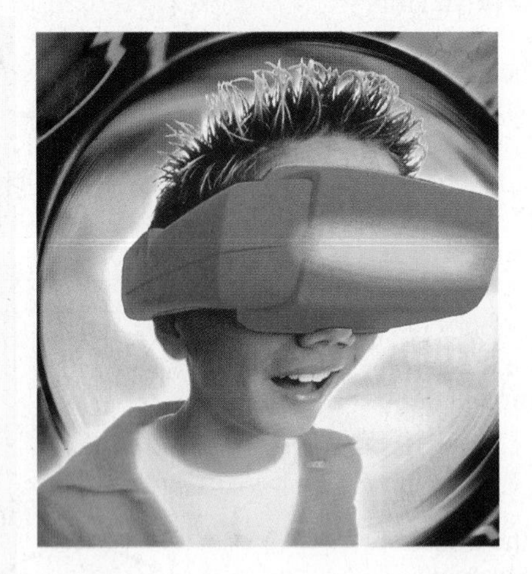

The time: the not-too-distant future. Your computer and its two-dimensional monitor have been replaced by a virtual reality system. You put on a headset and a virtual reality sensory suit. Suddenly you are skiing down a mountain. You feel the wind in your hair, the frost on your eyebrows, and the gentle heat of the sun on your face. Your heart races as you leap off cliffs, feeling the shudder through your ski boots as you race down the mountain.

Would you like to purchase a computer capable of helping you engage in virtual skiing? Luckily, your local computer store is having a sale right now. The models that are on sale cost either \$300 less than the regular price or 85% of the regular price. If x represents the computer's regular price, both discounts can be described with the following functions.

$$f(x) = x - 300 \qquad g(x) = 0.85x$$

The computer is on sale for \$300 less than its regular price.

The computer is on sale for 85% of its regular price.

At the store, you bargain with the salesperson. Eventually, she makes an offer you can't refuse: The sale price is 85% of the regular price followed by a \$300 reduction:

$$0.85x - 300$$

85% of the regular price

followed by a \$300 reduction

In terms of functions f and g, this offer can be obtained by taking the output of $g(x) = 0.85x$, namely $0.85x$, and using it as the input of f:

$$f(x) = x - 300$$

Replace x by $0.85x$, the output of $g(x) = 0.85x$.

$$f(0.85x) = 0.85x - 300.$$

Because $0.85x$ is $g(x)$, we can write this last equation as

$$f(g(x)) = 0.85x - 300.$$

We read this equation as "f of g of x is equal to $0.85x - 300$." We call $f(g(x))$ the **composition of the function f with g**, or a **composite function**. This composite function is written $f \circ g$. Thus,

$$(f \circ g)(x) = f(g(x)) = 0.85x - 300.$$

Like all functions, we can evaluate $f \circ g$ for a specified value of x in the function's domain. For example, here's how to find the value of this function at 1400:

$$(f \circ g)(x) = 0.85x - 300$$ This composite function describes the offer you cannot refuse.

Replace x by 1400.

$$(f \circ g)(1400) = 0.85(1400) - 300 = 1190 - 300 = 890.$$

This means that a computer that regularly sells for \$1400 is on sale for \$890 subject to both discounts.

In this section, we will focus on the composition of two functions. We will also study functions whose composition have a special relationship.

1 Form composite functions.

Before you run out to buy a new computer, let's generalize our discussion of the computer's double discount and define the composition of any two functions.

The Composition of Functions

The **composition of the function f with g** is denoted by $f \circ g$ and is defined by the equation

$$(f \circ g)(x) = f(g(x)).$$

The domain of the **composite function $f \circ g$** is the set of all x such that

1. x is in the domain of g and

2. $g(x)$ is in the domain of f.

The composition of f with g, $f \circ g$, is pictured as a machine with inputs and outputs in Figure 2.50. The diagram indicates that the output of g, or $g(x)$, becomes the input for "machine" f. If $g(x)$ is not in the domain of f, it cannot be input into machine f, and so $g(x)$ must be discarded.

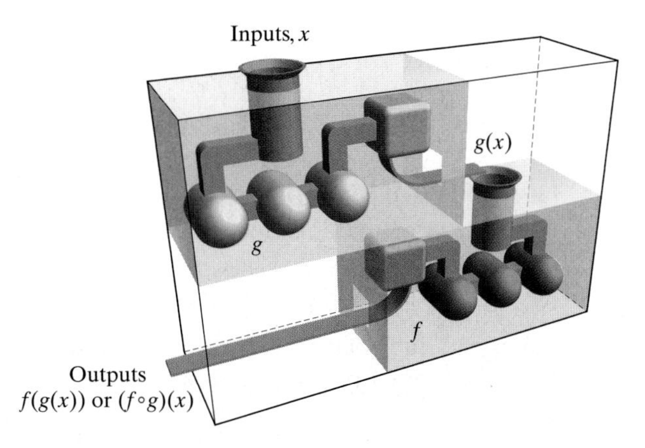

Figure 2.50 Inputting one function into a second function

EXAMPLE 1 Forming Composite Functions

Given $f(x) = 3x - 4$ and $g(x) = x^2 + 6$, find:

a. $(f \circ g)(x)$ b. $(g \circ f)(x)$.

Solution

a. We begin with $(f \circ g)(x)$, the composition of f with g. Because $(f \circ g)(x)$ means $f(g(x))$, we must replace each occurrence of x in the equation for f by $g(x)$.

$$f(x) = 3x - 4$$ This is the given equation for f.

Replace x by $g(x)$.

$$(f \circ g)(x) = f(g(x)) = 3g(x) - 4$$

$$= 3(x^2 + 6) - 4$$ Because $g(x) = x^2 + 6$, replace $g(x)$ with $x^2 + 6$

$$= 3x^2 + 18 - 4$$ Use the distributive property.

$$= 3x^2 + 14$$ Simplify.

Thus, $(f \circ g)(x) = 3x^2 + 14$.

b. Next, we find $(g \circ f)(x)$, the composition of g with f. Because $(g \circ f)(x)$ means $g(f(x))$, we must replace each occurrence of x in the equation for g by $f(x)$.

$$g(x) = x^2 + 6$$ This is the given equation for g.

Replace x by $f(x)$.

$$(g \circ f)(x) = g(f(x)) = (f(x))^2 + 6$$

$$= (3x - 4)^2 + 6$$ Because $f(x) = 3x - 4$, replace $f(x)$ with $3x - 4$.

$$= 9x^2 - 24x + 16 + 6$$ Use $(A - B)^2 = A^2 - 2AB + B^2$ to square $3x - 4$.

$$= 9x^2 - 24x + 22$$ Simplify.

Thus, $(g \circ f)(x) = 9x^2 - 24x + 22$. Notice that $(f \circ g)(x)$ is not the same function as $(g \circ f)(x)$.

Check Point 1 Given $f(x) = 5x + 6$ and $g(x) = x^2 - 1$, find:

a. $(f \circ g)(x)$ **b.** $(g \circ f)(x)$.

Inverse Functions

Here are two functions that describe situations related to the price of a computer, x:

$$f(x) = x - 300 \qquad g(x) = x + 300.$$

Function f subtracts \$300 from the computer's price and function g adds \$300 to the computer's price. Let's see what $f(g(x))$ does. Put $g(x)$ into f:

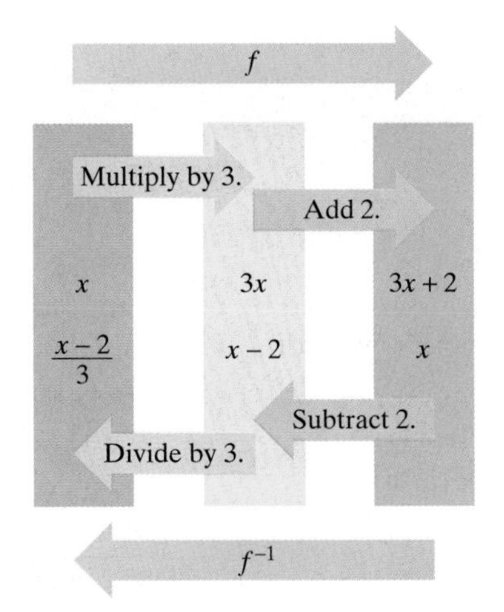

Figure 2.51 f^{-1} undoes the changes produced by f.

Because g is the inverse of f (and vice versa), we can use inverse notation and write

$$f(x) = 3x + 2 \quad \text{and} \quad f^{-1}(x) = \frac{x-2}{3}.$$

Notice how f^{-1} undoes the changes produced by f: f changes x by *multiplying* by 3 and *adding* 2, and f^{-1} undoes this by *subtracting* 2 and *dividing* by 3. This "undoing" process is illustrated in Figure 2.51.

Check Point 3 Show that each function is an inverse of the other:

$$f(x) = 4x - 7 \quad \text{and} \quad g(x) = \frac{x+7}{4}.$$

3 Find the inverse of a function.

Finding the Inverse of a Function

The definition of the inverse of a function tells us that the domain of f is equal to the range of f^{-1}, and vice versa. This means that if the function f is the set of ordered pairs (x, y), then the inverse of f is the set of ordered pairs (y, x). If a function is defined by an equation, we can obtain the equation for f^{-1}, the inverse of f, by interchanging the role of x and y in the equation for function f.

Study Tip

The procedure for finding a function's inverse uses a *switch-and-solve* strategy. Switch x and y, then solve for y.

Finding the Inverse of a Function

The equation for the inverse of a function f can be found as follows.

1. Replace $f(x)$ by y in the equation for $f(x)$.
2. Interchange x and y.
3. Solve for y. If this equation does not define y as a function of x, the function f does not have an inverse function and this procedure ends. If this equation does define y as a function of x, the function f has an inverse function.
4. If f has an inverse function, replace y in step 3 by $f^{-1}(x)$. We can verify our result by showing that $f(f^{-1}(x)) = x$ and $f^{-1}(f(x)) = x$.

EXAMPLE 4 Finding the Inverse of a Function

Find the inverse of $f(x) = 7x - 5$.

Solution

Step 1 Replace $f(x)$ by y:

$$y = 7x - 5$$

Step 2 Interchange x and y:

$$x = 7y - 5 \quad \text{This is the inverse function.}$$

Step 3 Solve for y:

$$x + 5 = 7y \quad \text{Add 5 to both sides.}$$

$$\frac{x+5}{7} = y \quad \text{Divide both sides by 7.}$$

Step 4 **Replace y by $f^{-1}(x)$:**

$$f^{-1}(x) = \frac{x+5}{7}$$ The equation is written with f^{-1} on the left.

Thus, $f(x) = 7x - 5$ and $f^{-1}(x) = \frac{x+5}{7}$.

The inverse function, f^{-1}, undoes the changes produced by f. f changes x by multiplying by 7 and subtracting 5. f^{-1} undoes this by adding 5 and dividing by 7.

Discovery

In Example 4, we found that if $f(x) = 7x - 5$, then

$$f^{-1}(x) = \frac{x+5}{7}.$$

Verify this result by showing that

$$f(f^{-1}(x)) = x$$

and

$$f^{-1}(f(x)) = x.$$

Check Point 4 Find the inverse of $f(x) = 2x + 7$.

EXAMPLE 5 Finding the Equation of the Inverse

Find the inverse of $f(x) = x^3 + 1$.

Solution

Step 1 **Replace $f(x)$ with y:** $y = x^3 + 1$

Step 2 **Interchange x and y:** $x = y^3 + 1$

Step 3 **Solve for y:**

$$x - 1 = y^3$$
$$\sqrt[3]{x-1} = \sqrt[3]{y^3}$$
$$\sqrt[3]{x-1} = y$$

Step 4 **Replace y with $f^{-1}(x)$:** $f^{-1}(x) = \sqrt[3]{x-1}$.

Thus, the inverse of $f(x) = x^3 + 1$ is $f^{-1}(x) = \sqrt[3]{x-1}$.

Check Point 5 Find the inverse of $f(x) = 4x^3 - 1$.

4 Use the horizontal line test to determine if a function has an inverse function.

The Horizontal Line Test and One-to-One Functions

Let's see what happens if we try to find the inverse of the standard quadratic function $f(x) = x^2$.

Step 1 **Replace $f(x)$ with y:** $y = x^2$.

Step 2 **Interchange x and y:** $x = y^2$.

Step 3 **Solve for y:** We apply the square root method to solve $y^2 = x$ for y. We obtain

$$y = \pm\sqrt{x}.$$

The $\pm$ in this last equation shows that for certain values of x (all positive real numbers), there are two values of y. Because this equation does not represent y as a function of x, the standard quadratic function does not have an inverse function.

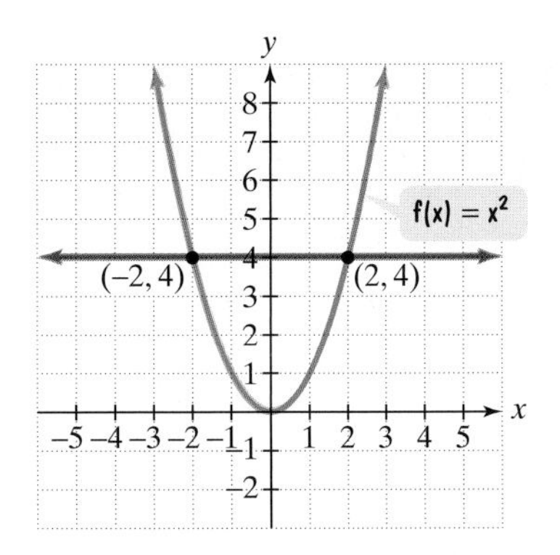

Figure 2.52 The horizontal line intersects the graph twice.

Discovery

How might you restrict the domain of $f(x) = x^2$, graphed in Figure 2.52, so that the remaining portion of the graph passes the horizontal line test?

Can we look at the graph of a function and tell if it represents a function with an inverse? Yes. The graph of the standard quadratic function is shown in Figure 2.52. Four units above the x-axis, a horizontal line is drawn. This line intersects the graph at two of its points, $(-2, 4)$ and $(2, 4)$. Because inverse functions have ordered pairs with the coordinates reversed, let's see what happens if we reverse these coordinates. We obtain $(4, -2)$ and $(4, 2)$. A function provides exactly one output for each input. However, the input 4 is associated with two outputs, -2 and 2. The points $(4, -2)$ and $(4, 2)$ do not define a function.

If any horizontal line, such as the one in Figure 2.52, intersects a graph at two or more points, these points will not define a function when their coordinates are reversed. This suggests the **horizontal line test** for inverse functions.

The Horizontal Line Test For Inverse Functions

A function f has an inverse that is a function, f^{-1}, if there is no horizontal line that intersects the graph of the function f at more than one point.

EXAMPLE 6 Applying the Horizontal Line Test

Which of the following graphs represent functions that have inverse functions?

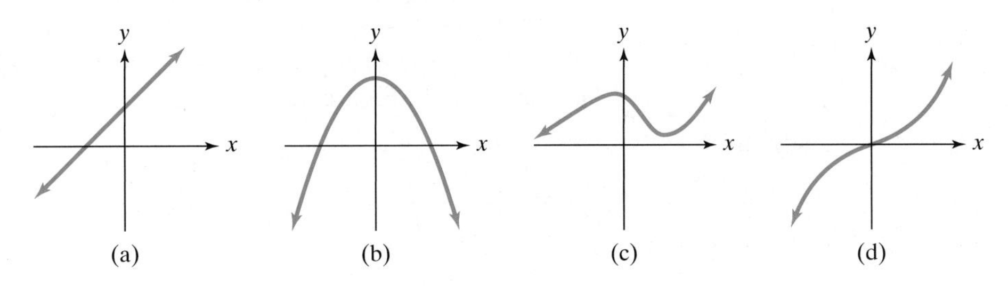

Solution Notice that horizontal lines can be drawn in parts (b) and (c) that intersect the graphs more than once. These graphs do not pass the horizontal line test. These are not the graphs of functions with inverse functions. By contrast, no horizontal line can be drawn in parts (a) and (d) that intersect the graphs more than once. These graphs pass the horizontal line test. Thus, the graphs in parts (a) and (d) represent functions that have inverse functions.

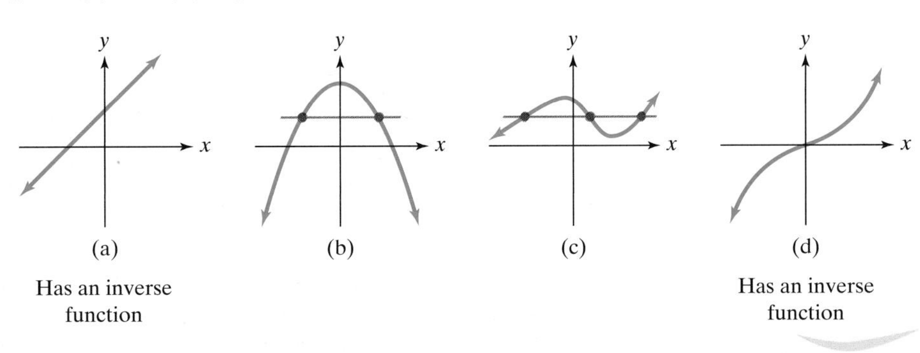

Check Point 6 Which of the following graphs represent functions that have inverse functions?

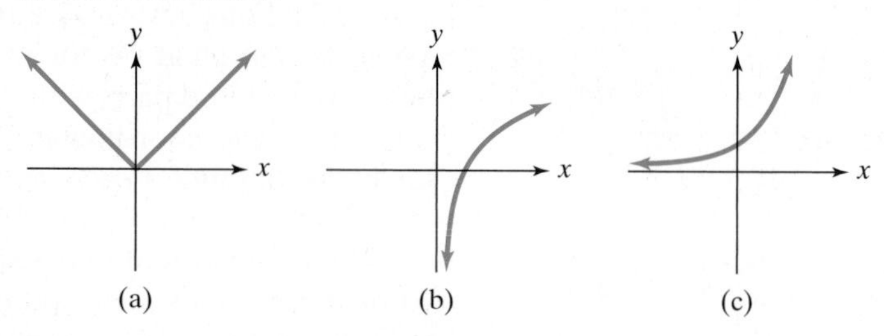

A function passes the horizontal line test when no two different ordered pairs have the same second component. This means that if $x_1 \neq x_2$, then $f(x_1) \neq f(x_2)$. Such a function is called a **one-to-one function**. Thus, a one-to-one function is a function in which no two different ordered pairs have the same second component. Only one-to-one functions have inverse functions. Any function that passes the horizontal line test is a one-to-one function. Any one-to-one function has a graph that passes the horizontal line test.

5 Use the graph of a one-to-one function to graph its inverse function.

Graphs of f and f^{-1}

There is a relationship between the graph of a one-to-one function f and its inverse f^{-1}. Because inverse functions have ordered pairs with the coordinates reversed, if the point (a, b) is on the graph of f, then the point (b, a) is on the graph of f^{-1}. The points (a, b) and (b, a) are symmetric with respect to the line $y = x$. Thus, **the graph of f^{-1} is a reflection of the graph of f about the line $y = x$.** This is illustrated in Figure 2.53.

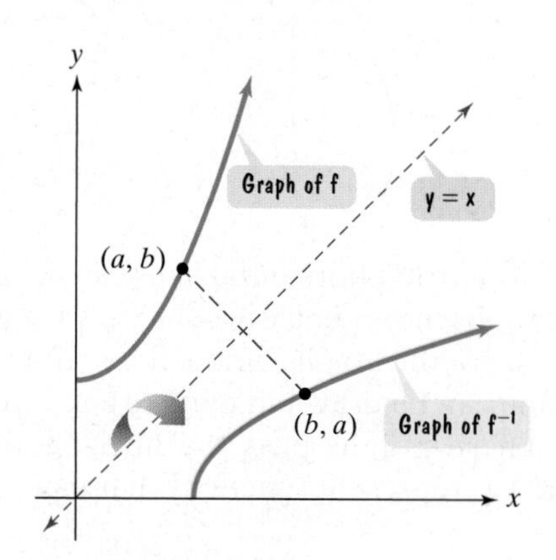

Figure 2.53 The graph of f^{-1} as a reflection of f about $y = x$

EXAMPLE 7 Graphing the Inverse Function

Use the graph of f in Figure 2.54 to draw the graph of its inverse function.

Solution We begin by noting that no horizontal line intersects the graph of f at more than one point, so f does have an inverse function. Because the points $(-3, -2)$, $(-1, 0)$, and $(4, 2)$ are on the graph of f, the graph of the inverse function, f^{-1}, has points with these ordered pairs reversed. Thus, $(-2, -3)$, $(0, -1)$, and $(2, 4)$ are on the graph of f^{-1}. We can use these points to graph f^{-1}. The graph of f^{-1} is shown in Figure 2.55. Note that the graph of f^{-1} is the reflection of the graph of f about the line $y = x$.

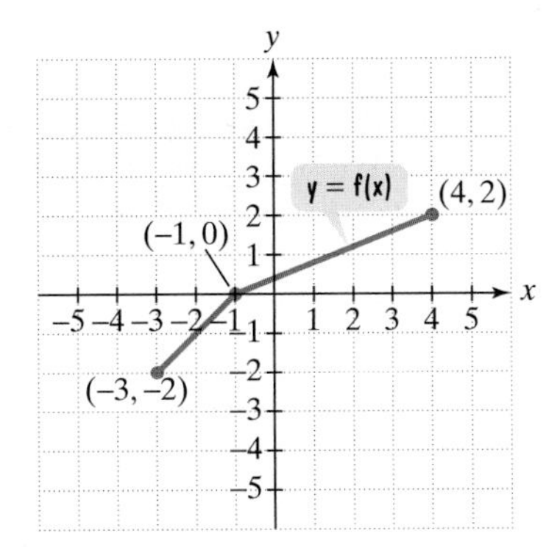

Figure 2.54

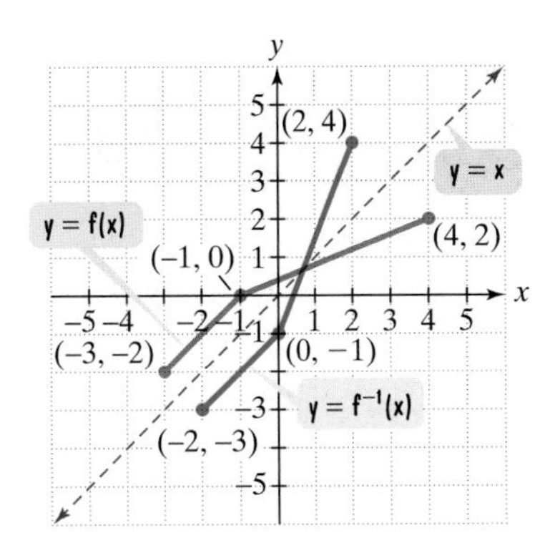

Figure 2.55 The graph of f and f^{-1}

Check Point 7 Use the graph of f in the figure below to draw the graph of its inverse function.

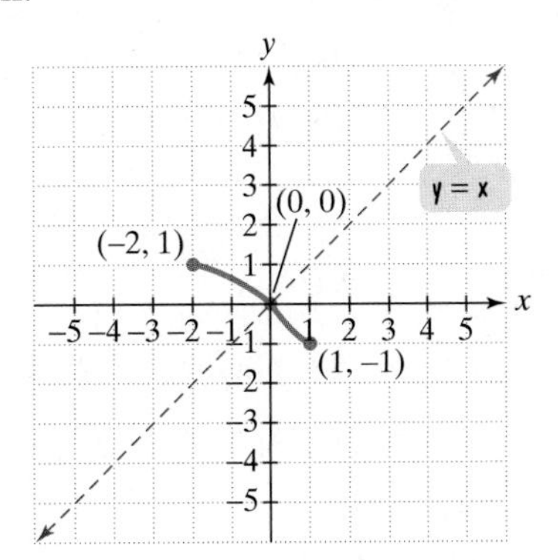

EXERCISE SET 2.6

Practice Exercises

In Exercises 1–14, find

a. $(f \circ g)(x)$;
b. $(g \circ f)(x)$;
c. $(f \circ g)(2)$.

1. $f(x) = 2x, \quad g(x) = x + 7$

2. $f(x) = 3x, \quad g(x) = x - 5$

3. $f(x) = x + 4, \quad g(x) = 2x + 1$

4. $f(x) = 5x + 2, \quad g(x) = 3x - 4$

5. $f(x) = 4x - 3, \quad g(x) = 5x^2 - 2$

6. $f(x) = 7x + 1, \quad g(x) = 2x^2 - 9$

7. $f(x) = x^2 + 2, \quad g(x) = x^2 - 2$

8. $f(x) = x^2 + 1, \quad g(x) = x^2 - 3$

9. $f(x) = \sqrt{x}, \quad g(x) = x - 1$

10. $f(x) = \sqrt{x}, \quad g(x) = x + 2$

11. $f(x) = 2x - 3, \quad g(x) = \dfrac{x + 3}{2}$

12. $f(x) = 6x - 3, \quad g(x) = \dfrac{x + 3}{6}$

13. $f(x) = \dfrac{1}{x}, \quad g(x) = \dfrac{1}{x}$

14. $f(x) = \dfrac{1}{x}, \quad g(x) = \dfrac{2}{x}$

In Exercises 15–24, find $f(g(x))$ and $g(f(x))$ and determine whether each pair of functions f and g are inverses of each other.

15. $f(x) = 4x$ and $g(x) = \dfrac{x}{4}$

16. $f(x) = 6x$ and $g(x) = \dfrac{x}{6}$

17. $f(x) = 3x + 8$ and $g(x) = \dfrac{x - 8}{3}$

18. $f(x) = 4x + 9$ and $g(x) = \dfrac{x - 9}{4}$

19. $f(x) = 5x - 9$ and $g(x) = \dfrac{x + 5}{9}$

20. $f(x) = 3x - 7$ and $g(x) = \dfrac{x + 3}{7}$

21. $f(x) = \dfrac{3}{x - 4}$ and $g(x) = \dfrac{3}{x} + 4$

22. $f(x) = \dfrac{2}{x - 5}$ and $g(x) = \dfrac{2}{x} + 5$

23. $f(x) = -x$ and $g(x) = -x$

24. $f(x) = \sqrt[3]{x - 4}$ and $g(x) = x^3 + 4$

The functions in Exercises 25–44 are all one-to-one. For each function:

- **a.** *Find an equation for $f^{-1}(x)$, the inverse function.*
- **b.** *Verify that your equation is correct by showing that $f(f^{-1}(x)) = x$ and $f^{-1}(f(x)) = x$.*

25. $f(x) = x + 3$

26. $f(x) = x + 5$

27. $f(x) = 2x$

28. $f(x) = 4x$

29. $f(x) = 2x + 3$

30. $f(x) = 3x - 1$

31. $f(x) = x^3 + 2$

32. $f(x) = x^3 - 1$

33. $f(x) = (x + 2)^3$

34. $f(x) = (x - 1)^3$

35. $f(x) = \dfrac{1}{x}$

36. $f(x) = \dfrac{2}{x}$

37. $f(x) = \sqrt{x}$

38. $f(x) = \sqrt[3]{x}$

39. $f(x) = x^2 + 1$, for $x \geq 0$

40. $f(x) = x^2 - 1$, for $x \geq 0$

41. $f(x) = \dfrac{2x + 1}{x - 3}$

42. $f(x) = \dfrac{2x - 3}{x + 1}$

43. $f(x) = \sqrt[3]{x - 4} + 3$

44. $f(x) = x^{3/5}$

Which graphs in Exercises 45–50 represent functions that have inverse functions?

45.

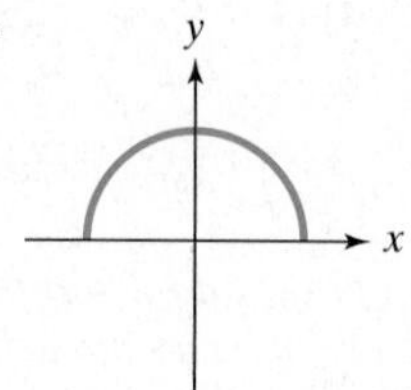

46.

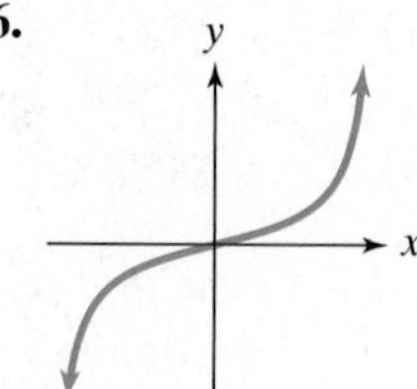

47.

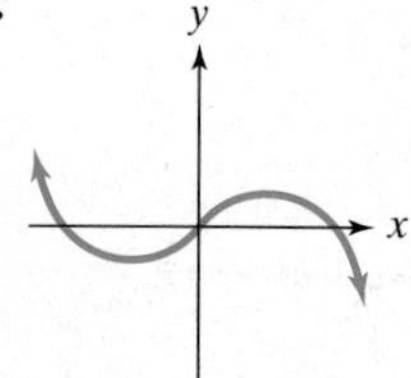

48.

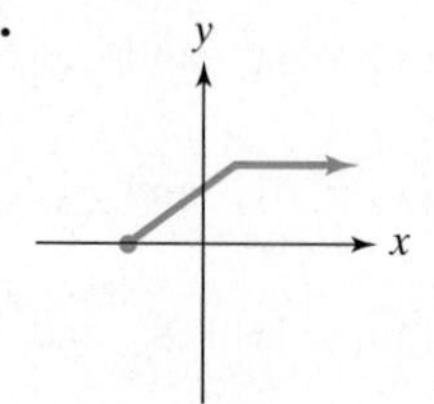

49.

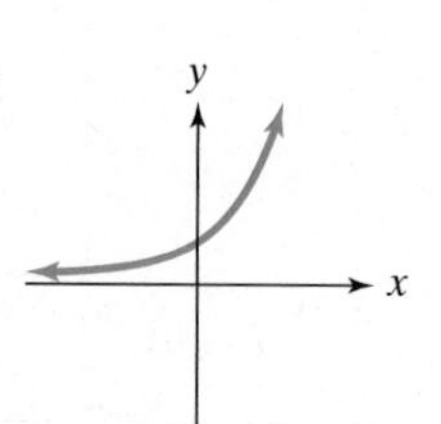

50.

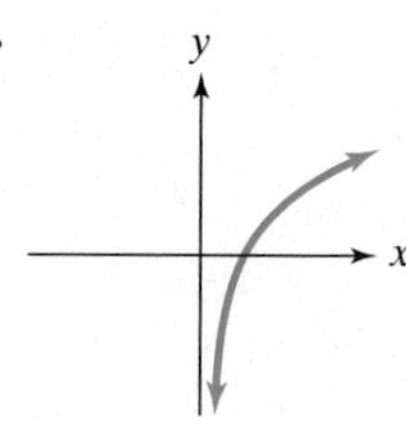

In Exercises 51–54, use the graph of f to draw the graph of its inverse function.

51.

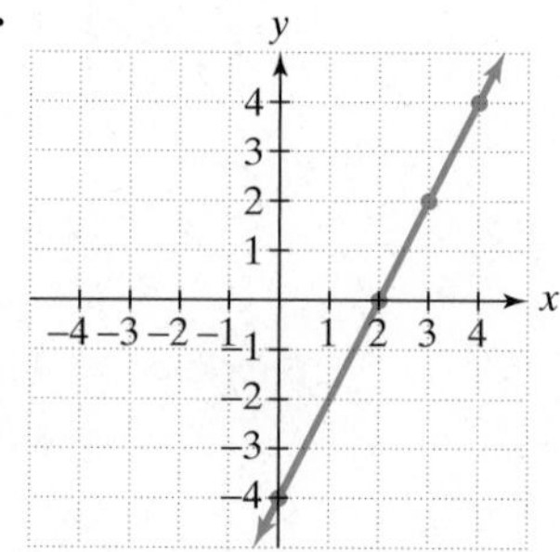

52.

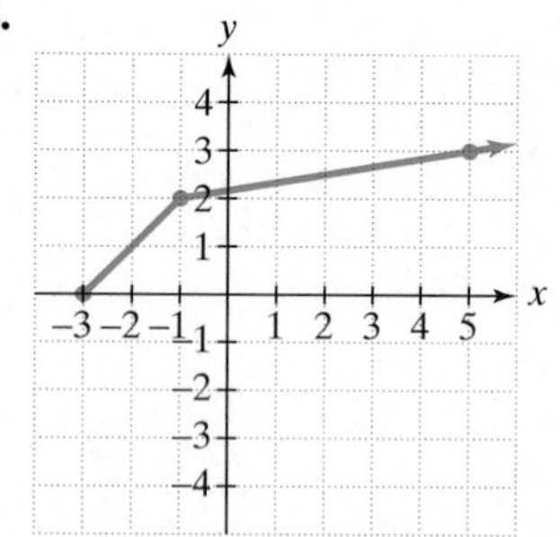

53.

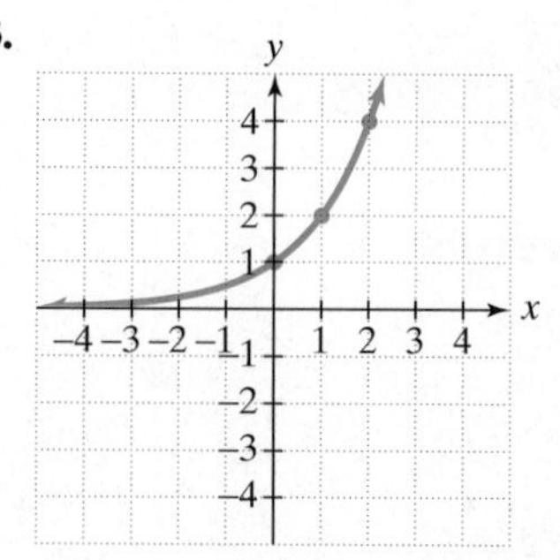

54.

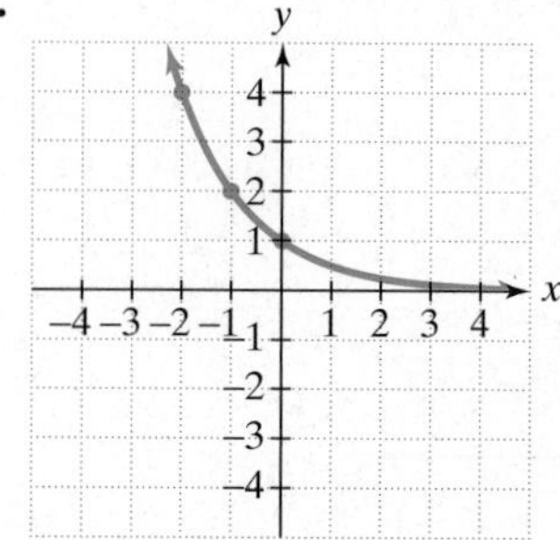

Application Exercises

55. The regular price of a computer is x dollars. Let $f(x) = x - 400$ and $g(x) = 0.75x$.

a. Describe what the functions f and g model in terms of the price of the computer.

b. Find $(f \circ g)(x)$ and describe what this models in terms of the price of the computer.

c. Repeat part (b) for $(g \circ f)(x)$.

d. Which composite function models the greater discount on the computer, $f \circ g$ or $g \circ f$? Explain.

e. Find f^{-1} and describe what this models in terms of the price of the computer.

56. The regular price of a pair of jeans is x dollars. Let $f(x) = x - 5$ and $g(x) = 0.6x$.

a. Describe what functions f and g model in terms of the price of the jeans.

b. Find $(f \circ g)(x)$ and describe what this models in terms of the price of the jeans.

c. Repeat part (b) for $(g \circ f)(x)$.

d. Which composite function models the greater discount on the jeans, $f \circ g$ or $g \circ f$? Explain.

e. Find f^{-1} and describe what this models in terms of the price of the jeans.

57. The graph represents the probability of two people in the same room sharing a birthday as a function of the number of people in the room. Call the function f.

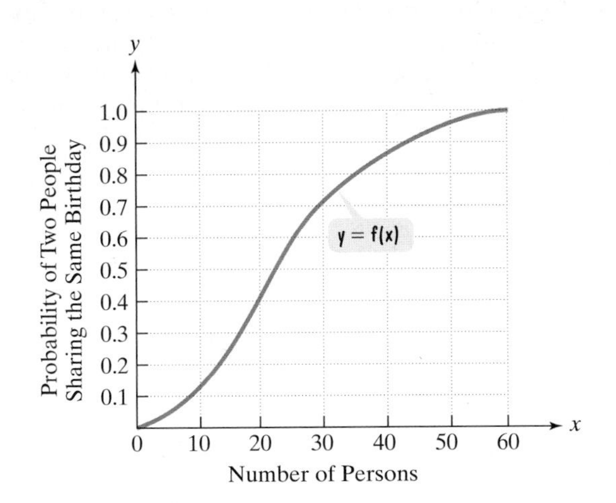

a. Explain why f has an inverse that is a function.

b. Describe in practical terms the meaning of $f^{-1}(0.25)$, $f^{-1}(0.5)$, and $f^{-1}(0.7)$.

58. The line graph shown at the top of the next column is based on data from the World Health Organization.

a. Explain why f has an inverse that is a function.

b. Describe in practical terms the meaning of $f^{-1}(20)$.

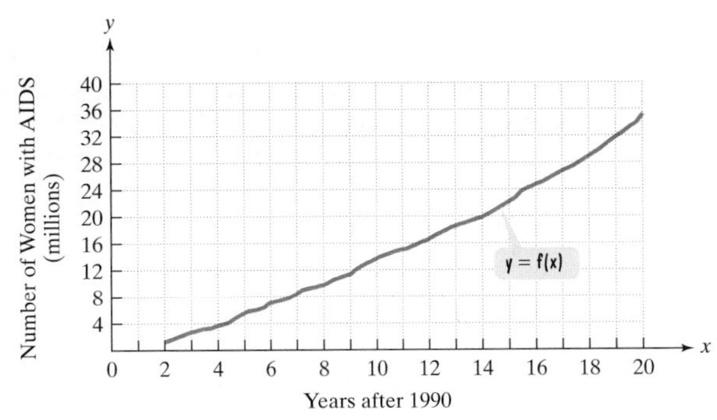

Source: Boston Globe

The graph shows the average age at which women in the United States marry for the first time over a 110-year period. Use the graph to solve Exercises 59–60.

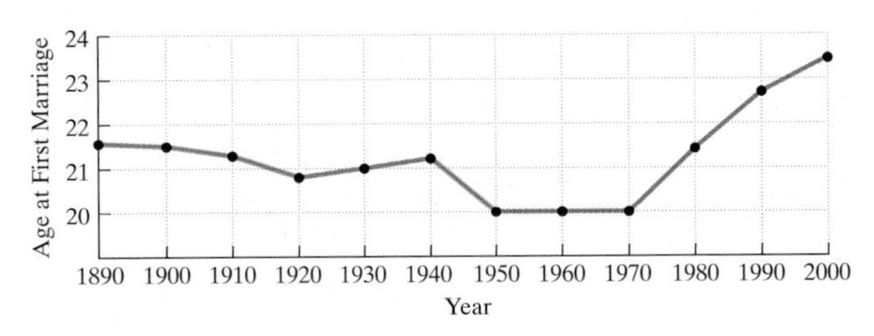

Source: U.S. Census Bureau

59. Does this graph have an inverse that is a function? What does this mean about the average age at which U.S. women marry during the period shown?

60. Identify two or more years in which U.S. women married for the first time at the same average age. What is this average age?

Writing in Mathematics

61. Describe a procedure for finding $(f \circ g)(x)$.

62. Explain how to determine if two functions are inverses of each other.

63. Describe how to find the inverse of a one-to-one function.

64. What is the horizontal line test and what does it indicate?

65. Describe how to use the graph of a one-to-one function to draw the graph of its inverse function.

66. How can a graphing utility be used to visually determine if two functions are inverses of each other?

67. Consider the following function:

(The Beatles, 20), (Elvis Presley, 18),

(Michael Jackson, 13)

(Mariah Carey, 13), (The Supremes, 12).

The domain is the set of the five recording artists with the most number 1 singles in the United States. The range is the set of the number of number 1 singles for each artist. Reverse each of the five ordered pairs. Is the resulting relation a function? Describe what this means in terms of whether or not the given function is one-to-one. (*Source: The Popular Music Database*)

Technology Exercises

In Exercises 68–76, use a graphing utility to graph the function. Use the graph to determine whether the function has an inverse that is a function (that is, whether the function is one-to-one).

68. $f(x) = x^2 - 1$

69. $f(x) = \sqrt[3]{2 - x}$

70. $f(x) = \dfrac{x^3}{2}$

71. $f(x) = \dfrac{x^4}{4}$

72. $f(x) = \text{int}(x - 2)$

73. $f(x) = |x - 2|$

74. $f(x) = (x - 1)^3$

75. $f(x) = -\sqrt{16 - x^2}$

76. $f(x) = x^3 + x + 1$

In Exercises 77–79, use a graphing utility to graph f and g in the same viewing rectangle. In addition, graph the line $y = x$ and visually determine if f and g are inverses.

77. $f(x) = 4x + 4, \quad g(x) = 0.25x - 1$

78. $f(x) = \dfrac{1}{x} + 2, \quad g(x) = \dfrac{1}{x - 2}$

79. $f(x) = \sqrt[3]{x} - 2, \quad g(x) = (x + 2)^3$

Critical Thinking Exercises

80. Which one of the following is true?

a. The inverse of $\{(1, 4), (2, 7)\}$ is $\{(2, 7), (1, 4)\}$.

b. The function $f(x) = 5$ is one-to-one.

c. If $f(x) = 3x$, then $f^{-1}(x) = \dfrac{1}{3x}$.

d. The domain of f is the same as the range of f^{-1}.

81. If $h(x) = \sqrt{3x^2 + 5}$, find functions f and g so that $h(x) = (f \circ g)(x)$.

82. If $f(x) = 3x$ and $g(x) = x + 5$, find $(f \circ g)^{-1}(x)$ and $(g^{-1} \circ f^{-1})(x)$.

83. Show that

$$f(x) = \frac{3x - 2}{5x - 3}$$

is its own inverse.

84. Consider the two functions defined by $f(x) = m_1x + b_1$ and $g(x) = m_2x + b_2$. Prove that the slope of the composite function of f with g is equal to the product of the slopes of the two functions.

Group Exercise

85. In Tom Stoppard's play *Arcadia*, the characters dream and talk about mathematics, including ideas involving graphing, composite functions, symmetry, and lack of symmetry in things that are tangled, mysterious, and unpredictable. Group members should rent and view the movie. Present a report on the ideas discussed by the characters that are related to concepts that we studied in this chapter. Bring in a copy of the video and show appropriate excerpts.

CHAPTER SUMMARY, REVIEW, AND TEST

Summary

2.1 Lines and Slope

a. The slope, m, of the line through (x_1, y_1) and (x_2, y_2) is $m = \dfrac{y_2 - y_1}{x_2 - x_1}$.

b. Equations of lines include point-slope form, $y - y_1 = m(x - x_1)$, slope-intercept form, $y = mx + b$, and general form, $Ax + By + C = 0$. The equation of a horizontal line is $y = b$; a vertical line is $x = a$.

2.2 Parallel and Perpendicular Lines; Circles

a. Parallel lines have equal slopes. Perpendicular lines have slopes that are negative reciprocals.

b. The standard form of the equation of a circle with center (h, k) and radius r is $(x - h)^2 + (y - k)^2 = r^2$.

c. The general form of the equation of a circle is $x^2 + y^2 + Dx + Ey + F = 0$.

d. To convert from the general form to the standard form of a circle's equation, complete the square on x and y.

2.3 Introduction to Functions

a. A relation is any set of ordered pairs. The set of first components is the domain and the set of second components is the range.

b. A function is a correspondence between two sets X (the domain) and Y (the range) that assigns to each element x in the domain exactly one element y in the range. If any element in a relation's domain corresponds to more than one element in the range, the relation is not a function.

c. Functions are usually given in terms of equations involving x and y, in which x is the independent variable and y is the dependent variable. If an equation is solved for y and more than one value of y can be obtained for a given x, then the equation does not define y as a function of x. If an equation defines a function, $f(x)$, the value of the function at x, often replaces y.

d. If a function f does not model data or verbal conditions, its domain is the largest set of real numbers for which the value of $f(x)$ is a real number. Exclude from the function's domain real numbers that cause division by zero and real numbers that result in an even root of a negative number.

2.4 Graphs of Functions

a. The graph of a function is the graph of its ordered pairs.

b. The vertical line test for functions: If any vertical line intersects a graph in more than one point, the graph does not define y as a function of x.

c. A function is increasing on intervals where its graph rises, decreasing on intervals where it falls, and constant on intervals where it neither rises nor falls. Precise definitions are given in the box on page 207.

d. The graph of an even function in which $f(-x) = f(x)$ is symmetric with respect to the y-axis. The graph of an odd function in which $f(-x) = -f(x)$ is symmetric with respect to the origin.

e. Table 2.3 on page 212 shows the graphs of the constant function, $f(x) = c$, the identity function, $f(x) = x$, the standard quadratic function, $f(x) = x^2$, the standard cubic function, $f(x) = x^3$, the square root function, $f(x) = \sqrt{x}$, and the absolute value function, $f(x) = |x|$. The table also lists characteristics of each function.

f. The graph of $f(x) = \text{int}(x)$, where $\text{int}(x)$ is the greatest integer that is less than or equal to x, has function values that form discontinuous steps, shown in Figure 2.32 on page 213. If $n \le x < n + 1$, where n is an integer, then $\text{int}(x) = n$.

2.5 Transformations and Combinations of Functions

a. Table 2.5 on page 228 summarizes how to graph a function using vertical shifts, $y = f(x) \pm c$, horizontal shifts, $y = f(x \pm c)$, reflections about the x-axis, $y = -f(x)$, reflections about the y-axis, $y = f(-x)$, vertical stretching, $y = cf(x)$, $c > 1$, and vertical shrinking, $y = cf(x), 0 < c < 1$.

b. A function involving more than one transformation can be graphed in the following order: (1) horizontal shifting; (2) vertical stretching or shrinking; (3) reflecting; (4) vertical shifting.

c. When functions are given as equations, they can be added, subtracted, multiplied, or divided by performing operations with the algebraic expressions that appear on the right side of the equations. Definitions for the sum $f + g$, the difference $f - g$, the product fg, and the quotient $\frac{f}{g}$ functions are given in the box on page 233.

2.6 Composite and Inverse Functions

a. The composition of functions f and g, $f \circ g$, is defined by $(f \circ g)(x) = f(g(x))$. The domain of the composite function $f \circ g$ is given in the box on page 239. This composite function is obtained by replacing each occurrence of x in the equation for f by $g(x)$.

b. If $f(g(x)) = x$ and $g(f(x)) = x$, function g is the inverse of function f, denoted f^{-1} and read "f inverse." Thus, to show that f and g are inverses of each other, one must show $f(g(x)) = x$ and $g(f(x)) = x$.

c. The procedure for finding a function's inverse uses a switch-and-solve strategy. Switch x and y, then solve for y. The procedure is given in the box on page 243.

d. The horizontal line test for inverse functions: A function f has an inverse that is a function, f^{-1}, if there is no horizontal line that intersects the graph of the function f at more than one point.

e. A one-to-one function is one in which no two different ordered pairs have the same second component. Only one-to-one functions have inverse functions.

f. If the point (a, b) is on the graph of f, then the point (b, a) is on the graph of f^{-1}. The graph of f^{-1} is a reflection of the graph of f about the line $y = x$.

Review Exercises

2.1

In Exercises 1–4, find the slope of the line passing through each pair of points or state that the slope is undefined. Then indicate whether the line through the points rises, falls, is horizontal, or is vertical.

1. $(3, 2)$ and $(5, 1)$

2. $(-1, -2)$ and $(-3, -4)$

3. $\left(-3, \frac{1}{4}\right)$ and $\left(6, \frac{1}{4}\right)$

4. $(-2, 5)$ and $(-2, 10)$

In Exercises 5–6, use the given conditions to write an equation for each line in point-slope form and slope-intercept form.

5. Passing through $(-3, 2)$ with a slope of -6

6. Passing through $(1, 6)$ and $(-1, 2)$

In Exercises 7–10, give the slope and y-intercept of each line whose equation is given. Then graph the line.

7. $y = \frac{2}{5}x - 1$ **8.** $y = -4x + 5$

9. $2x + 3y + 6 = 0$ **10.** $2y - 8 = 0$

11. In 1900, the typical surfboard was 16 feet long. Since then, they have become shorter and shorter. Here are two data measurements for a typical surfboard's length. (A scatter plot of all such data measurements through 1980 would show all data points on or near a straight line.)

x (Years since 1900)	y (Average Surfboard Length, in Feet)
0	16
30	12.1

Source: Bishop Museum

a. Write the point-slope form of the equation of the line on which these measurements fall.

b. Use the point-slope form of the equation to write the slope-intercept form of the equation.

c. Use the equation in part (b) to find average surfboard length in 1970 and 1980.

d. Does the equation in part (b) reasonably describe reality in 2000?

12. The scatter plot shows the number of minutes each that 16 people exercise per week and the number of headaches per month each person experiences.

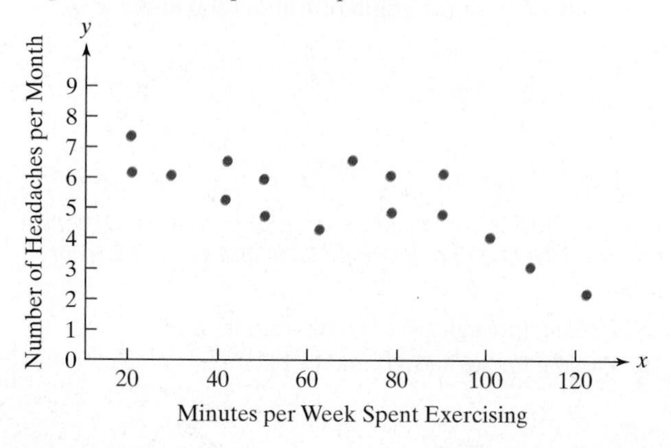

a. Draw a line that fits the data so that the spread of the data points around the line is as small as possible.

b. Use the coordinates of two points along your line to write its point-slope and slope-intercept equations.

c. Use the equation in part (b) to predict the number of headaches per month for a person exercising 130 minutes per week.

2.2

In Exercises 13–14, use the given conditions to write an equation for each line in point-slope form and slope-intercept form.

13. Passing through $(4, -7)$ and parallel to the line whose equation is $3x + y - 9 = 0$.

14. Passing through $(-3, 6)$ and perpendicular to the line whose equation is $y = \frac{1}{3}x + 4$.

In Exercises 15–16, write the standard form of the equation of the circle with the given center and radius.

15. Center $(0, 0), r = 3$ **16.** Center $(-2, 4), r = 6$

In Exercises 17–19, give the center and radius of each circle and graph its equation.

17. $x^2 + y^2 = 1$ **18.** $(x + 2)^2 + (y - 3)^2 = 9$

19. $x^2 + y^2 - 4x + 2y - 4 = 0$

2.3

In Exercises 20–22, determine whether each relation is a function. Give the domain and range for each relation.

20. $\{(2, 7), (3, 7), (5, 7)\}$ **21.** $\{(1, 10)\ (2, 500), (13, \pi)\}$

22. $\{(12, 13), (14, 15), (12, 19)\}$

In Exercises 23–25, determine whether each equation defines y as a function of x.

23. $2x + y = 8$ **24.** $3x^2 + y = 14$

25. $2x + y^2 = 6$

In Exercises 26–30, evaluate each function at the given values of the independent variable and simplify.

26. $f(x) = 5 - 7x$

a. $f(4)$ **b.** $f(x + 3)$ **c.** $f(-x)$

27. $g(x) = 3x^2 - 5x + 2$

a. $g(0)$ **b.** $g(-2)$

c. $g(x - 1)$ **d.** $g(-x)$

28. $f(x) = 4x - 3$

a. $f(a)$ **b.** $f(a + h)$

c. $\dfrac{f(a + h) - f(a)}{h}$, $h \neq 0$ **d.** $f(a) + f(h)$

29. $g(x) = \begin{cases} \sqrt{x - 4} & \text{if } x \geq 4 \\ 4 - x & \text{if } x < 4 \end{cases}$

a. $g(13)$ **b.** $g(0)$ **c.** $g(-3)$

30. $f(x) = \begin{cases} \dfrac{x^2 - 1}{x - 1} & \text{if } x \neq 1 \\ 12 & \text{if } x = 1 \end{cases}$

a. $f(-2)$ **b.** $f(1)$ **c.** $f(2)$

In Exercises 31–35, find the domain of each function.

31. $f(x) = x^2 + 6x - 3$

32. $g(x) = \dfrac{4}{x - 7}$

33. $h(x) = \sqrt{8 - 2x}$

34. $f(x) = \dfrac{x}{x^2 - 1}$

35. $g(x) = \dfrac{\sqrt{x - 2}}{x - 5}$

36. The function $f(x) = -0.46x^2 + 3.66x + 20.08$ models data from the U.S. Census Bureau regarding the number of participants in the Federal Food Stamp Program. The variable x represents the number of years after 1990 and $f(x)$ is the number of participants, in millions, in the program. Find and interpret $f(6)$.

2.4 255

Graph the functions in Exercises 37–38. Use the integer values of x given to the right of the function to obtain the ordered pairs. Use the graph to specify the function's domain and range.

37. $f(x) = x^2 - 4x + 4 \quad x = -1, 0, 1, 2, 3, 4$

38. $f(x) = |2 - x| \quad x = -1, 0, 1, 2, 3, 4$

In Exercises 39–41, use the graph to determine **a.** *the function's domain;* **b.** *the function's range;* **c.** *the x-intercepts, if any;* **d.** *the y-intercept, if any;* **e.** *intervals on which, the function is increasing, decreasing, or constant; and* **f.** *the function values indicated below the graphs.*

39.

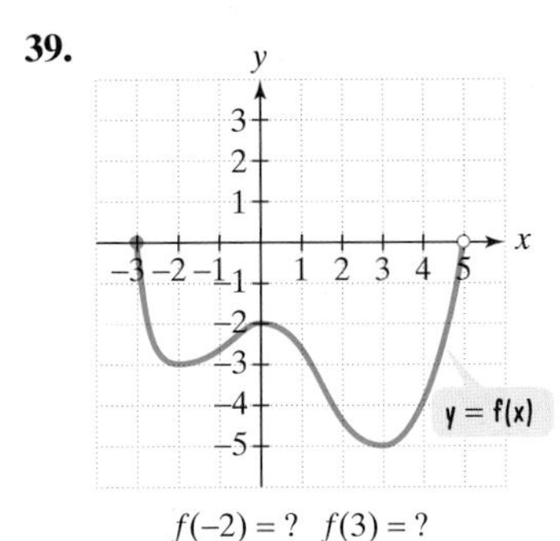

$f(-2) = ?\quad f(3) = ?$

40.

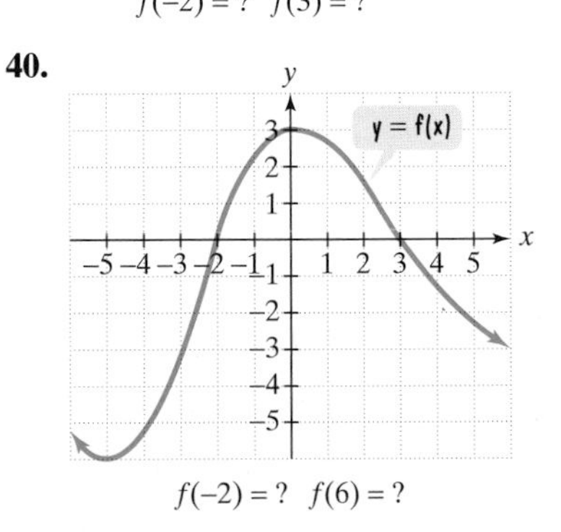

$f(-2) = ?\quad f(6) = ?$

41.

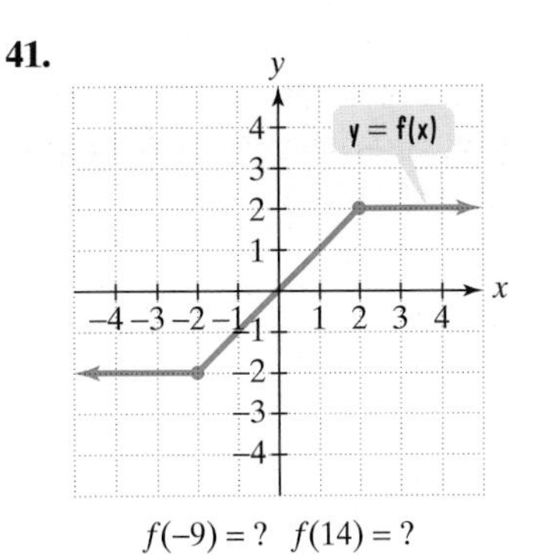

$f(-9) = ?\quad f(14) = ?$

In Exercises 42–45, use the vertical line test to identify graphs in which y is a function of x.

42. **43.**

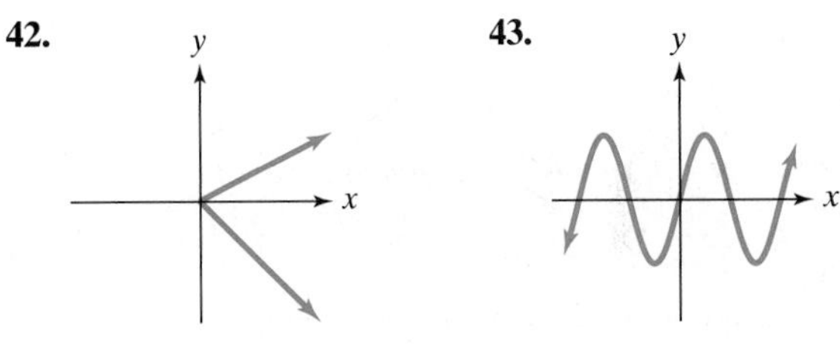

44. **45.**

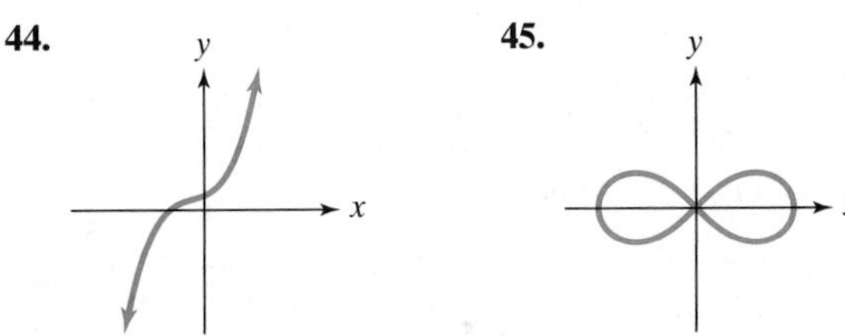

In Exercises 46–48, determine whether each function is even, odd, or neither. State each function's symmetry. If you are using a graphing utility, graph the function and verify its possible symmetry.

46. $f(x) = x^3 - 5x$

47. $f(x) = x^4 - 2x^2 + 1$

48. $f(x) = 2x\sqrt{1 - x^2}$

49. The graph shows the height (in meters) of a vulture as a function of its time (in seconds) in flight.

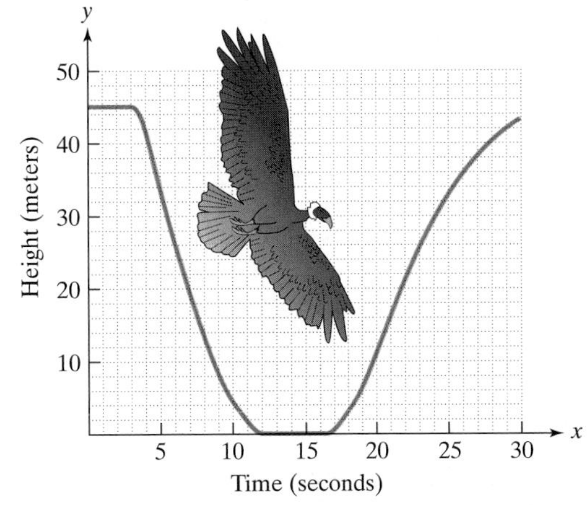

a. Is the vulture's height a function of time? Use the graph to explain why or why not.

b. On what interval is the function decreasing? Describe what this means in practical terms.

c. On what intervals is the function constant? What does this mean for each of these intervals?

d. On what interval is the function increasing? What does this mean?

50. A cargo service charges a flat fee of \$5 plus \$1.50 for each pound or fraction of a pound. Graph shipping cost (in dollars) as a function of weight (x, in pounds) for $0 < x \leq 5$.

2.5

In Exercises 51–53, begin by graphing the standard quadratic function, $f(x) = x^2$. Then use transformations of this graph to graph the given function.

51. $g(x) = x^2 + 2$ **52.** $h(x) = (x + 2)^2$

53. $r(x) = -(x + 1)^2$

In Exercises 54–56, begin by graphing the square root function, $f(x) = \sqrt{x}$. Then use transformations of this graph to graph the given function.

54. $g(x) = \sqrt{x + 3}$ **55.** $h(x) = \sqrt{3 - x}$

56. $r(x) = 2\sqrt{x + 2}$

In Exercises 57–59, begin by graphing the absolute value function, $f(x) = |x|$. Then use transformations of this graph to graph the given function.

57. $g(x) = |x + 2| - 3$ **58.** $h(x) = -|x - 1| + 1$

59. $r(x) = \frac{1}{2}|x + 2|$

In Exercises 60–62, begin by graphing the standard cubic function, $f(x) = x^3$. Then use transformations of this graph to graph the given function.

60. $g(x) = \frac{1}{2}(x - 1)^3$ **61.** $h(x) = -(x + 1)^3$

62. $r(x) = \frac{1}{4}x^3 - 1$

In Exercises 63–65, use the graph of the function f to sketch the graph of the given function g.

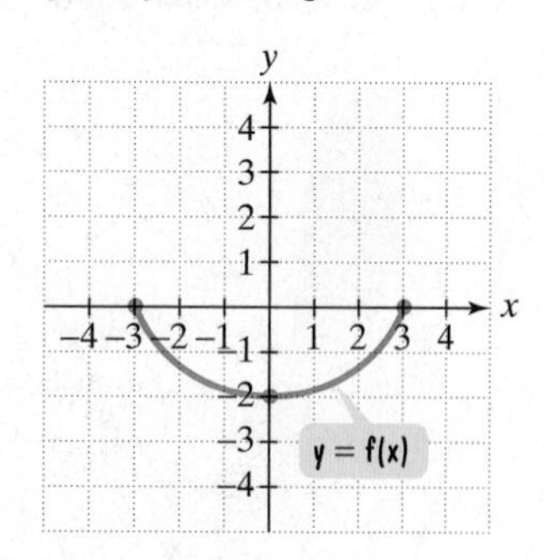

63. $g(x) = f(x + 2) + 3$ **64.** $g(x) = \frac{1}{2}f(x - 1)$

65. $g(x) = -2 + 2f(x + 2)$

In Exercises 66–68, find $f + g$, $f - g$, fg, and $\frac{f}{g}$. Determine the domain for each function.

66. $f(x) = 3x - 1,\quad g(x) = x - 5$

67. $f(x) = x^2 + x + 1,\quad g(x) = x^2 - 1$

68. $f(x) = \sqrt{x + 7},\quad g(x) = \sqrt{x - 2}$

2.6

In Exercises 69–70, find **a.** $(f \circ g)(x)$; **b.** $(g \circ f)(x)$; **c.** $(f \circ g)(3)$.

69. $f(x) = x^2 + 3,\quad g(x) = 4x - 1$

70. $f(x) = \sqrt{x},\quad g(x) = x + 1$

In Exercises 71–72, find $f(g(x))$ and $g(f(x))$ and determine whether each pair of functions f and g are inverses of each other.

71. $f(x) = \frac{3}{5}x + \frac{1}{2}$ and $g(x) = \frac{5}{3}x - 2$

72. $f(x) = 2 - 5x$ and $g(x) = \frac{2 - x}{5}$

The functions in Exercises 73–75 are all one-to-one. For each function:

a. *Find an equation of $f^{-1}(x)$, the inverse function.*

b. *Verify that your equation is correct by showing that $f(f^{-1}(x)) = x$ and $f^{-1}(f(x)) = x$.*

73. $f(x) = 4x - 3$ **74.** $f(x) = \sqrt{x + 2}$

75. $f(x) = 8x^3 + 1$

Which graphs in Exercises 76–79 represent functions that have inverse functions?

76.

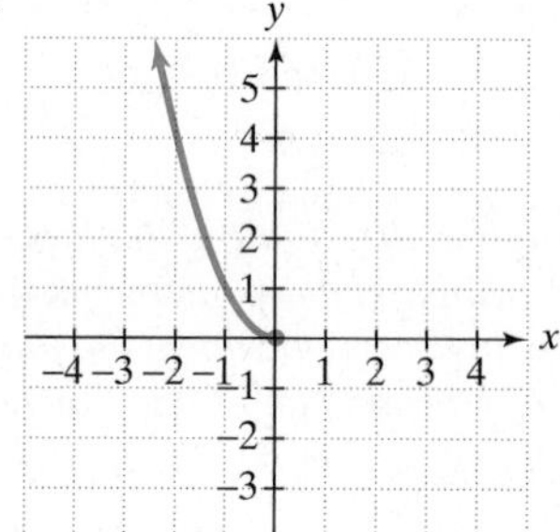

77.

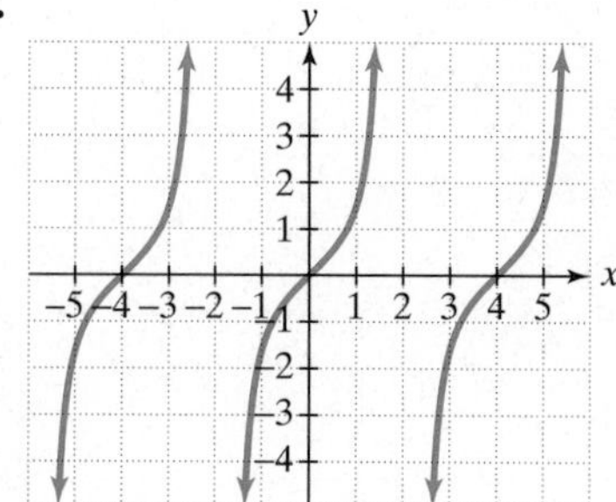

78.

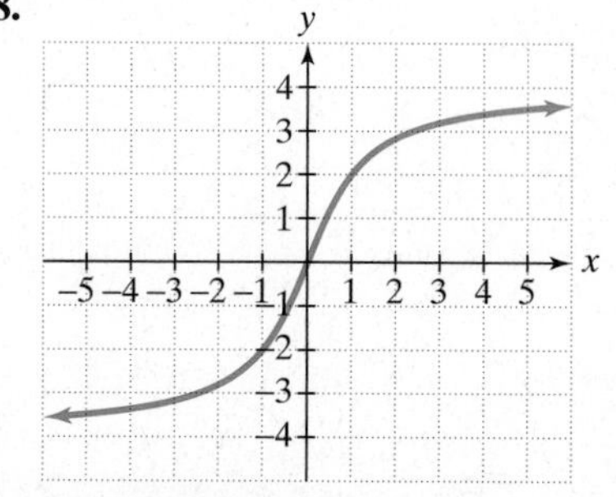

79.

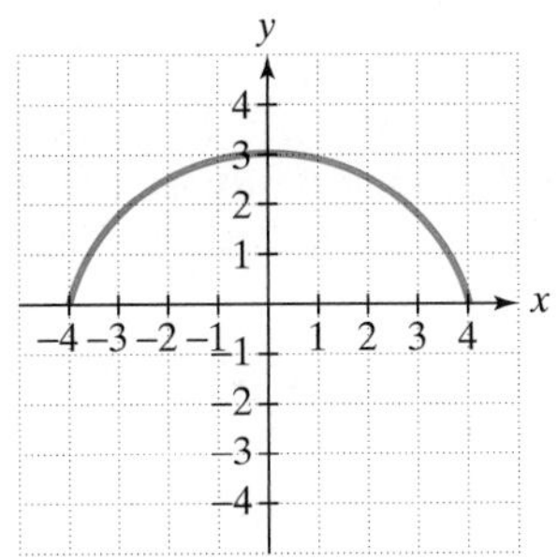

80. Use the graph of f in the figure shown to draw the graph of its inverse function.

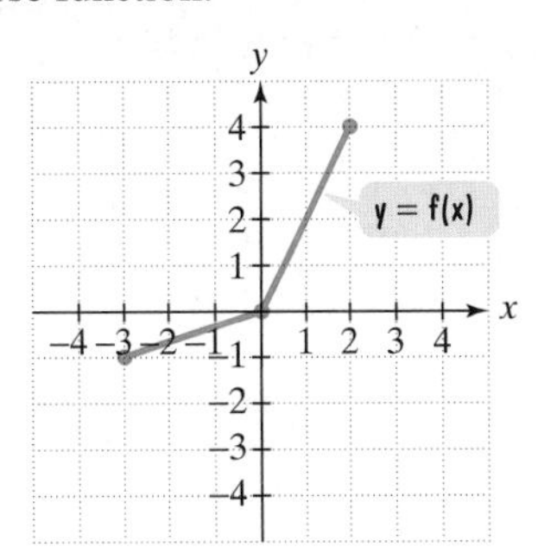

Chapter 2 Test

In Exercises 1–2, use the given conditions to write an equation for each line in point-slope form and slope-intercept form.

1. Passing through $(2, 1)$ and $(-1, -8)$
2. Passing through $(-4, 6)$ and perpendicular to the line whose equation is $y = -\frac{1}{4}x + 5$
3. The data points $(4, 401.1)$ and $(9, 475.6)$, shown and described in the table, fall on a straight line.

x (Number of Years after 1985)	y (Average Weekly Earnings of U.S. Workers)
4	401.1
9	475.6

   a. Write the equation of the line on which these measurements fall in point-slope form and slope-intercept form.
   b. Use the slope-intercept form of the equation to predict the average weekly earnings for U.S. workers for the year 2005.

4. Give the center and radius of the circle whose equation is $x^2 + y^2 + 4x - 6y - 3 = 0$ and graph the equation.
5. List by letter all relations that are not functions.
   a. $\{(7, 5), (8, 5), (9, 5)\}$
   b. $\{(5, 7), (5, 8), (5, 9)\}$
   c.

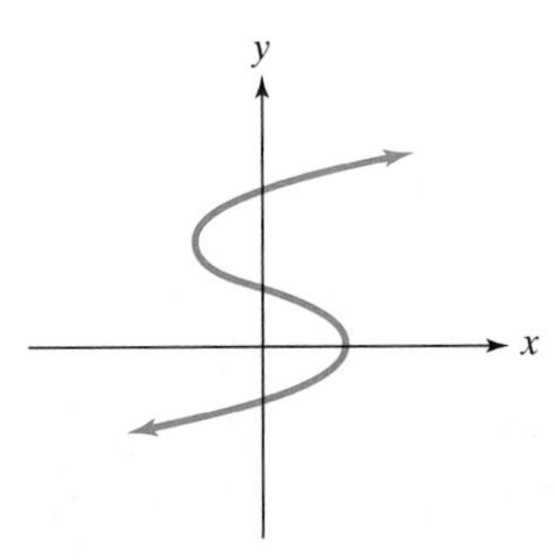

   d. $x^2 + y^2 = 100$
   e.

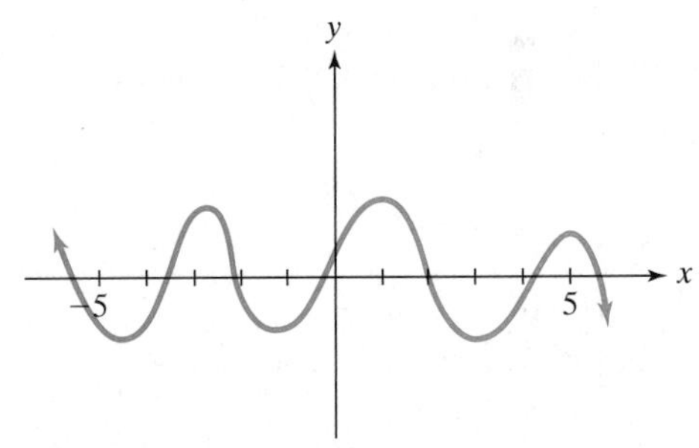

6. If $f(x) = x^2 - 2x + 5$, find $f(x - 1)$ and simplify.
7. If $g(x) = \begin{cases} \sqrt{x - 3} & \text{if } x \geq 3 \\ 3 - x & \text{if } x < 3 \end{cases}$, find $g(-1)$ and $g(7)$.
8. If $f(x) = \sqrt{12 - 3x}$, find the domain of f.
9. The function $f(x) = 0.79x^2 - 2x - 4$ models the number of board feet in a 16-foot log whose diameter averages x inches. Find and interpret $f(10)$.
10. Use the graph of function f to answer the following questions.

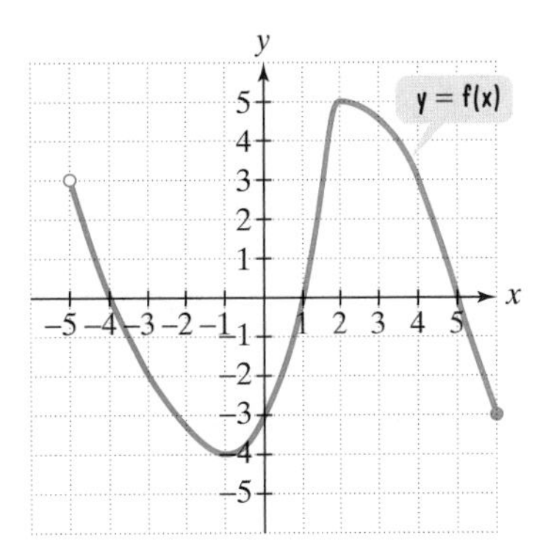

   a. What is $f(4) - f(-3)$?
   b. What is the domain of f?
   c. What is the range of f?
   d. On what interval or intervals is f increasing?
   e. On what interval or intervals is f decreasing?
   f. What are the x-intercepts?
   g. What is the y-intercept?

11. Determine whether $f(x) = x^4 - x^2$ is even, odd, or neither. Use your answer to explain why the graph in the figure shown cannot be the graph of f.

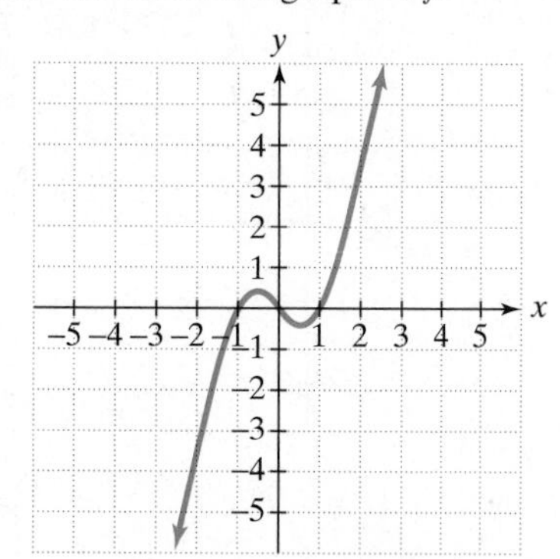

12. The figure shows how the graph of $h(x) = -2(x - 3)^2$ is obtained from the graph of $f(x) = x^2$. Describe this process, using the graph of g in your description.

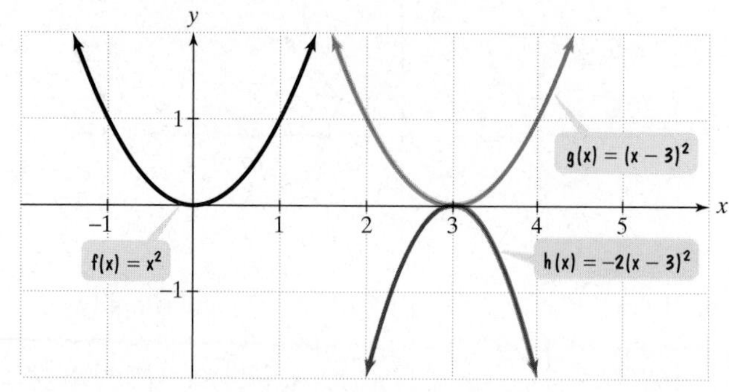

13. Begin by graphing the absolute value function, $f(x) = |x|$. Then use transformations of this graph to graph $g(x) = \frac{1}{2}|x + 1| + 3$.

If $f(x) = x^2 + 3x - 4$ and $g(x) = 5x - 2$, find each function or function value in Exercises 14–18.

14. $(f - g)(x)$

15. $\left(\frac{f}{g}\right)(x)$ and its domain

16. $(f \circ g)(x)$

17. $(g \circ f)(x)$

18. $f(g(2))$

19. If $f(x) = \sqrt{x - 2}$, find the equation for $f^{-1}(x)$. Then verify that your equation is correct by showing that $f(f^{-1}(x)) = x$ and $f^{-1}(f(x)) = x$.

20. A function f models the amount given to charity as a function of income. The graph of f is shown in the figure.

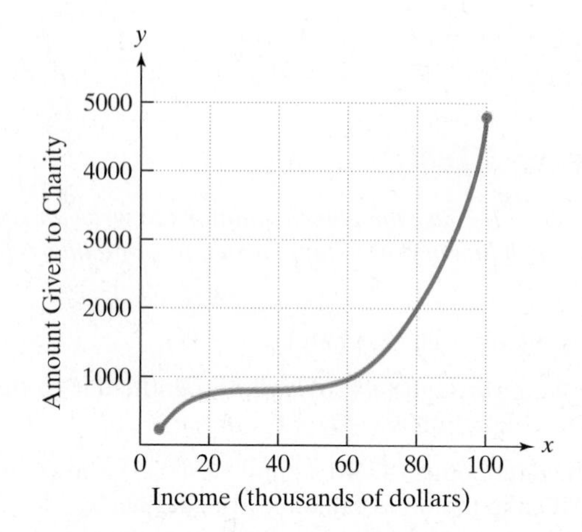

a. Explain why f has an inverse that is a function.
b. Find $f(80)$.
c. Describe in practical terms the meaning of $f^{-1}(2000)$.

21. Use a graphing utility to graph $f(x) = \frac{x^3}{3} + x^2 - 15x + 3$ in a $[-10, 10, 1]$ by $[-30, 70, 10]$ viewing rectangle. Use the graph to answer the following questions.
a. Is f one-to-one? Explain.
b. Is f even, odd, or neither? Explain.
c. What is the range of f?
d. On what interval or intervals is f increasing?
e. On what interval or intervals is f decreasing?

Cumulative Review Exercises (Chapters P–2)

Simplify each expression in Exercises 1 and 2.

1. $\dfrac{4x^2y}{2x^5y^{-3}}$

2. $\dfrac{5}{4\sqrt{2}}$

3. Factor: $x^3 - 4x^2 + 2x - 8$.

In Exercises 4 and 5, perform the operations and simplify.

4. $\dfrac{x - 3}{x + 4} + \dfrac{x}{x - 2}$

5. $\dfrac{4 + \frac{2}{x}}{4 - \frac{2}{x}}$

Solve each equation in Exercises 6–9.

6. $(x + 3)(x - 4) = 8$

7. $3(4x - 1) = 4 - 6(x - 3)$

8. $\sqrt{x + 2} = x$

9. $x^{2/3} - x^{1/3} - 6 = 0$

Solve each inequality in Exercises 10 and 11. Express the answer in interval notation.

10. $\dfrac{x}{2} - 3 \le \dfrac{x}{4} + 2$

11. $\dfrac{x + 3}{x - 2} \le 2$

12. Write the point-slope form and the slope-intercept form of the line passing through $(-2, 5)$ and perpendicular to the line whose equation is $y = -\frac{1}{4}x + \frac{1}{3}$.

13. Graph $f(x) = \sqrt{x}$ and then use transformations of this graph to graph $g(x) = \sqrt{x - 3} + 4$ in the same rectangular coordinate system.

14. If $f(x) = 2 + \sqrt{x - 3}$, find the equation for $f^{-1}(x)$.

15. If $f(x) = 3 - x^2$, find $f(x - 2)$ and simplify.

16. Solve for r: $G = \dfrac{a}{1 - r}$.

17. The length of a rectangular garden is 2 feet more than twice its width. If 22 feet of fencing is needed to enclose the garden, what are its dimensions?

18. With a 6% raise, you will earn $19,610 annually. What is your salary prior to this raise?

19. On the first five tests you have scores of 61, 95, 71, 83, and 80. The last test, a final exam, counts as two grades. What score do you need on the final in order to have an average score of 80?

20. If a rock is thrown from the ground with an initial velocity of 80 feet per second, then its height can be described by

$$f(x) = -16x^2 + 80x,$$

where x represents the number of seconds since the rock was thrown and $f(x)$ represents the rock's height, in feet. Find and interpret $f(3)$.

Polynomial and Rational Functions

One of the joys of your life is your dog, your very special buddy. Lately, however, you've noticed that your companion is slowing down a bit. He's now 8 years old and you wonder how this translates into human years. You remember something about every year of a dog's life being equal to seven years for a human. Is there a more accurate description?

There is a function that models the age in human years, $H(x)$, of a dog that is x years old:

$$H(x) = -0.001183x^4 + 0.05495x^3 - 0.8523x^2 + 9.054x + 6.748.$$

The function contains variables to powers that are whole numbers and is an example of a **polynomial function**. In this chapter, we study polynomial functions and functions that consist of quotients of polynomials, called **rational functions**.

SECTION 3.1 Quadratic Functions

Objectives

1. Recognize characteristics of parabolas.
2. Graph parabolas.
3. Solve problems involving minimizing or maximizing quadratic functions.

The wage gap is used to compare the status of women's earnings relative to men's. The wage gap is expressed as a percent and is calculated by dividing the median annual earnings for women by the median annual earnings for men. Based on data provided by the U.S. Women's Bureau, the function

$$f(x) = 0.022x^2 - 0.4x + 60.07$$

models women's earnings as a percentage of men's x years after 1960. For example, to calculate the wage gap in 1998, substitute 38 for x because 1998 is 38 years after 1960:

$$f(38) = 0.022(38)^2 - 0.4(38) + 60.07 \approx 76.6.$$

Thus, in 1998 women earned 76.6% as much as men. Since 1963, when the Equal Pay Act was signed, the wage gap between men and women has closed at a rate of less than half a penny per year.

The function $f(x) = 0.022x^2 - 0.4x + 60.07$ is an example of a *quadratic function*. A **quadratic function** is any function of the form

$$f(x) = ax^2 + bx + c$$

where a, b, and c are real numbers and $a \neq 0$. A quadratic function is a polynomial function whose highest power is 2. In this section we will study quadratic functions and their graphs.

1 Recognize characteristics of parabolas.

Graphs of Quadratic Functions

The graph of any quadratic function is called a **parabola**. Parabolas are shaped like cups, as shown in Figure 3.1 on page 260. If the coefficient of x^2 (the value of a in $ax^2 + bx + c$) is positive, the parabola opens upward. When the coefficient of x^2 is negative, the graph opens downward. The **vertex** (or turning point) of the parabola is the minimum point on the graph when it opens upward, and the maximum point on the graph when it opens downward.

Look at the unusual image on page 260 of the word "mirror." The artist, Scott Kim, has created the image so that the two halves of the whole are mirror images of each other. A parabola shares this kind of symmetry, in which a line through the vertex divides the figure in half. Parabolas are symmetric to this line,

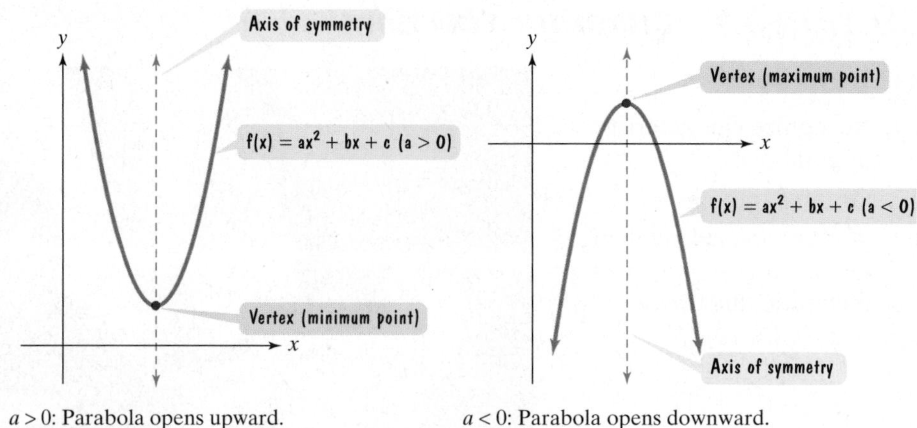

Figure 3.1 Characteristics of parabolas

$a > 0$: Parabola opens upward. $a < 0$: Parabola opens downward.

called the **axis of symmetry**. The movements of gymnasts, divers, and swimmers can approximate this symmetry.

Graphing Quadratic Functions in Standard Form

2 Graph parabolas.

In Section 2.5, we applied a series of transformations to the graph of $f(x) = x^2$. The graph of this function is a parabola. The vertex for this parabola is at (0, 0). In Figure 3.2(a), the graph of $f(x) = ax^2$ for $a > 0$ is shown in black; it opens *upward*. In Figure 3.2(b), the graph of $f(x) = ax^2$ for $a < 0$ is shown in black; it opens *downward*.

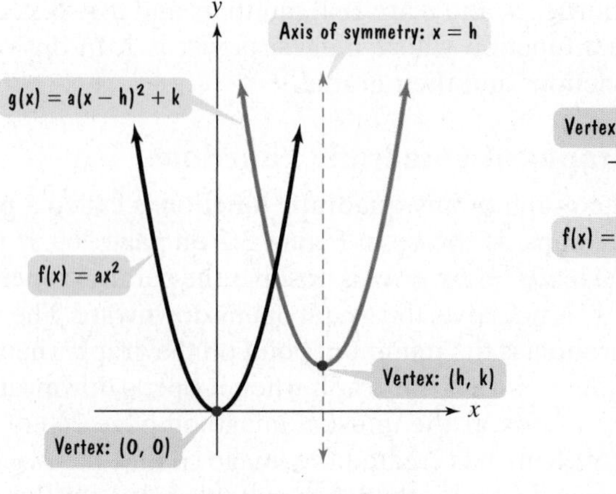

(a) $a > 0$: Parabola opens upward.

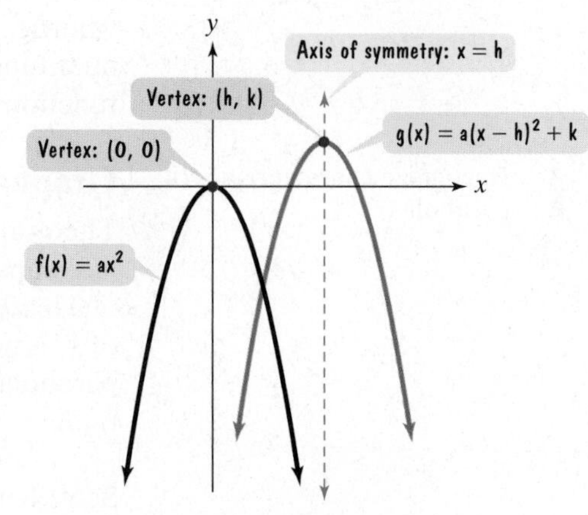

(b) $a < 0$: Parabola opens downward.

Figure 3.2 Transformations of $f(x) = ax^2$

Figure 3.2 also shows the graphs of $g(x) = a(x - h)^2 + k$ in blue. Compare these graphs to those of $f(x) = ax^2$. Observe that h determines the horizontal shift and k determines the vertical shift of the graph of $f(x) = ax^2$. Consequently, the vertex $(0, 0)$ on the graph of $f(x) = ax^2$ moves to the point (h, k) on the graph of $g(x) = a(x - h)^2 + k$. The axis of symmetry is the vertical line whose equation is $x = h$.

The form of the expression for g is convenient because it immediately identifies the vertex of the parabola as (h, k). This is the **standard form** of a quadratic function.

The Standard Form of a Quadratic Function

The quadratic function

$$f(x) = a(x - h)^2 + k, \qquad a \neq 0$$

is in **standard form**. The graph of f is a parabola whose vertex is the point (h, k). The parabola is symmetric to the line $x = h$. If $a > 0$, the parabola opens upward; if $a < 0$, the parabola opens downward.

The sign of a in $f(x) = a(x - h)^2 + k$ determines whether the parabola opens upward or downward. Furthermore, if $|a|$ is small, the parabola opens more widely than if $|a|$ is large. Here is a general procedure for graphing parabolas whose equations are in standard form.

Graphing Parabolas With Equations in Standard Form

To graph $f(x) = a(x - h)^2 + k$:

1. Determine whether the parabola opens upward or downward. If $a > 0$, it opens upward. If $a < 0$, it opens downward.
2. Determine the vertex of the parabola. The vertex is (h, k).
3. Find any x-intercepts by replacing $f(x)$ with 0. Solve the resulting quadratic equation for x.
4. Find the y-intercept by replacing x with 0.
5. Plot the intercepts and vertex. Connect these points with a smooth curve that is shaped like a cup.

EXAMPLE 1 Graphing a Parabola Whose Equation Is in Standard Form

Graph the quadratic function $f(x) = -2(x - 3)^2 + 8$.

Solution We can graph this function by following the steps in the preceding box. We begin by identifying values for a, h, and k.

Standard form $\quad f(x) = a(x - h)^2 + k$

$a = -2 \quad h = 3 \quad k = 8$

Given equation $\quad f(x) = -2(x - 3)^2 + 8$

Step 1 Determine how the parabola opens. Note that a, the coefficient of x^2, is -2. Thus, $a < 0$; this negative value tells us that the parabola opens downward.

Step 2 **Find the vertex.** The vertex of the parabola is at (h, k). Because $h = 3$ and $k = 8$, the parabola has its vertex at $(3, 8)$.

Step 3 **Find the x-intercepts.** Replace $f(x)$ with 0 in $f(x) = -2(x - 3)^2 + 8$.

$$0 = -2(x - 3)^2 + 8$$ Find x-intercepts, setting $f(x)$ equal to 0.

$$2(x - 3)^2 = 8$$ Solve for x. Add $2(x - 3)^2$ to both sides of the equation.

$$(x - 3)^2 = 4$$ Divide both sides by 2.

$$(x - 3) = \pm 2$$ Apply the square root method. If $(x - c)^2 = d$, then $x - c = \pm\sqrt{d}$.

$$x - 3 = -2 \quad \text{or} \quad x - 3 = 2$$ Express as two separate equations.

$$x = 1 \quad \text{or} \quad x = 5$$ Add 3 to both sides in each equation.

The x-intercepts are 1 and 5. The parabola passes through $(1, 0)$ and $(5, 0)$.

Step 4 **Find the y-intercept.** Replace x with 0 in $f(x) = -2(x - 3)^2 + 8$.

$$f(0) = -2(0 - 3)^2 + 8 = -2(-3)^2 + 8 = -2(9) + 8 = -10$$

The y-intercept is -10. The parabola passes through $(0, -10)$.

Step 5 **Graph the parabola.** With a vertex at $(3, 8)$, x-intercepts at 1 and 5, and a y-intercept at -10, the graph of f is shown in Figure 3.3. The axis of symmetry is the vertical line whose equation is $x = 3$.

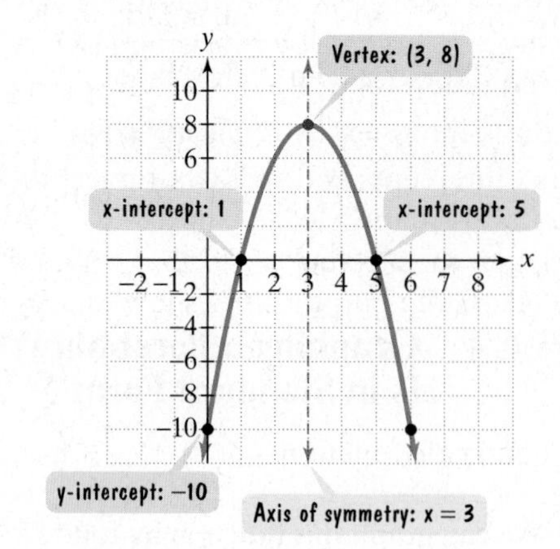

Figure 3.3 The graph of $f(x) = -2(x - 3)^2 + 8$

Check Point 1 Graph the quadratic function $f(x) = -(x - 1)^2 + 4$.

EXAMPLE 2 Graphing a Parabola Whose Equation Is in Standard Form

Graph the quadratic function $f(x) = (x + 3)^2 + 1$.

Solution We begin by finding values for a, h, and k.

Standard form	$f(x) = a(x - h)^2 + k$
Given equation	$f(x) = (x + 3)^2 + 1$
or	$f(x) = 1(x - (-3))^2 + 1$

$a = 1$ $h = -3$ $k = 1$

Step 1 Determine how the parabola opens. Note that a, the coefficient of x^2, is 1. Thus, $a > 0$; this positive value tells us that the parabola opens upward.

Step 2 Find the vertex. The vertex of the parabola is at (h, k). Because $h = -3$ and $k = 1$, the parabola has its vertex at $(-3, 1)$.

Step 3 Find the x-intercepts. Replace $f(x)$ with 0 in $f(x) = (x + 3)^2 + 1$. Because the vertex is at $(-3, 1)$, which lies above the x-axis, and the parabola opens upward, it appears that this parabola has no x-intercepts. We can verify this observation algebraically.

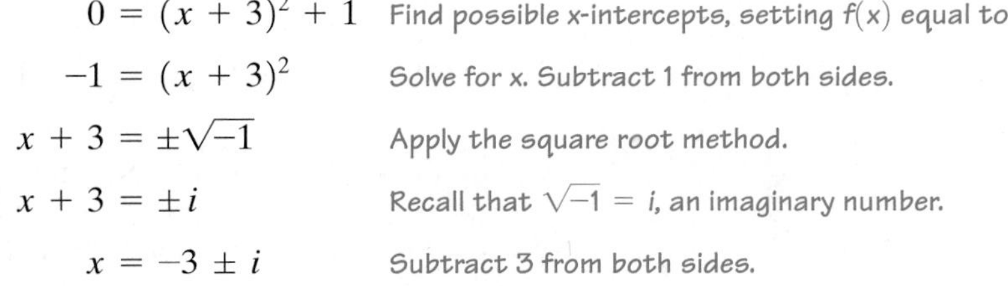

$0 = (x + 3)^2 + 1$	Find possible x-intercepts, setting f(x) equal to 0.
$-1 = (x + 3)^2$	Solve for x. Subtract 1 from both sides.
$x + 3 = \pm\sqrt{-1}$	Apply the square root method.
$x + 3 = \pm i$	Recall that $\sqrt{-1} = i$, an imaginary number.
$x = -3 \pm i$	Subtract 3 from both sides.

Because this equation has no real solutions, the parabola has no x-intercepts.

Step 4 Find the y-intercept. Replace x with 0 in $f(x) = (x + 3)^2 + 1$.

$$f(0) = (0 + 3)^2 + 1 = 3^2 + 1 = 9 + 1 = 10$$

The y-intercept is 10. The parabola passes through $(0, 10)$.

Step 5 Graph the parabola. With a vertex at $(-3, 1)$, no x-intercepts, and a y-intercept at 10, the graph of f is shown in Figure 3.4. The axis of symmetry is the vertical line whose equation is $x = -3$.

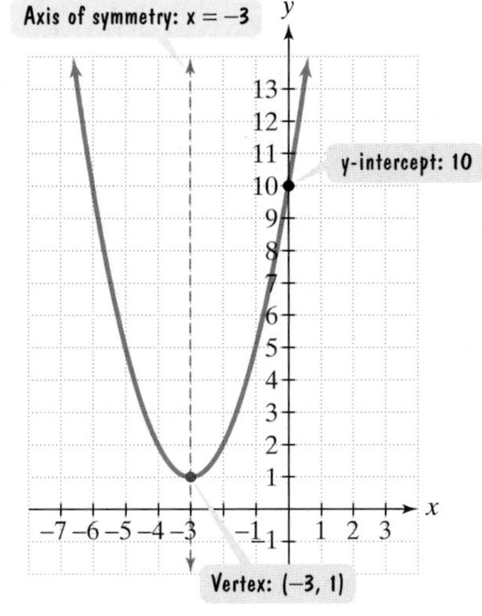

Figure 3.4 The graph of $g(x) = (x + 3)^2 + 1$

Check Point 2 Graph the quadratic function $f(x) = (x - 2)^2 + 1$.

Graphing Functions in the Form $f(x) = ax^2 + bx + c$

Quadratic functions are frequently expressed in the form $f(x) = ax^2 + bx + c$. How can we identify the vertex of a parabola whose equation is in this form? Completing the square provides the answer to this question.

$$f(x) = ax^2 + bx + c$$

$$= a\left(x^2 + \frac{b}{a}x\right) + c$$ Factor out a from $ax^2 + bx$.

$$= a\left(x^2 + \frac{b}{a}x + \frac{b^2}{4a^2}\right) + c - a\left(\frac{b^2}{4a^2}\right)$$

Complete the square by adding the square of half the coefficient of x.

By completing the square, we added $a \cdot \frac{b^2}{4a^2}$. To avoid changing the function's equation, we must subtract this term.

$$= a\left(x + \frac{b}{2a}\right)^2 + c - \frac{b^2}{4a}$$ Write the trinomial as the square of a binomial and simplify the constant term.

Compare the form of this equation with a quadratic function's standard form.

Standard form $\quad f(x) = a(x - h)^2 + k$

$h = -\frac{b}{2a} \qquad k = c - \frac{b^2}{4a}$

Equation under discussion $\quad f(x) = a\left(x - \left(-\frac{b}{2a}\right)\right)^2 + c - \frac{b^2}{4a}$

The important part of this observation is that h, the x-coordinate of the vertex, is $-\frac{b}{2a}$. The y-coordinate can be found by evaluating the function at $-\frac{b}{2a}$.

The Vertex of a Parabola Whose Equation Is $f(x) = ax^2 + bx + c$

Consider the parabola defined by the quadratic function $f(x) = ax^2 + bx + c$. The parabola's vertex is at $\left(-\frac{b}{2a}, f\left(-\frac{b}{2a}\right)\right)$.

We can apply our five-step procedure and graph parabolas in $f(x) = ax^2 + bx + c$ form. The only step that is different is how we determine the vertex.

EXAMPLE 3 Graphing a Parabola in $f(x) = ax^2 + bx + c$ Form

Graph the quadratic function $f(x) = -x^2 + 4x - 1$.

Solution

Step 1 Determine how the parabola opens. Note that a, the coefficient of x^2, is -1. Thus, $a < 0$; this negative value tells us that the parabola opens downward.

Step 2 Find the vertex. We know that the x-coordinate of the vertex is $x = -\frac{b}{2a}$. We identify a, b, and c in $f(x) = ax^2 + bx + c$.

$$f(x) = -x^2 + 4x - 1$$

$a = -1$ $b = 4$ $c = -1$

Substitute the values of a and b into the equation for the x-coordinate:

$$x = -\frac{b}{2a} = -\frac{4}{2(-1)} = \frac{-4}{-2} = 2.$$

The x-coordinate of the vertex is 2. We substitute 2 for x in the equation of the function to find the y-coordinate:

$$f(2) = -2^2 + 4 \cdot 2 - 1 = -4 + 8 - 1 = 3.$$

The vertex is at $(2, 3)$.

Step 3 Find the x-intercepts. Replace $f(x)$ with 0 in $f(x) = -x^2 + 4x - 1$. We obtain $0 = -x^2 + 4x - 1$ or $-x^2 + 4x - 1 = 0$. This equation cannot be solved by factoring. We will use the quadratic formula to solve it.

$$a = -1, \quad b = 4, \quad c = -1$$

$$x = \frac{-b \pm \sqrt{b^2 - 4ac}}{2a} = \frac{-4 \pm \sqrt{4^2 - 4(-1)(-1)}}{2(-1)} = \frac{-4 \pm \sqrt{16 - 4}}{-2}$$

$$x = \frac{-4 - \sqrt{12}}{-2} \approx 3.7 \quad \text{or} \quad x = \frac{-4 + \sqrt{12}}{-2} \approx 0.3$$

The x-intercepts are approximately 0.3 and 3.7. The parabola passes through $(0.3, 0)$ and $(3.7, 0)$.

Step 4 Find the y-intercept. Replace x with 0 in $f(x) = -x^2 + 4x - 1$.

$$f(0) = -0^2 + 4 \cdot 0 - 1 = -1$$

The y-intercept is -1. The parabola passes through $(0, -1)$.

Step 5 Graph the parabola. With a vertex at $(2, 3)$, x-intercepts at 0.3 and 3.7, and a y-intercept at -1, the graph of f is shown in Figure 3.5. The axis of symmetry is the vertical line whose equation is $x = 2$.

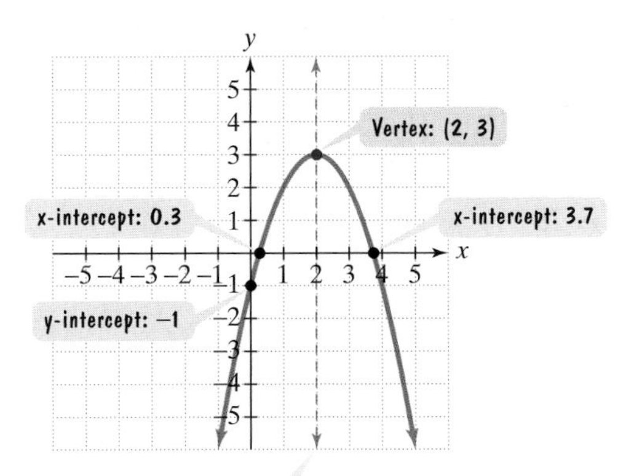

Figure 3.5 The graph of $f(x) = -x^2 + 4x - 1$

Check Point 3

Graph the quadratic function $f(x) = x^2 - 2x - 3$.

3 Solve problems involving minimizing or maximizing quadratic functions.

Applications of Quadratic Functions

When were women's earnings as a percentage of men's at the lowest? What is the age of a driver having the least number of car accidents? How much should a business spend on advertising to maximize its profits? The answers to these questions involve finding the maximum or minimum value of quadratic functions.

Consider the quadratic function $f(x) = ax^2 + bx + c$. If $a > 0$, the parabola opens upward and the vertex is its lowest point. If $a < 0$, the parabola opens downward and the vertex is its highest point. The x-coordinate of the vertex is $-\frac{b}{2a}$. Thus, we can find the minimum or maximum value of f by evaluating the quadratic function at $x = -\frac{b}{2a}$.

Minimum and Maximum: Quadratic Functions

Consider $f(x) = ax^2 + bx + c$.

1. If $a > 0$, then f has a minimum that occurs at $x = -\frac{b}{2a}$.

2. If $a < 0$, then f has a maximum that occurs at $x = -\frac{b}{2a}$.

EXAMPLE 4 An Application: The Wage Gap

The function

$$f(x) = 0.022x^2 - 0.4x + 60.07$$

models women's earnings as a percentage of men's x years after 1960. In which year was this percentage at a minimum? What was the percentage for that year?

Solution The quadratic function is in the form $f(x) = ax^2 + bx + c$ with $a = 0.022$ and $b = -0.4$. With $a > 0$, the function has a minimum when $x = -\frac{b}{2a}$.

$$x = -\frac{b}{2a} = -\frac{(-0.4)}{2(0.022)} \approx 9$$

This means that women's earnings as a percentage of men's were at their lowest approximately 9 years after 1960, or in 1969. The percentage for that year was

$$f(9) = 0.022(9)^2 - 0.4(9) + 60.07 \approx 58.$$

In 1969, women earned approximately 58% as much as men.

The function $f(x) = 0.4x^2 - 36x + 1000$ models the number of accidents, $f(x)$, per 50 million miles driven, as a function of a driver's age, x, in years, where $16 \le x \le 74$. What is the age of a driver having the least number of car accidents? What is the minimum number of car accidents per 50 million miles driven?

Modeling Data with Quadratic Functions

We've come a long way from the small nation of "embattled farmers" who launched the American Revolution. In the early years of our Republic, 95% of the population was involved in farming. Although U.S. agriculture is an integral part of the global economy, the number of U.S. farms has declined since the 1920s as individually owned family farms have been swallowed up by huge agribusinesses owned by corporations.

The graph in Figure 3.6 shows the number of farms in the United States from 1850 through 2010 (projected). Because the graph is shaped like a cup, with an increasing number of farms from 1850 to 1910 and a decreasing number of farms from 1910 to 2010, a quadratic function is an appropriate model for the data. You can use the statistical menu of a graphing utility to enter the data in Figure 3.6. We entered the data using

(number of decades after 1850, millions of U.S. farms).

Thus, we entered

$$(0, 2.3),\quad (2, 3.3),\quad (4, 5.1),\quad (6, 6.7),\quad (8, 6.4),$$
$$(10, 5.8),\quad (12, 3.6),\quad (14, 2.9),\quad (16, 2.3).$$

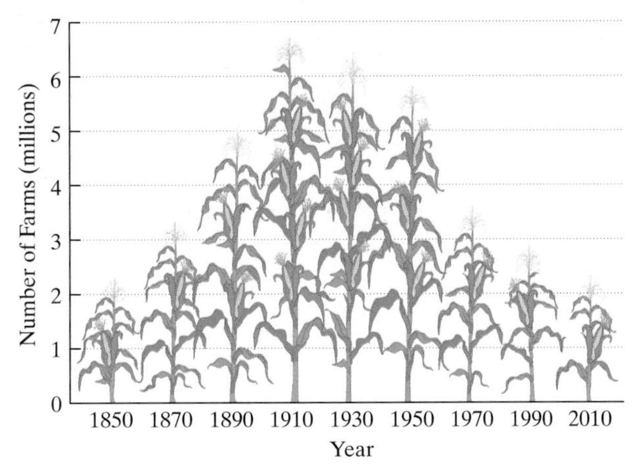

Source: U. S. Bureau of the Census

Figure 3.6 The number of U.S. farms is declining.

```
QuadReg
 y=ax²+bx+c
 a=-.0643668831
 b=.9873701299
 c=2.203636364
```

Figure 3.7 Executing the Quadratic Regression Program

Upon executing the QUADratic REGression program, we obtain the results shown in Figure 3.7. Thus, the quadratic function of best fit is

$$f(x) = -0.064x^2 + 0.99x + 2.2$$

where x represents the number of decades after 1850 and $f(x)$ represents the number of U.S. farms, in millions.

EXERCISE SET 3.1

Practice Exercises

In Exercises 1–4, the graph of a quadratic function is given. Write the function's equation, selecting from the following options.

$f(x) = (x + 1)^2 - 1$ $\quad g(x) = (x + 1)^2 + 1$

$h(x) = (x - 1)^2 + 1$ $\quad j(x) = (x - 1)^2 - 1$

1.

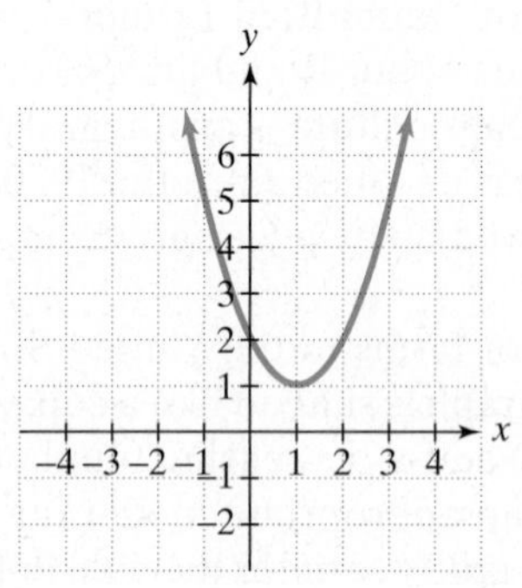

2.

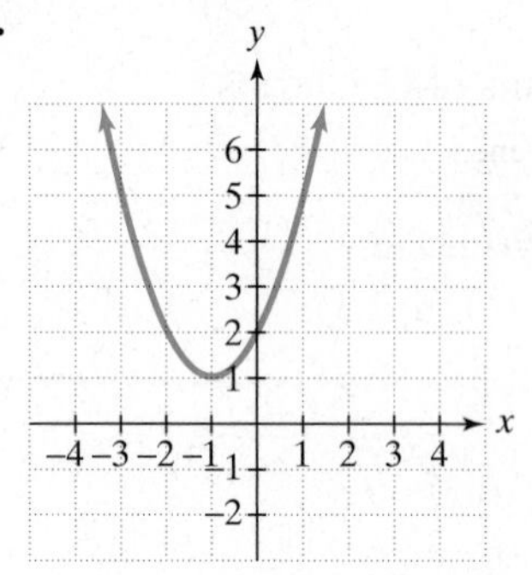

3.

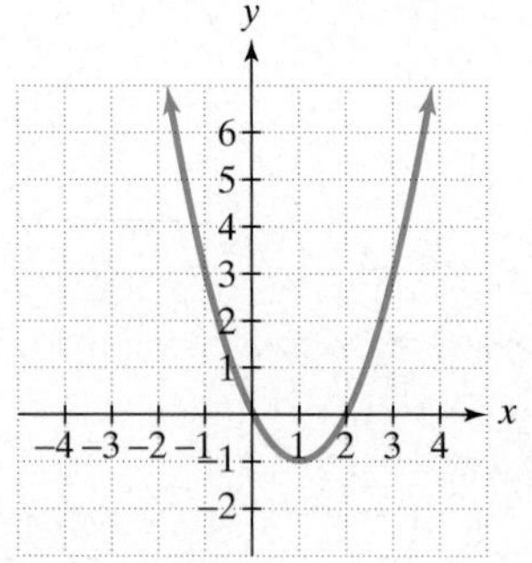

4.

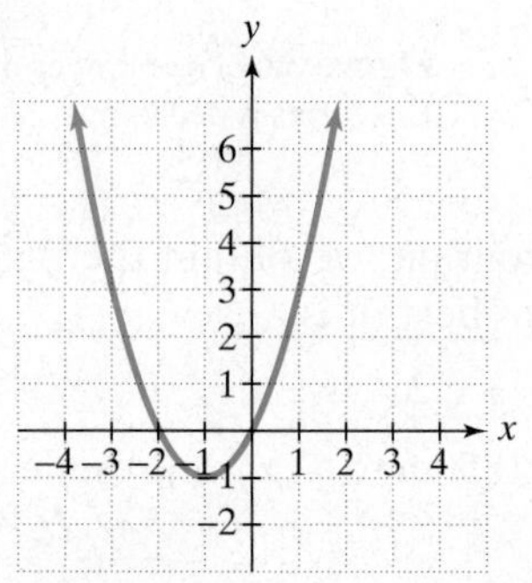

In Exercises 5–8, the graph of a quadratic function is given. Write the function's equation, selecting from the following options.

$f(x) = x^2 + 2x + 1$ $\quad g(x) = x^2 - 2x + 1$

$h(x) = x^2 - 1$ $\quad j(x) = -x^2 - 1$

5.

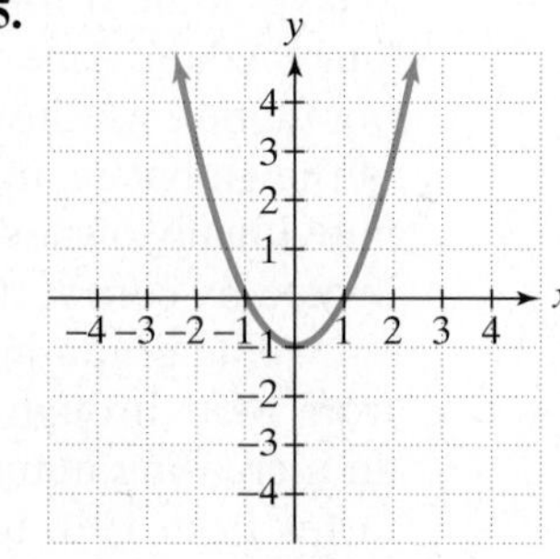

6.

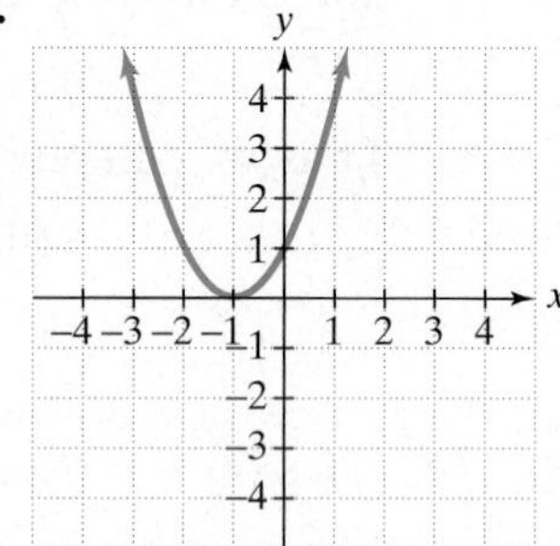

7.

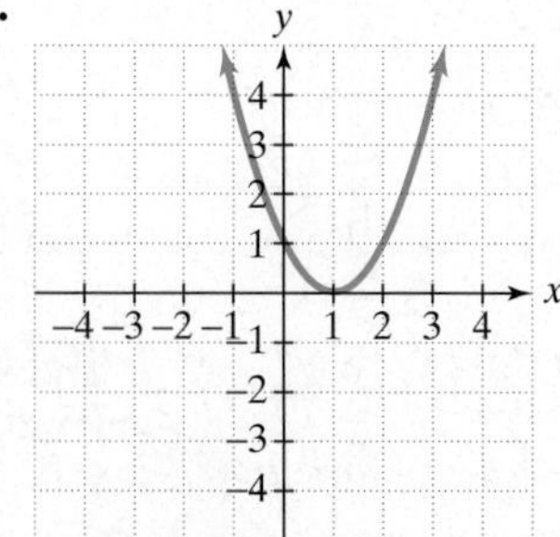

8.

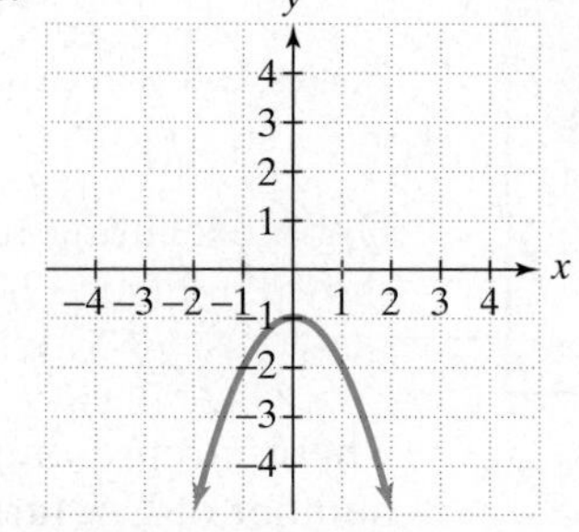

In Exercises 9–16, find the coordinates of the vertex for the parabola defined by the given quadratic function.

9. $f(x) = 2(x - 3)^2 + 1$ **10.** $f(x) = -3(x - 2)^2 + 12$
11. $f(x) = -2(x + 1)^2 + 5$ **12.** $f(x) = -2(x + 4)^2 - 8$
13. $f(x) = 2x^2 - 8x + 3$ **14.** $f(x) = 3x^2 - 12x + 1$
15. $f(x) = -x^2 - 2x + 8$ **16.** $f(x) = -2x^2 + 8x - 1$

In Exercises 17–34, use the vertex and intercepts to sketch the graph of each quadratic function. Give the equation for the parabola's axis of symmetry.

17. $f(x) = (x - 4)^2 - 1$ **18.** $f(x) = (x - 1)^2 - 2$
19. $f(x) = (x - 1)^2 + 2$ **20.** $f(x) = (x - 3)^2 + 2$
21. $y - 1 = (x - 3)^2$ **22.** $y - 3 = (x - 1)^2$
23. $f(x) = 2(x + 2)^2 - 1$ **24.** $f(x) = \frac{5}{4} - \left(x - \frac{1}{2}\right)^2$
25. $f(x) = 4 - (x - 1)^2$ **26.** $f(x) = 1 - (x - 3)^2$
27. $f(x) = x^2 - 2x - 3$ **28.** $f(x) = x^2 - 2x - 15$
29. $f(x) = x^2 + 3x - 10$ **30.** $f(x) = 2x^2 - 7x - 4$
31. $f(x) = 2x - x^2 + 3$ **32.** $f(x) = 5 - 4x - x^2$
33. $f(x) = 2x - x^2 - 2$ **34.** $f(x) = 6 - 4x + x^2$

In Exercises 35–40, determine, without graphing, whether the given quadratic function has a minimum value or a maximum value. Then find the coordinates of the minimum or the maximum point.

35. $f(x) = 3x^2 - 12x - 1$ **36.** $f(x) = 2x^2 - 8x - 3$
37. $f(x) = -4x^2 + 8x - 3$ **38.** $f(x) = -2x^2 - 12x + 3$
39. $f(x) = 5x^2 - 5x$ **40.** $f(x) = 6x^2 - 6x$

Application Exercises

41. The U.S. Center for Disease Control modeled the average annual per capita consumption C of cigarettes by Americans 18 and older as a function of time. The function is $C(t) = -3.1t^2 + 51.4t + 4024.5$, where t represents years after 1960. According to this function, in which year did cigarette consumption per capita reach a maximum? What was the consumption for that year?

Cigarette Consumption per U.S. Adult

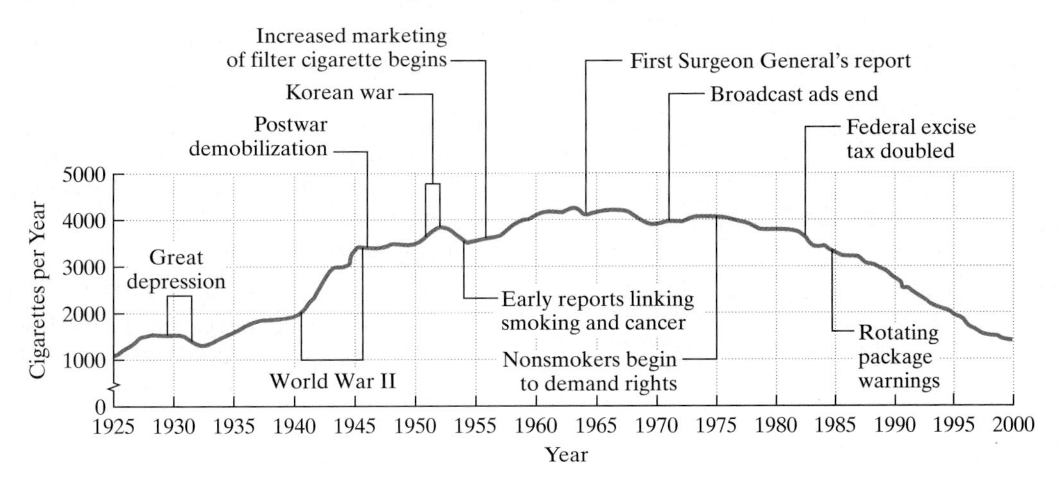

Source: U.S. Department of Health and Human Services

42. The function $R(x) = -0.0065x^2 + 0.23x + 8.47$ models the American marriage rate R (the number of marriages per 1000 population) x years after 1960. According to this function, in which year was the marriage rate the highest? What was the marriage rate for that year?

43. A person standing close to the edge of an 80-foot cliff throws a rock upward with an initial speed of 64 feet per second. The height of the rock above the water at the bottom of the cliff is a function of time, described by the quadratic function

$$y = -16x^2 + 64x + 80.$$

The variable x describes the number of seconds that the rock is in motion. The variable y describes the height of the rock, in feet, above the water at the bottom of the cliff. After how many seconds will the rock reach its maximum height above the water? How many feet above the water is the rock at that time?

44. There is a relationship between the amount of one's annual income, x, in thousands of dollars, and the percentage of this income, P, that one contributes to charities. This relationship is modeled by the quadratic function $P = 0.0014x^2 - 0.1529x + 5.855$, where $5 \le x \le 100$. What annual income corresponds to the minimum percentage given to charity? What is this minimum percentage?

45. Suppose that a quadratic function is used to model the data shown in the graph on page 270 using

(number of years after 1985,
number of U.S. children under 18
guilty of homicide using guns).

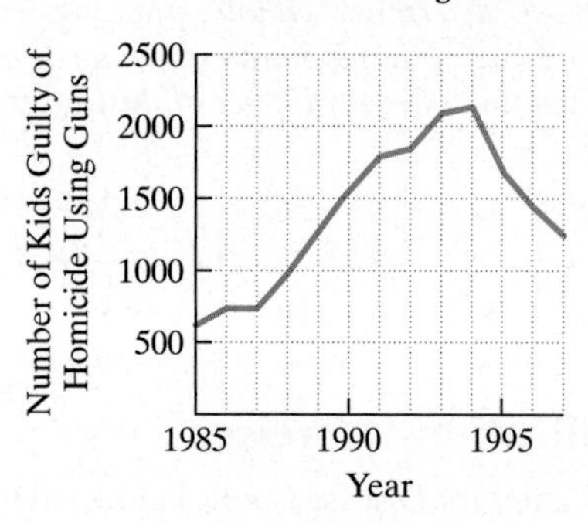

Source: National Center for Juvenile Justice

Determine, without obtaining the quadratic function of best fit, the approximate coordinates of the vertex for the function's graph.

46. Suppose that a quadratic function is used to model the data shown in the graph using

(number of years after 1971,
millions of students enrolled in U.S. schools).

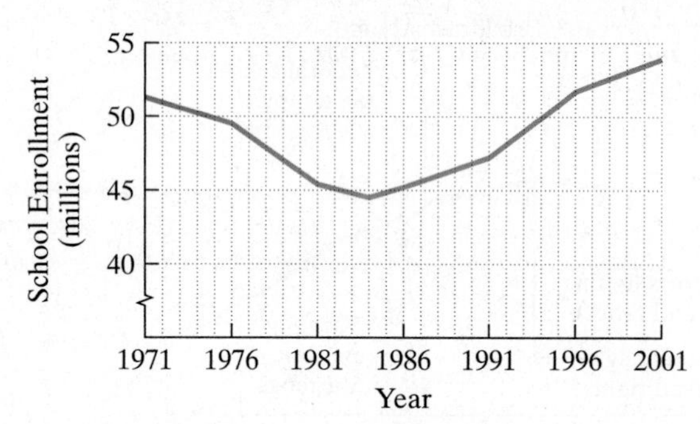

Source: National Education Association

Determine, without obtaining the quadratic function of best fit, the approximate coordinates of the vertex for the function's graph.

47. You have 120 feet of fencing to enclose a rectangular plot that borders on a river. If you do not fence the side along the river, find the length and width of the plot that will maximize the area. What is the largest area that can be enclosed?

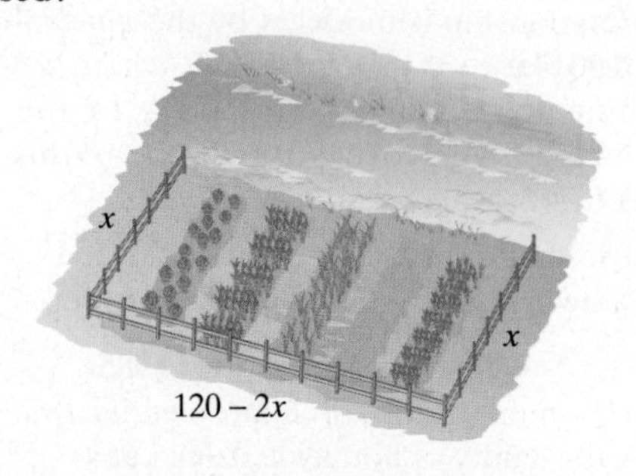

48. The figure shown indicates that you have 100 yards of fencing to enclose a rectangular area. Find the dimensions of the rectangle that maximize the enclosed area. What is the maximum area?

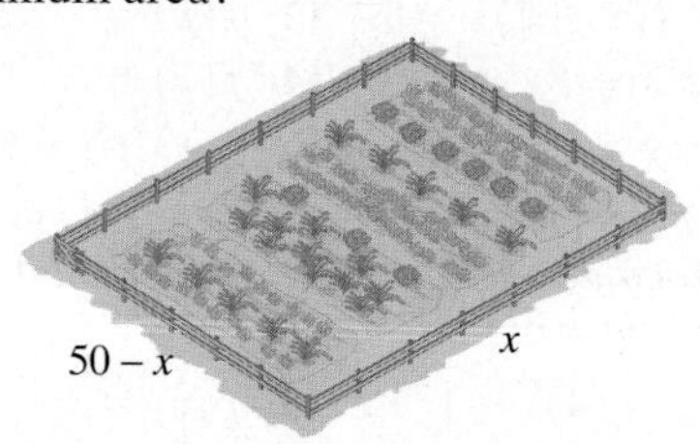

Writing in Mathematics

49. What is a quadratic function?

50. What is a parabola? Describe its shape.

51. Explain how to decide whether a parabola opens upward or downward.

52. Describe how to find a parabola's vertex if its equation is expressed in standard form. Give an example.

53. Describe how to find a parabola's vertex if its equation is in the form $f(x) = ax^2 + bx + c$. Use $f(x) = x^2 - 6x + 8$ as an example.

54. A parabola that opens upward has its vertex at $(1, 2)$. Describe as much as you can about the parabola based on this information. Include in your discussion the number of x-intercepts (if any) for the parabola.

55. The quadratic function

$$f(x) = -0.018x^2 + 1.93x - 25.34$$

describes the miles per gallon of a Ford Taurus driven at x miles per hour. Suppose that you own a Ford Taurus. Describe how you can use this function to save money.

Technology Exercises

56. Use a graphing utility to verify any five of your hand-drawn graphs in Exercises 17–34.

57. **a.** Use a graphing utility to graph $y = 2x^2 - 82x + 720$ in a standard viewing rectangle. What do you observe?

b. Find the coordinates of the vertex for the given quadratic function.

c. The answer to part (b) is $(20.5, -120.5)$. Because the leading coefficient of the given function (2) is positive, the vertex is a minimum point on the graph. Use this fact to help find a viewing rectangle that will give a relatively complete picture of the parabola. With an axis of symmetry at $x = 20.5$, the setting for x should extend past this, so try Xmin $= 0$ and Xmax $= 30$. The setting for y should include (and probably go below) the y-coordinate of the graph's minimum point, so try Ymin $= -130$. Experiment with Ymax until your utility shows the parabola's major features.

d. In general, explain how knowing the coordinates of a parabola's vertex can help determine a reasonable viewing rectangle on a graphing utility for obtaining a complete picture of the parabola.

In Exercises 58–61, find the vertex for each parabola. Then determine a reasonable viewing rectangle on your graphing utility and use it to graph the parabola.

58. $y = -0.25x^2 + 40x$

59. $y = -4x^2 + 20x + 160$

60. $y = 5x^2 + 40x + 600$

61. $y = 0.01x^2 + 0.6x + 100$

62. The quadratic function $f(x) = 0.013x^2 - 0.96x + 25.4$ describes the average yearly consumption of whole milk per person in the United States x years after 1970. The linear function $g(x) = 0.41x + 6.03$ describes the average yearly consumption of low-fat milk per person in the United States x years after 1970.

a. Use a graphing utility to graph each function in the same viewing rectangle for the years 1970 through 2000.

b. Use the graphs to describe the trend in consumption for both types of milk. What possible explanations are there for these consumption patterns?

63. The function $y = 0.011x^2 - 0.097x + 4.1$ models the number of people in the United States, y, in millions, holding more than one job x years after 1970. Use graphing utility to graph the function in a $[0, 20, 1]$ by $[3, 6, 1]$ viewing rectangle. TRACE along the curve or use your utility's minimum value feature to approximate the coordinates of the parabola's vertex. Describe what this represents in practical terms.

64. The following data show fuel efficiency, in miles per gallon, for all U.S. automobiles in the indicated year.

x (Years after 1940)	y (Average Number of Miles per Gallon for U.S. Automobiles)
1940: 0	14.8
1950: 10	13.9
1960: 20	13.4
1970: 30	13.5
1980: 40	15.5
1986: 46	18.3

Source: Statistical Abstract of the United States

a. Use a graphing utility to draw a scatter plot of the data. Explain why a quadratic function is appropriate for modeling these data.

b. Use the quadratic regression feature to find the quadratic function that best fits the data.

c. Use the equation in part (b) to determine the worst year for automobile fuel efficiency. What was the average number of miles per gallon for that year?

d. Use a graphing utility to draw a scatter plot of the data and graph the quadratic function of best fit on the scatter plot.

Critical Thinking Exercises

65. Which one of the following is true?

a. No quadratic functions have a range of $(-\infty, \infty)$.

b. The vertex of the parabola described by $f(x) = 2(x - 5)^2 - 1$ is at $(5, 1)$.

c. The graph of $f(x) = -2(x + 4)^2 - 8$ has one y-intercept and two x-intercepts.

d. The maximum value of y for the quadratic function $f(x) = -x^2 + x + 1$ is 1.

66. What explanations can you offer for your answer to Exercise 41? Use a graphing utility to graph C. Do you agree with the long-term predictions made by the graph? Explain.

In Exercises 67–68, find the axis of symmetry for each parabola whose equation is given. Use the axis of symmetry to find a second point on the parabola whose y-coordinate is the same as the given point.

67. $f(x) = 3(x + 2)^2 - 5; \quad (-1, -2)$

68. $f(x) = (x - 3)^2 + 2; \quad (6, 11)$

69. A rancher has 1000 feet of fencing to construct six corrals, as shown in the figure. Find the dimensions that maximize the enclosed area. What is the maximum area?

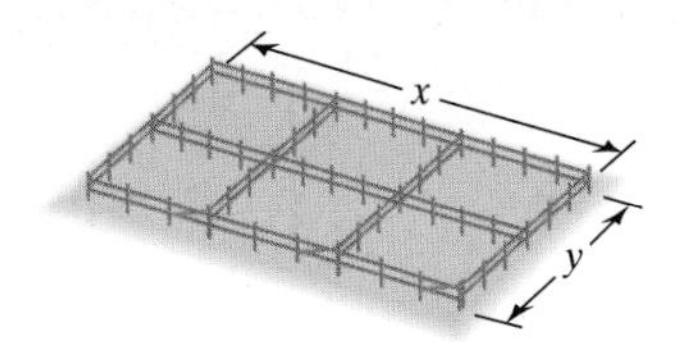

Group Exercise

70. Each group member should consult an almanac, newspaper, magazine, or the Internet to find data that can be modeled by a quadratic function. Group members should select the two sets of data that are most interesting and relevant. For each data set selected:

a. Use the quadratic regression feature of a graphing utility to find the quadratic function that best fits the data.

b. Use the equation of the quadratic function to make a prediction from the data. What circumstances might affect the accuracy of your prediction?

c. Use the equation of the quadratic function to write and solve a problem involving maximizing or minimizing the function.

SECTION 3.2 Polynomial Functions and Their Graphs

Objectives

1. Recognize characteristics of graphs of polynomial functions.
2. Determine end behavior.
3. Use factoring to find zeros of polynomial functions.
4. Identify the multiplicity of a zero.
5. Understand the relationship between degree and turning points.
6. Graph polynomial functions.

Magnified 6000 times, this color-scanned image shows a T-lymphocyte blood cell (green) infected with the HIV virus (red). Depletion of the number of T-cells causes destruction of the immune system.

In 1980, U.S. doctors diagnosed 41 cases of a rare form of cancer, Kaposi's sarcoma, that involved skin lesions, pneumonia, and severe immunological deficiencies. All cases involved gay men ranging in age from 26 to 51. By the end of 1998, approximately 680,000 Americans, straight and gay, male and female, old and young, were infected with the HIV virus.

Modeling AIDS-related data and making predictions about the epidemic's havoc is serious business. Changing circumstances and unforeseen events have resulted in models that are not particularly useful over long periods of time. For example, the function

$$f(x) = -143x^3 + 1810x^2 - 187x + 2331$$

models the number of AIDS cases diagnosed in the United States x years after 1983. The model was obtained using cases diagnosed from 1983 through 1991. Figure 3.8 shows the graph of f from 1983 through 1991 in a $[0, 8, 1]$ by $[0, 50{,}000, 5000]$ viewing rectangle. The function used to describe what was happening with new HIV infections over a limited period of time is an example of a **polynomial function**.

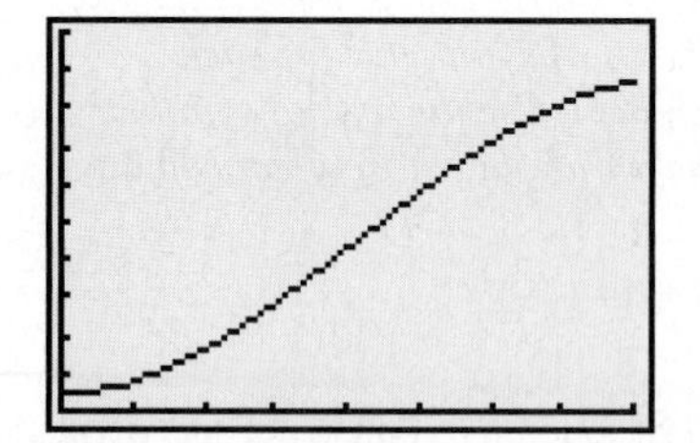

Figure 3.8 The graph of a function modeling the number of AIDS cases from 1983 through 1991

Definition of a Polynomial Function

Let n be a nonnegative integer and let $a_n, a_{n-1}, \dots, a_2, a_1, a_0$, be real numbers with $a_n \neq 0$. The function defined by

$$f(x) = a_n x^n + a_{n-1}x^{n-1} + \cdots + a_2 x^2 + a_1 x + a_0$$

is called a **polynomial function of x of degree n.** The number a_n, the coefficient of the variable to the highest power, is called the **leading coefficient**.

A constant function $f(x) = a$, where $a \neq 0$, is a polynomial function of degree 0. A linear function $f(x) = ax + b$, where $a \neq 0$, is a polynomial function of degree 1. A quadratic function $f(x) = ax^2 + bx + c$, where $a \neq 0$, is a polynomial function of degree 2. In this section, we focus on polynomial functions of degree 3 or higher.

1 Recognize characteristics of graphs of polynomial functions.

Smooth, Continuous Graphs

Polynomial functions of degree 2 or less have graphs that are either parabolas or lines. We can graph such functions by plotting points. We can also graph polynomial functions of degree 3 or higher by plotting points. However, the process is rather tedious: Many points must be plotted. It may be easier to use a graphing utility for such functions. Regardless of the graphing method you use, you will find an ability to recognize the basic features of polynomial functions helpful. For example, they may help you choose an appropriate viewing rectangle for a graphing utility.

Two important features of the graphs of polynomial functions are that they are *smooth* and *continuous*. By **smooth**, we mean that the graph contains only rounded curves with no sharp corners. By **continuous**, we mean that the graph has no breaks and can be drawn without lifting your pencil from the rectangular coordinate system. These ideas are illustrated in Figure 3.9.

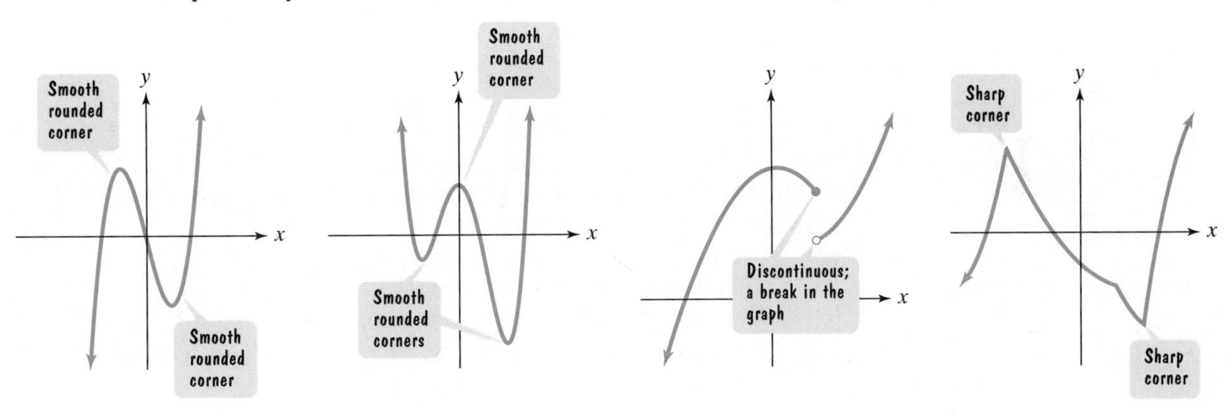

Figure 3.9 Recognizing graphs of polynomial functions

2 Determine end behavior.

End Behavior of Polynomial Functions

Figure 3.10 shows the graph of the function

$$f(x) = -143x^3 + 1810x^2 - 187x + 2331,$$

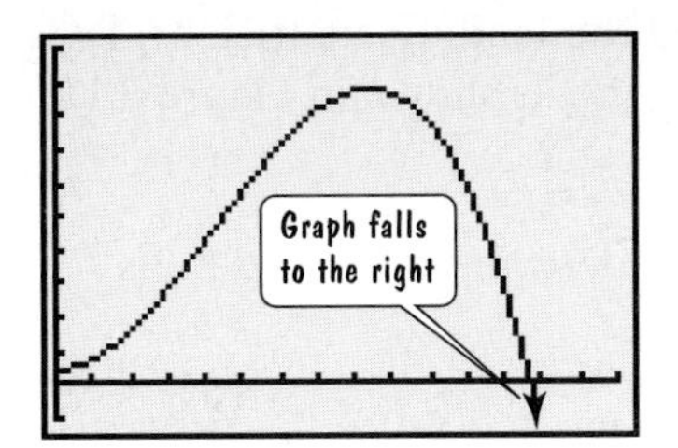

Figure 3.10 By extending the viewing rectangle, y is negative and the function no longer models the number of AIDS cases.

which models U.S. AIDS cases from 1983 through 1991. Look what happens to the graph when we extend the year up through 1998 with a $[0, 15, 1]$ by $[-5000, 50{,}000, 5000]$ viewing rectangle. By year 13 (1996), the values of y are negative and the function no longer models AIDS cases. We've added an arrow to the graph at the far right to emphasize that it continues to decrease without bound. It is this far-right *end behavior* of the graph that makes it inappropriate for modeling AIDS cases into the future.

The behavior of a graph of a function to the far left or the far right is called its **end behavior**. Although the graph of a polynomial function may have intervals where it increases or decreases, the graph will eventually rise or fall without bound as it moves far to the left or far to the right.

How can you determine whether the graph of a polynomial function goes up or down at each end? The end behavior of a polynomial function

$$f(x) = a_n x^n + a_{n-1}x^{n-1} + \cdots + a_1 x + a_0$$

depends upon the leading term $a_n x^n$. In particular, the sign of the leading coefficient a_n, and the degree, n, of the polynomial function reveal its end behavior. In terms of end behavior, only the term of highest degree counts, summarized by the **Leading Coefficient Test**.

The Leading Coefficient Test

As x increases or decreases without bound, the graph of the polynomial function

$$f(x) = a_n x^n + a_{n-1}x^{n-1} + a_{n-2}x^{n-2} + \cdots + a_1 x + a_0 \quad (a_n \neq 0)$$

eventually rises or falls. In particular,

1. For n odd:

If the leading coefficient is positive, the graph falls to the left and rises to the right.

$a_n > 0$

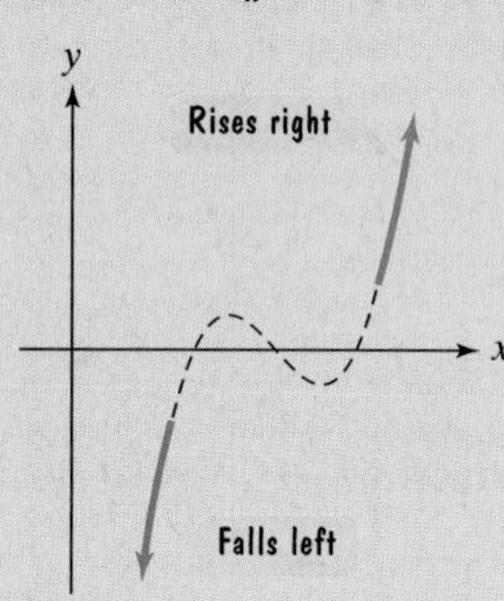

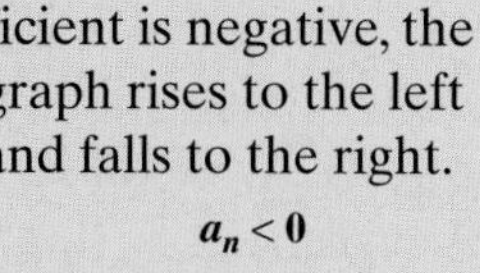
If the leading coefficient is negative, the graph rises to the left and falls to the right.

$a_n < 0$

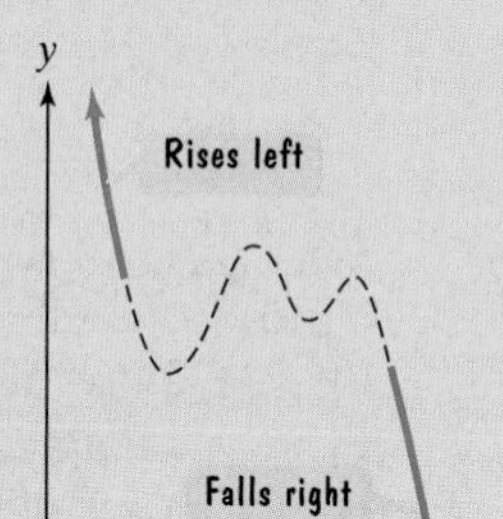

2. For n even:

If the leading coefficient is positive, the graph rises to the left and to the right.

$a_n > 0$

If the leading coefficient is negative, the graph falls to the left and to the right.

$a_n < 0$

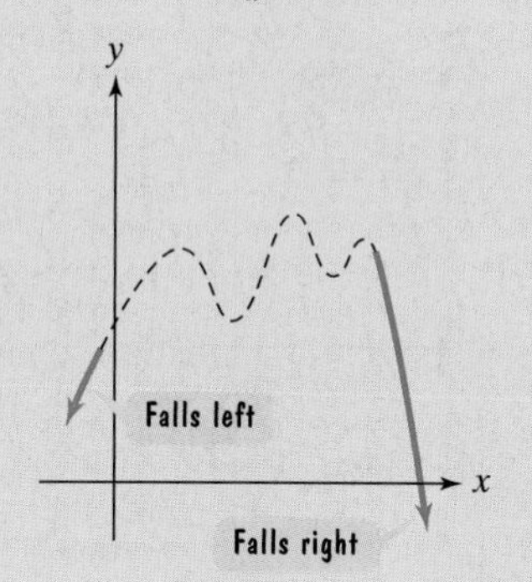

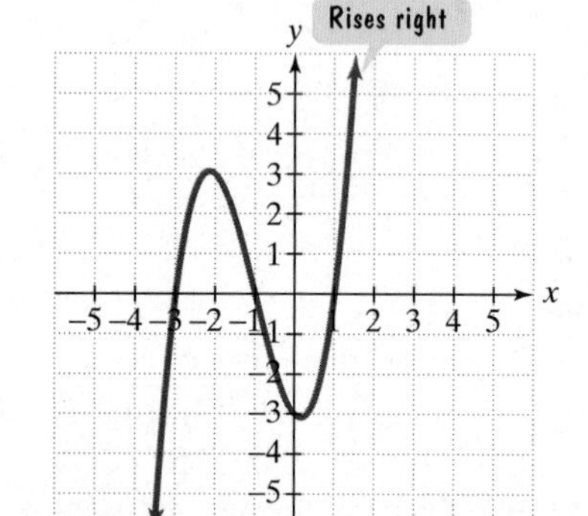

Figure 3.11 The graph of $f(x) = x^3 + 3x^2 - x - 3$

EXAMPLE 1 Using the Leading Coefficient Test

Use the Leading Coefficient Test to determine the end behavior of the graph of

$$f(x) = x^3 + 3x^2 - x - 3.$$

Solution Because the degree is odd ($n = 3$) and the leading coefficient, 1, is positive, the graph falls to the left and rises to the right, as shown in Figure 3.11.

Check Point 1 Use the Leading Coefficient Test to determine the end behavior of the graph of $f(x) = x^4 - 4x^2$.

EXAMPLE 2 Using the Leading Coefficient Test

Use end behavior to explain why

$$f(x) = -143x^3 + 1810x^2 - 187x + 2331$$

is only an appropriate model for AIDS cases for a limited time period.

Solution Because the degree is odd ($n = 3$) and the leading coefficient, -143, is negative, the graph rises to the left and falls to the right. The fact that it falls to the right indicates at some point the number of AIDS cases will be negative, an impossibility. No function with a graph that decreases without bound as x (time) increases can model nonnegative real-world phenomena over a long period of time.

Check Point 2

The polynomial function

$$f(x) = -0.27x^3 + 9.2x^2 - 102.9x + 400$$

models the ratio of students to computers in U.S. public schools x years after 1980. Use end behavior to determine whether this function could be an appropriate model for computers in the classroom well into the twenty-first century. Explain your answer.

If you use a graphing utility to graph a polynomial function, it is important to select a viewing rectangle that accurately reveals the graph's end behavior. If the viewing rectangle is too small, it may not accurately show end behavior.

EXAMPLE 3 Using the Leading Coefficient Test

The graph of $f(x) = -x^4 + 8x^3 + 4x^2 + 2$ was obtained with a graphing utility using a $[-8, 8, 1]$ by $[-10, 10, 1]$ viewing rectangle. The graph is shown in Figure 3.12(a). Does the graph show the end behavior of the function?

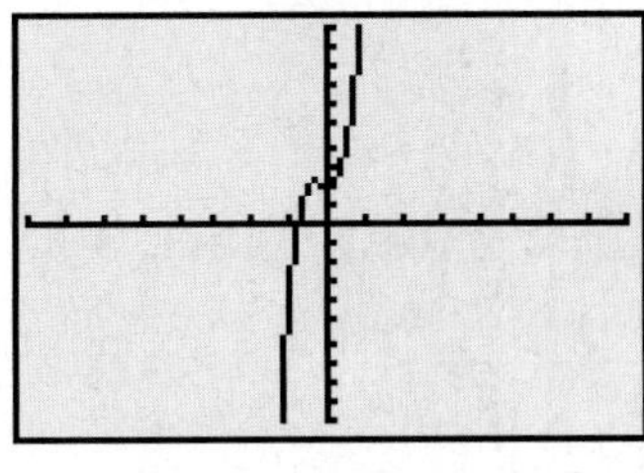

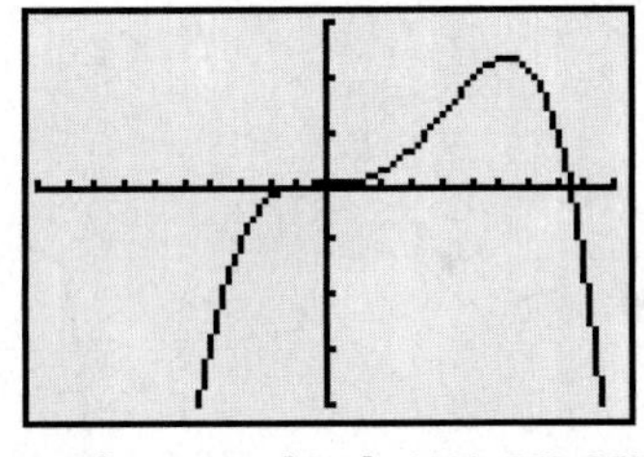

Figure 3.12 **(a)** $[-8, 8, 1]$ by $[-10, 10, 1]$ **(b)** $[-10, 10, 1]$ by $[-1000, 750, 250]$

Solution Note that the degree is even ($n = 4$) and the leading coefficient, -1, is negative. Thus, the Leading Coefficient Test indicates that the graph should fall to the left and the right. The graph in Figure 3.12(a) is falling to the left, but it is not falling to the right. Therefore, the graph is not complete enough to show end behavior. A more complete graph of the function is shown in a larger viewing rectangle in Figure 3.12(b).

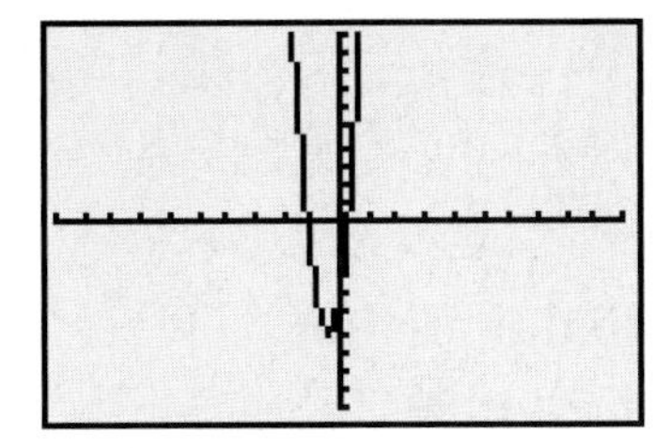

Figure 3.13

Check Point 3

The graph of $f(x) = x^3 + 13x^2 + 10x - 4$ is shown in a standard viewing rectangle in Figure 3.13. Use the Leading Coefficient Test to determine whether the graph shows the end behavior of the function. Explain your answer.

3 Use factoring to find zeros of polynomial functions.

Zeros of Polynomial Functions

If f is a polynomial function, then the values of x for which $f(x)$ is equal to 0 are called the **zeros** of f. These values of x are the **roots** of the polynomial equation $f(x) = 0$. Each real root of the polynomial equation appears as an x-intercept of the graph of the polynomial function.

EXAMPLE 4 Finding Zeros of a Polynomial Function

Find all zeros of $f(x) = x^3 + 3x^2 - x - 3$.

Solution By definition, the zeros are the values of x for which $f(x)$ is equal to 0. Thus, we set $f(x)$ equal to 0:

$$f(x) = x^3 + 3x^2 - x - 3 = 0.$$

We solve the polynomial equation $x^3 + 3x^2 - x - 3 = 0$ for x as follows:

$x^3 + 3x^2 - x - 3 = 0$ This is the equation needed to find the function's zeros.

$x^2(x + 3) - 1(x + 3) = 0$ Factor x^2 from the first two terms and -1 from the last two terms.

$(x + 3)(x^2 - 1) = 0$ A common factor of $x + 3$ is factored from the expression.

$x + 3 = 0$ or $x^2 - 1 = 0$ Set each factor equal to 0.

$x = -3$ $\quad x^2 = 1$ Solve for x.

$x = \pm 1$ Remember that if $x^2 = d$, then $x = \pm\sqrt{d}$.

The zeros of f are -3, -1, and 1. The graph of f in Figure 3.14 shows that each zero is an x-intercept.

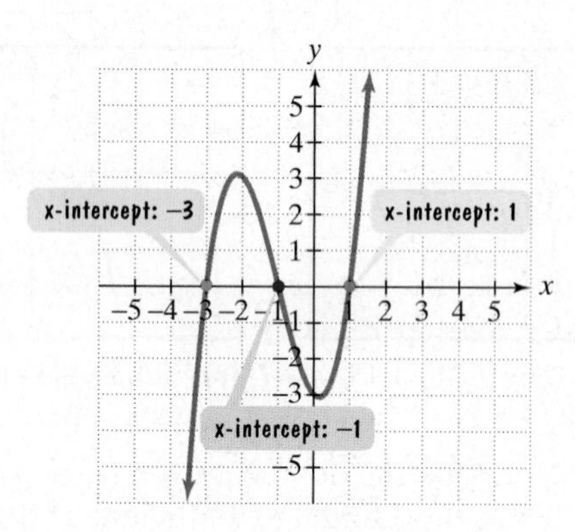

Figure 3.14 The real zeros of $f(x) = x^3 + 3x^2 - x - 3$ are the x-intercepts for the graph of f.

Check Point 4 Find all zeros of $f(x) = x^3 + 2x^2 - 4x - 8$.

EXAMPLE 5 Finding Zeros of a Polynomial Function

Find all zeros of $f(x) = -x^4 + 4x^3 - 4x^2$.

Solution We find the zeros of f by setting $f(x)$ equal to 0.

$-x^4 + 4x^3 - 4x^2 = 0$ We now have a polynomial equation.

$x^4 - 4x^3 + 4x^2 = 0$ Multiply both sides by -1. This step is optional.

$x^2(x^2 - 4x + 4) = 0$ Factor out x^2.

$x^2(x - 2)^2 = 0$ Factor completely.

$x^2 = 0$ or $(x - 2)^2 = 0$ Set each factor equal to 0.

$x = 0$ $\quad x = 2$ Solve for x.

The zeros of $f(x) = -x^4 + 4x^3 - 4x^2$ are 0 and 2. The graph of f, shown in Figure 3.15 has x-intercepts at 0 and 2.

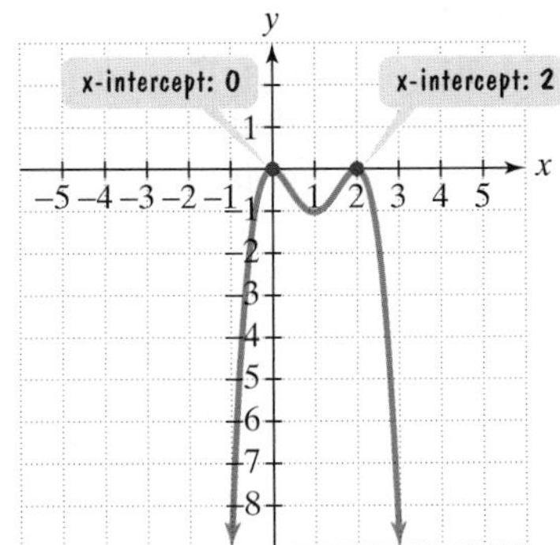

Figure 3.15 The zeros of $f(x) = -x^4 + 4x^3 - 4x^2$, namely 0 and 2, are the x-intercepts for the graph of f.

Check Point 5 Find all zeros of $f(x) = x^4 - 4x^2$.

In Example 5, we can use our factoring to express the function's equation as follows:

$$f(x) = -x^4 + 4x^3 - 4x^2 = -(x^4 - 4x^3 + 4x^2) = -x^2(x - 2)^2$$

The factor x occurs twice: $x^2 = x \cdot x$.

The factor $(x - 2)$ occurs twice: $(x - 2)^2 = (x - 2)(x - 2)$.

4 Identify the multiplicity of a zero.

Notice that each factor occurs twice. In factoring the equation for the polynomial function f, if the same factor $x - r$ occurs k times, but not $k + 1$ times, we call r a **repeated zero with multiplicity k**. For the polynomial

$$f(x) = -x^2(x - 2)^2,$$

0 and 2 are both repeated zeros with multiplicity 2. For the polynomial

$$f(x) = 4(x - 5)(x + 2)^3\left(x - \tfrac{1}{4}\right)^4,$$

5 is a zero with multiplicity 1, -2 is a repeated zero with multiplicity 3, and $\frac{1}{4}$ is a repeated zero with multiplicity 4.

The multiplicity of a zero tells us if the graph of a polynomial function touches the x-axis at the zero and turns around or crosses the x-axis at the zero. For example, look again at the graph of $f(x) = -x^4 + 4x^3 - 4x^2$ in Figure 3.15. Each zero, 0 and 2, is a repeated zero with multiplicity 2. The graph of f touches, but does not cross, the x-axis at each of these zeros of even multiplicity. By contrast, a graph crosses the x-axis at zeros of odd multiplicity.

Multiplicity and x-Intercepts

If r is a zero of even multiplicity, then the graph **touches** the x-axis and turns around at r. If r is a zero of odd multiplicity, then the graph **crosses** the x-axis at r. Regardless of whether a zero is even or odd, graphs tend to flatten out at zeros with multiplicity greater than one.

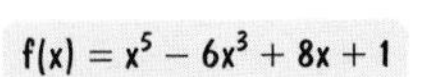

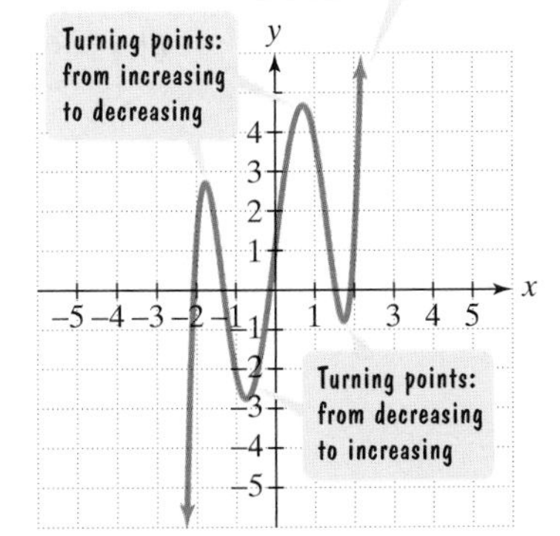

Figure 3.16 Graph with four turning points

5 Understand the relationship between degree and turning points.

Turning Points of Polynomial Functions

The graph of $f(x) = x^5 - 6x^3 + 8x + 1$ is shown in Figure 3.16. The graph has four smooth **turning points**. At each turning point, the graph changes direction from increasing to decreasing or vice versa. In calculus, these points are called **local maxima** or **local minima**. The given equation has 5 as its greatest exponent and is therefore a polynomial function of degree 5. Notice that the graph has four turning points. In general, **if f is a polynomial of degree n, then the graph of f has at most $n - 1$ turning points.**

6 Graph polynomial functions.

A Strategy for Graphing Polynomial Functions

Here's a general strategy for graphing a polynomial function. A graphing utility is a valuable complement to this strategy. Some of the steps listed in the following box will help you to select a viewing rectangle that shows the important parts of the graph.

Graphing a Polynomial Function

$$f(x) = a_n x^n + a_{n-1}x^{n-1} + a_{n-2}x^{n-2} + \cdots + a_1 x + a_0, \quad a_n \neq 0$$

1. Use the Leading Coefficient Test to determine the graph's end behavior.
2. Find x-intercepts by setting $f(x) = 0$ and solving the resulting polynomial equation. If there is an x-intercept at r as a result of $(x - r)^k$ in the complete factorization of $f(x)$, then:
   a. If k is even, the graph touches the x-axis at r and turns around.
   b. If k is odd, the graph crosses the x-axis at r.
   c. If $k > 1$, the graph flattens out at $(r, 0)$.
3. Find the y-intercept by setting x equal to 0 and computing $f(0)$.
4. Use symmetry, if applicable, to help draw the graph:
   a. y-axis symmetry: $f(-x) = f(x)$
   b. Origin symmetry: $f(-x) = -f(x)$
5. Use the fact that the maximum number of turning points of the graph is $n - 1$ to check whether it is drawn correctly.

EXAMPLE 6 Graphing a Polynomial Function

Graph: $f(x) = x^4 - 2x^2 + 1$.

Solution

Step 1 Determine end behavior. Because the degree is even ($n = 4$) and the leading coefficient, 1, is positive, the graph rises to the left and the right:

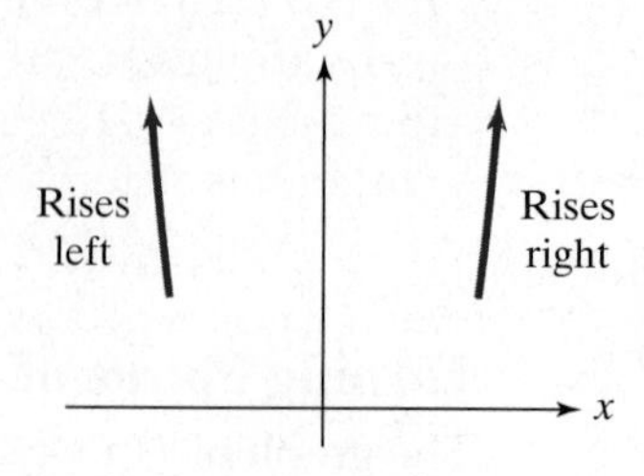

Step 2 Find x-intercepts (zeros of the function) by setting $f(x) = 0$.

$$x^4 - 2x^2 + 1 = 0$$

$$(x^2 - 1)(x^2 - 1) = 0 \quad \text{Factor.}$$

$$(x + 1)(x - 1)(x + 1)(x - 1) = 0 \quad \text{Factor completely.}$$

$$(x + 1)^2(x - 1)^2 = 0 \quad \text{Express the factoring in a more compact notation.}$$

$$(x + 1)^2 = 0 \quad \text{or} \quad (x - 1)^2 = 0 \quad \text{Set each factor equal to 0.}$$

$$x = -1 \qquad\qquad x = 1 \quad \text{Solve for x.}$$

We see that -1 and 1 are both repeated zeros with multiplicity 2. Because of the even multiplicity, the graph touches the x-axis at -1 and 1 and turns around. Furthermore, the graph tends to flatten out at these zeros with multiplicity greater than one:

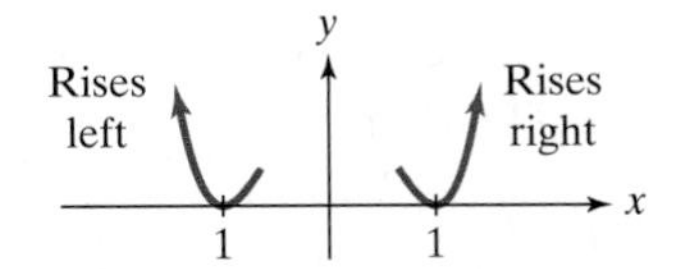

Step 3 Find the y-intercept by setting x equal to 0.

$$f(0) = 0^4 - 2 \cdot 0^2 + 1 = 1$$

There is a y-intercept at 1, so the graph passes through $(0, 1)$:

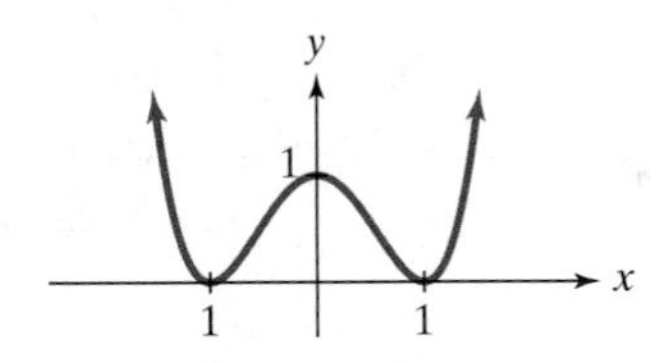

Step 4 Use possible symmetry to help draw the graph. Our partial graph suggests y-axis symmetry. Let's verify this by finding $f(-x)$.

$$f(x) = x^4 - 2x^2 + 1$$

Replace x with −x.

$$f(-x) = (-x)^4 - 2(-x)^2 + 1 = x^4 - 2x^2 + 1$$

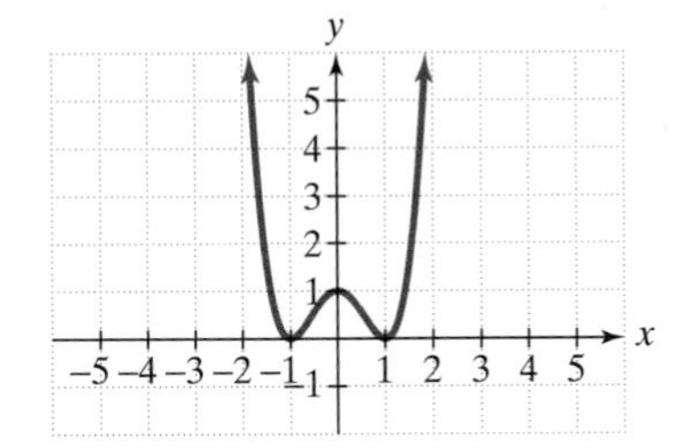

Figure 3.17 The graph of $f(x) = x^4 - 2x^2 + 1$

Because $f(-x) = f(x)$, the graph of f is symmetric with respect to the y-axis. Figure 3.17 shows the graph of $f(x) = x^4 - 2x^2 + 1$.

Step 5 Use the fact that the maximum number of turning points of the graph is $n - 1$ to check whether it is drawn correctly. Because $n = 4$, the maximum number of turning points is $4 - 1$, or 3. Because the graph in Figure 3.17 has three turning points, we have not violated the maximum number possible.

Check Point 6 Use the five-step strategy to graph $f(x) = x^3 - 3x^2$.

EXERCISE SET 3.2

Practice Exercises

In Exercises 1–10, determine which functions are polynomial functions. For those that are, identify the degree.

1. $f(x) = 5x^2 + 6x^3$
2. $f(x) = 7x^2 + 9x^4$
3. $g(x) = 7x^5 - \pi x^3 + \frac{1}{5}x$
4. $g(x) = 6x^7 + \pi x^5 + \frac{2}{3}x$
5. $h(x) = 7x^3 + 2x^2 + \frac{1}{x}$
6. $h(x) = 8x^3 - x^2 + \frac{2}{x}$
7. $f(x) = x^{1/2} - 3x^2 + 5$
8. $f(x) = x^{1/3} - 4x^2 + 7$
9. $f(x) = \frac{x^2 + 7}{x^3}$
10. $f(x) = \frac{x^2 + 7}{3}$

In Exercises 11–14, identify which graphs are not those of polynomial functions.

11.

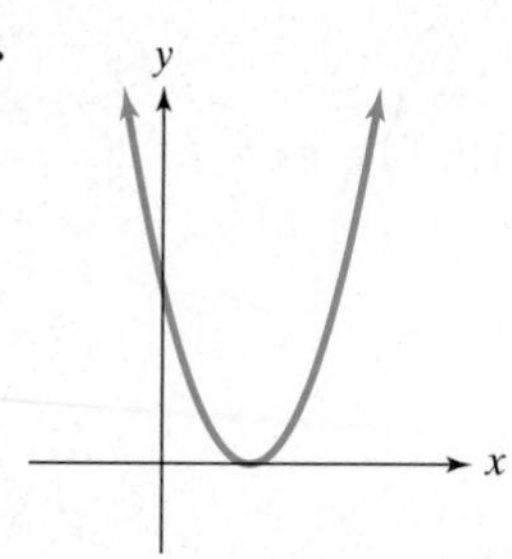

12.

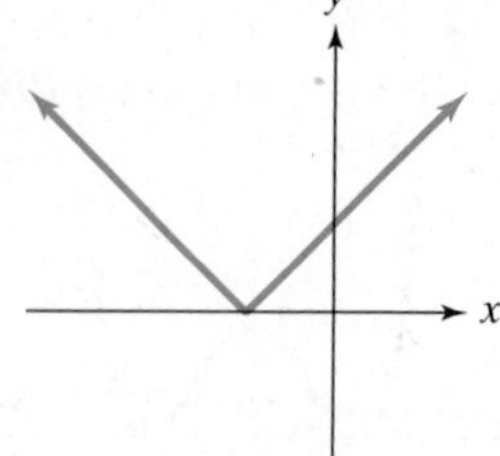

13.

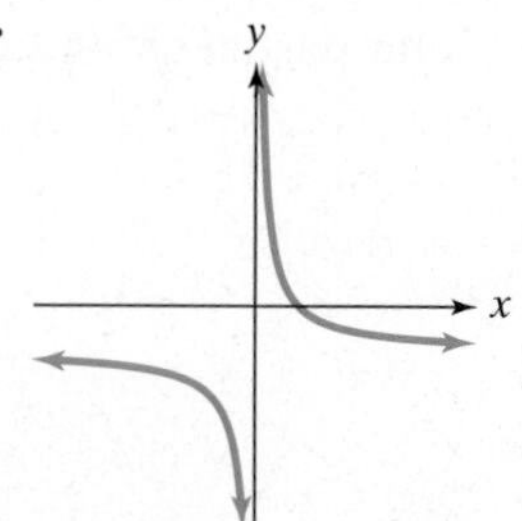

14.

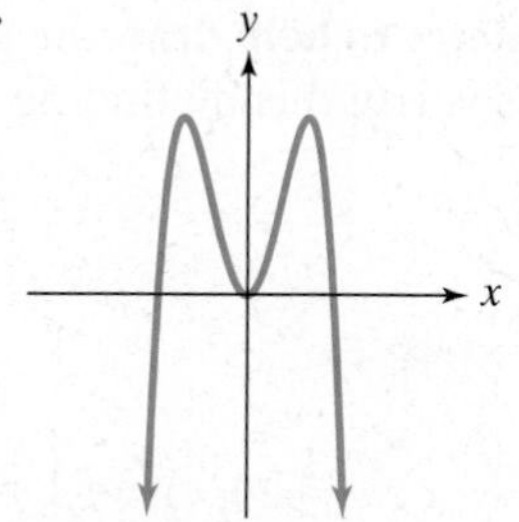

In Exercises 15–20, use end behavior to match the polynomial function with its graph. [The graphs are labeled (a) through (f).]

(a)

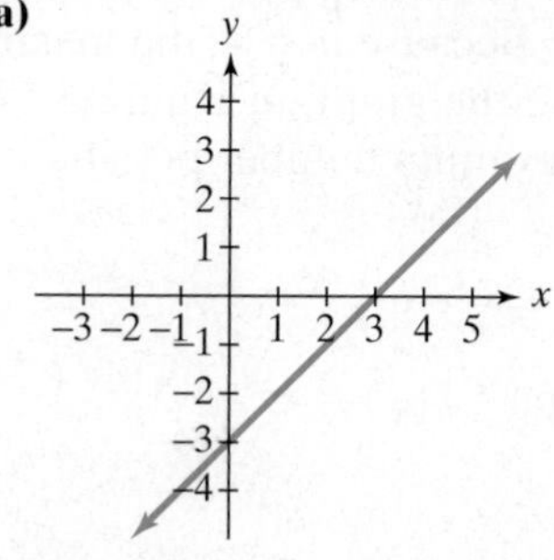

(b)

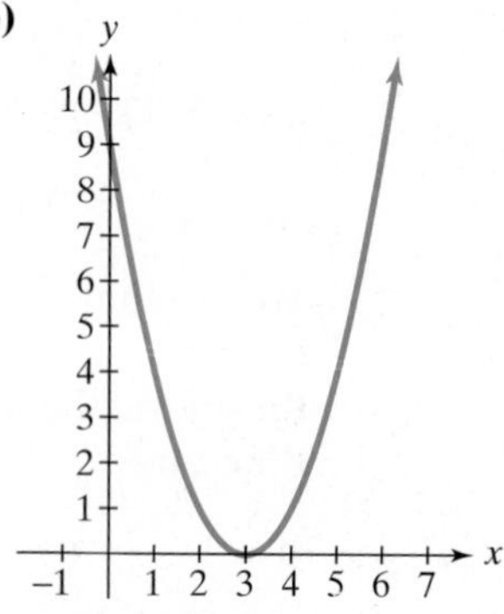

(c)

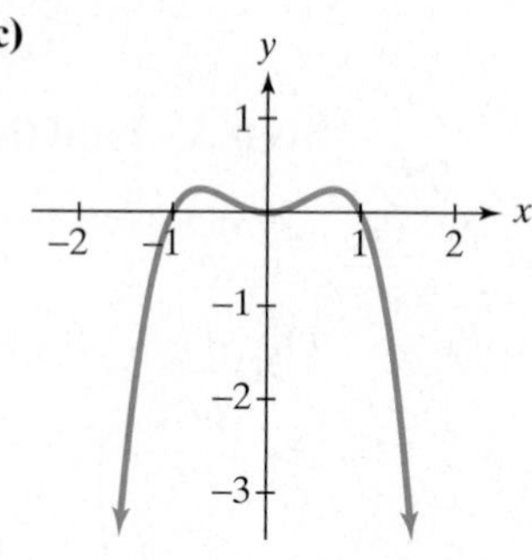

(d)

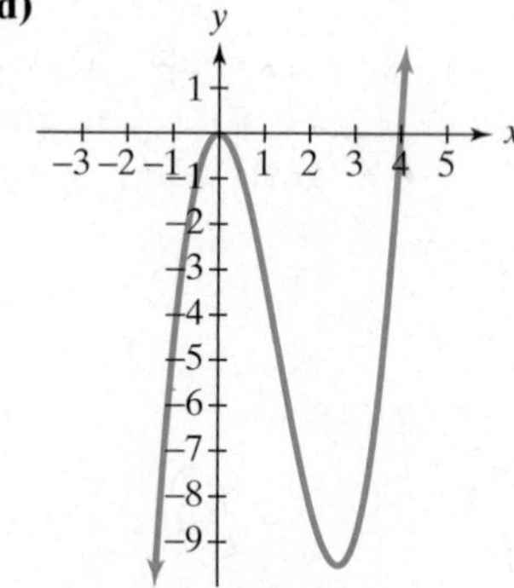

(e)

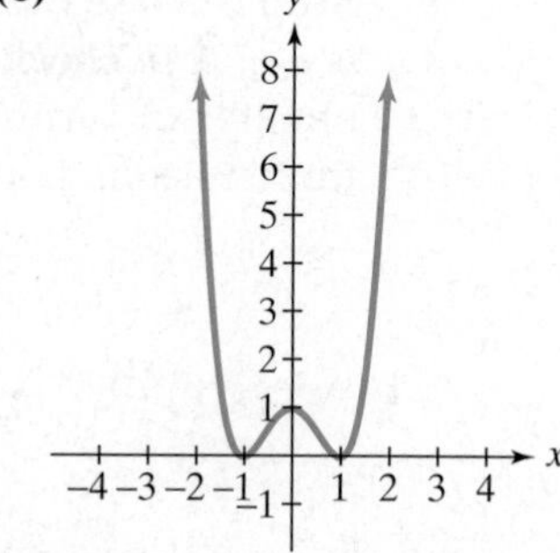

(f)

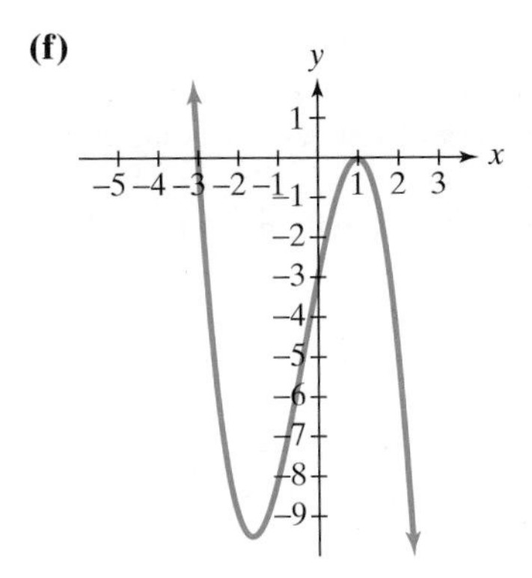

15. $f(x) = -x^4 + x^2$
16. $f(x) = x^3 - 4x^2$
17. $f(x) = (x - 3)^2$
18. $f(x) = -x^3 - x^2 + 5x - 3$
19. $f(x) = x - 3$
20. $f(x) = (x + 1)^2(x - 1)^2$

In Exercises 21–26, use the Leading Coefficient Test to determine the end behavior of the graph of the polynomial function.

21. $f(x) = 5x^3 + 7x^2 - x + 9$
22. $f(x) = 11x^3 - 6x^2 + x + 3$
23. $f(x) = 5x^4 + 7x^2 - x + 9$
24. $f(x) = 11x^4 - 6x^2 + x + 3$
25. $f(x) = -5x^4 + 7x^2 - x + 9$
26. $f(x) = -11x^4 - 6x^2 + x + 3$

In Exercises 27–34, find the zeros for each polynomial function and give the multiplicity for each zero. State whether the graph crosses the x-axis or touches the x-axis and turns around at each zero.

27. $f(x) = 2(x - 5)(x + 4)^2$
28. $f(x) = 3(x + 5)(x + 2)^2$
29. $f(x) = 4(x - 3)(x + 6)^3$
30. $f(x) = -3(x + \frac{1}{2})(x - 4)^3$
31. $f(x) = x^3 - 2x^2 + x$
32. $f(x) = x^3 + 4x^2 + 4x$
33. $f(x) = x^3 + 7x^2 - 4x - 28$
34. $f(x) = x^3 + 5x^2 - 9x - 45$

In Exercises 35–50,

a. *Use the Leading Coefficient Test to determine the graph's end behavior.*
b. *Find x-intercepts by setting $f(x) = 0$ and solving the resulting polynomial equation. State whether the graph crosses the x-axis or touches the x-axis and turns around at each intercept.*
c. *Find the y-intercept by setting x equal to 0 and computing $f(0)$.*
d. *Determine whether the graph has y-axis symmetry, origin symmetry, or neither.*
e. *If necessary, find a few additional points and graph the function. Use the fact that the maximum number of turning points of the graph is $n - 1$ to check whether it is drawn correctly.*

35. $f(x) = x^3 + 2x^2 - x - 2$
36. $f(x) = x^3 + x^2 - 4x - 4$
37. $f(x) = x^4 - 9x^2$
38. $f(x) = x^4 - x^2$
39. $f(x) = -x^4 + 16x^2$
40. $f(x) = -x^4 + 4x^2$
41. $f(x) = x^4 - 2x^3 + x^2$
42. $f(x) = x^4 - 6x^3 + 9x^2$
43. $f(x) = -2x^4 + 4x^3$
44. $f(x) = -2x^4 + 2x^3$
45. $f(x) = 6x^3 - 9x - x^5$
46. $f(x) = 6x - x^3 - x^5$
47. $f(x) = 3x^2 - x^3$
48. $f(x) = \frac{1}{2} - \frac{1}{2}x^4$
49. $f(x) = -3(x - 1)^2(x^2 - 4)$
50. $f(x) = -2(x - 4)^2(x^2 - 25)$

Application Exercises

51. A herd of 100 elk is introduced to a small island. The number of elk, $N(t)$, after t years is described by the polynomial function $N(t) = -t^4 + 21t^2 + 100$.
    a. Use the Leading Coefficient Test to determine the graph's end behavior to the right. What does this mean about what will eventually happen to the elk population?
    b. Graph the function.
    c. Graph only the portion of the function that serves as a realistic model for the elk population over time. When does the population become extinct?
52. The common cold is caused by a rhinovirus. After t days of invasion by the viral particles, there are N billion particles in our bodies, modeled by $N(t) = -\frac{3}{4}t^4 + 3t^3 + 5$. Use the Leading Coefficient Test to determine the graph's end behavior to the right. What does this mean about the number of viral particles in our bodies over time?
53. The following table shows the number of larceny thefts in the United States for the years 1988–1998, where 1 represents 1988, 2 represents 1989, and so on.

Year, x	Larceny Thefts, T, in thousands
1988, 1	7706
1989, 2	7872
1990, 3	7946
1991, 4	8142
1992, 5	7915
1993, 6	7821
1994, 7	7880
1995, 8	7998
1996, 9	7905
1997, 10	7744
1998, 11	7374

Using the polynomial regression feature of a graphing utility, the third-degree polynomial function of best fit for the data is

$$T(x) = -0.87x^3 + 0.35x^2 + 81.62x + 7684.94.$$

a. Use this function to predict the number of larceny thefts in 2005.

b. Will this function be useful in modeling the number of larceny thefts over an extended period of time? Explain your answer.

54. Suppose that a polynomial function is used to model the data shown in the graph using

(number of years after 1900, murder rate per 100,000 people).

Murders Per 100,000 People in the United States

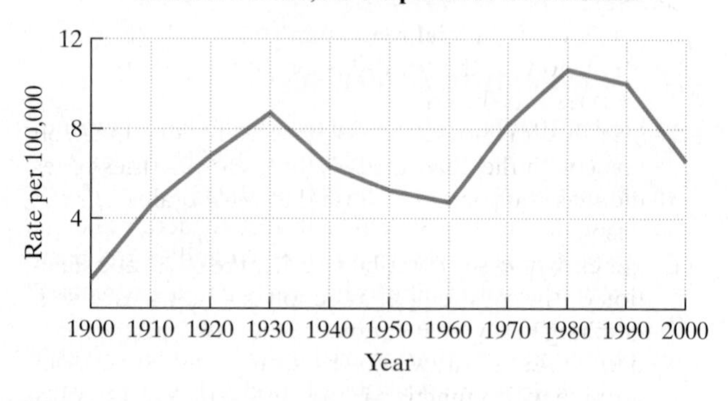

Source: National Center for Health Statistics

Determine the degree of the polynomial function of best fit. Should the leading coefficient be positive or negative? Explain your answers.

55. The polynomial function

$$H(x) = -0.001183x^4 + 0.05495x^3 - 0.8523x^2 + 9.054x + 6.748$$

models the age in human years, $H(x)$, of a dog that is x years old, where $x \geq 1$.

a. Use this function to find the equivalent age in human years for a 10-year-old dog.

b. If dogs lived as long as humans, would this function be useful in modeling the dog's equivalent age in human years? Explain your answer.

Writing in Mathematics

56. What is a polynomial function?

57. What do we mean when we describe the graph of a polynomial function as smooth and continuous?

58. What is meant by the end behavior of a polynomial function?

59. Explain how to use the Leading Coefficient Test to determine the end behavior of a polynomial function.

60. Why is a third-degree polynomial function with a negative leading coefficient not appropriate for modeling nonnegative real-world phenomena over a long period of time?

61. What are the zeros of a polynomial function and how are they found?

62. Explain the relationship between the multiplicity of a zero and whether or not the graph crosses or touches the x-axis at that zero.

63. Explain the relationship between the degree of a polynomial and the number of turning points on its graph.

64. Can the graph of a polynomial function have no x-intercepts? Explain.

65. Can the graph of a polynomial function have no y-intercept? Explain.

66. Describe a strategy for graphing a polynomial function. In your description, mention intercepts, the polynomial's degree, and turning points.

67. In a favorable habitat and without natural predators, a population of reindeer is introduced to an island preserve. The reindeer population t years after their introduction is modeled by the polynomial function $f(t) = -0.125t^5 + 3.125t^4 + 4000$. Discuss the growth and decline of the reindeer population. Describe the factors that might contribute to this population model.

Technology Exercises

68. Use a graphing utility to verify any five of the graphs that you drew by hand in Exercises 35–50.

Write a polynomial function that imitates the end behavior of each graph in Exercises 69–72. The dashed portions of the graphs indicate that you should focus only on imitating the left and right behavior of the graph and can be flexible about what occurs between the left and right ends. Then use your graphing utility to graph the polynomial function and verify that you imitated the end behavior shown in the given graph.

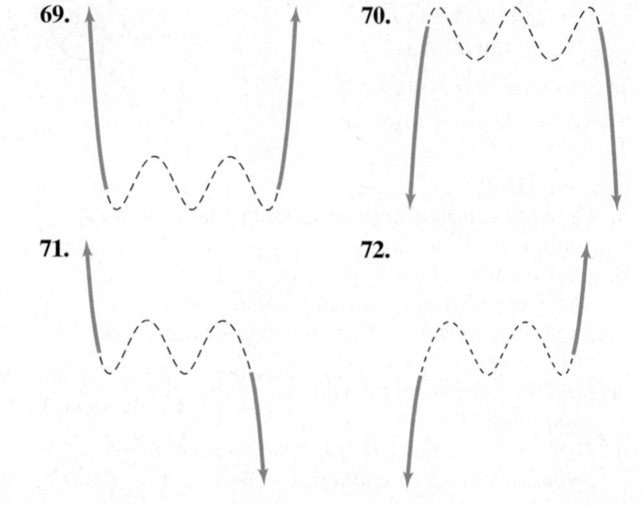

In Exercises 73–76, use a graphing utility with a viewing rectangle large enough to show end behavior to graph each polynomial function.

73. $f(x) = x^3 + 13x^2 + 10x - 4$

74. $f(x) = -2x^3 + 6x^2 + 3x - 1$

75. $f(x) = -x^4 + 8x^3 + 4x^2 + 2$

76. $f(x) = -x^5 + 5x^4 - 6x^3 + 2x + 20$

For Exercises 77–78, use a graphing utility to graph f and g in the same viewing rectangle. Then use the ZOOM OUT *feature to show that the end behavior of f and g is identical.*

77. $f(x) = x^3 - 6x + 1, \quad g(x) = x^3$

78. $f(x) = -x^4 + 2x^3 - 6x, \quad g(x) = -x^4$

Critical Thinking Exercises

79. Which one of the following is true?

a. If $f(x) = -x^3 + 4x$, then the graph of f falls to the left and to the right.

b. A mathematical model that is a polynomial of degree n whose leading term is $a_n x^n$, n odd and $a_n < 0$, is ideally suited to describe phenomena over unlimited periods of time.

c. There is more than one third-degree polynomial function with the same three x-intercepts.

d. The graph of a function with origin symmetry can rise to the left and to the right.

Use the descriptions in Exercises 80–81 to write an equation of a polynomial function with the given characteristics. Use a graphing utility to graph your function to see if you are correct. If not, modify the function's equation and repeat this process.

80. Crosses the x-axis at $-4, 0$, and 3; lies above the x-axis between -4 and 0; lies below the x-axis between 0 and 3

81. Touches the x-axis at 0 and crosses the x-axis at 2; lies below the x-axis between 0 and 2

Group Exercise

82. This exercise is based on the group's work in Exercise 70 of Exercise Set 3.1. For the two data sets that the group selected:

a. Use the polynomial regression feature of a graphing utility to find the third-degree polynomial function that best fits the data.

b. Use this function to repeat the predictions that you made with the quadratic function. How do these predictions compare with those that you obtained previously?

c. For each data set, describe whether the quadratic function or the third-degree function is a better fit. Use a graphing utility, a scatter plot of the data, and the function of best fit drawn on the scatter plot to help determine which function is the better fit.

SECTION 3.3 Dividing Polynomials; Remainder and Factor Theorems

Objectives

1. Use long division to divide polynomials.
2. Use synthetic division to divide polynomials.
3. Evaluate a polynomial using the Remainder Theorem.
4. Use the Factor Theorem to solve a polynomial equation.

For those of you who are dog lovers, you might still be thinking of the polynomial function that models the age in human years, $H(x)$, of a dog that is x years old, namely

$$H(x) = -0.001183x^4 + 0.05495x^3 - 0.8523x^2 + 9.054x + 6.748.$$

Suppose that you are in your twenties, say 25. What is Fido's equivalent age? To answer this question, we must substitute 25 for $H(x)$ and solve the resulting polynomial equation for x:

$$25 = -0.001183x^4 + 0.05495x^3 - 0.8523x^2 + 9.054x + 6.748.$$

How can we solve such an equation? You might begin by subtracting 25 from both sides to obtain zero on one side. But then what? The factoring that we used in the previous section will not work in this situation.

In Sections 3.4 and 3.5, we will present techniques for solving certain kinds of polynomial equations. These techniques will further enhance your ability to manipulate algebraically the formulas that model your world. Because these techniques are based on understanding polynomial division, in this section we look at two methods for dividing polynomials.

1 Use long division to divide polynomials.

Long Division of Polynomials and the Division Algorithm

We begin by looking at division by a polynomial containing more than one term, such as

$$x + 3\overline{)x^2 + 10x + 21}$$

Divisor has two terms.

Dividend has three terms.

When a divisor has more than one term, the four steps used to divide whole numbers—**divide**, **multiply**, **subtract**, **bring down the next term**—form the repetitive procedure for polynomial long division.

EXAMPLE 1 Long Division of Polynomials

Divide $x^2 + 10x + 21$ by $x + 3$.

Solution The following steps illustrate how polynomial division is very similar to numerical division.

$$x + 3\overline{)x^2 + 10x + 21}$$

Arrange the terms of the dividend $(x^2 + 10x + 21)$ and the divisor $(x + 3)$ in descending powers of x.

$$\begin{array}{r} x \\ x + 3\overline{)x^2 + 10x + 21} \end{array}$$

Divide x^2 (the first term in the dividend) by x (the first term in the divisor): $\frac{x^2}{x} = x$. Align like terms.

$$\begin{array}{r} x \\ x + 3\overline{)x^2 + 10x + 21} \\ \underline{x^2 + 3x} \end{array}$$

(times; equals)

Multiply each term in the divisor $(x + 3)$ by x, aligning terms of the product under like terms in the dividend.

$$\begin{array}{r} x \\ x + 3\overline{)x^2 + 10x + 21} \\ \underline{x^2 + 3x} \\ 7x \end{array}$$

Subtract $x^2 + 3x$ from $x^2 + 10x$ by changing the sign of each term in the lower expression and adding.

$$\begin{array}{r} x \\ x + 3\overline{)x^2 + 10x + 21} \\ \underline{x^2 + 3x} \quad \downarrow \\ 7x + 21 \end{array}$$

Bring down 21 from the original dividend and add algebraically to form a new dividend.

$$\begin{array}{r} x + 7 \\ x+3\overline{)x^2 + 10x + 21} \\ \underline{x^2 + 3x} \quad \downarrow \\ 7x + 21 \end{array}$$

Find the second term of the quotient. **Divide** the first term of $7x + 21$ by x, the first term of the divisor: $\frac{7x}{x} = 7$.

times

$$\begin{array}{r} x + 7 \\ x+3\overline{)x^2 + 10x + 21} \\ \underline{x^2 + 3x} \\ 7x + 21 \\ \underline{7x + 21} \\ 0 \end{array}$$

equals

Multiply the divisor $(x + 3)$ by 7, aligning under like terms in the new dividend. Then **subtract** to obtain the remainder of 0.

Remainder

The quotient is $x + 7$. Because the remainder is 0, we can conclude that $x + 3$ is a factor of $x^2 + 10x + 21$ and

$$x^2 + 10x + 21 = (x + 3)(x + 7).$$

Check Point 1 Divide $x^2 + 14x + 45$ by $x + 9$.

Before considering additional examples, let's summarize the general procedure for dividing one polynomial by another.

Long Division of Polynomials

1. **Arrange the terms** of both the dividend and the divisor in descending powers of any variable.
2. **Divide** the first term in the dividend by the first term in the divisor. The result is the first term of the quotient.
3. **Multiply** every term in the divisor by the first term in the quotient. Write the resulting product beneath the dividend with like terms lined up.
4. **Subtract** the product from the dividend.
5. **Bring down** the next term in the original dividend and write it next to the remainder to form a new dividend.
6. Use this new expression as the dividend and repeat this process until the remainder can no longer be divided. This will occur when the degree of the remainder (the highest exponent on a variable in the remainder) is less than the degree of the divisor.

In our next long division, we will obtain a nonzero remainder.

EXAMPLE 2 Long Division of Polynomials

Divide $4 - 5x - x^2 + 6x^3$ by $3x - 2$.

Solution We begin by writing the divisor and dividend in descending powers of x.

Is It Hot in Here, or Is It Just Me?

In the 1980s, a rising trend in global surface temperature was observed and the term "global warming" was coined. Scientists are more convinced than ever that burning coal, oil, and gas results in a buildup of gases and particles that trap heat and raise the planet's temperature. The average increase in global surface temperature, in degrees Centigrade, x years after 1980 can be modeled by the polynomial function

$$T(x) = \frac{21}{5{,}000{,}000}x^3 - \frac{127}{1{,}000{,}000}x^2 + \frac{1293}{50{,}000}x.$$

Use your graphing utility to graph the function in a [0, 60, 3] by [0, 2, 0.1] viewing rectangle. (Place parentheses around each fractional coefficient when you enter the equation.) What do you observe about global warming through the year 2040?

Multiply.

$$\begin{array}{r} 2x^2 \\ 3x - 2\overline{)6x^3 - x^2 - 5x + 4} \\ \ominus\, 6x^3 \oplus 4x^2 \\ \hline 3x^2 - 5x \end{array}$$

Divide: $\frac{6x^3}{3x} = 2x^2$.
Multiply: $2x^2(3x - 2) = 6x^3 - 4x^2$.
Subtract $6x^3 - 4x^2$ from $6x^3 - x^2$ and bring down $-5x$.

Now we divide $3x^2$ by $3x$ to obtain x, multiply x and the divisor, and subtract.

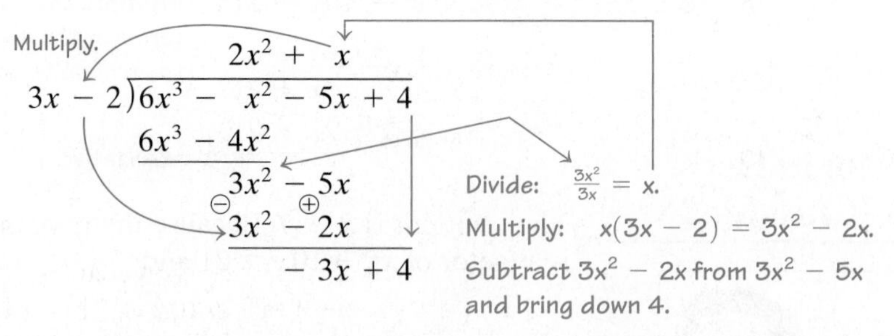

Now we divide $-3x$ by $3x$ to obtain -1, multiply -1 and the divisor, and subtract.

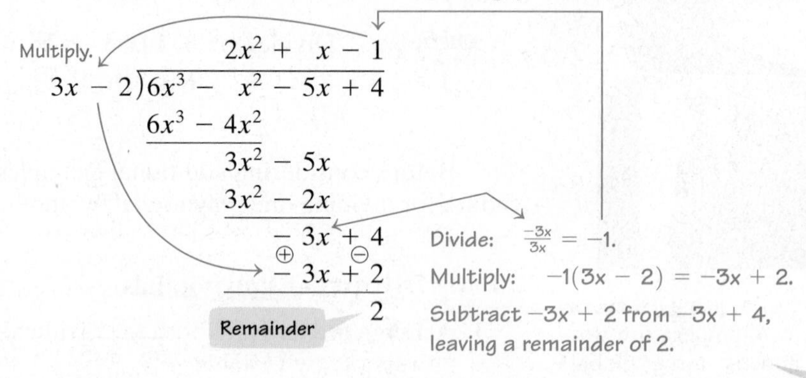

In Example 2, the quotient is $2x^2 + x - 1$ and the remainder is 2. This can be written in fractional form as follows:

Dividend — Quotient — Remainder

$$\frac{6x^3 - x^2 - 5x + 4}{3x - 2} = 2x^2 + x - 1 + \frac{2}{3x - 2}$$

Divisor — Divisor

Multiplying both sides of this equation by $3x - 2$ results in the following equation:

$$6x^3 - x^2 - 5x + 4 = (3x - 2)(2x^2 + x - 1) + 2$$

Dividend — Divisor — Quotient — Remainder

Polynomial long division is checked by multiplying the divisor with the quotient and then adding the remainder. This should give the dividend. The process illustrates the **Division Algorithm**.

The Division Algorithm

If $f(x)$ and $d(x)$ are polynomials, with $d(x) \neq 0$, and the degree of $d(x)$ is less than or equal to the degree of $f(x)$, then there exist unique polynomials $q(x)$ and $r(x)$ such that

$$\underset{\text{Dividend}}{f(x)} = \underset{\text{Divisor}}{d(x)} \cdot \underset{\text{Quotient}}{q(x)} + \underset{\text{Remainder}}{r(x)}.$$

The remainder, $r(x)$, equals 0 or it is of degree less than the degree of $d(x)$. If $r(x) = 0$, we say that $d(x)$ **divides evenly** into $f(x)$ and that $d(x)$ and $q(x)$ are **factors** of $f(x)$.

Check Point 2 Divide $7 - 11x - 3x^2 + 2x^3$ by $x - 3$. Use the remainder to express your result in fractional form.

If a power of x is missing in either a dividend or a divisor, add that power of x with a coefficient of 0 and then divide. In this way, like terms will be aligned as you carry out the long division.

EXAMPLE 3 Long Division of Polynomials

Divide $6x^4 + 5x^3 + 3x - 5$ by $3x^2 - 2x$.

Solution We write the dividend, $6x^4 + 5x^3 + 3x - 5$, as $6x^4 + 5x^3 + 0x^2 + 3x - 5$ so as to keep all like terms aligned.

Multiply.

$$\begin{array}{r} 2x^2 + 3x + 2 \\ 3x^2 - 2x \overline{)6x^4 + 5x^3 + 0x^2 + 3x - 5} \\ \ominus\, 6x^4 \overset{\oplus}{-} 4x^3 \qquad\qquad\qquad \\ \hline 9x^3 + 0x^2 \qquad\qquad \\ \ominus\, 9x^3 \overset{\oplus}{-} 6x^2 \qquad\qquad \\ \hline 6x^2 + 3x \qquad \\ \ominus\, 6x^2 \overset{\oplus}{-} 4x \qquad \\ \hline 7x - 5 \end{array}$$

$2x^2(3x^2 - 2x)$

$3x(3x^2 - 2x)$

$2(3x^2 - 2x)$

The division process is finished because the degree of $7x - 5$, which is 1, is less than the degree of the divisor $3x^2 - 2x$, which is 2. The answer is

$$\frac{6x^4 + 5x^3 + 3x - 5}{3x^2 - 2x} = 2x^2 + 3x + 2 + \frac{7x - 5}{3x^2 - 2x}.$$

Check Point 3 Divide $2x^4 + 3x^3 - 7x - 10$ by $x^2 - 2x$.

2 Use synthetic division to divide polynomials.

Dividing Polynomials Using Synthetic Division

We can use **synthetic division** to divide polynomials if the divisor is of the form $x - c$. This method provides a quotient more quickly than long division. Let's compare the two methods showing $x^3 + 4x^2 - 5x + 5$ divided by $x - 3$.

Long Division

Quotient

$$\begin{array}{r} x^2 + 7x + 16 \\ x - 3\overline{)x^3 + 4x^2 - 5x + 5} \\ \ominus x^3 - 3x^2 \qquad\qquad \\ \hline 7x^2 - 5x \qquad \\ \ominus 7x^2 - 21x \qquad \\ \hline 16x + 5 \\ \ominus 16x - 48 \\ \hline 53 \end{array}$$

Divisor $x - c$; $c = 3$

Dividend

Remainder

Synthetic Division

$$\begin{array}{r|rrrr} 3 & 1 & 4 & -5 & 5 \\ & & 3 & 21 & 48 \\ \hline & 1 & 7 & 16 & 53 \end{array}$$

Notice the relationship between the polynomials in the long division process and the numbers that appear in synthetic division.

These are the coefficients of the dividend $x^3 + 4x^2 - 5x + 5$.

The divisor is $x - 3$. This is 3, or c in $x - c$.

$$\begin{array}{r|rrrr} 3 & 1 & 4 & -5 & 5 \\ & & 3 & 21 & 48 \\ \hline & 1 & 7 & 16 & 53 \end{array}$$

These are the coefficients of the quotient $x^2 + 7x + 16$.

This is the remainder.

Now let's look at the steps involved in synthetic division.

Synthetic Division

To divide a polynomial by $x - c$,

	Example
1. Arrange polynomials in descending powers, with a 0 coefficient for any missing term.	$x - 3\overline{)x^3 + 4x^2 - 5x + 5}$
2. Write c for the divisor, $x - c$. To the right, write the coefficients of the dividend.	$\underline{3\rvert}\ \ 1\ \ 4\ \ -5\ \ 5$
3. Write the leading coefficient of the dividend on the bottom row.	$\underline{3\rvert}\ \ 1\ \ 4\ \ -5\ \ 5$; ↓ Bring down 1.; bottom row: 1
4. Multiply c (in this case, 3) times the value just written on the bottom row. Write the product in the next column in the second row.	$\underline{3\rvert}\ \ 1\ \ 4\ \ -5\ \ 5$; second row: 3; bottom row: 1; Multiply by 3.
5. Add the values in this new column, writing the sum in the bottom row.	$\underline{3\rvert}\ \ 1\ \ 4\ \ -5\ \ 5$; second row: 3 (Add.); bottom row: 1 7

6. Repeat this series of multiplications and additions until all columns are filled in.

$$\begin{array}{r|rrrr} 3 & 1 & 4 & -5 & 5 \\ & & 3 & 21 & \\ \hline & 1 & 7 & 16 & \end{array}$$

Add. Multiply by 3.

$$\begin{array}{r|rrrr} 3 & 1 & 4 & -5 & 5 \\ & & 3 & 21 & 48 \\ \hline & 1 & 7 & 16 & 53 \end{array}$$

Add. Multiply by 3.

7. Use the numbers in the last row to write the quotient and remainder in fractional form. **The degree of the first term of the quotient is one less than the degree of the first term of the dividend.** The final value in this row is the remainder.

Written from the last row of the synthetic division

$$x - 3\overline{)x^3 + 4x^2 - 5x + 5} \quad 1x^2 + 7x + 16 + \frac{53}{x - 3}$$

EXAMPLE 4 Using Synthetic Division

Use synthetic division to divide $5x^3 + 6x + 8$ by $x + 2$.

Solution The divisor must be in the form $x - c$. Thus, we write $x + 2$ as $x - (-2)$. This means that $c = -2$. Writing a 0 coefficient for the missing x^2-term in the dividend, we can express the division as follows:

$$x - (-2)\overline{)5x^3 + 0x^2 + 6x + 8}.$$

Now we are ready to set up the problem so that we can use synthetic division.

Use the coefficients of the dividend $5x^3 + 0x^2 + 6x + 8$ in descending powers of x.

This is c in $x-(-2)$.

$$\begin{array}{r|rrrr} -2 & 5 & 0 & 6 & 8 \end{array}$$

We begin the synthetic division process by bringing down 5. This is followed by a series of multiplications and additions.

1. Bring down 5.

$$\begin{array}{r|rrrr} -2 & 5 & 0 & 6 & 8 \\ & & & & \\ \hline & 5 & & & \end{array}$$

2. Multiply: $-2(5) = -10$.

$$\begin{array}{r|rrrr} -2 & 5 & 0 & 6 & 8 \\ & & -10 & & \\ \hline & 5 & & & \end{array}$$

Multiply by -2.

3. Add: $0 + (-10) = -10$.

$$\begin{array}{r|rrrr} -2 & 5 & 0 & 6 & 8 \\ & & -10 & & \\ \hline & 5 & -10 & & \end{array}$$

Add.

4. Multiply: $-2(-10) = 20$.

$$\begin{array}{r|rrrr} -2 & 5 & 0 & 6 & 8 \\ & & -10 & 20 & \\ \hline & 5 & -10 & & \end{array}$$

Multiply by -2.

5. Add: $6 + 20 = 26$.

$$\begin{array}{r|rrrr} -2 & 5 & 0 & 6 & 8 \\ & & -10 & 20 & \\ \hline & 5 & -10 & 26 & \end{array}$$

Add.

6. Multiply: $-2(26) = -52$.

$$\begin{array}{r|rrrr} -2 & 5 & 0 & 6 & 8 \\ & & -10 & 20 & -52 \\ \hline & 5 & -10 & 26 & \end{array}$$

Multiply by -2.

7. Add: $8 + (-52) = -44$.

$$\begin{array}{r|rrrr} -2 & 5 & 0 & 6 & 8 \\ & & -10 & 20 & -52 \\ \hline & 5 & -10 & 26 & -44 \end{array}$$

Add.

The numbers in the last row represent the coefficients of the quotient and the remainder. The degree of the first term of the quotient is one less than that of the dividend. Because the degree of the dividend is 3, the degree of the quotient is 2. This means that the 5 in the last row represents $5x^2$.

$$\begin{array}{r|rrrr} -2 & 5 & 0 & 6 & 8 \\ & & -10 & 20 & -52 \\ \hline & 5 & -10 & 26 & -44 \end{array}$$

The quotient is $5x^2 - 10x + 26$.

The remainder is -44.

Thus,

$$x + 2 \overline{)5x^3 + 6x + 8} = 5x^2 - 10x + 26 - \frac{44}{x+2}$$

Check Point 4 Use synthetic division to divide $x^3 - 7x - 6$ by $x + 2$.

3 Evaluate a polynomial using the Remainder Theorem.

The Remainder Theorem

Let's consider the Division Algorithm when the dividend, $f(x)$, is divided by $x - c$. In this case, the remainder must be a constant because its degree is less than one, the degree of $x - c$.

$f(x) = d(x)q(x) + r(x)$ This is the Division Algorithm.

Dividend Divisor Quotient Remainder

$f(x) = (x - c)q(x) + r$ The divisor is $x - c$. Call the constant remainder r.

Now let's evaluate f at c.

$f(c) = (c - c)q(c) + r$ Find $f(c)$, setting $x = c$. This will give an expression for r.

$f(c) = 0 \cdot q(c) + r$ $c - c = 0$ and $0 \cdot q(c) = 0$.

$f(c) = r$ On the right, $0 + r = r$.

What does this last equation mean? If a polynomial is divided by $x - c$, the remainder is the value of the polynomial at c. This result is called the **Remainder Theorem.**

The Remainder Theorem

If the polynomial $f(x)$ is divided by $x - c$, then the remainder is $f(c)$.

Example 5 shows how we can use the Remainder Theorem to evaluate a polynomial function at 2. Rather than substituting 2 for x, we divide the function by $x - 2$. The remainder is $f(2)$.

EXAMPLE 5 Using the Remainder Theorem to Evaluate a Polynomial Function

Given $f(x) = x^3 - 4x^2 + 5x + 3$, use the Remainder Theorem to find $f(2)$.

Solution By the Remainder Theorem, if $f(x)$ is divided by $x - 2$, then the remainder is $f(2)$. We'll use synthetic division to divide.

$$\begin{array}{r|rrrr} 2 & 1 & -4 & 5 & 3 \\ & & 2 & -4 & 2 \\ \hline & 1 & -2 & 1 & 5 \end{array}$$

Remainder

The remainder, 5, is the value of $f(2)$. Thus, $f(2) = 5$. We can verify that this is correct by evaluating $f(2)$ directly. Using $f(x) = x^3 - 4x^2 + 5x + 3$, we obtain

$$f(2) = 2^3 - 4 \cdot 2^2 + 5 \cdot 2 + 3 = 8 - 16 + 10 + 3 = 5.$$

Check Point 5 Given $f(x) = 3x^3 + 4x^2 - 5x + 3$, use the Remainder Theorem to find $f(-4)$.

4 Use the Factor Theorem to solve a polynomial equation.

The Factor Theorem

Let's look again at the Division Algorithm when the divisor is of the form $x - c$.

$$f(x) = (x - c)q(x) + r$$

Dividend | Divisor | Quotient | Constant remainder

By the Remainder Theorem, the remainder r is $f(c)$, so we can substitute $f(c)$ for r:

$$f(x) = (x - c)q(x) + f(c).$$

Notice that if $f(c) = 0$, then

$$f(x) = (x - c)q(x)$$

so that $x - c$ is a factor of $f(x)$. This means that for the polynomial function $f(x)$, if $f(c) = 0$, then $x - c$ is a factor of $f(x)$.

Let's reverse directions and see what happens if $x - c$ is a factor of $f(x)$. This means that

$$f(x) = (x - c)q(x).$$

If we replace x with c, we obtain

$$f(c) = (c - c)q(c) = 0.$$

Thus, if $x - c$ is a factor of $f(x)$, then $f(c) = 0$.

We have proved a result known as the **Factor Theorem**.

The Factor Theorem

Let $f(x)$ be a polynomial.

a. If $f(c) = 0$, then $x - c$ is a factor of $f(x)$.

b. If $x - c$ is a factor of $f(x)$, then $f(c) = 0$.

The example that follows shows how the Factor Theorem can be used to solve a polynomial equation.

EXAMPLE 6 Using the Factor Theorem

Solve the equation $2x^3 - 3x^2 - 11x + 6 = 0$ given that 3 is a zero of $f(x) = 2x^3 - 3x^2 - 11x + 6$.

Solution We are given that $f(3) = 0$. The Factor Theorem tells us that $x - 3$ is a factor of $f(x)$. We'll use synthetic division to divide $f(x)$ by $x - 3$.

$$\begin{array}{r|rrrr} 3 & 2 & -3 & -11 & 6 \\ & & 6 & 9 & -6 \\ \hline & 2 & 3 & -2 & 0 \end{array}$$

$$x - 3\overline{)2x^3 - 3x^2 - 11x + 6} \quad \text{quotient: } 2x^2 + 3x - 2$$

Equivalently,

$$2x^3 - 3x^2 - 11x + 6 = (x - 3)(2x^2 + 3x - 2)$$

Technology

Because the solution set of

$$2x^3 - 3x^2 - 11x + 6 = 0$$

is $\{-2, \frac{1}{2}, 3\}$, this implies that the polynomial function

$$f(x) = 2x^3 - 3x^2 - 11x + 6$$

has x-intercepts (or zeros) at $-2, \frac{1}{2}$, and 3. This is verified by the graph of f.

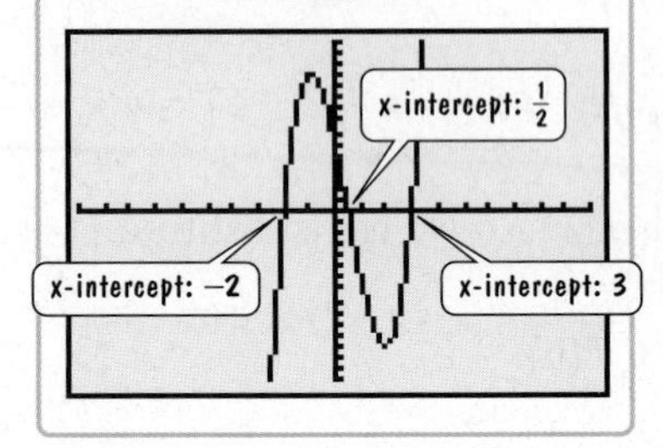

Now we can solve the polynomial equation.

$2x^3 - 3x^2 - 11x + 6 = 0$ This is the given equation.

$(x - 3)(2x^2 + 3x - 2) = 0$ Factor using the result from the synthetic division.

$(x - 3)(2x - 1)(x + 2) = 0$ Factor the trinomial.

$x - 3 = 0$ or $2x - 1 = 0$ or $x + 2 = 0$ Set each factor equal to 0.

$x = 3$ $\quad x = \frac{1}{2}$ $\quad x = -2$ Solve for x.

The solution set is $\{-2, \frac{1}{2}, 3\}$.

By the Factor Theorem, the following statements are useful in solving polynomial equations:

1. If $f(x)$ is divided by $x - c$ and the remainder is zero, then c is a zero of f and c is a root of the polynomial equation $f(x) = 0$.
2. If $f(x)$ is divided by $x - c$ and the remainder is zero, then $x - c$ is a factor of $f(x)$.

Check Point 6 Solve the equation $15x^3 + 14x^2 - 3x - 2 = 0$ given that -1 is a zero of $f(x) = 15x^3 + 14x^2 - 3x - 2$.

EXERCISE SET 3.3

Practice Exercises

In Exercises 1–16, divide by long division.

1. $(x^2 + 8x + 15) \div (x + 5)$
2. $(x^2 + 3x - 10) \div (x - 2)$
3. $(x^3 + 5x^2 + 7x + 2) \div (x + 2)$
4. $(x^3 - 2x^2 - 5x + 6) \div (x - 3)$
5. $(6x^3 + 7x^2 + 12x - 5) \div (3x - 1)$
6. $(6x^3 + 17x^2 + 27x + 20) \div (3x + 4)$
7. $(12x^2 + x - 4) \div (3x - 2)$
8. $(4x^2 - 8x + 6) \div (2x - 1)$
9. $\dfrac{2x^3 + 7x^2 + 9x - 20}{x + 3}$
10. $\dfrac{3x^2 - 2x + 5}{x - 3}$
11. $\dfrac{4x^4 - 4x^2 + 6x}{x - 4}$
12. $\dfrac{x^4 - 81}{x - 3}$
13. $\dfrac{6x^3 + 13x^2 - 11x - 15}{3x^2 - x - 3}$
14. $\dfrac{x^4 + 2x^3 - 4x^2 - 5x - 6}{x^2 + x - 2}$
15. $\dfrac{18x^4 + 9x^3 + 3x^2}{3x^2 + 1}$
16. $\dfrac{2x^5 - 8x^4 + 2x^3 + x^2}{2x^3 + 1}$

In Exercises 17–32, divide by synthetic division.

17. $(2x^2 + x - 10) \div (x - 2)$
18. $(x^2 + x - 2) \div (x - 1)$
19. $(3x^2 + 7x - 20) \div (x + 5)$
20. $(5x^2 - 12x - 8) \div (x + 3)$
21. $(4x^3 - 3x^2 + 3x - 1) \div (x - 1)$
22. $(5x^3 - 6x^2 + 3x + 11) \div (x - 2)$
23. $(6x^5 - 2x^3 + 4x^2 - 3x + 1) \div (x - 2)$
24. $(x^5 + 4x^4 - 3x^2 + 2x + 3) \div (x - 3)$
25. $(x^2 - 5x - 5x^3 + x^4) \div (5 + x)$
26. $(x^2 - 6x - 6x^3 + x^4) \div (6 + x)$
27. $\dfrac{x^5 + x^3 - 2}{x - 1}$
28. $\dfrac{x^7 + x^5 - 10x^3 + 12}{x + 2}$
29. $\dfrac{x^4 - 256}{x - 4}$
30. $\dfrac{x^7 - 128}{x - 2}$
31. $\dfrac{2x^5 - 3x^4 + x^3 - x^2 + 2x - 1}{x + 2}$
32. $\dfrac{x^5 - 2x^4 - x^3 + 3x^2 - x + 1}{x - 2}$
33. Given $f(x) = 2x^3 - 11x^2 + 7x - 5$, use the Remainder Theorem to find $f(4)$.
34. Given $f(x) = x^3 - 7x^2 + 5x - 6$, use the Remainder Theorem to find $f(3)$.
35. Given $f(x) = 7x^4 - 3x^3 + 6x + 9$, use the Remainder Theorem to find $f(-5)$.
36. Given $f(x) = 3x^4 + 6x^3 - 2x + 4$, use the Remainder Theorem to find $f(-4)$.
37. Use synthetic division to divide $f(x) = x^3 - 4x^2 + x + 6$ by $x + 1$. Use the result to find all zeros of f.
38. Use synthetic division to divide $f(x) = x^3 - 2x^2 - x + 2$ by $x + 1$. Use the result to find all zeros of f.
39. Solve the equation $2x^3 - 5x^2 + x + 2 = 0$ given that 2 is a zero of $f(x) = 2x^3 - 5x^2 + x + 2$.
40. Solve the equation $2x^3 - 3x^2 - 11x + 6 = 0$ given that -2 is a zero of $f(x) = 2x^3 - 3x^2 - 11x + 6$.
41. Solve the equation $12x^3 + 16x^2 - 5x - 3 = 0$ given that $-\frac{3}{2}$ is a root.
42. Solve the equation $3x^3 + 7x^2 - 22x - 8 = 0$ given that $-\frac{1}{3}$ is a root.

Application Exercises

43. A rectangle with length $2x + 5$ inches has an area of $2x^4 + 15x^3 + 7x^2 - 135x - 225$ square inches. Write a polynomial that represents its width.
44. If you travel a distance of $x^3 + 3x^2 + 5x + 3$ miles at a rate of $x + 1$ miles per hour, write a polynomial that represents the number of hours you traveled.
45. Two people are 25 years old and 20 years old, respectively. In x years from now, their ages can be represented by $x + 25$ and $x + 20$.

    a. Use long division to find $\dfrac{x + 25}{x + 20}$, the ratio of the older person's age in x years to the younger person's age in x years.

    b. Complete the following table.

x	0	5	10	25	50	75
$\dfrac{x + 25}{x + 20}$						

    c. Describe what is happening to the ratio $\dfrac{x + 25}{x + 20}$ as x increases. How can this be verified using the result of the long division in part (a)?

Writing in Mathematics

46. Explain how to perform long division of polynomials. Use $2x^3 - 3x^2 - 11x + 7$ divided by $x - 3$ in your explanation.
47. In your own words, state the Division Algorithm.
48. How can the Division Algorithm be used to check the quotient and remainder in a long division problem?
49. Explain how to perform synthetic division. Use the division problem in Exercise 46 to support your explanation.
50. State the Remainder Theorem.

51. Explain how the Remainder Theorem can be used to find $f(-6)$ if $f(x) = x^4 + 7x^3 + 8x^2 + 11x + 5$. What advantage is there to using the Remainder Theorem in this situation rather than evaluating $f(-6)$ directly?

52. How can the Factor Theorem be used to determine if $x - 1$ is a factor of $x^3 - 2x^2 - 11x + 12$?

53. If you know that -2 is a zero of
$$f(x) = x^3 + 7x^2 + 4x - 12,$$
explain how to solve the equation
$$x^3 + 7x^2 + 4x - 12 = 0.$$

Technology Exercises

In Exercises 54–57, use a graphing utility to graph the function on each side of the given equation. If the graphs coincide, this verifies that the expressions are equivalent and the division has been performed correctly. If the graphs do not coincide, correct the expression on the right by performing the division. Then use your graphing utility to verify your result.

54. $\dfrac{x^4 + 6x^3 + 6x^2 - 10x - 3}{x^2 + 2x - 3} = x^2 + 4x + 1,$
$x \neq -3, \quad x \neq 1$

55. $\dfrac{2x^3 - 3x^2 - 3x + 4}{x - 1} = 2x^2 - x + 4, \quad x \neq 1$

56. $\dfrac{3x^4 + 4x^3 - 32x^2 - 5x - 20}{x + 4} = 3x^3 + 8x^2 - 5,$
$x \neq -4$

57. $\dfrac{10x^3 - 26x^2 + 17x - 13}{5x - 3} = 2x^2 - 4x + 1 - \dfrac{10}{5x - 3},$
$x \neq \frac{3}{5}$

Critical Thinking Exercises

58. Which one of the following is true?

a. If a trinomial in x of degree 6 is divided by a trinomial in x of degree 3, the degree of the quotient is 2.

b. Synthetic division could not be used to find the quotient of $10x^3 - 6x^2 + 4x - 1$ and $x - \frac{1}{2}$.

c. Any problem that can be done by synthetic division can also be done by the method for long division of polynomials.

d. If a polynomial long-division problem results in a remainder that is a whole number, then the divisor is a factor of the dividend.

59. Find k so that $4x + 3$ is a factor of
$$20x^3 + 23x^2 - 10x + k.$$

60. When $2x^2 - 7x + 9$ is divided by a polynomial, the quotient is $2x - 3$ and the remainder is 3. Find the polynomial.

61. Find the quotient of $x^{3n} + 1$ and $x^n + 1$.

62. Synthetic division is a process for dividing a polynomial by $x - c$. The coefficient of x is 1. How might synthetic division be used if you are dividing by $2x - 4$?

SECTION 3.4 *Zeros of Polynomial Functions*

Objectives

1. Use the Rational Zero Theorem to find possible rational zeros.
2. Find zeros of a polynomial function.
3. Solve polynomial equations.
4. Use Descartes's Rule of Signs.

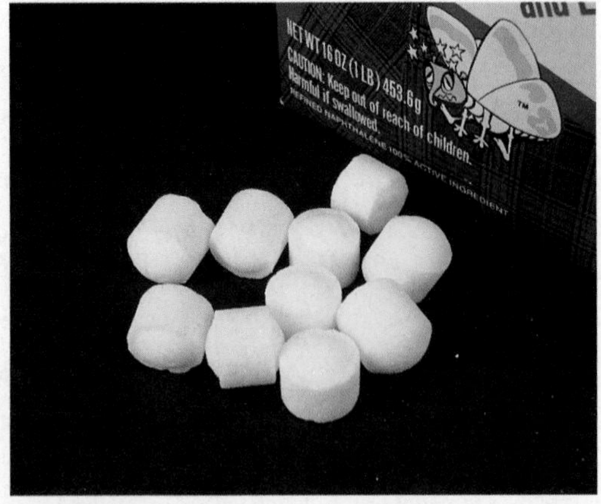

The solution to a multitude of moths?

A moth has moved into your closet. She appeared in your bedroom at night, but somehow her relatively stout body escaped your clutches. Within a few weeks swarms of moths in your tattered wardrobe suggest that Mama Moth was in the family way. There must be at least 200 critters nesting in every crevice of your clothing.

Two hundred plus moth-tykes from one female moth; is this possible? Indeed it is. The number of eggs, N, in a female moth is a function of her abdominal width, W, in millimeters, modeled by

$$N = 14W^3 - 17W^2 - 16W + 34$$

for $1.5 \le W \le 3.5$. Because there are 200 moths feasting on your favorite sweaters, Mama's abdominal width can be estimated by finding the roots of the polynomial equation

$$14W^3 - 17W^2 - 16W + 34 = 200.$$

With mathematics present even in your quickly disappearing attire, we move from rags to polynomial equations. The process of solving such equations begins with listing possibilities for Mama Moth's abdominal width. To do this, we turn to a theorem that plays an important role in finding zeros of polynomial functions.

1 Use the Rational Zero Theorem to find possible rational zeros.

The Rational Zero Theorem

The Rational Zero Theorem gives a list of possible rational zeros of a polynomial function. Equivalently, the theorem gives all possible rational roots of a polynomial equation. Not every number in the list will be a zero of the function, but every rational zero of the polynomial function will appear somewhere in the list.

The Rational Zero Theorem

If $f(x) = a_n x^n + a_{n-1}x^{n-1} + \cdots + a_1 x + a_0$ has *integer* coefficients and $\frac{p}{q}$ (where $\frac{p}{q}$ is reduced) is a rational zero, then p is a factor of the constant term a_0 and q is a factor of the leading coefficient a_n.

You can explore the "why" behind the Rational Zero Theorem in Exercise 64 of Exercise Set 3.4. For now, let's see if we can figure out what the theorem tells us about possible rational zeros. In order to use the theorem, list all the integers that are factors of the constant term, a_0. Then list all the integers that are factors of the leading coefficient, a_n. Finally list all possible rational zeros:

$$\text{Possible rational zeros} = \frac{\text{Factors of the constant term}}{\text{Factors of the leading coefficient}}.$$

EXAMPLE 1 Using the Rational Zero Theorem

List all possible rational zeros of $f(x) = -x^4 + 4x^2 + 4$.

Solution The constant term is 4. We list all of its factors: $\pm1, \pm2, \pm4$. The leading coefficient is -1. Its factors are ±1.

Factors of the constant term: $\pm1, \quad \pm2 \quad \pm4$

Factors of the leading coefficient: ±1

Because

$$\text{Possible rational zeros} = \frac{\text{Factors of the constant term}}{\text{Factors of the leading coefficient}},$$

we must take each number in the first row, $\pm1, \pm2, \pm4$, and divide by each number in the second row, ±1.

$$\text{Possible rational zeros} = \frac{\text{Factors of } 4}{\text{Factors of } -1} = \frac{\pm 1, \pm 2, \pm 4}{\pm 1} = \pm 1, \quad \pm 2, \quad \pm 4$$

Divide ± 1 by ± 1. Divide ± 2 by ± 1. Divide ± 4 by ± 1.

There are six possible rational zeros. The graph of $f(x) = -x^4 + 4x^2 + 4$ is shown in Figure 3.18. The x-intercepts are -2 and 2. Thus, -2 and 2 are the actual rational zeros.

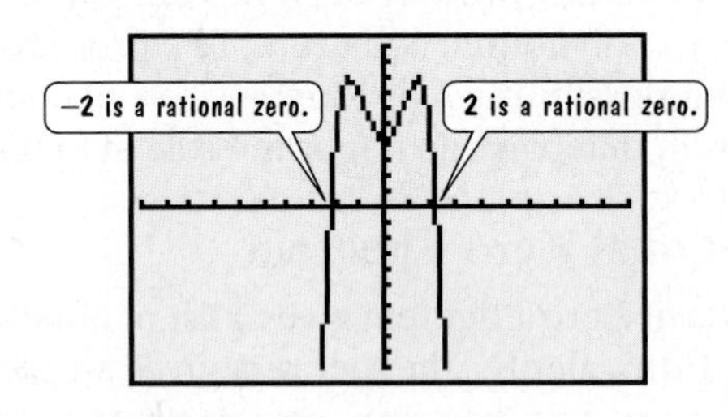

Figure 3.18 The graph of $f(x) = -x^4 + 4x^2 + 4$ shows that -2 and 2 are rational zeros.

Check Point 1 List all possible rational zeros of $f(x) = x^3 + 2x^2 - 5x - 6$.

EXAMPLE 2 Using the Rational Zero Theorem

List all possible rational zeros of $f(x) = 15x^3 + 14x^2 - 3x - 2$.

Solution The constant term is -2 and the leading coefficient is 15.

$$\text{Possible rational zeros} = \frac{\text{Factors of the constant term, } -2}{\text{Factors of the leading coefficient, } 15}$$

$$= \frac{\pm 1, \pm 2}{\pm 1, \pm 3, \pm 5, \pm 15}$$

$$= \pm 1, \quad \pm 2, \quad \pm \tfrac{1}{3}, \quad \pm \tfrac{2}{3}, \quad \pm \tfrac{1}{5}, \quad \pm \tfrac{2}{5}, \quad \pm \tfrac{1}{15}, \quad \pm \tfrac{2}{15}$$

Divide ± 1 and ± 2 by ± 1. Divide ± 1 and ± 2 by ± 3. Divide ± 1 and ± 2 by ± 5. Divide ± 1 and ± 2 by ± 15.

There are 16 possible rational zeros. The actual solution set to $15x^3 + 14x^2 - 3x - 2 = 0$ is $\{-1, -\frac{1}{3}, \frac{2}{5}\}$, which contains 3 of the 16 possible zeros.

Check Point 2 Find all possible rational zeros of $f(x) = 4x^5 + 12x^4 - x - 3$.

2 Find zeros of a polynomial function.

How do we determine which (if any) of the possible rational zeros are rational zeros of the polynomial function? To find the first rational zero, we can use a trial-and-error process involving synthetic division. [Recall that if $f(x)$ is divided by $x - c$ and the remainder is zero, then c is a zero of f.] After we identify the first rational zero, we use the result of the synthetic division to factor the original polynomial. Then we set each factor equal to zero to identify any additional rational zeros.

EXAMPLE 3 Finding Zeros of a Polynomial Function

Find all rational zeros of $f(x) = x^3 + 2x^2 - 5x - 6$.

Solution We begin by listing all possible rational zeros.

Possible rational zeros

$$= \frac{\text{Factors of the constant term, } -6}{\text{Factors of the leading coefficient, } 1} = \frac{\pm 1, \pm 2, \pm 3, \pm 6}{\pm 1} = \pm 1, \pm 2, \pm 3, \pm 6$$

Divide the eight numbers in the numerator by ± 1.

Now we will use synthetic division to see if we can find a rational root among the possible rational zeros ± 1, ± 2, ± 3, ± 6. Keep in mind that if $f(x)$ is divided by $x - c$ and the remainder is zero, then c is a zero of f. Let's start by testing 1. If 1 is not a rational zero, then we will test other possible rational zeros.

Test 1

Coefficients of $f(x) = x^3 + 2x^2 - 5x - 6$

Possible rational zero

$$\begin{array}{r|rrrr} 1 & 1 & 2 & -5 & -6 \\ & & 1 & 3 & -2 \\ \hline & 1 & 3 & -2 & -8 \end{array}$$

The nonzero remainder shows that 1 is not a zero.

Test 2

Coefficients of $f(x) = x^3 + 2x^2 - 5x - 6$

Possible rational zero

$$\begin{array}{r|rrrr} 2 & 1 & 2 & -5 & -6 \\ & & 2 & 8 & 6 \\ \hline & 1 & 4 & 3 & 0 \end{array}$$

The zero remainder shows that 2 is a zero.

The zero remainder tells us that 2 is a zero of the polynomial function $f(x) = x^3 + 2x^2 - 5x - 6$. Equivalently, 2 is a solution, or root, of the polynomial equation $x^3 + 2x^2 - 5x - 6 = 0$. Thus, $x - 2$ is a factor of the polynomial.

$x^3 + 2x^2 - 5x - 6 = 0$ Finding the zeros of $f(x) = x^3 + 2x^2 - 5x - 6$ is the same as finding the roots of this equation.

$(x - 2)(x^2 + 4x + 3) = 0$ Factor using the result from the synthetic division.

$(x - 2)(x + 3)(x + 1) = 0$ Factor completely.

$x - 2 = 0$ or $x + 3 = 0$ or $x + 1 = 0$ Set each factor equal to zero.

$x = 2$ $\quad x = -3$ $\quad x = -1$ Solve for x.

The solution set is $\{-3, -1, 2\}$. The rational zeros of f are -3, -1, and 2.

Check Point 3 Find all rational zeros of $f(x) = x^3 + 8x^2 + 11x - 20$.

Our work in Example 3 involved solving a third-degree equation. We found one factor by synthetic division and factored the remaining quadratic factor using the FOIL method. If the degree of a polynomial function or equation is 4 or higher, it is often necessary to find more than one linear factor by synthetic division.

One way to speed up the process of finding the first zero is to graph the function. Any x-intercept is a zero.

3 Solve polynomial equations.

EXAMPLE 4 Solving a Polynomial Equation

Solve: $x^4 - 6x^2 - 8x + 24 = 0$.

Solution Recall that we refer to the zeros of a polynomial function and the roots of a polynomial equation. Because we are given an equation, we will use the word "roots," rather than "zeros," in the solution process. We begin by listing all possible rational roots.

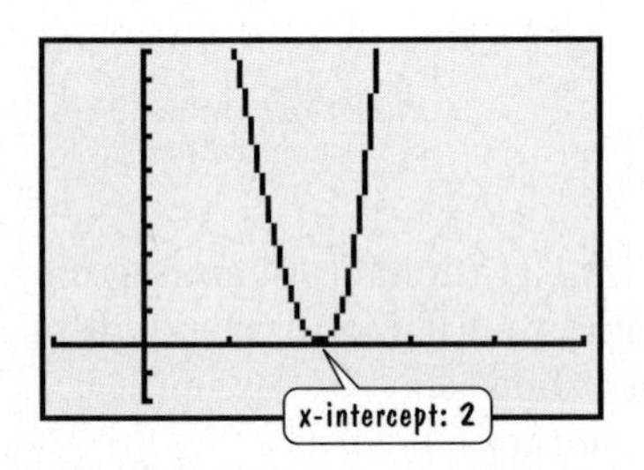

Figure 3.19 The graph of $f(x) = x^4 - 6x^2 - 8x + 24$ in a $[-1, 5, 1]$ by $[-2, 10, 1]$ viewing rectangle

Possible rational roots

$$= \frac{\text{Factors of the constant term, 24}}{\text{Factors of the leading coefficient, 1}}$$

$$= \frac{\pm1, \pm2, \pm3, \pm4, \pm6, \pm8, \pm12, \pm24}{\pm1} = \pm1, \pm2, \pm3, \pm4, \pm6, \pm8, \pm12, \pm24$$

The graph of $f(x) = x^4 - 6x^2 - 8x + 24$ is shown in Figure 3.19. Because the x-intercept is 2, we will test 2 by synthetic division and show that it is a root of the given equation.

$$\begin{array}{r|rrrrr} 2 & 1 & 0 & -6 & -8 & 24 \\ & & 2 & 4 & -4 & -24 \\ \hline & 1 & 2 & -2 & -12 & 0 \end{array}$$

Careful! $x^4 - 6x^2 - 8x + 24 = x^4 + 0x^3 - 6x^2 - 8x + 24$

The zero remainder indicates that 2 is a root of $x^4 - 6x^2 - 8x + 24 = 0$.

Now we can rewrite the given equation in factored form.

$x^4 - 6x^2 - 8x + 24 = 0$ — This is the given equation.

$(x - 2)(x^3 + 2x^2 - 2x - 12) = 0$ — This is the result obtained from the synthetic division.

$x - 2 = 0$ or $x^3 + 2x^2 - 2x - 12 = 0$ — Set each factor equal to 0.

We can use the same approach to look for rational roots of the polynomial equation $x^3 + 2x^2 - 2x - 12 = 0$, listing all possible rational roots. However, take a second look at the graph in Figure 3.19. Because the graph turns around at 2, this means that 2 is a root of even multiplicity. Thus, 2 must also be a root of $x^3 + 2x^2 - 2x - 12 = 0$, confirmed by the following synthetic division.

$$\begin{array}{r|rrrr} 2 & 1 & 2 & -2 & -12 \\ & & 2 & 8 & 12 \\ \hline & 1 & 4 & 6 & 0 \end{array}$$

These are the coefficients of $x^3 + 2x^2 - 2x - 12 = 0$.

The zero remainder indicates that 2 is a root of $x^3 + 2x^2 - 2x - 12 = 0$.

Now we can solve the original equation as follows:

$x^4 - 6x^2 - 8x + 24 = 0$ — This is the given equation.

$(x - 2)(x^3 + 2x^2 - 2x - 12) = 0$ — This was obtained from the first synthetic division.

$(x - 2)(x - 2)(x^2 + 4x + 6) = 0$ — This was obtained from the second synthetic division.

$x - 2 = 0$ or $x - 2 = 0$ or $x^2 + 4x + 6 = 0$ — Set each factor equal to 0.

$x = 2$ $\quad$ $x = 2$ $\quad$ $x^2 + 4x + 6 = 0$ — Solve.

We can use the quadratic formula to solve $x^2 + 4x + 6 = 0$.

$$x = \frac{-b \pm \sqrt{b^2 - 4ac}}{2a}$$ We use the quadratic formula because $x^2 + 4x + 6$ cannot be factored.

$$= \frac{-4 \pm \sqrt{4^2 - 4(1)(6)}}{2(1)}$$ Let $a = 1$, $b = 4$, and $c = 6$.

$$= \frac{-4 \pm \sqrt{-8}}{2}$$ Multiply and subtract under the radical.

$$= \frac{-4 \pm 2i\sqrt{2}}{2}$$ $\sqrt{-8} = \sqrt{4(2)(-1)} = 2i\sqrt{2}$

$$= -2 \pm i\sqrt{2}$$ Simplify.

The solution set of the original equation is $\{2, -2 - i\sqrt{2}, -2 + i\sqrt{2}\}$.

In Example 4, 2 is a repeated root of the equation with multiplicity 2. The example illustrates two general properties.

Properties of Polynomial Equations

1. If a polynomial equation is of degree n, then counting multiple roots separately, the equation has n roots.
2. If $a + bi$ is a root of a polynomial equation $(b \neq 0)$, then the nonreal complex number $a - bi$ is also a root. Nonreal complex roots, if they exist, occur in conjugate pairs.

These ideas will be developed in more detail in the next section.

Check Point 4 Solve: $x^4 - 6x^3 + 22x^2 - 30x + 13 = 0$.

4 Use Descartes's Rule of Signs.

Descartes's Rule of Signs

Because an nth-degree polynomial equation might have roots that are imaginary numbers, we should note that such an equation can have *at most n* real roots. **Descartes's Rule of Signs** provides even more specific information about the number of real zeros that a polynomial can have. The rule is based on considering *variations in sign* between consecutive coefficients. For example, the function

$$f(x) = 3x^7 - 2x^5 - x^4 + 7x^2 + x - 3$$

has three sign changes.

An equation can have as many true [positive] roots as it contains changes of sign, from plus to minus or from minus to plus.... René Descartes (1596–1650) in *La Géométrie* (1637)

Descartes's Rule of Signs

Let $f(x) = a_nx^n + a_{n-1}x^{n-1} + \cdots + a_2x^2 + a_1x + a_0$ be a polynomial with real coefficients.

1. The number of *positive real zeros* of f is either equal to the number of sign changes of $f(x)$ or is less than that number by an even integer. If

there is only one variation in sign, there is exactly one positive real zero.

2. The number of *negative real zeros* of f is either equal to the number of sign changes of $f(-x)$ or is less than that number by an even integer. If $f(-x)$ has only one variation in sign, then f has exactly one negative real zero.

EXAMPLE 5 Using Descartes's Rule of Signs

Determine the possible number of positive and negative real zeros of $f(x) = x^3 + 2x^2 + 5x + 4$.

Solution

1. To find possibilities for positive real zeros, count the number of sign changes in the equation for $f(x)$. Because all the terms are positive, there are no variations in sign. Thus, there are no positive real zeros.
2. To find possibilities for negative real zeros, count the number of sign changes in the equation for $f(-x)$. We obtain this equation by replacing x with $-x$ in the given function.

$$f(x) = x^3 + 2x^2 + 5x + 4$$ This is the given polynomial function.

Replace x with −x.

$$f(-x) = (-x)^3 + 2(-x)^2 + 5(-x) + 4$$
$$= -x^3 + 2x^2 - 5x + 4$$

Now count the sign changes.

$$f(-x) = -x^3 + 2x^2 - 5x + 4$$

1 2 3

There are three variations in sign. The number of negative real zeros of f is either equal to the number of sign changes, 3, or is less than this number by an even integer. This means that there are either 3 negative real zeros or $3 - 2 = 1$ negative real zero.

What do the results of Example 5 mean in terms of solving

$$x^3 + 2x^2 + 5x + 4 = 0?$$

Without using Descartes's Rule of Signs, we list possible rational roots as follows:

Possible rational roots

$$= \frac{\text{Factors of the constant term, 4}}{\text{Factors of the leading coefficient, 1}} = \frac{\pm 1, \pm 2, \pm 4}{\pm 1} = \pm 1, \quad \pm 2, \quad \pm 4.$$

However, Descartes's Rule of Signs informed us that $f(x) = x^3 + 2x^2 + 5x + 4$ has no positive real zeros. Thus, the polynomial equation $x^3 + 2x^2 + 5x + 4 = 0$ has no positive real roots. This means that we can eliminate the positive numbers from our list of possible rational roots. Possible rational roots include only $-1, -2$, and -4. We can use synthetic division and test two of the three possible rational roots as follows:

$$\begin{array}{r|rrrr} -1 & 1 & 2 & 5 & 4 \\ & & -1 & -1 & -4 \\ \hline & 1 & 1 & 4 & 0 \end{array} \qquad \begin{array}{r|rrrr} -2 & 1 & 2 & 5 & 4 \\ & & -2 & 0 & -10 \\ \hline & 1 & 0 & 5 & -6 \end{array}$$

The zero remainder shows that −1 is a root.

The nonzero remainder shows that −2 is not a root.

We do not need to test the third possible rational root using synthetic division. Based on our work in Example 5, we know that there are either one or three negative roots. If -2 is not a root and it is one of three possible negative roots, the polynomial equation cannot have three negative roots. Therefore, there is only one negative root, -1. The equation is of degree 3 and will have a total of three solutions. Two of the solutions will be nonreal complex numbers in a conjugate pair. (Verify this by completing the solution process.)

Check Point 5

Determine the possible number of positive and negative real zeros of $f(x) = x^4 - 14x^3 + 71x^2 - 154x + 120$.

EXERCISE SET 3.4

Practice Exercises

In Exercises 1–8, use the Rational Zero Theorem to list all possible rational zeros for each given function.

1. $f(x) = x^3 + x^2 - 4x - 4$
2. $f(x) = x^3 + 3x^2 - 6x - 8$
3. $f(x) = 3x^4 - 11x^3 - x^2 + 19x + 6$
4. $f(x) = 2x^4 + 3x^3 - 11x^2 - 9x + 15$
5. $f(x) = 4x^4 - x^3 + 5x^2 - 2x - 6$
6. $f(x) = 3x^4 - 11x^3 - 3x^2 - 6x + 8$
7. $f(x) = x^5 - x^4 - 7x^3 + 7x^2 - 12x - 12$
8. $f(x) = 4x^5 - 8x^4 - x + 2$

In Exercises 9–14,

a. *List all possible rational zeros.*

b. *Use synthetic division to test the possible rational zeros and find an actual zero.*

c. *Use the zero from part (b) to find all the zeros of the polynomial function.*

9. $f(x) = x^3 + x^2 - 4x - 4$
10. $f(x) = x^3 - 2x^2 - 11x + 12$
11. $f(x) = 2x^3 - 3x^2 - 11x + 6$
12. $f(x) = 2x^3 - 5x^2 + x + 2$
13. $f(x) = 3x^3 + 7x^2 - 22x - 8$
14. $f(x) = 3x^3 + 8x^2 - 15x + 4$

In Exercises 15–22,

a. *List all possible rational roots.*

b. *Use synthetic division to test the possible rational roots and find an actual root.*

c. *Use the root from part (b) and solve the equation.*

15. $x^3 - 2x^2 - 11x + 12 = 0$
16. $x^3 - 2x^2 - 7x - 4 = 0$
17. $x^3 - 10x - 12 = 0$
18. $x^3 - 5x^2 + 17x - 13 = 0$
19. $6x^3 + 25x^2 - 24x + 5 = 0$
20. $2x^3 - 5x^2 - 6x + 4 = 0$
21. $x^4 - 2x^3 - 5x^2 + 8x + 4 = 0$
22. $x^4 - 2x^2 - 16x - 15 = 0$

In Exercises 23–28, use Descartes's Rule of Signs to determine the possible number of positive and negative real zeros for each given function.

23. $f(x) = x^3 + 2x^2 + 5x + 4$
24. $f(x) = x^3 + 7x^2 + x + 7$
25. $f(x) = 5x^3 - 3x^2 + 3x - 1$
26. $f(x) = -2x^3 + x^2 - x + 7$
27. $f(x) = 2x^4 - 5x^3 - x^2 - 6x + 4$
28. $f(x) = 4x^4 - x^3 + 5x^2 - 2x - 6$

In Exercises 29–40, find all zeros of the polynomial function or solve the given polynomial equation. Use the Rational Zero Theorem and Descartes's Rule of Signs as an aid in obtaining the first zero or the first root.

29. $f(x) = x^3 - 4x^2 - 7x + 10$

30. $f(x) = x^3 + 12x^2 + 21x + 10$

31. $2x^3 - x^2 - 9x - 4 = 0$

32. $3x^3 - 8x^2 - 8x + 8 = 0$

33. $x^4 - 3x^3 - 20x^2 - 24x - 8 = 0$

34. $x^4 - x^3 + 2x^2 - 4x - 8 = 0$

35. $f(x) = 3x^4 - 11x^3 - x^2 + 19x + 6$

36. $f(x) = 2x^4 + 3x^3 - 11x^2 - 9x + 15$

37. $4x^4 - x^3 + 5x^2 - 2x - 6 = 0$

38. $3x^4 - 11x^3 - 3x^2 - 6x + 8 = 0$

39. $2x^5 + 7x^4 - 18x^2 - 8x + 8 = 0$

40. $4x^5 + 12x^4 - 41x^3 - 99x^2 + 10x + 24 = 0$

Application Exercises

41. Suppose that a polynomial function f is used to model the data shown in the graph using

(number of years after 1993, thousands of deaths at the workplace).

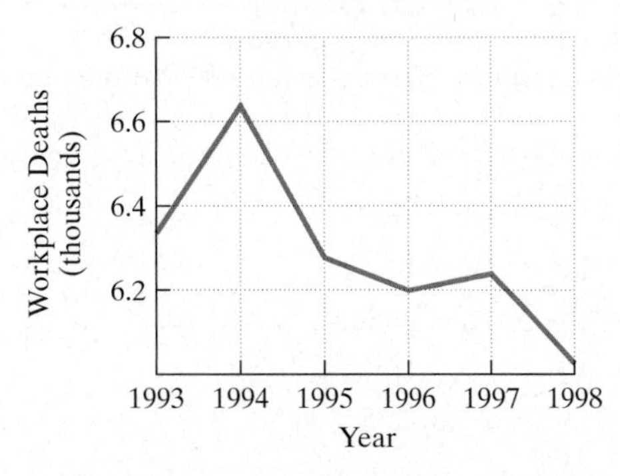

Source: F. B. I.

a. Use the graph to solve the polynomial equation $f(x) = 6.2$.

b. Describe the degree and the leading coefficient of the function f that can be used to model the data in the graph.

42. Suppose that a polynomial function f is used to model the data shown in the graph at the top of the next column using

(number of years after 1995, average cost of a computer in thousands of dollars).

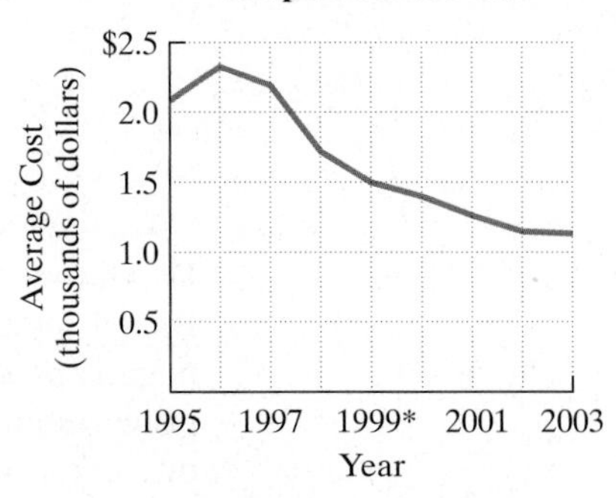

*Costs are projected from 1999 to 2003
Source: National Science Foundation

Use the graph to solve the polynomial equation $f(x) = 1.5$.

43. The number of eggs, N, in a female moth is a function of her abdominal width, W, in millimeters, modeled by $N = 14W^3 - 17W^2 - 16W + 34$, for $1.5 \leq W \leq 3.5$. What is the abdominal width when there are 211 eggs?

44. The concentration of a drug, in parts per million, in a patient's blood x hours after the drug is administered is given by the function

$$f(x) = -x^4 + 12x^3 - 58x^2 + 132x.$$

How many hours after the drug is administered will it be eliminated from the bloodstream?

45. The width of a rectangular box is twice the height and the length is 7 inches more than the height. If the volume is 72 cubic inches, find the dimensions of the box.

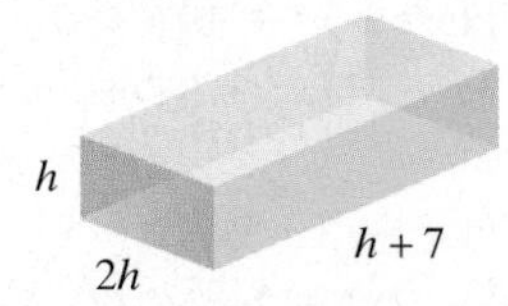

46. A box with an open top is formed by cutting squares out of the corners of a rectangular piece of cardboard 10 inches by 8 inches and then folding up the sides. If x represents the length of the side of the square cut from each corner of the rectangle, what size square must be cut if the volume of the box is to be 48 cubic inches?

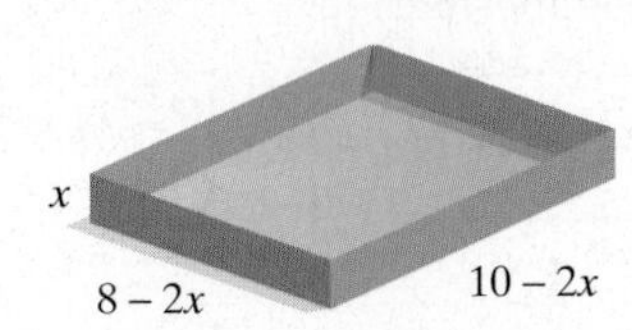

Writing in Mathematics

47. Describe how to find the possible rational zeros of a polynomial function.

48. Describe how to use Descartes's Rule of signs to determine the possible number of positive real zeros of a polynomial function.

49. Describe how to use Descartes's rule of signs to determine the possible number of negative roots of a polynomial equation.

50. Why must every polynomial equation of degree 3 have at least one real root?

51. Explain why the equation $x^4 + 6x^2 + 2 = 0$ has no rational roots.

52. Suppose $\frac{3}{4}$ is a root of a polynomial equation. What does this tell us about the leading coefficient and the constant term in the equation?

53. The number of AIDS cases in the United States for the years 1983 through 1990 is approximated by the function

$$f(x) = -143x^3 + 1810x^2 - 187x + 2331$$

where x represents the number of years after 1983. Use the Rational Zero Theorem to explain why, according to this formula, 14,199 cases could not have occurred 5 years after 1983.

Technology Exercises

The equations in Exercises 54–57 have real roots that are rational. Use the Rational Zero Theorem to list all possible rational roots. Then graph the polynomial function in the given viewing rectangle to determine which possible rational roots are actual roots of the equation.

54. $2x^3 - 15x^2 + 22x + 15 = 0$; $[-1, 6, 1]$ by $[-50, 50, 1]$

55. $6x^3 - 19x^2 + 16x - 4 = 0$; $[0, 2, 1]$ by $[-3, 2, 1]$

56. $2x^4 + 7x^3 - 4x^2 - 27x - 18 = 0$; $[-4, 3, 1]$ by $[-45, 45, 1]$

57. $4x^4 + 4x^3 + 7x^2 - x - 2 = 0$; $[-2, 2, 1]$ by $[-5, 5, 1]$

58. Use Descartes's Rule of Signs to determine the possible number of positive and negative real zeros of $f(x) = 3x^4 + 5x^2 + 2$. What does this mean in terms of the graph of f? Verify your result by using a graphing utility to graph f.

59. Use Descartes's Rule of Signs to determine the possible number of positive and negative real zeros of $f(x) = x^5 - x^4 + x^3 - x^2 + x - 8$. Verify your result by using a graphing utility to graph f.

60. Make up a number of polynomial functions of odd degree and graph each function. Is it possible for the graph to have no real zeros? Explain. Try doing the same thing for polynomial functions of even degree. Now is it possible to have no real zeros?

Critical Thinking Exercises

61. Which one of the following is true?

a. The equation $x^3 + 5x^2 + 6x + 1 = 0$ has one positive real root.

b. Descartes's Rule of Signs gives the exact number of positive and negative real roots for a polynomial equation.

c. Every polynomial equation of degree 3 has at least one rational root.

d. None of the above is true.

62. Give an example of a polynomial equation that has no real roots. Describe how you obtained the equation.

63. If the volume of the solid shown in the figure is 208 cubic inches, find the value of x.

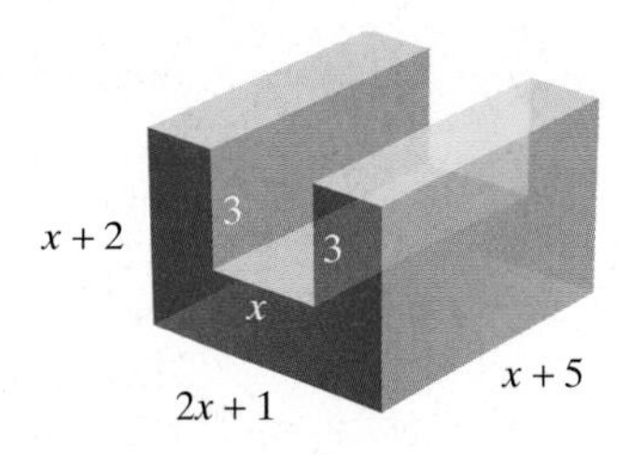

64. In this exercise, we lead you through the steps involved in the proof of the Rational Zero Theorem. Consider the polynomial equation

$$a_n x^n + a_{n-1}x^{n-1} + a_{n-2}x^{n-2} + \cdots + a_1 x + a_0 = 0$$

where $\frac{p}{q}$ is a rational root reduced to lowest terms.

a. Substitute $\frac{p}{q}$ for x in the equation and show that the equation can be written as

$$a_n p^n + a_{n-1}p^{n-1}q + a_{n-2}p^{n-2}q^2 + \cdots + a_1 pq^{n-1} = -a_0 q^n.$$

b. Why is p a factor of the left side of the equation?

c. Because p divides the left side, it must also divide the right side. However, because $\frac{p}{q}$ is reduced to lowest terms, p cannot divide q. Thus, p and q have no common factors other than -1 and 1. Because p does divide the right side and it is not a factor of q^n, what can you conclude?

d. Rewrite the equation from part (a) with all terms containing q on the left and the term that does not have a factor of q on the right. Use an argument that parallels parts (b) and (c) to conclude that q is a factor of a_n.

SECTION 3.5 More On Zeros of Polynomial Functions

Objectives

1. Find bounds for the roots of a polynomial equation.
2. Approximate real zeros.
3. Use conjugate roots to solve a polynomial equation.
4. Use the Linear Factorization Theorem to factor a polynomial.
5. Find polynomials with given zeros.

You stole my formula!

Tartaglia's Secret Formula for One Solution of $x^3 + mx = n$

$$x = \sqrt[3]{\sqrt{\left(\frac{n}{2}\right)^2 + \left(\frac{m}{3}\right)^3} + \frac{n}{2}} - \sqrt[3]{\sqrt{\left(\frac{n}{2}\right)^2 + \left(\frac{m}{3}\right)^3} - \frac{n}{2}}$$

Popularizers of mathematics are sharing bizarre stories that are giving math a secure place in popular culture. One episode, able to compete with the wildest fare served up by television talk shows and the tabloids, involves three Italian mathematicians and, of all things, zeros of polynomial functions.

Tartaglia (1499–1557), poor and starving, has found a formula that gives a root for a third-degree polynomial equation. Cardano (1501–1576) begs Tartaglia to reveal the secret formula, wheedling it from him with the promise he will find the impoverished Tartalia a patron. Then Cardano publishes his famous work *Ars Magna*, in which he presents Tartaglia's formula as his own. Cardano uses his most talented student, Ferrari (1522–1565), who derived a formula for a root of a fourth-degree polynomial equation, to falsely accuse Tartaglia of plagiarism. The dispute becomes violent and Tartaglia is fortunate to escape alive.

The noise from this "You Stole My Formula" episode is quieted by the work of French mathematician Evariste Galois (1811–1832). Galois proved that there is no general formula for finding roots of polynomial equations of degree 5 or higher. There are, of course, methods for finding roots. In this section, we continue our study of methods for finding zeros of polynomial functions.

1 Find bounds for the roots of a polynomial equation.

Upper and Lower Bounds for Roots

The **Upper and Lower Bound Theorem** helps us rule out many of a polynomial equation's possible rational roots. Figure 3.20 illustrates that a is a **lower bound** and b is an **upper bound** for the roots of $f(x) = 0$ because every real root c of the equation satisfies $a \le c \le b$.

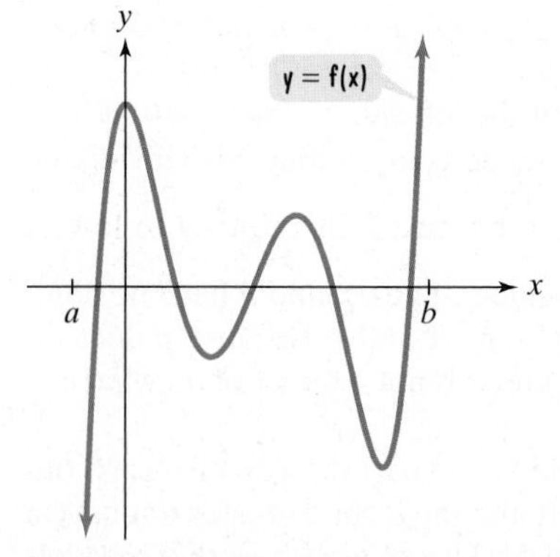

Figure 3.20 b is an upper bound and a is a lower bound for the real roots of $f(x) = 0$.

The Upper and Lower Bound Theorem

Let $f(x)$ be a polynomial with real coefficients and a positive leading coefficient, and let a and b be nonzero real numbers.

1. Divide $f(x)$ by $x - b$ (where $b > 0$) using synthetic division. If the last row containing the quotient and remainder has no negative numbers, then b is an **upper bound** for the real roots of $f(x) = 0$.
2. Divide $f(x)$ by $x - a$ (where $a < 0$) using synthetic division. If the last row containing the quotient and remainder has numbers that alternate in sign (zero entries count as positive or negative), then a is a **lower bound** for the real roots of $f(x) = 0$.

EXAMPLE 1 Finding Bounds for the Roots

Show that all the real roots of the equation $8x^3 + 10x^2 - 39x + 9 = 0$ lie between -3 and 2.

Solution We begin by showing that 2 is an upper bound. Divide the polynomial by $x - 2$. If all the numbers in the bottom row of the synthetic division are nonnegative, then 2 is an upper bound.

$$\begin{array}{r|rrrr} 2 & 8 & 10 & -39 & 9 \\ & & 16 & 52 & 26 \\ \hline & 8 & 26 & 13 & 35 \end{array}$$

All numbers in this row are nonnegative.

The nonnegative entries in the last row verify that 2 is an upper bound. Next, we show that -3 is a lower bound. Divide the polynomial by $x - (-3)$, or $x + 3$. If the numbers in the bottom row of the synthetic division alternate in sign, then -3 is a lower bound. Remember that the number zero can be considered positive or negative.

$$\begin{array}{r|rrrr} -3 & 8 & 10 & -39 & 9 \\ & & -24 & 42 & -9 \\ \hline & 8 & -14 & 3 & 0 \end{array}$$

Counting 0 as negative, the signs alternate: $+, -, +, -$.

By the Upper and Lower Bound Theorem, the alternating signs in the last row indicate that -3 is a lower bound for the roots. (The zero remainder indicates that -3 is also a root.)

Check Point 1 Show that all the real roots of the equation $2x^3 + 11x^2 - 7x - 6 = 0$ lie between -7 and 2.

How might the Upper and Lower Bound Theorem be helpful in solving a polynomial equation? Consider the equation

$$x^4 + 3x^3 - 27x^2 + 3x - 28 = 0.$$

With a leading coefficient of 1 and a constant term of -28, the possible rational roots are

$$\pm 1, \quad \pm 2, \quad \pm 4, \quad \pm 7, \quad \pm 14, \quad \pm 28.$$

We begin testing for an actual root using synthetic division. The following divisions indicate that 1 and 2 are not roots because of the nonzero remainders. However, something interesting happens when testing 4.

$$\begin{array}{r|rrrrr} 1 & 1 & 3 & -27 & 3 & -28 \\ & & 1 & 4 & -23 & -20 \\ \hline & 1 & 4 & -23 & -20 & -48 \end{array} \qquad \begin{array}{r|rrrrr} 2 & 1 & 3 & -27 & 3 & -28 \\ & & 2 & 10 & -34 & -62 \\ \hline & 1 & 5 & -17 & -31 & -90 \end{array}$$

$$\begin{array}{r|rrrrr} 4 & 1 & 3 & -27 & 3 & -28 \\ & & 4 & 28 & 4 & 28 \\ \hline & 1 & 7 & 1 & 7 & 0 \end{array}$$

Nonnegative numbers

4 is a root of $f(x) = 0$ because the remainder is 0.

4 is an upper bound for the roots of $f(x) = 0$.

Notice that 4 is both a root and an upper bound for the roots. Should you take the time to use synthetic division and test 7, 14, and 28? There is no need to do this because all three numbers exceed 4, the upper bound for the roots. Thus, 7, 14, and 28 cannot be roots of the equation.

Technology

The Upper and Lower Bound Theorem and your knowledge of polynomial functions can help you to find a reasonable range setting when using your graphing utility. Consider

$$f(x) = x^4 + 3x^3 - 27x^2 + 3x - 28.$$

Based on our discussion, 4 is a zero and an upper bound for the zeros. We can also use synthetic division to show that -7 is a zero and a lower bound for the zeros. We can use these lower and upper bounds to determine Xmin and Xmax. We'll go one unit to the left and to the right of these bounds and use $[-8, 5, 1]$. Now, how do we determine Ymin and Ymax? Let's see what kinds of values of y we obtain when we evaluate the function between -8 and 5. Using synthetic division, direct substitution, or the table feature of some graphing utilities, we have $f(-6) = -370$, $f(-5) = -468$, $f(0) = -28$, and $f(3) = -100$. These evaluations suggest that we can use -500 for Ymin and 100 for Ymax. The graph of $f(x) = x^4 + 3x^3 - 27x^2 + 3x - 28$ is shown in a $[-8, 5, 1]$ by $[-500, 100, 20]$ viewing rectangle in Figure 3.21. Because the degree is even ($n = 4$) and the leading coefficient, 1, is positive, the graph should rise to the left and right. This is precisely what occurs in Figure 3.21. Our work in obtaining this complete graph is an excellent illustration of the fact that technology complements human knowledge and is not intended to replace it.

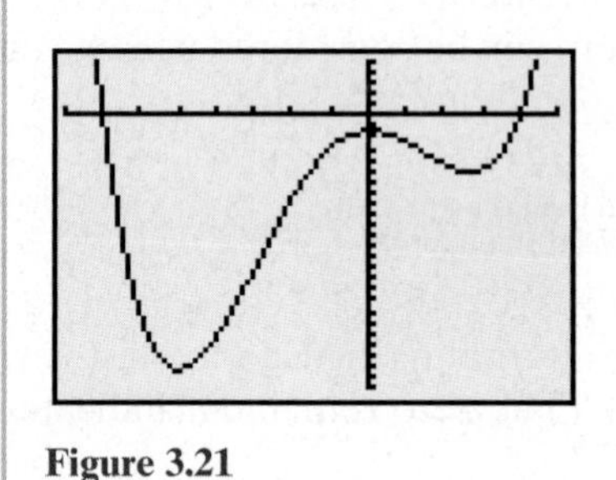

Figure 3.21

2 Approximate real zeros.

The Intermediate Value Theorem

We can find decimal approximations for real zeros of polynomial functions using a graphing utility. The **Intermediate Value Theorem** tells us of the existence of real zeros and how to approximate them. The idea behind the theorem is illustrated in Figure 3.22. The figure shows that if $(a, f(a))$ lies below the x-axis and $(b, f(b))$ lies above the x-axis, the smooth, continuous graph of a polynomial function f must cross the x-axis at some value c between a and b. This value is a real zero for the function.

These observations are summarized in the **Intermediate Value Theorem**.

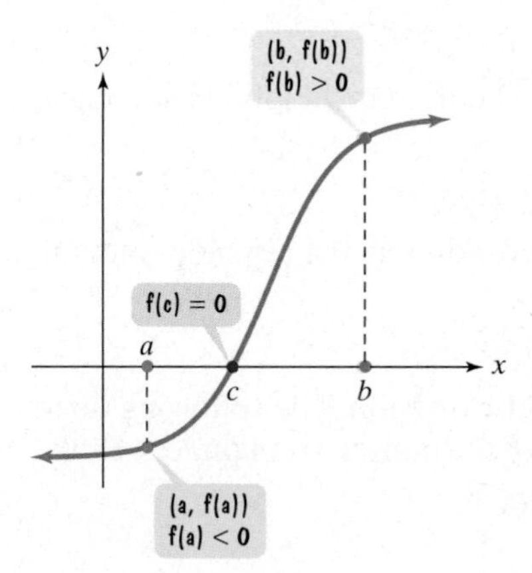

Figure 3.22 The graph must cross the x-axis at some value between a and b.

The Intermediate Value Theorem for Polynomials

Let $f(x)$ be a polynomial function with real coefficients. If $f(a)$ and $f(b)$ have opposite signs, then there is at least one value of c between a and b for which $f(c) = 0$. Equivalently, the equation $f(x) = 0$ has at least one real root between a and b.

EXAMPLE 2 Approximating a Real Zero

a. Show that the polynomial function $f(x) = x^3 - 2x - 5$ has a real zero between 2 and 3.

b. Use the Intermediate Value Theorem to find an approximation for this real zero to the nearest tenth.

Solution

a. Let us evaluate $f(x)$ at 2 and 3. If $f(2)$ and $f(3)$ have opposite signs, then there is a real zero between 2 and 3. Using $f(x) = x^3 - 2x - 5$, we obtain

$$f(2) = 2^3 - 2 \cdot 2 - 5 = 8 - 4 - 5 = -1$$

$f(2)$ is negative.

and

$$f(3) = 3^3 - 2 \cdot 3 - 5 = 27 - 6 - 5 = 16.$$

$f(3)$ is positive.

This sign change shows that the polynomial function has a real zero between 2 and 3.

b. A numerical approach is to evaluate f at successive tenths between 2 and 3, looking for a sign change. This sign change will place the real zero between a pair of successive tenths.

x	$f(x) = x^3 - 2x - 5$
2	$f(2) = 2^3 - 2(2) - 5 \quad = -1$
2.1	$f(2.1) = (2.1)^3 - 2(2.1) - 5 = 0.061$

Sign change

The sign change indicates that f has a real zero between 2 and 2.1. We now follow a similar procedure to locate the real zero between successive hundredths. We divide the interval $[2, 2.1]$ into ten equal subintervals. Then we evaluate f at each endpoint and look for a sign change.

$f(2.00) = -1$ | $f(2.06) = -0.378184$
$f(2.01) = -0.899399$ | $f(2.07) = -0.270257$
$f(2.02) = -0.797592$ | $f(2.08) = -0.161088$
$f(2.03) = -0.694573$ | $f(2.09) = -0.050671$
$f(2.04) = -0.590336$ | $f(2.1) = 0.061$
$f(2.05) = -0.484875$

Sign change

The sign change indicates that f has a real zero between 2.09 and 2.1. Correct to the nearest tenth, the zero is 2.1.

Technology

The following graph was obtained by entering

$$y = x^3 - 2x - 5$$

and

$$y = 0$$

and using the intersection feature in a $[-3, 3, 1]$ by $[-10, 10, 1]$ viewing rectangle. Correct to the nearest thousandth, the function's real zero is 2.095.

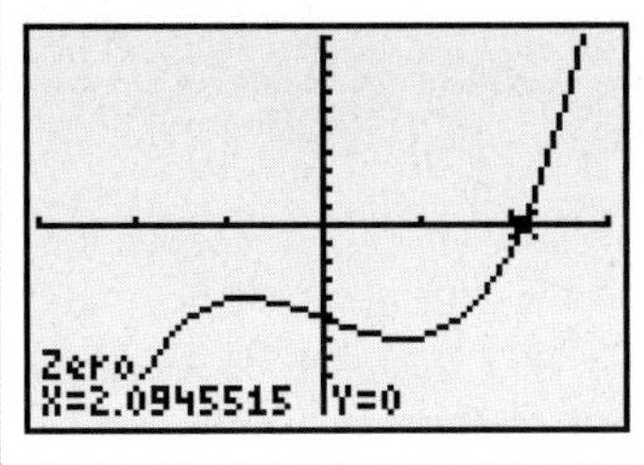

Check Point 2 Show that the polynomial function $f(x) = 3x^3 - 10x + 9$ has a real zero between -3 and -2.

3 Use conjugate roots to solve a polynomial equation.

The Fundamental Theorem of Algebra

We have seen that if a polynomial equation is of degree n, then counting multiple roots separately, the equation has n roots. Some of these roots may be nonreal complex numbers that occur in conjugate pairs, such as $2 + i$ and $2 - i$.

EXAMPLE 3 Using Conjugate Roots to Solve a Polynomial Equation

Solve $x^4 - 4x^3 + 3x^2 + 8x - 10 = 0$ given that $2 + i$ is a root.

Solution The degree of the given equation is 4. This means that there are four roots. One of the roots is $2 + i$. Because complex nonreal roots come in conjugate pairs, we know that $2 - i$ is a second root. By the Factor Theorem, both

$$[x - (2 + i)] \quad \text{and} \quad [x - (2 - i)]$$

are factors of the given polynomial. We multiply these known factors.

$$[x - (2 + i)][x - (2 - i)]$$

F O I L

$$= x^2 - x(2 - i) - x(2 + i) + (2 + i)(2 - i)$$ Multiply using the FOIL method.

$$= x^2 - 2x + ix - 2x - ix + (4 - i^2)$$ Continue multiplying.

$$= x^2 - 2x + ix - 2x - ix + [4 - (-1)]$$ Simplify using $i^2 = -1$.

$$= x^2 - 4x + 5$$ Combine like terms.

At this point we have only two of the four possible roots, $2 + i$ and $2 - i$. We can find the other two roots by factoring the given equation. We have found that $x^2 - 4x + 5$ is one of the factors. We can find the other factor(s) by dividing $x^2 - 4x + 5$ into the polynomial on the left side of the given equation.

$$\begin{array}{r} x^2 \quad\quad\quad\quad\quad - 2 \\ x^2 - 4x + 5 \overline{) x^4 - 4x^3 + 3x^2 + 8x - 10} \\ \underline{x^4 - 4x^3 + 5x^2} \quad\quad\quad\quad\quad \\ -2x^2 + 8x - 10 \\ \underline{-2x^2 + 8x - 10} \\ 0 \end{array}$$

The zero remainder confirms that $x^2 - 4x + 5$ is a factor.

Technology

The graph of $f(x) = x^4 - 4x^3 + 3x^2 + 8x - 10$ is shown in a $[-4, 4, 1]$ by $[-15, 5, 1]$ viewing rectangle. The real roots of $f(x) = 0$, the equation in Example 3, are $-\sqrt{2}$ and $\sqrt{2}$. These appear as x-intercepts at approximately -1.4 and 1.4.

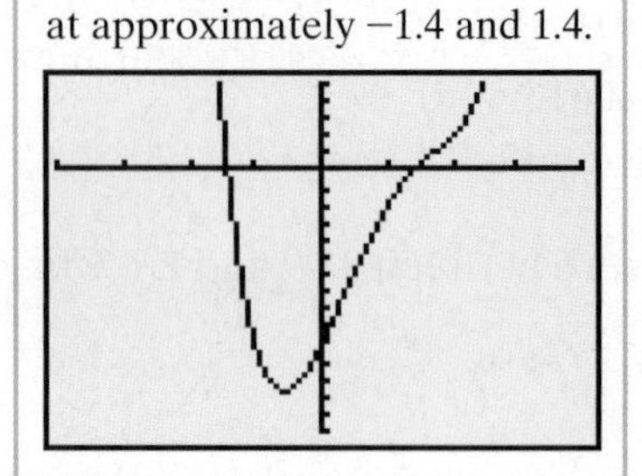

We can now solve the given equation.

$$x^4 - 4x^3 + 3x^2 + 8x - 10 = 0$$ This is the original equation.

$$(x^2 - 4x + 5)(x^2 - 2) = 0$$ Factor using the result of the polynomial long division.

$$x^2 - 4x + 5 = 0 \quad \text{or} \quad x^2 - 2 = 0$$ Set each factor equal to 0.

$$x = 2 \pm i \quad\quad\quad x = \pm\sqrt{2}$$ Solve for x. We know the roots for the first equation, $x^2 - 4x + 5 = 0$, by our previous analysis.

The solution set is $\{2 \pm i, \pm\sqrt{2}\}$.

Check Point 3 Solve $x^4 - 8x^3 + 64x - 105 = 0$ given that $2 - i$ is a root.

The fact that a polynomial equation of degree n has n roots is a consequence of a theorem proved in 1799 by a 22-year-old student named Carl Friedrich Gauss in his doctoral dissertation. His result is called the **Fundamental Theorem of Algebra**.

The Fundamental Theorem of Algebra

If $f(x)$ is a polynomial of degree n, where $n \geq 1$, then the equation $f(x) = 0$ has at least one complex root.

Suppose, for example, that $f(x) = 0$ represents a polynomial equation of degree n. By the Fundamental Theorem of Algebra, we know that this equation has at least one complex root; we'll call it c_1. By the Factor Theorem, we know that $x - c_1$ is a factor of $f(x)$. Therefore, we obtain

$$(x - c_1)q_1(x) = 0 \quad \text{The degree of the polynomial } q_1(x) \text{ is } n - 1.$$

$$x - c_1 = 0 \quad \text{or} \quad q_1(x) = 0 \quad \text{Set each factor equal to 0.}$$

If the degree of $q_1(x)$ is at least 1, by the Fundamental Theorem of Algebra the equation $q_1(x) = 0$ has at least one complex root. We'll call it c_2. The Factor Theorem gives us

$$q_1(x) = 0 \quad \text{The degree of } q_1(x) \text{ is } n - 1.$$

$$(x - c_2)q_2(x) = 0 \quad \text{The degree of } q_2(x) \text{ is } n - 2.$$

$$x - c_2 = 0 \quad \text{or} \quad q_2(x) = 0. \quad \text{Set each factor equal to 0.}$$

Let's see what we have up to this point, and then continue the process.

$$f(x) = 0 \quad \text{This is the original polynomial equation of degree } n.$$

$$(x - c_1)q_1(x) = 0 \quad \text{This is the result from our first application of the Fundamental Theorem.}$$

$$(x - c_1)(x - c_2)q_2(x) = 0 \quad \text{This is the result from our second application of the Fundamental Theorem.}$$

By continuing this process, we will obtain the product of n linear factors. Setting each of these linear factors equal to zero results in n complex roots. Thus, if $f(x)$ is a polynomial of degree n, where $n \geq 1$, then $f(x) = 0$ has exactly n roots, where roots are counted according to their multiplicity.

4 Use the Linear Factorization Theorem to factor a polynomial.

The Linear Factorization Theorem

In Example 3, we found that $x^4 - 4x^3 + 3x^2 + 8x - 10 = 0$ has $\{2 \pm i, \pm\sqrt{2}\}$ as a solution set. The polynomial can be factored over the complex nonreal numbers as follows:

These are the four zeros.

$$f(x) = x^4 - 4x^3 + 3x^2 + 8x - 10$$
$$= [x - (2 + i)][x - (2 - i)](x + \sqrt{2})(x - \sqrt{2})$$

These are four linear factors.

This fourth-degree polynomial has four linear factors. Just as an nth-degree polynomial equation has n roots, an nth-degree polynomial has n linear factors. This is formally stated as the **Linear Factorization Theorem**.

The Linear Factorization Theorem

If $f(x) = a_n x^n + a_{n-1}x^{n-1} + \cdots + a_1 x + a_0$, where $n \geq 1$ and $a_n \neq 0$, then

$$f(x) = a_n(x - c_1)(x - c_2)\cdots(x - c_n)$$

where $c_1, c_2, \ldots, c_n$ are complex numbers (possibly real and not necessarily distinct). In words: An nth-degree polynomial can be expressed as the product of n linear factors.

The Linear Factorization Theorem involves factors somewhat different than those you are used to seeing. For example, the polynomial $x^2 - 3$ is irreducible over the rational numbers. However, it can be factored over the real numbers as follows:

$$x^2 - 3 = (x + \sqrt{3})(x - \sqrt{3}).$$ Use $a^2 - b^2 = (a + b)(a - b)$ with $a = x$ and $b = \sqrt{3}$.

Study Tip

The sum of squares, irreducible over the real numbers, can be factored over the complex nonreal numbers as

$$a^2 + b^2 = (a + bi)(a - bi).$$

The polynomial $x^2 + 1$ is irreducible over the real numbers, but reducible over the complex nonreal numbers.

$$x^2 + 1 = (x + i)(x - i)$$

EXAMPLE 4 Factoring a Polynomial

Factor $x^4 - 3x^2 - 28$:

a. As the product of factors that are irreducible over the rational numbers.
b. As the product of factors that are irreducible over the real numbers.
c. In completely factored form involving complex nonreal numbers.

Solution

a. $x^4 - 3x^2 - 28 = (x^2 - 7)(x^2 + 4)$ Both quadratic factors are irreducible over the rational numbers.

b. $= (x + \sqrt{7})(x - \sqrt{7})(x^2 + 4)$ The third factor is still irreducible over the real numbers.

c. $= (x + \sqrt{7})(x - \sqrt{7})(x + 2i)(x - 2i)$ This is the completely factored form using complex nonreal numbers.

Factor $x^4 - 4x^2 - 5$ as the product of factors that are irreducible over **a.** the rational numbers; **b.** the real numbers; **c.** the complex nonreal numbers.

5 Find polynomials with given zeros.

Reversing Things: Finding Polynomials When the Zeros Are Given

Many of our problems involving polynomial functions and polynomial equations dealt with the process of finding zeros and roots. The Linear Factorization Theorem enables us to reverse this process, finding a polynomial function when the zeros are given.

EXAMPLE 5 Finding a Polynomial Function with Given Zeros

Find a fourth-degree polynomial function $f(x)$ with real coefficients that has $-2, 2$, and i as zeros and such that $f(3) = -150$.

Solution Because i is a zero and the polynomial has real coefficients, the conjugate must also be a zero. We can now use the Linear Factorization Theorem.

$f(x) = a_n(x - c_1)(x - c_2)(x - c_3)(x - c_4)$ This is the linear factorization for a fourth-degree polynomial.

$= a_n(x + 2)(x - 2)(x - i)(x + i)$ Use the given zeros: $c_1 = -2$, $c_2 = 2$, $c_3 = i$, and, from above, $c_4 = -i$.

Technology

The graph of $f(x) = -3x^4 + 9x^2 + 12$, shown in a $[-3, 3, 1]$ by $[-200, 20, 20]$ viewing rectangle, verifies that -2 and 2 are real zeros. By tracing along the curve, we can check that $f(3) = -150$.

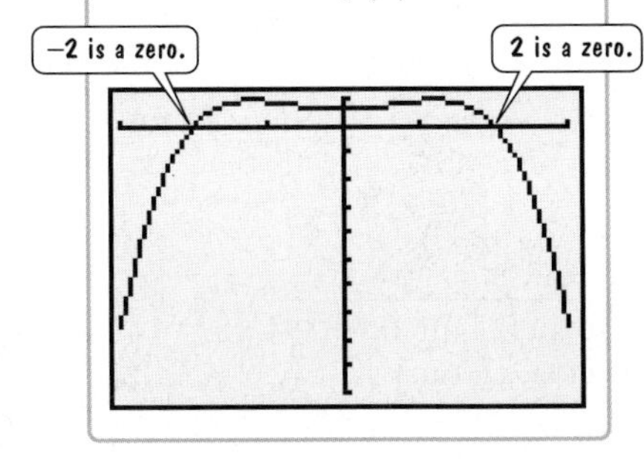

$$= a_n(x^2 - 4)(x^2 + 1) \quad \text{Multiply.}$$

$$f(x) = a_n(x^4 - 3x^2 - 4) \quad \text{Complete the multiplication.}$$

$$f(3) = a_n(3^4 - 3 \cdot 3^2 - 4) = -150 \quad \text{To find } a_n\text{, use the fact that } f(3) = -150.$$

$$a_n(81 - 27 - 4) = -150 \quad \text{Solve for } a_n.$$

$$50a_n = -150$$

$$a_n = -3$$

Substituting -3 for a_n in the formula for $f(x)$, we obtain

$$f(x) = -3(x^4 - 3x^2 - 4).$$

Equivalently,

$$f(x) = -3x^4 + 9x^2 + 12.$$

Check Point 5 Find a third-degree polynomial function $f(x)$ with real coefficients that has -3 and i as zeros and such that $f(1) = 8$.

EXERCISE SET 3.5

Practice Exercises

Use the Upper and Lower Bound Theorem to solve Exercises 1–6.

1. Show that all the real roots of the equation $x^4 - 5x^3 + 11x^2 + 33x - 18 = 0$ lie between -4 and 7.
2. Show that all the real roots of the equation $x^4 + 11x^3 - 12x^2 + 6 = 0$ lie between -13 and 1.
3. Show that all the real roots of the equation $2x^3 + 5x^2 - 8x - 7 = 0$ lie between -4 and 2.
4. Show that all the real roots of the equation $2x^5 - 13x^3 + 2x - 5 = 0$ lie between -3 and 3.
5. Consider the equation $x^4 + 3x^3 + 2x^2 - 5x + 12 = 0$.
   a. List all possible rational roots.
   b. Determine whether 1 is a root using synthetic division. What two conclusions can you draw?
   c. Based on part (b), what possible rational roots can you eliminate?
   d. Determine whether -3 is a root using synthetic division. What two conclusions can you draw?
   e. Based on part (d), what possible rational roots can you eliminate?
6. Consider the equation $2x^5 + 5x^4 - 8x^3 - 14x^2 + 6x + 9 = 0$.
   a. List all possible rational roots.
   b. Determine whether $\frac{3}{2}$ is a root using synthetic division. What two conclusions can you draw?
   c. Based on part (b), what possible rational roots can you eliminate?
   d. Determine whether -3 is a root using synthetic division. What two conclusions can you draw?
   e. Based on part (d), what possible rational roots can you eliminate?

In Exercises 7–14, show that each polynomial has a real zero between the given integers. Then use the Intermediate Value Theorem to find an approximation for this zero to the nearest tenth.

7. $f(x) = x^3 - x - 1$; between 1 and 2
8. $f(x) = x^3 - 4x^2 + 2$; between 0 and 1
9. $f(x) = 2x^4 - 4x^2 + 1$; between -1 and 0
10. $f(x) = x^4 + 6x^3 - 18x^2$; between 2 and 3
11. $f(x) = x^3 + x^2 - 2x + 1$; between -3 and -2
12. $f(x) = x^5 - x^3 - 1$; between 1 and 2
13. $f(x) = 3x^3 - 10x + 9$; between -3 and -2
14. $f(x) = 3x^3 - 8x^2 + x + 2$; between 2 and 3

In Exercises 15–22, use the given root to find the solution set of the polynomial equation.

15. $x^3 - 2x^2 + 4x - 8 = 0$; $-2i$
16. $x^4 + 13x^2 + 36 = 0$; $3i$
17. $3x^3 - 7x^2 + 8x - 2 = 0$; $1 + i$
18. $x^3 - 7x^2 + 16x - 10 = 0$; $3 + i$
19. $x^4 - 6x^2 + 25 = 0$; $2 - i$
20. $x^4 - x^3 - 9x^2 + 29x - 60 = 0$; $1 + 2i$

21. $x^4 - 8x^3 + 64x - 105 = 0;\ 2 - i$

22. $4x^4 - 28x^3 + 129x^2 - 130x + 125 = 0;\ 3 - 4i$

In Exercises 23–28, factor each polynomial:

a. *as the product of factors that are irreducible over the rational numbers*

b. *as the product of factors that are irreducible over the real numbers*

c. *in completely factored form involving complex nonreal numbers*

23. $x^4 - x^2 - 20$

24. $x^4 + 6x^2 - 27$

25. $x^4 + x^2 - 6$

26. $x^4 - 9x^2 - 22$

27. $x^4 - 2x^3 + x^2 - 8x - 12$
(*Hint*: One factor is $x^2 + 4$.)

28. $x^4 - 4x^3 + 14x^2 - 36x + 45$
(*Hint*: One factor is $x^2 + 9$.)

In Exercises 29–36, find an nth-degree polynomial function with real coefficients satisfying the given conditions. If you are using a graphing utility, use it to graph the function and verify the real zeros and the given function value.

29. $n = 3$; 1 and $5i$ are zeros; $f(-1) = -104$

30. $n = 3$; 4 and $2i$ are zeros; $f(-1) = -50$

31. $n = 3$; -5 and $4 + 3i$ are zeros; $f(2) = 91$

32. $n = 3$; 6 and $-5 + 2i$ are zeros; $f(2) = -636$

33. $n = 4$; i and $3i$ are zeros; $f(-1) = 20$

34. $n = 4$; $-2, -\frac{1}{2}$, and i are zeros; $f(1) = 18$

35. $n = 4$; $-2, 5$, and $3 + 2i$ are zeros; $f(1) = -96$

36. $n = 4$; $-4, \frac{1}{3}$, and $2 + 3i$ are zeros; $f(1) = 100$

In Exercises 37–44, find all the zeros of the function and write the polynomial as a product of linear factors.

37. $f(x) = x^3 - x^2 + 25x - 25$

38. $f(x) = x^3 - 10x^2 + 33x - 34$

39. $f(x) = x^3 - 8x^2 + 25x - 26$

40. $f(x) = x^3 - 8x^2 + 17x - 4$

41. $f(x) = x^4 + 37x^2 + 36$

42. $f(x) = x^4 + 8x^3 + 9x^2 - 10x + 100$

43. $f(x) = 16x^4 + 36x^3 + 16x^2 + x - 30$

44. $f(x) = 2x^4 - x^3 + 7x^2 - 4x - 4$

Application Exercises

We have seen the polynomial function

$$H(x) = -0.001183x^4 + 0.05495x^3 - 0.8523x^2 + 9.054x + 6.748$$

that models the age in human years, $H(x)$, of a dog that is x years old, where $x \geq 1$. Although the coefficients make it difficult to solve equations algebraically using this function, a graph of the function makes approximate solutions possible. Use the graph shown to solve Exercises 45–46.

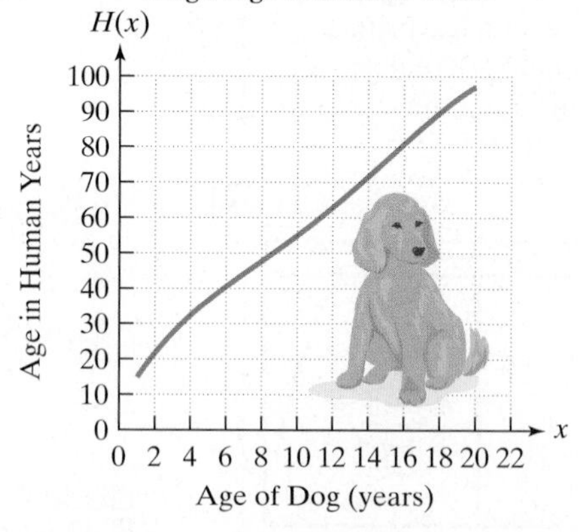

45. If you are 25, what is the equivalent age for dogs?

46. If you are 30, what is the equivalent age for dogs?

47. Set up an equation to answer the question in either Exercise 45 or 46. Bring all terms to one side and obtain zero on the other side. What are some of the difficulties involved in solving this equation? Explain how the Intermediate Value Theorem can be used to verify the approximate solution that you obtained from the graph.

The bar graph shows the cost of Medicare, in billions of dollars, projected through 2005. Using the regression feature of a graphing utility, these data can be modeled by

a linear function, $f(x) = 27x + 163$;

a quadratic function, $g(x) = 1.2x^2 + 15.2x + 181.4$;

a third-degree polynomial function,
$h(x) = 0.08x^3 - 0.06x^2 + 20.08x + 178.32$.

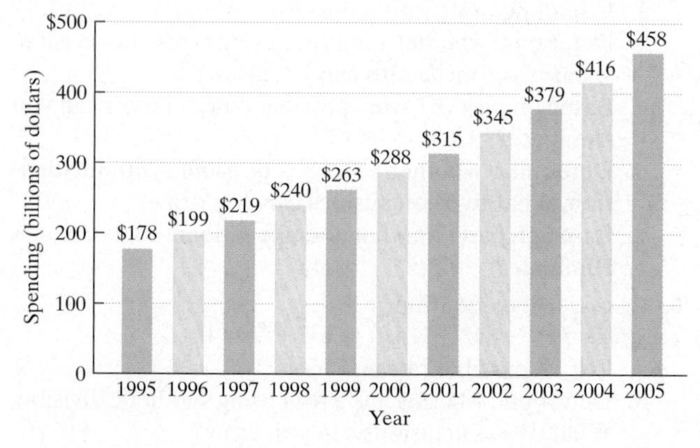

Source: Congressional Budget Office

For each of these functions, x represents the number of years after 1995 and the function value represents Medicare spending, in billions of dollars, for that year.

48. The graph indicates that Medicare spending will reach \$379 billion in 2003. Substitute 379 for $f(x)$, $g(x)$, and $h(x)$ in each of the three models. Then solve each resulting equation, if possible, to find how many years after 1995 spending will reach \$379 billion. Which of the three functions is the best model for 2003?

49. The graph indicates that Medicare spending will reach \$458 billion in 2005. Substitute 458 for $f(x)$, $g(x)$, and $h(x)$ in each of the three models. Then solve each resulting equation, if possible, to find how many years after 1995 spending will reach \$458 billion. Which of the three functions is the best model for 2005?

Writing in Mathematics

50. When testing a number using synthetic division, how do you know if it is an upper bound for the real roots?

51. When testing a number using synthetic division, how do you know if it is a lower bound for the real roots?

52. How do you show that a polynomial function has a real zero between two given numbers?

53. How does the linear factorization of $f(x)$, that is,

$$f(x) = a_n(x - c_1)(x - c_2)\cdots(x - c_n),$$

show that a polynomial equation of degree n has n roots?

Technology Exercises

54. Show that -1 is a lower bound of $f(x) = x^3 - 53x^2 + 103x - 51$. Show that 60 is an upper bound. Use this information and a graphing utility to draw a relatively complete graph of f.

For Exercises 55–56, use a graphing utility to determine upper and lower bounds for the zeros of f. Does synthetic division verify your observations?

55. $f(x) = 2x^3 + x^2 - 14x - 7$

56. $f(x) = 2x^4 - 7x^3 - 5x^2 + 28x - 12$

57. The function $f(x) = -0.00002x^3 + 0.008x^2 - 0.3x + 6.95$ models the number of annual physician visits, f, by a person of age x.

a. Graph the function for meaningful values of x and discuss what the graph reveals in terms of the variables described by the model.

b. Use the polynomial root-finding capability of your graphing utility to find the age (to the nearest year) for the group that averages 13.43 annual physician visits.

c. Verify part (b) using the graph of f.

Use a graphing utility to obtain a complete graph for each polynomial function in Exercises 58–61. Then determine the number of real zeros and the number of nonreal complex zeros for each function.

58. $f(x) = x^3 - 6x - 9$

59. $f(x) = 3x^5 - 2x^4 + 6x^3 - 4x^2 - 24x + 16$

60. $f(x) = 3x^4 + 4x^3 - 7x^2 - 2x - 3$

61. $f(x) = x^6 - 64$

Critical Thinking Exercises

In Exercises 62–64, what is the smallest degree that each polynomial could have?

62.

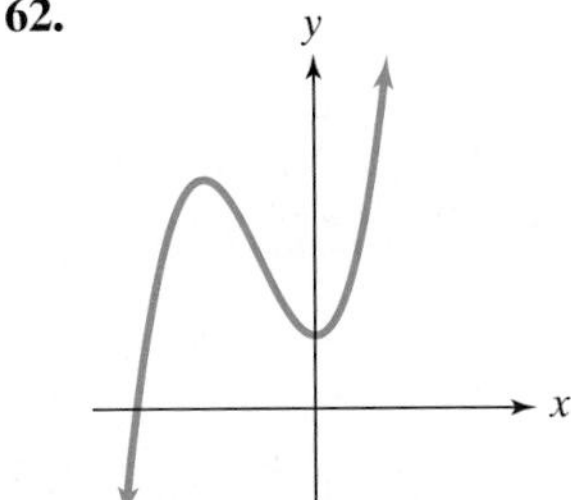

63.

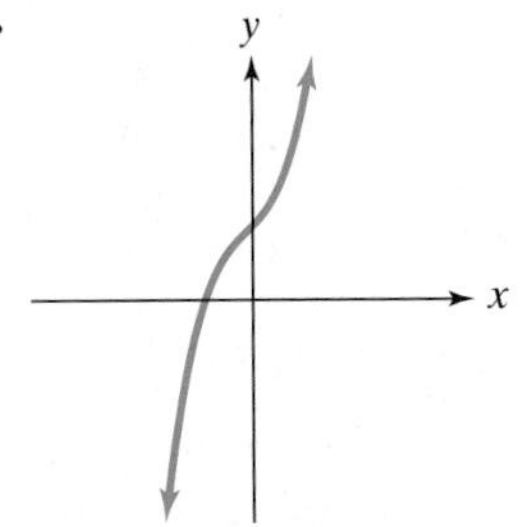

64.

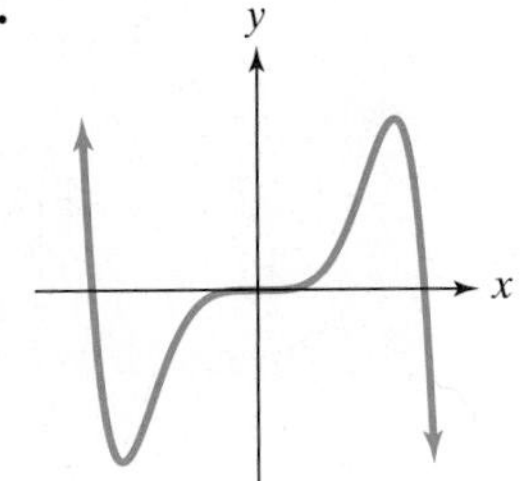

65. Explain why nonreal complex zeros are gained or lost in pairs in terms of graphs of polynomial functions.

66. Explain why a polynomial function of degree 20 cannot cross the x-axis exactly once.

67. Give an example of a function that is not subject to the Intermediate Value Theorem.

Group Exercise

68. The graphs on page 314 show costs for private and public four-year colleges projected through the year 2017. According to these projections, your daughter's college education at a private four-year school could cost about \$250,000. This activity involves forming and using models from these data. Group members should begin by deciding whether to work with data for private or public colleges.

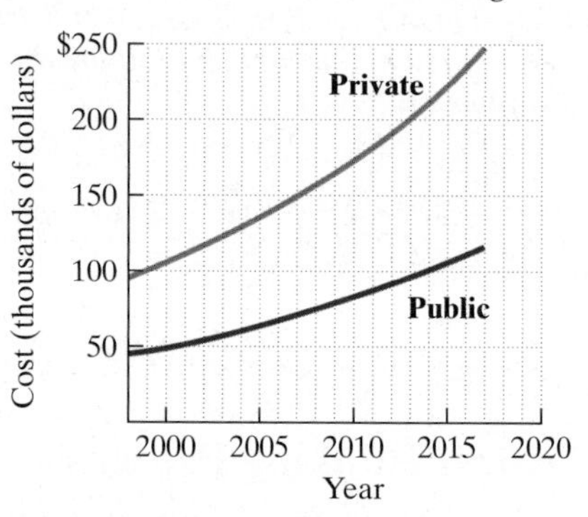

Source: U.S. Department of Education

a. Let $x = 0$ correspond to 1998, $x = 1$ to 1999, $x = 2$ to 2000, and so on up through $x = 19$ for 2017. Group members should use the chosen line graph to obtain a reasonable estimate for the cost of an education, y, in thousands of dollars, for each x.

b. Use the regression feature of a graphing utility to model the data for the cost of a four-year college x years after 1998 using a linear function, a quadratic function, and a third-degree polynomial function.

c. Use these functions to write and solve a problem similar to Exercise 48 or 49.

d. Use these functions to make predictions well into the future. Which function, if any, seems to be most reasonable in its predicted cost? Of course, these are only predictions, subject to unforeseeable events. What events might render each of these models relatively useless over long periods of time?

SECTION 3.6 Rational Functions and Their Graphs

Objectives

1. Find the domain of rational functions.
2. Use arrow notation.
3. Identify vertical asymptotes.
4. Identify horizontal asymptotes.
5. Graph rational functions.
6. Identify slant asymptotes.
7. Solve applied problems involving rational functions.

Technology is now promising to bring light, fast, and beautiful wheelchairs to millions of disabled people. The cost of manufacturing these radically different wheelchairs can be modeled by rational functions. In this section we will see how graphs of these functions illustrate that low prices are possible with high production levels, urgently needed in this situation. There are more than half a billion people with disabilities in developing countries; an estimated 20 million need wheelchairs right now.

1 Find the domain of rational functions.

Rational Functions

Rational functions are quotients of polynomial functions. This means that rational functions can be expressed as

$$f(x) = \frac{p(x)}{q(x)}$$

where $p(x)$ and $q(x)$ are polynomial functions and $q(x) \neq 0$. The **domain** of a rational function is the set of all real numbers except the x-values that make the denominator zero. For example, the domain of the rational function

$$f(x) = \frac{x^2 + 7x + 9}{x(x - 2)(x + 5)}$$

This is $p(x)$.

This is $q(x)$.

is the set of all real numbers except 0, 2, and −5.

EXAMPLE 1 Finding the Domain of a Rational Function

Find the domain of each rational function.

a. $f(x) = \frac{x^2 - 9}{x - 3}$ b. $g(x) = \frac{x}{x^2 - 9}$ c. $h(x) = \frac{x + 3}{x^2 + 9}$

Solution Rational functions contain division. Because division by 0 is undefined, we must exclude from the domain of each function values of x that cause the polynomial function in the denominator to be 0.

a. The denominator of $f(x) = \frac{x^2 - 9}{x - 3}$ is 0 if $x = 3$. Thus, x cannot equal 3. The domain of f consists of all real numbers except 3, written $\{x \mid x \neq 3\}$.

b. The denominator of $g(x) = \frac{x}{x^2 - 9}$ is 0 if $x = -3$ or $x = 3$. Thus, the domain of g consists of all real numbers except −3 and 3, written $\{x \mid x \neq -3, x \neq 3\}$.

c. No real numbers cause the denominator of $h(x) = \frac{x + 3}{x^2 + 9}$ to equal 0. The domain of h consists of all real numbers.

Check Point 1 Find the domain of each rational function.

a. $f(x) = \frac{x^2 - 25}{x - 5}$ **b.** $g(x) = \frac{x}{x^2 - 25}$ **c.** $h(x) = \frac{x + 5}{x^2 + 25}$

2 Use arrow notation.

The most basic rational function is the **reciprocal function**, defined by $f(x) = \frac{1}{x}$. The denominator of the reciprocal function is zero when $x = 0$, so the domain of f is the set of all real numbers except for 0.

Let's look at the behavior of f near the excluded value 0. We start by evaluating $f(x)$ to the left of 0.

x approaches 0 from the left.

x	−1	−0.5	−0.1	−0.01	−0.001
$f(x) = \frac{1}{x}$	−1	−2	−10	−100	−1000

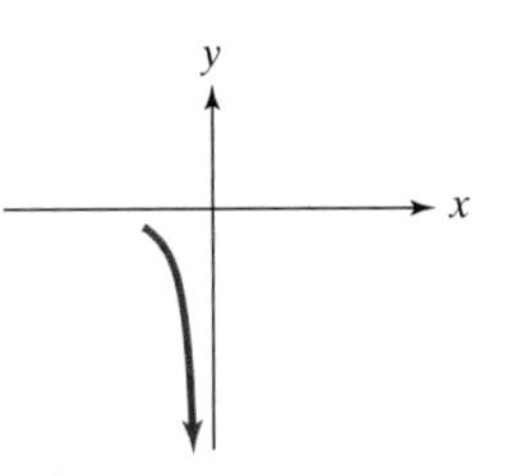

Mathematically, we say that "x approaches 0 from the left." From the table and the accompanying graph, it appears that as x approaches 0 from the left, the function values, $f(x)$, decrease without bound. We say that "$f(x)$ approaches negative infinity." We use a special arrow notation to describe this situation symbolically:

$$f(x) \to -\infty \quad \text{as} \quad x \to 0^-$$

f(x) approaches negative infinity (that is, the graph falls) as x approaches 0 from the left.

Observe that the minus ($-$) superscript on the 0 $(x \to 0^-)$ is read "from the left."

Next, we evaluate $f(x)$ to the right of 0.

x approaches 0 from the right.

x	0.001	0.01	0.1	0.5	1
$f(x) = \frac{1}{x}$	1000	100	10	2	1

Mathematically, we say that "x approaches 0 from the right." From the table and the accompanying graph, it appears that as x approaches 0 from the right, the function values, $f(x)$, increase without bound. We say that "$f(x)$ approaches infinity." We again use a special arrow notation to describe this situation symbolically:

$$f(x) \to \infty \quad \text{as} \quad x \to 0^+$$

f(x) approaches infinity (that is, the graph rises) as x approaches 0 from the right.

Observe that the plus ($+$) superscript on the 0 $(x \to 0^+)$ is read "from the right."

Now let's see what happens to the function values, $f(x)$, as x gets farther away from the origin. The following tables suggest what happens to $f(x)$ as x increases or decreases without bound.

x increases without bound:

x	1	10	100	1000
$f(x) = \frac{1}{x}$	1	0.1	0.01	0.001

x decreases without bound:

x	-1	-10	-100	-1000
$f(x) = \frac{1}{x}$	-1	-0.1	-0.01	-0.001

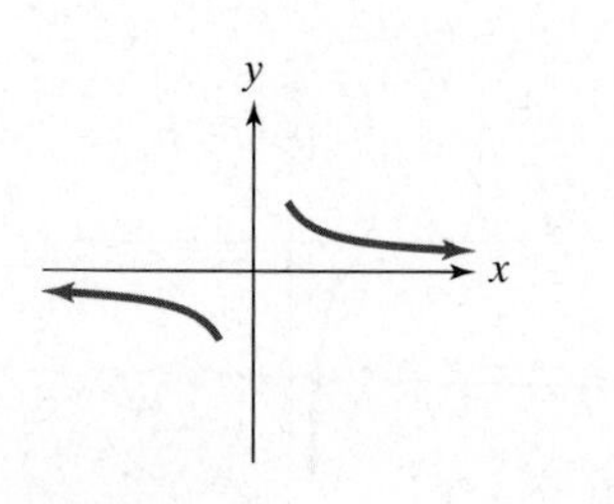

Figure 3.23
$f(x)$ approaches 0 as x increases or decreases without bound

Figure 3.23 illustrates the end behavior of $f(x) = \frac{1}{x}$ as x increases or decreases without bound. The function values, $f(x)$, are getting progressively closer to 0. This means that the graph of f is approaching the horizontal line $y = 0$ (that is, the x-axis) as x increases or decreases without bound. We use the arrow notation to describe this situation.

$$f(x) \to 0 \quad \text{as} \quad x \to \infty \qquad \text{and} \qquad f(x) \to 0 \quad \text{as} \quad x \to -\infty$$

f(x) approaches 0 as x increases without bound.

f(x) approaches 0 as x decreases without bound.

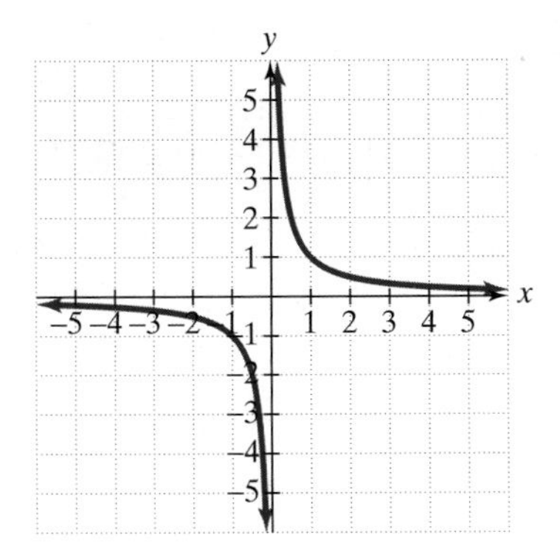

Figure 3.24 The graph of the reciprocal function $f(x) = \frac{1}{x}$

Thus, as x approaches infinity ($x \to \infty$) or as x approaches negative infinity ($x \to -\infty$), the function values are approaching zero: $f(x) \to 0$.

The graph of the reciprocal function $f(x) = \frac{1}{x}$ is shown in Figure 3.24. Unlike the graph of a polynomial function, the graph of the reciprocal function has a break in it and is composed of two distinct branches.

The arrow notation used throughout our discussion of the reciprocal function is summarized in the following box.

Arrow Notation

Symbol	Meaning
$x \to a^+$	x approaches a from the right.
$x \to a^-$	x approaches a from the left.
$x \to \infty$	x approaches infinity; that is, x increases without bound.
$x \to -\infty$	x approaches negative infinity; that is, x decreases without bound.

Vertical Asymptotes of Rational Functions

3 Identify vertical asymptotes.

Look again at the graph of $f(x) = \frac{1}{x}$. The curve approaches, but does not touch, the y-axis. The y-axis, or $x = 0$, is said to be a **vertical asymptote** of the graph. A rational function may have no vertical asymptotes, one vertical asymptote, or several vertical asymptotes. The graph of a rational function never intersects a vertical asymptote. We will use dashed lines to show asymptotes.

Definition of a Vertical Asymptote

The line $x = a$ is a vertical asymptote of the graph of a function f if $f(x)$ increases or decreases without bound as x approaches a.

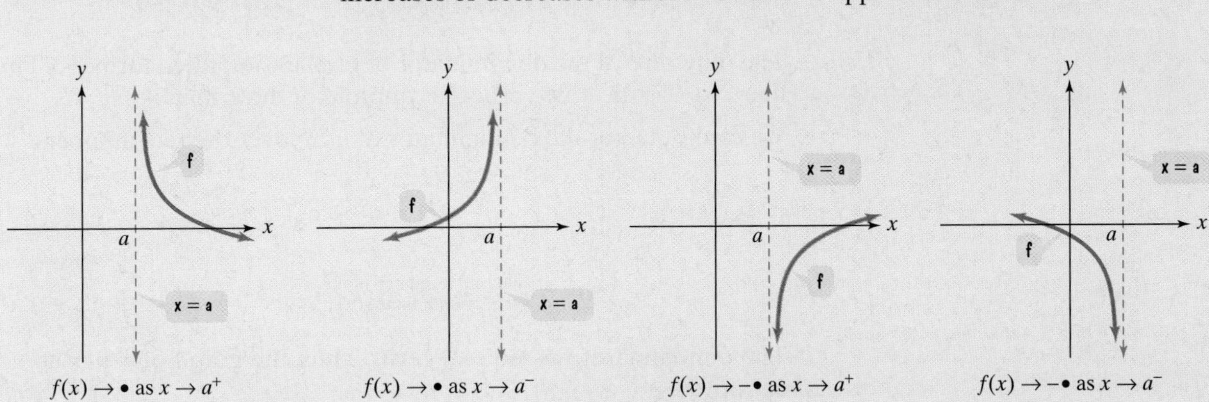

Thus, $f(x) \to \infty$ or $f(x) \to -\infty$ as x approaches a from either the left or the right.

If the graph of a rational function has vertical asymptotes, they can be located by using the following theorem.

Locating Vertical Asymptotes

If $f(x) = \dfrac{p(x)}{q(x)}$ is a rational function in which $p(x)$ and $q(x)$ have no common factors and a is a zero of $q(x)$, the denominator, then $x = a$ is a vertical asymptote of the graph of f.

EXAMPLE 2 Finding the Vertical Asymptotes of a Rational Function

Find the vertical asymptotes, if any, of the graph of each rational function.

a. $f(x) = \dfrac{x}{x^2 - 9}$ b. $g(x) = \dfrac{x+3}{x^2 - 9}$ c. $h(x) = \dfrac{x+3}{x^2 + 9}$

Solution Factoring is usually helpful in identifying zeros of denominators.

a.
$$f(x) = \frac{x}{x^2 - 9} = \frac{x}{(x+3)(x-3)}$$

This factor is 0 if $x = -3$. This factor is 0 if $x = 3$.

There are no common factors in the numerator and the denominator. The zeros of the denominator are -3 and 3. Thus, the lines $x = -3$ and $x = 3$ are the vertical asymptotes for the graph of f.

b. We will use factoring to see if there are common factors.

$$g(x) = \frac{x+3}{x^2 - 9} = \frac{(x+3)}{(x+3)(x-3)} = \frac{1}{x-3}$$

There is a common factor, $x + 3$, so simplify. This denominator is 0 if $x = 3$.

The only zero of the denominator of $g(x)$ in simplified form is 3. Thus, the line $x = 3$ is the only vertical asymptote of the graph of g.

c. We cannot factor the denominator of $h(x)$ over the real numbers.

$$h(x) = \frac{x+3}{x^2 + 9}$$

No real numbers make this denominator 0.

The denominator has no real zeros. Thus, the graph of h has no vertical asymptotes.

Check Point 2 Find the vertical asymptotes, if any, of the graph of each rational function.

a. $f(x) = \dfrac{x}{x^2 - 1}$ **b.** $g(x) = \dfrac{x-1}{x^2 - 1}$ **c.** $h(x) = \dfrac{x-1}{x^2 + 1}$

A value where the denominator of a function is zero does not necessarily result in a vertical asymptote. There is a hole corresponding to $x = a$, and not a vertical asymptote, in the graph of a function under the following conditions: The value a causes the denominator to be zero, but there is a reduced form of the function's equation in which a does not cause the denominator to be zero.

Consider, for example, the function

$$f(x) = \frac{x^2 - 9}{x - 3}.$$

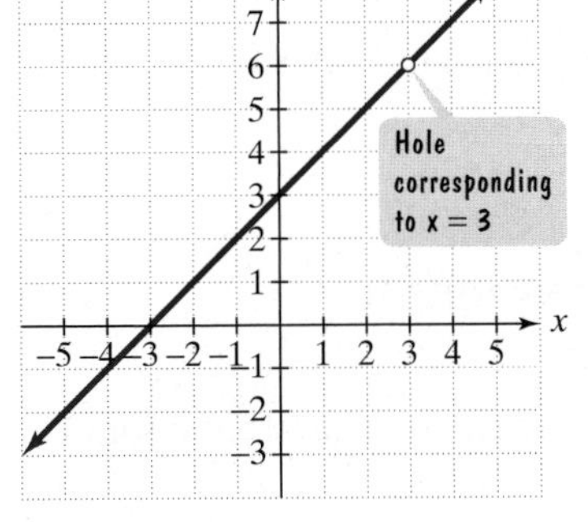

Figure 3.25

Because the denominator is zero when $x = 3$, the function's domain is all real numbers except 3. However, there is a reduced form of the equation in which 3 does not cause the denominator to be zero:

$$f(x) = \frac{x^2 - 9}{x - 3} = \frac{(x + 3)\cancel{(x - 3)}}{\cancel{x - 3}} = x + 3, \quad x \neq 3$$

Denominator is zero at x = 3.

In this reduced form, 3 does not result in a zero denominator.

Figure 3.25 shows that the graph has a hole corresponding to $x = 3$. Graphing utilities do not show this feature of the graph.

4 Identify horizontal asymptotes.

Horizontal Asymptotes of Rational Functions

Figure 3.24 shows the graph of the reciprocal function $f(x) = \frac{1}{x}$. As $x \to \infty$ and as $x \to -\infty$, the function values are approaching 0: $f(x) \to 0$. The line $y = 0$ (that is, the x-axis) is a **horizontal asymptote** of the graph. Many, but not all, rational functions have horizontal asymptotes.

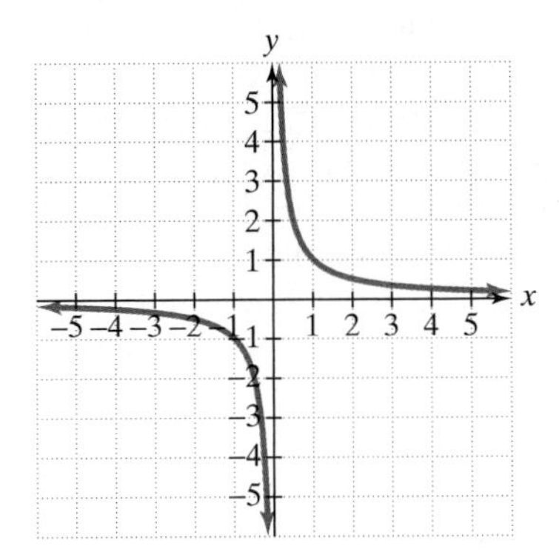

Figure 3.24 The graph of the reciprocal function $f(x) = \frac{1}{x}$, repeated

Definition of a Horizontal Asymptote

The line $y = b$ is a horizontal asymptote of the graph of a function f if $f(x)$ approaches b as x increases or decreases without bound.

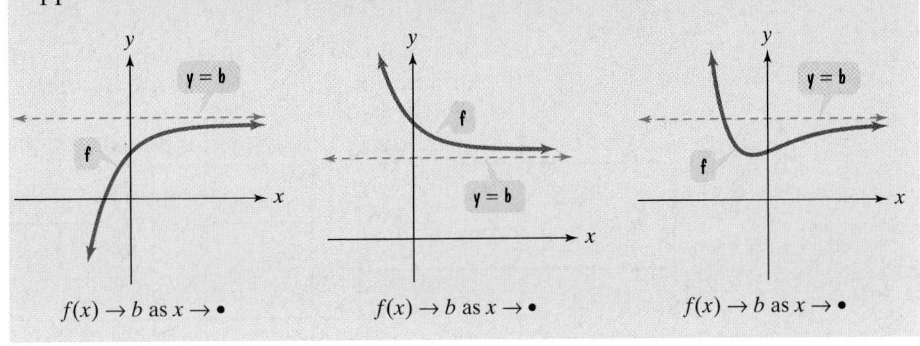

Recall that a rational function may have several vertical asymptotes. By contrast, it can have at most one horizontal asymptote. Although a graph can never intersect a vertical asymptote, it may cross its horizontal asymptote.

If the graph of a rational function has a horizontal asymptote, it can be located by using the following theorem.

Locating Horizontal Asymptotes

Let f be the rational function given by

$$f(x) = \frac{a_n x^n + a_{n-1}x^{n-1} + \cdots + a_1 x + a_0}{b_m x^m + b_{m-1}x^{m-1} + \cdots + b_1 x + b_0}, \quad a_n \neq 0, b_m \neq 0.$$

The degree of the numerator is n. The degree of the denominator is m.

1. If $n < m$, the x-axis is the horizontal asymptote of the graph of f.
2. If $n = m$, the line $y = \frac{a_n}{b_m}$ is the horizontal asymptote of the graph of f.
3. If $n > m$, the graph of f has no horizontal asymptote.

EXAMPLE 3 Finding the Horizontal Asymptote of a Rational Function

Find the horizontal asymptote, if any, of the graph of each rational function.

a. $f(x) = \frac{4x}{2x^2 + 1}$ b. $g(x) = \frac{4x^2}{2x^2 + 1}$ c. $h(x) = \frac{4x^3}{2x^2 + 1}$

Solution

a. $f(x) = \frac{4x}{2x^2 + 1}$

The degree of the numerator, 1, is less than the degree of the denominator, 2. Thus, the graph of f has the x-axis as a horizontal asymptote [see Figure 3.26(a)]. The equation of the horizontal asymptote is $y = 0$.

b. $g(x) = \frac{4x^2}{2x^2 + 1}$

The degree of the numerator, 2, is equal to the degree of the denominator, 2. The leading coefficients of the numerator and denominator, 4 and 2, are used to obtain the equation of the horizontal asymptote. The equation of the horizontal asymptote is $y = \frac{4}{2}$ or $y = 2$ [see Figure 3.26(b)].

c. $h(x) = \frac{4x^3}{2x^2 + 1}$

The degree of the numerator, 3, is greater than the degree of the denominator, 2. Thus, the graph of h has no horizontal asymptote [see Figure 3.26(c)].

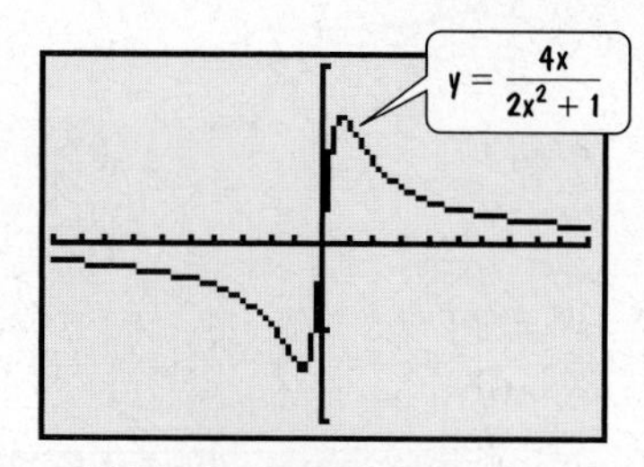

(a) The horizontal asymptote of the graph is $y = 0$.

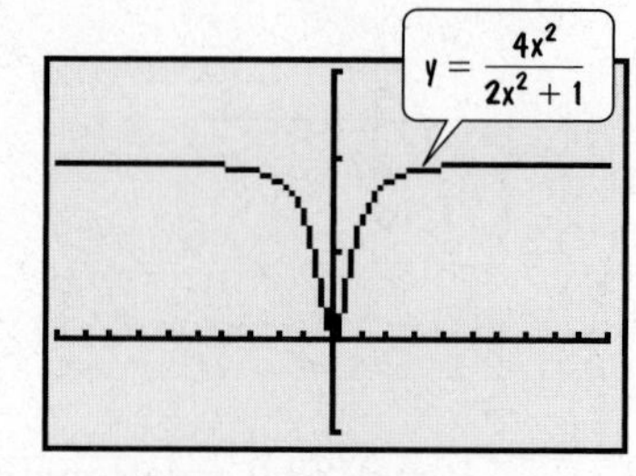

(b) The horizontal asymptote of the graph is $y = 2$.

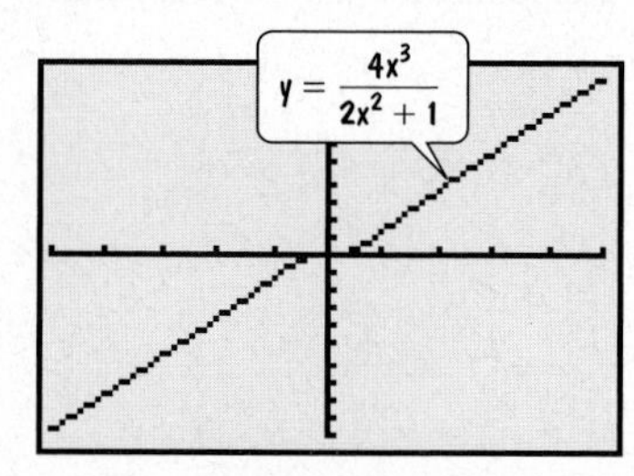

(c) The graph has no horizontal asymptote.

Figure 3.26

EXAMPLE 5 Graphing a Rational Function

Graph: $f(x) = \dfrac{3x^2}{x^2 - 4}$.

Solution

Step 1 Determine symmetry: $f(-x) = \dfrac{3(-x)^2}{(-x)^2 - 4} = \dfrac{3x^2}{x^2 - 4} = f(x)$: Symmetric with respect to the y-axis.

Step 2 Find the y-intercept: $f(0) = \dfrac{3 \cdot 0^2}{0^2 - 4} = \dfrac{0}{-4} = 0$: y-intercept is 0.

Step 3 Find the x-intercept: $3x^2 = 0$, so $x = 0$: x-intercepts is 0.

Step 4 Find the vertical asymptotes: Set $q(x) = 0$.

$$x^2 - 4 = 0 \quad \text{Set the denominator equal to 0.}$$
$$x^2 = 4$$
$$x = \pm 2$$

Vertical asymptotes: $x = -2$ and $x = 2$

Step 5 Find the horizontal asymptote: $y = \frac{3}{1} = 3$.

Step 6 Plot points between and beyond the x-intercept and the vertical asymptotes. With an x-intercept at 0 and vertical asymptotes at $x = -2$ and $x = 2$, we evaluate the function at $-3, -1, 1, 3$, and 4.

x	-3	-1	1	3	4
$f(x) = \dfrac{3x^2}{x^2 - 4}$	$\dfrac{27}{5}$	-1	-1	$\dfrac{27}{5}$	4

Figure 3.29 shows these points, the y-intercept, the x-intercept, and the asymptotes.

Step 7 Graph the function. The graph of $f(x) = \dfrac{3x^2}{x^2 - 4}$ is shown in Figure 3.30. The y-axis symmetry is now obvious.

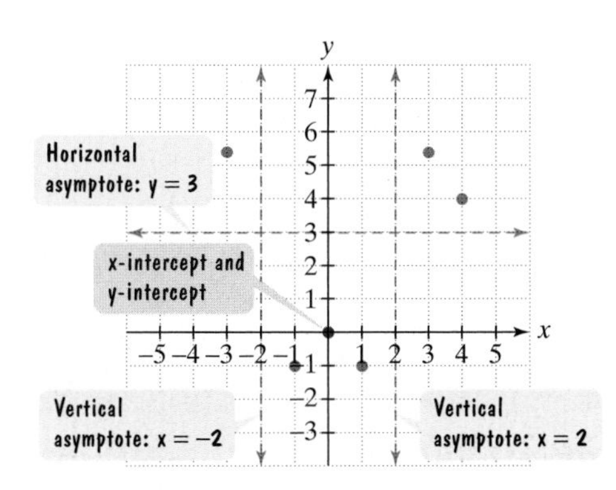

Figure 3.29 Preparing to graph $f(x) = \dfrac{3x^2}{x^2 - 4}$

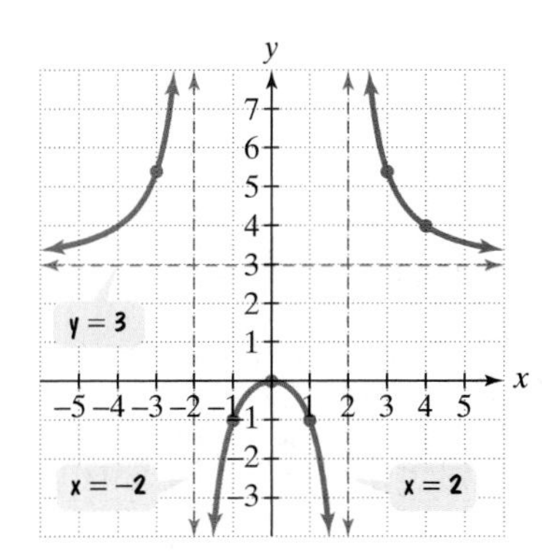

Figure 3.30 The graph of $f(x) = \dfrac{3x^2}{x^2 - 4}$

Technology

The graph of $y = \dfrac{3x^2}{x^2 - 4}$ generated by a graphing utility verifies that our hand-drawn graph is correct.

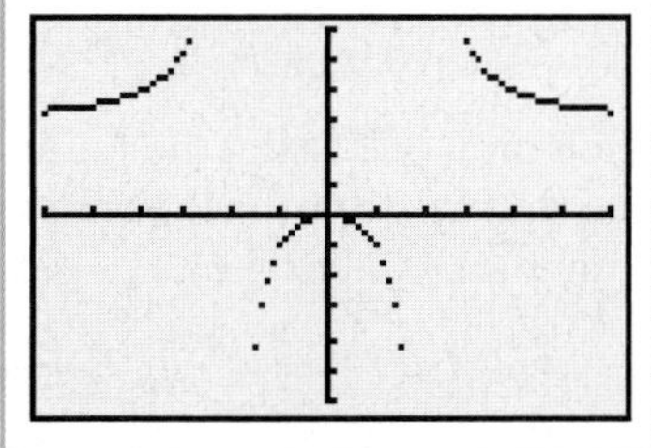

Check Point 5 Graph: $f(x) = \dfrac{2x^2}{x^2 - 9}$.

Example 6 illustrates that not every rational function has vertical and horizontal asymptotes.

EXAMPLE 6 Graphing a Rational Function

Graph: $f(x) = \dfrac{x^4}{x^2 + 1}$.

Solution

Step 1 Determine symmetry: $f(-x) = \dfrac{(-x)^4}{(-x)^2 + 1} = \dfrac{x^4}{x^2 + 1} = f(x)$: Symmetric with respect to the y-axis.

Step 2 Find the y-intercept: $f(0) = \dfrac{0^4}{0^2 + 1} = \dfrac{0}{1} = 0$: y-intercept is 0.

Step 3 Find the x-intercept: $x^4 = 0$, so $x = 0$: x-intercept is 0.

Step 4 Find the vertical asymptote: Set $q(x) = 0$.

$$x^2 + 1 = 0 \quad \text{Set the denominator equal to 0.}$$

$$x^2 = -1$$

Although this equation has imaginary roots ($x = \pm i$), there are no real roots. Thus, there is no vertical asymptote.

Step 5 Find the horizontal asymptote: Because the degree of the numerator, 4, is greater than the degree of the denominator, 2, there is no horizontal asymptote.

Step 6 Plot points between and beyond the x-intercept and the vertical asymptotes. With an x-intercept at 0 and no vertical asymptotes, we evaluate the function at −2, −1, 1, and 2.

x	−2	−1	1	2
$f(x) = \dfrac{x^4}{x^2 + 1}$	$\dfrac{16}{5}$	$\dfrac{1}{2}$	$\dfrac{1}{2}$	$\dfrac{16}{5}$

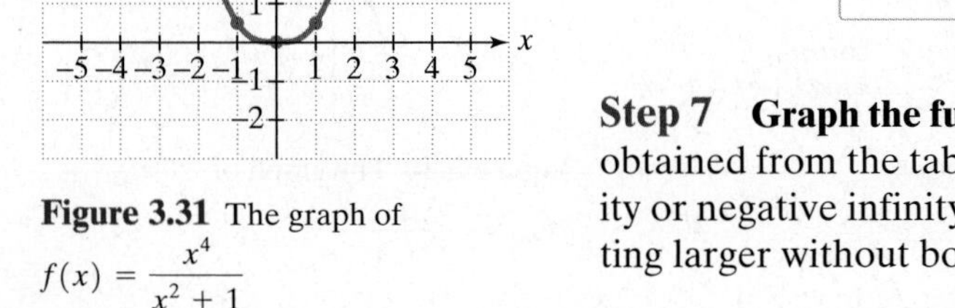

Figure 3.31 The graph of $f(x) = \dfrac{x^4}{x^2 + 1}$

Step 7 Graph the function. Figure 3.31 shows the graph of f using the points obtained from the table and y-axis symmetry. Notice that as x approaches infinity or negative infinity ($x \to \infty$ or $x \to -\infty$), the function values, $f(x)$, are getting larger without bound $[f(x) \to \infty]$.

Check Point 6 Graph: $f(x) = \dfrac{x^4}{x^2 + 2}$.

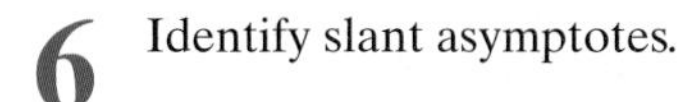

6 Identify slant asymptotes.

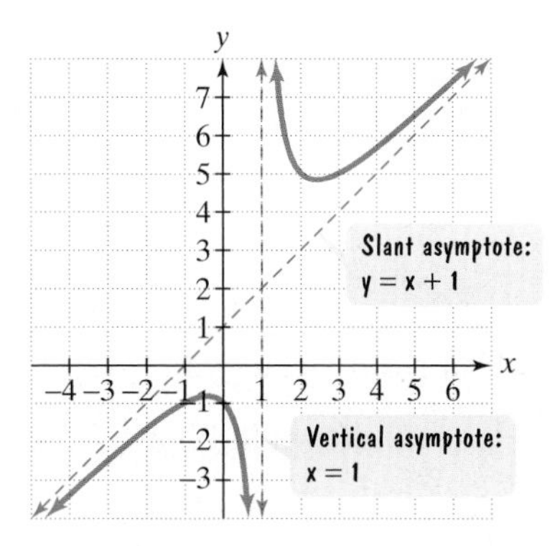

Figure 3.32 The graph of $f(x) = \dfrac{x^2 + 1}{x - 1}$ with a slant asymptote

Slant Asymptotes

Examine the graph of

$$f(x) = \frac{x^2 + 1}{x - 1}$$

shown in Figure 3.32. Note that the degree of the numerator, 2, is greater than the degree of the denominator, 1. Thus, the graph of this function has no horizontal asymptote. However, the graph has a **slant asymptote**, $y = x + 1$.

The graph of a rational function has a slant asymptote if the degree of the numerator is one more than the degree of the denominator. The equation of the slant asymptote can be found by division. For example, to find the slant asymptote for the graph of $f(x) = \dfrac{x^2 + 1}{x - 1}$, divide $x - 1$ into $x^2 + 1$:

$$\begin{array}{r|rrr} 1 & 1 & 0 & 1 \\ & & 1 & 1 \\ \hline & 1 & 1 & 2 \end{array} \qquad \begin{array}{r} 1x + 1 + \dfrac{2}{x - 1} \\ x - 1 \overline{)\, x^2 + 0x + 1} \end{array}$$

Remainder

Observe that

$$f(x) = \frac{x^2 + 1}{x - 1} = \underbrace{x + 1}_{\text{Slant asymptote: } y = x + 1} + \frac{2}{x - 1}$$

If $|x| \to \infty$, the value of $\dfrac{2}{x - 1}$ is approximately 0. Thus, when $|x|$ is large, the function is very close to $y = x + 1 + 0$. This means that as $x \to \infty$ or as $x \to -\infty$, the graph of f gets closer and closer to the line whose equation is $y = x + 1$. The line $y = x + 1$ is a slant asymptote of the graph.

In general, if $f(x) = \dfrac{p(x)}{q(x)}$ and the degree of p is one greater than the degree of q, find the slant asymptote by dividing $q(x)$ into $p(x)$. The division will take the form

$$\frac{p(x)}{q(x)} = mx + b + \frac{\text{remainder}}{q(x)}.$$

Slant asymptote: $y = mx + b$

The equation of the slant asymptote is $y = mx + b$.

EXAMPLE 7 Finding the Slant Asymptote of a Rational Function

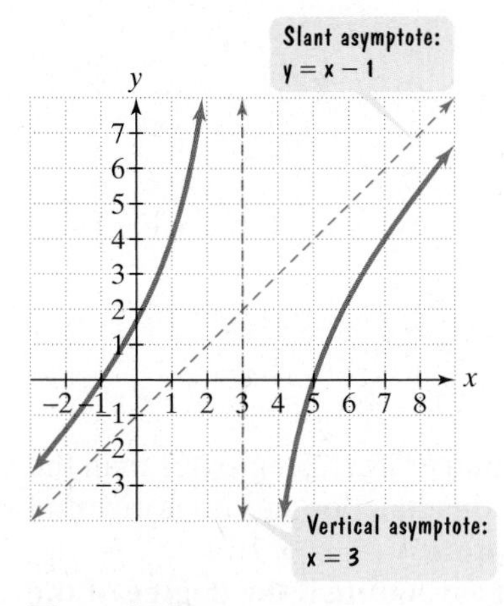

Figure 3.33 The graph of $f(x) = \dfrac{x^2 - 4x - 5}{x - 3}$

Find the slant asymptote of $f(x) = \dfrac{x^2 - 4x - 5}{x - 3}$.

Solution Because the degree of the numerator, 2, is exactly one more than the degree of the denominator, 1, the graph of f has a slant asymptote. To find the equation of the slant asymptote, divide $x - 3$ into $x^2 - 4x - 5$:

$$\begin{array}{r|rrr} 3 & 1 & -4 & -5 \\ & & 3 & -3 \\ \hline & 1 & -1 & -8 \end{array} \qquad \begin{array}{r} 1x - 1 - \dfrac{8}{x-3} \\ x - 3\,\overline{)\,x^2 - 4x - 5} \end{array}$$

Remainder

The equation of the slant asymptote is $y = x - 1$. Using our strategy for graphing rational functions, the graph of $f(x) = \dfrac{x^2 - 4x - 5}{x - 3}$ is shown in Figure 3.33.

Find the slant asymptote of $f(x) = \dfrac{2x^2 - 5x + 7}{x - 2}$.

7 Solve applied problems involving rational functions.

Applications

There are numerous examples of asymptotic behavior in functions that describe real-world phenomena.

EXAMPLE 8 Average Cost of Producing a Wheelchair

A company that manufactures wheelchairs has costs given by the function

$$C(x) = 400x + 500{,}000$$

where x is the number of wheelchairs produced per month and $C(x)$ is measured in dollars. The average cost per wheelchair for the company is given by

$$\overline{C}(x) = \frac{400x + 500{,}000}{x}.$$

a. Find and interpret $\overline{C}(1000)$, $\overline{C}(10{,}000)$, and $\overline{C}(100{,}000)$.

b. What is the horizontal asymptote for the average cost function, $\overline{C}(x)$? Describe what this represents for the company.

Solution

a. $\overline{C}(1000) = \dfrac{400(1000) + 500{,}000}{1000} = 900$

The average cost per wheelchair of producing 1000 wheelchairs per month is \$900.00.

$\overline{C}(10{,}000) = \dfrac{400(10{,}000) + 500{,}000}{10{,}000} = 450$

The average cost per wheelchair of producing 10,000 wheelchairs per month is \$450.

$$\bar{C}(100{,}000) = \frac{400(100{,}000) + 500{,}000}{100{,}000} = 405$$

The average cost per wheelchair of producing 100,000 wheelchairs per month is \$405. Notice that with higher production levels, the cost of producing each wheelchair decreases.

b. We are given the average cost function

$$\bar{C}(x) = \frac{400x + 500{,}000}{x}$$

in which the degree of the numerator, 1, is equal to the degree of the denominator, 1. The leading coefficients of the numerator and denominator, 400 and 1, are used to obtain the equation of the horizontal asymptote. The equation of the horizontal asymptote is

$$y = \frac{400}{1} \quad \text{or} \quad y = 400.$$

The horizontal asymptote is shown in Figure 3.34. This means that the more wheelchairs produced per month, the closer the average cost per wheelchair for the company comes to \$400. The least possible cost per wheelchair is approaching \$400. Competitively low prices take place with high production levels, posing a major problem for small businesses.

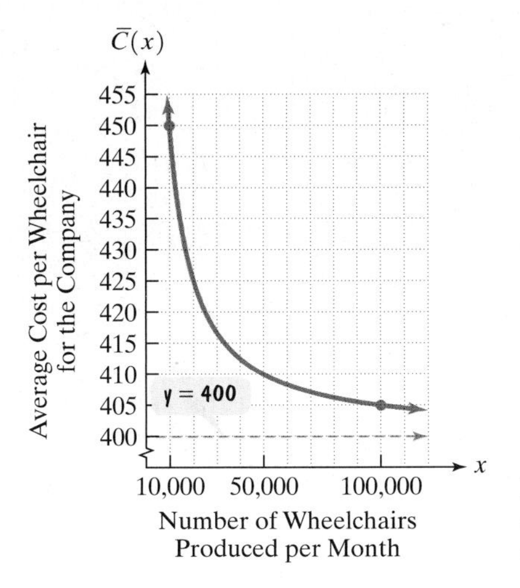

Figure 3.34 As production level increases, the average cost per wheelchair approaches \$400.

Check Point 8

A company that manufactures running shoes has costs given by the function $C(x) = 30x + 300{,}000$, where x is the number of pairs of shoes produced per week and $C(x)$ is measured in dollars. The average cost per pair for the company is given by

$$\bar{C}(x) = \frac{30x + 300{,}000}{x}.$$

a. Find and interpret $\bar{C}(1000)$, $\bar{C}(10{,}000)$, and $\bar{C}(100{,}000)$.

b. What is the horizontal asymptote for the average cost function, $\bar{C}(x)$? Describe what this represents for the company.

EXERCISE SET 3.6

Practice Exercises

In Exercises 1–8, find the domain of each rational function.

1. $f(x) = \dfrac{5x}{x - 4}$ **2.** $f(x) = \dfrac{7x}{x - 8}$

3. $g(x) = \dfrac{3x^2}{(x - 5)(x + 4)}$ **4.** $g(x) = \dfrac{2x^2}{(x - 2)(x + 6)}$

5. $h(x) = \dfrac{x + 7}{x^2 - 49}$ **6.** $h(x) = \dfrac{x + 8}{x^2 - 64}$

7. $f(x) = \dfrac{x + 7}{x^2 + 49}$ **8.** $f(x) = \dfrac{x + 8}{x^2 + 64}$

Use the graph of the rational function in the figure shown to complete each statement in Exercises 9–14.

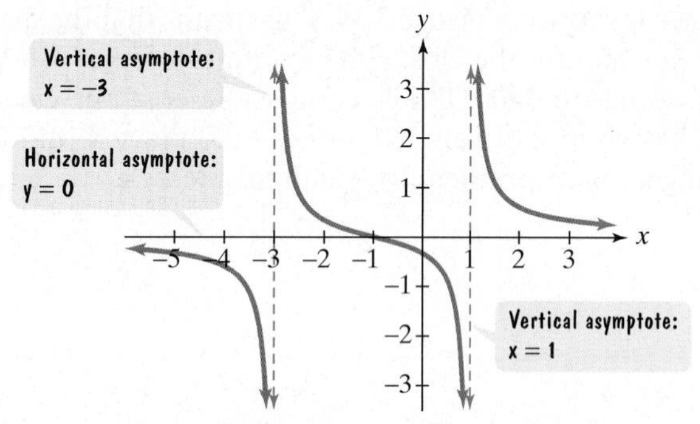

9. As $x \to -3^-$, $f(x) \to$ ______.

10. As $x \to -3^+$, $f(x) \to$ ______.

11. As $x \to 1^-$, $f(x) \to$ ______.

12. As $x \to 1^+$, $f(x) \to$ ______.

13. As $x \to -\infty$, $f(x) \to$ ______.

14. As $x \to \infty$, $f(x) \to$ ______.

Use the graph of the rational function in the figure shown to complete each statement in Exercises 15–20.

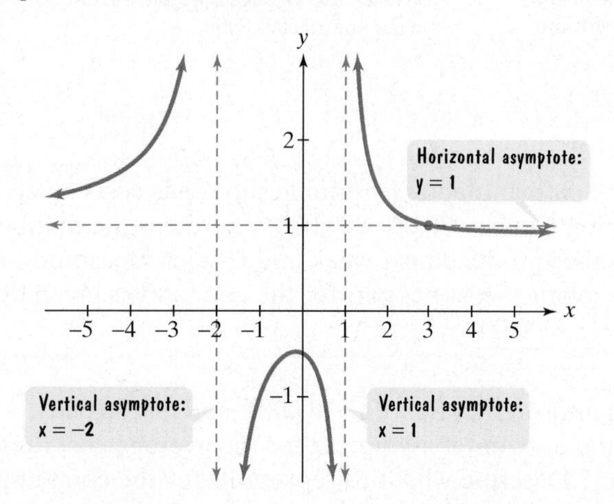

15. As $x \to 1^+$, $f(x) \to$ ______.

16. As $x \to 1^-$, $f(x) \to$ ______.

17. As $x \to -2^+$, $f(x) \to$ ______.

18. As $x \to -2^-$, $f(x) \to$ ______.

19. As $x \to \infty$, $f(x) \to$ ______.

20. As $x \to -\infty$, $f(x) \to$ ______.

In Exercises 21–28, find the vertical asymptotes, if any, of the graph of each rational function.

21. $f(x) = \dfrac{x}{x + 4}$ **22.** $f(x) = \dfrac{x}{x - 3}$

23. $g(x) = \dfrac{x + 3}{x(x + 4)}$ **24.** $g(x) = \dfrac{x + 3}{x(x - 3)}$

25. $h(x) = \dfrac{x}{x(x + 4)}$ **26.** $h(x) = \dfrac{x}{x(x - 3)}$

27. $r(x) = \dfrac{x}{x^2 + 4}$ **28.** $r(x) = \dfrac{x}{x^2 + 3}$

In Exercises 29–36, find the horizontal asymptote, if any, of the graph of each rational function.

29. $f(x) = \dfrac{12x}{3x^2 + 1}$ **30.** $f(x) = \dfrac{15x}{3x^2 + 1}$

31. $g(x) = \dfrac{12x^2}{3x^2 + 1}$ **32.** $g(x) = \dfrac{15x^2}{3x^2 + 1}$

33. $h(x) = \dfrac{12x^3}{3x^2 + 1}$ **34.** $h(x) = \dfrac{15x^3}{3x^2 + 1}$

35. $f(x) = \dfrac{-2x + 1}{3x + 5}$ **36.** $f(x) = \dfrac{-3x + 7}{5x - 2}$

In Exercises 37–58, follow the seven steps on page 321 to graph each rational function.

37. $f(x) = \dfrac{4x}{x - 2}$ **38.** $f(x) = \dfrac{3x}{x - 1}$

39. $f(x) = \dfrac{2x}{x^2 - 4}$ **40.** $f(x) = \dfrac{4x}{x^2 - 1}$

41. $f(x) = \dfrac{2x^2}{x^2 - 1}$ **42.** $f(x) = \dfrac{4x^2}{x^2 - 9}$

43. $f(x) = \dfrac{-x}{x + 1}$ **44.** $f(x) = \dfrac{-3x}{x + 2}$

45. $f(x) = -\dfrac{1}{x^2 - 4}$ **46.** $f(x) = -\dfrac{2}{x^2 - 1}$

47. $f(x) = \dfrac{2}{x^2 + x - 2}$ **48.** $f(x) = \dfrac{-2}{x^2 - x - 2}$

49. $f(x) = \dfrac{2x^2}{x^2 + 4}$ **50.** $f(x) = \dfrac{4x^2}{x^2 + 1}$

51. $f(x) = \dfrac{x + 2}{x^2 + x - 6}$

52. $f(x) = \dfrac{x - 4}{x^2 - x - 6}$

53. $f(x) = \dfrac{x^4}{x^2 + 2}$

54. $f(x) = \dfrac{2x^4}{x^2 + 1}$

55. $f(x) = \dfrac{x^2 + x - 12}{x^2 - 4}$

56. $f(x) = \dfrac{x^2}{x^2 + x - 6}$

57. $f(x) = \dfrac{3x^2 + x - 4}{2x^2 - 5x}$

58. $f(x) = \dfrac{x^2 - 4x + 3}{(x + 1)^2}$

In Exercises 59–66, **a.** *Find the slant asymptote of the graph of each rational function and* **b.** *Follow the seven-step strategy and use the slant asymptote to graph each rational function.*

59. $f(x) = \dfrac{x^2 - 1}{x}$

60. $f(x) = \dfrac{x^2 - 4}{x}$

61. $f(x) = \dfrac{x^2 + 1}{x}$

62. $f(x) = \dfrac{x^2 + 4}{x}$

63. $f(x) = \dfrac{x^2 + x - 6}{x - 3}$

64. $f(x) = \dfrac{x^2 - x + 1}{x - 1}$

65. $f(x) = \dfrac{x^3 + 1}{x^2 + 2x}$

66. $f(x) = \dfrac{x^3 - 1}{x^2 - 9}$

Application Exercises

67. A company that manufactures small canoes has costs given by the function $C(x) = 20x + 20{,}000$, where x is the number of canoes manufactured and $C(x)$ is measured in dollars. The average cost to manufacture each canoe is given by

$$\bar{C}(x) = \frac{20x + 20{,}000}{x}.$$

a. Find the average cost per canoe when $x = 100, 1000,$ 10,000, and 100,000.
b. What is the horizontal asymptote for the function $\bar{C}$, and what does it represent?

68. A company that manufactures bicycles has costs given by the function $C(x) = 100x + 100{,}000$, where x is the number of bicycles manufactured and $C(x)$ is measured in dollars. The average cost to manufacture each bicycle is given by

$$\bar{C}(x) = \frac{100x + 100{,}000}{x}.$$

a. Find and interpret $\bar{C}(500)$, $\bar{C}(1000)$, $\bar{C}(2000)$, and $\bar{C}(4000)$.
b. What is the horizontal asymptote for the function $\bar{C}$? Describe what this means in practical terms.

69. The cost, in dollars, of removing p percent of the air pollutants in the smokestack emission of a utility company that burns coal to generate electricity is given by

$$C(p) = \frac{60{,}000p}{100 - p}.$$

a. Current law requires that the company remove 80% of the pollutants from its smokestack emissions. A new law before the legislature would require increasing this amount by 5%. How much will it cost to remove another 5% of the pollutants?
b. Does this function indicate the possibility of removing 100% of the pollutants? Explain.

70. The rational function

$$C(x) = \frac{130x}{100 - x}, \quad 0 \le x < 100,$$

describes the cost C, in millions of dollars, to inoculate x% of the population against a particular strain of flu.
a. Find and interpret $C(80) - C(40)$.
b. Graph the function.
c. Describe the practical meaning of the observation that $x = 100$ is an asymptote.

71. The temperature, F, in degrees Fahrenheit, of a dessert placed in an icebox for t hours is modeled by

$$F(t) = \frac{80}{t^2 + 4t + 1}.$$

a. Find and interpret $F(0)$.
b. Find the temperature of the dessert after 1 hour, 2 hours, 3 hours, 4 hours, and 5 hours.
c. What is the equation of the horizontal asymptote associated with this function? Describe what this means in terms of the dessert's temperature over time.
d. Graph the function.

72. The function $f(x) = \dfrac{72{,}900}{100x^2 + 729}$ models the percentage of people in the United States who are unemployed as a function of years of education, x.
a. Find and interpret $f(0)$.
b. Find and interpret $f(20)$.
c. Is there an education level that leads to guaranteed employment? If not, how is this indicated by the equation of the horizontal asymptote associated with this function?
d. Graph the function.

73. Rational functions are often used to model how much we remember over time. In an experiment on memory, students in a language class are asked to memorize 40 vocabulary words in Latin, a language with which the students are not familiar. After studying the words for one day, the class is tested each day thereafter to see how many words they remember. The class average is taken and the results are graphed as shown on page 330.

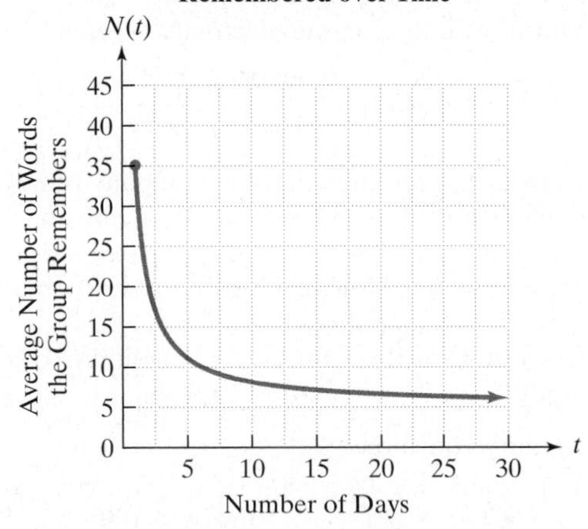

a. Use the graph to find a reasonable estimate of the number of Latin words remembered after 1 day, 5 days, and 15 days.

b. The function that models the number of Latin words remembered by the students after t days is given by

$$N(t) = \frac{5t + 30}{t}, \quad \text{where } t \geq 1.$$

Find $N(1)$, $N(5)$, and $N(15)$, comparing these values with your estimates from part (a).

c. What does the graph indicate about the number of Latin words remembered by the group over time?

d. Use the function in part (b) to find the horizontal asymptote for the graph. Describe what this horizontal asymptote means in terms of the variables modeled in this situation.

74. A drug is injected into a patient and the concentration of the drug in the bloodstream is monitored. The drug's concentration, $C(t)$, in milligrams per liter, after t hours is modeled by

$$C(t) = \frac{5t}{t^2 + 1}.$$

The graph of this rational function, obtained with a graphing utility, is shown in the figure.

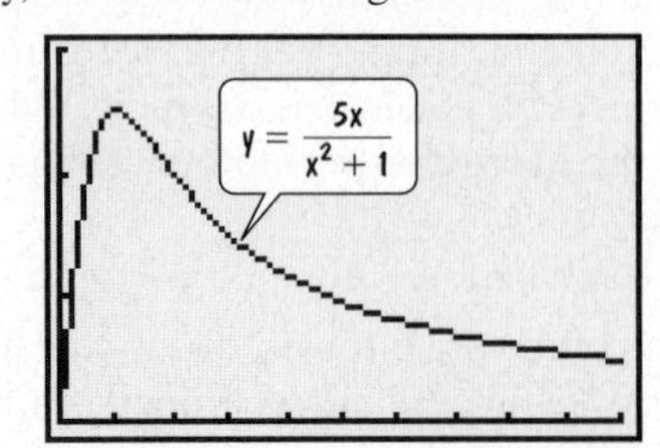

a. Use the graph to obtain a reasonable estimate of the drug's concentration after 3 hours. Then verify this estimate algebraically.

b. Use the function's equation to find the horizontal asymptote for the graph. Describe what this means about the drug's concentration in the patient's bloodstream as time increases.

Writing in Mathematics

75. What is a rational function?

76. Use everyday language to describe the graph of a rational function f such that $f(x) \to 3$ as $x \to -\infty$.

77. Use everyday language to describe the behavior of a graph near its vertical asymptote if $f(x) \to \infty$ as $x \to -2^-$ and $f(x) \to -\infty$ as $x \to -2^+$.

78. If you are given the equation of a rational function, explain how to find the vertical asymptotes, if any, of the function's graph.

79. If you are given the equation of a rational function, explain how to find the horizontal asymptote, if any, of the function's graph.

80. Describe how to graph a rational function.

81. If you are given the equation of a rational function, how can you tell if the graph has a slant asymptote? If it does, how do you find its equation?

82. Is every rational function a polynomial function? Why or why not? Does a true statement result if the two adjectives rational and polynomial are reversed? Explain.

83. The function $f(x) = \frac{5000x}{x^2 + 36}$ describes the population density, $f(x)$, in people per square mile in a large city x miles from the city's center. Describe what eventually happens to the population density as the distance from the city's center increases.

Technology Exercises

84. Use a graphing utility to verify any five of your hand-drawn graphs in Exercises 37–66.

85. Use a graphing utility to verify your hand-drawn graphs in Exercises 71–72.

86. Use a graphing utility to graph $y = \frac{1}{x}$, $y = \frac{1}{x^3}$, and $\frac{1}{x^5}$ in the same viewing rectangle. For odd values of n, how does changing n affect the graph of $y = \frac{1}{x^n}$?

87. Use a graphing utility to graph $y = \frac{1}{x^2}$, $y = \frac{1}{x^4}$, and $y = \frac{1}{x^6}$ in the same viewing rectangle. For even values of n, how does changing n affect the graph of $y = \frac{1}{x^n}$?

88. A grocery store sells 4000 cases of canned soup per year. By averaging costs to purchase soup and pay storage costs, the owner has determined that if x cases are ordered at a time, the inventory cost will be

$$C(x) = \frac{10{,}000}{x} + 3x.$$

a. Use a graphing utility to graph the inventory function.
b. Use the ZOOM and TRACE features or the minimum function feature of your graphing utility to approximate the number of cases that should be ordered to minimize inventory cost. What is the minimum cost?

89. Use a graphing utility to graph

$$f(x) = \frac{x^2 - 4x + 3}{x - 2} \quad \text{and} \quad g(x) = \frac{x^2 - 5x + 6}{x - 2}.$$

What differences do you observe between the graph of f and g? How do you account for these differences?

90. Use a graphing utility to graph

$$f(x) = \frac{2|x - 1|}{x + 2}.$$

How does the situation with horizontal asymptotes differ from the other functions graphed throughout this section?

Critical Thinking Exercises

91. Which one of the following is true?
a. The graph of a rational function cannot have both a vertical and a horizontal asymptote.
b. It is not possible to have a rational function whose graph has no y-intercept.
c. The graph of a rational function can have three horizontal asymptotes.
d. The graph of a rational function can never cross a vertical asymptote.

92. Which one of the following is true?
a. The function $f(x) = \frac{1}{\sqrt{x} - 3}$ is a rational function.
b. The x-axis is a horizontal asymptote for the graph of $f(x) = \frac{4x - 1}{x + 3}$.
c. The number of televisions that a company can produce per week after t weeks of production is given by

$$N(t) = \frac{3000t^2 + 30{,}000t}{t^2 + 10t + 25}.$$

Using this model, the company will eventually be able to produce 30,000 televisions in a single week.
d. None of the given statements is true.

In Exercises 93–96, write the equation of a rational function $f(x) = \frac{p(x)}{q(x)}$ having the indicated properties, in which the degrees of p and q are as small as possible. More than one correct function may be possible. Graph your function using a graphing utility to verify that it has the required properties.

93. f has a vertical asymptote given by $x = 3$, a horizontal asymptote $y = 0$, y-intercept $= -1$, and no x-intercept.

94. f has vertical asymptotes given by $x = -2$ and $x = 2$, a horizontal asymptote $y = 2$, y-intercept $= \frac{9}{2}$, x-intercepts of -3 and 3, and y-axis symmetry.

95. f has a vertical asymptote given by $x = 1$, a slant asymptote whose equation is $y = x$, y-intercept $= 2$, and x-intercepts of -1 and 2.

96. f has no vertical, horizontal, or slant asymptotes, and no x-intercepts.

Group Exercise

97. Group members make up the sales team for a company that makes computer video games. It has been determined that the rational function

$$f(x) = \frac{200x}{x^2 + 100}$$

models the monthly sales, in thousands of games, of a new video game as a function of the number of months, x, after the game is introduced. The figure shows the graph of the function. What are the team's recommendations to the company in terms of how long the video game should be on the market before another new video game is introduced? What other factors might members want to take into account in terms of the recommendations? What will eventually happen to sales, and how is this indicated by the graph? What does this have to do with a horizontal asymptote? What could the company do to change the behavior of this function and continue generating sales? Would this be cost effective?

Monthly Sales of a New Video Game

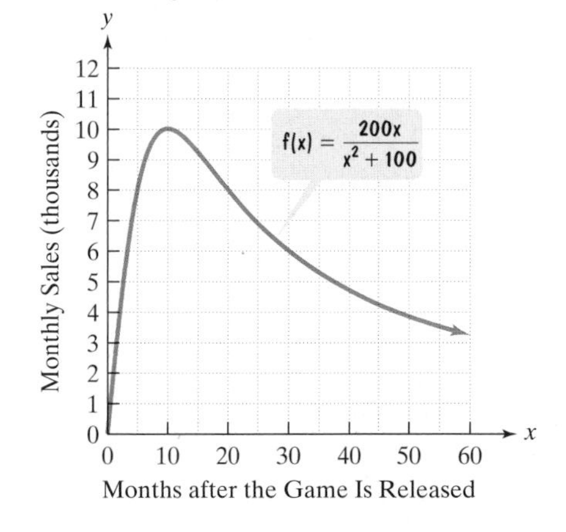

SECTION 3.7 Modeling Using Variation

Objectives

1. Solve direct variation problems.
2. Solve inverse variation problems.
3. Solve combined variation problems.
4. Solve problems involving joint variation.

Have you ever wondered how telecommunication companies estimate the number of phone calls expected per day between two cities? The formula

$$N = \frac{400P_1P_2}{d^2}$$

shows that the daily number of phone calls, N, increases as the populations of the cities, P_1 and P_2, in thousands, increase and decreases as the distance, d, between the cities increases.

Certain formulas occur so frequently in applied situations that they are given special names. Variation formulas show how one quantity changes in relation to other quantities. Quantities can vary *directly*, *inversely*, or *jointly*. In this section, we look at situations that can be modeled by each of these kinds of variation. And think of this: The next time you get one of those "all-circuits-are-busy" messages, you will be able to use a variation formula to estimate how many other callers you're competing with for those precious 5-cent minutes.

1 Solve direct variation problems.

Direct Variation

Because light travels faster than sound, during a thunderstorm we see lightning before we hear thunder. The formula

$$d = 1080t$$

describes the distance, in feet, of the storm's center if it takes t seconds to hear thunder after seeing lightning. Thus,

If $t = 1$, $d = 1080 \cdot 1 = 1080$: If it takes 1 second to hear thunder, the storm's center is 1080 feet away.

If $t = 2$, $d = 1080 \cdot 2 = 2160$: If it takes 2 seconds to hear thunder, the storm's center is 2160 feet away.

If $t = 3$, $d = 1080 \cdot 3 = 3240$: If it takes 3 seconds to hear thunder, the storm's center is 3240 feet away.

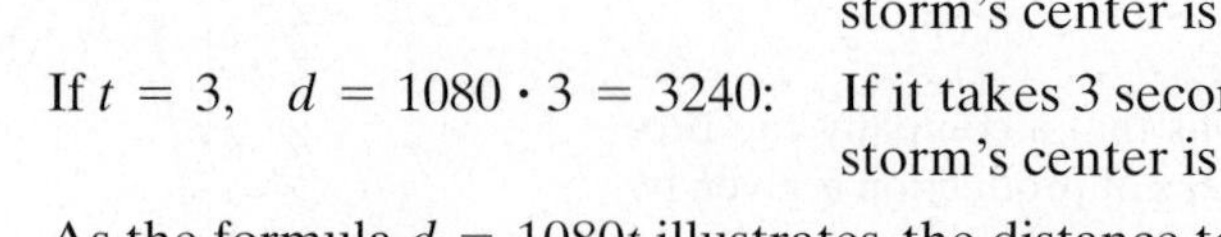

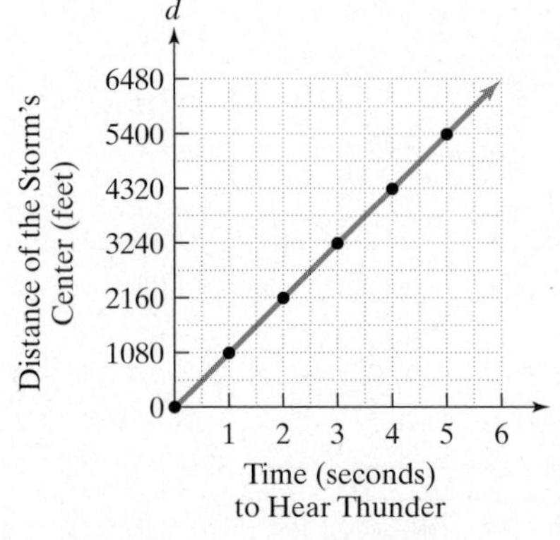

The graph of $d = 1080t$. Distance to a storm's center varies directly as the time it takes to hear thunder.

As the formula $d = 1080t$ illustrates, the distance to the storm's center is a constant multiple of how long it takes to hear the thunder. When the time is doubled, the storm's distance is doubled; when the time is tripled, the storm's distance is tripled; and so on. Because of this, the distance is said to **vary directly** as the time. The **equation of variation** is

$$d = 1080t.$$

Generalizing, we obtain the following statement.

Direct Variation

If a situation is described by an equation in the form

$$y = kx$$

where k is a constant, we say that ***y* varies directly as *x***. The number k is called the **constant of variation**.

EXAMPLE 1 Writing a Direct Variation Equation

A person's salary, S, varies directly as the number of hours worked, h.

a. Write an equation that expresses this relationship.
b. Margarita earns \$18 per hour. Substitute 18 for k, the constant of variation, in the equation in part (a) and write the equation for Margarita's salary.

Solution

a. We know that y varies directly as x is expressed as

$$y = kx.$$

By changing letters, we can write an equation that describes the following English statement: Salary, S, varies directly as the number of hours worked, h.

$$S = kh$$

b. Substituting 18 for k in the direct variation equation gives

$$S = 18h.$$

This equation describes Margarita's salary in terms of the number of hours she works. For example, if she works 10 hours, we can substitute 10 for h and determine her salary:

$$S = 18(10) = 180.$$

Her salary for working 10 hours is \$180. Notice that, as the number of hours worked increases, the salary increases.

Check Point 1

A person's hair length, L, in inches, varies directly as the number of years it has been growing, N.

a. Write an equation that expresses this relationship.
b. The longest moustache on record was grown by Kalyan Sain of India. His moustache grew 4 inches each year. Substitute 4 for k, the constant of variation, in the equation in part (a) and write the equation for the length of Sain's moustache.
c. Sain grew his moustache for 17 years. Substitute 17 for N in the equation from part (b) and find its length.

In Example 1 and Checkpoint 1, the constants of variation were given. If the constant of variation is not given, we can find it by substituting given values in the variation formula and solving for k. Example 2 shows how this is done.

EXAMPLE 2 Finding k, the Constant of Variation

Height, H, varies directly as foot length, F.

a. Write an equation that expresses this relationship.
b. Photographs of large footprints were published in 1951. Some speculated that these footprints were made by the Abominable Snowman. Each footprint was 23 inches long. The Abominable Snowman's height was determined to be 154.1 inches. (This is 12 feet, 10.1 inches, so it might not be a pleasant experience to run into this critter on a mellow hike through the woods!) Use $H = 154.1$ and $F = 23$ to find the constant of variation.

Solution

a. We know that y varies directly as x is expressed as

$$y = kx.$$

By changing letters, we can write an equation that describes the following English statement: Height, H, varies directly as foot length, F.

$$H = kF$$

b. The Abominable Snowman's height is 154.1 inches, and foot length is 23 inches. Substitute 154.1 for H and 23 for F in the direct variation equation.

$$H = kF$$

$$154.1 = k \cdot 23$$

Solve for k, the constant of variation, by dividing both sides of the equation by 23:

$$\frac{154.1}{23} = \frac{k \cdot 23}{23}$$

$$6.7 = k$$

Thus, the constant of variation is 6.7.

In Example 2, now that we know the constant of variation ($k = 6.7$), we can rewrite $H = kF$ using this constant. The equation of variation is

$$H = 6.7F.$$

We can use this equation to find other values. For example, if your foot length is 10 inches, your height is

$$H = 6.7(10) = 67,$$

or approximately 67 inches.

Check Point 2

The weight, W, of an aluminum canoe varies directly as its length, L.

a. Write an equation that expresses this relationship.
b. A 6-foot canoe weighs 75 pounds. Substitute 75 for W and 6 for L in the equation from part (a) and find k, the constant of variation.
c. Substitute the value of k into your equation in part (a) and write the equation that describes the weight of this type of canoe in terms of its length.
d. Use the equation from part (c) to find the weight of a 16-foot canoe of this type.

Our work up to this point provides a step-by-step procedure for solving variation problems. This procedure applies to direct variation problems as well as to the other kinds of variation problems that we will discuss.

Solving Variation Problems

1. Write an equation that describes the given English statement.
2. Substitute the given pair of values into the equation in step 1 and find the value of k.
3. Substitute the value of k into the equation in step 1.
4. Use the equation from step 3 to answer the problem's question.

EXAMPLE 3 Solving a Direct Variation Problem

The amount of garbage, G, varies directly as the population, P. Allegheny County, Pennsylvania, has a population of 1.3 million and creates 26 million pounds of garbage each week. Find the weekly garbage produced by New York City with a population of 7.3 million.

Solution

Step 1 Write an equation. We know that y varies directly as x is expressed as

$$y = kx.$$

By changing letters, we can write an equation that describes the following English statement: Garbage production, G, varies directly as the population, P.

$$G = kP$$

Step 2 Use the given values to find k. Allegheny County has a population of 1.3 million and creates 26 million pounds of garbage weekly. Substitute 26 for G and 1.3 for P in the direct variation equation. Then solve for k.

$$G = kP$$

$$26 = k \cdot 1.3$$

$$\frac{26}{1.3} = \frac{k \cdot 1.3}{1.3} \quad \text{Divide both sides by 1.3.}$$

$$20 = k \quad \text{Simplify.}$$

Step 3 Substitute the value of k into the equation.

$$G = kP \quad \text{Use the equation from step 1.}$$

$$G = 20P \quad \text{Replace } k \text{, the constant of variation, with 20.}$$

Step 4 Answer the problem's question. New York City has a population of 7.3 million. To find its weekly garbage production, substitute 7.3 for P in $G = 20P$ and solve for G.

$$G = 20P \quad \text{Use the equation from step 3.}$$

$$G = 20(7.3) \quad \text{Substitute 7.3 for } P.$$

$$G = 146$$

The weekly garbage produced by New York City weighs approximately 146 million pounds.

Check Point 3 The pressure, P, of water on an object below the surface varies directly as its distance, D, below the surface. If a submarine experiences a pressure of 25 pounds per square inch 60 feet below the surface, how much pressure will it experience 330 feet below the surface?

The direct variation equation $y = kx$ is a linear function. If $k > 0$, then the slope of the line is positive. Consequently, as x increases, y also increases.

A direct variation situation can involve variables to higher powers. For example, y can vary directly as x^2 $(y = kx^2)$ or as x^3 $(y = kx^3)$.

Direct Variation With Powers

y* varies directly as the *n*th power of *x if there exists some nonzero constant k such that

$$y = kx^n.$$

Direct variation with powers is modeled by polynomial functions. In our next example, the graph of the variation equation is the familiar parabola.

EXAMPLE 4 Solving a Direct Variation Problem

The distance, s, that a body falls from rest varies directly as the square of the time, t, of the fall. If skydivers fall 64 feet in 2 seconds, how far will they fall in 4.5 seconds?

Solution

Step 1 Write an equation. We know that y varies directly as the square of x is expressed as

$$y = kx^2.$$

By changing letters, we can write an equation that describes the following English statement: Distance, s, varies directly as the square of time, t, of the fall.

$$s = kt^2$$

Step 2 Use the given values to find *k*. Skydivers fall 64 feet in 2 seconds. Substitute 64 for s and 2 for t in the direct variation equation. Then solve for k.

$$s = kt^2$$

$$64 = k \cdot 2^2$$

$$64 = 4k$$

$$\frac{64}{4} = \frac{4k}{4}$$

$$16 = k$$

Step 3 Substitute the value of *k* into the equation.

$s = kt^2$ Use the equation from step 1.

$s = 16t^2$ Replace k, the constant of variation, with 16.

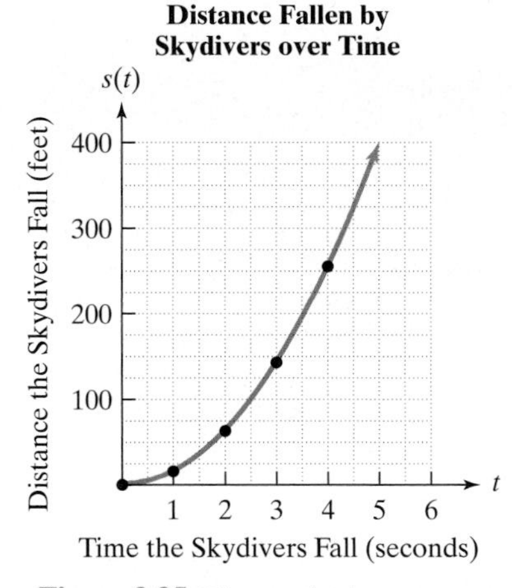

Figure 3.35 The graph of $s(t) = 16t^2$

Step 4 Answer the problem's question. How far will the skydivers fall in 4.5 seconds? Substitute 4.5 for t in $s = 16t^2$ and solve for s.

$$s = 16(4.5)^2 = 16(20.25) = 324$$

Thus, in 4.5 seconds, skydivers will fall 324 feet.

We can express the variation equation from Example 4 in function notation, writing

$$s(t) = 16t^2.$$

The distance that a body falls from rest is a function of the time, t, of the fall. The parabola that is the graph of this quadratic function is shown in Figure 3.35. The graph increases rapidly from left to right, showing the effects of the acceleration of gravity.

Check Point 4 The distance required to stop a car varies directly as the square of its speed. If 200 feet are required to stop a car traveling 60 miles per hour, how many feet are required to stop a car traveling 100 miles per hour?

2 Solve inverse variation problems.

Inverse Variation

The distance from Atlanta, Georgia, to Orlando, Florida, is 450 miles. The time that it takes to drive from Atlanta to Orlando depends on the rate at which one drives and is given by

$$\text{Time} = \frac{450}{\text{Rate}}.$$

For example, if you average 45 miles per hour, the time for the drive is

$$\text{Time} = \frac{450}{45} = 10,$$

or 10 hours. If you ignore speed limits and average 75 miles per hour, the time for the drive is

$$\text{Time} = \frac{450}{75} = 6,$$

or 6 hours. As your rate (or speed) increases, the time for the trip decreases and vice versa.

We can express the time for the Atlanta–Orlando trip using t for time and r for rate:

$$t = \frac{450}{r}.$$

This equation is an example of an **inverse variation** equation. Time, t, **varies inversely** as rate, r. When two quantities vary inversely, one quantity increases as the other decreases, and vice versa.

Generalizing, we obtain the following statement.

Inverse Variation

If a situation is described by an equation in the form

$$y = \frac{k}{x}$$

where k is a constant, we say that **y varies inversely as x.** The number k is called the **constant of variation.**

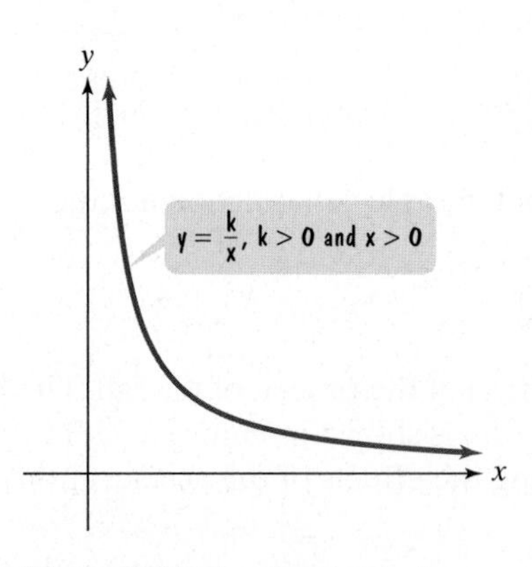

Figure 3.36 The graph of the inverse variation equation

Notice that the inverse variation equation

$$y = \frac{k}{x}, \quad \text{or} \quad f(x) = \frac{k}{x},$$

is a rational function. For $k > 0$ and $x > 0$, the graph of the function takes on the shape shown in Figure 3.36.

We use the same procedure to solve inverse variation problems as we did to solve direct variation problems. Example 5 illustrates this procedure.

EXAMPLE 5 Solving an Inverse Variation Problem

To continue making money, the number of new songs, S, a rock band needs to record each year varies inversely as the number of years, N, the band has been recording. After 4 years of recording, a band needs to record 15 new songs per year to be profitable. After 6 years, how many new songs will the band need to record in order to make a profit in the seventh year?

Solution

Step 1 Write an equation. We know that y varies inversely as x is expressed as

$$y = \frac{k}{x}.$$

By changing letters, we can write an equation that describes the following English statement: The number of new songs each year, S, varies inversely as the number of years, N.

$$S = \frac{k}{N}$$

Step 2 Use the given values to find k. After 4 years of recording, the band needs to record 15 new songs. Substitute 15 for S and 4 for N in the inverse variation equation. Then solve for k.

$$S = \frac{k}{N}$$

$$15 = \frac{k}{4}$$

$$15 \cdot 4 = \frac{k}{4} \cdot 4 \quad \text{Multiply both sides by 4.}$$

$$60 = k \quad \text{Simplify.}$$

Step 3 Substitute the value of k into the equation.

$$S = \frac{k}{N} \quad \text{Use the equation from step 1.}$$

$$S = \frac{60}{N} \quad \text{Replace } k\text{, the constant of variation, with 60.}$$

Step 4 Answer the problem's question. We need to find how many new songs the band will need to record after 6 years to make a profit in the seventh year. Substitute 6 for N in the equation above and solve for S.

$$S = \frac{60}{N} = \frac{60}{6} = 10$$

The band will need to record 10 new songs after 6 years.

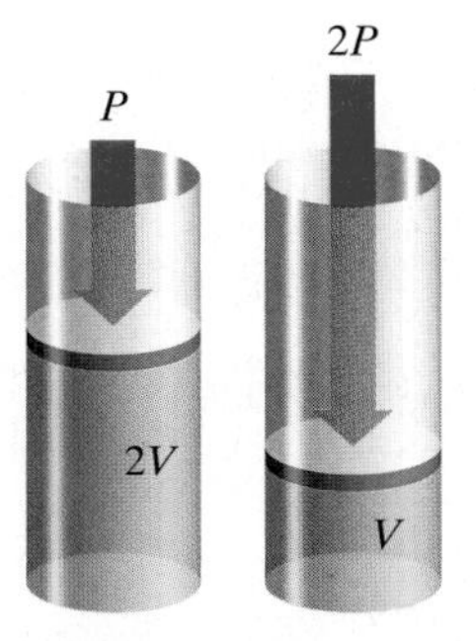

Doubling the pressure halves the volume.

Check Point 5 When you use a spray can and press the valve at the top, you decrease the pressure of the gas in the can. This decrease of pressure causes the volume of the gas in the can to increase. Because the gas needs more room than is provided in the can, it expands in spray form through the small hole near the valve. In general, if the temperature is constant, the pressure, P, of a gas in a container varies inversely as the volume, V, of the container. The pressure of a gas sample in a container whose volume is 8 cubic inches is 12 pounds per square inch. If the sample expands to a volume of 22 cubic inches, what is the new pressure of the gas?

3 Solve combined variation problems.

Combined Variation

In a **combined variation** situation, direct and inverse variation occur at the same time. For example, as the advertising budget, A, of a company increases, its monthly sales, S, also increase. Monthly sales vary directly as the advertising budget:

$$S = kA.$$

By contrast, as the price of the company's product, P, increases, its monthly sales, S, decrease. Monthly sales vary inversely as the price of the product:

$$S = \frac{k}{P}.$$

We can combine these two variation equations into one combined equation:

$$S = \frac{kA}{P}.$$

The following example illustrates the application of combined variation.

EXAMPLE 6 Solving a Combined Variation Problem

The owners of Rollerblades Now determine that the monthly sales, S, of its skates vary directly as its advertising budget, A, and inversely as the price of the skates, P. When \$60,000 is spent on advertising and the price of the skates is \$40, the monthly sales are 12,000 pairs of rollerblades.

a. Write an equation of variation that describes this situation.

b. Determine monthly sales if the amount of the advertising budget is increased to \$70,000.

Solution

a. Write an equation.

$$S = \frac{kA}{P}$$ Translate "sales vary directly as the advertising budget and inversely as the skates' price."

Use the given values to find k.

$$12{,}000 = \frac{k(60{,}000)}{40}$$ When \$60,000 is spent on advertising ($A = 60{,}000$) and the price is \$40 ($P = 40$), monthly sales are 12,000 units ($S = 12{,}000$).

$$12{,}000 = k \cdot 1500$$ Divide 60,000 by 40.

$$\frac{12{,}000}{1500} = \frac{k \cdot 1500}{1500}$$ Divide both sides of the equation by 1500.

$$8 = k$$ Simplify.

Therefore, the equation of variation that describes monthly sales is

$$S = \frac{8A}{P}.$$

b. The advertising budget is increased to \$70,000, so $A = 70{,}000$. The skates' price is still \$40, so $P = 40$.

$$S = \frac{8A}{P}$$ This is the equation from part (a).

$$S = \frac{8(70{,}000)}{40}$$ Substitute 70,000 for A and 40 for P.

$$S = 14{,}000$$

With a \$70,000 advertising budget and \$40 price, the company can expect to sell 14,000 pairs of rollerblades in a month (up from 12,000).

Check Point 6 The number of minutes needed to solve an exercise set of variation problems varies directly as the number of problems and inversely as the number of people working to solve the problems. It takes 4 people 32 minutes to solve 16 problems. How many minutes will it take 8 people to solve 24 problems?

4 Solve problems involving joint variation.

Joint Variation

Joint variation is a variation in which a variable varies directly as the product of two or more other variables. Thus, the equation $y = kxz$ is read "y varies jointly as x and z."

Joint variation plays a critical role in Isaac Newton's formula for gravitation:

$$F = G\frac{m_1 m_2}{d^2}$$

The formula states that the force of gravitation, F, between two bodies varies jointly as the product of their masses, m_1 and m_2, and inversely as the square of the distance between them, d. (G is the gravitational constant.) The formula indicates that gravitational force exists between any two objects in the universe, increasing as the distance between the bodies decreases. One practical result is that

the pull of the moon on the oceans is greater on the side of Earth closer to the moon. This gravitational imbalance is what produces tides.

EXAMPLE 7 Modeling Centrifugal Force

The centrifugal force, C, of a body moving in a circle varies jointly with the radius of the circular path, r, and the body's mass, m, and inversely with the square of the time, t, it takes to move about one full circle. A 6-gram body moving in a circle with radius 100 centimeters at a rate of 1 revolution in 2 seconds has a centrifugal force of 6000 dynes. Find the centrifugal force of an 18-gram body moving in a circle with radius 100 centimeters at a rate of 1 revolution in 3 seconds.

Solution

$$C = \frac{krm}{t^2}$$ Translate "Centrifugal force, C, varies jointly with radius, r, and mass, m, and inversely with the square of time, t."

$$6000 = \frac{k(100)(6)}{2^2}$$ If $r = 100$, $m = 6$, and $t = 2$, then $C = 6000$.

$$40 = k$$ Solve for k.

$$C = \frac{40rm}{t^2}$$ Substitute 40 for k in the model for centrifugal force.

$$= \frac{40(100)(18)}{3^2}$$ Find C when $r = 100$, $m = 18$, and $t = 3$.

$$= 8000$$

The centrifugal force is 8000 dynes.

Check Point 7 The volume of a cone, V, varies jointly as its height, h, and the square of its radius r. A cone with a radius measuring 6 feet and a height measuring 10 feet has a volume of 120π cubic feet. Find the volume of a cone having a radius of 12 feet and a height of 2 feet.

EXERCISE SET 3.7

Practice Exercises

In Exercises 1–12, write an equation that expresses each relationship. Use k as the constant of variation.

1. g varies directly as h.
2. v varies directly as r.
3. a varies directly as the square of b.
4. s varies directly as the cube of v.
5. r varies inversely as t.
6. w varies inversely as l.
7. a varies inversely as the cube of b.
8. y varies inversely as the square root of x.
9. r varies directly as s and inversely as v.
10. a varies directly as d and inversely as g.
11. s varies jointly as g and the square of t.
12. V varies jointly as h and the square of r.

In Exercises 13–22, determine the constant of variation for each stated condition.

13. y varies directly as x, and $y = 75$ when $x = 3$.
14. y varies directly as x, and $y = 55$ when $x = 11$.
15. y varies directly as x^2, and $y = 45$ when $x = 3$.
16. y varies directly as x^2, and $y = 72$ when $x = 6$.
17. W varies inversely as r, and $W = 500$ when $r = 10$.
18. T varies inversely as n, and $T = 7$ when $n = 12$.
19. A varies directly as B and inversely as C, and $A = 9$ when $B = 12$ and $C = 4$.
20. D varies directly as E and inversely as F, and $D = 6$ when $E = 12$ and $F = 10$.

21. a varies jointly as b and c, and $a = 72$ when $b = 18$ and $c = 2$.

22. z varies jointly as w and y, and $z = 38$ when $w = 38$ and $y = 2$.

Use the four-step procedure for solving variation problems given on page 335 to solve Exercises 23–30.

23. y varies directly as x. $y = 35$ when $x = 5$. Find y when $x = 12$.

24. y varies directly as x. $y = 55$ when $x = 5$. Find y when $x = 13$.

25. y varies inversely as x. $y = 10$ when $x = 5$. Find y when $x = 2$.

26. y varies inversely as x. $y = 5$ when $x = 3$. Find y when $x = 9$.

27. y varies directly as x and inversely as the square of z. $y = 20$ when $x = 50$ and $z = 5$. Find y when $x = 3$ and $z = 6$.

28. a varies directly as b and inversely as the square of c. $a = 7$ when $b = 9$ and $c = 6$. Find a when $b = 4$ and $c = 8$.

29. y varies jointly as x and z. $y = 25$ when $x = 2$ and $z = 5$. Find y when $x = 8$ and $z = 12$.

30. C varies jointly as A and T. $C = 175$ when $A = 2100$ and $T = 4$. Find C when $A = 2400$ and $T = 6$.

Application Exercises

31. A person's fingernail length, L, in inches, varies directly as the number of weeks it has been growing, W.

a. Write an equation that expresses this relationship.

b. Fingernails grow at a rate of about 0.02 inch per week. Substitute 0.02 for k, the constant of variation, in the equation in part (a) and write the equation for fingernail length.

c. Substitute 52 for W to determine your fingernail length at the end of one year if for some bizarre reason you decided not to cut them and they did not break.

32. A person's salary, S, varies directly as the number of hours worked, h.

a. Write an equation that expresses this relationship.

b. For a 40-hour work week, Gloria earned \$1400. Substitute 1400 for S and 40 for h in the equation from part (a) and find k, the constant of variation.

c. Substitute the value of k into your equation in part (a) and write the equation that describes Gloria's salary in terms of the number of hours she works.

d. Use the equation from part (c) to find Gloria's salary for 25 hours of work.

Use the four-step procedure for solving variation problems given on page 335 to solve Exercises 33–49.

33. The cost, C, of an airplane ticket varies directly as the number of miles, M, in the trip. A 3000-mile trip costs \$400. What is the cost of a 450-mile trip?

34. An object's weight on the moon, M, varies directly as its weight on Earth, E. A person who weighs 55 kilograms on Earth weights 8.8 kilograms on the moon. What is the moon weight of a person who weighs 90 kilograms on Earth?

35. The Mach number is a measurement of speed named after the man who suggested it, Ernst Mach (1838–1916). The speed of an aircraft varies directly as its Mach number. Shown here are two aircraft. Use the figures for the Concord to determine the Blackbird's speed.

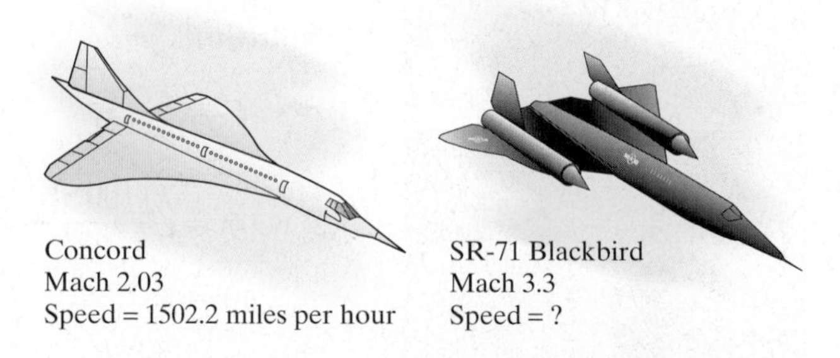

Concord
Mach 2.03
Speed = 1502.2 miles per hour

SR-71 Blackbird
Mach 3.3
Speed = ?

36. Do you still own records, or are you strictly a CD person? Record owners claim that the quality of sound on good vinyl surpasses that of a CD, although this is up for debate. This, however, is not debatable: The number of revolutions a record makes as it is being played varies directly as the time that it is on the turntable. A record that lasted 3 minutes made 135 revolutions. If a record takes 2.4 minutes to play, how many revolutions does it make?

37. If all men had identical body types, their weight would vary directly as the cube of their height. Shown is Robert Wadlow, who reached a record height of 8 feet 11 inches (107 inches) before his death at age 22. If a man who is 5 feet 10 inches tall (70 inches) with the same body type as Mr. Wadlow weighs 170 pounds, what was Robert Wadlow's weight shortly before his death?

38. The distance that an object falls varies directly as the square of the time it has been falling. An object falls 144 feet in 3 seconds. Find how far it will fall in 7 seconds.

39. The time that it takes you to get to campus varies inversely as your driving rate. Averaging 20 miles per hour in terrible traffic, it takes you 1.5 hours to get to campus. How long would the trip take averaging 60 miles per hour?

40. The weight that can be supported by a 2-inch by 4-inch piece of pine (called a 2-by-4) varies inversely as its length. A 10-foot 2-by-4 pine can support 500 pounds. What weight can be supported by a 125-foot 2-by-4 pine?

41. The volume of a gas in a container at a constant temperature varies inversely as the pressure. If the volume is 32 cubic centimeters at a pressure of 8 pounds, find the pressure when the volume is 40 cubic centimeters.

42. The current in a circuit varies inversely as the resistance. The current is 20 amperes when the resistance is 5 ohms. Find the current for a resistance of 16 ohms.

43. A person's body-mass index is used to assess levels of fatness, with an index from 20 to 26 considered in the desirable range. The index varies directly as one's weight, in pounds, and inversely as one's height, in inches. A person who weighs 150 pounds and is 70 inches tall has an index of 21. What is the body-mass index of a person who weighs 240 pounds and is 74 inches tall? Because the index is rounded to the nearest whole number, do so and then determine if this person's level of fatness is in the desirable range.

44. The volume of a gas varies directly as its temperature and inversely as its pressure. At a temperature of 100 Kelvin and a pressure of 15 kilograms per square meter, the gas occupies a volume of 20 cubic meters. Find the volume at a temperature of 150 Kelvin and a pressure of 30 kilograms per square meter.

45. The intensity of illumination on a surface varies inversely as the square of the distance of the light source from the surface. The illumination from a source is 25 foot-candles at a distance of 4 feet. What is the illumination when the distance is 6 feet?

46. The gravitational force with which Earth attracts an object varies inversely with the square of the distance from the center of Earth. A gravitational force of 160 pounds acts on an object 400 miles from Earth's center. Find the force of attraction on an object 6000 miles from the center of Earth.

47. Kinetic energy varies jointly as the mass and the square of the velocity. A mass of 8 grams and velocity of 3 centimeters per second has a kinetic energy of 36 ergs. Find the kinetic energy for a mass of 4 grams and velocity of 6 centimeters per second.

48. The electrical resistance of a wire varies directly as its length and inversely as the square of its diameter. A wire of 720 feet with $\frac{1}{4}$-inch diameter has a resistance of $1\frac{1}{2}$ ohms. Find the resistance for 960 feet of the same kind of wire if its diameter is doubled.

49. The average number of phone calls between two cities in a day varies jointly as the product of their populations and inversely as the square of the distance between them. The population of Minneapolis is 2538 thousand and the population of Cincinnati is 1818 thousand. Separated by 108 miles, the average number of telephone calls per day between the two cities is 158,233. Find the average number of telephone calls per day between Orlando, Florida (population 1225 thousand) and Seattle, Washington (population 2970 thousand), two cities that are 3403 miles apart.

Writing in Mathematics

50. What does it mean if two quantities vary directly?

51. In your own words, explain how to solve a variation problem.

52. What does it mean if two quantities vary inversely?

53. Explain what is meant by combined variation. Give an example with your explanation.

54. Explain what is meant by joint variation. Give an example with your explanation.

In Exercises 55–56, describe in words the variation shown by the given equation.

55. $z = \dfrac{k\sqrt{x}}{y^2}$

56. $z = kx^2\sqrt{y}$

57. We have seen that the daily number of phone calls between two cities varies jointly as their populations and inversely as the square of the distance between them. This model, used by telecommunication companies to estimate the line capacities needed among various cities, is called the *gravity model*. Compare the model to Newton's formula for gravitation on page 340 and describe why the name *gravity model* is appropriate.

Technology Exercise

58. Use a graphing utility to graph any three of the variation equations in Exercises 33–42. Then TRACE along each curve and identify the point that corresponds to the problem's solution.

Critical Thinking Exercises

59. In a hurricane, the wind pressure varies directly as the square of the wind velocity. If wind pressure is a measure of a hurricane's destructive capacity, what happens to this destructive power when the wind speed doubles?

60. The illumination from a light source varies inversely as the square of the distance from the light source. If you raise a lamp from 15 inches to 30 inches over your desk, what happens to the illumination?

61. The heat generated by a stove element varies directly as the square of the voltage and inversely as the resistance. If the voltage remains constant, what needs to be done to triple the amount of heat generated?

62. Galileo's telescope brought about revolutionary changes in astronomy. A comparable leap in our ability to observe the universe took place as a result of the Hubble Space Telescope. The space telescope can see stars and galaxies whose brightness is $\frac{1}{50}$ of the faintest objects now observable using ground-based telescopes. Use the fact that the brightness of a point source, such as a star, varies inversely as the square of its distance from an observer to show that the space telescope can see about seven times farther than a ground-based telescope.

Group Exercise

63. Begin by deciding on a product that interests the group because you are now in charge of advertising this product. Members were told that the demand for the product varies directly as the amount spent on advertising and inversely as the price of the product. However, as more money is spent on advertising, the price of your product rises. Under what conditions would members recommend an increased expense in advertising? Once you've determined what your product is, write formulas for the given conditions and experiment with hypothetical numbers. What other factors might you take into consideration in terms of your recommendation? How do these factor affect the demand for your product?

CHAPTER SUMMARY, REVIEW, AND TEST

Summary

3.1 Quadratic Functions

a. A quadratic function is of the form $f(x) = ax^2 + bx + c, a \neq 0$.

b. The standard form of a quadratic function is $f(x) = a(x - h)^2 + k, a \neq 0$.

c. The graph of a quadratic function is a parabola. The vertex is (h, k) or $\left(-\frac{b}{2a}, f\left(-\frac{b}{2a}\right)\right)$. A procedure for graphing a parabola is given in the box on page 261.

3.2 Polynomial Functions and Their Graphs

a. Polynomial Function of x of Degree n: $f(x) = a_n x^n + a_{n-1}x^{n-1} + \cdots + a_2x^2 + a_1x + a_0, \quad a_n \neq 0$

b. The graphs of polynomial functions are smooth and continuous.

c. The end behavior of the graph of a polynomial function depends on the leading term, given by the Leading Coefficient Test in the box on page 274.

d. The values of x for which $f(x)$ is equal to 0 are the zeros of the polynomial function f. These values are the roots of the polynomial equation $f(x) = 0$.

e. If $x - r$ occurs k times in a polynomial function's factorization, r is a repeated zero with multiplicity k. If k is even, the graph touches the x-axis at r; if odd, it crosses the x-axis at r.

f. If f is a polynomial of degree n, the graph of f has at most $n - 1$ turning points.

g. A strategy for graphing a polynomial function is given in the box on page 278.

3.3 Dividing Polynomials; Remainder and Factor Theorems

a. Long division of polynomials is performed by dividing, multiplying, subtracting, bringing down the next term, and repeating this process until the degree of the remainder is less than the degree of the divisor. The details are given in the box on page 285.

b. The Division Algorithm: $f(x) = d(x)q(x) + r(x)$. The dividend is the product of the divisor and the quotient plus the remainder.

c. Synthetic division is used to divide a polynomial by $x - c$. The details are given in the box on pages 288–289.

d. The Remainder Theorem: If the polynomial $f(x)$ is divided by $x - c$, then the remainder is $f(c)$.

e. The Factor Theorem: If a polynomial function $f(x)$ is divided by $x - c$ and the remainder is zero, c is a zero of f and a root of $f(x) = 0$. If c is a zero of f or a root of $f(x) = 0$, then $x - c$ is a factor of $f(x)$.

3.4 Zeros of Polynomial Functions

a. The Rational Zero Theorem states that possible rational zeros of a polynomial function $= \dfrac{\text{Factors of the constant term}}{\text{Factors of the leading coefficient}}$. The theorem is stated in the box on page 295.

b. Descartes's Rule of Signs: The number of positive real zeros of f equals the number of sign changes of $f(x)$ or is less than that number by an even integer. The number of negative real zeros of f applies a similar statement to $f(-x)$.

3.5 More on Zeros of Polynomial Functions

a. The Upper and Lower Bound Theorem: The number $b > 0$ is an upper bound for the real roots of $f(x) = 0$ if synthetic division of $f(x)$ by $x - b$ results in no negative numbers. The number $a < 0$ is a lower bound if synthetic division by $x - a$ results in numbers that alternate in sign, counting zero entries as positive or negative.

b. The Intermediate Value Theorem: If $f(a)$ and $f(b)$ have opposite signs, there is at least one value of c between a and b for which $f(c) = 0$.

c. Number of roots: If $f(x)$ is a polynomial of degree $n \geq 1$, then, counting multiple roots separately, the equation $f(x) = 0$ has n roots.

d. If $a + bi$ is a root of $f(x) = 0$, then $a - bi$ is also a root.

e. The Linear Factorization Theorem: An nth-degree polynomial can be expressed as the product of n linear factors. Thus,

$$f(x) = a_n(x - c_1)(x - c_2)\cdots(x - c_n).$$

3.6 Rational Functions and Their Graphs

a. Rational function: $f(x) = \dfrac{p(x)}{q(x)}$; $p(x)$ and $q(x)$ are polynomial functions and $q(x) \neq 0$. The domain of f is the set of all real numbers excluding values of x that make $q(x)$ zero.

b. Arrow notation is summarized in the box on page 317.

c. The line $x = a$ is a vertical asymptote of the graph of f if $f(x)$ increases or decreases without bound as x approaches a. Vertical asymptotes are identified using the location theorem in the box on page 318.

d. The line $y = b$ is a horizontal asymptote of the graph of f if $f(x)$ approaches b as x increases or decreases without bound. Horizontal asymptotes are identified using the location theorem in the box on page 320.

e. A strategy for graphing rational functions is given in the box on page 321.

f. The graph of a rational functions has a slant asymptote when the degree of the numerator is one more than the degree of the denominator. The equation of the slant asymptote is found using division and ignoring the remainder term.

3.7 Modeling Using Variation

a.

English Statement	**Equation**
y varies directly as x.	$y = kx$
y varies directly as x^n.	$y = kx^n$
y varies inversely as x.	$y = \dfrac{k}{x}$
y varies inversely as x^n.	$y = \dfrac{k}{x^n}$
y varies jointly as x and z.	$y = kxz$

b. A procedure for solving variation problems is given in the box on page 335.

Review Exercises

3.1

In Exercises 1–4, use the vertex and intercepts to sketch the graph of each quadratic function. Give the equation for the parabola's axis of symmetry.

1. $f(x) = -2(x - 1)^2 + 3$ **2.** $f(x) = (x + 4)^2 - 2$

3. $f(x) = -x^2 + 2x + 3$ **4.** $f(x) = 2x^2 - 4x - 6$

5. The function $f(x) = 104.5x^2 - 1501.5x + 6016$ describes the death rate per year per 100,000 males, $f(x)$, for U.S. men who average x hours of sleep each night. How many hours of sleep, to the nearest tenth of an hour, corresponds to the minimum death rate? What is this minimum death rate, to the nearest whole number?

6. A person standing close to the edge on the top of an 80-foot building throws a ball vertically upward with an initial velocity of 64 feet per second. The function $s(t) = -16t^2 + 64t + 80$ describes the ball's height above the ground, $s(t)$, in feet, t seconds after it is thrown. After how many seconds does the ball reach its maximum height? What is the maximum height?

3.2

In Exercises 7–10, use end behavior and, if necessary, zeros to match each polynomial function with its graph on page 346.

7. $f(x) = -x^3 + 12x^2 - x$

8. $g(x) = x^6 - 6x^4 + 9x^2$

9. $h(x) = x^5 - 5x^3 + 4x$

10. $r(x) = x^3 + 1$

a.

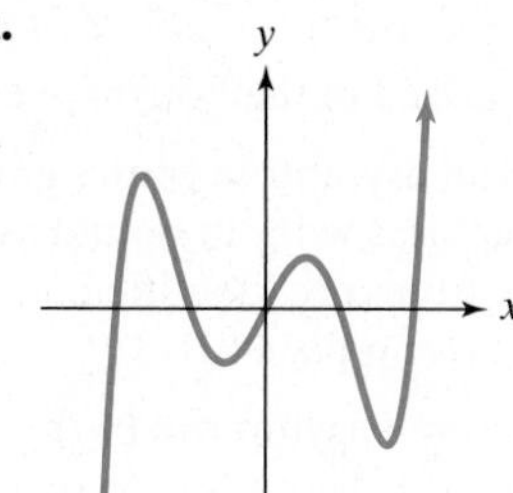

b.

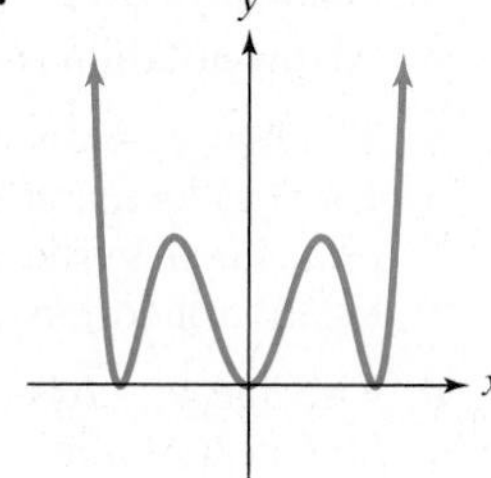

c.

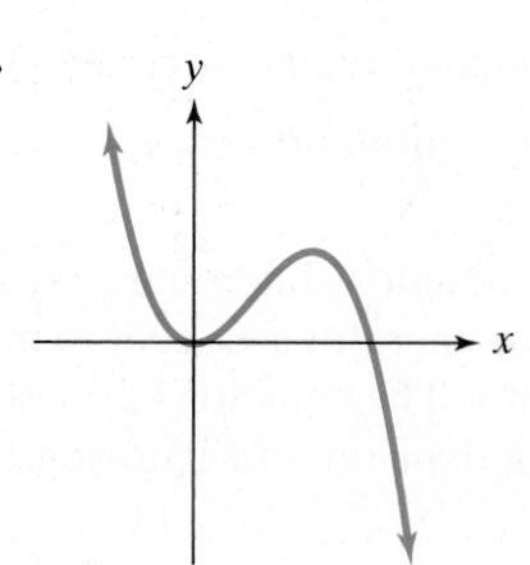

d.

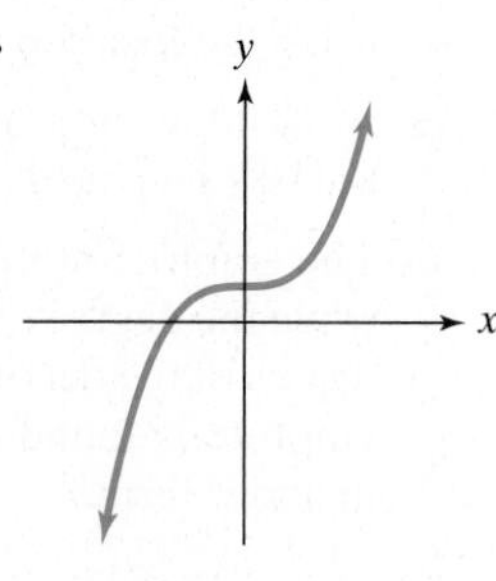

11. The function $f(x) = -0.0013x^3 + 0.78x^2 - 1.43x + 18.1$ models the percentage of U.S. families below the poverty level x years after 1960. Use end behavior to explain why the model is valid only for a limited period of time.

12. Despite a combination of drugs used to inhibit the growth of the HIV virus, a patient dies as a result of the virus overwhelming his body. Could the function $N(t) = -\frac{3}{4}t^4 + 3t^3 + 5$ model the number of viral particles, in billions, in this patient's body over time? Use the graph's end behavior to the right to answer the question. Explain your answer.

In Exercises 13–14, find the zeros for each polynomial function and give the multiplicity of each zero. State whether the graph crosses or touches the x-axis at each zero.

13. $f(x) = -2(x - 1)(x + 2)^2(x + 5)^3$

14. $f(x) = x^3 - 5x^2 - 25x + 125$

In Exercises 15–20,

a. *Use the Leading Coefficient Test to determine the graph's end behavior.*

b. *Determine whether the graph has y-axis symmetry, origin symmetry, or neither.*

c. *Graph the function.*

15. $f(x) = x^3 - x^2 - 9x + 9$
16. $f(x) = 4x - x^3$
17. $f(x) = 2x^3 + 3x^2 - 8x - 12$
18. $f(x) = -x^4 + 25x^2$
19. $f(x) = -x^4 + 6x^3 - 9x^2$
20. $f(x) = 3x^4 - 15x^3$

3.3

In Exercises 21–23, divide by long division.

21. $(4x^3 - 3x^2 - 2x + 1) \div (x + 1)$

22. $(10x^3 - 26x^2 + 17x - 13) \div (5x - 3)$

23. $(4x^4 + 6x^3 + 3x - 1) \div (2x^2 + 1)$

In Exercises 24–25, divide by synthetic division.

24. $(3x^4 + 11x^3 - 20x^2 + 7x + 35) \div (x + 5)$

25. $(3x^4 - 2x^2 - 10x) \div (x - 2)$

26. Given $f(x) = 2x^3 - 7x^2 + 9x - 3$, use the Remainder Theorem to find $f(-13)$.

27. Use synthetic division to divide $f(x) = 2x^3 + x^2 - 13x + 6$ by $x - 2$. Use the result to find all zeros of f.

28. Solve the equation $x^3 - 17x + 4 = 0$ given that 4 is a root.

3.4

In Exercises 29–30, use the Rational Zero Theorem to list all possible rational zeros for each given function.

29. $f(x) = x^4 - 6x^3 + 14x^2 - 14x + 5$

30. $f(x) = 3x^5 - 2x^4 - 15x^3 + 10x^2 + 12x - 8$

In Exercises 31–32, use Descartes's Rule of Signs to determine the possible number of positive and negative real zeros for each given function.

31. $f(x) = 3x^4 - 2x^3 - 8x + 5$

32. $f(x) = 2x^5 - 3x^3 - 5x^2 + 3x - 1$

33. Use Descartes's Rule of Signs to explain why $2x^4 + 6x^2 + 8 = 0$ has no real roots.

For Exercises 34–39,

a. *List all possible rational roots or rational zeros.*

b. *Use Descartes's Rule of Signs to determine the possible number of positive and negative real roots or real zeros.*

c. *Use synthetic division to test the possible rational roots or zeros and find an actual root or zero.*

d. *Use the root or zero from part (c) to find all the zeros or roots.*

34. $f(x) = x^3 + 3x^2 - 4$

35. $f(x) = 6x^3 + x^2 - 4x + 1$

36. $8x^3 - 36x^2 + 46x - 15 = 0$

37. $x^4 - x^3 - 7x^2 + x + 6 = 0$

38. $4x^4 + 7x^2 - 2 = 0$

39. $f(x) = 2x^4 + x^3 - 9x^2 - 4x + 4$

3.5

40. Show that all real roots of the equation
$$2x^4 - 7x^3 - 5x^2 + 28x - 12 = 0$$
lie between -2 and 6. Use this result to list all possible rational roots.

41. Consider the equation $2x^4 - x^3 - 5x^2 + 10x + 12 = 0$.

a. List all possible rational roots.

b. Determine whether 2 is a root using synthetic division. In terms of bounds, what can you conclude?

c. Determine whether -2 is a root using synthetic division. In terms of bounds, what can you conclude?

d. Use the results of parts (b) and (c) to discard some of the possible rational roots from part (a). Now what are the possible rational roots?

In Exercises 42–43, show that the polynomial has a zero between the given integers. Then use the Intermediate Value

Theorem to find an approximation for this zero to the nearest tenth.

42. $f(x) = x^3 - 2x - 1$; between 1 and 2

43. $f(x) = 3x^3 + 2x^2 - 8x + 7$; between -3 and -2

In Exercises 44–46, use the given root to find the solution set of the polynomial equation.

44. $4x^3 - 47x^2 + 232x + 61 = 0$; $6 + 5i$

45. $x^4 - 4x^3 + 16x^2 - 24x + 20 = 0$; $1 - 3i$

46. $2x^4 - 17x^3 + 137x^2 - 57x - 65 = 0$; $4 + 7i$

In Exercises 47–49, find an nth-degree polynomial function with real coefficients satisfying the given conditions. If you are using a graphing utility, graph the function and verify the real zeros and the given function value.

47. $n = 3$; 2 and $2 - 3i$ are zeros; $f(1) = -10$

48. $n = 4$; i is a zero; -3 is a zero of multiplicity 2; $f(-1) = 16$

49. $n = 4$; $-2, 3$, and $1 + 3i$ are zeros; $f(2) = -40$

In Exercises 50–51, find all the zeros of each polynomial function and write the polynomial as a product of linear factors.

50. $f(x) = 2x^4 + 3x^3 + 3x - 2$

51. $g(x) = x^4 - 6x^3 + x^2 + 24x + 16$

In Exercises 52–55, graphs of fifth-degree polynomial functions are shown. In each case, specify the number of real zeros and the number of nonreal complex zeros. Indicate whether there are any real zeros with multiplicity other than 1.

52.

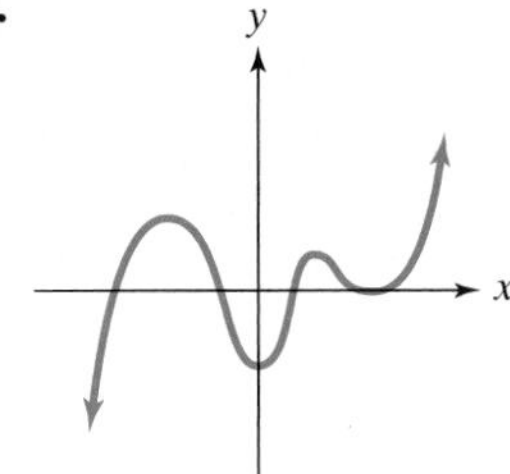

53.

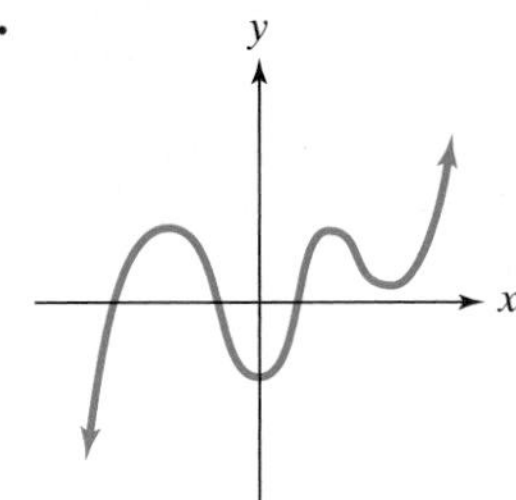

54.

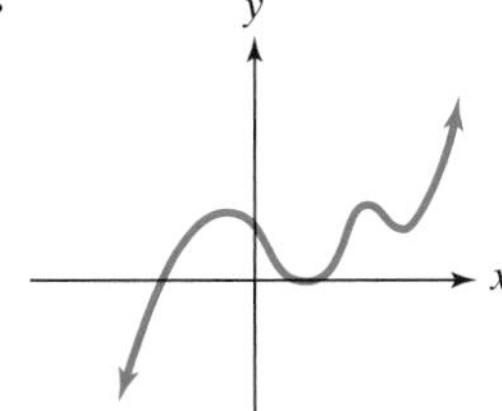

55.

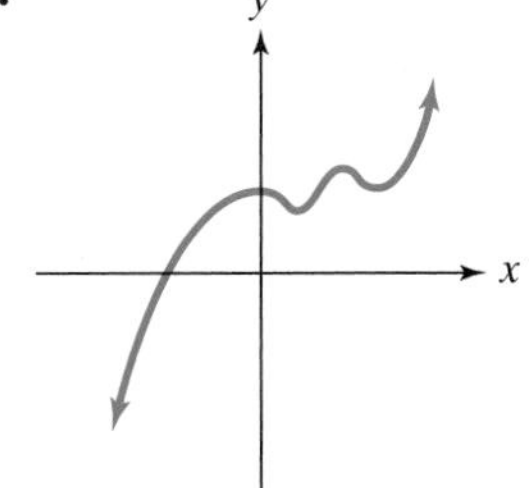

3.6

In Exercises 56–63, find the vertical asymptotes, if any, the horizontal asymptote, if there is one, and the slant asymptote, if there is one, of the graph of each rational function. Then graph the rational function.

56. $f(x) = \dfrac{2x}{x^2 - 9}$

57. $g(x) = \dfrac{2x - 4}{x + 3}$

58. $h(x) = \dfrac{x^2 - 3x - 4}{x^2 - x - 6}$

59. $r(x) = \dfrac{x^2 + 4x + 3}{(x + 2)^2}$

60. $y = \dfrac{x^2}{x + 1}$

61. $y = \dfrac{x^2 + 2x - 3}{x - 3}$

62. $f(x) = \dfrac{-2x^3}{x^2 + 1}$

63. $g(x) = \dfrac{4x^2 - 16x + 16}{2x - 3}$

64. A company that manufactures graphing calculators has costs given by the function $C(x) = 25x + 50{,}000$, where x is the number of calculators manufactured and $C(x)$ is measured in dollars. The average cost to manufacture each calculator is given by

$$\bar{C}(x) = \frac{25x + 50{,}000}{x}.$$

a. Find and interpret $\bar{C}(50)$, $\bar{C}(100)$, $\bar{C}(1000)$, and $\bar{C}(100{,}000)$.

b. What is the horizontal asymptote for this function, and what does it represent?

65. In Silicon Valley, California, a government agency ordered computer-related companies to contribute to a monetary pool to clean up underground water supplies. (The companies had stored toxic chemicals in leaking underground containers.) The rational function

$$C(x) = \frac{200x}{100 - x}$$

models the cost, $C(x)$, in tens of thousands of dollars, for removing x percent of the contaminants.

a. Find and interpret $C(90) - C(50)$.

b. What is the equation for the vertical asymptote? What does this mean in terms of the variables given by the function?

Exercises 66–67 involve rational functions that model the given situations. In each case, find the horizontal asymptote as $x \to \infty$ and then describe what this means in practical terms.

66. $F(x) = \dfrac{30(4 + 5x)}{1 + 0.05x}$; the number of fish, F, in thousands, after x weeks in a lake that was stocked with 120,000 fish.

67. $P(x) = \dfrac{72{,}900}{100x^2 + 729}$; the percentage rate, P, of U.S. unemployment for groups with x years of education.

68. In a get-tough drug policy, a politician promises to spend whatever it takes to seize all illegal drugs as they enter the country. If the cost of this venture is

$$C(p) = \frac{Ap}{100 - p}$$

where A is a positive constant, C is expressed in millions of dollars, and p is the percentage of illegal drugs seized, use this function to evaluate the politician's promise.

3.7

Solve the variation problems in Exercises 69–74.

69. An electric bill varies directly as the amount of electricity used. The bill for 1400 kilowatts of electricity is \$98. What is the bill for 2200 kilowatts of electricity?

70. The distance that a body falls from rest varies directly as the square of the time of the fall. If skydivers fall 144 feet in 3 seconds, how far will they fall in 10 seconds?

71. The time it takes to drive a certain distance varies inversely as the rate of travel. If it takes 4 hours at 50 miles per hour to drive the distance, how long will it take at 40 miles per hour?

72. The loudness of a stereo speaker, measured in decibels, varies inversely as the square of your distance from the speaker. When you are 8 feet from the speaker, the loudness is 28 decibels. What is the loudness when you are 4 feet from the speaker?

73. The time required to assemble computers varies directly as the number of computers assembled and inversely as the number of workers. If 30 computers can be assembled by 6 workers in 10 hours, how long would it take 5 workers to assemble 40 computers?

74. The volume of a pyramid varies jointly as its height and the area of its base. A pyramid with a height of 15 feet and a base with an area of 35 square feet has a volume of 175 cubic feet. Find the volume of a pyramid with a height of 20 feet and a base with an area of 120 square feet.

Chapter 3 Test

In Exercises 1–2, use the vertex and intercepts to sketch the graph of each quadratic function. Give the equation for the parabola's axis of symmetry.

1. $f(x) = (x + 1)^2 + 4$ **2.** $f(x) = x^2 - 2x - 3$

3. Determine, without graphing, whether the quadratic function $f(x) = -2x^2 + 12x - 16$ has a minimum value or a maximum value. Then find the coordinates of the minimum or the maximum point.

4. The function $f(x) = -x^2 + 46x - 360$ models the daily profit, $f(x)$, in hundreds of dollars, for a company that manufactures x VCRs daily. How many VCRs should be manufactured each day to maximize profit? What is the maximum daily profit?

5. Consider the function $f(x) = x^3 - 5x^2 - 4x + 20$.

a. Use factoring to find all zeros of f.

b. Use the Leading Coefficient Test and the zeros of f to graph the function.

6. Use end behavior to explain why the graph cannot be the graph of $f(x) = x^5 - x$. Then use intercepts to explain why the graph cannot represent $f(x) = x^5 - x$.

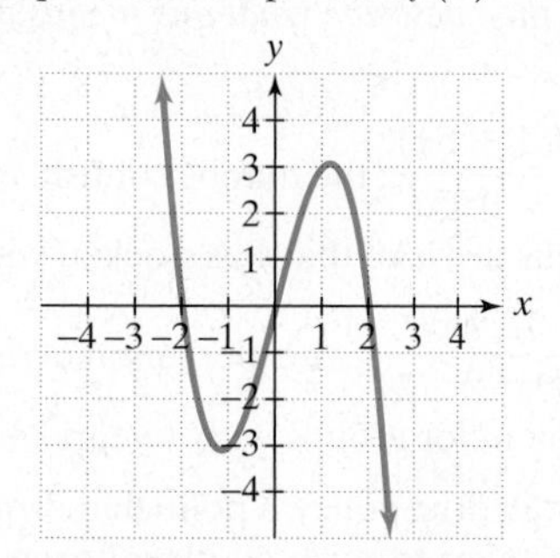

7. The graph of $f(x) = 6x^3 - 19x^2 + 16x - 4$ is shown in the figure at the top of the next column.

a. Based on the graph of f, find the root of the equation $6x^3 - 19x^2 + 16x - 4 = 0$ that is an integer.

b. Use synthetic division to find the other two roots of $6x^3 - 19x^2 + 16x - 4 = 0$.

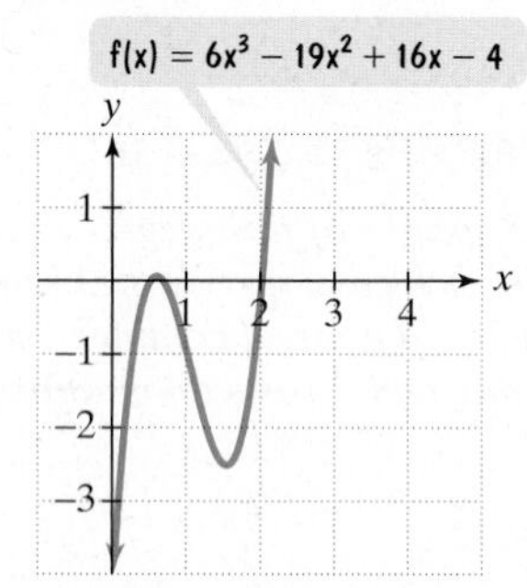

8. Use the Rational Zero Theorem to list all possible rational zeros of $f(x) = 2x^3 + 11x^2 - 7x - 6$.

9. Use Descartes's Rule of Signs to determine the possible number of positive and negative real zeros of $f(x) = 3x^5 - 2x^4 - 2x^2 + x - 1$.

10. Solve: $x^3 + 6x^2 - x - 30 = 0$.

11. Consider the function whose equation is given by $f(x) = 2x^4 - x^3 - 13x^2 + 5x + 15$.

a. List all possible rational zeros.

b. Use the graph of f in the figure shown and synthetic division to find all zeros of the function.

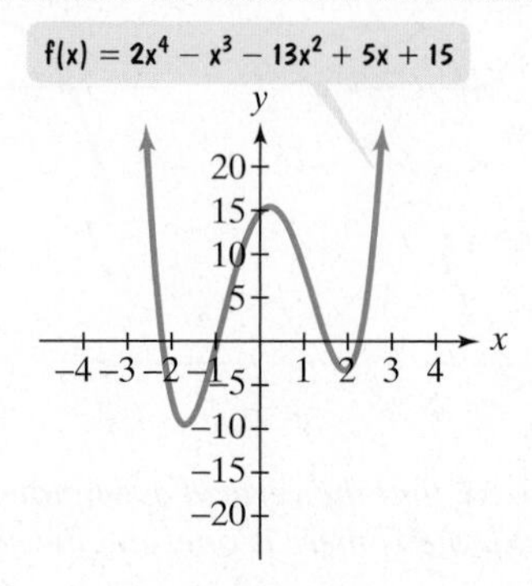

12. Use the graph of $f(x) = 3x^4 + 4x^3 - 7x^2 - 2x - 3$ in the figure shown to find the smallest positive integer that is an upper bound and the largest negative integer that is a lower bound for the real roots of

$$3x^4 + 4x^3 - 7x^2 - 2x - 3 = 0.$$

Then use synthetic division to show that all the real roots of the equation lie between these integers.

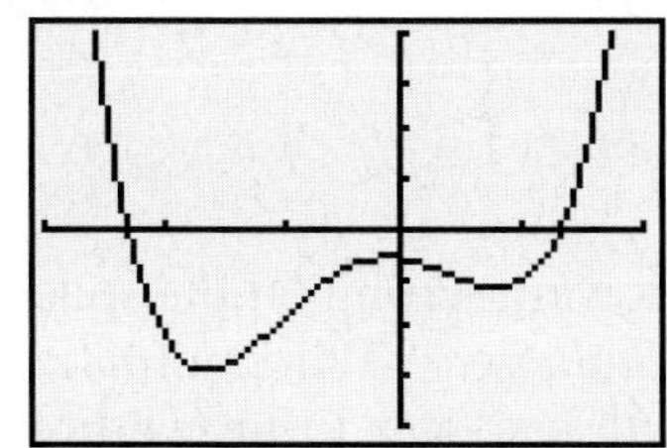

13. Solve $x^4 - 7x^3 + 18x^2 - 22x + 12 = 0$ given that $1 - i$ is a root.

14. Use the graph of $f(x) = x^3 + 3x^2 - 4$ in the figure shown to factor $x^3 + 3x^2 - 4$.

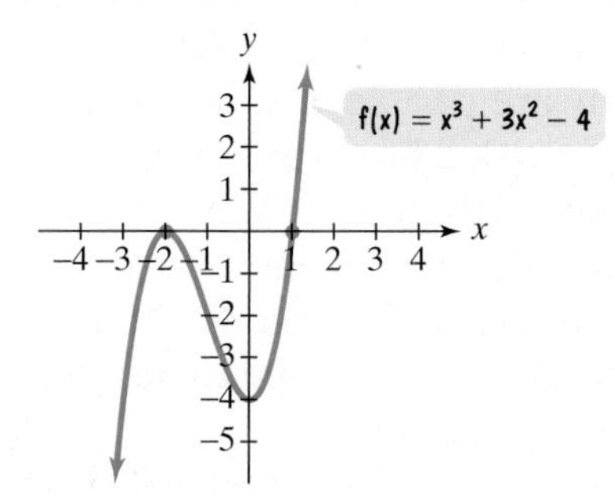

In Exercises 15–18, find the domain of each rational function and graph the function.

15. $f(x) = \dfrac{x}{x^2 - 16}$ 16. $f(x) = \dfrac{x^2 - 9}{x - 2}$

17. $f(x) = \dfrac{x + 1}{x^2 + 2x - 3}$ 18. $f(x) = \dfrac{4x^2}{x^2 + 3}$

19. A number of deer are placed into a newly acquired habitat. The deer population over time is modeled by a rational function whose graph is shown in the figure. Use the graph to answer each of the following questions.
   a. How many deer were introduced into the habitat?
   b. What is the population after 10 years?
   c. What is the equation of the horizontal asymptote shown in the figure? What does this mean in terms of the deer population?

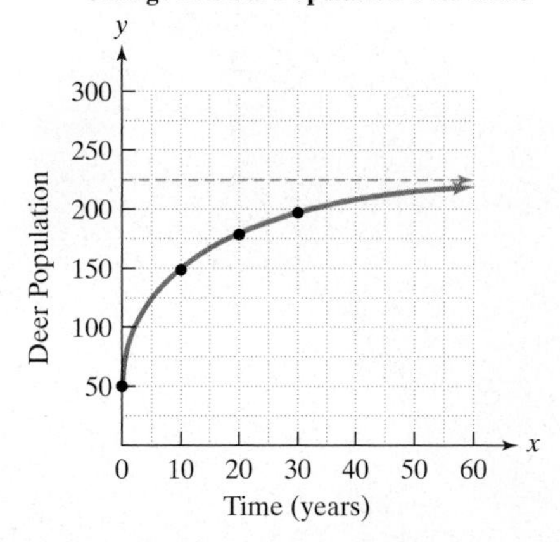

20. The intensity of light received at a source varies inversely as the square of the distance from the source. A particular light has an intensity of 20 foot-candles at 15 feet. What is the light's intensity at 10 feet?

Cumulative Review Exercises (Chapters P–3)

Simplify each expression in Exercises 1–3.

1. $\dfrac{1}{2 - \sqrt{3}}$
2. $3(x^2 - 3x + 1) - 2(3x^2 + x - 4)$
3. $3\sqrt{8} + 5\sqrt{50} - 4\sqrt{32}$
4. Factor completely: $x^7 - x^5$.

Solve each equation in Exercises 5–8.

5. $|2x - 1| = 3$ 6. $3x^2 - 5x + 1 = 0$

7. $9 + \dfrac{3}{x} = \dfrac{2}{x^2}$ 8. $x^3 + 2x^2 - 5x - 6 = 0$

Solve each inequality in Exercises 9–10. Express the answer in interval notation.

9. $|2x - 5| > 3$ 10. $3x^2 > 2x + 5$

11. Give the center and radius. Then graph the equation $x^2 + y^2 - 2x + 4y - 4 = 0$.

12. Solve for t: $V = C(1 - t)$.

13. If $f(x) = \sqrt{45 - 9x}$, find the domain of f.

If $f(x) = x^2 + 2x - 5$ and $g(x) = 4x - 1$, find each function or function value in Exercises 14–16.

14. $(f - g)(x)$ 15. $(f \circ g)(x)$

16. $g(f(-3))$

17. Consider the function $f(x) = x^3 - 4x^2 - x + 4$.
   a. Use factoring to find all zeros of f.
   b. Use the Leading Coefficient Test and the zeros of f to graph the function.

Graph each function in Exercises 18–20.

18. $f(x) = x^2 + 2x - 8$ 19. $f(x) = x^2(x - 3)$

20. $f(x) = \dfrac{x - 1}{x - 2}$

Exponential and Logarithmic Functions

Chapter 4

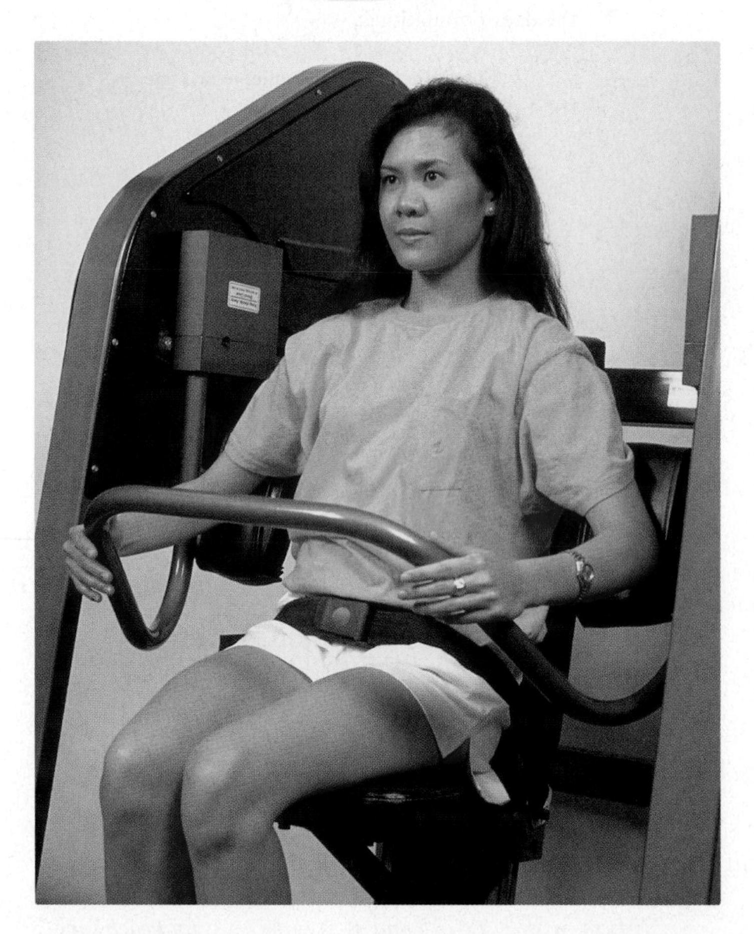

What went wrong on the space shuttle *Challenger*? Will population growth lead to a future without comfort or individual choice? Can I put aside a small amount of money and have millions for early retirement? Why did I feel I was walking too slowly on my visit to New York City? Why are people in California at far more risk from drunk drivers than from earthquakes? What is the difference between earthquakes measuring 6 and 7 on the Richter scale? And what can I hope to accomplish in weightlifting?

The functions that you will be learning about in this chapter will provide you with the mathematics for answering these questions. You will see how these remarkable functions enable us to predict the future and rediscover the past.

You've recently taken up weightlifting, recording the maximum number of pounds you can lift at the end of each week. At first your weight limit increases rapidly, but now you notice that this growth is beginning to level off. You wonder about a function that would serve as a mathematical model to predict the number of pounds you can lift as you continue the sport.

SECTION 4.1 Exponential Functions

Objectives

1. Evaluate exponential functions.
2. Graph exponential functions.
3. Evaluate functions with base e.
4. Use compound interest formulas.

The space shuttle *Challenger* exploded approximately 73 seconds into flight on January 28, 1986. The tragedy involved damage to O-rings, which were used to seal the connections between different sections of the shuttle engines. The number of O-rings damaged increases dramatically as Fahrenheit temperature falls.

The function

$$f(x) = 13.49(0.967)^x - 1$$

models the number of O-rings expected to fail when the temperature is x°F. Can you see how this function is different from polynomial functions? The variable x is in the exponent. Functions whose equations contain a variable in the exponent are called **exponential functions**. Many real-life situations, including population growth, growth of epidemics, radioactive decay, and other changes that involve rapid increase or decrease, can be described using exponential functions.

Definition of the Exponential Function

The **exponential function f with base b** is defined by

$$f(x) = b^x \quad \text{or} \quad y = b^x$$

where b is a positive constant other than $1 (b > 0$ and $b \neq 1)$ and x is any real number.

Here are some examples of exponential functions.

$$f(x) = 2^x \qquad g(x) = 10^x \qquad h(x) = 3^{x+1}$$

Base is 2. Base is 10. Base is 3.

Each of these functions has a constant base and a variable exponent. By contrast, the following functions are not exponential.

$$F(x) = x^2 \qquad G(x) = 1^x \qquad H(x) = x^x$$

Variable is the base and not the exponent. The base of an exponential function must be a positive constant other than 1. Variable is both the base and the exponent.

Why is $G(x) = 1^x$ not classified as an exponential function? The number 1 raised to any power is 1. Thus, the function G can be written as $G(x) = 1$, which is a constant function.

1 Evaluate exponential functions.

You will need a calculator to evaluate exponential expressions. Most scientific calculators have an $\boxed{x^y}$ key. Graphing calculators have a $\boxed{\wedge}$ key. To evaluate expressions of the form b^x, enter the base b, press $\boxed{x^y}$ or $\boxed{\wedge}$, enter the exponent x, and finally press $\boxed{=}$ or $\boxed{\text{ENTER}}$.

EXAMPLE 1 Evaluating an Exponential Function

The exponential function $f(x) = 13.49(0.967)^x - 1$ describes the number of O-rings expected to fail, $f(x)$, when the temperature is x°F. On the morning the *Challenger* was launched, the temperature was 31°F, colder than any previous experience. Find the number of O-rings expected to fail at this temperature.

Solution Because the temperature was 31°F, substitute 31 for x and evaluate the function at 31.

$$f(x) = 13.49(0.967)^x - 1 \quad \text{This is the given function.}$$

$$f(31) = 13.49(0.967)^{31} - 1 \quad \text{Substitute 31 for } x.$$

Use a scientific or graphing calculator to evaluate $(0.967)^{31}$. Press the following keys on your calculator to do this:

Scientific calculator: .967 $\boxed{x^y}$ 31 $\boxed{=}$

Graphing calculator: .967 $\boxed{\wedge}$ 31 $\boxed{\text{ENTER}}$

The display should be approximately .353362693426. Multiplying this number by 13.49 and subtracting 1, we obtain

$$f(31) = 13.49(0.967)^{31} - 1 \approx 4.$$

Thus, four O-rings are expected to fail at a temperature of 31°F.

Check Point 1 Use the function in Example 1 to find the number of O-rings expected to fail at a temperature of 60°F.

2 Graph exponential functions.

Graphing Exponential Functions

We are familiar with expressions involving b^x where x is a rational number. For example,

$$b^{1.7} = b^{17/10} = \sqrt[10]{b^{17}} \quad \text{and} \quad b^{1.73} = b^{173/100} = \sqrt[100]{b^{173}}.$$

However, note that the definition of $f(x) = b^x$ includes all real numbers for the domain x. You may wonder what b^x means when x is an irrational number, such as $b^{\sqrt{3}}$ or b^{π}. Using the nonrepeating and nonterminating approximation 1.73205 for $\sqrt{3}$, we can think of $b^{\sqrt{3}}$ as the value that has the successively closer approximations

$$b^{1.7}, b^{1.73}, b^{1.732}, b^{1.73205}, \ldots.$$

In this way, we can graph the exponential function with no holes, or points of discontinuity, at the irrational domain values.

EXAMPLE 2 Graphing an Exponential Function

Graph: $f(x) = 2^x$.

Solution We begin by setting up a table of coordinates.

x	$f(x) = 2^x$
-3	$f(-3) = 2^{-3} = \frac{1}{8}$
-2	$f(-2) = 2^{-2} = \frac{1}{4}$
-1	$f(-1) = 2^{-1} = \frac{1}{2}$
0	$f(0) = 2^0 = 1$
1	$f(1) = 2^1 = 2$
2	$f(2) = 2^2 = 4$
3	$f(3) = 2^3 = 8$

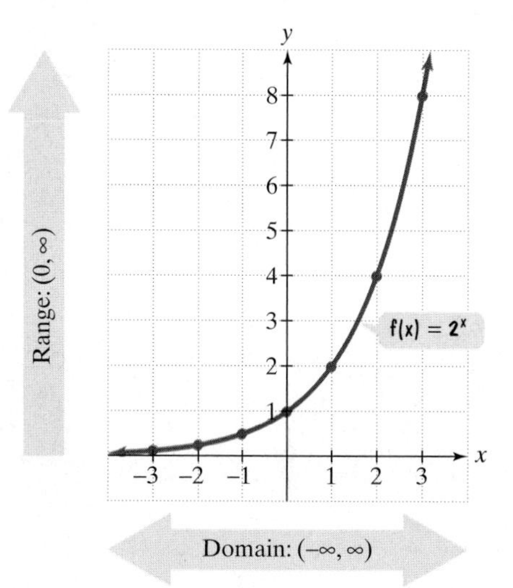

Figure 4.1 The graph of $f(x) = 2^x$

We plot these points, connecting them with a continuous curve. Figure 4.1 shows the graph of $f(x) = 2^x$. Observe that the graph approaches but never touches the negative portion of the x-axis. Thus, the x-axis is a horizontal asymptote. The range is all positive real numbers. Although we used integers for x in our table of coordinates, you can use a calculator to find additional points. For example, $f(0.3) = 2^{0.3} \approx 1.231$, $f(0.95) = 2^{0.95} \approx 1.932$. The points $(0.3, 1.231)$ and $(0.95, 1.932)$ fit the graph.

Check Point 2 Graph $f(x) = 3^x$.

Four exponential functions have been graphed in Figure 4.2. Compare the graphs of functions where $b > 1$ to those where $b < 1$. When $b > 1$, the value of y increases as the value of x increases. When $b < 1$, the value of y decreases as the value of x increases. Notice that all four graphs pass through $(0, 1)$.

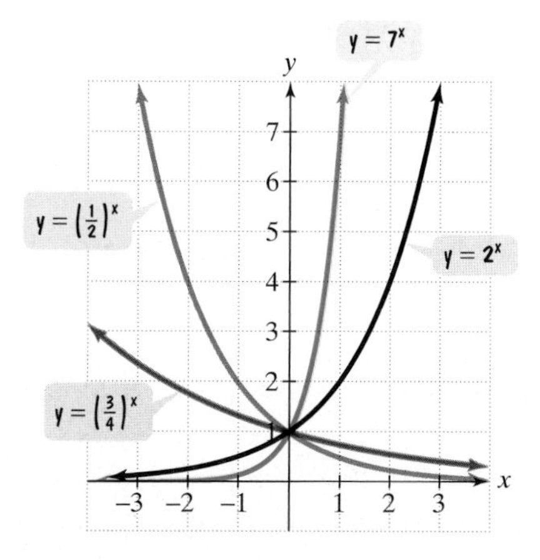

Figure 4.2 Graphs of four exponential functions

These graphs illustrate the following general characteristics of exponential functions.

Characteristics of Exponential Functions

1. The domain of $f(x) = b^x$ consists of all real numbers. The range of $f(x) = b^x$ consists of all positive real numbers.
2. The graphs of all exponential functions pass through the point $(0, 1)$ because $f(0) = b^0 = 1 (b \neq 0)$.
3. If $b > 1$, $f(x) = b^x$ has a graph that goes up to the right and is an increasing function.
4. If $0 < b < 1$, $f(x) = b^x$ has a graph that goes down to the right and is a decreasing function.
5. $f(x) = b^x$ is one-to-one and has an inverse that is a function.
6. The graph of $f(x) = b^x$ approaches but does not cross the x-axis. The x-axis is a horizontal asymptote.

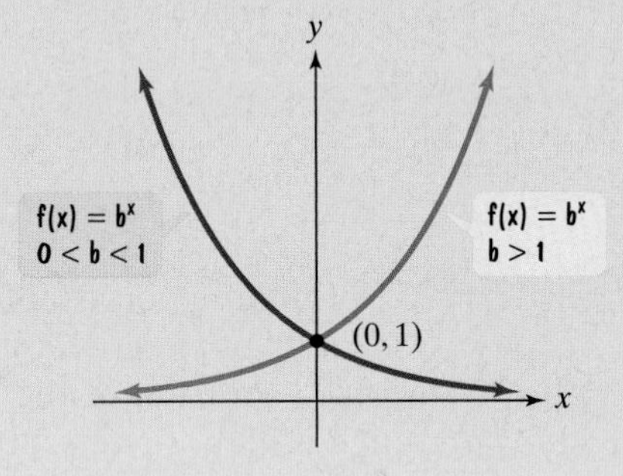

Transformations of Exponential Functions The graphs of exponential functions can be translated vertically or horizontally, reflected, stretched, or shrunk. We use the ideas of Section 2.5 to do so, as summarized in Table 4.1.

Table 4.1 Transformations Involving Exponential Functions

Transformation	Equation	Description
Horizontal translation	$g(x) = b^{x+c}$	• Shifts the graph of $f(x) = b^x$ to the left c units if $c > 0$. • Shifts the graph of $f(x) = b^x$ to the right c units if $c < 0$.
Vertical stretching or shrinking	$g(x) = cb^x$	Multiplying y-coordinates of $f(x) = b^x$ by c, • Stretches the graph of $f(x) = b^x$ if $c > 1$. • Shrinks the graph of $f(x) = b^x$ if $0 < c < 1$.
Reflecting	$g(x) = -b^x$ $g(x) = b^{-x}$	• Reflects the graph of $f(x) = b^x$ about the x-axis. • Reflects the graph of $f(x) = b^x$ about the y-axis.
Vertical translation	$g(x) = b^x + c$	• Shifts the graph of $f(x) = b^x$ upward c units if $c > 0$. • Shifts the graph of $f(x) = b^x$ downward c units if $c < 0$.

Using the information in Table 4.1 and a table of coordinates, you will obtain relatively accurate graphs that can be verified using a graphing utility.

EXAMPLE 3 Transformations Involving Exponential Functions

Use the graph of $f(x) = 3^x$ to obtain the graph of $g(x) = 3^{x+1}$.

Solution Examine Table 4.1. Note that the function $g(x) = 3^{x+1}$ has the general form $g(x) = b^{x+c}$, where $c = 1$. Because $c > 0$, we graph $g(x) = 3^{x+1}$ by shifting the graph of $f(x) = 3^x$ *one* unit to the *left*. We construct a table showing some of the coordinates for f and g. The graphs of f and g are shown in Figure 4.3.

x	$f(x) = 3^x$	$g(x) = 3^{x+1}$
-2	$3^{-2} = \frac{1}{9}$	$3^{-2+1} = 3^{-1} = \frac{1}{3}$
-1	$3^{-1} = \frac{1}{3}$	$3^{-1+1} = 3^0 = 1$
0	$3^0 = 1$	$3^{0+1} = 3^1 = 3$
1	$3^1 = 3$	$3^{1+1} = 3^2 = 9$
2	$3^2 = 9$	$3^{2+1} = 3^3 = 27$

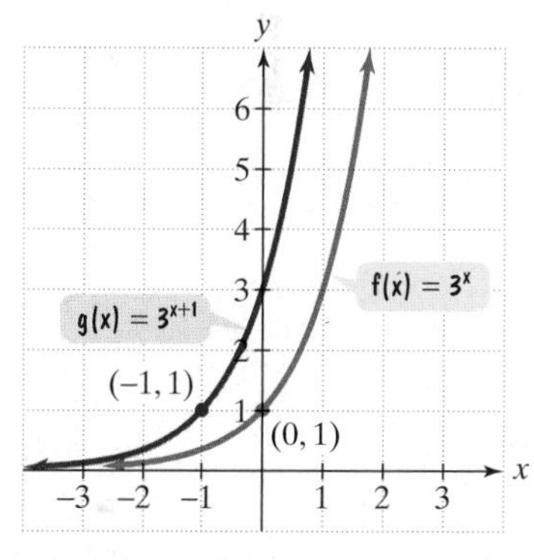

Figure 4.3 The graph of $g(x) = 3^{x+1}$ shifts the graph of $f(x) = 3^x$ one unit to the left.

Check Point 3 Use the graph of $f(x) = 3^x$ to obtain the graph of $g(x) = 3^{x-1}$.

If an exponential function is translated upward or downward, the horizontal asymptote is shifted by the amount of the vertical shift.

EXAMPLE 4 Transformations Involving Exponential Functions

Use the graph of $f(x) = 2^x$ to obtain the graph of $g(x) = 2^x - 3$.

Solution Examine Table 4.1. Note that the function $g(x) = 2^x - 3$ has the general form $g(x) = b^x + c$, where $c = -3$. Because $c < 0$, we graph $g(x) = 2^x - 3$ by shifting the graph of $f(x) = 2^x$ *down three* units. We construct a table showing some of the coordinates for f and g. The graphs of f and g are shown in Figure 4.4 on page 356. Notice that the horizontal asymptote for f, the x-axis, is shifted down three units for the horizontal asymptote for g. Thus, $y = -3$ is the horizontal asymptote for g.

x	$f(x) = 2^x$	$g(x) = 2^x - 3$
−2	$\frac{1}{4}$	$\frac{1}{4} - 3 = -2\frac{3}{4}$
−1	$\frac{1}{2}$	$\frac{1}{2} - 3 = -2\frac{1}{2}$
0	1	$1 - 3 = -2$
1	2	$2 - 3 = -1$
2	4	$4 - 3 = 1$

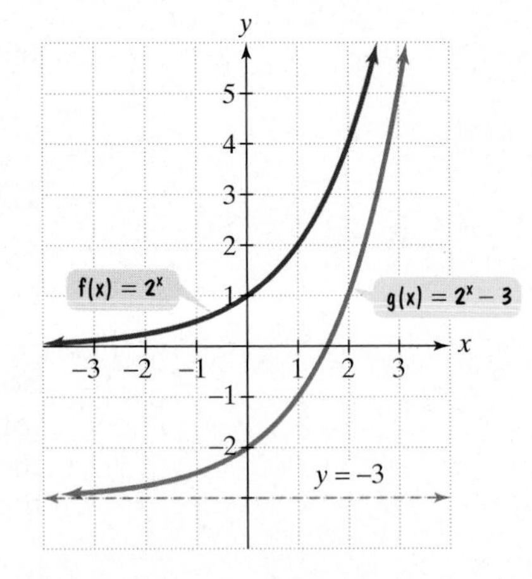

Figure 4.4 The graph of $g(x) = 2^x - 3$ shifts the graph of $f(x) = 2^x$ down three units.

Check Point 4 Use the graph of $f(x) = 2^x$ to obtain the graph of $g(x) = 2^x + 1$.

3 Evaluate functions with base e.

The Natural Base e

An irrational number, symbolized by the letter e, appears as the base in many applied exponential functions. This irrational number is approximately equal to 2.72. More accurately,

$$e \approx 2.71828\ldots.$$

The number e is called the **natural base**. The function $f(x) = e^x$ is called the **natural exponential function**.

Use a scientific or graphing calculator with an $\boxed{e^x}$ key to evaluate e to various powers. For example, to find e^2, press the following keys on most calculators:

Scientific calculator: $2\ \boxed{e^x}$

Graphing calculator: $\boxed{e^x}\ 2\ \boxed{\text{ENTER}}$

The display is approximately 7.389.

$$e^2 \approx 7.389$$

The number e lies between 2 and 3. Because $2^2 = 4$ and $3^2 = 9$, it makes sense that e^2, approximately 7.389, lies between 4 and 9.

Because $2 < e < 3$, the graph of $y = e^x$ is between the graphs of $y = 2^x$ and $y = 3^x$, shown in Figure 4.5.

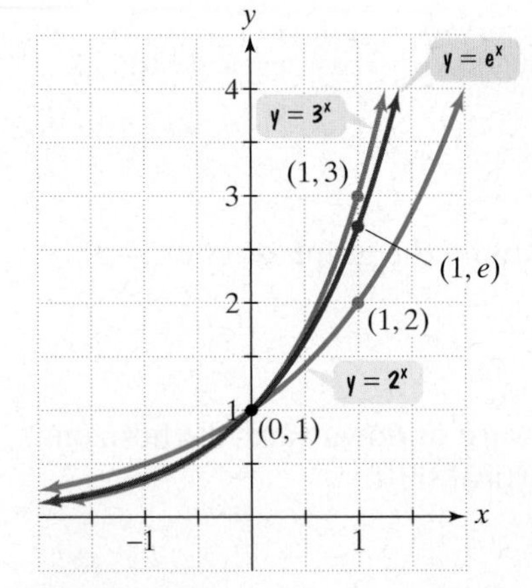

Figure 4.5 Graphs of three exponential functions

EXAMPLE 5 World Population

In a report entitled *Resources and Man*, the U.S. National Academy of Sciences concluded that a world population of 10 billion "is close to (if not above) the maximum that an intensely managed world might hope to support with some degree of comfort and individual choice." At the time the report was issued in 1969,

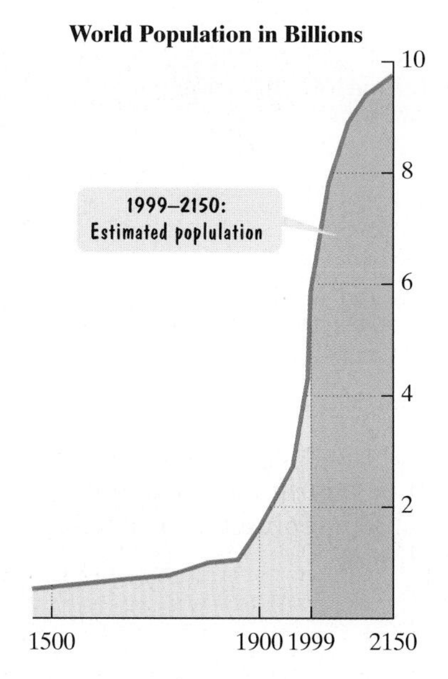

Source: U.N. Population Division

the world population was approximately 3.6 billion, with a growth rate of 2% per year. The function

$$f(x) = 3.6e^{0.02x}$$

describes world population, $f(x)$, in billions, x years after 1969. Use the function to find world population in the year 2020. Is there cause for alarm?

Solution Because 2020 is 51 years after 1969, we substitute 51 for x:

$$f(51) = 3.6e^{0.02(51)}.$$

Although this computation can be done on your calculator in one step, we will break it down into smaller steps so that you can clearly see how we use the $\boxed{e^x}$ key. First find 0.02(51):

$$0.02(51) = 1.02.$$

Now find $e^{1.02}$:

Scientific calculator: $1.02\ \boxed{e^x}$

Graphing calculator: $\boxed{e^x}\ 1.02\ \boxed{\text{ENTER}}$

The display is approximately 2.7731948. Multiplying this number by 3.6, we obtain

$$f(51) = 3.6e^{0.02(51)} = 3.6e^{1.02} \approx 9.98.$$

This indicates that world population in the year 2020 will be approximately 9.98 billion. Because this number is quite close to 10 billion, the given function suggests that there may be cause for alarm.

World population in 1999 was 6 billion, but the growth rate was no longer 2%. It had slowed down to 1.3%. Using this current growth rate, exponential functions now predict a world population of 7.6 billion in the year 2020. Experts think the population may stabilize at 10 billion after 2200 if the deceleration in growth rate continues.

Check Point 5 The function $f(x) = 6e^{0.013x}$ describes world population, $f(x)$, in billions, x years after 1999 subject to a growth rate of 1.3% annually. Use the function to find world population in 2050.

4 Use compound interest formulas.

Compound Interest

We all want a wonderful life with fulfilling work, good health, and loving relationships. And let's be honest: Financial security wouldn't hurt! Achieving this goal depends on understanding how money in savings accounts grows in remarkable ways as a result of *compound interest*. **Compound interest** is interest computed on your original investment as well as on any accumulated interest.

Suppose a sum of money called the **principal**, P, is invested at an annual percentage rate r, in decimal form, compounded once per year. Because the interest is added to the principal at year's end, the accumulated value A is

$$A = P + Pr = P(1 + r).$$

The accumulated amount of money follows this pattern of multiplying the previous principal by $(1 + r)$ for each successive year, as indicated in Table 4.2.

Table 4.2

Time in Years	Accumulated Value after Each Compounding
0	$A = P$
1	$A = P(1 + r)$
2	$A = P(1 + r)(1 + r) = P(1 + r)^2$
3	$A = P(1 + r)^2(1 + r) = P(1 + r)^3$
4	$A = P(1 + r)^3(1 + r) = P(1 + r)^4$
⋮	⋮
t	$A = P(1 + r)^t$

n	$\left(1 + \frac{1}{n}\right)^n$
1	2
2	2.25
5	2.48832
10	2.59374246
100	2.704813829
1,000	2.716923932
10,000	2.718145927
100,000	2.718268237
1,000,000	2.718280469
1,000,000,000	2.718281827

As n takes on increasingly large values, the expression $\left(1 + \frac{1}{n}\right)^n$ approaches e.

If money invested at a specified rate of interest is compounded more than once a year, then the formula $A = P(1 + r)^t$ can be adjusted to take into account the number of compounding periods in a year. If n represents the number of compounding periods in a year, the formula becomes

$$A = P\left(1 + \frac{r}{n}\right)^{nt}.$$

Some banks use **continuous compounding**, where the number of compounding periods increases infinitely (compounding interest every trillionth of a second, every quadrillionth of a second, etc.). As n, the number of compounding periods in a year, increases without bound, the expression $\left(1 + \frac{1}{n}\right)^n$ approaches e. As a result, the formula for continuous compounding is $A = Pe^{rt}$. Although continuous compounding sounds terrific, it yields only a fraction of a percent more interest over a year than daily compounding.

Formulas for Compound Interest

After t years, the balance A in an account with principal P and annual interest rate r (in decimal form) is given by the following formulas:

1. For n compoundings per year: $A = P\left(1 + \frac{r}{n}\right)^{nt}$
2. For continuous compounding: $A = Pe^{rt}$

EXAMPLE 6 Choosing Between Investments

You want to invest \$8000 for 6 years, and you have a choice between two accounts. The first pays 7% per year, compounded monthly. The second pays 6.85% per year, compounded continuously. Which is the better investment?

Solution The better investment is the one with the greater balance in the account after 6 years. Let's begin with the account with monthly compounding. We use the compound interest model with $P = 8000$, $r = 7\% = 0.07$, $n = 12$ (monthly compounding, means 12 compoundings per year), and $t = 6$.

$$A = P\left(1 + \frac{r}{n}\right)^{nt} = 8000\left(1 + \frac{0.07}{12}\right)^{12 \cdot 6} \approx 12{,}160.84$$

The balance in this account after 6 years is \$12,160.84. For the second investment option, we use the model for continuous compounding with $P = 8000$, $r = 6.85\% = 0.0685$, and $t = 6$.

$$A = Pe^{rt} = 8000e^{0.0685(6)} \approx 12{,}066.60$$

The balance in this account after 6 years is $12,066.60, slightly less than the previous amount. Thus, the better investment is the 7% monthly compounding option.

A sum of $10,000 is invested at an annual rate of 8%. Find the balance in the account after 5 years subject to **a.** quarterly compounding and **b.** continuous compounding.

EXERCISE SET 4.1

Practice Exercises

In Exercises 1–10, approximate each number using a calculator. Round your answer to three decimal places.

1. $2^{3.4}$ **2.** $3^{2.4}$ **3.** $3^{\sqrt{5}}$ **4.** $5^{\sqrt{3}}$ **5.** $4^{-1.5}$

6. $6^{-1.2}$ **7.** $e^{2.3}$ **8.** $e^{3.4}$ **9.** $e^{-0.95}$ **10.** $e^{-0.75}$

In Exercises 11–18, graph each function by making a table of coordinates. If applicable, use a graphing utility to confirm your hand-drawn graph.

11. $f(x) = 4^x$ **12.** $f(x) = 5^x$

13. $g(x) = \left(\frac{3}{2}\right)^x$ **14.** $g(x) = \left(\frac{4}{3}\right)^x$

15. $h(x) = \left(\frac{1}{2}\right)^x$ **16.** $h(x) = \left(\frac{1}{3}\right)^x$

17. $f(x) = (0.6)^x$ **18.** $f(x) = (0.8)^x$

In Exercises 19–24, the graph of an exponential function is given. Select the function for each graph from the following options:

$$f(x) = 3^x,\ g(x) = 3^{x-1},\ h(x) = 3^x - 1,$$
$$F(x) = -3^x,\ G(x) = 3^{-x},\ H(x) = -3^{-x}.$$

19.

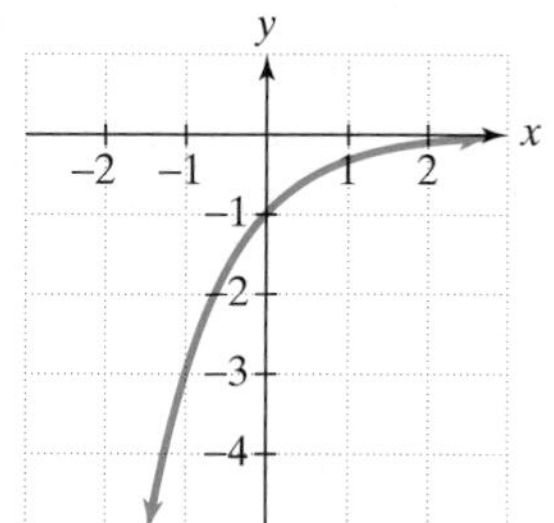

20.

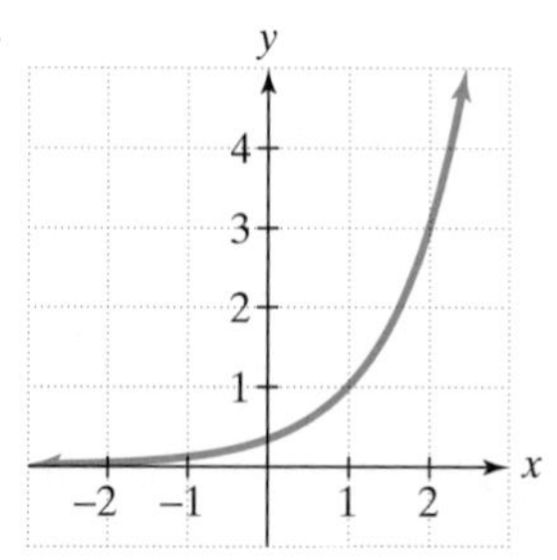

21.

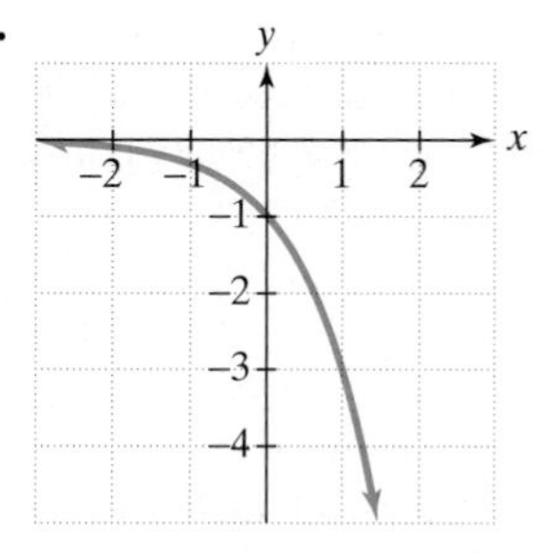

22.

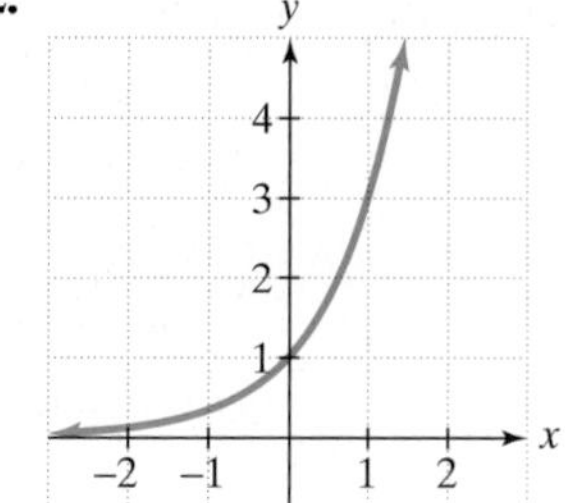

23.

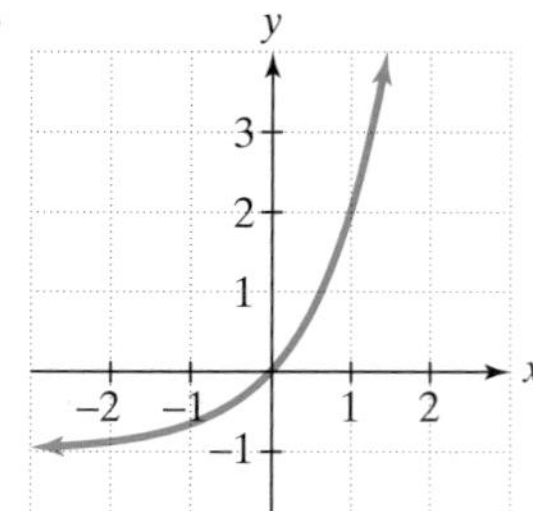

24.

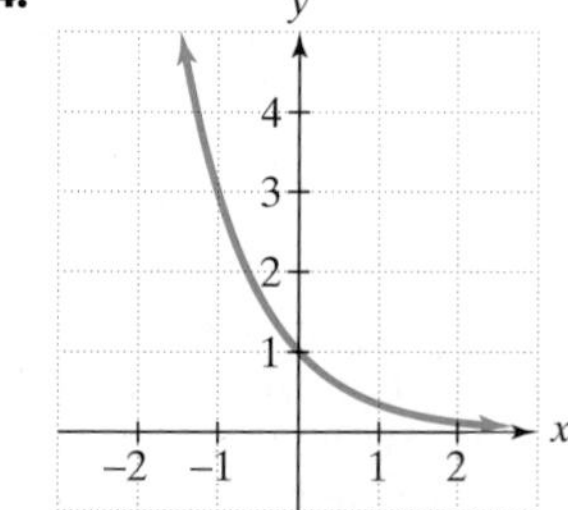

In Exercises 25–34, begin by graphing $f(x) = 2^x$. Then use transformations of this graph and a table of coordinates to graph the given function.

25. $g(x) = 2^{x+1}$
26. $g(x) = 2^{x+2}$
27. $g(x) = 2^x - 1$
28. $g(x) = 2^x + 2$
29. $h(x) = 2^{x+1} - 1$
30. $h(x) = 2^{x+2} - 1$
31. $g(x) = -2^x$
32. $g(x) = 2^{-x}$
33. $g(x) = 2 \cdot 2^x$
34. $g(x) = \frac{1}{2} \cdot 2^x$

Use the compound interest formulas $A = P\left(1 + \frac{r}{n}\right)^{nt}$ and $A = Pe^{rt}$ to solve Exercises 35–38.

35. Find the accumulated value of an investment of \$10,000 for 5 years at an interest rate of 5.5% if the money is **a.** compounded semiannually; **b.** compounded monthly; **c.** compounded continuously.

36. Find the accumulated value of an investment of \$5000 for 10 years at an interest rate of 6.5% if the money is **a.** compounded semiannually; **b.** compounded monthly; **c.** compounded continuously.

37. Suppose that you have \$12,000 to invest. Which investment yields the greatest return over 3 years: 7% compounded monthly or 6.85% compounded continuously?

38. Suppose that you have \$6000 to invest. Which investment yields the greatest return over 4 years: 8.25% compounded quarterly or 8.3% compounded semiannually?

Application Exercises

Use a calculator with an $\boxed{x^y}$ key or a $\boxed{\wedge}$ key to solve Exercises 39–46.

39. The exponential function $f(x) = 67.38(1.026)^x$ describes the population of Mexico, $f(x)$, in millions, x years after 1980.

a. Substitute 0 for x and, without using a calculator, find Mexico's population in 1980.
b. Substitute 27 for x and use your calculator to find Mexico's population in the year 2007 as predicted by this function.
c. Find Mexico's population in the year 2034 as predicted by this function.
d. Find Mexico's population in the year 2061 as predicted by this function.
e. What appears to be happening to Mexico's population every 27 years?

40. The 1986 explosion at the Chernobyl nuclear power plant in the former Soviet Union sent about 1000 kilograms of radioactive cesium–137 into the atmosphere. The function $f(x) = 1000(0.5)^{x/30}$ describes the amount, $f(x)$, in kilograms, of cesium-137 remaining in Chernobyl x years after 1986. If even 100 kilograms of cesium-137 remain in Chernobyl's atmosphere, the area is considered unsafe for human habitation. Find $f(80)$ and determine if Chernobyl will be safe for human habitation by 2066.

The function

$$f(x) = \frac{0.9}{1 + 271(0.885)^x}$$

models the fraction of people x years old with some coronary heart disease. Use this function to solve Exercises 41–42.

41. Evaluate $f(25)$ and describe what this means in practical terms.

42. Evaluate $f(70)$ and describe what this means in practical terms.

The formula $S = C(1 + r)^t$ models inflation, where C = the value today, r = the annual inflation rate, and S = the inflated value t years from now. Use this formula to solve Exercises 43–44.

43. If the inflation rate is 6%, how much will a house now worth \$65,000 be worth in 10 years?

44. If the inflation rate is 3%, how much will a house now worth \$110,000 be worth in 5 years?

45. A decimal approximation for $\sqrt{3}$ is 1.7320508. Use a calculator to find $2^{1.7}$, $2^{1.73}$, $2^{1.732}$, $2^{1.73205}$, and $2^{1.7320508}$. Now find $2^{\sqrt{3}}$. What do you observe?

46. A decimal approximation for π is 3.141593. Use a calculator to find 2^3, $2^{3.1}$, $2^{3.14}$, $2^{3.141}$, $2^{3.1415}$, $2^{3.14159}$, and $2^{3.141593}$. Now find 2^π. What do you observe?

Use a calculator with an $\boxed{e^x}$ key to solve Exercises 47–51. The function $f(x) = 24{,}000e^{0.21x}$ describes the number of AIDS cases in the United States among intravenous drug users x years after 1989. Use this function to solve Exercises 47–48.

47. Evaluate $f(11)$ and describe what this means in practical terms.

48. Evaluate $f(31)$ and describe what this means in practical terms.

49. In college, we study large volumes of information—information that, unfortunately, we do not often retain for very long. The function

$$f(x) = 80e^{-0.5x} + 20$$

describes the percentage of information, $f(x)$, that a particular person remembers x weeks after learning the information.

a. Substitute 0 for x and, without using a calculator, find the percentage of information remembered at the moment it is first learned.
b. Substitute 1 for x and find the percentage of information that is remembered after 1 week.
c. Find the percentage of information that is remembered after 4 weeks.
d. Find the percentage of information that is remembered after one year (52 weeks).

50. In 1626, Peter Minuit convinced the Wappinger Indians to sell him Manhattan Island for \$24. If the Native Americans had put the \$24 into a bank account paying 5% interest, how much would the investment be worth in the year 2000 if interest were compounded
a. monthly? **b.** continuously?

51. The function

$$N(t) = \frac{30{,}000}{1 + 20e^{-1.5t}}$$

describes the number of people, $N(t)$, who become ill with influenza t weeks after its initial outbreak in a town with 30,000 inhabitants. The horizontal asymptote in the graph indicates that there is a limit to the epidemic's growth.
a. How many people became ill with the flu when the epidemic began? (When the epidemic began, $t = 0$.)
b. How many people were ill by the end of the third week?
c. Why can't the spread of an epidemic simply grow indefinitely? What does the horizontal asymptote shown in the graph indicate about the limiting size of the population that becomes ill?

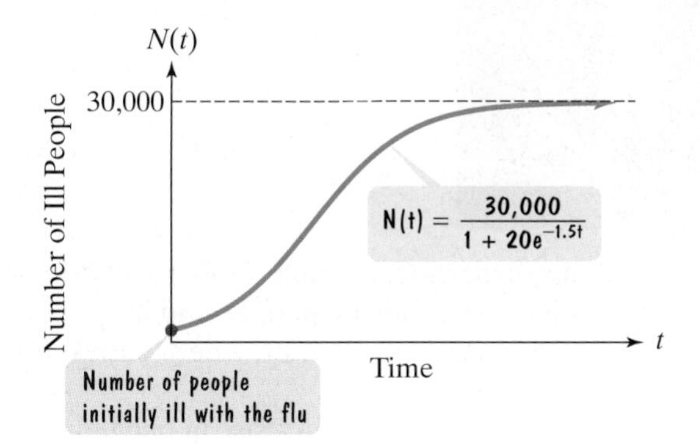

Writing in Mathematics

52. What is an exponential function?

53. What is the natural exponential function?

54. Use a calculator to evaluate $\left(1 + \frac{1}{x}\right)^x$ for $x = 10, 100,$ 1000, 10,000, 100,000, and 1,000,000. Describe what happens to the expression as x increases.

55. Write an example similar to Example 6 on page 358 in which continuous compounding at a slightly lower yearly interest rate is a better investment than compounding n times per year.

56. Describe how you could use the graph of $f(x) = 2^x$ to obtain a decimal approximation for $\sqrt{2}$.

57. The exponential function $y = 2^x$ is one-to-one and has an inverse function. Try finding the inverse function by exchanging x and y and solving for y. Describe the difficulty that you encounter in this process. What is needed to overcome this problem?

58. In 1999, world population was 6 billion with an annual growth rate of 1.3%. Discuss two factors that would cause this growth rate to slow down over the next ten years.

Technology Exercises

59. Graph $y = 13.49(0.967)^x - 1$, the function for the number of O-rings expected to fail at x°F, in a $[0, 90, 10]$ by $[0, 20, 5]$ viewing rectangle. If NASA engineers had used this function and its graph, is it likely they would have allowed the *Challenger* to be launched when the temperature was 31°F? Explain.

60. The student–teacher ratio in U.S. elementary and secondary schools can be modeled by $y = 25.34\,(0.987)^x$, where x represents the number of years since 1959 and y represents the student–teacher ratio. Graph the function in a $[1, 40, 1]$ by $[0, 26, 1]$ viewing rectangle. When did the student–teacher ratio become less than 21 students per teacher?

61. You have \$10,000 to invest. One bank pays 5% interest compounded quarterly and the other pays 4.5% interest compounded monthly.
a. Use the formula for compound interest to write a function for the balance in each account at any time t.
b. Use a graphing utility to graph both functions in an appropriate viewing rectangle. Based on the graphs, which bank offers the better return on your money?

62. a. Graph $y = e^x$ and $y = 1 + x + \frac{x^2}{2}$ in the same viewing rectangle.
b. Graph $y = e^x$ and $y = 1 + x + \frac{x^2}{2} + \frac{x^3}{6}$ in the same viewing rectangle.
c. Graph $y = e^x$ and $y = 1 + x + \frac{x^2}{2} + \frac{x^3}{6} + \frac{x^4}{24}$ in the same viewing rectangle.
d. Describe what you observe in parts (a)–(c). Try generalizing this observation.

Critical Thinking Exercises

63. Which one of the following is true?
a. As the number of compounding periods increases on a fixed investment, the amount of money in the account over a fixed interval of time will increase without bound.
b. The functions $f(x) = 3^{-x}$ and $g(x) = -3^x$ have the same graph.
c. $e = 2.718$.
d. The functions $f(x) = \left(\frac{1}{3}\right)^x$ and $g(x) = 3^{-x}$ have the same graph.

64. The graphs labeled (a)–(d) in the figure represent $y = 3^x$, $y = 5^x$, $y = \left(\frac{1}{3}\right)^x$, and $y = \left(\frac{1}{5}\right)^x$, but not necessarily in that order. Which is which? Describe the process that enables you to make this decision.

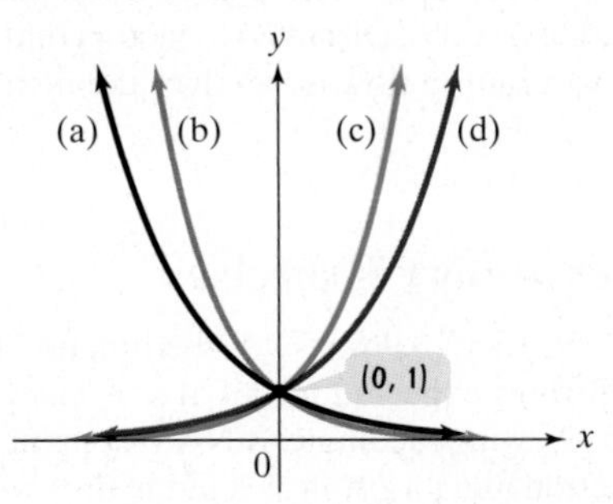

65. The hyperbolic cosine and hyperbolic sine functions are defined by

$$\cosh x = \frac{e^x + e^{-x}}{2} \quad \text{and} \quad \sinh x = \frac{e^x - e^{-x}}{2}.$$

Prove that $(\cosh x)^2 - (\sinh x)^2 = 1$.

4.2 Logarithmic Functions

Objectives

1. Change from logarithmic to exponential form.
2. Change from exponential to logarithmic form.
3. Evaluate logarithms.
4. Use basic logarithmic properties.
5. Graph logarithmic functions.
6. Find the domain of a logarithmic function.
7. Use common logarithms.
8. Use natural logarithms.

The earthquake that ripped through northern California on October 17, 1989, measured 7.1 on the Richter scale, killed more than 60 people, and injured more than 2400. Shown here is San Francisco's Marina district, where shock waves tossed houses off their foundations and into the street.

The Richter scale is misleading because for each increase in one unit on the scale, there is a tenfold increase in the intensity of an earthquake. In this section our focus is on the inverse of the exponential function, called the logarithmic function. The logarithmic function will help you to understand diverse phenomena, including earthquake intensity, human memory, and the pace of life in large cities.

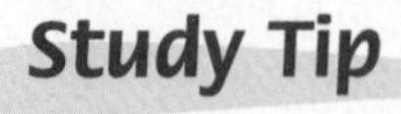

In case you need to review inverse functions, they are discussed on pages 246–247. The horizontal line test appears on page 245.

The Definition of Logarithmic Functions

No horizontal line can be drawn that intersects the graph of an exponential function at more than one point. This means that the exponential function is one-to-one and has an inverse. The inverse function of the exponential function with base b is called the **logarithmic function with base *b*.**

Definition of the Logarithmic Function

For $x > 0$ and $b > 0$, $b \neq 1$,

$$y = \log_b x \quad \text{is equivalent to } b^y = x.$$

The function $f(x) = \log_b x$ is the **logarithmic function with base *b*.**

The equations

$$y = \log_b x \text{ and } b^y = x$$

are different ways of expressing the same thing. The first equation is in **logarithmic form** and the second equivalent equation is in **exponential form**.

Notice that a **logarithm**, y, **is an exponent**. You should learn the location of the base and exponent in each form.

Location of Base and Exponent in Exponential and Logarithmic Forms

Exponent | Exponent

Logarithmic Form: $y = \log_b x$ | Exponential Form: $b^y = x$

Base | Base

1 Change from logarithmic to exponential form.

EXAMPLE 1 Changing From Logarithmic to Exponential Form

Write each equation in its equivalent exponential form.

a. $2 = \log_5 x$ **b.** $3 = \log_b 64$ **c.** $\log_3 7 = y$

Solution We use the fact that $y = \log_b x$ means $b^y = x$.

a. $2 = \log_5 x$ means $5^2 = x$. **b.** $3 = \log_b 64$ means $b^3 = 64$.

Logarithms are exponents. | Logarithms are exponents.

c. $\log_3 7 = y$ or $y = \log_3 7$ means $3^y = 7$.

Check Point 1 Write each equation in its equivalent exponential form.

a. $3 = \log_7 x$ **b.** $2 = \log_b 25$ **c.** $\log_4 26 = y$

2 Change from exponential to logarithmic form.

EXAMPLE 2 Changing From Exponential to Logarithmic Form

Write each equation in its equivalent logarithmic form.

a. $12^2 = x$ **b.** $b^3 = 8$ **c.** $e^y = 9$

Solution We use the fact that $b^y = x$ means $y = \log_b x$.

a. $12^2 = x$ means $2 = \log_{12} x$. **b.** $b^3 = 8$ means $3 = \log_b 8$.

Exponents are logarithms. | Exponents are logarithms.

c. $e^y = 9$ means $y = \log_e 9$.

Check Point 2 Write each equation in its equivalent logarithmic form.

a. $2^5 = x$ **b.** $b^3 = 27$ **c.** $e^y = 33$

3 Evaluate logarithms.

Remembering that logarithms are exponents makes it possible to evaluate some logarithms by inspection. The logarithm of x with base b, $\log_b x$, is the exponent to which b must be raised to get x. For example, suppose we want to evaluate $\log_2 32$. We ask, 2 to what power gives 32? Because $2^5 = 32$, $\log_2 32 = 5$.

EXAMPLE 3 Evaluating Logarithms

Evaluate:

a. $\log_2 16$ **b.** $\log_3 9$ **c.** $\log_{25} 5$

Solution

Logarithmic Expression	Question Needed for Evaluation	Logarithmic Expression Evaluated
a. $\log_2 16$	2 to what power gives 16?	$\log_2 16 = 4$ because $2^4 = 16$.
b. $\log_3 9$	3 to what power gives 9?	$\log_3 9 = 2$ because $3^2 = 9$.
c. $\log_{25} 5$	25 to what power gives 5?	$\log_{25} 5 = \frac{1}{2}$ because $25^{1/2} = \sqrt{25} = 5$.

Check Point 3 Evaluate:

a. $\log_{10} 100$ **b.** $\log_3 3$ **c.** $\log_{36} 6$

4 Use basic logarithmic properties.

Basic Logarithmic Properties

Because logarithms are exponents, they have properties that can be verified using properties of exponents.

Basic Logarithmic Properties Involving One

1. $\log_b b = 1$ because 1 is the exponent to which b must be raised to obtain b. $(b^1 = b)$
2. $\log_b 1 = 0$ because 0 is the exponent to which b must be raised to obtain 1. $(b^0 = 1)$

EXAMPLE 4 Using Properties of Logarithms

Evaluate:

a. $\log_7 7$ **b.** $\log_5 1$

Solution

a. Because $\log_b b = 1$, we conclude $\log_7 7 = 1$.

b. Because $\log_b 1 = 0$, we conclude $\log_5 1 = 0$.

Check Point 4

Evaluate:

a. $\log_9 9$ **b.** $\log_8 1$

The inverse of the exponential function is the logarithmic function. Thus, if $f(x) = b^x$, then $f^{-1}(x) = \log_b x$. In Chapter 2, we saw how inverse functions "undo" one another. In particular,

$$f(f^{-1}(x)) = x \text{ and } f^{-1}(f(x)) = x.$$

Applying these relationships to exponential and logarithmic functions, we obtain the following **inverse properties of logarithms**.

Inverse Properties of Logarithms

For $b > 0$ and $b \neq 1$,

$\log_b b^x = x$ The logarithm with base b of b raised to a power equals that power.

$b^{\log_b x} = x$ b raised to the logarithm with base b of a number equals that number.

EXAMPLE 5 Using Inverse Properties of Logarithms

Evaluate:

a. $\log_4 4^5$ **b.** $6^{\log_6 9}$.

Solution

a. Because $\log_b b^x = x$, we conclude $\log_4 4^5 = 5$.

b. Because $b^{\log_b x} = x$, we conclude $6^{\log_6 9} = 9$.

Check Point 5

Evaluate:

a. $\log_7 7^8$ **b.** $3^{\log_3 17}$

5 Graph logarithmic functions.

Graphs of Logarithmic Functions

How do we graph logarithmic functions? We use the fact that the logarithmic function is the inverse of the exponential function. This means that the logarithmic function reverses the coordinates of the exponential function. It also means that the graph of the logarithmic function is a reflection of the graph of the exponential function about the line $y = x$.

EXAMPLE 6 Graphs of Exponential and Logarithmic Functions

Graph $f(x) = 2^x$ and $g(x) = \log_2 x$ in the same rectangular coordinate system.

Solution We first set up a table of coordinates for $f(x) = 2^x$. Reversing, these coordinates gives the coordinates for the inverse function $g(x) = \log_2 x$.

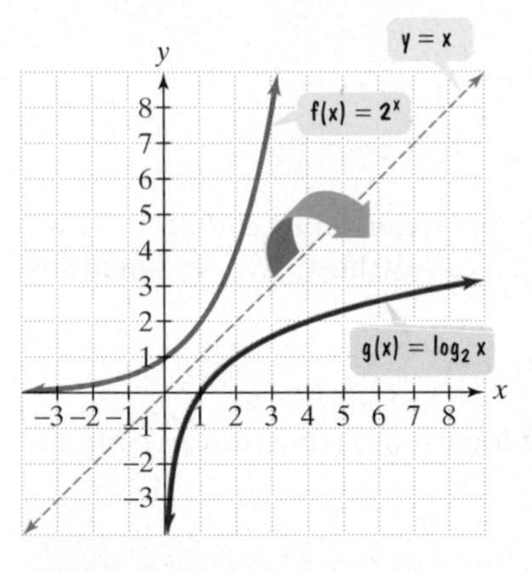

Figure 4.6 The graphs of $f(x) = 2^x$ and its inverse function

x	-2	-1	0	1	2	3
$f(x) = 2^x$	$\frac{1}{4}$	$\frac{1}{2}$	1	2	4	8

Reverse coordinates.

x	$\frac{1}{4}$	$\frac{1}{2}$	1	2	4	8
$g(x) = \log_2 x$	-2	-1	0	1	2	3

We now plot the ordered pairs in both tables, connecting them with smooth curves. Figure 4.6 shows the graphs of $f(x) = 2^x$ and its inverse function $g(x) = \log_2 x$. The graph of the inverse can also be drawn by reflecting the graph of $f(x) = 2^x$ about the line $y = x$.

Check Point 6 Graph $f(x) = 3^x$ and $g(x) = \log_3 x$ in the same rectangular coordinate system.

Figure 4.7 illustrates the relationship between the graph of the exponential function, shown in blue and its inverse, the logarithmic function, shown in red, for bases greater than 1 and for bases between 0 and 1.

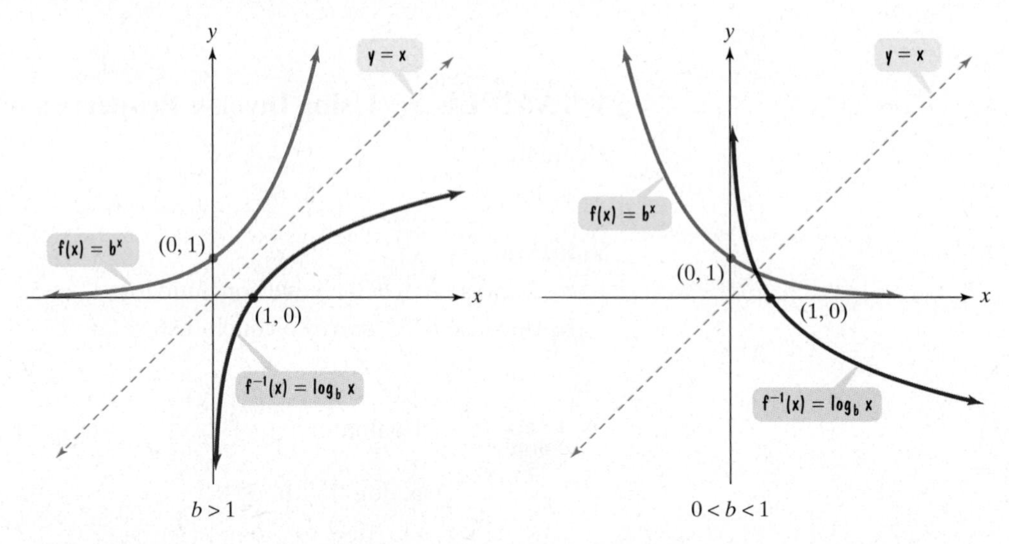

Figure 4.7 Graphs of exponential and logarithmic functions

Characteristics of the Graphs of Logarithmic Functions

- The x-intercept is 1. There is no y-intercept.
- The y-axis is a vertical asymptote.
- If $b > 1$, the function is increasing. If $0 < b < 1$, the function is decreasing.
- The graph is smooth and continuous. It has no sharp corners or gaps.

The graphs of logarithmic functions can be translated vertically or horizontally, reflected, stretched, or shrunk. We use the ideas of Section 2.5 to do so, as summarized in Table 4.3.

Table 4.3 Transformations Involving Logarithmic Functions

Transformation	Equation	Description
Horizontal translation	$g(x) = \log_b(x + c)$	• Shifts the graph of $f(x) = \log_b x$ to the left c units if $c > 0$. Vertical asymptote: $x = -c$. • Shifts the graph of $f(x) = \log_b x$ to the right c units if $c < 0$. Vertical asymptote: $x = -c$.
Vertical stretching or shrinking	$g(x) = c \log_b x$	Multiplying y-coordinates of $f(x) = \log_b x$ by c, • Stretches the graph of $f(x) = \log_b x$ if $c > 1$. • Shrinks the graph of $f(x) = \log_b x$ if $0 < c < 1$.
Reflecting	$g(x) = -\log_b x$ $g(x) = \log_b(-x)$	• Reflects the graph of $f(x) = \log_b x$ about the x-axis. • Reflects the graph of $f(x) = \log_b x$ about the y-axis.
Vertical translation	$g(x) = c + \log_b x$	• Shifts the graph of $f(x) = \log_b x$ upward c units if $c > 0$. • Shifts the graph of $f(x) = \log_b x$ downward c units if $c < 0$.

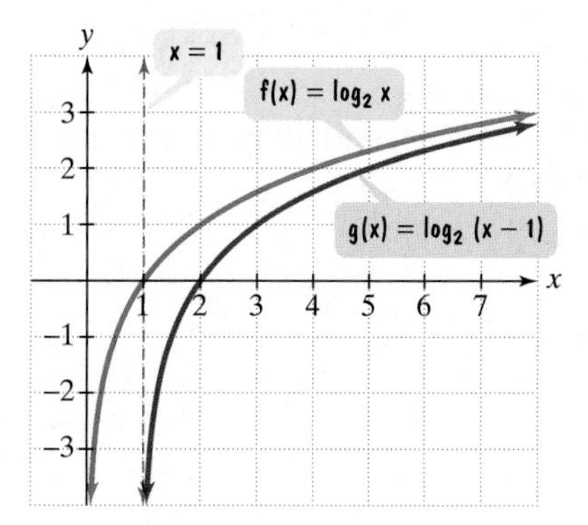

Figure 4.8 Shifting $f(x) = \log_2 x$ one unit to the right

For example, Figure 4.8 illustrates that the graph of $g(x) = \log_2(x - 1)$ is the graph of $f(x) = \log_2 x$ moved one unit to the right. If a logarithmic function is translated to the left or to the right, both the x-intercept and the vertical asymptote are shifted by the amount of the horizontal shift. In Figure 4.8, the x-intercept of f is 1. Because g is shifted one unit to the right, its x-intercept is 2. Also observe that the vertical asymptote for f, the y-axis, is shifted one unit to the right for the vertical asymptote for g. Thus, $x = 1$ is the vertical asymptote for g.

Here are some other examples of transformations of graphs of logarithmic functions.

- The graph of $g(x) = 3 + \log_4 x$ is the graph of $f(x) = \log_4 x$ moved up three unit, shown in Figure 4.9
- The graph of $h(x) = -\log_2 x$ is the graph of $f(x) = \log_2 x$ reflected about the x-axis, shown in Figure 4.10
- The graph of $r(x) = \log_2(-x)$ is the graph of $f(x) = \log_2 x$ reflected about the y-axis, shown in Figure 4.11.

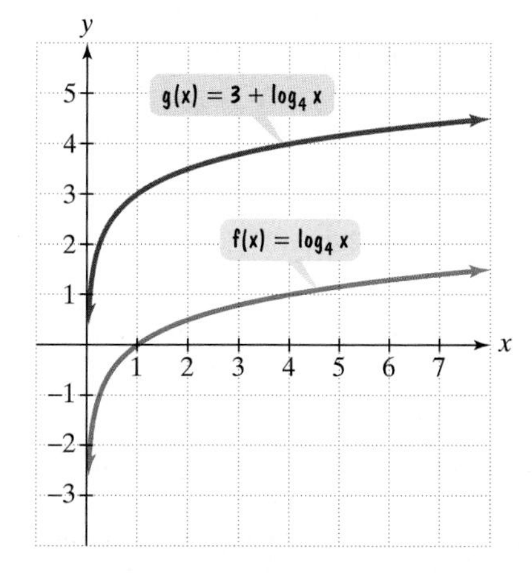

Figure 4.9 Shifting vertically up three units

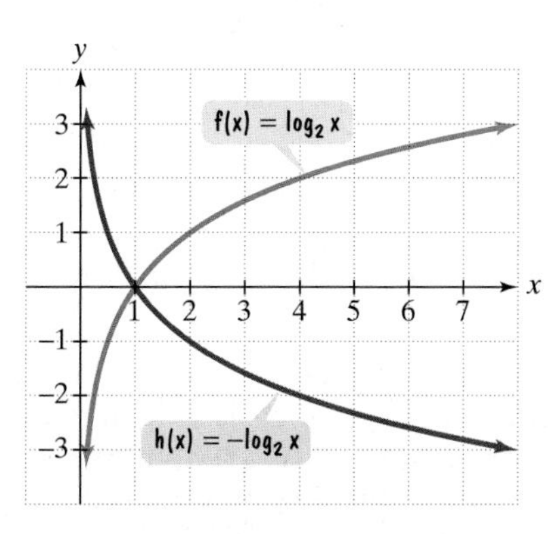

Figure 4.10 Reflection about the x-axis

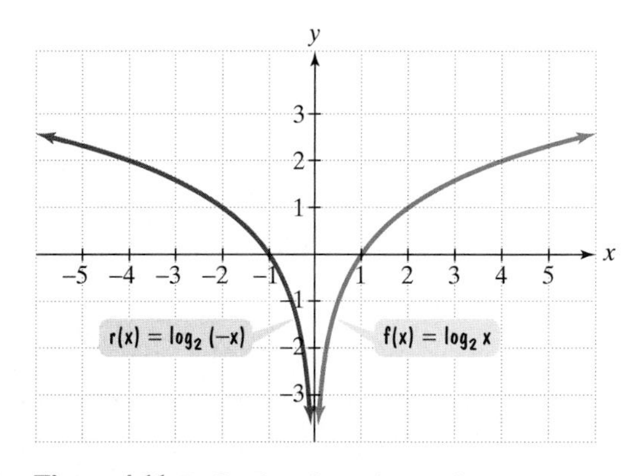

Figure 4.11 Reflection about the y-axis

6 Find the domain of a logarithmic function.

The Domain of a Logarithmic Function

In Section 4.1 we learned that the domain of an exponential function includes all real numbers and its range is the set of positive real numbers. Because the logarithmic function reverses the domain and the range of the exponential function, the **domain of a logarithmic function is the set of all positive real numbers.** Thus, $\log_2 8$ is defined because the value of x in the logarithmic expression, 8, is greater than zero and therefore is included in the domain of the logarithmic function $f(x) = \log_2 x$. However, $\log_2 0$ and $\log_2(-8)$ are not defined because 0 and -8 are not positive real numbers and therefore are excluded from the domain of the logarithmic function $f(x) = \log_2 x$. In general, the domain of $f(x) = \log_b(x + c)$ consists of all x for which $x + c > 0$.

EXAMPLE 7 Finding the Domain of a Logarithmic Function

Find the domain of $g(x) = \log_4(x + 3)$.

Solution The domain of g consists of all x for which $x + 3 > 0$. Solving this inequality for x, we obtain $x > -3$. Thus, the domain of g is $(-3, \infty)$. This is illustrated in Figure 4.12. The vertical asymptote is $x = -3$, and all points on the graph of g have x-coordinates that are greater than -3.

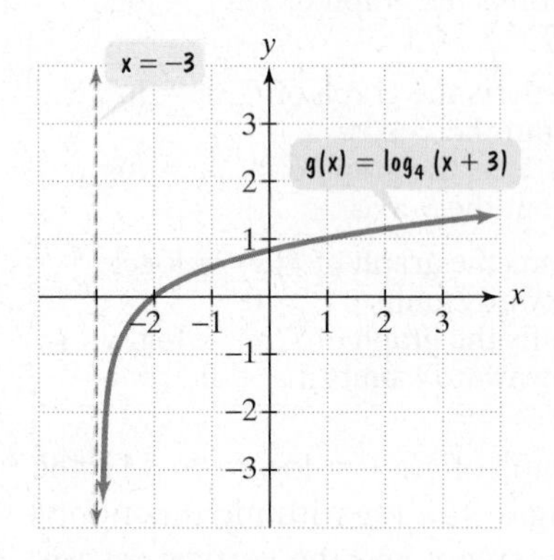

Figure 4.12 The domain of $g(x) = \log_4(x + 3)$ is $(-3, \infty)$.

Check Point 7 Find the domain of $h(x) = \log_4(x - 5)$.

7 Use common logarithms.

Common Logarithms

The logarithmic function with base 10 is called the **common logarithmic function.** The function $f(x) = \log_{10} x$ is usually expressed as $f(x) = \log x$. A calculator with a [LOG] key can be used to evaluate common logarithms. Here are some examples:

Logarithm	Graphing Calculator Keystrokes	Display (or Approximate Display)
$\log 1000$	[LOG] 1000 [ENTER]	3
$\log \frac{5}{2}$	[LOG] [(] 5 [÷] 2 [)] [ENTER]	0.39794
$\frac{\log 5}{\log 2}$	[LOG] 5 [÷] [LOG] 2 [ENTER]	2.32192
$\log(-3)$	[LOG] [(−)] 3 [ENTER]	[ERROR]

The error message given by many graphing calculators for $\log(-3)$ is a reminder that the domain of every logarithmic function, including the common logarithmic function, is the set of positive real numbers.

Many real-life phenomena start with rapid growth, and then the growth begins to level off. This type of behavior can be modeled by logarithmic functions.

EXAMPLE 8 Modeling Height of Children

The percentage of adult height attained by a boy who is x years old can be modeled by

$$f(x) = 29 + 48.8\log(x + 1)$$

where x represents the boy's age and $f(x)$ represents the percentage of his adult height. Approximately what percentage of his adult height is a boy at age eight?

Solution We substitute the boy's age, 8, for x and evaluate the function at 8.

$f(x) = 29 + 48.8\log(x + 1)$ This is the given function.

$f(8) = 29 + 48.8\log(8 + 1)$ Substitute 8 for x.

$= 29 + 48.8\log 9$ Graphing calculator keystrokes:

≈ 76 29 [+] 48.8 [×] [LOG] 9 [ENTER]

Thus, an 8-year-old boy is approximately 76% of his adult height.

Check Point 8 Use the function in Example 8 to answer this question: Approximately what percentage of his adult height is a boy at age 10?

The basic properties of logarithms that were listed earlier in this section can be applied to common logarithms.

Properties of Common Logarithms

General Properties	Common Logarithms
1. $\log_b 1 = 0$	**1.** $\log 1 = 0$
2. $\log_b b = 1$	**2.** $\log 10 = 1$
3. $\log_b b^x = x$	**3.** $\log 10^x = x$
4. $b^{\log_b x} = x$	**4.** $10^{\log x} = x$

The property $\log 10^x = x$ can be used to evaluate common logarithms involving powers of 10. For example,

$$\log 100 = \log 10^2 = 2, \quad \log 1000 = \log 10^3 = 3, \quad \log 10^{7.1} = 7.1.$$

EXAMPLE 9 Earthquake Intensity

The magnitude R on the Richter scale of an earthquake of intensity I is given by

$$R = \log\frac{I}{I_0}$$

where I_0 is the intensity of a barely felt zero-level earthquake. The earthquake that destroyed San Francisco in 1906 was $10^{8.3}$ times as intense as a zero-level earthquake. What was its magnitude on the Richter scale?

Solution Because the earthquake was $10^{8.3}$ times as intense as a zero-level earthquake, the intensity I is $10^{8.3}I_0$.

$$R = \log \frac{I}{I_0}$$ This is the formula for magnitude on the Richter scale.

$$R = \log \frac{10^{8.3} I_0}{I_0}$$ Substitute $10^{8.3} I_0$ for I.

$$= \log 10^{8.3}$$ Simplify.

$$= 8.3$$ Use the property $\log 10^x = x$.

San Francisco's 1906 earthquake registered 8.3 on the Richter scale.

Check Point 9 Use the formula in Example 9 to solve this problem. If an earthquake is 10,000 times as intense as a zero-level quake ($I = 10{,}000 I_0$), what is its magnitude on the Richter scale?

8 Use natural logarithms.

Natural Logarithms

The logarithmic function with base e is called the **natural logarithmic function**. The function $f(x) = \log_e x$ is usually expressed as $f(x) = \ln x$, read "el en of x." A calculator with an LN key can be used to evaluate natural logarithms.

Like the domain of all logarithmic functions, the domain of the natural logarithmic function is the set of all positive real numbers. Thus, the domain of $f(x) = \ln(x + c)$ consists of all x for which $x + c > 0$.

EXAMPLE 10 Finding Domains of Natural Logarithmic Functions

Find the domain of each function.

a. $f(x) = \ln(3 - x)$ **b.** $g(x) = \ln(x - 3)^2$

Solution

a. The domain of f consists of all x for which $3 - x > 0$. Solving this inequality for x, we obtain $x < 3$. Thus, the domain of f is $(-\infty, 3)$. This is verified by the graph in Figure 4.13.

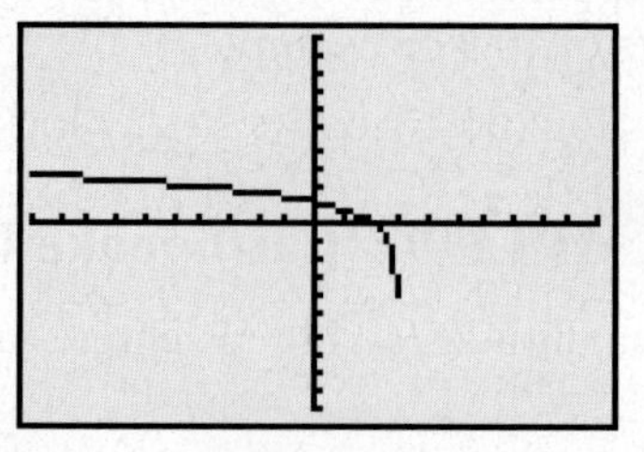

Figure 4.13 The domain of $f(x) = \ln(3 - x)$ is $(-\infty, 3)$.

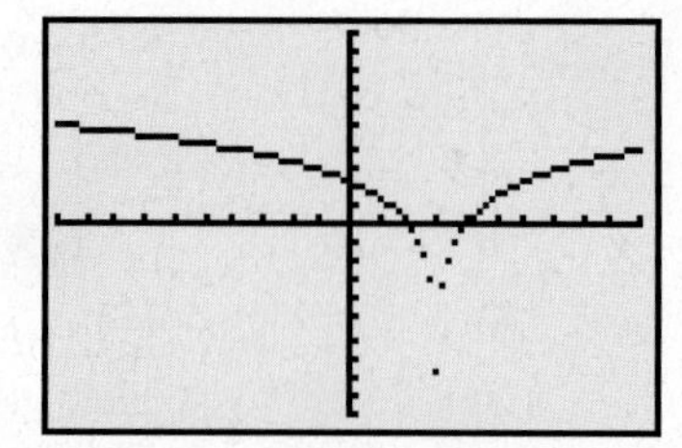

Figure 4.14 3 is excluded from the domain of $g(x) = \ln(x - 3)^2$.

b. The domain of g consists of all x for which $(x - 3)^2 > 0$. It follows that the domain of g is all real numbers except 3. This is shown by the graph in Figure 4.14. If it is not obvious that 3 is excluded from the domain, try using a dot format.

Check Point 10 Find the domain of each function.

a. $f(x) = \ln(4 - x)$ **b.** $g(x) = \ln x^2$

The basic properties of logarithms that were listed earlier in this section can be applied to natural logarithms.

Properties of Natural Logarithms

General Properties	Natural Logarithms
1. $\log_b 1 = 0$	**1.** $\ln 1 = 0$
2. $\log_b b = 1$	**2.** $\ln e = 1$
3. $\log_b b^x = x$	**3.** $\ln e^x = x$
4. $b^{\log_b x} = x$	**4.** $e^{\ln x} = x$

The property $\ln e^x = x$ can be used to evaluate natural logarithms involving powers of e. For example,

$$\ln e^2 = 2, \quad \ln e^3 = 3, \quad \ln e^{7.1} = 7.1, \quad \text{and} \quad \ln \frac{1}{e} = \ln e^{-1} = -1.$$

EXAMPLE 11 Using Inverse Properties

Use inverse properties to simplify:

a. $\ln e^{7x}$ **b.** $e^{\ln 4x^2}$

Solution

a. Because $\ln e^x = x$, we conclude that $\ln e^{7x} = 7x$.

b. Because $e^{\ln x} = x$, we conclude $e^{\ln 4x^2} = 4x^2$.

Check Point 11 Use inverse properties to simplify:

a. $\ln e^{25x}$ **b.** $e^{\ln \sqrt{x}}$

EXAMPLE 12 Walking Speed and City population

As the population of a city increases, the pace of life also increases. The formula

$$W = 0.35 \ln P + 2.74$$

models average walking speed, W, in feet per second, for a resident of a city whose population is P thousand. Find the average walking speed for people living in New York City with a population of 7323 thousand.

Solution We use the formula and substitute 7323 for P, the population in thousands.

$$W = 0.35 \ln P + 2.74 \quad \text{This is the given formula.}$$

$$W = 0.35 \ln 7323 + 2.74 \quad \text{Substitute 7323 for } P.$$

$$\approx 5.9 \quad \text{Graphing calculator keystrokes: } 0.35 \ \boxed{\times} \ \boxed{\text{LN}} \ 7323 \ \boxed{+} \ 2.74 \ \boxed{\text{ENTER}}$$

The average walking speed in New York City is approximately 5.9 feet per second.

Use the formula $W = 0.35 \ln P + 2.74$ to find the average walking speed in Jackson, Mississippi with a population of 197 thousand.

EXERCISE SET 4.2

Practice Exercises

In Exercises 1–8, write each equation in its equivalent exponential form.

1. $4 = \log_2 16$
2. $6 = \log_2 64$
3. $2 = \log_3 x$
4. $2 = \log_9 x$
5. $5 = \log_b 32$
6. $3 = \log_b 27$
7. $\log_6 216 = y$
8. $\log_5 125 = y$

In Exercises 9–20, write each equation in its equivalent logarithmic form.

9. $2^3 = 8$
10. $5^4 = 625$
11. $2^{-4} = \frac{1}{16}$
12. $5^{-3} = \frac{1}{125}$
13. $\sqrt[3]{8} = 2$
14. $\sqrt[3]{64} = 4$
15. $13^2 = x$
16. $15^2 = x$
17. $b^3 = 1000$
18. $b^3 = 343$
19. $7^y = 200$
20. $8^y = 300$

In Exercises 21–38, evaluate each expression without using a calculator.

21. $\log_4 16$
22. $\log_7 49$
23. $\log_2 64$
24. $\log_3 27$
25. $\log_7 \sqrt{7}$
26. $\log_6 \sqrt{6}$
27. $\log_2 \frac{1}{8}$
28. $\log_3 \frac{1}{9}$
29. $\log_{64} 8$
30. $\log_{81} 9$
31. $\log_5 5$
32. $\log_{11} 11$
33. $\log_4 1$
34. $\log_6 1$
35. $\log_5 5^7$
36. $\log_4 4^6$
37. $8^{\log_8 19}$
38. $7^{\log_7 23}$
39. Graph $f(x) = 4^x$ and $g(x) = \log_4 x$ in the same rectangular coordinate system.
40. Graph $f(x) = 5^x$ and $g(x) = \log_5 x$ in the same rectangular coordinate system.
41. Graph $f(x) = (\frac{1}{2})^x$ and $g(x) = \log_{1/2} x$ in the same rectangular coordinate system.
42. Graph $f(x) = (\frac{1}{4})^x$ and $g(x) = \log_{1/4} x$ in the same rectangular coordinate system.

In Exercises 43–48, the graph of a logarithmic function is given. Select the function for each graph from the following options.

$$f(x) = \log_3 x, \ g(x) = \log_3(x - 1), \ h(x) = \log_3 x - 1,$$

$$F(x) = -\log_3 x, \ G(x) = \log_3(-x), \ H(x) = 1 - \log_3 x$$

43.

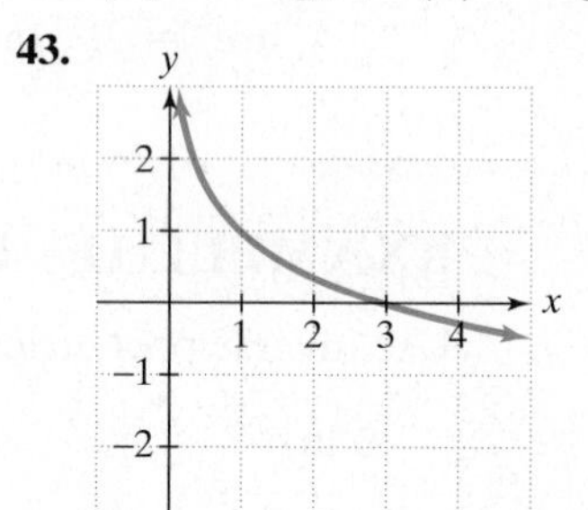

44.

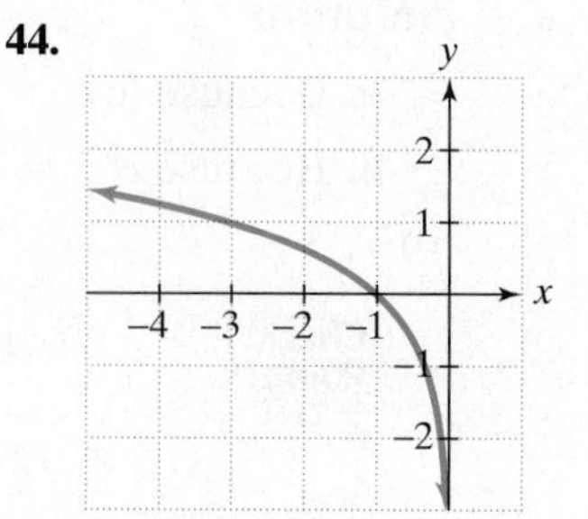

45.

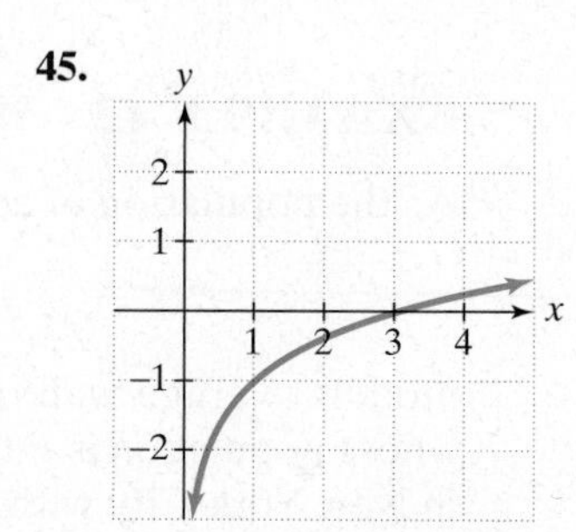

46.

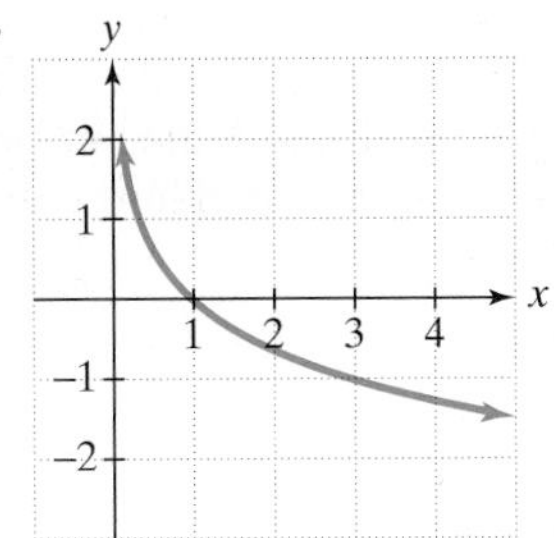

47.

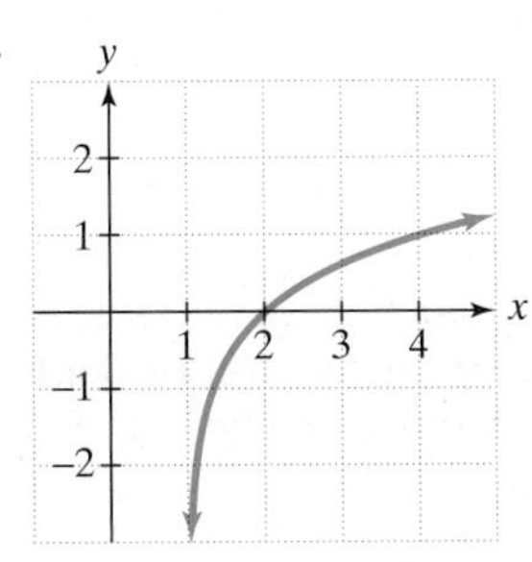

48.

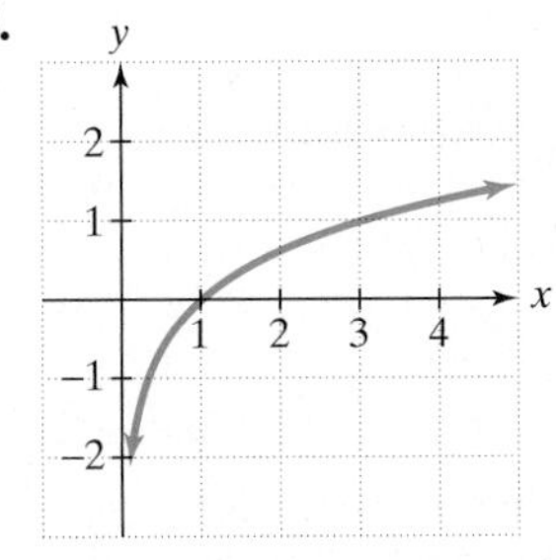

In Exercises 49–54, begin by graphing $f(x) = \log_2 x$. Then use transformations of this graph to graph the given function. What is the graph's x-intercept? What is the vertical asymptote?

49. $g(x) = \log_2(x + 1)$

50. $g(x) = \log_2(x + 2)$

51. $h(x) = 1 + \log_2 x$

52. $h(x) = 2 + \log_2 x$

53. $g(x) = \frac{1}{2}\log_2 x$

54. $g(x) = -2\log_2 x$

In Exercises 55–60, find the domain of each logarithmic function.

55. $f(x) = \log_5(x + 4)$

56. $f(x) = \log_5(x + 6)$

57. $f(x) = \log(2 - x)$

58. $f(x) = \log(7 - x)$

59. $f(x) = \ln(x - 2)^2$

60. $f(x) = \ln(x - 7)^2$

In Exercises 61–74, evaluate each expression without using a calculator.

61. $\log 100$

62. $\log 1000$

63. $\log 10^7$

64. $\log 10^8$

65. $10^{\log 33}$

66. $10^{\log 53}$

67. $\ln 1$

68. $\ln e$

69. $\ln e^6$

70. $\ln e^7$

71. $\ln \frac{1}{e^6}$

72. $\ln \frac{1}{e^7}$

73. $e^{\ln 125}$

74. $e^{\ln 300}$

In Exercises 75–80, use inverse properties of logarithms to simplify each expression.

75. $\ln e^{9x}$

76. $\ln e^{13x}$

77. $e^{\ln 5x^2}$

78. $e^{\ln 7x^2}$

79. $10^{\log \sqrt{x}}$

80. $10^{\log \sqrt[3]{x}}$

Application Exercises

The percentage of adult height attained by a girl who is x years old can be modeled by

$$f(x) = 62 + 35\log(x - 4)$$

where x represents the girl's age (from 5 to 15) and $f(x)$ represents the percentage of her adult height. Use the formula to solve Exercises 81–82.

81. Approximately what percentage of her adult height is a girl at age 13?

82. Approximately what percentage of her adult height is a girl at age ten?

83. The annual amount that we spend to attend sporting events can be modeled by

$$f(x) = 2.05 + 1.3\ln x$$

where x represents the number of years since 1984 and $f(x)$ represents the total annual expenditures for admission to spectator sports, in billions of dollars. In 2000, approximately how much was spent on admission to spectator sports?

84. The percentage of U.S. households with cable television can be modeled by

$$f(x) = 18.32 + 15.94\ln x$$

where x represents the number of years since 1979 and $f(x)$ represents the percentage of U.S. households with cable television. What percentage of U.S. households had cable television in 1990?

The loudness level of a sound, D, in decibels, is given by the formula

$$D = 10\log(10^{12}I)$$

where I is the intensity of the sound, in watts per meter^2. Decibel levels range from 0, a barely audible sound, to 160, a sound resulting in a ruptured eardrum. Use the formula to solve Exercises 85–86.

85. The sound of a blue whale can be heard 500 miles away, reaching an intensity of 6.3×10^6 watts per meter^2. Determine the decibel level of this sound. At close range, can the sound of a blue whale rupture the human eardrum?

86. What is the decibel level of a normal conversation, 3.2×10^{-6} watts per meter^2?

87. Students in a psychology class took a final examination. As part of an experiment to see how much of the course content they remembered over time, they took equivalent forms of the exam in monthly intervals thereafter. The average score for the group, $f(t)$, after t months was modeled by the function

$$f(t) = 88 - 15\ln(t + 1), \qquad 0 \le t \le 12.$$

a. What was the average score on the original exam?
b. What was the average score after 2 months? 4 months? 6 months? 8 months? 10 months? one year?
c. Sketch the graph of f (either by hand or with a graphing utility). Describe what the graph indicates in terms of the material retained by the students.

Writing in Mathematics

88. Describe the relationship between an equation in logarithmic form and an equivalent equation in exponential form.

89. What question can be asked to help evaluate $\log_3 81$?

90. Explain why the logarithm of 1 with base b is 0.

91. Describe the following property using words: $\log_b b^x = x$.

92. Explain how to use the graph of $f(x) = 2^x$ to obtain the graph of $g(x) = \log_2 x$.

93. Explain how to find the domain of a logarithmic function.

94. New York City is one of the world's great walking cities. Use the formula in Example 12 on page 371 to describe what frequently happens to tourists exploring the city by foot.

95. Logarithmic models are well suited to phenomena in which growth is initially rapid but then begins to level off. Describe something that is changing over time that can be modeled using a logarithmic function.

96. Suppose that a girl is 4′ 6″ at age 10. Explain how to use the function in Exercises 81–82 to determine how tall she can expect to be as an adult.

Technology Exercises

In Exercises 97–100, graph f and g in the same viewing rectangle. Then describe the relationship of the graph of g to the graph of f.

97. $f(x) = \ln x,\ g(x) = \ln(x + 3)$

98. $f(x) = \ln x,\ g(x) = \ln x + 3$

99. $f(x) = \log x,\ g(x) = -\log x$

100. $f(x) = \log x,\ g(x) = \log(x - 2) + 1$

101. Students in a mathematics class took a final examination. They took equivalent forms of the exam in monthly intervals thereafter. The average score, $f(t)$, for the group after t months was modeled by the human memory function $f(t) = 75 - 10\log(t + 1)$, where $0 \le t \le 12$. Use a graphing utility to graph the function. Then determine how many months will elapse before the average score falls below 65.

102. Graph f and g in the same viewing rectangle.
a. $f(x) = \ln(3x),\ g(x) = \ln 3 + \ln x$
b. $f(x) = \log(5x^2),\ g(x) = \log 5 + \log x^2$
c. $f(x) = \ln(2x^3),\ g(x) = \ln 2 + \ln x^3$
d. Describe what you observe in parts (a)–(c). Generalize this observation by writing an equivalent expression for $\log_b(MN)$, where $M > 0$ and $N > 0$.
e. Complete this statement: The logarithm of a product is equal to ________________.

103. Graph each of the following functions in the same viewing rectangle and then place the functions in order from the one that increases most slowly to the one that increases most rapidly.

$$y = x,\ y = \sqrt{x},\ y = e^x,\ y = \ln x,\ y = x^x,\ y = x^2$$

Critical Thinking Exercises

104. Which one of the following is true?
a. $\dfrac{\log_2 8}{\log_2 4} = \dfrac{8}{4}$
b. $\log(-100) = -2$.
c. The domain of $f(x) = \log_2 x$ is $(-\infty, \infty)$.
d. $\log_b x$ is the exponent to which b must be raised to obtain x.

105. Without using a calculator, find the exact value of

$$\frac{\log_3 81 - \log_\pi 1}{\log_{2\sqrt{2}} 8 - \log 0.001}.$$

106. Solve for x: $\log_4[\log_3(\log_2 x)] = 0$.

107. Without using a calculator, determine which is the greater number: $\log_4 60$ or $\log_3 40$.

Group Exercise

108. This group exercise involves exploring the way we grow. Group members should create a graph for the function that models the percentage of adult height attained by a boy who is x years old, $f(x) = 29 + 48.8\log(x + 1)$. Let $x = 1, 2, 3, \ldots, 12$, find function values, and connect the resulting points with a smooth curve. Then create a function that models the percentage of adult height attained by a girl who is x years old, $g(x) = 62 + 35\log(x - 4)$. Let $x = 5, 6, 7, \ldots, 15$, find function values, and connect the resulting points by a smooth curve. Group members should then discuss similarities and differences in the growth patterns for boys and girls based on the graphs.

SECTION 4.3 Properties of Logarithms

Objectives

1. Use the product rule.
2. Use the quotient rule.
3. Use the power rule.
4. Expand logarithmic expressions.
5. Condense logarithmic expressions.
6. Use the change-of-base property.

We all learn new things in different ways. In this section, we consider important properties of logarithms. What would be the most effective way for you to learn about these properties? Would it be helpful to use your graphing utility and discover one of these properties for yourself? To do so, work Exercise 102 in Exercise Set 4.2 before continuing. Would the properties become more meaningful if you could see exactly where they come from? If so, you will find details of the proofs of many of these properties in the appendix. The remainder of our work in this chapter will be based on the properties of logarithms that you learn in this section.

1 Use the product rule.

The Product Rule

Properties of exponents correspond to properties of logarithms. For example, when we multiply with the same base, we add exponents:

$$b^M \cdot b^N = b^{M+N}.$$

This property of exponents, coupled with an awareness that a logarithm is an exponent, suggests the following property, called the **product rule**.

> **The Product Rule**
>
> Let b, M, and N be positive real numbers with $b \neq 1$.
>
> $$\log_b(MN) = \log_b M + \log_b N$$
>
> The logarithm of a product is the sum of the logarithms.

When we use the product rule to write a single logarithm as the sum of two logarithms, we say that we are **expanding a logarithmic expression**. For example, we can use the product rule to expand $\ln(4x)$:

$$\ln(4x) = \ln 4 + \ln x$$

The logarithm of a product | is | the sum of the logarithms.

EXAMPLE 1 Using the Product Rule

Use the product rule to expand

a. $\log_4(7 \cdot 9)$ **b.** $\log(10x)$

Solution

a. $\log_4(7 \cdot 9) = \log_4 7 + \log_4 9$ The logarithm of a product is the sum of the logarithms.

b. $\log(10x) = \log 10 + \log x$ The logarithm of a product is the sum of the logarithms. These are common logarithms with base 10 understood.

$= 1 + \log x$ Because $\log_b b = 1$, then $\log_{10} 10 = 1$.

Check Point 1 Use the product rule to expand

a. $\log_6(10 \cdot 9)$ **b.** $\log(100x)$

The Quotient Rule

2 Use the quotient rule.

When we divide with the same base, we subtract exponents:

$$\frac{b^M}{b^N} = b^{M-N}.$$

This property suggests the following property of logarithms, called the **quotient rule**.

The Quotient Rule

Let b, M, and N be positive real numbers with $b \neq 1$.

$$\log_b\left(\frac{M}{N}\right) = \log_b M - \log_b N$$

The logarithm of a quotient is the difference of the logarithms.

When we use the quotient rule to write a single logarithm as the difference of two logarithms, we say that we are **expanding a logarithmic expression**. For example, we can use the quotient rule to expand $\log \frac{x}{2}$:

$$\log \frac{x}{2} = \log x - \log 2$$

The logarithm of a quotient | is | the difference of the logarithms.

EXAMPLE 2 Using the Quotient Rule

Use the quotient rule to expand

a. $\log_7\left(\frac{14}{x}\right)$ **b.** $\ln\left(\frac{e^3}{7}\right)$

Solution

a. $\log_7\left(\frac{14}{x}\right) = \log_7 14 - \log_7 x$ The logarithm of a quotient is the difference of the logarithms.

b. $\ln\left(\frac{e^3}{7}\right) = \ln e^3 - \ln 7$ The logarithm of a quotient is the difference of the logarithms. These are natural logarithms with base e understood.

$= 3 - \ln 7$ Because $\ln e^x = x$, then $\ln e^3 = 3$.

Check Point 2 Use the quotient rule to expand

a. $\log_8\left(\frac{23}{x}\right)$ **b.** $\ln\left(\frac{e^5}{11}\right)$

3 Use the power rule.

The Power Rule

When an exponential expression is raised to a power, we multiply exponents:

$$(b^M)^p = b^{Mp}.$$

This property suggests the following property of logarithms, called the **power rule**.

The Power Rule

Let b, M, and N be positive real numbers with $b \neq 1$, and let p be any real number.

$$\log_b M^p = p \log_b M$$

The logarithm of a number with an exponent is the product of the exponent and the logarithm of that number.

When we use the power rule to "pull the exponent to the front," we say that we are **expanding a logarithmic expression**. For example, we can use the power rule to expand $\ln x^2$:

$$\ln x^2 = 2 \ln x.$$

The logarithm of a number with an exponent | is | the product of the exponent and the logarithm of that number.

Figure 4.15 shows the graphs of $y = \ln x^2$ and $y = 2 \ln x$. Are $\ln x^2$ and $2 \ln x$ the same? The graphs illustrate that $y = \ln x^2$ and $y = 2 \ln x$ have different domains. The graphs are only the same if $x > 0$. Thus, we should write

$$\ln x^2 = 2 \ln x \text{ for } x > 0.$$

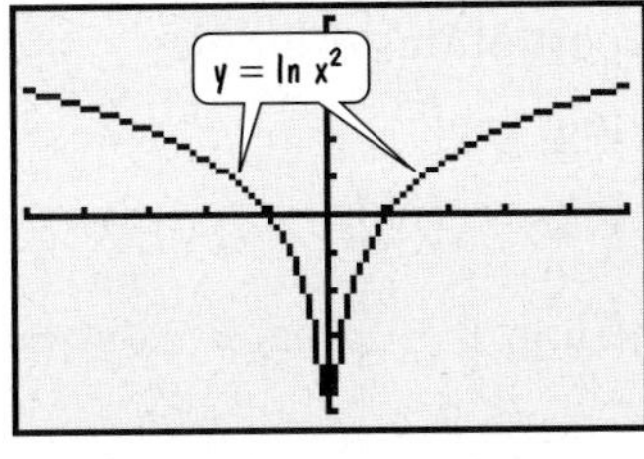

Domain: $(-\infty, 0)$ or $(0, \infty)$

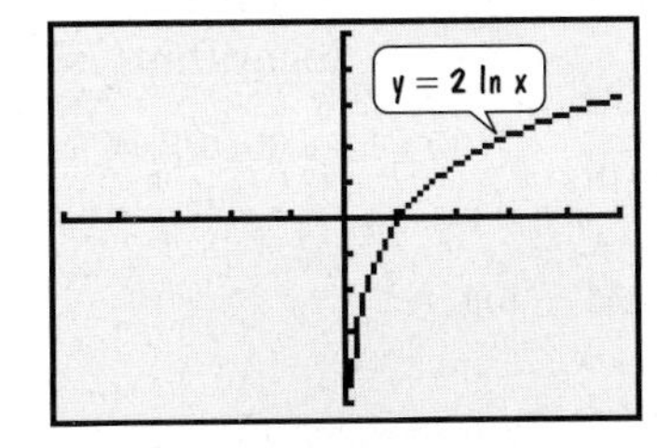

Domain: $(0, \infty)$

Figure 4.15 $\ln x^2$ and $2 \ln x$ have different domains.

When expanding a logarithmic expression, you might want to determine whether the rewriting has changed the domain of the expression.

EXAMPLE 3 Using the Power Rule

Use the power rule to expand

4 Expand logarithmic expressions.

a. $\log_5 7^4$ **b.** $\ln \sqrt{x}$

Solution

a. $\log_5 7^4 = 4 \log_5 7$ The logarithm of a number with an exponent is the exponent times the logarithm of the number.

b. $\ln \sqrt{x} = \ln x^{1/2}$ Rewrite the radical using a rational exponent.

$= \frac{1}{2} \ln x$ Use the power rule to bring the exponent to the front.

Check Point 3 Use the power rule to expand

a. $\log_6 8^9$ **b.** $\ln \sqrt[3]{x}$

Study Tip

The graphs show

$$y_1 = \ln(x + 3)$$

and $y_2 = \ln x + \ln 3$. The graphs are not the same. The graph of y_1 is the graph of the natural logarithmic function shifted 3 units to the left. By contrast, the graph of y_2 is the graph of the natural logarithmic function shifted upward by $\ln 3$, or about 1.1 units. Thus we see that

$$\ln(x + 3) \neq \ln x + \ln 3.$$

In general,

$$\log_b(M + N) \neq \log_b M + \log_b N.$$

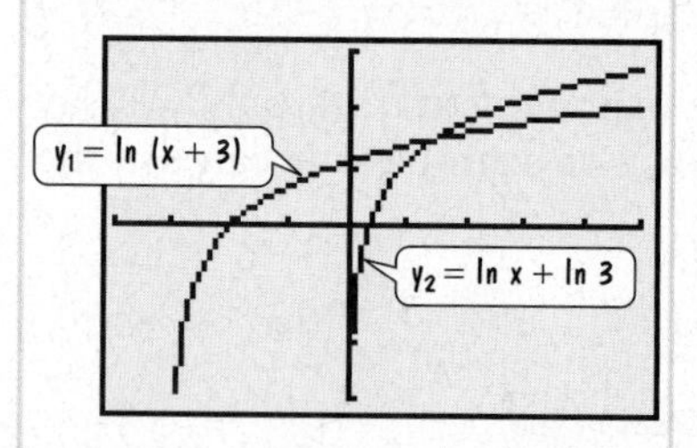

Try to avoid the following errors.

INCORRECT

$\log_b(M + N) = \log_b M + \log_b N$

$\log_b(M - N) = \log_b M - \log_b N$

$\log_b(M \cdot N) = \log_b M \cdot \log_b N$

$\log_b\left(\frac{M}{N}\right) = \frac{\log_b M}{\log_b N}$

$\frac{\log_b M}{\log_b N} = \log_b M - \log_b N$

Expanding Logarithmic Expressions

It is sometimes necessary to use more than one property of logarithms when you expand a logarithmic expression. Properties for expanding logarithmic expressions are as follows:

Properties for Expanding Logarithmic Expressions

1. $\log_b(MN) = \log_b M + \log_b N$ Product rule
2. $\log_b\left(\frac{M}{N}\right) = \log_b M - \log_b N$ Quotient rule
3. $\log_b M^p = p \log_b M$ Power rule

In all cases, $M > 0$ and $N > 0$.

EXAMPLE 4 Expanding Logarithmic Expressions

Use logarithmic properties to expand each expression as much as possible.

a. $\log_b x^2\sqrt{y}$ **b.** $\log_6\left(\frac{\sqrt[3]{x}}{36y^4}\right)$

Solution We will have to use two or more of the properties for expanding logarithms in each part of this example.

a. $\log_b x^2\sqrt{y} = \log_b x^2 y^{1/2}$ Use exponential notation.

$= \log_b x^2 + \log_b y^{1/2}$ Use the product rule.

$= 2 \log_b x + \frac{1}{2} \log_b y$ Use the power rule.

b. $\log_6\left(\frac{\sqrt[3]{x}}{36y^4}\right) = \log_6 \frac{x^{1/3}}{36y^4}$ Use exponential notation.

$= \log_6 x^{1/3} - \log_6 36y^4$ Use the quotient rule.

$= \log_6 x^{1/3} - (\log_6 36 + \log_6 y^4)$ Use the product rule on $\log_6 36y^4$.

$= \frac{1}{3}\log_6 x - (\log_6 36 + 4\log_6 y)$ Use the power rule.

$= \frac{1}{3}\log_6 x - \log_6 36 - 4\log_6 y$ Apply the distributive property.

$= \frac{1}{3}\log_6 x - 2 - 4\log_6 y$ $\log_6 36 = 2$ because 2 is the power to which we must raise 6 to get 36. $(6^2 = 36)$

Check Point 4 Use logarithmic properties to expand each expression as much as possible.

a. $\log_b x^4\sqrt[3]{y}$ **b.** $\log_5 \frac{\sqrt{x}}{25y^3}$

5 Condense logarithmic expressions.

Condensing Logarithmic Expressions

To **condense a logarithmic expression**, we write the sum or difference of two or more logarithmic expressions as a single logarithmic expression. We use the properties of logarithms to do so.

Properties for Condensing Logarithmic Expressions

1. $\log_b M + \log_b N = \log_b(MN)$ Product rule
2. $\log_b M - \log_b N = \log_b\left(\frac{M}{N}\right)$ Quotient rule
3. $p\log_b M = \log_b M^p$ Power rule

In all cases, $M > 0$ and $N > 0$.

EXAMPLE 5 Condensing Logarithmic Expressions

Write as a single logarithm:

a. $\log_4 2 + \log_4 32$ **b.** $\log(4x - 3) - \log x$

Solution

a. $\log_4 2 + \log_4 32 = \log_4(2 \cdot 32)$ Use the product rule.

$= \log_4 64$ We now have a single logarithm. However, we can simplify.

$= 3$ $\log_4 64 = 3$ because $4^3 = 64$.

b. $\log(4x - 3) - \log x = \log \frac{4x - 3}{x}$ Use the quotient rule.

Check Point 5 Write as a single logarithm:

a. $\log 25 + \log 4$ **b.** $\log(7x + 6) - \log x$

Coefficients of logarithms must be 1 before you can condense them using the product and quotient rules. For example, to condense

$$2 \ln x + \ln(x + 1),$$

the coefficient of the first term must be 1. We use the power rule to rewrite the coefficient as an exponent:

1. Make the number in front an exponent.

$$2 \ln x + \ln(x + 1) = \ln x^2 + \ln(x + 1) = \ln x^2(x + 1)$$

2. Use the product rule. The sum of logarithms with coefficients 1 is the logarithm of the product.

EXAMPLE 6 Condensing Logarithmic Expressions

a. $\frac{1}{2} \log x + 4 \log(x - 1)$ **b.** $3 \ln(x + 7) - \ln x$

Solution

a. $\frac{1}{2} \log x + 4 \log(x - 1)$

$= \log x^{1/2} + \log(x - 1)^4$ Use the power rule so that all coefficients are 1.

$= \log x^{1/2}(x - 1)^4$ Use the product rule.

b. $3 \ln (x + 7) - \ln x$

$= \ln (x + 7)^3 - \ln x$ Use the power rule so that all coefficients are 1.

$= \ln \dfrac{(x + 7)^3}{x}$ Use the quotient rule.

Check Point 6 Write as a single logarithm:

a. $2 \ln x + \frac{1}{3} \ln(x + 5)$ **b.** $2 \log(x - 3) - \log x$

6 Use the change-of-base property.

The Change-of-Base Property

We have seen that calculators give the values of both common logarithms (base 10) and natural logarithms (base e). To find a logarithm with any other base, we can use the following change-of-base property.

The Change-of-Base Property

For any logarithmic bases a and b, and any positive number M,

$$\log_b M = \frac{\log_a M}{\log_a b}.$$

The logarithm of M with base b is equal to the logarithm of M with any new base divided by the logarithm of b with that new base.

In the change-of-base property, base b is the base of the original logarithm. Base a is a new base that we introduce. Thus, the change-of-base property allows us to change from base b to *any* new base a, as long as the newly introduced base is a positive number not equal to 1.

The change-of-base property is used to write a logarithm in terms of quantities that can be evaluated with a calculator. Because calculators contain keys for common (base 10) and natural (base e) logarithms, we will frequently introduce base 10 or base e.

Change-of-Base Property

$$\log_b M = \frac{\log_a M}{\log_a b}$$

a is the new introduced base.

Introducing Common Logarithms

$$\log_b M = \frac{\log_{10} M}{\log_{10} b}$$

10 is the new introduced base.

Introducing Natural Logarithms

$$\log_b M = \frac{\log_e M}{\log_e b}$$

e is the new introduced base.

Using the notations for common logarithms and natural logarithms, we have the following results.

Discovery

Find a reasonable estimate of $\log_5 140$ to the nearest whole number. 5 to what power is 140? Compare your estimate to the value obtained in Example 7.

The Change-of-Base Property: Introducing Common and Natural Logarithms

Introducing Common Logarithms

$$\log_b M = \frac{\log M}{\log b}$$

Introducing Natural Logarithms

$$\log_b M = \frac{\ln M}{\ln b}$$

EXAMPLE 7 Changing Base to Common Logarithms

Use common logarithms to evaluate $\log_5 140$.

Solution Because $\log_b M = \dfrac{\log M}{\log b}$,

$$\log_5 140 = \frac{\log 140}{\log 5} \approx 3.07.$$

Use a calculator: [LOG] 140 [÷] [LOG] 5 [ENTER]

This means that $\log_5 140 \approx 3.07$.

Check Point 7 Use common logarithms to evaluate $\log_7 2506$.

EXAMPLE 8 Changing Base to Natural Logarithms

Use natural logarithms to evaluate $\log_5 140$.

Solution Because $\log_b M = \dfrac{\ln M}{\ln b}$,

$$\log_5 140 = \frac{\ln 140}{\ln 5} \approx 3.07.$$

Use a calculator: [LN] 140 [÷] [LN] 5 [ENTER]

We have again shown that $\log_5 140 \approx 3.07$.

Check Point 8 Use natural logarithms to evaluate $\log_7 2506$.

We can use the change-of-base property to graph logarithmic functions with bases other than 10 or e on a graphing utility.

EXAMPLE 9 Using a Graphing Utility to Graph Logarithmic Functions

Graph $y = \log_2 x$ and $y = \log_{20} x$ in the same viewing rectangle.

Solution Because $\log_2 x = \dfrac{\ln x}{\ln 2}$ and $\log_{20} x = \dfrac{\ln x}{\ln 20}$ the functions are entered as

$y_1 =$ [LN] x [÷] [LN] 2 [ENTER]

and $y_2 =$ [LN] x [÷] [LN] 20 [ENTER].

Using a $[0, 10, 1] \times [-3, 3, 1]$ viewing rectangle, the graphs are shown in Figure 4.16.

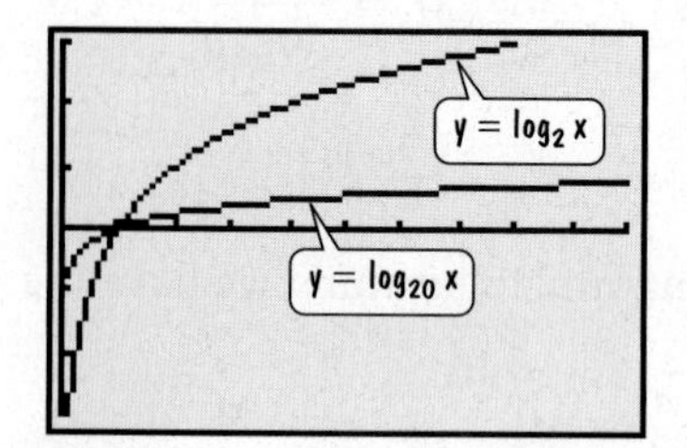

Figure 4.16 Using the change-of-base property to graph logarithmic functions

Check Point 9 Graph $y = \log_3 x$ and $y = \log_{15} x$ in the same viewing rectangle.

EXERCISE SET 4.3

Practice Exercises

In Exercises 1–32, use properties of logarithms to expand each logarithmic expression as much as possible. Where possible, evaluate logarithmic expressions without using a calculator.

1. $\log_5(12 \cdot 3)$
2. $\log_8(13 \cdot 9)$
3. $\log_7(7x)$
4. $\log_9(9x)$
5. $\log(1000x)$
6. $\log(10{,}000x)$
7. $\log_7\left(\frac{7}{x}\right)$
8. $\log_9\left(\frac{9}{x}\right)$
9. $\log\left(\frac{x}{100}\right)$
10. $\log\left(\frac{x}{1000}\right)$
11. $\log_4\left(\frac{64}{y}\right)$
12. $\log_5\left(\frac{125}{y}\right)$
13. $\ln\left(\frac{e^2}{5}\right)$
14. $\ln\left(\frac{e^4}{8}\right)$
15. $\log_b x^3$
16. $\log_b x^7$
17. $\log N^{-6}$
18. $\log M^{-8}$
19. $\ln \sqrt[5]{x}$
20. $\ln \sqrt[7]{x}$
21. $\log_b x^2 y$
22. $\log_b xy^3$
23. $\log_4\left(\frac{\sqrt{x}}{64}\right)$
24. $\log_5\left(\frac{\sqrt{x}}{25}\right)$
25. $\log_6\left(\frac{36}{\sqrt{x+1}}\right)$
26. $\log_8\left(\frac{64}{\sqrt{x+1}}\right)$
27. $\log_b \frac{x^2 y}{z^2}$
28. $\log_b \frac{x^3 y}{z^2}$
29. $\log \sqrt{100x}$
30. $\ln \sqrt{ex}$
31. $\log \sqrt[3]{\frac{x}{y}}$
32. $\log \sqrt[5]{\frac{x}{y}}$

In Exercises 33–52, use properties of logarithms to condense each logarithmic expression. Write the expression as a single logarithm whose coefficient is 1. Where possible, evaluate logarithmic expressions.

33. $\log 5 + \log 2$
34. $\log 250 + \log 4$
35. $\ln x + \ln 7$
36. $\ln x + \ln 3$
37. $\log_2 96 - \log_2 3$
38. $\log_3 405 - \log_3 5$
39. $\log(2x + 5) - \log x$
40. $\log(3x + 7) - \log x$
41. $\log x + 3 \log y$
42. $\log x + 7 \log y$
43. $\frac{1}{2} \ln x + \ln y$
44. $\frac{1}{3} \ln x + \ln y$

45. $2\log_b x + 3\log_b y$
46. $5\log_b x + 6\log_b y$
47. $5\ln x - 2\ln y$
48. $7\ln x - 3\ln y$
49. $3\ln x - \frac{1}{3}\ln y$
50. $2\ln x - \frac{1}{2}\ln y$
51. $4\ln(x + 6) - 3\ln x$
52. $8\ln(x + 9) - 4\ln x$

In Exercises 53–60, use common logarithms or natural logarithms and a calculator to evaluate to four decimal places.

53. $\log_5 13$
54. $\log_6 17$
55. $\log_{14} 87.5$
56. $\log_{16} 57.2$
57. $\log_{0.1} 17$
58. $\log_{0.3} 19$
59. $\log_\pi 63$
60. $\log_\pi 400$

Application Exercises

61. The loudness level of a sound can be expressed by comparing the sound's intensity to the intensity of a sound barely audible to the human ear. The formula

$$D = 10(\log I - \log I_0)$$

describes the loudness level of a sound, D, in decibels, where I is the intensity of the sound, in watts per meter^2, and I_0 is the intensity of a sound barely audible to the human ear.
  a. Express the formula so that the expression in parentheses is written as a single logarithm.
  b. Use the form of the formula from part (a) to answer this question. If a sound has an intensity 100 times the intensity of a softer sound, how much larger on the decibel scale is the loudness level of the more intense sound?

62. The formula

$$t = \frac{1}{c}[\ln A - \ln(A - N)]$$

describes the time, t, in weeks, that it takes to achieve mastery of a portion of a task, where A is the maximum learning possible, N is the portion of the learning that is to be achieved, and c is a constant used to measure an individual's learning style.
  a. Express the formula so that the expression in brackets is written as a single logarithm.
  b. The formula is also used to determine how long it will take chimpanzees and apes to master a task. For example, a typical chimpanzee learning sign language can master a maximum of 65 signs. Use the form of the formula from part (a) to answer this question. How many weeks will it take a chimpanzee to master 30 signs if c for that chimp is 0.03?

Writing in Mathematics

63. Describe the product rule for logarithms and give an example.
64. Describe the quotient rule for logarithms and give an example.
65. Describe the power rule for logarithms and give an example.
66. Without showing the details, explain how to condense $\ln x - 2\ln(x + 1)$.
67. Describe the change-of-base property and give an example.
68. Explain how to use your calculator to find $\log_{14} 283$.
69. You overhear a student talking about a property of logarithms in which division becomes subtraction. Explain what the student means by this.
70. Find $\ln 2$ using a calculator. Then calculate each of the following: $1 - \frac{1}{2}$; $1 - \frac{1}{2} + \frac{1}{3}$; $1 - \frac{1}{2} + \frac{1}{3} - \frac{1}{4}$; $1 - \frac{1}{2} + \frac{1}{3} - \frac{1}{4} + \frac{1}{5}$; .... Describe what you observe.

Technology Exercises

71. a. Use a graphing utility (and the change-of-base property) to graph $y = \log_3 x$.
  b. Graph $y = 2 + \log_3 x$, $y = \log_3(x + 2)$, and $y = -\log_3 x$ in the same viewing rectangle as $y = \log_3 x$. Then describe the change or changes that need to be made to the graph of $y = \log_3 x$ to obtain each of these three graphs.
72. Graph $y = \log x$, $y = \log(10x)$, and $y = \log(0.1x)$ in the same viewing rectangle. Describe the relationship among the three graphs. What logarithmic property accounts for this relationship?
73. Use a graphing utility and the change-of-base property to graph $y = \log_3 x$, $y = \log_{25} x$, and $y = \log_{100} x$ in the same viewing rectangle.
  a. Which graph is on the top in the interval $(0, 1)$? Which is on the bottom?
  b. Which graph is on the top in the interval $(1, \infty)$? Which is on the bottom?
  c. Generalize by writing a statement about which graph is on top, which is on the bottom, and in which intervals, using $y = \log_b x$ where $b > 1$.

Disprove each statement in Exercises 74–78 by
  a. *letting y equal a positive constant of your choice.*
  b. *using a graphing utility to graph the function on each side of the equal sign. The two functions should have different graphs, showing that the equation is not true in general.*

74. $\log(x + y) = \log x + \log y$
75. $\log\frac{x}{y} = \frac{\log x}{\log y}$
76. $\ln(x - y) = \ln x - \ln y$
77. $\ln(xy) = (\ln x)(\ln y)$
78. $\frac{\ln x}{\ln y} = \ln x - \ln y$

Critical Thinking Exercises

79. Which one of the following is true?

a. $\frac{\log_7 49}{\log_7 7} = \log_7 49 - \log_7 7$

b. $\log_b(x^3 + y^3) = 3\log_b x + 3\log_b y$

c. $\log_b(xy)^5 = (\log_b x + \log_b y)^5$

d. $\ln\sqrt{2} = \frac{\ln 2}{2}$

80. Use the change-of-base property to prove that

$$\log e = \frac{1}{\ln 10}$$

81. If $\log 3 = A$ and $\log 7 = B$, find $\log_7 9$ in terms of A and B.

82. Write as a single term that does not contain a logarithm:

$$e^{\ln 8x^5 - \ln 2x^2}.$$

SECTION 4.4 Exponential and Logarithmic Equations

Objectives

1. Solve exponential equations.
2. Solve logarithmic equations.
3. Solve applied problems involving exponential and logarithmic equations.

Is an early retirement awaiting you?

You inherited $30,000. You'd like to put aside $25,000 and eventually have over half a million dollars for early retirement. Is this possible? In this section you will see how techniques for solving equations with variable exponents provide an answer to this question.

1 Solve exponential equations.

Exponential Equations

An **exponential equation** is an equation containing a variable in an exponent. Examples of exponential equations include

$$4^x = 15 \quad \text{and} \quad 40e^{0.6x} = 240.$$

Logarithms are extremely useful in solving such equations. The solution begins with isolating the exponential expression and taking the natural logarithm on both sides. Why can we do this? All logarithmic functions are one-to-one—that is, no two different ordered pairs have the same second component. Thus, if M and N are positive real numbers and $M = N$, then $\log_b M = \log_b N$.

Using Natural Logarithms to Solve Exponential Equations

1. Isolate the exponential expression.
2. Take the natural logarithm on both sides of the equation.
3. Simplify using one of the following properties:

$$\ln b^x = x \ln b \quad \text{or} \quad \ln e^x = x.$$

4. Solve for the variable.

EXAMPLE 1 Solving an Exponential Equation

Solve: $4^x = 15$.

Solution Because the exponential expression, 4^x, is already isolated on the left, we begin by taking the natural logarithm on both sides of the equation.

$$4^x = 15$$ This is the given equation.

$$\ln 4^x = \ln 15$$ Take the natural logarithm on both sides.

$$x \ln 4 = \ln 15$$ Use the power rule and bring the variable exponent to the front: $\ln b^x = x \ln b$.

$$x = \frac{\ln 15}{\ln 4}$$ Solve for x by dividing both sides by $\ln 4$.

We now have an exact value for x. We use the exact value for x in the equation's solution set. Thus, the equation's solution set is $\left\{\frac{\ln 15}{\ln 4}\right\}$. We can obtain a decimal approximation by using a calculator:

[LN] 15 [÷] [LN] 4 [ENTER]

Using these keystrokes, $x \approx 1.95$. Because $4^2 = 16$, it seems reasonable that the solution to $4^x = 15$ is approximately 1.95.

Check Point 1 Solve: $5^x = 134$. Find the solution set and then use a calculator to obtain a decimal approximation to two decimal places for the solution.

EXAMPLE 2 Solving an Exponential Equation

Solve: $40e^{0.6x} = 240$.

Solution We begin by dividing both sides by 40 to isolate the exponential expression, $e^{0.6x}$. Then we take the natural logarithm on both sides of the equation.

$$40e^{0.6x} = 240$$ This is the given equation.

$$e^{0.6x} = 6$$ Isolate the exponential factor by dividing both sides by 40.

$$\ln e^{0.6x} = \ln 6$$ Take the natural logarithm on both sides.

$$0.6x = \ln 6$$ Use the inverse property $\ln e^x = x$ on the left.

$$x = \frac{\ln 6}{0.6} \approx 2.99$$ Divide both sides by 0.6.

Thus, the solution of the equation is $\frac{\ln 6}{0.6} \approx 2.99$. Try checking this approximate solution in the original equation, verifying that $\left\{\frac{\ln 6}{0.6}\right\}$ is the solution set.

Check Point 2 Solve: $7e^{2x} = 63$. Find the solution set and then use a calculator to obtain a decimal approximation to two decimal places for the solution.

EXAMPLE 3 Solving an Exponential Equation

Solve: $5^{4x-7} - 3 = 10$

Solution We begin by adding 3 to both sides to isolate the exponential expression, 5^{4x-7}. Then we take the natural logarithm on both sides of the equation.

$$5^{4x-7} - 3 = 10 \quad \text{This is the given equation.}$$

$$5^{4x-7} = 13 \quad \text{Add 3 to both sides.}$$

$$\ln 5^{4x-7} = \ln 13 \quad \text{Take the natural logarithm on both sides.}$$

$$(4x - 7)\ln 5 = \ln 13 \quad \text{Use the power rule to bring the exponent to the front: } \ln b^x = x \ln b.$$

$$4x \ln 5 - 7\ln 5 = \ln 13 \quad \text{Use the distributive property and distribute } \ln 5 \text{ to both terms in parentheses.}$$

$$4x \ln 5 = \ln 13 + 7\ln 5 \quad \text{Isolate the variable term by adding } 7\ln 5 \text{ to both sides.}$$

$$x = \frac{\ln 13 + 7\ln 5}{4\ln 5} \quad \text{Isolate } x \text{ by dividing both sides by } 4\ln 5.$$

The solution set is $\left\{\frac{\ln 13 + 7\ln 5}{4\ln 5}\right\}$, approximately 2.15.

Check Point 3 Solve: $6^{3x-4} - 7 = 2081$. Find the solution set and then use a calculator to obtain a decimal approximation to two decimal places for the solution.

Technology

Shown below is the graph of $y = e^{2x} - 4e^x + 3$. There are two x-intercepts, one at 0 and one at approximately 1.099. These intercepts verify our algebraic solution.

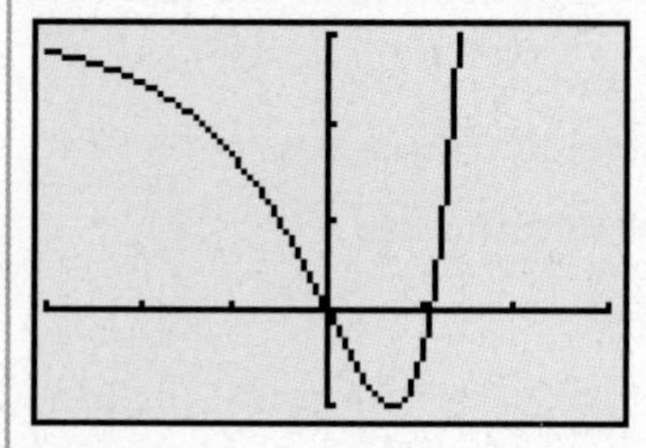

EXAMPLE 4 Solving an Exponential Equation

Solve: $e^{2x} - 4e^x + 3 = 0$.

Solution The given equation is quadratic in form. If $t = e^x$, the equation can be expressed as $t^2 - 4t + 3 = 0$. Because this equation can be solved by factoring, we factor to isolate the exponential term.

$$e^{2x} - 4e^x + 3 = 0 \quad \text{This is the given equation.}$$

$$(e^x - 3)(e^x - 1) = 0 \quad \text{Factor on the left. Notice that if } t = e^x, \; t^2 - 4t + 3 = (t-3)(t-1).$$

$$e^x - 3 = 0 \quad \text{or} \quad e^x - 1 = 0 \quad \text{Set each factor equal to 0.}$$

$$e^x = 3 \qquad e^x = 1 \quad \text{Solve for } e^x.$$

$\ln e^x = \ln 3 \qquad x = 0$ Take the natural logarithm on both sides of the first equation. The equation on the right can be solved by inspection.

$x = \ln 3$ $\ln e^x = x$

The solution set is $\{0, \ln 3\}$. The solutions are 0 and (approximately) 1.099.

Check Point 4 Solve: $e^{2x} - 8e^x + 7 = 0$. Find the solution set and then use a calculator to obtain a decimal approximation to two decimal places, if necessary.

2 Solve logarithmic equations.

Logarithmic Equations

A **logarithmic equation** is an equation containing a variable in a logarithmic expression. Examples of logarithmic equations include

$$\log_4(x+3) = 2 \quad \text{and} \quad \ln 2x = 3.$$

If a logarithmic equation is in the form $\log_b x = c$, we can solve the equation by rewriting it in its equivalent exponential form $b^c = x$. Example 5 illustrates how this is done.

Technology

The graphs of

$y_1 = \log_4(x+3)$ and $y_2 = 2$

have an intersection point whose x-coordinate is 13. This verifies that $\{13\}$ is the solution set for $\log_4(x+3) = 2$.

Note:
Because

$$\log_b x = \frac{\ln x}{\ln b}$$

(change-of-base property),

we entered y_1 using

$$y_1 = \log_4(x+3) = \frac{\ln(x+3)}{\ln 4}$$

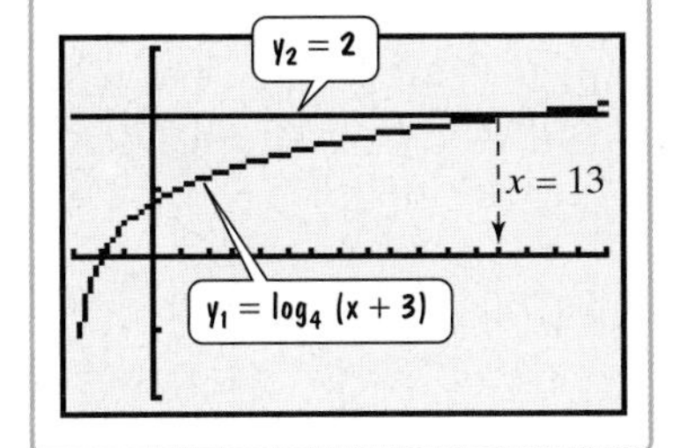

EXAMPLE 5 Solving a Logarithmic Equation

Solve: $\log_4(x+3) = 2$.

Solution We first rewrite the equation as an equivalent equation in exponential form using the fact that $\log_b x = c$ means $b^c = x$.

$$\log_4(x+3) = 2 \quad \text{means} \quad 4^2 = x + 3$$

Logarithms are exponents.

Now we solve the equivalent equation for x.

$4^2 = x + 3$ This is the equivalent equation.

$16 = x + 3$ Square 4.

$13 = x$ Subtract 3 from both sides.

Check

$\log_4(x+3) = 2$ This is the given logarithmic equation.

$\log_4(13+3) \stackrel{?}{=} 2$ Substitute 13 for x.

$\log_4 16 \stackrel{?}{=} 2$

$2 = 2$ ✓ $\log_4 16 = 2$ because $4^2 = 16$.

This true statement indicates that the solution set is $\{13\}$.

Check Point 5 Solve: $\log_2(x-4) = 3$.

Logarithmic expressions are defined only for logarithms of positive real numbers. Always check proposed solutions of a logarithmic equation in the original equation. Exclude from the solution set any proposed solution that produces the logarithm of a negative number or the logarithm of 0.

In order to rewrite the logarithmic equation $\log_b x = c$ in the equivalent exponential form $b^c = x$, we need a single logarithm whose coefficient is one. It is sometimes necessary to use properties of logarithms to condense logarithms into a single logarithm. In the next example we use the product rule for logarithms to obtain a single logarithmic expression on the left side.

EXAMPLE 6 Using the Product Rule to Solve a Logarithmic Equation

Solve: $\log_2 x + \log_2(x - 7) = 3$.

Solution

$\log_2 x + \log_2(x - 7) = 3$ — This is the given equation.

$\log_2 x(x - 7) = 3$ — Use the product rule to obtain a single logarithm: $\log_b M + \log_b N = \log_b(MN)$.

$2^3 = x(x - 7)$ — $\log_b x = c$ means $b^c = x$.

$8 = x^2 - 7x$ — Apply the distributive property on the right.

$0 = x^2 - 7x - 8$ — Set the equation equal to 0.

$0 = (x - 8)(x + 1)$ — Factor.

$x - 8 = 0$ or $x + 1 = 0$ — Set each factor equal to 0.

$x = 8$ $\quad$ $x = -1$ — Solve for x.

Check

Checking 8:

$$\log_2 x + \log_2(x - 7) = 3$$
$$\log_2 8 + \log_2(8 - 7) \stackrel{?}{=} 3$$
$$\log_2 8 + \log_2 1 \stackrel{?}{=} 3$$
$$3 + 0 \stackrel{?}{=} 3$$
$$3 = 3 \checkmark$$

Checking −1:

$$\log_2 x + \log_2(x - 7) = 3$$
$$\log_2(-1) + \log_2(-1 - 7) \stackrel{?}{=} 3$$

The number −1 does not check. Negative numbers do not have logarithms.

The solution set is $\{8\}$.

Check Point 6 Solve: $\log x + \log(x - 3) = 1$.

Equations involving natural logarithms can be solved using the inverse property $e^{\ln x} = x$. For example, to solve

$$\ln x = 5$$

we write both sides of the equation as exponents on base e:

$$e^{\ln x} = e^5$$

This is called **exponentiating both sides** of the equation. Using the inverse property $e^{\ln x} = x$, we simplify the left side of the equation and obtain the solution:

$$x = e^5.$$

EXAMPLE 7 Solving an Equation with a Natural Logarithm

Solve: $3 \ln 2x = 12$.

Solution

$3 \ln 2x = 12$ This is the given equation.

$\ln 2x = 4$ Divide both sides by 3.

$e^{\ln 2x} = e^4$ Exponentiate both sides.

$2x = e^4$ Use the inverse property to simplify the left side: $e^{\ln \square} = \square$.

$x = \frac{e^4}{2} \approx 27.30$ Divide both sides by 2.

Check

$3 \ln 2x = 12$ This is the given logarithmic equation.

$3 \ln 2\left(\frac{e^4}{2}\right) \stackrel{?}{=} 12$ Substitute $\frac{e^4}{2}$ for x.

$3 \ln e^4 \stackrel{?}{=} 12$ Simplify: $\frac{\cancel{2}}{1} \cdot \frac{e^4}{\cancel{2}} = e^4$.

$3 \cdot 4 \stackrel{?}{=} 12$ Because $\ln e^x = x$, we conclude $\ln e^4 = 4$.

$12 = 12$ ✓

This true statement indicates that the solution set is $\left\{\frac{e^4}{2}\right\}$.

Check Point 7 Solve: $4 \ln 3x = 8$.

3 Solve applied problems involving exponential and logarithmic equations.

Applications

Our first applied example provides a mathematical perspective on the old slogan "Alcohol and driving don't mix." In California, where 38% of fatal traffic crashes involve drinking drivers, it is illegal to drive with a blood alcohol concentration of 0.08 or higher. At these levels, drivers may be arrested and charged with driving under the influence.

EXAMPLE 8 Alcohol and Risk of a Car Accident

Medical research indicates that the risk of having a car accident increases exponentially as the concentration of alcohol in the blood increases. The risk is modeled by

$$R = 6e^{12.77x}$$

where x is the blood alcohol concentration and R, given as a percent, is the risk of having a car accident. What blood alcohol concentration corresponds to a 17% risk of a car accident?

Solution For a risk of 17%, we let $R = 17$ in the equation and solve for x, the blood alcohol concentration.

$$R = 6e^{12.77x}$$ This is the given equation.

$$6e^{12.77x} = 17$$ Substitute 17 for R and (optional) reverse the two sides of the equation.

$$e^{12.77x} = \frac{17}{6}$$ Isolate the exponential factor by dividing both sides by 6.

$$\ln e^{12.77x} = \ln\left(\frac{17}{6}\right)$$ Take the natural logarithm on both sides.

$$12.77x = \ln\left(\frac{17}{6}\right)$$ Use the inverse property $\ln e^x = x$ on the left.

$$x = \frac{\ln\left(\frac{17}{6}\right)}{12.77} \approx 0.08$$ Divide both sides by 12.77.

For a blood alcohol concentration of 0.08, the risk of a car accident is 17%. In many states, it is illegal to drive at this blood alcohol concentration.

Check Point 8 Use the formula in Example 8 to solve this problem. What blood alcohol concentration corresponds to a 7% risk of a car accident? (In many states, drivers under the age of 21 can lose their license for driving at this level.)

Playing Doubles: Interest Rates and Doubling Time

One way to calculate what your savings will be worth at some point in the future is to consider doubling time. Shown below is how long it takes for your money to double at different annual interest rates subject to continuous compounding.

Annual Interest Rate	Years to Double
5%	13.9 years
7%	9.9 years
9%	7.7 years
11%	6.3 years

Of course, the first problem is collecting some money to invest. The second problem is finding a reasonably safe investment with a return of 9% or more.

Suppose that you inherit \$30,000. Is it possible to invest \$25,000 and have over half a million dollars for early retirement? Our next example illustrates the power of compound interest.

EXAMPLE 9 Revisiting the Formula for Compound Interest

The formula

$$A = P\left(1 + \frac{r}{n}\right)^{nt}$$

describes the accumulated value A of a sum of money P, the principal, after t years at annual percentage rate r (in decimal form) compounded n times a year. How long will it take \$25,000 to grow to \$500,000 at 9% annual interest compounded monthly?

Solution

$$A = P\left(1 + \frac{r}{n}\right)^{nt}$$ This is the given formula.

$$500{,}000 = 25{,}000\left(1 + \frac{0.09}{12}\right)^{12t}$$ A (the desired accumulated value) = \$500,000, P (the principal) = \$25,000, r (the interest rate) = 9% = 0.09, and $n = 12$ (monthly compounding).

Our goal is to solve the equation for t. Let's reverse the two sides of the equation and then simplify within parentheses.

$$25{,}000\left(1 + \frac{0.09}{12}\right)^{12t} = 500{,}000$$

$$25{,}000(1 + 0.0075)^{12t} = 500{,}000 \qquad \text{Divide within parentheses: } \frac{0.09}{12} = 0.0075.$$

$$25{,}000(1.0075)^{12t} = 500{,}000 \qquad \text{Add within parentheses.}$$

$$(1.0075)^{12t} = 20 \qquad \text{Divide both sides by 25,000.}$$

$$\ln(1.0075)^{12t} = \ln 20 \qquad \text{Take the natural logarithm on both sides.}$$

$$12t\ln(1.0075) = \ln 20 \qquad \text{Use the power rule to bring the exponent to the front: } \ln b^x = x\ln b.$$

$$t = \frac{\ln 20}{12\ln 1.0075} \qquad \text{Solve for } t\text{, dividing both sides by } 12\ln 1.0075.$$

$$\approx 33.4 \qquad \text{Use a calculator.}$$

After approximately 33.4 years, the \$25,000 will grow to an accumulated value of \$500,000. If you set aside the money at age 20, you can begin enjoying a life of leisure at about age 53.

Check Point 9 How long, to the nearest tenth of a year, will it take \$1000 to grow to \$3600 at 8% annual interest compounded quarterly?

Yogi Berra, catcher and renowned hitter for the New York Yankees (1946–1963), said it best: "Prediction is very hard, especially when it's about the future." At the start of the twenty-first century, we are plagued by questions about the environment. Will we run out of gas? How hot will it get? Will there be neighborhoods where the air is pristine? Can we make garbage disappear? Will there be any wilderness left? Which wild animals will become extinct? These concerns have led to the growth of the environmental industry in the United States.

EXAMPLE 10 The Growth of the Environmental Industry

The formula

$$N = 461.87 + 299.4\ln x$$

models the thousands of workers, N, in the environmental industry in the United States x years after 1979. By which year will there be 1,500,000, or 1500 thousand, U.S. workers in the environmental industry?

Solution We substitute 1500 for N and solve for x, the number of years after 1979.

$$N = 461.87 + 299.4\ln x \qquad \text{This is the given formula.}$$

$$461.87 + 299.4\ln x = 1500 \qquad \text{Substitute 1500 for } N \text{ and reverse the two sides of the equation.}$$

Our goal is to isolate $\ln x$. We can then find x by exponentiating both sides of the equation, using the inverse property $e^{\ln x} = x$.

$$\begin{aligned} 299.4 \ln x &= 1038.13 && \text{Subtract 461.87 from both sides.} \\ \ln x &= \frac{1038.13}{299.4} && \text{Divide both sides by 299.4.} \\ e^{\ln x} &= e^{1038.13/299.4} && \text{Exponentiate both sides.} \\ x &= e^{1038.13/299.4} && e^{\ln x} = x \\ &\approx 32 && \text{Use a calculator.} \end{aligned}$$

Approximately 32 years after 1979, in the year 2011, there will be 1.5 million U.S. workers in the environmental industry.

Use the formula in Example 10 to find by what year there will be two million, or 2000 thousand, U.S. workers in the environmental industry.

EXERCISE SET 4.4

Practice Exercises

Solve each exponential equation in Exercises 1–22. Express the solution set in terms of natural logarithms. Then use a calculator to obtain a decimal approximation, correct to two decimal places, for the solution.

1. $10^x = 3.91$ **2.** $10^x = 8.07$

3. $e^x = 5.7$ **4.** $e^x = 0.83$

5. $5^x = 17$ **6.** $19^x = 143$

7. $5e^x = 23$ **8.** $9e^x = 107$

9. $3e^{5x} = 1977$ **10.** $4e^{7x} = 10{,}273$

11. $e^{1-5x} = 793$ **12.** $e^{1-8x} = 7957$

13. $e^{5x-3} - 2 = 10{,}476$ **14.** $e^{4x-5} - 7 = 11{,}243$

15. $7^{x+2} = 410$ **16.** $5^{x-3} = 137$

17. $7^{0.3x} = 813$ **18.** $3^{x/7} = 0.2$

19. $e^{2x} - 3e^x + 2 = 0$ **20.** $e^{2x} - 2e^x - 3 = 0$

21. $e^{4x} + 5e^{2x} - 24 = 0$ **22.** $e^{4x} - 3e^{2x} - 18 = 0$

Solve each logarithmic equation in Exercises 23–36. Be sure to reject any value of x that produces the logarithm of a negative number or the logarithm of 0.

23. $\log_3 x = 4$ **24.** $\log_5 x = 3$

25. $\log_4(x + 5) = 3$ **26.** $\log_5(x - 7) = 2$

27. $\log_3(x - 4) = -3$ **28.** $\log_7(x + 2) = -2$

29. $\log_4(3x + 2) = 3$ **30.** $\log_2(4x + 1) = 5$

31. $\log_5 x + \log_5(4x - 1) = 1$

32. $\log_6(x + 5) + \log_6 x = 2$

33. $\log_3(x - 5) + \log_3(x + 3) = 2$

34. $\log_2(x - 1) + \log_2(x + 1) = 3$

35. $\log_2(x + 2) - \log_2(x - 5) = 3$

36. $\log_4(x + 2) - \log_4(x - 1) = 1$

Exercises 37–44 involve equations with natural logarithms. Solve each equation by isolating the natural logarithm and exponentiating both sides. Express the answer in terms of e. Then use a calculator to obtain a decimal approximation, correct to two decimal places, for the solution.

37. $\ln x = 2$ **38.** $\ln x = 3$

39. $5 \ln 2x = 20$ **40.** $6 \ln 2x = 30$

41. $6 + 2 \ln x = 5$ **42.** $7 + 3 \ln x = 6$

43. $\ln \sqrt{x + 3} = 1$ **44.** $\ln \sqrt{x + 4} = 1$

Application Exercises

Use the formula $R = 6e^{12.77x}$, where x is the blood alcohol concentration and R, given as a percent, is the risk of having a car accident, to solve Exercises 45–46.

45. What blood alcohol concentration corresponds to certainty, or a 100% risk, of a car accident?

46. What blood alcohol concentration corresponds to a 50% risk of a car accident?

47. The formula $A = 18.2e^{0.001t}$ models the population of New York State, in millions, t years after 1994.

a. What was the population of New York in 1994?

b. When will the population of New York reach 18.5 million?

48. The formula $A = 14e^{0.168t}$ models the population of Florida, in millions, t years after 1994.

a. What was the population of Florida in 1994?

b. When will the population of Florida reach 18.5 million?

In Exercises 49–52, complete the table for a savings account subjected to n compoundings yearly $\left(A = P\left(1 + \frac{r}{n}\right)^{nt}\right)$.

	Amount Invested	Number of Compounding Periods	Annual Interest Rate	Accumulated Amount	Time t in Years
49.	\$12,500	4	5.75%	\$20,000	
50.	\$7250	12	6.5%	\$15,000	
51.	\$1000	360		\$1400	2
52.	\$5000	360		\$9000	4

In Exercises 53–56, complete the table for a savings account subjected to continuous compounding $(A = Pe^{rt})$.

	Amount Invested	Annual Interest Rate	Accumulated Amount	Time t in Years
53.	\$8000	8%	Double the amount invested	
54.	\$8000		\$12,000	2
55.	\$2350		Triple the amount invested	7
56.	\$17,425	4.25%	\$25,000	

57. The formula $C = 15{,}557 + 5259 \ln x$ models the average cost of a new car x years after 1989. When will the average cost of a new car be \$25,000?

58. The formula $C = 280 \ln(A + 1) + 1925$ models the number of calories, C, consumed each day by a person who owns A acres of land in a developing country, where $0 \le A \le 4$. How many acres of land are owned by a person who consumes 2200 calories daily in a developing country? (Source: Grigg, D. *The World Food Problem.* Oxford: Blackwell Publishers, 1993.)

The formula $P = 95 - 30 \log_2 x$ models the percentage, P, of students who could recall the important features of a classroom lecture as a function of time, where x represents the number of days that have elapsed since the lecture was given. The figure shows the graph of the formula. Use this information to solve Exercises 59–60.

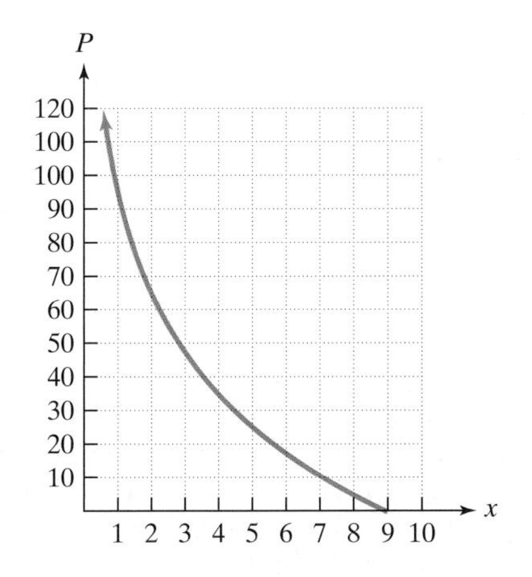

59. After how many days do only half the students recall the important features of the classroom lecture? (Let $P = 50$ and solve for x.) Can you approximately locate the point on the graph that conveys this information?

60. After how many days have all students forgotten the important features of the classroom lecture? (Let $P = 0$ and solve for x.) Can you approximately locate the point on the graph that conveys this information?

The pH of a solution ranges from 0 to 14. An acid solution has a pH less than 7. Pure water is neutral and has a pH of 7. Normal, unpolluted rain has a pH of about 5.6. The pH of a solution is given by

$$\text{pH} = -\log x$$

where x represents the concentration of the hydrogen ions in the solution in moles per liter. Use the formula to solve Exercises 61–62.

61. An environmental concern involves the destructive effects of acid rain. The most acidic rainfall ever had a pH of 2.4. What was the hydrogen ion concentration? Express the answer as a power of 10, and then round to the nearest thousandth.

62. The figure on page 394 shows very acidic rain in the northeast United States. What is the hydrogen ion concentration of rainfall with a pH of 4.2? Express the answer as a power of 10, and then round to the nearest hundred-thousandth.

Acid Rain Over Canada and the United States

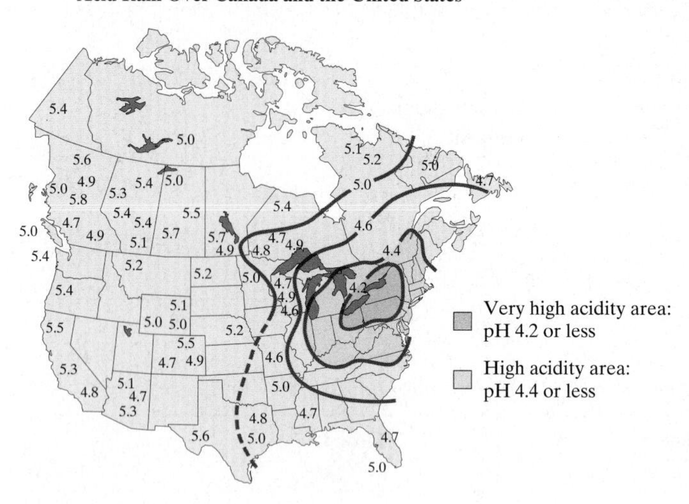

Source: National Atmospheric Program

Writing in Mathematics

63. Explain how to solve an exponential equation. Use $3^x = 140$ in your explanation.

64. Explain how to solve a logarithmic equation. Use $\log_3(x - 1) = 4$ in your explanation.

65. In many states, a 17% risk of a car accident with a blood alcohol concentration of 0.08 is the lowest level for charging a motorist with driving under the influence. Do you agree with the 17% risk as a cutoff percentage, or do you feel that the percentage should be lower or higher? Explain your answer. What blood alcohol concentration corresponds to what you believe is an appropriate percentage?

66. Have you purchased a new or used car recently? If so, describe if the formula in Exercise 58 accurately models what you paid for your car. If there is a big difference between the figure given by the formula and the amount that you paid, how can you explain this difference?

Technology Exercises

In Exercises 67–74, use your graphing utility to graph each side of the equation in the same viewing rectangle. Then use the x-coordinate of the intersection point to find the equation's solution set. Verify this value by direct substitution into the equation.

67. $2^{x+1} = 8$

68. $3^{x+1} = 9$

69. $\log_3(4x - 7) = 2$

70. $\log_3(3x - 2) = 2$

71. $\log(x + 3) + \log x = 1$

72. $\log(x - 15) + \log x = 2$

73. $3^x = 2x + 3$

74. $5^x = 3x + 4$

Hurricanes are one of nature's most destructive forces. These low-pressure areas often have diameters of over 500 miles. The function $f(x) = 0.48\ln(x + 1) + 27$ models the barometric air pressure, $f(x)$, in inches of mercury, at a distance of x miles from the eye of a hurricane. Use this function to solve Exercises 75–76.

75. Graph the function in a $[0, 500, 50]$ by $[27, 30, 1]$ viewing rectangle. What does the shape of the graph indicate about barometric air pressure as the distance from the eye increases?

76. Use an equation to answer this question: How far from the eye of a hurricane is the barometric air pressure 29 inches of mercury? Use the TRACE and ZOOM features or the intersect command of your graphing utility to verify your answer.

77. The formula $P = 145e^{-0.092t}$ models a runner's pulse, P, in beats per minute, t minutes after a race, where $0 \le t \le 15$. Graph the formula using a graphing utility. TRACE along the graph and determine after how many minutes the runner's pulse will be 70 beats per minute. Verify your observation algebraically.

78. The formula $W = 2600(1 - 0.51e^{-0.075t})^3$ models the weight, W, in kilograms, of a female African elephant at age t years. (1 kilogram $\approx$ 2.2 pounds) Use a graphing utility to graph the formula. Then TRACE along the curve to estimate the age of an adult female elephant weighing 1800 kilograms.

Critical Thinking Exercises

79. Which one of the following is true?

a. If $\log(x + 3) = 2$, then $e^2 = x + 3$.

b. If $\log(7x + 3) - \log(2x + 5) = 4$, then in exponential form $10^4 = (7x + 3) - (2x + 5)$.

c. If $x = \frac{1}{k}\ln y$, then $y = e^{kx}$.

d. Examples of exponential equations include $10^x = 5.71$, $e^x = 0.72$, and $x^{10} = 5.71$.

80. If \$4000 is deposited into an account paying 3% interest compounded annually and at the same time \$2000 is deposited into an account paying 5% interest compounded annually, after how long will the two accounts have the same balance?

Solve each equation in Exercises 81–83. Check each proposed solution by direct substitution or with a graphing utility.

81. $(\ln x)^2 = \ln x^2$

82. $(\log x)(2\log x + 1) = 6$

83. $\ln(\ln x) = 0$

Group Exercise

84. Research applications of logarithmic functions as mathematical models and plan a seminar based on your group's research. Each group member should research one of the following areas or any other area of interest: pH (acidity of solutions), intensity of sound (decibels), brightness of stars, consumption of natural resources, human memory, progress over time in a sport, profit over time. For the area that you select, explain how logarithmic functions are used and provide examples.

SECTION 4.5 Modeling with Exponential and Logarithmic Functions

Objectives

1. Model exponential growth and decay.
2. Use logistic growth models.
3. Model data with exponential and logarithmic functions.
4. Express an exponential model in base e.

The most casual cruise on the Internet shows how people disagree when it comes to making predictions about the effects of the world's growing population. Some argue that there is a recent slowdown in the growth rate, economies remain robust, and famines in Biafra and Ethiopia are aberrations rather than signs of the future. Others say that the 6 billion people on Earth is twice as many as can be supported in middle-class comfort, and the world is running out of arable land and fresh water. Debates about entities that are growing exponentially can be approached mathematically: We can create functions that model data and use these functions to make predictions. In this section we will show you how this is done.

1 Model exponential growth and decay.

Exponential Growth and Decay

One of algebra's many applications is to predict the behavior of variables. This can be done with **exponential growth** and **decay models**. With exponential growth and decay, quantities grow or decay at a rate directly proportional to their size. Populations that are growing exponentially grow extremely rapidly as they get larger because there are more adults to have offspring. For example, the **growth rate** for world population is 1.3%, or 0.013. This means that each year world population is 1.3% more than what it was in the previous year. In 1999, world population was 6 billion. Thus, we compute the world population in 2000 as follows:

$$6 \text{ billion} + 1.3\% \text{ of } 6 \text{ billion} = 6 + (0.013)(6) = 6.078.$$

This computation suggests that 6.078 billion people will populate the world in 2000. The 0.078 billion represents an increase of 78 million people from 1999 to 2000, the equivalent of the population of Germany. Using 1.3% as the annual growth rate, world population for 2001 is found in a similar manner:

$$6.078 \text{ billion} + 1.3\% \text{ of } 6.078 \text{ billion} = 6.078 + (0.013)(6.078) \approx 6.157.$$

This computation suggests that approximately 6.157 billion people will populate the world in 2001.

The explosive growth of world population may remind you of the growth of money in an account subject to compound interest. Just as the growth rate for world population is multiplied by the the population plus any increase in the population, a compound interest rate is multiplied by your original investment plus any accumulated interest. The balance in an account subject to continuous compounding and world population are special cases of an *exponential growth model*.

Exponential Growth and Decay Models

The mathematical model for **exponential growth** or **decay** is given by

$$f(t) = A_0e^{kt} \quad \text{or} \quad A = A_0e^{kt}.$$

- **If $k > 0$, the function models the amount or size of a *growing* entity.** A_0 is the original amount or size of the growing entity at time $t = 0$, A is the amount at time t, and k is a constant representing the growth rate.
- **If $k < 0$, the function models the amount or size of a *decaying* entity.** A_0 is the original amount or size of the decaying entity at time $t = 0$, A is the amount at time t, and k is a constant representing the decay rate.

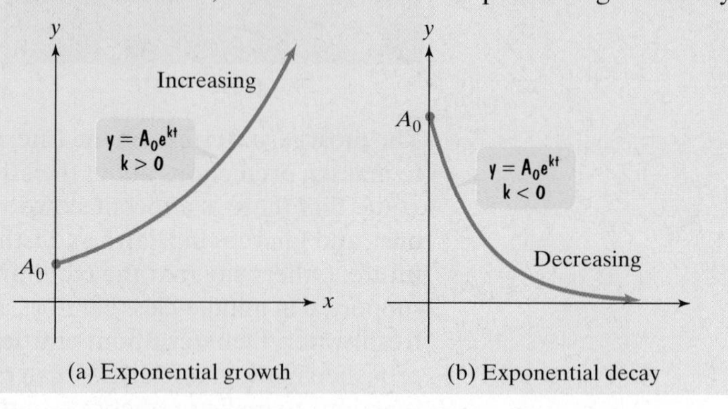

(a) Exponential growth (b) Exponential decay

Sometimes we need to use given data to determine k, the rate of growth or decay. After we compute the value of k, we can use the formula $A = A_0e^{kt}$ to make predictions. This idea is illustrated in our first two examples.

EXAMPLE 1 Modeling Mexico City's Growth

The graph in Figure 4.17 shows the growth of the Mexico City metropolitan area from 1970 through 2000. In 1970, the population of Mexico City was 9.4 million. By 1990, it had grown to 20.2 million.

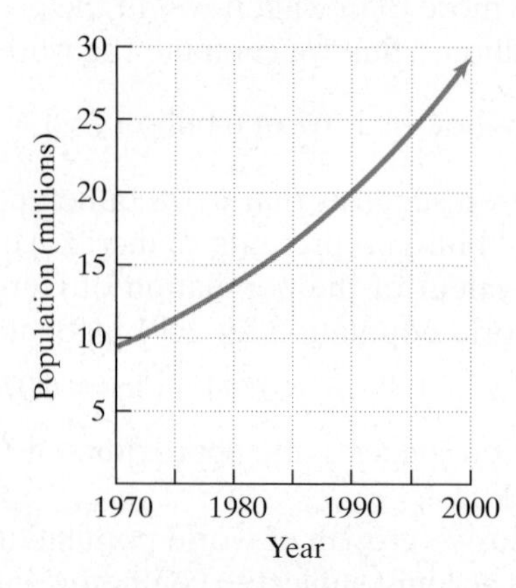

Figure 4.17 Mexico City's population has grown exponentially.

a. Find the exponential growth function that models the data.

b. By what year will the population reach 40 million?

Solution

a. We use the exponential growth model

$$A = A_0 e^{kt}$$

in which t is the number of years since 1970. This means that 1970 corresponds to $t = 0$. At that time there were 9.4 million inhabitants, so we substitute 9.4 for A_0 in the growth model.

$$A = 9.4e^{kt}$$

We are given that there were 20.2 million inhabitants in 1990. Because 1990 is 20 years after 1970, when $t = 20$ the value of A is 20.2. Substituting these numbers into the growth model will enable us to find k, the growth rate. We know that $k > 0$ because the problem involves growth.

$$A = 9.4e^{kt}$$ Use the growth model with $A_0 = 9.4$.

$$20.2 = 9.4e^{k \cdot 20}$$ When $t = 20$, $A = 20.2$. Substitute these numbers into the model.

$$e^{20k} = \frac{20.2}{9.4}$$ Isolate the exponential factor by dividing both sides by 9.4. We also reversed the sides.

$$\ln e^{20k} = \ln \frac{20.2}{9.4}$$ Take the natural logarithm on both sides.

$$20k = \ln \frac{20.2}{9.4}$$ Simplify the left side using $\ln e^x = x$.

$$k = \frac{\ln \frac{20.2}{9.4}}{20} \approx 0.038$$ Divide both sides by 20 and solve for k.

We substitute 0.038 for k in the growth model to obtain the exponential growth function for Mexico City. It is

$$A = 9.4e^{0.038t}$$

where t is measured in years since 1970.

b. To find the year in which the population will reach 40 million, we substitute 40 for A in the model from part (a) and solve for t.

$$A = 9.4e^{0.038t}$$ This is the model from part (a).

$$40 = 9.4e^{0.038t}$$ Substitute 40 for A.

$$e^{0.038t} = \frac{40}{9.4}$$ Divide both sides by 9.4.

$$\ln e^{0.038t} = \ln \frac{40}{9.4}$$ Take the natural logarithm on both sides.

$$0.038t = \ln \frac{40}{9.4}$$ Simplify on the left using $\ln e^x = x$.

$$t = \frac{\ln \frac{40}{9.4}}{0.038} \approx 38$$ Solve for t by dividing both sides by 0.038.

Because 38 is the number of years after 1970, the model indicates that the population of Mexico City will reach 40 million by 1970 + 38, or in the year 2008.

Check Point 1 In 1980, the population of Africa was 491 million and by 1990 it had grown to 643 million.

a. Use the exponential growth model $A = A_0e^{kt}$, in which t is the number of years since 1980, to find the exponential growth function that models the data.

b. By what year will Africa's population reach 1000 million, or one billion?

Carbon Dating and Artistic Development

The artistic community was electrified by the discovery in 1995 of spectacular cave paintings in a limestone cavern in France. Carbon dating of the charcoal from the site showed that the images, created by artists of remarkable talent, were 30,000 years old, making them the oldest cave paintings ever found. The artists seemed to have used the cavern's natural contours to heighten a sense of perspective. The quality of the painting suggests that the art of early humans did not mature steadily from primitive to sophisticated in any simple linear fashion.

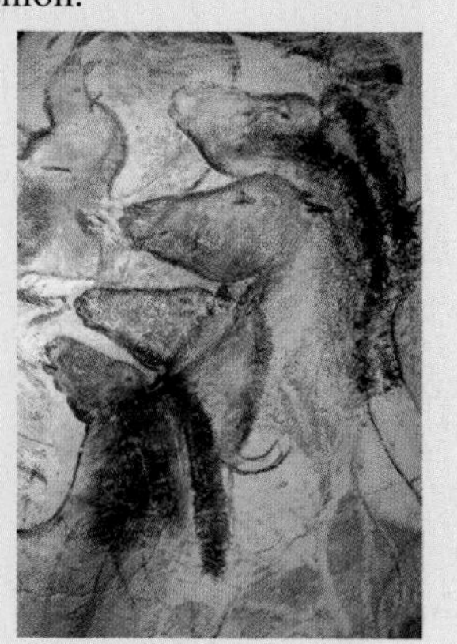

Our next example involves exponential decay and its use in determining the age of fossils and artifacts. The method is based on considering the percentage of carbon-14 remaining in the fossil or artifact. Carbon-14 decays exponentially with a *half-life* of approximately 5715 years. The **half-life** of a substance is the time required for half of a given sample to disintegrate. Thus, after 5715 years a given amount of carbon-14 will have decayed to half the original amount. Carbon dating is useful for artifacts or fossils up to 80,000 years old. Older objects do not have enough carbon-14 left to date age accurately.

EXAMPLE 2 Carbon-14 Dating: The Dead Sea Scrolls

a. Use the fact that after 5715 years a given amount of carbon-14 will have decayed to half the original amount to find the exponential decay model for carbon-14.

b. In 1947, earthenware jars containing what are known as the Dead Sea Scrolls were found by an Arab Bedouin herdsman. Analysis indicated that the scroll wrappings contained 76% of their original carbon-14. Estimate the age of the Dead Sea Scrolls.

Solution We begin with the exponential decay model $A = A_0e^{kt}$. We know that $k < 0$ because the problem involves the decay of carbon-14. After 5715 years ($t = 5715$), the amount of carbon-14 present, A, is half the original amount A_0. Thus we can substitute $\frac{A_0}{2}$ for A in the exponential decay model. This will enable us to find k, the decay rate.

a. $\frac{A_0}{2} = A_0e^{k5715}$ After 5715 years ($t = 5715$), $A = \frac{A_0}{2}$ (because the amount present, A, is half the original amount, A_0).

$\frac{1}{2} = e^{5715k}$ Divide both sides of the equation by A_0.

$\ln\frac{1}{2} = \ln e^{5715k}$ Take the natural logarithm of both sides.

$\ln\frac{1}{2} = 5715k$ $\ln e^x = x$

$k = \frac{\ln\frac{1}{2}}{5715} \approx -0.000121$ Solve for k.

Substituting for k in the decay model, the model for carbon-14 is $A = A_0e^{-0.000121t}$.

b.

$A = A_0 e^{-0.000121t}$	This is the decay model for carbon-14.
$0.76A_0 = A_0 e^{-0.000121t}$	A, the amount present, is 76% of the original amount, so $A = 0.76A_0$.
$0.76 = e^{-0.000121t}$	Divide both sides of the equation by A_0.
$\ln 0.76 = \ln e^{-0.000121t}$	Take the natural logarithm on both sides.
$\ln 0.76 = -0.000121t$	$\ln e^x = x$
$t = \dfrac{\ln 0.76}{-0.000121} \approx 2268$	Solve for t.

The Dead Sea Scrolls are approximately 2268 years old plus the number of years between 1947 and the current year.

Check Point 2 Strontium-90 is a waste product from nuclear reactors. As a consequence of fallout from atmospheric nuclear tests, we all have a measurable amount of strontium-90 in our bones.

a. Use the fact that after 28 years a given amount of strontium-90 will have decayed to half the original amount to find the exponential decay model for strontium-90.

b. Suppose that a nuclear accident occurs and releases 60 grams of strontium-90 into the atmosphere. How long will it take for strontium-90 to decay to a level of 10 grams?

2 Use logistic growth models.

Logistic Growth Models

From population growth to the spread of an epidemic, nothing on Earth can grow exponentially indefinitely. Growth is always limited. This is shown in Figure 4.18 by the horizontal asymptote. The **logistic growth model** is an exponential function used to model situations in which growth is limited.

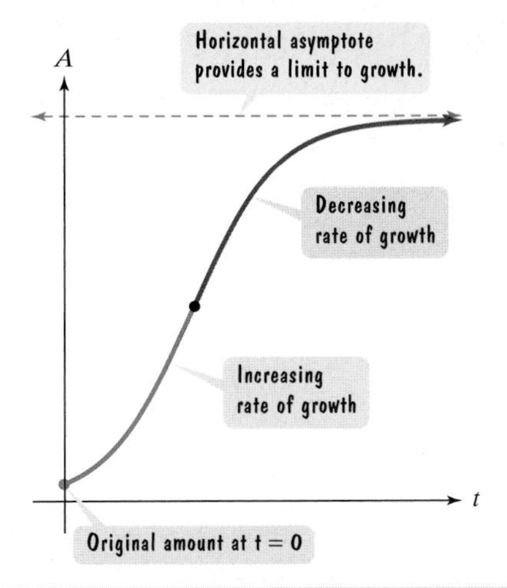

Figure 4.18 The logistic growth curve has a horizontal asymptote that limits the growth of A over time.

Logistic Growth Model

The mathematical model for limited logistic growth is given by

$$f(t) = \frac{c}{1 + ae^{-bt}} \quad \text{or} \quad A = \frac{c}{1 + ae^{-bt}}$$

where a, b, and c are constants with $c > 0$ and $b > 0$.

As time increases ($t \to \infty$), the expression ae^{-bt} in the model approaches 0 and A gets closer and closer to c. This means that $y = c$ is a horizontal asymptote for the graph of the function. Thus, the value of A can never exceed c and c represents the limiting size that A can attain.

EXAMPLE 3 Modeling the Spread of the Flu

The function

$$f(t) = \frac{30{,}000}{1 + 20e^{-1.5t}}$$

describes the number of people, $f(t)$, who have become ill with influenza t weeks after its initial outbreak in a town with 30,000 inhabitants.

a. How many people became ill with the flu when the epidemic began?
b. How many people were ill by the end of the fourth week?
c. What is the limiting size of $f(t)$, the population that becomes ill?

Solution

a. The time at the beginning of the flu epidemic is $t = 0$. Thus, we can find the number of people who were ill at the beginning of the epidemic by substituting 0 for t.

$$f(t) = \frac{30{,}000}{1 + 20e^{-1.5t}} \quad \text{This is the given logistic growth function.}$$

$$f(0) = \frac{30{,}000}{1 + 20e^{-1.5(0)}} \quad \text{When the epidemic began, } t = 0.$$

$$= \frac{30{,}000}{1 + 20} \quad e^{-1.5(0)} = e^0 = 1$$

$$\approx 1429$$

Approximately 1429 people were ill when the epidemic began.

b. We find the number of people who were ill at the end of the fourth week by substituting 4 for t in the logistic growth function.

$$f(t) = \frac{30{,}000}{1 + 20e^{-1.5t}} \quad \text{Use the given logistic growth function.}$$

$$f(4) = \frac{30{,}000}{1 + 20e^{-1.5(4)}} \quad \text{To find the number of people ill by the end of week four, let } t = 4.$$

$$= 28{,}583 \quad \text{Use a calculator.}$$

Approximately 28,583 people were ill by the end of the fourth week. Compared with the number of people who were ill initially, this illustrates the virulence of the epidemic.

c. Recall that in the logistic growth model, $f(t) = \frac{c}{1 + ae^{-bt}}$, the constant c represents the limiting size that $f(t)$ can attain. Thus, the number in the numerator, 30,000, is the limiting size of the population that becomes ill.

Technology

The graph of the logistic growth function for the flu epidemic

$$y = \frac{30{,}000}{1 + 20e^{-1.5x}}$$

can be obtained using a graphing calculator. We started x at 0 and ended at 10. This takes us to week 10. (In Example 3, we found that by week 4 approximately 28,583 people were ill.) We also know that 30,000 is the limiting size, so we took values of y up to 30,000. Using a [0, 10, 1] by [0, 30,000, 3000] viewing rectangle, the graph of the logistic growth function is shown below.

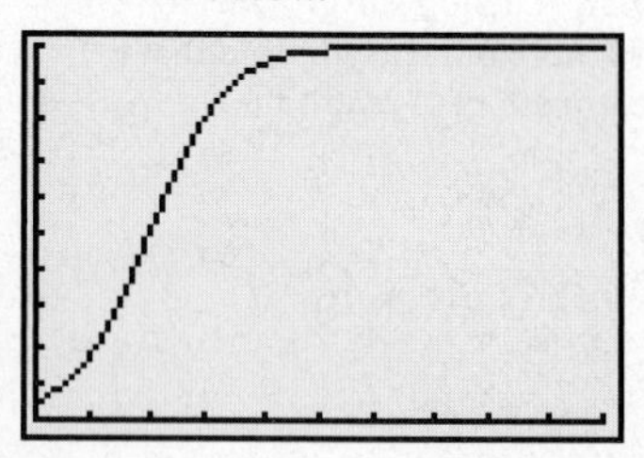

Check Point 3

In a learning theory project, psychologists discovered that

$$f(t) = \frac{0.8}{1 + e^{-0.2t}}$$

is a model for describing the proportion of correct responses after t learning trials.

a. Find the proportion of correct responses prior to learning trials taking place.

b. Find the proportion of correct responses after 10 learning trials.

c. What is the limiting size of $f(t)$, the proportion of correct responses as continued learning trials take place?

3 Model data with exponential and logarithmic functions.

The Art of Modeling

Throughout this chapter, we have been working with models that were given. However, we can create functions that model data by observing patterns in scatter plots. Figure 4.19 shows scatter plots for data that are exponential or logarithmic.

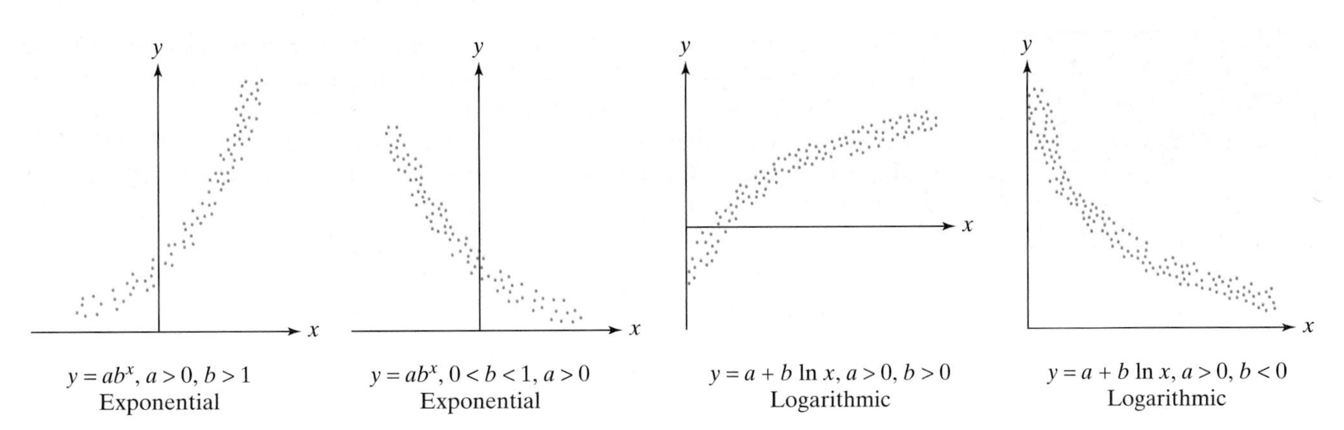

Figure 4.19 Scatter plots for exponential or logarithmic models

Graphing utilities can be used to find the equation of a function that is derived from data. For example, earlier in the chapter we encountered a function that modeled the size of a city and the average walking speed, in feet per second, of pedestrians. The function was derived from the data in Table 4.4 on page 402. The scatter plot is shown in Figure 4.20 on page 402.

Because the data in this scatter plot increase rapidly at first and then begin to level off a bit, the shape suggests that a logarithmic model might be a good choice. A graphing utility fits the data in Table 4.4 to a logarithmic model of the form $y = a + b \ln x$ by using the Logarithmic REGression option

Table 4.4

x, Population (thousands)	*y*, Walking Speed (feet per second)
5.5	3.3
14	3.7
71	4.3
138	4.4
342	4.8

Source: Mark and Helen Bornstein, "The Pace of Life"

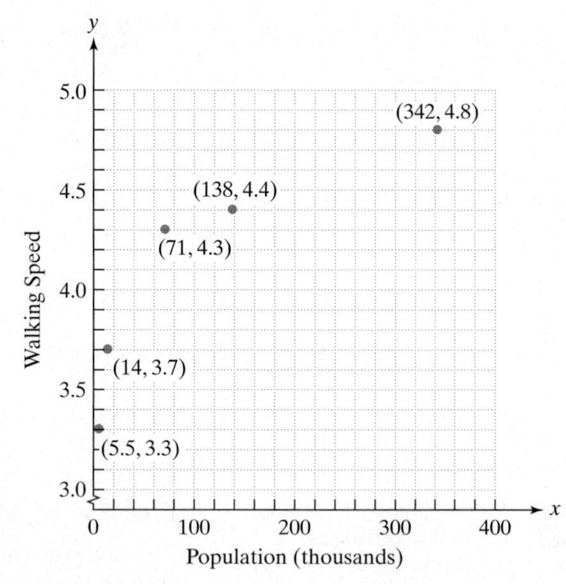

Figure 4.20 Scatter plot for data in Table 4.4

(see Figure 4.21). From the figure, we see that the logarithmic model of the data, with numbers rounded to three decimal places, is

$$y = 2.735 + 0.352 \ln x.$$

```
LnReg
 y=a+blnx
 a=2.734605225
 b=.3524772878
 r=.9958996083
```

Figure 4.21 A logarithmic model for the data in Table 4.4.

The number r that appears in Figure 4.21 is called the **correlation coefficient** and is a measure of how well the model fits the data. The value of r is such that $-1 \leq r \leq 1$. A positive r means that as the x-values increase, so do the y-values. A negative r means that as the x-values increase, the y-values decrease. **The closer that r is to -1 or 1, the better the model fits the data.** Because r is approximately 0.996, the model

$$y = 2.735 + 0.352 \ln x$$

fits the data very well.

Now let's look at data whose scatter plot suggests an exponential model. The data in Table 4.5 indicate world population for six years. The scatter plot is shown in Figure 4.22.

Table 4.5

x, Year	*y*, World Population (billions)
1950	2.6
1960	3.1
1970	3.7
1980	4.5
1989	5.3
1999	6.0

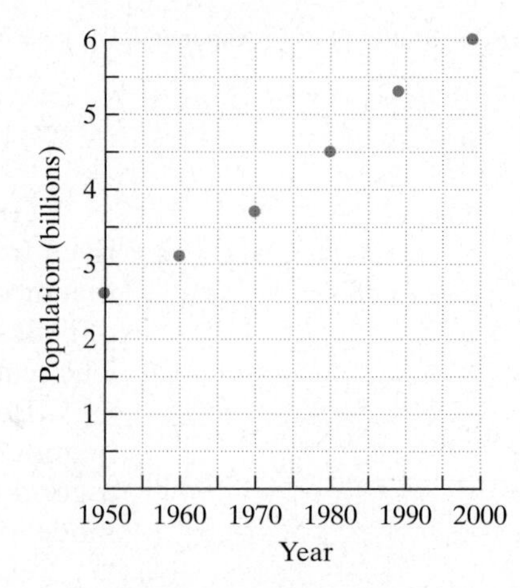

Figure 4.22 A scatter plot for data in Table 4.5.

The Future of World Population

In the future, a new world order looms. The percentage of world population in less developed countries will increase, North America's will remain relatively stable, and Europe's will decrease.

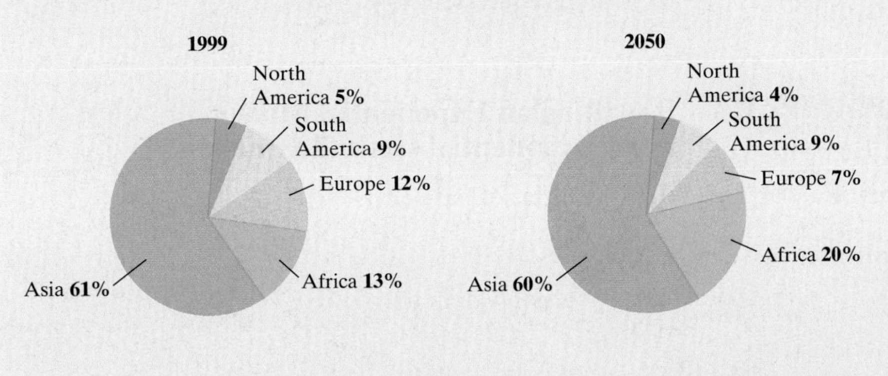

Most Populous Countries, 2050

1. India	1529 million
2. China	1478 million
3. U.S.	349 million
4. Pakistan	345 million
5. Indonesia	312 million
6. Nigeria	244 million
7. Brazil	244 million
8. Bangladesh	212 million
9. Ethiopia	169 million
10. Congo	160 million

Source: United Nations Population Fund

Because the data in this scatter plot have a rapidly increasing pattern, the shape suggests that an exponential model might be a good choice. (You might also want to try a linear model.) If you go with the exponential option, you will use a graphing utility's Exponential REGression option. With this feature, a graphing utility fits the data to an exponential model of the form $y = ab^x$.

When computing an exponential model of the form $y = ab^x$, a graphing utility rewrites the equation using logarithms. Because the domain of the logarithmic function is the set of positive numbers, **zero must not be a value for** $\boldsymbol{x}$. What does this mean in terms of our data for world population that starts in the year 1950? We must start values of x after 0. Thus, we'll assign x to represent the number of years since 1949. This gives us the data shown in Table 4.6. Using the Exponential REGression option, we obtain the equation in Figure 4.23.

Table 4.6

x, Numbers of Years after 1949	y, World Population (billions)
1 (1950)	2.6
11 (1960)	3.1
21 (1970)	3.7
31 (1980)	4.5
40 (1989)	5.3
50 (1999)	6.0

```
ExpReg
 y=a*b^x
 a=2.569837845
 b=1.017658868
 r=.9983590746
```

Figure 4.23 An exponential model for the data in Table 4.6

From Figure 4.23, we see that the exponential model of the data for world population x years after 1949, with numbers rounded to three decimal places, is

$$y = 2.570(1.018)^x.$$

The correlation coefficient, r, is close to 1, indicating that the model fits the data very well.

4 Express an exponential model in base e.

Because $b = e^{\ln b}$, we can rewrite any model in the form $y = ab^x$ in terms of base e.

Expressing an Exponential Model in Base e

$$y = ab^x \text{ is equivalent to } y = ae^{(\ln b)\cdot x}.$$

EXAMPLE 4 Rewriting an Exponential Model as an Exponential Growth Function

Rewrite $y = 2.57(1.018)^x$ in terms of base e.

Solution

$y = ab^x$ is equivalent to $y = ae^{(\ln b)\cdot x}$.

$$y = 2.57(1.018)^x \quad \text{is equivalent to} \quad y = 2.57e^{(\ln 1.018)\cdot x}.$$

Using $\ln 1.018 \approx 0.018$, the exponential growth model for world population x years after 1949 is

$$y = 2.57e^{0.018x}.$$

In Example 4, we can replace y by A and x by t so that the model has the same letters as those in the exponential growth model $A = A_0e^{kt}$.

$A = A_0 e^{kt}$ This is the exponential growth model.

$A = 2.57e^{0.018t}$ This is the model for world population.

The value of k, 0.018, indicates a growth rate of 1.8%. Although this is an excellent model for the data, we must be careful about making projections about world population using this growth function. Why? World population growth rate is now 1.3%, not 1.8%, so our model will overestimate future populations.

Check Point 4 Rewrite $y = 4(7.8)^x$ in terms of base e. Express the answer in terms of a natural logarithm, and then round to three decimal places.

When using a graphing utility to model data, begin with a scatter plot, drawn either by hand or with the graphing utility, to obtain a general picture for the shape of the data. It might be difficult to determine what model best fits the data—linear, logarithmic, exponential, quadratic, or something else. If necessary, use your graphing utility to fit several models to the data. The best model is the one that yields the value r, the correlation coefficient, closest to 1 or -1. Finding a proper fit for data can be almost as much art as it is mathematics. In this era of technology, the process of creating models that best fit data is one that involves more decision making than computation.

EXERCISE SET 4.5

Practice and Application Exercises

The exponential growth model $A = 208e^{0.008t}$ *describes the population of the United States, in millions, t years after 1970. Use this model to solve Exercises 1–4.*

1. What was the population of the United States in 1970?

2. By what percentage is the population of the United States increasing each year?

3. When will the U.S. population be 300 million?

4. When will the U.S. population be 350 million?

India is currently one of the world's fastest-growing countries. By 2040, the population of India will be larger than the population of China; by 2050, nearly one-third of the world's population will live in these two countries alone. The exponential growth model $A = 574e^{0.026t}$ *describes the population of India, in millions, t years after 1974. Use this model to solve Exercises 5–8.*

5. By what percentage is the population of India increasing each year?

6. What was the population of India in 1974?

7. When will India's population be 1624 million?

8. When will India's population be 2732 million?

The value of houses in a neighborhood follows a pattern of exponential growth. In the year 2000, you purchased a house in this neighborhood. The value of your house, in thousands of dollars, t years after 2000 is given by the exponential growth model $V = 140e^{0.068t}$. *Use this model to solve Exercises 9–12.*

9. What did you pay for your house?

10. By what percentage is the price of houses in your neighborhood increasing each year?

11. When will your house be worth $200,000?

12. When will your house be worth $300,000?

13. Through the end of 1991, 200,000 cases of AIDS had been reported to the Centers for Disease Control in the United States. By the end of 1998, the number had grown to 680,000. The exponential growth function $A = 200e^{kt}$ describes the thousands of AIDS cases in the United States t years after 1991. Use the fact that 7 years after 1991 there were 680 thousand cases to find k to three decimal places. Then write the exponential growth function. According to your model, by what percentage is the number of AIDS cases in the United States increasing each year?

14. In 1980, China's population was 983 million; in 1990, it was 1154 million. The exponential growth function $A = 983e^{kt}$ describes the population of China, in millions, t years after 1980. Use the fact that 10 years after 1980 the population was 1154 million to find k to three decimal places. Then write the exponential growth function. According to your model, by what percentage is the population of China increasing each year?

An artifact originally had 16 grams of carbon-14 present. The decay model $A = 16e^{-0.000121t}$ *describes the amount of carbon-14 present after t years. Use this model to solve Exercises 15–16.*

15. How many grams of carbon-14 will be present in 5715 years?

16. How many grams of carbon-14 will be present in 11,430 years?

17. The half-life of the radioactive element krypton-91 is 10 seconds. If 16 grams of krypton-91 are initially present, how many grams are present after 10 seconds? 20 seconds? 30 seconds? 40 seconds? 50 seconds?

18. The half-life of the radioactive element plutonium-239 is 25,000 years. If 16 grams of plotonium-239 are initially present how many grams are present after 25,000 years? 50,000 years? 75,000 years? 100,000 years? 125,000 years?

Use the exponential decay model for carbon-14, $A = A_0e^{-0.000121t}$, *to solve Exercises 19–20.*

19. Prehistoric cave paintings were discovered in a cave in France. The paint contained 15% of the original carbon-14. Estimate the age of the paintings.

20. Skeletons were found at a construction site in San Francisco in 1989. The skeletons contained 88% of the expected amount of carbon-14 found in a living person. In 1989, how old were the skeletons?

21. The August 1978 issue of *National Geographic* described the 1964 find of dinosaur bones of a newly discovered dinosaur weighing 170 pounds, measuring 9 feet, with a 6-inch claw on one toe of each hind foot. The age of the dinosaur was estimated using potassium-40 dating of rocks surrounding the bones.

a. Potassium-40 decays exponentially with a half-life of approximately 1.31 billion years. Use the fact that after 1.31 billion years a given amount of potassium-40 will have decayed to half the original amount to show that the decay model for potassium-40 is given by $A = A_0e^{-0.52912t}$, where t is in billions of years.

b. Analysis of the rocks surrounding the dinosaur bones indicated that 94.5% of the original amount of potassium-40 was still present. Let $A = 0.945A_0$ in the model in part (a) and estimate the age of the bones of the dinosaur.

22. A bird species in danger of extinction has a population that is decreasing exponentially $(A = A_0e^{kt})$. Five years ago the population was at 1400 and today only 1000 of the birds are alive. Once the population drops below 100, the situation will be irreversible. When will this happen?

23. Use the exponential growth model, $A = A_0e^{kt}$, to show that the time it takes a population to double (to grow from A_0 to $2A_0$) is given by $t = \frac{\ln 2}{k}$.

24. Use the exponential growth model, $A = A_0e^{kt}$, to show that the time it takes a population to triple (to grow from A_0 to $3A_0$) is given by $t = \frac{\ln 3}{k}$.

Use the formula $t = \frac{\ln 2}{k}$ that gives the time for a population with a growth rate k to double to solve Exercises 25–26. Express each answer to the nearest whole year.

25. China is growing at a rate of 1.1% per year. How long will it take China to double its population?

26. Japan is growing at a rate of 0.3% per year. How long will it take Japan to double its population?

27. The logistic growth function

$$f(t) = \frac{100{,}000}{1 + 5000e^{-t}}$$

describes the number of people who have become ill with influenza t weeks after its initial outbreak in a particular community.

a. How many people became ill with the flu when the epidemic began?

b. How many people were ill by the end of the fourth week?

c. What is the limiting size of the population that becomes ill?

28. The logistic growth function

$$f(t) = \frac{500}{1 + 83.3e^{-0.162t}}$$

describes the population of an endangered species of birds t years after they are introduced to a non-threatening habitat.

a. How many birds were initially introduced to the habitat?

b. How many birds are expected in the habitat after 10 years?

c. What is the limiting size of the bird population that the habitat will sustain?

The logistic growth function

$$P(x) = \frac{0.9}{1 + 271e^{-0.122x}}$$

models the probability that an American who is x years old has some coronary heart disease. Use the function to solve Exercises 29–32.

29. What is the probability that a 20-year-old has some coronary heart disease?

30. What is the probability that an 80-year-old has some coronary heart disease?

31. At what age is the probability of some coronary heart disease 0.5?

32. At what age is the probability of some coronary heart disease 0.7?

In Exercises 33–36, rewrite the equation in terms of base e. Express the answer in terms of a natural logarithm, and then round to three decimal places.

33. $y = 100(4.6)^x$

34. $y = 1000(7.3)^x$

35. $y = 2.5(0.7)^x$

36. $y = 4.5(0.6)^x$

Writing in Mathematics

37. Nigeria has a growth rate of 0.031 or 3.1%. Describe what this means.

38. How can you tell if an exponential model describes exponential growth or exponential decay?

39. Suppose that a population that is growing exponentially increases from 800,000 people in 1997 to 1,000,000 people in 2000. Without showing the details, describe how to obtain the exponential growth function that models this data.

40. What is the half-life of a substance?

41. Describe a difference between exponential growth and logistic growth.

42. Describe the shape of a scatter plot that suggests modeling the data with an exponential function.

43. You take up weightlifting and record the maximum number of pounds you can lift at the end of each week. You start off with rapid growth in terms of the weight you can lift from week to week, but then the growth begins to level off. Describe how to obtain a function that models the number of pounds you can lift at the end of each week. How can you use this function to predict what might happen if you continue the sport?

44. Would you prefer that your salary be modeled exponentially or logarithmically? Explain your answer.

45. One problem with all exponential growth models is that nothing can grow exponentially forever. Describe factors that might limit the size of a population.

Technology Exercises

The consumer price index measures changes in prices over time. The consumer price index for the bars shown in the graph indicates that what cost $1.00 in 1969 (the reference year) cost about $1.16 in 1970, $1.61 in 1975, $2.49 in 1980, $3.22 in 1985, $3.91 in 1990, $4.57 in 1995, and $4.96 in 1999. Use the data in the table to solve Exercises 46–50.

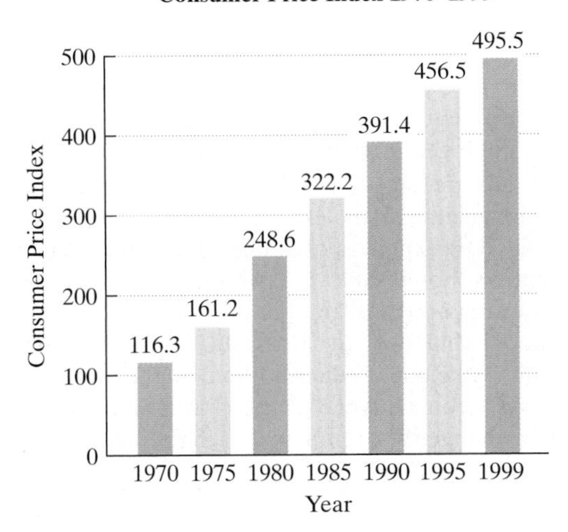

Source: U. S. Department of Labor

x, Number of Years, after 1969	y, Consumer Price Index
1	116.3
6	161.2
11	248.8
16	322.2
21	391.4
26	456.5
30	495.5

46. Use your graphing utility's Exponential REGression option to obtain a model of the form $y = ab^x$ that fits the data. How well does the correlation coefficient, r, indicate that the model fits the data?

47. Use your graphing utility's Logarithmic REGression option to obtain a model of the form $y = a + b \ln x$ that fits the data. How well does the correlation coefficient, r, indicate that the model fits the data?

48. Use your graphing utility's Linear REGression option to obtain a model of the form $y = ax + b$ that fits the data. How well does the correlation coefficient, r, indicate that the model fits the data?

49. Use your graphing utility's Power REGression option to obtain a model of the form $y = ax^b$ that fits the data. How well does the correlation coefficient, r, indicate that the model fits the data?

50. Use the value of r in Exercises 46–49 to select the model of best fit. Use this model to predict the consumer price index in 2007. How much will what cost $1.00 in 1969 cost in 2007?

51. In Exercises 29–32 you worked with the logistic growth function

$$P(x) = \frac{0.9}{1 + 271e^{-0.122x}}$$

that models the probability that an American who is x years old has some coronary heart disease. Use your graphing utility to graph the function in a $[0, 100, 10]$ by $[0, 1, 0.1]$ viewing rectangle. Describe as specifically as possible what the logistic curve indicates about aging and the probability of coronary heart disease.

In Exercises 52–53, use the data and a graphing utility to find the model that best fits the given data. Then use the model to make a prediction about what might occur in the future.

52.

x, Number of Years after 1984	y, Millions of Computers in Use in the U.S.
1	21.5
4	40.8
5	47.6
7	62.0
8	68.2
9	76.5
10	85.8
11	96.2
16	160.5

Source: Computer Industry Almanac

53.

x, Number of Years after 1989	y, Millions of CDs Sold in the U.S.
1	286.5
2	333.3
3	407.5
4	495.4
5	662.1
6	722.9
7	778.9
8	753.1
9	847.0

Source: Recording Industry Association of America

Critical Thinking Exercises

54. The World Health Organization makes predictions about the number of AIDS cases based on a compromise between a linear model and an exponential growth model. Explain why the World Health Organization does this.

55. Over a period of time, a hot object cools to the temperature of the surrounding air. This is described mathematically by

$$T = C + (T_0 - C)e^{-kt}$$

where t is the time it takes for an object to cool from temperature T_0 to temperature T, C is the surrounding air temperature, and k is a positive constant that is associated with the cooling object. A cake removed from the oven has a temperature of 210°F and is left to cool in a room that has a temperature of 70°F. After 30 minutes, the temperature of the cake is 140°F. What is the temperature of the cake after 40 minutes?

Group Exercises

56. This activity is intended for three or four people who would like to take up weightlifting. Each person in the group should record the maximum number of pounds that he or she can lift at the end of each week for the first 10 consecutive weeks. Use the Logarithmic REGression option of a graphing utility to obtain a model showing the amount of weight that group members can lift from week 1 to week 10. Graph each of the models in the same viewing rectangle to observe similarities and differences among weight-growth patterns of each member. Use the functions to predict the amount of weight that group members will be able to lift in the future. If the group continues to work out together, check the accuracy of these predictions.

57. Each group member should consult an almanac, newspaper, magazine, or the Internet to find data that can be modeled by exponential or logarithmic functions. Group members should select the two sets of data that are most interesting and relevant. For each data set selected, find a model that best fits the data. Each group member should make one prediction based on the model and then discuss a consequence of this prediction. What factors might change the accuracy of each prediction?

CHAPTER SUMMARY, REVIEW, AND TEST

Summary

4.1 Exponential Functions

a. The exponential function with base b is defined by $f(x) = b^x$, where $b > 0$ and $b \neq 1$.

b. Characteristics of exponential functions and graphs for $0 < b < 1$ and $b > 1$ are shown in the box on page 354.

c. Transformations involving exponential functions are summarized in Table 4.1 on page 354.

d. The natural exponential function: $f(x) = e^x$. The irrational number e is called the natural base, where $e \approx 2.7183$.

e. Formulas for compound interest: After t years, the balance A in an account with principal P and annual interest rate r (in decimal form) is given by one of the following formulas:

1. For n compoundings per year: $A = P\left(1 + \frac{r}{n}\right)^{nt}$
2. For continuous compounding: $A = Pe^{rt}$

4.2 Logarithmic Functions

a. Definition of the logarithmic function: For $x > 0$ and $b > 0$, $b \neq 1$, $y = \log_b x$ is equivalent to $b^y = x$. The function $f(x) = \log_b x$ is the logarithmic function with base b. This function is the inverse function of the exponential function with base b.

b. Graphs of logarithmic functions for $b > 1$ and $0 < b < 1$ are shown in Figure 4.7 on page 366. Characteristics of the graphs are summarized in the box that follows the figure.

c. Transformations involving logarithmic functions are summarized in Table 4.3 on page 367.

d. The domain of a logarithmic function is the set of all positive real numbers. The domain of $f(x) = \log_b(x + c)$ consists of all x for which $x + c > 0$.

e. Common and natural logarithms: $f(x) = \log x$ means $f(x) = \log_{10} x$ and is the common logarithmic function. $f(x) = \ln x$ means $f(x) = \log_e x$ and is the natural logarithmic function.

f. Basic Logarithmic Properties

Base b ($b > 0, b \neq 1$)	Base 10 (Common Logarithms)	Base e (Natural Logarithms)
$\log_b 1 = 0$	$\log 1 = 0$	$\ln 1 = 0$
$\log_b b = 1$	$\log 10 = 1$	$\ln e = 1$
$\log_b b^x = x$	$\log 10^x = x$	$\ln e^x = x$
$b^{\log_b x} = x$	$10^{\log x} = x$	$e^{\ln x} = x$

4.3 Properties of Logarithms

a. *The Product Rule:* $\log_b(MN) = \log_b M + \log_b N$

b. *The Quotient Rule:* $\log_b\left(\frac{M}{N}\right) = \log_b M - \log_b N$

c. *The Power Rule:* $\log_b M^p = p \log_b M$

d. *The Change-of Base Property:*

The General Property	Introducing Common Logarithms	Introducing Natural Logarithms
$\log_b M = \frac{\log_a M}{\log_a b}$	$\log_b M = \frac{\log M}{\log b}$	$\log_b M = \frac{\ln M}{\ln b}$

4.4 Exponential and Logarithmic Equations

a. An exponential equation is an equation containing a variable in an exponent. The solution procedure involves isolating the exponential expression and taking the natural logarithm on both sides. The box on page 385 provides the details.

b. A logarithmic equation is an equation containing a variable in a logarithmic expression. Logarithmic equations in the form $\log_b x = c$ can be solved by rewriting as $b^c = x$.

c. When checking logarithmic equations, reject proposed solutions that produce the logarithm of a negative number or the logarithm of 0 in the original equation.

d. Equations involving natural logarithms are solved by isolating the natural logarithm with coefficient 1 on one side and exponentiating both sides. Simplify using $e^{\ln x} = x$.

4.5 Modeling with Exponential and Logarithmic Functions

a. Exponential growth and decay models are given by $A = A_0 e^{kt}$ in which t represents time, A_0 is the amount present at $t = 0$, and A is the amount present at time t. If $k > 0$, the model describes growth and k is the growth rate. If $k < 0$, the model describes decay and k is the decay rate.

b. The logistic growth model, given by $A = \frac{c}{1 + ae^{-bt}}$, describes situations in which growth is limited. $y = c$ is a horizontal asymptote for the graph and growth, A, can never exceed c.

c. Scatter plots for exponential and logarithmic models are shown in Figure 4.19 on page 401. When using a graphing utility to model data, the closer that the correlation coefficient r is to -1 or 1, the better the model fits the data.

d. Expressing an Exponential Model in Base e: $y = ab^x$ is equivalent to $y = ae^{(\ln b)\cdot x}$.

Review Exercises

4.1

In Exercises 1–4, the graph of an exponential function is given. Select the function for each graph from the following options:

$$f(x) = 4^x,\ g(x) = 4^{-x},$$

$$h(x) = -4^{-x},\ r(x) = -4^{-x} + 3.$$

1.

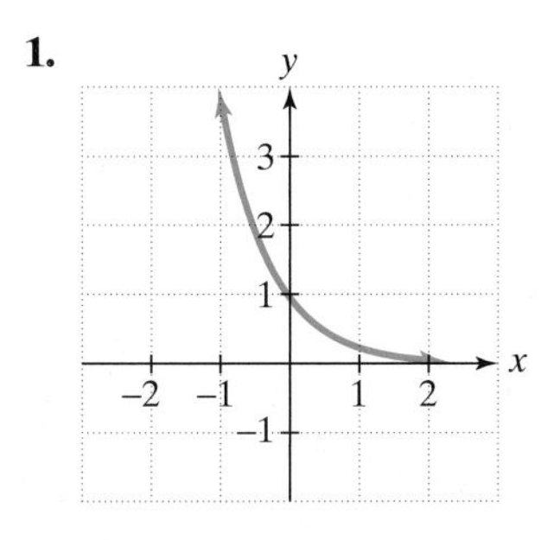

2.

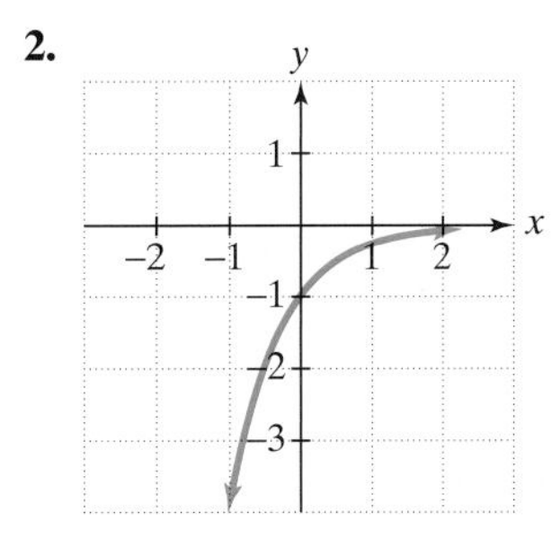

3.

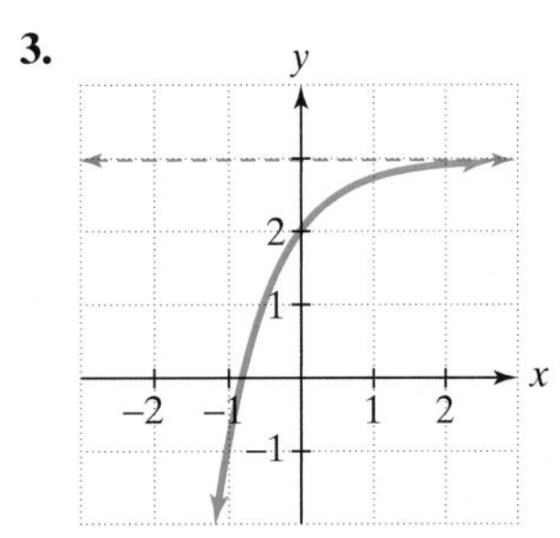

4.

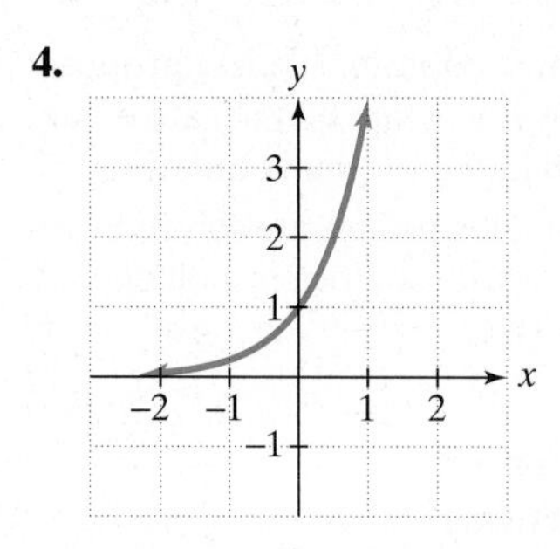

In Exercises 5–8, sketch by hand the graphs of the two functions in the same rectangular coordinate system. Use a table of coordinates to sketch the first function and transformations of this function plus a table of coordinates to graph the second function.

5. $f(x) = 2^x$ and $g(x) = 2^{x-1}$

6. $f(x) = 3^x$ and $g(x) = 3^x - 1$

7. $f(x) = 3^x$ and $g(x) = -3^x$

8. $f(x) = \left(\frac{1}{2}\right)^x$ and $g(x) = \left(\frac{1}{2}\right)^{-x}$

Use the compound interest formulas to solve Exercises 9–10.

9. Suppose that you have \$5000 to invest. Which investment yields the greater return over 5 years: 5.5% compounded semiannually or 5.25% compounded monthly?

10. Suppose that you have \$14,000 to invest. Which investment yields the greater return over 10 years: 7% compounded monthly or 6.85% compounded continuously?

11. A cup of coffee is taken out of a microwave oven and placed in a room. The temperature, T, in degrees Fahrenheit, of the coffee after t minutes is modeled by the function $T = 70 + 130e^{-0.04855t}$. The graph of the function is shown in the figure. Use the graph to answer each of the following questions.

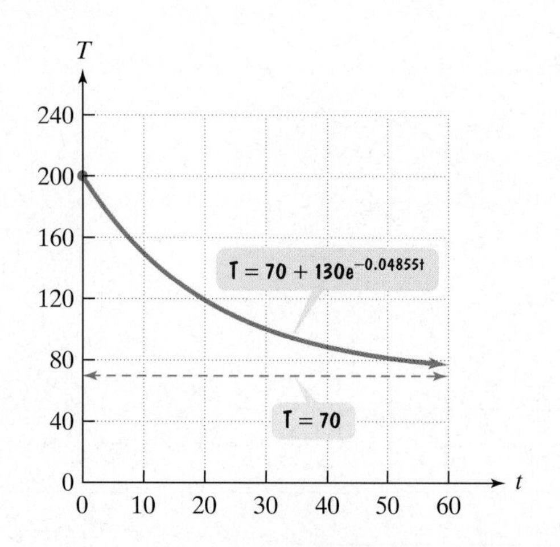

a. What was the temperature of the coffee when it was first taken out of the microwave?

b. What is a reasonable estimate of the temperature of the coffee after 20 minutes? Use your calculator to verify this estimate.

c. What is the limit of the temperature to which the coffee will cool? What does this tell you about the temperature of the room?

4.2

In Exercises 12–14, write each equation in its equivalent exponential form.

12. $\frac{1}{2} = \log_{49} 7$ **13.** $3 = \log_4 x$ **14.** $\log_3 81 = y$

In Exercises 15–17, write each equation in its equivalent logarithmic form.

15. $6^3 = 216$ **16.** $b^4 = 625$ **17.** $13^y = 874$

In Exercises 18–25, evaluate each expression without using a calculator. If evaluation is not possible, state the reason.

18. $\log_4 64$ **19.** $\log_5 \frac{1}{25}$ **20.** $\log_3(-9)$

21. $\log_{16} 4$ **22.** $\log_{17} 17$ **23.** $\log_3 3^8$

24. $\ln e^5$ **25.** $\log_3(\log_8 8)$

26. Graph $f(x) = 2^x$ and $g(x) = \log_2 x$ in the same rectangular coordinate system.

27. Graph $f(x) = \left(\frac{1}{3}\right)^x$ and $g(x) = \log_{1/3} x$ in the same rectangular coordinate system.

In Exercises 28–31, the graph of a logarithmic function is given. Select the function for each graph from the following options:

$$f(x) = \log x,\ g(x) = \log(-x),$$

$$h(x) = \log(2 - x),\ r(x) = 1 + \log(2 - x).$$

28.

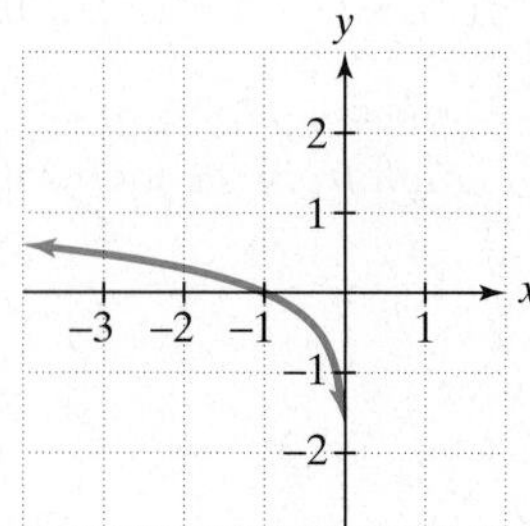

29.

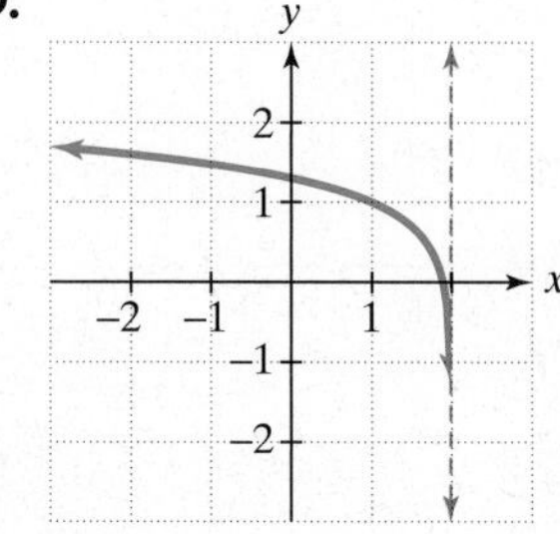

30.

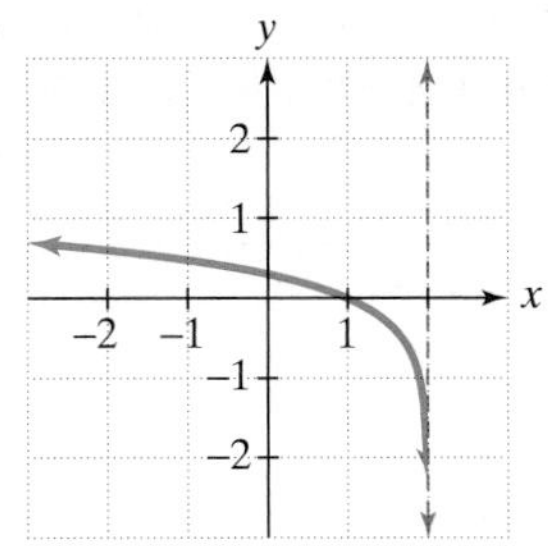

31.

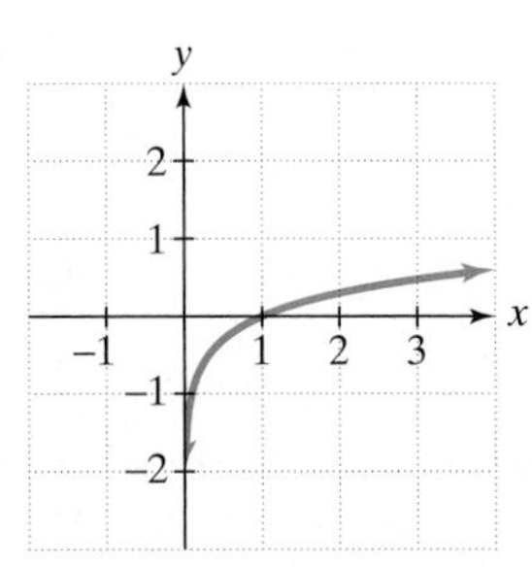

In Exercises 32–34, begin by graphing $f(x) = \log_2 x$. Then use transformations of this graph to graph the given function. What is the graph's x-intercept? What is the vertical asymptote?

32. $g(x) = \log_2(x - 2)$ **33.** $h(x) = -1 + \log_2 x$
34. $r(x) = \log_2(-x)$

In Exercises 35–37, find the domain of each logarithmic function.

35. $f(x) = \log_8(x + 5)$ **36.** $f(x) = \log(3 - x)$
37. $f(x) = \ln(x - 1)^2$

In Exercises 38–40, use inverse properties of logarithms to simplify each expression.

38. $\ln e^{6x}$ **39.** $e^{\ln \sqrt{x}}$ **40.** $10^{\log 4x^2}$

41. On the Richter scale, the magnitude, R, of an earthquake of intensity I is given by $R = \log \frac{I}{I_0}$, where I_0 is the intensity of a barely felt zero-level earthquake. If the intensity of an earthquake is $1000I_0$, what is its magnitude on the Richter scale?

42. Students in a psychology class took a final examination. As part of an experiment to see how much of the course content they remembered over time, they took equivalent forms of the exam in monthly intervals thereafter. The average score, $f(t)$, for the group after t months was modeled by the function $f(t) = 76 - 18\log(t + 1)$, where $0 \le t \le 12$.

a. What was the average score when the exam was first given?

b. What was the average score after 2 months? 4 months? 6 months? 8 months? one year?

c. Use the results from parts (a) and (b) to graph f. Describe what the shape of the graph indicates in terms of the material retained by the students.

43. The formula

$$t = \frac{1}{c}\ln\left(\frac{A}{A - N}\right)$$

describes the time, t, in weeks, that it takes to achieve mastery of a portion of a task. In the formula, A represents maximum learning possible, N is the portion of the learning that is to be achieved, and c is a constant used to measure an individual's learning style. A 50-year-old man decides to start running as a way to maintain good health. He feels that the maximum rate he could ever hope to achieve is 12 miles per hour. How many weeks will it take before the man can run 5 miles per hour if $c = 0.06$ for this person?

4.3

In Exercises 44–47, use properties of logarithms to expand each logarithmic expression as much as possible. Where possible, evaluate logarithmic expressions without using a calculator.

44. $\log_6(36x^3)$ **45.** $\log_4 \frac{\sqrt{x}}{64}$

46. $\log_2 \frac{xy^2}{64}$ **47.** $\ln \sqrt[3]{\frac{x}{e}}$

In Exercises 48–51, use properties of logarithms to condense each logarithmic expression. Write the expression as a single logarithm whose coefficient is 1.

48. $\log_b 7 + \log_b 3$ **49.** $\log 3 - 3\log x$

50. $3\ln x + 4\ln y$ **51.** $\frac{1}{2}\ln x - \ln y$

In Exercises 52–53, use common logarithms or natural logarithms and a calculator to evaluate to four decimal places.

52. $\log_6 72{,}348$ **53.** $\log_4 0.863$

4.4

Solve each exponential equation in Exercises 54–58. Express the answer in terms of natural logarithms. Then use a calculator to obtain a decimal approximation, correct to the nearest thousandth, for the solution.

54. $8^x = 12{,}143$ **55.** $9e^{5x} = 1269$

56. $e^{12-5x} - 7 = 123$ **57.** $5^{4x+2} = 37{,}500$

58. $e^{2x} - e^x - 6 = 0$

Solve each logarithmic equation in Exercises 59–63.

59. $\log_4(3x - 5) = 3$

60. $\log_2(x + 3) + \log_2(x - 3) = 4$

61. $\log_3(x - 1) - \log_3(x + 2) = 2$

62. $\ln x = -1$

63. $3 + 4\ln 2x = 15$

64. The formula $A = 10.1e^{0.005t}$ models the population of Los Angeles, California, in millions, t years after 1992. If the growth rate continues into the future, when will the population reach 13 million?

65. The amount of carbon dioxide in the atmosphere, measured in parts per million, has been increasing as a result of the burning of oil and coal. The buildup of gases and particles trap heat and raise the planet's temperature, a phenomenon called the greenhouse effect. Carbon dioxide accounts for about half of the warming. The formula $A = 364(1.005)^t$ projects carbon dioxide concentration, A, in parts per million, t years after 2000. Using the projections given by the formula, when will the carbon dioxide concentration be double the preindustrial level of 280 parts per million?

66. The formula $C = 15{,}557 + 5259\ln x$ models the average cost of a new car x years after 1989. When will the average cost of a new car be $30,000?

67. Use the formula for compound interest with n compoundings each year to solve this problem. How long, to the nearest tenth of a year, will it take $12,500 to grow to $20,000 at 6.5% annual interest compounded quarterly?

Use the formula for continuous compounding to solve Exercises 68–69.

68. How long, to the nearest tenth of a year, will it take $50,000 to triple in value at 7.5% annual interest compounded continuously?

69. What interest rate is required for an investment subject to continuous compounding to triple in 5 years?

4.5

70. According to the U.S. Bureau of the Census, in 1980 there were 14.6 million residents of Hispanic origin living in the United States. By 1997, the number had increased to 29.3 million. The exponential growth function $A = 14.6e^{kt}$ describes the U.S. Hispanic population, in millions, t years after 1980.

a. Find k, correct to three decimal places.

b. Use the resulting model to project the Hispanic resident population in 2005.

c. In what year will the Hispanic resident population reach 50 million?

71. Use the exponential decay model for carbon-14, $A = A_0e^{-0.000121t}$, to solve this exercise. Prehistoric cave paintings were discovered in the Lascaux cave in France. The paint contained 15% of the original carbon-14. Estimate the age of the paintings at the time of the discovery.

72. Europe's Great Plague of 1666 devastated Eyam, England. There were 261 people in the village; only 83 survived. The logistic growth function

$$f(t) = \frac{171}{1 + 18.6e^{-0.0747t}}$$

models the number of people in Eyam who were infected t days after the outbreak. (Source: Raggett, G. "Modeling the Eyam Plague." *The Institute of Mathematics and Its Application* 18:221–226.)

a. How many people were infected when the outbreak began?

b. How many people were infected after 45 days?

c. According to the model, what is the limiting size of Eyam's population that can become infected? With 83 survivors among 261 people, does this mean that the size of the infected population surpassed the limit set by the model? Explain your answer.

In Exercises 73–74, rewrite the equation in terms of base e. Express the answer in terms of a natural logarithm, and then round to three decimal places.

73. $y = 73(2.6)^x$

74. $y = 6.5(0.43)^x$

75. The figure shows world population projections through the year 2150. The data are from the United Nations Family Planning Program and are based on optimistic or pessimistic expectations for successful control of human population growth. Suppose that you are interested in modeling these data using exponential, logarithmic, linear, and quadratic functions. Which function would you use to model each of the projections? Explain your choices. For the choice corresponding to a quadratic model, would your formula involve one with a positive or negative leading coefficient? Explain.

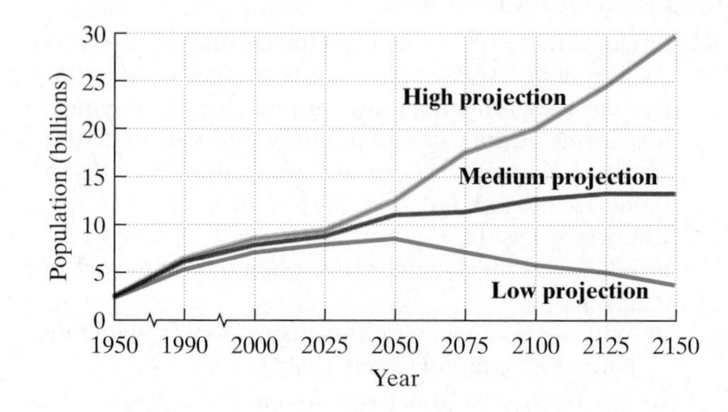

76. The figure shows the number of people in the United States age 65 and over, with projected figures for the year 2000 and beyond.

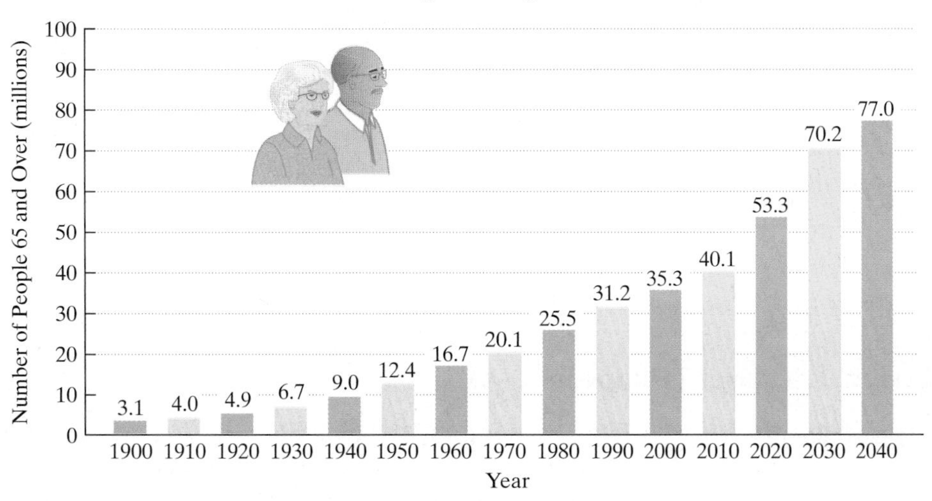

Source: U. S. Bureau of the Census

Let x represent the number of years since 1899 and let y represent the U.S. population, in millions. Use your graphing utility to find the model that best fits the data in the bar graph. Then use the model to find the U.S. population age 65 and over in 2050.

Chapter 4 Test

1. Graph $f(x) = 2^x$ and $g(x) = 2^{x+1}$ in the same rectangular coordinate system.

2. Graph $f(x) = \log_2 x$ and $g(x) = \log_2(x - 1)$ in the same rectangular coordinate system.

3. Write in exponential form: $\log_5 125 = 3$.

4. Write in logarithmic form: $\sqrt{36} = 6$.

5. Find the domain of $f(x) = \ln(3 - x)$.

In Exercises 6–7, use properties of logarithms to expand each logarithmic expression as much as possible. Where possible, evaluate logarithmic expressions without using a calculator.

6. $\log_4(64x^5)$ **7.** $\log_3 \dfrac{\sqrt[3]{x}}{81}$

In Exercises 8–9, write each expression as a single logarithm.

8. $6 \log x + 2 \log y$ **9.** $\ln 7 - 3 \ln x$

10. Use a calculator to evaluate $\log_{15} 71$ to four decimal places.

In Exercises 11–16, solve each equation.

11. $5^x = 1.4$ **12.** $400e^{0.005x} = 1600$

13. $e^{2x} - 6e^x + 5 = 0$ **14.** $\log_6(4x - 1) = 3$

15. $\log x + \log(x + 15) = 2$ **16.** $2 \ln 3x = 8$

17. Suppose you have \$3000 to invest. Which investment yields the greater return over 10 years: 6.5% compounded semianually or 6% compounded continuously? How much more (to the nearest dollar) is yielded by the better investment?

18. On the decibel scale, the loudness of a sound, in decibels, is given by $D = 10 \log \dfrac{I}{I_0}$, where I is the intensity of the sound, in watts per meter^2, and I_0 is the intensity of a sound barely audible to the human ear. If the intensity of a sound is $10^{12} I_0$, what is its loudness in decibels? (Such a sound is potentially damaging to the ear.)

19. The percentage of married men in the United States who are employed is modeled by $P = 89.18e^{-0.004t}$. The model indicates that P% of married men were employed t years after 1959.

a. What percentage of married men were employed in 1959?

b. Is the percentage of married men who are employed increasing or decreasing? Explain.

c. In what year were 77% of U.S. married men employed?

20. The 1980 population of Europe was 484 million; in 1990, it was 509 million. Write the exponential growth function that describes the population of Europe, in millions, t years after 1980.

21. Use the exponential decay model for carbon-14, $A = A_0e^{-0.000121t}$, to solve this exercise. Bones of a prehistoric man were discovered and contained 5% of the original amount of carbon-14. How long ago did the man die?

22. The logistic growth function

$$f(t) = \frac{140}{1 + 9e^{-0.165t}}$$

describes the population of an endangered species of elk t years after they were introduced to a nonthreatening habitat.

a. How many elk were initially introduced to the habitat?
b. How many elk are expected in the habitat after 10 years?
c. What is the limiting size of the elk population that the habitat will sustain?

Cumulative Review Exercises (Chapters 1–4)

Solve each equation in Exercises 1–5.

1. $|3x - 4| = 2$

2. $\sqrt{2x - 5} - \sqrt{x - 3} = 1$

3. $x^4 + x^3 - 3x^2 - x + 2 = 0$

4. $e^{5x} - 32 = 96$

5. $\log_2(x + 5) + \log_2(x - 1) = 4$

Solve each inequality in Exercises 6–7. Express the answer in interval notation.

6. $14 - 5x \geq -6$ **7.** $|2x - 4| \leq 2$

8. Write the point-slope form and the slope-intercept form of the line passing through $(1, 3)$ and $(3, -3)$.

9. If $f(x) = x^2$ and $g(x) = x + 2$, find $(f \circ g)(x)$ and $(g \circ f)(x)$.

10. If $f(x) = 2x - 7$, find $f^{-1}(x)$.

11. Divide $x^3 + 5x^2 + 3x - 10$ by $x + 2$.

12. Use the Rational Zero Theorem to list all possible rational zeros for $f(x) = 4x^3 - 7x - 3$.

13. The value of y varies directly as the square of x. If $x = 3$ when $y = 12$, find y when $x = 15$.

14. Solve $x^3 - 4x^2 + 6x - 4 = 0$ given that $1 + i$ is a root.

In Exercises 15–18, graph each equation.

15. $(x - 3)^2 + (y + 2)^2 = 4$ **16.** $f(x) = (x - 2)^2 - 1$

17. $f(x) = \dfrac{x^2 - 1}{x^2 - 4}$ **18.** $f(x) = (x - 2)^2(x + 1)$

19. You are paid time-and-a-half for each hour worked over 40 hours a week. Last week you worked 50 hours and earned $660. What is your normal hourly salary?

20. The formula $F = 1 - k\ln(t + 1)$ models the fraction of people, F, who remember all the words in a list of nonsense words t hours after memorizing the list. After 3 hours only half the people could remember all the words. Determine the value of k and then predict the fraction of people in the group who will remember all the words after 6 hours.

Systems of Equations and Inequalities

Chapter 5

Most things in life depend on many variables. Temperature and precipitation are two variables that have a critical effect on whether regions are forests, grasslands, or deserts. Airlines deal with numerous variables during weather disruptions at large connecting airports. They must solve the problem of putting their operation back together again to minimize the cost of the disruption and passenger inconvenience. In this chapter, forests, grasslands, and airline service are viewed in the same way—situations with several variables. You will learn methods for modeling and solving problems in these situations.

A major weather disruption delayed your flight for hours, but you finally made it. You are in Yosemite National Park in California, surrounded by evergreen forests, alpine meadows, and sheer walls of granite. Soaring cliffs, plunging waterfalls, gigantic trees, rugged canyons, mountains and valleys stand in stark contrast to the angry chaos at the airport. This is so different from where you live and attend college, a region in which grasslands predominate.

SECTION 5.1 Systems of Linear Equations in Two Variables

Objectives

1. Decide whether an ordered pair is a solution of a linear system.
2. Solve linear systems by substitution.
3. Solve linear systems by addition.
4. Identify systems that do not have exactly one ordered-pair solution.
5. Solve problems using systems of linear equations.

Key West residents Brian Goss (left), George Wallace, and Michael Mooney (right) hold on to each other as they battle 90 mph winds along Houseboat Row in Key West, Fla., on Friday, Sept. 25, 1998. The three had sought shelter behind a Key West hotel as Hurricane Georges descended on the Florida Keys but were forced to seek other shelter when the storm conditions became too rough. Hundreds of people were killed by the storm when it swept through the Caribbean.

Real-world problems often involve solving thousands of equations, sometimes containing a million variables. Problems ranging from scheduling airline flights to controlling traffic flow to routing phone calls over the nation's communication network often require solutions in a matter of moments. AT&T's domestic long distance network involves 800,000 variables! Meteorologists describing atmospheric conditions surrounding a hurricane must solve problems involving thousands of equations rapidly and efficiently. The difference between a two-hour warning and a two-day warning is a life-and-death issue for thousands of people in the path of one of nature's most destructive forces.

Although we will not be solving 800,000 equations with 800,000 variables, we will turn our attention to two equations with two variables, such as

$$2x - 3y = -4$$
$$2x + y = 4.$$

The methods that we consider for solving such problems provide the foundation for solving far more complex systems with many variables.

1 Decide whether an ordered pair is a solution of a linear system.

Systems of Linear Equations and Their Solutions

We have seen that all equations in the form $Ax + By = C$ are straight lines when graphed. Two such equations, such as those listed above, are called a **system of linear equations**. A **solution to a system of linear equations** is an ordered pair that satisfies all equations in the system. For example, (3, 4) satisfies the system

$$x + y = 7 \quad (3 + 4 \text{ is, indeed, } 7.)$$
$$x - y = -1 \quad (3 - 4 \text{ is, indeed, } -1.)$$

Thus, (3, 4) satisfies both equations and is a solution of the system. The solution can be described by saying that $x = 3$ and $y = 4$. The solution can also be described using set notation. The solution set to the system is $\{(3, 4)\}$—that is, the set consisting of the ordered pair (3, 4).

A system of linear equations can have exactly one solution, no solution, or infinitely many solutions. We will focus on systems with exactly one solution.

EXAMPLE 1 Determining Whether an Ordered Pair Is a Solution of a System

Determine whether (4, −1) is a solution of the system

$$x + 2y = 2$$
$$x - 2y = 6.$$

Solution Because 4 is the x-coordinate and −1 is the y-coordinate of (4, −1), we replace x by 4 and y by −1.

$$\begin{aligned} x + 2y &= 2 \\ 4 + 2(-1) &\stackrel{?}{=} 2 \\ 4 + (-2) &\stackrel{?}{=} 2 \\ 2 &= 2 \quad \text{true} \end{aligned} \qquad \begin{aligned} x - 2y &= 6 \\ 4 - 2(-1) &\stackrel{?}{=} 6 \\ 4 - (-2) &\stackrel{?}{=} 6 \\ 4 + 2 &\stackrel{?}{=} 6 \\ 6 &= 6 \quad \text{true} \end{aligned}$$

The pair (4, −1) satisfies both equations: It makes each equation true. Thus, the pair is a solution of the system. The solution set to the system is $\{(4, -1)\}$.

The solution to a system of linear equations can be found by graphing both of the equations in the same rectangular coordinate system. For a system with one solution, the **coordinates of the point of intersection give the system's solution**. For example, the system in Example 1 is graphed in Figure 5.1. The solution of the system, (4, −1), corresponds to the point of intersection of the lines.

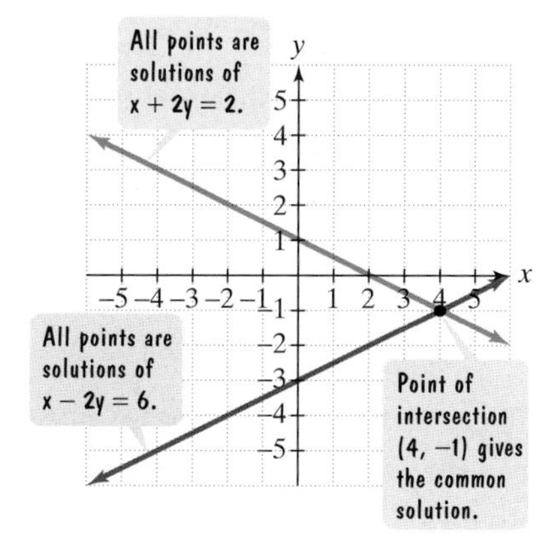

Figure 5.1 Visualizing a system's solution.

Check Point 1 Determine whether (1, 2) is a solution of the system

$$2x - 3y = -4$$
$$2x + y = 4.$$

2 Solve linear systems by substitution.

Eliminating a Variable Using the Substitution Method

Finding the solution to a linear system by graphing equations may not be easy to do. For example, a solution of $\left(-\frac{2}{3}, \frac{157}{29}\right)$ would be difficult to "see" as an intersection point on a graph.

Let's consider a method that does not depend on finding a system's solution visually: the substitution method. This method involves converting the system to one equation in one variable by an appropriate substitution.

EXAMPLE 2 Solving a System by Substitution

Solve by the substitution method:

$$y = -x - 1$$
$$4x - 3y = 24.$$

Solution

Step 1 Solve either of the equations for one variable in terms of the other. This step has already been done for us. The first equation, $y = -x - 1$, has y solved in terms of x.

Step 2 Substitute the expression from step 1 into the other equation. We substitute the expression $-x - 1$ for y in the other equation:

$$y = \boxed{-x - 1} \qquad 4x - 3\boxed{y} = 24 \quad \text{Substitute } -x - 1 \text{ for } y.$$

This gives us an equation in one variable, namely

$$4x - 3(-x - 1) = 24.$$

The variable y has been eliminated.

Step 3 Solve the resulting equation containing one variable.

$$4x - 3(-x - 1) = 24 \quad \text{This is the equation containing one variable.}$$
$$4x + 3x + 3 = 24 \quad \text{Apply the distributive property.}$$
$$7x + 3 = 24 \quad \text{Combine like terms.}$$
$$7x = 21 \quad \text{Subtract 3 from both sides.}$$
$$x = 3 \quad \text{Divide both sides by 7.}$$

Step 4 Back-substitute the obtained value into the equation from step 1. We now know that the x-coordinate of the solution is 3. To find the y-coordinate, we back-substitute the x-value into the equation from step 1,

$$y = -x - 1.$$

Substitute 3 for x.

$$y = -3 - 1 = -4$$

With $x = 3$ and $y = -4$, the proposed solution is $(3, -4)$.

Step 5 Check the proposed solution in both of the system's given equations. Replace x with 3 and y with -4.

$$y = -x - 1 \qquad\qquad 4x - 3y = 24$$
$$-4 \stackrel{?}{=} -3 - 1 \qquad\qquad 4(3) - 3(-4) \stackrel{?}{=} 24$$
$$-4 = -4 \text{ true} \qquad\qquad 12 + 12 \stackrel{?}{=} 24$$
$$24 = 24 \text{ true}$$

Technology

A graphing utility can be used to solve the system in Example 2. Graph each equation and use the intersection feature. The utility displays the solution $(3, -4)$ as $x = 3$, $y = -4$.

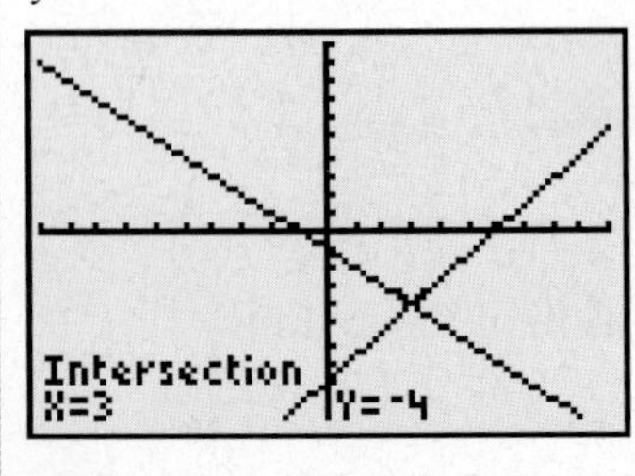

The pair $(3, -4)$ satisfies both equations. The system's solution set is $\{(3, -4)\}$.

Check Point 2 Solve by the substitution method:

$$\begin{aligned} y &= 5x - 13 \\ 2x + 3y &= 12. \end{aligned}$$

Before considering additional examples, let's summarize the steps used in the substitution method.

Study Tip

In step 1, if possible, solve for a variable whose coefficient is 1 or -1 to avoid working with fractions.

Solving Linear Systems by Substitution

1. Solve either of the equations for one variable in terms of the other. (If one of the equations is already in this form, you can skip this step.)
2. Substitute the expression found in step 1 into the other equation. This will result in an equation in one variable.
3. Solve the equation obtained in step 2.
4. Back-substitute the value found in step 3 into the equation from step 1. Simplify and find the value of the remaining variable.
5. Check the proposed solution in both of the system's given equations.

EXAMPLE 3 Solving a System by Substitution

Solve by the substitution method:

$$\begin{aligned} 5x - 4y &= 9 \\ x - 2y &= -3. \end{aligned}$$

Solution

Step 1 Solve either of the equations for one variable in terms of the other. We begin by isolating one of the variables in either of the equations. By solving for x in the second equation, which has a coefficient of 1, we can avoid fractions.

$x - 2y = -3$ This is the second equation in the given system.

$x = 2y - 3$ Solve for x by adding 2y to both sides.

Step 2 Substitute the expression from step 1 into the other equation. We substitute $2y - 3$ for x in the first equation.

$$x = \boxed{2y - 3} \qquad 5\boxed{x} - 4y = 9$$

This gives us an equation in one variable, namely

$$5(2y - 3) - 4y = 9.$$

The variable x has been eliminated.

Step 3 Solve the resulting equation containing one variable.

$5(2y - 3) - 4y = 9$ This is the equation containing one variable.

$10y - 15 - 4y = 9$ Apply the distributive property.

$6y - 15 = 9$ Combine like terms.

$6y = 24$ Add 15 to both sides.

$y = 4$ Divide both sides by 6.

Step 4 Back-substitute the obtained value into the equation from step 1. Now that we have the y-coordinate of the solution, we back-substitute 4 for y in the equation $x = 2y - 3$.

$$x = 2y - 3 \quad \text{Use the equation obtained in step 1.}$$
$$x = 2(4) - 3 \quad \text{Substitute 4 for } y.$$
$$x = 8 - 3 \quad \text{Multiply.}$$
$$x = 5 \quad \text{Subtract.}$$

With $x = 5$ and $y = 4$, the proposed solution is $(5, 4)$.

Step 5 Check. Take a moment to show that $(5, 4)$ satisfies both given equations. The solution set is $\{(5, 4)\}$.

Study Tip

Get into the habit of checking ordered-pair solutions in *both* equations of the system.

Check Point 3 Solve by the substitution method:

$$3x + 2y = -1$$
$$x - y = 3.$$

3 Solve linear systems by addition.

Eliminating a Variable Using the Addition Method

The substitution method is most useful if one of the given equations has an isolated variable. A second, and frequently the easiest, method for solving a linear system is the addition method. Like the substitution method, the addition method involves eliminating a variable and ultimately solving an equation containing only one variable. However, this time we eliminate a variable by adding the equations.

For example, consider the following equations:

$$3x - 4y = 11$$
$$-3x + 2y = -7.$$

When we add these two equations, the x-terms are eliminated. This occurs because the coefficients of the x-terms, 3 and -3, are opposites (additive inverses) of each other:

$$3x - 4y = 11$$
$$\underline{-3x + 2y = -7}$$
$$\text{Add:} \quad -2y = 4$$
$$y = -2 \quad \text{Solve for } y\text{, dividing both sides by } -2.$$

Now we can back-substitute -2 for y into one of the original equations to find x. It does not matter which equation you use; you will obtain the same value for x in either case. If we use either equation, we can show that $x = 1$ and the solution $(1, -2)$ satisfies both equations in the system.

When we use the addition method, we want to obtain two equations whose sum is an equation containing only one variable. The key step is to obtain, for one of the variables, coefficients that differ only in sign. In order to do this, we may need to multiply one or both equations by some nonzero number so that the coefficients of one of the variables, x or y, become opposites. Then when the two equations are added, this variable is eliminated. Let's see exactly how this works by considering Example 4.

EXAMPLE 4 Solving a System by the Addition Method

Solve by the addition method:

$$3x + 2y = 48$$
$$9x - 8y = -24.$$

Solution We must rewrite one or both equations in equivalent forms so that the coefficients of the same variable (either x or y) are opposites of each other. Consider the terms in x in each equation, that is, $3x$ and $9x$. To eliminate x, we can multiply each term of the first equation by -3 and then add the equations.

$$3x + 2y = 48 \xrightarrow{\text{Multiply by } -3.} -9x - 6y = -144$$
$$9x - 8y = -24 \xrightarrow{\text{No change}} 9x - 8y = -24$$
$$\text{Add:} \quad -14y = -168$$
$$y = 12 \quad \text{Solve for } y\text{, dividing both sides by } -14.$$

Thus, $y = 12$. We back-substitute this value into either one of the given equations. We'll use the first one.

$3x + 2y = 48$	This the first equation in the given system.
$3x + 2(12) = 48$	Substitute 12 for y.
$3x + 24 = 48$	Multiply.
$3x = 24$	Subtract 24 from both sides.
$x = 8$	Divide both sides by 3.

The solution $(8, 12)$ can be shown to satisfy both equations in the system. Consequently, the solution set is $\{(8, 12)\}$.

Solving Linear Systems by Addition

1. If necessary, rewrite both equations in the form $Ax + By = C$.
2. If necessary, multiply either equation or both equations by appropriate nonzero numbers so that the sum of the x-coefficients or the sum of the y-coefficients is 0.
3. Add the equations in step 2. The sum is an equation in one variable.
4. Solve the equation from step 3.
5. Back-substitute the value obtained in step 4 into either of the given equations and solve for the other variable.
6. Check the solution in both of the original equations.

Check Point 4 Solve by the addition method:

$$4x + 5y = 3$$
$$2x - 3y = 7.$$

Some linear systems have solutions that are not integers. If the value of one variable turns out to be a "messy" fraction, back-substitution might lead to cumbersome arithmetic. If this happens, you can return to the original system and use addition to find the value of the other variable.

EXAMPLE 5 Solving a System by the Addition Method

Solve by the addition method:

$$2x = 7y - 17$$
$$5y = 17 - 3x.$$

Solution

Step 1 Rewrite both equations in the form $Ax + By = C$. We first arrange the system so that variable terms appear on the left and constants appear on the right. We obtain

$$2x - 7y = -17 \quad \text{Subtract } 7y \text{ from both sides of the first equation.}$$
$$3x + 5y = 17 \quad \text{Add } 3x \text{ to both sides of the second equation.}$$

Step 2 If necessary, multiply either equation or both equations by appropriate numbers so that the sum of the *x*-coefficients or the sum of the *y*-coefficients is 0. We can eliminate x or y. Let's eliminate x by multiplying the first equation by 3 and the second equation by -2.

$$2x - 7y = -17 \xrightarrow{\text{Multiply by 3.}} 3 \cdot 2x - 3 \cdot 7y = 3(-17) \longrightarrow 6x - 21y = -51$$
$$3x + 5y = 17 \xrightarrow{\text{Multiply by } -2.} -2 \cdot 3x + (-2) \cdot 5y = -2(17) \longrightarrow -6x - 10y = -34$$

Steps 3 and 4 Add the equations and solve for the remaining variable.

$$\begin{aligned} 6x - 21y &= -51 \\ -6x - 10y &= -34 \\ \text{Add:} \quad -31y &= -85 \\ \frac{-31y}{-31} &= \frac{-85}{-31} \quad \text{Divide both sides by } -31. \\ y &= \frac{85}{31} \quad \text{Simplify.} \end{aligned}$$

Step 5 Back-substitute and find the value for the other variable. Back-substitution of $\frac{85}{31}$ for y into either of the given equations results in cumbersome arithmetic. Instead, let's use the addition method on the given system in the form $Ax + By = C$ to find the value for x. Thus, we eliminate y by multiplying the first equation by 5 and the second equation by 7.

$$2x - 7y = -17 \xrightarrow{\text{Multiply by 5.}} 10x - 35y = -85$$
$$3x + 5y = 17 \xrightarrow{\text{Multiply by 7.}} 21x + 35y = 119$$
$$\text{Add:} \quad 31x = 34$$
$$x = \frac{34}{31} \quad \text{Divide both sides by 31.}$$

Step 6 Check. For this system, a calculator is helpful in showing the solution $\left(\frac{34}{31}, \frac{85}{31}\right)$ satisfies both equations. Consequently, the solution set is $\left\{\left(\frac{34}{31}, \frac{85}{31}\right)\right\}$.

Check Point 5 Solve by the addition method:

$$4x = 5 + 2y$$
$$3y = 4 - 2x.$$

4 Identify systems that do not have exactly one ordered-pair solution.

Linear Systems Having No Solution or Infinitely Many Solutions

We have seen that a system of linear equations in two variables represents a pair of lines. The lines either intersect, are parallel, or are identical. Thus, there are three possibilities for the number of solutions to a system of two linear equations.

The Number of Solutions to a System of Two Linear Equations

The number of solutions to a system of two linear equations in two variables is given by one of the following. (See Figure 5.2.)

Number of Solutions	What This Means Graphically
Exactly one ordered-pair solution	The two lines intersect at one point.
No solution	The two lines are parallel.
Infinitely many solutions	The two lines are identical.

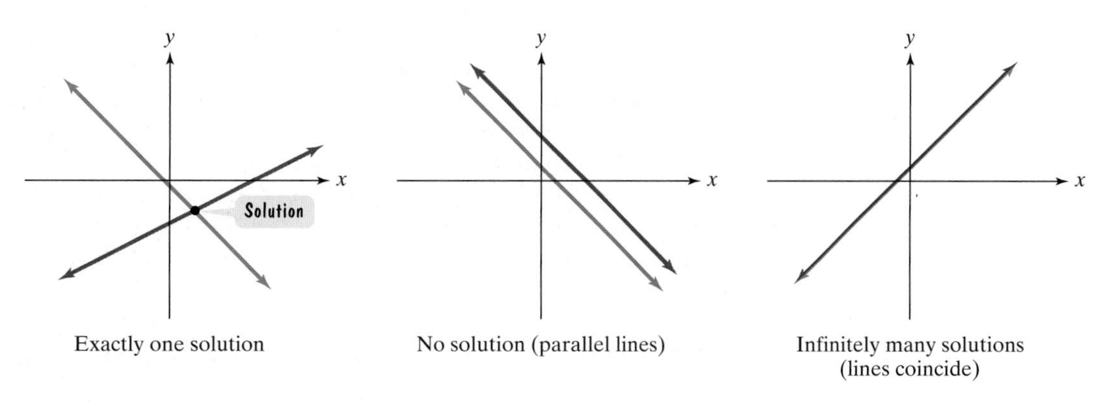

Figure 5.2 Possible graphs for a system of two linear equations in two variables

A linear system with no solution is called an **inconsistent system**. If you attempt to solve such a system by substitution or addition, you will eliminate both variables. A false statement such as $0 = 17$ will be the result.

EXAMPLE 6 A System with No Solution

Solve the system:

$$4x + 6y = 12.$$
$$6x + 9y = 12.$$

Solution Because no variable is isolated, we will use the addition method. To obtain coefficients of x that differ only in sign, we multiply the first equation by 3 and multiply the second equation by -2.

$$\begin{aligned} 4x + 6y &= 12 \xrightarrow{\text{Multiply by 3.}} & 12x + 18y &= 36 \\ 6x + 9y &= 12 \xrightarrow{\text{Multiply by } -2.} & -12x - 18y &= -24 \\ & \text{Add:} & 0 &= 12 \end{aligned}$$

There are no values of x and y for which $0 = 12$.

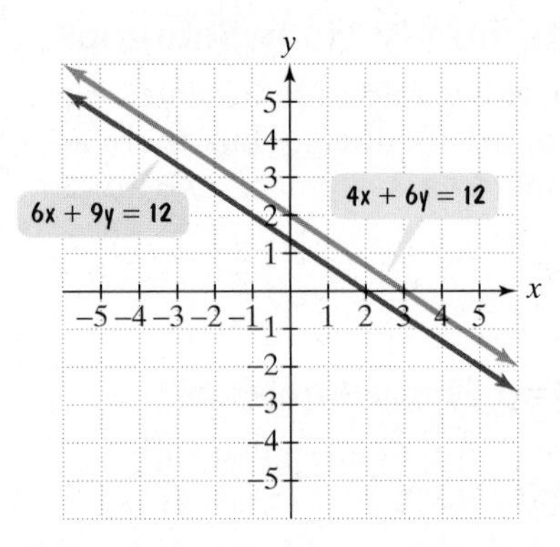

Figure 5.3 The graph of an inconsistent system

The false statement $0 = 12$ indicates that the system is inconsistent and has no solution. The solution set is the empty set, $\varnothing$.

The lines corresponding to the two equations in Example 6 are shown in Figure 5.3. The lines are parallel and have no point of intersection.

Discovery

Show that the graphs of $4x + 6y = 12$ and $6x + 9y = 12$ must be parallel lines by solving each equation for y. What is the slope and y-intercept for each line? What does this mean? If a linear system is inconsistent, what must be true about the slopes and y-intercepts for the system's graphs?

Check Point 6 Solve the system:

$$x + 2y = 4$$
$$3x + 6y = 13.$$

A linear system that has at least one solution is called a **consistent system**. Lines that intersect and lines that coincide both represent consistent systems. If the lines coincide, then the consistent system has infinitely many solutions, represented by every point on the line.

The equations in a linear system with infinitely many solutions are called **dependent**. If you attempt to solve such a system by substitution or addition, you will eliminate both variables. However, a true statement such as $0 = 0$ will be the result.

EXAMPLE 7 A System with Infinitely Many Solutions

Solve the system:

$$y = 3 - 2x$$
$$4x + 2y = 6.$$

Solution Because the variable y is isolated in the first equation, we can use the substitution method. We substitute the expression for y in the other equation.

$y = \boxed{3 - 2x} \quad 4x + 2\boxed{y} = 6$ Substitute $3 - 2x$ for y.

$4x + 2y = 6$ This is the second equation in the given system.

$4x + 2(3 - 2x) = 6$ Substitute $3 - 2x$ for y.

$4x + 6 - 4x = 6$ Apply the distributive property.

$6 = 6$ Simplify. This statement is true for all values of x and y.

In our final step, both variables have been eliminated, and the resulting statement $6 = 6$ is true. This true statement indicates that the system has infinitely many solutions. The solution set consists of all points (x, y) lying on the line $y = 3 - 2x$, as shown in Figure 5.4.

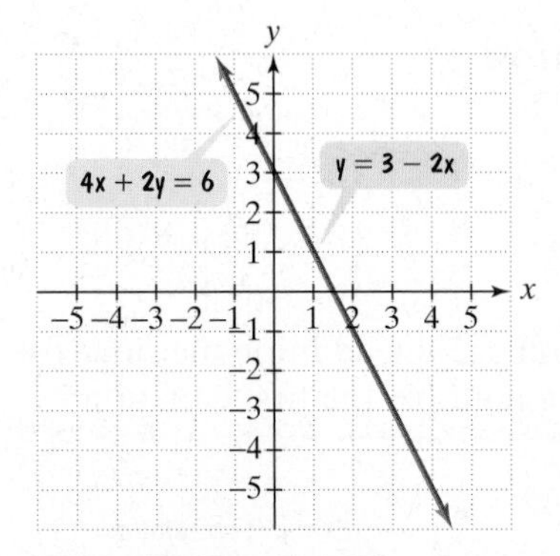

Figure 5.4 The graph of a system with infinitely many solutions

We express the solution set for the system in one of two equivalent ways:

$\{(x, y) | y = 3 - 2x\}$ The set of all ordered pairs (x, y) such that $y = 3 - 2x$.

or $\{(x, y) | 4x + 2y = 6\}$ The set of all ordered pairs (x, y) such that $4x + 2y = 6$.

Check Point 7 Solve the system:

$$y = 4x - 4$$
$$8x - 2y = 8.$$

5 Solve problems using systems of linear equations.

Applications

An important application of systems of equations arises in connection with supply and demand. As the price of a product increases, the demand for that product decreases. However, at higher prices suppliers are willing to produce greater quantities of the product.

EXAMPLE 8 Supply and Demand Models

A chain of video stores specializes in cult films. The weekly demand and supply models for *The Rocky Horror Picture Show* are given by

$$N = -13p + 760 \quad \text{Demand model}$$
$$N = 2p + 430 \quad \text{Supply model}$$

in which p is the price of the video and N is the number of copies of the video sold or supplied each week to the chain of stores.

a. How many copies of the video can be sold and supplied at $18 per copy?

b. Find the price at which supply and demand are equal. At this price, how many copies of *Rocky Horror* can be supplied and sold each week?

Solution

a. To find how many copies of the video can be sold and supplied at $18 per copy, we substitute 18 for p in the demand and supply models.

Demand Model	**Supply Model**
$N = -13p + 760$	$N = 2p + 430$
Substitute 18 for p.	Substitute 18 for p.
$N = -13 \cdot 18 + 760 = 526$	$N = 2 \cdot 18 + 430 = 466$

At $18 per video, the chain can sell 526 copies of *Rocky Horror* in a week. The manufacturer is willing to supply 466 copies per week. This will result in a shortage of copies of the video. Under these conditions, the retail chain is likely to raise the price of the video.

b. We can find the price at which supply and demand are equal by solving the demand-supply linear system. We will use substitution, substituting $-13p + 760$ for N in the second equation.

$$N = \boxed{-13p + 760} \qquad \boxed{N} = 2p + 430$$ Substitute $-13p + 760$ for N.

$$-13p + 760 = 2p + 430$$ The resulting equation contains only one variable.

$$-15p + 760 = 430$$ Subtract $2p$ from both sides.

$$-15p = -330$$ Subtract 760 from both sides.

$$p = 22$$ Divide both sides by -15.

The price at which supply and demand are equal is \$22 per video. To find the value of N, the number of videos supplied and sold weekly at this price, we back-substitute 22 for p into either the demand or the supply model. We'll use both models to make sure we get the same number in each case.

Demand Model

$N = -13p + 760$

Substitute 22 for p.

$N = -13 \cdot 22 + 760 = 474$

Supply Model

$N = 2p + 430$

Substitute 22 for p.

$N = 2 \cdot 22 + 430 = 474$

At a price of \$22, 474 units of the video can be supplied and sold weekly. The intersection point, (22, 474), is shown in Figure 5.5.

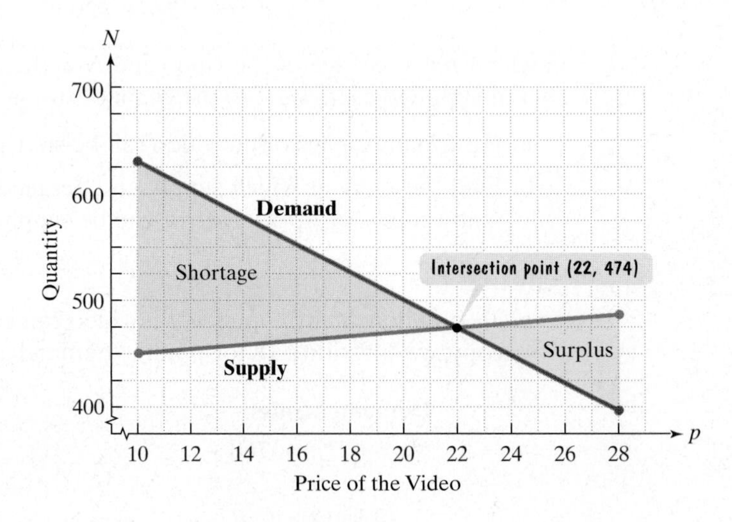

Figure 5.5 Priced at \$22, 474 copies of the video can be supplied and sold weekly.

Check Point 8

The demand for a product is modeled by $N = -20p + 1000$ and the supply for the product by $N = 5p + 250$. In these models, p is the price of the product and N is the number supplied or sold weekly. At what price will supply equal demand? At that price, how many units of the product will be supplied and sold each week?

EXERCISE SET 5.1

Practice Exercises

In Exercises 1–4, determine whether the given ordered pair is a solution of the system.

1. $(2, 3)$
$x + 3y = 11$
$x - 5y = -13$

2. $(-3, 5)$
$9x + 7y = 8$
$8x - 9y = -69$

3. $(2, 5)$
$2x + 3y = 17$
$x + 4y = 16$

4. $(8, 5)$
$5x - 4y = 20$
$3y = 2x + 1$

In Exercises 5–16, solve each system by the substitution method.

5. $x + y = 4$
$y = 3x$

6. $x + y = 6$
$y = 2x$

7. $x + 3y = 8$
$y = 2x - 9$

8. $2x - 3y = -13$
$y = 2x + 7$

9. $x + 3y = 5$
$4x + 5y = 13$

10. $x + 2y = 5$
$2x - y = -15$

11. $2x - y = -5$
$x + 5y = 14$

12. $2x + 3y = 11$
$x - 4y = 0$

13. $2x - y = 3$
$5x - 2y = 10$

14. $-x + 3y = 10$
$2x + 8y = -6$

15. $x + 8y = 6$
$2x + 4y = -3$

16. $-4x + y = -11$
$2x - 3y = 5$

In Exercises 17–28, solve each system by the addition method.

17. $x + y = 1$
$x - y = 3$

18. $x + y = 6$
$x - y = -2$

19. $2x + 3y = 6$
$2x - 3y = 6$

20. $3x + 2y = 14$
$3x - 2y = 10$

21. $x + 2y = 2$
$-4x + 3y = 25$

22. $2x - 7y = 2$
$3x + y = -20$

23. $4x + 3y = 15$
$2x - 5y = 1$

24. $3x - 7y = 13$
$6x + 5y = 7$

25. $3x - 4y = 11$
$2x + 3y = -4$

26. $2x + 3y = -16$
$5x - 10y = 30$

27. $3x = 4y + 1$
$3y = 1 - 4x$

28. $5x = 6y + 40$
$2y = 8 - 3x$

In Exercises 29–36, solve by the method of your choice. Identify systems with no solution and systems with infinitely many solutions, using set notation to express their solution sets.

29. $x = 9 - 2y$
$x + 2y = 13$

30. $6x + 2y = 7$
$y = 2 - 3x$

31. $y = 3x - 5$
$21x - 35 = 7y$

32. $9x - 3y = 12$
$y = 3x - 4$

33. $3x - 2y = -5$
$4x + y = 8$

34. $2x + 5y = -4$
$3x - y = 11$

35. $x + 3y = 2$
$3x + 9y = 6$

36. $4x - 2y = 2$
$2x - y = 1$

In Exercises 37–40, let x represent one number and let y represent the other number. Use the given conditions to write a system of equations. Solve the system and find the numbers.

37. The sum of two numbers is 7. If one number is subtracted from the other, their difference is -1. Find the numbers.

38. The sum of two numbers is 2. If one number is subtracted from the other, their difference is 8. Find the numbers.

39. Three times a first number decreased by a second number is 1. The first number increased by twice the second number is 12. Find the numbers.

40. The sum of three times a first number and twice a second number is 8. If the second number is subtracted from twice the first number, the result is 3. Find the numbers.

Application Exercises

41. At a price of p dollars per ticket, the number of tickets to a rock concert that can be sold is given by the demand model $N = -25p + 7500$. At a price of p dollars per ticket, the number of tickets that the concert's promoters are willing to make available is given by the supply model $N = 5p + 6000$.

a. How many tickets can be sold and supplied for $40 per ticket?

b. Find the ticket price at which supply and demand are equal. At this price, how many tickets will be supplied and sold?

42. The weekly demand and supply models for a particular brand of scientific calculator for a chain of stores are given by the demand model $N = -53p + 1600$, and the supply model $N = 75p + 320$. In these models, p is the price of the calculator and N is the number of calculators sold or supplied each week to the stores.

a. How many calculators can be sold and supplied at \$12 per calculator?

b. Find the price at which supply and demand are equal. At this price, how many calculators of this type can be supplied and sold each week?

A business breaks even when the cost for running the business is equal to the money taken in by the business. In Exercises 43–44, determine how many units must be sold so that a business breaks even, experiencing neither loss nor profit.

43. A gasoline station has weekly costs and revenue (the money taken in by the station) that are functions of the number of gallons of gasoline purchased and sold. If x gallons are purchased and sold, weekly costs are given by $C(x) = 1.2x + 1080$ and weekly revenue by $R(x) = 1.6x$. How many gallons of gasoline must be sold weekly for the station to break even?

44. An artist has monthly costs and revenue (the money taken in by the artist) that are functions of the number of ceramic pieces produced and sold. If x ceramic pieces are produced and sold, monthly costs are given by $C(x) = 4x + 2000$ and monthly revenue by $R(x) = 9x$. How many ceramic pieces must be sold monthly for the artist to break even?

Use a system of linear equations to solve Exercises 45–48.

45. The verdict is in: After years of research, the nation's health experts agree that high cholesterol in the blood is a major contributor to heart disease. Cholesterol intake should be limited to 300 mg or less each day. Fast foods provide a cholesterol carnival. Two McDonald's Quarter Pounders and three Burger King Whoppers with cheese contain 520 mg of cholesterol. Three Quarter Pounders and one Whopper with cheese exceed the suggested daily cholesterol intake by 53 mg. Determine the cholesterol content in each item.

46. How do the Quarter Pounder and Whopper with cheese measure up in the calorie department? Actually, not too well. Two Quarter Pounders and three Whoppers with cheese provide 2607 calories. Even one of each provide enough calories to bring tears to Jenny Craig's eyes—9 calories in excess of what is allowed on a 1000 calorie-a-day diet. Find the caloric content of each item.

47. The graph at the top of the next column makes Super Bowl Sunday look like a day of snack food binging in the United States. The number of pounds of guacamole consumed is ten times the difference between the number of pounds of potato and tortilla chips eaten on the same day. On Super Bowl Sunday Americans also eat a total quantity of potato and tortilla chips that exceeds popcorn consumption by 7.3 million pounds. How many millions of pounds of potato chips and tortilla chips are consumed on Super Bowl Sunday?

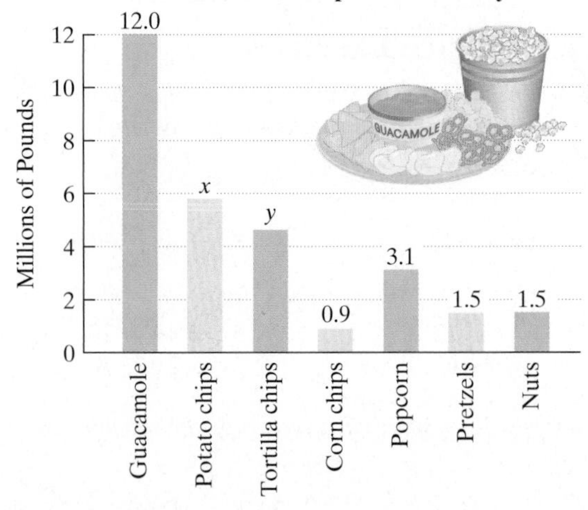

Source: Association of American Snack Foods

48. The bar graph indicates countries in which ten or more languages have become extinct. The number of extinct languages in Brazil is 7.5 times the difference between the number in the United States and Colombia. The number of extinct languages in the United States and Colombia combined exceeds the number in Australia by 24. How many languages have become extinct in the United States and Colombia?

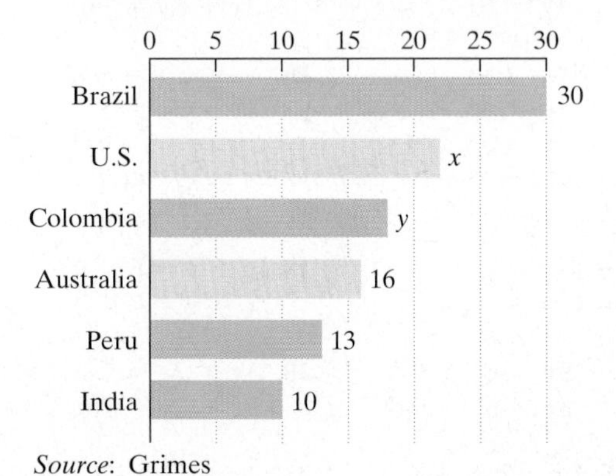

Source: Grimes

49. The June 7, 1999 issue of *Newsweek* presents statistics showing progress African Americans have made in education, health, and finance. Infant mortality for blacks is decreasing at a faster rate than it is for whites, shown by the graphs on page 429. Infant mortality for blacks can be modeled by $M = -0.41x + 22$ and for whites by $M = -0.18x + 10$. In both models, x is the number of years since 1980 and M is infant mortality, measured in deaths per 1000 live births. Use these models to project when infant mortality for blacks and whites will be the same. What is infant mortality for both groups at that time?

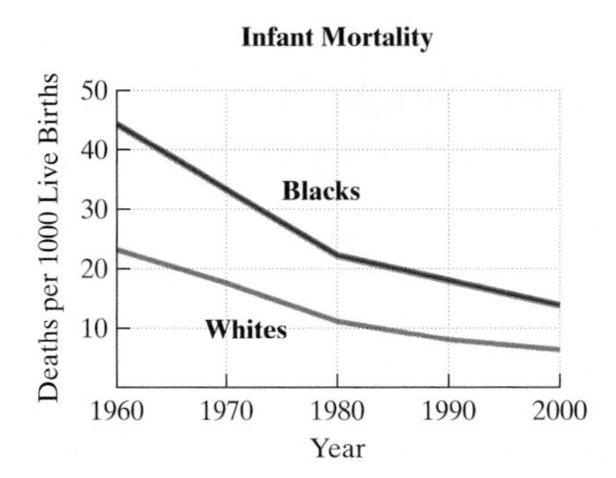

Source: National Center for Health Statistics

50. The equation $x + 10y = 2120$ models deaths from gunfire in the United States, y, in deaths per hundred thousand Americans, in year x. The equation $7x + 8y = 14{,}065$ models deaths from car accidents in the United States, y, in deaths per hundred thousand Americans, in year x. Solve the linear system formed by the two models. Then describe what the solution means in terms of the variables in the given models.

Writing in Mathematics

51. What is a system of linear equations? Provide an example with your description.

52. What is the solution to a system of linear equations?

53. Explain how to solve a system of equations using the substitution method. Use $y = 3 - 3x$ and $3x + 4y = 6$ to illustrate your explanation.

54. Explain how to solve a system of equations using the addition method. Use $3x + 5y = -2$ and $2x + 3y = 0$ to illustrate your explanation.

55. When is it easier to use the addition method rather than the substitution method when solving a system of equations?

56. When using the addition or substitution method, how can you tell if a system of linear equations has infinitely many solutions? What is the relationship between the graphs of the two equations?

57. When using the addition or substitution method, how can you tell if a system of linear equations has no solution? What is the relationship between the graphs of the two equations?

58. The law of supply and demand states that, in a free market economy, a commodity tends to be sold at its equilibrium price. At this price, the amount that the seller will supply is the same amount that the consumer will buy. Explain how systems of equations can be used to determine the equilibrium price.

59. The graphs at the top of the next column show median weekly earnings of full-time wage and salary workers 25 years and older, by education attainment. Which graphs look like they might intersect sometime after 1997? Describe how to use algebra to model the data and determine the year in which the groups might have the same weekly earnings.

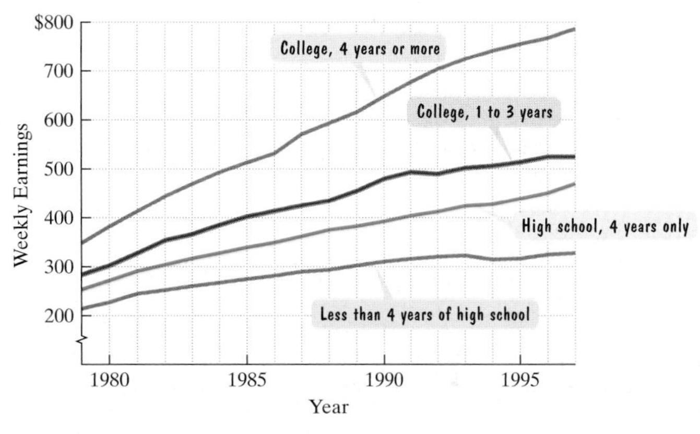

Source: U.S. Bureau of Labor Statistics

Technology Exercises

60. Verify your solutions to any five exercises from 5 through 36 by using a graphing utility to graph the two equations in the system in the same viewing rectangle. Then use the intersection feature to display the solution.

61. Some graphing utilities can give the solution to a linear system of equations. (Consult your manual for details.) This capability is usually accessed with the [SIMULT] (simultaneous equations) feature. First, you will enter 2, for two equations in two variables. With each equation in $Ax + By = C$ form, you will then enter the coefficients for x and y and the constant term, one equation at a time. After entering all six numbers, press [SOLVE]. The solution will be displayed on the screen. (The x-value may be displayed as $x_1 =$ and the y-value as $x_2 =$.) Use this capability to verify the solution to any five of the exercises you solved in the practice exercises of this exercise set. Describe what happens when you use your graphing utility on a system with no solution or infinitely many solutions.

Critical Thinking Exercises

62. Write a system of equations having $\{(-2, 7)\}$ as a solution set. (More than one system is possible.)

63. Solve the system for x and y in terms of a_1, b_1, c_1, a_2, b_2, and c_2:

$$a_1x + b_1y = c_1$$
$$a_2x + b_2y = c_2.$$

64. Two identical twins can only be recognized by the characteristic that one always tells the truth and the other always lies. One twin tells you of a lucky number pair: "When I multiply my first lucky number by 3 and my second lucky number by 6, the addition of the resulting numbers produces a sum of 12. When I add my first lucky number and twice my second lucky number, the sum is 5." Which twin is talking?

65. A marching band has 52 members, and there are 24 in the pom-pom squad. They wish to form several hexagons and squares like those diagrammed below. Can it be done with no people left over?

Hexagon with pom–pom person in center

B B

B P B

B B

Square with band member in center

P P

B

P P

B = Band Member

P = Pom-pom Person

Group Exercise

66. The group should write four different word problems that can be solved using a system of linear equations in two variables. All of the problems should be on different topics. Select from the following topics: a number problem (see Exercises 37–40); a problem using supply and demand models (see Exercises 41–42); a problem involving a business breaking even (see Exercises 43–44); a problem based on two missing numbers in a graph (see Exercises 47–48; you'll need to find an interesting graph); a problem involving linear modeling, finding the year when the quantity modeled will be the same for two groups (see Exercise 49). Of course, you can also base the problem on any topic of interest, but remember—only one problem per topic. The group should turn in the four problems and their algebraic solutions.

SECTION 5.2 Systems of Linear Equations in Three Variables

Objectives

1. Verify the solution of a linear system in three variables.
2. Solve linear systems in three variables.
3. Solve problems using systems in three variables.

All animals sleep, but the length of time they sleep varies widely: Cattle sleep for only a few minutes at a time. We humans seem to need more sleep than other animals, up to eight hours a day. Without enough sleep, we have difficulty concentrating, make mistakes in routine tasks, lose energy, and feel bad-tempered. There is a relationship between hours of sleep and death rate per year per 100,000 people. How many hours of sleep will put you in the group with the minimum death rate? In this section you will learn how to solve linear systems with more than two variables in order to answer this question.

1 Verify the solution of a linear system in three variables.

Systems of Linear Equations in Three Variables and Their Solutions

An equation such as $x + 2y - 3z = 9$ is called a **linear equation in three variables**. In general, any equation of the form

$$Ax + By + Cz = D$$

where $A, B, C,$ and D are real numbers such that $A, B,$ and C are not all 0, is a linear equation in the variables x, y, and z. The graph of this linear equation in three variables is a plane in three-dimensional space.

The process of solving a system of three linear equations in three variables is geometrically equivalent to finding the point of intersection (assuming that there is one) of three planes in space (see Figure 5.6). A **solution** to a system of linear equations in three variables is an ordered triple of real numbers that satisfies all equations of the system. The **solution set** of the system is the set of all its solutions.

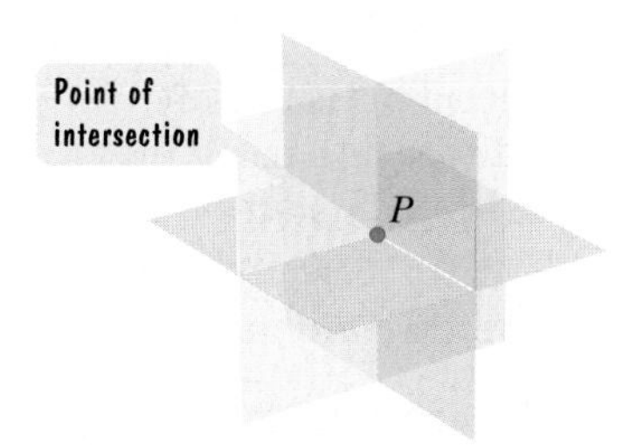

Figure 5.6

EXAMPLE 1 Determining Whether an Ordered Triple Satisfies a System

Show that the ordered triple $(-1, 2, -2)$ is a solution of the system:

$$\begin{aligned} x + 2y - 3z &= 9 \\ 2x - y + 2z &= -8 \\ -x + 3y - 4z &= 15. \end{aligned}$$

Solution Because -1 is the x-coordinate, 2 is the y-coordinate, and -2 is the z-coordinate of $(-1, 2, -2)$, we replace x by -1, y by 2, and z by -2 in each of the three equations.

$$\begin{aligned} x + 2y - 3z &= 9 \\ -1 + 2(2) - 3(-2) &\stackrel{?}{=} 9 \\ -1 + 4 + 6 &\stackrel{?}{=} 9 \\ 9 &= 9 \quad \text{true} \end{aligned}$$

$$\begin{aligned} 2x - y + 2z &= -8 \\ 2(-1) - 2 + 2(-2) &\stackrel{?}{=} -8 \\ -2 - 2 - 4 &\stackrel{?}{=} -8 \\ -8 &= -8 \quad \text{true} \end{aligned}$$

$$\begin{aligned} -x + 3y - 4z &= 15 \\ -(-1) + 3(2) - 4(-2) &\stackrel{?}{=} 15 \\ 1 + 6 + 8 &\stackrel{?}{=} 15 \\ 15 &= 15 \quad \text{true} \end{aligned}$$

The ordered triple $(-1, 2, -2)$ satisfies the three equations: It makes each equation true. Thus, the ordered triple is a solution of the system. The solution set to the system is $\{(-1, 2, -2)\}$.

Check Point 1 Show that the ordered triple $(-1, -4, 5)$ is a solution of the system:

$$\begin{aligned} x - 2y + 3z &= 22 \\ 2x - 3y - z &= 5 \\ 3x + y - 5z &= -32. \end{aligned}$$

2 Solve linear systems in three variables.

Solving Systems of Linear Equations in Three Variables by Eliminating Variables

The method for solving a system of linear equations in three variables is similar to that used on systems of linear equations in two variables. We use addition to eliminate any variable, reducing the system to two equations in two variables.

Once we obtain a system of two equations in two variables, we use addition or substitution to eliminate a variable. The result is a single equation in one variable. We solve this equation to get the value of the remaining variable. Other variable values are found by back-substitution.

Solving Linear Systems in Three Variables by Eliminating Variables

1. Reduce the system to two equations in two variables. This is usually accomplished by taking two different pairs of equations and using the addition method to eliminate the same variable from each pair.
2. Solve the resulting system of two equations in two variables using addition or substitution. The result is an equation in one variable that gives the value of that variable.
3. Back-substitute the value of the variable found in step 2 into either of the equations in two variables to find the value of the second variable.
4. Use the values of the two variables from steps 2 and 3 to find the value of the third variable by back-substituting into one of the original equations.
5. Check the proposed solution in each of the original equations.

EXAMPLE 2 Solving a System in Three Variables

Solve the system:

$$\begin{aligned} 5x - 2y - 4z &= 3 && \text{Equation 1} \\ 3x + 3y + 2z &= -3 && \text{Equation 2} \\ -2x + 5y + 3z &= 3. && \text{Equation 3} \end{aligned}$$

Solution There are many ways to proceed. Because our initial goal is to reduce the system to two equations in two variables, **the central idea is to take two different pairs of equations and eliminate the same variable from each pair.**

Step 1 Reduce the system to two equations in two variables. We choose any two equations and use the addition method to eliminate a variable. Let's eliminate z from Equations 1 and 2. We do so by multiplying Equation 2 by 2. Then we add equations.

$$\begin{aligned} &(\text{Equation 1})\ 5x - 2y - 4z = 3 \xrightarrow{\text{No change}} & 5x - 2y - 4z &= 3 \\ &(\text{Equation 2})\ 3x + 3y + 2z = -3 \xrightarrow{\text{Multiply by 2.}} & 6x + 6y + 4z &= -6 \\ &\text{Add:} & 11x + 4y \quad &= -3 \quad \text{Equation 4} \end{aligned}$$

Now we must eliminate the *same* variable from another pair of equations. We can eliminate z from Equations 2 and 3. First, we multiply Equation 2 by −3. Next, we multiply Equation 3 by 2. Finally, we add equations.

$$\begin{aligned} &(\text{Equation 2})\ 3x + 3y + 2z = -3 \xrightarrow{\text{Multiply by } -3.} & -9x - 9y - 6z &= 9 \\ &(\text{Equation 3})\ -2x + 5y + 3z = 3 \xrightarrow{\text{Multiply by 2.}} & -4x + 10y + 6z &= 6 \\ &\text{Add:} & -13x + y \quad &= 15 \quad \text{Equation 5} \end{aligned}$$

Equations 4 and 5 give us a system of two equations in two variables.

Step 2 Solve the resulting system of two equations in two variables. We will use the addition method to solve Equations 4 and 5 for x and y. To do so, we multiply Equation 5 on both sides by -4 and add this to Equation 4.

$$\begin{array}{llcl} \text{(Equation 4)} & 11x + 4y = -3 & \xrightarrow{\text{No change}} & 11x + 4y = -3 \\ \text{(Equation 5)} & -13x + y = 15 & \xrightarrow{\text{Multiply by } -4.} & 52x - 4y = -60 \\ & & \text{Add:} & 63x = -63 \\ & & & x = -1 \quad \text{Divide both sides by 63.} \end{array}$$

Step 3 Use back-substitution in one of the equations in two variables to find the value of the second variable. We back-substitute -1 for x in either Equation 4 or 5 to find the value of y.

$$\begin{aligned} -13x + y &= 15 && \text{Equation 5} \\ -13(-1) + y &= 15 && \text{Substitute } -1 \text{ for } x. \\ 13 + y &= 15 && \text{Multiply.} \\ y &= 2 && \text{Subtract 13 from both sides.} \end{aligned}$$

Step 4 Back-substitute the values found for two variables into one of the original equations to find the value of the third variable. We can now use any one of the original equations and back-substitute the values of x and y to find the value for z. We will use Equation 2.

$$\begin{aligned} 3x + 3y + 2z &= -3 && \text{Equation 2} \\ 3(-1) + 3(2) + 2z &= -3 && \text{Substitute } -1 \text{ for } x \text{ and 2 for } y. \\ 3 + 2z &= -3 && \text{Multiply and then add.} \\ 2z &= -6 && \text{Subtract 3 from both sides.} \\ z &= -3 && \text{Divide both sides by 2.} \end{aligned}$$

With $x = -1$, $y = 2$, and $z = -3$, the proposed solution is the ordered triple $(-1, 2, -3)$.

Step 5 Check. Check the proposed solution, $(-1, 2, -3)$, by substituting the values for x, y, and z into each of the three original equations. These substitutions yield three true statements. Thus, the solution set is $\{(-1, 2, -3)\}$.

Check Point 2 Solve the system:

$$\begin{aligned} x + 4y - z &= 20 \\ 3x + 2y + z &= 8 \\ 2x - 3y + 2z &= -16. \end{aligned}$$

In some examples, one of the variables is already eliminated from an original equation. In this case, the same variable should be eliminated from the other two equations, thereby making it possible to omit one of the elimination steps. We illustrate this idea in Example 3.

EXAMPLE 3 Solving a System of Equations with a Missing Term

Solve the system:

$$\begin{aligned} x + \quad\;\; z &= 8 \quad \text{Equation 1} \\ x + y + 2z &= 17 \quad \text{Equation 2} \\ x + 2y + z &= 16 \quad \text{Equation 3} \end{aligned}$$

Solution

Step 1 Reduce the system to two equations in two variables. Because Equation 1 contains only x and z, we could eliminate y from Equations 2 and 3. This will give us two equations in x and z. To eliminate y from Equations 2 and 3, we multiply Equation 2 by -2 and add Equation 3.

$$\begin{aligned} &(\text{Equation 2})\; x + y + 2z = 17 \xrightarrow{\text{Multiply by } -2.} & -2x - 2y - 4z &= -34 \\ &(\text{Equation 3})\; x + 2y + z = 16 \xrightarrow{\text{No change}} & x + 2y + z &= 16 \\ &\text{Add:} & -x \qquad\quad -3z &= -18 \quad \text{Equation 4} \end{aligned}$$

Equation 4 and the given Equation 1 provide us with a system of two equations in two variables.

Step 2 Solve the resulting system of two equations in two variables. We will solve Equations 1 and 4 for x and z.

$$\begin{aligned} x + z &= 8 \quad \text{Equation 1} \\ -x - 3z &= -18 \quad \text{Equation 4} \\ \text{Add:} \quad -2z &= -10 \\ z &= 5 \quad \text{Divide both sides by } -2. \end{aligned}$$

Step 3 Use back-substitution in one of the equations in two variables to find the value of the second variable. To find x, we back-substitute 5 for z in either Equation 1 or 4. We will use Equation 1.

$$\begin{aligned} x + z &= 8 \quad \text{Equation 1} \\ x + 5 &= 8 \quad \text{Substitute 5 for } z. \\ x &= 3 \quad \text{Subtract 5 from both sides.} \end{aligned}$$

Step 4 Back-substitute the values found for two variables into one of the original equations to find the value of the third variable. To find y, we back-substitute 3 for x and 5 for z into Equation 2 or 3. We can't use Equation 1 because y is missing in this equation. We will use Equation 2.

$$\begin{aligned} x + y + 2z &= 17 \quad \text{Equation 2} \\ 3 + y + 2(5) &= 17 \quad \text{Substitute 3 for } x \text{ and 5 for } z. \\ y + 13 &= 17 \quad \text{Multiply and add.} \\ y &= 4 \quad \text{Subtract 13 from both sides.} \end{aligned}$$

We found that $z = 5$, $x = 3$, and $y = 4$. Thus, the proposed solution is the ordered triple $(3, 4, 5)$.

Step 5 Check. Substituting 3 for x, 4 for y, and 5 for z into each of the three original equations yields three true statements. Consequently, the solution set is $\{(3, 4, 5)\}$.

Check Point 3 Solve the system:

$$\begin{aligned} 2y - z &= 7 \\ x + 2y + z &= 17 \\ 2x - 3y + 2z &= -1 \end{aligned}$$

A system of linear equations in three variables represents three planes. The three planes need not intersect at one point. The planes may have no common point of intersection and represent an inconsistent system with no solution. By contrast, the planes may coincide or intersect along a line. In these cases, the planes have infinitely many points in common and represent systems with infinitely many solutions. Systems of linear equations in three variables that are inconsistent or that contain dependent equations will be discussed in Chapter 6.

3 Solve problems using systems in three variables.

Applications

Systems of equations may allow us to find models for data without using a graphing utility. Quadratic functions of the form $y = ax^2 + bx + c$ often model situations in which values of y are decreasing and then increasing, suggesting the cuplike shape of a parabola.

EXAMPLE 4 Modeling Data Relating Sleep and Death Rate

In a study relating sleep and death rate, the following data were obtained. Use the function $y = ax^2 + bx + c$ to model the data.

x (Average Number of Hours of Sleep)	y (Death Rate per Year Per 100,000 Males)
4	1682
7	626
9	967

Solution We need to find values for a, b, and c. We can do so by solving a system of three linear equations in a, b, and c. We obtain the three equations by using the values of x and y from the data as follows:

$\mathbf{y = ax^2 + bx + c}$ — Use the quadratic function to model the data.

When $x = 4, y = 1682$: $1682 = a \cdot 4^2 + b \cdot 4 + c$ or $16a + 4b + c = 1682$

When $x = 7, y = 626$: $626 = a \cdot 7^2 + b \cdot 7 + c$ or $49a + 7b + c = 626$

When $x = 9, y = 967$: $967 = a \cdot 9^2 + b \cdot 9 + c$ or $81a + 9b + c = 967$.

The easiest way to solve this system is to eliminate c from two pairs of equations, obtaining two equations in a and b. Solving this system gives $a = 104.5$, $b = -1501.5$, and $c = 6016$. We now substitute the values for a, b, and c into $y = ax^2 + bx + c$. The function that models the given data is

$$y = 104.5x^2 - 1501.5x + 6016.$$

Discovery

Use the x-coordinate of a parabola's vertex, $x = -\frac{b}{2a}$, and the function on the right to find the hours of sleep that minimize the death rate. Round to the nearest tenth of an hour. What is the minimum death rate per year per 100,000 males?

We can use the model that we obtained in Example 4 to find the death rate of males who average, say, 6 hours of sleep. Substitute 6 for x:

$$y = 104.5(6)^2 - 1501.5(6) + 6016 = 769.$$

According to the model, the death rate for males who average 6 hours of sleep is 769 deaths per 100,000 males.

Check Point 4 Find the quadratic function $y = ax^2 + bx + c$ whose graph passes through the points $(1, 4)$, $(2, 1)$, and $(3, 4)$.

EXERCISE SET 5.2

Practice Exercises

In Exercises 1–4, determine if the given ordered triple is a solution of the system.

1. $\begin{aligned} x + y + z &= 4 \\ x - 2y - z &= 1 \\ 2x - y - 2 &= -1 \end{aligned}$ $(2, -1, 3)$

2. $\begin{aligned} x + y + z &= 0 \\ x + 2y - 3z &= 5 \\ 3x + 4y + 2z &= -1 \end{aligned}$ $(5, -3, -2)$

3. $\begin{aligned} x - 2y &= 2 \\ 2x + 3y &= 11 \\ y - 4z &= -7 \end{aligned}$ $(4, 1, 2)$

4. $\begin{aligned} x - 2z &= -5 \\ y - 3z &= -3 \\ 2x - z &= -4 \end{aligned}$ $(-1, 3, 2)$

Solve each system in Exercises 5–18.

5. $\begin{aligned} x + y + 2z &= 11 \\ x + y + 3z &= 14 \\ x + 2y - z &= 5 \end{aligned}$

6. $\begin{aligned} 2x + y - 2z &= -1 \\ 3x - 3y - z &= 5 \\ x - 2y + 3z &= 6 \end{aligned}$

7. $\begin{aligned} 4x - y + 2z &= 11 \\ x + 2y - z &= -1 \\ 2x + 2y - 3z &= -1 \end{aligned}$

8. $\begin{aligned} x - y + 3z &= 8 \\ 3x + y - 2z &= -2 \\ 2x + 4y + z &= 0 \end{aligned}$

9. $\begin{aligned} 3x + 5y + 2z &= 0 \\ 12x - 15y + 4z &= 12 \\ 6x - 25y - 8z &= 8 \end{aligned}$

10. $\begin{aligned} 2x + 3y + 7z &= 13 \\ 3x + 2y - 5z &= -22 \\ 5x + 7y - 3z &= -28 \end{aligned}$

11. $\begin{aligned} 2x - 4y + 3z &= 17 \\ x + 2y - z &= 0 \\ 4x - y - z &= 6 \end{aligned}$

12. $\begin{aligned} x + z &= 3 \\ x + 2y - z &= 1 \\ 2x - y + z &= 3 \end{aligned}$

13. $\begin{aligned} 2x + y &= 2 \\ x + y - z &= 4 \\ 3x + 2y + z &= 0 \end{aligned}$

14. $\begin{aligned} x + 3y + 5z &= 20 \\ y - 4z &= -16 \\ 3x - 2y + 9z &= 36 \end{aligned}$

15. $\begin{aligned} x + y &= -4 \\ y - z &= 1 \\ 2x + y + 3z &= -21 \end{aligned}$

16. $\begin{aligned} x + y &= 4 \\ x + z &= 4 \\ y + z &= 4 \end{aligned}$

17. $\begin{aligned} 3(2x + y) + 5z &= -1 \\ 2(x - 3y + 4z) &= -9 \\ 4(1 + x) &= -3(z - 3y) \end{aligned}$

18. $\begin{aligned} 7z - 3 &= 2(x - 3y) \\ 5y + 3z - 7 &= 4x \\ 4 + 5z &= 3(2x - y) \end{aligned}$

In Exercises 19–20, let x represent the first number, y the second number, and z the third number. Use the given conditions to write a system of equations. Solve the system and find the numbers.

19. The sum of three numbers is 16. The sum of twice the first number, 3 times the second number, and 4 times the third number is 46. The difference between 5 times the first number and the second number is 31. Find the three numbers.

20. Three numbers are unknown. Three times the first number plus the second number plus twice the third number is 5. If 3 times the second number is subtracted from the sum of the first number and 3 times the third number, the result is 2. If the third number is subtracted from 2 times the first number and 3 times the second number, the result is 1. Find the numbers.

In Exercises 21–24, find the quadratic function $y = ax^2 + bx + c$ whose graph passes through the given points.

21. $(-1, 6), (1, 4), (2, 9)$

22. $(-2, 7), (1, -2), (2, 3)$

23. $(-1, -4), (1, -2), (2, 5)$

24. $(1, 3), (3, -1), (4, 0)$

Application Exercises

25. The bar graph at the top of the next page shows the average starting salaries for the five top-paying fields for college graduates. If we add the average starting salaries for college graduates who are chemical, mechanical, and electrical engineers, the total is \$121,421. The difference between the starting salaries for chemical and mechanical engineers is \$2906. The difference between the starting salaries for mechanical engineers and electrical engineers is \$1041. Find the average starting salaries for chemical, mechanical, and electrical engineers.

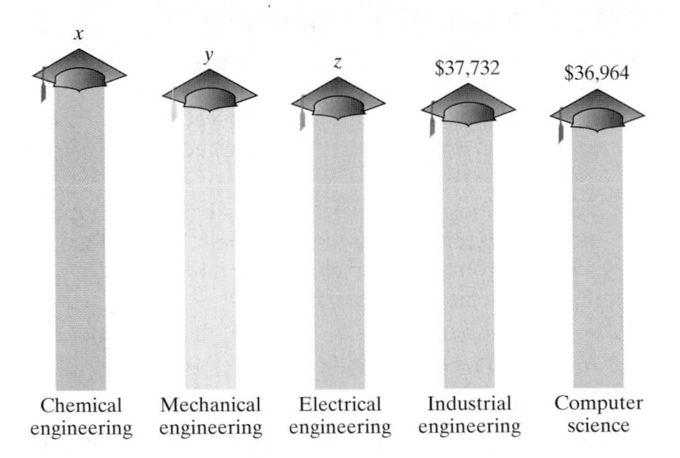

Source: Michigan State University

26. The table shows a list of the most frequently spoken languages in the United States, not counting English. Yiddish, Thai, and Persian are spoken by 621 thousand people in the United States. The difference between the number of people who speak Yiddish and the number who speak Thai is 7 thousand. The difference between the number of people who speak Thai and the number who speak Persian is 4 thousand. Find the thousands of people in the United States who speak Yiddish, Thai, and Persian.

Languages Spoken in the United States

Language	Number of Speakers
1. Spanish	17,339,000
2. French	1,702,000
3. German	1,547,000
4. Italian	1,309,000
5. Chinese	1,249,000
6. Tagalog	843,000
7. Polish	723,000
8. Korean	626,000
9. Vietnamese	507,000
10. Portuguese	430,000
11. Japanese	428,000
12. Greek	388,000
13. Arabic	355,000
14. Hindi, Urdu, & related languages	331,000
15. Russian	242,000
16. Yiddish	x
17. Thai	y
18. Persian	z

Source: Bureau of the census

27. The equation $y = \frac{1}{2}Ax^2 + Bx + C$ gives the relationship between the number of feet a car travels once the brakes are applied, y, and the number of seconds the car is in motion after the brakes are applied, x. A research firm discovered that when a car was in motion for 1 second after the brakes were applied, the car traveled 46 feet. (When $x = 1$, $y = 46$.) Similarly, it was found that when x was 2, y was 84, and when x was 3, y was 114. Use these values to find the constants A, B, and C in the equation. What is the value for y when $x = 6$? Describe what this means.

28. A ball is thrown directly upward from the top of a building. The position function

$$s = \frac{1}{2}at^2 + v_0t + s_0$$

describes the ball's height, s, in feet, after t seconds. Find the values of a, v_0, and s_0 if $s = 224$ at $t = 1$, $s = 176$ at $t = 3$, and $s = 104$ at $t = 4$. What is the value for s when $t = 5$? Describe what this means.

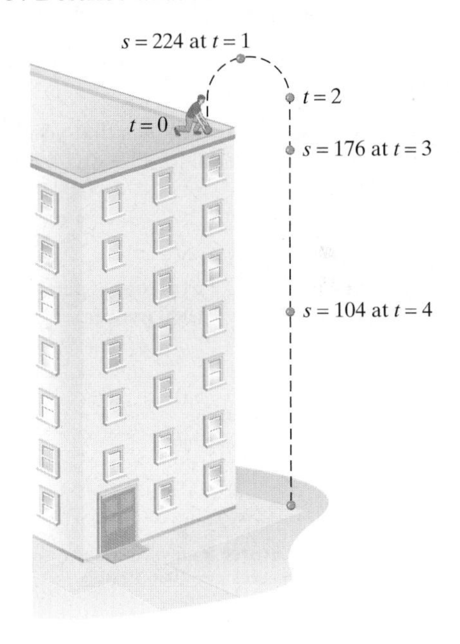

Use a system of linear equations in three variables to solve Exercises 29–32.

29. At a college production of *Evita*, 400 tickets were sold. The ticket prices were \$8, \$10, and \$12, and the total income from ticket sales was \$3700. How many tickets of each type were sold if the combined number of \$8 and \$10 tickets sold was 7 times the number of \$12 tickets sold?

30. A certain brand of razor blades comes in packages of 6, 12, and 24 blades, costing \$2, \$3, and \$4 per package, respectively. A store sold 12 packages containing a total of 162 razor blades and took in \$35. How many packages of each type were sold?

31. A person invested \$6700 for one year, part at 8%, part at 10%, and the remainder at 12%. The total annual income from these investments was \$716. The amount of money invested at 12% was \$300 more than the amount invested at 8% and 10% combined. Find the amount invested at each rate.

32. A person invested \$17,000 for one year, part at 10%, part at 12%, and the remainder at 15%. The total annual income from these investments was \$2110. The amount of money invested at 12% was \$1000 less than the amount invested at 10% and 15% combined. Find the amount invested at each rate.

Writing in Mathematics

33. What is a system of linear equations in three variables?

34. How do you determine whether a given ordered triple is a solution of a system in three variables?

35. Describe in general terms how to solve a system in three variables.

36. Describe how to use the techniques that you learned in this section to obtain a model for U.S. divorce rates from 1970 to 1997.

x (years after 1970)	y (divorces per 1000 people)
0	3.5
15	5.0
27	4.3

U.S. Divorce Rates: Number of Divorces per 1000 People

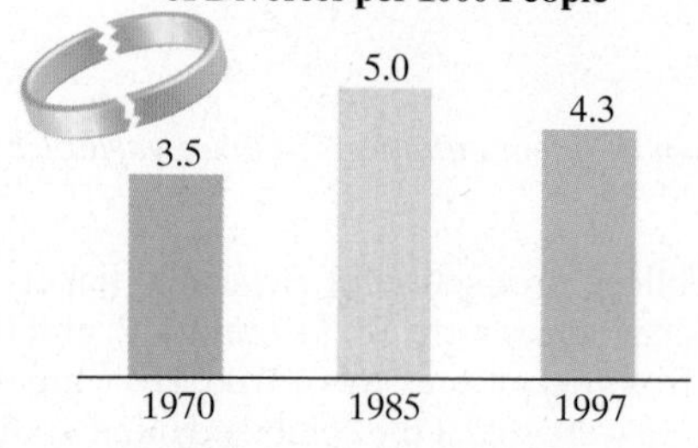

Source: U.S. Census Bureau

Technology Exercises

37. Does your graphing utility have a feature that allows you to solve linear systems by entering coefficients and constant terms? If so, use this feature to verify the solutions to any five exercises that you worked by hand from Exercises 5–16.

38. Verify your results in Exercises 21–24 by using a graphing utility to graph the resulting parabola. Trace along the curve and convince yourself that the three points given in the exercise lie on the parabola.

39. Some graphing utilities will do three-dimensional graphing. For example, on the TI-92, press MODE, go to GRAPH, press the arrow to the right, select 3D, then ENTER. When you display the Y = screen, you will see the equations are functions of x and y. Thus, you must solve each of a linear system's equations for z before entering the equation. For example,

$$x + y + z = 19$$

is solved for z, giving

$$z = 19 - x - y.$$

(Consult your manual.) If your utility does three-dimensional graphing, graph five of the systems in Exercises 5–16 and trace along the planes to find their common point of intersection.

Critical Thinking Exercises

40. Describe how the system

$$\begin{aligned} x + y - z - 2w &= -8 \\ x - 2y + 3z + w &= 18 \\ 2x + 2y + 2z - 2w &= 10 \\ 2x + y - z + w &= 3 \end{aligned}$$

could be solved. Is it likely that in the near future a graphing utility will be available to provide a geometric solution (using intersecting graphs) to this system? Explain.

41. A modernistic painting consists of triangles, rectangles, and pentagons, all drawn so as to not overlap or share sides. Within each rectangle are drawn 2 red roses, and each pentagon contains 5 carnations. How many triangles, rectangles, and pentagons appear in the painting if the painting contains a total of 40 geometric figures, 153 sides of geometric figures, and 72 flowers?

SECTION 5.3 Partial Fractions

Objective

1. Find the partial fraction decomposition of a rational expression.

The rising and setting of the sun suggest the obvious: Things change over time. Calculus is the study of rates of change, allowing the motion of the rising sun to be measured by "freezing the frame" at one instant in time. If you are given a function, calculus reveals its rate of change at any "frozen" instant. In this section, you will learn an algebraic technique used in calculus to find a function if its rate of change is known.

The Idea Behind Partial Fraction Decomposition

Systems of linear equations can be used to reverse the process of adding and subtracting rational expressions—for example,

$$\frac{3}{x-4} - \frac{2}{x+2} = \frac{3(x+2) - 2(x-4)}{(x-4)(x+2)}$$

$$= \frac{3x+6-2x+8}{(x-4)(x+2)} = \frac{x+14}{(x-4)(x+2)}.$$

In order to reverse this process, we must show that

$$\frac{x+14}{(x-4)(x+2)} = \frac{3}{x-4} - \frac{2}{x+2} \quad \text{or} \quad \frac{3}{x-4} + \frac{-2}{x+2}.$$

Each of the two fractions on the right is called a **partial fraction**. The sum of these fractions is called the **partial fraction decomposition** of the rational expression on the left-hand side.

Partial fraction decompositions can be written for rational expressions of the form $\frac{P(x)}{Q(x)}$, where P and Q have no common factors and the highest power in the numerator is less than the highest power in the denominator. In this section, we will show you how to write the partial fraction decompositions for each of the following rational expressions:

$$\frac{9x^2 - 9x + 6}{(2x-1)(x+2)(x-2)}$$

$P(x) = 9x^2 - 9x + 6$; highest power $= 2$

$Q(x) = (2x-1)(x+2)(x-2)$; multiplying factors, highest power $= 3$

$$\frac{5x^3 - 3x^2 + 7x - 3}{(x^2+1)^2}$$

$P(x) = 5x^3 - 3x^2 + 7x - 3$; highest power $= 3$

$Q(x) = (x^2+1)^2$; squaring this expression, highest power $= 4$

1 Find the partial fraction decomposition of a rational expression.

The Steps in Partial Fraction Decomposition

The partial fraction decomposition of a rational expression depends on the factors of the denominator. We consider four cases involving different kinds of factors in the denominator.

Case 1: The Partial Fraction Decomposition of a Rational Expression with Distinct Linear Factors in the Denominator If the denominator has a linear factor of the form $ax + b$, then the partial fraction decomposition will contain a term of the form

$$\frac{A}{ax+b}.$$

Constant

Linear factor

Each distinct linear factor in the denominator produces a partial fraction of the form *constant over linear factor*. For example,

$$\frac{9x^2 - 9x + 6}{(2x-1)(x+2)(x-2)} = \frac{A}{2x-1} + \frac{B}{x+2} + \frac{C}{x-2}.$$

We write a constant over each linear factor in the denominator.

The form of the partial fraction decomposition for a rational expression with distinct linear factors in the denominator is

$$\frac{P(x)}{(a_1x + b_1)(a_2x + b_2)(a_3x + b_3)\cdots(a_nx + b_n)}$$
$$= \frac{A_1}{a_1x + b_1} + \frac{A_2}{a_2x + b_2} + \frac{A_3}{a_3x + b_3} + \cdots + \frac{A_n}{a_nx + b_n}.$$

EXAMPLE 1 Partial Fraction Decomposition with Distinct Linear Factors

Find the partial fraction decomposition of

$$\frac{x+14}{(x-4)(x+2)}.$$

Solution We begin by setting up the partial fraction decomposition with the unknown constants. Write a constant over each of the two distinct linear factors in the denominator.

$$\frac{x+14}{(x-4)(x+2)} = \frac{A}{x-4} + \frac{B}{x+2}$$

Our goal is to find A and B. We do this by multiplying both sides of the equation by the least common denominator.

$$(x-4)(x+2)\frac{x+14}{(x-4)(x+2)} = (x-4)(x+2)\left(\frac{A}{x-4} + \frac{B}{x+2}\right)$$

We use the distributive property on the right side.

$$\cancel{(x-4)(x+2)}\frac{x+14}{\cancel{(x-4)(x+2)}}$$
$$= \cancel{(x-4)}(x+2)\frac{A}{\cancel{(x-4)}} + (x-4)\cancel{(x+2)}\frac{B}{\cancel{(x+2)}}$$

Dividing out common factors in numerators and denominators, we obtain

$$x + 14 = A(x + 2) + B(x - 4).$$

To find values for A and B that make both sides equal, we'll express the sides in exactly the same form by writing the variable x-terms and then writing the constant terms. Apply the distributive property on the right side.

$$x + 14 = Ax + 2A + Bx - 4B$$

$$x + 14 = Ax + Bx + 2A - 4B$$

$$1x + 14 = (A + B)x + (2A - 4B)$$

As shown by the arrows, if two polynomials are equal, coefficients of like powers of x must be equal $(A + B = 1)$ and their constant terms must be equal $(2A - 4B = 14)$. Consequently, A and B satisfy the following two equations.

$$A + B = 1$$

$$2A - 4B = 14$$

We can use the addition method to solve this linear system in two variables. By multiplying the first equation by -2 and adding equations, we obtain $A = 3$ and $B = -2$. Thus,

$$\frac{x + 14}{(x - 4)(x + 2)} = \frac{A}{x - 4} + \frac{B}{x + 2} = \frac{3}{x - 4} + \frac{-2}{x + 2}\left(\text{or } \frac{3}{x - 4} - \frac{2}{x + 2}\right).$$

Steps in Partial Fraction Decomposition

1. Set up the partial fraction decomposition with the unknown constants A, B, C, etc., in the numerator of the decomposition.
2. Multiply both sides of the resulting equation by the least common denominator.
3. Simplify the right-hand side of the equation.
4. Write both sides in descending powers, equate coefficients of like powers of x, and equate constant terms.
5. Solve the resulting linear system for A, B, C, etc.
6. Substitute the values for A, B, C, etc., into the equation in step 1 and write the partial fraction decomposition.

Check Point 1 Find the partial fraction decomposition of $\dfrac{5x - 1}{(x - 3)(x + 4)}$.

Case 2: The Partial Fraction Decomposition of a Rational Expression with Linear Factors in the Denominator, Some of Which Are Repeated
Suppose that $(ax + b)^n$ is a factor of the denominator. This means that the linear

factor $ax + b$ is repeated n times. When this occurs, the partial fraction decomposition will contain the following sum of n fractions.

$$\frac{P(x)}{(ax+b)^n} = \frac{A_1}{ax+b} + \frac{A_2}{(ax+b)^2} + \frac{A_3}{(ax+b)^3} + \cdots + \frac{A_n}{(ax+b)^n}$$

Include one fraction with a constant numerator for each power of $ax + b$.

EXAMPLE 2 Partial Fraction Decomposition with Repeated Linear Factors

Find the partial fraction decomposition of $\frac{x-18}{x(x-3)^2}$.

Solution

Step 1 Set up the partial fraction decomposition with the unknown constants. Because the linear factor $x - 3$ is repeated twice, we must include one fraction with a constant numerator for each power of $x - 3$.

$$\frac{x-18}{x(x-3)^2} = \frac{A}{x} + \frac{B}{x-3} + \frac{C}{(x-3)^2}$$

Step 2 Multiply both sides of the resulting equation by the least common denominator. We clear fractions, multiplying both sides by $x(x-3)^2$, the least common denominator.

$$x(x-3)^2\left[\frac{x-18}{x(x-3)^2}\right] = x(x-3)^2\left[\frac{A}{x} + \frac{B}{x-3} + \frac{C}{(x-3)^2}\right]$$

We use the distributive property on the right side.

$$\cancel{x(x-3)^2} \cdot \frac{x-18}{\cancel{x(x-3)^2}}$$

$$= \cancel{x}(x-3)^2 \cdot \frac{A}{\cancel{x}} + x(x-3)^2 \cdot \frac{B}{(x-3)} + x\cancel{(x-3)^2} \cdot \frac{C}{\cancel{(x-3)^2}}$$

Dividing out common factors in numerators and denominators, we obtain

$$x - 18 = A(x-3)^2 + Bx(x-3) + Cx.$$

Step 3 Simplify the right side of the equation. Square $x - 3$. Then apply the distributive property.

$$x - 18 = A(x^2 - 6x + 9) + Bx(x-3) + Cx$$

Square $x - 3$ using $(A - B)^2 = A^2 - 2AB + B^2$.

$$x - 18 = Ax^2 - 6Ax + 9A + Bx^2 - 3Bx + Cx$$

Apply the distributive property.

Step 4 Write both sides in descending powers, equate coefficients of like powers of x, and equate constant terms. The left side, $x - 18$, is in descending powers of x: $x - 18x^0$. We will write the right side in descending powers of x.

$$x - 18 = Ax^2 + Bx^2 - 6Ax - 3Bx + Cx + 9A$$

Express both sides in the same form.

$$0x^2 + 1x - 18 = (A + B)x^2 + (-6A - 3B + C)x + 9A$$

Equating coefficients of like powers of x and constant terms results in the following system of linear equations.

$$A + B = 0$$
$$-6A - 3B + C = 1$$
$$9A = -18$$

Step 5 Solve the resulting system for A, B, and C. Dividing both sides of the last equation by 9, we obtain $A = -2$. Substituting -2 for A in the first equation, $A + B = 0$, gives $-2 + B = 0$ or $B = 2$. We find C by substituting -2 for A and 2 for B in the middle equation, $-6A - 3B + C = 1$. We obtain $C = -5$.

Step 6 Substitute the values of A, B, and C and write the partial fraction decomposition. With $A = -2$, $B = 2$, and $C = -5$, the required partial fraction decomposition is

$$\frac{x - 18}{x(x - 3)^2} = \frac{A}{x} + \frac{B}{x - 3} + \frac{C}{(x - 3)^2} = -\frac{2}{x} + \frac{2}{x - 3} - \frac{5}{(x - 3)^2}.$$

Check Point 2 Find the partial fraction decomposition of $\frac{x + 2}{x(x - 1)^2}$.

Case 3: The Partial Fraction Decomposition of a Rational Expression with Prime, Nonrepeated Quadratic Factors in the Denominator Suppose that $ax^2 + bx + c$ is a factor of the denominator and that this quadratic factor cannot be factored into linear factors with real coefficients. Under these conditions, the partial fraction decomposition will contain a term of the form

$$\frac{Ax + B}{ax^2 + bx + c}.$$

Linear numerator

Quadratic factor

Each distinct prime quadratic factor in the denominator produces a partial fraction of the form *linear numerator over quadratic factor*. For example,

$$\frac{3x^2 + 17x + 14}{(x - 2)(x^2 + 2x + 4)} = \frac{A}{x - 2} + \frac{Bx + C}{x^2 + 2x + 4}.$$

We write a constant over the linear factor in the denominator.

We write a linear numerator over the prime quadratic factor in the denominator.

Our next example illustrates how a linear system in three variables is used to determine values for A, B, and C.

EXAMPLE 3 Partial Fraction Decomposition

Find the partial fraction decomposition of

$$\frac{3x^2 + 17x + 14}{(x - 2)(x^2 + 2x + 4)}.$$

Solution

Step 1 Set up the partial fraction decomposition with the unknown constants. We put a constant (A) over the linear factor and a linear expression $(Bx + C)$ over the prime quadratic factor.

$$\frac{3x^2 + 17x + 14}{(x - 2)(x^2 + 2x + 4)} = \frac{A}{x - 2} + \frac{Bx + C}{x^2 + 2x + 4}$$

Step 2 Multiply both sides of the resulting equation by the least common denominator. We clear fractions, multiplying both sides by $(x - 2)(x^2 + 2x + 4)$, the least common denominator.

$$(x - 2)(x^2 + 2x + 4)\left[\frac{3x^2 + 17x + 14}{(x - 2)(x^2 + 2x + 4)}\right] = (x - 2)(x^2 + 2x + 4)\left[\frac{A}{x - 2} + \frac{Bx + C}{x^2 + 2x + 4}\right]$$

We use the distributive property on the right side.

$$(x - 2)(x^2 + 2x + 4) \cdot \frac{3x^2 + 17x + 14}{(x - 2)(x^2 + 2x + 4)}$$

$$= (x - 2)(x^2 + 2x + 4) \cdot \frac{A}{x - 2} + (x - 2)(x^2 + 2x + 4) \cdot \frac{Bx + C}{x^2 + 2x + 4}$$

Dividing out common factors in numerators and denominators, we obtain

$$3x^2 + 17x + 14 = A(x^2 + 2x + 4) + (Bx + C)(x - 2).$$

Step 3 Simplify the right side of the equation. We multiply on the right side by distributing A over each term in parentheses and multiplying $(Bx + C)(x - 2)$ using the FOIL method.

$$3x^2 + 17x + 14 = Ax^2 + 2Ax + 4A + Bx^2 - 2Bx + Cx - 2C$$

Step 4 Write both sides in descending powers, equate coefficients of like powers of x, and equate constant terms. The left side, $3x^2 + 17x + 14$, is in descending powers of x. We write the right side in descending powers of x

$$3x^2 + 17x + 14 = Ax^2 + Bx^2 + 2Ax - 2Bx + Cx + 4A - 2C$$

and express both sides in the same form.

$$3x^2 + 17x + 14 = (A + B)x^2 + (2A - 2B + C)x + (4A - 2C)$$

Equating coefficients of like powers of x and constant terms results in the following system of linear equations.

$$A + B = 3$$
$$2A - 2B + C = 17$$
$$4A - 2C = 14$$

Step 5 Solve the resulting system for A, B, and C. Because the first equation involves A and B, we can obtain another equation in A and B by eliminating C from the second and third equations. Multiply the second equation by 2 and add equations. Solving in this manner, we obtain $A = 5$, $B = -2$, and $C = 3$.

Step 6 Substitute the values of A, B, and C and write the partial fraction decomposition. With $A = 5$, $B = -2$, and $C = 3$, the required partial fraction decomposition is

$$\frac{3x^2 + 17x + 14}{(x-2)(x^2+2x+4)} = \frac{A}{x-2} + \frac{Bx + C}{x^2 + 2x + 4} = \frac{5}{x-2} + \frac{-2x+3}{x^2+2x+4}.$$

Check Point 3 Find the partial fraction decomposition of

$$\frac{8x^2 + 12x - 20}{(x+3)(x^2+x+2)}.$$

Case 4: The Partial Fraction Decomposition of a Rational Expression with a Prime, Repeated Quadratic Factor in the Denominator Suppose that $(ax^2 + bx + c)^n$ is a factor of the denominator and that $ax^2 + bx + c$ cannot be factored further. This means that the quadratic factor $ax^2 + bx + c$ is repeated n times. When this occurs, the partial fraction decomposition will contain a linear numerator for each power of $ax^2 + bx + c$.

$$\frac{P(x)}{(ax^2+bx+c)^n} = \frac{A_1x + B_1}{ax^2+bx+c} + \frac{A_2x + B_2}{(ax^2+bx+c)^2} + \frac{A_3x + B_3}{(ax^2+bx+c)^3} + \cdots + \frac{A_nx + B_n}{(ax^2+bx+c)^n}$$

Include one fraction with a linear numerator for each power of $ax^2 + bx + c$.

EXAMPLE 4 Partial Fraction Decomposition with a Repeated Quadratic Factor

Find the partial fraction decomposition of

$$\frac{5x^3 - 3x^2 + 7x - 3}{(x^2+1)^2}.$$

Solution

Step 1 Set up the partial fraction decomposition with the unknown constants. Because the quadratic factor $x^2 + 1$ is repeated twice, we must include one fraction with a linear numerator for each power of $x^2 + 1$.

$$\frac{5x^3 - 3x^2 + 7x - 3}{(x^2+1)^2} = \frac{Ax + B}{x^2+1} + \frac{Cx + D}{(x^2+1)^2}$$

Step 2 Multiply both sides of the resulting equation by the least common denominator. We clear fractions, multiplying both sides by $(x^2 + 1)^2$, the least common denominator.

$$(x^2+1)^2\left[\frac{5x^3 - 3x^2 + 7x - 3}{(x^2+1)^2}\right] = (x^2+1)^2\left[\frac{Ax + B}{x^2+1} + \frac{Cx + D}{(x^2+1)^2}\right]$$

Now we multiply and simplify.

$$5x^3 - 3x^2 + 7x - 3 = (x^2+1)(Ax + B) + Cx + D$$

Step 3 Simplify the right side of the equation. We multiply $(x^2 + 1)(Ax + B)$ using the FOIL method.

$$5x^3 - 3x^2 + 7x - 3 = Ax^3 + Bx^2 + Ax + B + Cx + D$$

Step 4 Write both sides in descending powers, equate coefficients of like powers of x, and equate constant terms.

$$5x^3 - 3x^2 + 7x - 3 = Ax^3 + Bx^2 + Ax + Cx + B + D$$

$$5x^3 - 3x^2 + 7x - 3 = Ax^3 + Bx^2 + (A + C)x + (B + D)$$

Equating coefficients of like powers of x and constant terms results in the following system of linear equations.

$$A = 5$$

$$B = -3$$

$$A + C = 7 \quad \text{With } A = 5\text{, we immediately obtain } C = 2.$$

$$B + D = -3 \quad \text{With } B = -3\text{, we immediately obtain } D = 0.$$

Step 5 Solve the resulting system for A, B, C, and D. Based on our observations in step 4, $A = 5$, $B = -3$, $C = 2$, and $D = 0$.

Step 6 Substitute the values of A, B, C, and D and write the partial fraction decomposition.

$$\frac{5x^3 - 3x^2 + 7x - 3}{(x^2 + 1)^2} = \frac{Ax + B}{x^2 + 1} + \frac{Cx + D}{(x^2 + 1)^2} = \frac{5x - 3}{x^2 + 1} + \frac{2x}{(x^2 + 1)^2}$$

Check Point 4 Find the partial fraction decomposition of $\dfrac{2x^3 + x + 3}{(x^2 + 1)^2}$.

EXERCISE SET 5.3

Practice Exercises

In Exercises 1–8, write the form of the partial fraction decomposition of the rational expression. It is not necessary to solve for the constants.

1. $\dfrac{11x - 10}{(x - 2)(x + 1)}$
2. $\dfrac{5x + 7}{(x - 1)(x + 3)}$
3. $\dfrac{6x^2 - 14x - 27}{(x + 2)(x - 3)^2}$
4. $\dfrac{3x + 16}{(x + 1)(x - 2)^2}$
5. $\dfrac{5x^2 - 6x + 7}{(x - 1)(x^2 + 1)}$
6. $\dfrac{5x^2 - 9x + 19}{(x - 4)(x^2 + 5)}$
7. $\dfrac{x^3 + x^2}{(x^2 + 4)^2}$
8. $\dfrac{7x^2 - 9x + 3}{(x^2 + 7)^2}$

In Exercises 9–38, write the partial fraction decomposition of each rational expression.

9. $\dfrac{x}{(x - 3)(x - 2)}$
10. $\dfrac{1}{x(x - 1)}$
11. $\dfrac{3x + 50}{(x - 9)(x + 2)}$
12. $\dfrac{5x - 1}{(x - 2)(x + 1)}$
13. $\dfrac{7x - 4}{x^2 - x - 12}$
14. $\dfrac{9x + 21}{x^2 + 2x - 15}$
15. $\dfrac{4x^2 + 13x - 9}{x(x - 1)(x + 3)}$
16. $\dfrac{4x^2 - 5x - 15}{x(x + 1)(x - 5)}$
17. $\dfrac{4x^2 - 7x - 3}{x^3 - x}$
18. $\dfrac{2x^2 - 18x - 12}{x^3 - 4x}$

19. $\dfrac{6x - 11}{(x - 1)^2}$

20. $\dfrac{x}{(x + 1)^2}$

21. $\dfrac{x^2 - 6x + 3}{(x - 2)^3}$

22. $\dfrac{2x^2 + 8x + 3}{(x + 1)^3}$

23. $\dfrac{x^2 + 2x + 7}{x(x - 1)^2}$

24. $\dfrac{3x^2 + 49}{x(x + 7)^2}$

25. $\dfrac{5x^2 + 21x + 4}{(x + 1)^2(x - 3)}$

26. $\dfrac{x}{(x + 2)^2(x + 1)}$

27. $\dfrac{5x^2 - 6x + 7}{(x - 1)(x^2 + 1)}$

28. $\dfrac{5x^2 - 9x + 19}{(x - 4)(x^2 + 5)}$

29. $\dfrac{5x^2 + 6x + 3}{(x + 1)(x^2 + 2x + 2)}$

30. $\dfrac{9x + 2}{(x - 2)(x^2 + 2x + 2)}$

31. $\dfrac{6x^2 - x + 1}{x^3 + x^2 + x + 1}$

32. $\dfrac{3x^2 - 2x + 8}{x^3 + 2x^2 + 4x + 8}$

33. $\dfrac{x^3 + x^2 + 2}{(x^2 + 2)^2}$

34. $\dfrac{x^2 + 2x + 3}{(x^2 + 4)^2}$

35. $\dfrac{x^3 - 4x^2 + 9x - 5}{(x^2 - 2x + 3)^2}$

36. $\dfrac{3x^3 - 6x^2 + 7x - 2}{(x^2 - 2x + 2)^2}$

37. $\dfrac{4x^2 + 3x + 14}{x^3 - 8}$

38. $\dfrac{2x + 4}{x^3 - 1}$

Application Exercises

39. Find the partial fraction decomposition for $\dfrac{1}{x(x + 1)}$ and use the result to find the following sum:

$$\frac{1}{1 \cdot 2} + \frac{1}{2 \cdot 3} + \frac{1}{3 \cdot 4} + \cdots + \frac{1}{99 \cdot 100}.$$

40. Find the partial fraction decomposition for $\dfrac{2}{x(x + 2)}$ and use the result to find the following sum:

$$\frac{2}{1 \cdot 3} + \frac{2}{3 \cdot 5} + \frac{2}{5 \cdot 7} + \cdots + \frac{2}{99 \cdot 101}.$$

Writing in Mathematics

41. Explain what is meant by the partial fraction decomposition of a rational expression.

42. Explain how to find the partial fraction decomposition of a rational expression with distinct linear factors in the denominator.

43. Explain how to find the partial fraction decomposition of a rational expression with a repeated linear factor in the denominator.

44. Explain how to find the partial fraction decomposition of a rational expression with a prime quadratic factor in the denominator.

45. Explain how to find the partial fraction decomposition of a rational expression with a repeated, prime quadratic factor in the denominator.

46. How can you verify your result for the partial fraction decomposition for a given rational expression without using a graphing utility?

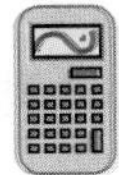

Technology Exercises

47. A graphing utility can be used to check the partial fraction decomposition for a given rational expression. Graph y_1 = *the given rational expression* and y_2 = *its partial fraction decomposition* on the same screen. If the graphs are identical, the decomposition is correct. Use this method to verify any five of the decompositions that you obtained in Exercises 9–38.

48. As you worked Exercise 47, did you find that it took a while to determine the range setting that showed a graph for the rational function and its decomposition? Suggest another method for showing that $y_1 = y_2$ using your graphing utility. Use this method to check the results of the same five decompositions you worked with in Exercise 47.

Critical Thinking Exercises

49. If a, b, and c are constants, find the partial fraction decomposition of

$$\frac{ax + b}{(x - c)^2}.$$

50. Find the partial fraction decomposition of

$$\frac{4x^2 + 5x - 9}{x^3 - 6x - 9}.$$

SECTION 5.4 Systems of Nonlinear Equations in Two Variables

Objectives

1. Recognize systems of nonlinear equations in two variables.
2. Solve nonlinear systems by substitution.
3. Solve nonlinear systems by addition.
4. Solve problems using systems of nonlinear equations.

Scientists debate the probability that a "doomsday rock" will collide with Earth. It has been estimated that an asteroid, a tiny planet that revolves around the sun, crashes into Earth about once every 250,000 years, and that such a collision would have disastrous results. In 1908 a small fragment struck Siberia, leveling thousands of acres of trees. One theory about the extinction of dinosaurs 65 million years ago involves Earth's collision with a large asteroid and the resulting drastic changes in Earth's climate.

Understanding the path of Earth and the path of a comet is essential to detecting threatening space debris. Orbits about the sun are not described by linear equations in the form $Ax + By = C$. The ability to solve systems that do not contain linear equations provides NASA scientists watching for troublesome asteroids with a possible collision point with Earth's orbit.

1 Recognize systems of nonlinear equations in two variables.

Systems of Nonlinear Equations and Their Solutions

A **system of** two **nonlinear equations** in two variables contains at least one equation that cannot be expressed in the form $Ax + By = C$. Here are two examples:

$$x^2 = 2y + 10$$
$$3x - y = 9$$

Not in the form $Ax + By = C$. The term x^2 is not linear.

$$y = x^2 + 3$$
$$x^2 + y^2 = 9$$

Neither equation is in the form $Ax + By = C$. The terms x^2 and y^2 are not linear.

A **solution** to a nonlinear system in two variables is an ordered pair of real numbers that satisfies all equations in the system. The **solution set** to the system is the set of all such ordered pairs. As with linear systems in two variables, the solution to a nonlinear system (if there is one) corresponds to the intersection point(s) of the graphs of the equations in the system. Unlike linear systems, the graphs can be circles, parabolas, or anything other than two lines. We will solve nonlinear systems using the substitution method and the addition method.

2 Solve nonlinear systems by substitution.

Eliminating a Variable Using the Substitution Method

The substitution method involves converting a nonlinear system to one equation in one variable by an appropriate substitution. The steps in the solution process are exactly the same as those used to solve a linear system by substitution. However, when you obtain an equation in one variable, this equation will not be linear. In our first example, this equation is quadratic.

EXAMPLE 1 Solving a Nonlinear System by the Substitution Method

Solve by the substitution method:

$$x^2 = 2y + 10 \quad \text{The graph is a parabola.}$$
$$3x - y = 9. \quad \text{The graph is a line.}$$

Solution

Step 1 Solve one of the equations for one variable in terms of the other. We begin by isolating one of the variables raised to the first power in either of the equations. By solving for y in the second equation, which has a coefficient of -1, we can avoid fractions.

$$3x - y = 9 \quad \text{This is the second equation in the given system.}$$
$$3x = y + 9 \quad \text{Add } y \text{ to both sides.}$$
$$3x - 9 = y \quad \text{Subtract 9 from both sides.}$$

Step 2 Substitute the expression from step 1 into the other equation. We substitute $3x - 9$ for y in the first equation.

$$y = \boxed{3x - 9} \qquad x^2 = 2\boxed{y} + 10$$

This gives us an equation in one variable, namely

$$x^2 = 2(3x - 9) + 10.$$

The variable y has been eliminated.

Step 3 Solve the resulting equation containing one variable.

$$x^2 = 2(3x - 9) + 10 \quad \text{This is the equation containing one variable.}$$
$$x^2 = 6x - 18 + 10 \quad \text{Use the distributive property.}$$
$$x^2 = 6x - 8 \quad \text{Combine numerical terms on the right.}$$
$$x^2 - 6x + 8 = 0 \quad \text{Move all terms to one side and set the quadratic equation equal to 0.}$$
$$(x - 4)(x - 2) = 0 \quad \text{Factor.}$$
$$x - 4 = 0 \quad \text{or} \quad x - 2 = 0 \quad \text{Set each factor equal to 0.}$$
$$x = 4 \quad \text{or} \quad x = 2 \quad \text{Solve for } x.$$

Step 4 Back-substitute the obtained values into the equation from step 1. Now that we have the x-coordinates of the solutions, we back-substitute 4 for x and 2 for x in the equation $y = 3x - 9$.

If x is 4, $y = 3(4) - 9 = 3$, so $(4, 3)$ is a solution.

If x is 2, $y = 3(2) - 9 = -3$, so $(2, -3)$ is a solution.

Step 5 Check the proposed solutions in both of the system's given equations. We begin by checking $(4, 3)$. Replace x with 4 and y with 3.

$$x^2 = 2y + 10 \qquad 3x - y = 9 \quad \text{These are the given equations.}$$
$$4^2 \stackrel{?}{=} 2(3) + 10 \qquad 3(4) - 3 \stackrel{?}{=} 9 \quad \text{Let } x = 4 \text{ and } y = 3.$$
$$16 \stackrel{?}{=} 6 + 10 \qquad 12 - 3 \stackrel{?}{=} 9 \quad \text{Simplify.}$$
$$16 = 16 \checkmark \qquad 9 = 9 \checkmark \quad \text{True statements result.}$$

The ordered pair $(4, 3)$ satisfies both equations. Thus, $(4, 3)$ is a solution to the system.

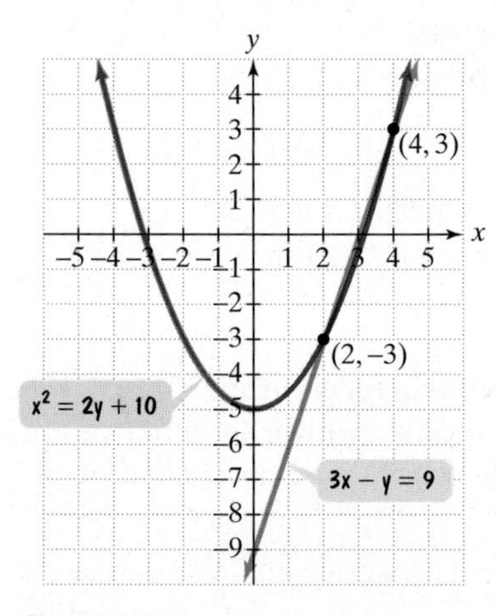

Figure 5.7 Points of intersection illustrate the nonlinear system's solutions.

Now let's check $(2, -3)$. Replace x with 2 and y with -3 in both given equations.

$x^2 = 2y + 10$	$3x - y = 9$	These are the given equations.
$2^2 \stackrel{?}{=} 2(-3) + 10$	$3(2) - (-3) \stackrel{?}{=} 9$	Let $x = 2$ and $y = -3$.
$4 \stackrel{?}{=} -6 + 10$	$6 + 3 \stackrel{?}{=} 9$	Simplify.
$4 = 4$ ✓	$9 = 9$ ✓	True statements result.

The ordered pair $(2, -3)$ also satisfies both equations and is a solution to the system. The solution set is $\{(4, 3), (2, -3)\}$. Figure 5.7 shows the graphs of the equations in the system and the solutions as intersection points.

Check Point 1 Solve by the substitution method:

$$x^2 = y - 1$$
$$4x - y = -1.$$

Study Tip

Recall from Chapter 2 that

$$(x - h)^2 + (y - k)^2 = r^2$$

describes a circle with center (h, k) and radius r.

EXAMPLE 2 Solving a Nonlinear System by the Substitution Method

Solve by the substitution method:

$$x - y = 3 \quad \text{The graph is a line.}$$
$$(x - 2)^2 + (y + 3)^2 = 4. \quad \text{The graph is a circle.}$$

Solution Graphically, we are finding the intersection of a line and a circle whose center is at $(2, -3)$ and whose radius measures 2.

Step 1 Solve one of the equations for one variable in terms of the other. We will solve for x in the linear equation – that is, the first equation. (We could also solve for y.)

$$x - y = 3 \quad \text{This is the first equation in the given system.}$$
$$x = y + 3 \quad \text{Add } y \text{ to both sides.}$$

Step 2 Substitute the expression from step 1 into the other equation. We substitute $y + 3$ for x in the second equation.

$$x = \boxed{y + 3} \qquad (\boxed{x} - 2)^2 + (y + 3)^2 = 4$$

This gives an equation in one variable, namely

$$(y + 3 - 2)^2 + (y + 3)^2 = 4.$$

The variable x has been eliminated.

Step 3 Solve the resulting equation containing one variable.

$(y + 3 - 2)^2 + (y + 3)^2 = 4$	This is the equation containing one variable.
$(y + 1)^2 + (y + 3)^2 = 4$	Combine numerical terms in the first parentheses.
$y^2 + 2y + 1 + y^2 + 6y + 9 = 4$	Use the formula $(A + B)^2 = A^2 + 2AB + B^2$ to square $y + 1$ and $y + 3$.
$2y^2 + 8y + 10 = 4$	Combine like terms on the left.
$2y^2 + 8y + 6 = 0$	Subtract 4 from both sides and set the quadratic equation equal to 0.

$$y^2 + 4y + 3 = 0 \quad \text{Simplify by dividing both sides by 2.}$$

$$(y + 3)(y + 1) = 0 \quad \text{Factor.}$$

$$y + 3 = 0 \quad \text{or} \quad y + 1 = 0 \quad \text{Set each factor equal to 0.}$$

$$y = -3 \quad \text{or} \quad y = -1 \quad \text{Solve for } y.$$

Step 4 Back-substitute the obtained values into the equation from step 1. Now that we have the y-coordinates of the solutions, we back-substitute -3 for y and -1 for y in the equation $x = y + 3$.

If $y = -3$: $x = -3 + 3 = 0$, so $(0, -3)$ is a solution.

If $y = -1$: $x = -1 + 3 = 2$, so $(2, -1)$ is a solution.

Step 5 Check the proposed solution in both of the system's given equations. Take a moment to show that each ordered pair satisfies both equations. The solution set of the given system is $\{(0, -3), (2, -1)\}$.

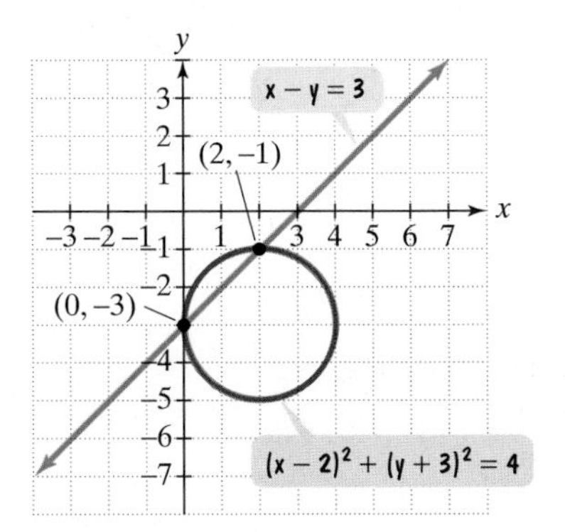

Figure 5.8 Points of intersection illustrate the nonlinear system's solutions.

Figure 5.8 shows the graphs of the equations in the system and the solutions as intersection points.

Check Point 2 Solve by the substitution method:

$$x + 2y = 0$$
$$(x - 1)^2 + (y - 1)^2 = 5.$$

3 Solve nonlinear systems by addition.

Eliminating a Variable Using the Addition Method

In solving linear systems with two variables, we learned that the addition method works well when each equation is in the form $Ax + By = C$. For nonlinear systems, the addition method can be used when each equation is in the form $Ax^2 + By^2 = C$. If necessary, we will multiply either equation or both equations by appropriate numbers so that the coefficients of x^2 or y^2 will have a sum of 0. We then add equations. The sum will be an equation in one variable.

EXAMPLE 3 Solving a Nonlinear System by the Addition Method

Solve the system

$$4x^2 + y^2 = 13 \quad \text{Equation 1}$$
$$x^2 + y^2 = 10. \quad \text{Equation 2}$$

Solution We can use the same steps that we did when we solved linear systems by the addition method.

Step 1 Write both equations in the form $Ax^2 + By^2 = C$. Both equations are already in this form, so we can skip this step.

Step 2 If necessary, multiply either equation or both equations by appropriate numbers so that the sum of the x^2-coefficients or the sum of the y^2-coefficients is 0. We can eliminate y^2 by multiplying Equation 2 by -1.

$$4x^2 + y^2 = 13 \xrightarrow{\text{No change}} 4x^2 + y^2 = 13$$
$$x^2 + y^2 = 10 \xrightarrow{\text{Multiply by } -1.} -x^2 - y^2 = -10$$

Steps 3 and 4 **Add equations and solve for the remaining variable.**

$$\begin{aligned} 4x^2 + y^2 &= 13 \\ -x^2 - y^2 &= -10 \\ \text{Add: } 3x^2 &= 3 \\ x^2 &= 1 \quad \text{Divide both sides by 3.} \\ x &= \pm 1 \quad \text{Use the square root method: If } x^2 = c \text{, then } x = \pm\sqrt{c}. \end{aligned}$$

Step 5 **Back-substitute and find the values for the other variables.** We must back-substitute each value of x into either one of the original equations. Let's use $x^2 + y^2 = 10$, Equation 2. If $x = 1$,

$$\begin{aligned} 1^2 + y^2 &= 10 \quad \text{Replace x with 1 in Equation 2.} \\ y^2 &= 9 \quad \text{Subtract 1 from both sides.} \\ y &= \pm 3 \quad \text{Apply the square root method.} \end{aligned}$$

$(1, 3)$ and $(1, -3)$ are solutions. If $x = -1$,

$$\begin{aligned} (-1)^2 + y^2 &= 10 \quad \text{Replace x with } -1 \text{ in Equation 2.} \\ y^2 &= 9 \quad \text{The steps are the same as before.} \\ y &= \pm 3 \end{aligned}$$

$(-1, 3)$ and $(-1, -3)$ are solutions.

Step 6 **Check.** Take a moment to show that each of the four ordered pairs satisfies Equation 1 and Equation 2. The solution set of the given system is $\{(1, 3), (1, -3), (-1, 3), (-1, -3)\}$.

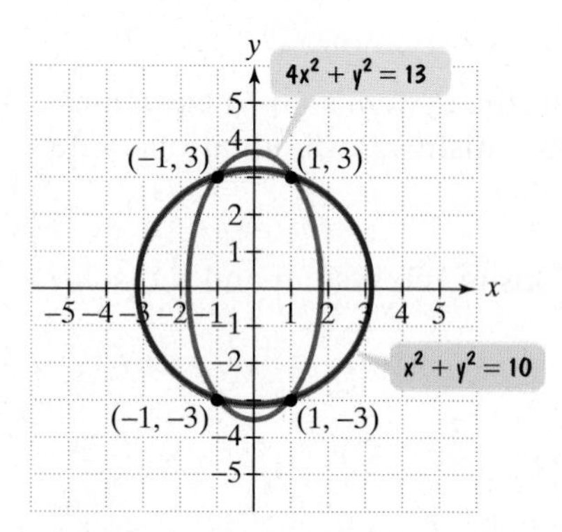

Figure 5.9 A system with four solutions

Figure 5.9 shows the graphs of the equations in the system and the solutions as intersection points.

Study Tip

When solving nonlinear systems, extra solutions may be introduced that do not satisfy both equations in the system. Therefore, you should get into the habit of checking all proposed pairs in each of the system's two equations.

Check Point 3 Solve the system:

$$\begin{aligned} 3x^2 + 2y^2 &= 35 \\ 4x^2 + 3y^2 &= 48. \end{aligned}$$

In solving nonlinear systems, we include only ordered pairs with real numbers in the solution set. We have seen that each of these ordered pairs corresponds to a point of intersection of the system's graphs.

EXAMPLE 4 Solving a Nonlinear System by the Addition Method

Solve the system:

$$\begin{aligned} y &= x^2 + 3 \quad \text{Equation 1 (The graph is a parabola.)} \\ x^2 + y^2 &= 9. \quad \text{Equation 2 (The graph is a circle.)} \end{aligned}$$

Solution We could use substitution because Equation 1 has y expressed in terms of x, but this would result in a fourth-degree equation. However, we can rewrite Equation 1 by subtracting x^2 from both sides and adding the equations to eliminate the x^2-terms.

$$\begin{aligned} -x^2 + y &= 3 \quad \text{Subtract } x^2 \text{ from both sides of Equation 1.} \\ x^2 + y^2 &= 9 \quad \text{This is Equation 2.} \\ y + y^2 &= 12 \quad \text{Add the equations.} \end{aligned}$$

We now solve this quadratic equation.

$$y + y^2 = 12$$

$$y^2 + y - 12 = 0 \quad \text{Subtract 12 from both sides and set the quadratic equation equal to 0.}$$

$$(y + 4)(y - 3) = 0 \quad \text{Factor.}$$

$$y + 4 = 0 \quad \text{or} \quad y - 3 = 0 \quad \text{Set each factor equal to 0.}$$

$$y = -4 \quad \text{or} \quad y = 3 \quad \text{Solve for } y.$$

To complete the solution, we must back-substitute each value of y into either one of the original equations. We will use $y = x^2 + 3$, Equation 1. First, we substitute -4 for y.

$$-4 = x^2 + 3$$

$$-7 = x^2 \quad \text{Subtract 3 from both sides.}$$

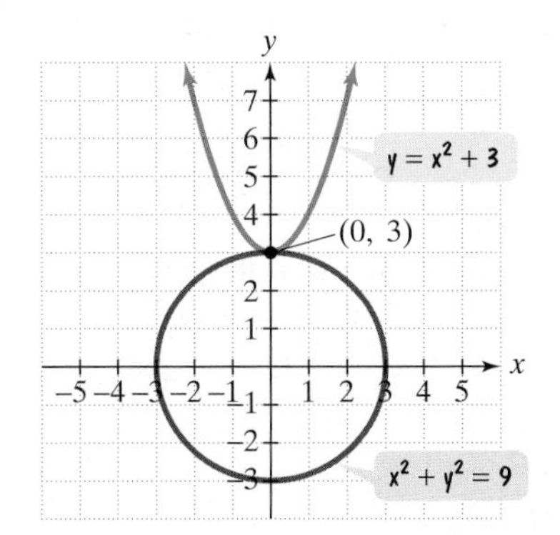

Figure 5.10 A system with one real solution

Because the square of a real number cannot be negative, the equation $x^2 = -7$ does not have real-number solutions. Thus, we move on to our other value for y, 3, and substitute this value into Equation 1.

$$y = x^2 + 3 \quad \text{This is Equation 1.}$$

$$3 = x^2 + 3 \quad \text{Back-substitute 3 for } y.$$

$$0 = x^2 \quad \text{Subtract 3 from both sides.}$$

$$0 = x \quad \text{Solve for } x.$$

We showed that if $y = 3$, then $x = 0$. Thus, $(0, 3)$ is the solution. Take a moment to show that $(0, 3)$ satisfies Equation 1 and Equation 2. The solution set of the given system is $\{(0, 3)\}$. Figure 5.10 shows the system's graphs and the solution as an intersection point.

Check Point 4 Solve the system:

$$y = x^2 + 5$$
$$x^2 + y^2 = 25.$$

4 Solve problems using systems of nonlinear equations.

Applications

Many geometric problems can be modeled and solved by the use of nonlinear systems of equations. We will use our step-by-step strategy for solving problems using mathematical models that are created from verbal models.

EXAMPLE 5 An Application of a Nonlinear System

You have 36 yards of fencing to build the enclosure in Figure 5.11. Some of this fencing is to be used to build an internal divider. If you'd like to enclose 54 square yards, what are the dimensions of the enclosure?

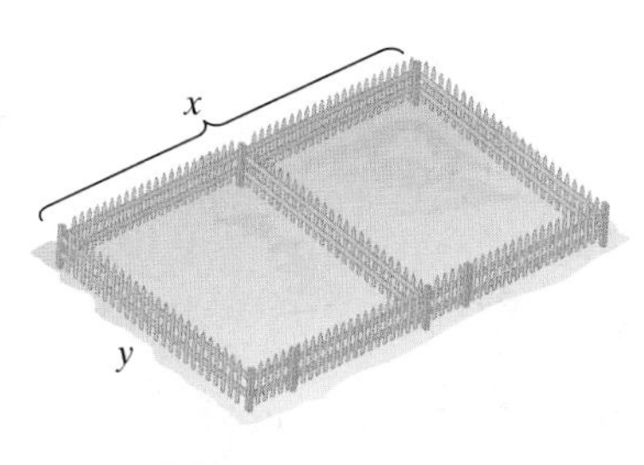

Figure 5.11 Building an enclosure

Solution

Step 1 Use variables to represent unknown quantities. Let x = the enclosure's length and y = the enclosure's width. These variables are shown in Figure 5.11.

Step 2 Write a system of equations describing the problem's conditions. The first condition is that you have 36 yards of fencing.

Fencing along both lengths	plus	Fencing along both widths	plus	Fencing for the internal divider	equals	36 yards.
$2x$	$+$	$2y$	$+$	y	$=$	36

Adding like terms, we can express the equation that models the verbal conditions for the fencing as $2x + 3y = 36$.

The second condition is that you'd like to enclose 54 square yards. The rectangle's area, the product of its length and its width, must be 54 square yards.

Length	times	width	is	54 square yards.
x	$\cdot$	y	$=$	54

Step 3 Solve the system and answer the problem's question. We must solve the system

$$2x + 3y = 36 \quad \text{Equation 1}$$
$$xy = 54. \quad \text{Equation 2}$$

We will use substitution. Because Equation 1 has no coefficients of 1 or -1, we will solve Equation 2 for y. Dividing both sides of $xy = 54$ by x, we obtain

$$y = \frac{54}{x}.$$

Now we substitute $\frac{54}{x}$ for y in Equation 1 and solve for x.

$$2x + 3y = 36 \quad \text{This is Equation 1.}$$
$$2x + 3 \cdot \frac{54}{x} = 36 \quad \text{Substitute } \frac{54}{x} \text{ for } y.$$
$$2x + \frac{162}{x} = 36 \quad \text{Multiply.}$$
$$x\left(2x + \frac{162}{x}\right) = 36 \cdot x \quad \text{Clear fractions by multiplying both sides by } x.$$
$$2x^2 + 162 = 36x \quad \text{Use the distributive property on the left side.}$$
$$2x^2 - 36x + 162 = 0 \quad \text{Subtract } 36x \text{ from both sides and set the quadratic equation equal to 0.}$$
$$x^2 - 18x + 81 = 0 \quad \text{Simplify by dividing both sides by 2.}$$
$$(x - 9)^2 = 0 \quad \text{Factor using } A^2 - 2AB + B^2 = (A - B)^2.$$
$$x - 9 = 0 \quad \text{Set the factor equal to zero.}$$
$$x = 9 \quad \text{Solve for } x.$$

We back-substitute this value of x into $y = \frac{54}{x}$.

$$\text{If } x = 9, \quad y = \tfrac{54}{9} = 6.$$

This means that the dimensions of the enclosure are 9 yards by 6 yards.

Step 4 Check the proposed solution in the original wording of the problem. With a length of 9 yards and a width of 6 yards, take a moment to check that this results in 36 yards of fencing and an area of 54 square yards.

Check Point 5 Find the length and width of a rectangle whose perimeter is 20 feet and whose area is 21 square feet.

EXERCISE SET 5.4

Practice Exercises

In Exercises 1–18, solve each system by the substitution method.

1. $x + y = 2$
   $y = x^2 - 4$

2. $x - y = -1$
   $y = x^2 + 1$

3. $x - y = -1$
   $y = x^2 + 2x - 3$

4. $2x + y = -5$
   $y = x^2 + 6x + 7$

5. $y = x^2 - 4x - 10$
   $y = -x^2 - 2x + 14$

6. $y = x^2 + 4x + 5$
   $y = x^2 + 2x - 1$

7. $x^2 + y^2 = 25$
   $x - y = 1$

8. $x^2 + y^2 = 5$
   $3x - y = 5$

9. $xy = 6$
   $2x - y = 1$

10. $xy = -12$
    $x - 2y + 14 = 0$

11. $y^2 = x^2 - 9$
    $2y = x - 3$

12. $x^2 + y = 4$
    $2x + y = 1$

13. $xy = 3$
    $x^2 + y^2 = 10$

14. $xy = 4$
    $x^2 + y^2 = 8$

15. $x + y = 1$
    $x^2 + xy - y^2 = -5$

16. $x + y = -3$
    $x^2 + 2y^2 = 12y + 18$

17. $x + y = 1$
    $(x - 1)^2 + (y + 2)^2 = 10$

18. $2x + y = 4$
    $(x + 1)^2 + (y - 2)^2 = 4$

In Exercises 19–28, solve each system by the addition method.

19. $x^2 + y^2 = 13$
    $x^2 - y^2 = 5$

20. $4x^2 - y^2 = 4$
    $4x^2 + y^2 = 4$

21. $x^2 - 4y^2 = -7$
    $3x^2 + y^2 = 31$

22. $3x^2 - 2y^2 = -5$
    $2x^2 - y^2 = -2$

23. $3x^2 + 4y^2 - 16 = 0$
    $2x^2 - 3y^2 - 5 = 0$

24. $32x^2 + 2y^2 - 50 = 0$
    $x^2 - y^2 - 10 = 0$

25. $x^2 + y^2 = 25$
    $(x - 8)^2 + y^2 = 41$

26. $x^2 + y^2 = 5$
    $x^2 + (y - 8)^2 = 41$

27. $y^2 - x = 4$
    $x^2 + y^2 = 4$

28. $x^2 - 2y = 8$
    $x^2 + y^2 = 16$

In Exercises 29–42, solve each system by the method of your choice.

29. $3x^2 + 4y^2 = 16$
    $2x^2 - 3y^2 = 5$

30. $x + y^2 = 4$
    $x^2 + y^2 = 16$

31. $2x^2 + y^2 = 18$
    $xy = 4$

32. $x^2 + 4y^2 = 20$
    $xy = 4$

33. $x^2 + 4y^2 = 20$
    $x + 2y = 6$

34. $3x^2 - 2y^2 = 1$
    $4x - y = 3$

35. $x^3 + y = 0$
    $x^2 - y = 0$

36. $x^3 + y = 0$
    $2x^2 - y = 0$

37. $x^2 + (y - 2)^2 = 4$
    $x^2 - 2y = 0$

38. $x^2 - y^2 - 4x + 6y - 4 = 0$
    $x^2 + y^2 - 4x - 6y + 12 = 0$

39. $y = (x + 3)^2$
    $x + 2y = -2$

40. $(x - 1)^2 + (y + 1)^2 = 5$
    $2x - y = 3$

41. $x^2 + y^2 + 3y = 22$
    $2x + y = -1$

42. $2x - y = -3$
    $x^2 + y^2 - 4x = 0$

In Exercises 43–46, let x represent one number and let y represent the other number. Use the given conditions to write a system of nonlinear equations. Solve the system and find the numbers.

43. The sum of two numbers is 10 and their product is 24. Find the numbers.

44. The sum of two numbers is 20 and their product is 96. Find the numbers.

45. The difference between the squares of two numbers is 3. Twice the square of the first number increased by the square of the second number is 9. Find the numbers.

46. The difference between the squares of two numbers is 5. Twice the square of the second number subtracted from three times the square of the first number is 19. Find the numbers.

Application Exercises

47. A planet's orbit follows a path described by $16x^2 + 4y^2 = 64$. A comet follows the parabolic path $y = x^2 - 4$. Where might the comet intersect the orbiting planet?

48. A system for tracking ships indicates that a ship lies on a path described by $16y^2 - x^2 = 16$. The process is repeated and the ship is found to lie on a path described by $9x^2 - 4y^2 = 36$. If it is known that the ship is located in the first quadrant of the coordinate system, determine its exact location.

49. Find the length and width of a rectangle whose perimeter is 36 feet and whose area is 77 square feet.

50. Find the length and width of a rectangle whose perimeter is 40 feet and whose area is 96 square feet.

Use the formula for the area of a rectangle and the Pythagorean Theorem to solve Exercises 51–52.

51. A small television has a picture with a diagonal measure of 10 inches and a viewing area of 48 square inches. Find the length and width of the screen.

52. The area of a rug is 108 square feet, and the length of its diagonal is 15 feet. Find the length and width of the rug.

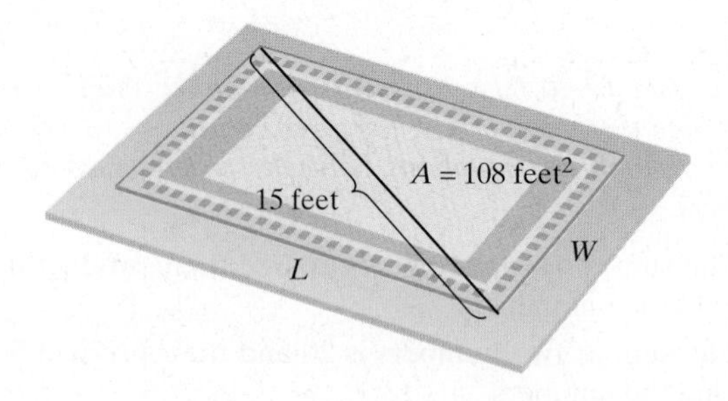

53. The figure at the top of the next column shows a square floor plan with a smaller square area that will accommodate a combination fountain and pool. The floor with the fountain-pool area removed has an area of 21 square meters and a perimeter of 24 meters. Find the dimensions of the floor and the dimensions of the square that will accommodate the pool.

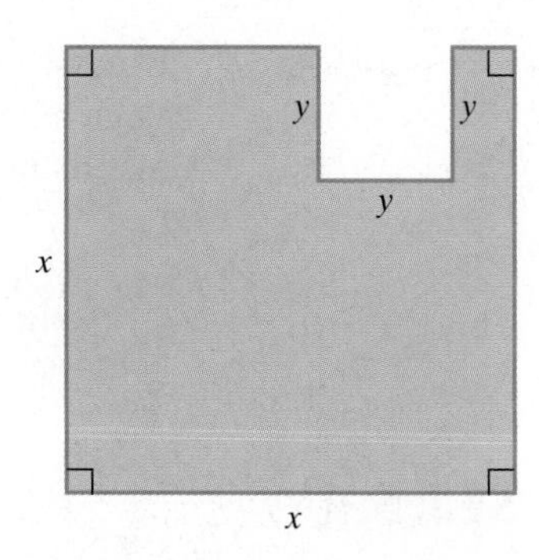

54. The area of the rectangular piece of cardboard shown on the left is 216 square inches. The cardboard is used to make an open box by cutting a 2-inch square from each corner and turning up the sides. If the box is to have a volume of 224 cubic inches, find the length and width of the cardboard that must be used.

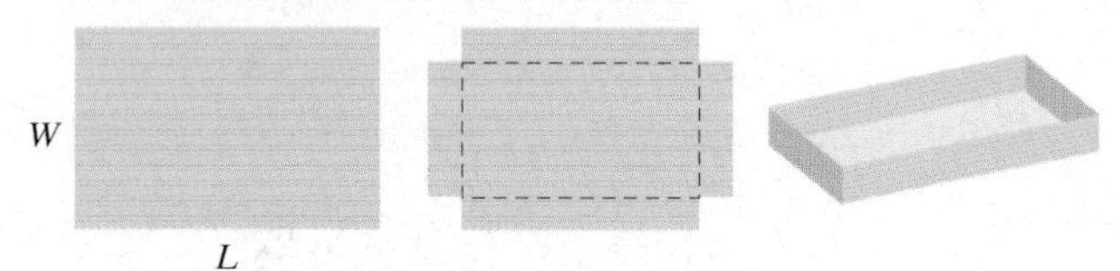

Writing in Mathematics

55. What is a system of nonlinear equations? Provide an example with your description.

56. Explain how to solve a nonlinear system using the substitution method. Use $x^2 + y^2 = 9$ and $2x - y = 3$ to illustrate your explanation.

57. Explain how to solve a nonlinear system using the addition method. Use $x^2 - y^2 = 5$ and $3x^2 - 2y^2 = 19$ to illustrate your explanation.

58. The daily demand and supply models for a carrot cake supplied by a bakery to a convenience store are given by the demand model $N = 40 - 3p$ and the supply model $N = \frac{p^2}{10}$, in which p is the price of the cake and N is the number of cakes sold or supplied each day to the convenience store. Explain how to determine the price at which supply and demand are equal. Then describe how to find how many carrot cakes can be supplied and sold each day at this price.

Technology Exercises

59. Verify your solutions to any five exercises from 1 through 42 by using a graphing utility to graph the two equations in the system in the same viewing rectangle. Then use the trace or intersection feature to verify the solutions.

60. Write a system of equations, one equation whose graph is a line and the other whose graph is a parabola, that has no ordered pairs that are real numbers in its solution set. Graph the equations using a graphing utility and verify that you are correct.

Critical Thinking Exercises

61. Which one of the following is true?

a. A system of two equations in two variables whose graphs represent a circle and a line can have four real solutions.

b. A system of two equations in two variables whose graphs represent a parabola and a circle can have four real solutions.

c. A system of two equations in two variables whose graphs represent two circles must have at least two real solutions.

d. A system of two equations in two variables whose graphs represent a parabola and a circle cannot have only one real solution.

62. The points of intersection of the graphs of $xy = 20$ and $x^2 + y^2 = 41$ are joined to form a rectangle. Find the area of the rectangle.

63. Find a and b in this figure.

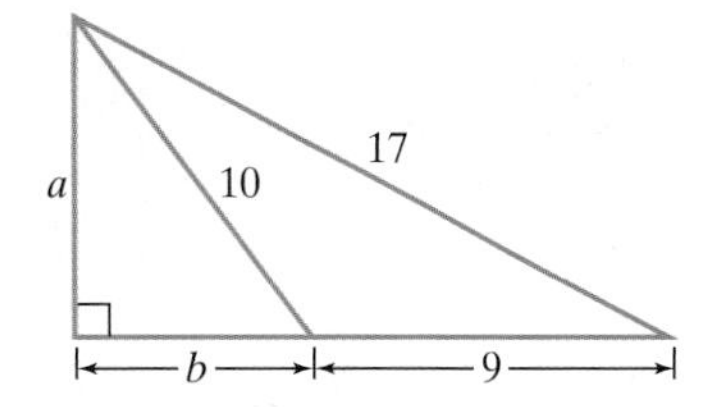

Solve the systems in Exercises 64–65.

64. $\log_y x = 3$
$\log_y (4x) = 5$

65. $\log x^2 = y + 3$
$\log x = y - 1$

SECTION 5.5 Systems of Inequalities

Objectives

1. Graph a linear inequality in two variables.
2. Graph a nonlinear inequality in two variables.
3. Graph a system of inequalities.
4. Solve applied problems involving systems of inequalities.

Had a good workout lately? If so, could you tell if you were overdoing it or not pushing yourself hard enough? In this section, we will use systems of inequalities in two variables to help you establish a target zone for your workouts.

1 Graph a linear inequality in two variables.

Graphing a Linear Inequality in Two Variables

We have seen that equations in the form $Ax + By = C$ are straight lines when graphed. If we change the = sign to $>, <, \geq$, or $\leq$, we obtain a **linear inequality in two variables**. Some examples of linear inequalities in two variables are $x + y > 2$, $3x - 5y \leq 15$, and $2x - y < 4$.

Figure 5.12 shows the graph of the linear equation $x + y = 2$. The line divides the points in the rectangular coordinate system into three sets. First, there is the set of points along the line, satisfying $x + y = 2$. Next, there is the set of points in the green region above the line. Points in the green region satisfy the linear inequality $x + y > 2$. Finally, there is the set of points in the pink region below the line. Points in the pink region satisfy the linear inequality $x + y < 2$.

Half-plane $x + y > 2$

Half-plane $x + y < 2$

Line $x + y = 2$

Figure 5.12 All points on the line satisfy $x + y = 2$. All points in the green half-plane above the line satisfy $x + y > 2$. All points in the pink half-plane below the line satisfy $x + y < 2$.

A **half-plane** is the set of all the points on one side of a line. In Figure 5.12, the green region is a half-plane. The pink region is also a half-plane. A half-plane is the solution set of a linear inequality that involves $>$ or $<$. The solution set of an inequality that involves $\geq$ or $\leq$ is a half-plane and a line. A solid line is used to show that the line is part of the solution set. A dashed line is used to show that a line is not part of a solution set.

Graphing a Linear Inequality in Two Variables

1. Replace the inequality symbol with an equal sign and graph the corresponding linear equation. Draw a solid line if the original inequality contains a $\leq$ or $\geq$ symbol. Draw a dashed line if the original inequality contains a $<$ or $>$ symbol.
2. Choose a test point in one of the half-planes that is not on the line. Substitute the coordinates of the test point into the inequality.
3. If a true statement results, shade the half-plane containing this test point. If a false statement results, shade the half-plane not containing this test point.

EXAMPLE 1 Graphing a Linear Inequality in Two Variables

Graph: $3x - 5y < 15$.

Solution

Step 1 Replace the inequality symbol by = and graph the linear equation. We need to graph $3x - 5y = 15$. We can use intercepts to graph this line.

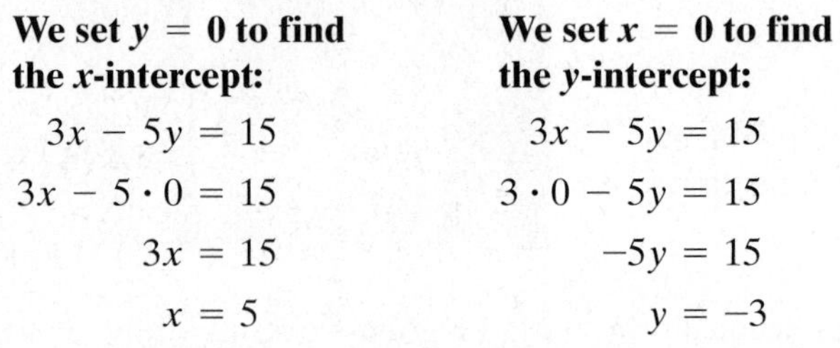

We set $y = 0$ to find the x-intercept:	We set $x = 0$ to find the y-intercept:
$3x - 5y = 15$	$3x - 5y = 15$
$3x - 5 \cdot 0 = 15$	$3 \cdot 0 - 5y = 15$
$3x = 15$	$-5y = 15$
$x = 5$	$y = -3$

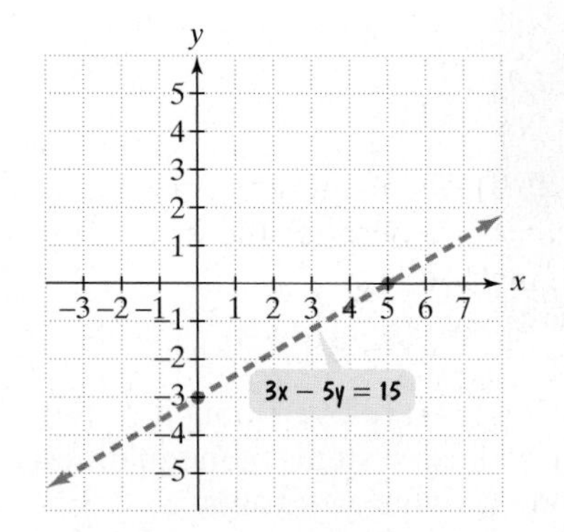

Figure 5.13 Preparing to graph $3x - 5y < 15$

The x-intercept is 5, so the line passes through $(5, 0)$. The y-intercept is -3, so the line passes through $(0, -3)$. The graph is indicated by a dashed line because the inequality $3x - 5y < 15$ contains a $<$ symbol, rather than $\leq$. The graph of the line is shown in Figure 5.13.

Step 2 Choose a test point in one of the half-planes that is not on the line. Substitute its coordinates into the inequality. The line $3x - 5y = 15$ divides the plane into three parts—the line itself and two half-planes. The points in one half-plane satisfy $3x - 5y > 15$. The points in the other half-plane satisfy $3x - 5y < 15$. We need to find which half-plane is the solution. To do so, we test a point from either half-plane. The origin, $(0, 0)$, is the easiest point to test.

$$3x - 5y < 15 \quad \text{This is the given inequality.}$$

$$\text{Is } 3 \cdot 0 - 5 \cdot 0 < 15? \quad \text{Test } (0, 0) \text{ by substituting 0 for } x \text{ and 0 for } y.$$

$$0 - 0 < 15$$

$$0 < 15, \text{ true}$$

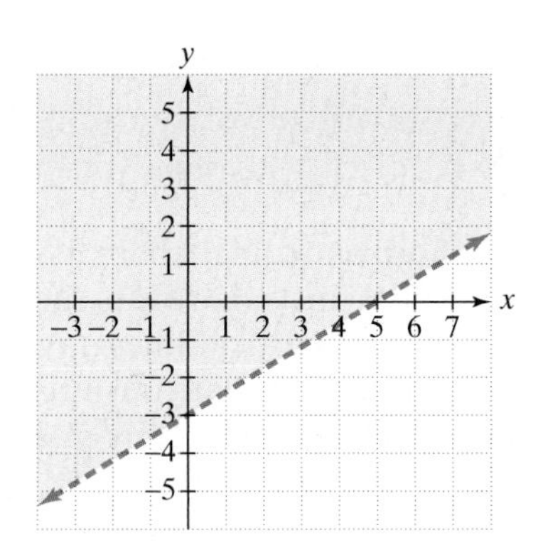

Figure 5.14 The graph of $3x - 5y < 15$

Step 3 If a true statement results, shade the half-plane containing the test point. Because 0 is less than 15, the test point $(0, 0)$ is part of the solution set. All the points on the same side of the line $3x - 5y = 15$ as the point $(0, 0)$ are members of the solution set. The solution set is the half-plane that contains the point $(0, 0)$, indicated by shading this half-plane. The graph is shown using green shading and a dashed blue line in Figure 5.14.

Check Point 1 Graph: $2x - 4y < 8$.

When graphing a linear inequality, test a point that lies in one of the half-planes and *not on the line dividing the half-planes.* The test point $(0, 0)$ is convenient because it is easy to calculate when 0 is substituted for each variable. However, if $(0, 0)$ lies on the dividing line and not in a half-plane, a different test point must be selected.

EXAMPLE 2 Graphing a Linear Inequality

Graph: $y \leq \frac{2}{3}x$.

Solution

Step 1 Replace the inequality symbol by = and graph the linear equation. We need to graph $y = \frac{2}{3}x$. We can use the slope and the y-intercept to graph this line.

$$y = \frac{2}{3}x + 0$$

Slope $= \frac{2}{3} = \frac{\text{rise}}{\text{run}}$ y-intercept = 0

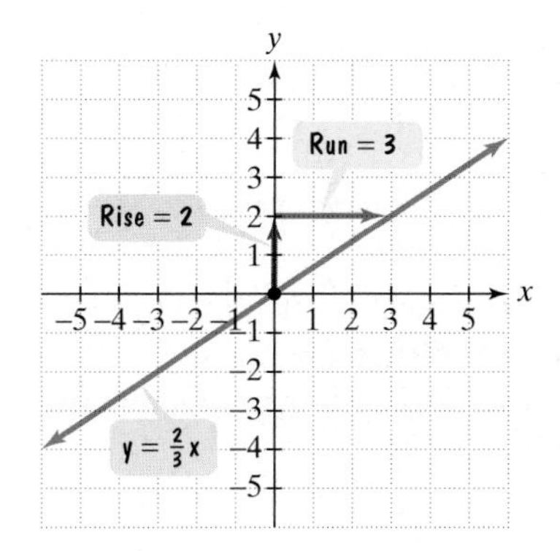

Figure 5.15 Preparing to graph $y \leq \frac{2}{3}x$

The y-intercept is 0 and the line passes through $(0, 0)$. Using the y-intercept and the slope, the line is shown in Figure 5.15 as a solid line because the inequality $y \leq \frac{2}{3}x$ contains a $\leq$ symbol, in which equality is included.

Step 2 Choose a test point in one of the half-planes that is not on the line. Substitute its coordinates into the inequality. We cannot use $(0, 0)$ as a test point because it lies on the line and not in a half-plane. Let's use $(1, 1)$, which lies in the half-plane above the line.

$$y \leq \frac{2}{3}x \quad \text{This is the given inequality.}$$

$$\text{Is } 1 \leq \frac{2}{3} \cdot 1? \quad \text{Test } (1, 1) \text{ by substituting 1 for } x \text{ and 1 for } y.$$

$$1 \leq \frac{2}{3}, \text{ false}$$

Step 3 If a false statement results, shade the half-plane not containing the test point. Because 1 is not less than or equal to $\frac{2}{3}$, the test point $(1, 1)$ is not part of the solution set. Thus, the half-plane below the solid line $y = \frac{2}{3}x$ is part of the solution set. The solution set is the line and the half-plane that does not contain the point $(1, 1)$, indicated by shading this half-plane. The graph is shown using green shading and a blue line in Figure 5.16.

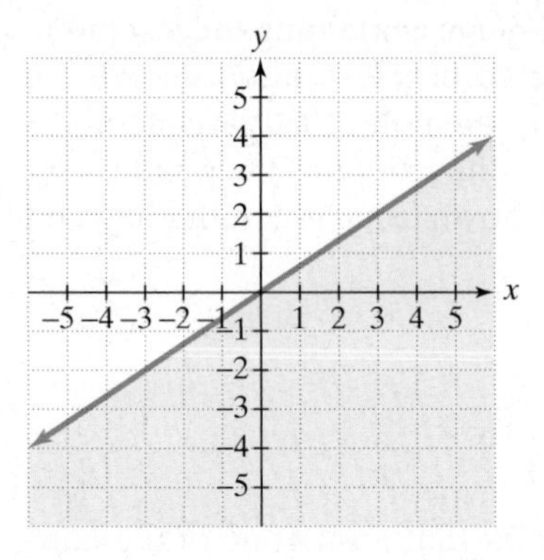

Figure 5.16 The graph of $y \le \frac{2}{3}x$

Check Point 2 Graph: $y \ge \frac{1}{2}x$.

In Chapter 1, we learned that $y = b$ graphs as a horizontal line, where b is the y-intercept. Similarly, the graph of $x = a$ is a vertical line, where a is the x-intercept. Half-planes can be separated by horizontal or vertical lines. For example, Figure 5.17 shows the graph of $y \le 2$. Because $(0, 0)$ satisfies this inequality ($0 \le 2$ is true), the graph consists of the half-plane below the line $y = 2$ and the line. Similarly, Figure 5.18 shows the graph of $x < 4$.

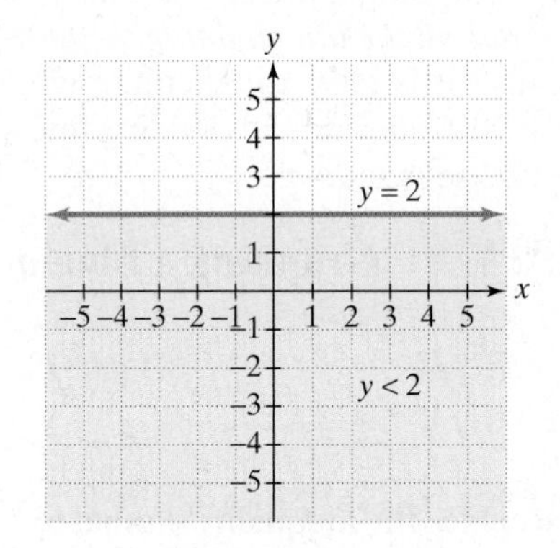

Figure 5.17 The graph of $y \le 2$

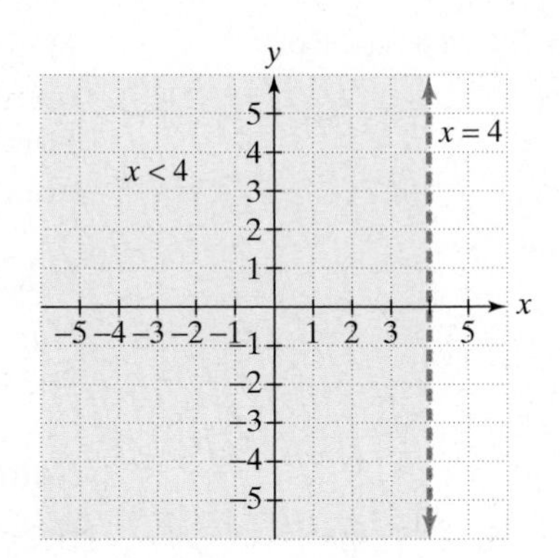

Figure 5.18 The graph of $x < 4$

2 Graph a nonlinear inequality in two variables.

Graphing a Nonlinear Inequality in Two Variables

Example 3 illustrates that a nonlinear inequality in two variables is graphed in the same way that we graph a linear inequality.

EXAMPLE 3 Graphing a Nonlinear Inequality in Two Variables

Graph: $x^2 + y^2 \le 9$.

Solution

Step 1 Replace the inequality symbol by = and graph the nonlinear equation. We need to graph $x^2 + y^2 = 9$. The graph is a circle of radius 3 with its center at the origin. The graph is shown in Figure 5.19 as a solid circle because equality is included in the $\le$ symbol.

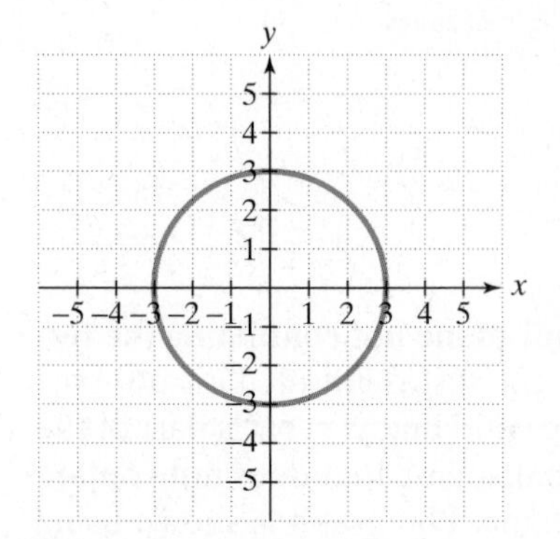

Figure 5.19 Preparing to graph $x^2 + y^2 \le 9$

Step 2 Choose a test point in one of the regions that is not on the circle. Substitute its coordinates into the inequality. The circle divides the plane into three parts—the circle itself, the region inside the circle, and the region outside the circle. We need to determine whether the region inside or outside the circle is the solution. To do so, we will use the test point $(0, 0)$ from inside the circle.

$x^2 + y^2 \le 9$ This is the given inequality.

Is $0^2 + 0^2 \le 9$?   Test $(0, 0)$ by substituting 0 for x and 0 for y.

$0 + 0 \le 9$

$0 \le 9$, true

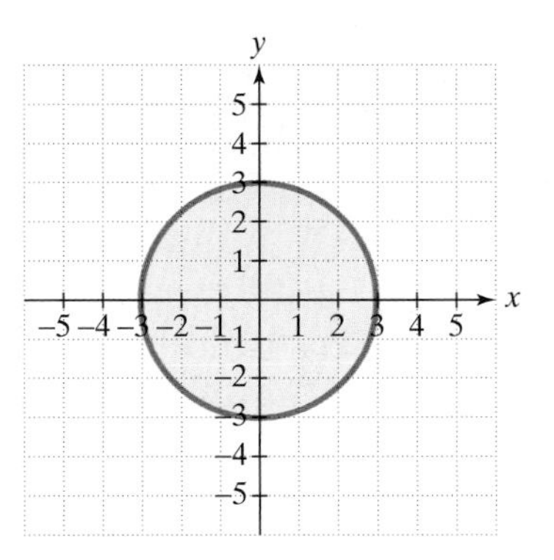

Figure 5.20 The graph of $x^2 + y^2 \le 9$

Step 3 If a true statement results, shade the region containing the test point. The true statement tells us that all the points inside the circle satisfy $x^2 + y^2 \le 9$. The graph is shown using green shading and a solid blue circle in Figure 5.20.

Check Point 3 Graph: $x^2 + y^2 \ge 16$.

3 Graph a system of inequalities.

Systems of Inequalities in Two Variables

The **solution set of a system of inequalities** in two variables x and y is the set of all points (x, y) that satisfy each inequality in the system. The **graph of a system of inequalities** in two variables is the graph of the system's solution set. Thus, to graph a system of inequalities in two variables, begin by graphing each individual inequality in the same rectangular coordinate system. Then find the region, if there is one, that is common to every graph in the system.

EXAMPLE 4 Graphing a System of Linear Inequalities

Graph the solution set:

$$2x - y < 4$$
$$x + y \ge -1.$$

Solution We begin by graphing $2x - y < 4$. Because the inequality contains a $<$ symbol, rather than $\le$, we graph $2x - y = 4$ as a dashed line. (If $x = 0$, then $y = -4$, and if $y = 0$, then $x = 2$. The x-intercept is 2 and the y-intercept is -4.) Because $(0, 0)$ makes the inequality $2x - y < 4$ true, we shade the half-plane containing $(0, 0)$, shown in yellow in Figure 5.21.

Now we graph $x + y \ge -1$ in the same rectangular coordinate system. Because the inequality contains a $\ge$ symbol, in which equality is included, we graph $x + y = -1$ as a solid line. (If $x = 0$, then $y = -1$, and if $y = 0$, then $x = -1$. The x-intercept and y-intercept are both -1.) Because $(0, 0)$ makes the inequality true, we shade the half-plane containing $(0, 0)$. This is shown in Figure 5.22 using green vertical shading. The solution set of the system is shown graphically by the intersection (the overlap) of the two half-planes. This is shown in Figure 5.22 as the region in which the yellow shading and the green vertical shading overlap. The solution of the system is shown again in Figure 5.23.

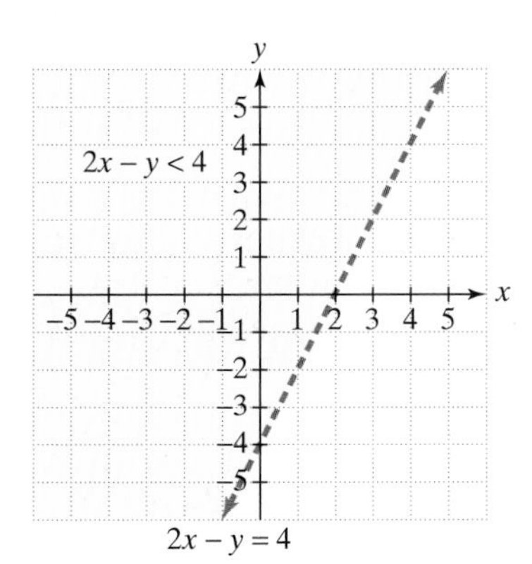

Figure 5.21 The graph of $2x - y < 4$

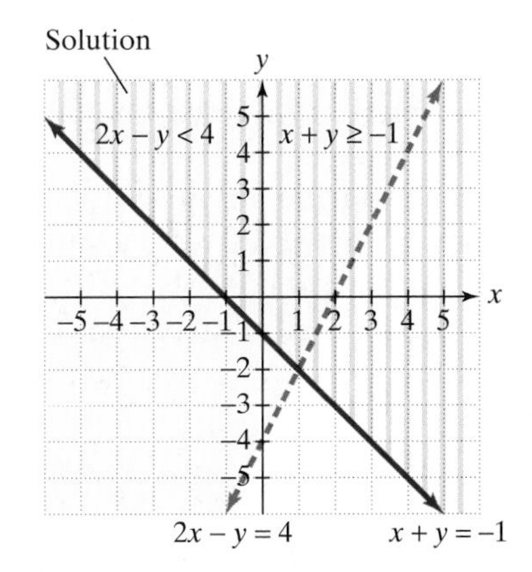

Figure 5.22 Adding the graph of $x + y \ge -1$

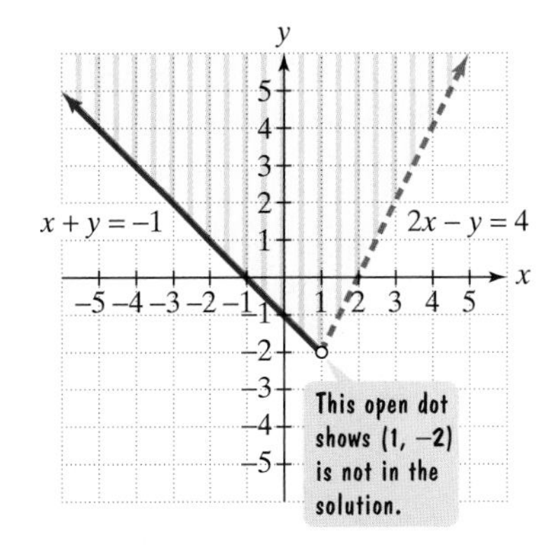

Figure 5.23 The graph of $2x - y < 4$ and $x + y \ge -1$

Check Point 4 Graph the solution set:

$$x + 2y > 4$$
$$2x - 3y \leq -6.$$

EXAMPLE 5 Graphing a System of Inequalities

Graph the solution set:

$$y \geq x^2 - 4$$
$$x - y \geq 2.$$

Solution We begin by graphing $y \geq x^2 - 4$. Because equality is included in $\geq$, we graph $y = x^2 - 4$ as a solid parabola. Because (0, 0) makes the inequality $y \geq x^2 - 4$ true (we obtain $0 \geq -4$), we shade the interior portion of the parabola containing (0, 0), shown in yellow in Figure 5.24.

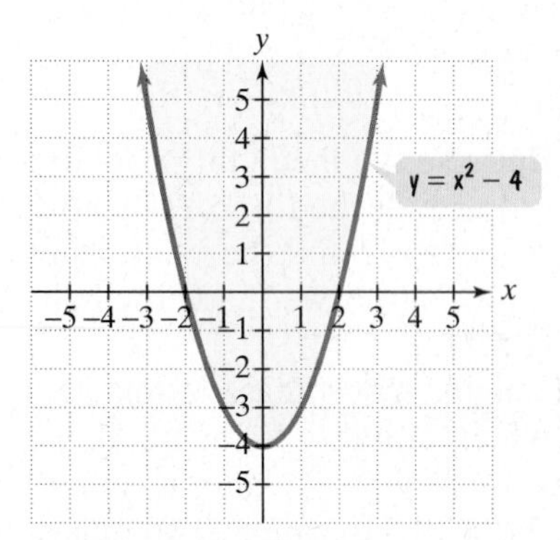

Figure 5.24 The graph of $y \geq x^2 - 4$

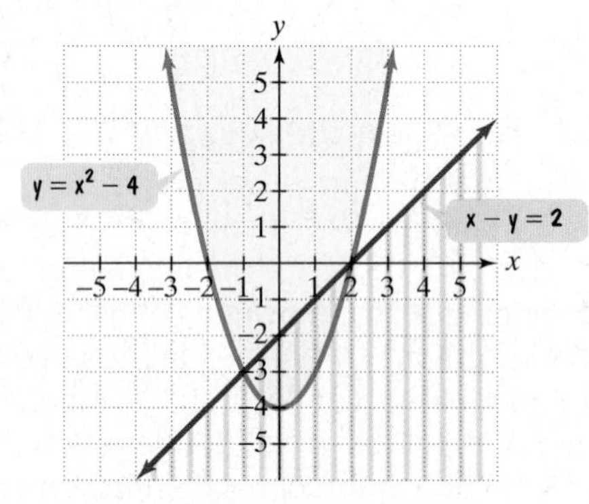

Figure 5.25 Adding the graph of $x - y \geq 2$

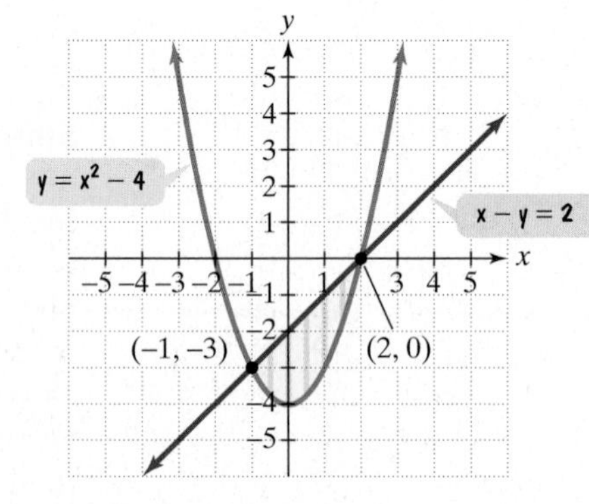

Figure 5.26 The graph of $y \geq x^2 - 4$ and $x - y \geq 2$

Now we graph $x - y \geq 2$ in the same rectangular coordinate system. First we graph the line $x - y = 2$ using its x-intercept, 2, and its y-intercept, -2. Because (0, 0) makes the inequality $x - y \geq 2$ false (we obtain $0 \geq 2$), we shade the half-plane below the line. This is shown in Figure 5.25 using green vertical shading.

The solution of the system is shown in Figure 5.25 by the intersection (the overlap) of the solid yellow and green vertical shadings. The graph of the system's solution set consists of the region enclosed by the parabola and the line. To find the points of intersection of the parabola and the line, use the substitution method to solve the nonlinear system

$$y = x^2 - 4$$
$$x - y = 2.$$

Take a moment to show that the solutions are $(-1, -3)$ and $(2, 0)$, as shown in Figure 5.26.

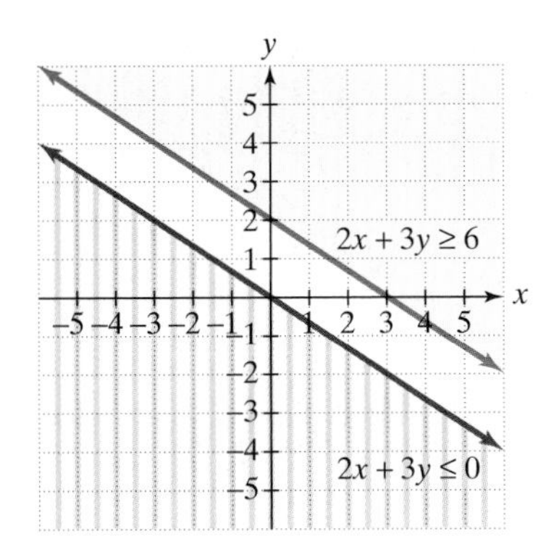

Figure 5.27 A system of inequalities with no solution.

Check Point 5 Graph the solution set:

$$y \geq x^2 - 4$$
$$x + y \leq 2.$$

A system of inequalities has no solution if there are no points in the rectangular coordinate system that simultaneously satisfy each inequality in the system. For example, the system

$$2x + 3y \geq 6$$
$$2x + 3y \leq 0$$

whose graph is shown in Figure 5.27 has no overlapping region. Thus, the system has no solution.

EXAMPLE 6 Graphing a System of Inequalities

Graph the solution set:

$$x - y < 2$$
$$-2 \leq x < 4$$
$$y < 3.$$

Solution We begin by graphing $x - y < 2$, the first given inequality. The line $x - y = 2$ has an x-intercept of 2 and a y-intercept of -2. The test point $(0, 0)$ makes the inequality $x - y < 2$ true, and its graph is shown in Figure 5.28.

Now let's consider the second given inequality $-2 \leq x < 4$. Replacing the inequality symbols by $=$, we obtain $x = -2$ and $x = 4$, graphed as vertical lines. The line of $x = 4$ is not included. Using $(0, 0)$ as a test point and substituting the x-coordinate, 0, into $-2 \leq x < 4$, we obtain the true statement $-2 \leq 0 < 4$. We therefore shade the region between the vertical lines. We've added this region to Figure 5.28, intersecting the region between the vertical lines with the yellow region in Figure 5.28. The resulting region is shown in yellow and green vertical shading in Figure 5.29.

Finally, let's consider the third given inequality, $y < 3$. Replacing the inequality symbol by $=$, we obtain $y = 3$, which graphs as a horizontal line. Because $(0, 0)$ satisfies $y < 3$ ($0 < 3$ is true), the graph consists of the half-plane below the line $y = 3$. We've added this half-plane to the region in Figure 5.29, intersecting the half-plane with this region. The resulting region is shown in yellow and green vertical shading in Figure 5.30. This region represents the graph of the solution set of the given system.

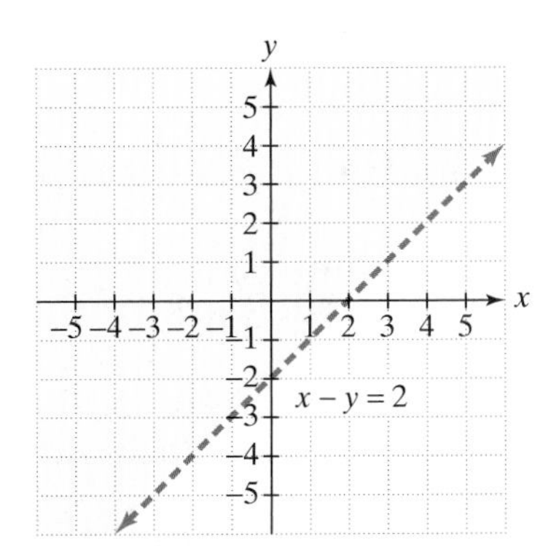

Figure 5.28 The graph of $x - y < 2$

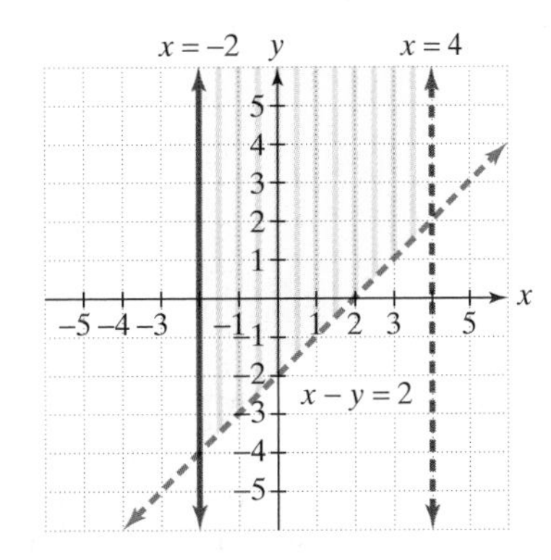

Figure 5.29 The graph of $x - y < 2$ and $-2 \leq x < 4$

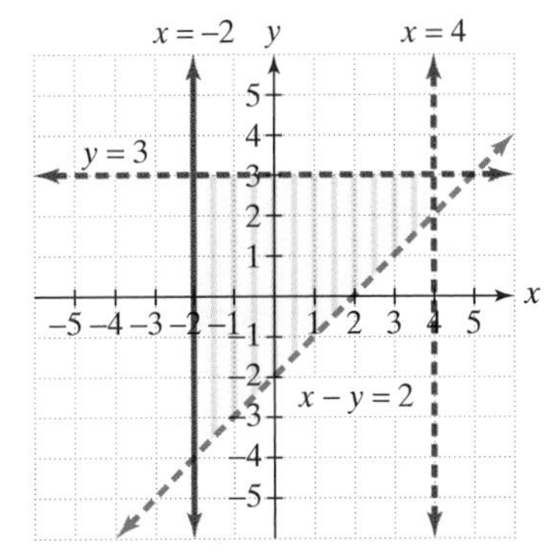

Figure 5.30 The graph of $x - y < 2$ and $-2 \leq x < 4$ and $y < 3$

Check Point 6 Graph the solution set:

$$\begin{aligned} x + y &< 2 \\ -2 \leq x &< 1 \\ y &> -3. \end{aligned}$$

4 Solve applied problems involving systems of inequalities.

Applications

Now we are ready to use a system of inequalities to establish a target zone for your workouts.

EXAMPLE 7 Inequalities and Aerobic Exercise

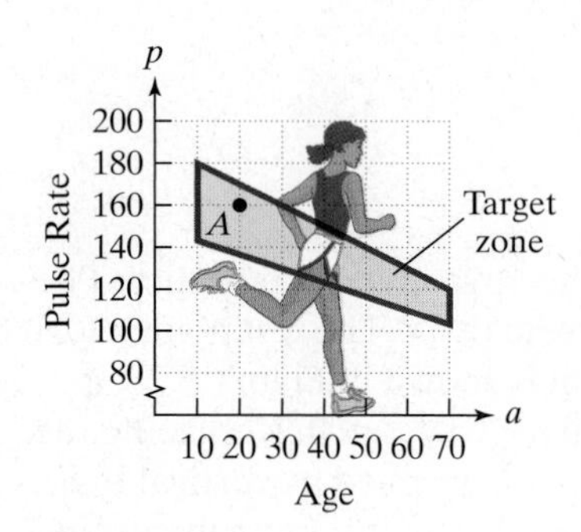

Figure 5.31

For people between ages 10 and 70, inclusive, the target zone for aerobic exercise is given by the following system of inequalities in which a represents one's age and p is one's pulse rate.

$$2a + 3p \geq 450$$
$$a + p \leq 190$$

The graph of this target zone is shown in Figure 5.31. Find your age. The line segments on the top and bottom of the shaded region indicate upper and lower limits for your pulse rate, in beats per minute, when engaging in aerobic exercise.

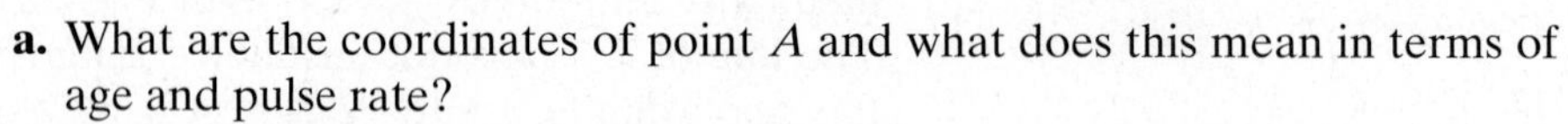

a. What are the coordinates of point A and what does this mean in terms of age and pulse rate?

b. Show that the coordinates of point A satisfy each inequality in the system.

Solution

a. Point A has coordinates (20, 160). This means that a pulse rate of 160 beats per minute is within the target zone for a 20-year-old person engaged in aerobic exercise.

b. We can show that (20, 160) satisfies each inequality by substituting 20 for a and 160 for p.

$$2a + 3p \geq 450 \qquad\qquad a + p \leq 190$$

$$\text{Is } 2(20) + 3(160) \geq 450? \qquad\qquad \text{Is } 20 + 160 \leq 190?$$

$$40 + 480 \geq 450 \qquad\qquad 180 \leq 190, \text{ true}$$

$$520 \geq 450, \text{ true}$$

The pair (20, 160) makes each inequality true, so it satisfies each inequality in the system.

Check Point 7 Identify a point other than A in the target zone in Figure 5.31.

a. What are the coordinates of this point and what does this mean in terms of age and pulse rate?

b. Show that the coordinates of the point satisfy each inequality in the system in Example 7.

EXERCISE SET 5.5

Practice Exercises

In Exercises 1–22, graph each inequality.

1. $x + 2y \le 8$
2. $3x - 6y \le 12$
3. $x - 2y > 10$
4. $2x - y > 4$
5. $y \le \frac{1}{3}x$
6. $y \le \frac{1}{4}x$
7. $y > 2x - 1$
8. $y > 3x + 2$
9. $x \le 1$
10. $x \le -3$
11. $y > 1$
12. $y > -3$
13. $x^2 + y^2 \le 1$
14. $x^2 + y^2 \le 4$
15. $x^2 + y^2 > 25$
16. $x^2 + y^2 > 36$
17. $y < x^2 - 1$
18. $y < x^2 - 9$
19. $y \ge x^2 - 9$
20. $y \ge x^2 - 1$
21. $y > 2^x$
22. $y \le 3^x$

In Exercises 23–52, graph the solution set of each system of inequalities or indicate that the system has no solution.

23. $3x + 6y \le 6$, $2x + y \le 8$
24. $x - y \ge 4$, $x + y \le 6$
25. $2x - 5y \le 10$, $3x - 2y > 6$
26. $2x - y \le 4$, $3x + 2y > -6$
27. $y > 2x - 3$, $y < -x + 6$
28. $y < -2x + 4$, $y < x - 4$
29. $x + 2y \le 4$, $y \ge x - 3$
30. $x + y \le 4$, $y \ge 2x - 4$
31. $x \le 2$, $y \ge -1$
32. $x \le 3$, $y \le -1$
33. $-2 \le x < 5$
34. $-2 < y \le 5$
35. $x - y \le 1$, $x \ge 2$
36. $4x - 5y \ge -20$, $x \ge -3$
37. $x + y > 4$, $x + y < -1$
38. $x + y > 3$, $x + y < -2$
39. $x + y > 4$, $x + y > -1$
40. $x + y > 3$, $x + y > -2$
41. $y \ge x^2 - 1$, $x - y \ge -1$
42. $y \ge x^2 - 4$, $x - y \ge 2$
43. $x^2 + y^2 \le 16$, $x + y > 2$
44. $x^2 + y^2 \le 4$, $x + y > 1$
45. $x^2 + y^2 > 1$, $x^2 + y^2 < 4$
46. $x^2 + y^2 > 1$, $x^2 + y^2 < 9$
47. $x - y \le 2$, $x \ge -2$, $y \le 3$
48. $3x + y \le 6$, $x \ge -2$, $y \le 4$
49. $x \ge 0$, $y \ge 0$, $2x + 5y \le 10$, $3x + 4y \le 12$
50. $x \ge 0$, $y \ge 0$, $2x + y \le 4$, $2x - 3y \le 6$
51. $3x + y \le 6$, $2x - y \le -1$, $x \ge -2$, $y \le 4$
52. $2x + y \le 6$, $x + y \ge 2$, $1 \le x \le 2$, $y \le 3$

Application Exercises

53. Use Figure 5.31 on page 464 to solve this exercise.
    a. Find a pulse rate that lies within the target zone for a person your age engaged in aerobic exercise.
    b. Express your answer in part (a) as an ordered pair. Show that the coordinates of this ordered pair satisfy each inequality in the system.

The shaded region in the figure shows recommended weight and height combinations based on information from the Department of Agriculture. Use this region to solve Exercises 54–57.

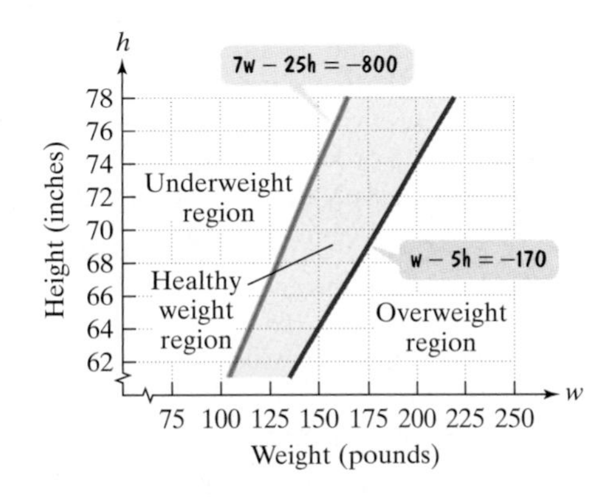

54. Is a person who is 70 inches tall weighing 175 pounds within the healthy weight region?

55. Is a person who is 64 inches tall weighing 105 pounds within the healthy weight region?

56. Estimate the recommended weight range for a person who is 6 feet tall.

57. Write a system of linear inequalities that describes the region for recommended weight and height combinations.

Temperature and precipitation affect whether or not trees and forests can grow. At certain levels of precipitation and temperature, only grasslands and deserts will exist. The figure shows three kinds of regions—deserts, grasslands, and forests—that result from various ranges of temperature and precipitation. Use the figure to solve Exercises 58–60.

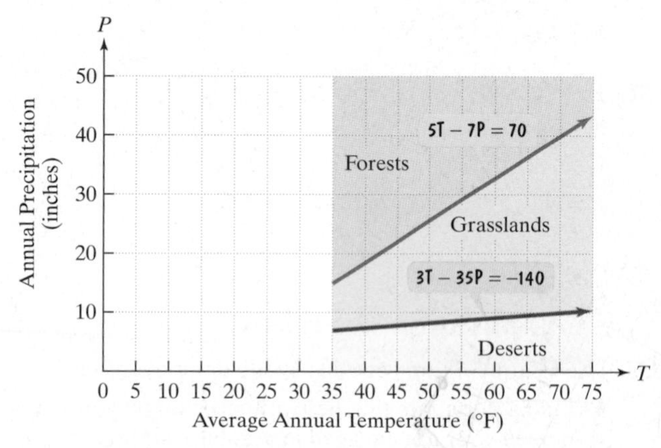

Source: A. Miller and J. Thompson, *Elements of Meterology*

58. Write a system of inequalities that describe where deserts occur.

59. Write a system of inequalities that describe where forests occur.

60. Write a system of inequalities that describe where grasslands occur.

61. Many elevators have a capacity of 2000 pounds. If a child averages 50 pounds and an adult 150 pounds, write an inequality that describes when x children and y adults will cause the elevator to be overloaded. Graph the inequality. Select an ordered pair satisfying the inequality. Describe what this means in practical terms.

62. Suppose a patient is not allowed to have more than 330 milligrams of cholesterol from a diet of eggs and meat. Each egg provides 165 milligrams of cholesterol and each ounce of meat provides 110 milligrams of cholesterol. Thus, $165x + 110y \leq 330$, where x is the number of eggs and y the number of ounces of meat. Graph the inequality in the first quadrant. Give the coordinates of any two points in the solution set. Describe what each set of coordinates means in terms of the variables in the problem.

63. A person with $15,000 plans to place the money in two investments. One investment is high risk, high yield; the other is low risk, low yield. At least $2000 is to be placed in the high-risk investment. Furthermore, the amount invested at low risk should be at least three times the amount invested at high risk. Find and graph a system of inequalities that describes all possibilities for placing the money in the high- and low-risk investments.

64. Promoters of a rock concert must sell at least 25,000 tickets priced at $35 and $50 per ticket. Furthermore, the promoters must take in at least $1,025,000 in ticket sales. Find and graph a system of inequalities that describes all possibilities for selling the $35 tickets and the $50 tickets.

Writing in Mathematics

65. What is a half-plane?

66. What does a dashed line mean in the graph of an inequality?

67. Explain how to graph $2x - 3y < 6$.

68. Compare the graphs of $3x - 2y > 6$ and $3x - 2y \leq 6$. Discuss similarities and differences between the graphs.

69. Describe how to solve a system of inequalities.

70. Look at the shaded region showing recommended weight and height combinations in the figure for Exercises 54–57. Describe why a system of inequalities, rather than an equation, is better suited to give the recommended combinations.

Technology Exercises

Graphing utilities can be used to shade regions in the rectangular coordinate system, thereby graphing an inequality in two variables. Read the section of the user's manual for your graphing utility that describes how to shade a region. Then use your graphing utility to graph the inequalities in Exercises 71–74.

71. $y \leq 4x + 4$

72. $y \geq \frac{2}{3}x - 2$

73. $y \geq x^2 - 4$

74. $y \geq \frac{1}{2}x^2 - 2$

75. Does your graphing utility have any limitations in terms of graphing inequalities? If so, what are they?

76. Use a graphing utility with a SHADE feature to verify any five of the graphs that you drew by hand in Exercises 1–22.

77. Use a graphing utility with a SHADE feature to verify any five of the graphs that you drew by hand for the systems in Exercises 23–52.

Critical Thinking Exercises

78. Write a system of inequalities that has no solution.

79. Write a system of inequalities that describes the shaded region in the figure at the top of the next column.

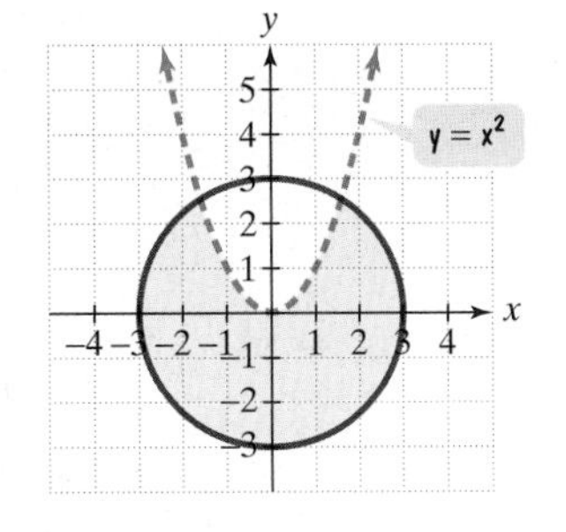

80. Sketch the graph of the solution set for the following system of inequalities:

$$y \geq nx + b \quad (n < 0, b > 0)$$
$$y \leq mx + b \quad (m > 0, b > 0).$$

81. Sketch the graph of the solution set for the following system of inequalities:

$$|x + y| \leq 3$$
$$|y| \leq 2.$$

SECTION 5.6 *Linear Programming*

Objectives

1. Write an objective function describing a quantity that must be maximized or minimized.
2. Use inequalities to describe limitations in a situation.
3. Use linear programming to solve problems.

West Berlin children at Tempelhof airport watch fleets of U.S. airplanes bringing in supplies to circumvent the Russian blockade. The airlift began June 28, 1948 and continued for 15 months.

The Berlin Airlift (1948–1949) was an operation by the United States and Great Britain. It was a response to military action by the former Soviet Union: The Soviet troops closed all roads and rail lines between West Germany and Berlin, cutting off supply routes to the city. The Allies used a mathematical technique developed during World War II to maximize the amount of supplies transported. During the 15-month airlift, 278,228 flights provided basic necessities to blockaded Berlin, saving one of the world's great cities.

In this section, we will look at an important application of systems of linear inequalities. Such systems arise in **linear programming**, a method for solving problems in which a particular quantity that must be maximized or minimized is limited. Linear programming is one of the most widely used tools in management science. It helps businesses allocate resources to manufacture products in a way that will maximize profit. Linear programming accounts for more than 50% and perhaps as much as 90% of all computing time used for management decisions in business. The Allies used linear programming to save Berlin.

1 Write an objective function describing a quantity that must be maximized or minimized.

Objective Functions in Linear Programming

Many problems involve quantities that must be maximized or minimized. Businesses are interested in maximizing profit. An operation in which bottled water and medical kits are shipped to earthquake victims needs to maximize the number of victims helped by this shipment. An **objective function** is an algebraic expression in two or more variables describing a quantity that must be maximized or minimized.

EXAMPLE 1 Writing an Objective Function

Bottled water and medical supplies are to be shipped to victims of an earthquake by plane. Each container of bottled water will serve 10 people and each medical kit will aid 6 people. If x represents the number of bottles of water to be shipped and y represents the number of medical kits, write the objective function that describes the number of people that can be helped.

Solution Because each bottle of water serves 10 people and each medical kit aids 6 people, we have

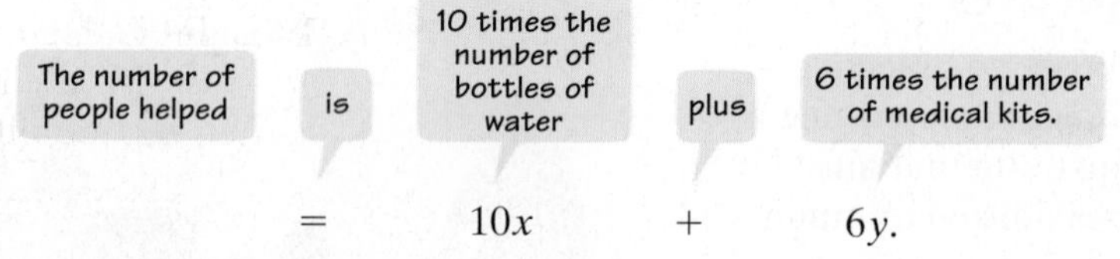

$$= \quad 10x \quad + \quad 6y.$$

Using z to represent the objective function, we have

$$z = 10x + 6y.$$

Unlike the functions that we have seen so far, the objective function is an equation in three variables. For a value of x and a value of y, there is one and only one value of z. Thus, z is a function of x and y.

A company manufactures bookshelves and desks for computers. Let x represent the number of bookshelves manufactured daily and y the number of desks manufactured daily. The company's profits are \$25 per bookshelf and \$55 per desk. Write the objective function that describes the company's total daily profit, z, from x bookshelves and y desks. (Checkpoints 1 through 4 are related to this situation, so keep track of your answers.)

2 Use inequalities to describe limitations in a situation.

Constraints in Linear Programming

Ideally, the number of earthquake victims helped in Example 1 should increase without restriction so that every victim receives water and medical kits. However, the planes that ship these supplies are subject to weight and volume restrictions. In linear programming problems, such restrictions are called **constraints**. Each constraint is expressed as a linear inequality. The list of constraints forms a system of linear inequalities.

EXAMPLE 2 Writing a Constraint

Each plane can carry no more than 80,000 pounds. The bottled water weighs 20 pounds per container and each medical kit weighs 10 pounds. If x represents the number of bottles of water to be shipped and y represents the number of medical kits, write an inequality that describes this constraint.

Solution Because each plane can carry no more than 80,000 pounds, we have

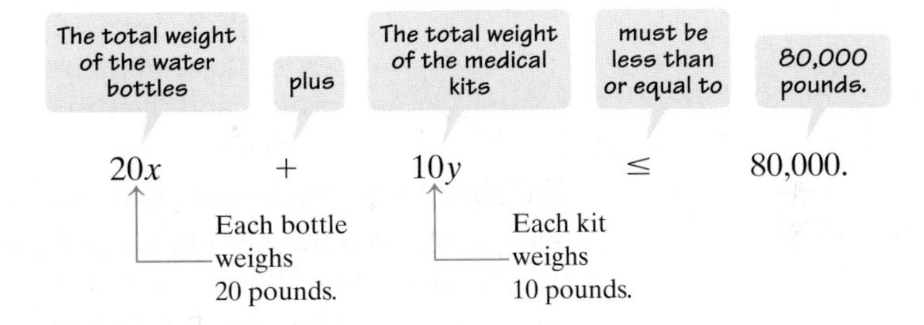

The plane's weight constraint is described by the inequality

$$20x + 10y \leq 80{,}000.$$

Check Point 2 To maintain high quality, the company in Checkpoint 1 should not manufacture more than 80 bookshelves and desks per day. Write an inequality that describes this constraint.

In addition to a weight constraint on its cargo, each plane has a limited amount of space in which to carry supplies. Example 3 demonstrates how to express this constraint.

EXAMPLE 3 Writing a Constraint

Planes can carry a total volume for supplies that does not exceed 6000 cubic feet. Each water bottle is 1 cubic foot and each medical kit also has a volume of 1 cubic foot. With x still representing the number of water bottles and y the number of medical kits, write an inequality that describes this second constraint.

Solution Because each plane can carry a volume of supplies that does not exceed 6000 cubic feet, we have

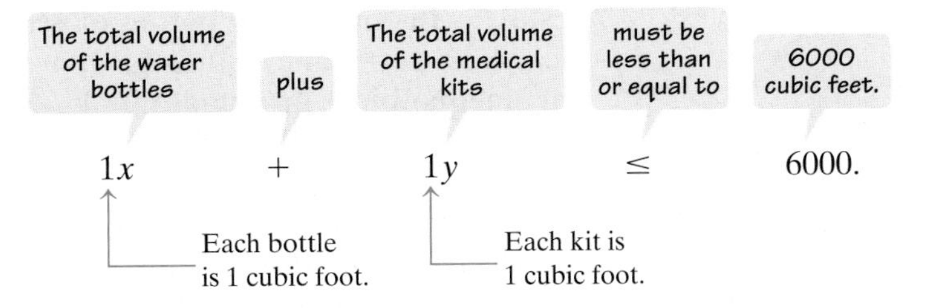

The plane's volume constraint is described by the inequality $x + y \leq 6000$.

In summary, here's what we have described in this aid-to-earthquake-victims situation:

$$z = 10x + 6y$$ This is the objective function describing the number of people helped with x bottles of water and y medical kits.

$$20x + 10y \leq 80{,}000$$
$$x + y \leq 6000.$$
These are the constraints based on each plane's weight and volume limitations.

Check Point 3 To meet customer demand, the company in Checkpoint 1 must manufacture between 30 and 80 bookshelves per day. Furthermore, the company must manufacture at least 10 and no more than 30 desks per day. Write an inequality that describes each of these sentences. Then summarize what you have described about this company by writing the objective function for its profits, and the three constraints.

3 Use linear programming to solve problems.

Solving Problems with Linear Programming

The problem in the earthquake situation described previously is to maximize the number of victims who can be helped, subject to the planes' weight and volume constraints. The process of solving this problem is called linear programming, based on a theorem that was proven during World War II.

Solving a Linear Programming Problem

Let $z = ax + by$ be an objective function that depends on x and y. Furthermore, z is subject to a number of constraints on x and y. If a maximum or minimum value of z exists, it can be determined as follows:

1. Graph the system of inequalities representing the constraints.
2. Find the value of the objective function at each corner, or **vertex**, of the graphed region. The maximum and minimum of the objective function occur at one or more of the corner points.

EXAMPLE 4 Solving a Linear Programming Problem

Determine how many bottles of water and how many medical kits should be sent on each plane to maximize the number of earthquake victims who can be helped.

Solution We must maximize $z = 10x + 6y$ subject to the constraints

$$20x + 10y \leq 80{,}000$$
$$x + y \leq 6000.$$

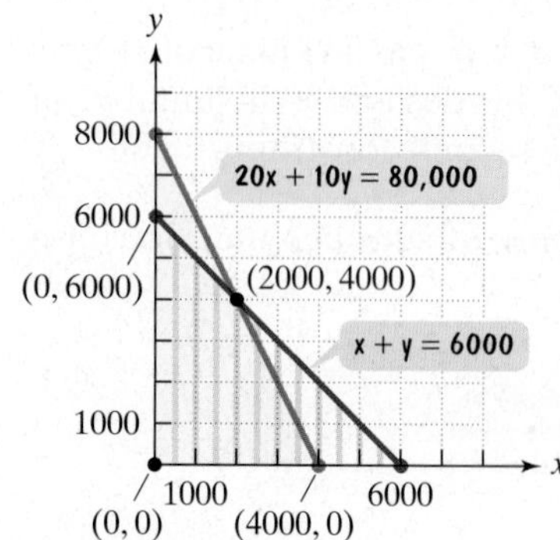

Figure 5.32 The region in quadrant I representing the constraints

$$20x + 10y \leq 80{,}000$$
$$x + y \leq 6000$$

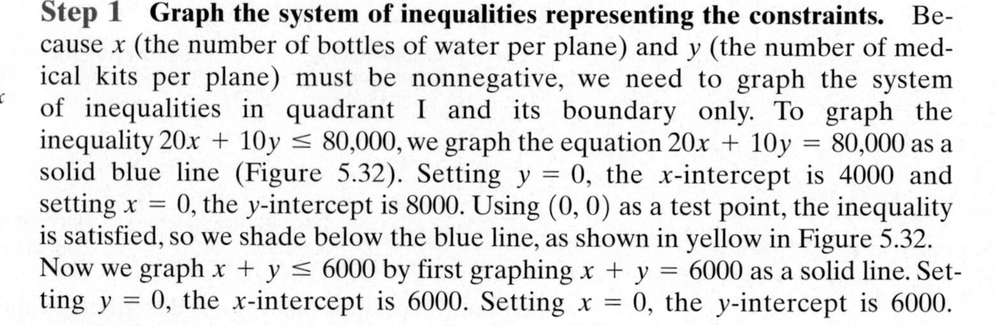

Step 1 Graph the system of inequalities representing the constraints. Because x (the number of bottles of water per plane) and y (the number of medical kits per plane) must be nonnegative, we need to graph the system of inequalities in quadrant I and its boundary only. To graph the inequality $20x + 10y \leq 80{,}000$, we graph the equation $20x + 10y = 80{,}000$ as a solid blue line (Figure 5.32). Setting $y = 0$, the x-intercept is 4000 and setting $x = 0$, the y-intercept is 8000. Using $(0, 0)$ as a test point, the inequality is satisfied, so we shade below the blue line, as shown in yellow in Figure 5.32. Now we graph $x + y \leq 6000$ by first graphing $x + y = 6000$ as a solid line. Setting $y = 0$, the x-intercept is 6000. Setting $x = 0$, the y-intercept is 6000.

Using (0, 0) as a test point, the inequality is satisfied, so we shade below the red line, as shown using green vertical shading in Figure 5.32.

We use the addition method to find where the lines $20x + 10y = 80{,}000$ and $x + y = 6000$ intersect.

$$\begin{array}{rcl} 20x + 10y = 80{,}000 & \xrightarrow{\text{No change}} & 20x + 10y = 80{,}000 \\ x + y = 6000 & \xrightarrow{\text{Multiply by } -10.} & -10x - 10y = -60{,}000 \\ & \text{Add:} & 10x = 20{,}000 \\ & & x = 2000 \end{array}$$

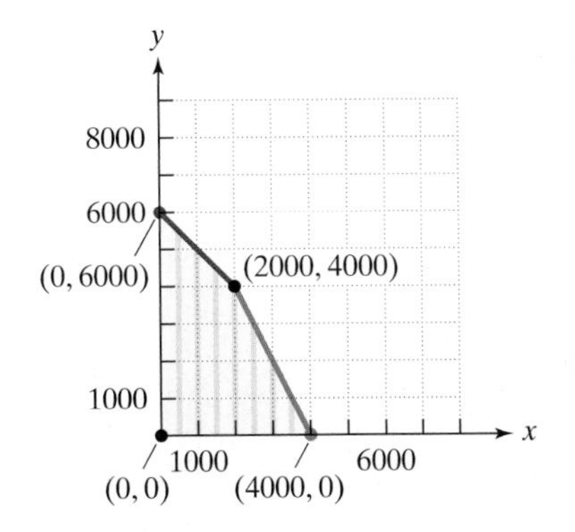

Figure 5.33

Back-substituting 2000 for x in $x + y = 6000$, we find $y = 4000$, so the intersection point is (2000, 4000).

The system of inequalities representing the constraints is shown by the region in which the yellow shading and the green vertical shading overlap in Figure 5.32. The graph of the system of inequalities is shown again in Figure 5.33. The red and blue line segments are included in the graph.

Step 2 Find the value of the objective function at each corner of the graphed region. The maximum and minimum of the objective function occur at one or more of the corner points. We must evaluate the objective function, $z = 10x + 6y$, at the four corners of the region in Figure 5.33.

Corner (x, y)	**Objective Function** $z = 10x + 6y$
(0, 0)	$z = 10(0) + 6(0) = 0$
(4000, 0)	$z = 10(4000) + 6(0) = 40{,}000$
(2000, 4000)	$z = 10(2000) + 6(4000) = 44{,}000$ ← maximum
(0, 6000)	$z = 10(0) + 6(6000) = 36{,}000$

Thus, the maximum value of z is 44,000 and this occurs when $x = 2000$ and $y = 4000$. In practical terms, this means that the maximum number of earthquake victims who can be helped with each plane shipment is 44,000. This can be accomplished by sending 2000 water bottles and 4000 medical kits per plane.

For the company in Checkpoints 1–3, how many bookshelves and how many desks should be manufactured per day to obtain maximum profit? What is the maximum daily profit?

EXAMPLE 5 Solving a Linear Programming Problem

Find the maximum value of the objective function

$$z = 2x + y$$

subject to the constraints:

$$x \ge 0,\ y \ge 0$$
$$x + 2y \le 5$$
$$x - y \le 2.$$

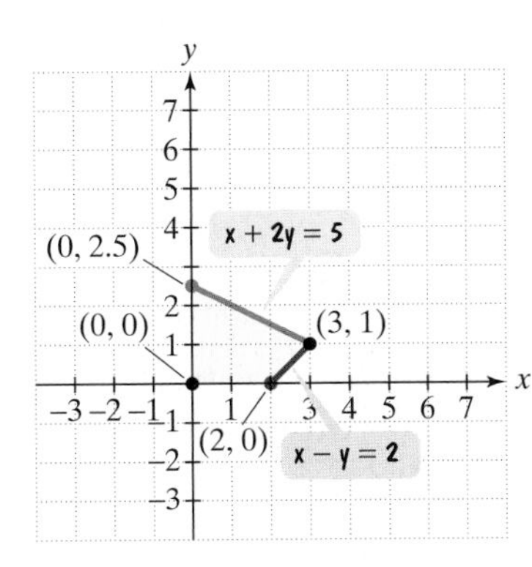

Figure 5.34 The graph of $x + 2y \le 5$ and $x - y \le 2$ in quadrant I

Solution We begin by graphing the region in quadrant I ($x \ge 0, y \ge 0$) formed by the constraints. The graph is shown in Figure 5.34.

Now we evaluate the objective function at the four vertices of this region.

Objective function: $z = 2x + y$

At $(0, 0)$: $z = 2 \cdot 0 + 0 = 0$

At $(2, 0)$: $z = 2 \cdot 2 + 0 = 4$

At $(3, 1)$: $z = 2 \cdot 3 + 1 = 7$ Maximum value of z

At $(0, 2.5)$: $z = 2 \cdot 0 + 2.5 = 2.5$

Thus, the maximum value of z is 7, and this occurs when $x = 3$ and $y = 1$.

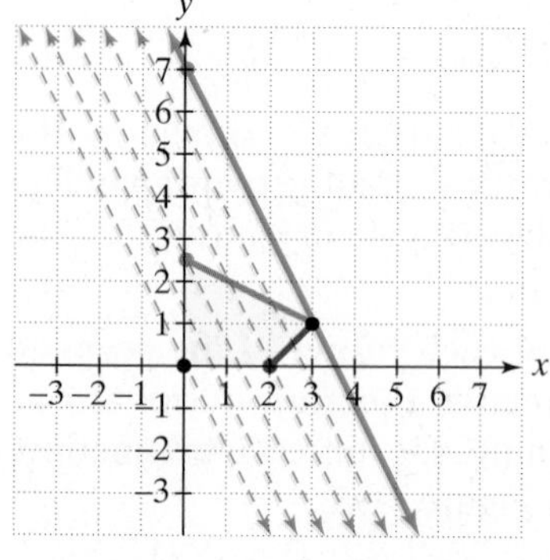

Figure 5.35 The line with slope -2 with the greatest y-intercept that intersects the shaded region passes through one of its vertices.

We can see why the objective function in Example 5 has a maximum value that occurs at a vertex by solving the equation for y.

$z = 2x + y$ This is the objective function of Example 5.

$y = -2x + z$ Solve for y. Recall that the slope-intercept form of a line is $y = mx + b$.

Slope $= -2$ y-intercept $= z$

In this form, z represents the y-intercept of the objective function. The equation describes infinitely many parallel lines, each with a slope of -2. The process in linear programming involves finding the maximum z-value for all lines that intersect the region determined by the constraints. Of all the lines whose slope is -2, we're looking for the one with the greatest y-intercept that intersects the given region. As we see in Figure 5.35, such a line will pass through one (or possibly more) of the vertices of the region.

Check Point 5 Find the maximum value of the objective function $z = 3x + 5y$ subject to the constraints $x \geq 0, y \geq 0, x + y \geq 1, x + y \leq 6$.

Faster and Faster

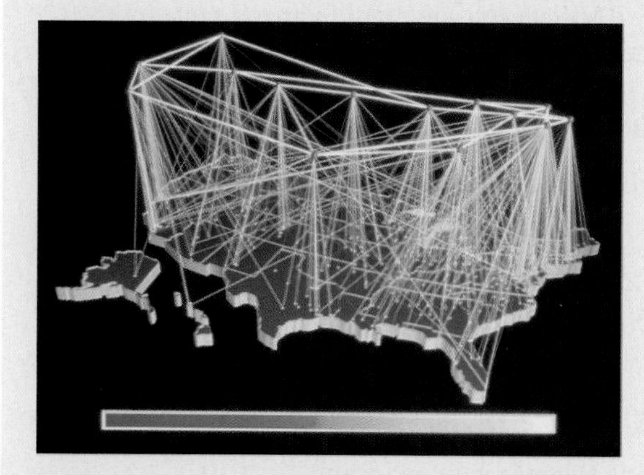

The network of computer linkages in the United States is growing exponentially.

The problems we solve nowadays have thousands of equations, sometimes a million variables. One of the things that still amazes me is to see a program run on the computer—and to see the answer come out. If we think of the number of combinations of different solutions that we're trying to choose the best of, it's akin to the stars in the heavens. Yet we solve them in a matter of moments. This, to me, is staggering. Not that we can solve them—but that we can solve them so rapidly and efficiently.

—George Dantzig
Inventor of a linear programming method

Problems in linear programming can involve objective functions with thousands of variables subject to thousands of constraints. Several nongeometric linear programming methods are available on software for solving such problems. And we continue to search for faster and faster linear programming methods. This area of applied mathematics has a direct impact on the efficiency and profitability of numerous industries, including telephone and computer communications, and the airlines.

EXERCISE SET 5.6

Practice Exercises

In Exercises 1–4, find the value of the objective function at each corner of the graphed region. What is the maximum value of the objective function? What is the minimum value of the objective function?

1. Objective Function $\quad z = 5x + 6y$

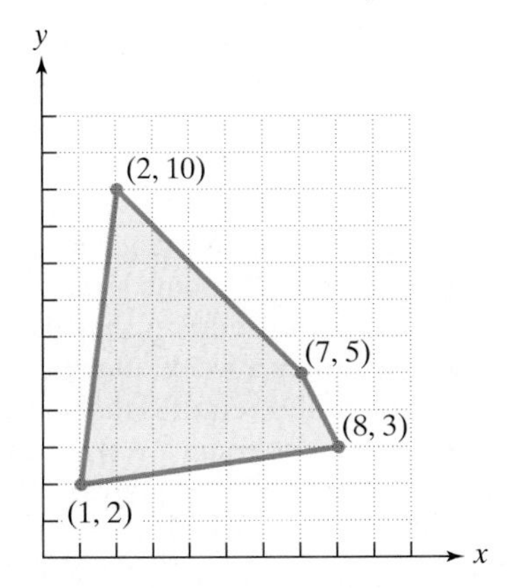

2. Objective Function $\quad z = 3x + 2y$

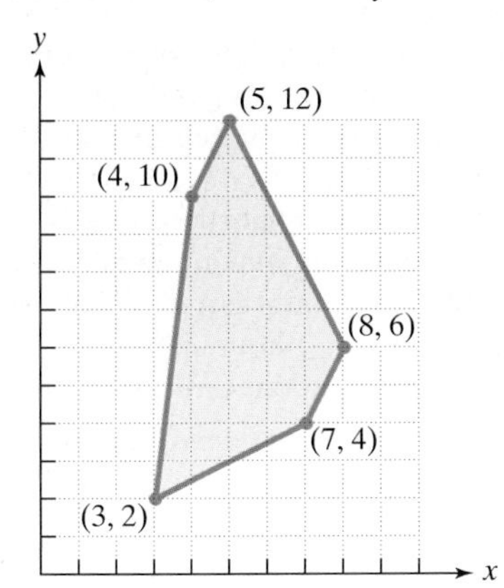

3. Objective Function $\quad z = 40x + 50y$

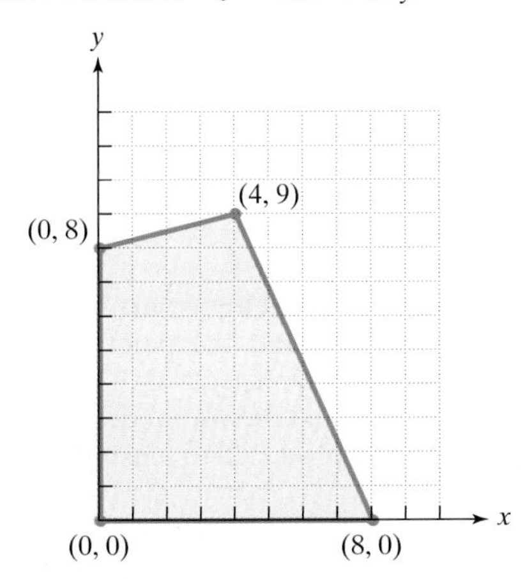

4. Objective Function $\quad z = 30x + 45y$

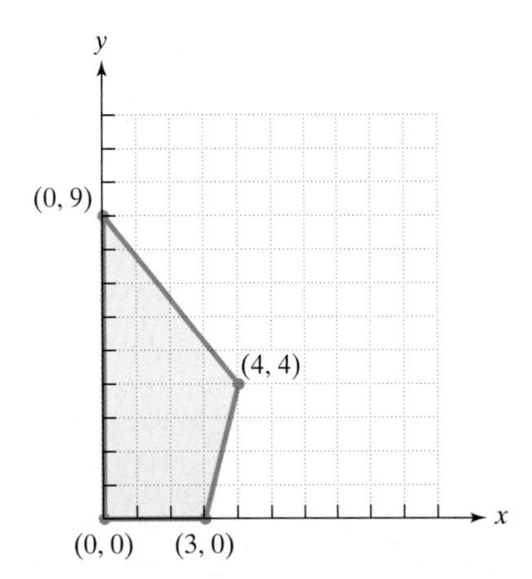

In Exercises 5–14, an objective function and a system of linear inequalities representing constraints are given.

a. *Graph the system of inequalities representing the constraints.*

b. *Find the value of the objective function at each corner of the graphed region.*

c. *Use the values in part (b) to determine the maximum value of the objective function and the values of x and y for which the maximum occurs.*

5. Objective Function $\quad z = 2x + 3y$

Constraints $\quad x \geq 0, y \geq 0$

$3x + y \leq 6$

$2x + 3y \leq 12$

6. Objective Function $\quad z = 3x + 2y$

Constraints $\quad x \geq 0, y \geq 0$

$2x + y \leq 8$

$x + y \geq 4$

7. Objective Function $\quad z = 4x + y$

Constraints $\quad x \geq 0, y \geq 0$

$2x + 3y \leq 12$

$x + y \geq 3$

8. Objective Function $\quad z = x + 6y$

Constraints $\quad x \geq 0, y \geq 0$

$2x + y \leq 10$

$x - 2y \geq -10$

9. Objective Function $\quad z = 3x - 2y$

Constraints $\quad 1 \leq x \leq 5$

$y \geq 2$

$x - y \geq -3$

10. Objective Function $z = 5x - 2y$
Constraints
$0 \le x \le 5$
$0 \le y \le 3$
$x + y \ge 2$

11. Objective Function $z = 4x + 2y$
Constraints
$x \ge 0, y \ge 0$
$2x + 3y \le 12$
$3x + 2y \le 12$
$x + y \ge 2$

12. Objective Function $z = 2x + 4y$
Constraints
$x \ge 0, y \ge 0$
$x + 3y \ge 6$
$x + y \ge 3$
$x + y \le 9$

13. Objective Function $z = 10x + 12y$
Constraints
$x \ge 0, y \ge 0$
$x + y \le 7$
$2x + y \le 10$
$2x + 3y \le 18$

14. Objective Function $z = 5x + 6y$
Constraints
$x \ge 0, y \ge 0$
$2x + y \ge 10$
$x + 2y \ge 10$
$x + y \le 10$

Application Exercises

15. A television manufacturer makes console and wide-screen televisions. The profit per unit is \$125 for the console televisions and \$200 for the wide-screen televisions.

a. Let

x = the number of consoles manufactured in a month and

y = the number of wide-screens manufactured in a month.

Write the objective function that describes the total monthly profit.

b. The manufacturer is bound by the following constraints:

1. Equipment in the factory allows for making at most 450 console televisions in one month.
2. Equipment in the factory allows for making at most 200 wide-screen televisions in one month.
3. The cost to the manufacturer per unit is \$600 for the console televisions and \$900 for the wide-screen televisions. Total monthly costs cannot exceed \$360,000.

Write a system of three inequalities that describes these constraints.

c. Graph the system of inequalities in part (b). Use only the first quadrant and its boundary, because x and y must both be nonnegative.

d. Evaluate the objective function for total monthly profit at each of the five vertices of the graphed region. (The vertices should occur at (0, 0), (0, 200), (300, 200), (450, 100), and (450, 0).)

e. Complete the missing portions of this statement: The television manufacturer will make the greatest profit by manufacturing ___ console televisions each month and ___ wide-screen televisions each month. The maximum monthly profit is \$ ___.

16. a. A student earns \$10 per hour for tutoring and \$7 per hour as a teacher's aid. Let x = the number of hours each week spent tutoring, and y = the number of hours each week spent as a teacher's aid. Write the objective function that describes total weekly earnings.

b. The student is bound by the following constraints:

- To have enough time for studies, the student can work no more than 20 hours a week.
- The tutoring center requires that each tutor spend at least three hours a week tutoring.
- The tutoring center requires that each tutor spend no more than eight hours a week tutoring.

Write a system of three inequalities that describes these constraints.

c. Graph the system of inequalities in part (b). Use only the first quadrant and its boundary, because x and y are nonnegative.

d. Evaluate the objective function for total weekly earnings at each of the four vertices of the graphed region. (The vertices should occur at (3, 0), (8, 0), (3, 17), and (8, 12).)

e. Complete the missing portions of this statement: The student can earn the maximum amount per week by tutoring for ___ hours per week and working as a teacher's aid for ___ hours per week. The maximum amount that the student can earn each week is \$ ___.

Use the two steps for solving a linear programming problem, given in the box on page 470, to solve the problems in Exercises 17–23.

17. A manufacturer produces two models of mountain bicycles. The times (in hours) required for assembling and painting each model are given in the following table.

	Model *A*	Model *B*
Assembling	5	4
Painting	2	3

The maximum total weekly hours available in the assembly department and the paint department are 200 hours and 108 hours, respectively. The profits per unit are \$25 for model *A* and \$15 for model *B*. How many of each type should be produced to maximize profit?

18. A large institution is preparing lunch menus containing foods A and B. The specifications for the two foods are given in the following table.

Food	Units of Fat per Ounce	Units of Carbohydrates per Ounce	Units of Protein per Ounce
A	1	2	1
B	1	1	1

Each lunch must provide at least 6 units of fat per serving, no more than 7 units of protein, and at least 10 units of carbohydrates. The institution can purchase food A for $0.12 per ounce and food B for $0.08 per ounce. How many ounces of each food should a serving contain to meet the dietary requirements at the least cost?

19. Food and clothing are shipped to victims of a natural disaster. Each carton of food will feed 5 people, while each carton of clothing will help 6 people. Each 30-cubic-foot box of food weighs 50 pounds and each 20-cubic-foot box of clothing weighs 5 pounds. The commercial carriers transporting food and clothing are bound by the following constraints:

1. The total weight per carrier cannot exceed 18,000 pounds.
2. The total volume must be less than 12,000 cubic feet.

How many cartons of food and clothing should be sent with each plane shipment to maximize the number of people who can be helped?

20. On June 24, 1948, the former Soviet Union blocked all land and water routes through East Germany to Berlin. A gigantic airlift was organized using American and British planes to supply food, clothing, and other supplies to the more than 2 million people in West Berlin. The cargo capacity was 30,000 cubic feet for an American plane and 20,000 cubic feet for a British plane. To break the Soviet blockade, the Western Allies had to maximize cargo capacity, but were subject to the following restrictions:

1. No more than 44 planes could be used.
2. The larger American planes required 16 personnel per flight, double that of the requirement for the British planes. The total number of personnel available could not exceed 512.
3. The cost of an American flight was $9000 and the cost of a British flight was $5000. Total weekly costs could not exceed $300,000.

Find the number of American and British planes that were used to maximize cargo capacity.

21. A theater is presenting a program on drinking and driving for students and their parents. The proceeds will be donated to a local alcohol information center. Admission is $2.00 for parents and $1.00 for students. However, the situation has two constraints: The theater can hold no more than 150 people and every two parents must bring at least one student. How many parents and students should attend to raise the maximum amount of money?

22. You are about to take a test that contains computation problems worth 6 points each and word problems worth 10 points each. You can do a computation problem in 2 minutes and a word problem in 4 minutes. You have 40 minutes to take the test and may answer no more than 12 problems. Assuming you answer all the problems attempted correctly, how many of each type of problem must you do to maximize your score? What is the maximum score?

23. In 1978, a ruling by the Civil Aeronautics Board allowed Federal Express to purchase larger aircraft. Federal Express's options included 20 Boeing 727s that United Airlines was retiring and/or the French-built Dassault Fanjet Falcon 20. To aid in their decision, executives at Federal Express analyzed the following data:

	Boeing 727	Falcon 20
Direct Operating Cost	$1400 per hour	$500 per hour
Payload	42,000 pounds	6000 pounds

Federal Express was faced with the following constraints:

1. Hourly operating cost was limited to $35,000.
2. Total payload had to be at least 672,000 pounds.
3. Only twenty 727s were available.

Given the constraints, how many of each kind of aircraft should Federal Express have purchased to maximize the number of aircraft?

Writing in Mathematics

24. What kinds of problems are solved using the linear programming method?

25. What is an objective function in a linear programming problem?

26. What is a constraint in a linear programming problem? How is a constraint represented?

27. In your own words, describe how to solve a linear programming problem.

28. Describe a situation in your life in which you would really like to maximize something, but you are limited by at least two constraints. Can linear programming be used in this situation? Explain your answer.

Technology Exercises

In Exercises 29–32, use a graphing utility to sketch the region determined by the constraints. Then determine the maximum value of the objective function subject to the contraints.

29. Objective Function $z = 6x + 8y$
Constraints $x \geq 0, y \geq 0$
$x + 2y \leq 6$

30. Objective Function $z = 30x + 20y$
Constraints $x \geq 0, y \geq 0$
$2x + y \leq 14$
$3x + y \leq 18$

31. Objective Function $z = 9x + 14y$
Constraints $x \geq 0, y \geq 0$
$2x + y \leq 10$
$2x + 3y \leq 18$

32. Objective Function $z = 10x + 3y$
Constraints $0 \leq x \leq 10, \quad y \geq 0$
$4x + 5y \leq 60$
$4x - 5y \geq -20$

Critical Thinking Exercises

33. Suppose that you inherit $10,000. The will states how you must invest the money. Some (or all) of the money must be invested in stocks and bonds. The requirements are that at least $3000 be invested in bonds, with expected returns of $0.08 per dollar, and at least $2000 be invested in stocks, with expected returns of $0.12 per dollar. Because the stocks are medium risk, the final stipulation requires that the investment in bonds should never be less than the investment in stocks. How should the money be invested so as to maximize your expected returns?

34. Consider the objective function $z = Ax + By$ ($A > 0$ and $B > 0$) subject to the following constraints: $2x + 3y \leq 9$, $x - y \leq 2$, $x \geq 0$, and $y \geq 0$. Prove that the objective function will have the same maximum value at the vertices $(3, 1)$ and $(0, 3)$ if $A = \frac{2}{3}B$.

Group Exercises

35. Group members should choose a particular field of interest. Research how linear programming is used to solve problems in that field. If possible, investigate the solution of a specific practical problem. Present a report on your findings, including the contributions of George Dantzig, Narendra Karmarkar, and L.G. Khachion to linear programming.

36. Members of the group should interview a business executive who is in charge of deciding the product mix for a business. How are production policy decisions made? Are other methods used in conjunction with linear programming? What are these methods? What sort of academic background, particularly in mathematics, does this executive have? Present a group report addressing these questions, emphasizing the role of linear programming for the business.

CHAPTER SUMMARY, REVIEW, AND TEST

Summary

5.1 Systems of Linear Equations in Two Variables

a. Two equations in the form $Ax + By = C$ are called a system of linear equations. A solution to the system is an ordered pair that satisfies both equations in the system.

b. Linear systems in two variables can be solved by eliminating a variable, using the substitution method (see the box on page 419) or the addition method (see the box on page 421).

c. Some linear systems have no solution and are called inconsistent systems; others have infinitely many solutions. The equations in a linear system with infinitely many solutions are called dependent. For details, see the box on page 423.

5.2 Systems of Linear Equations in Three Variables

a. Three equations in the form $Ax + By + Cz = D$ are called a system of linear equations in three variables. A solution to the system is an ordered triple that satisfies all three equations in the system.

b. A system of linear equations in three variables can be solved by eliminating variables. Use the addition method to eliminate any variable, reducing the system to two equations in two variables. Use substitution or the addition method to solve the resulting system in two variables. Details are found in the box on page 432.

5.3 Partial Fraction Decomposition

a. Partial fraction decomposition is used on rational expressions in which the numerator and denominator have no common factors and the highest power in the numerator is less than the highest power in the denominator. The steps in partial fraction decomposition are given in the box on page 441.

b. Include one partial fraction with a constant numerator for each distinct linear factor in the denominator. Include one partial fraction with a constant numerator for each power of a repeated linear factor in the denominator.

c. Include one partial fraction with a linear numerator for each distinct prime quadratic factor in the denominator. Include one partial fraction with a linear numerator for each power of a prime, repeated quadratic factor in the denominator.

5.4 Systems of Nonlinear Equations in Two Variables

a. A system of two nonlinear equations in two variables contains at least one equation that cannot be expressed as $Ax + By = C$.

b. Nonlinear systems of equations can be solved algebraically by eliminating all occurrences of one of the variables by the substitution and addition methods.

5.5 Systems of Inequalities

a. A linear inequality in two variables can be written in the form $Ax + By > C$, $Ax + By \geq C$, $Ax + By < C$, or $Ax + By \leq C$.

b. The procedure for graphing a linear inequality in two variables is given in the box on page 458. A nonlinear inequality in two variables is graphed using the same procedure.

c. To graph the solution set to a system of inequalities, graph each inequality in the system in the same rectangular coordinate system. Then find the region, if there is one, that is common to every graph in the system.

5.6 Linear Programming

a. An objective function is an algebraic expression in three variables describing a quantity that must be maximized or minimized.

b. Constraints are restrictions, expressed as linear inequalities.

c. Steps for solving a linear programming problem are given in the box on page 470.

Review Exercises

5.1

In Exercises 1–5, solve by the method of your choice. Identify systems with no solution and systems with infinitely many solutions, using set notation to express their solution sets.

1. $y = 4x + 1$
$3x + 2y = 13$

2. $x + 4y = 14$
$2x - y = 1$

3. $5x + 3y = 1$
$3x + 4y = -6$

4. $2y - 6x = 7$
$3x - y = 9$

5. $4x - 8y = 16$
$3x - 6y = 12$

6. Can the graphing-utility-generated screen be the solution for the system

$$x + y = 2$$
$$2x + y = -5?$$

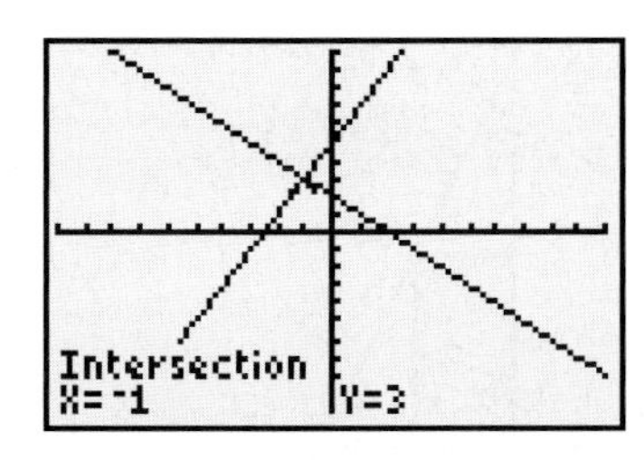

Explain.

7. Health experts agree that cholesterol intake should be limited to 300 mg or less each day. Three ounces of shrimp and 2 ounces of scallops contain 156 mg of cholesterol. Five ounces of shrimp and 3 ounces of scallops contain 45 mg of cholesterol less than the suggested maximum daily intake. Determine the cholesterol content in an ounce of each item.

8. The calorie-nutrient information for an apple and an avocado is given in the table. How many of each should be eaten to get exactly 1000 calories and 100 grams of carbohydrates?

	One Apple	One Avocado
Calories	100	350
Carbohydrates (grams)	24	14

9. The weekly demand and supply models for the video *Titanic* at a chain of stores that sells videos are given by the demand model $N = -60p + 1000$ and the supply model $N = 4p + 200$, in which p is the price of the video and N is the number of videos sold or supplied each week to the chain of stores. Find the price at which supply and demand are equal. At this price, how many copies of *Titanic* can be supplied and sold each week?

5.2

Solve each system in Exercises 10–11.

10. $2x - y + z = 1$
$3x - 3y + 4z = 5$
$4x - 2y + 3z = 4$

11. $x + 2y - z = 5$
$2x - y + 3x = 0$
$2y + z = 1$

12. Find the quadratic function $y = ax^2 + bx + c$ whose graph passes through the points $(1, 4)$, $(3, 20)$, and $(-2, 25)$.

13. The graph shows a low savings rate in the United States compared to that of many industrialized countries. The combined rate for Japan, Germany, and France is 45%. The savings rate in Japan exceeds that for Germany by 1% and is 12% less than twice that for France. Find the savings rates for Japan, Germany, and France.

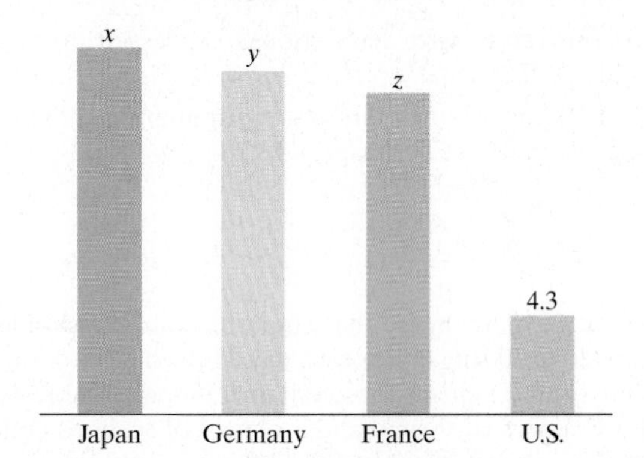

Source: Office of Management and Budget

14. Describe how to obtain a model for the millions of Americans living in poverty by using the ordered pairs for 1994, 1996, and 1997 in the graph.

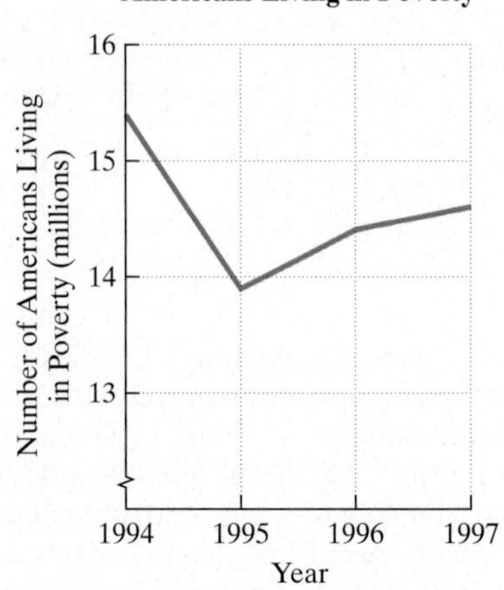

Source: U.S. Census Bureau

5.3

In Exercises 15–23, write the partial fraction decomposition of each rational expression.

15. $\dfrac{x}{(x - 3)(x + 2)}$

16. $\dfrac{11x - 2}{x^2 - x - 12}$

17. $\dfrac{4x^2 - 3x - 4}{x(x + 2)(x - 1)}$

18. $\dfrac{2x + 1}{(x - 2)^2}$

19. $\dfrac{2x - 6}{(x - 1)(x - 2)^2}$

20. $\dfrac{3x}{(x - 2)(x^2 + 1)}$

21. $\dfrac{7x^2 - 7x + 23}{(x - 3)(x^2 + 4)}$

22. $\dfrac{x^3}{(x^2 + 4)^2}$

23. $\dfrac{4x^3 + 5x^2 + 7x - 1}{(x^2 + x + 1)^2}$

5.4

In Exercises 24–34, solve each system by the method of your choice.

24. $5y = x^2 - 1$
$x - y = 1$

25. $y = x^2 + 2x + 1$
$x + y = 1$

26. $x^2 + y^2 = 2$
$x + y = 0$

27. $2x^2 + y^2 = 24$
$x^2 + y^2 = 15$

28. $xy - 4 = 0$
$y - x = 0$

29. $y^2 = 4x$
$x - 2y + 3 = 0$

30. $x^2 + y^2 = 10$
$y = x + 2$

31. $xy = 1$
$y = 2x + 1$

32. $x + y + 1 = 0$
$x^2 + y^2 + 6y - x = -5$

33. $x^2 + y^2 = 13$
$x^2 - y = 7$

34. $2x^2 + 3y^2 = 21$
$3x^2 - 4y^2 = 23$

35. The perimeter of a rectangle is 26 meters, and its area is 40 square meters. Find its dimensions.

36. Find the coordinates of all points (x, y) that lie on the line whose equation is $2x + y = 8$, so that the area of the rectangle shown in the figure is 6 square units.

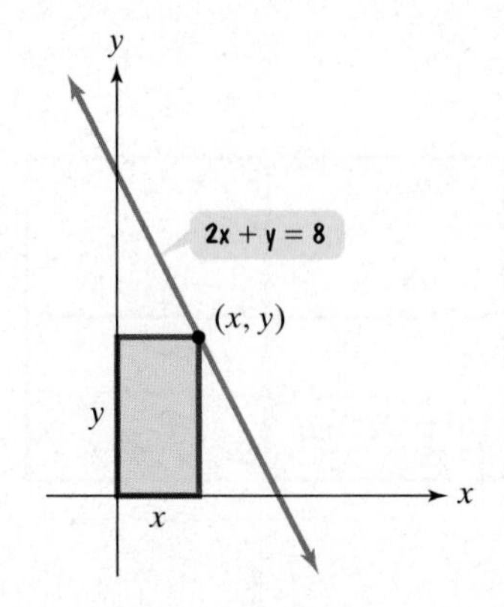

37. Two adjoining square fields with an area of 2900 square feet are to be enclosed with 240 feet of fencing. The situation is represented in the figure. Find the length of each side where a variable appears.

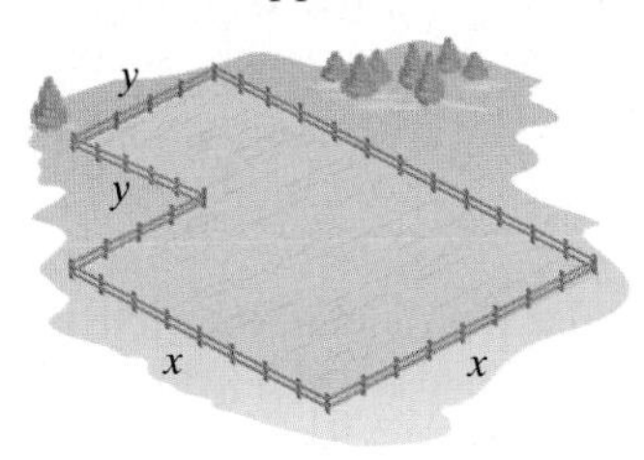

5.5

In Exercises 38–44, graph each inequality.

38. $3x - 4y > 12$

39. $y \leq -\frac{1}{2}x + 2$

40. $x < -2$

41. $y \geq 3$

42. $x^2 + y^2 > 4$

43. $y \leq x^2 - 1$

44. $y \leq 2^x$

In Exercises 45–54, graph the solution set of each system of inequalities or indicate that the system has no solution.

45. $3x + 2y \geq 6$
$2x + y \geq 6$

46. $2x - y \geq 4$
$x + 2y < 2$

47. $y < x$
$y \leq 2$

48. $y \leq x$
$2x + 5y \leq 10$

49. $0 \leq x \leq 3$
$y > 2$

50. $2x + y < 4$
$2x + y > 6$

51. $x^2 + y^2 \leq 16$
$x + y < 2$

52. $x^2 + y^2 \leq 9$
$y < -3x + 1$

53. $y > x^2$
$x + y < 6$
$y < x + 6$

54. $x \geq 0, y \geq 0$
$2x + 3y \leq 12$
$3x + y \leq 6$

5.6

55. Find the value of the objective function $z = 2x + 3y$ at each corner of the graphed region shown. What is the maximum value of the objective function? What is the minimum value of the objective function?

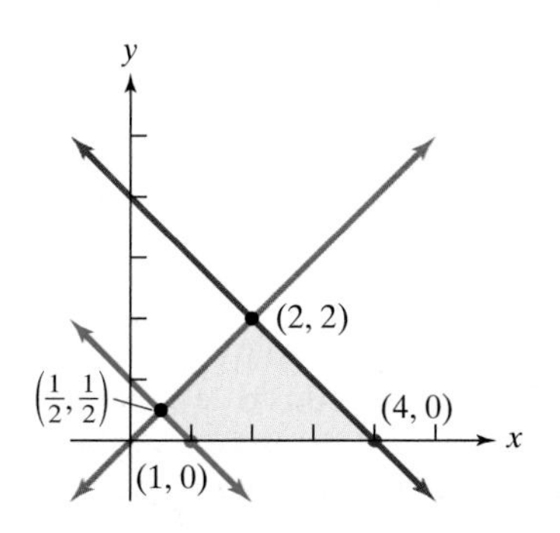

In Exercises 56–58, graph the region determined by the constraints. Then find the maximum value of the given objective function, subject to the constraints.

56. Objective Function $z = 2x + 3y$
Constraints $x \geq 0, y \geq 0,$
$x + y \leq 8$
$3x + 2y \geq 6$

57. Objective Function $z = x + 4y$
Contraints $0 \leq x \leq 5, 0 \leq y \leq 7$
$x + y \geq 3$

58. Objective Function $z = 5x + 6y$
Constraints $x \geq 0, y \geq 0$
$y \leq x$
$2x + y \leq 12$
$2x + 3y \geq 6$

59. A paper manufacturing company converts wood pulp to writing paper and newsprint. The profit on a unit of writing paper is \$500 and the profit on a unit of newsprint is \$350.

a. Let x represent the number of units of writing paper produced daily. Let y represent the number of units of newsprint produced daily. Write the objective function that models total daily profit.

b. The manufacturer is bound by the following constraints:

1. Equipment in the factory allows for making at most 200 units of paper (writing paper and newsprint) in a day.
2. Regular customers require at least 10 units of writing paper and at least 80 units of newsprint daily.

Write a system of inequalities that models these constraints.

c. Graph the inequalities in part (b). Use only the first quadrant, because x and y must both be positive. (*Suggestion:* Let each unit along the x- and y-axes represent 20.)

d. Evaluate the objective profit function at each of the three vertices of the graphed region.

e. Complete the missing portions of this statement: The company will make the greatest profit by producing ___ units of writing paper and ___ units of newsprint each day. The maximum daily profit is \$ ___.

60. A manufacturer of lightweight tents makes two models whose specifications are given in the following table.

	Cutting Time per Tent	Assembly Time per Tent
Model *A*	0.9 hour	0.8 hour
Model *B*	1.8 hours	1.2 hours

On a monthly basis, the manufacturer has no more than 864 hours of labor available in the cutting department and at most 672 hours in the assembly division. The profits come to \$25 per tent for model *A* and \$40 per tent for model *B*. How many of each should be manufactured monthly to maximize the profit?

Chapter 5 Test

In Exercises 1–5, solve the system.

1. $x = y + 4$
$3x + 7y = -18$

2. $2x + 5y = -2$
$3x - 4y = 20$

3. $x + y + z = 6$
$3x + 4y - 7z = 1$
$2x - y + 3z = 5$

4. $x^2 + y^2 = 25$
$x + y = 1$

5. $2x^2 - 5y^2 = -2$
$3x^2 + 2y^2 = 35$

6. Find the partial fraction decomposition for
$$\frac{x}{(x + 1)(x^2 + 9)}.$$

In Exercises 7–10, graph the solution set of each inequality or system of inequalities.

7. $x - 2y < 8$

8. $x \geq 0, y \geq 0$
$3x + y \leq 9$
$2x + 3y \geq 6$

9. $x^2 + y^2 > 1$
$x^2 + y^2 < 4$

10. $y \leq 1 - x^2$
$x^2 + y^2 \leq 9$

11. Find the maximum value of the objective function $z = 3x + 5y$ subject to the following constraints: $x \geq 0$, $y \geq 0, x + y \leq 6, x \geq 2$.

12. A theater sells all orchestra seats at one price and all mezzanine seats at another price. One person purchased 4 orchestra tickets and 3 mezzanine tickets for a total of $134.00. A second person purchased 5 orchestra tickets and 2 mezzanine tickets for $143.00. What is the price of one orchestra ticket and one mezzanine ticket?

13. The demand and supply models for a product are given, respectively, by $N = 1000 - 20p$ and $N = 250 + 5p$. At what price will supply equal demand? At that price, how many units of the product will be supplied and sold?

14. Find the quadratic function $y = ax^2 + bx + c$ whose graph passes through the points $(-1, -2)$, $(2, 1)$, and $(-2, 1)$.

15. The rectangular plot of land shown in the figure is to be fenced along three sides using 39 feet of fencing. No fencing is to be placed along the river's edge. The area of the plot is 180 square feet. What are its dimensions?

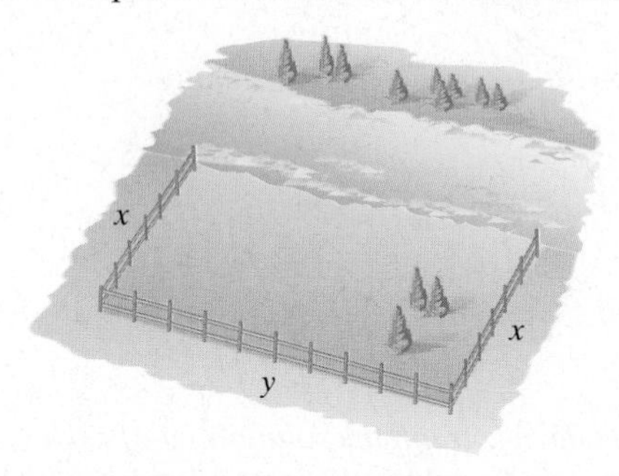

16. A manufacturer makes two types of jet skis, regular and deluxe. The profit on a regular jet ski is $200 and the profit on the deluxe model is $250. To meet customer demand, the company must manufacture at least 50 regular jet skis per week and at least 75 deluxe models. To maintain high quality, the total number of both models of jet skis manufactured by the company should not exceed 150 per week. How many jet skis of each type should be manufactured per week to obtain maximum profit? What is the maximum weekly profit?

Cumulative Review Exercises (Chapters 1–5)

Solve each equation or inequality in Exercises 1–8.

1. $\sqrt{x^2 - 3x} = 2x - 6$

2. $4x^2 = 8x - 7$

3. $\left|\frac{x}{3} + 2\right| < 4$

4. $\frac{x + 5}{x - 1} > 2$

5. $2x^3 + x^2 - 13x + 6 = 0$

6. $6x - 3(5x + 2) = 4(1 - x)$

7. $\log(x + 3) + \log x = 1$

8. $3^{x+2} = 11$

In Exercises 9–12, graph each equation, function, or inequality in the rectangular coordinate system.

9. $f(x) = (x + 2)^2 - 4$

10. $2x - 3y \leq 6$

11. $y = 3^{x-2}$

12. $f(x) = \frac{x^2 - x - 6}{x + 1}$

13. Expand and simplify: $\log_2(8x^5)$.

14. What interest rate is required for an investment of $6000 subject to continuous compounding to grow to $18,000 in 10 years?

15. If $f(x) = 7x - 3$, find $f^{-1}(x)$.

16. If $f(x) = 7x - 3$ and $g(x) = 3x - 7$, find $g(f(x))$.

17. Explain why $x^2 + y^2 = 4$ does not represent y as a function of x.

18. Solve the system:
$$3x - y = -2$$
$$2x^2 - y = 0.$$

19. The length of a rectangle is 1 meter more than twice the width. If the rectangle's area is 36 square meters, find its dimensions.

20. The function $f(x) = 0.1x^2 - 3x + 22$ describes the distance, $f(x)$, in feet, needed for an airplane to land when its initial landing speed is x feet per second. Find and interpret $f(90)$. Will there be a problem if 550 feet of runway is available? Explain.

Matrices and Determinants

You are being drawn deeper into cyberspace, spending more time online each week. With constantly improving high-resolution images, cyberspace is reshaping your life by nourishing shared enthusiasms. The people who built your computer talk of "bandwidth out the wazoo" that will give you the visual experience, in high-definition 3-D format, of being in the same room with a person who is actually in another city.

Jaron Lanier, who first used the term "virtual reality," is chief scientist for the "tele-immersion" project, which explores the impact of massive bandwidth and computing power. Rectangular arrays of numbers, called *matrices*, play a central role in representing computer images and in the forthcoming technology of tele-immersion. In this chapter, we study matrices and their applications. We begin with solving linear systems using matrices, which leads to a discussion of how computers might unjam traffic and give us a gridlock-free future.

SECTION 6.1 *Matrix Solutions to Linear Systems*

Objectives

1. Write the augmented matrix for a linear system.
2. Perform matrix row operations.
3. Use matrices and Gaussian elimination to solve systems.
4. Use matrices and Gauss-Jordan elimination to solve systems.

Yes, we overindulged, but it was delicious. Anyway, a few hours of moderate activity and we'll just burn off those extra calories. The following chart should help. We see that the number of calories burned per hour depends on our weight. Four hours of tennis and we'll be as good as new!

How Fast You Burn Off Calories

	Weight (pounds)					
	110	**132**	**154**	**176**	**187**	**209**
Activity	**Calories Burned per Hour**					
Housework	175	210	245	285	300	320
Cycling	190	215	245	270	280	295
Tennis	335	380	425	470	495	520
Watching TV	60	70	80	85	90	95

The 24 numbers inside the red brackets are arranged in four rows and six columns. This rectangular array of 24 numbers, arranged in rows and columns and placed in brackets, is an example of a **matrix** (plural: **matrices**). The numbers inside the brackets are called **elements** of the matrix. Matrices are used to display information and to solve systems of linear equations. Because systems involving two equations in two variables can easily be solved by substitution or addition, we will focus on matrix solutions to linear systems in three or more variables.

1 Write the augmented matrix for a linear system.

Solving Linear Systems by Using Matrices

A matrix gives us a shortened way of writing a system of equations. The first step in solving a system of linear equations using matrices is to write the augmented matrix. An **augmented matrix** has a vertical bar separating the columns of the matrix into two groups. The coefficients of each variable are placed to the left of the vertical line, and the constants are placed to the right. If any variable is missing, its coefficient is 0. Here are two examples.

System of Linear Equations	Augmented Matrix
$\begin{aligned} 3x + y + 2z &= 31 \\ x + y + 2z &= 19 \\ x + 3y + 2z &= 25 \end{aligned}$	$\left[\begin{array}{ccc\|c} 3 & 1 & 2 & 31 \\ 1 & 1 & 2 & 19 \\ 1 & 3 & 2 & 25 \end{array}\right]$
$\begin{aligned} x + 2y - 5z &= -19 \\ y + 3z &= 9 \\ z &= 4 \end{aligned}$	$\left[\begin{array}{ccc\|c} 1 & 2 & -5 & -19 \\ 0 & 1 & 3 & 9 \\ 0 & 0 & 1 & 4 \end{array}\right]$

Notice how the second matrix contains 1s down the diagonal from upper left to lower right and 0s below the 1s. This arrangement makes it easy to find the solution of the system of equations, as Example 1 shows.

EXAMPLE 1 Solving a System Using a Matrix

Write the solution set for a system of equations represented by the matrix

$$\left[\begin{array}{ccc|c} 1 & 2 & -5 & -19 \\ 0 & 1 & 3 & 9 \\ 0 & 0 & 1 & 4 \end{array}\right].$$

Solution The system represented by the given matrix is

$$\left[\begin{array}{ccc|c} 1 & 2 & -5 & -19 \\ 0 & 1 & 3 & 9 \\ 0 & 0 & 1 & 4 \end{array}\right] \rightarrow \begin{aligned} 1x + 2y - 5z &= -19 \\ 0x + 1y + 3z &= 9 \\ 0x + 0y + 1z &= 4 \end{aligned}.$$

This system can be simplified as follows.

$$\begin{aligned} x + 2y - 5z &= -19 && \text{Equation 1} \\ y + 3z &= 9 && \text{Equation 2} \\ z &= 4 && \text{Equation 3} \end{aligned}$$

The value of z is known. We can find y by back-substitution.

$$\begin{aligned} y + 3z &= 9 && \text{Equation 2} \\ y + 3(4) &= 9 && \text{Substitute 4 for } z. \\ y + 12 &= 9 && \text{Multiply.} \\ y &= -3 && \text{Subtract 12 from both sides.} \end{aligned}$$

With values for y and z, we can now use back-substitution to find x.

$$\begin{aligned} x + 2y - 5z &= -19 && \text{Equation 1} \\ x + 2(-3) - 5(4) &= -19 && \text{Substitute } -3 \text{ for } y \text{ and 4 for } z. \\ x - 6 - 20 &= -19 && \text{Multiply.} \\ x - 26 &= -19 && \text{Add.} \\ x &= 7 && \text{Add 26 to both sides.} \end{aligned}$$

We see that $x = 7$, $y = -3$, and $z = 4$. The solution set for the system is $\{(7, -3, 4)\}$.

Check Point 1 Write the solution set for a system of equations represented by the matrix

$$\left[\begin{array}{ccc|c} 1 & -1 & 1 & 8 \\ 0 & 1 & -12 & -15 \\ 0 & 0 & 1 & 1 \end{array}\right].$$

Our goal in solving a linear system using matrices is to produce a matrix similar to the one in Example 1. In general, the matrix will be of the form

$$\left[\begin{array}{ccc|c} 1 & a & b & c \\ 0 & 1 & d & e \\ 0 & 0 & 1 & f \end{array}\right],$$

where a through f represent real numbers. The third row of this matrix gives us the value of one variable. The other variables can then be found by back-substitution.

2 Perform matrix row operations.

A matrix with 1s down the main diagonal and 0s below the 1s is said to be in **triangular form**. How do we produce a matrix in this form? We use **row operations** on the augmented matrix. These row operations are just like what you did when solving a linear system by the addition method. The difference is that we no longer write the variables, usually represented by x, y, and z.

Matrix Row Operations

These row operations produce matrices that lead to systems with the same solution set as the original system.

1. Two rows of a matrix may be interchanged. This is the same as interchanging two equations in the linear system.
2. The elements in any row may be multiplied by a nonzero number. This is the same as multiplying both sides of an equation by a nonzero number.
3. The elements in any row may be multiplied by a nonzero number, and these products may be added to the corresponding elements in any other row. This is the same as multiplying both sides of an equation by a nonzero number and then adding equations to eliminate a variable.

Two matrices are **row equivalent** if one can be obtained from the other by a sequence of row operations.

Study Tip

When performing the row operation

$$kR_i + R_j$$

you use row i to find the products. However, **elements in row i do not change. It is the elements in row j that change:** Add k times the elements in row i to the corresponding elements in row j. Replace elements in row j by these sums.

Each matrix row operation in the preceding box can be expressed symbolically as follows:

1. Interchange the elements in the ith and jth rows: $R_i \leftrightarrow R_j$.
2. Multiply each element in the ith row by k: kR_i.
3. Add k times the elements in row i to the corresponding elements in row j: $kR_i + R_j$.

EXAMPLE 2 Performing Matrix Row operations

Use the matrix

$$\left[\begin{array}{ccc|c} 3 & 18 & -12 & 21 \\ 1 & 2 & -3 & 5 \\ -2 & -3 & 4 & -6 \end{array}\right]$$

and perform each indicated row operation:

a. $R_1 \leftrightarrow R_2$ **b.** $\frac{1}{3}R_1$ **c.** $2R_2 + R_3$

Solution

a. The notation $R_1 \leftrightarrow R_2$ means to interchange the elements in row 1 and row 2. This results in the row-equivalent matrix.

$$\left[\begin{array}{rrr|r} 1 & 2 & -3 & 5 \\ 3 & 18 & -12 & 21 \\ -2 & -3 & 4 & -6 \end{array}\right].$$

This was row 2; now it's row 1.

This was row 1; now it's row 2.

b. The notation $\frac{1}{3}R_1$ means to multiply each element in row 1 by $\frac{1}{3}$. This results in the row-equivalent matrix

$$\left[\begin{array}{cccc|c} \frac{1}{3}(3) & \frac{1}{3}(18) & \frac{1}{3}(-12) & & \frac{1}{3}(21) \\ 1 & 2 & -3 & & 5 \\ -2 & -3 & 4 & & -6 \end{array}\right] = \left[\begin{array}{rrr|r} 1 & 6 & -4 & 7 \\ 1 & 2 & -3 & 5 \\ -2 & -3 & 4 & -6 \end{array}\right].$$

c. The notation $2R_2 + R_3$ means to add 2 times the elements in row 2 to the corresponding elements in row 3. Replace the elements in row 3 by these sums. First, we find 2 times the elements in row 2:

$$2(1) \text{ or } 2, \quad 2(2) \text{ or } 4, \quad 2(-3) \text{ or } -6, \quad 2(5) \text{ or } 10.$$

Now we add these products to the corresponding elements in row 3. Although we use row 2 to find the products, row 2 does not change. It is the elements in row 3 that change, resulting in the row-equivalent matrix

$$\left[\begin{array}{ccc|c} 3 & 18 & -12 & 21 \\ 1 & 2 & -3 & 5 \\ -2+2=0 & -3+4=1 & 4+(-6)=-2 & -6+10=4 \end{array}\right]$$

$$= \left[\begin{array}{rrr|r} 3 & 18 & -12 & 21 \\ 1 & 2 & -3 & 5 \\ 0 & 1 & -2 & 4 \end{array}\right].$$

Check Point 2

Use the matrix

$$\left[\begin{array}{rrr|r} 4 & 12 & -20 & 8 \\ 1 & 6 & -3 & 7 \\ -3 & -2 & 1 & -9 \end{array}\right]$$

and perform each indicated row operation:

a. $R_1 \leftrightarrow R_2$ **b.** $\frac{1}{4}R_1$ **c.** $3R_2 + R_3$

3 Use matrices and Gaussian elimination to solve systems.

The process that we use to solve linear systems using matrix row operations is called **Gaussian elimination,** after the German mathematician Carl Friedrich Gauss (1777–1835). Here are the steps used in Gaussian elimination.

Solving Linear Systems Using Gaussian Elimination

1. Write the augmented matrix for the system.
2. Use matrix row operations to simplify the matrix to one with 1s down the diagonal from upper left to lower right, and 0s below the 1s.
3. Write the system of linear equations corresponding to the matrix in step 2, and use back-substitution to find the system's solution.

EXAMPLE 3 Gaussian Elimination with Back-Substitution

Use matrices to solve the system

$$\begin{aligned} 3x + y + 2z &= 31 \\ x + y + 2z &= 19 \\ x + 3y + 2z &= 25. \end{aligned}$$

Solution

Step 1 Write the augmented matrix for the system.

Linear System	Augmented Matrix
$3x + y + 2z = 31$	$\left[\begin{array}{ccc|c} 3 & 1 & 2 & 31 \\ 1 & 1 & 2 & 19 \\ 1 & 3 & 2 & 25 \end{array}\right]$
$x + y + 2z = 19$	
$x + 3y + 2z = 25$	

Study Tip

Start with the augmented matrix.

$$\left[\begin{array}{ccc|c} * & * & * & * \\ * & * & * & * \\ * & * & * & * \end{array}\right]$$

Get a one in upper left-hand corner.

$$\left[\begin{array}{ccc|c} 1 & * & * & * \\ * & * & * & * \\ * & * & * & * \end{array}\right]$$

Get zeros in first column beneath the one.

$$\left[\begin{array}{ccc|c} 1 & * & * & * \\ 0 & * & * & * \\ 0 & * & * & * \end{array}\right]$$

Get a one in the second row/second column position.

$$\left[\begin{array}{ccc|c} 1 & * & * & * \\ 0 & 1 & * & * \\ 0 & * & * & * \end{array}\right]$$

Get zero below the one in the second column.

$$\left[\begin{array}{ccc|c} 1 & * & * & * \\ 0 & 1 & * & * \\ 0 & 0 & * & * \end{array}\right]$$

Get a one in the third row/third column position.

$$\left[\begin{array}{ccc|c} 1 & * & * & * \\ 0 & 1 & * & * \\ 0 & 0 & 1 & * \end{array}\right]$$

Step 2 Use matrix row operations to simplify the matrix to one with 1s down the diagonal from upper left to lower right, and 0s below the 1s. Our goal is to obtain a matrix of the form

$$\left[\begin{array}{ccc|c} 1 & a & b & c \\ 0 & 1 & d & e \\ 0 & 0 & 1 & f \end{array}\right].$$

Our first step in achieving this goal is to get 1 in the top position of the first column.

We want 1 in this position.

$$\left[\begin{array}{ccc|c} 3 & 1 & 2 & 31 \\ 1 & 1 & 2 & 19 \\ 1 & 3 & 2 & 25 \end{array}\right]$$

To get 1 in this position, we interchange rows 1 and 2. (We could also interchange rows 1 and 3 to attain our goal.)

$$\left[\begin{array}{ccc|c} 1 & 1 & 2 & 19 \\ 3 & 1 & 2 & 31 \\ 1 & 3 & 2 & 25 \end{array}\right]$$

This was row 2; now it's row 1.

This was row 1; now it's row 2.

Now we want to get 0s below the 1 in the first column.

We want 0 in these positions.

$$\left[\begin{array}{ccc|c} 1 & 1 & 2 & 19 \\ 3 & 1 & 2 & 31 \\ 1 & 3 & 2 & 25 \end{array}\right]$$

Let's first get a 0 where there is now a 3. If we multiply the top row of numbers by -3 and add these products to the second row of numbers, we will get 0 in this position. The top row of numbers multiplied by -3 gives

$$-3(1) \text{ or } -3, \quad -3(1) \text{ or } -3, \quad -3(2) \text{ or } -6, \quad -3(19) \text{ or } -57.$$

Now add these products to the corresponding numbers in row 2. Notice that although we use row 1 to find the products, row 1 does not change.

$$\left[\begin{array}{ccc|c} 1 & 1 & 2 & 19 \\ 3+(-3) & 1+(-3) & 2+(-6) & 31+(-57) \\ 1 & 3 & 2 & 25 \end{array}\right] = \left[\begin{array}{ccc|c} 1 & 1 & 2 & 19 \\ 0 & -2 & -4 & -26 \\ 1 & 3 & 2 & 25 \end{array}\right]$$

We want 0 in this position.

We are not yet done with the first column. The voice balloon shows that we want to get another 0 in this column. If we multiply the top row of numbers by −1 and add these products to the third row of numbers, we will get 0 in this position. The top row of numbers multiplied by −1 gives

$$-1(1) \text{ or } -1, \quad -1(1) \text{ or } -1, \quad -1(2) \text{ or } -2, \quad -1(19) \text{ or } -19.$$

Now add these products to the corresponding numbers in row 3.

$$\left[\begin{array}{ccc|c} 1 & 1 & 2 & 19 \\ 0 & -2 & -4 & -26 \\ 1+(-1)=0 & 3+(-1)=2 & 2+(-2)=0 & 25+(-19)=6 \end{array}\right] = \left[\begin{array}{ccc|c} 1 & 1 & 2 & 19 \\ 0 & -2 & -4 & -26 \\ 0 & 2 & 0 & 6 \end{array}\right]$$

We move on to the second column. We want 1 in the second row, second column.

We want 1 in this position.

$$\left[\begin{array}{ccc|c} 1 & 1 & 2 & 19 \\ 0 & -2 & -4 & -26 \\ 0 & 2 & 0 & 6 \end{array}\right]$$

To get 1 in the desired position, we multiply −2 by its reciprocal, $-\frac{1}{2}$. Therefore, we multiply all the numbers in the second row by $-\frac{1}{2}$ to get

$$\left[\begin{array}{ccc|c} 1 & 1 & 2 & 19 \\ -\frac{1}{2}(0) & -\frac{1}{2}(-2) & -\frac{1}{2}(-4) & -\frac{1}{2}(-26) \\ 0 & 2 & 0 & 6 \end{array}\right] = \left[\begin{array}{ccc|c} 1 & 1 & 2 & 19 \\ 0 & 1 & 2 & 13 \\ 0 & 2 & 0 & 6 \end{array}\right].$$

We want 0 in this position.

We are not yet done with the second column. The voice balloon shows that we want to get a 0 where there is now a 2. If we multiply the second row of numbers by −2 and add these products to the third row of numbers, we will get 0 in this position. The second row of numbers multiplied by −2 gives

$$-2(0) \text{ or } 0, \quad -2(1) \text{ or } -2, \quad -2(2) \text{ or } -4, \quad -2(13) \text{ or } -26.$$

Now add these products to the corresponding numbers in row 3.

$$\left[\begin{array}{ccc|c} 1 & 1 & 2 & 19 \\ 0 & 1 & 2 & 13 \\ 0+0 & 2+(-2) & 0+(-4) & 6+(-26) \end{array}\right] = \left[\begin{array}{ccc|c} 1 & 1 & 2 & 19 \\ 0 & 1 & 2 & 13 \\ 0 & 0 & -4 & -20 \end{array}\right]$$

We move on to the third column. We want 1 in the third row, third column.

We want 1 in this position.

$$\left[\begin{array}{ccc|c} 1 & 1 & 2 & 19 \\ 0 & 1 & 2 & 13 \\ 0 & 0 & -4 & -20 \end{array}\right]$$

To get 1 in the desired position, we multiply −4 by its reciprocal, $-\frac{1}{4}$. Therefore, we multiply all the numbers in the third row by $-\frac{1}{4}$ to get

$$\left[\begin{array}{ccc|c} 1 & 1 & 2 & 19 \\ 0 & 1 & 2 & 13 \\ -\frac{1}{4}(0) & -\frac{1}{4}(0) & -\frac{1}{4}(-4) & -\frac{1}{4}(-20) \end{array}\right] = \left[\begin{array}{ccc|c} 1 & 1 & 2 & 19 \\ 0 & 1 & 2 & 13 \\ 0 & 0 & 1 & 5 \end{array}\right].$$

We now have the desired matrix with 1s down the diagonal and 0s below the 1s.

Step 3 Write the system of linear equations corresponding to the matrix in step 2, and use back-substitution to find the system's solution. The system represented by the matrix in step 2 is

$$\left[\begin{array}{ccc|c} 1 & 1 & 2 & 19 \\ 0 & 1 & 2 & 13 \\ 0 & 0 & 1 & 5 \end{array}\right] \rightarrow \begin{aligned} 1x + 1y + 2z &= 19 \\ 0x + 1y + 2z &= 13 \\ 0x + 0y + 1z &= 5 \end{aligned} \quad \text{or} \quad \begin{aligned} x + y + 2z &= 19 \\ y + 2z &= 13. \\ z &= 5 \end{aligned}$$

We immediately see that the value for z is 5. To find y, we back-substitute 5 for z in the second equation.

$$y + 2z = 13 \quad \text{Equation 2}$$

$$y + 2(5) = 13 \quad \text{Substitute 5 for } z.$$

$$y = 3 \quad \text{Solve for } y.$$

Finally, back-substitute 3 for y and 5 for z in the first equation:

$$x + y + 2z = 19 \quad \text{Equation 1}$$

$$x + 3 + 2(5) = 19 \quad \text{Substitute 3 for } y \text{ and 5 for } z.$$

$$x + 13 = 19 \quad \text{Multiply and add.}$$

$$x = 6 \quad \text{Subtract 13 both sides.}$$

The solution set for the original system is $\{(6, 3, 5)\}$.

Check Point 3 Use matrices to solve the system

$$\begin{aligned} 2x + y + 2z &= 18 \\ x - y + 2z &= 9 \\ x + 2y - z &= 6. \end{aligned}$$

Modern supercomputers are capable of solving systems with more than 600,000 variables. The augmented matrices for such systems are huge, but the solution using matrices is exactly like what we did in Example 2. Work with the augmented matrix, one column at a time. First, get 1 in the desired position. Then get 0s below the 1. Let's see how this works for a linear system involving four equations in four variables.

EXAMPLE 4 Gaussian Elimination with Back-Substitution

Use matrices to solve the system

$$\begin{aligned} 2x + y + 3z - w &= 6 \\ x - y + 2z - 2w &= -1 \\ x - y - z + w &= -4 \\ -x + 2y - 2z - w &= -7. \end{aligned}$$

Solution

Step 1 Write the augmented matrix for the system.

Linear System

$$\begin{aligned} 2x + y + 3z - w &= 6 \\ x - y + 2z - 2w &= -1 \\ x - y - z + w &= -4 \\ -x + 2y - 2z - w &= -7 \end{aligned}$$

Augmented Matrix

$$\left[\begin{array}{rrrr|r} 2 & 1 & 3 & -1 & 6 \\ 1 & -1 & 2 & -2 & -1 \\ 1 & -1 & -1 & 1 & -4 \\ -1 & 2 & -2 & -1 & -7 \end{array}\right]$$

Step 2 Use matrix row operations to simplify the matrix to one with 1s down the diagonal from upper left to lower right, and 0s below the 1s. Our first step in achieving this goal is to get 1 in the top position of the first column. To do this, we interchange rows 1 and 2.

$$\left[\begin{array}{rrrr|r} 1 & -1 & 2 & -2 & -1 \\ 2 & 1 & 3 & -1 & 6 \\ 1 & -1 & -1 & 1 & -4 \\ -1 & 2 & -2 & -1 & -7 \end{array}\right]$$

We want 0s in these positions.

This was row 2; now it's row 1.

This was row 1; now it's row 2.

Now we want 0s below the 1 in the first column. To get the first 0, multiply the top row of numbers by -2 and add these products to the second row of numbers. To get the second 0, multiply the top row of numbers by -1 and add these products to the third row of numbers. To get the third 0, multiply the top row of numbers by 1 and add these products to the fourth row of numbers. (Equivalently, add corresponding numbers in rows 1 and 4.) Performing these operations, we obtain the following matrix.

$$\left[\begin{array}{rrrr|r} 1 & -1 & 2 & -2 & -1 \\ 0 & 3 & -1 & 3 & 8 \\ 0 & 0 & -3 & 3 & -3 \\ 0 & 1 & 0 & -3 & -8 \end{array}\right]$$

We want 1 in this position.

Use the previous matrix and:
Replace row 2 by $-2R_1 + R_2$.
Replace row 3 by $-1R_1 + R_3$.
Replace row 4 by $1R_1 + R_4$.

We move on to the second column. We can obtain 1 in the desired position by multiplying the numbers in the second row by $\frac{1}{3}$, the reciprocal of 3.

$$\left[\begin{array}{cccc|c} 1 & -1 & 2 & -2 & -1 \\ \frac{1}{3}(0) & \frac{1}{3}(3) & \frac{1}{3}(-1) & \frac{1}{3}(3) & \frac{1}{3}(8) \\ 0 & 0 & -3 & 3 & -3 \\ 0 & 1 & 0 & -3 & -8 \end{array}\right] = \left[\begin{array}{rrrr|r} 1 & -1 & 2 & -2 & -1 \\ 0 & 1 & -\frac{1}{3} & 1 & \frac{8}{3} \\ 0 & 0 & -3 & 3 & -3 \\ 0 & 1 & 0 & -3 & -8 \end{array}\right] \quad \frac{1}{3}R_2$$

We want 0s in these positions. The top position already has a 0.

Now we want 0s below the 1 in the second column. The top position already has a 0. To obtain a 0 on the bottom, we multiply the second row by -1 and add the product to the corresponding numbers of the last row. (What would happen if we added rows 1 and 4?) Performing these operations, we obtain the following matrix.

$$\left[\begin{array}{cccc|c} 1 & -1 & 2 & -2 & -1 \\ 0 & 1 & -\frac{1}{3} & 1 & \frac{8}{3} \\ 0 & 0 & -3 & 3 & -3 \\ 0 & 0 & \frac{1}{3} & -4 & -\frac{32}{3} \end{array}\right]$$

We want 1 in this position.

Replace row 4 in the previous matrix by $-1R_2 + R_4$.

We move on to the third column. We can obtain 1 in the desired position by multiplying the numbers in the third row by $-\frac{1}{3}$, the reciprocal of -3.

$$\left[\begin{array}{cccc|c} 1 & -1 & 2 & -2 & -1 \\ 0 & 1 & -\frac{1}{3} & 1 & \frac{8}{3} \\ -\frac{1}{3}(0) & -\frac{1}{3}(0) & -\frac{1}{3}(-3) & -\frac{1}{3}(3) & -\frac{1}{3}(-3) \\ 0 & 0 & \frac{1}{3} & -4 & -\frac{32}{3} \end{array}\right] = \left[\begin{array}{cccc|c} 1 & -1 & 2 & -2 & -1 \\ 0 & 1 & -\frac{1}{3} & 1 & \frac{8}{3} \\ 0 & 0 & 1 & -1 & 1 \\ 0 & 0 & \frac{1}{3} & -4 & -\frac{32}{3} \end{array}\right]$$

$-\frac{1}{3}R_3$

We want 0 in this position.

Now we want 0 below the 1 in the third column. If we multiply the third row of numbers by $-\frac{1}{3}$ and add these products to the fourth row of numbers, we will get 0 in this position. Performing these operations, we obtain the following matrix.

$$\left[\begin{array}{cccc|c} 1 & -1 & 2 & -2 & -1 \\ 0 & 1 & -\frac{1}{3} & 1 & \frac{8}{3} \\ 0 & 0 & 1 & -1 & 1 \\ 0 & 0 & 0 & -\frac{11}{3} & -11 \end{array}\right]$$

We want 1 in this position.

Replace row 4 in the previous matrix by $-\frac{1}{3}R_3 + R_4$.

We move on to the fourth column. Because we want 1s down the main diagonal, we want 1 where there is now $-\frac{11}{3}$. We can obtain 1 in this position by multiplying the numbers in the fourth row by $-\frac{3}{11}$.

$$\left[\begin{array}{cccc|c} 1 & -1 & 2 & -2 & -1 \\ 0 & 1 & -\frac{1}{3} & 1 & \frac{8}{3} \\ 0 & 0 & 1 & -1 & 1 \\ -\frac{3}{11}(0) & -\frac{3}{11}(0) & -\frac{3}{11}(0) & -\frac{3}{11}(-\frac{11}{3}) & -\frac{3}{11}(-11) \end{array}\right]$$

$$= \left[\begin{array}{cccc|c} 1 & -1 & 2 & -2 & -1 \\ 0 & 1 & -\frac{1}{3} & 1 & \frac{8}{3} \\ 0 & 0 & 1 & -1 & 1 \\ 0 & 0 & 0 & 1 & 3 \end{array}\right]$$

$-\frac{3}{11}R_4$

We now have the desired matrix with 1s down the diagonal and 0s below the 1s.

Step 3 Write the system of linear equations corresponding to the matrix in step 2, and use back-substitution to find the system's solution. The system represented by the matrix in step 2 is

$$\left[\begin{array}{cccc|c} 1 & -1 & 2 & -2 & -1 \\ 0 & 1 & -\frac{1}{3} & 1 & \frac{8}{3} \\ 0 & 0 & 1 & -1 & 1 \\ 0 & 0 & 0 & 1 & 3 \end{array}\right] \rightarrow \begin{aligned} 1x - 1y + 2z - 2w &= -1 \\ 0x + 1y - \tfrac{1}{3}z + 1w &= \tfrac{8}{3} \\ 0x + 0y + 1z - 1w &= 1 \\ 0x + 0y + 0z + 1w &= 3 \end{aligned} \quad \text{or} \quad \begin{aligned} x - y + 2z - 2w &= -1 \\ y - \tfrac{1}{3}z + w &= \tfrac{8}{3} \\ z - w &= 1 \\ w &= 3 \end{aligned}$$

We immediately see that the value for w is 3. We can now use back-substitution to find the values for z, y, and x.

$w = 3$	$z - w = 1$	$y - \frac{1}{3}z + w = \frac{8}{3}$	$x - y + 2z - 2w = -1$
	$z - 3 = 1$	$y - \frac{1}{3}(4) + 3 = \frac{8}{3}$	$x - 1 + 2(4) - 2(3) = -1$
	$z = 4$	$y + \frac{5}{3} = \frac{8}{3}$	$x - 1 + 8 - 6 = -1$
		$y = 1$	$x + 1 = -1$
			$x = -2$

Let's agree to write the solution set for the system in the order in which the variables for the given system appeared, from left to right, namely (x, y, z, w). Thus, the solution set is $\{(-2, 1, 4, 3)\}$. We can verify this solution set by substituting the value for each variable into the original system of equations.

Check Point 4 Use matrices to solve the system

$$\begin{aligned} x - 3y - 2z + w &= -3 \\ 2x - 7y - z + 2w &= 1 \\ 3x - 7y - 3z + 3w &= -5 \\ 5x + y + 4z - 2w &= 18. \end{aligned}$$

4 Use matrices and Gauss-Jordan elimination to solve systems.

Gauss-Jordan Elimination

Using Gaussian elimination, we obtain a matrix with 1s down the main diagonal and 0s below the 1s. A second method, called **Gauss-Jordan elimination**, after Carl Friedrich Gauss and Wilhelm Jordan (1842–1899), continues the process until a matrix with 1s down the main diagonal from left to right and 0s in every position *above and below* each 1 is found. For a system of linear equations in three variables, x, y, and z, we try to get the augmented matrix into the form

$$\left[\begin{array}{ccc|c} 1 & 0 & 0 & a \\ 0 & 1 & 0 & b \\ 0 & 0 & 1 & c \end{array}\right].$$

Based on this matrix, we conclude that $x = a$, $y = b$, and $z = c$.

EXAMPLE 5 Using Gauss-Jordan Elimination

Use Gauss-Jordan elimination to solve the system

$$\begin{aligned} 3x + y + 2z &= 31 \\ x + y + 2z &= 19 \\ x + 3y + 2z &= 25. \end{aligned}$$

Solution In Example 3, we used Gaussian elimination to obtain the following matrix:

$$\left[\begin{array}{ccc|c} 1 & 1 & 2 & 19 \\ 0 & 1 & 2 & 13 \\ 0 & 0 & 1 & 5 \end{array}\right].$$

Study Tip

The advantage to Gauss-Jordan elimination is that from the augmented matrix we can simply read the solution. The disadvantage is that we must continue row operations in the augmented matrix from the Gaussian elimination process, and it's fairly easy to make computational errors.

To find a solution using Gauss-Jordan elimination, we need to work with this matrix and convert the boxed numbers to 0s. Thus, we will apply matrix row operations to get 0s *above the 1s* in the main diagonal. To get 0 in row 1, column 2 (where there is now a 1), we multiply each number in the second row by -1 and add these products to the corresponding numbers in the first row. Performing these operations, we obtain the following matrix.

$$\left[\begin{array}{ccc|c} 1 & 0 & 0 & 6 \\ 0 & 1 & 2 & 13 \\ 0 & 0 & 1 & 5 \end{array}\right]$$

We want 0s in these positions. The top position already has a 0.

We want 0s above the 1 in the third column. The top position already has a 0. To obtain a 0 where there is now 2, we multiply each number in the bottom row by -2 and add these products to the corresponding numbers in the second row. Performing these operations, we obtain the following matrix.

$$\left[\begin{array}{ccc|c} 1 & 0 & 0 & 6 \\ 0 & 1 & 0 & 3 \\ 0 & 0 & 1 & 5 \end{array}\right]$$

Replace row 2 in the previous matrix by $-2R_3 + R_2$.

This last matrix corresponds to

$$x = 6, \quad y = 3, \quad z = 5.$$

As we found in Example 3, the solution set is $\{(6, 3, 5)\}$.

Check Point 5 Solve the system in Checkpoint 2 using Gauss-Jordan elimination. Begin by working with the matrix that you obtained in Checkpoint 2.

EXERCISE SET 6.1

Practice Exercises

In Exercises 1–8, write the augmented matrix for each system of linear equations.

1. $2x + y + 2z = 2$, $3x - 5y - z = 4$, $x - 2y - 3z = -6$

2. $3x - 2y + 5z = 31$, $x + 3y - 3z = -12$, $-2x - 5y + 3z = 11$

3. $x - y + z = 8$, $y - 12z = -15$, $z = 1$

4. $x - 2y + 3z = 9$, $y + 3z = 5$, $z = 2$

5. $5x - 2y - 3z = 0$, $x + y = 5$, $2x - 3z = 4$

6. $x - 2y + z = 10$, $3x + y = 5$, $7x + 2z = 2$

7. $2x + 5y - 3z + w = 2$, $3y + z = 4$, $x - y + 5z = 9$, $5x - 5y - 2z = 1$

8. $4x + 7y - 8z + w = 3$, $5y + z = 5$, $x - y - z = 17$, $2x - 2y + 11z = 4$

In Exercises 9–12, write the system of linear equations represented by the augmented matrix. Use x, y, z, and, if necessary, w for the variables.

9. $\left[\begin{array}{ccc|c} 5 & 0 & 3 & -11 \\ 0 & 1 & -4 & 12 \\ 7 & 2 & 0 & 3 \end{array}\right]$

10. $\left[\begin{array}{ccc|c} 7 & 0 & 4 & -13 \\ 0 & 1 & -5 & 11 \\ 2 & 7 & 0 & 6 \end{array}\right]$

11. $\left[\begin{array}{cccc|c} 1 & 1 & 4 & 1 & 3 \\ -1 & 1 & -1 & 0 & 7 \\ 2 & 0 & 0 & 5 & 11 \\ 0 & 0 & 12 & 4 & 5 \end{array}\right]$

12. $\left[\begin{array}{cccc|c} 4 & 1 & 5 & 1 & 6 \\ 1 & -1 & 0 & -1 & 8 \\ 3 & 0 & 0 & 7 & 4 \\ 0 & 0 & 11 & 5 & 3 \end{array}\right]$

In Exercises 13–18, write the system of linear equations represented by the augmented matrix. Use x, y, z, and, if necessary, w for the variables. Once the system is written, use back-substitution to find its solution.

13. $\left[\begin{array}{ccc|c} 1 & 0 & -4 & 5 \\ 0 & 1 & -12 & 13 \\ 0 & 0 & 1 & -\frac{1}{2} \end{array}\right]$

14. $\left[\begin{array}{ccc|c} 1 & 2 & 1 & 0 \\ 0 & 1 & 0 & -2 \\ 0 & 0 & 1 & 3 \end{array}\right]$

15. $\left[\begin{array}{ccc|c} 1 & \frac{1}{2} & 1 & \frac{11}{2} \\ 0 & 1 & \frac{3}{2} & 7 \\ 0 & 0 & 1 & 4 \end{array}\right]$

16. $\left[\begin{array}{ccc|c} 1 & 1 & 0 & 3 \\ 0 & 1 & \frac{3}{2} & -2 \\ 0 & 0 & 1 & 0 \end{array}\right]$

17. $\left[\begin{array}{cccc|c} 1 & -1 & 1 & 1 & 3 \\ 0 & 1 & -2 & -1 & 0 \\ 0 & 0 & 1 & 6 & 17 \\ 0 & 0 & 0 & 1 & 3 \end{array}\right]$

18. $\left[\begin{array}{cccc|c} 1 & 2 & -1 & 0 & 2 \\ 0 & 1 & 1 & -2 & -3 \\ 0 & 0 & 1 & -1 & -2 \\ 0 & 0 & 0 & 1 & 3 \end{array}\right]$

In Exercises 19–24, perform each matrix row operation and write the new matrix.

19. $\left[\begin{array}{ccc|c} 2 & -6 & 4 & 10 \\ 1 & 5 & -5 & 0 \\ 3 & 0 & 4 & 7 \end{array}\right] \quad \frac{1}{2}R_1$

20. $\left[\begin{array}{ccc|c} 3 & -12 & 6 & 9 \\ 1 & -4 & 4 & 0 \\ 2 & 0 & 7 & 4 \end{array}\right] \quad \frac{1}{3}R_1$

21. $\left[\begin{array}{ccc|c} 1 & -3 & 2 & 0 \\ 3 & 1 & -1 & 7 \\ 2 & -2 & 1 & 3 \end{array}\right] \quad -3R_1 + R_2$

22. $\left[\begin{array}{ccc|c} 1 & -1 & 5 & -6 \\ 3 & 3 & -1 & 10 \\ 1 & 3 & 2 & 5 \end{array}\right] \quad -3R_1 + R_2$

23. $\left[\begin{array}{cccc|c} 1 & -1 & 1 & 1 & 3 \\ 0 & 1 & -2 & -1 & 0 \\ 2 & 0 & 3 & 4 & 11 \\ 5 & 1 & 2 & 4 & 6 \end{array}\right] \quad \begin{array}{l} -2R_1 + R_3 \\ -5R_1 + R_4 \end{array}$

24. $\left[\begin{array}{cccc|c} 1 & -5 & 2 & -2 & 4 \\ 0 & 1 & -3 & -1 & 0 \\ 3 & 0 & 2 & -1 & 6 \\ -4 & 1 & 4 & 2 & -3 \end{array}\right] \quad \begin{array}{l} -3R_1 + R_3 \\ 4R_1 + R_4 \end{array}$

In Exercises 25–26, a few steps in the process of simplifying the given matrix to one with 1s down the diagonal from upper left to lower right, and 0s below the 1s, are shown. Fill in the missing numbers in the steps that are shown.

25. $\left[\begin{array}{ccc|c} 1 & -1 & 1 & 8 \\ 2 & 3 & -1 & -2 \\ 3 & -2 & -9 & 9 \end{array}\right] \rightarrow \left[\begin{array}{ccc|c} 1 & -1 & 1 & 8 \\ 0 & 5 & \square & \square \\ 0 & 1 & \square & \square \end{array}\right]$

$\rightarrow \left[\begin{array}{ccc|c} 1 & -1 & 1 & 8 \\ 0 & 1 & \square & \square \\ 0 & 1 & \square & \square \end{array}\right]$

26. $\left[\begin{array}{ccc|c} 1 & -2 & 3 & 4 \\ 2 & 1 & -4 & 3 \\ -3 & 4 & -1 & -2 \end{array}\right] \rightarrow \left[\begin{array}{ccc|c} 1 & -2 & 3 & 4 \\ 0 & 5 & \square & \square \\ 0 & -2 & \square & \square \end{array}\right]$

$\rightarrow \left[\begin{array}{ccc|c} 1 & -2 & 3 & 4 \\ 0 & 1 & \square & \square \\ 0 & -2 & \square & \square \end{array}\right]$

In Exercises 27–40, solve each system of equations using matrices. Use Gaussian elimination with back-substitution or Gauss-Jordan elimination.

27. $\begin{aligned} x + y - z &= -2 \\ 2x - y + z &= 5 \\ -x + 2y + 2z &= 1 \end{aligned}$

28. $\begin{aligned} x - 2y - z &= 2 \\ 2x - y + z &= 4 \\ -x + y - 2z &= -4 \end{aligned}$

29. $\begin{aligned} x + 3y &= 0 \\ x + y + z &= 1 \\ 3x - y - z &= 11 \end{aligned}$

30. $\begin{aligned} 3y - z &= -1 \\ x + 5y - z &= -4 \\ -3x + 6y + 2z &= 11 \end{aligned}$

31. $\begin{aligned} 2x + 2y + 7z &= -1 \\ 2x + y + 2z &= 2 \\ 4x + 6y + z &= 15 \end{aligned}$

32. $\begin{aligned} 3x + 2y + 3z &= 3 \\ 4x - 5y + 7z &= 1 \\ 2x + 3y - 2z &= 6 \end{aligned}$

33. $\begin{aligned} x + y + z + w &= 4 \\ 2x + y - 2z - w &= 0 \\ x - 2y - z - 2w &= -2 \\ 3x + 2y + z + 3w &= 4 \end{aligned}$

34. $\begin{aligned} x + y + z + w &= 5 \\ x + 2y - z - 2w &= -1 \\ x - 3y - 3z - w &= -1 \\ 2x - y + 2z - w &= -2 \end{aligned}$

35. $\begin{aligned} 3x - 4y + z + w &= 9 \\ x + y - z - w &= 0 \\ 2x + y + 4z - 2w &= 3 \\ -x + 2y + z - 3w &= 3 \end{aligned}$

36. $\begin{aligned} 2x + z - 3w &= 8 \\ x - y + 4w &= -10 \\ 3x + 5y - z - w &= 20 \\ x + y - z - w &= 6 \end{aligned}$

37. $\begin{aligned} 2x + 3y - z - w &= -3 \\ 2x - y - 3z + 2w &= -5 \\ x - y + z - w &= -4 \\ 3x - 2y + z + w &= 0 \end{aligned}$

38. $\begin{aligned} 2x - y - z + w &= 4 \\ x + 3y - 2z - 3w &= 6 \\ x - y + z - w &= 2 \\ -x + 2y - z - w &= -1 \end{aligned}$

39. $\begin{aligned} 2x_1 - 2x_2 + 3x_3 - x_4 &= 12 \\ x_1 + 2x_2 - x_3 + 2x_4 - x_5 &= -7 \\ x_1 + x_3 - x_4 - 5x_5 &= 5 \\ -x_1 + x_2 - x_3 - 2x_4 - 3x_5 &= 0 \\ x_1 - x_2 - x_4 + x_5 &= 4 \end{aligned}$

40. $\begin{aligned} 2x - 2z + 4w - 4v &= -6 \\ -x - y - z - w - u - v &= -12 \\ x + y - z - w &= -2 \\ y - z + u - v &= -1 \\ x - y + z - w + u - v &= 0 \\ 3y - z + v &= 4 \end{aligned}$

Application Exercises

41. The table shows the number of inmates in federal and state prisons in the United States for three selected years.

x (Number of Years after 1980)	1	5	10
y (Number of Inmates, in thousands)	344	480	740

a. Use the quadratic function $y = ax^2 + bx + c$ to model the data. Solve the system of linear equations involving a, b, and c using matrices.

b. Predict the number of inmates in the year 2010.

c. List one factor that would change the accuracy of this model for the year 2010.

42. A football is kicked straight upward. The position function

$$s = \frac{1}{2}at^2 + v_0 t + s_0$$

describes the ball's height, s, in feet, after t seconds. Use the points labeled in the graph to find the values of a, v_0, and s_0. Solve the system of linear equations involving a, v_0, and s_0 using matrices. What is the value for s when $t = 7$? Describe what this means.

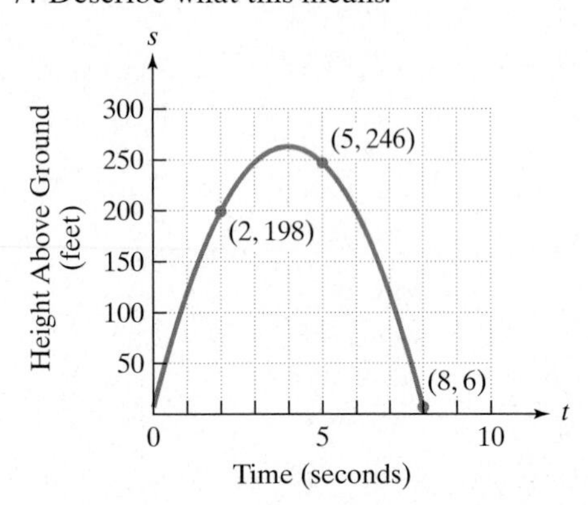

Write a system of linear equations in three variables to solve Exercises 43–46. Then use matrices to solve the system.

43. The circle graph indicates the ages of the 40 million online users in the United States. The percentage of online users in the youngest (under 30) and oldest (50 and over) age groups combined exceeds the percentage in the 30–49 age group by 2%. If the percentage of users in the oldest age group is doubled, it is 3% less than the percentage of users in the youngest age group. Find the percentage of online users in each of the three age groups.

Age of U.S. Online Users

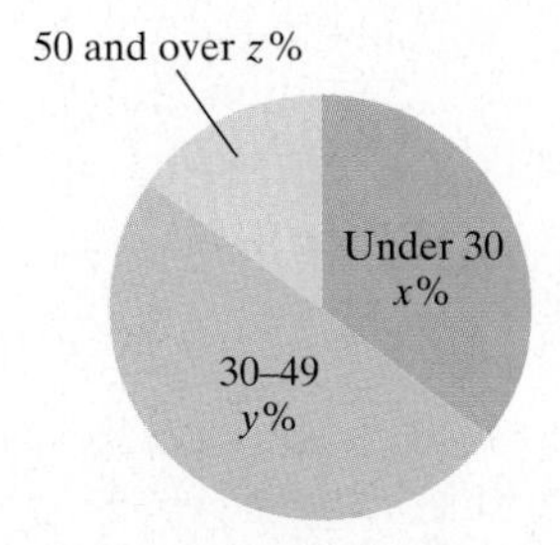

Source: U.S. Census Bureau

44. The circle graph indicates computers in use for the United States and the rest of the world. The percentage of the world's computers in Europe and Japan combined is 13% less than the percentage of the world's computers in the United States. If the percentage of the world's computers in Europe is doubled, it is only 3% more than the percentage of the world's computers in the United States. Find the percentage of the world's computers in the United States, Europe, and Japan.

Percentage of the World's Computers: U.S. and the World

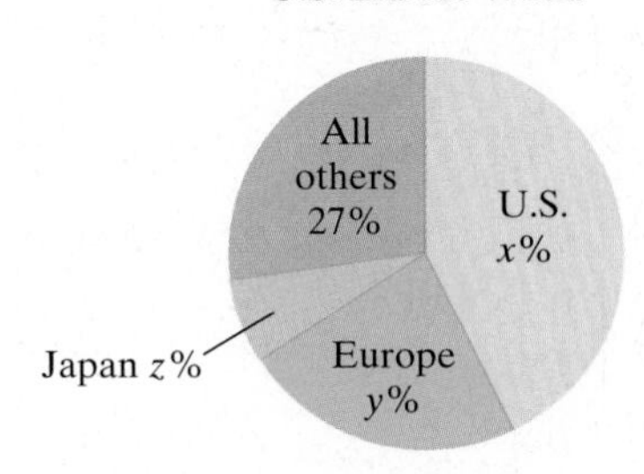

Source: Jupiter Communications

45. Three foods have the following nutritional content per ounce.

	Calories	Protein (in grams)	Vitamin C (in milligrams)
Food *A*	40	5	30
Food *B*	200	2	10
Food *C*	400	4	300

If a meal consisting of the three foods allows exactly 660 calories, 25 grams of protein, and 425 milligrams of vitamin C, how many ounces of each kind of food should be used?

46. A furniture company produces three types of desks: a children's model, an office model, and a deluxe model. Each desk is manufactured in three stages: cutting, construction, and finishing. The time requirements for each model and manufacturing stage are given in the following table.

	Children's model	Office model	Deluxe model
Cutting	2 hr	3 hr	2 hr
Construction	2 hr	1 hr	3 hr
Finishing	1 hr	1 hr	2 hr

Each week the company has available a maximum of 100 hours for cutting, 100 hours for construction, and 65 hours for finishing. If all available time must be used, how many of each type of desk should be produced each week?

Writing in Mathematics

47. What is a matrix?

48. Describe what is meant by the augmented matrix of a system of linear equations.

49. In your own words, describe each of the three matrix row operations. Give an example with each of the operations.

50. Describe how to use row operations and matrices to solve a system of linear equations.

51. What is the difference between Gaussian elimination and Gauss-Jordan elimination?

52. The graphs show the percentage of recorded music on CDs, cassettes, and LPs from 1981–2001. For this time period, which of these three forms of recorded music would you model using a quadratic function? Explain your answer.

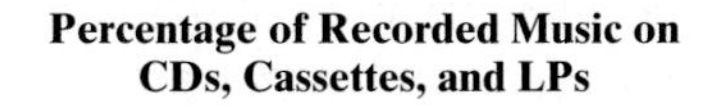

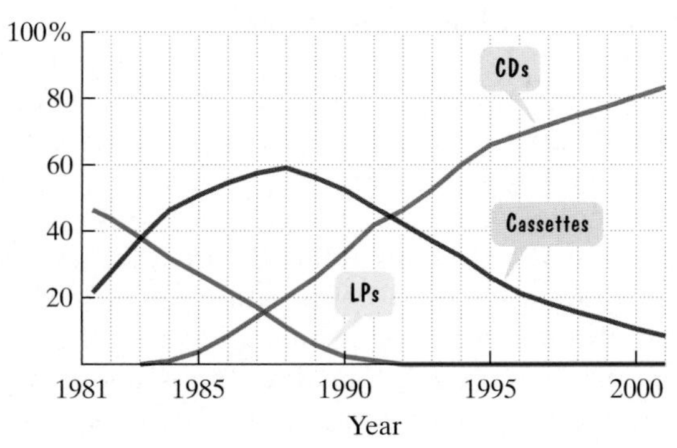

Source: Recording Industry Association of America

53. In Exercise 52, assume that you plan to obtain the quadratic model by hand. Explain how to use the graph for the form that you selected to find a, b, and c in $y = ax^2 + bx + c$, where x represents years since 1981 and y represents the percentage of recorded music on this form. Describe the role that matrices play in the process of obtaining the model.

Technology Exercises

54. Most graphing utilities can perform row operations on matrices. Consult the owner's manual for your graphing utility to learn proper keystrokes for performing these operations. Then duplicate the row operations of any three exercises that you solved from Exercises 19–24.

55. The final augmented matrix that we obtain when using Gaussian elimination is said to be in **row-echelon form**. For systems of linear equations with unique solutions, this form results when each entry in the main diagonal is 1 and all entries below the main diagonal are 0s. Some graphing utilities can transform a matrix to row-echelon form. Consult the owner's manual for your graphing utility. If your utility has this capability, enter the augmented matrix and obtain the final matrices of Example 3 on page 486 and Example 4 on page 488. Then use this capability to solve any five of the systems in Exercises 27–40.

56. The final augmented matrix that we obtain when using Gauss-Jordan elimination is said to be in **reduced row-echelon form**. For systems of linear equations with unique solutions, this form results when each entry on the main diagonal is 1 and all entries below and above that main diagonal are 0s. Some graphing utilities can transform a matrix to reduced row-echelon form. Consult the owner's manual for your graphing utility. If your utility has this capability, obtain the final matrix of Example 5 on pages 491–492 beginning with the augmented matrix for Example 3 on page 486. Then use this capability to solve any five of the systems in Exercises 27–40.

Critical Thinking Exercises

57. Find a cubic function whose graph passes through the points $(0, -3)$, $(1, 5)$, $(-1, -7)$, and $(-2, -13)$. (*Hint:* Use the equation $y = ax^3 + bx^2 + cx + d$.)

58. The table shows the daily production level and profit for a business.

x (Number of units Produced Daily)	30	50	100
y (Daily Profit)	\$5900	\$7500	\$4500

Use the quadratic function $y = ax^2 + bx + c$ to determine the number of units that should be produced each day for maximum profit. What is the maximum daily profit?

Group Exercise

59. In Chapter 5, you learned how to fit a quadratic function of the form $y = ax^2 + bx + c$ to data without using the regression feature of a graphing utility (see pages 435–436). Each group member should find an interesting data set. Group members should select the two sets of data that are most interesting and relevant.

a. For one of the data sets selected, use the function $y = ax^3 + bx^2 + cx + d$ and four ordered pairs of values (x, y) to find the cubic function that models the data. Use matrices or a graphing utility to solve the resulting system in four variables for a, b, c, and d.

b. For the other data set selected, fit a higher-degree polynomial function to the data. Use a graphing utility to solve the resulting system in five or more variables.

SECTION 6.2 Inconsistent and Dependent Systems and Their Applications

Objectives

1. Apply Gaussian elimination to systems without unique solutions.
2. Apply Gaussian elimination to systems with differing numbers of variables and equations.
3. Solve problems involving systems without unique solutions.

Traffic jams getting you down? Powerful computers, able to solve systems with hundreds of thousands of variables in a single bound, may promise a gridlock-free future. The computer in your car could be linked to a central computer that manages traffic flow by controlling traffic lights, rerouting you away from traffic congestion, issuing weather reports, and selecting the best route to your destination. New technologies could eventually drive your car at a steady 75 miles per hour along automated highways as you comfortably nap. In this section, we look at the role of linear systems without unique solutions in a future free of traffic jams.

1 Apply Gaussian elimination to systems without unique solutions.

Linear systems can have one solution, no solutions, or infinitely many solutions. We can use Gaussian elimination on systems with three or more variables to determine how many solutions such systems may have. In the case of systems with no solutions or infinitely many solutions, it is impossible to rewrite the augmented matrix in the desired form with 1s down the main diagonal and 0s below the 1s. Let's see what this means by looking at a system that has no solutions.

EXAMPLE 1 A System With No Solutions

Use Gaussian elimination to solve the system

$$\begin{aligned} x - y - 2z &= 2 \\ 2x - 3y + 6z &= 5 \\ 3x - 4y + 4z &= 12. \end{aligned}$$

Solution

Step 1 Write the augmented matrix for the system.

Linear System

$$\begin{aligned} x - y - 2z &= 2 \\ 2x - 3y + 6z &= 5 \\ 3x - 4y + 4z &= 12 \end{aligned}$$

Augmented Matrix

$$\left[\begin{array}{rrr|r} 1 & -1 & -2 & 2 \\ 2 & -3 & 6 & 5 \\ 3 & -4 & 4 & 12 \end{array}\right]$$

Discovery

Use the addition method to solve Example 1. Describe what happens. Why does this mean that there is no solution?

Step 2 Attempt to simplify the matrix to one with 1s down the diagonal and 0s below the 1s. Notice that the augmented matrix already has a 1 in the top position of the first column. Now we want 0s below the 1. To get the first 0, multiply row 1 by −2 and add these products to row 2. To get the second 0, multiply row 1 by −3 and add these products to row 3. Performing these operations, we obtain the following matrix.

$$\left[\begin{array}{ccc|c} 1 & -1 & -2 & 2 \\ 0 & -1 & 10 & 1 \\ 0 & -1 & 10 & 6 \end{array}\right]$$

We want 1 in this position.

Use the previous matrix and:
Replace row 2 by $-2R_1 + R_2$.
Replace row 3 by $-3R_1 + R_3$.

Moving on to the second column, we obtain 1 in the desired position by multiplying row 2 by −1.

$$\left[\begin{array}{ccc|c} 1 & -1 & -2 & 2 \\ 0(-1) & -1(-1) & 10(-1) & 1(-1) \\ 0 & -1 & 10 & 6 \end{array}\right] = \left[\begin{array}{ccc|c} 1 & -1 & -2 & 2 \\ 0 & 1 & -10 & -1 \\ 0 & -1 & 10 & 6 \end{array}\right]$$

$-1R_2$

We want 0 in this position.

Now we want a 0 below the 1 in column 2. To get the 0, multiply row 2 by 1 and add these products to row 3. (Equivalently, add row 2 to row 3.) We obtain the following matrix.

$$\left[\begin{array}{ccc|c} 1 & -1 & -2 & 2 \\ 0 & 1 & -10 & -1 \\ 0 & 0 & 0 & 5 \end{array}\right]$$

Replace row 3 in the previous matrix by $1R_2 + R_3$.

It is impossible to convert this last matrix to the desired form of 1s down the main diagonal. If we translate the last row back into equation form, we get

$$0x + 0y + 0z = 5,$$

which is false. Regardless of which values we select for x, y, and z, the last equation can never be a true statement. Consequently, the system has no solution. The solution set is ∅, the empty set.

Use Gaussian elimination to solve the system

$$\begin{aligned} x - 2y - z &= -5 \\ 2x - 3y - z &= 0 \\ 3x - 4y - z &= 1. \end{aligned}$$

Recall that the graph of a system of three linear equations in three variables consists of three planes. When these planes intersect in a single point, the system has precisely one ordered-triple solution. When the planes have no point in common, the system has no solution, like the one in Example 1. Figure 6.1 illustrates some of the geometric possibilities for these inconsistent systems.

Now let's see what happens when we apply Gaussian elimination to a system with infinitely many solutions. Representing the solution set for these systems can be a bit tricky.

Three planes are parallel with no common intersection point.

Two planes are parallel with no common intersection point.

Planes intersect two at a time. There is no intersection point common to all three planes.

Figure 6.1 Three planes may have no common point of intersection.

EXAMPLE 2 A System with an Infinite Number of Solutions

Use Gaussian elimination to solve the following system:

$$\begin{aligned} 3x - 4y + 4z &= 7 \\ x - y - 2z &= 2 \\ 2x - 3y + 6z &= 5. \end{aligned}$$

Solution As always, we start with the augmented matrix.

$$\left[\begin{array}{rrr|r} 3 & -4 & 4 & 7 \\ 1 & -1 & -2 & 2 \\ 2 & -3 & 6 & 5 \end{array}\right] \xrightarrow[\text{1 and 2.}]{\substack{R_1 \leftrightarrow R_2 \\ \text{Reverse rows}}} \left[\begin{array}{rrr|r} 1 & -1 & -2 & 2 \\ 3 & -4 & 4 & 7 \\ 2 & -3 & 6 & 5 \end{array}\right] \xrightarrow{\substack{\text{Replace row 2} \\ \text{by } -3R_1 + R_2. \\ \text{Replace row 3} \\ \text{by } -2R_1 + R_3.}}$$

$$\left[\begin{array}{rrr|r} 1 & -1 & -2 & 2 \\ 0 & -1 & 10 & 1 \\ 0 & -1 & 10 & 1 \end{array}\right] \xrightarrow[\text{2 by } -1.]{\substack{-1R_2 \\ \text{Multiply row}}} \left[\begin{array}{rrr|r} 1 & -1 & -2 & 2 \\ 0 & 1 & -10 & -1 \\ 0 & -1 & 10 & 1 \end{array}\right] \xrightarrow{\substack{\text{Replace row 3} \\ \text{by } 1R_2 + R_3.}}$$

$$\left[\begin{array}{rrr|r} 1 & -1 & -2 & 2 \\ 0 & 1 & -10 & -1 \\ 0 & 0 & 0 & 0 \end{array}\right]$$

If we translate row 3 of the matrix into equation form, we obtain

$$0x + 0y + 0z = 0$$

or

$$0 = 0.$$

This equation results in a true statement regardless of which values we select for x, y, and z. Consequently, the equation $0x + 0y + 0z = 0$ is *dependent* on the other two equations in the system in the sense that it adds no new information about the variables. Thus, we can drop it from the system, which can now be expressed in the form

$$\left[\begin{array}{rrr|r} 1 & -1 & -2 & 2 \\ 0 & 1 & -10 & -1 \end{array}\right].$$

The original system is equivalent to the system

$$\begin{aligned} x - y - 2z &= 2 \\ y - 10z &= -1. \end{aligned}$$

Although neither of these equations gives a value for z, we can use them to express x and y in terms of z. From the last equation we obtain

$$y = 10z - 1. \quad \text{Add } 10z \text{ to both sides and isolate } y.$$

Back-substituting for y into the previous equation, we can find x in terms of z.

$x - y - 2z = 2$ — This is the first equation obtained from the final matrix.

$x - (10z - 1) - 2z = 2$ — Because $y = 10z - 1$, substitute $10z - 1$ for y.

$x - 10z + 1 - 2z = 2$ — Apply the distributive property.

$x - 12z + 1 = 2$ — Combine like terms.

$x = 12z + 1$ — Solve for x in terms of z.

Because no value is determined for z, we can find a solution to the system by letting z equal any real number and then using the above equations to obtain x and y. For example, if $z = 1$, then

$$x = 12z + 1 = 12(1) + 1 = 13 \text{ and}$$
$$y = 10z - 1 = 10(1) - 1 = 9.$$

Consequently, $(13, 9, 1)$ is a solution to the system. On the other hand, if we let $z = -1$, then

$$x = 12z + 1 = 12(-1) + 1 = -11 \text{ and}$$
$$y = 10z - 1 = 10(-1) - 1 = -11.$$

Thus, $(-11, -11, -1)$ is another solution to the system. Finally, letting $z = t$ (or any letter of our choice), the solutions to the system are all of the form

$$x = 12t + 1, \qquad y = 10t - 1, \qquad z = t,$$

where t is a real number. Therefore, every ordered triple that is of the form $(12t + 1, 10t - 1, t)$, where t is a real number, is a solution of the system. The solution set of the system with dependent equations can be written as $\{(12t + 1, 10t - 1, t)\}$.

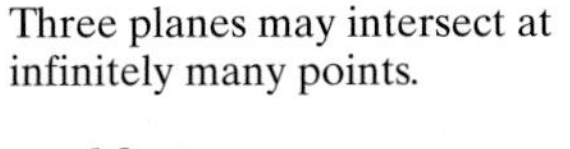
Three planes may intersect at infinitely many points.

Figure 6.2

We have seen that when three planes have no point in common, the corresponding system has no solution. When the system has infinitely many solutions, like the one in Example 2, the three planes intersect in more than one point. Figure 6.2 illustrates one geometric possibility for systems with dependent equations.

Check Point 2 Use Gaussian elimination to solve the following system:

$$\begin{aligned} x - 2y - z &= 5 \\ 2x - 5y + 3z &= 6 \\ x - 3y + 4z &= 1. \end{aligned}$$

2 Apply Gaussian elimination to systems with differing numbers of variables and equations.

Nonsquare Systems

Up to this point, we have encountered only *square* systems in which the number of equations is equal to the number of variables. In a **nonsquare system**, the number of variables differs from the number of equations.

EXAMPLE 3 A System with Fewer Equations Than Variables

Use Gaussian elimination to solve the system

$$\begin{aligned} 3x + 7y + 6z &= 26 \\ x + 2y + z &= 8. \end{aligned}$$

Solution We begin with the augmented matrix.

$$\left[\begin{array}{ccc|c} 3 & 7 & 6 & 26 \\ 1 & 2 & 1 & 8 \end{array}\right] \xrightarrow{R_1 \leftrightarrow R_2} \left[\begin{array}{ccc|c} 1 & 2 & 1 & 8 \\ 3 & 7 & 6 & 26 \end{array}\right] \xrightarrow[\text{by } -3R_1 + R_2.]{\text{Replace row 2}} \left[\begin{array}{ccc|c} 1 & 2 & 1 & 8 \\ 0 & 1 & 3 & 2 \end{array}\right]$$

Because we now have 1s down the diagonal that begins with the upper-left entry and a 0 below this 1, we translate the matrix back into equation form.

$$x + 2y + z = 8 \qquad \text{Equation 1}$$
$$y + 3z = 2 \qquad \text{Equation 2}$$

Discovery

Let $t = 1$ for the solution set

$$\{(5t + 4, -3t + 2, t)\}.$$

What solution do you obtain? Substitute these three values in the two equations in Example 3 and show that each equation is satisfied. Repeat this process for another two values for t.

We can let z equal any real number and use back-substitution to express x and y in terms of z.

Equation 2	**Equation 1**
$y + 3z = 2$	$x + 2y + z = 8$
$y = -3z + 2$	$x + 2(-3z + 2) + z = 8$
	$x - 6z + 4 + z = 8$
	$x - 5z + 4 = 8$
	$x = 5z + 4$

With $z = t$, the ordered solution (x, y, z) enables us to express the system's solution set as

$$\{(5t + 4, -3t + 2, t)\}$$

where t is any real number.

Check Point 3 Use Gaussian elimination to solve the system

$$x + 2y + 3z = 70$$
$$x + y + z = 60.$$

3 Solve problems involving systems without unique solutions.

Applications

How will computers be programmed to control traffic flow and avoid congestion? They will be required to solve systems continually based on the following premise: If traffic is to keep moving, during any period of time the number of cars entering an intersection must equal the number of cars leaving that intersection. Let's see what this means by looking at the intersections of four one-way city streets.

EXAMPLE 4 Traffic Control

Figure 6.3, shows the intersections of four one-way streets. As you study the figure, notice, that 300 cars per hour want to enter intersection I_1 from the north on 27th Avenue. Also, 200 cars per hour want to head east from intersection I_2 on Palm Drive. The letters x, y, z, and w stand for the number of cars passing between the intersections.

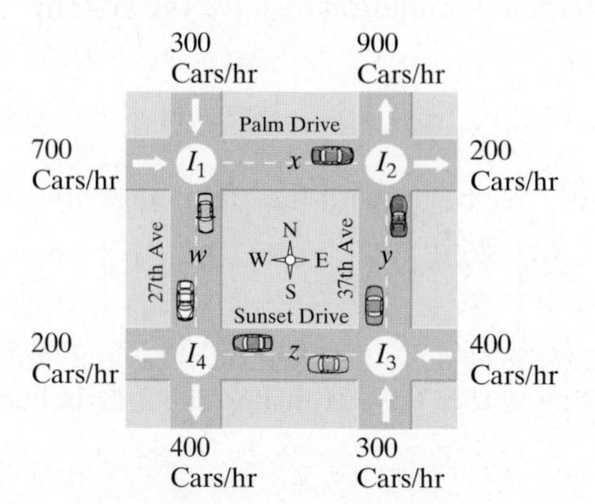

Figure 6.3 The intersections of four one-way streets

a. If the traffic is to keep moving, at each intersection the number of cars entering per hour must equal the number of cars leaving per hour. Use this idea to set up a linear system of equations involving x, y, z, and w.

b. Use Gaussian elimination to solve the system.

c. If construction on 27th Avenue limits w to 50 cars per hour, how many cars per hour must pass between the other intersections to keep traffic flowing?

Automated Highways

New technologies are making automated highways a reality. Experiments are taking place with cars that can steer, accelerate, and brake by themselves. A computer in the car picks up signals from magnets set in the road. Commuters can sit back, play with their laptop computers, read the newspaper, and enjoy the journey.

Solution

a. Set up the system by considering one intersection at a time, referring to Figure 4.3.

For Intersection I_1: Because $300 + 700 = 1000$ cars enter I_1, and $x + w$ cars leave the intersection, then $x + w = 1000$.

For Intersection I_2: Because $x + y$ cars enter the intersection, and $200 + 900 = 1100$ cars leave I_2, then $x + y = 1100$.

For Intersection I_3: Figure 6.3 indicates that $300 + 400 = 700$ cars enter and $y + z$ leave, so $y + z = 700$.

For Intersection I_4: With $z + w$ cars entering and $200 + 400 = 600$ cars exiting, traffic will keep flowing if $z + w = 600$.

The system of equations that describes this situation is given by

$$\begin{aligned} x + w &= 1000 \\ x + y &= 1100 \\ y + z &= 700 \\ z + w &= 600. \end{aligned}$$

b. To solve this system using Gaussian elimination, we begin with the augmented matrix.

System of Linear Equations (showing missing variables with 0 coefficients)	Augmented Matrix
$1x + 0y + 0z + 1w = 1000$	$\left[\begin{array}{cccc\|c} 1 & 0 & 0 & 1 & 1000 \end{array}\right]$
$1x + 1y + 0z + 0w = 1100$	$\left[\begin{array}{cccc\|c} 1 & 1 & 0 & 0 & 1100 \end{array}\right]$
$0x + 1y + 1z + 0w = 700$	$\left[\begin{array}{cccc\|c} 0 & 1 & 1 & 0 & 700 \end{array}\right]$
$0x + 0y + 1z + 1w = 600$	$\left[\begin{array}{cccc\|c} 0 & 0 & 1 & 1 & 600 \end{array}\right]$

We can now use row operations to obtain the matrix

$$\left[\begin{array}{cccc|c} 1 & 0 & 0 & 1 & 1000 \\ 0 & 1 & 0 & -1 & 100 \\ 0 & 0 & 1 & 1 & 600 \\ 0 & 0 & 0 & 0 & 0 \end{array}\right].$$

$x + w = 1000$

$y - w = 100$

$z + w = 600$

The last row of the matrix shows that the system in the voice balloons has dependent equations and infinitely many solutions. To write the solution set containing these infinitely many solutions, let w equal any real number. Use the three equations in the voice balloons to express x, y, and z in terms of w: $x = 1000 - w$, $y = 100 + w$, and $z = 600 - w$.
With $w = t$, the ordered solution (x, y, z, w) enables us to express the system's solution set as

$$\{(1000 - t, 100 + t, 600 - t, t)\}.$$

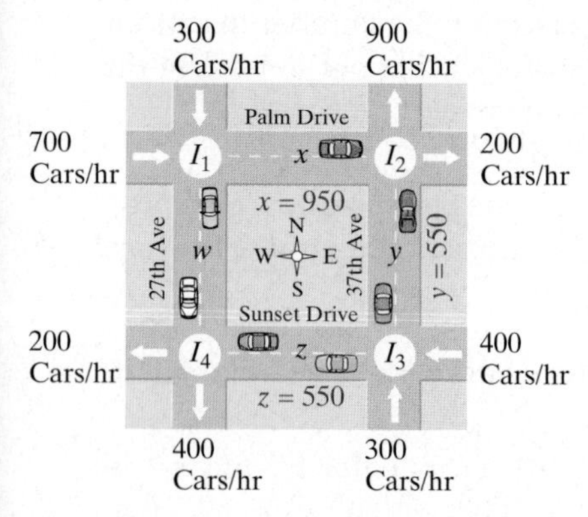

Figure 6.4 With w limited to 50 cars per hour, values for x, y, and z are determined.

c. We are given that construction limits w to 50 cars per hour. Because $w = t$, we replace 50 for t in the system's ordered solution:

$$(1000 - t, 100 + t, 600 - t, t) \qquad \text{Use the system's solution.}$$

$$= (1000 - 50, 100 + 50, 600 - 50, 50) \qquad t = 50$$

$$= (950, 150, 550, 50)$$

Thus, $x = 950$, $y = 150$, and $z = 550$. (See Figure 6.4.) With construction on 27th Avenue, this means that to keep traffic flowing, 950 cars per hour must be routed between I_1 and I_2, 150 per hour between I_3 and I_2, and 550 per hour between I_3 and I_4.

Figure 6.5 shows a system of four one-way streets. The numbers in the figure denote the number of cars per minute that travel in the direction shown.

a. Use the requirement that the number of cars entering each of the intersections per minute must equal the number of cars leaving per minute to set up a system of equations in x, y, z, and w.

b. Use Gaussian elimination to solve the system.

c. If construction limits w to 10 cars per minute, how many cars per minute must pass between the other intersections to keep traffic flowing?

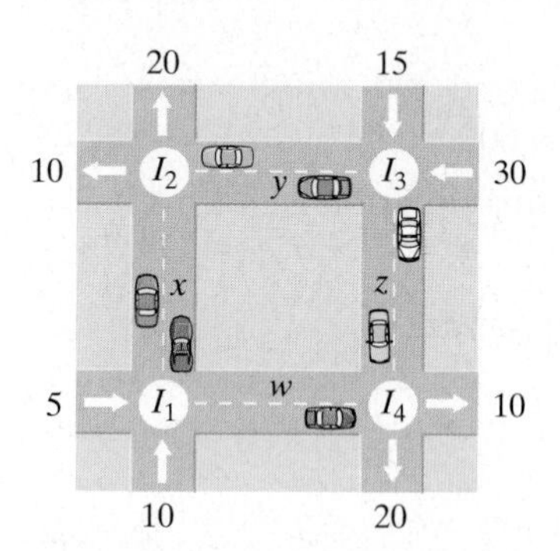

Figure 6.5

EXERCISE SET 6.2

Practice Exercises

In Exercises 1–24, use Gaussian elimination to find the complete solution to each system of equations, or show that none exists.

1. $\begin{aligned} 5x + 12y + z &= 10 \\ 2x + 5y + 2z &= -1 \\ x + 2y - 3z &= 5 \end{aligned}$

2. $\begin{aligned} 2x - 4y + z &= 3 \\ x - 3y + z &= 5 \\ 3x - 7y + 2z &= 12 \end{aligned}$

3. $\begin{aligned} 5x + 8y - 6z &= 14 \\ 3x + 4y - 2z &= 8 \\ x + 2y - 2z &= 3 \end{aligned}$

4. $\begin{aligned} 5x - 11y + 6z &= 12 \\ -x + 3y - 2z &= -4 \\ 3x - 5y + 2z &= 4 \end{aligned}$

5. $\begin{aligned} 3x + 4y + 2z &= 3 \\ 4x - 2y - 8z &= -4 \\ x + y - z &= 3 \end{aligned}$

6. $\begin{aligned} 2x - y - z &= 0 \\ x + 2y + z &= 3 \\ 3x + 4y + 2z &= 8 \end{aligned}$

7. $\begin{aligned} 8x + 5y + 11z &= 30 \\ -x - 4y + 2z &= 3 \\ 2x - y + 5z &= 12 \end{aligned}$

8. $\begin{aligned} x + y - 10z &= -4 \\ x - 7z &= -5 \\ 3x + 5y - 36z &= -10 \end{aligned}$

9. $\begin{aligned} x - 2y - z - 3w &= -9 \\ x + y - z &= 0 \\ 3x + 4y + w &= 6 \\ 2y - 2z + w &= 3 \end{aligned}$

10. $\begin{aligned} 2x + y - 2z - w &= 3 \\ x - 2y + z + w &= 4 \\ -x - 8y + 7z + 5w &= 13 \\ 3x + y - 2z + 2w &= 6 \end{aligned}$

11. $\begin{aligned} 2x + y - z &= 3 \\ x - 3y + 2z &= -4 \\ 3x + y - 3z + w &= 1 \\ x + 2y - 4z - w &= -2 \end{aligned}$

12. $\begin{aligned} 2x - y + 3z + w &= 0 \\ 3x + 2y + 4z - w &= 0 \\ 5x - 2y - 2z - w &= 0 \\ 2x + 3y - 7z - 5w &= 0 \end{aligned}$

13.
$$\begin{aligned} x - 3y + z - 4w &= 4 \\ -2x + y + 2z &= -2 \\ 3x - 2y + z - 6w &= 2 \\ -x + 3y + 2z - w &= -6 \end{aligned}$$

14.
$$\begin{aligned} 3x + 2y - z + 2w &= -12 \\ 4x - y + z + 2w &= 1 \\ x + y + z + w &= -2 \\ -2x + 3y + 2z - 3w &= 10 \end{aligned}$$

15.
$$\begin{aligned} 2x + y - z &= 2 \\ 3x + 3y - 2z &= 3 \end{aligned}$$

16.
$$\begin{aligned} 3x + 2y - z &= 5 \\ x + 2y - z &= 1 \end{aligned}$$

17.
$$\begin{aligned} x + 2y + 3z &= 5 \\ y - 5z &= 0 \end{aligned}$$

18.
$$\begin{aligned} 3x - y + 4z &= 8 \\ y + 2z &= 1 \end{aligned}$$

19.
$$\begin{aligned} x + y - 2z &= 2 \\ 3x - y - 6z &= -7 \end{aligned}$$

20.
$$\begin{aligned} -2x - 5y + 10z &= 19 \\ x + 2y - 4z &= 12 \end{aligned}$$

21.
$$\begin{aligned} x + y - z + w &= -2 \\ 2x - y + 2z - w &= 7 \\ -x + 2y + z + 2w &= -1 \end{aligned}$$

22.
$$\begin{aligned} 2x - 3y + 4z + w &= 7 \\ x - y + 3z - 5w &= 10 \\ 3x + y - 2z - 2w &= 6 \end{aligned}$$

23.
$$\begin{aligned} x + 2y + 3z - w &= 7 \\ 2y - 3z + w &= 4 \\ x - 4y + z &= 3 \end{aligned}$$

24.
$$\begin{aligned} x - y + w &= 0 \\ x - 4y + z + 2w &= 0 \\ 3x - z + 2w &= 0 \end{aligned}$$

Application Exercises

The figure for Exercises 25–28 shows the intersection of three one-way streets. To keep traffic moving, the number of cars per minute entering an intersection must equal the number exiting that intersection. For intersection I_1, $x + 10$ cars enter and $y + 14$ cars exit per minute. Thus, $x + 10 = y + 14$.

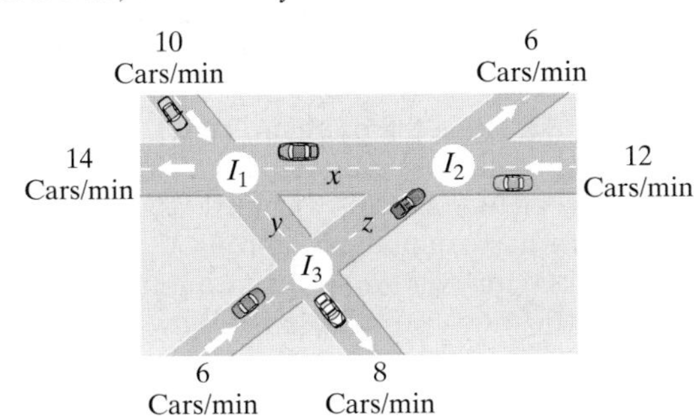

25. Write an equation for intersection I_2 that keeps traffic moving.

26. Write an equation for intersection I_3 that keeps traffic moving.

27. Use Gaussian elimination to solve the system formed by the equation given prior to Exercise 25 and the two equations that you obtained in Exercises 25–26.

28. Use your ordered solution obtained in Exercise 27 to solve this exercise. If construction limits z to 4 cars per minute, how many cars per minute must pass between the other intersections to keep traffic flowing?

29. The figure shows the intersection of four one-way streets.

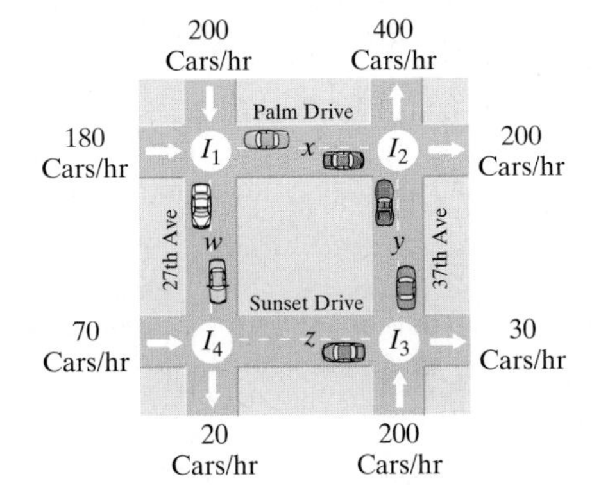

a. Set up a system of equations that keep traffic moving.
b. Use Gaussian elimination to solve the system.
c. If construction limits w to 50 cars per hour, how many cars per hour must pass between the other intersections to keep traffic moving?

30. The vitamin content per ounce for three foods is given in the following table.

	Milligrams per Ounce		
	Thiamin	**Riboflavin**	**Niacin**
Food A	3	7	1
Food B	1	5	3
Food C	3	8	2

a. Use matrices to show that no combination of these foods can provide exactly 14 mg of thiamin, 32 mg of riboflavin, and 9 mg of niacin.
b. Use matrices to describe in practical terms what happens if the riboflavin requirement is increased by 5 mg and the other requirements stay the same.

31. Three foods have the following nutritional content per ounce.

	Units per Ounce		
	Vitamin A	**Iron**	**Calcium**
Food 1	20	20	10
Food 2	30	10	10
Food 3	10	10	30

a. A diet must consist precisely of 220 units of vitamin A, 180 units of iron, and 340 units of calcium. However, the dietician runs out of Food 1. Use a matrix approach to show that under these conditions the dietary requirements cannot be met.

b. Now suppose that all three foods are available, but due to problems with vitamin A for pregnant women, a hospital dietician no longer wants to include this vitamin in the diet. Use matrices to give two possible ways to meet the iron and calcium requirements with the three foods.

32. A company that manufactures products A, B, and C does both manufacturing and testing. The hours needed to manufacture and test each product are shown in the table.

	Hours Needed Weekly to Manufacture	Hours Needed Weekly to Test
Product A	7	2
Product B	6	2
Product C	3	1

The company has exactly 67 hours per week available for manufacturing and 20 hours per week available for testing. Give two different combinations for the number of products that can be manufactured and tested weekly.

Writing in Mathematics

33. Describe what happens when Gaussian elimination is used to solve an inconsistent system.

34. Describe what happens when Gaussian elimination is used to solve a system with dependent equations.

35. In solving a system of dependent equations in three variables, one student simply said that there are infinitely many solutions. A second student expressed the solution set as $\{(4t + 3, 5t - 1, t)\}$. Which is the better form of expressing the solution set and why?

Technology Exercise

36. a. The figure at the top of the next column shows the intersections of a number of one-way streets. The numbers given represent traffic flow at a peak period (from 4 P.M. to 5:30 P.M.). Use the figure to write a linear system of six equations in seven variables based on the idea that at each intersection the number of cars entering must equal the number of cars leaving.

b. Use a graphing utility with matrix capabilities to find the complete solution to the system.

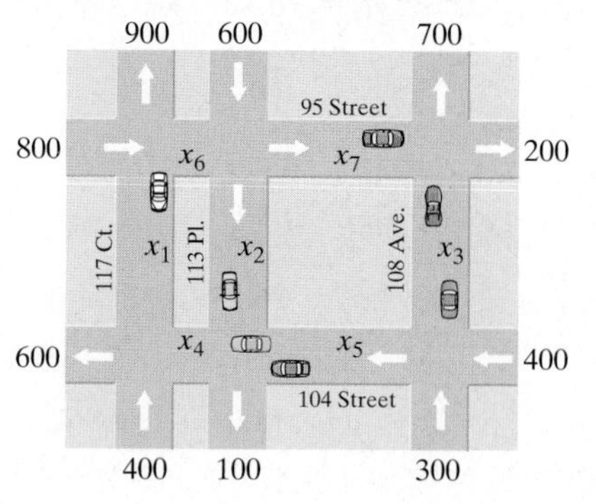

Critical Thinking Exercise

37. Consider the linear system

$$\begin{aligned} x + 3y + z &= a^2 \\ 2x + 5y + 2az &= 0 \\ x + y + a^2z &= -9. \end{aligned}$$

For what values of a will the system be inconsistent?

Group Exercise

38. Before beginning this exercise, the group needs to read and solve Exercise 36.

a. A political group is planning a demonstration on 95th Street between 113th Place and 117th Court for 5 P.M. Wednesday. The problem becomes one of minimizing traffic flow on 95th Street (between 113th and 117th) without causing traffic tie-ups on other streets. One possible solution is to close off traffic on 95th Street between 113th and 117th (let $x_6 = 0$). What can group members conclude about x_7 under these conditions?

b. Working with a matrix allows us to simplify the problem caused by the political demonstration, but it did not actually solve the problem. There are an infinite number of solutions; each value of x_7 we choose gives us a new picture. We also assumed x_6 was equal to 0; changing that assumption would also lead to different solutions. With your group, design another solution to the traffic flow problem caused by the political demonstration.

SECTION 6.3 Matrix Operations and Their Applications

Objectives

1. Use matrix notation.
2. Understand what is meant by equal matrices.
3. Add and subtract matrices.
4. Perform scalar multiplication.
5. Multiply matrices.
6. Describe applied situations with matrix operations.

Turn on your computer and read your e-mail or write a paper. When you need to do research, use the Internet to browse through art museums and photography exhibits. When you need a break, load a flight simulator program and fly through a photorealistic computer world. As different as these experiences may be, they all share one thing—you're looking at images based on matrices. Matrices have applications in numerous fields, including the new technology of digital photography in which pictures are represented by numbers rather than film. In this section, we turn our attention to matrix algebra and some of its applications.

1 Use matrix notation.

Notations for Matrices

We have seen that an array of numbers, arranged in rows and columns and placed in brackets, is called a matrix. We can represent the matrix in two different ways.

- A capital letter, such as A, B, or C, can denote a matrix.
- A lowercase letter enclosed in brackets, such as that shown below, can denote a matrix.

$$A = [a_{ij}]$$

Matrix A with elements a_{ij}

A general element in matrix A is denoted by a_{ij}. This refers to the element in the ith row and jth column. For example, a_{32} is the element of A located in the third row, second column.

A matrix of **order** $m \times n$ has m rows and n columns. If $m = n$, a matrix has the same number of rows as columns and is called a **square matrix**.

EXAMPLE 1 Matrix Notation

Let

$$A = \begin{bmatrix} 3 & 2 & 0 \\ -4 & -5 & -\frac{1}{5} \end{bmatrix}.$$

a. What is the order of A?

b. If $A = [a_{ij}]$, identify a_{23} and a_{12}.

$$A = \begin{bmatrix} 3 & 2 & 0 \\ -4 & -5 & -\frac{1}{5} \end{bmatrix}$$

Matrix A, shown again, to avoid turning back a page

Solution

a. The matrix has 2 rows and 3 columns, so it is of order 2×3.

b. The element a_{23} is in the second row and third column. Thus, $a_{23} = -\frac{1}{5}$. The element a_{12} is in the first row and second column, and consequently $a_{12} = 2$.

Check Point 1 Let

$$A = \begin{bmatrix} 5 & -2 \\ -3 & \pi \\ 1 & 6 \end{bmatrix}.$$

a. What is the order of A?

b. Identify a_{12} and a_{31}.

2 Understand what is meant by equal matrices.

Equality of Matrices

Two matrices are **equal** if and only if they have the same order and corresponding elements are equal.

Definition of Equality of Matrices

Two matrices A and B are **equal** if and only if they have the same order $m \times n$ and $a_{ij} = b_{ij}$ for $i = 1, 2, \ldots, m$ and $j = 1, 2, \ldots, n$.

For example, if $A = \begin{bmatrix} x & y+1 \\ z & 6 \end{bmatrix}$ and $B = \begin{bmatrix} 1 & 5 \\ 3 & 6 \end{bmatrix}$, then $A = B$ if and only if $x = 1$, $y + 1 = 5$ (so $y = 4$), and $z = 3$.

3 Add and subtract matrices.

Matrix Addition and Subtraction

Table 6.1 shows that matrices of the same order can be added or subtracted by simply adding or subtracting corresponding elements.

Table 6.1 Adding and subtracting matrices (Let $A = [a_{ij}]$ and $B = [b_{ij}]$ be matrices of order $m \times n$.)

Definition	The Definition in Words	Example
Matrix Addition $A + B = [a_{ij} + b_{ij}]$	Matrices of the same order are added by adding the elements in corresponding positions.	$\begin{bmatrix} 1 & -2 \\ 3 & 5 \end{bmatrix} + \begin{bmatrix} -1 & 6 \\ 0 & 4 \end{bmatrix} = \begin{bmatrix} 1+(-1) & -2+6 \\ 3+0 & 5+4 \end{bmatrix} = \begin{bmatrix} 0 & 4 \\ 3 & 9 \end{bmatrix}$
Matrix Subtraction $A - B = [a_{ij} - b_{ij}]$	Matrices of the same order are subtracted by subtracting the elements in corresponding positions.	$\begin{bmatrix} 1 & -2 \\ 3 & 5 \end{bmatrix} - \begin{bmatrix} -1 & 6 \\ 0 & 4 \end{bmatrix} = \begin{bmatrix} 1-(-1) & -2-6 \\ 3-0 & 5-4 \end{bmatrix} = \begin{bmatrix} 2 & -8 \\ 3 & 1 \end{bmatrix}$

The sum or difference of two matrices of different orders is undefined. For example, consider the matrices

$$A = \begin{bmatrix} 0 & 3 \\ 4 & 3 \end{bmatrix} \quad \text{and} \quad B = \begin{bmatrix} 1 & 9 \\ 4 & 5 \\ 2 & 3 \end{bmatrix}.$$

The order of A is 2×2; the order of B is 3×2. These matrices are of different orders and cannot be added or subtracted.

Technology

Graphing utilities can add and subtract matrices. Enter the matrices and use a keystroke sequence similar to

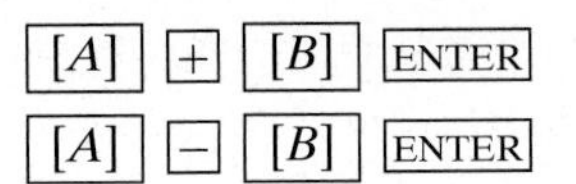

Consult your manual and verify the results in Example 2.

EXAMPLE 2 Adding and Subtracting Matrices

Perform the indicated matrix operations:

a. $\begin{bmatrix} 0 & 5 & 3 \\ -2 & 6 & -8 \end{bmatrix} + \begin{bmatrix} -2 & 3 & 5 \\ 7 & -9 & 6 \end{bmatrix}$

b. $\begin{bmatrix} -6 & 7 \\ 2 & -3 \end{bmatrix} - \begin{bmatrix} -5 & 6 \\ 0 & -4 \end{bmatrix}$.

Solution

a. $\begin{bmatrix} 0 & 5 & 3 \\ -2 & 6 & -8 \end{bmatrix} + \begin{bmatrix} -2 & 3 & 5 \\ 7 & -9 & 6 \end{bmatrix}$

$= \begin{bmatrix} 0 + (-2) & 5 + 3 & 3 + 5 \\ -2 + 7 & 6 + (-9) & -8 + 6 \end{bmatrix}$ Add the corresponding elements in the 2×3 matrices.

$= \begin{bmatrix} -2 & 8 & 8 \\ 5 & -3 & -2 \end{bmatrix}$ Simplify.

b. $\begin{bmatrix} -6 & 7 \\ 2 & -3 \end{bmatrix} - \begin{bmatrix} -5 & 6 \\ 0 & -4 \end{bmatrix}$

$= \begin{bmatrix} -6 - (-5) & 7 - 6 \\ 2 - 0 & -3 - (-4) \end{bmatrix}$ Subtract the corresponding elements in the 2×2 matrices.

$= \begin{bmatrix} -1 & 1 \\ 2 & 1 \end{bmatrix}$ Simplify.

Check Point 2 Perform the indicated matrix operations:

a. $\begin{bmatrix} -4 & 3 \\ 7 & -6 \end{bmatrix} + \begin{bmatrix} 6 & -3 \\ 2 & -4 \end{bmatrix}$ **b.** $\begin{bmatrix} 5 & 4 \\ -3 & 7 \\ 0 & 1 \end{bmatrix} - \begin{bmatrix} -4 & 8 \\ 6 & 0 \\ -5 & 3 \end{bmatrix}$

A matrix whose elements are all equal to 0 is called a **zero matrix**. If A is an $m \times n$ matrix and 0 is an $m \times n$ zero matrix, then $A + 0 = A$. For example,

$$\begin{bmatrix} -5 & 2 \\ 3 & 6 \end{bmatrix} + \begin{bmatrix} 0 & 0 \\ 0 & 0 \end{bmatrix} = \begin{bmatrix} -5 & 2 \\ 3 & 6 \end{bmatrix}.$$

An $m \times n$ zero matrix is called the **additive identity** for $m \times n$ matrices.

For any matrix A, the **additive inverse** of A, written $-A$, is the matrix of the same order of A such that every element of $-A$ is the opposite of the corresponding element of A. Because corresponding elements are added in matrix addition, $A + (-A)$ is a zero matrix. For example,

$$\begin{bmatrix} -5 & 2 \\ 3 & 6 \end{bmatrix} + \begin{bmatrix} 5 & -2 \\ -3 & -6 \end{bmatrix} = \begin{bmatrix} 0 & 0 \\ 0 & 0 \end{bmatrix}.$$

Properties of matrix addition are similar to properties involved with adding real numbers.

Properties of Matrix Addition

If A, B, and C are $m \times n$ matrices and 0 is an $m \times n$ zero matrix, then the following properties are true.

1. $A + B = B + A$ — Commutative Property of Addition
2. $(A + B) + C = A + (B + C)$ — Associative Property of Addition
3. $A + 0 = 0 + A = A$ — Additive Identity Property
4. $A + (-A) = (-A) + A = 0$ — Additive Inverse Property

4 Perform scalar multiplication.

Scalar Multiplication

A matrix of order 1×1, such as [6], contains only one entry. To distinguish this matrix from the number 6, we refer to 6 as a **scalar**. In general, in our work with matrices, we will refer to real numbers as scalars.

To multiply a matrix A by a scalar c, we multiply each entry in A by c. For example,

$$4\begin{bmatrix} 2 & 5 \\ -3 & 0 \end{bmatrix} = \begin{bmatrix} 4(2) & 4(5) \\ 4(-3) & 4(0) \end{bmatrix} = \begin{bmatrix} 8 & 20 \\ -12 & 0 \end{bmatrix}.$$

Scalar   Matrix

Definition of Scalar Multiplication

If $A = [a_{ij}]$ is a matrix of order $m \times n$ and c is a scalar, then the matrix cA is the $m \times n$ matrix given by

$$cA = [ca_{ij}].$$

This matrix is obtained by multiplying each element of A by the real number c. We call cA a **scalar multiple** of A.

EXAMPLE 3 Scalar Multiplication

If $A = \begin{bmatrix} -1 & 4 \\ 3 & 0 \end{bmatrix}$ and $B = \begin{bmatrix} 2 & -3 \\ 5 & -6 \end{bmatrix}$, find **a.** $-5B$ **b.** $2A + 3B$.

Solution

a. $-5B = -5\begin{bmatrix} 2 & -3 \\ 5 & -6 \end{bmatrix} = \begin{bmatrix} -5(2) & -5(-3) \\ -5(5) & -5(-6) \end{bmatrix} = \begin{bmatrix} -10 & 15 \\ -25 & 30 \end{bmatrix}$

Multiply each element in B by -5.

b. $2A + 3B = 2\begin{bmatrix} -1 & 4 \\ 3 & 0 \end{bmatrix} + 3\begin{bmatrix} 2 & -3 \\ 5 & -6 \end{bmatrix}$

$= \begin{bmatrix} 2(-1) & 2(4) \\ 2(3) & 2(0) \end{bmatrix} + \begin{bmatrix} 3(2) & 3(-3) \\ 3(5) & 3(-6) \end{bmatrix}$

Multiply each element in A by 2.

Multiply each element in B by 3.

$= \begin{bmatrix} -2 & 8 \\ 6 & 0 \end{bmatrix} + \begin{bmatrix} 6 & -9 \\ 15 & -18 \end{bmatrix} = \begin{bmatrix} -2+6 & 8+(-9) \\ 6+15 & 0+(-18) \end{bmatrix}$

Perform the addition of these 2×2 matrices by adding corresponding elements.

$= \begin{bmatrix} 4 & -1 \\ 21 & -18 \end{bmatrix}$

Check Point 3

If $A = \begin{bmatrix} -4 & 1 \\ 3 & 0 \end{bmatrix}$ and $B = \begin{bmatrix} -1 & -2 \\ 8 & 5 \end{bmatrix}$, find

a. $-6B$ **b.** $3A + 2B$.

Discovery

Verify each of the four properties listed in the box using

$A = \begin{bmatrix} 2 & -4 \\ -5 & 3 \end{bmatrix}$,

$B = \begin{bmatrix} 4 & 0 \\ 1 & -6 \end{bmatrix}$,

$c = 4$, and $d = 2$.

Properties of Scalar Multiplication

If A and B are $m \times n$ matrices, and c and d are scalars, then the following properties are true.

1. $(cd)A = c(dA)$ — Associative Property of Scalar Multiplication
2. $1A = A$ — Scalar Identity Property
3. $c(A + B) = cA + cB$ — Distributive Property
4. $(c + d)A = cA + dA$ — Distributive Property

5 Multiply matrices.

Matrix Multiplication

We do not multiply two matrices by multiplying the corresponding entries of matrices. Instead, we must think of matrix multiplication as *row-by-column multiplication*. To better understand how this works, let's begin with the definition of matrix multiplication for matrices of order 2×2.

Definition of Matrix Multiplication: 2 × 2 matrices

Row 1 of A × Column 1 of B

Row 1 of A × Column 2 of B

$$AB = \begin{bmatrix} a & b \\ c & d \end{bmatrix}\begin{bmatrix} e & f \\ g & h \end{bmatrix} = \begin{bmatrix} ae + bg & af + bh \\ ce + dg & cf + dh \end{bmatrix}$$

Row 2 of A × Column 1 of B

Row 2 of A × Column 2 of B

Notice that we obtain the element in the ith row and jth column in AB by performing computations with elements in the ith row of A and the jth column of B. For example, we obtain the element in the first row and first column of AB by performing computations with elements in the first row of A and the first column of B.

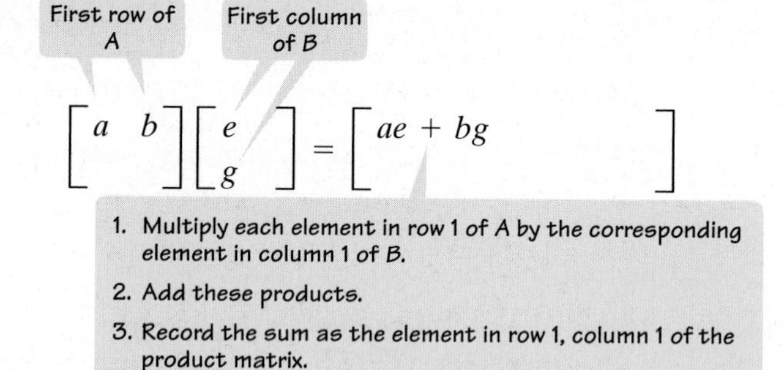

Corresponding elements

Corresponding elements

Figure 6.6 Finding corresponding elements when multiplying matrices

You may wonder how to find the corresponding elements in step 1 in the voice balloon. The element at the far left of row 1 corresponds to the element at the top of column 1. The second element from the left of row 1 corresponds to the second element from the top of column 1. This is illustrated in Figure 6.6.

EXAMPLE 4 Multiplying Matrices

Find AB, given

$$A = \begin{bmatrix} 2 & 3 \\ 4 & 7 \end{bmatrix} \quad \text{and} \quad B = \begin{bmatrix} 0 & 1 \\ 5 & 6 \end{bmatrix}.$$

Solution We will perform a row-by-column computation.

$$AB = \begin{bmatrix} 2 & 3 \\ 4 & 7 \end{bmatrix}\begin{bmatrix} 0 & 1 \\ 5 & 6 \end{bmatrix}$$

Row 1 of A × Column 1 of B

Row 1 of A × Column 2 of B

$$= \begin{bmatrix} 2(0) + 3(5) & 2(1) + 3(6) \\ 4(0) + 7(5) & 4(1) + 7(6) \end{bmatrix} = \begin{bmatrix} 15 & 20 \\ 35 & 46 \end{bmatrix}$$

Row 2 of A × Column 1 of B

Row 2 of A × Column 2 of B

Check Point 4 Find AB, given $A = \begin{bmatrix} 1 & 3 \\ 2 & 5 \end{bmatrix}$ and $B = \begin{bmatrix} 4 & 6 \\ 1 & 0 \end{bmatrix}$.

We can generalize the process of Example 4 to multiplying an $m \times n$ matrix and an $n \times p$ matrix. **For the product of two matrices to be defined, the number of columns of the first matrix must equal the number of rows of the second matrix.**

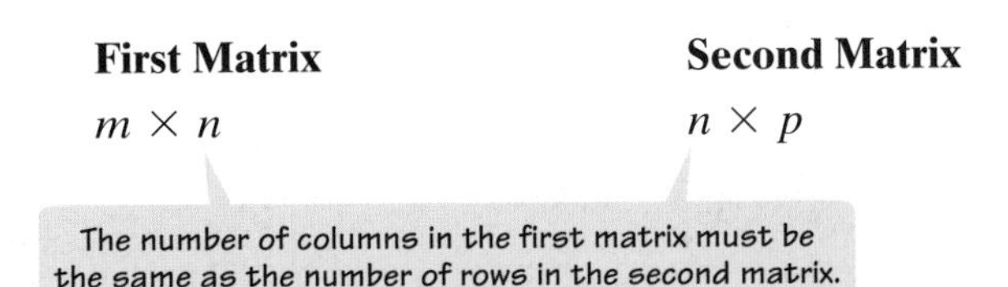

Study Tip

The following diagram illustrates the first sentence in the box defining matrix multiplication. The diagram is helpful in determining the order of the product AB.

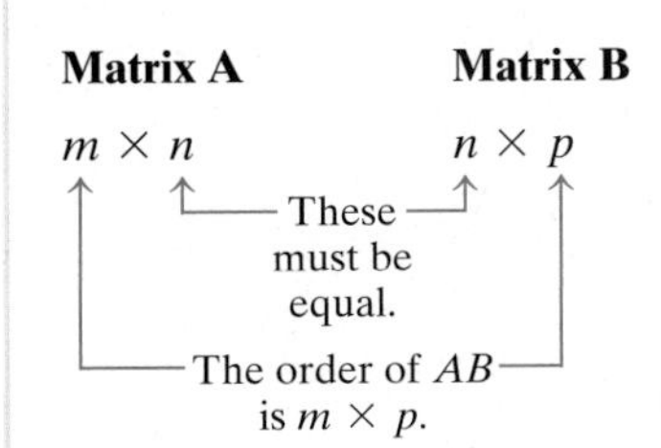

Definition of Matrix Multiplication

The **product** of an $m \times n$ matrix, A, and an $n \times p$ matrix, B, is an $m \times p$ matrix, AB, whose elements are found as follows. The element in the ith row and jth column of AB is found by multiplying the each element in the ith row of A by the corresponding element in the jth column of B and adding the products.

To find a product AB, each row of A must have the same number of elements as each column of B. We obtain p_{ij}, the element in the ith row and jth column in AB, by performing computations with elements in the ith row of A and the jth column of B:

$$\begin{bmatrix} & & \\ * & * & * \\ & & \end{bmatrix} \begin{bmatrix} & * & \\ & * & \\ & * & \end{bmatrix} = \begin{bmatrix} & & \\ & p_{ij} & \\ & & \end{bmatrix}$$

ith row of A — jth column of B — Element in the ith row and jth column of AB

When multiplying corresponding elements, keep in mind that the element at the far left of row i corresponds to the element at the top of column j. The element second from the left of row i corresponds to the element second from the top of column j. Likewise, the element third from the left of row i corresponds to the element third from the top of column j, and so on.

EXAMPLE 5 Multiplying Matrices

Matrices A and B are defined as follows.

$$A = \begin{bmatrix} 1 & 2 & 3 \end{bmatrix} \qquad B = \begin{bmatrix} 4 \\ 5 \\ 6 \end{bmatrix}$$

Find **a.** AB and **b.** BA.

Solution

a. Matrix A is a 1×3 matrix and matrix B is a 3×1 matrix. Thus, the product is a 1×1 matrix.

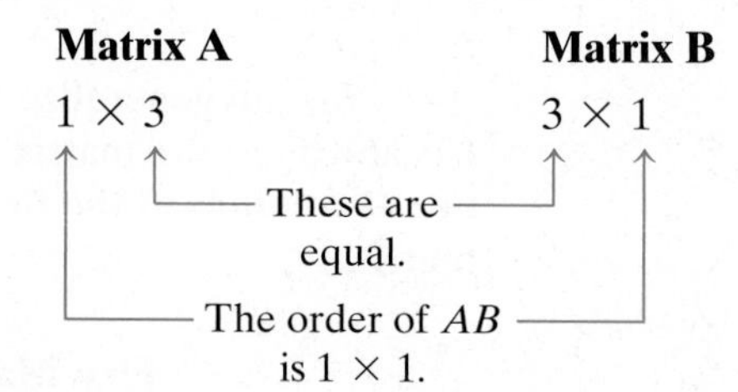

$$AB = [1 \quad 2 \quad 3]\begin{bmatrix} 4 \\ 5 \\ 6 \end{bmatrix}$$ We will perform a row-by-column computation.

$$= [(1)(4) + (2)(5) + (3)(6)]$$ Multiply elements in row 1 of A by corresponding elements in column 1 of B and add the products.

$$= [4 + 10 + 18]$$

$$= [32]$$

b. Matrix B is a 3×1 matrix and matrix A is a 1×3 matrix. Thus, the product BA is a 3×3 matrix.

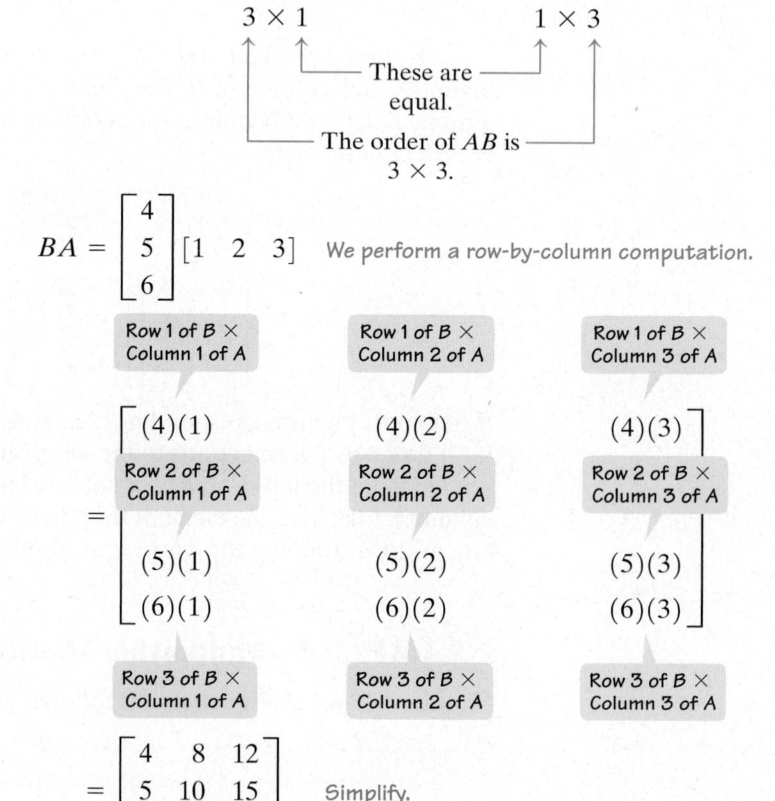

$$BA = \begin{bmatrix} 4 \\ 5 \\ 6 \end{bmatrix}[1 \quad 2 \quad 3]$$ We perform a row-by-column computation.

$$= \begin{bmatrix} (4)(1) & (4)(2) & (4)(3) \\ (5)(1) & (5)(2) & (5)(3) \\ (6)(1) & (6)(2) & (6)(3) \end{bmatrix}$$

$$= \begin{bmatrix} 4 & 8 & 12 \\ 5 & 10 & 15 \\ 6 & 12 & 18 \end{bmatrix}$$ Simplify.

Arthur Cayley

The Granger Collection

Matrices were first studied intensively by the English mathematician Arthur Cayley (1821–1895). Before reaching the age of 25, he published 25 papers, setting a pattern of prolific creativity that lasted throughout his life. Cayley was a lawyer, painter, mountaineer, and Cambridge professor whose greatest invention was that of matrices and matrix theory. Cayley's matrix algebra, especially the noncommutativity of multiplication ($AB \neq BA$), opened up a new area of mathematics called abstract algebra.

In Example 5, notice that AB and BA are different matrices. For most matrices $AB \neq BA$. Because **matrix multiplication is not commutative**, be careful about the order in which matrices appear when performing this operation.

Check Point 5 If $A = \begin{bmatrix} 2 & 0 & 4 \end{bmatrix}$ and $B = \begin{bmatrix} 1 \\ 3 \\ 7 \end{bmatrix}$, find AB and BA.

EXAMPLE 6 Multiplying Matrices

Where possible, find each product:

a. $\begin{bmatrix} 4 & 2 \\ 1 & 3 \end{bmatrix}\begin{bmatrix} 1 & 2 & 3 & 4 \\ 0 & 2 & -1 & 6 \end{bmatrix}$ **b.** $\begin{bmatrix} 1 & 2 & 3 & 4 \\ 0 & 2 & -1 & 6 \end{bmatrix}\begin{bmatrix} 4 & 2 \\ 1 & 3 \end{bmatrix}$

Solution

a. The first matrix is a 2×2 matrix and the second is a 2×4 matrix. The product will be a 2×4 matrix.

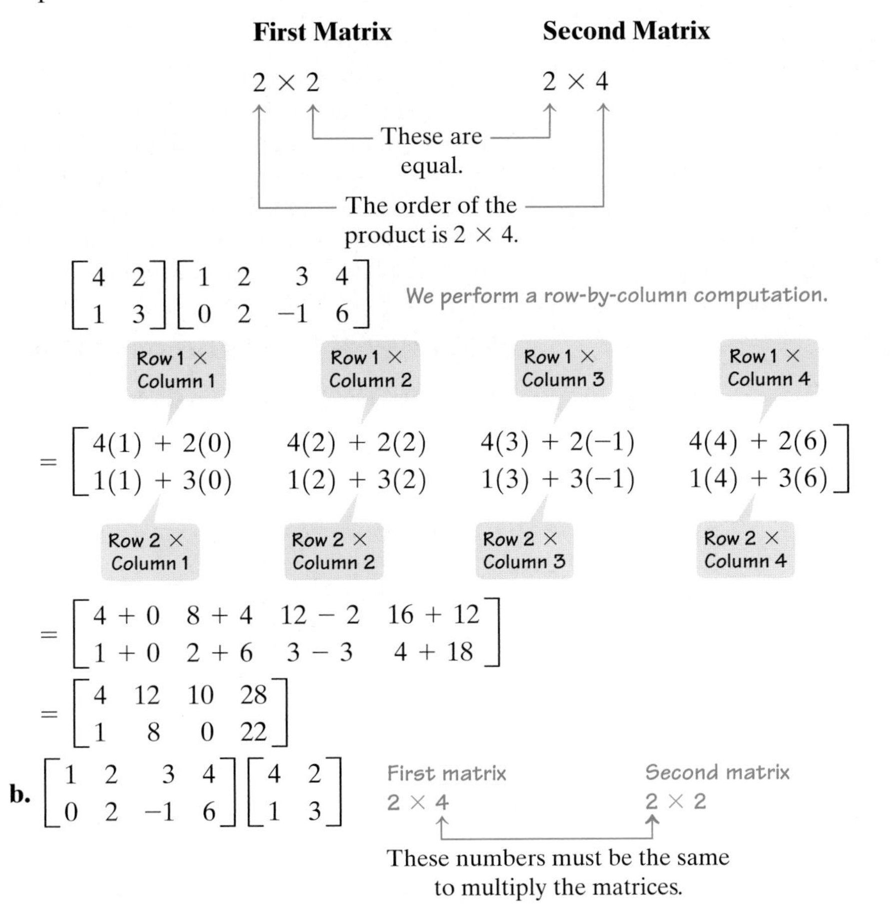

The number of columns in the first matrix does not equal the number of rows in the second matrix. Thus, the product of these two matrices is undefined.

Check Point 6 Where possible, find each product.

a. $\begin{bmatrix} 1 & 3 \\ 0 & 2 \end{bmatrix}\begin{bmatrix} 2 & 3 & -1 & 6 \\ 0 & 5 & 4 & 1 \end{bmatrix}$ **b.** $\begin{bmatrix} 2 & 3 & -1 & 6 \\ 0 & 5 & 4 & 1 \end{bmatrix}\begin{bmatrix} 1 & 3 \\ 0 & 2 \end{bmatrix}$

Although matrix multiplication is not commutative, it does obey many of the properties of real numbers.

Discovery

Verify the properties listed in the box using

$$A = \begin{bmatrix} 3 & 2 \\ -1 & 4 \end{bmatrix}$$

$$B = \begin{bmatrix} 1 & 0 \\ 3 & 2 \end{bmatrix}$$

$$C = \begin{bmatrix} 1 & 2 \\ -1 & 1 \end{bmatrix}$$

and $c = 3$.

Properties of Matrix Multiplication

If A, B, and C are matrices and c is a scalar, then the following properties are true. (Assume the order of each matrix is such that all operations in these properties are defined.)

1. $(AB)C = A(BC)$ — Associative Property of Matrix Multiplication
2. $A(B + C) = AB + AC$; $(A + B)C = AC + BC$ — Distributive Properties of Matrix Multiplication
3. $c(AB) = (cA)B$ — Associative Property of Scalar Multiplication

6 Describe applied situations with matrix operations.

Applications

All of the still images that you see on the Web have been created or manipulated on a computer in a digital format—made up of hundreds of thousands, or even millions, of tiny squares called **pixels**. Pixels are created by dividing an image into a grid. The computer can change the brightness of every square or pixel in this grid. A digital camera captures photos in this digital format. Also, you can scan pictures to convert them into digital format. Example 7 illustrates the role that matrices play in this new technology.

EXAMPLE 7 Matrices and Digital Photography

The letter T in Figure 6.7 is shown using 9 pixels in a 3×3 grid. The colors possible in the grid are shown in Figure 6.8. Each color is represented by a specific number: 0, 1, 2, or 3.

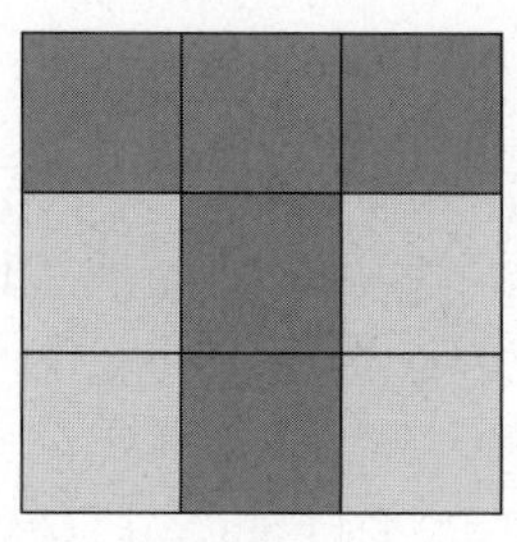

Figure 6.7 The letter T

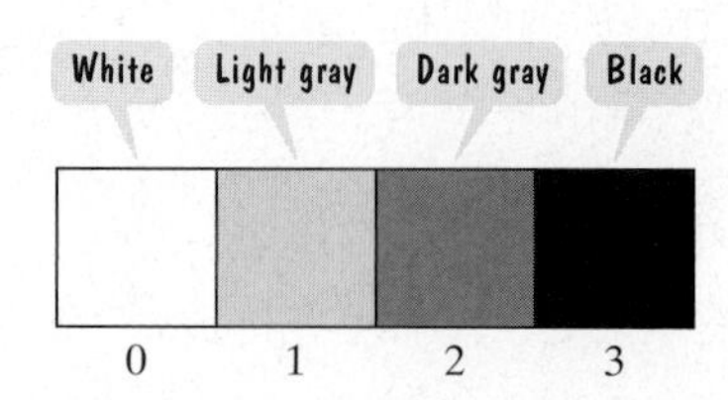

Figure 6.8 Color levels

a. Find a matrix that represents a digital photograph of this letter T.

b. Increase the contrast of the letter T by changing the dark gray to black and the light gray to white. Use matrix addition to accomplish this.

Solution

a. Look at the T and the background in Figure 6.7. Because the T is dark gray and the background is light gray, a digital photograph of Figure 6.7 can be represented by the matrix

$$\begin{bmatrix} 2 & 2 & 2 \\ 1 & 2 & 1 \\ 1 & 2 & 1 \end{bmatrix}.$$

b. We can make the T black by increasing each 2 in the above matrix to 3. We can make the background white by decreasing each 1 in the matrix to 0. This is accomplished using the following matrix addition.

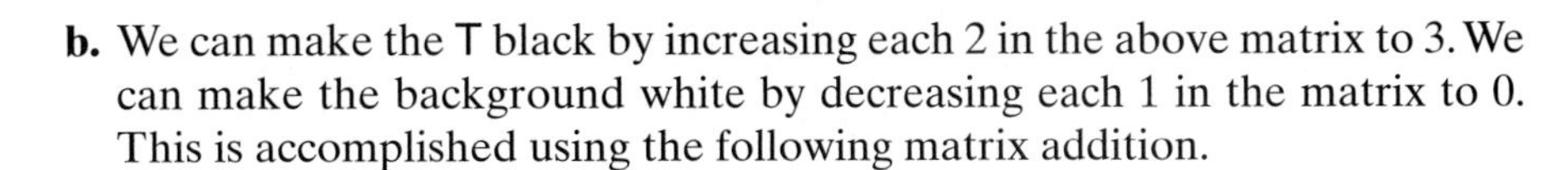

$$\begin{bmatrix} 2 & 2 & 2 \\ 1 & 2 & 1 \\ 1 & 2 & 1 \end{bmatrix} + \begin{bmatrix} 1 & 1 & 1 \\ -1 & 1 & -1 \\ -1 & 1 & -1 \end{bmatrix} = \begin{bmatrix} 3 & 3 & 3 \\ 0 & 3 & 0 \\ 0 & 3 & 0 \end{bmatrix}$$

The picture corresponding to the matrix sum to the right of the equal sign is shown in Figure 6.9.

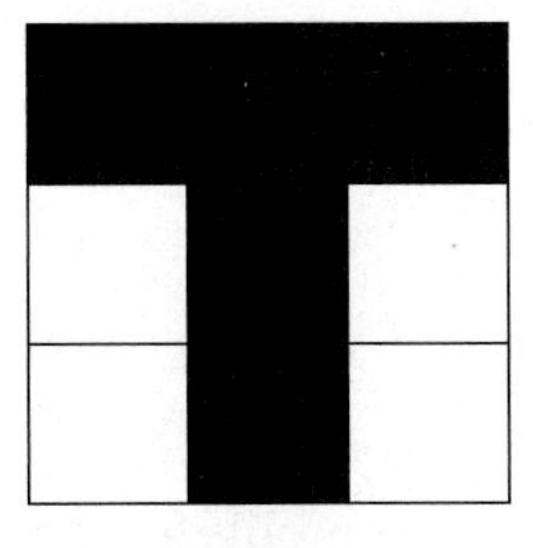

Figure 6.9 Changing contrast: the letter T

Check Point 7 Change the contrast of the letter T in Figure 6.7 by making the T light gray and the background black. Use matrix addition to accomplish this.

Images of Space

Photographs sent back from space use matrices with thousands of pixels. Each pixel is assigned a number from 0 to 63 representing its color—0 for pure white and 63 for pure black. In the image of Saturn shown here, matrix operations provide false colors that emphasize the banding of the planet's upper atmosphere.

EXAMPLE 8 Applying Matrix Multiplication

At a certain gas station, the number of gallons of regular, unleaded, and super unleaded gas sold on Monday, Tuesday, and Wednesday of a particular week is given by the following matrix.

	Regular	Unleaded	Super Unleaded
Monday	240	300	160
Tuesday	200	280	180
Wednesday	260	310	200

$= A$

A second matrix gives the selling price per gallon and the profit per gallon for the three types of gas sold by the station.

	Selling price per Gallon	Profit per Gallon
Regular	1.15	0.15
Unleaded	1.20	0.17
Super Unleaded	1.25	0.19

$= B$

a. Calculate the product AB.

b. What is the gas station's profit for Monday through Wednesday?

Solution

a. $$AB = \begin{bmatrix} 240 & 300 & 160 \\ 200 & 280 & 180 \\ 260 & 310 & 200 \end{bmatrix} \begin{bmatrix} 1.15 & 0.15 \\ 1.20 & 0.17 \\ 1.25 & 0.19 \end{bmatrix}$$

$$= \begin{bmatrix} 240(1.15) + 300(1.20) + 160(1.25) & 240(0.15) + 300(0.17) + 160(0.19) \\ 200(1.15) + 280(1.20) + 180(1.25) & 200(0.15) + 280(0.17) + 180(0.19) \\ 260(1.15) + 310(1.20) + 200(1.25) & 260(0.15) + 310(0.17) + 200(0.19) \end{bmatrix}$$

Perform to row-by-column multiplications.

$$= \begin{bmatrix} 836 & 117.40 \\ 791 & 111.80 \\ 921 & 129.70 \end{bmatrix}$$ Multiply and add as indicated.

b. The entries in the second column of the product matrix represent profits for Monday, Tuesday, and Wednesday, respectively. The gas station's profit for Monday through Wednesday is \$117.40 + \$111.80 + \$129.70 or \$358.90.

Use the product matrix in Example 8a to answer this question. What are the gas station's total sales for Monday, Tuesday, and Wednesday?

EXERCISE SET 6.3

Practice Exercises

In Exercises 1–4,

a. *Give the order of each matrix.*

b. *If $A = [a_{ij}]$, identify a_{32} and a_{23} or explain why identification is not possible.*

1. $\begin{bmatrix} 4 & -7 & 5 \\ -6 & 8 & -1 \end{bmatrix}$

2. $\begin{bmatrix} -6 & 4 & -1 \\ -9 & 0 & \frac{1}{2} \end{bmatrix}$

3. $\begin{bmatrix} 1 & -5 & \pi & e \\ 0 & 7 & -6 & -\pi \\ -2 & \frac{1}{2} & 11 & -\frac{1}{5} \end{bmatrix}$

4. $\begin{bmatrix} -4 & 1 & 3 & -5 \\ 2 & -1 & \pi & 0 \\ 1 & 0 & -e & \frac{1}{5} \end{bmatrix}$

In Exercises 5–8, find values for the variables so that the matrices in each exercise are equal.

5. $\begin{bmatrix} x \\ 4 \end{bmatrix} = \begin{bmatrix} 6 \\ y \end{bmatrix}$

6. $\begin{bmatrix} x \\ 7 \end{bmatrix} = \begin{bmatrix} 11 \\ y \end{bmatrix}$

7. $\begin{bmatrix} x & 2y \\ z & 9 \end{bmatrix} = \begin{bmatrix} 4 & 12 \\ 3 & 9 \end{bmatrix}$

8. $\begin{bmatrix} x & y+3 \\ 2z & 8 \end{bmatrix} = \begin{bmatrix} 12 & 5 \\ 6 & 8 \end{bmatrix}$

In Exercises 9–16, find

a. $A + B$ **b.** $A - B$

c. $-4A$ **d.** $3A + 2B$

9. $A = \begin{bmatrix} 4 & 1 \\ 3 & 2 \end{bmatrix}$, $B = \begin{bmatrix} 5 & 9 \\ 0 & 7 \end{bmatrix}$

10. $A = \begin{bmatrix} -2 & 3 \\ 0 & 1 \end{bmatrix}$, $B = \begin{bmatrix} 8 & 1 \\ 5 & 4 \end{bmatrix}$

11. $A = \begin{bmatrix} 1 & 3 \\ 3 & 4 \\ 5 & 6 \end{bmatrix}$, $B = \begin{bmatrix} 2 & -1 \\ 3 & -2 \\ 0 & 1 \end{bmatrix}$

12. $A = \begin{bmatrix} 3 & 1 & 1 \\ -1 & 2 & 5 \end{bmatrix}$, $B = \begin{bmatrix} 2 & -3 & 6 \\ -3 & 1 & -4 \end{bmatrix}$

13. $A = \begin{bmatrix} 2 \\ -4 \\ 1 \end{bmatrix}$, $B = \begin{bmatrix} -5 \\ 3 \\ -1 \end{bmatrix}$

14. $A = [6 \;\; 2 \;\; -3]$, $B = [4 \;\; -2 \;\; 3]$

15. $A = \begin{bmatrix} 2 & -10 & -2 \\ 14 & 12 & 10 \\ 4 & -2 & 2 \end{bmatrix}$, $B = \begin{bmatrix} 6 & 10 & -2 \\ 0 & -12 & -4 \\ -5 & 2 & -2 \end{bmatrix}$

16. $A = \begin{bmatrix} 6 & -3 & 5 \\ 6 & 0 & -2 \\ -4 & 2 & -1 \end{bmatrix}$, $B = \begin{bmatrix} -3 & 5 & 1 \\ -1 & 2 & -6 \\ 2 & 0 & 4 \end{bmatrix}$

In Exercises 17–26, find (if possible)
a. AB *and* **b.** BA.

17. $A = \begin{bmatrix} 1 & 3 \\ 5 & 3 \end{bmatrix}$, $B = \begin{bmatrix} 3 & -2 \\ -1 & 6 \end{bmatrix}$

18. $A = \begin{bmatrix} 3 & -2 \\ 1 & 5 \end{bmatrix}$, $B = \begin{bmatrix} 0 & 0 \\ 5 & -6 \end{bmatrix}$

19. $A = [1 \quad 2 \quad 3 \quad 4]$, $B = \begin{bmatrix} 1 \\ 2 \\ 3 \\ 4 \end{bmatrix}$

20. $A = \begin{bmatrix} -1 \\ -2 \\ -3 \end{bmatrix}$, $B = [1 \quad 2 \quad 3]$

21. $A = \begin{bmatrix} 1 & -1 & 4 \\ 4 & -1 & 3 \\ 2 & 0 & -2 \end{bmatrix}$, $B = \begin{bmatrix} 1 & 1 & 0 \\ 1 & 2 & 4 \\ 1 & -1 & 3 \end{bmatrix}$

22. $A = \begin{bmatrix} 1 & -1 & 1 \\ 5 & 0 & -2 \\ 3 & -2 & 2 \end{bmatrix}$, $B = \begin{bmatrix} 1 & 1 & 0 \\ 1 & -4 & 5 \\ 3 & -1 & 2 \end{bmatrix}$

23. $A = \begin{bmatrix} 4 & 2 \\ 6 & 1 \\ 3 & 5 \end{bmatrix}$, $B = \begin{bmatrix} 2 & 3 & 4 \\ -1 & -2 & 0 \end{bmatrix}$

24. $A = \begin{bmatrix} 2 & 4 \\ 3 & 1 \\ 4 & 2 \end{bmatrix}$, $B = \begin{bmatrix} 3 & 2 & 0 \\ -1 & -3 & 5 \end{bmatrix}$

25. $A = \begin{bmatrix} 2 & -3 & 1 & -1 \\ 1 & 1 & -2 & 1 \end{bmatrix}$, $B = \begin{bmatrix} 1 & 2 \\ -1 & 1 \\ 5 & 4 \\ 10 & 5 \end{bmatrix}$

26. $A = \begin{bmatrix} 2 & -1 & 3 & 2 \\ 1 & 0 & -2 & 1 \end{bmatrix}$, $B = \begin{bmatrix} -1 & 2 \\ 1 & 1 \\ 3 & -4 \\ 6 & 5 \end{bmatrix}$

In Exercises 27–34, perform the indicated matrix operations given that A, B, and C are defined as follows. If an operation is not defined, state the reason.

$$A = \begin{bmatrix} 4 & 0 \\ -3 & 5 \\ 0 & 1 \end{bmatrix} \qquad B = \begin{bmatrix} 5 & 1 \\ -2 & -2 \end{bmatrix} \qquad C = \begin{bmatrix} 1 & -1 \\ -1 & 1 \end{bmatrix}$$

27. $4B - 3C$ **28.** $5C - 2B$
29. $BC + CB$ **30.** $A(B + C)$
31. $A - C$ **32.** $B - A$
33. $A(BC)$ **34.** $A(CB)$

Application Exercises

The + sign in the figure is shown using 9 pixels in a 3 × 3 grid. The color levels are given to the right of the figure. Use the matrix $\begin{bmatrix} 1 & 3 & 1 \\ 3 & 3 & 3 \\ 1 & 3 & 1 \end{bmatrix}$ *that represents a digital photograph of the + sign to solve Exercises 35–38.*

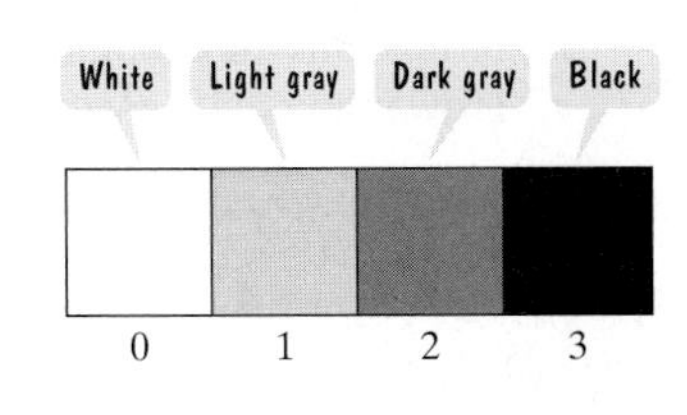

35. Adjust the contrast by changing the black to dark gray and the light gray to white. Use matrix addition to accomplish this.

36. Adjust the contrast by changing the black to dark gray and the light gray to black. Use matrix addition to accomplish this.

37. Adjust the contrast by changing the black to light gray and the light gray to dark gray. Use matrix addition to accomplish this.

38. Adjust the contrast by leaving the black alone and changing the light gray to white. Use matrix addition to accomplish this.

39. a. Write a 3 × 3 matrix A that represents a digital photograph of the number 1 in black on a white background.
b. Find a matrix B so that $A + B$ darkens only the white background to light gray.

40. a. Write a 3 × 3 matrix A that represents a digital photograph of the letter T in light gray on a white background.
b. Find a matrix B so that $A + B$ darkens only the letter T from light gray to black.

41. A virus strikes a college campus. Students are either sick, well, or carriers of the virus. The percentages of people in each category are given by the following matrix, which we'll call A.

	Freshman	Sophomore	Junior	Senior
Well	15%	25%	20%	10%
Sick	35%	40%	35%	70%
Carrier	50%	35%	45%	20%

The student population is distributed by class and gender as given by the following matrix, which we'll call B.

	Male	Female
Freshman	820	640
Sophomore	950	1020
Junior	680	720
Senior	930	910

a. Calculate the product AB.
b. How many sick females are there?
c. How many male carriers are there?

42. In a certain county, the proportion of voters in each age group registered as Republicans, Democrats, or Independents is given by the following matrix, which we'll call A.

	Age 18–30	31–50	Over 50
Republicans	0.4	0.30	0.70
Democrats	0.30	0.60	0.25
Independents	0.30	0.10	0.05

The distribution, by age and gender, of this county's voting population is given by the following matrix, which we'll call B.

		Male	Female
	18–30	6000	8000
Age	**31–50**	12,000	14,000
	Over 50	14,000	16,000

a. Calculate the product AB.
b. How many female Democrats are there?
c. How many male Republicans are there?

43. The final grade in a particular course is determined by grades on the midterm and final. The grades for five students and the two grading systems are modeled by the following matrices. Call the first matrix A and the second B.

	Midterm	Final
Student 1	76	92
Student 2	74	84
Student 3	94	86
Student 4	84	62
Student 5	58	80

	System 1	System 2
Midterm	0.5	0.3
Final	0.5	0.7

a. Describe the grading system that is represented by matrix B.
b. Compute the matrix AB and assign each of the five students a final course grade first using system 1 and then using system 2. ($89.5 - 100 = A$, $79.5 - 89.4 = B$, $69.5 - 79.4 = C$, $59.5 - 69.4 = D$, below $59.5 = F$)

44. In the matrices shown below, a 1 represents a yes, and a 0 represents a no. The first matrix, A, describes whether or not three colleges in a state university system offer degrees in each program.

	Programs: Liberal Arts	Engineering	Education
College 1	1	1	0
College 2	1	1	1
College 3	0	1	0

Each program requires that certain math courses be completed, indicated by the following matrix called B.

	General College Math	Intermediate Algebra	College Algebra	Trigonometry	Calculus
Liberal Arts	1	1	0	0	0
Engineering	0	0	1	1	1
Education	1	1	1	0	0

Find the product AB. Explain how this helps the college decide which courses to offer.

Writing in Mathematics

45. What is meant by the order of a matrix? Give an example with your explanation.

46. What does a_{ij} mean?

47. What are equal matrices?

48. How are matrices added?

49. Describe how to subtract matrices.

50. Describe matrices that cannot be added or subtracted.

51. Describe how to perform scalar multiplication. Provide an example with your description.

52. Describe how to multiply matrices.

53. Describe when the multiplication of two matrices is not defined.

54. If two matrices can be multiplied, describe how to determine the order of the product.

55. Low-resolution digital photographs use 262,144 pixels in a 512×512 grid. If you enlarge a low-resolution digital photograph enough, describe what will happen.

Technology Exercise

56. Use the matrix feature of a graphing utility to verify each of your answers to Exercises 27–34.

Critical Thinking Exercises

57. Find two matrices A and B such that $AB = BA$.

58. Consider a square matrix such that each element that is not on the main diagonal is zero. Experiment with such matrices (call each matrix A) by finding AA. Then write a sentence or two describing a method for multiplying this kind of matrix by itself.

59. If $AB = -BA$, then A and B are said to be anticommutative. Are $A = \begin{bmatrix} 0 & -1 \\ 1 & 0 \end{bmatrix}$ and $B = \begin{bmatrix} 1 & 0 \\ 0 & -1 \end{bmatrix}$ anticommutative?

Group Exercise

60. The interesting and useful applications of matrix theory are nearly unlimited. Applications of matrices range from representing digital photographs to predicting long-range trends in the stock market. Members of the group should research an application of matrices that they find intriguing. The group should then present a seminar to the class about this application.

SECTION 6.4 Multiplicative Inverses of Matrices and Matrix Equations

Objectives

1. Find the multiplicative inverse of a square matrix.
2. Use inverses to solve matrix equations.
3. Encode and decode messages.

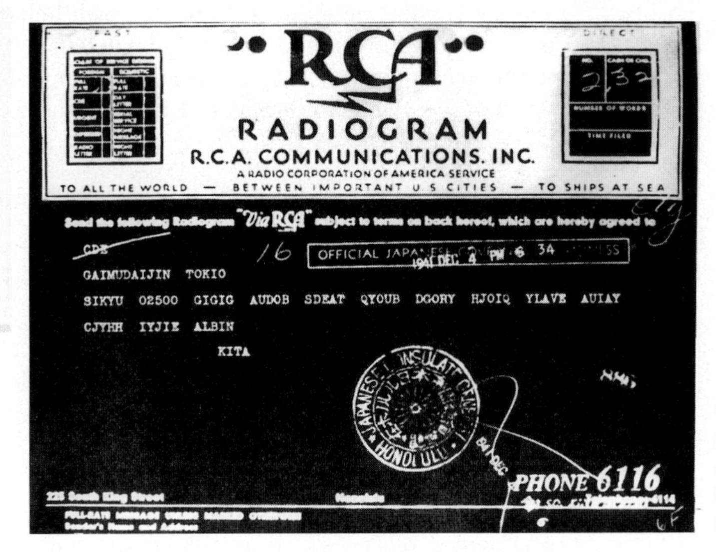

This 1941 RCA radiogram shows an encoded message from the Japanese government.

In 1939, Britain's secret service hired top chess players, mathematicians, and other masters of logic to break the code used by the Nazis in communications between headquarters and troops. The project, which employed over 10,000 people, broke the code less than a year later, providing the Allies with information about Nazi troop movements throughout World War II.

Messages must often be sent in such a way that the real meaning is hidden from everyone but the sender and the recipient. In this section, we will look at the role that matrices and their inverses play in this process.

The Multiplicative Identity Matrix

For the real numbers, we know that 1 is the multiplicative identity because $a \cdot 1 = 1 \cdot a = a$. Is there a similar property for matrix multiplication? That is, is there a matrix I such that $AI = A$ and $IA = A$? The answer is yes. A square matrix with 1s down the main diagonal and 0s elsewhere does not change the elements in a matrix when it multiplies that matrix. In the case of 2×2 matrices,

$$\begin{bmatrix} a_{11} & a_{12} \\ a_{21} & a_{22} \end{bmatrix}\begin{bmatrix} 1 & 0 \\ 0 & 1 \end{bmatrix} = \begin{bmatrix} a_{11} & a_{12} \\ a_{21} & a_{22} \end{bmatrix}$$

The elements in the matrix do not change.

and $$\begin{bmatrix} 1 & 0 \\ 0 & 1 \end{bmatrix}\begin{bmatrix} a_{11} & a_{12} \\ a_{21} & a_{22} \end{bmatrix} = \begin{bmatrix} a_{11} & a_{12} \\ a_{21} & a_{22} \end{bmatrix}.$$

The elements in the matrix do not change.

An $n \times n$ square matrix whose main diagonal elements are 1s, while all other elements are 0s, is called the **multiplicative identity matrix of order *n***, designated by I_n. For example.

$$I_2 = \begin{bmatrix} 1 & 0 \\ 0 & 1 \end{bmatrix}, \quad I_3 = \begin{bmatrix} 1 & 0 & 0 \\ 0 & 1 & 0 \\ 0 & 0 & 1 \end{bmatrix},$$

and so on.

1 Find the multiplicative inverse of a square matrix.

The Multiplicative Inverse of a Matrix

The multiplicative identity matrix, I_n, will help us to define a new concept: the multiplicative inverse of a matrix. To do so, let's consider a similar concept, the multiplicative inverse of a nonzero number, a. Recall that the multiplicative inverse of a is $\frac{1}{a}$. The multiplicative inverse has the following property:

$$a \cdot \frac{1}{a} = 1 \quad \text{and} \quad \frac{1}{a} \cdot a = 1.$$

We can define the multiplicative inverse of a square matrix in a similar manner.

Definition of the Multiplicative Inverse of a Square Matrix

Let A be an $n \times n$ matrix. If there exists an $n \times n$ matrix A^{-1} (read: "A inverse") such that

$$AA^{-1} = I_n \quad \text{and} \quad A^{-1}A = I_n,$$

then A^{-1} is the **multiplicative inverse** of A.

We have seen that matrix multiplication is not commutative. Thus, to show that matrix B is the multiplicative inverse of matrix A, find both AB and BA. If B is the multiplicative inverse of A, both products (AB and BA) will be the multiplicative identity matrix, I_n.

EXAMPLE 1 The Multiplicative Inverse of a Matrix

Show that B is the multiplicative inverse of A, where

$$A = \begin{bmatrix} -1 & 3 \\ 2 & -5 \end{bmatrix} \quad \text{and} \quad B = \begin{bmatrix} 5 & 3 \\ 2 & 1 \end{bmatrix}.$$

Solution To show that B is the multiplicative inverse of A, we must find the products AB and BA. If B is the multiplicative inverse of A, then AB will be the multiplicative identity matrix and BA will be the multiplicative identity matrix. Because A and B are 2×2 matrices, $n = 2$. Thus, we denote the multiplicative identity matrix as I_2; it is also a 2×2 matrix. We must show that

- $AB = I_2 = \begin{bmatrix} 1 & 0 \\ 0 & 1 \end{bmatrix}$.
- $BA = I_2 = \begin{bmatrix} 1 & 0 \\ 0 & 1 \end{bmatrix}$.

$$AB = \begin{bmatrix} -1 & 3 \\ 2 & -5 \end{bmatrix}\begin{bmatrix} 5 & 3 \\ 2 & 1 \end{bmatrix}$$

$$= \begin{bmatrix} -1(5) + 3(2) & -1(3) + 3(1) \\ 2(5) + (-5)(2) & 2(3) + (-5)(1) \end{bmatrix} = \begin{bmatrix} 1 & 0 \\ 0 & 1 \end{bmatrix}$$

$$BA = \begin{bmatrix} 5 & 3 \\ 2 & 1 \end{bmatrix}\begin{bmatrix} -1 & 3 \\ 2 & -5 \end{bmatrix}$$

$$= \begin{bmatrix} 5(-1) + 3(2) & 5(3) + 3(-5) \\ 2(-1) + 1(2) & 2(3) + 1(-5) \end{bmatrix} = \begin{bmatrix} 1 & 0 \\ 0 & 1 \end{bmatrix}$$

Both products give the multiplicative identity matrix. Thus, B is the multiplicative inverse of A and we can designate B as $A^{-1} = \begin{bmatrix} 5 & 3 \\ 2 & 1 \end{bmatrix}$.

Check Point 1 Show that B is the multiplicative inverse of A, where

$$A = \begin{bmatrix} 2 & 1 \\ 1 & 1 \end{bmatrix} \quad \text{and} \quad B = \begin{bmatrix} 1 & -1 \\ -1 & 2 \end{bmatrix}.$$

One method for finding the multiplicative inverse of a matrix A is to begin by denoting the elements in A^{-1} with variables. Using the equation $AA^{-1} = I_n$ we can find a value for each element in the multiplicative inverse that was represented by a variable. Example 2 shows how this is done.

EXAMPLE 2 Finding the Multiplicative Inverse of a Matrix

Find the multiplicative inverse of

$$A = \begin{bmatrix} 2 & 1 \\ 5 & 3 \end{bmatrix}.$$

Solution Let us denote the multiplicative inverse by

$$A^{-1} = \begin{bmatrix} x & y \\ z & w \end{bmatrix}.$$

Because A is a 2×2 matrix, we use the equation $AA^{-1} = I_2$ to find values for x, y, z, and w.

$$\underbrace{\begin{bmatrix} 2 & 1 \\ 5 & 3 \end{bmatrix}}_{A}\underbrace{\begin{bmatrix} x & y \\ z & w \end{bmatrix}}_{A^{-1}} = \underbrace{\begin{bmatrix} 1 & 0 \\ 0 & 1 \end{bmatrix}}_{I_2}$$

$$\begin{bmatrix} 2x + z & 2y + w \\ 5x + 3z & 5y + 3w \end{bmatrix} = \begin{bmatrix} 1 & 0 \\ 0 & 1 \end{bmatrix}$$

Use row-by-column matrix multiplication on the left.

We now equate corresponding elements to obtain the following two systems of linear equations.

$$\begin{aligned} 2x + z &= 1 \\ 5x + 3z &= 0 \end{aligned} \qquad \text{and} \qquad \begin{aligned} 2y + w &= 0 \\ 5y + 3w &= 1 \end{aligned}$$

Each of these systems can be solved using the addition method.

$$\begin{aligned} 2x + z &= 1 \xrightarrow{\text{Multiply by } -3.} & -6x - 3z &= -3 \\ 5x + 3z &= 0 \xrightarrow{\text{No change}} & 5x + 3z &= 0 \\ & \text{Add:} & -x &= -3 \\ & & x &= 3 \\ & \text{Use back-substitution.} & z &= -5 \end{aligned}$$

$$\begin{aligned} 2y + w &= 0 \xrightarrow{\text{Multiply by } -3.} & -6y - 3w &= 0 \\ 5y + 3w &= 1 \xrightarrow{\text{No change}} & 5y + 3w &= 1 \\ & \text{Add:} & -y &= 1 \\ & & y &= -1 \\ & \text{Use back-substitution.} & w &= 2 \end{aligned}$$

Using these values, we have

$$A^{-1} = \begin{bmatrix} x & y \\ z & w \end{bmatrix} = \begin{bmatrix} 3 & -1 \\ -5 & 2 \end{bmatrix}.$$

Discovery

Verify that the inverse matrix found in Example 2 is correct. Use matrix multiplication to show that

$AA^{-1} = I_2$ and $A^{-1}A = I_2$,

where

$$I_2 = \begin{bmatrix} 1 & 0 \\ 0 & 1 \end{bmatrix}.$$

Check Point 2 Find the multiplicative inverse of $A = \begin{bmatrix} 5 & 7 \\ 2 & 3 \end{bmatrix}$.

Only square matrices of order $n \times n$ have multiplicative inverses, but not every square matrix possesses a multiplicative inverse. For example, suppose that you apply the procedure of Example 2 to $A = \begin{bmatrix} -6 & 4 \\ -3 & 2 \end{bmatrix}$:

This is A. — This represents A^{-1}. — This is the multiplicative identity matrix.

$$\begin{bmatrix} -6 & 4 \\ -3 & 2 \end{bmatrix} \begin{bmatrix} x & y \\ z & w \end{bmatrix} = \begin{bmatrix} 1 & 0 \\ 0 & 1 \end{bmatrix}.$$

Multiplying matrices on the left and equating corresponding elements results in inconsistent systems with no solutions. There are no values for x, y, z, and w. This shows that matrix A does not have a multiplicative inverse.

A nonsquare matrix, one with a different number of rows than columns, cannot have a multiplicative inverse. If A is an $m \times n$ matrix and B is an $n \times m$ matrix ($n \neq m$), then the products AB and BA are of different orders. This means that they could not be equal to each other, so that AB and BA could not both equal the multiplicative identity matrix.

If a square matrix has a multiplicative inverse, that inverse is unique. This means that the square matrix has no more than one inverse. If a square matrix has a multiplicative inverse, it is said to be **invertible**.

Study Tip

To find the matrix that appears as the second factor for the inverse of

$$A = \begin{bmatrix} a & b \\ c & d \end{bmatrix}.$$

- Reverse a and d, the numbers in the main diagonal.
- Negate b and c, the numbers in the other diagonal.

A Quick Method for Finding the Multiplicative Inverse of a 2 × 2 Matrix

The following rule enables us to calculate the multiplicative inverse, if there is one, of a 2 × 2 matrix.

Multiplicative Inverse of a 2 × 2 Matrix

If $A = \begin{bmatrix} a & b \\ c & d \end{bmatrix}$, then $A^{-1} = \dfrac{1}{ad - bc}\begin{bmatrix} d & -b \\ -c & a \end{bmatrix}$.

The matrix A is invertible if and only if $ad - bc \neq 0$. If $ad - bc = 0$, then A does not have a multiplicative inverse.

EXAMPLE 3 Using the Quick Method to Find Multiplicative Inverses

Find the multiplicative inverse of

$$A = \begin{bmatrix} -1 & -2 \\ 3 & 4 \end{bmatrix}.$$

Study Tip

When using the formula to find the multiplicative inverse, start by computing $ad - bc$. If the computed value is 0, there is no need to continue. The given matrix does not have a multiplicative inverse.

Solution

$$A = \begin{bmatrix} -1 & -2 \\ 3 & 4 \end{bmatrix}$$

(Labels: a = −1, b = −2, c = 3, d = 4.) This is the given matrix. We've designated the elements a, b, c, and d.

$$A^{-1} = \frac{1}{ad - bc}\begin{bmatrix} d & -b \\ -c & a \end{bmatrix}$$

This is the formula for the inverse of $\begin{bmatrix} a & b \\ c & d \end{bmatrix}$.

$$= \frac{1}{(-1)(4) - (-2)(3)}\begin{bmatrix} 4 & -(-2) \\ -3 & -1 \end{bmatrix}$$

Apply the formula with $a = -1$, $b = -2$, $c = 3$, and $d = 4$.

$$= \frac{1}{2}\begin{bmatrix} 4 & 2 \\ -3 & -1 \end{bmatrix}$$

Simplify.

$$= \begin{bmatrix} 2 & 1 \\ -\frac{3}{2} & -\frac{1}{2} \end{bmatrix}$$

Perform the scalar multiplication by multiplying each element in the matrix by $\frac{1}{2}$.

The inverse of $A = \begin{bmatrix} -1 & -2 \\ 3 & 4 \end{bmatrix}$ is $A^{-1} = \begin{bmatrix} 2 & 1 \\ -\frac{3}{2} & -\frac{1}{2} \end{bmatrix}$.

We can verify this result by showing that $AA^{-1} = I_2$ and $A^{-1}A = I_2$.

Check Point 3 Find the multiplicative inverse of

$$A = \begin{bmatrix} 3 & -2 \\ -1 & 1 \end{bmatrix}.$$

Finding Multiplicative Inverses of $n \times n$ Matrices with n Greater Than 2

To find the multiplicative inverse of a 3×3 invertible matrix, we begin by denoting the elements in the multiplicative inverse with variables. Here is an example:

$$\begin{bmatrix} -1 & -1 & -1 \\ 4 & 5 & 0 \\ 0 & 1 & -3 \end{bmatrix} \begin{bmatrix} x_1 & x_2 & x_3 \\ y_1 & y_2 & y_3 \\ z_1 & z_2 & z_3 \end{bmatrix} = \begin{bmatrix} 1 & 0 & 0 \\ 0 & 1 & 0 \\ 0 & 0 & 1 \end{bmatrix}.$$

This is matrix A whose inverse we wish to find.

This represents A^{-1}.

This is the multiplicative identity matrix, I_3.

We multiply the matrices on the left, using the row-by-column definition of matrix multiplication.

$$\begin{bmatrix} -x_1 - y_1 - z_1 & -x_2 - y_2 - z_2 & -x_3 - y_3 - z_3 \\ 4x_1 + 5y_1 + 0z_1 & 4x_2 + 5y_2 + 0z_2 & 4x_3 + 5y_3 + 0z_3 \\ 0x_1 + 1y_1 - 3z_1 & 0x_2 + 1y_2 - 3z_2 & 0x_3 + 1y_3 - 3z_3 \end{bmatrix} = \begin{bmatrix} 1 & 0 & 0 \\ 0 & 1 & 0 \\ 0 & 0 & 1 \end{bmatrix}$$

We now equate corresponding entries to obtain the following three systems of linear equations.

$$\begin{aligned} -x_1 - y_1 - z_1 &= 1 \\ 4x_1 + 5y_1 + 0z_1 &= 0 \\ 0x_1 + y_1 - 3z_1 &= 0 \end{aligned} \qquad \begin{aligned} -x_2 - y_2 - z_2 &= 0 \\ 4x_2 + 5y_2 + 0z_2 &= 1 \\ 0x_2 + y_2 - 3z_2 &= 0 \end{aligned} \qquad \begin{aligned} -x_3 - y_3 - z_3 &= 0 \\ 4x_3 + 5y_3 + 0z_3 &= 0 \\ 0x_3 + y_3 - 3z_3 &= 1 \end{aligned}$$

Notice that the variables on the left of the equal sign have the same coefficients in each system. We can use Gauss-Jordan elimination to solve all three at once. Form an augmented matrix that contains the coefficients of the three systems to the left of the vertical line and the constants for the systems to the right.

$$\left[\begin{array}{ccc|ccc} -1 & -1 & -1 & 1 & 0 & 0 \\ 4 & 5 & 0 & 0 & 1 & 0 \\ 0 & 1 & -3 & 0 & 0 & 1 \end{array}\right]$$

Coefficients of the three systems

Constants on the right in each of the three systems

To solve all three systems using Gauss-Jordan elimination, we must obtain $\begin{bmatrix} 1 & 0 & 0 \\ 0 & 1 & 0 \\ 0 & 0 & 1 \end{bmatrix}$ to the left of the vertical line. Use matrix row operations, working one column at a time. Obtain 1 in the required position. Then obtain 0s in the other two positions. Using these operations, we obtain the matrix

$$\left[\begin{array}{ccc|ccc} 1 & 0 & 0 & 15 & 4 & -5 \\ 0 & 1 & 0 & -12 & -3 & 4 \\ 0 & 0 & 1 & -4 & -1 & 1 \end{array}\right].$$

This augmented matrix provides the solutions to the three systems of equations. They are given by

$$\left[\begin{array}{ccc|c} 1 & 0 & 0 & 15 \\ 0 & 1 & 0 & -12 \\ 0 & 0 & 1 & -4 \end{array}\right] \qquad \begin{aligned} x_1 &= 15 \\ y_1 &= -12 \\ z_1 &= -4 \end{aligned}$$

and

$$\left[\begin{array}{ccc|c} 1 & 0 & 0 & 4 \\ 0 & 1 & 0 & -3 \\ 0 & 0 & 1 & -1 \end{array}\right] \qquad \begin{aligned} x_2 &= 4 \\ y_2 &= -3 \\ z_2 &= -1 \end{aligned}$$

and

$$\left[\begin{array}{ccc|c} 1 & 0 & 0 & -5 \\ 0 & 1 & 0 & 4 \\ 0 & 0 & 1 & 1 \end{array}\right] \qquad \begin{aligned} x_3 &= -5 \\ y_3 &= 4 \\ z_3 &= 1 \end{aligned}$$

The inverse matrix is

$$\begin{bmatrix} x_1 & x_2 & x_3 \\ y_1 & y_2 & y_3 \\ z_1 & z_2 & z_3 \end{bmatrix} = \begin{bmatrix} 15 & 4 & -5 \\ -12 & -3 & 4 \\ -4 & -1 & 1 \end{bmatrix}.$$

Technology

You can use a graphing utility to find the inverse of

$$A = \begin{bmatrix} -1 & -1 & -1 \\ 4 & 5 & 0 \\ 0 & 1 & -3 \end{bmatrix}.$$

Enter MATRIX [A] and then use the inverse key. The display, $[A]^{-1}$, should be

$$\begin{bmatrix} 15 & 4 & -5 \\ -12 & -3 & 4 \\ -4 & -1 & 1 \end{bmatrix}.$$

Take a second look at the matrix obtained at the point where Gauss-Jordan elimination was completed. Notice that the 3×3 matrix to the right of the vertical bar is the multiplicative inverse of A. Also notice that the multiplicative identity matrix, I_3 is the matrix that appears to the left of the vertical bar.

$$\left[\begin{array}{ccc|ccc} 1 & 0 & 0 & 15 & 4 & -5 \\ 0 & 1 & 0 & -12 & -3 & 4 \\ 0 & 0 & 1 & -4 & -1 & 1 \end{array}\right]$$

This is the multiplicative identity, I_3.

This is the multiplicative inverse of A.

The observations in the voice balloons and the procedures followed above give us a general method for finding the multiplicative inverse of an invertible matrix.

Study Tip

Because we have a quick method for finding the multiplicative inverse of a 2×2 matrix, the procedure on the right is recommended for matrices of order 3×3 or greater when a graphing utility is not being used.

Procedure for Finding the Multiplicative Inverse of an Invertible Matrix

To find A^{-1} for any $n \times n$ matrix A for which A^{-1} exists:

1. Form the augmented matrix $[A|I]$, where I is the multiplicative identity matrix of the same order as the given matrix A.
2. Perform row transformations on $[A|I]$ to obtain a matrix of the form $[I|B]$. This is equivalent to using Gauss-Jordan elimination to change A into the identity matrix.
3. Matrix B is A^{-1}.
4. Verify the result by showing that $AA^{-1} = I$ and $A^{-1}A = I$.

EXAMPLE 4 Finding the Multiplicative Inverse of a 3 × 3 Matrix

Find the multiplicative inverse of

$$A = \begin{bmatrix} 1 & -1 & 1 \\ 0 & -2 & 1 \\ -2 & -3 & 0 \end{bmatrix}.$$

Solution

Step 1 Form the augmented matrix $[A \mid I_3]$.

$$\left[\begin{array}{ccc|ccc} 1 & -1 & 1 & 1 & 0 & 0 \\ 0 & -2 & 1 & 0 & 1 & 0 \\ -2 & -3 & 0 & 0 & 0 & 1 \end{array}\right]$$

This is matrix A.

This is I_3, the multiplicative identity matrix, with 1s down the main diagonal and 0s elsewhere.

Step 2 Perform row transformations on $[A \mid I_3]$ to obtain a matrix of the form $[I_3 \mid B]$. We want 1s down the main diagonal to the left of the vertical dividing line and 0s elsewhere.

$$\left[\begin{array}{ccc|ccc} 1 & -1 & 1 & 1 & 0 & 0 \\ 0 & -2 & 1 & 0 & 1 & 0 \\ -2 & -3 & 0 & 0 & 0 & 1 \end{array}\right] \xrightarrow[\text{by } 2R_1 + R_3]{\text{Replace row 3}} \left[\begin{array}{ccc|ccc} 1 & -1 & 1 & 1 & 0 & 0 \\ 0 & -2 & 1 & 0 & 1 & 0 \\ 0 & -5 & 2 & 2 & 0 & 1 \end{array}\right] \xrightarrow{-\frac{1}{2}R_2}$$

$$\left[\begin{array}{ccc|ccc} 1 & -1 & 1 & 1 & 0 & 0 \\ 0 & 1 & -\frac{1}{2} & 0 & -\frac{1}{2} & 0 \\ 0 & -5 & 2 & 2 & 0 & 1 \end{array}\right] \xrightarrow[\text{Replace row 3 by } 5R_2 + R_3.]{\text{Replace row 1 by } 1R_2 + R_1.} \left[\begin{array}{ccc|ccc} 1 & 0 & \frac{1}{2} & 1 & -\frac{1}{2} & 0 \\ 0 & 1 & -\frac{1}{2} & 0 & -\frac{1}{2} & 0 \\ 0 & 0 & -\frac{1}{2} & 2 & -\frac{5}{2} & 1 \end{array}\right] \xrightarrow{-2R_3}$$

$$\left[\begin{array}{ccc|ccc} 1 & 0 & \frac{1}{2} & 1 & -\frac{1}{2} & 0 \\ 0 & 1 & -\frac{1}{2} & 0 & -\frac{1}{2} & 0 \\ 0 & 0 & 1 & -4 & 5 & -2 \end{array}\right] \xrightarrow[\text{Replace row 2 by } \frac{1}{2}R_3 + R_2.]{\text{Replace row 1 by } -\frac{1}{2}R_3 + R_1.} \left[\begin{array}{ccc|ccc} 1 & 0 & 0 & 3 & -3 & 1 \\ 0 & 1 & 0 & -2 & 2 & -1 \\ 0 & 0 & 1 & -4 & 5 & -2 \end{array}\right]$$

This is the multiplicative identity, I_3.

This is the multiplicative inverse of A.

Step 3 Matrix B is A^{-1}. The matrix just shown is in the form $[I_3 \mid B]$. The multiplicative identity matrix is on the left of the vertical bar. Matrix B, the multiplicative inverse of A, is on the right. Thus, the multiplicative inverse of A is

$$A^{-1} = \begin{bmatrix} 3 & -3 & 1 \\ -2 & 2 & -1 \\ -4 & 5 & -2 \end{bmatrix}.$$

Step 4 Verify the result by showing that $AA^{-1} = I_3$ and $A^{-1}A = I_3$. Try confirming the result by multiplying A and A^{-1} to obtain I_3. Do you obtain I_3 if you reverse the order of the multiplication?

We have seen that not all square matrices have multiplicative inverses. If the row transformations in step 2 result in all zeros in a row or column to the left of the vertical line, the given matrix does not have a multiplicative inverse.

Check Point 4 Find the multiplicative inverse of

$$A = \begin{bmatrix} 1 & 0 & 2 \\ -1 & 2 & 3 \\ 1 & -1 & 0 \end{bmatrix}.$$

Summary: Finding Multiplicative Inverses for Invertible Matrices

Use a graphing utility with matrix capabilities,

or

a. If the matrix is 2×2: The inverse of $A = \begin{bmatrix} a & b \\ c & d \end{bmatrix}$ is

$$A^{-1} = \frac{1}{ad - bc}\begin{bmatrix} d & -b \\ -c & a \end{bmatrix}.$$

b. If the matrix A is $n \times n$ where $n > 2$: Use the procedure on page 525. Form $[A|I]$ and use row transformations to obtain $[I|B]$. $A^{-1} = [B]$.

2 Use inverses to solve matrix equations.

Solving Systems of Equations Using Multiplicative Inverses of Matrices

Matrix multiplication can be used to represent a system of linear equations.

Linear System

$$a_1x + b_1y + c_1z = d_1$$
$$a_2x + b_2y + c_2z = d_2$$
$$a_3x + b_3y + c_3z = d_3$$

Matrix Form of the System

$$\begin{bmatrix} a_1 & b_1 & c_1 \\ a_2 & b_2 & c_2 \\ a_3 & b_3 & c_3 \end{bmatrix}\begin{bmatrix} x \\ y \\ z \end{bmatrix} = \begin{bmatrix} d_1 \\ d_2 \\ d_3 \end{bmatrix}$$

This matrix contains the system's coefficients.

This matrix contains the system's variables.

This matrix contains the system's constants.

You can work with the matrix form on the right and obtain the form of the linear system on the left. To do so, perform the matrix multiplication on the left side of the matrix equation. Then equate the corresponding elements.

The matrix equation

$$\underbrace{\begin{bmatrix} a_1 & b_1 & c_1 \\ a_2 & b_2 & c_2 \\ a_3 & b_3 & c_3 \end{bmatrix}}_{A}\underbrace{\begin{bmatrix} x \\ y \\ z \end{bmatrix}}_{X} = \underbrace{\begin{bmatrix} d_1 \\ d_2 \\ d_3 \end{bmatrix}}_{B}$$

is abbreviated as $AX = B$, where A is the **coefficient matrix** of the system, and X and B are matrices containing one column, called **column matrices**. The matrix B is called the **constant matrix**.

Here is a specific example of a linear system and its matrix form.

Linear System

$$\begin{aligned} x - y + z &= 2 \\ -2y + z &= 2 \\ -2x - 3y \quad &= \tfrac{1}{2} \end{aligned}$$

Matrix Form

Coefficients

$$\begin{bmatrix} 1 & -1 & 1 \\ 0 & -2 & 1 \\ -2 & -3 & 0 \end{bmatrix} \begin{bmatrix} x \\ y \\ z \end{bmatrix} = \begin{bmatrix} 2 \\ 2 \\ \tfrac{1}{2} \end{bmatrix}$$

Constants

A, the coefficient matrix · X = B, the constant matrix

The matrix equation $AX = B$ can be solved using A^{-1} if it exists.

$AX = B$ — This is the matrix equation.

$A^{-1}AX = A^{-1}B$ — Multiply both sides by A^{-1}. Because matrix multiplication is not commutative, put A^{-1} in the same left position on both sides.

$I_nX = A^{-1}B$ — The multiplicative inverse property tells us that $A^{-1}A = I_n$.

$X = A^{-1}B$ — Because I_n is the multiplicative identity, $I_nX = X$.

We see that if $AX = B$, then $X = A^{-1}B$.

Solving a System Using A^{-1}

If $AX = B$ has a unique solution, $X = A^{-1}B$. To solve a linear system of equations, multiply A^{-1} and B to find X.

EXAMPLE 5 Using the Inverse of a Matrix to Solve a System

Solve the system by using A^{-1}, the inverse of the coefficient matrix.

$$\begin{aligned} x - y + z &= 2 \\ -2y + z &= 2 \\ -2x - 3y \quad &= \tfrac{1}{2} \end{aligned}$$

Solution The linear system can be written as

$$\begin{bmatrix} 1 & -1 & 1 \\ 0 & -2 & 1 \\ -2 & -3 & 0 \end{bmatrix} \begin{bmatrix} x \\ y \\ z \end{bmatrix} = \begin{bmatrix} 2 \\ 2 \\ \tfrac{1}{2} \end{bmatrix}.$$

A X B

The solution is given by $X = A^{-1}B$. Consequently, we must find A^{-1}. We found the inverse of matrix A in Example 4. Using this result,

$$X = A^{-1}B = \begin{bmatrix} 3 & -3 & 1 \\ -2 & 2 & -1 \\ -4 & 5 & -2 \end{bmatrix} \begin{bmatrix} 2 \\ 2 \\ \tfrac{1}{2} \end{bmatrix} = \begin{bmatrix} 3 \cdot 2 + (-3) \cdot 2 + 1 \cdot \tfrac{1}{2} \\ -2 \cdot 2 + 2 \cdot 2 + (-1) \cdot \tfrac{1}{2} \\ -4 \cdot 2 + 5 \cdot 2 + (-2) \cdot \tfrac{1}{2} \end{bmatrix} = \begin{bmatrix} \tfrac{1}{2} \\ -\tfrac{1}{2} \\ 1 \end{bmatrix}$$

Thus, $x = \tfrac{1}{2}$, $y = -\tfrac{1}{2}$, and $z = 1$. The solution set is $\{(\tfrac{1}{2}, -\tfrac{1}{2}, 1)\}$.

Check Point 5 Solve the system by using A^{-1}, the inverse of the coefficient matrix that you found in Checkpoint 4.

$$\begin{aligned} x \quad\quad + 2z &= 6 \\ -x + 2y + 3z &= -5 \\ x - y \quad\quad &= 6 \end{aligned}$$

3 Encode and decode messages.

Applications of Matrix Inverses to Coding

A **cryptogram** is a message written so that no one other than the intended recipient can understand it. To encode a message, we begin by assigning a number to each letter in the alphabet: $A = 1, B = 2, C = 3, \ldots, Z = 26$, and a space $= 0$. For example, the numerical equivalent of the word MATH is 13, 1, 20, 8. The numerical equivalent of the message is then converted into a matrix. Finally, an invertible matrix can be used to convert the message into code. The multiplicative inverse of this matrix can be used to decode the message.

Encoding a Word or Message

1. Express the word or message numerically.
2. List the numbers in step 1 by columns and form a square matrix. If you do not have enough numbers to form a square matrix, put zeros in any remaining spaces in the last column.
3. Select any square invertible matrix, called the **coding matrix**, the same size as the matrix in step 2. Multiply the coding matrix by the square matrix that expresses the message numerically. The resulting matrix is the **coded matrix**.
4. Use the numbers, by columns, from the coded matrix in step 3 to write the encoded message.

EXAMPLE 6 Encoding a Word

Use matrices to encode the word MATH.

Solution

Step 1 Express the word numerically. As shown previously, the numerical equivalent of MATH is 13, 1, 20, 8.

Step 2 List the numbers in step 1 by columns and form a square matrix. The 2×2 matrix is

$$\begin{bmatrix} 13 & 20 \\ 1 & 8 \end{bmatrix}.$$

Step 3 Multiply the matrix in step 2 by a square invertible matrix. We will use $\begin{bmatrix} -2 & -3 \\ 3 & 4 \end{bmatrix}$ as the coding matrix.

$$\begin{bmatrix} -2 & -3 \\ 3 & 4 \end{bmatrix}\begin{bmatrix} 13 & 20 \\ 1 & 8 \end{bmatrix} = \begin{bmatrix} -2(13) - 3(1) & -2(20) - 3(8) \\ 3(13) + 4(1) & 3(20) + 4(8) \end{bmatrix}$$

$$= \begin{bmatrix} -29 & -64 \\ 43 & 92 \end{bmatrix}$$

Coding matrix; Numerical representation of MATH; Coded matrix

Step 4 Use the numbers, by columns, from the coded matrix in step 3 to write the encoded message. The encoded message is −29, 43, −64, 92.

Check Point 6 Use the coding matrix in Example 6, $\begin{bmatrix} -2 & -3 \\ 3 & 4 \end{bmatrix}$, to encode the word BASE.

The inverse of a coding matrix can be used to decode a word or message that was encoded.

Decoding a Word or Message That Was Encoded

1. Find the multiplicative inverse of the coding matrix.
2. Multiply the multiplicative inverse of the coding matrix and the coded matrix.
3. Express the numbers, by columns, from the matrix in step 2 as letters.

EXAMPLE 7 Decoding a Word

Decode −29, 43, −64, 92 from Example 6.

Solution

Step 1 Find the inverse of the coding matrix. The coding matrix in Example 6 was $\begin{bmatrix} -2 & -3 \\ 3 & 4 \end{bmatrix}$. We use the formula for the multiplicative inverse of a 2×2 matrix to find the multiplicative inverse of this matrix. It is $\begin{bmatrix} 4 & 3 \\ -3 & -2 \end{bmatrix}$.

Step 2 Multiply the multiplicative inverse of the coding matrix and the coded matrix.

$$\begin{bmatrix} 4 & 3 \\ -3 & -2 \end{bmatrix}\begin{bmatrix} -29 & -64 \\ 43 & 92 \end{bmatrix} = \begin{bmatrix} 4(-29) + 3(43) & 4(-64) + 3(92) \\ -3(-29) - 2(43) & -3(-64) - 2(92) \end{bmatrix}$$

$$= \begin{bmatrix} 13 & 20 \\ 1 & 8 \end{bmatrix}$$

Multiplicative inverse of the coding matrix

Coded matrix

Step 3 Express the numbers, by columns, from the matrix in step 2 as letters. The numbers are 13, 1, 20, and 8. Using letters, the decoded message is MATH.

Check Point 7 Decode the word that you encoded in Checkpoint 6.

Decoding is simple for an authorized receiver who knows the coding matrix. Because any invertible matrix can be used for the coding matrix, decoding a cryptogram for an unauthorized receiver who does not know this matrix is extremely difficult.

EXERCISE SET 6.4

Practice Exercises

In Exercises 1–12 find the products AB and BA to determine whether B is the multiplicative inverse of A.

1. $A = \begin{bmatrix} 4 & -3 \\ -5 & 4 \end{bmatrix}$, $B = \begin{bmatrix} 4 & 3 \\ 5 & 4 \end{bmatrix}$

2. $A = \begin{bmatrix} -2 & -1 \\ -1 & 1 \end{bmatrix}$, $B = \begin{bmatrix} 1 & 1 \\ 1 & 2 \end{bmatrix}$

3. $A = \begin{bmatrix} -4 & 0 \\ 1 & 3 \end{bmatrix}$, $B = \begin{bmatrix} -2 & 4 \\ 0 & 1 \end{bmatrix}$

4. $A = \begin{bmatrix} -2 & 4 \\ 1 & -2 \end{bmatrix}$, $B = \begin{bmatrix} 1 & 2 \\ -1 & -2 \end{bmatrix}$

5. $A = \begin{bmatrix} -2 & 1 \\ \frac{3}{2} & -\frac{1}{2} \end{bmatrix}$, $B = \begin{bmatrix} 1 & 2 \\ 3 & 4 \end{bmatrix}$

6. $A = \begin{bmatrix} 4 & 5 \\ 2 & 3 \end{bmatrix}$, $B = \begin{bmatrix} \frac{3}{2} & -\frac{5}{2} \\ -1 & 2 \end{bmatrix}$

7. $A = \begin{bmatrix} 0 & 1 & 0 \\ 0 & 0 & 1 \\ 1 & 0 & 0 \end{bmatrix}$ $B = \begin{bmatrix} 0 & 0 & 1 \\ 1 & 0 & 0 \\ 0 & 1 & 0 \end{bmatrix}$

8. $A = \begin{bmatrix} -2 & 1 & -1 \\ -5 & 2 & -1 \\ 3 & -1 & 1 \end{bmatrix}$ $B = \begin{bmatrix} 1 & 0 & 1 \\ 2 & 1 & 3 \\ -1 & 1 & 1 \end{bmatrix}$

9. $A = \begin{bmatrix} 1 & 2 & 3 \\ 1 & 3 & 4 \\ 1 & 4 & 3 \end{bmatrix}$ $B = \begin{bmatrix} \frac{7}{2} & -3 & \frac{1}{2} \\ -\frac{1}{2} & 0 & \frac{1}{2} \\ -\frac{1}{2} & 1 & -\frac{1}{2} \end{bmatrix}$

10. $A = \begin{bmatrix} 0 & 2 & 0 \\ 3 & 3 & 2 \\ 2 & 5 & 1 \end{bmatrix}$ $B = \begin{bmatrix} -3.5 & -1 & 2 \\ 0.5 & 0 & 0 \\ 4.5 & 2 & -3 \end{bmatrix}$

11. $A = \begin{bmatrix} 0 & 0 & -2 & 1 \\ -1 & 0 & 1 & 1 \\ 0 & 1 & -1 & 0 \\ 1 & 0 & 0 & -1 \end{bmatrix}$, $B = \begin{bmatrix} 1 & 2 & 0 & 3 \\ 0 & 1 & 1 & 1 \\ 0 & 1 & 0 & 1 \\ 1 & 2 & 0 & 2 \end{bmatrix}$

12. $A = \begin{bmatrix} 1 & -2 & 1 & 0 \\ 0 & 1 & -2 & 1 \\ 0 & 0 & 1 & -2 \\ 0 & 0 & 0 & 1 \end{bmatrix}$, $B = \begin{bmatrix} 1 & 2 & 3 & 4 \\ 0 & 1 & 2 & 3 \\ 0 & 0 & 1 & 2 \\ 0 & 0 & 0 & 1 \end{bmatrix}$

In Exercises 13–18, use the fact that if $A = \begin{bmatrix} a & b \\ c & d \end{bmatrix}$, *then* $A^{-1} = \frac{1}{ad - bc}\begin{bmatrix} d & -b \\ -c & a \end{bmatrix}$ *to find the inverse of each matrix, if possible. Check that* $AA^{-1} = I_2$ *and* $A^{-1}A = I_2$.

13. $A = \begin{bmatrix} 2 & 3 \\ -1 & 2 \end{bmatrix}$

14. $A = \begin{bmatrix} 0 & 3 \\ 4 & -2 \end{bmatrix}$

15. $A = \begin{bmatrix} 3 & -1 \\ -4 & 2 \end{bmatrix}$

16. $A = \begin{bmatrix} 2 & -6 \\ 1 & -2 \end{bmatrix}$

17. $A = \begin{bmatrix} 10 & -2 \\ -5 & 1 \end{bmatrix}$

18. $A = \begin{bmatrix} 6 & -3 \\ -2 & 1 \end{bmatrix}$

In Exercises 19–24, find A^{-1} *by forming* $[A|I]$ *and then using row transformations to obtain* $[I|B]$, *where* $A^{-1} = [B]$. *Check that* $AA^{-1} = I$ *and* $A^{-1}A = I$.

19. $A = \begin{bmatrix} 2 & 2 & -1 \\ 0 & 3 & -1 \\ -1 & -2 & 1 \end{bmatrix}$

20. $A = \begin{bmatrix} 1 & -1 & 1 \\ 0 & 2 & -1 \\ 2 & 3 & 0 \end{bmatrix}$

21. $A = \begin{bmatrix} 5 & 0 & 2 \\ 2 & 2 & 1 \\ -3 & 1 & -1 \end{bmatrix}$

22. $A = \begin{bmatrix} 3 & 2 & 6 \\ 1 & 1 & 2 \\ 2 & 2 & 5 \end{bmatrix}$

23. $A = \begin{bmatrix} 1 & 0 & 0 & 0 \\ 0 & -1 & 0 & 0 \\ 0 & 0 & 3 & 0 \\ 1 & 0 & 0 & 1 \end{bmatrix}$

24. $A = \begin{bmatrix} 2 & 0 & 0 & 1 \\ 0 & 1 & 0 & 0 \\ 0 & 0 & -1 & 0 \\ 0 & 0 & 0 & 2 \end{bmatrix}$

In Exercises 25–28, write each linear system as a matrix equation in the form $AX = B$, *where A is the coefficient matrix and B is the constant matrix.*

25. $6x + 5y = 13$
$5x + 4y = 10$

26. $7x + 5y = 23$
$3x + 2y = 10$

27. $x + 3y + 4z = -3$
$x + 2y + 3z = -2$
$x + 4y + 3z = -6$

28. $x + 4y - z = 3$
$x + 3y - 2z = 5$
$2x + 7y - 5z = 12$

In Exercises 29–32, write each matrix equation as a system of linear equations without matrices.

29. $\begin{bmatrix} 4 & -7 \\ 2 & -3 \end{bmatrix}\begin{bmatrix} x \\ y \end{bmatrix} = \begin{bmatrix} -3 \\ 1 \end{bmatrix}$

30. $\begin{bmatrix} 3 & 0 \\ -3 & 1 \end{bmatrix}\begin{bmatrix} x \\ y \end{bmatrix} = \begin{bmatrix} 6 \\ -7 \end{bmatrix}$

31. $\begin{bmatrix} 2 & 0 & -1 \\ 0 & 3 & 0 \\ 1 & 1 & 0 \end{bmatrix}\begin{bmatrix} x \\ y \\ z \end{bmatrix} = \begin{bmatrix} 6 \\ 9 \\ 5 \end{bmatrix}$

32. $\begin{bmatrix} -1 & 0 & 1 \\ 0 & -1 & 0 \\ 0 & 1 & 1 \end{bmatrix}\begin{bmatrix} x \\ y \\ z \end{bmatrix} = \begin{bmatrix} -4 \\ 2 \\ 4 \end{bmatrix}$

In Exercises 33–38,

a. *Write each linear system as a matrix equation in the form $AX = B$.*

b. *Solve the system using the inverse of the coefficient matrix.*

33. $\begin{aligned} 2x + 6y + 6z &= 8 \\ 2x + 7y + 6z &= 10 \\ 2x + 7y + 7z &= 9 \end{aligned}$ The inverse of $\begin{bmatrix} 2 & 6 & 6 \\ 2 & 7 & 6 \\ 2 & 7 & 7 \end{bmatrix}$ is $\begin{bmatrix} \frac{7}{2} & 0 & -3 \\ -1 & 1 & 0 \\ 0 & -1 & 1 \end{bmatrix}$.

34. $\begin{aligned} x + 2y + 5z &= 2 \\ 2x + 3y + 8z &= 3 \\ -x + y + 2z &= 3 \end{aligned}$ The inverse of $\begin{bmatrix} 1 & 2 & 5 \\ 2 & 3 & 8 \\ -1 & 1 & 2 \end{bmatrix}$ is $\begin{bmatrix} 2 & -1 & -1 \\ 12 & -7 & -2 \\ -5 & 3 & 1 \end{bmatrix}$.

35. $\begin{aligned} x - y + z &= 8 \\ 2y - z &= -7 \\ 2x + 3y &= 1 \end{aligned}$ The inverse of $\begin{bmatrix} 1 & -1 & 1 \\ 0 & 2 & -1 \\ 2 & 3 & 0 \end{bmatrix}$ is $\begin{bmatrix} 3 & 3 & -1 \\ -2 & -2 & 1 \\ -4 & -5 & 2 \end{bmatrix}$.

36. $\begin{aligned} x - 6y + 3z &= 11 \\ 2x - 7y + 3z &= 14 \\ 4x - 12y + 5z &= 25 \end{aligned}$ The inverse of $\begin{bmatrix} 1 & -6 & 3 \\ 2 & -7 & 3 \\ 4 & -12 & 5 \end{bmatrix}$ is $\begin{bmatrix} 1 & -6 & 3 \\ 2 & -7 & 3 \\ 4 & -12 & 5 \end{bmatrix}$.

37. $\begin{aligned} x - y + 2z \quad &= -3 \\ y - z + w &= 4 \\ -x + y - z + 2w &= 2 \\ -y + z - 2w &= -4 \end{aligned}$

The inverse of $\begin{bmatrix} 1 & -1 & 2 & 0 \\ 0 & 1 & -1 & 1 \\ -1 & 1 & -1 & 2 \\ 0 & -1 & 1 & -2 \end{bmatrix}$ is $\begin{bmatrix} 0 & 0 & -1 & -1 \\ 1 & 4 & 1 & 3 \\ 1 & 2 & 1 & 2 \\ 0 & -1 & 0 & -1 \end{bmatrix}$.

38. $\begin{aligned} 2x + z + w &= 6 \\ 3x + w &= 9 \\ -x + y - 2z + w &= 4 \\ 4x - y + z &= 6 \end{aligned}$

The inverse of $\begin{bmatrix} 2 & 0 & 1 & 1 \\ 3 & 0 & 0 & 1 \\ -1 & 1 & -2 & 1 \\ 4 & -1 & 1 & 0 \end{bmatrix}$ is $\begin{bmatrix} -1 & 2 & -1 & -1 \\ -4 & 9 & -5 & -6 \\ 0 & 1 & -1 & -1 \\ 3 & -5 & 3 & 3 \end{bmatrix}$.

Application Exercises

In Exercises 39–40, use the coding matrix

$A = \begin{bmatrix} 4 & -1 \\ -3 & 1 \end{bmatrix}$ *and its inverse* $A^{-1} = \begin{bmatrix} 1 & 1 \\ 3 & 4 \end{bmatrix}$ *to encode and then decode the given message.*

39. HELP

40. LOVE

In Exercises 41–42, use the coding matrix

$A = \begin{bmatrix} 1 & -1 & 0 \\ 3 & 0 & 2 \\ -1 & 0 & -1 \end{bmatrix}$ *and its inverse*

$A^{-1} = \begin{bmatrix} 0 & 1 & 2 \\ -1 & 1 & 2 \\ 0 & -1 & -3 \end{bmatrix}$ *to write a cryptogram for each message. Check your result by decoding the cryptogram.*

41. S E N D _ C A S H

19 5 14 4 0 3 1 19 8

Use $\begin{bmatrix} 19 & 4 & 1 \\ 5 & 0 & 19 \\ 14 & 3 & 8 \end{bmatrix}$.

42. S T A Y _ W E L L

19 20 1 25 0 23 5 12 12

Use $\begin{bmatrix} 19 & 25 & 5 \\ 20 & 0 & 12 \\ 1 & 23 & 12 \end{bmatrix}$.

Writing in Mathematics

43. What is the multiplicative identity matrix?

44. If you are given two matrices, A and B, explain how to determine if B is the multiplicative inverse of A.

45. Explain why a matrix that does not have the same number of rows and columns cannot have a multiplicative inverse.

46. Explain how to find the multiplicative inverse for a 2×2 invertible matrix.

47. Explain how to find the multiplicative inverse for a 3×3 invertible matrix.

48. Explain how to write a linear system of three equations in three variables as a matrix equation.

49. Explain how to solve the matrix equation $AX = B$.

50. What is a cryptogram?

51. It's January 1, and you've written down your major goal for the year. You do not want those closest to you to see what you've written in case you do not accomplish your objective. Consequently, you decide to use a coding matrix to encode your goal. Explain how this can be accomplished.

52. A year has passed since Exercise 51. (Time flies when you're solving exercises in algebra books.) It's been a terrific year and so many wonderful things have happened that you can't remember your goal from a year ago. You consult your personal journal and you find the encoded message and the coding matrix. How can you use these to find your original goal?

Technology Exercises

In Exercises 53–58, use a graphing utility to find the multiplicative inverse of each matrix. Check that the displayed inverse is correct.

53. $\begin{bmatrix} 3 & -1 \\ -2 & 1 \end{bmatrix}$

54. $\begin{bmatrix} -4 & 1 \\ 6 & -2 \end{bmatrix}$

55. $\begin{bmatrix} -2 & 1 & -1 \\ -5 & 2 & -1 \\ 3 & -1 & 1 \end{bmatrix}$

56. $\begin{bmatrix} 1 & 1 & -1 \\ -3 & 2 & -1 \\ 3 & -3 & 2 \end{bmatrix}$

57. $\begin{bmatrix} 7 & -3 & 0 & 2 \\ -2 & 1 & 0 & -1 \\ 4 & 0 & 1 & -2 \\ -1 & 1 & 0 & -1 \end{bmatrix}$

58. $\begin{bmatrix} 1 & 2 & 0 & 0 \\ 0 & 0 & 1 & 0 \\ 1 & 3 & 0 & 1 \\ 4 & 0 & 0 & 2 \end{bmatrix}$

In Exercises 59–64, write each system in the form $AX = B$. Then solve the system by entering A and B into your graphing utility and computing $A^{-1}B$.

59. $\begin{aligned} x - y + z &= -6 \\ 4x + 2y + z &= 9 \\ 4x - 2y + z &= -3 \end{aligned}$

60. $\begin{aligned} y + 2z &= 0 \\ -x + y &= 1 \\ 2x - y + z &= -1 \end{aligned}$

61. $\begin{aligned} 3x - 2y + z &= -2 \\ 4x - 5y + 3z &= -9 \\ 2x - y + 5z &= -5 \end{aligned}$

62. $\begin{aligned} x - y &= 1 \\ 6x + y + 20z &= 14 \\ y + 3z &= 1 \end{aligned}$

63. $\begin{aligned} x - 3z + v &= -3 \\ y + w &= -1 \\ z + v &= 7 \\ x + y - z + 4w &= -8 \\ x + y + z + w + v &= 8 \end{aligned}$

64. $\begin{aligned} x + y + z + w &= 4 \\ x + 3y - 2z + 2w &= 7 \\ 2x + 2y + z + w &= 3 \\ x - y + 2z + 3w &= 5 \end{aligned}$

In Exercises 65–66, use a coding matrix A of your choice. Use a graphing utility to find the multiplicative inverse of your coding matrix. Write a cryptogram for each message. Check your result by decoding the cryptogram. Use your graphing utility to perform all necessary matrix multiplications.

65. A R R I V E D _ S A F E L Y
1 18 18 9 22 5 4 0 19 1 6 5 12 25

66. A R T _ E N R I C H E S
1 18 20 0 5 14 18 9 3 8 5 19

Critical Thinking Exercises

67. Which one of the following is true?
a. Some nonsquare matrices have inverses.
b. All square 2×2 matrices have inverses because there is a formula for finding these inverses.
c. Two 2×2 invertible matrices can have a matrix sum that is not invertible.
d. To solve the matrix equation $AX = B$ for X, multiply A and the inverse of B.

68. Which one of the following is true?
a. $(AB)^{-1} = A^{-1}B^{-1}$, assuming A, B, and AB, are invertible.
b. $(A + B)^{-1} = A^{-1} + B^{-1}$, assuming A, B, and $A + B$ are invertible.
c. $\begin{bmatrix} 1 & -3 \\ -1 & 3 \end{bmatrix}$ is an invertible matrix.
d. None of the above is true.

69. Give an example of a 2×2 matrix that is its own inverse.

70. If $A = \begin{bmatrix} 3 & 5 \\ 2 & 4 \end{bmatrix}$, find $\left(A^{-1}\right)^{-1}$.

71. Find values of a for which the following matrix is not invertible.

$$\begin{bmatrix} 1 & a + 1 \\ a - 2 & 4 \end{bmatrix}$$

Group Exercise

72. Each person in the group should work with one partner. Send a coded word or message to each other by giving your partner the coded matrix and the coding matrix that you selected. Once messages are sent, each person should decode the message received.

SECTION 6.5 Determinants and Cramer's Rule

Objectives

1. Evaluate a second-order determinant.
2. Solve a linear system of equations in two variables using Cramer's rule.
3. Evaluate a third-order determinant.
4. Solve a linear system of equations in three variables using Cramer's rule.
5. Use determinants to identify inconsistent systems and systems with dependent equations.
6. Evaluate higher-order determinants.

A portion of Charles Babbage's unrealized Difference Engine

As cyberspace absorbs more and more of our work, play, shopping, and socializing, where will it all end? Which activities will still be offline in 2025?

Our technologically transformed lives can be traced back to the English inventor Charles Babbage (1792–1871). Babbage knew of a method for solving linear systems called Cramer's rule, in honor of the Swiss geometer Gabriel Cramer (1704–1752). Cramer's rule was simple, but involved numerous multiplications for large systems. Babbage designed a machine, called the "difference engine," that consisted of toothed wheels on shafts for performing these multiplications. Despite the fact that only one-seventh of the functions ever worked, Babbage's invention demonstrated how complex calculations could be handled mechanically. In 1944, scientists at IBM used the lessons of the the difference engine to create the world's first computer.

Those who invented computers hoped to relegate the drudgery of repeated computation to a machine. In this section, we look at a method for solving linear systems that played a critical role in this process. The method uses arrays of numbers called *determinants*. As with matrix methods, solutions are obtained by writing down the coefficients and constants of a linear system and performing operations with them.

1 Evaluate a second-order determinant.

The Determinant of a 2 × 2 Matrix

Associated with every square matrix is a real number called its **determinant**. The determinant for a 2 × 2 square matrix is defined as follows.

Definition of the Determinant of a 2 × 2 Matrix

The determinant of the matrix $\begin{bmatrix} a_1 & b_1 \\ a_2 & b_2 \end{bmatrix}$ is denoted by $\begin{vmatrix} a_1 & b_1 \\ a_2 & b_2 \end{vmatrix}$ and is defined by

$$\begin{vmatrix} a_1 & b_1 \\ a_2 & b_2 \end{vmatrix} = a_1b_2 - a_2b_1.$$

We also say that the **value** of the **second-order determinant** $\begin{vmatrix} a_1 & b_1 \\ a_2 & b_2 \end{vmatrix}$ is $a_1b_2 - a_2b_1$.

Study Tip

To evaluate a determinant, find the difference of the product of the two diagonals.

$$\begin{vmatrix} a_1 & b_1 \\ a_2 & b_2 \end{vmatrix} = a_1b_2 - a_2b_1$$

Example 1 illustrates that the determinant of a matrix may be positive or negative. The determinant can also have 0 as its value.

EXAMPLE 1 Evaluating the Determinant of a 2 × 2 Matrix

Evaluate the determinant of:

a. $\begin{bmatrix} 5 & 6 \\ 7 & 3 \end{bmatrix}$ **b.** $\begin{bmatrix} 2 & 4 \\ -3 & -5 \end{bmatrix}$.

Discovery

Write and then evaluate three determinants, one whose value is positive, one whose value is negative, and one whose value is 0.

Solution We multiply and subtract as indicated.

a. $\begin{vmatrix} 5 & 6 \\ 7 & 3 \end{vmatrix} = 5 \cdot 3 - 7 \cdot 6 = 15 - 42 = -27$ The value of the second-order determinant is −27.

b. $\begin{vmatrix} 2 & 4 \\ -3 & -5 \end{vmatrix} = 2(-5) - (-3)(4) = -10 + 12 = 2$ The value of the second-order determinant is 2.

Check Point 1 Evaluate the determinant of:

a. $\begin{bmatrix} 10 & 9 \\ 6 & 5 \end{bmatrix}$ **b.** $\begin{bmatrix} 4 & 3 \\ -5 & -8 \end{bmatrix}$.

2 Solve a linear system of equations in two variables using Cramer's rule.

Solving Linear Systems of Equations in Two Variables Using Determinants

Determinants can be used to solve a linear system in two variables. In general, such a system appears as

$$a_1x + b_1y = c_1$$
$$a_2x + b_2y = c_2.$$

Let's first solve this system for x using the addition method. We can solve for x by eliminating y from the equations. Multiply the first equation by b_2 and the second equation by $-b_1$. Then add the two equations:

$$\begin{aligned} a_1x + b_1y &= c_1 \xrightarrow{\text{Multiply by } b_2.} & a_1b_2x + b_1b_2y &= c_1b_2 \\ a_2x + b_2y &= c_2 \xrightarrow{\text{Multiply by } -b_1.} & -a_2b_1x - b_1b_2y &= -c_2b_1 \\ & \text{Add:} & (a_1b_2 - a_2b_1)x &= c_1b_2 - c_2b_1 \\ & & x &= \frac{c_1b_2 - c_2b_1}{a_1b_2 - a_2b_1} \end{aligned}$$

Because

$$\begin{vmatrix} c_1 & b_1 \\ c_2 & b_2 \end{vmatrix} = c_1b_2 - c_2b_1 \quad \text{and} \quad \begin{vmatrix} a_1 & b_1 \\ a_2 & b_2 \end{vmatrix} = a_1b_2 - a_2b_1$$

we can express our answer for x as the quotient of two determinants:

$$x = \frac{\begin{vmatrix} c_1 & b_1 \\ c_2 & b_2 \end{vmatrix}}{\begin{vmatrix} a_1 & b_1 \\ a_2 & b_2 \end{vmatrix}}.$$

In a similar way, we could use the addition method to solve our system for y, again expressing y as the quotient of two determinants. This method of using determinants to solve the linear system, called **Cramer's rule**, is summarized in the box.

Solving a Linear System in Two Variables Using Determinants

Cramer's Rule

If

$$a_1x + b_1y = c_1$$
$$a_2x + b_2y = c_2$$

then

$$x = \frac{\begin{vmatrix} c_1 & b_1 \\ c_2 & b_2 \end{vmatrix}}{\begin{vmatrix} a_1 & b_1 \\ a_2 & b_2 \end{vmatrix}} \quad \text{and} \quad y = \frac{\begin{vmatrix} a_1 & c_1 \\ a_2 & c_2 \end{vmatrix}}{\begin{vmatrix} a_1 & b_1 \\ a_2 & b_2 \end{vmatrix}}$$

where

$$\begin{vmatrix} a_1 & b_1 \\ a_2 & b_2 \end{vmatrix} \neq 0.$$

Here are some helpful tips when solving

$$a_1x + b_1y = c_1$$
$$a_2x + b_2y = c_2$$

using determinants.

1. Three different determinants are used to find x and y. The determinants in the denominators for x and y are identical. The determinants in the numerators for x and y differ. In abbreviated notation, we write

$$x = \frac{D_x}{D} \quad \text{and} \quad y = \frac{D_y}{D} \quad \text{where } D \neq 0.$$

2. The elements of D, the determinant in the denominator, are the coefficients of the variables in the system.

$$D = \begin{vmatrix} a_1 & b_1 \\ a_2 & b_2 \end{vmatrix}$$

3. D_x, the determinant in the numerator of x, is obtained by replacing the x-coefficients, a_1, a_2, in D with the constants on the right side of the equations, c_1, c_2.

$$D = \begin{vmatrix} a_1 & b_1 \\ a_2 & b_2 \end{vmatrix} \quad \text{and} \quad D_x = \begin{vmatrix} c_1 & b_1 \\ c_2 & b_2 \end{vmatrix}$$

Replace the column with a_1 and a_2 with the constants c_1 and c_2 to get D_x.

4. D_y, the determinant in the numerator for y, is obtained by replacing the y-coefficients, b_1, b_2 in D with the constants on the right side of the equations, c_1, c_2.

$$D = \begin{vmatrix} a_1 & b_1 \\ a_2 & b_2 \end{vmatrix} \quad \text{and} \quad D_y = \begin{vmatrix} a_1 & c_1 \\ a_2 & c_2 \end{vmatrix}$$

Replace the column with b_1 and b_2 with the constants c_1 and c_2 to get D_y.

Example 2 illustrates the use of Cramer's rule.

EXAMPLE 2 Using Cramer's Rule to Solve a Linear System

Use Cramer's rule to solve the system:

$$5x - 4y = 2$$
$$6x - 5y = 1.$$

Solution Because

$$x = \frac{D_x}{D} \quad \text{and} \quad y = \frac{D_y}{D},$$

we will set up and evaluate the three determinants D, D_x, and D_y.

1. D, the determinant in both denominators, consists of the x- and y-coefficients.

$$D = \begin{vmatrix} 5 & -4 \\ 6 & -5 \end{vmatrix} = (5)(-5) - (6)(-4) = -25 + 24 = -1$$

Because this determinant is not zero, we continue to use Cramer's rule to solve the system.

2. D_x, the determinant in the numerator for x, is obtained by replacing the x-coefficients in D, 5 and 6, by the constants on the right side of the equation, 2 and 1.

$$D_x = \begin{vmatrix} 2 & -4 \\ 1 & -5 \end{vmatrix} = (2)(-5) - (1)(-4) = -10 + 4 = -6$$

3. D_y, the determinant in the numerator for y, is obtained by replacing the y-coefficients in D, -4 and -5, by the constants on the right side of the equation, 2 and 1.

$$D_y = \begin{vmatrix} 5 & 2 \\ 6 & 1 \end{vmatrix} = (5)(1) - (6)(2) = 5 - 12 = -7$$

4. Thus,

$$x = \frac{D_x}{D} = \frac{-6}{-1} = 6 \quad \text{and} \quad y = \frac{D_y}{D} = \frac{-7}{-1} = 7.$$

As always, the solution (6, 7) can be checked by substituting these values into the original equations. The solution set is $\{(6, 7)\}$.

Check Point 2 Use Cramer's rule to solve the system:

$$5x + 4y = 12$$
$$3x - 6y = 24.$$

3 Evaluate a third-order determinant.

The Determinant of a 3 × 3 Matrix

Associated with every square matrix is a real number called its determinant. The determinant for a 3×3 matrix is defined as follows.

Definition of a Third-Order Determinant

$$\begin{vmatrix} a_1 & b_1 & c_1 \\ a_2 & b_2 & c_2 \\ a_3 & b_3 & c_3 \end{vmatrix} = a_1b_2c_3 + b_1c_2a_3 + c_1a_2b_3 - a_3b_2c_1 - b_3c_2a_1 - c_3a_2b_1$$

The six terms and the three factors in each term in this complicated evaluation formula can be rearranged, and then we can apply the distributive property. We obtain

$$\begin{aligned} &a_1b_2c_3 - a_1b_3c_2 - a_2b_1c_3 + a_2b_3c_1 + a_3b_1c_2 - a_3b_2c_1 \\ &= a_1(b_2c_3 - b_3c_2) - a_2(b_1c_3 - b_3c_1) + a_3(b_1c_2 - b_2c_1) \\ &= a_1\begin{vmatrix} b_2 & c_2 \\ b_3 & c_3 \end{vmatrix} - a_2\begin{vmatrix} b_1 & c_1 \\ b_3 & c_3 \end{vmatrix} + a_3\begin{vmatrix} b_1 & c_1 \\ b_2 & c_2 \end{vmatrix}. \end{aligned}$$

You can evaluate each of the second-order determinants and obtain the three expressions in parentheses in the second step.

In summary, we now have arranged the definition of a third-order determinant as follows.

Definition of the Determinant of a 3 × 3 Matrix

A third-order determinant is defined by

Subtract. Add.

$$\begin{vmatrix} a_1 & b_1 & c_1 \\ a_2 & b_2 & c_2 \\ a_3 & b_3 & c_3 \end{vmatrix} = a_1\begin{vmatrix} b_2 & c_2 \\ b_3 & c_3 \end{vmatrix} - a_2\begin{vmatrix} b_1 & c_1 \\ b_3 & c_3 \end{vmatrix} + a_3\begin{vmatrix} b_1 & c_1 \\ b_2 & c_2 \end{vmatrix}.$$

The a's on the right come from the first column.

Here are some tips that may be helpful when evaluating the determinant of a 3 × 3 matrix.

1. Each of the three terms in the definition contains two factors—a numerical factor and a second-order determinant.
2. The numerical factor in each term is an element from the first column of the third-order determinant.
3. The minus sign precedes the second term.
4. The second-order determinant that appears in each term is obtained by crossing out the row and the column containing the numerical factor.

$$a_1\begin{vmatrix} b_2 & c_2 \\ b_3 & c_3 \end{vmatrix} - a_2\begin{vmatrix} b_1 & c_1 \\ b_3 & c_3 \end{vmatrix} + a_3\begin{vmatrix} b_1 & c_1 \\ b_2 & c_2 \end{vmatrix}$$

$$\downarrow \qquad\qquad \downarrow \qquad\qquad \downarrow$$

$$\begin{vmatrix} \not{a_1} & \not{b_1} & \not{c_1} \\ \not{a_2} & b_2 & c_2 \\ \not{a_3} & b_3 & c_3 \end{vmatrix} \begin{vmatrix} \not{a_1} & b_1 & c_1 \\ \not{a_2} & \not{b_2} & \not{c_2} \\ \not{a_3} & b_3 & c_3 \end{vmatrix} \begin{vmatrix} \not{a_1} & b_1 & c_1 \\ \not{a_2} & b_2 & c_2 \\ \not{a_3} & \not{b_3} & \not{c_3} \end{vmatrix}$$

The **minor** of an element is the determinant that remains after deleting the row and column of that element. For this reason, we call this method **expansion by minors.**

EXAMPLE 3 Evaluating the Determinant of a 3 × 3 Matrix

Evaluate the determinant of

$$\begin{bmatrix} 4 & 1 & 0 \\ -9 & 3 & 4 \\ -3 & 8 & 1 \end{bmatrix}.$$

Solution We know that each of the three terms in the determinant contains a numerical factor and a second-order determinant. The numerical factors are from the first column of the determinant of the given matrix. They are highlighted in the following matrix:

$$\begin{vmatrix} 4 & 1 & 0 \\ -9 & 3 & 4 \\ -3 & 8 & 1 \end{vmatrix}$$

We find the minor for each numerical factor by deleting the row and column of that element:

$$\begin{bmatrix} 4 & 1 & 0 \\ -9 & 3 & 4 \\ -3 & 8 & 1 \end{bmatrix} \begin{bmatrix} 4 & 1 & 0 \\ -9 & 3 & 4 \\ -3 & 8 & 1 \end{bmatrix} \begin{bmatrix} 4 & 1 & 0 \\ -9 & 3 & 4 \\ -3 & 8 & 1 \end{bmatrix}$$

The minor for 4 is $\begin{vmatrix} 3 & 4 \\ 8 & 1 \end{vmatrix}$.

The minor for −9 is $\begin{vmatrix} 1 & 0 \\ 8 & 1 \end{vmatrix}$.

The minor for −3 is $\begin{vmatrix} 1 & 0 \\ 3 & 4 \end{vmatrix}$.

Now we have three numerical factors, 4, −9, and −3, and three second-order determinants. We multiply each numerical factor by its second-order determinant to find the three terms of the third-order determinant:

$$4\begin{vmatrix} 3 & 4 \\ 8 & 1 \end{vmatrix}, \quad -9\begin{vmatrix} 1 & 0 \\ 8 & 1 \end{vmatrix}, \quad -3\begin{vmatrix} 1 & 0 \\ 3 & 4 \end{vmatrix}.$$

Based on the preceding definition, we subtract the second term from the first term and add the third term:

Don't forget to supply the minus sign.

$$\begin{vmatrix} 4 & 1 & 0 \\ -9 & 3 & 4 \\ -3 & 8 & 1 \end{vmatrix} = 4\begin{vmatrix} 3 & 4 \\ 8 & 1 \end{vmatrix} - (-9)\begin{vmatrix} 1 & 0 \\ 8 & 1 \end{vmatrix} - 3\begin{vmatrix} 1 & 0 \\ 3 & 4 \end{vmatrix}$$

$$= 4(3 \cdot 1 - 8 \cdot 4) + 9(1 \cdot 1 - 8 \cdot 0) - 3(1 \cdot 4 - 3 \cdot 0)$$

$$= 4(3 - 32) + 9(1 - 0) - 3(4 - 0)$$

Evaluate the three second-order determinants.

$$= 4(-29) + 9(1) - 3(4)$$

$$= -119$$

Technology

Verify the result of Example 3 by using your graphing utility to enter the given matrix as

$$A = \begin{bmatrix} 4 & 1 & 0 \\ -9 & 3 & 4 \\ -3 & 8 & 1 \end{bmatrix}.$$

Then enter

det [A] ENTER.

The result should be −119.

Check Point 3 Evaluate the determinant of

$$\begin{bmatrix} 2 & 1 & 7 \\ -5 & 6 & 0 \\ -4 & 3 & 1 \end{bmatrix}.$$

The six terms in the definition of a third-order determinant can be rearranged and factored in a variety of ways. Thus, it is possible to expand a determinant by minors about any row or any column. *Minus signs must be supplied preceding any element appearing in a position where the sum of its row and its column is an odd number.* For example, expanding about the elements in column 2 gives us

$$\begin{vmatrix} a_1 & b_1 & c_1 \\ a_2 & b_2 & c_2 \\ a_3 & b_3 & c_3 \end{vmatrix} = -b_1\begin{vmatrix} a_2 & c_2 \\ a_3 & c_3 \end{vmatrix} + b_2\begin{vmatrix} a_1 & c_1 \\ a_3 & c_3 \end{vmatrix} - b_3\begin{vmatrix} a_1 & c_1 \\ a_2 & c_2 \end{vmatrix}.$$

Minus sign is supplied because b_1 appears in row 1 and column 2; $1+2=3$, an odd number.

Minus sign is supplied because b_3 appears in row 3 and column 2; $3+2=5$, an odd number.

Expanding by minors about column 3, we obtain

$$\begin{vmatrix} a_1 & b_1 & c_1 \\ a_2 & b_2 & c_2 \\ a_3 & b_3 & c_3 \end{vmatrix} = c_1\begin{vmatrix} a_2 & b_2 \\ a_3 & b_3 \end{vmatrix} - c_2\begin{vmatrix} a_1 & b_1 \\ a_3 & b_3 \end{vmatrix} + c_3\begin{vmatrix} a_1 & b_1 \\ a_2 & b_2 \end{vmatrix}.$$

Minus sign must be supplied because c_2 appears in row 2 and column 3; $2+3=5$, an odd number.

Study Tip

Keep in mind that you can expand a determinant by minors about any row or column. Use alternating plus and minus signs to precede the numerical factors of the minors according to the following sign array:

$$\begin{vmatrix} + & - & + \\ - & + & - \\ + & - & + \end{vmatrix}.$$

When evaluating a 3×3 determinant using expansion by minors, you can expand about any row or column. To simplify the arithmetic, if a row or column contains one or more 0s, expand about that row or column.

EXAMPLE 4 Evaluating a Third-Order Determinant

Evaluate:

$$\begin{vmatrix} 9 & 5 & 0 \\ -2 & -3 & 0 \\ 1 & 4 & 2 \end{vmatrix}.$$

Solution Note that the last column has two 0s. We will expand the determinant about the elements in that column.

$$\begin{vmatrix} 9 & 5 & 0 \\ -2 & -3 & 0 \\ 1 & 4 & 2 \end{vmatrix} = 0\begin{vmatrix} -2 & -3 \\ 1 & 4 \end{vmatrix} - 0\begin{vmatrix} 9 & 5 \\ 1 & 4 \end{vmatrix} + 2\begin{vmatrix} 9 & 5 \\ -2 & -3 \end{vmatrix}$$

$$= 0 - 0 + 2[9(-3) - (-2) \cdot 5]$$

Evaluate the second-order determinant whose numerical factor is not 0.

$$= 2(-27 + 10)$$

$$= 2(-17)$$

$$= -34$$

Check Point 4 Evaluate:

$$\begin{vmatrix} 6 & 4 & 0 \\ -3 & -5 & 3 \\ 1 & 2 & 0 \end{vmatrix}.$$

4 Solve a linear system of equations in three variables using Cramer's rule.

Solving Linear Systems of Equations in Three Variables Using Determinants

Cramer's rule can be applied to solving systems of linear equations in three variables. The determinants in the numerator and denominator of all variables are third-order determinants.

Solving Three Equations in Three Variables Using Determinants

Cramer's Rule

If

$$\begin{aligned} a_1x + b_1y + c_1z &= d_1 \\ a_2x + b_2y + c_2z &= d_2 \\ a_3x + b_3y + c_3z &= d_3 \end{aligned}$$

then

$$x = \frac{D_x}{D}, \; y = \frac{D_y}{D}, \text{ and } z = \frac{D_z}{D}.$$

These four third-order determinants are given by

$$D = \begin{vmatrix} a_1 & b_1 & c_1 \\ a_2 & b_2 & c_2 \\ a_3 & b_3 & c_3 \end{vmatrix}$$ These are the coefficients of the variables x, y, and z. $D \neq 0$.

$$D_x = \begin{vmatrix} d_1 & b_1 & c_1 \\ d_2 & b_2 & c_2 \\ d_3 & b_3 & c_3 \end{vmatrix}$$ Replace x-coefficients in D with the **constants at the right of the three equations.**

$$D_y = \begin{vmatrix} a_1 & d_1 & c_1 \\ a_2 & d_2 & c_2 \\ a_3 & d_3 & c_3 \end{vmatrix}$$ Replace y-coefficients in D with the **constants at the right of the three equations.**

$$D_z = \begin{vmatrix} a_1 & b_1 & d_1 \\ a_2 & b_2 & d_2 \\ a_3 & b_3 & d_3 \end{vmatrix}$$ Replace z-coefficients in D with the **constants at the right of the three equations.**

EXAMPLE 5 Using Cramer's Rule to Solve a Linear System in Three Variables

Use Cramer's rule to solve:

$$\begin{aligned} x + 2y - z &= -4 \\ x + 4y - 2z &= -6 \\ 2x + 3y + z &= 3. \end{aligned}$$

Solution Because

$$x = \frac{D_x}{D}, \quad y = \frac{D_y}{D}, \quad \text{and} \quad z = \frac{D_z}{D},$$

we need to set up and evaluate four determinants.

$$\begin{aligned} x + 2y - z &= -4 \\ x + 4y - 2z &= -6 \\ 2x + 3y + z &= 3 \end{aligned}$$

The linear system is shown again so that you do not need to turn back a page.

Step 1 Set up the determinants.

1. D, the determinant in all three denominators, consists of the x-, y-, and z-coefficients.

$$D = \begin{vmatrix} 1 & 2 & -1 \\ 1 & 4 & -2 \\ 2 & 3 & 1 \end{vmatrix}$$

2. D_x, the determinant in the numerator for x, is obtained by replacing the x-coefficients in D, 1, 1, and 2, with the constants on the right side of the equation, −4, −6, and 3.

$$D_x = \begin{vmatrix} -4 & 2 & -1 \\ -6 & 4 & -2 \\ 3 & 3 & 1 \end{vmatrix}$$

3. D_y, the determinant in the numerator for y, is obtained by replacing the y-coefficients in D, 2, 4, and 3, with the constants on the right side of the equation, −4, −6, and 3.

$$D_y = \begin{vmatrix} 1 & -4 & -1 \\ 1 & -6 & -2 \\ 2 & 3 & 1 \end{vmatrix}$$

4. D_z, the determinant in the numerator for z, is obtained by replacing the z-coefficients in D, −1, −2, and 1, with the constants on the right side of the equation, −4, −6, and 3.

$$D_z = \begin{vmatrix} 1 & 2 & -4 \\ 1 & 4 & -6 \\ 2 & 3 & 3 \end{vmatrix}$$

Step 2 Evaluate the four determinants.

$$D = \begin{vmatrix} 1 & 2 & -1 \\ 1 & 4 & -2 \\ 2 & 3 & 1 \end{vmatrix} = 1\begin{vmatrix} 4 & -2 \\ 3 & 1 \end{vmatrix} - 1\begin{vmatrix} 2 & -1 \\ 3 & 1 \end{vmatrix} + 2\begin{vmatrix} 2 & -1 \\ 4 & -2 \end{vmatrix}$$

$$= 1(4 + 6) - 1(2 + 3) + 2(-4 + 4)$$

$$= 1(10) - 1(5) + 2(0) = 5$$

Using the same technique to evaluate each determinant, we obtain

$$D_x = -10, \quad D_y = 5, \quad \text{and} \quad D_z = 20.$$

Step 3 Substitute these four values and solve the system.

$$x = \frac{D_x}{D} = \frac{-10}{5} = -2$$

$$y = \frac{D_y}{D} = \frac{5}{5} = 1$$

$$z = \frac{D_z}{D} = \frac{20}{5} = 4$$

The solution $(-2, 1, 4)$ can be checked by substitution into the original three equations. The solution set is $\{(-2, 1, 4)\}$.

Check Point 5 Use Cramer's rule to solve the system:

$$\begin{aligned} 3x - 2y + z &= 16 \\ 2x + 3y - z &= -9. \\ x + 4y + 3z &= 2 \end{aligned}$$

5 Use determinants to identify inconsistent systems and systems with dependent equations.

Cramer's Rule with Inconsistent and Dependent Systems

If D, the determinant in the denominator, is 0, the variables described by the quotient of determinants are not real numbers. However, when $D = 0$, this indicates that the system is inconsistent or contains dependent equations. This gives rise to the following two situations.

Discovery

Write a system of two equations that is inconsistent. Now use determinants and the result boxed on the right to verify that this is truly an inconsistent system. Repeat the same process for a system with two dependent equations.

Determinants: Inconsistent and Dependent-Systems

1. If $D = 0$ and at least one of the determinants in the numerator is not 0, then the system is inconsistent. The solution set is $\varnothing$.
2. If $D = 0$ and all the determinants in the numerators are 0, then the equations in the system are dependent.

Although we have focused on applying determinants to solve linear systems, they have other applications, some of which we consider in the exercise set that follows.

6 Evaluate higher-order determinants.

The Determinant of Any $n \times n$ Matrix

A determinant with n rows and n columns is said to be an ***n*th-order determinant**. The value of an nth-order determinant $(n > 2)$ can be found in terms of determinants of order $n - 1$. For example, we found the value of a third-order determinant in terms of determinants of order 2.

We can generalize this idea for fourth-order determinants and higher. We have seen that the **minor** of the element a_{ij} is the determinant obtained by deleting the ith row and the jth column in the given array of numbers. The **cofactor** of the element a_{ij} is $(-1)^{i+j}$ times the minor of the a_{ij}th entry. If the sum of the row and column $(i + j)$ is even, the cofactor is the same as the minor. If the sum of the row and column $(i + j)$ is odd, the cofactor is the opposite of the minor.

Let's see what this means in the case of a fourth-order determinant.

EXAMPLE 6 Evaluating the Determinant of a 4 × 4 Matrix

Evaluate the determinant of

$$A = \begin{bmatrix} 1 & -2 & 3 & 0 \\ -1 & 1 & 0 & 2 \\ 0 & 2 & 0 & -3 \\ 2 & 3 & -4 & 1 \end{bmatrix}.$$

Solution

$$|A| = \begin{vmatrix} 1 & -2 & 3 & 0 \\ -1 & 1 & 0 & 2 \\ 0 & 2 & 0 & -3 \\ 2 & 3 & -4 & 1 \end{vmatrix}$$

With two 0s in the third column, we will expand along the third column.

$$= (-1)^{1+3}3\begin{vmatrix} -1 & 1 & 2 \\ 0 & 2 & -3 \\ 2 & 3 & 1 \end{vmatrix} + (-1)^{4+3}(-4)\begin{vmatrix} 1 & -2 & 0 \\ -1 & 1 & 2 \\ 0 & 2 & -3 \end{vmatrix}$$

3 is in row 1, column 3.

−4 is in row 4, column 3.

$$= 3\begin{vmatrix} -1 & 1 & 2 \\ 0 & 2 & -3 \\ 2 & 3 & 1 \end{vmatrix} + 4\begin{vmatrix} 1 & -2 & 0 \\ -1 & 1 & 2 \\ 0 & 2 & -3 \end{vmatrix}$$

The determinant that follows 3 is obtained by crossing out the row and the column (row 1, column 3) in the original determinant. The minor for −4 is obtained in the same manner.

Evaluate the two third-order determinants to get

$$|A| = 3(-25) + 4(-1) = -79.$$

Why Modern Software Packages for Solving Linear Systems Do Not Use Cramer's Rule

The fastest supercomputers can perform one trillion (10^{12}) multiplications per second. To solve a linear system with a "mere" 20 equations using Cramer's rule requires over 5×10^{19} multiplications. This would take a supercomputer more than 590 days. The cost for this venture? At $2200 per hour, typical of the costs for supercomputer time, the computations required by Cramer's rule would cost more than $30 million!

Check Point 6 Evaluate the determinant of

$$A = \begin{bmatrix} 0 & 4 & 0 & -3 \\ -1 & 1 & 5 & 2 \\ 1 & -2 & 0 & 6 \\ 3 & 0 & 0 & 1 \end{bmatrix}.$$

If a linear system has n equations, Cramer's rule requires you to compute $n + 1$ determinants of nth order. The excessive number of calculations required to perform Cramer's rule for systems with four or more equations makes it an inefficient method for solving large systems.

EXERCISE SET 6.5

Practice Exercises

Evaluate each determinant in Exercises 1–10.

1. $\begin{vmatrix} 5 & 7 \\ 2 & 3 \end{vmatrix}$

2. $\begin{vmatrix} 4 & 8 \\ 5 & 6 \end{vmatrix}$

3. $\begin{vmatrix} -4 & 1 \\ 5 & 6 \end{vmatrix}$

4. $\begin{vmatrix} 7 & 9 \\ -2 & -5 \end{vmatrix}$

5. $\begin{vmatrix} -7 & 14 \\ 2 & -4 \end{vmatrix}$

6. $\begin{vmatrix} 1 & -3 \\ -8 & 2 \end{vmatrix}$

7. $\begin{vmatrix} -5 & -1 \\ -2 & -7 \end{vmatrix}$

8. $\begin{vmatrix} \frac{1}{5} & \frac{1}{6} \\ -6 & 5 \end{vmatrix}$

9. $\begin{vmatrix} \frac{1}{2} & \frac{1}{2} \\ \frac{1}{8} & -\frac{3}{4} \end{vmatrix}$

10. $\begin{vmatrix} \frac{2}{3} & \frac{1}{3} \\ -\frac{1}{2} & \frac{3}{4} \end{vmatrix}$

For Exercises 11–26, use Cramer's rule to solve each system or to determine that the system is inconsistent or contains dependent equations.

11. $x + y = 7$
    $x - y = 3$

12. $2x + y = 3$
    $x - y = 3$

13. $12x + 3y = 15$
$2x - 3y = 13$

14. $x - 2y = 5$
$5x - y = -2$

15. $4x - 5y = 17$
$2x + 3y = 3$

16. $3x + 2y = 2$
$2x + 2y = 3$

17. $x + 2y = 3$
$5x + 10y = 15$

18. $2x - 9y = 5$
$3x - 3y = 11$

19. $3x - 4y = 4$
$2x + 2y = 12$

20. $3x = 7y + 1$
$2x = 3y - 1$

21. $2x = 3y + 2$
$5x = 51 - 4y$

22. $x + 2y - 3 = 0$
$12 = 8y + 4x$

23. $3x = 2 - 3y$
$2y = 3 - 2x$

24. $y = -4x + 2$
$2x = 3y + 8$

25. $4y = 16 - 3x$
$5x = 12 - 3y$

26. $2x = 7 + 3y$
$4x - 6y = 3$

Evaluate each determinant in Exercises 27–32.

27. $\begin{vmatrix} 3 & 0 & 0 \\ 2 & 1 & -5 \\ 2 & 5 & -1 \end{vmatrix}$

28. $\begin{vmatrix} 4 & 0 & 0 \\ 3 & -1 & 4 \\ 2 & -3 & 5 \end{vmatrix}$

29. $\begin{vmatrix} 3 & 1 & 0 \\ -3 & 4 & 0 \\ -1 & 3 & -5 \end{vmatrix}$

30. $\begin{vmatrix} 2 & -4 & 2 \\ -1 & 0 & 5 \\ 3 & 0 & 4 \end{vmatrix}$

31. $\begin{vmatrix} 1 & 1 & 1 \\ 2 & 2 & 2 \\ -3 & 4 & -5 \end{vmatrix}$

32. $\begin{vmatrix} 1 & 2 & 3 \\ 2 & 2 & -3 \\ 3 & 2 & 1 \end{vmatrix}$

In Exercises 33–40, use Cramer's rule to solve each system.

33. $x + y + z = 0$
$2x - y + z = -1$
$-x + 3y - z = -8$

34. $x - y + 2z = 3$
$2x + 3y + z = 9$
$-x - y + 3z = 11$

35. $4x - 5y - 6z = -1$
$x - 2y - 5z = -12$
$2x - y = 7$

36. $x - 3y + z = -2$
$x + 2y = 8$
$2x - y = 1$

37. $x + y + z = 4$
$x - 2y + z = 7$
$x + 3y + 2z = 4$

38. $2x + 2y + 3z = 10$
$4x - y + z = -5$
$5x - 2y + 6z = 1$

39. $x + 2z = 4$
$2y - z = 5$
$2x + 3y = 13$

40. $3x + 2z = 4$
$5x - y = -4$
$4y + 3z = 22$

Evaluate each determinant in Exercises 41–44.

41. $\begin{vmatrix} 4 & 2 & 8 & -7 \\ -2 & 0 & 4 & 1 \\ 5 & 0 & 0 & 5 \\ 4 & 0 & 0 & -1 \end{vmatrix}$

42. $\begin{vmatrix} 3 & -1 & 1 & 2 \\ -2 & 0 & 0 & 0 \\ 2 & -1 & -2 & 3 \\ 1 & 4 & 2 & 3 \end{vmatrix}$

43. $\begin{vmatrix} -2 & -3 & 3 & 5 \\ 1 & -4 & 0 & 0 \\ 1 & 2 & 2 & -3 \\ 2 & 0 & 1 & 1 \end{vmatrix}$

44. $\begin{vmatrix} 1 & -3 & 2 & 0 \\ -3 & -1 & 0 & -2 \\ 2 & 1 & 3 & 1 \\ 2 & 0 & -2 & 0 \end{vmatrix}$

Application Exercises

Determinants are used to find the area of a triangle whose vertices are given by three points in a rectangular coordinate system. The area of a triangle with vertices (x_1, y_1), (x_2, y_2), and (x_3, y_3) is

$$\text{Area} = \pm\frac{1}{2}\begin{vmatrix} x_1 & y_1 & 1 \\ x_2 & y_2 & 1 \\ x_3 & y_3 & 1 \end{vmatrix}$$

where the symbol ($\pm$) indicates that the appropriate sign should be chosen to yield a positive area. Use this information to work Exercises 45–46.

45. **a.** Use determinants to find the area of the triangle whose vertices are $(3, -5)$, $(2, 6)$, and $(-3, 5)$.
b. Graph the triangle in part (a) and then confirm your answer by using the formula for a triangle's area, $A = \frac{1}{2}bh$.

46. Find the area of the triangle whose vertices are $(1, 1)$, $(-2, -3)$, and $(11, -3)$.

Determinants are used to show that three points lie on the same line (are collinear). If

$$\begin{vmatrix} x_1 & y_1 & 1 \\ x_2 & y_2 & 1 \\ x_3 & y_3 & 1 \end{vmatrix} = 0$$

then the points (x_1, y_1), (x_2, y_2), and (x_3, y_3) are collinear. If the determinant does not equal 0, then the points are not collinear. Use this information to work Exercises 47–48.

47. Are the points $(3, -1)$, $(0, -3)$, and $(12, 5)$ collinear?

48. Are the points $(-4, -6)$, $(1, 0)$, and $(11, 12)$ collinear?

Determinants are used to write an equation of a line passing through two points. An equation of the line passing through the distinct points (x_1, y_1) and (x_2, y_2) is given by

$$\begin{vmatrix} x & y & 1 \\ x_1 & y_1 & 1 \\ x_2 & y_2 & 1 \end{vmatrix} = 0.$$

Use this information to work Exercises 49–50.

49. Use the determinant to write an equation for the line passing through $(3, -5)$ and $(-2, 6)$. Then expand the determinant, expressing the line's equation in slope-intercept form.

50. Use the determinant to write an equation for the line passing through $(-1, 3)$ and $(2, 4)$. Then expand the determinant, expressing the line's equation in slope-intercept form.

Writing in Mathematics

51. Explain how to evaluate a second-order determinant.

52. Describe the determinants D_x and D_y in terms of the coefficients and constants in a system of two equations in two variables.

53. Explain how to evaluate a third-order determinant.

54. When expanding a determinant by minors, when is it necessary to supply minus signs?

55. Without going into too much detail, describe how to solve a linear system in three variables using Cramer's rule.

56. In applying Cramer's rule, what does it mean if $D = 0$?

57. The process of solving a linear system in three variables using Cramer's rule can involve tedious computation. Is there a way of speeding up this process, perhaps using Cramer's rule to find the value for only one of the variables? Describe how this process might work, presenting a specific example with your description. Remember that your goal is still to find the value for each variable in the system.

58. If you could use only one method to solve linear systems in three variables, which method would you select? Explain why this is so.

Technology Exercises

59. Use the feature of your graphing utility that evaluates the determinant of a square matrix to verify any five of the determinants that you evaluated by hand in Exercises 1–10, 27–32, or 41–44.

In Exercises 60–61, use a graphing utility to evaluate the determinant for the given matrix.

60. $\begin{bmatrix} 3 & -2 & -1 & 4 \\ -5 & 1 & 2 & 7 \\ 2 & 4 & 5 & 0 \\ -1 & 3 & -6 & 5 \end{bmatrix}$

61. $\begin{bmatrix} 8 & 2 & 6 & -1 & 0 \\ 2 & 0 & -3 & 4 & 7 \\ 2 & 1 & -3 & 6 & -5 \\ -1 & 2 & 1 & 5 & -1 \\ 4 & 5 & -2 & 3 & -8 \end{bmatrix}$

62. What is the fastest method for solving a linear system with your graphing utility?

Critical Thinking Exercises

63. a. Evaluate: $\begin{vmatrix} a & a \\ 0 & a \end{vmatrix}$.

b. Evaluate: $\begin{vmatrix} a & a & a \\ 0 & a & a \\ 0 & 0 & a \end{vmatrix}$.

c. Evaluate: $\begin{vmatrix} a & a & a & a \\ 0 & a & a & a \\ 0 & 0 & a & a \\ 0 & 0 & 0 & a \end{vmatrix}$.

d. Describe the pattern in the given determinants.

e. Describe the pattern in the evaluations.

64. Evaluate: $\begin{vmatrix} 2 & 0 & 0 & 0 & 0 \\ 0 & 3 & 0 & 0 & 0 \\ 0 & 0 & 2 & 0 & 0 \\ 0 & 0 & 0 & 1 & 0 \\ 0 & 0 & 0 & 0 & 4 \end{vmatrix}$.

65. What happens to the value of a second-order determinant if the two columns are interchanged?

66. Consider the system

$$a_1x + b_1y = c_1$$
$$a_2x + b_2y = c_2.$$

Use Cramer's rule to prove that if the first equation of the system is replaced by the sum of the two equations, the resulting system has the same solution as the original system.

Group Exercise

67. We have seen that determinants can be used to solve linear equations, give areas of triangles in rectangular coordinates, and determine equations of lines. Not impressed with these applications? Members of the group should research an application of determinants that they find intriguing. The group should then present a seminar to the class about this application.

CHAPTER SUMMARY, REVIEW, AND TEST

Summary

6.1 Matrix Solution to Linear Systems

a. Matrix row operations are described in the box on page 484.

b. To solve a linear system using Gaussian elimination, begin with the system's augmented matrix. Use matrix operations to get 1s down the main diagonal and 0s below the 1s. Details are in the box on page 485.

c. To solve a linear system using Gauss-Jordan elimination, use the procedure of Gaussian elimination, but obtain 0s above and below the 1s in the main diagonal.

6.2 Inconsistent and Dependent Systems

a. If Gaussian elimination results in a matrix with a row containing all 0s to the left of the vertical line and a nonzero number to the right, the system has no solution (is inconsistent).

b. If Gaussian elimination results in a matrix with a row with all 0s, the system has an infinite number of solutions (contains dependent equations).

6.3 Matrix Operations

a. Two matrices are equal if and only if they have the same order and corresponding elements are equal.

b. Matrix Addition and Subtraction: Matrices of the same order are added or subtracted by adding or subtracting corresponding elements. Properties of matrix addition are given in the box on page 508.

c. Scalar Multiplication: If A is a matrix and c is a scalar, then cA is the matrix formed by multiplying each element in A by c. Properties of scalar multiplication are given in the box on page 509.

d. Matrix Multiplication: The product of an $m \times n$ matrix A and an $n \times p$ matrix B is an $m \times p$ matrix AB. The element in the ith row and jth column of AB is found by multiplying each element in the ith row of A by the corresponding element in the jth column of B and adding the products. Matrix multiplication is not commutative: $AB \neq BA$. Properties of matrix multiplication are given in the box on page 514.

6.4 Multiplicative Inverses of Matrices; Matrix Equations

a. The multiplicative identity matrix I_n is an $n \times n$ matrix with 1s down the main diagonal and 0s elsewhere.

b. Let A be an $n \times n$ square matrix. If there is a square matrix A^{-1} such that $AA^{-1} = I_n$ and $A^{-1}A = I_n$, then A^{-1} is the multiplicative inverse of A.

c. If a square matrix has a multiplicative inverse, it is invertible. Methods for finding multiplicative inverses for invertible matrices, including a formula for 2×2 matrices, are given in the box on page 527.

d. Linear systems can be represented by matrix equations $AX = B$ in which A is the coefficient matrix and B is the constant matrix. If $AX = B$ has a unique solution, then $X = A^{-1}B$.

6.5 Determinants and Cramer's Rule

a. Value of a Second-Order Determinant:

$$\begin{vmatrix} a_1 & b_1 \\ a_2 & b_2 \end{vmatrix} = a_1b_2 - a_2b_1$$

b. Cramer's rule for solving linear systems in two variables uses three second-order determinants and is stated in the box on page 536.

c. To evaluate an nth-order determinant, where $n > 2$,

1. Select a row or column about which to expand.
2. For each element a_{ij} in the row or column, multiply by $(-1)^{i+j}$ times the determinant obtained by deleting the ith row and the jth column in the given array of numbers.
3. The value of the determinant is the sum of the products found in step 2.

d. Cramer's rule for solving linear systems in three variables uses four third-order determinants and is stated in the box on page 541.

e. Cramer's rule with inconsistent and dependent systems is summarized by the two situations in the box on page 543.

Review Exercises

6.1

In Exercises 1–2, write the system of linear equations represented by the augmented matrix. Use x, y, z, and, if necessary, w for the variables. Once the system is written, use back-substitution to find its solution.

1. $\left[\begin{array}{ccc|c} 1 & 1 & 3 & 12 \\ 0 & 1 & -2 & -4 \\ 0 & 0 & 1 & 3 \end{array}\right]$

2. $\left[\begin{array}{cccc|c} 1 & 0 & -2 & 2 & 1 \\ 0 & 1 & 1 & -1 & 0 \\ 0 & 0 & 1 & -\frac{7}{3} & -\frac{1}{3} \\ 0 & 0 & 0 & 1 & 1 \end{array}\right]$

In Exercises 3–4, perform each matrix row operation and write the new matrix.

3. $\left[\begin{array}{ccc|c} 1 & 2 & 2 & 2 \\ 0 & 1 & -1 & 2 \\ 0 & 5 & 4 & 1 \end{array}\right]$ Multiply row 2 by −5 and add to corresponding entries in row 3.

4. $\left[\begin{array}{ccc|c} 2 & -2 & 1 & -1 \\ 1 & 2 & -1 & 2 \\ 6 & 4 & 3 & 5 \end{array}\right]$ Multiply row 1 by $\frac{1}{2}$.

In Exercises 5–7, solve each system of equations using matrices. Use Gaussian elimination with back-substitution or Gauss-Jordan elimination.

5.
$$\begin{aligned} x + 2y + 3z &= -5 \\ 2x + y + z &= 1 \\ x + y - z &= 8 \end{aligned}$$

6.
$$\begin{aligned} x - 2y + z &= 0 \\ y - 3z &= -1 \\ 2y + 5z &= -2 \end{aligned}$$

7.
$$\begin{aligned} 3x_1 + 5x_2 - 8x_3 + 5x_4 &= -8 \\ x_1 + 2x_2 - 3x_3 + x_4 &= -7 \\ 2x_1 + 3x_2 - 7x_3 + 3x_4 &= -11 \\ 4x_1 + 8x_2 - 10x_3 + 7x_4 &= -10 \end{aligned}$$

8. The table shows the pollutants in the air in a city on a typical summer day.

x (Hours after 6 A.M.)	y (Amount of Pollutants in the Air, in parts per million)
2	98
4	138
10	162

a. Use the function $y = ax^2 + bx + c$ to model the data. Use either Gaussian elimination with back-substitution or Gauss-Jordan elimination to find the values for a, b, and c.

b. Use the function to find the time of day at which the city's air pollution level is at a maximum. What is the maximum level?

6.2

In Exercises 9–12, use Gaussian elimination to find the complete solution to each system, or show that none exists.

9.
$$\begin{aligned} 2x - 3y + z &= 1 \\ x - 2y + 3z &= 2 \\ 3x - 4y - z &= 1 \end{aligned}$$

10.
$$\begin{aligned} x - 3y + z &= 1 \\ -2x + y + 3z &= -7 \\ x - 4y + 2z &= 0 \end{aligned}$$

11.
$$\begin{aligned} x_1 + 4x_2 + 3x_3 - 6x_4 &= 5 \\ x_1 + 3x_2 + x_3 - 4x_4 &= 3 \\ 2x_1 + 8x_2 + 7x_3 - 5x_4 &= 11 \\ 2x_1 + 5x_2 - 6x_4 &= 4 \end{aligned}$$

12.
$$\begin{aligned} 2x + 3y - 5z &= 15 \\ x + 2y - z &= 4 \end{aligned}$$

13. The figure shows the intersections of three one-way streets. The numbers given represent traffic flow in cars per hour at a peak period (from 4 P.M. to 6 P.M.).

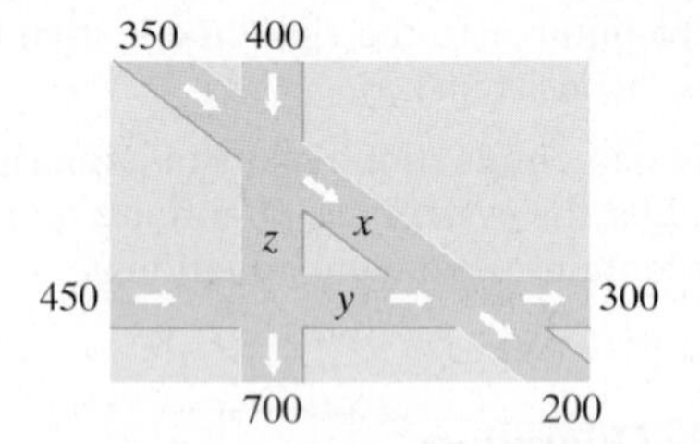

a. Use the idea that the number of cars entering each intersection per hour must equal the number of cars leaving per hour to set up a linear system of equations involving x, y, and z.

b. Use Gaussian elimination to solve the system.

c. If construction limits the value of z to 400, how many cars per hour must pass between the other intersections to keep traffic flowing?

6.3

14. Find values for x, y, and z so that the following matrices are equal:

$$\begin{bmatrix} 2x & y+7 \\ z & 4 \end{bmatrix} = \begin{bmatrix} -10 & 13 \\ 6 & 4 \end{bmatrix}.$$

In Exercises 15–28, perform the indicated matrix operations given that A, B, C, and D are defined as follows. If an operation is not defined, state the reason.

$$A = \begin{bmatrix} 2 & -1 & 2 \\ 5 & 3 & -1 \end{bmatrix}, \quad B = \begin{bmatrix} 0 & -2 \\ 3 & 2 \\ 1 & -5 \end{bmatrix},$$

$$C = \begin{bmatrix} 1 & 2 & 3 \\ -1 & 1 & 2 \\ -1 & 2 & 1 \end{bmatrix}, \quad \text{and} \quad D = \begin{bmatrix} -2 & 3 & 1 \\ 3 & -2 & 4 \end{bmatrix}$$

15. $A + D$

16. $2B$

17. $D - A$

18. $B + C$

19. $3A + 2D$

20. $-2A + 4D$

21. $-5(A + D)$

22. AB

23. BA

24. BD

25. DB

26. $AB - BA$

27. $(A - D)C$

28. $B(AC)$

In Exercises 29–30, use nine pixels in a 3×3 grid and the color levels shown.

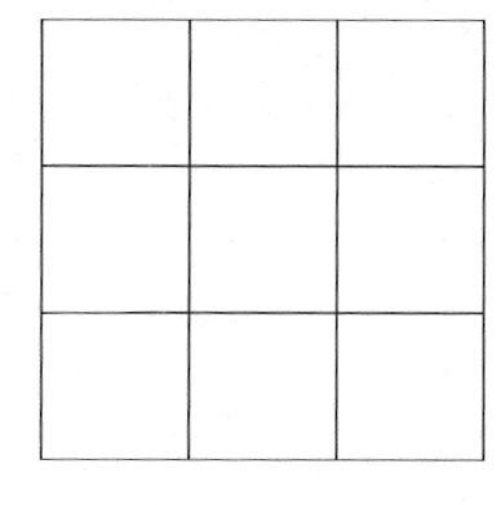

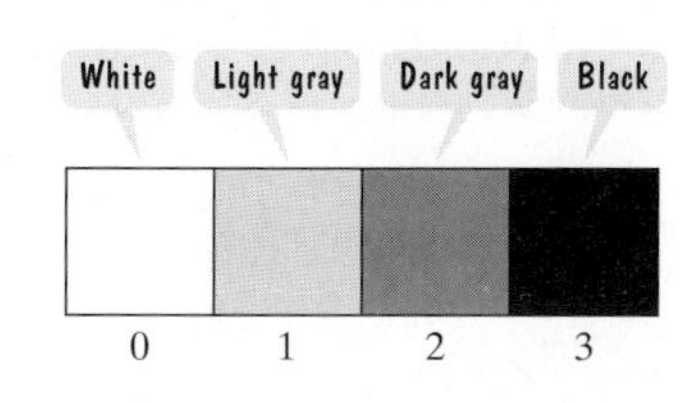

29. Write a 3×3 matrix that represents a digital photograph of the letter L in dark gray on a light gray background.

30. Find a matrix B so that $A + B$ increases the contrast of the letter L by changing the dark gray to black and the light gray to white.

31. An automobile dealership sells three models of cars at its three outlets. The inventory of models at each store is given by the following matrix.

	Model X	Model Y	Model Z	
Outlet 1	12	7	6	
Outlet 2	20	8	10	$= A$
Outlet 3	7	2	3	

The next matrix gives the wholesale and retail prices for each model.

	Wholesale Price	Retail Price	
Model X	16,000	19,000	
Model Y	12,000	15,000	$= B$
Model Z	14,000	18,500	

a. Calculate the product AB.

b. Describe what the matrix AB represents and interpret the elements.

c. What is the wholesale value of the cars at outlet 1?

d. What is the retail value of the cars at outlet 2?

e. If outlet 3 sells all of the inventory in matrix A, what is the profit for that branch of the dealership?

6.4

In Exercises 32–33, find the products AB and BA to determine whether B is the multiplicative inverse of A.

32. $A = \begin{bmatrix} 2 & 7 \\ 1 & 4 \end{bmatrix}, \quad B = \begin{bmatrix} 4 & -7 \\ -1 & 3 \end{bmatrix}$

33. $A = \begin{bmatrix} 1 & 0 & 0 \\ 0 & 2 & -7 \\ 0 & -1 & 4 \end{bmatrix}, \quad B = \begin{bmatrix} 1 & 0 & 0 \\ 0 & 4 & 7 \\ 0 & 1 & 2 \end{bmatrix}$

In Exercises 34–37, find A^{-1}. Check that $AA^{-1} = I$ and $A^{-1}A = I$.

34. $A = \begin{bmatrix} 1 & -1 \\ -2 & 3 \end{bmatrix}$

35. $A = \begin{bmatrix} 0 & 1 \\ 5 & 3 \end{bmatrix}$

36. $A = \begin{bmatrix} 1 & 0 & -2 \\ 2 & 1 & 0 \\ 1 & 0 & -3 \end{bmatrix}$

37. $A = \begin{bmatrix} 1 & 3 & -2 \\ 4 & 13 & -7 \\ 5 & 16 & -8 \end{bmatrix}$

In Exercises 38–39,

a. *Write each linear system as a matrix equation in the form $AX = B$.*

b. *Solve the system using the inverse of the coefficient matrix.*

38. $\begin{aligned} x + y + 2z &= 7 \\ y + 3z &= -2 \\ 3x \quad - 2z &= 0 \end{aligned}$

The inverse of $\begin{bmatrix} 1 & 1 & 2 \\ 0 & 1 & 3 \\ 3 & 0 & -2 \end{bmatrix}$ is $\begin{bmatrix} -2 & 2 & 1 \\ 9 & -8 & -3 \\ -3 & 3 & 1 \end{bmatrix}$.

39. $\begin{aligned} x - y + 2z &= 12 \\ y - z &= -5 \\ x \quad + 2z &= 10 \end{aligned}$

The inverse of $\begin{bmatrix} 1 & -1 & 2 \\ 0 & 1 & -1 \\ 1 & 0 & 2 \end{bmatrix}$ is $\begin{bmatrix} 2 & 2 & -1 \\ -1 & 0 & 1 \\ -1 & -1 & 1 \end{bmatrix}$.

40. Use the coding-matrix, $A = \begin{bmatrix} 3 & 2 \\ 4 & 3 \end{bmatrix}$ and its inverse $A^{-1} = \begin{bmatrix} 3 & -2 \\ -4 & 3 \end{bmatrix}$ to encode and then decode the word RULE.

6.5

In Exercises 41–46, evaluate each determinant.

41. $\begin{vmatrix} 3 & 2 \\ -1 & 5 \end{vmatrix}$

42. $\begin{vmatrix} -2 & -3 \\ -4 & -8 \end{vmatrix}$

43. $\begin{vmatrix} 2 & 4 & -3 \\ 1 & -1 & 5 \\ -2 & 4 & 0 \end{vmatrix}$

44. $\begin{vmatrix} 4 & 7 & 0 \\ -5 & 6 & 0 \\ 3 & 2 & -4 \end{vmatrix}$

45. $\begin{vmatrix} 1 & 1 & 0 & 2 \\ 0 & 3 & 2 & 1 \\ 0 & -2 & 4 & 0 \\ 0 & 3 & 0 & 1 \end{vmatrix}$

46. $\begin{vmatrix} 2 & 2 & 2 & 2 \\ 0 & 2 & 2 & 2 \\ 0 & 0 & 2 & 2 \\ 0 & 0 & 0 & 2 \end{vmatrix}$

In Exercises 47–50, use Cramer's rule to solve each system.

47. $\begin{aligned} x - 2y &= 8 \\ 3x + 2y &= -1 \end{aligned}$

48. $\begin{aligned} 7x + 2y &= 0 \\ 2x + y &= -3 \end{aligned}$

49. $\begin{aligned} x + 2y + 2z &= 5 \\ 2x + 4y + 7z &= 19 \\ -2x - 5y - 2z &= 8 \end{aligned}$

50. $\begin{aligned} 2x + y &= -4 \\ y - 2z &= 0 \\ 3x - 2z &= -11 \end{aligned}$

51. Use the quadratic function $y = ax^2 + bx + c$ to model the following data:

x (Age of a Driver)	y (Average Number of Automobile Accidents per Day in the United States)
20	400
40	150
60	400

Use Cramer's rule to determine values for a, b, and c. Then use the model to write a statement about the average number of automobile accidents in which 30-year-olds and 50-year-olds are involved daily.

Chapter 6 Test

In Exercises 1–2, solve each system of equations using matrices.

1. $\begin{aligned} x + 2y - z &= -3 \\ 2x - 4y + z &= -7 \\ -2x + 2y - 3z &= 4 \end{aligned}$

2. $\begin{aligned} x - 2y + z &= 2 \\ 2x - y - z &= 1 \end{aligned}$

In Exercises 3–6, let

$$A = \begin{bmatrix} 3 & 1 \\ 1 & 0 \\ 2 & 1 \end{bmatrix}, \quad B = \begin{bmatrix} 1 & -1 \\ 2 & 1 \end{bmatrix}, \text{ and } C = \begin{bmatrix} 1 & 2 \\ -1 & 3 \end{bmatrix}.$$

Carry out the indicated operations.

3. $2B + 3C$

4. AB

5. C^{-1}

6. $BC - 3B$

7. If $A = \begin{bmatrix} 1 & 2 & 2 \\ 2 & 3 & 3 \\ 1 & -1 & -2 \end{bmatrix}$ and $B = \begin{bmatrix} -3 & 2 & 0 \\ 7 & -4 & 1 \\ -5 & 3 & -1 \end{bmatrix}$, show that B is the inverse of A.

8. Consider the system

$$\begin{aligned} 3x + 5y &= 9 \\ 2x - 3y &= -13. \end{aligned}$$

a. Express the system in the form $AX = B$, where A, X, and B are appropriate matrices.

b. Find A^{-1}, the inverse of the coefficient matrix.

c. Use A^{-1} to solve the given system.

9. Evaluate: $\begin{vmatrix} 4 & -1 & 3 \\ 0 & 5 & -1 \\ 5 & 2 & 4 \end{vmatrix}$.

10. Solve for x only using Cramer's rule:

$$\begin{aligned} 3x + y - 2z &= -3 \\ 2x + 7y + 3z &= 9 \\ 4x - 3y - z &= 7. \end{aligned}$$

Cumulative Review Exercises (Chapters 1–6)

Solve each equation or inequality in Exercises 1–6.

1. $2x^2 = 4 - x$

2. $5x + 8 \le 7(1 + x)$

3. $\sqrt{2x + 4} - \sqrt{x + 3} - 1 = 0$

4. $3x^3 + 8x^2 - 15x + 4 = 0$

5. $e^{2x} - 14e^x + 45 = 0$

6. $\log_3 x + \log_3(x + 2) = 1$

7. Use matrices to solve this system.

$$\begin{aligned} x - y + z &= 17 \\ 2x + 3y + z &= 8 \\ -4x + y + 5z &= -2 \end{aligned}$$

8. Solve for y using Cramer's rule.

$$\begin{aligned} x - 2y + z &= 7 \\ 2x + y - z &= 0 \\ 3x + 2y - 2z &= -2 \end{aligned}$$

9. If $f(x) = \sqrt{4x - 7}$, find $f^{-1}(x)$.

10. Graph: $f(x) = \dfrac{x}{x^2 - 16}$.

11. Use the graph of $f(x) = 4x^4 - 4x^3 - 25x^2 + x + 6$ shown in the figure to factor the polynomial completely.

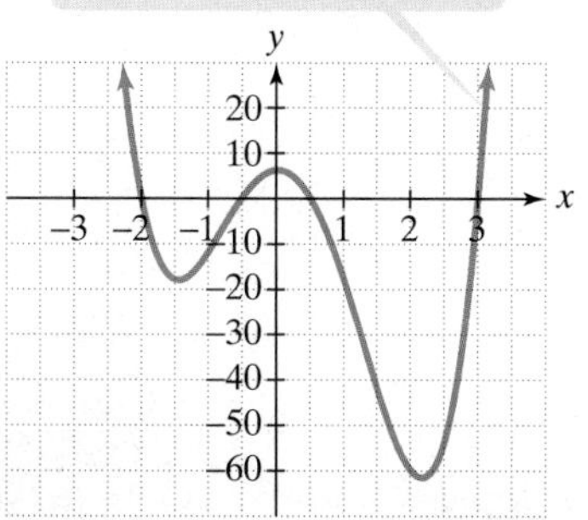

12. Graph $y = \log_2 x$ and $y = \log_2(x + 1)$ in the same rectangular coordinate system.

13. Use the exponential decay model $A = A_0e^{kt}$ to solve this problem. A radioactive substance has a half-life of 40 days. There are initially 900 grams of the substance.

a. Find the decay model for this substance.

b. How much of the substance will remain after 10 days?

14. Multiply the matrices: $\begin{bmatrix} 1 & -1 & 0 \\ 2 & 1 & 3 \end{bmatrix}\begin{bmatrix} 4 & -1 \\ 2 & 0 \\ 1 & 1 \end{bmatrix}$.

15. Find the partial fraction decomposition of

$$\frac{3x^2 + 17x - 38}{(x - 3)(x - 2)(x + 2)}.$$

In Exercises 16–19, graph each equation, function, or inequality in the rectangular coordinate system.

16. $y = -\frac{2}{3}x - 1$

17. $3x - 5y < 15$

18. $f(x) = x^2 - 2x - 3$

19. $(x - 1)^2 + (y + 1)^2 = 9$

20. Use synthetic division to divide $x^3 - 6x + 4$ by $x - 2$.

Conic Sections

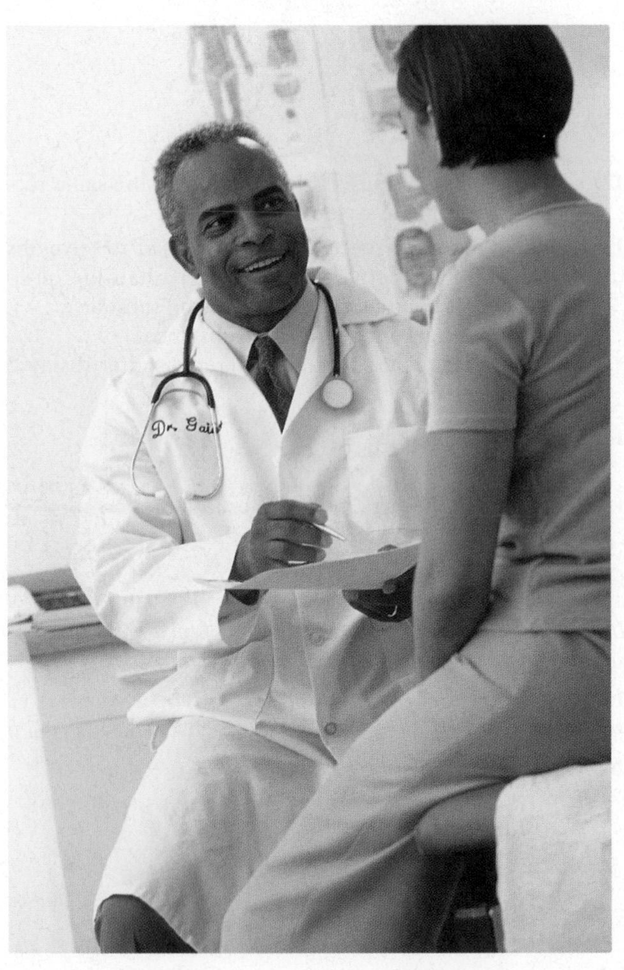

From ripples in water to the path on which humanity journeys through space, certain curves occur naturally throughout the universe. Over 2000 years ago the ancient Greeks studied these curves, called *conic sections*, without regard to their immediate usefulness simply because the study elicited ideas that were exciting, challenging, and interesting. The ancient Greeks could not have imagined the applications of these curves in the twenty-first century. Overwhelmed by the choices on satellite television? Blame it on a conic section! In this chapter, we use the rectangular coordinate system to study the conic sections and the mathematics behind their surprising applications.

One minute you're in class, enjoying the lecture. Then a sharp pain radiates down your side. The next minute you're being diagnosed with, of all things, a kidney stone. It took your cousin six weeks to recover from kidney stone surgery, but your doctor assures you there is nothing to worry about. A new procedure, based on a curve that looks like the cross section of a football, will dissolve the stone painlessly and let you return to class in a day or two. How can this be?

SECTION 7.1 The Ellipse

Objectives

1. Graph ellipses centered at the origin.
2. Write equations of ellipses in standard form.
3. Graph ellipses not centered at the origin.
4. Solve applied problems involving ellipses.

You took on a summer job driving a truck, delivering books that were ordered online. You're an avid reader, so just being around books sounded appealing. However, now you're feeling a bit shaky driving the truck for the first time. It's 10 feet wide and 9 feet high; compared to your compact car, it feels like you're behind the wheel of a tank. Up ahead you see a sign at the semielliptical entrance to a tunnel: Caution! Tunnel is 10 Feet High at Center Peak. Then you see another sign: Caution! Tunnel is 40 Feet Wide. Will your truck clear the opening of the tunnel's archway?

The mathematics of your world is present in the movements of planets, bridge and tunnel construction, navigational systems used to keep track of a ship's location, manufacture of lenses for telescopes, and even a procedure for disintegrating kidney stones. The mathematics behind these applications involves conic sections. **Conic sections** are curves that result from the intersection of a right circular cone and a plane. Figure 7.1 illustrates the four conic sections: the circle, the ellipse, the parabola, and the hyperbola.

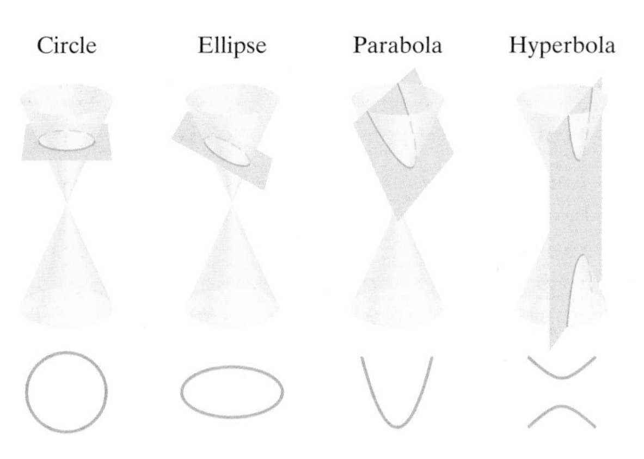

Figure 7.1 Obtaining the conic sections by intersecting a plane and a cone

In this section, we study the symmetric oval-shaped curve known as the ellipse. We will use a geometric definition for an ellipse to derive its equations. With these equations, we will determine if your delivery truck will clear the tunnel's entrance.

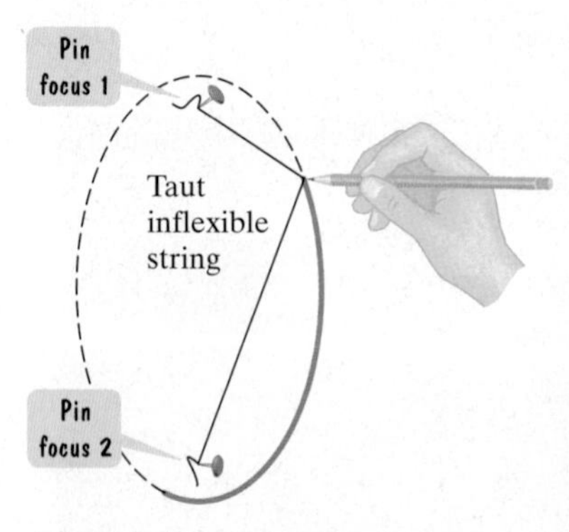

Figure 7.2 Drawing an ellipse

Definition of an Ellipse

Figure 7.2 illustrates how to draw an ellipse. Place pins at two fixed points, each of which is called a focus (plural: foci). If the ends of a fixed length of string are fastened to the pins and we draw the string taut with a pencil, the path traced by the pencil will be an ellipse. Notice that the sum of the distances of the pencil point from the foci remains constant because the length of the string is fixed. This procedure for drawing an ellipse illustrates its geometric definition.

Definition of an Ellipse

An **ellipse** is the set of all points in a plane the sum of whose distances from two fixed points, F_1 and F_2, is constant (see Figure 7.3). These two fixed points are called the **foci** (plural of **focus**). The midpoint of the segment connecting the foci is the **center** of the ellipse.

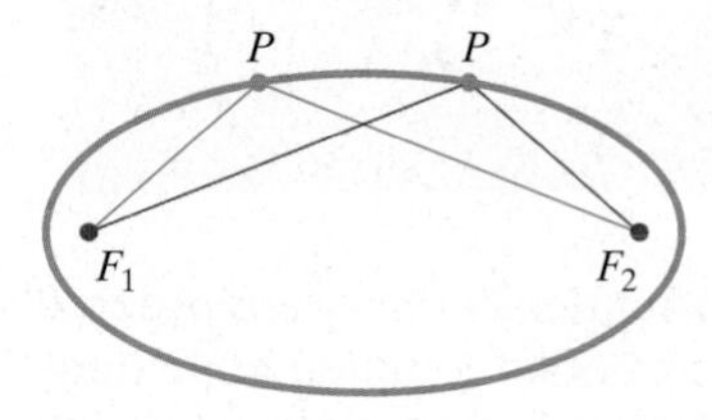

Figure 7.3

Figure 7.4 illustrates that an ellipse can be elongated horizontally or vertically. The line through the foci intersects the ellipse at two points, called the **vertices** (singular: **vertex**). The line segment that joins the vertices is the **major axis**. Notice that the midpoint of the major axis is the center of the ellipse. The line segment whose endpoints are on the ellipse that is perpendicular to the major axis at the center is the **minor axis** of the ellipse.

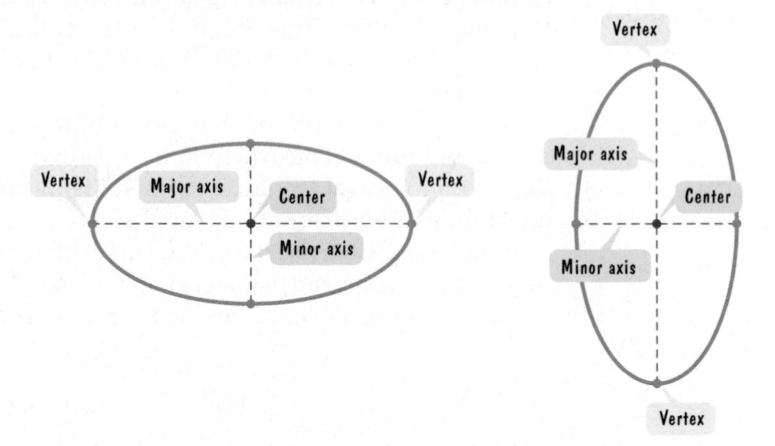

Figure 7.4 Horizontal and vertical elongations of an ellipse

Standard Form of the Equation of an Ellipse

The rectangular coordinate system gives us a unique way of describing an ellipse. It enables us to translate an ellipse's geometric definition into an algebraic equation.

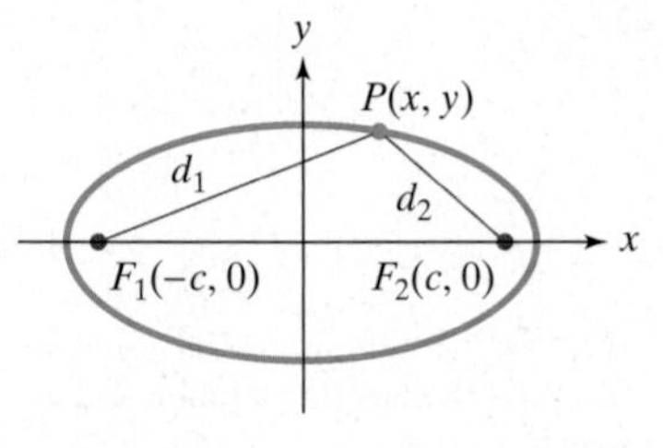

Figure 7.5

We start with Figure 7.5 to obtain an ellipse's equation. We've placed an ellipse that is elongated horizontally into a rectangular coordinate system. The foci are on the x-axis at $(-c, 0)$ and $(c, 0)$, as in Figure 7.5. In this way, the center of the ellipse is at the origin. We let (x, y) represent the coordinates of any point on the ellipse.

What does the definition of an ellipse tell us about the point (x, y) in Figure 7.5? For any point (x, y) on the ellipse, the sum of the distances to the two foci, $d_1 + d_2$, must be constant. We denote this constant by $2a$. Thus, the point (x, y) is on the ellipse if and only if

Discovery

Perform the algebra mentioned on the right by eliminating radicals and obtaining the equation shown.

$$d_1 + d_2 = 2a.$$

$$\sqrt{(x+c)^2 + y^2} + \sqrt{(x-c)^2 + y^2} = 2a \quad \text{Use the distance formula.}$$

After eliminating radicals and simplifying, we obtain

$$(a^2 - c^2)x^2 + a^2y^2 = a^2(a^2 - c^2).$$

Look at the triangle in Figure 7.5. Notice that the distance from F_1 to $F_2 < d_1 + d_2$. Equivalently, $2c < 2a$ and $c < a$. Consequently, $a^2 - c^2 > 0$. For convenience, let $b^2 = a^2 - c^2$. Substituting b^2 for $a^2 - c^2$ in the preceding equation, we obtain

$$b^2x^2 + a^2y^2 = a^2b^2$$

$$\frac{b^2x^2}{a^2b^2} + \frac{a^2y^2}{a^2b^2} = \frac{a^2b^2}{a^2b^2} \quad \text{Divide both sides by } a^2b^2.$$

$$\frac{x^2}{a^2} + \frac{y^2}{b^2} = 1 \quad \text{Simplify.}$$

This last equation is the **standard form of the equation of an ellipse.** There are two such equations, one for a horizontal major axis and one for a vertical major axis.

Standard Forms of the Equations of an Ellipse

The **standard form of the equation of an ellipse** with center at the origin, and major and minor axes of lengths $2a$ and $2b$ (where a and b are positive, and $a^2 > b^2$) is

$$\frac{x^2}{a^2} + \frac{y^2}{b^2} = 1 \quad \text{or} \quad \frac{x^2}{b^2} + \frac{y^2}{a^2} = 1.$$

Figure 7.6 illustrates that the vertices are on the major axis, a units from the center. The foci are are on the major axis, c units from the center. For both equations, $b^2 = a^2 - c^2$.

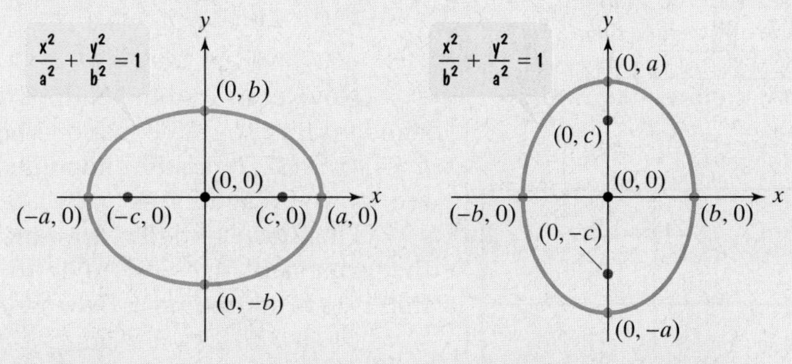

Figure 7.6 **(a)** Major axis is horizontal with length $2a$. **(b)** Major axis is vertical with length $2a$.

Using the Standard Form of the Equation of an Ellipse

We can use the standard form of an ellipse's equation to graph the ellipse. Although the definition of the ellipse is given in terms of its foci, the foci are not part of the graph. A complete graph of an ellipse can be obtained without graphing the foci.

1 Graph ellipses centered at the origin.

EXAMPLE 1 Graphing an Ellipse Centered at the Origin

Graph and locate the foci: $\frac{x^2}{9} + \frac{y^2}{4} = 1.$

Solution The given equation is the standard form of an ellipse's equation with $a^2 = 9$ and $b^2 = 4$.

$$\frac{x^2}{9} + \frac{y^2}{4} = 1$$

$a^2 = 9$. This is the larger of the two numbers in the denominator.

$b^2 = 4$. This is the smaller of the two numbers in the denominator.

Because the denominator of the x^2 term is greater than the denominator of the y^2 term, the major axis is horizontal. Based on the standard form of the equation, we know the vertices are $(-a, 0)$ and $(a, 0)$. Because $a^2 = 9$, $a = 3$. Thus, the vertices are $(-3, 0)$ and $(3, 0)$, shown in Figure 7.7.

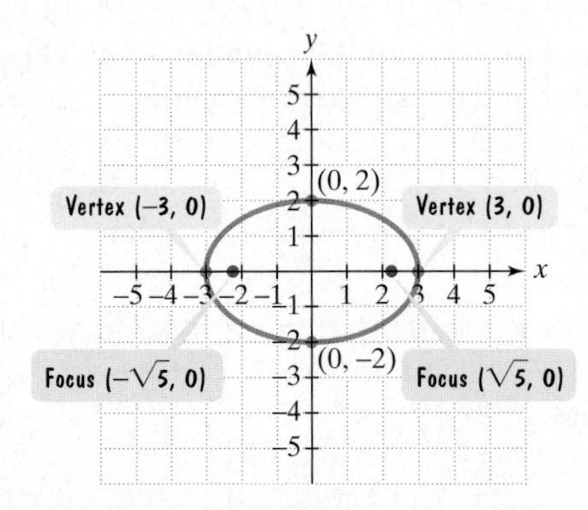

Figure 7.7 The graph of $\frac{x^2}{9} + \frac{y^2}{4} = 1$

Now let us find the endpoints of the vertical minor axis. According to the standard form of the equation, these endpoints are $(0, -b)$ and $(0, b)$. Because $b^2 = 4$, $b = 2$. Thus, the endpoints of the minor axis are $(0, -2)$ and $(0, 2)$. They are shown in Figure 7.7.

Finally, we find the foci, which are located at $(-c, 0)$ and $(c, 0)$. We can use the formula $b^2 = a^2 - c^2$ to do so. We know that $a^2 = 9$ and $b^2 = 4$; we need to find c^2 in order to find c. Because $b^2 = a^2 - c^2$, we obtain

$$c^2 = a^2 - b^2 = 9 - 4 = 5.$$

Because $c^2 = 5, c = \sqrt{5}$. The foci, $(-c, 0)$ and $(c, 0)$, are located at $(-\sqrt{5}, 0)$ and $(\sqrt{5}, 0)$. They are shown in Figure 7.7.

Technology

We graph $\frac{x^2}{9} + \frac{y^2}{4} = 1$ with a graphing utility by solving for y and defining two functions.

$$\frac{y^2}{4} = 1 - \frac{x^2}{9}$$

$$y^2 = 4\left(1 - \frac{x^2}{9}\right)$$

$$y = \pm 2\sqrt{1 - \frac{x^2}{9}}$$

Enter

$y_1 = 2$ [√] (1 [−] x [^] 2 [÷] 9)

and

$$y_2 = -y_1.$$

To see the true shape of the ellipse, use the [ZOOM SQUARE] feature so that one unit on the x-axis is the same length as one unit on the y-axis.

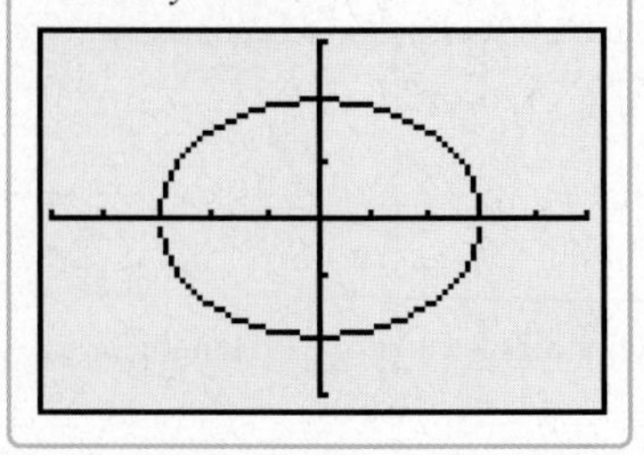

You can sketch the ellipse in Figure 7.7 by locating endpoints on the major and minor axes.

$$\frac{x^2}{3^2} + \frac{y^2}{2^2} = 1$$

Endpoints of the major axis are 3 units to the right and left of the center.

Endpoints of the minor axis are 2 units up and down from the center.

Check Point 1 Graph and locate the foci: $\frac{x^2}{36} + \frac{y^2}{9} = 1.$

EXAMPLE 2 Graphing an Ellipse Centered at the Origin

Graph and locate the foci: $25x^2 + 16y^2 = 400$.

Solution We begin by expressing the equation in standard form. Because we want 1 on the right side, we divide both sides by 400.

$$\frac{25x^2}{400} + \frac{16y^2}{400} = \frac{400}{400}$$

$$\frac{x^2}{16} + \frac{y^2}{25} = 1$$

$b^2 = 16$. This is the smaller of the two numbers in the denominator.

$a^2 = 25$. This is the larger of the two numbers in the denominator.

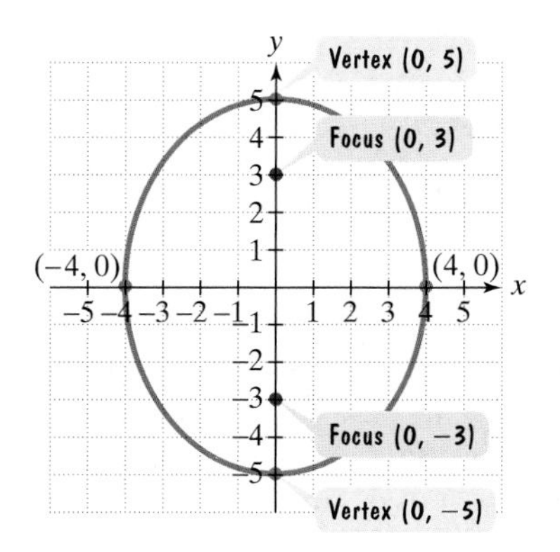

Figure 7.8 The graph of $\frac{x^2}{16} + \frac{y^2}{25} = 1$

The equation is the standard form of an ellipse's equation with $a^2 = 25$ and $b^2 = 16$. Because the denominator of the y^2 term is greater than the denominator of the x^2 term, the major axis is vertical. Based on the standard form of the equation, we know the vertices are $(0, -a)$ and $(0, a)$. Because $a^2 = 25$, $a = 5$. Thus, the vertices are $(0, -5)$ and $(0, 5)$, shown in Figure 7.8.

Now let us find the endpoints of the horizontal minor axis. According to the standard form of the equation, these endpoints are $(-b, 0)$ and $(b, 0)$. Because $b^2 = 16$, $b = 4$. Thus, the endpoints of the minor axis are $(-4, 0)$ and $(4, 0)$. They are shown in Figure 7.8.

Finally, we find the foci, which are located at $(0, -c)$ and $(0, c)$. We can use the formula $b^2 = a^2 - c^2$ to do so. We know that $a^2 = 25$ and $b^2 = 16$; we need to find c^2 in order to find c. Because $b^2 = a^2 - c^2$, we obtain

$$c^2 = a^2 - b^2 = 25 - 16 = 9.$$

Because $c^2 = 9$, $c = 3$. The foci, $(0, -c)$ and $(0, c)$, are located at $(0, -3)$ and $(0, 3)$. They are shown in Figure 7.8. You can sketch the ellipse in Figure 7.8 by locating endpoints on the major and minor axes:

$$\frac{x^2}{4^2} + \frac{y^2}{5^2} = 1.$$

Endpoints of the minor axis are 4 units to the right and left of the center.

Endpoints of the major axis are 5 units up and down from the center.

Check Point 2 Graph and locate the foci: $16x^2 + 9y^2 = 144$.

2 Write equations of ellipses in standard form.

In Examples 1 and 2, we used the equation of an ellipse to find its foci and vertices. In the next example, we reverse this procedure.

EXAMPLE 3 Finding the Equation of an Ellipse from Its Foci and Vertices

Find the standard form of the equation of an ellipse with foci at $(-1, 0)$ and $(1, 0)$ and vertices $(-2, 0)$ and $(2, 0)$.

Solution Because the foci are located at $(-1, 0)$ and $(1, 0)$, on the x-axis, the major axis is horizontal. The center of the ellipse is midway between the foci, located at $(0, 0)$. Thus, the form of the equation is

$$\frac{x^2}{a^2} + \frac{y^2}{b^2} = 1.$$

We need to determine the values for a^2 and b^2. The distance from the center $(0, 0)$ to either vertex, $(-2, 0)$ or $(2, 0)$, is 2. Thus, $a = 2$.

$$\frac{x^2}{2^2} + \frac{y^2}{b^2} = 1 \quad \text{or} \quad \frac{x^2}{4} + \frac{y^2}{b^2} = 1$$

We must still find b^2. The distance from the center $(0, 0)$ to either focus, $(-1, 0)$ or $(1, 0)$, is 1, so $c = 1$. Because $b^2 = a^2 - c^2$, we have

$$b^2 = 2^2 - 1^2 = 4 - 1 = 3.$$

Substituting 3 for b^2 in the last equation gives us the standard form of the ellipse's equation. The equation is

$$\frac{x^2}{4} + \frac{y^2}{3} = 1.$$

Check Point 3 Find the standard form of the equation of an ellipse with foci at $(-2, 0)$ and $(2, 0)$ and vertices $(-3, 0)$ and $(3, 0)$.

3 Graph ellipses not centered at the origin.

Translations of Ellipses

Despite the fact that an ellipse is not the graph of a function, its graph can be translated in the same manner as that of a function. Figure 7.9 illustrates that the graphs of

$$\frac{(x-h)^2}{a^2}+\frac{(y-k)^2}{b^2}=1 \quad \text{and} \quad \frac{x^2}{a^2}+\frac{y^2}{b^2}=1$$

have the same size and shape. However, the graph of the first equation is centered at (h, k) rather than at the origin.

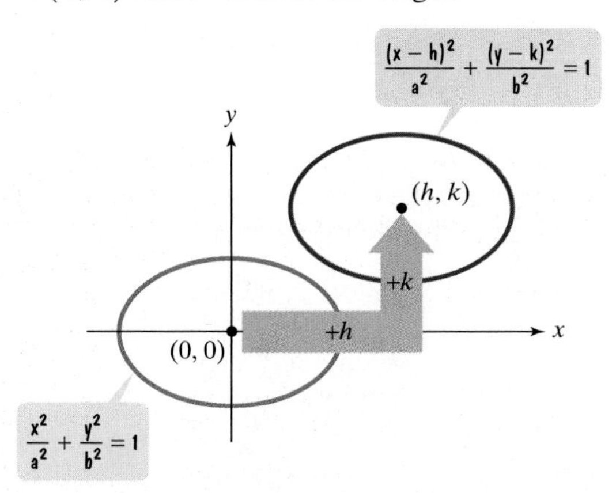

Figure 7.9 Translating an ellipse's graph

Table 7.1 gives the standard forms of equations of ellipses centered at (h, k). Figure 7.10 shows their graphs.

Table 7.1 Standard Forms of Equations of Ellipses Centered at (h, k)

Equation	Center	Major Axis	Foci	Vertices
$\frac{(x-h)^2}{a^2}+\frac{(y-k)^2}{b^2}=1,$ $a^2>b^2$ and $b^2=a^2-c^2$	(h, k)	Parallel to the x-axis, horizontal	$(h-c, k)$ $(h+c, k)$	$(h-a, k)$ $(h+a, k)$
$\frac{(x-h)^2}{b^2}+\frac{(y-k)^2}{a^2}=1,$ $a^2>b^2$ and $b^2=a^2-c^2$	(h, k)	Parallel to the y-axis, vertical	$(h, k-c)$ $(h, k+c)$	$(h, k-a)$ $(h, k+a)$

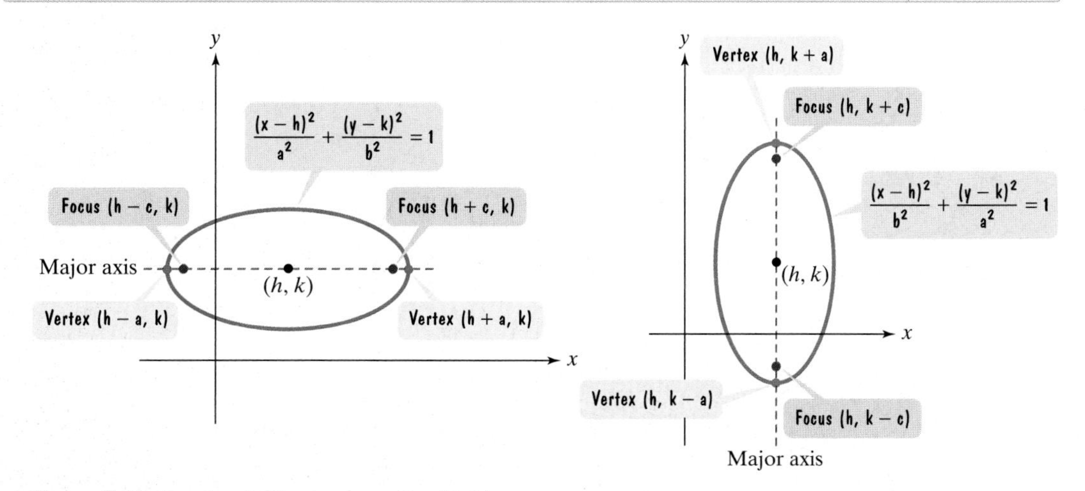

Figure 7.10 Graphs of ellipses centered at (h, k)

EXAMPLE 4 Graphing an Ellipse Centered at (h, k)

Graph: $\dfrac{(x-1)^2}{4} + \dfrac{(y+2)^2}{9} = 1$. Where are the foci located?

Solution In order to graph the ellipse, we need to know its center (h, k). In the standard forms of equations centered at (h, k), h is the number subtracted from x and k is the number subtracted from y.

$$\frac{(x-1)^2}{4} + \frac{(y-(-2))^2}{9} = 1$$

This is $(x-h)^2$ with $h = 1$.

This is $(y-k)^2$ with $k = -2$.

We see that $h = 1$ and $k = -2$. Thus, the center of the ellipse, (h, k), is $(1, -2)$. We can graph the ellipse by locating endpoints on the major and minor axes. To do this, we must identify a^2 and b^2.

$$\frac{(x-1)^2}{4} + \frac{(y+2)^2}{9} = 1$$

$b^2 = 4$. This is the smaller of the two numbers in the denominator.

$a^2 = 9$. This is the larger of the two numbers in the denominator.

The larger number is under the expression involving y. This means that the major axis is vertical and parallel to the y-axis. Because $a^2 = 9$, $a = 3$ and the vertices lie three units above and below the center. Also, because $b^2 = 4$, $b = 2$ and the endpoints of the minor axis lie two units to the right and left of the center. We categorize these observations as follows:

Center	Vertices	Endpoints of Minor Axis
$(1, -2)$	$(1, -2 + 3) = (1, 1)$	$(1 + 2, -2) = (3, -2)$
	$(1, -2 - 3) = (1, -5)$	$(1 - 2, -2) = (-1, -2)$

Using the center and these four points, we can sketch the ellipse shown in Figure 7.11.

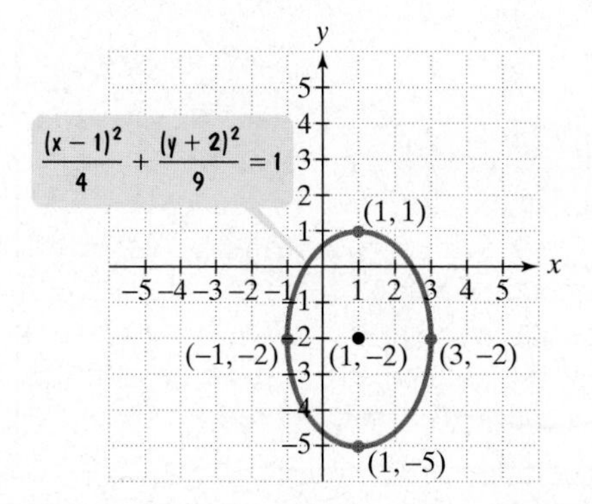

Figure 7.11 The graph of an ellipse centered at $(1, -2)$

With $b^2 = a^2 - c^2$, we have $4 = 9 - c^2$, and $c^2 = 5$. So the foci are located $\sqrt{5}$ units above and below the center, at $(1, -2 + \sqrt{5})$ and $(1, -2 - \sqrt{5})$.

Check Point 4 Graph: $\frac{(x+1)^2}{9} + \frac{(y-2)^2}{4} = 1$. Where are the foci located?

In some cases, it is necessary to convert the equation of an ellipse to standard form by completing the square on x and y. For example, suppose that we wish to graph the ellipse whose equation is

$$9x^2 + 4y^2 - 18x + 16y - 11 = 0.$$

Because we plan to complete the square on both x and y, we need to rearrange terms so that

- x terms are arranged in descending order.
- y terms are arranged in descending order.
- the constant term appears on the right.

$$9x^2 + 4y^2 - 18x + 16y - 11 = 0$$ This is the given equation.

$$(9x^2 - 18x) + (4y^2 + 16y) = 11$$ Group terms and add 11 to both sides.

$$9(x^2 - 2x + \square) + 4(y^2 + 4y + \square) = 11$$ To complete the square, coefficients of x^2 and y^2 must be 1. Factor out 9 and 4, respectively.

$$9(x^2 - 2x + 1) + 4(y^2 + 4y + 4) = 11 + 9 + 16$$ Complete each square by adding the square of half the coefficient of x and y, respectively.

$$9(x-1)^2 + 4(y+2)^2 = 36$$ Factor.

$$\frac{9(x-1)^2}{36} + \frac{4(y+2)^2}{36} = \frac{36}{36}$$ Divide both sides by 36.

$$\frac{(x-1)^2}{4} + \frac{(y+2)^2}{9} = 1$$ Simplify.

Study Tip

When completing the square, remember that changes made on the left side of the equation must also be made on the right side of the equation.

The equation is now in standard form. This is precisely the form of the equation that we graphed in Example 4.

4 Solve applied problems involving ellipses.

Applications

Ellipses have many applications. German scientist Johannes Kepler (1571–1630) showed that the planets in our solar system move in elliptical orbits, with the sun at a focus. Earth satellites also travel in elliptical orbits, with Earth at a focus.

One intriguing aspect of the ellipse is that a ray of light or a sound wave from one focus will be reflected from the ellipse exactly to the other focus. A whispering gallery is an elliptical room with an elliptical, dome-shaped ceiling. People standing at the foci can whisper and hear each other quite clearly, while persons in other locations in the room cannot hear them. Statuary Hall in the U.S. Capitol Building is elliptical. President John Quincy Adams, while a member of the House of Representatives, was aware of this acoustical phenomenon. He situated his desk at a focal point of the elliptical ceiling, easily eavesdropping on the private conversations of other House members located near the other focus.

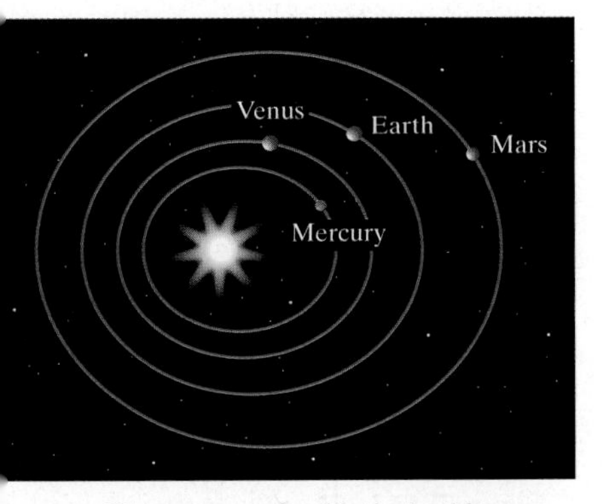

Planets move in elliptical orbits.

The elliptical reflection principle is used in a procedure for disintegrating kidney stones. The patient is placed within a device that is elliptical in shape. The patient is at one focus, while ultrasound waves from the other focus hit the walls and are reflected to the kidney stone. The convergence of the ultrasound waves at the kidney stone causes vibrations that shatter it into fragments. The small pieces can then be passed painlessly through the patient's system. The patient recovers in days, as opposed to up to six weeks if surgery is used instead.

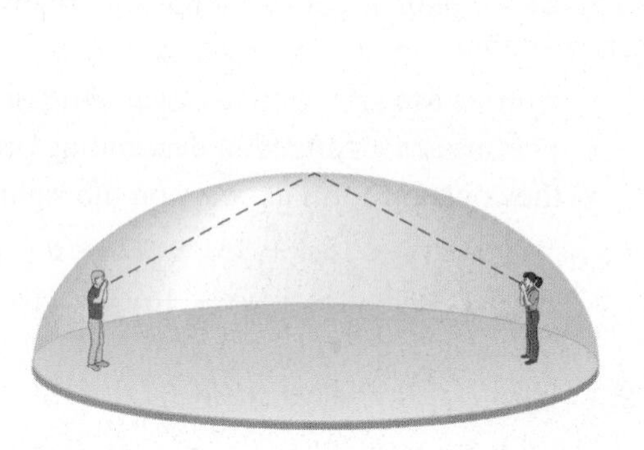

Whispering in an elliptical dome

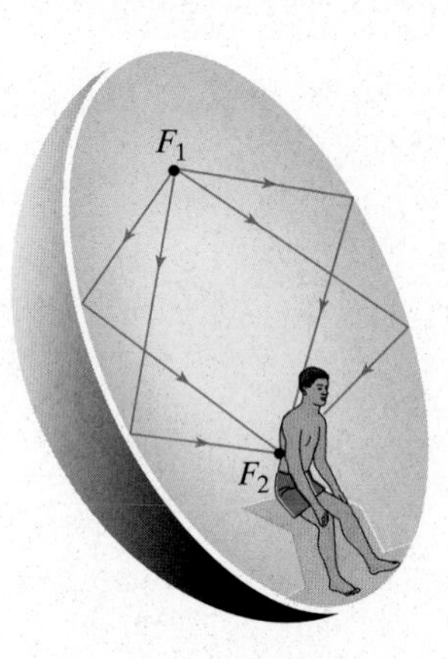

Disintegrating kidney stones

Ellipses are often used for supporting arches of bridges and in tunnel construction. This application forms the basis of our next example.

EXAMPLE 5 An Application Involving an Ellipse

A semielliptical archway over a one-way road has a height of 10 feet and a width of 40 feet (see Figure 7.12). Your truck has a width of 10 feet and a height of 9 feet. Will your truck clear the opening of the archway?

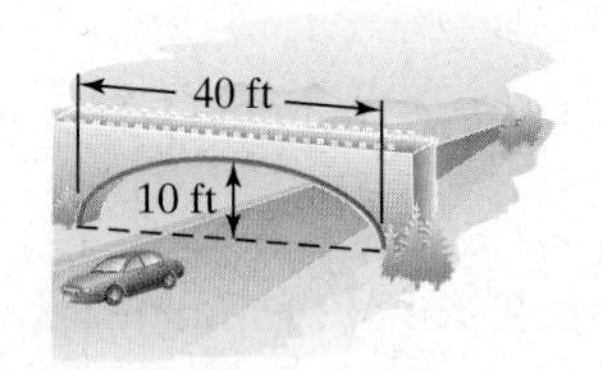

Figure 7.12 A semielliptical archway

Solution To determine the clearance, we must find the height of the archway 5 feet from the center. If that height is 9 feet or less, the truck will not clear the opening.

In Figure 7.13, we've constructed a coordinate system with the x-axis on the ground and the origin at the center of the archway. Also shown is the truck, whose height is 9 feet.

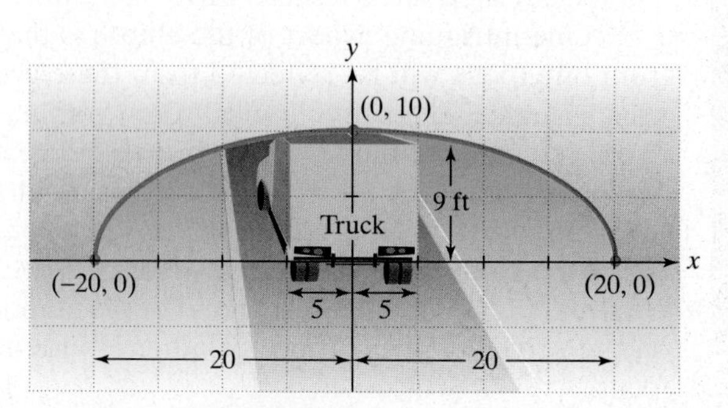

Figure 7.13

Halley's Comet

Halley's Comet has an elliptical orbit with the sun at one focus. The comet returns every 76.3 years. The first recorded sighting was in 239 B.C. It was last seen in 1986. At that time, spacecraft went close to the comet, measuring its nucleus to be 7 miles long and 4 miles wide. By 2024, Halley's Comet will have reached the farthest point in its elliptical orbit before returning to be next visible from Earth in 2062.

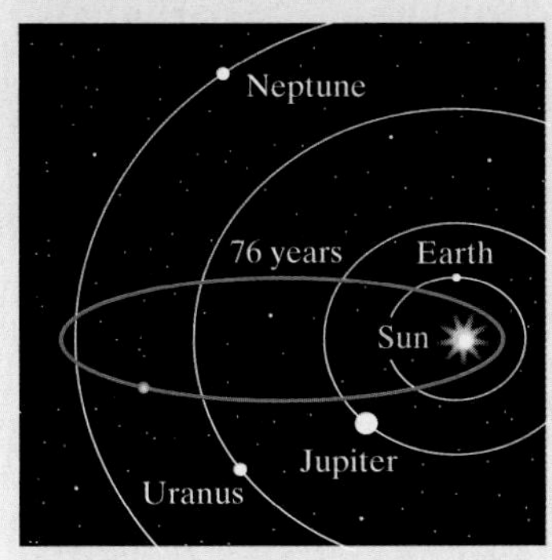

The elliptical orbit of Halley's Comet

Using the equation $\frac{x^2}{a^2} + \frac{y^2}{b^2} = 1$, we can express the equation of the blue archway in Figure 7.13 as $\frac{x^2}{20^2} + \frac{y^2}{10^2} = 1$ or $\frac{x^2}{400} + \frac{y^2}{100} = 1$.

As shown in Figure 7.13, the edge of the 10-foot-wide truck corresponds to $x = 5$. We find the height of the archway 5 feet from the center by substituting 5 for x and solving for y.

$$\frac{5^2}{400} + \frac{y^2}{100} = 1 \quad \text{Substitute 5 for } x.$$

$$\frac{25}{400} + \frac{y^2}{100} = 1$$

$$\frac{1}{16} + \frac{y^2}{100} = 1$$

$$1600\left(\frac{1}{16} + \frac{y^2}{100}\right) = 1600(1) \quad \text{Clear fractions by multiplying both sides by 1600.}$$

$$100 + 16y^2 = 1600 \quad \text{Use the distributive property and simplify.}$$

$$16y^2 = 1500 \quad \text{Subtract 100 from both sides.}$$

$$y^2 = \frac{1500}{16} \quad \text{Divide both sides by 16.}$$

$$y = \sqrt{\frac{1500}{16}} \quad \text{Take only the positive square root. The archway is above the x-axis and y is nonnegative.}$$

$$\approx 9.68$$

Thus, the height of the archway 5 feet from the center is approximately 9.68 feet. Because your truck's height is 9 feet, there is enough room for the truck to clear the archway.

Check Point 5 Will a truck that is 12 feet wide and has a height of 9 feet clear the opening of the archway described in Example 5?

EXERCISE SET 7.1

Practice Exercises

In Exercises 1–16, graph each ellipse and locate the foci.

1. $\frac{x^2}{16} + \frac{y^2}{4} = 1$

2. $\frac{x^2}{25} + \frac{y^2}{16} = 1$

3. $\frac{x^2}{9} + \frac{y^2}{36} = 1$

4. $\frac{x^2}{16} + \frac{y^2}{49} = 1$

5. $\frac{x^2}{25} + \frac{y^2}{64} = 1$

6. $\frac{x^2}{49} + \frac{y^2}{36} = 1$

7. $\frac{x^2}{49} + \frac{y^2}{81} = 1$

8. $\frac{x^2}{64} + \frac{y^2}{100} = 1$

9. $25x^2 + 4y^2 = 100$

10. $9x^2 + 4y^2 = 36$

11. $4x^2 + 16y^2 = 64$

12. $16x^2 + 9y^2 = 144$

13. $25x^2 + 9y^2 = 225$

14. $4x^2 + 25y^2 = 100$

15. $x^2 + 2y^2 = 8$

16. $12x^2 + 4y^2 = 36$

In Exercises 17–20, find the standard form of the equation of each ellipse and give the location of its foci.

17.

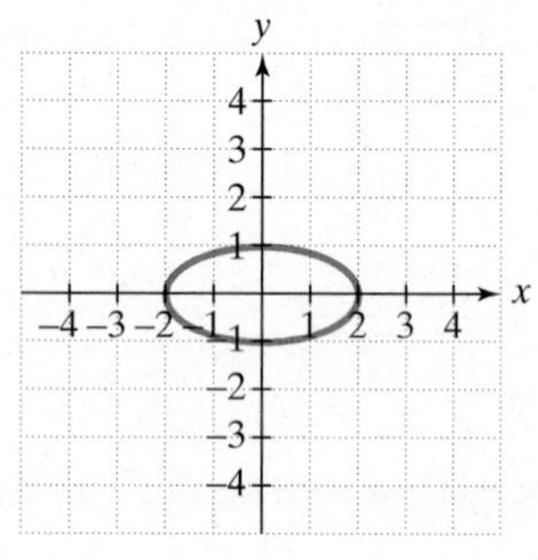

18.

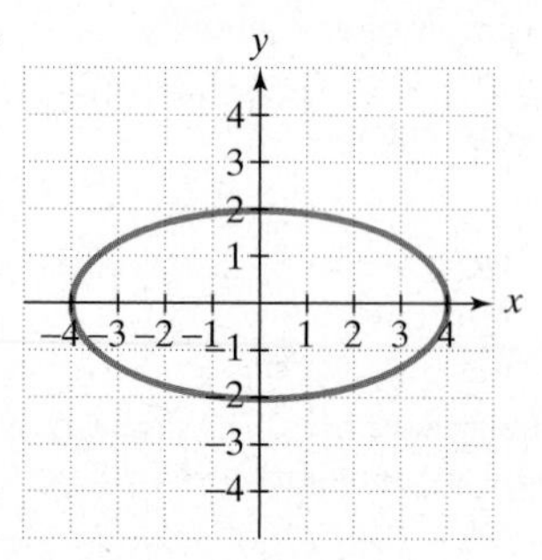

19.

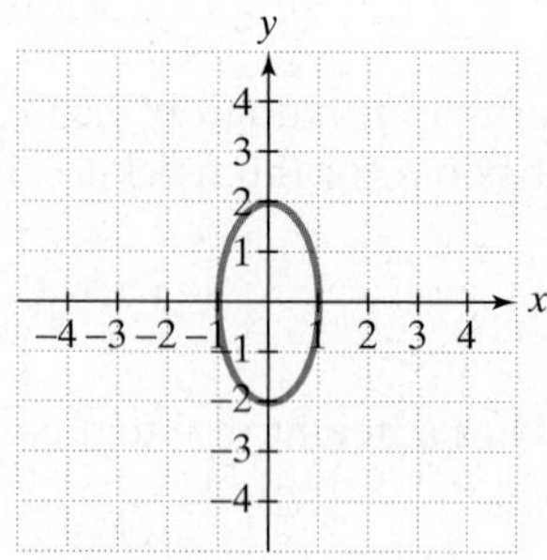

20.

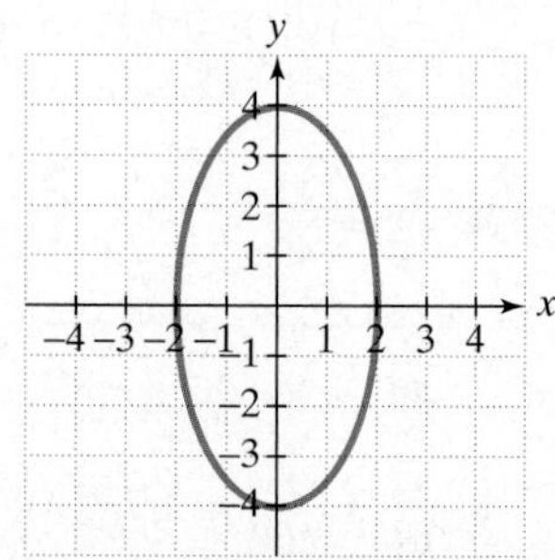

In Exercises 21–30, find the standard form of the equation of each ellipse centered at the origin satisfying the given conditions.

21. Foci: $(-5, 0), (5, 0)$; vertices: $(-8, 0), (8, 0)$

22. Foci: $(-2, 0), (2, 0)$; vertices: $(-6, 0), (6, 0)$

23. Foci: $(0, -4), (0, 4)$; vertices: $(0, -7), (0, 7)$

24. Foci: $(0, -3), (0, 3)$; vertices: $(0, -4), (0, 4)$

25. Foci: $(-2, 0), (2, 0)$; y-intercepts: -3 and 3

26. Foci: $(0, -2), (0, 2)$; x-intercepts: -2 and 2

27. Major axis horizontal with length 8; length of minor axis $= 4$

28. Major axis horizontal with length 12; length of minor axis $= 6$

29. Major axis vertical with length 10; length of minor axis $= 4$

30. Major axis vertical with length 20; length of minor axis $= 10$

In Exercises 31–42 graph each ellipse and give the location of its foci.

31. $\dfrac{(x-2)^2}{9} + \dfrac{(y-1)^2}{4} = 1$

32. $\dfrac{(x-1)^2}{16} + \dfrac{(y+2)^2}{9} = 1$

33. $(x+3)^2 + 4(y-2)^2 = 16$

34. $(x-3)^2 + 9(y+2)^2 = 18$

35. $\dfrac{(x-4)^2}{9} + \dfrac{(y+2)^2}{25} = 1$

36. $\dfrac{(x-3)^2}{9} + \dfrac{(y+1)^2}{16} = 1$

37. $\dfrac{x^2}{25} + \dfrac{(y-2)^2}{36} = 1$

38. $\dfrac{(x-4)^2}{4} + \dfrac{y^2}{25} = 1$

39. $\dfrac{(x+3)^2}{9} + (y-2)^2 = 1$

40. $\dfrac{(x+2)^2}{16} + (y-3)^2 = 1$

41. $9(x-1)^2 + 4(y+3)^2 = 36$

42. $36(x+4)^2 + (y+3)^2 = 36$

In Exercises 43–48, convert each equation to standard form by completing the square on x and y. Then graph the ellipse and give the location of its foci.

43. $9x^2 + 25y^2 - 36x + 50y - 164 = 0$

44. $4x^2 + 9y^2 - 32x + 36y + 64 = 0$

45. $9x^2 + 16y^2 - 18x + 64y - 71 = 0$

46. $x^2 + 4y^2 + 10x - 8y + 13 = 0$

47. $4x^2 + y^2 + 16x - 6y - 39 = 0$

48. $4x^2 + 25y^2 - 24x + 100y + 36 = 0$

Application Exercises

49. Will a truck that is 8 feet wide carrying a load that reaches 7 feet above the ground clear the semielliptical arch on the one-way road that passes under the bridge shown in the figure?

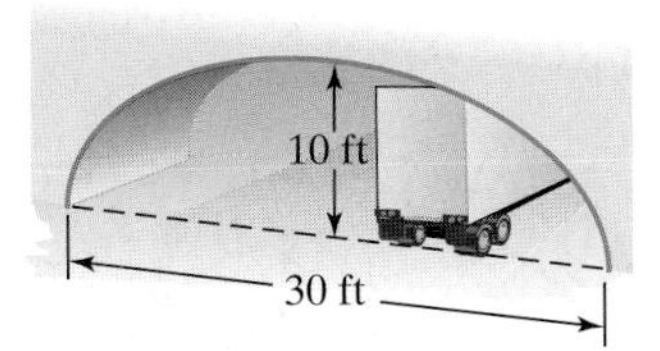

50. A semielliptic archway has a height of 20 feet and a width of 50 feet, as shown in the figure. Can a truck 14 feet high and 10 feet wide drive under the archway without going into the other lane?

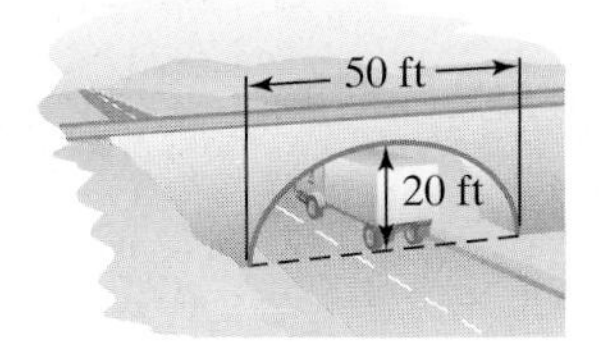

51. The elliptical ceiling in Statuary Hall in the U.S. Capitol Building is 96 feet long and 23 feet tall.

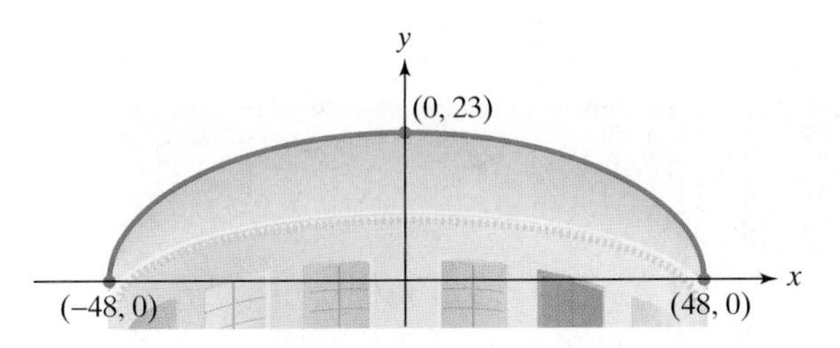

a. Using the rectangular coordinate system in the figure shown, write the standard form of the equation of the elliptical ceiling.

b. John Quincy Adams discovered that he could overhear the conversations of opposing party leaders near the left side of the chamber if he situated his desk at the focus at the right side of the chamber. How far from the center of the ellipse along the major axis did Adams situate his desk? (Round to the nearest foot.)

52. If an elliptical whispering room has a height of 30 feet and a width of 100 feet, where should two people stand if they would like to whisper back and forth and be heard?

Writing in Mathematics

53. What is an ellipse?

54. Describe how to graph $\frac{x^2}{25} + \frac{y^2}{16} = 1$.

55. Describe how to locate the foci for $\frac{x^2}{25} + \frac{y^2}{16} = 1$.

56. Describe one similarity and one difference between the graphs of $\frac{x^2}{25} + \frac{y^2}{16} = 1$ and $\frac{x^2}{16} + \frac{y^2}{25} = 1$.

57. Describe one similarity and one difference between the graphs of $\frac{x^2}{25} + \frac{y^2}{16} = 1$ and $\frac{(x-1)^2}{25} + \frac{(y-1)^2}{16} = 1$.

58. An elliptipool is an elliptical pool table with only one pocket. A pool shark places a ball on the table, hits it in what appears to be a random direction, and yet it bounces off the edge, falling directly into the pocket. Explain why this happens.

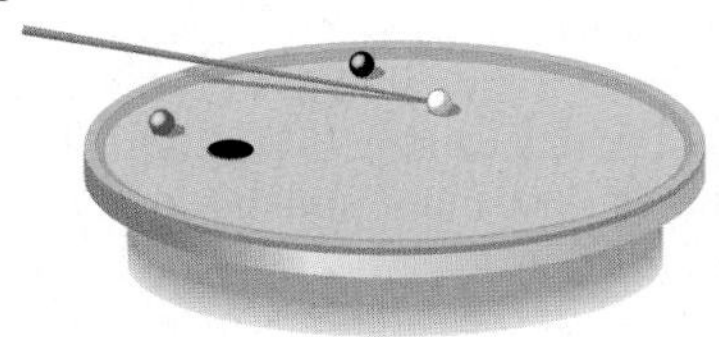

Technology Exercises

59. Use a graphing utility to graph any five of the ellipses that you graphed by hand in Exercises 1–16.

60. Use a graphing utility to graph any three of the ellipses that you graphed by hand in Exercises 31–42. First solve the given equation for y by using the square root method. Enter each of the two resulting equations to produce each half of the ellipse.

61. Use a graphing utility to graph any one of the ellipses that you graphed by hand in Exercises 43–48. Write the equation as a quadratic equation in y and use the quadratic formula to solve for y. Enter each of the two resulting equations to produce each half of the ellipse.

62. Write an equation for the path of each of the following elliptical orbits. Then use a graphing utility to graph the two ellipses in the same viewing rectangle. Can you see why early astronomers had difficulty detecting that these orbits are ellipses rather than circles?

Earth's orbit:	Length of major axis: 186 million miles
	Length of minor axis: 185.8 million miles
Mars's orbit:	Length of major axis: 283.5 million miles
	Length of minor axis: 278.5 million miles

Critical Thinking Exercises

63. Find the standard form of the equation of an ellipse with vertices at $(0, -6)$ and $(0, 6)$, passing through $(2, -4)$.

64. An Earth satellite has an elliptical orbit described by

$$\frac{x^2}{(5000)^2} + \frac{y^2}{(4750)^2} = 1.$$

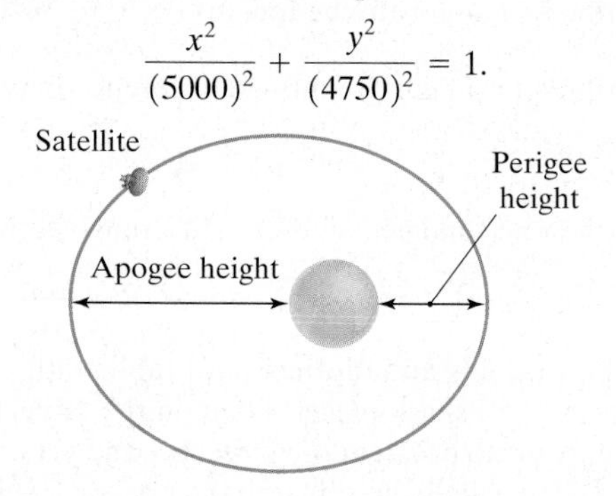

(All units are in miles.) The coordinates of the center of Earth are $(16, 0)$.

a. The perigee of the satellite's orbit is the point that is nearest Earth's center. If the radius of Earth is approximately 4000 miles, find the distance of the perigee above Earth's surface.

b. The apogee of the satellite's orbit is the point that is the greatest distance from Earth's center. Find the distance of the apogee above Earth's surface.

65. The equation of the red ellipse in the following figure is

$$\frac{x^2}{25} + \frac{y^2}{9} = 1.$$

Write the equation for each circle shown in the figure.

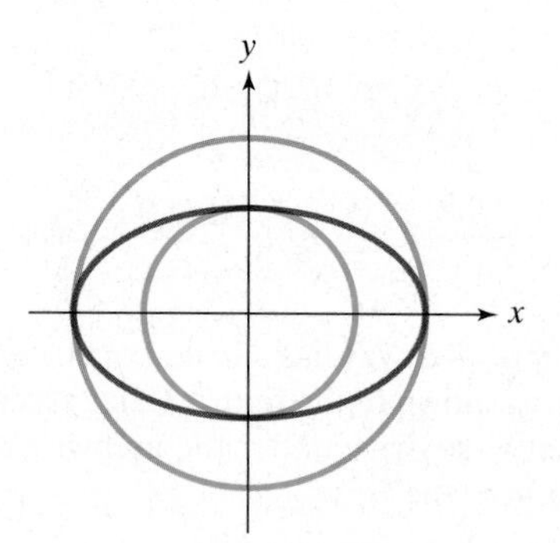

66. What happens to the shape of the graph of $\frac{x^2}{a^2} + \frac{y^2}{b^2} = 1$ as $\frac{c}{a}$ is close to zero?

SECTION 7.2 *The Hyperbola*

Objectives

1. Locate a hyperbola's vertices and foci.
2. Write equations of hyperbolas in standard form.
3. Graph hyperbolas centered at the origin.
4. Graph hyperbolas not centered at the origin.
5. Solve applied problems involving hyperbolas.

St. Mary's Cathedral

Conic sections are often used to create unusual architectural designs. The top of St. Mary's Cathedral in San Francisco is a 2135-cubic-foot dome with walls rising 200 feet above the floor and supported by four massive concrete pylons that extend 94 feet into the ground. Cross sections of the roof are parabolas and hyperbolas. In this section, we study the curve with two parts known as the hyperbola.

Definition of a Hyperbola

Figure 7.14 Casting hyperbolic shadows

Figure 7.14 shows a cylindrical lampshade casting two shadows on a wall. These shadows indicate the distinguishing feature of hyperbolas: Their graphs contain two disjoint parts called **branches**. Although each branch might look like a parabola, its shape is actually quite different.

The definition of a hyperbola is similar to that of the ellipse. For the ellipse, the *sum* of the distances to the foci is a constant. By contrast, for a hyperbola the *difference* of the distances to the foci is a constant.

> **Definition of a Hyperbola**
>
> A **hyperbola** is the set of points in a plane the difference of whose distances from two fixed points (called foci) is a constant.

Figure 7.15 illustrates the two branches of a hyperbola's graph. The line through the foci intersects the hyperbola at two points, called the **vertices**. The line segment that joins the vertices is the **transverse axis**. The midpoint of the transverse axis is the **center** of the hyperbola. Notice that the center lies midway between the vertices, as well as midway between the foci.

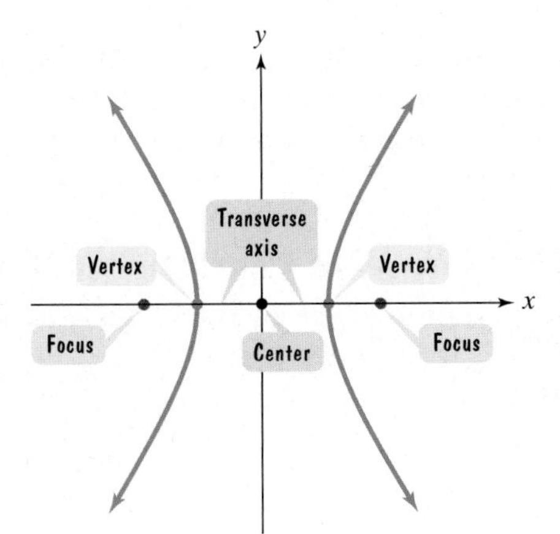

Figure 7.15 The two branches of a hyperbola

Standard Form of the Equation of a Hyperbola

The rectangular coordinate system enables us to translate a hyperbola's geometric definition into an algebraic equation. Figure 7.16 is our starting point for obtaining an equation. We place the foci on the x-axis at the points $(-c, 0)$ and $(c, 0)$. Note that the center of this hyperbola is at the origin. We let (x, y) represent the coordinates of any point on the hyperbola.

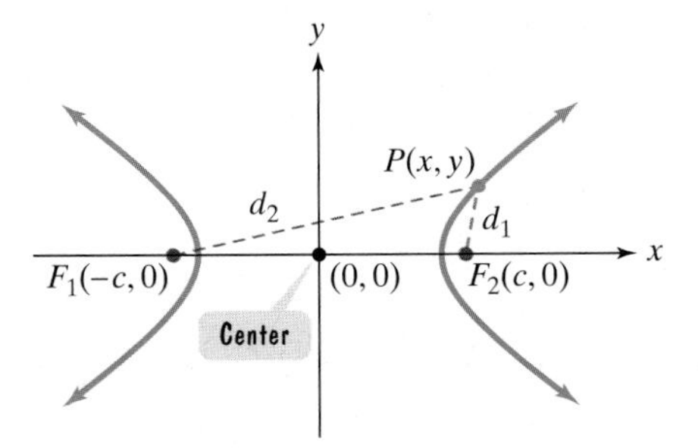

Figure 7.16

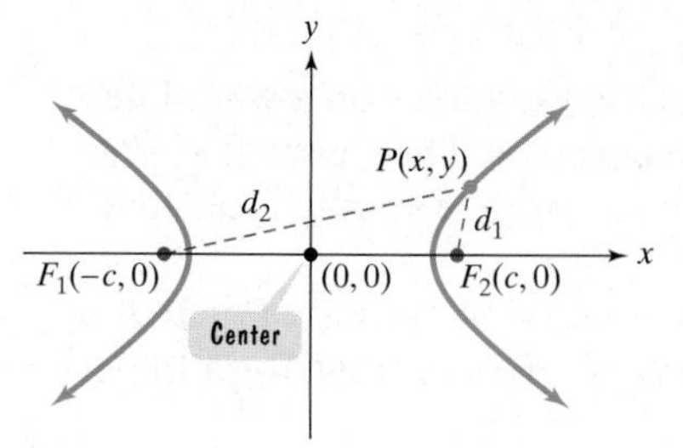

Figure 7.16, shown again so that you do not have to turn back a page

What does the definition of a hyperbola tell us about the point (x, y) in Figure 7.16? For any point (x, y) on the hyperbola, the absolute value of the difference of the distances from the two foci, $|d_2 - d_1|$, must be constant. We denote this constant by $2a$, just as we did for the ellipse. Thus, the point (x, y) is on the hyperbola if and only if

$$|d_2 - d_1| = 2a$$

$$\left|\sqrt{(x+c)^2 + (y-0)^2} - \sqrt{(x-c)^2 + (y-0)^2}\right| = 2a \quad \text{Use the distance formula.}$$

After eliminating radicals and simplifying, we obtain

$$(c^2 - a^2)x^2 - a^2y^2 = a^2(c^2 - a^2).$$

For convenience, let $b^2 = c^2 - a^2$. Substituting b^2 for $c^2 - a^2$ in the preceding equation, we obtain

$$b^2x^2 - a^2y^2 = a^2b^2$$

$$\frac{b^2x^2}{a^2b^2} - \frac{a^2y^2}{a^2b^2} = \frac{a^2b^2}{a^2b^2} \quad \text{Divide both sides by } a^2b^2.$$

$$\frac{x^2}{a^2} - \frac{y^2}{b^2} = 1 \quad \text{Simplify.}$$

This last equation is called the **standard form of the equation of a hyperbola.** There are two such equations. The first is for a hyperbola in which the transverse axis lies on the x-axis. The second is for a hyperbola in which the transverse axis lies on the y-axis.

Standard Forms of the Equations of a Hyperbola

The **standard form of the equation of a hyperbola** with center at the origin is

$$\frac{x^2}{a^2} - \frac{y^2}{b^2} = 1 \quad \text{or} \quad \frac{y^2}{a^2} - \frac{x^2}{b^2} = 1.$$

Figure 7.17 illustrates that for the equation on the left, the transverse axis lies on the x-axis. For the equation on the right, the transverse axis lies on the y-axis. The vertices are a units from the center and the foci are c units from the center. For both equations, $b^2 = c^2 - a^2$.

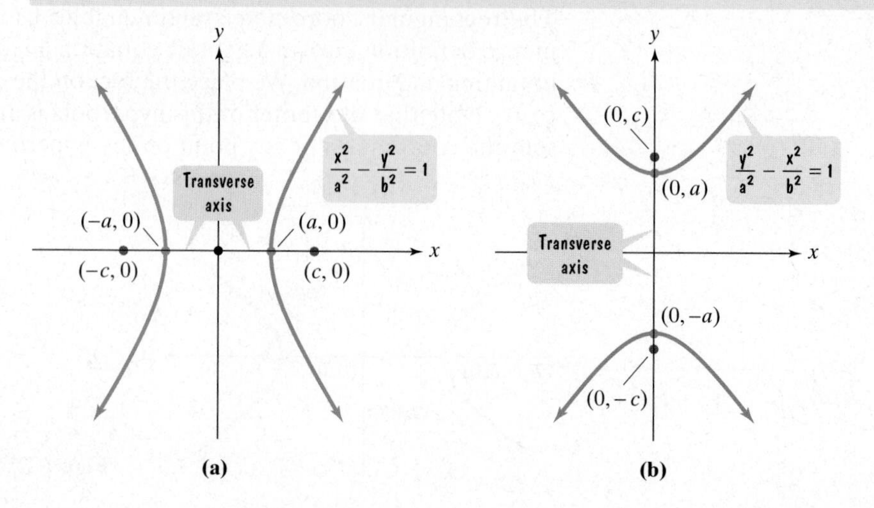

Figure 7.17 **(a)** Transverse axis lies on the x-axis. **(b)** Transverse axis lies on the y-axis.

1 Locate a hyperbola's vertices and foci.

Using the Standard Form of the Equation of a Hyperbola

We can use the standard form of the equation of a hyperbola to find its vertices and locate its foci. Because the vertices are a units from the center, begin by identifying a^2 in the equation. In the standard form of a hyperbola's equation, **a^2 is the number under the variable whose term is preceded by a plus sign** $(+)$. If the x^2 term is preceded by a plus sign, the transverse axis lies along the x-axis. Thus, the vertices are a units to the right and left of the origin. If the y^2 term is preceded by a plus sign, the transverse axis lies along the y-axis. Thus, the vertices are a units above and below the origin.

We know that the foci are c units from the center. The substitution that we used to derive the hyperbola's equation, $b^2 = c^2 - a^2$, is needed to locate the foci when a^2 and b^2 are known. To find c^2, and then c, we will use an equivalent form of $b^2 = c^2 - a^2$, namely $c^2 = a^2 + b^2$.

EXAMPLE 1 Finding Vertices and Foci from a Hyperbola's Equation

Find the vertices and locate the foci for each of the following hyperbolas with the given equation.

a. $\frac{x^2}{16} - \frac{y^2}{9} = 1$ **b.** $\frac{y^2}{9} - \frac{x^2}{16} = 1$

Solution Both equations are in standard form. We begin by identifying a^2 and b^2 in each equation.

a. The first equation is in the form $\frac{x^2}{a^2} - \frac{y^2}{b^2} = 1$.

$$\frac{x^2}{16} - \frac{y^2}{9} = 1$$

$a^2 = 16$. This is the number in the denominator of the term preceded by a plus sign.

$b^2 = 9$. This is the number in the denominator of the term preceded by a minus sign.

Because the x^2 term is preceded by a plus sign, the transverse axis lies along the x-axis. Thus, the vertices are a units to the *right* and *left* of the origin. Based on the standard form of the equation, we know the vertices are $(-a, 0)$ and $(a, 0)$. Because $a^2 = 16$, $a = 4$. Thus, the vertices are $(-4, 0)$ and $(4, 0)$, shown in Figure 7.18.

We use $c^2 = a^2 + b^2$ to find the foci, which are located at $(-c, 0)$ and $(c, 0)$. We know that $a^2 = 16$ and $b^2 = 9$; we need to find c^2 in order to find c.

$$c^2 = a^2 + b^2 = 16 + 9 = 25$$

Because $c^2 = 25$, $c = 5$. The foci are located at $(-5, 0)$ and $(5, 0)$. They are shown in Figure 7.18.

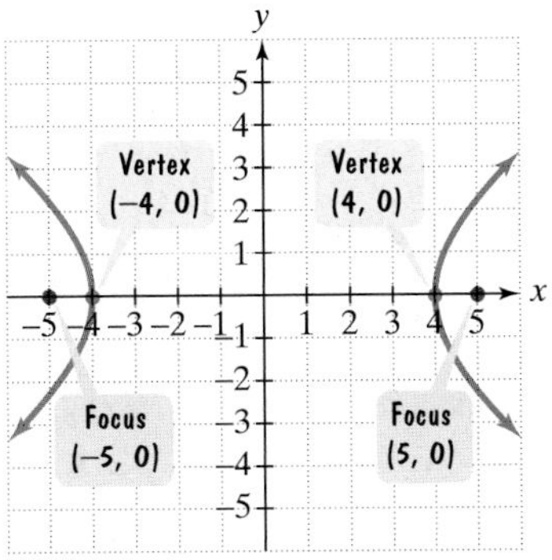

Figure 7.18 The graph of $\frac{x^2}{16} - \frac{y^2}{9} = 1$

b. The second given equation is in the form $\frac{y^2}{a^2} - \frac{x^2}{b^2} = 1$.

$$\frac{y^2}{9} - \frac{x^2}{16} = 1$$

$a^2 = 9$. This is the number in the denominator of the term preceded by a plus sign.

$b^2 = 16$. This is the number in the denominator of the term preceded by a minus sign.

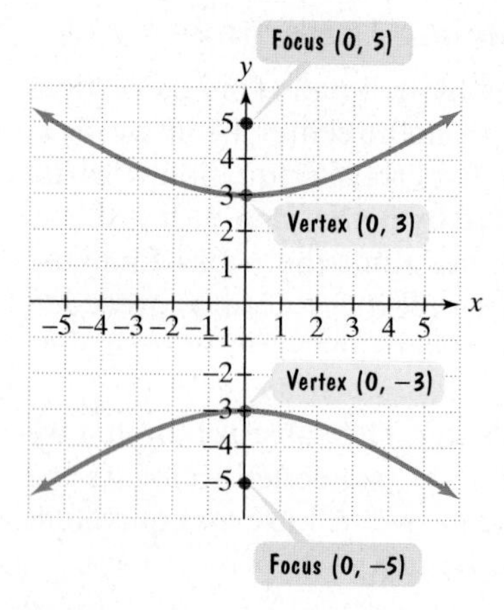

Figure 7.19 The graph of $\frac{y^2}{9} - \frac{x^2}{16} = 1$

Because the y^2 term is preceded by a plus sign, the transverse axis lies along the y-axis. Thus, the vertices are a units *above* and *below* the origin. Based on the standard form of the equation, we know the vertices are $(0, -a)$ and $(0, a)$. Because $a^2 = 9$, $a = 3$. Thus, the vertices are $(0, -3)$ and $(0, 3)$, shown in Figure 7.19.

We use $c^2 = a^2 + b^2$ to find the foci, which are located at $(0, -c)$ and $(0, c)$.

$$c^2 = a^2 + b^2 = 9 + 16 = 25$$

Because $c^2 = 25$, $c = 5$. The foci are located at $(0, -5)$ and $(0, 5)$. They are shown in Figure 7.19.

Check Point 1 Find the vertices and locate the foci for each of the following hyperbolas with the given equation.

a. $\frac{x^2}{25} - \frac{y^2}{16} = 1$ **b.** $\frac{y^2}{25} - \frac{x^2}{16} = 1$

2 Write equations of hyperbolas in standard form.

In Example 1, we used equations of hyperbolas to find their foci and vertices. In the next example, we reverse this procedure.

EXAMPLE 2 Finding the Equation of a Hyperbola from Its Foci and Vertices

Find the standard form of the equation of a hyperbola with foci at $(0, -3)$ and $(0, 3)$ and vertices $(0, -2)$ and $(0, 2)$, shown in Figure 7.20.

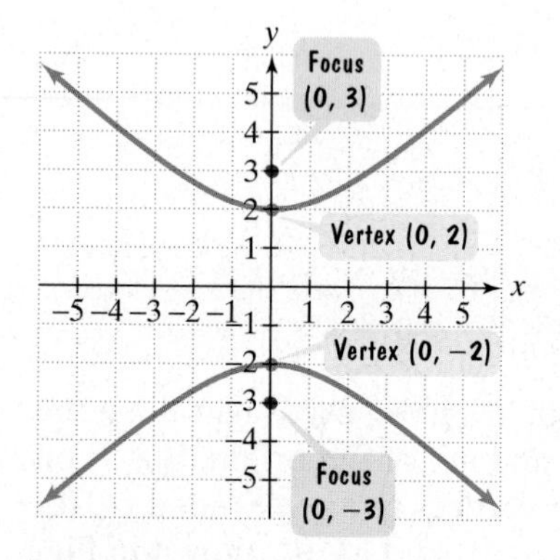

Figure 7.20

Solution Because the foci are located at $(0, -3)$ and $(0, 3)$, on the y-axis, the transverse axis lies on the y-axis. The center of the hyperbola is midway between the foci, located at $(0, 0)$. Thus, the form of the equation is

$$\frac{y^2}{a^2} - \frac{x^2}{b^2} = 1.$$

We need to determine the values for a^2 and b^2. The distance from the center $(0, 0)$ to either vertex, $(0, -2)$ or $(0, 2)$, is 2, so $a = 2$.

$$\frac{y^2}{2^2} - \frac{x^2}{b^2} = 1 \quad \text{or} \quad \frac{y^2}{4} - \frac{x^2}{b^2} = 1$$

We must still find b^2. The distance from the center, $(0, 0)$, to either focus, $(0, -3)$ or $(0, 3)$, is 3. Thus, $c = 3$. Because $b^2 = c^2 - a^2$, we have

$$b^2 = 3^2 - 2^2 = 9 - 4 = 5.$$

Substituting 5 for b^2 in the last equation gives us the standard form of the hyperbola's equation. The equation is

$$\frac{y^2}{4} - \frac{x^2}{5} = 1.$$

Check Point 2 Find the standard form of the equation of a hyperbola with foci at $(0, -5)$ and $(0, 5)$ and vertices $(0, -3)$ and $(0, 3)$.

The Asymptotes of a Hyperbola

As x and y get larger, the two branches of the graph of a hyperbola approach a pair of intersecting straight lines called **asymptotes.** The asymptotes pass through the center of the hyperbola and are helpful in graphing hyperbolas.

Figure 7.21 shows the asymptotes for the graphs of hyperbolas centered at the origin. The asymptotes pass through the corners of a rectangle. Note that the dimensions of this rectangle are $2a$ by $2b$. The line segment of length $2b$ is the **conjugate axis** of the hyperbola and is perpendicular to the transverse axis.

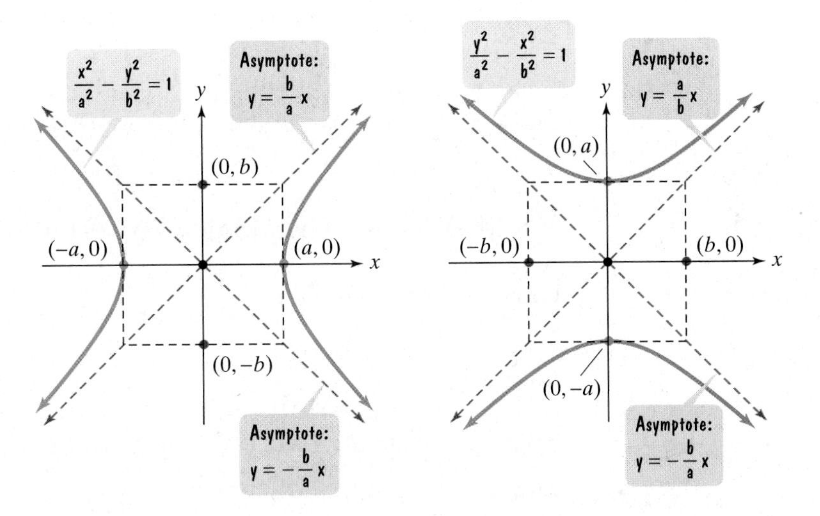

Figure 7.21 Asymptotes of a hyperbola

The Asymptotes of a Hyperbola Centered at the Origin

The hyperbola $\frac{x^2}{a^2} - \frac{y^2}{b^2} = 1$ with a horizontal transverse axis has the two asymptotes

$$y = \frac{b}{a}x \quad \text{and} \quad y = -\frac{b}{a}x.$$

The hyperbola $\frac{y^2}{a^2} - \frac{x^2}{b^2} = 1$ with a vertical transverse axis has the two asymptotes

$$y = \frac{a}{b}x \quad \text{and} \quad y = -\frac{a}{b}x.$$

Why are $y = \pm\frac{b}{a}x$ the asymptotes for a hyperbola whose transverse axis is horizontal? The proof can be found in the appendix.

3 Graph hyperbolas centered at the origin.

Graphing Hyperbolas Centered at the Origin

Hyperbolas are graphed using vertices and asymptotes.

Graphing Hyperbolas

1. Locate the vertices.
2. Draw the rectangle centered at the origin with sides parallel to the axes, crossing one axis at $\pm a$ and the other at $\pm b$.
3. Draw the diagonals of this rectangle and extend them to obtain the asymptotes.
4. Draw the two branches of the hyperbola by starting at each vertex and approaching the asymptotes.

The rectangle in step 2 and the asymptotes in step 3 are drawn using dashed lines to show that they are not part of the hyperbola.

EXAMPLE 3 Graphing a Hyperbola

Graph and locate the foci: $\frac{x^2}{25} - \frac{y^2}{16} = 1.$

Solution

Step 1 Locate the vertices. The given equation is in the form $\frac{x^2}{a^2} - \frac{y^2}{b^2} = 1$, with $a^2 = 25$ and $b^2 = 16$.

$$\frac{x^2}{25} - \frac{y^2}{16} = 1$$

$a^2 = 25$ $b^2 = 16$

Based on the standard form of the equation with the transverse axis on the x-axis, we know that the vertices are $(-a, 0)$ and $(a, 0)$. Because $a^2 = 25$, $a = 5$. Thus, the vertices are $(-5, 0)$ and $(5, 0)$, shown in Figure 7.22.

Step 2 Draw a rectangle. Because $a^2 = 25$ and $b^2 = 16$, $a = 5$ and $b = 4$. We construct a rectangle to find the asymptotes, using -5 and 5 on the x-axis (the ver-

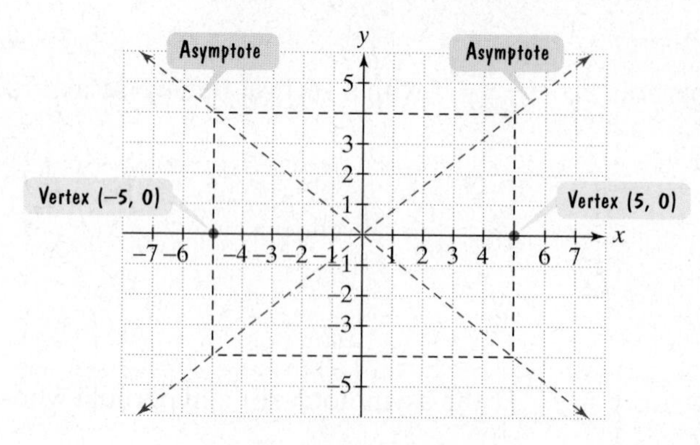

Figure 7.22 Preparing to graph $\frac{x^2}{25} - \frac{y^2}{16} = 1$

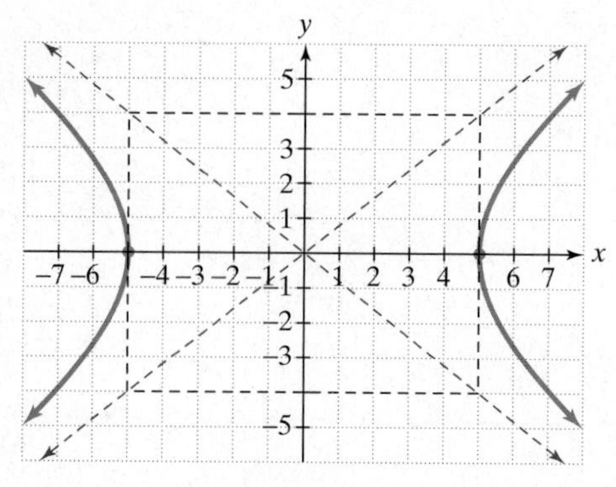

Figure 7.23 The graph of $\frac{x^2}{25} - \frac{y^2}{16} = 1$

Technology

Graph $\frac{x^2}{25} - \frac{y^2}{16} = 1$ by solving for y:

$$y_1 = \frac{\sqrt{16x^2 - 400}}{5}$$

$$y_2 = -\frac{\sqrt{16x^2 - 400}}{5} = -y_1.$$

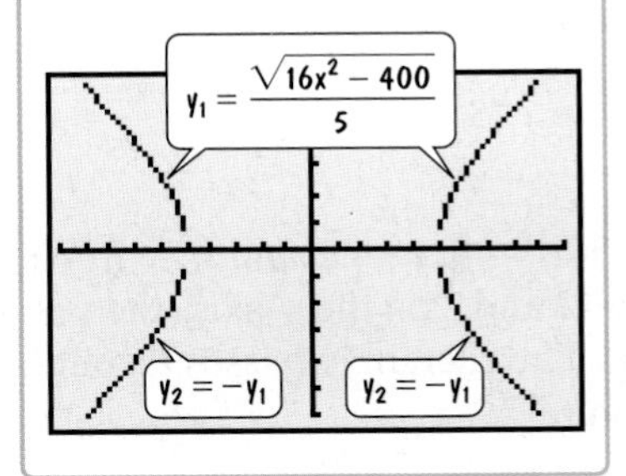

tices are located here) and −4 and 4 on the y-axis. The rectangle passes through these four points, shown using dashed lines in Figure 7.22.

Step 3 Draw extended diagonals for the rectangle to obtain the asymptotes. We draw dashed lines through the opposite corners of the rectangle, shown in Figure 7.22, to obtain the graph of the asymptotes. Based on the standard form of the hyperbola's equation, the equations for these asymptotes are

$$y = \pm\frac{b}{a}x \quad \text{or} \quad y = \pm\frac{4}{5}x.$$

Step 4 Draw the two branches of the hyperbola by starting at each vertex and approaching the asymptotes. The hyperbola is shown in Figure 7.23.

The foci are located at $(-c, 0)$ and $(c, 0)$. We find c using $c^2 = a^2 + b^2$.

$$c^2 = 25 + 16 = 41$$

Because $c^2 = 41$, $c = \sqrt{41}$. The foci are located at $(-\sqrt{41}, 0)$ and $(\sqrt{41}, 0)$, approximately $(-6.4, 0)$ and $(6.4, 0)$.

Check Point 3 Graph and locate the foci: $\frac{x^2}{36} - \frac{y^2}{9} = 1$.

EXAMPLE 4 Graphing a Hyperbola

Graph and locate the foci: $9y^2 - 4x^2 = 36$.

Solution We begin by writing the equation in standard form. The right side should be 1, so we divide both sides by 36.

$$\frac{9y^2}{36} - \frac{4x^2}{36} = \frac{36}{36}$$

$$\frac{y^2}{4} - \frac{x^2}{9} = 1 \quad \text{Simplify. The right side is now 1.}$$

Now we are ready to use our four-step procedure for graphing hyperbolas.

Step 1 Locate the vertices. The equation that we obtained is in the form $\frac{y^2}{a^2} - \frac{x^2}{b^2} = 1$, with $a^2 = 4$ and $b^2 = 9$.

$$\frac{y^2}{4} - \frac{x^2}{9} = 1$$

$a^2 = 4$ $b^2 = 9$

Based on the standard form of the equation with the transverse axis on the y-axis, we know that the vertices are $(0, -a)$ and $(0, a)$. Because $a^2 = 4$, $a = 2$. Thus, the vertices are $(0, -2)$ and $(0, 2)$, shown in Figure 7.24.

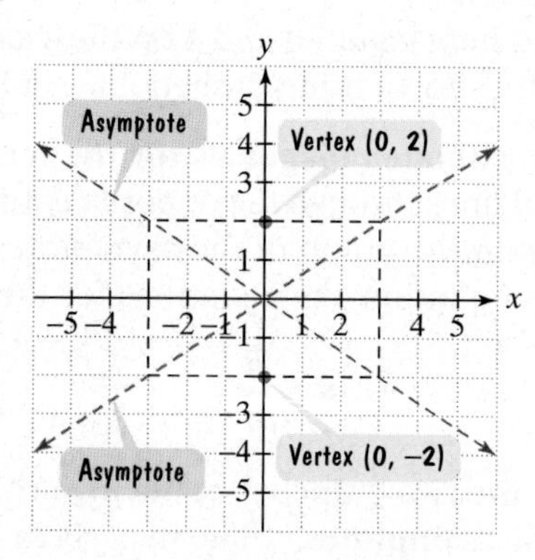

Figure 7.24 Preparing to graph $\frac{y^2}{4} - \frac{x^2}{9} = 1$

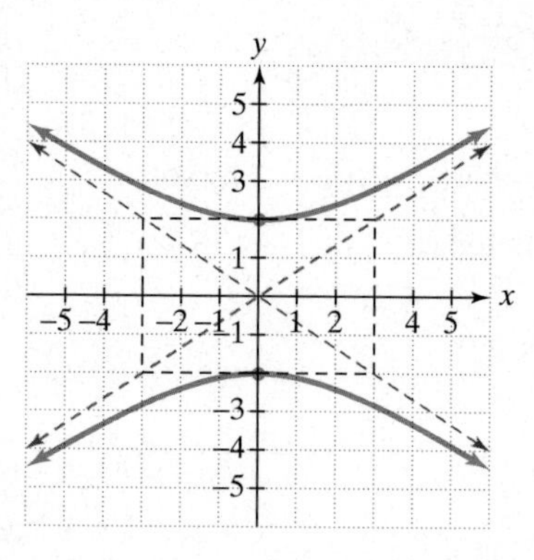

Figure 7.25 The graph of $\frac{y^2}{4} - \frac{x^2}{9} = 1$

Step 2 Draw a rectangle. Because $a^2 = 4$ and $b^2 = 9$, $a = 2$ and $b = 3$. We construct a rectangle to find the asymptotes, using -2 and 2 on the y-axis (the vertices are located here) and -3 and 3 on the x-axis. The rectangle passes through these four points, shown using dashed lines in Figure 7.24.

Step 3 Draw extended diagonals of the rectangle to obtain the asymptotes. We draw dashed lines through the opposite corners of the rectangle, shown in Figure 7.24, to obtain the graph of the asymptotes. Based on the standard form of the hyperbola's equation, the equations of these asymptotes are

$$y = \pm\frac{a}{b}x \quad \text{or} \quad y = \pm\frac{2}{3}x.$$

Step 4 Draw the two branches of the hyperbola by starting at each vertex and approaching the asymptotes. The hyperbola is shown in Figure 7.25.

The foci are located at $(0, -c)$ and $(0, c)$. We find c using $c^2 = a^2 + b^2$.

$$c^2 = 4 + 9 = 13$$

Because $c^2 = 13$, $c = \sqrt{13}$. The foci are located at $(0, -\sqrt{13})$ and $(0, \sqrt{13})$, approximately $(0, -3.6)$ and $(0, 3.6)$.

Check Point 4 Graph and locate the foci: $y^2 - 4x^2 = 4$.

4 Graph hyperbolas not centered at the origin.

Translations of Hyperbolas

The graph of a hyperbola can be centered at (h, k) rather than at the origin. Horizontal and vertical translations are accomplished by replacing x with $x - h$ and y with $y - k$ in the standard form of the hyperbola's equation.

Table 7.2 gives the standard forms of equations of hyperbolas centered at (h, k). Figure 7.26 shows their graphs.

Table 7.2 Standard Forms of Equations of Hyperbolas Centered at (h, k)

Equation	Center	Transverse Axis	Foci	Vertices
$\frac{(x-h)^2}{a^2} - \frac{(y-k)^2}{b^2} = 1,$ $b^2 = c^2 - a^2$	(h, k)	Parallel to x-axis; horizontal	$(h-c, k)$ $(h+c, k)$	$(h-a, k)$ $(h+a, k)$
$\frac{(y-k)^2}{a^2} - \frac{(x-h)^2}{b^2} = 1,$ $b^2 = c^2 - a^2$	(h, k)	Parallel to y-axis; vertical	$(h, k-c)$ $(h, k+c)$	$(h, k-a)$ $(h, k+a)$

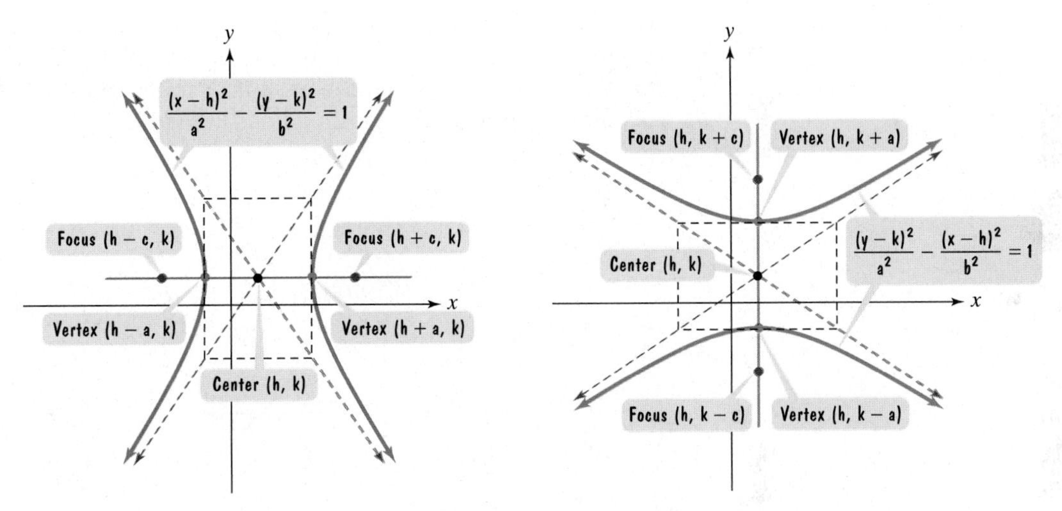

Figure 7.26 Graphs of hyperbolas centered at (h, k).

EXAMPLE 5 Graphing a Hyperbola Centered at (h, k)

Graph: $\frac{(x-2)^2}{16} - \frac{(y-3)^2}{9} = 1$. Where are the foci located?

Solution In order to graph the hyperbola, we need to know its center (h, k). In the standard forms of equations centered at (h, k), h is the number subtracted from x and k is the number subtracted from y.

$$\frac{(x-2)^2}{16} - \frac{(y-3)^2}{9} = 1$$

This is $(x-h)^2$, with $h = 2$.

This is $(y-k)^2$, with $k = 3$.

We see that $h = 2$ and $k = 3$. Thus, the center of hyperbola, (h, k), is $(2, 3)$. We can graph the hyperbola by using vertices, asymptotes, and our four-step graphing procedure.

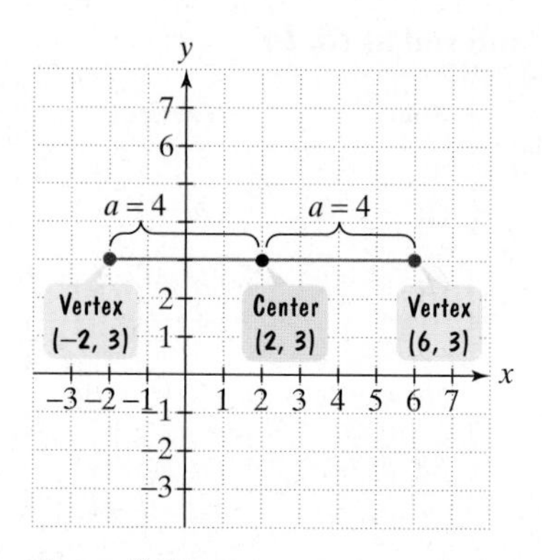

Figure 7.27 Locating a hyperbola's center and vertices

Step 1 Locate the vertices. To do this, we must identify a^2.

$$\frac{(x-2)^2}{16} - \frac{(y-3)^2}{9} = 1 \quad \text{The form of this equation is } \frac{(x-h)^2}{a^2} - \frac{(y-k)^2}{b^2} = 1.$$

$a^2 = 16$ $\quad b^2 = 9$

Based on the standard form of the equation with a horizontal transverse axis, the vertices are a units to the right and left of the center. Because $a^2 = 16$, $a = 4$. This means that the vertices are 4 units to the right and left of the center, (2, 3). Four units to the right of (2, 3) puts one vertex at $(2 + 4, 3)$, or (6, 3). Four units to the left of (2, 3) puts the other vertex at $(2 - 4, 3)$, or $(-2, 3)$. The vertices are shown in Figure 7.27.

Step 2 Draw a rectangle. Because $a^2 = 16$ and $b^2 = 9$, $a = 4$ and $b = 3$. The rectangle passes through points that are 4 units to the right and left of the center (the vertices are located here) and 3 units above and below the center. The rectangle is shown using dashed lines in Figure 7.28.

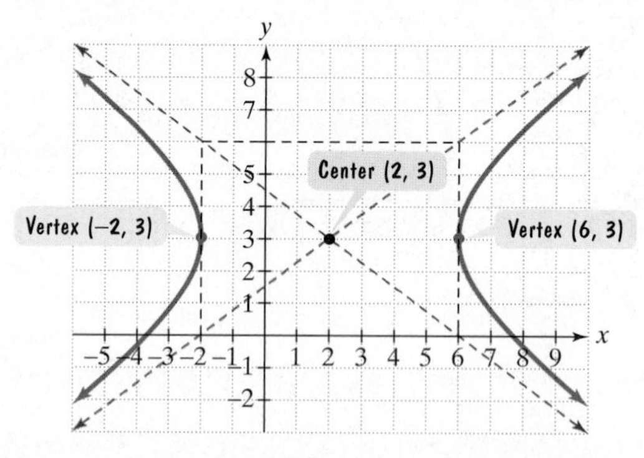

Figure 7.28 The graph of $\frac{(x-2)^2}{16} - \frac{(y-3)^2}{9} = 1$

Step 3 Draw extended diagonals of the rectangle to obtain the asymptotes. We draw dashed lines through the opposite corners of the rectangle, shown in Figure 7.28, to obtain the graph of the asymptotes. The equations of the asymptotes of the unshifted hyperbola $\frac{x^2}{16} - \frac{y^2}{9} = 1$ are $y = \pm\frac{b}{a}x$, or $y = \pm\frac{3}{4}x$. Thus, the asymptotes for the hyperbola that is shifted two units to the right and three units up, namely

$$\frac{(x-2)^2}{16} - \frac{(y-3)^2}{9} = 1$$

have equations that can be expressed as

$$y - 3 = \pm\frac{3}{4}(x - 2).$$

Step 4 Draw the two branches of the hyperbola by starting at each vertex and approaching the asymptotes. The hyperbola is shown in Figure 7.28.

The foci are located c units to the right and left of the center. We find c using $c^2 = a^2 + b^2$.

$$c^2 = 16 + 9 = 25$$

Because $c^2 = 25$, $c = 5$. This means that the foci are 5 units to the right and left of the center, (2, 3). Five units to the right of (2, 3) puts one focus at $(2 + 5, 3)$, or $(7, 3)$. Five units to the left of $(2, 3)$ puts the other focus at $(2 - 5, 3)$, or $(-3, 3)$.

Check Point 5 Graph and locate the foci: $\dfrac{(x-3)^2}{4} - \dfrac{(y-1)^2}{1} = 1.$

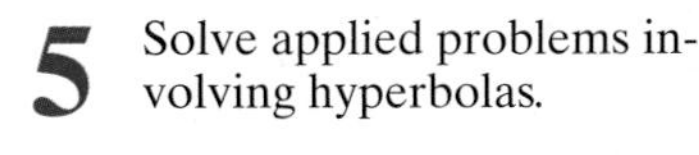

5 Solve applied problems involving hyperbolas.

Applications

Hyperbolas have many applications. When a jet flies at a speed greater than the speed of sound, the shock wave that is created is heard as a sonic boom. The wave has the shape of a cone. The shape formed as the cone hits the ground is one branch of a hyperbola.

Halley's Comet, a permanent part of our solar system, travels around the sun in an elliptical orbit. Other comets pass through the solar system only once, following a hyperbolic path with the sun as a focus.

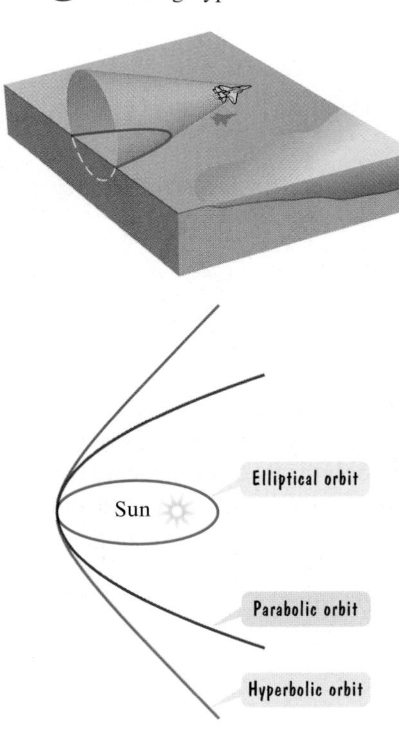

Hyperbolas are of practical importance in fields ranging from architecture to navigation. Cooling towers used in the design for nuclear power plants have cross sections that are both ellipses and hyperbolas. Three-dimensional solids whose cross sections are hyperbolas are used in some rather unique architectural creations, including the TWA building at Kennedy Airport and the St. Louis Science Center Planetarium.

EXAMPLE 6 An Application Involving Hyperbolas

An explosion is recorded by two microphones that are 2 miles apart. Microphone M_1 received the sound 4 seconds before microphone M_2. Assuming sound travels at 1100 feet per second, determine the possible locations of the explosion relative to the location of the microphones.

Solution We begin by putting the microphones in a coordinate system. Because 1 mile $=$ 5280 feet, we place M_1 5280 feet on a horizontal axis to the right of the origin and M_2 5280 feet on a horizontal axis to the left of the origin. Figure 7.29 illustrates that the two microphones are 2 miles apart.

We know that M_2 received the sound 4 seconds after M_1. Because sound travels at 1100 feet per second, the difference between the distance from P to M_1 and the distance from P to M_2 is 4400 feet. The set of all points P (or locations

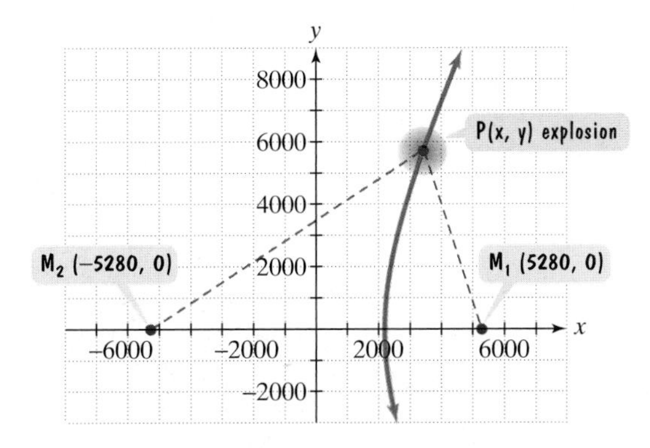

Figure 7.29 Locating an explosion on the branch of a hyperbola

Where Exactly Am I?

The hyperbola is the basis for the navigational system LORAN (for long-range navigation), used by a ship or aircraft to determine its location. The measured time-of-arrival difference between signals transmitted from two ground stations determines the hyperbola on which the ship or aircraft is located. The process is then repeated by taking a similar time-difference reading from a second pair of stations, determining a second hyperbola. The point of intersection of the two hyperbolas is the location of the ship or aircraft.

LORAN will eventually be replaced by the Global Positioning System. Using 24 satellites that orbit at 11,000 miles above Earth, the system is able to show you your exact position on Earth anytime, in any weather, anywhere.

of the explosion) satisfying these conditions fits the definition of a hyperbola, with microphones M_1 and M_2 at the foci.

$$\frac{x^2}{a^2} - \frac{y^2}{b^2} = 1$$ Use the standard form of the hyperbola's equation. $P(x, y)$, the explosion point, lies on this hyperbola. We must find a^2 and b^2.

The difference between the distances, represented by $2a$ in the derivation of the hyperbola's equation, is 4400 feet. Thus, $2a = 4400$ and $a = 2200$.

$$\frac{x^2}{(2200)^2} - \frac{y^2}{b^2} = 1$$ Substitute 2200 for a.

Because $c = 5280$ and $a = 2200$, then $b^2 = c^2 - a^2 = 5280^2 - 2200^2 = 23{,}038{,}400$.

$$\frac{x^2}{4{,}840{,}000} - \frac{y^2}{23{,}038{,}400} = 1$$ Substitute 23,038,400 for b^2.

We can conclude that the explosion occurred somewhere on the right branch (the branch closest to M_1) of the hyperbola given by

$$\frac{x^2}{4{,}840{,}000} - \frac{y^2}{23{,}038{,}400} = 1.$$

In Example 6, we determined that the explosion occurred somewhere along one branch of a hyperbola, but not exactly where on the hyperbola. If, however, we had received the sound from another pair of microphones, we could locate the sound along a branch of another hyperbola. The exact location of the explosion would be the point where the two hyperbolas intersect.

Check Point 6 Rework Example 6 if Microphone M_1 receives the sound 3 seconds before Microphone M_2.

EXERCISE SET 7.2

Practice Exercises

In Exercises 1–4, find the vertices and locate the foci of each hyperbola with the given equation. Then match each equation to one of the graphs that are shown and labeled (a)–(d).

1. $\frac{x^2}{4} - \frac{y^2}{1} = 1$
2. $\frac{x^2}{1} - \frac{y^2}{4} = 1$
3. $\frac{y^2}{4} - \frac{x^2}{1} = 1$
4. $\frac{y^2}{1} - \frac{x^2}{4} = 1$

a.

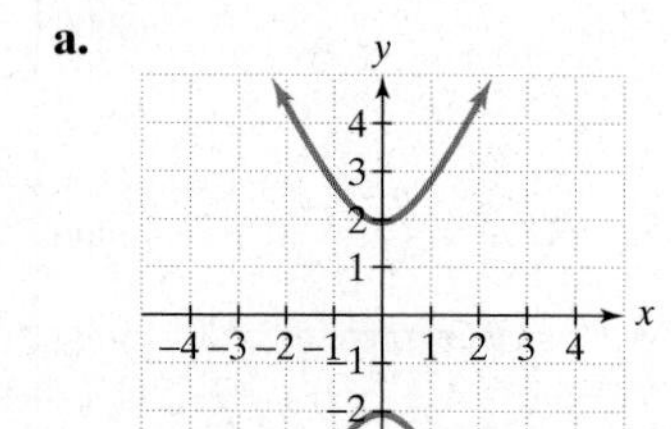

b.

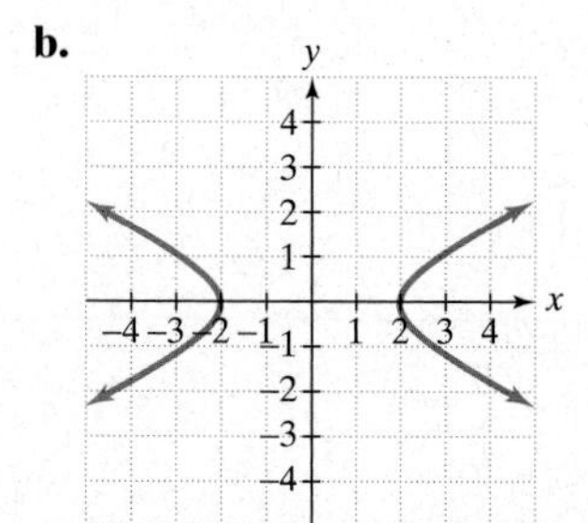

c.

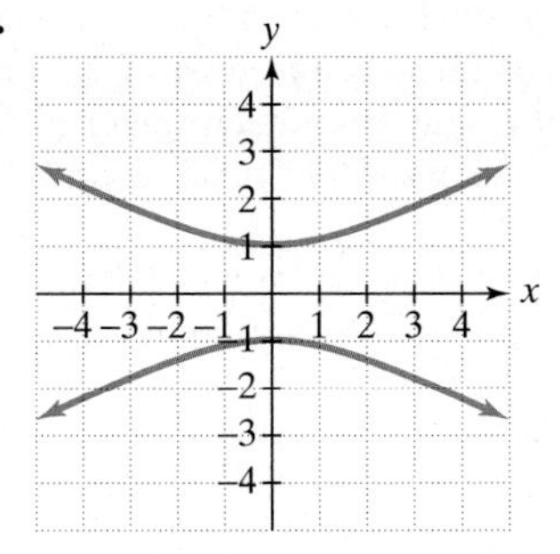

d.

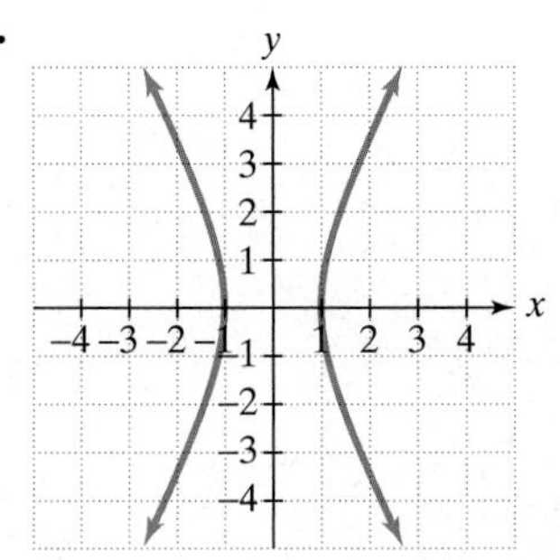

In Exercises 5–8, find the standard form of the equation of each hyperbola centered at the origin satisfying the given conditions.

5. Foci: $(0, -3), (0, 3)$; vertices: $(0, -1), (0, 1)$

6. Foci: $(0, -6), (0, 6)$; vertices: $(0, -2), (0, 2)$

7. Foci: $(-4, 0), (4, 0)$; vertices: $(-3, 0), (3, 0)$

8. Foci: $(-7, 0), (7, 0)$; vertices: $(-5, 0), (5, 0)$

In Exercises 9–22, use vertices and asymptotes to graph each hyperbola. Locate the foci.

9. $\dfrac{x^2}{9} - \dfrac{y^2}{25} = 1$

10. $\dfrac{x^2}{16} - \dfrac{y^2}{25} = 1$

11. $\dfrac{x^2}{100} - \dfrac{y^2}{64} = 1$

12. $\dfrac{x^2}{144} - \dfrac{y^2}{81} = 1$

13. $\dfrac{y^2}{16} - \dfrac{x^2}{36} = 1$

14. $\dfrac{y^2}{25} - \dfrac{x^2}{64} = 1$

15. $\dfrac{y^2}{36} - \dfrac{x^2}{25} = 1$

16. $\dfrac{y^2}{100} - \dfrac{x^2}{49} = 1$

17. $9x^2 - 4y^2 = 36$

18. $4x^2 - 25y^2 = 100$

19. $9y^2 - 25x^2 = 225$

20. $16y^2 - 9x^2 = 144$

21. $4x^2 = 4 + y^2$

22. $25y^2 = 225 + 9x^2$

In Exercises 23–26, find the standard form of the equation of each hyperbola.

23.

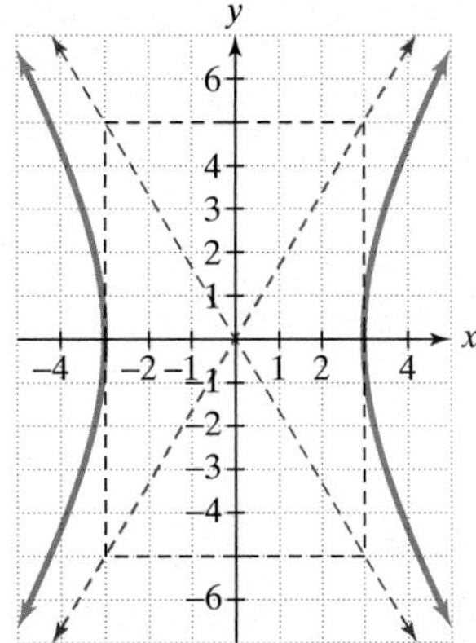

24.

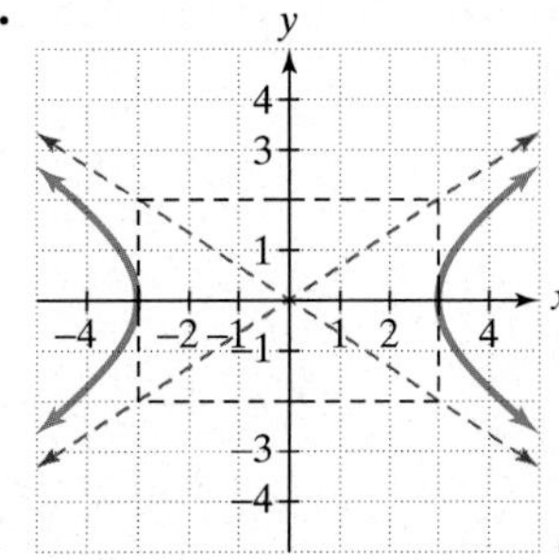

25.

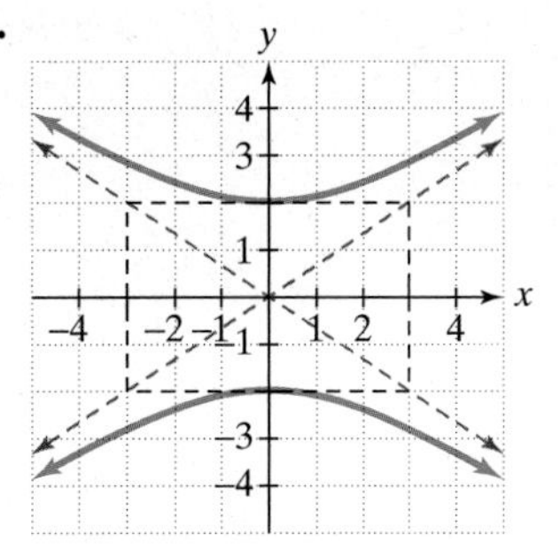

26.

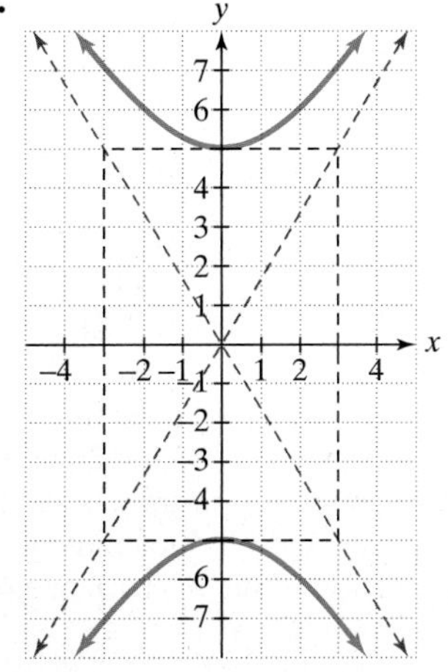

In Exercises 27–36, use the center, vertices, and asymptotes to graph each hyperbola. Locate the foci.

27. $\frac{(x+4)^2}{9} - \frac{(y+3)^2}{16} = 1$

28. $\frac{(x+2)^2}{9} - \frac{(y-1)^2}{25} = 1$

29. $\frac{(x+3)^2}{25} - \frac{y^2}{16} = 1$

30. $\frac{(x+2)^2}{9} - \frac{y^2}{25} = 1$

31. $\frac{(y+2)^2}{4} - \frac{(x-1)^2}{16} = 1$

32. $\frac{(y-2)^2}{36} - \frac{(x+1)^2}{49} = 1$

33. $(x-3)^2 - 4(y+3)^2 = 4$

34. $(x+3)^2 - 9(y-4)^2 = 9$

35. $(x-1)^2 - (y-2)^2 = 4$

36. $(y-2)^2 - (x+3)^2 = 4$

In Exercises 37–44, convert each equation to standard form by completing the square on x and y. Then graph the hyperbola and give the location of its foci.

37. $x^2 - y^2 - 2x - 4y - 4 = 0$

38. $4x^2 - y^2 + 32x + 6y + 39 = 0$

39. $16x^2 - y^2 + 64x - 2y + 67 = 0$

40. $9y^2 - 4x^2 - 18y + 24x - 63 = 0$

41. $4x^2 - 9y^2 - 16x + 54y - 101 = 0$

42. $4x^2 - 9y^2 + 8x - 18y - 6 = 0$

43. $4x^2 - 25y^2 - 32x + 164 = 0$

44. $9x^2 - 16y^2 - 36x - 64y + 116 = 0$

Application Exercises

45. An explosion is recorded by two microphones that are 1 mile apart. Microphone M_1 received the sound 2 seconds before microphone M_2. Assuming sound travels at 1100 feet per second, determine the possible locations of the explosion relative to the location of the microphones.

46. Radio towers A and B, 200 kilometers apart, are situated along the coast, with A located due west of B. Simultaneous radio signals are sent from each tower to a ship, with the signal from B received 500 microseconds before the signal from A.

a. Assuming that the radio signals travel 300 meters per microsecond, determine the equation of the hyperbola on which the ship is located.

b. If the ship lies due north of tower B, how far out at sea is it?

47. An architect designs two houses that are shaped and positioned like a part of the branches of the hyperbola whose equation is $625y^2 - 400x^2 = 250{,}000$, where x and y are in yards. How far apart are the houses at their closest point?

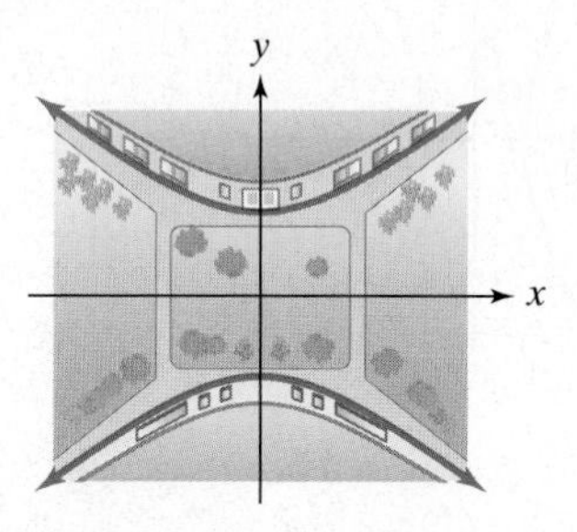

48. Scattering experiments, in which moving particles are deflected by various forces, led to the concept of the nucleus of an atom. In 1911, the physicist Ernest Rutherford (1871–1937) discovered that when alpha particles are directed toward the nuclei of gold atoms, they are eventually deflected along hyperbolic paths, illustrated in the figure. If a particle gets as close as 3 units to the nucleus along a hyperbolic path with an asymptote given by $y = \frac{1}{2}x$, what is the equation of its path?

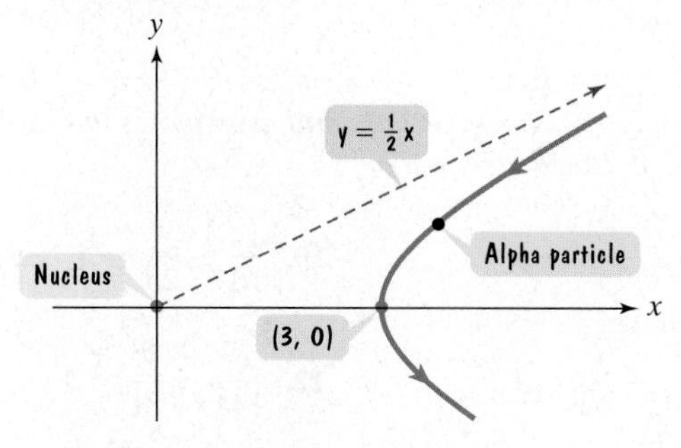

Writing in Mathematics

49. What is a hyperbola?

50. Describe how to graph $\frac{x^2}{9} - \frac{y^2}{1} = 1$.

51. Describe how to locate the foci of the graph of $\frac{x^2}{9} - \frac{y^2}{1} = 1$.

52. Describe one similarity and one difference between the graphs of $\frac{x^2}{9} - \frac{y^2}{1} = 1$ and $\frac{y^2}{9} - \frac{x^2}{1} = 1$.

53. Describe one similarity and one difference between the graphs of $\frac{x^2}{9} - \frac{y^2}{1} = 1$ and $\frac{(x-3)^2}{9} - \frac{(y+3)^2}{1} = 1$.

54. How can you distinguish an ellipse from a hyperbola by looking at their equations?

55. In 1992, a NASA team began a project called Spaceguard Survey, calling for an international watch for comets that might collide with Earth. Why is it more difficult to detect a possible "doomsday comet" with a hyperbolic orbit than one with an elliptical orbit?

Technology Exercises

56. Use a graphing utility to graph any five of the hyperbolas that you graphed by hand in Exercises 9–22.

57. Use a graphing utility to graph any three of the hyperbolas that you graphed by hand in Exercises 27–36. First solve the given equation for y by using the square root method. Enter each of the two resulting equations to produce each branch of the hyperbola.

58. Use a graphing utility to graph any one of the hyperbolas that you graphed by hand in Exercises 37–44. Write the equation as a quadratic equation in y and use the quadratic formula to solve for y. Enter each of the two resulting equations to produce each branch of the hyperbola.

59. Use a graphing utility to graph $\frac{x^2}{4} - \frac{y^2}{9} = 0$. Is the graph a hyperbola? In general, what is the graph of $\frac{x^2}{a^2} - \frac{y^2}{b^2} = 0$?

60. Graph $\frac{x^2}{a^2} - \frac{y^2}{b^2} = 1$ and $\frac{x^2}{a^2} - \frac{y^2}{b^2} = -1$ in the same viewing rectangle for values of a^2 and b^2 of your choice. Describe the relationship between the two graphs.

61. Write $4x^2 - 6xy + 2y^2 - 3x + 10y - 6 = 0$ as a quadratic equation in y and then use the quadratic formula to express y in terms of x. Graph the resulting two equations using a graphing utility and a $[-50, 70, 10]$ by $[-30, 50, 10]$ viewing rectangle. What effect does the xy-term have on the graph of the resulting hyperbola? What problems would you encounter if you attempted to write the given equation in standard form by completing the square?

62. Graph $\frac{x^2}{16} - \frac{y^2}{9} = 1$ and $\frac{x|x|}{16} - \frac{y|y|}{9} = 1$ in the same viewing rectangle. Explain why the graphs are not the same.

Critical Thinking Exercises

63. Which one of the following is true?

a. If one branch of a hyperbola is removed from a graph, then the branch that remains must define y as a function of x.

b. All points on the asymptotes of a hyperbola also satisfy the hyperbola's equation.

c. The graph of $\frac{x^2}{9} - \frac{y^2}{4} = 1$ does not intersect the line $y = -\frac{2}{3}x$.

d. Two different hyperbolas can never share the same asymptotes.

64. What happens to the shape of the graph of $\frac{x^2}{a^2} - \frac{y^2}{b^2} = 1$ as $\frac{c}{a}$ gets larger and larger?

65. Find the standard form of the equation of the hyperbola with vertices $(5, -6)$ and $(5, 6)$, and passing through $(0, 9)$.

66. Find the equation of a hyperbola whose asymptotes are perpendicular.

SECTION 7.3 *The Parabola*

Objectives

1. Graph parabolas with vertices at the origin.
2. Write equations of parabolas in standard form.
3. Graph parabolas with vertices not at the origin.
4. Solve applied problems involving parabolas.

At first glance, this image looks like columns of smoke rising from a fire into a starry sky. Those are, indeed, stars in the background, but you are not looking at ordinary smoke columns. These stand almost 6 trillion miles high and are 7000 light-years from Earth—more than 400 million times as far away as the sun.

This NASA photograph is one of a series of stunning images captured from the ends of the universe by the Hubble Space Telescope. The image shows infant star systems the size of our solar system emerging from the gas and dust that shrouded their creation. Using a parabolic mirror that is 94.5 inches in diameter, the Hubble is providing answers to many of the profound mysteries of the cosmos: How big and how old is the universe? How did the galaxies come to exist? Do other Earth-like planets orbit other sun-like stars?

In Chapter 3, we studied parabolas, viewing them as graphs of the quadratic function $y = ax^2 + bx + c$. In this section, we will use a geometric definition of a parabola to derive its equation. We will also consider applications of parabolas, including parabolic shapes that gather distant rays of light and focus them into spectacular images.

Definition of a Parabola

The definitions of ellipses and hyperbolas involved two fixed points, the foci. By contrast, the definition of a parabola is based on one point and a line.

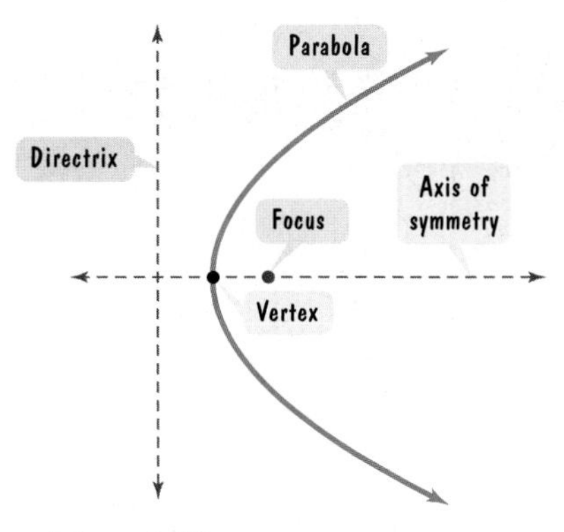

Figure 7.30

Definition of a Parabola

A **parabola** is the set of all points in a plane that are equidistant from a fixed line (the **directrix**) and a fixed point (the **focus**) that is not on the line (see Figure 7.30).

In Figure 7.30, find the line passing through the focus and perpendicular to the directrix. This is the **axis of symmetry** of the parabola. The point of intersection of the parabola with its axis of symmetry is called the **vertex**. Notice that the vertex is midway between the focus and the directrix.

Standard Form of the Equation of a Parabola

The rectangular coordinate system enables us to translate a parabola's geometric definition into an algebraic equation. Figure 7.31 is our starting point for obtaining an equation. We place the focus on the x-axis at the point $(p, 0)$. The directrix has an equation given by $x = -p$. The vertex, located midway between the focus and the directrix, is at the origin.

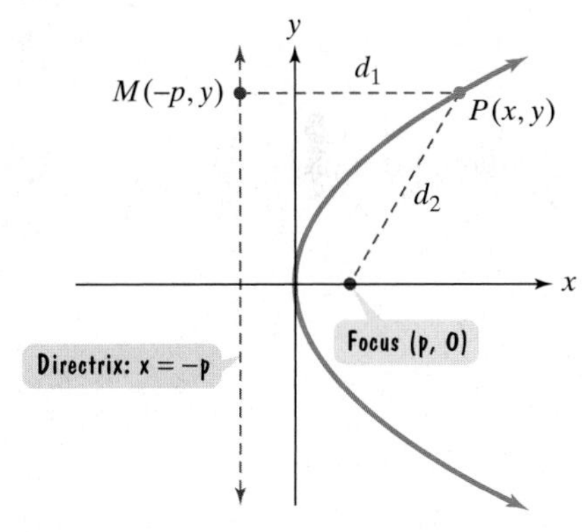

Figure 7.31

What does the definition of a parabola tell us about the point (x, y) in Figure 7.31? For any point (x, y) on the parabola, the distance d_1 to the directrix is equal to the distance d_2 to the focus. Thus, the point (x, y) is on the parabola if and only if

$$d_1 = d_2$$

$$\sqrt{(x+p)^2 + (y-y)^2} = \sqrt{(x-p)^2 + (y-0)^2} \quad \text{Use the distance formula.}$$

$$(x+p)^2 = (x-p)^2 + y^2 \quad \text{Square both sides of the equation.}$$

$$x^2 + 2px + p^2 = x^2 - 2px + p^2 + y^2 \quad \text{Square } x + p \text{ and } x - p.$$

$$2px = -2px + y^2 \quad \text{Subtract } x^2 + p^2 \text{ from both sides of the equation.}$$

$$y^2 = 4px \quad \text{Solve for } y^2.$$

This last equation is called the **standard form of the equation of a parabola.** There are two such equations, one for a focus on the x-axis and one for a focus on the y-axis.

Standard Forms of the Equations of a Parabola

The **standard form of the equation of a parabola** with vertex at the origin is

$$y^2 = 4px \quad \text{or} \quad x^2 = 4py.$$

Figure 7.32 illustrates that for the equation on the left, the focus is on the x-axis, which is the axis of symmetry. For the equation on the right, the focus is on the y-axis, which is the axis of symmetry.

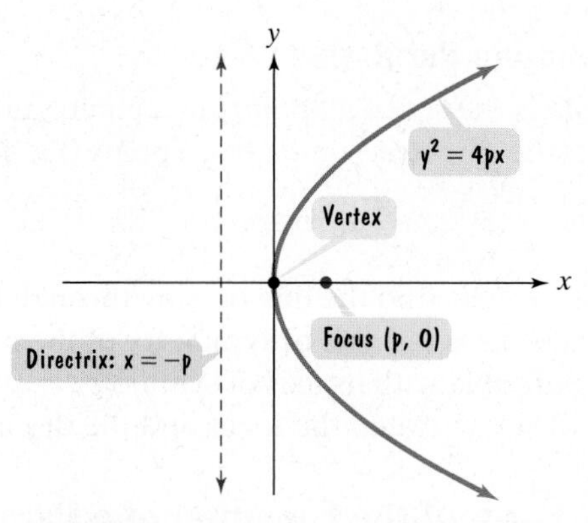

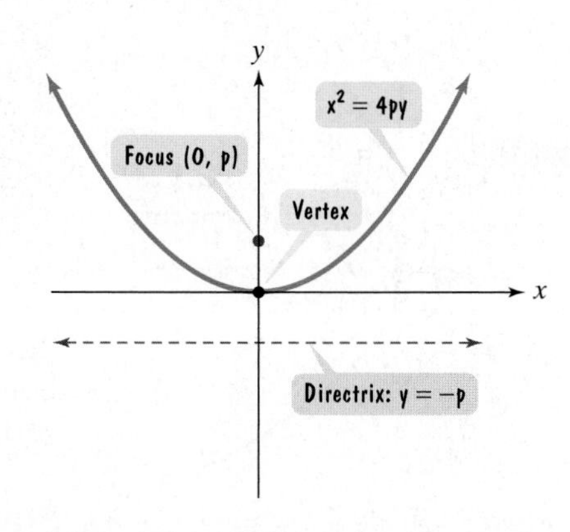

Figure 7.32 **(a)** Parabola with the x-axis as the axis of symmetry **(b)** Parabola with the y-axis as the axis of symmetry

1 Graph parabolas with vertices at the origin.

Using the Standard Form of the Equation of a Parabola

We can use the standard form of the equation of a parabola to find its focus and directrix. Remember that the focus is located on the axis corresponding to the variable in the equation that is *not* squared.

$$y^2 = 4px \qquad x^2 = 4py$$

x is not squared. Focus is on the x-axis at $(p, 0)$.

y is not squared. Focus is on the y-axis at $(0, p)$.

Although the definition of a parabola is given in terms of its focus and its directrix, the focus and directrix are not part of the graph. The vertex, located at the origin, is a point on the graph of $y^2 = 4px$ and $x^2 = 4py$. You can find two additional points on the parabola by assigning a value to x or y that makes y^2 or x^2 a perfect square. For example, consider $y^2 = 8x$. A value of x that makes the right side a perfect square is 2. If $x = 2$, then $y^2 = 8(2)$, or 16. Because $y^2 = 16$, $y = \pm 4$. Thus, the parabola passes through the points $(2, 4)$ and $(2, -4)$. The parabola can be graphed by connecting the vertex, $(0, 0)$, to each of these points with a smooth curve.

EXAMPLE 1 Finding the Focus and Directrix of a Parabola

Find the focus and directrix of the parabola given by $y^2 = 12x$. Then graph the parabola.

Solution The given equation is in the standard form $y^2 = 4px$, so $4p = 12$.

$$y^2 = 12x$$

This is $4p$.

We can find both the focus and the directrix by finding p.

$4p = 12$ Remember that the focus is at $(p, 0)$ and the directrix is given by $x = -p$.

$p = 3$ Divide both sides by 4.

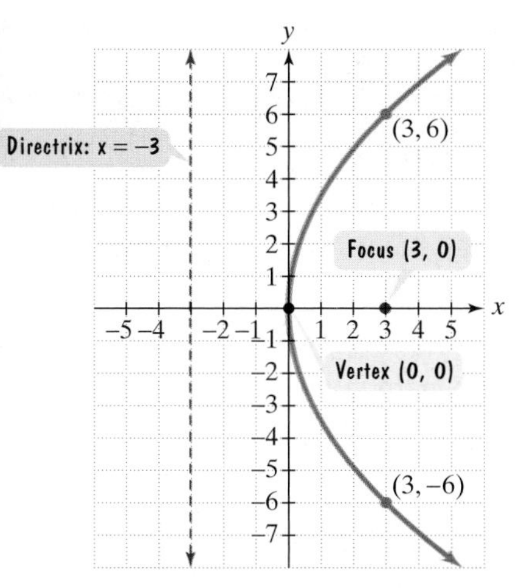

Figure 7.33 The graph of $y^2 = 12x$

Using this value for p, we obtain

$$\text{Focus:} \quad (p, 0) = (3, 0)$$
$$\text{Directrix:} \quad x = -p;\ x = -3.$$

Observe that the vertex, midway between the focus and the directrix, is at the origin. To graph $y^2 = 12x$, we assign x a value that makes the right side a perfect square. If $x = 3$, then $y^2 = 12(3)$ or $y^2 = 36$. Because $y = \pm 6$, the parabola passes through the points $(3, 6)$ and $(3, -6)$. The graph is sketched in Figure 7.33.

Check Point 1 Find the focus and directrix of the parabola given by $y^2 = 8x$. Then graph the parabola.

Parabolas with vertices at the origin can open to the right, left, upward, or downward. The graph of $y^2 = 4px$ opens to the right if $p > 0$ or left if $p < 0$. For example, Figure 7.34 shows that $y^2 = x$ opens to the right and $y^2 = -x$ opens to the left. The graph of $x^2 = 4py$ opens upward if $p > 0$ or downward if $p < 0$. Figure 7.34 shows that $x^2 = y$ opens upward and $x^2 = -y$ opens downward.

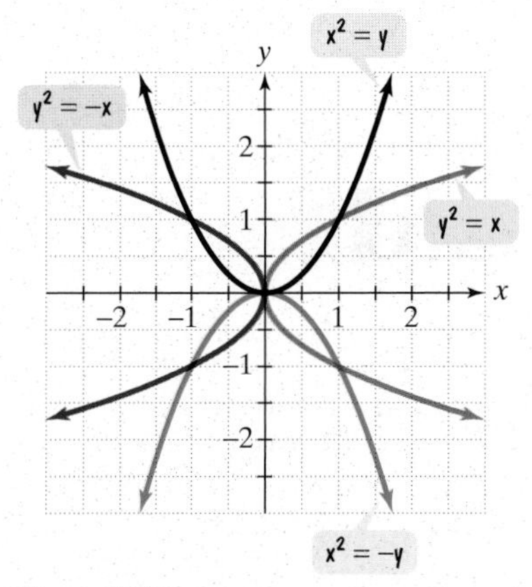

Figure 7.34

EXAMPLE 2 Finding the Focus and Directrix of a Parabola

Find the focus and directrix of the parabola given by $x^2 = -8y$. Then graph the parabola.

Solution The given equation is in the standard form $x^2 = 4py$, so $4p = -8$.

$$x^2 = -8y$$

This is $4p$.

We can find both the focus and the directrix by finding p.

$$4p = -8$$
$$p = -2$$

The focus, on the y-axis, is at $(0, p)$ and the directrix is given by $y = -p$.

Because $p < 0$, the parabola opens downward. Using this value for p, we obtain

$$\text{Focus:} \quad (0, p) = (0, -2)$$
$$\text{Directrix:} \quad y = -p;\ y = 2.$$

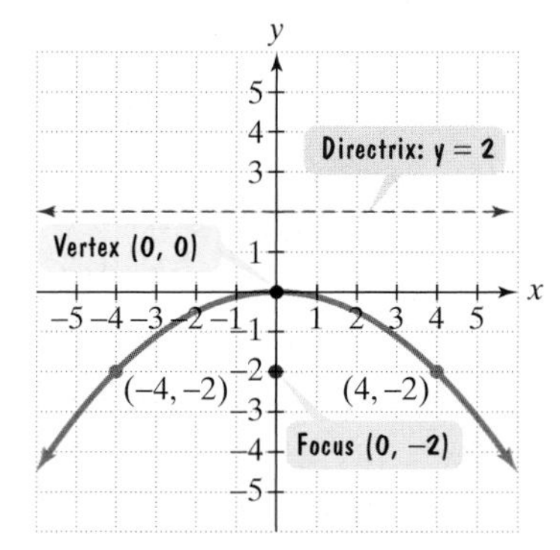

Figure 7.35 The graph of $x^2 = -8y$

To graph $x^2 = -8y$, we assign y a value that makes the right side a perfect square. If $y = -2$, then $x^2 = -8(-2) = 16$, so $x = \pm 4$. The parabola passes through the points $(4, -2)$ and $(-4, -2)$. The graph is sketched in Figure 7.35.

Check Point 2 Find the focus and directrix of the parabola given by $x^2 = -12y$. Then graph the parabola.

2 Write equations of parabolas in standard form.

In Examples 1 and 2, we used the equation of a parabola to find its focus and directrix. In the next example, we reverse this procedure.

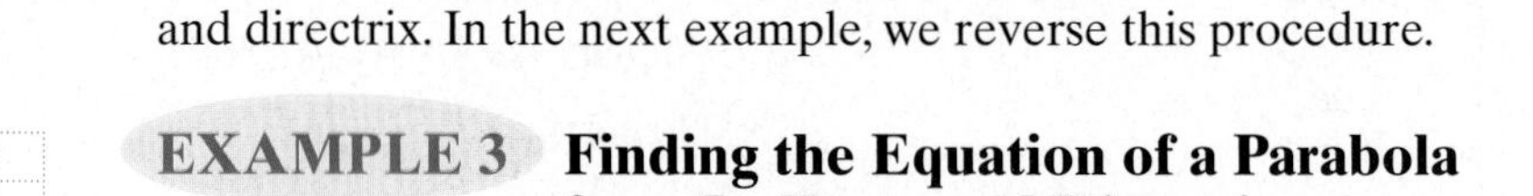

EXAMPLE 3 Finding the Equation of a Parabola from Its Focus and Directrix

Find the standard form of the equation of a parabola with focus $(5, 0)$ and directrix $x = -5$, shown in Figure 7.36.

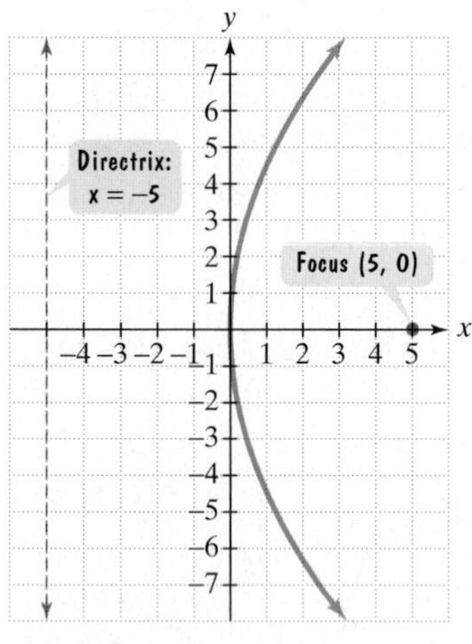

Figure 7.36

Solution The focus is $(5, 0)$. Thus, the focus is on the x-axis. We use the standard form of the equation in which x is not squared, namely $y^2 = 4px$.

We need to determine the value of p. Recall that the focus, located at $(p, 0)$, is p units from the vertex, $(0, 0)$. Thus, if the focus is $(5, 0)$, then $p = 5$. We substitute 5 for p into $y^2 = 4px$ to obtain the standard form of the equation of the parabola. The equation is

$$y^2 = 4 \cdot 5x \qquad \text{or} \qquad y^2 = 20x.$$

Check Point 3 Find the standard form of the equation of a parabola with focus $(8, 0)$ and directrix $x = -8$.

3 Graph parabolas with vertices not at the origin.

Translations of Parabolas

The graph of a parabola can have its vertex at (h, k) rather than at the origin. Horizontal and vertical translations are accomplished by replacing x with $x - h$ and y with $y - k$ in the standard form of the parabola's equation.

Table 7.3 gives the standard forms of equations of parabolas with vertex at (h, k). Figure 7.37 shows their graphs.

Table 7.3 Standard Forms of Equations of Parabolas with Vertex at (h, k)

Equation	Vertex	Axis of Symmetry	Focus	Directrix	Description
$(y - k)^2 = 4p(x - h)$	(h, k)	Horizontal	$(h + p, k)$	$x = h - p$	If $p > 0$, opens to right. If $p < 0$, opens to left.
$(x - h)^2 = 4p(y - k)$	(h, k)	Vertical	$(h, k + p)$	$y = k - p$	If $p > 0$, opens up. If $p < 0$, opens down.

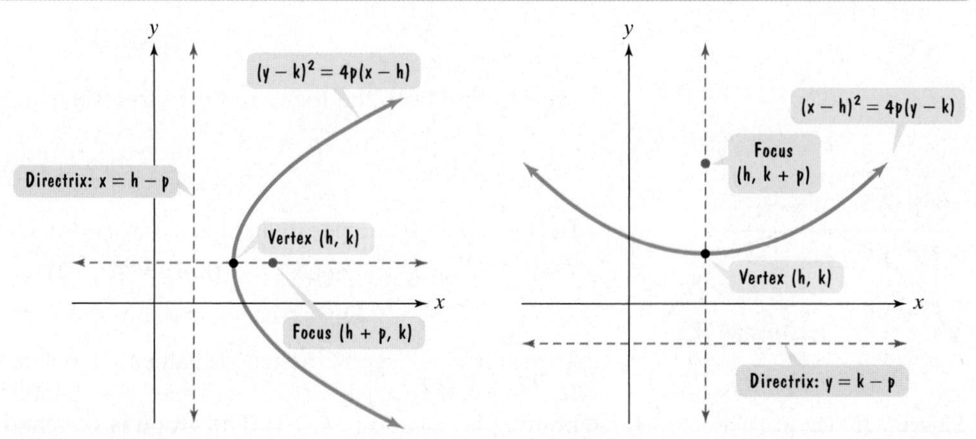

Figure 7.37 Graphs of parabolas with vertex at (h, k)

The two parabolas shown in Figure 7.37 illustrate standard forms of equations for $p > 0$. If $p < 0$, a parabola with a horizontal axis of symmetry will open to the left and the focus will lie to the left of the directrix. If $p < 0$, a parabola with a vertical axis of symmetry will open downward and the focus will lie below the directix.

EXAMPLE 4 Graphing a Parabola with Vertex at (h, k)

Find the vertex, focus, and directrix of the parabola given by

$$(x - 3)^2 = 8(y + 1).$$

Then graph the parabola.

Solution In order to find the focus and directrix, we need to know the vertex. In the standard forms of equations with vertex at (h, k), h is the number subtracted from x, and k is the number subtracted from y.

$$(x - 3)^2 = 8(y - (-1))$$

This is $(x - h)^2$, with $h = 3$.

This is $y - k$, with $k = -1$.

We see that $h = 3$ and $k = -1$. Thus, the vertex of the parabola is $(h, k) = (3, -1)$.

Now that we have the vertex, we can find both the focus and directrix by finding p.

$(x - 3)^2 = 8(y + 1)$ The equation is in the standard form $(x - h)^2 = 4p(y - k)$.

This is $4p$.

Because $4p = 8$, $p = 2$. Based on the standard form of the equation, the axis of symmetry is vertical. With a positive value for p and a vertical axis of symmetry, the parabola opens upward. Because $p = 2$, the focus is located 2 units above the vertex, $(3, -1)$. Likewise, the directrix is located 2 units below the vertex.

Focus: $(h, k + p) = (3, -1 + 2) = (3, 1)$

The vertex, (h, k), is $(3, -1)$.

The focus is 2 units above the vertex, $(3, -1)$.

Directrix:

$$y = k - p$$
$$y = -1 - 2 = -3$$

The directrix is 2 units below the vertex, $(3, -1)$.

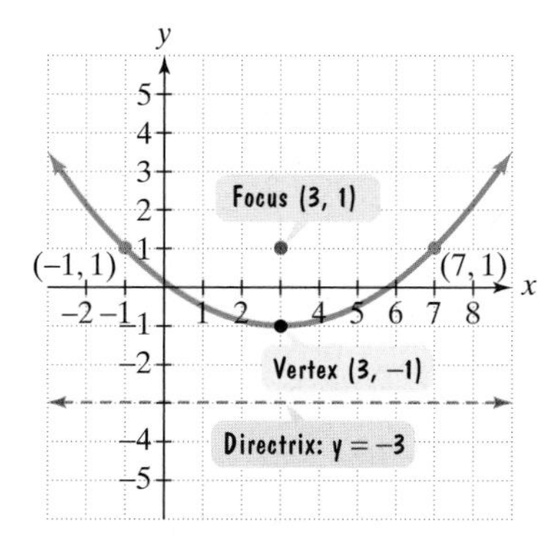

Figure 7.38 The graph of $(x - 3)^2 = 8(y + 1)$

Thus, the focus is $(3, 1)$ and the directrix is $y = -3$. They are shown in Figure 7.38.

To graph $(x - 3)^2 = 8(y + 1)$, we assign y a value that makes the right side of the equation a perfect square. If $y = 1$, the right side is $8(1 + 1)$ or 16, a perfect square. We let $y = 1$ and solve for x to obtain points on the parabola.

$(x - 3)^2 = 8(1 + 1)$ Substitute 1 for y in $(x - 3)^2 = 8(y + 1)$.

$(x - 3)^2 = 16$ Simplify.

$x - 3 = \pm\sqrt{16}$ Apply the square root method.

$x - 3 = 4$ or $x - 3 = -4$ Write $\sqrt{16}$ as 4 and express as two separate equations.

$x = 7$ $\quad x = -1$ Solve for x by adding 3 to both sides.

Because we obtained these values of x for $y = 1$, the parabola passes through the points $(7, 1)$ and $(-1, 1)$. Passing a smooth curve through the vertex and each of these points, we sketch the parabola shown in Figure 7.38.

Check Point 4 Find the vertex, focus, and directrix of the parabola given by $(x - 2)^2 = 4(y + 1)$. Then graph the parabola.

In some cases, we need to convert the equation of a parabola to standard form by completing the square on x or y, whichever variable is squared. Let's see how this is done.

EXAMPLE 5 Graphing a Parabola with Vertex at (h, k)

Find the vertex, focus, and directrix of the parabola given by

$$y^2 + 2y + 12x - 23 = 0.$$

Then graph the parabola.

Solution We convert the given equation to standard form by completing the square on the variable y. We isolate the terms involving y on the left side.

$$y^2 + 2y + 12x - 23 = 0$$ This is the given equation.

$$y^2 + 2y = -12x + 23$$ Isolate the terms involving y.

$$y^2 + 2y + 1 = -12x + 23 + 1$$ Complete the square by adding the square of half the coefficient of y.

$$(y + 1)^2 = -12x + 24$$

To express this equation in the standard form $(y - k)^2 = 4p(x - h)$, we factor -12 on the right. The standard form of the parabola's equation is

$$(y + 1)^2 = -12(x - 2).$$

We use this form to identify the vertex, (h, k), and the value for p needed to locate the focus and the directrix.

$$(y - (-1))^2 = -12(x - 2)$$ The equation is in the standard form $(y - k)^2 = 4p(x - h)$.

This is $(y - k)^2$, with $k = -1$. This is $4p$. This is $x - h$, with $h = 2$.

We see that $h = 2$ and $k = -1$. Thus, the vertex of the parabola is $(h, k) = (2, -1)$. Because $4p = -12$, $p = -3$. Based on the standard form of the equation, the axis of symmetry is horizontal. With a negative value for p and a horizontal axis of symmetry, the parabola opens to the left. We locate the focus and directrix as follows.

Focus: $$(h + p, k) = (2 + (-3), -1) = (-1, -1)$$

The vertex, (h, k), is $(2, -1)$. $h = 2$ $p = -3$ $k = -1$

Directrix:

$$x = h - p$$

$$x = 2 - (-3) = 5$$

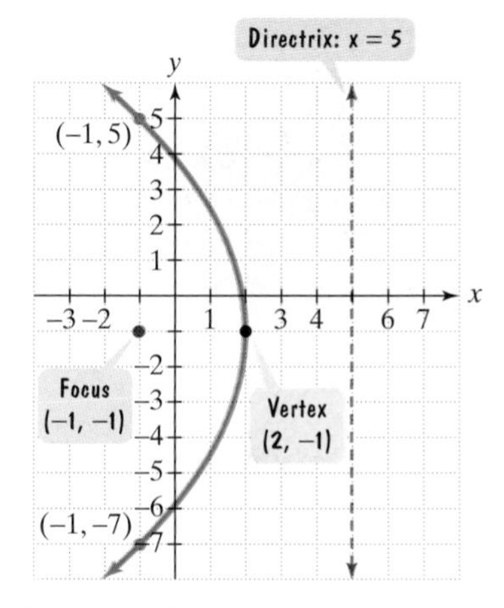

Figure 7.39 The graph of $(y + 1)^2 = -12(x - 2)$

Thus, the focus is $(-1, -1)$ and the directrix is $x = 5$. They are shown in Figure 7.39.

To graph $(y + 1)^2 = -12(x - 2)$, we assign x a value that makes the right side of the equation a perfect square. If $x = -1$, the right side is $-12(-1 - 2) = -12(-3) = 36$, a perfect square. We will let $x = -1$ and solve for y to obtain points on the parabola.

$$(y + 1)^2 = -12(-1 - 2)$$ Substitute -1 for x in $(y + 1)^2 = -12(x - 2)$.

$$(y + 1)^2 = 36$$ Simplify: $-12(-1 - 2) = -12(-3) = 36$.

$$y + 1 = \pm\sqrt{36}$$ Apply the square root method.

$$y + 1 = 6 \quad \text{or} \quad y + 1 = -6$$ Write $\sqrt{36}$ as 6 and express as two separate equations.

$$y = 5 \qquad\qquad y = -7$$ Solve for y by subtracting 1 from both sides.

Because we obtained these values of y for $x = -1$, the parabola passes through the points $(-1, 5)$ and $(-1, -7)$. Passing a smooth curve through the vertex and these two points, we sketch the parabola shown in Figure 7.39.

Find the vertex, focus, and directrix of the parabola given by $y^2 + 2y + 4x - 7 = 0$. Then graph the parabola.

4 Solve applied problems involving parabolas.

Applications

Parabolas have many applications. Cables hung between structures to form suspension bridges form parabolas. Arches constructed of steel and concrete, whose main purpose is strength, are usually parabolic in shape.

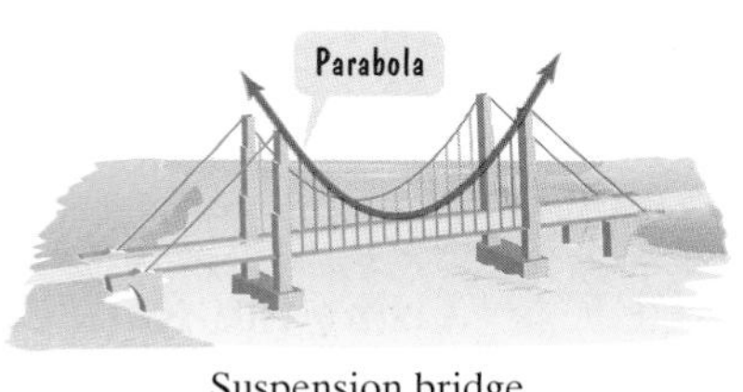

Suspension bridge

Arch bridge

Figure 7.40 Multiflash photo showing the parabolic path of a ball thrown into the air

We have seen that comets in our solar system travel in orbits that are ellipses and hyperbolas. Some comets also follow parabolic paths. Only comets with elliptical orbits, such as Halley's Comet, return to our part of the galaxy.

A projectile, such as a baseball thrown directly upward, moves along a parabolic path, illustrated in Figure 7.40.

If a parabola is rotated about its axis of symmetry, a parabolic surface is formed. Figure 7.41 (a) shows how a parabolic surface can be used to reflect light.

The Hubble Space Telescope

The Hubble Space Telescope

For decades, astronomers hoped to create an observatory above the atmosphere that would provide an unobscured view of the universe. This dream came true with the 1990 launching of the Hubble Space Telescope. The telescope initially had blurred vision due to problems with its parabolic mirror. The mirror had been ground two millionths of a meter smaller than design specifications. In 1993, astronauts from the Space Shuttle *Endeavor* equipped the telescope with optics to correct the blurred vision. "A small change for a mirror, a giant leap for astronomy," Christopher J. Burrows of the Space Telescope Science Institute said when clear images from the ends of the universe were presented to the public after the repair mission.

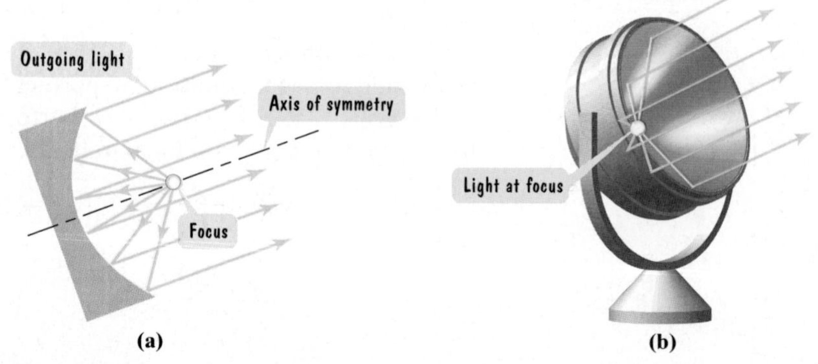

Figure 7.41 **(a)** Parabolic surface reflecting light **(b)** Light from the focus is reflected parallel to the axis of symmetry.

Light originates at the focus. Note how the light is reflected by the parabolic surface, so that the outgoing light is parallel to the axis of symmetry. The reflective properties of parabolic surfaces are used in the design of searchlights [Figure 7.41(b)], automobile headlights, and parabolic microphones.

Figure 7.42(a) shows how a parabolic surface can be used to reflect *incoming* light. Note that light rays strike the surface and are reflected *to the focus*. This principle is used in the design of reflecting telescopes, radar, and television satellite dishes. Reflecting telescopes magnify the light from distant stars by reflecting the light from these bodies to the focus of a parabolic mirror [Figure 7.42(b)].

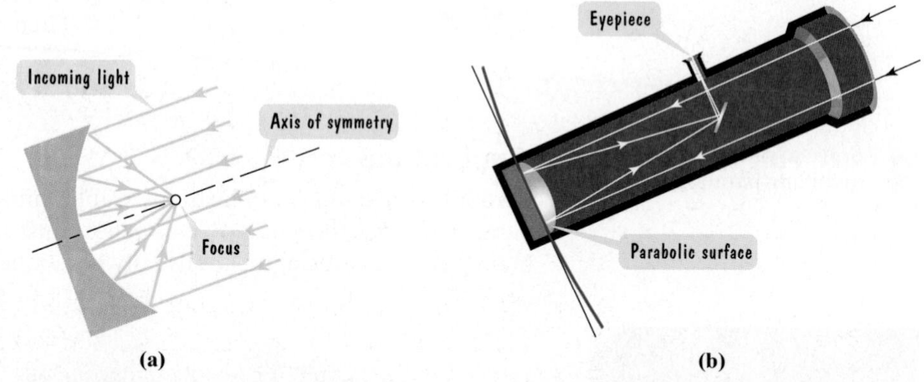

Figure 7.42 **(a)** Parabolic surface reflecting incoming light **(b)** Incoming light rays are reflected to the focus.

EXAMPLE 6 Using the Reflection Property of Parabolas

An engineer is designing a flashlight using a parabolic reflecting mirror and a light source, shown in Figure 7.43. The casting has a diameter of 4 inches and a depth of 2 inches. What is the equation of the parabola used to shape the mirror? At what point should the light source be placed relative to the mirror's vertex?

Solution We position the parabola with its vertex at the origin and opening upward (Figure 7.44). Thus, the focus is on the y-axis, located at $(0, p)$. We use the standard form of the equation in which y is not squared, namely $x^2 = 4py$. We need to find p. Because $(2, 2)$ lies on the parabola, we let $x = 2$ and $y = 2$ in $x^2 = 4py$.

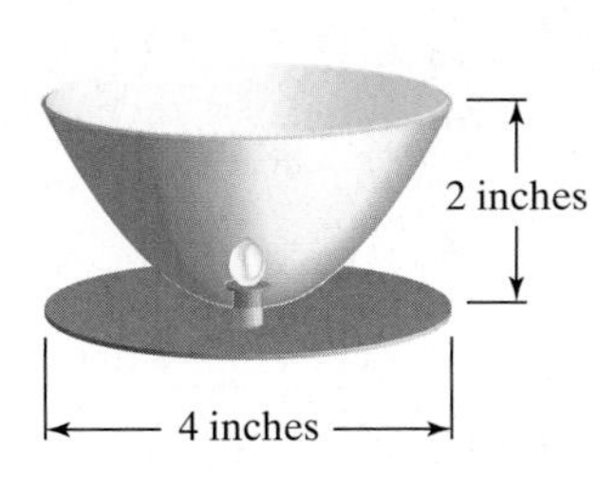

Figure 7.43 Designing a flashlight

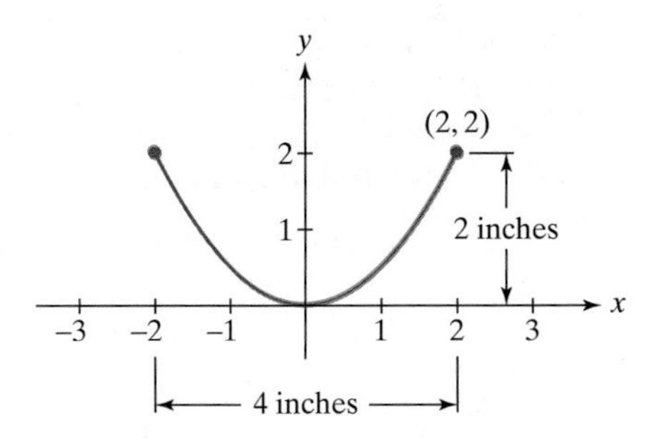

Figure 7.44

$2^2 = 4p \cdot 2$ Substitute 2 for x and 2 for y in $x^2 = 4py$.

$4 = 8p$ Simplify.

$p = \frac{1}{2}$ Divide both sides of the equation by 8 and reduce the resulting fraction.

We substitute $\frac{1}{2}$ for p in $x^2 = 4py$ to obtain the standard form of the equation of the parabola. The equation of the parabola used to shape the mirror is

$$x^2 = 4 \cdot \tfrac{1}{2}y \quad \text{or} \quad x^2 = 2y.$$

The light source should be placed at the focus $(0, p)$. Because $p = \frac{1}{2}$, the light should be placed at $\left(0, \frac{1}{2}\right)$, or $\frac{1}{2}$ inch above the vertex.

Check Point 6 In Example 6, suppose that the casting has a diameter of 6 inches and a depth of 4 inches. What is the equation of the parabola used to shape the mirror? At what point should the light source be placed relative to the mirror's vertex?

Degenerate Conic Sections

We opened the chapter by noting that conic sections are curves that result from the intersection of a cone and a plane. However, these intersections might not result in a conic section. Three degenerate cases occur when the cutting plane passes through the vertex. These **degenerate conic sections** are a point, a line, and a pair of intersecting lines, illustrated in Figure 7.45.

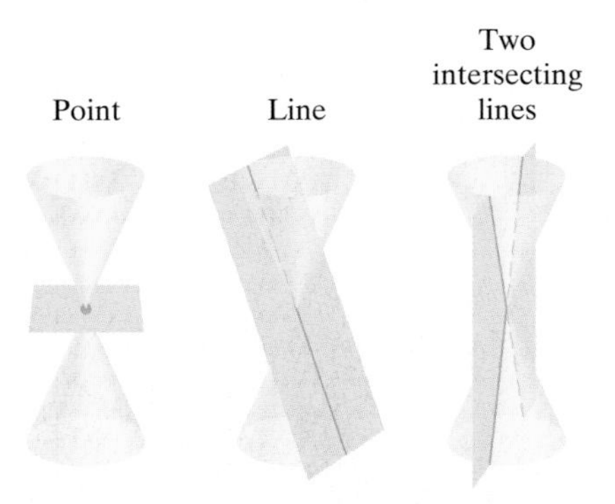

Figure 7.45 Degenerate conics

EXERCISE SET 7.3

Practice Exercises

In Exercises 1–4, find the focus and directrix of each parabola with the given equation. Then match each equation to one of the graphs that are shown and labeled (a)–(d).

1. $y^2 = 4x$ **2.** $x^2 = 4y$

3. $x^2 = -4y$ **4.** $y^2 = -4x$

a.

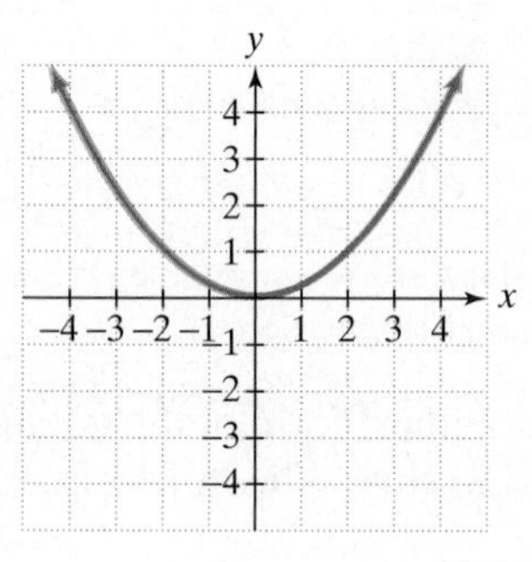

b.

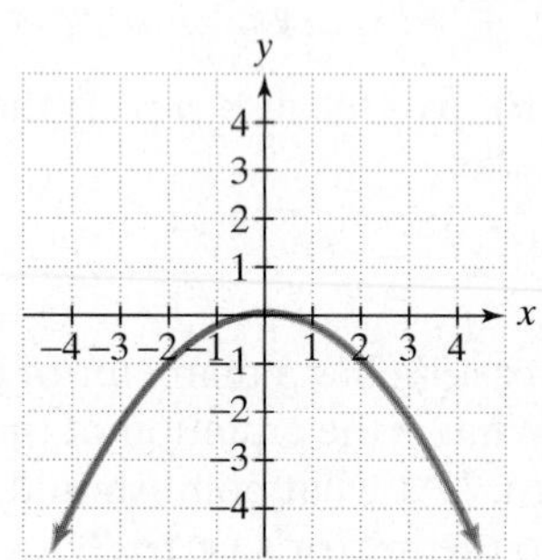

c.

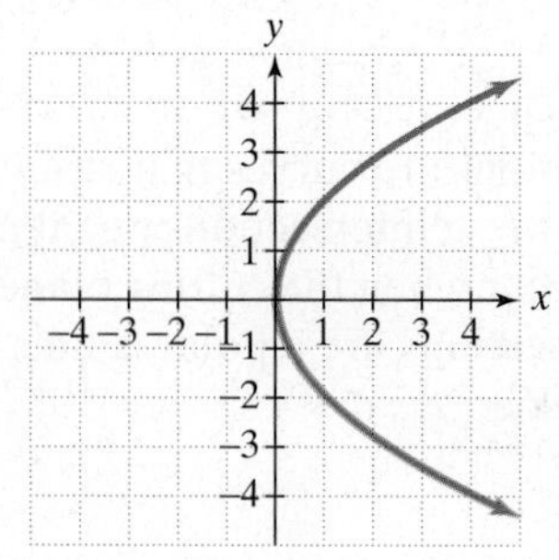

d.

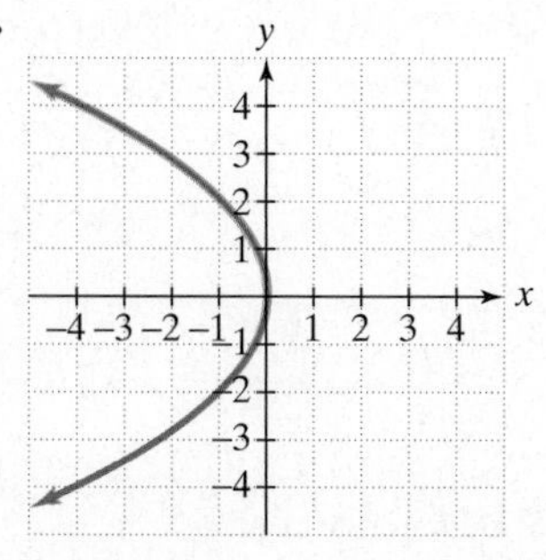

In Exercises 5–14, find the focus and directrix of the parabola with the given equation. Then graph the parabola.

5. $y^2 = 16x$ **6.** $y^2 = 4x$

7. $y^2 = -8x$ **8.** $y^2 = -12x$

9. $x^2 = 12y$ **10.** $x^2 = 8y$

11. $x^2 = -16y$ **12.** $x^2 = -20y$

13. $y^2 - 6x = 0$ **14.** $x^2 - 6y = 0$

In Exercises 15–22, find the standard form of the equation of each parabola with vertex at the origin satisfying the given conditions.

15. Focus: $(7, 0)$; Directrix: $x = -7$

16. Focus: $(9, 0)$; Directrix: $x = -9$

17. Focus: $(-5, 0)$; Directrix: $x = 5$

18. Focus: $(-10, 0)$; Directrix: $x = 10$

19. Focus: $(0, 15)$; Directrix: $y = -15$

20. Focus: $(0, 20)$; Directrix: $y = -20$

21. Focus: $(0, -25)$; Directrix: $y = 25$

22. Focus: $(0, -15)$; Directrix: $y = 15$

In Exercises 23–26, find the vertex, focus, and directrix of each parabola with the given equation. Then match each equation to one of the graphs that are shown and labeled (a)–(d).

23. $(y - 1)^2 = 4(x - 1)$ **24.** $(x + 1)^2 = 4(y + 1)$

25. $(x + 1)^2 = -4(y + 1)$ **26.** $(y - 1)^2 = -4(x - 1)$

a.

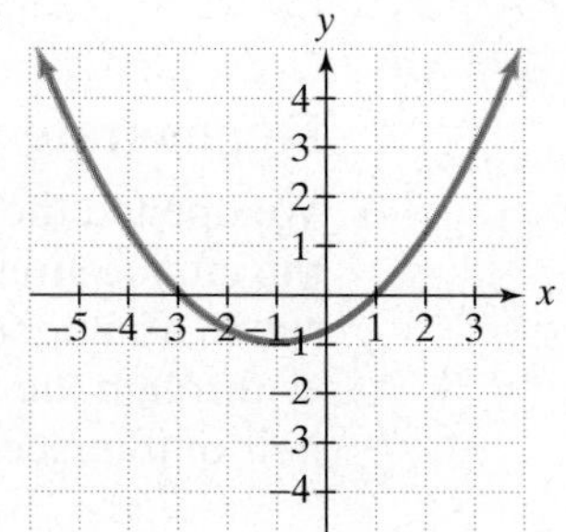

b.

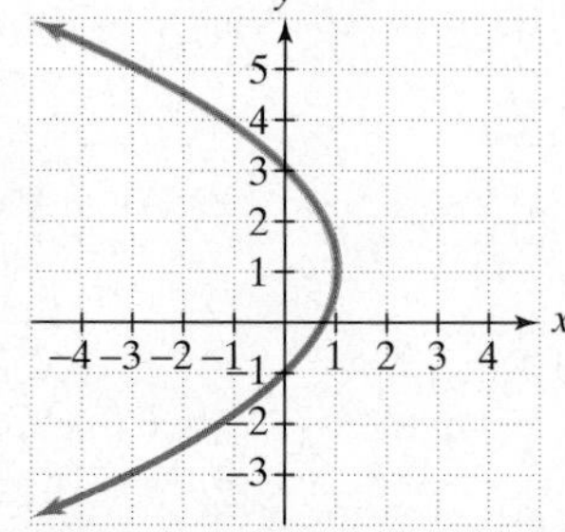

c.

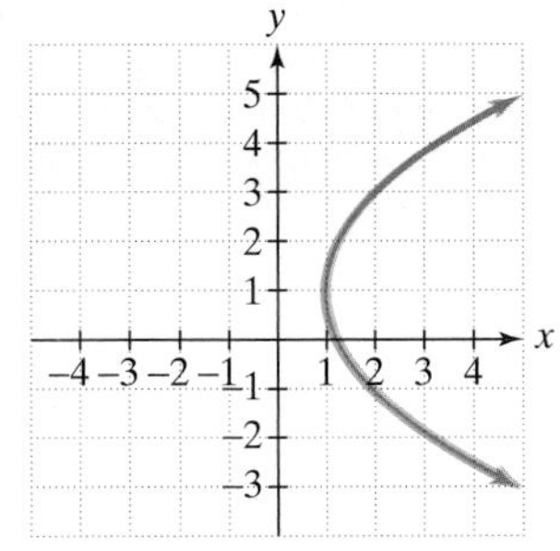

d.

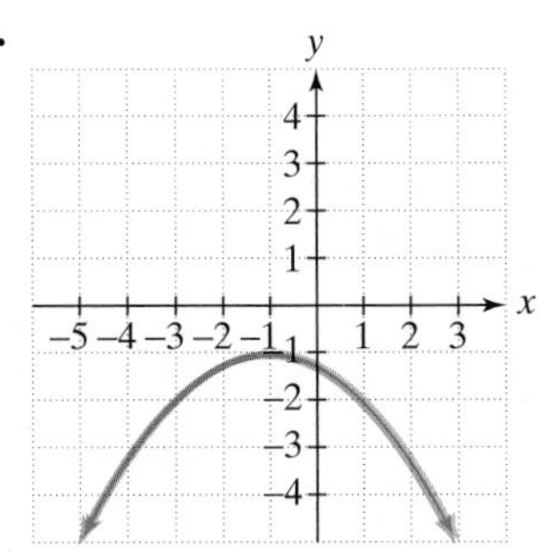

In Exercises 27–34, find the vertex, focus, and directrix of each parabola with the given equation. Then graph the parabola.

27. $(x - 2)^2 = 8(y - 1)$ **28.** $(x + 2)^2 = 4(y + 1)$

29. $(x + 1)^2 = -8(y + 1)$ **30.** $(x + 2)^2 = -8(y + 2)$

31. $(y + 3)^2 = 12(x + 1)$ **32.** $(y + 4)^2 = 12(x + 2)$

33. $(y + 1)^2 = -8x$ **34.** $(y - 1)^2 = -8x$

In Exercises 35–40, convert each equation to standard form by completing the square on x or y. Then find the vertex, focus, and directrix of the parabola. Finally, graph the parabola.

35. $x^2 - 2x - 4y + 9 = 0$ **36.** $x^2 + 6x + 8y + 1 = 0$

37. $y^2 - 2y + 12x - 35 = 0$ **38.** $y^2 - 2y - 8x + 1 = 0$

39. $x^2 + 6x - 4y + 1 = 0$ **40.** $x^2 + 8x - 4y + 8 = 0$

Application Exercises

41. The reflector of a flashlight is in the shape of a parabolic surface. The casting has a diameter of 4 inches and a depth of 1 inch. How far from the vertex should the light bulb be placed?

42. The reflector of a flashlight is in the shape of a parabolic surface. The casting has a diameter of 8 inches and a depth of 1 inch. How far from the vertex should the light bulb be placed?

43. A satellite dish, like the one shown at the top of the next column, is in the shape of a parabolic surface. Signals coming from a satellite strike the surface of the dish and are reflected to the focus, where the receiver is located. The satellite dish shown has a diameter of 12 feet and a depth of 2 feet. How far from the base of the dish should the receiver be placed?

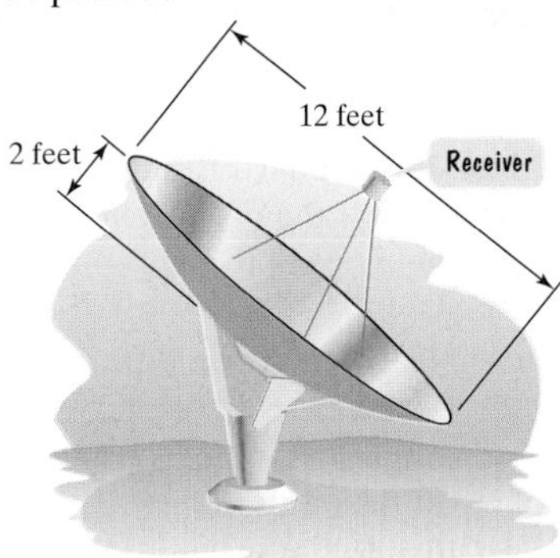

44. In Exercise 43, if the diameter of the dish is halved and the depth stays the same, how far from the base of the smaller dish should the receiver be placed?

45. The towers of the Golden Gate Bridge connecting San Francisco to Marin County are 1280 meters apart and rise 160 meters above the road. The cable between the towers has the shape of a parabola, and the cable just touches the sides of the road midway between the towers. What is the height of the cable 200 meters from a tower?

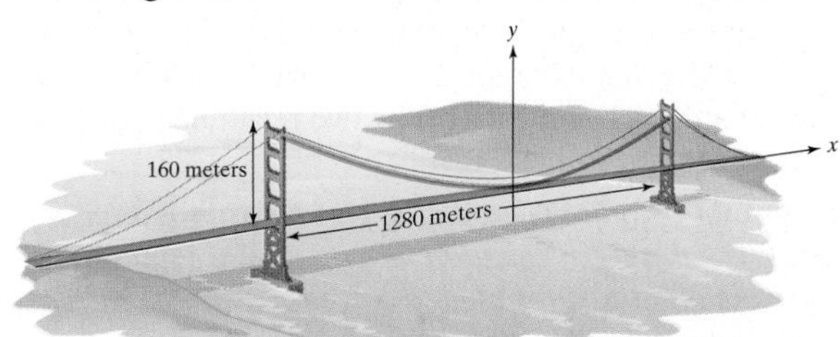

46. The towers of a suspension bridge are 800 feet apart and rise 160 feet above the road. The cable between the towers has the shape of a parabola, and the cable just touches the sides of the road midway between the towers. What is the height of the cable 100 feet from a tower?

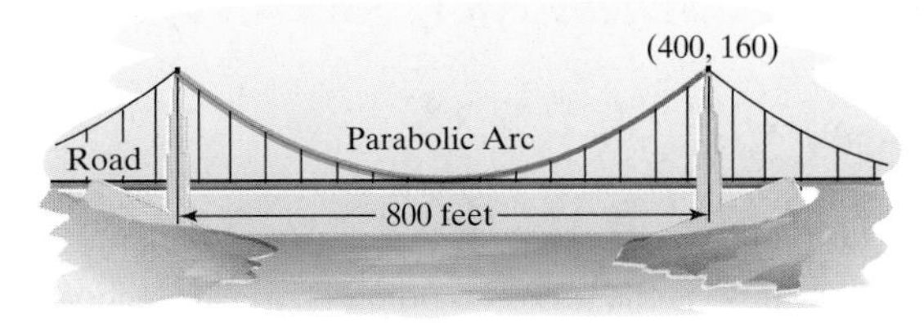

47. The parabolic arch shown in the figure is 50 feet above the water at the center and 200 feet wide at the base. Will a boat that is 30 feet tall clear the arch 30 feet from the center?

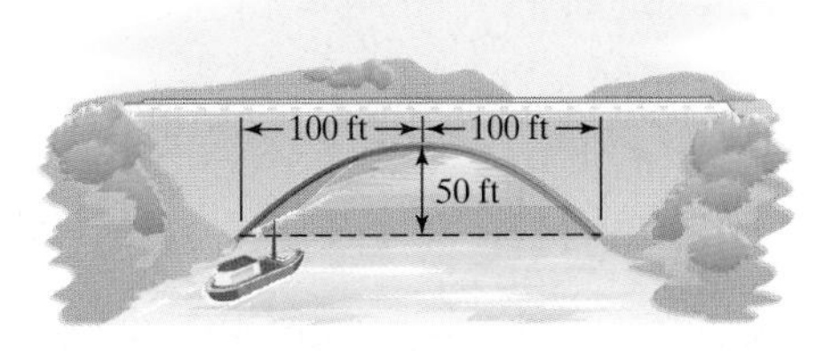

Writing in Mathematics

48. What is a parabola?

49. Explain how to use $y^2 = 8x$ to find the parabola's focus and directrix.

50. If you are given the standard form of the equation of a parabola with vertex at the origin, explain how to determine if the parabola opens to the right, left, upward, or downward.

51. Describe one similarity and one difference between the graphs of $y^2 = 4x$ and $(y - 1)^2 = 4(x - 1)$.

52. How can you distinguish parabolas from other conic sections by looking at their equations?

53. Look at the satellite dish shown in Exercise 43. Why must the receiver for a shallow dish be farther from the base of the dish than for a deeper dish of the same diameter?

Technology Exercises

54. Use a graphing utility to graph any five of the parabolas that you graphed by hand in Exercises 5–14.

55. Use a graphing utility to graph any three of the parabolas that you graphed by hand in Exercises 27–34. First solve the given equation for y, possibly using the square root method. Enter each of the two resulting equations to produce the complete graph.

Use a graphing utility to graph the parabolas in Exercises 56–57. Write the given equation as a quadratic equation in y and use the quadratic formula to solve for y. Enter each of the equations to produce the complete graph.

56. $y^2 + 2y - 6x + 13 = 0$

57. $y^2 + 10y - x + 25 = 0$

In Exercises 58–59, write each equation as a quadratic equation in y and then use the quadratic formula to express y in terms of x. Graph the resulting two equations using a graphing utility. What effect does the xy-term have on the graph of the resulting parabola?

58. $16x^2 - 24xy + 9y^2 - 60x - 80y + 100 = 0$

59. $x^2 + 2\sqrt{3}xy + 3y^2 + 8\sqrt{3}x - 8y + 32 = 0$

Critical Thinking Exercises

60. Which one of the following is true?

a. The parabola whose equation is $x = 2y - y^2 + 5$ opens to the right.

b. If the parabola whose equation is $x = ay^2 + by + c$ has its vertex at $(3, 2)$ and $a > 0$, then it has no y-intercepts.

c. Some parabolas that open to the right have equations that define y as a function of x.

d. The graph of $x = a(y - k) + h$ is a parabola with vertex at (h, k).

61. A satellite dish in the shape of a parabolic surface has a diameter of 20 feet. If the receiver is to be placed 6 feet from the base, how deep should the dish be?

62. Write the standard form of the equation of a parabola whose points are equidistant from $y = 4$ and $(-1, 0)$.

Group Exercise

63. Consult the research department of your library or the Internet to find an example of architecture that incorporates one or more conic sections in its design. Share this example with other group members. Explain precisely how conic sections are used. Do conic sections enhance the appeal of the architecture? In what ways?

CHAPTER SUMMARY, REVIEW, AND TEST

Summary

7.1 The Ellipse

a. An ellipse is the set of all points in a plane the sum of whose distances from two fixed points, the foci, is constant.

b. Standard forms of the equations of an ellipse with center at the origin are $\frac{x^2}{a^2} + \frac{y^2}{b^2} = 1$ [foci: $(-c, 0), (c, 0)$] and $\frac{x^2}{b^2} + \frac{y^2}{a^2} = 1$ [foci: $(0, -c)$, $(0, c)$], where $b^2 = a^2 - c^2$. See the box on page 555 and Figure 7.6.

c. Standard forms of the equations of an ellipse centered at (h, k) are $\frac{(x-h)^2}{a^2} + \frac{(y-k)^2}{b^2} = 1$ and $\frac{(x-h)^2}{b^2} + \frac{(y-k)^2}{a^2} = 1$. See Table 7.1 on page 559 and Figure 7.10.

7.2 The Hyperbola

a. A hyperbola is the set of all points in a plane the difference of whose distances from two fixed points, the foci, is constant.

b. Standard forms of the equations of a hyperbola with center at the origin are $\frac{x^2}{a^2} - \frac{y^2}{b^2} = 1$ [foci: $(-c, 0)$, $(c, 0)$] and $\frac{y^2}{a^2} - \frac{x^2}{b^2} = 1$ [foci: $(0, -c)$, $(0, c)$], where $b^2 = c^2 - a^2$. See the box on page 568 and Figure 7.17.

c. Asymptotes for $\frac{x^2}{a^2} - \frac{y^2}{b^2} = 1$ are $y = \pm\frac{b}{a}x$. Asymptotes for $\frac{y^2}{a^2} - \frac{x^2}{b^2} = 1$ are $y = \pm\frac{a}{b}x$.

d. A procedure for graphing hyperbolas is given in the box on page 572.

e. Standard forms of the equations of a hyperbola centered at (h, k) are $\frac{(x-h)^2}{a^2} - \frac{(y-k)^2}{b^2} = 1$ and $\frac{(y-k)^2}{a^2} - \frac{(x-h)^2}{b^2} = 1$. See Table 7.2 on page 575 and Figure 7.26.

7.3 The Parabola

a. A parabola is the set of all points in a plane that are equidistant from a fixed line, the directrix, and a fixed point, the focus.

b. Standard forms of the equations of parabolas with vertex at the origin are $y^2 = 4px$ [focus: $(p, 0)$] and $x^2 = 4py$ [focus: $(0, p)$]. See the box on page 583 and Figure 7.32 on page 584.

c. Standard forms of the equations of a parabola with vertex at (h, k) are $(y - k)^2 = 4p(x - h)$ and $(x - h)^2 = 4p(y - k)$. See Table 7.3 on page 586 and Figure 7.37.

Review Exercises

7.1

In Exercises 1–8, graph each ellipse and locate the foci.

1. $\frac{x^2}{36} + \frac{y^2}{25} = 1$

2. $\frac{y^2}{25} + \frac{x^2}{16} = 1$

3. $4x^2 + y^2 = 16$

4. $4x^2 + 9y^2 = 36$

5. $\frac{(x-1)^2}{16} + \frac{(y+2)^2}{9} = 1$

6. $\frac{(x+1)^2}{9} + \frac{(y-2)^2}{16} = 1$

7. $4x^2 + 9y^2 + 24x - 36y + 36 = 0$

8. $9x^2 + 4y^2 - 18x + 8y - 23 = 0$

In Exercises 9–11, find the standard form of the equation of each ellipse centered at the origin satisfying the given conditions.

9. Foci: $(-4, 0)$, $(4, 0)$; Vertices: $(-5, 0)$, $(5, 0)$

10. Foci: $(0, -3)$, $(0, 3)$; Vertices: $(0, -6)$, $(0, 6)$

11. Major axis horizontal with length 12; length of minor axis = 4

12. A semielliptical arch supports a bridge that spans a river 20 yards wide. The center of the arch is 6 yards above the river's center. Write an equation for the ellipse so that the x-axis coincides with the water level and the y-axis passes through the center of the arch.

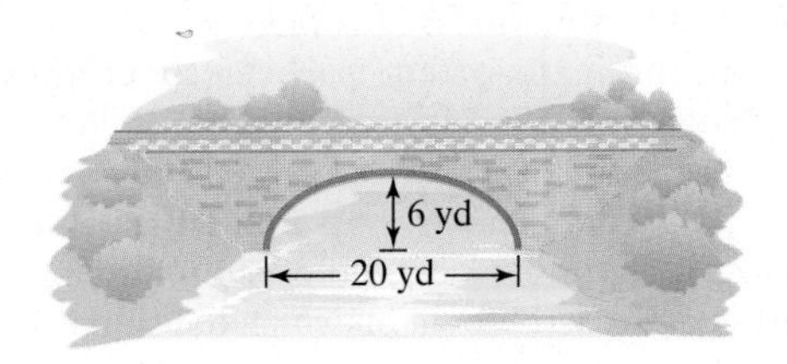

13. A semielliptic archway has a height of 15 feet at the center and a width of 50 feet, as shown in the figure. The 50-foot width consists of a two-lane road. Can a truck that is 12 feet high and 14 feet wide drive under the archway without going into the other lane?

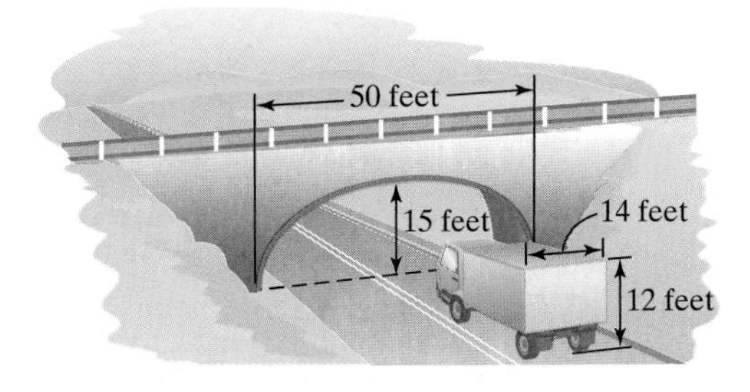

14. An elliptical pool table has a ball placed at each focus. If one ball is hit toward the side of the table, explain what will occur.

7.2

In Exercises 15–22, graph each hyperbola and locate the foci.

15. $\frac{x^2}{16} - y^2 = 1$

16. $\frac{y^2}{16} - x^2 = 1$

17. $9x^2 - 16y^2 = 144$

18. $4y^2 - x^2 = 16$

19. $\frac{(x-2)^2}{25} - \frac{(y+3)^2}{16} = 1$

20. $\frac{(y+2)^2}{25} - \frac{(x-3)^2}{16} = 1$

21. $y^2 - 4y - 4x^2 + 8x - 4 = 0$

22. $x^2 - y^2 - 2x - 2y - 1 = 0$

In Exercises 23–24, find the standard form of the equation of each hyperbola centered at the origin satisfying the given conditions.

23. Foci: $(0, -4)$, $(0, 4)$; vertices: $(0, -2)$, $(0, 2)$

24. Foci: $(-8, 0)$, $(8, 0)$; vertices: $(-3, 0)$, $(3, 0)$

25. Explain why it is not possible for a hyperbola to have foci at $(0, -2)$ and $(0, 2)$ and vertices at $(0, -3)$ and $(0, 3)$.

26. Radio tower M_2 is located 200 miles due west of radio tower M_1. The situation is illustrated in the figure shown, where a coordinate system has been superimposed. Simultaneous radio signals are sent from each tower to a ship, with the signal from M_2 received 500 microseconds before the signal from M_1. Assuming that radio signals travel at 0.186 miles per microsecond, determine the equation of the hyperbola on which the ship is located.

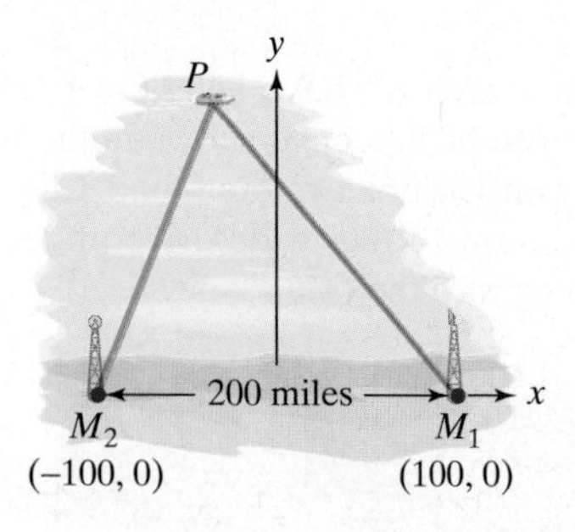

7.3

In Exercises 27–33, find the vertex, focus, and directrix of each parabola with the given equation. Then graph the parabola.

27. $y^2 = 8x$

28. $x^2 + 16y = 0$

29. $(y - 2)^2 = -16x$

30. $(x - 4)^2 = 4(y + 1)$

31. $x^2 + 4y = 4$

32. $y^2 - 4x - 10y + 21 = 0$

33. $x^2 - 4x - 2y = 0$

In Exercises 34–35, find the standard form of the equation of each parabola with vertex at the origin satisfying the given conditions.

34. Focus: $(12, 0)$; Directrix: $x = -12$

35. Focus: $(0, -11)$; Directrix: $y = 11$

36. An engineer is designing headlight units for automobiles. The unit has a parabolic surface with a diameter of 12 inches and a depth of 3 inches. The situation is illustrated in the figure, where a coordinate system has been superimposed. What is the equation of the parabola in this system? Where should the light source be placed? Describe this placement relative to the vertex.

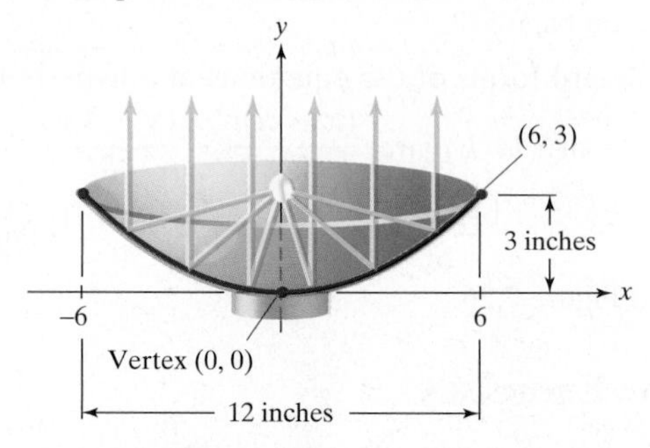

37. The George Washington Bridge spans the Hudson River from New York to New Jersey. Its two towers are 3500 feet apart and rise 316 feet above the road. The cable between the towers has the shape of a parabola, and the cable just touches the sides of the road midway between the towers. What is the height of the cable 1000 feet from a tower?

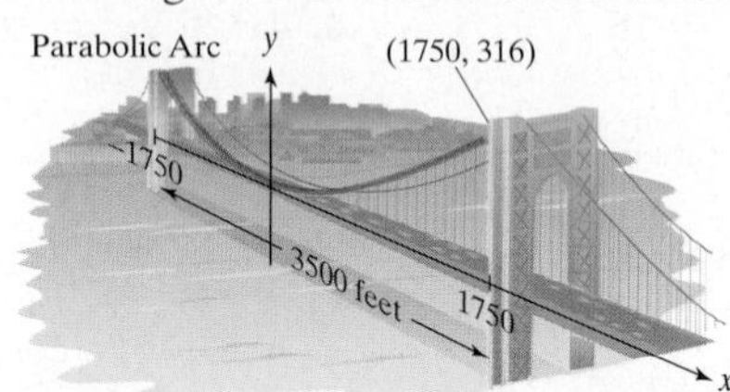

38. The giant satellite dish in the figure shown is in the shape of a parabolic surface. Signals strike the surface and are reflected to the focus, where the receiver is located. The diameter of the dish is 300 feet and its depth is 44 feet. How far, to the nearest foot, from the base of the dish should the receiver be placed?

Chapter 7 Test

In Exercises 1–5, graph the conic section with the given equation. For ellipses and hyperbolas, find the foci. For parabolas, find the vertex, focus, and directrix.

1. $9x^2 - 4y^2 = 36$ **2.** $x^2 = -8y$

3. $\frac{(x+2)^2}{25} + \frac{(y-5)^2}{9} = 1$

4. $4x^2 - y^2 + 8x + 2y + 7 = 0$

5. $(x+5)^2 = 8(y-1)$

In Exercises 6–8, find the standard form of the equation of the conic section satisfying the given conditions.

6. Ellipse; Foci: $(-7, 0)$, $(7, 0)$; Vertices: $(-10, 0)$, $(10, 0)$

7. Hyperbola; Foci: $(0, -10)$, $(0, 10)$; Vertices: $(0, -7)$, $(0, 7)$

8. Parabola; Focus: $(50, 0)$; Directrix: $x = -50$

9. A sound whispered at one focus of a whispering gallery can be heard at the other focus. The figure shows a whispering gallery whose cross section is a semielliptical arch with a height of 24 feet and a width of 80 feet. How far from the room's center should two people stand so that they can whisper back and forth and be heard?

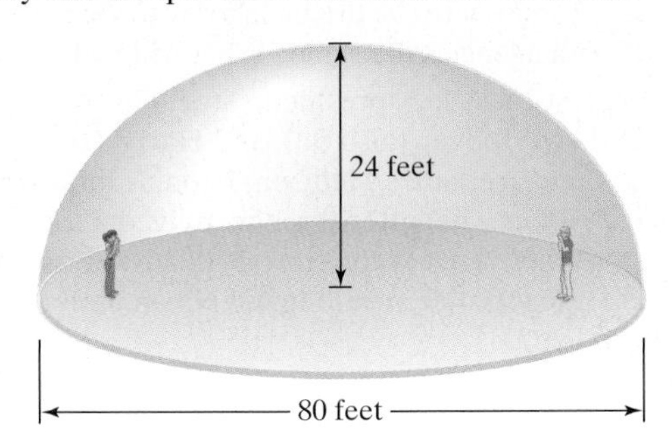

10. An engineer is designing headlight units for cars. The unit shown in the figure has a parabolic surface with a diameter of 6 inches and a depth of 3 inches.

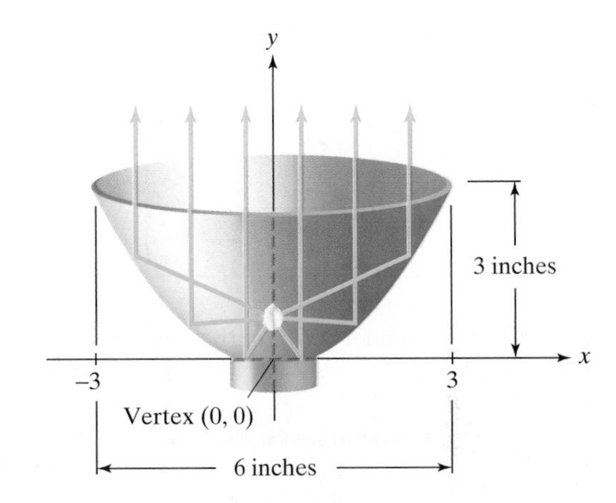

a. Using the coordinate system that has been positioned on the unit, find the parabola's equation.

b. If the light source is located at the focus, describe its placement relative to the vertex.

Cumulative Review Exercises (Chapters 1–7)

Solve each equation or inequality in Exercises 1–7.

1. $2(x-3) + 5x = 8(x-1)$

2. $-3(2x-4) > 2(6x-12)$

3. $x - 5 = \sqrt{x+7}$ **4.** $(x-2)^2 = 20$

5. $|2x-1| \geq 7$ **6.** $3x^3 + 4x^2 - 7x + 2 = 0$

7. $\log_2(x+1) + \log_2(x-1) = 3$

Solve each system in Exercises 8–10

8. $3x + 4y = 2$
$2x + 5y = -1$

9. $2x^2 - y^2 = -8$
$x - y = 6$

10. (Use matrices.)
$x - y + z = 17$
$-4x + y + 5z = -2$
$2x + 3y + z = 8$

In Exercises 11–13, graph each equation, function, or system in the rectangular coordinate system.

11. $f(x) = (x-1)^2 - 4$ **12.** $\frac{x^2}{9} + \frac{y^2}{4} = 1$

13. $5x + y \leq 10$
$y \geq \frac{1}{4}x + 2$

14. a. List all possible rational roots of

$$32x^3 - 52x^2 + 17x + 3 = 0.$$

b. The graph of $f(x) = 32x^3 - 52x^2 + 17x + 3$ is shown in the figure. Use the graph of f and synthetic division to solve the equation in part (a).

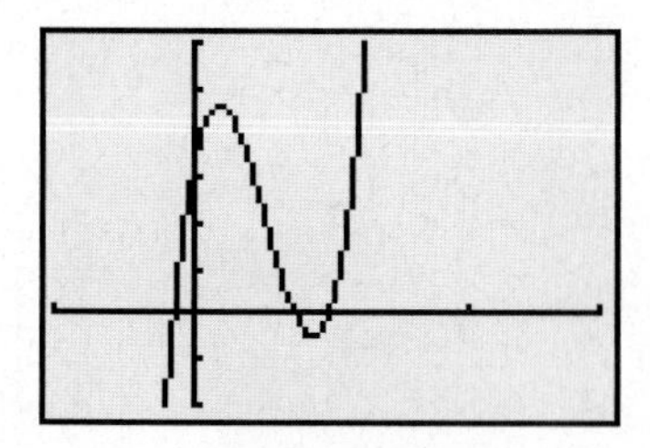

15. The graph shows gender ratios in the United States, with future projections.

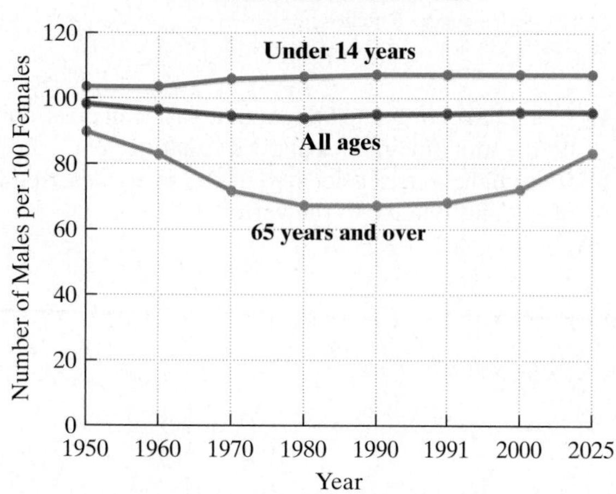

Source: U.S. Census Bureau

For males age 65 and over, shown by the blue graph:

a. In what time interval is the number of males per 100 females constant?

b. In what time interval is the number of males per 100 females increasing?

c. In what time interval is the number of males per 100 females decreasing?

For all ages, shown by the red graph:

d. Write a constant function $f(x)$ that approximately models the data shown for x in the interval [1950, 2025].

e. What is misleading about the scale on the horizontal axis?

16. If $f(x) = x^2 - 4$ and $g(x) = x + 2$, find $(g \circ f)(x)$.

17. Expand using logarithmic properties. Where possible, evaluate logarithmic expressions.

$$\log_5 \frac{x^3\sqrt{y}}{125}$$

18. Write the slope-intercept form of the equation of the line passing through $(1, -4)$ and $(-5, 8)$.

19. Rent-a-Truck charges a daily rental rate for a truck of \$39 plus \$0.16 a mile. A competing agency, Ace Truck Rentals, charges \$25 a day plus \$0.24 a mile for the same truck. How many miles must be driven in a day to make the daily cost of both agencies the same? What will be the cost?

20. The longest-lived U.S. presidents are John Adams (age 90), Herbert Hoover (also 90), and Harry Truman (88). Behind them are James Madison, Thomas Jefferson, and Richard Nixon. The latter three men lived a total of 249 years, and their ages at the time of death form consecutive odd integers. For how long did Nixon, Jefferson, and Madison live?

Sequences, Induction, and Probability

We often save for the future by investing small amounts at periodic intervals. To understand how our savings accumulate, we need to understand properties of lists of numbers that are related to each other by a rule. Such lists are called *sequences*. Learning about properties of sequences will show you how to make your financial goals a reality. Your knowledge of sequences will enable you to inform your college roommate of the best of the three appealing offers.

Something incredible has happened. Your college roommate, a gifted athlete, has been given a six-year contract with a professional baseball team. He will be playing against the likes of Mark McGwire and Sammy Sosa. Management offers him three options. One is a beginning salary of $1,700,000 with annual increases of $70,000 per year starting in the second year. A second option is $1,700,000 the first year with an annual increase of 2% per year beginning in the second year. The third offer involves less money the first year—$1,500,000—but there is an annual increase of 9% yearly after that. Which option offers the most money over the six-year contract?

SECTION 8.1 Sequences and Summation Notation

Objectives

1. Find particular terms of a sequence from the general term.
2. Use recursion formulas.
3. Use factorial notation.
4. Use summation notation.

Sequences

Many creations in nature involve intricate mathematical designs, including a variety of spirals. For example, the arrangement of the individual florets in the head of a sunflower forms spirals. In some species, there are 21 spirals in the clockwise direction and 34 in the counterclockwise direction. The precise numbers depend on the species of sunflower: 21 and 34, or 34 and 55, or 55 and 89, or even 89 and 144.

This observation becomes even more interesting when we consider a sequence of numbers investigated by Leonardo of Pisa, also known as Fibonacci, an Italian mathematician of the thirteenth century. The **Fibonacci sequence** of numbers is an infinite sequence that begins as follows:

$$1, 1, 2, 3, 5, 8, 13, 21, 34, 55, 89, 144, 233....$$

The first two terms are 1. Every term thereafter is the sum of the two preceding terms. For example, the third term, 2, is the sum of the first and second terms: $1 + 1 = 2$. The fourth term, 3, is the sum of the second and third terms: $1 + 2 = 3$, and so on. Did you know that the number of spirals in a daisy or a sunflower, 21 and 34, are two Fibonacci numbers? The number of spirals in a pine cone, 8 and 13, and a pineapple, 8 and 13, are also Fibonacci numbers.

We can think of the Fibonacci sequence as a function. The terms of the sequence

$$1, 1, 2, 3, 5, 8, 13, 21, 34, 55, 89, 144, 233, \ldots$$

are the range values for a function whose domain is the set of positive integers.

$$\begin{array}{llllllll} \text{Domain:} & 1, & 2, & 3, & 4, & 5, & 6, & 7, \ldots \\ & \downarrow & \downarrow & \downarrow & \downarrow & \downarrow & \downarrow & \downarrow \\ \text{Range:} & 1, & 1, & 2, & 3, & 5, & 8, & 13, \ldots \end{array}$$

Thus, $f(1) = 1, f(2) = 1, f(3) = 2, f(4) = 3, f(5) = 5, f(6) = 8, f(7) = 13$, and so on.

The letter a with a subscript is used to represent function values of a sequence, rather than the usual function notation. The subscripts make up the domain of the sequence, and they identify the location of a term. Thus, a_1 represents the first term of the sequence, a_2 represents the second term, a_3 the third term, and so on. This notation is shown for the first six terms of the Fibonacci sequence:

$$\begin{array}{cccccc} 1, & 1, & 2, & 3, & 5, & 8. \\ a_1 = 1 & a_2 = 1 & a_3 = 2 & a_4 = 3 & a_5 = 5 & a_6 = 8 \end{array}$$

Fibonacci Numbers on the Piano Keyboard

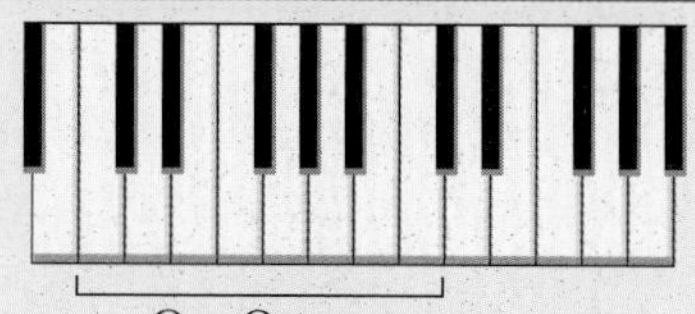

Numbers in the Fibonacci sequence can be found in an octave on the piano keyboard. The octave contains 2 black keys in one cluster, 3 black keys in another cluster, 5 black keys, 8 white keys, and a total of 13 keys altogether. The numbers 2, 3, 5, 8, and 13 are the third through seventh terms of the Fibonacci sequence.

The notation a_n represents the nth term, or **general term**, of a sequence. The entire sequence is represented by $\{a_n\}$.

Definition of a Sequence

An **infinite sequence** $\{a_n\}$ is a function whose domain is the set of positive integers. The function values, or **terms**, of the sequence are represented by

$$a_1, a_2, a_3, a_4, \ldots, a_n, \ldots.$$

Sequences whose domains consist only of the first n positive integers are called **finite sequences**.

1 Find particular terms of a sequence from the general term.

EXAMPLE 1 Writing Terms of a Sequence from the General Term

Write the first four terms of the sequence whose nth term, or general term, is given.

a. $a_n = 3n + 4$ **b.** $a_n = \dfrac{(-1)^n}{3^n - 1}$

Solution

a. We need to find the first four terms of the sequence whose general term is $a_n = 3n + 4$. To do so, we replace n in the formula by 1, 2, 3, and 4.

a_1, 1st term: $3 \cdot 1 + 4 = 3 + 4 = 7$

a_2, 2nd term: $3 \cdot 2 + 4 = 6 + 4 = 10$

a_3, 3rd term: $3 \cdot 3 + 4 = 9 + 4 = 13$

a_4, 4th term: $3 \cdot 4 + 4 = 12 + 4 = 16$

The first four terms are 7, 10, 13, and 16. The sequence defined by $a_n = 3n + 4$ can be written as

$$7,\ 10,\ 13,\ \ldots,\ 3n + 4,\ \ldots.$$

Study Tip

The factor $(-1)^n$ in the general term of a sequence causes the signs of the terms to alternate between positive and negative, depending on whether n is even or odd.

b. We need to find the first four terms of the sequence whose general term is $a_n = \dfrac{(-1)^n}{3^n - 1}$. To do so, we replace each occurrence of n in the formula by 1, 2, 3, and 4.

a_1, 1st term: $\dfrac{(-1)^1}{3^1 - 1} = \dfrac{-1}{3 - 1} = -\dfrac{1}{2}$

a_2, 2nd term: $\dfrac{(-1)^2}{3^2 - 1} = \dfrac{1}{9 - 1} = \dfrac{1}{8}$

a_3, 3rd term: $\dfrac{(-1)^3}{3^3 - 1} = \dfrac{-1}{27 - 1} = -\dfrac{1}{26}$

a_4, 4th term: $\dfrac{(-1)^4}{3^4 - 1} = \dfrac{1}{81 - 1} = \dfrac{1}{80}$

The first four terms are $-\frac{1}{2}, \frac{1}{8}, -\frac{1}{26}, \frac{1}{80}$. The sequence defined by $\dfrac{(-1)^n}{3^n - 1}$ can be written as

$$-\frac{1}{2}, \frac{1}{8}, -\frac{1}{26}, \ldots, \frac{(-1)^n}{3^n - 1}, \ldots.$$

Technology

Graphing utilities can write the terms of a sequence and graph them. For example, to find the first six terms of $\{a_n\} = \left\{\frac{1}{n}\right\}$, enter

General term — Stop at a_6.

SEQ $(1 \div x, x, 1, 6, 1)$.

Variable used in general term — Start at a_1. — The "step" from a_1 to a_2, a_2 to a_3, etc., is 1.

The first few terms of the sequence are shown in the viewing rectangle. By pressing the right arrow key to scroll right, you can see the remaining terms.

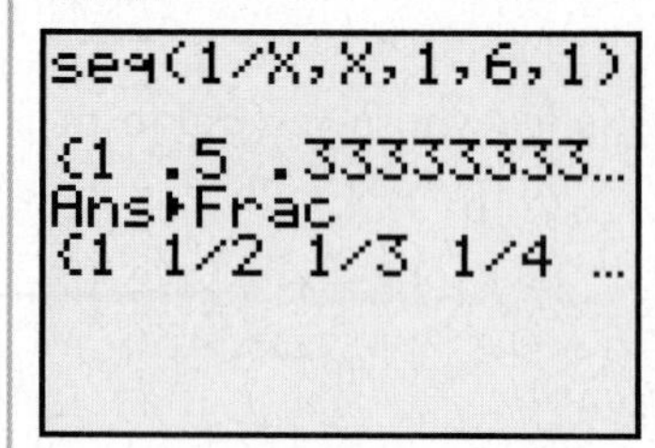

Check Point 1 Write the first four terms of the sequence whose nth term, or general term, is given.

a. $a_n = 2n + 5$ **b.** $a_n = \dfrac{(-1)^n}{2^n + 1}$

Although sequences are usually named with the letter a, any lowercase letter can be used. For example, the first four terms of the sequence $\{b_n\} = \{(\frac{1}{2})^n\}$ are $b_1 = \frac{1}{2}$, $b_2 = \frac{1}{4}$, $b_3 = \frac{1}{8}$, and $b_4 = \frac{1}{16}$.

Because a sequence is a function whose domain is the set of positive integers, the **graph of a sequence** is a set of discrete points. For example, consider the sequence whose general term is $a_n = \frac{1}{n}$. How does the graph of this sequence differ from the graph of the function $f(x) = \frac{1}{x}$? The graph of $f(x) = \frac{1}{x}$ is shown in Figure 8.1(a) for positive values of x. To obtain the graph of the sequence $\{a_n\} = \{\frac{1}{n}\}$, remove all the points from the graph of f except those whose x-coordinates are positive integers. Thus, we remove all points except $(1, 1)$, $(2, \frac{1}{2})$, $(3, \frac{1}{3})$, $(4, \frac{1}{4})$, and so on. The remaining points are the graph of the sequence $\{a_n\} = \{\frac{1}{n}\}$, shown in Figure 8.1(b). Notice that the horizontal axis is labeled n and the vertical axis a_n.

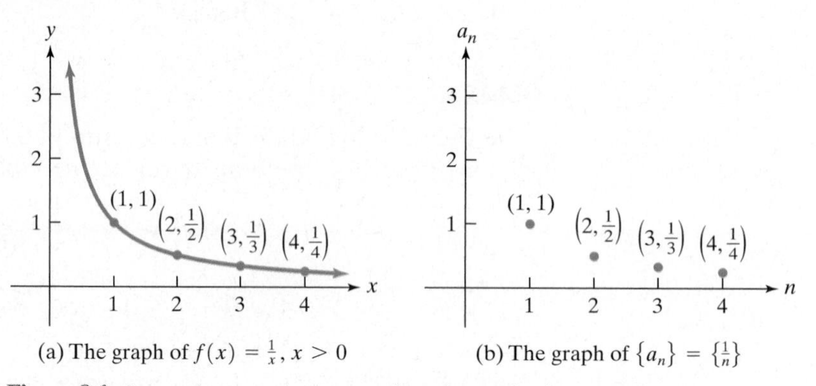

Figure 8.1 Comparing a continuous graph to the graph of a sequence

2 Use recursion formulas.

Recursion Formulas

In Example 1, the formulas used for the nth term of a sequence expressed the term as a function of n, the number of the term. Sequences can also be defined using **recursion formulas.** A recursion formula defines the nth term of a sequence as a function of the previous term. Our next example illustrates that if the first term of a sequence is known, then the recursion formula can be used to determine the remaining terms.

EXAMPLE 2 Using a Recursion Formula

Find the first four terms of the sequence in which $a_1 = 5$ and $a_n = 3a_{n-1} + 2$ for $n \geq 2$.

Solution

$a_1 = 5$ — This is the given first term.

$a_2 = 3a_1 + 2$ — Use $a_n = 3a_{n-1} + 2$, with $n = 2$. Thus, $a_2 = 3a_{2-1} + 2 = 3a_1 + 2$.

$= 3(5) + 2 = 17$ — Substitute 5 for a_1.

$a_3 = 3a_2 + 2$ — Again use $a_n = 3a_{n-1} + 2$, with $n = 3$.

$= 3(17) + 2 = 53$ — Substitute 17 for a_2.

$a_4 = 3a_3 + 2$ — Notice that a_4 is defined in terms of a_3. We used $a_n = 3a_{n-1} + 2$, with $n = 4$.

$= 3(53) + 2 = 161$ — Use the value of a_3, the third term, obtained from above.

The first four terms are 5, 17, 53, and 161.

Check Point 2 Find the first four terms of the sequence in which $a_1 = 3$ and $a_n = 2a_{n-1} + 5$ for $n \geq 2$.

3 Use factorial notation.

Factorial Notation

Products of consecutive positive integers occur quite often in sequences. These products can be expressed in a special notation, called **factorial notation**.

Factorial Notation

If n is a positive integer, the notation $n!$ (read "n factorial") is the product of all positive integers from n down through 1.

$$n! = n(n-1)(n-2)\ldots(3)(2)(1)$$

0! (zero factorial), by definition, is 1.

$$0! = 1$$

Factorials from 0 through 20

$n!$	Value
0!	1
1!	1
2!	2
3!	6
4!	24
5!	120
6!	720
7!	5040
8!	40,320
9!	362,880
10!	3,628,800
11!	39,916,800
12!	479,001,600
13!	6,227,020,800
14!	87,178,291,200
15!	1,307,674,368,000
16!	20,922,789,888,000
17!	355,687,428,096,000
18!	6,402,373,705,728,000
19!	121,645,100,408,832,000
20!	2,432,902,008,176,640,000

As n increases, $n!$ grows very rapidly. Factorial growth is more explosive than exponential growth discussed in Chapter 4.

The values of $n!$ for the first six positive integers are

$$1! = 1$$
$$2! = 2 \cdot 1 = 2$$
$$3! = 3 \cdot 2 \cdot 1 = 6$$
$$4! = 4 \cdot 3 \cdot 2 \cdot 1 = 24$$
$$5! = 5 \cdot 4 \cdot 3 \cdot 2 \cdot 1 = 120$$
$$6! = 6 \cdot 5 \cdot 4 \cdot 3 \cdot 2 \cdot 1 = 720.$$

Factorials affect only the number or variable that they follow unless grouping symbols appear. For example,

$$2 \cdot 3! = 2(3 \cdot 2 \cdot 1) = 2 \cdot 6 = 12$$

whereas

$$(2 \cdot 3)! = 6! = 6 \cdot 5 \cdot 4 \cdot 3 \cdot 2 \cdot 1 = 720.$$

In this sense, factorials are similar to exponents.

EXAMPLE 3 Finding Terms of a Sequence Involving Factorials

Write the first four terms of the sequence whose nth term is

$$a_n = \frac{2^n}{(n-1)!}.$$

Technology

Graphing utilities have factorial keys. To find 5 factorial, enter

5 [!] [ENTER].

Because $n!$ becomes quite large as n increases, your utility will display these larger values in scientific notation.

Solution We need to find the first four terms of the sequence. To do so, we replace each n in the formula by 1, 2, 3, and 4.

a_1, 1st term: $$\frac{2^1}{(1-1)!} = \frac{2}{0!} = \frac{2}{1} = 2$$

a_2, 2nd term: $$\frac{2^2}{(2-1)!} = \frac{4}{1!} = \frac{4}{1} = 4$$

a_3, 3rd term: $$\frac{2^3}{(3-1)!} = \frac{8}{2!} = \frac{8}{2 \cdot 1} = 4$$

a_4, 4th term: $$\frac{2^4}{(4-1)!} = \frac{16}{3!} = \frac{16}{3 \cdot 2 \cdot 1} = \frac{16}{6} = \frac{8}{3}$$

The first four terms are $2, 4, 4, \frac{8}{3}$.

Check Point 3 Write the first four terms of the sequence whose nth term is

$$a_n = \frac{20}{(n+1)!}.$$

When evaluating fractions with factorials in the numerator and the denominator, try to reduce the fraction before performing the multiplications. For example, consider $\frac{26!}{21!}$. Rather than write out 26! as the product of all integers from 26 down to 1, we can express 26! as

$$26! = 26 \cdot 25 \cdot 24 \cdot 23 \cdot 22 \cdot 21!.$$

In this way, we can divide both the numerator and the denominator by the common factor, 21!.

$$\frac{26!}{21!} = \frac{26 \cdot 25 \cdot 24 \cdot 23 \cdot 22 \cdot \cancel{21!}}{\cancel{21!}} = 26 \cdot 25 \cdot 24 \cdot 23 \cdot 22 = 7{,}893{,}600$$

EXAMPLE 4 Evaluating Fractions with Factorials

Evaluate each factorial expression.

a. $\frac{10!}{2!8!}$ **b.** $\frac{(n+1)!}{n!}$

Solution

a. $$\frac{10!}{2!\,8!} = \frac{10 \cdot 9 \cdot \cancel{8!}}{2 \cdot 1 \cdot \cancel{8!}} = \frac{90}{2} = 45$$

b. $$\frac{(n+1)!}{n!} = \frac{(n+1) \cdot \cancel{n!}}{\cancel{n!}} = n + 1$$

Check Point 4 Evaluate each factorial expression.

a. $\frac{14!}{2!12!}$ **b.** $\frac{n!}{(n-1)!}$

4 Use summation notation.

Summation Notation

It is sometimes useful to find the sum of the first n terms of a sequence. For example, consider the number of AIDS cases diagnosed in the United States from 1991 to 1997, shown in Table 8.1.

Table 8.1 AIDS Cases Diagnosed in the United States, 1991–1997

Year	1991	1992	1993	1994	1995	1996	1997
Cases Diagnosed	60,124	79,054	79,049	71,209	66,233	54,656	31,153

Source: U.S. Department of Health and Human Services

We can let a_n represent the number of AIDS cases diagnosed in year n, where $n = 1$ corresponds to 1991, $n = 2$ to 1992, $n = 3$ to 1993, and so on. The terms of the finite sequence in Table 8.1 are given as follows.

60,124 79,054, 79,049, 71,209, 66,233, 54,656, 31,153

a_1 a_2 a_3 a_4 a_5 a_6 a_7

Why might we want to add the terms of this sequence? We do this to find the number of AIDS cases diagnosed from 1991 to 1997. Thus,

$$\begin{aligned} &a_1 + a_2 + a_3 + a_4 + a_5 + a_6 + a_7 \\ &= 60{,}124 + 79{,}054, + 79{,}049, + 71{,}209, + 66{,}233, + 54{,}656, + 31{,}153 \\ &= 441{,}478. \end{aligned}$$

We see that there were 441,478 AIDS cases diagnosed in the United States from 1991 to 1997.

There is a compact notation for expressing the sum of the first n terms of a sequence. For example, rather than write

$$a_1 + a_2 + a_3 + a_4 + a_5 + a_6 + a_7,$$

we can use **summation notation** to express the sum as

$$a_1 + a_2 + a_3 + a_4 + a_5 + a_6 + a_7 = \sum_{i=1}^{7} a_i.$$

We read the expression on the right as "the sum as i goes from 1 to 7 of a_i." The letter i is called the **index of summation** and is not related to the use of i to represent $\sqrt{-1}$.

You can think of the symbol Σ (the uppercase Greek letter sigma) as an instruction to add up terms of a sequence.

Summation Notation

The sum of the first n terms of a sequence is represented by the **summation notation**

$$\sum_{i=1}^{n} a_i = a_1 + a_2 + a_3 + a_4 + \cdots + a_n$$

where i is the **index of summation**, n is the **upper limit of summation**, and 1 is the **lower limit of summation.**

Any letter can be used for the index of summation. The letters i, j, and k are used commonly. Furthermore, the lower limit of summation can be an integer other than 1.

When we write out a sum that is given in summation notation, we are **expanding the summation notation.** Example 5 shows how to do this.

EXAMPLE 5 Using Summation Notation

Expand and evaluate the sum:

a. $\sum_{i=1}^{6}(i^2 + 1)$ **b.** $\sum_{k=4}^{7}[(-2)^k - 5]$ **c.** $\sum_{i=1}^{5} 3$

Technology

Graphing utilities can calculate the sum of a sequence. For example, to find the sum of the sequence in Example 5a, enter

SUM SEQ $(x^2 + 1, x, 1, 6, 1)$.

Then press ENTER; 97 should be displayed. Use this capability to verify Example 5b.

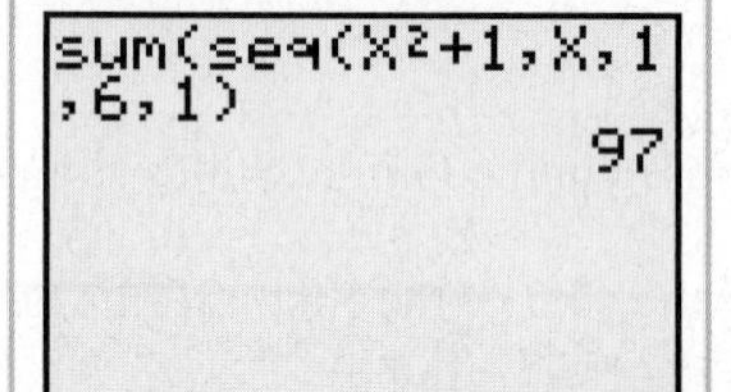

Solution

a. We must replace i in the expression $i^2 + 1$ with all consecutive integers from 1 to 6 inclusively. Then we add.

$$\sum_{i=1}^{6}(i^2 + 1) = (1^2 + 1) + (2^2 + 1) + (3^2 + 1) + (4^2 + 1) + (5^2 + 1) + (6^2 + 1)$$
$$= 2 + 5 + 10 + 17 + 26 + 37$$
$$= 97$$

b. This time the index of summation is k. First we evaluate $(-2)^k - 5$ for all consecutive integers from 4 through 7 inclusively. Then we add.

$$\sum_{k=4}^{7}[(-2)^k - 5] = [(-2)^4 - 5] + [(-2)^5 - 5] + [(-2)^6 - 5] + [(-2)^7 - 5]$$
$$= (16 - 5) + (-32 - 5) + (64 - 5) + (-128 - 5)$$
$$= 11 + (-37) + 59 + (-133)$$
$$= -100$$

c. To find $\sum_{i=1}^{5} 3$, we observe that every term of the sum is 3. The notation $i = 1$ through 5 indicates that we must add the first five 3s from a sequence in which every term is 3.

$$\sum_{i=1}^{5} 3 = 3 + 3 + 3 + 3 + 3 = 15$$

Check Point 5 Expand and evaluate the sum:

a. $\sum_{i=1}^{6} 2i^2$ **b.** $\sum_{k=3}^{5}(2^k - 3)$ **c.** $\sum_{i=1}^{5} 4$

For a given sum, we can vary the upper and lower limits of summation as well as the letter used for the index of summation. By doing so, we can produce different-looking summation notations for the same sum. For example, the sum of the squares of the first four integers can be expressed in a number of equivalent ways:

$$\sum_{i=1}^{4} i^2 = 1^2 + 2^2 + 3^2 + 4^2 = 30$$

$$\sum_{i=0}^{3} (i+1)^2 = (0+1)^2 + (1+1)^2 + (2+1)^2 + (3+1)^2$$

$$= 1^2 + 2^2 + 3^2 + 4^2 = 30$$

$$\sum_{k=2}^{5} (k-1)^2 = (2-1)^2 + (3-1)^2 + (4-1)^2 + (5-1)^2$$

$$= 1^2 + 2^2 + 3^2 + 4^2 = 30.$$

EXAMPLE 6 Writing Sums in Summation Notation

Express each sum using summation notation.

a. $1^3 + 2^3 + 3^3 + \cdots + 7^3$ **b.** $1 + \frac{1}{3} + \frac{1}{9} + \frac{1}{27} + \cdots + \frac{1}{3^{n-1}}$

Solution In each case, we will use 1 as the lower limit of summation and i for the index of summation.

a. The sum $1^3 + 2^3 + 3^3 + \cdots + 7^3$ has seven terms, each of the form i^3, starting at $i = 1$ and ending at $i = 7$. Thus,

$$1^3 + 2^3 + 3^3 + \cdots + 7^3 = \sum_{i=1}^{7} i^3.$$

b. The sum

$$1 + \frac{1}{3} + \frac{1}{9} + \frac{1}{27} + \cdots + \frac{1}{3^{n-1}}$$

has n terms, each of the form $\frac{1}{3^{i-1}}$, starting at $i = 1$ and ending at $i = n$.

Thus,

$$1 + \frac{1}{3} + \frac{1}{9} + \frac{1}{27} + \cdots + \frac{1}{3^{n-1}} = \sum_{i=1}^{n} \frac{1}{3^{i-1}}.$$

Check Point 6 Express each sum using summation notation.

a. $1^2 + 2^2 + 3^2 + \cdots + 9^2$ **b.** $1 + \frac{1}{2} + \frac{1}{4} + \frac{1}{8} + \cdots + \frac{1}{2^{n-1}}$

Table 8.2 contains some important properties of sums expressed in summation notation.

Table 8.2 Properties of Sums

Property	Example
1. $\sum_{i=1}^{n} ca_i = c\sum_{i=1}^{n} a_i$, c any real number	$\sum_{i=1}^{4} 3i^2 = 3 \cdot 1^2 + 3 \cdot 2^2 + 3 \cdot 3^2 + 3 \cdot 4^2$ $3\sum_{i=1}^{4} i^2 = 3(1^2 + 2^2 + 3^2 + 4^2) = 3 \cdot 1^2 + 3 \cdot 2^2 + 3 \cdot 3^2 + 3 \cdot 4^2$ Conclusion: $\sum_{i=1}^{4} 3i^2 = 3\sum_{i=1}^{4} i^2$
2. $\sum_{i=1}^{n} (a_i + b_i) = \sum_{i=1}^{n} a_i + \sum_{i=1}^{n} b_i$	$\sum_{i=1}^{4} (i + i^2) = (1 + 1^2) + (2 + 2^2) + (3 + 3^2) + (4 + 4^2)$ $\sum_{i=1}^{4} i + \sum_{i=1}^{4} i^2 = (1 + 2 + 3 + 4) + (1^2 + 2^2 + 3^2 + 4^2)$ $= (1 + 1^2) + (2 + 2^2) + (3 + 3^2) + (4 + 4^2)$ Conclusion: $\sum_{i=1}^{4} (i + i^2) = \sum_{i=1}^{4} i + \sum_{i=1}^{4} i^2$
3. $\sum_{i=1}^{n} (a_i - b_i) = \sum_{i=1}^{n} a_i - \sum_{i=1}^{n} b_i$	$\sum_{i=3}^{5} (i^2 - i^3) = (3^2 - 3^3) + (4^2 - 4^3) + (5^2 - 5^3)$ $\sum_{i=3}^{5} i^2 - \sum_{i=3}^{5} i^3 = (3^2 + 4^2 + 5^2) - (3^3 + 4^3 + 5^3)$ $= (3^2 - 3^3) + (4^2 - 4^3) + (5^2 - 5^3)$ Conclusion: $\sum_{i=3}^{5} (i^2 - i^3) = \sum_{i=3}^{5} i^2 - \sum_{i=3}^{5} i^3$

EXERCISE SET 8.1

Practice Exercises

In Exercises 1–12, write the first four terms of each sequence whose general term is given.

1. $a_n = 3n + 2$
2. $a_n = 4n - 1$
3. $a_n = 3^n$
4. $a_n = \left(\frac{1}{3}\right)^n$
5. $a_n = (-3)^n$
6. $a_n = \left(-\frac{1}{3}\right)^n$
7. $a_n = (-1)^n(n + 3)$
8. $a_n = (-1)^{n+1}(n + 4)$
9. $a_n = \frac{2n}{n + 4}$
10. $a_n = \frac{3n}{n + 5}$
11. $a_n = \frac{(-1)^{n+1}}{2^n - 1}$
12. $a_n = \frac{(-1)^{n+1}}{2^n + 1}$

The sequences in Exercises 13–18 are defined using recursion formulas. Write the first four terms of each sequence.

13. $a_1 = 7$ and $a_n = a_{n-1} + 5$ for $n \geq 2$
14. $a_1 = 12$ and $a_n = a_{n-1} + 4$ for $n \geq 2$
15. $a_1 = 3$ and $a_n = 4a_{n-1}$ for $n \geq 2$
16. $a_1 = 2$ and $a_n = 5a_{n-1}$ for $n \geq 2$
17. $a_1 = 4$ and $a_n = 2a_{n-1} + 3$ for $n \geq 2$
18. $a_1 = 5$ and $a_n = 3a_{n-1} - 1$ for $n \geq 2$

In Exercises 19–22, the general term of a sequence is given and involves a factorial. Write the first four terms of each sequence.

19. $a_n = \frac{n^2}{n!}$
20. $a_n = \frac{(n + 1)!}{n^2}$
21. $a_n = 2(n + 1)!$
22. $a_n = -2(n - 1)!$

In Exercises 23–28, evaluate each factorial expression.

23. $\frac{17!}{15!}$
24. $\frac{18!}{16!}$
25. $\frac{16!}{2!14!}$
26. $\frac{20!}{2!18!}$

27. $\dfrac{(n+2)!}{n!}$

28. $\dfrac{(2n+1)!}{(2n)!}$

In Exercises 29–42, find each indicated sum.

29. $\sum_{i=1}^{6} 5i$

30. $\sum_{i=1}^{6} 7i$

31. $\sum_{i=1}^{4} 2i^2$

32. $\sum_{i=1}^{5} i^3$

33. $\sum_{k=1}^{5} k(k+4)$

34. $\sum_{k=1}^{4} (k-3)(k+2)$

35. $\sum_{i=1}^{4} \left(-\frac{1}{2}\right)^i$

36. $\sum_{i=2}^{4} \left(-\frac{1}{3}\right)^i$

37. $\sum_{i=5}^{9} 11$

38. $\sum_{i=3}^{7} 12$

39. $\sum_{i=0}^{4} \dfrac{(-1)^i}{i!}$

40. $\sum_{i=0}^{4} \dfrac{(-1)^{i+1}}{(i+1)!}$

41. $\sum_{i=1}^{5} \dfrac{i!}{(i-1)!}$

42. $\sum_{i=1}^{5} \dfrac{(i+2)!}{i!}$

In Exercises 43–54, express each sum using summation notation. Use 1 as the lower limit of summation and i for the index of summation.

43. $1^2 + 2^2 + 3^2 + \cdots + 15^2$

44. $1^4 + 2^4 + 3^4 + \cdots + 12^4$

45. $2 + 2^2 + 2^3 + \cdots + 2^{11}$

46. $5 + 5^2 + 5^3 + \cdots + 5^{12}$

47. $1 + 2 + 3 + \cdots + 30$

48. $1 + 2 + 3 + \cdots + 40$

49. $\frac{1}{2} + \frac{2}{3} + \frac{3}{4} + \cdots + \frac{14}{14+1}$

50. $\frac{1}{3} + \frac{2}{4} + \frac{3}{5} + \cdots + \frac{16}{16+2}$

51. $4 + \frac{4^2}{2} + \frac{4^3}{3} + \cdots + \frac{4^n}{n}$

52. $\frac{1}{9} + \frac{2}{9^2} + \frac{3}{9^3} + \cdots + \frac{n}{9^n}$

53. $1 + 3 + 5 + \cdots + (2n-1)$

54. $a + ar + ar^2 + \cdots + ar^{n-1}$

In Exercises 55–60, express each sum using summation notation. Use a lower limit of summation of your choice and k for the index of summation.

55. $5 + 7 + 9 + 11 + \cdots + 31$

56. $6 + 8 + 10 + 12 + \cdots + 32$

57. $a + ar + ar^2 + \cdots + ar^{12}$

58. $a + ar + ar^2 + \cdots + ar^{14}$

59. $a + (a+d) + (a+2d) + \cdots + (a+nd)$

60. $(a+d) + (a+d^2) + \cdots + (a+d^n)$

Application Exercises

61. The bar graph shows the number of children home-educated in the United States. Let a_n represent the number of children, in thousands, home-educated in year n, where $n = 2$ corresponds to 1992, $n = 3$ to 1993, and so on.

Number of Children Home-Educated in the U.S.

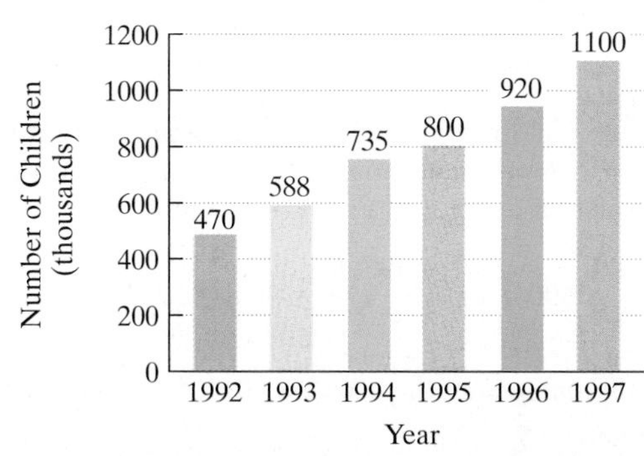

Source: National Home Education Research Institute

a. Find $\sum_{i=2}^{7} a_i$. What does this represent?

b. Find $\frac{1}{6}\sum_{i=2}^{7} a_i$. What does this represent?

62. The bar graph shows the number of business failures in the United States. Let a_n represent the number of business failures in year n, where $n = 0$ corresponds to 1990, $n = 1$ to 1991, $n = 2$ to 1992, and so on.

Number of Business Failures in the U.S.

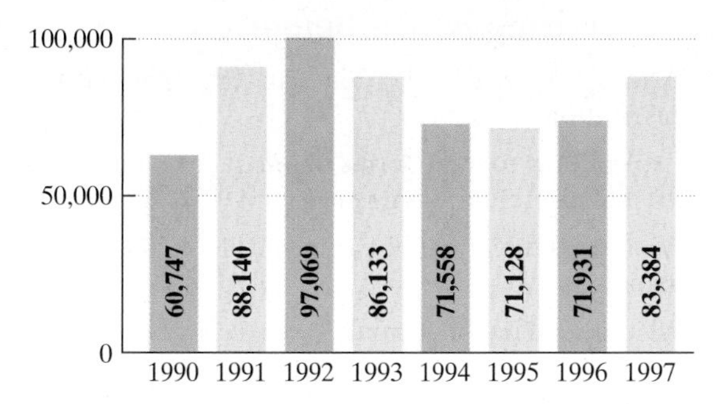

Source: Dun & Bradstreet

a. Find $\sum_{i=0}^{7} a_i$. What does this represent?

b. Find $\frac{1}{8}\sum_{i=0}^{7} a_i$. What does this represent?

63. The finite sequence whose general term is

$$a_n = 0.16n^2 - 1.04n + 7.39$$

where $n = 1, 2, 3, \ldots, 8$ models the total number of dollars, in billions, that Americans spent on recreational boating from 1991 through 1998. Find and interpret

$$\sum_{i=1}^{5} a_i.$$

64. The finite sequence whose general term is

$$a_n = 2.54e^{-0.09n}$$

where $n = 0, 1, 2, \ldots, 9$ models the number of new foreign cars sold in the United States, in millions, from 1990 through 1999. Find and interpret

$$\sum_{i=0}^{4} a_i.$$

65. A deposit of \$6000 is made in an account that earns 6% interest compounded quarterly. The balance in the account after n quarters is given by the sequence

$$a_n = 6000\left(1 + \frac{0.06}{4}\right)^n, \quad n = 1, 2, 3, \ldots.$$

Find the balance in the account after five years by computing a_{20}.

66. A deposit of \$10,000 is made in an account that earns 8% interest compounded quarterly. The balance in the account after n quarters is given by the sequence

$$a_n = 10{,}000\left(1 + \frac{0.08}{4}\right)^n, \quad n = 1, 2, 3, \ldots.$$

Find the balance in the account after six years by computing a_{24}.

Writing in Mathematics

67. What is a sequence? Give an example with your description.

68. Explain how to write terms of a sequence if the formula for the general term is given.

69. What does the graph of a sequence look like? How is it obtained?

70. What is a recursion formula?

71. Explain how to find $n!$ if n is a positive integer.

72. Explain the best way to evaluate $\frac{900!}{899!}$ without a calculator.

73. What is the meaning of the symbol Σ? Give an example with your description.

74. You buy a new car for \$24,000. At the end of n years, the value of your car is given by the sequence

$$a_n = 24{,}000\left(\frac{3}{4}\right)^n, \quad n = 1, 2, 3, \ldots.$$

Find a_5 and write a sentence explaining what this value represents. Describe the nth term of the sequence in terms of the value of your car at the end of each year.

Technology Exercises

In Exercises 75–79, use the factorial key of a graphing utility to evaluate each expression.

75. $\frac{200!}{198!}$ **76.** $\left(\frac{300}{20}\right)!$ **77.** $\frac{20!}{300}$

78. $\frac{20!}{(20-3)!}$ **79.** $\frac{54!}{(54-3)!\,3!}$

80. Use the [SEQ] (sequence) capability of a graphing utility to verify the terms of the sequences you obtained for any five sequences from Exercises 1–12 or 19–22.

81. Use the [SUM] [SEQ] (sum of the sequence) capability of a graphing utility to verify any five of the sums you obtained in Exercises 29–42.

82. As n increases, the terms of the sequence

$$a_n = \left(1 + \frac{1}{n}\right)^n$$

get closer and closer to the number e (where $e \approx 2.7183$). Use a calculator to find a_{10}, a_{100}, a_{1000}, $a_{10{,}000}$, and $a_{100{,}000}$, comparing these terms to the decimal approximation for e.

Many graphing utilities have a sequence-graphing mode that plots the terms of a sequence as points on a rectangular coordinate system. Consult your manual; if your graphing utility has this capability, use it to graph each of the sequences in Exercises 83–86. What appears to be happening to the terms of each sequence as n gets larger?

83. $a_n = \frac{n}{n+1}$ $\quad n{:}[0, 10, 1] \times a_n{:}[0, 1, 0.1]$

84. $a_n = \frac{100}{n}$ $\quad n{:}[0, 1000, 100] \times a_n{:}[0, 1, 0.1]$

85. $a_n = \frac{2n^2 + 5n - 7}{n^3}$ $\quad n{:}[0, 10, 1] \times a_n{:}[0, 2, 0.2]$

86. $a_n = \frac{3n^4 + n - 1}{5n^4 + 2n^2 + 1}$ $\quad n{:}[0, 10, 1] \times a_n{:}[0, 1, 0.1]$

Critical Thinking Exercises

87. Which one of the following is true?

a. $\frac{n!}{(n-1)!} = \frac{1}{n-1}$

b. The Fibonacci sequence 1, 1, 2, 3, 5, 8, 13, 21, 34, 55, 89, 144, ... can be defined recursively using $a_0 = 1, a_1 = 1, a_n = a_{n-2} + a_{n-1}$, where $n \geq 2$.

c. $\sum_{i=1}^{2} (-1)^i 2^i = 0$

d. $\sum_{i=1}^{2} a_i b_i = \sum_{i=1}^{2} a_i \sum_{i=1}^{2} b_i$

88. Write the first five terms of the sequence whose first term is 9 and whose general term is

$$a_n = \begin{cases} \dfrac{a_{n-1}}{2} & \text{if } a_{n-1} \text{ is even} \\ 3a_{n-1} + 5 & \text{if } a_{n-1} \text{ is odd.} \end{cases}$$

Group Exercise

89. Enough curiosities involving the Fibonacci sequence exist to warrant a flourishing Fibonacci Association, which publishes a quarterly journal. Do some research on the Fibonacci sequence by consulting the Internet or the research department of your library, and find one property that interests you. After doing this research, get together with your group to share these intriguing properties.

SECTION 8.2 Arithmetic Sequences

Objectives

1. Find the common difference for an arithmetic sequence.
2. Write terms of an arithmetic sequence.
3. Use the formula for the general term of an arithmetic sequence.
4. Use the formula for the sum of the first n terms of an arithmetic sequence.

Your grandmother and her financial counselor are looking at options in case nursing home care is needed in the future. The good news is that your grandmother's total assets are $350,000. The bad news is that yearly nursing home costs average $49,730, increasing by $1800 each year. In this section, we will see how sequences can be used to describe your grandmother's situation and help her to identify realistic options.

Arithmetic Sequences

A mathematical model for the average annual salaries of major league baseball players generates the following data.

Year	1991	1992	1993	1994	1995	1996	1997	1998
Salary	801,000	892,000	983,000	1,074,000	1,165,000	1,256,000	1,347,000	1,438,000

From 1991 to 1992, salaries increased by $892,000 − $801,000 = $91,000. From 1992 to 1993, salaries increased by $983,000 − $892,000 = $91,000. If we make these computations for each year, we find that the yearly salary increase is $91,000. The sequence of annual salaries shows that each term after the first, 801,000, differs from the preceding term by a constant amount, namely 91,000. The sequence of annual salaries

$$801{,}000,\ 892{,}000,\ 983{,}000,\ 1{,}074{,}000,\ 1{,}165{,}000,\ 1{,}256{,}000, \ldots$$

is an example of an **arithmetic sequence**.

Definition of an Arithmetic Sequence

An **arithmetic sequence** is a sequence in which each term after the first differs from the preceding term by a constant amount. The difference between consecutive terms is called the **common difference** of the sequence.

1 Find the common difference of an arithmetic sequence.

The common difference, d, is found by subtracting any term from the term that directly follows it. In the following examples, the common difference is found by subtracting the first term from the second term, $a_2 - a_1$.

Arithmetic sequence	Common difference
801,000, 892,000, 983,000, 1,074,000,...	$d = 892{,}000 - 801{,}000 = 91{,}000$
2, 6, 10, 14, 18,...	$d = 6 - 2 = 4$
−2, −7, −12, −17,...	$d = -7 - (-2) = -5$

If the first term of an arithmetic sequence is a_1, each term after the first is obtained by adding d, the common difference, to the previous term. This can be expressed recursively as follows:

$$a_n = a_{n-1} + d$$

Add d to the term in any position to get the next term.

To use this recursion formula, we must be given the first term.

2 Write the terms of an arithmetic sequence.

EXAMPLE 1 Writing the Terms of an Arithmetic Sequence Using the First Term and the Common Difference

The recursion formula $a_n = a_{n-1} - 24$ models the thousands of Air Force personnel on active duty for each year starting with 1986. In 1986, there were 624 thousand personnel on active duty. Find the first five terms of the arithmetic sequence in which $a_1 = 624$ and $a_n = a_{n-1} - 24$.

Solution The recursion formula $a_n = a_{n-1} - 24$ indicates that each term after the first is obtained by adding −24 to the previous term. Thus, each year there are 24 thousand fewer personnel on active duty in the Air Force than in the previous year.

$a_1 = 624$ This is given.

$a_2 = a_1 - 24 = 624 - 24 = 600$ Use $a_n = a_{n-1} - 24$ with $n = 2$.

$a_3 = a_2 - 24 = 600 - 24 = 576$ Use $a_n = a_{n-1} - 24$ with $n = 3$.

$a_4 = a_3 - 24 = 576 - 24 = 552$ Use $a_n = a_{n-1} - 24$ with $n = 4$.

$a_5 = a_4 - 24 = 552 - 24 = 528$ Use $a_n = a_{n-1} - 24$ with $n = 5$.

The first five terms are

624, 600, 576, 552, and 528.

Check Point 1 Find the first five terms of the arithmetic sequence in which $a_1 = 100$ and $a_n = a_{n-1} - 30$.

3 Use the formula for the general term of an arithmetic sequence.

The General Term of an Arithmetic Sequence

Consider an arithmetic sequence whose first term is a_1 and whose common difference is d. We are looking for a formula for the general term, a_n. Let's begin by writing the first six terms. The first term is a_1. The second term is $a_1 + d$. The third term is $a_1 + d + d$, or $a_1 + 2d$. Thus, we start with a_1 and add d to each successive term. The first six terms are

$$a_1, \quad a_1 + d, \quad a_1 + 2d, \quad a_1 + 3d, \quad a_1 + 4d, \quad a_1 + 5d.$$

a_1, first term; a_2, second term; a_3, third term; a_4, fourth term; a_5, fifth term; a_6, sixth term

Compare the coefficient of d and the subscript of a denoting the term number. Can you see that the coefficient of d is 1 less than the subscript of a denoting the term number?

$$a_3\text{: third term} = a_1 + 2d \qquad a_4\text{: fourth term} = a_1 + 3d$$

2 is one less than 3. 3 is one less than 4.

Thus, the formula for the nth term is

$$a_n\text{: }n\text{th term} = a_1 + (n - 1)d.$$

$n - 1$ is one less than n.

General Term of an Arithmetic Sequence

The nth term (the general term) of an arithmetic sequence with first term a_1 and common difference d is

$$a_n = a_1 + (n - 1)d.$$

EXAMPLE 2 Using the Formula for the General Term of an Arithmetic Sequence

Find the eighth term of the arithmetic sequence whose first term is 4 and whose common difference is −7.

Solution To find the eighth term, a_8, we replace n in the formula with 8, a_1 with 4, and d with −7.

$$a_n = a_1 + (n - 1)d$$
$$a_8 = 4 + (8 - 1)(-7) = 4 + 7(-7) = 4 + (-49) = -45$$

The eighth term is −45. We can check this result by writing the first eight terms of the sequence:

$$4, -3, -10, -17, -24, -31, -38, -45.$$

Check Point 2 Find the ninth term of the arithmetic sequence whose first term is 6 and whose common difference is −5.

EXAMPLE 3 Using an Arithmetic Sequence to Model Teachers' Earnings

According to the National Education Association, teachers in the United States earned an average of \$21,700 per year in 1984. This amount has increased by approximately \$1472 yearly.

a. Write a formula for the nth term of the arithmetic sequence that describes teachers' average earnings n years after 1983.

b. How much will U.S. teachers earn by the year 2005?

Solution

a. We can express teachers' earnings by the following arithmetic sequence:

$$21{,}700, \quad 23{,}172, \quad 24{,}644, \quad 26{,}116, \ldots$$

a_1: earnings in 1984, 1 year after 1983

a_2: earnings in 1985, 2 years after 1983

a_3: earnings in 1986, 3 years after 1983

a_4: earnings in 1987, 4 years after 1983

In this sequence a_1, the first term, represents the amount teachers earned in 1984. Each subsequent year this amount increases by \$1472, so $d = 1472$. We use the formula for the general term of an arithmetic sequence to write the nth term of the sequence that describes teachers' earnings n years after 1983.

$a_n = a_1 + (n - 1)d$ — This is the formula for the general term of an arithmetic sequence.

$a_n = 21{,}700 + (n - 1)1472$ — $a_1 = 21{,}700$ and $d = 1472$.

$a_n = 21{,}700 + 1472n - 1472$ — Distribute 1472 to each term in parentheses.

$a_n = 1472n + 20{,}228$ — Simplify.

Thus, teachers' earnings n years after 1983 can be described by $a_n = 1472n + 20{,}228$.

b. Now we need to find teachers' earnings in 2005. The year 2005 is 22 years after 1983: That is, $2005 - 1983 = 22$. Thus, $n = 22$. We substitute 22 for n in $a_n = 1472n + 20{,}228$.

$$a_{22} = 1472 \cdot 22 + 20{,}228 = 52{,}612$$

The 22nd term of the sequence is 52,612. Therefore, U.S. teachers are predicted to earn an average of \$52,612 by the year 2005.

Check Point 3 According to the U.S. Bureau of Economic Analysis, U.S. travelers spent \$12,808 million in other countries in 1984. This amount has increased by approximately \$2350 million yearly.

a. Write a formula for the nth term of the arithmetic sequence that describes what U.S. travelers spend in other countries n years after 1983.

b. How much will U.S. travelers spend in other countries by the year 2010?

4 Use the formula for the sum of the first n terms of an arithmetic sequence.

The Sum of the First n Terms of an Arithmetic Sequence

The sum of the first n terms of an arithmetic sequence, denoted by S_n, can be found without having to add up all the terms. Let

$$S_n = a_1 + a_2 + a_3 + \cdots + a_n$$

be the sum of the first n terms of an arithmetic sequence. Because d is the common difference between terms, S_n can be written forward and backward as follows.

Forward: Start with the first term. Keep adding d.

Backward: Start with the last term. Keep subtracting d.

$$
\begin{array}{rcl}
S_n &=& a_1 + (a_1 + d) + (a_1 + 2d) + \cdots + a_n \\
S_n &=& a_n + (a_n - d) + (a_n - 2d) + \cdots + a_1 \\
\hline
2S_n &=& (a_1 + a_n) + (a_1 + a_n) + (a_1 + a_n) + \cdots + (a_1 + a_n)
\end{array}
$$

Add the two equations.

Because there are n sums of $(a_1 + a_n)$ on the right side, we can express this side as $n(a_1 + a_n)$. Thus, the last equation can be simplified:

$$2S_n = n(a_1 + a_n)$$

$$S_n = \frac{n}{2}(a_1 + a_n)$$

Solve for S_n, dividing both sides by 2.

We have proved the following result.

The Sum of the First *n* Terms of an Arithmetic Sequence

The sum, S_n, of the first n terms of an arithmetic sequence is given by

$$S_n = \frac{n}{2}(a_1 + a_n)$$

in which a_1 is the first term and a_n is the nth term.

To find the sum of the terms of an arithmetic sequence, we need to know the first term, a_1, the last term, a_n, and the number of terms, n. The following examples illustrate how to use this formula.

EXAMPLE 4 Finding the Sum of *n* Terms of an Arithmetic Sequence

Find the sum of the first 100 terms of the arithmetic sequence: 1, 3, 5, 7, ...

Solution We are finding the sum of the first 100 odd numbers. To find the sum of the first 100 terms, S_{100}, we replace n in the formula with 100.

$$S_n = \frac{n}{2}(a_1 + a_n)$$

$$S_{100} = \frac{100}{2}(a_1 + a_{100})$$

The first term, a_1, is 1.

We must find a_{100}, the 100th term.

We use the formula for the general term of a sequence to find a_{100}. The common difference, d, of 1, 3, 5, 7, ..., is 2.

$$a_n = a_1 + (n - 1)d$$

This is the formula for the nth term of an arithmetic sequence. Use it to find the 100th term.

$$a_{100} = 1 + (100 - 1) \cdot 2$$

Substitute 100 for n, 2 for d, and 1 (the first term) for a_1.

$$= 1 + 99 \cdot 2$$

$$= 199$$

Now we are ready to find the sum of the first 100 terms of $1, 3, 5, 7, \ldots, 199$.

$$S_n = \frac{n}{2}(a_1 + a_n)$$ Use the formula for the sum of the first n terms of an arithmetic sequence. Let $n = 100$, $a_1 = 1$, and $a_{100} = 199$.

$$S_{100} = \frac{100}{2}(1 + 199) = 50(200) = 10{,}000$$

The sum of the first 100 odd numbers is 10,000.

Check Point 4 Find the sum of the first 15 terms of the arithmetic sequence: $3, 6, 9, 12, \ldots$.

Technology

To find

$$\sum_{i=1}^{25}(5i - 9)$$

on a graphing utility, enter: SUM SEQ $(5x - 9, x, 1, 25, 1)$. Then press ENTER.

```
sum(seq(5X-9,X,1
,25,1)
            1400
```

EXAMPLE 5 Using S_n to Evaluate a Summation

Find the following sum: $\sum_{i=1}^{25}(5i - 9)$.

Solution

$$\sum_{i=1}^{25}(5i - 9) = (5 \cdot 1 - 9) + (5 \cdot 2 - 9) + (5 \cdot 3 - 9) + \cdots + (5 \cdot 25 - 9)$$
$$= -4 \quad + 1 \quad + 6 \quad + \cdots + 116$$

By evaluating the first three terms and the last term, we see that $a_1 = -4$; d, the common difference, is $1 - (-4)$ or 5; and a_{25}, the last term, is 116.

$$S_n = \frac{n}{2}(a_1 + a_n)$$ Use the formula for the sum of the first n terms of an arithmetic sequence. Let $n = 25$, $a_1 = -4$, and $a_{25} = 116$.

$$S_{25} = \frac{25}{2}(-4 + 116) = \frac{25}{2}(112) = 1400.$$

Thus,

$$\sum_{i=1}^{25}(5i - 9) = 1400.$$

Check Point 5 Find the following sum: $\sum_{i=1}^{30}(6i - 11)$.

EXAMPLE 6 Modeling Total Nursing Home Costs over a Six-Year Period

Your grandmother has assets of \$350,000. One option that she is considering involves nursing home care for a six-year period beginning in 2001. The model

$$a_n = 1800n + 49{,}730$$

describes yearly nursing home costs n years after 2000. Does your grandmother have enough to pay for the facility?

Solution We must find the sum of an arithmetic sequence. The first term of the sequence corresponds to nursing home costs in the year 2001. The last term corresponds to nursing home costs in the year 2006. Because the model describes costs n years after 2000, $n = 1$ describes the year 2001 and $n = 6$ describes the year 2006.

$a_n = 1800n + 49{,}730$ This is the given formula for the general term of the sequence.

$a_1 = 1800 \cdot 1 + 49{,}730 = 51{,}530$ Find a_1 by replacing n by 1.

$a_6 = 1800 \cdot 6 + 49{,}730 = 60{,}530$ Find a_6 by replacing n by 6.

The first year the facility will cost \$51,530. By year six, the facility will cost \$60,530. Now we must find the sum of these costs for all six years. We focus on the sum of the first six terms of the arithmetic sequence

$$51{,}530, \quad 53{,}330, \quad \ldots, \quad 60{,}530.$$

a_1 a_2 a_6

We find this sum using the formula for the sum of the first n terms of an arithmetic sequence. We are adding 6 terms: $n = 6$. The first term is 51,530: $a_1 = 51{,}530$. The last term—that is, the sixth term—is 60,530: $a_6 = 60{,}530$.

$$S_n = \frac{n}{2}(a_1 + a_n)$$

$$S_6 = \frac{6}{2}(51{,}530 + 60{,}530) = 3(112{,}060) = 336{,}180$$

Total nursing home costs for your grandmother are predicted to be \$336,180. Because your grandmother's assets are \$350,000, she has enough to pay for the facility.

In Example 6, how much would it cost for nursing home care for a ten-year period beginning in 2001?

EXERCISE SET 8.2

Practice Exercises

In Exercises 1–14, write the first six terms of each arithmetic sequence.

1. $a_1 = 200, d = 20$
2. $a_1 = 300, d = 50$
3. $a_1 = -7, d = 4$
4. $a_1 = -8, d = 5$
5. $a_1 = 300, d = -90$
6. $a_1 = 200, d = -60$
7. $a_1 = \frac{5}{2}, d = -\frac{1}{2}$
8. $a_1 = \frac{3}{4}, d = -\frac{1}{4}$
9. $a_n = a_{n-1} + 6, a_1 = -9$
10. $a_n = a_{n-1} + 4, a_1 = -7$
11. $a_n = a_{n-1} - 10, a_1 = 30$
12. $a_n = a_{n-1} - 20, a_1 = 50$
13. $a_n = a_{n-1} - 0.4, a_1 = 1.6$
14. $a_n = a_{n-1} - 0.3, a_1 = -1.7$

In Exercises 15–22, find the indicated term of the arithmetic sequence with first term, a_1, and common difference, d.

15. Find a_6 when $a_1 = 13, d = 4$.
16. Find a_{16} when $a_1 = 9, d = 2$.
17. Find a_{50} when $a_1 = 7, d = 5$.
18. Find a_{60} when $a_1 = 8, d = 6$.
19. Find a_{200} when $a_1 = -40, d = 5$.
20. Find a_{150} when $a_1 = -60, d = 5$.
21. Find a_{60} when $a_1 = 35, d = -3$.
22. Find a_{70} when $a_1 = -32, d = 4$.

In Exercises 23–34, write a formula for the general term (the nth term) of each arithmetic sequence. Do not use a recursion formula. Then use the formula for a_n to find a_{20}, the 20th term of the sequence.

23. $1, 5, 9, 13, \ldots$
24. $2, 7, 12, 17, \ldots$
25. $7, 3, -1, -5, \ldots$
26. $6, 1, -4, -9, \ldots$
27. $a_1 = 9, d = 2$
28. $a_1 = 6, d = 3$
29. $a_1 = -20, d = -4$
30. $a_1 = -70, d = -5$
31. $a_n = a_{n-1} + 3, a_1 = 4$
32. $a_n = a_{n-1} + 5, a_1 = 6$
33. $a_n = a_{n-1} - 10, a_1 = 30$
34. $a_n = a_{n-1} - 12, a_1 = 24$

35. Find the sum of the first 20 terms of the arithmetic sequence: 4, 10, 16, 22, ...

36. Find the sum of the first 25 terms of the arithmetic sequence: 7, 19, 31, 43, ...

37. Find the sum of the first 50 terms of the arithmetic sequence: −10, −6, −2, 2, ...

38. Find the sum of the first 50 terms of the arithmetic sequence: −15, −9, −3, 3, ...

39. Find $1 + 2 + 3 + 4 + \dots + 100$, the sum of the first 100 natural numbers.

40. Find $2 + 4 + 6 + 8 + \dots + 200$, the sum of the first 100 positive even integers.

41. Find the sum of the first 60 positive even integers.

42. Find the sum of the first 80 positive even integers.

43. Find the sum of the even integers between 21 and 45.

44. Find the sum of the odd integers between 30 and 54.

For Exercises 45–50, write out the first three terms and the last term. Then use the formula for the sum of the first n terms of an arithmetic sequence to find the indicated sum.

45. $\sum_{i=1}^{17} (5i + 3)$

46. $\sum_{i=1}^{20} (6i - 4)$

47. $\sum_{i=1}^{30} (-3i + 5)$

48. $\sum_{i=1}^{40} (-2i + 6)$

49. $\sum_{i=1}^{100} 4i$

50. $\sum_{i=1}^{50} -4i$

Application Exercises

51. According to the U.S. Bureau of Labor Statistics, in 1990 there were 126,424 thousand employees in the United States. This number has increased by approximately 1265 thousand employees each year.

a. Write the general term for the arithmetic sequence modeling the thousands of employees in the United States n years after 1989.

b. How many thousands of employees will there be by the year 2005?

52. According to the National Center for Education Statistics, the total enrollment in U.S. public elementary and secondary schools in 1985 was 39.05 million. Enrollment has increased by approximately 0.45 million each year.

a. Write the general term for the arithmetic sequence modeling the millions of students enrolled in U.S. public elementary and secondary schools n years after 1984.

b. How many millions of students will be enrolled by the year 2005?

53. Company A pays \$24,000 yearly with raises of \$1600 per year. Company B pays \$28,000 yearly with raises of \$1000 per year. Which company will pay more in year 10? How much more?

54. Company A pays \$23,000 yearly with raises of \$1200 per year. Company B pays \$26,000 yearly with raises of \$800 per year. Which company will pay more in year 10? How much more?

55. According to the Environmental Protection Agency, in 1960 the United States recovered 3.78 million tons of solid waste. Due primarily to recycling programs, this amount has increased by approximately 0.576 million ton each year.

a. Write the general term for the arithmetic sequence modeling the amount of solid waste recovered in the United States n years after 1959.

b. What is the total amount of solid waste recovered from 1960 through 2000?

56. According to the Environmental Protection Agency, in 1960 the United States generated 87.1 million tons of solid waste. This amount has increased by approximately 3.14 million tons each year.

a. Write the general term for the arithmetic sequence modeling the amount of solid waste generated in the United States n years after 1959.

b. What is the total amount of solid waste generated from 1960 through 2000?

57. A company offers a starting yearly salary of \$33,000 with raises of \$2500 per year. Find the total salary over a ten-year period.

58. You are considering two job offers. Company A will start you at \$19,000 a year and guarantee you a raise of \$2600 per year. Company B will start you at a higher salary, \$27,000 a year, but will only guarantee a raise of \$1200 per year. Find the total salary that each company will pay you over a ten-year period. Which company pays the greater total amount?

59. A theater has 30 seats in the first row, 32 seats in the second row, increasing by 2 seats each row for a total of 26 rows. How many seats are there in the theater?

60. A section in a stadium has 20 seats in the first row, 23 seats in the second row, increasing by 3 seats each row for a total of 38 rows. How many seats are in this section of the stadium?

Writing in Mathematics

61. What is an arithmetic sequence? Give an example with your explanation.

62. What is the common difference in an arithmetic sequence?

63. Explain how to find the general term of an arithmetic sequence.

64. Explain how to find the sum of the first n terms of an arithmetic sequence without having to add up all the terms.

65. Teachers' earnings n years after 1983 can be described by $a_n = 1472n + 20{,}228$. According to this model, what will teachers earn in 2083? Describe two possible circumstances that would render this predicted salary incorrect.

Technology Exercises

66. Use the SEQ (sequence) capability of a graphing utility and the formula you obtained for a_n to verify the value you found for a_{20} in any five exercises from Exercises 23–34.

67. Use the capability of a graphing utility to calculate the sum of a sequence to verify any five of your answers to Exercises 45–50.

Critical Thinking Exercises

68. Give examples of two different arithmetic sequences whose fourth term, a_4, is 10.

69. In the sequence 21,700, 23,172, 24,644, 26,116, . . . , which term is 314,628?

70. A *degree-day* is a unit used to measure the fuel requirements of buildings. By definition, each degree that the average daily temperature is below 65°F is 1 degree-day. For example, a temperature of 42°F constitutes 23 degree-days. If the average temperature on January 1 was 42°F and fell 2°F for each subsequent day up to and including January 10, how many degree-days are included from January 1 to January 10?

71. Show that the sum of the first n positive odd integers,

$$1 + 3 + 5 + \cdots + (2n - 1),$$

is n^2.

Group Exercise

72. Members of your group have been hired by the Environmental Protection Agency to write a report on whether we are making significant progress in recovering solid waste. Use the models from Exercises 55 and 56 as the basis for your report. A graph of each model from 1960 through 2000 would be helpful. What percentage of solid waste generated is actually recovered on a year-to-year basis? Be as creative as you want in your report and then draw conclusions. The group should write up the report and perhaps even include suggestions as to how we might improve recycling progress.

SECTION 8.3 *Geometric Sequences*

Objectives

1. Find the common ratio of a geometric sequence.
2. Write terms of a geometric sequence.
3. Use the formula for the general term of a geometric sequence.
4. Use the formula for the sum of the first n terms of a geometric sequence.
5. Find the value of an annuity.
6. Use the formula for the sum of an infinite geometric series.

Here we are at the closing moments of a job interview. You're shaking hands with the manager. You managed to answer all the tough questions without losing your poise, and now you've been offered a job. As a matter of fact, your qualifications are so terrific that you've been offered two jobs—one just the day before, with a rival company in the same field! One company offers \$30,000 the first year, with increases of 6% per year for four years after that. The other offers \$32,000 the first year, with annual increases of 3% per year after that. Over a five-year period, which is the better offer?

If salary raises amount to a certain percent each year, the yearly salaries over time form a geometric sequence. In this section, we investigate geometric sequences and their properties. After studying the section, you will be in a position to decide which job offer to accept: you will know which company will pay you more over five years.

Geometric Sequences

Figure 8.2 shows a sequence in which the number of squares is increasing. From left to right, the number of squares is 1, 5, 25, 125, and 625. In this sequence, each term after the first, 1, is obtained by multiplying the preceding term by a constant amount, namely 5. This sequence of increasing number of squares is an example of a *geometric sequence.*

Figure 8.2 A geometric sequence of squares

Definition of a Geometric Sequence

A **geometric sequence** is a sequence in which each term after the first is obtained by multiplying the preceding term by a fixed nonzero constant. The amount by which we multiply each time is called the **common ratio** of the sequence.

1 Find the common ratio of a geometric sequence.

The common ratio, r, is found by dividing any term after the first term by the term that directly precedes it. In the following examples, the common ratio is found by dividing the second term by the first term, $\frac{a_2}{a_1}$.

Geometric sequence	Common ratio
$1, 5, 25, 125, 625, \ldots$	$r = \frac{5}{1} = 5$
$4, 8, 16, 32, 64, \ldots$	$r = \frac{8}{4} = 2$
$6, -12, 24, -48, 96, \ldots$	$r = \frac{-12}{6} = -2$
$9, -3, 1, -\frac{1}{3}, \frac{1}{9}, \ldots$	$r = \frac{-3}{9} = -\frac{1}{3}$

Study Tip

When the common ratio of a geometric sequence is negative, the signs of the terms alternate.

2 Write terms of a geometric sequence.

How do we write out the terms of a geometric sequence when the first term and the common ratio are known? We multiply the first term by the common ratio to get the second term, multiply the second term by the common ratio to get the third term, and so on.

EXAMPLE 1 Writing the Terms of a Geometric Sequence

Write the first six terms of the geometric sequence with first term 6 and common ratio $\frac{1}{3}$.

Solution The first term is 6. The second term is $6 \cdot \frac{1}{3}$, or 2. The third term is $2 \cdot \frac{1}{3}$, or $\frac{2}{3}$. The fourth term is $\frac{2}{3} \cdot \frac{1}{3}$, or $\frac{2}{9}$, and so on. The first six terms are

$$6, 2, \tfrac{2}{3}, \tfrac{2}{9}, \tfrac{2}{27}, \tfrac{2}{81}.$$

Check Point 1 Write the first six terms of the geometric sequence with first term 12 and common ratio $\frac{1}{2}$.

3 Use the formula for the general term of a geometric sequence.

The General Term of a Geometric Sequence

Consider a geometric sequence whose first term is a_1, and whose common ratio is r. We are looking for a formula for the general term, a_n. Let's begin by writing the first six terms. The first term is a_1. The second term is $a_1 r$. The third term is $a_1 r \cdot r$, or $a_1 r^2$. The fourth term is $a_1 r^2 \cdot r$, or $a_1 r^3$, and so on. Starting with a_1 and multiplying each successive term by r, the first six terms are

$$a_1, \quad a_1 r, \quad a_1 r^2, \quad a_1 r^3, \quad a_1 r^4, \quad a_1 r^5.$$

a_1, first term; a_2, second term; a_3, third term; a_4, fourth term; a_5, fifth term; a_6, sixth term

Compare the exponent on r and the subscript of a denoting the term number. Can you see that the exponent on r is 1 less than the subscript of a denoting the term number?

$$a_3\text{: third term} = a_1 r^2 \qquad a_4\text{: fourth term} = a_1 r^3$$

2 is one less than 3. 3 is one less than 4.

Thus, the formula for the nth term is

$$a_n = a_1 r^{n-1}.$$

$n-1$ is one less than n.

General Term of a Geometric Sequence

The nth term (the general term) of a geometric sequence with first term a_1 and common ratio r is

$$a_n = a_1 r^{n-1}.$$

Study Tip

Be careful with the order of operations when evaluating

$$a_1 r^{n-1}.$$

First find r^{n-1}. Then multiply the result by a_1.

EXAMPLE 2 Using the Formula for the General Term of a Geometric Sequence

Find the eighth term of the geometric sequence whose first term is -4 and whose common ratio is -2.

Solution To find the eighth term, a_8, we replace n in the formula with 8, a_1 with -4, and r with -2.

$$a_n = a_1 r^{n-1}$$
$$a_8 = -4(-2)^{8-1} = -4(-2)^7 = -4(-128) = 512$$

The eighth term is 512. We can check this result by writing the first eight terms of the sequence:

$$-4, 8, -16, 32, -64, 128, -256, 512.$$

Check Point 2 Find the seventh term of the geometric sequence whose first term is 5 and whose common ratio is -3.

In Chapter 4, we studied exponential functions of the form $f(x) = b^x$ and the explosive exponential growth of world population. In our next example, we consider Florida's geometric population growth. Because **a geometric sequence is an exponential function whose domain is the set of positive integers,** geometric and exponential growth mean the same thing. (By contrast, an arithmetic sequence is a *linear function* whose domain is the set of positive integers.)

Geometric Population Growth

Economist Thomas Malthus (1766–1834) predicted that population growth would increase as a geometric sequence and food production would increase as an arithmetic sequence. He concluded that eventually population would exceed food production. If two sequences, one geometric and one arithmetic, are increasing, the geometric sequence will eventually overtake the arithmetic sequence, regardless of any head start that the arithmetic sequence might initially have.

EXAMPLE 3 Geometric Population Growth

The population of Florida from 1980 through 1987 is shown in the following table.

Year	1980	1981	1982	1983	1984	1985	1986	1987
Population in millions	9.75	10.03	10.32	10.62	10.93	11.25	11.58	11.92

a. Show that the population is increasing geometrically.

b. Write the general term for the geometric sequence describing population growth for Florida n years after 1979.

c. Estimate Florida's population, in millions, for the year 2000.

Solution

a. First, we divide the population for each year by the population in the preceding year.

$$\frac{10.03}{9.75} \approx 1.029, \quad \frac{10.32}{10.03} \approx 1.029, \quad \frac{10.62}{10.32} \approx 1.029$$

Continuing in this manner, we will keep getting approximately 1.029. This means that the population is increasing geometrically with $r \approx 1.029$. In this situation, the common ratio is the growth rate, indicating that the population of Florida in any year shown in the table is approximately 1.029 times the population the year before.

b. The sequence of Florida's population growth is

$$9.75, 10.03, 10.32, 10.62, 10.93, 11.25, 11.58, 11.92, \ldots.$$

Because the population is increasing geometrically, we can find the general term of this sequence using

$$a_n = a_1 r^{n-1}.$$

In this sequence, $a_1 = 9.75$ and r [from part (a)] ≈ 1.029. We substitute these values into the formula for the general term. This gives the general term for the geometric sequence describing Florida's population n years after 1979.

$$a_n = 9.75(1.029)^{n-1}$$

c. We can use the formula for the general term, a_n, in part (b) to estimate Florida's population for the year 2000. The year 2000 is 21 years after 1979—that is, $2000 - 1979 = 21$. Thus, $n = 21$. We substitute 21 for n in $a_n = 9.75(1.029)^{n-1}$.

$$a_{21} = 9.75(1.029)^{21-1} = 9.75(1.029)^{20} \approx 17.27$$

The formula predicts that Florida will have a population of approximately 17.27 million in the year 2000.

Check Point 3 Write the general term for the geometric sequence

$$3, 6, 12, 24, 48, \ldots.$$

Then use the formula for the general term to find the eighth term.

4 Use the formula for the sum of the first n terms of a geometric sequence.

The Sum of the First n Terms of a Geometric Sequence

The sum of the first n terms of a geometric sequence, denoted by S_n, can be found without having to add up all the terms. Recall that the first n terms of a geometric sequence are

$$a_1, a_1r, a_1r^2, \ldots, a_1r^{n-2}, a_1r^{n-1}.$$

We proceed as follows:

$$S_n = a_1 + a_1r + a_1r^2 + \cdots + a_1r^{n-2} + a_1r^{n-1}$$

S_n is the sum of the first n terms of the sequence.

$$rS_n = a_1r + a_1r^2 + a_1r^3 + \cdots + a_1r^{n-1} + a_1r^n$$

Multiply both sides of the equation by r.

$$S_n - rS_n = a_1 - a_1r^n$$

Subtract the second equation from the first equation.

$$S_n(1 - r) = a_1(1 - r^n)$$

Factor out S_n on the left and a_1 on the right.

$$S_n = \frac{a_1(1 - r^n)}{1 - r}$$

Solve for S_n by dividing both sides by $1 - r$ (assuming that $r \neq 1$).

We have proved the following result.

Study Tip

If the common ratio is 1, the geometric sequence is

$$a_1, a_1, a_1, a_1, \ldots.$$

The sum of the first n terms of this sequence is na_1:

$$S_n = \underbrace{a_1 + a_1 + a_1 + \cdots + a_1}_{\text{There are } n \text{ terms.}} = na_1.$$

The Sum of the First n Terms of a Geometric Sequence

The sum, S_n, of the first n terms of a geometric sequence is given by

$$S_n = \frac{a_1(1 - r^n)}{1 - r}$$

in which a_1 is the first term and r is the common ratio ($r \neq 1$).

To find the sum of the terms of a geometric sequence, we need to know the first term, a_1, the common ratio, r, and the number of terms, n. The following examples illustrate how to use this formula.

EXAMPLE 4 Finding the Sum of n Terms of a Geometric Sequence

Find the sum of the first 18 terms of the geometric sequence: 2, −8, 32, −128, … .

Solution To find the sum of the first 18 terms, S_{18}, we replace n in the formula with 18.

$$S_n = \frac{a_1(1 - r^n)}{1 - r}$$

$$S_{18} = \frac{a_1(1 - r^{18})}{1 - r}$$

The first term, a_1, is 2. We must find r, the common ratio.

We can find the common ratio by dividing the second term by the first term.

$$r = \frac{a_2}{a_1} = \frac{-8}{2} = -4$$

Now we are ready to find the sum of the first 18 terms of 2, −8, 32, −128, … .

$$S_n = \frac{a_1(1 - r^n)}{1 - r}$$ Use the formula for the sum of the first n terms of a geometric sequence.

$$S_{18} = \frac{2(1 - (-4)^{18})}{1 - (-4)}$$ a_1 (the first term) $= 2$, $r = -4$, and $n = 18$ because we want the sum of the first 18 terms.

$$= -27{,}487{,}790{,}694$$ Use a calculator.

The sum of the first 18 terms is −27,487,790,694.

Check Point 4 Find the sum of the first nine terms of the geometric sequence: 2, −6, 18, −54, … .

EXAMPLE 5 Using S_n to Evaluate a Summation

Find the following sum: $\sum_{i=1}^{10} 6 \cdot 2^i$

Solution Let's write out a few terms in the sum.

$$\sum_{i=1}^{10} 6 \cdot 2^i = 6 \cdot 2 + 6 \cdot 2^2 + 6 \cdot 2^3 + \cdots + 6 \cdot 2^{10}$$

Can you see that each term after the first is obtained by multiplying the preceding term by 2? To find the sum of the 10 terms ($n = 10$), we need to know the first term, a_1, and the common ratio, r. The first term is $6 \cdot 2$ or 12: $a_1 = 12$. The common ratio is 2.

$$S_n = \frac{a_1(1 - r^n)}{1 - r}$$ Use the formula for the sum of the first n terms of a geometric sequence.

$$S_{10} = \frac{12(1 - 2^{10})}{1 - 2}$$ a_1 (the first term) $= 12$, $r = 2$, and $n = 10$ because we are adding ten terms.

$$= 12{,}276$$ Use a calculator.

Technology

To find

$$\sum_{i=1}^{10} 6 \cdot 2^i$$

on a graphing utility, enter

SUM SEQ $(6 \times 2^x, x, 1, 10, 1)$.

Then press ENTER.

```
sum(seq(6*2^X,X,
1,10,1)
           12276
```

Thus,

$$\sum_{i=1}^{10} 6 \cdot 2^i = 12{,}276$$

Check Point 5 Find the following sum: $\sum_{i=1}^{8} 2 \cdot 3^i$.

Some of the exercises in the previous exercise set involved situations in which salaries increase by a fixed amount each year. A more realistic situation is one in which salary raises increase by a certain percent each year. Example 6 shows how such a situation can be described using a geometric series.

EXAMPLE 6 Computing a Lifetime Salary

A union contract specifies that each worker will receive a 5% pay increase each year for the next 30 years. One worker is paid \$20,000 the first year. What is this person's total lifetime salary over a 30-year period?

Solution The salary for the first year is \$20,000. With a 5% raise, the second-year salary is computed as follows:

$$\text{Salary for year 2} = 20{,}000 + 20{,}000(0.05) = 20{,}000(1.05).$$

Each year, the salary is 1.05 times what it was in the previous year. Thus, the salary for year 3 is 1.05 times 20,000(1.05), or $20{,}000(1.05)^2$. The salaries for the first five years are given in the table.

Yearly Salaries					
Year 1	**Year 2**	**Year 3**	**Year 4**	**Year 5**	...
20,000	20,000(1.05)	$20{,}000(1.05)^2$	$20{,}000(1.05)^3$	$20{,}000(1.05)^4$	...

The numbers in the second row form a geometric sequence with $a_1 = 20{,}000$ and $r = 1.05$. To find the total salary over 30 years, we use the formula for the sum of the first n terms of a geometric sequence, with $n = 30$.

$$S_n = \frac{a_1(1 - r^n)}{1 - r}$$

$$S_{30} = \frac{20{,}000(1 - (1.05)^{30})}{1 - 1.05}$$

$$= \frac{20{,}000(1 - (1.05)^{30})}{-0.05}$$

$$\approx 1{,}328{,}777$$

Total salary over 30 years

Use a calculator.

The total salary over the 30-year period is approximately \$1,328,777.

Check Point 6 A job pays a salary of \$30,000 the first year. During the next 29 years, the salary increases by 6% each year. What is the total lifetime salary over the 30-year period?

5 Find the value of an annuity.

Annuities

The compound interest formula

$$A = P(1 + r)^t$$

gives the future value, A, after + years, when a fixed amount of money, P, the principal, is deposited in an account that pays an annual interest rate r (in decimal form) compounded once a year. However, money is often invested in small amounts at periodic intervals. For example, to save for retirement, you might decide to place \$1000 into an Individual Retirement Account (IRA) at the end of each year until you retire. An **annuity** is a sequence of equal payments made at equal time periods. An IRA is an example of an annuity.

Suppose P dollars is deposited into an account at the end of each year. The account pays an annual interest rate, r, compounded annually. At the end of the first year, the account contains P dollars. At the end of the second year, P dollars is deposited again. At the time of this deposit, the first deposit has received interest earned during the second year. The **value of the annuity** is the sum of all deposits made plus all interest paid. Thus, the value of the annuity after two years is

$$P + P(1 + r).$$

Deposit of P dollars at end of second year

First-year deposit of P dollars with interest earned for a year

The value of the annuity after three years is

$$P \quad + \quad P(1 + r) \quad + \quad P(1 + r)^2.$$

Deposit of P dollars at end of third year

Second-year deposit of P dollars with interest earned for a year

First-year deposit of P dollars with interest earned over two years

The value of the annuity after t years is

$$P + P(1 + r) + P(1 + r)^2 + P(1 + r)^3 + \cdots + P(1 + r)^{t-1}$$

Deposit of P dollars at end of year t

First-year deposit of P dollars with interest earned over t − 1 years

This is a geometric series with first term P and common ratio $1 + r$. We use the formula

$$S_n = \frac{a_1(1 - r^n)}{1 - r}$$

to find the sum of the terms:

$$S_n = \frac{P(1 - (1 + r)^t)}{1 - (1 + r)} = \frac{P(1 - (1 + r)^t)}{-r} = P\frac{(1 + r)^t - 1}{r}.$$

This formula gives the value of an annuity after t years if interest is compounded once a year. We can adjust the formula to find the value of an annuity if equal payments are made at the end of each of n yearly compounding periods.

Value of an Annuity: Interest Compounded n Times per Year

If P is the deposit made at the end of each compounding period for an annuity at r percent annual interest compounded n times per year, the value, A, of the annuity after t years is

$$A = P\frac{\left(1+\frac{r}{n}\right)^{nt}-1}{\frac{r}{n}}.$$

EXAMPLE 7 Determining the Value of an Annuity

To save for retirement, you decide to deposit \$1000 into an IRA at the end of each year for the next 30 years. If the interest rate is 10% per year compounded annually, find the value of the IRA after 30 years.

Solution The annuity involves 30 year-end deposits of $P = \$1000$. The interest rate is 10%: $r = 0.10$. Because the deposits are made once a year and the interest is compounded once a year, $n = 1$. The number of years is 30: $t = 30$. We replace the variables in the formula for the value of an annuity with these numbers.

$$A = P\frac{\left(1+\frac{r}{n}\right)^{nt}-1}{\frac{r}{n}}$$

$$A = 1000\frac{\left(1+\frac{0.10}{1}\right)^{1\cdot 30}-1}{\frac{0.10}{1}} \approx 164{,}494$$

The value of the IRA at the end of 30 years is approximately \$164,494.

Check Point 7 If \$3000 is deposited into an IRA at the end of each year for 40 years and the interest rate is 10% per year compounded annually, find the value of the IRA after 40 years.

6 Use the formula for the sum of an infinite geometric series.

Geometric Series

An infinite sum of the form

$$a_1 + a_1r + a_1r^2 + a_1r^3 + \cdots + a_1r^{n-1} + \cdots$$

with first term a_1 and common ratio r is called an **infinite geometric series**. How can we determine which infinite geometric series have sums and which do not? We look at what happens to r^n as n gets larger in the formula for the sum of the first n terms of this series, namely

$$S_n = \frac{a_1(1-r^n)}{1-r}.$$

If r is any number between -1 and 1, that is, $-1 < r < 1$, the term r^n approaches 0 as n gets larger. For example, consider what happens to r^n for $r = \frac{1}{2}$:

$$\left(\tfrac{1}{2}\right)^1 = \tfrac{1}{2} \quad \left(\tfrac{1}{2}\right)^2 = \tfrac{1}{4} \quad \left(\tfrac{1}{2}\right)^3 = \tfrac{1}{8} \quad \left(\tfrac{1}{2}\right)^4 = \tfrac{1}{16} \quad \left(\tfrac{1}{2}\right)^5 = \tfrac{1}{32} \quad \left(\tfrac{1}{2}\right)^6 = \tfrac{1}{64}$$

These numbers are approaching 0 as n gets larger.

Take another look at the formula for the sum of the first n terms of a geometric sequence.

$$S_n = \frac{a_1(1 - r^n)}{1 - r}$$

If $-1 < r < 1$, r^n approaches 0 as n approaches infinity ($n \to \infty$).

Let us replace r^n with 0 in the formula for S_n. This change gives us a formula for the sum of infinite geometric series with common ratios between -1 and 1.

The Sum of an Infinite Geometric Series

If $-1 < r < 1$ (equivalently, $|r| < 1$), then the sum of the infinite geometric series

$$a_1 + a_1 r + a_1 r^2 + a_1 r^3 + \cdots$$

in which a_1 is the first term and r is the common ratio is given by

$$S = \frac{a_1}{1 - r}.$$

If $|r| \geq 1$, the infinite series does not have a sum.

To use the formula for the sum of an infinite geometric series, we need to know the first term and the common ratio. For example, consider

$$\tfrac{1}{2} + \tfrac{1}{4} + \tfrac{1}{8} + \tfrac{1}{16} + \tfrac{1}{32} + \cdots.$$

First term, a_1, is $\frac{1}{2}$.

Common ratio, r, is $\frac{a_2}{a_1}$.

$$r = \frac{1}{4} \div \frac{1}{2} = \frac{1}{4} \cdot 2 = \frac{1}{2}$$

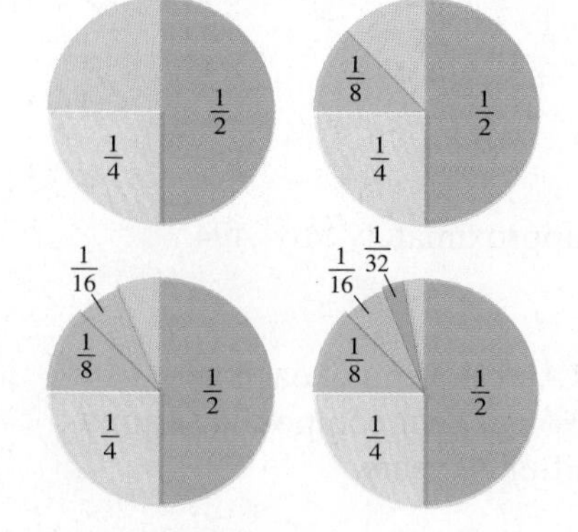

Figure 8.3 The sum $\frac{1}{2} + \frac{1}{4} + \frac{1}{8} + \frac{1}{16} + \frac{1}{32} + \cdots$ is approaching 1.

With $r = \frac{1}{2}$, the condition that $|r| < 1$ is met, so the infinite geometric series has a sum given by $S = \dfrac{a_1}{1 - r}$. The sum of the series is found as follows:

$$\tfrac{1}{2} + \tfrac{1}{4} + \tfrac{1}{8} + \tfrac{1}{16} + \tfrac{1}{32} + \cdots = \frac{a_1}{1 - r} = \frac{\frac{1}{2}}{1 - \frac{1}{2}} = \frac{\frac{1}{2}}{\frac{1}{2}} = 1.$$

Thus, the sum of the infinite geometric series is 1. Notice how this is illustrated in Figure 8.3. As more terms are included, the sum is approaching the area of one complete circle.

EXAMPLE 8 Finding the Sum of an Infinite Geometric Series

Find the sum of the infinite geometric series: $\frac{3}{8} - \frac{3}{16} + \frac{3}{32} - \frac{3}{64} + \cdots$.

Solution Before finding the sum, we must find the common ratio.

$$r = \frac{a_2}{a_1} = \frac{-\frac{3}{16}}{\frac{3}{8}} = -\frac{3}{16} \cdot \frac{8}{3} = -\frac{1}{2}$$

Because $r = -\frac{1}{2}$, the condition that $|r| < 1$ is met. Thus, the infinite geometric series has a sum.

$$S = \frac{a_1}{1 - r}$$

This is the formula for the sum of an infinite geometric series. Let $a_1 = \frac{3}{8}$ and $r = -\frac{1}{2}$.

$$= \frac{\frac{3}{8}}{1 - \left(-\frac{1}{2}\right)} = \frac{\frac{3}{8}}{\frac{3}{2}} = \frac{3}{8} \cdot \frac{2}{3} = \frac{1}{4}$$

Thus, the sum of this infinite geometric series is $\frac{1}{4}$. Put in an informal way, as we continue to add more and more terms, the sum is approximately $\frac{1}{4}$.

Check Point 8 Find the sum of the infinite geometric series: $3 + 2 + \frac{4}{3} + \frac{8}{9} + \cdots$.

We can use the formula for the sum of an infinite series to express a repeating decimal as a fraction in lowest terms.

EXAMPLE 9 Writing a Repeating Decimal as a Fraction

Express $0.\overline{7}$ as a fraction in lowest terms.

Solution

$$0.\overline{7} = 0.7777\ldots = \frac{7}{10} + \frac{7}{100} + \frac{7}{1000} + \frac{7}{10{,}000} + \cdots$$

Observe that $0.\overline{7}$ is an infinite geometric series with first term $\frac{7}{10}$ and common ratio $\frac{1}{10}$. Because $r = \frac{1}{10}$, the condition that $|r| < 1$ is met. Thus, we can use our formula to find the sum. Therefore,

$$0.\overline{7} = \frac{a_1}{1 - r} = \frac{\frac{7}{10}}{1 - \frac{1}{10}} = \frac{\frac{7}{10}}{\frac{9}{10}} = \frac{7}{10} \cdot \frac{10}{9} = \frac{7}{9}.$$

An equivalent fraction for $0.\overline{7}$ is $\frac{7}{9}$.

Check Point 9 Express $0.\overline{9}$ as a fraction in lowest terms.

Infinite geometric series have many applications, as illustrated in Example 10.

EXAMPLE 10 Tax Rebates and the Multiplier Effect

A tax rebate that returns a certain amount of money to taxpayers can have a total effect on the economy that is many times this amount. In economics, this phenomenon is called the **multiplier effect**. Suppose, for example, that the government reduces taxes so that each consumer has \$2000 more income. The government assumes that each person will spend 70% of this (= \$1400). The individuals and businesses receiving this \$1400 in turn spend 70% of it (= \$980),

creating extra income for other people to spend, and so on. Determine the total amount spent on consumer goods from the initial $2000 tax rebate.

Solution The total amount spent is given by the infinite geometric series

$$1400 + 980 + 686 + \cdots.$$

The first term is 1400: $a_1 = 1400$. The common ratio is 70%, or 0.7: $r = 0.7$. Because $r = 0.7$, the condition that $|r| < 1$ is met. Thus, we can use our formula to find the sum. Therefore,

$$1400 + 980 + 686 + \cdots = \frac{a_1}{1 - r} = \frac{1400}{1 - 0.7} \approx 4667.$$

This means that the total amount spent on consumer goods from the initial $2000 rebate is approximately $4667.

Check Point 10 Rework Example 10 and determine the total amount spent on consumer goods with a $1000 tax rebate and 80% spending down the line.

EXERCISE SET 8.3

Practice Exercises

In Exercises 1–8, write the first five terms of each geometric sequence.

1. $a_1 = 5, \quad r = 3$

2. $a_1 = 4, \quad r = 3$

3. $a_1 = 20, \quad r = \frac{1}{2}$

4. $a_1 = 24, \quad r = \frac{1}{3}$

5. $a_n = -4a_{n-1}, \quad a_1 = 10$

6. $a_n = -3a_{n-1}, \quad a_1 = 10$

7. $a_n = -5a_{n-1}, \quad a_1 = -6$

8. $a_n = -6a_{n-1}, \quad a_1 = -2$

In Exercises 9–16, find the indicated term of the geometric sequence with first term, a_1, and common ratio, r.

9. Find a_8 when $a_1 = 6, r = 2$.

10. Find a_8 when $a_1 = 5, r = 3$.

11. Find a_{12} when $a_1 = 5, r = -2$.

12. Find a_{12} when $a_1 = 4, r = -2$.

13. Find a_{40} when $a_1 = 1000, r = -\frac{1}{2}$.

14. Find a_{30} when $a_1 = 8000, r = -\frac{1}{2}$.

15. Find a_8 when $a_1 = 1{,}000{,}000, r = 0.1$.

16. Find a_8 when $a_1 = 40{,}000, r = 0.1$.

In Exercises 17–24, write a formula for the general term (the nth term) of each geometric sequence. Then use the formula for a_n to find a_7, the seventh term of the sequence.

17. $3, 12, 48, 192, \ldots$

18. $3, 15, 75, 375, \ldots$

19. $18, 6, 2, \frac{2}{3}, \ldots$

20. $12, 6, 3, \frac{3}{2}, \ldots$

21. $1.5, -3, 6, -12, \ldots$

22. $5, -1, \frac{1}{5}, -\frac{1}{25}, \ldots$

23. $0.0004, -0.004, 0.04, -0.4, \ldots$

24. $0.0007, -0.007, 0.07, -0.7, \ldots$

Use the formula for the sum of the first n terms of a geometric sequence to solve Exercises 25–36.

25. Find the sum of the first 12 terms of the geometric sequence: $2, 6, 18, 54, \ldots$.

26. Find the sum of the first 12 terms of the geometric sequence: $3, 6, 12, 24, \ldots$.

27. Find the sum of the first 11 terms of the geometric sequence: $3, -6, 12, -24, \ldots$.

28. Find the sum of the first 11 terms of the geometric sequence: $4, -12, 36, -108, \ldots$.

29. Find the sum of the first 14 terms of the geometric sequence: $-\frac{3}{2}, 3, -6, 12, \ldots$.

30. Find the sum of the first 14 terms of the geometric sequence: $-\frac{1}{24}, \frac{1}{12}, -\frac{1}{6}, \frac{1}{3}, \ldots$.

In Exercises 31–36, find the indicated sum.

31. $\sum_{i=1}^{8} 3^i$

32. $\sum_{i=1}^{6} 4^i$

33. $\sum_{i=1}^{10} 5 \cdot 2^i$ **34.** $\sum_{i=1}^{7} 4(-3)^i$

35. $\sum_{i=1}^{6} \left(\frac{1}{2}\right)^{i+1}$ **36.** $\sum_{i=1}^{6} \left(\frac{1}{3}\right)^{i+1}$

In Exercises 37–44, find the sum of each infinite geometric series.

37. $1 + \frac{1}{3} + \frac{1}{9} + \frac{1}{27} + \cdots$ **38.** $1 + \frac{1}{4} + \frac{1}{16} + \frac{1}{64} + \cdots$

39. $3 + \frac{3}{4} + \frac{3}{4^2} + \frac{3}{4^3} + \cdots$ **40.** $5 + \frac{5}{6} + \frac{5}{6^2} + \frac{5}{6^3} + \cdots$

41. $1 - \frac{1}{2} + \frac{1}{4} - \frac{1}{8} + \cdots$ **42.** $3 - 1 + \frac{1}{3} - \frac{1}{9} + \cdots$

43. $\sum_{i=1}^{\infty} 8(-0.3)^{i-1}$ **44.** $\sum_{i=1}^{\infty} 12(-0.7)^{i-1}$

In Exercises 45–50, express each repeating decimal as a fraction in lowest terms.

45. $0.\overline{5} = \frac{5}{10} + \frac{5}{100} + \frac{5}{1000} + \frac{5}{10{,}000} + \cdots$

46. $0.\overline{1} = \frac{1}{10} + \frac{1}{100} + \frac{1}{1000} + \frac{1}{10{,}000} + \cdots$

47. $0.\overline{47} = \frac{47}{100} + \frac{47}{10{,}000} + \frac{47}{1{,}000{,}000} + \cdots$

48. $0.\overline{83} = \frac{83}{100} + \frac{83}{10{,}000} + \frac{83}{1{,}000{,}000} + \cdots$

49. $0.\overline{257}$ **50.** $0.\overline{529}$

In Exercises 51–56, the general term of a sequence is given. Determine whether the sequence is arithmetic, geometric, or neither. If the sequence is arithmetic, find the common difference; if it is geometric, find the common ratio.

51. $a_n = n + 5$ **52.** $a_n = n - 3$

53. $a_n = 2^n$ **54.** $a_n = \left(\frac{1}{2}\right)^n$

55. $a_n = n^2 + 5$ **56.** $a_n = n^2 - 3$

Application Exercises

Use the formula for the general term (the nth term) of a geometric sequence to solve Exercises 57–60.

In Exercises 57–58, suppose you save \$1 the first day of a month, \$2 the second day, \$4 the third day, and so on. That is, each day you save twice as much as you did the day before.

57. What will you put aside for savings on the fifteenth day of the month?

58. What will you put aside for savings on the thirtieth day of the month?

59. A professional baseball player signs a contract with a beginning salary of \$3,000,000 for the first year with an annual increase of 4% per year beginning in the second year. That is, beginning in year 2, the athlete's salary will be 1.04 times what it was in the previous year. What is the athlete's salary for year 7 of the contract?

60. You are offered a job that pays \$30,000 for the first year with an annual increase of 5% per year beginning in the second year. That is, beginning in year 2, your salary will be 1.05 times what it was in the previous year. What can you expect to earn in your sixth year on the job?

61. The population of Iraq from 1995 through 1998 is shown in the following table.

Year	1995	1996	1997	1998
Population in millions	20.60	21.36	22.19	23.02

Source: U.N. Population Division

a. Divide the population for each year by the population in the preceding year. Round to two decimal places and show that Iraq's population is increasing geometrically.

b. Write the general term of the geometric sequence describing population growth for Iraq n years after 1994.

c. Estimate Iraq's population, in millions, for the year 2005.

62. The population of China from 1995 through 1998 is shown in the following table.

Year	1995	1996	1997	1998
Population in millions	1218.80	1232.21	1245.76	1259.46

Source: U.N. Population Division

a. Divide the population for each year by the population in the preceding year. Round to two decimal places and show that China's population is increasing geometrically.

b. Write the general term of the geometric sequence describing population growth for China n years after 1994.

c. Estimate China's population, in millions, for the year 2005.

Use the formula for the sum of the first n terms of a geometric sequence to solve Exercises 63–68.

In Exercises 63–64, you save \$1 the first day of a month, \$2 the second day, \$4 the third day, continuing to double your savings each day.

63. What will your total savings be for the first 15 days?

64. What will your total savings be for the first 30 days?

65. A job pays a salary of \$24,000 the first year. During the next 19 years, the salary increases by 5% each year. What is the total lifetime salary over the 20-year period?

66. You are investigating two employment opportunities. Company A offers \$30,000 the first year. During the next four years, the salary is guaranteed to increase by 6% per year. Company B offers \$32,000 the first year, with guaranteed annual increases of 3% per year after that. Which company offers the better total salary for a five-year contract? By how much?

67. A pendulum swings through an arc of 20 inches. On each successive swing, the length of the arc is 90% of the previous length.

$$20, \quad 0.9(20), \quad 0.9^2(20), \quad 0.9^3(20), \quad \ldots$$

1st swing, 2nd swing, 3rd swing, 4th swing

After 10 swings, what is the total length of the distance the pendulum has swung?

68. A pendulum swings through an arc of 16 inches. On each successive swing, the length of the arc is 96% of the previous length.

$$16, \quad 0.96(16), \quad (0.96)^2(16), \quad (0.96)^3(16), \quad \ldots$$

1st swing, 2nd swing, 3rd swing, 4th swing

After 10 swings, what is the total length of the distance the pendulum has swung?

Use the formula for the value of an annuity to solve Exercises 69–72.

69. To save for retirement, you decide to deposit \$2500 into an IRA at the end of each year for the next 40 years. If the interest rate is 9% per year compounded annually, find the value of the IRA after 40 years.

70. You decide to deposit \$100 at the end of each month into an account paying 8% interest compounded monthly to save for your child's education. How much will you save over 16 years?

71. You contribute \$600 at the end of each quarter to a Tax Sheltered Annuity (TSA) paying 8% annual interest compounded quarterly. Find the value of the TSA after 18 years.

72. To save for a new home, you invest \$500 per month in a mutual fund with an annual rate of return of 10% compounded monthly. How much will you have saved after four years?

Use the formula for the sum of an infinite geometric series to solve Exercises 73–75.

73. A new factory in a small town has an annual payroll of \$6 million. It is expected that 60% of this money will be spent in the town by factory personnel. The people in the town who receive this money are expected to spend 60% of what they receive in the town, and so on. What is the total of all this spending, called the total economic impact of the factory, on the town each year?

74. How much additional spending will be generated by a \$10 billion tax rebate if 60% of all income is spent?

75. If the shading process shown in the figure is continued indefinitely, what fractional part of the largest square is eventually shaded?

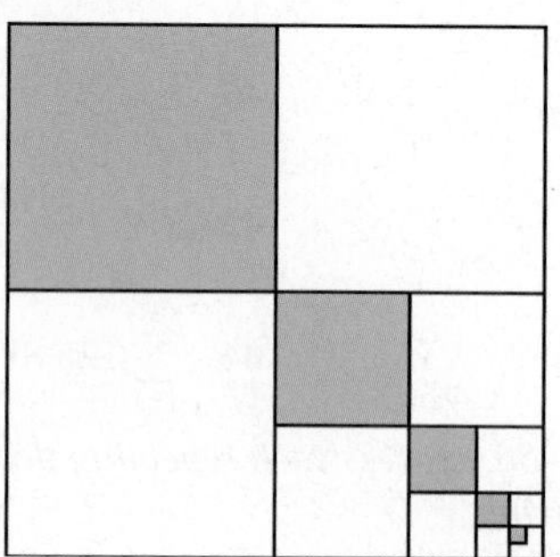

Writing in Mathematics

76. What is a geometric sequence? Give an example with your explanation.

77. What is the common ratio in a geometric sequence?

78. Explain how to find the general term of a geometric sequence.

79. Explain how to find the sum of the first n terms of a geometric sequence without having to add up all the terms.

80. What is an annuity?

81. What is the difference between a geometric sequence and an infinite geometric series?

82. How do you determine if an infinite geometric series has a sum? Explain how to find the sum of an infinite geometric series.

83. Would you rather have \$10,000,000 and a brand new BMW or 1¢ today, 2¢ tomorrow, 4¢ on day 3, 8¢ on day 4, 16¢ on day 5, and so on, for 30 days? Explain.

84. For the first 30 days of a flu outbreak, the number of students on your campus who become ill is increasing. Which is worse: The number of students with the flu is increasing arithmetically or is increasing geometrically? Explain your answer.

Technology Exercises

85. Use the SEQ (sequence) capability of a graphing utility and the formula you obtained for a_n to verify the value you found for a_7 in any three exercises from Exercises 17–24.

86. Use the capability of a graphing utility to calculate the sum of a sequence to verify any three of your answers to Exercises 31–36.

In Exercises 87–88, use a graphing utility to graph the function. Determine the horizontal asymptote for the graph of f and discuss its relationship to the sum of the given series.

	Function	Series
87.	$f(x) = \dfrac{2\left[1 - \left(\frac{1}{3}\right)^x\right]}{1 - \frac{1}{3}}$	$2 + 2(\frac{1}{3}) + 2(\frac{1}{3})^2 + 2(\frac{1}{3})^3 + \cdots$
88.	$f(x) = \dfrac{4[1 - (0.6)^x]}{1 - 0.6}$	$4 + 4(0.6) + 4(0.6)^2 + 4(0.6)^3 + \cdots$

Critical Thinking Exercises

89. Which one of the following is true?

a. The sequence 2, 6, 24, 120, ... is an example of a geometric sequence.

b. The sum of the geometric series $\frac{1}{2} + \frac{1}{4} + \frac{1}{8} + \cdots + \frac{1}{512}$ can only be estimated without knowing precisely what terms occur between $\frac{1}{8}$ and $\frac{1}{512}$.

c. $10 - 5 + \frac{5}{2} - \frac{5}{4} + \cdots = \dfrac{10}{1 - \frac{1}{2}}$

d. If the nth term of a geometric sequence is $a_n = 3(0.5)^{n-1}$, the common ratio is $\frac{1}{2}$.

90. In a pest-eradication program, sterilized male flies are released into the general population each day. Ninety percent of those flies will survive a given day. How many flies should be released each day if the long-range goal of the program is to keep 20,000 sterilized flies in the population?

91. You are now 25 years old and would like to retire at age 55 with a retirement fund of \$1,000,000. How much should you deposit at the end of each month for the next 30 years in an IRA paying 10% annual interest compounded monthly to achieve your goal?

Group Exercise

92. Group members serve as a financial team analyzing the three options given to the professional baseball player described in the chapter opener. As a group, determine which option provides the most amount of money over the six-year contract and which provides the least. Describe one advantage and one disadvantage to each option.

SECTION 8.4 *Mathematical Induction*

Objectives

1. Understand the principle of mathematical induction.
2. Prove statements using mathematical induction.

After ten years of work, Princeton University's Andrew Wiles proved Fermat's Last Theorem.

Pierre de Fermat (1601–1665) was a lawyer who enjoyed studying mathematics. In a margin of one of his books he claimed that no positive integers satisfy

$$x^n + y^n = z^n$$

if n is an integer greater than or equal to 3.

If $n = 2$, we can find positive integers satisfying the equation:

$$3^2 + 4^2 = 5^2.$$

However, Fermat claimed that no positive integers satisfy

$$x^3 + y^3 = z^3, \quad x^4 + y^4 = z^4, \quad x^5 + y^5 = z^5,$$

and so on. Fermat claimed to have a proof of his conjecture, but added, "The margin of my book is too narrow to write it down." Some believe that he never had a proof and intended to frustrate his colleagues.

In 1994, 40-year-old Princeton math professor Andrew Wiles proved Fermat's Last Theorem using a principle called *mathematical induction*. In this section, you will learn how to use this powerful method to prove statements about the positive integers.

1 Understand the principle of mathematical induction.

The Principle of Mathematical Induction

How do we prove statements using mathematical induction? Let's consider an example. We will prove a statement that appears to give a correct formula for the sum of the first n positive integers.

$$S_n: 1 + 2 + 3 + \cdots + n = \frac{n(n+1)}{2}$$

We can verify this statement for, say, the first four positive integers.

If $n = 1$, the statement S_1 is

Take the first term on the left.

$$1 \stackrel{?}{=} \frac{1(1+1)}{2}$$

Substitute 1 for n on the right.

$$1 \stackrel{?}{=} \frac{1 \cdot 2}{2}$$

$$1 = 1 \checkmark.$$

This true statement shows that S_1 is true.

If $n = 2$, the statement S_2 is

Add the first two terms on the left.

$$1 + 2 \stackrel{?}{=} \frac{2(2+1)}{2}$$

Substitute 2 for n on the right.

$$3 \stackrel{?}{=} \frac{2 \cdot 3}{2}$$

$$3 = 3 \checkmark.$$

This true statement shows S_2 is true.

If $n = 3$, the statement S_3 is

Add the first three terms on the left.

$$1 + 2 + 3 \stackrel{?}{=} \frac{3(3+1)}{2}$$

Substitute 3 for n on the right.

$$6 \stackrel{?}{=} \frac{3 \cdot 4}{2}$$

$$6 = 6 \checkmark.$$

This true statement shows S_3 is true.

Finally, if $n = 4$, the statement S_4 is

Add the first four terms on the left.

$$1 + 2 + 3 + 4 \stackrel{?}{=} \frac{4(4+1)}{2}$$

Substitute 4 for n on the right.

$$10 \stackrel{?}{=} \frac{4 \cdot 5}{2}$$

$$10 = 10 \checkmark.$$

This true statement shows S_4 is true.

This approach does *not* prove that the given statement S_n is true for every positive integer n. The fact that the formula produces true statements for $n = 1$, 2, 3, and 4 does not guarantee that it is valid for all positive integers n. Thus, we need to be able to verify the truth of S_n without verifying the statement for each and every one of the positive integers.

A legitimate proof of the given statement S_n involves a technique called **mathematical induction**.

The Principle of Mathematical Induction

Let S_n be a statement involving the positive integer n. If

1. S_1 is true, and
2. the truth of the statement S_k implies the truth of the statement S_{k+1}, for every positive integer k,

then the statement S_n is true for all positive integers n.

Figure 8.4 Falling dominoes illustrate the principle of mathematical induction.

The principle of mathematical induction can be illustrated using an unending line of dominoes, as shown in Figure 8.4. If the first domino is pushed over, it knocks down the next, which knocks down the next, and so on, in a chain reaction. To topple all the dominoes in the infinite sequence, two conditions must be satisfied:

1. The first domino must be knocked down.
2. If the domino in position k is knocked down, then the domino in position $k + 1$ must be knocked down.

If the second condition is not satisfied, it does not follow that all the dominoes will topple. For example, suppose the dominoes are spaced far enough apart so that a falling domino does not push over the next domino in the line.

The domino analogy provides the two steps that are required in a proof by mathematical induction.

The Steps in a Proof by Mathematical Induction

Let S_n be a statement involving the positive integer n. To prove that S_n is true for all positive integers n requires two steps.

Step 1 Show that S_1 is true.

Step 2 Show that if S_k is assumed to be true, then S_{k+1} is also true, for every positive integer k.

Notice that to prove S_n, we work only with the statements S_1, S_k, and S_{k+1}. Our first example provides practice in writing these statements.

EXAMPLE 1 Writing S_1, S_k, and S_{k+1}

For the given statement S_n, write the three statements S_1, S_k, and S_{k+1}.

a. S_n: $1 + 2 + 3 + \cdots + n = \dfrac{n(n+1)}{2}$

b. S_n: $1^2 + 2^2 + 3^2 + \cdots + n^2 = \dfrac{n(n+1)(2n+1)}{6}$

Solution

a. We begin with

$$S_n: 1 + 2 + 3 + \cdots + n = \frac{n(n + 1)}{2}.$$

Write S_1 by taking the first term on the left and replacing n with 1 on the right.

$$S_1: 1 = \frac{1(1 + 1)}{2}$$

Write S_k by taking the sum of the first k terms on the left and replacing n with k on the right.

$$S_k: 1 + 2 + 3 + \cdots + k = \frac{k(k + 1)}{2}$$

Write S_{k+1} by taking the sum of the first $k + 1$ terms on the left and replacing n with $k + 1$ on the right.

$$S_{k+1}: 1 + 2 + 3 + \cdots + (k + 1) = \frac{(k + 1)[(k + 1) + 1]}{2}$$

$$S_{k+1}: 1 + 2 + 3 + \cdots + (k + 1) = \frac{(k + 1)(k + 2)}{2}$$

Simplify on the right.

b. We begin with

$$S_n: 1^2 + 2^2 + 3^2 + \cdots + n^2 = \frac{n(n + 1)(2n + 1)}{6}.$$

Write S_1 by taking the first term on the left and replacing n with 1 on the right.

$$S_1: 1^2 = \frac{1(1 + 1)(2 \cdot 1 + 1)}{6}$$

Write S_k by taking the sum of the first k terms on the left and replacing n with k on the right.

$$S_k: 1^2 + 2^2 + 3^2 + \cdots + k^2 = \frac{k(k + 1)(2k + 1)}{6}$$

Write S_{k+1} by taking the sum of the first $k + 1$ terms on the left and replacing n with $k + 1$ on the right.

$$S_{k+1}: 1^2 + 2^2 + 3^2 + \cdots + (k + 1)^2 = \frac{(k + 1)[(k + 1) + 1][2(k + 1) + 1]}{6}$$

$$S_{k+1}: 1^2 + 2^2 + 3^2 + \cdots + (k + 1)^2 = \frac{(k + 1)(k + 2)(2k + 3)}{6}$$

Simplify on the right.

For the given statement S_n, write the three statements S_1, S_k, and S_{k+1}.

a. $2 + 4 + 6 + \cdots + 2n = n(n + 1)$

b. $1^3 + 2^3 + 3^3 + \cdots + n^3 = \dfrac{n^2(n + 1)^2}{4}$

Always simplify S_{k+1} before trying to use mathematical induction to prove that S_n is true. For example, consider

$$S_n\colon\ 1^2 + 3^2 + 5^2 + \cdots + (2n-1)^2 = \frac{n(2n-1)(2n+1)}{3}.$$

Begin by writing S_{k+1} as follows:

$$S_{k+1}\colon\ 1^2 + 3^2 + 5^2 + \cdots + [2(k+1)-1]^2 = \frac{(k+1)[2(k+1)-1][2(k+1)+1]}{3}.$$

The sum of the first $k + 1$ terms

Replace n by $k + 1$ on the right side of S_n.

Now simplify the algebra.

$$S_{k+1}\colon\ 1^2 + 3^2 + 5^2 + \cdots + (2k+2-1)^2 = \frac{(k+1)(2k+2-1)(2k+2+1)}{3}$$

$$S_{k+1}\colon\ 1^2 + 3^2 + 5^2 + \cdots + (2k+1)^2 = \frac{(k+1)(2k+1)(2k+3)}{3}$$

2 Prove statements using mathematical induction.

Proving Statements about Positive Integers Using Mathematical Induction

Now that we know how to find S_1, S_k, and S_{k+1}, let's see how we can use these statements to carry out the two steps in a proof by mathematical induction. In Examples 2 and 3, we will use the statements S_1, S_k, and S_{k+1} to prove each of the statements S_n that we worked with in Example 1.

EXAMPLE 2 Proving a Formula by Mathematical Induction

Use mathematical induction to prove that

$$1 + 2 + 3 + \cdots + n = \frac{n(n+1)}{2}$$

for all positive integers n.

Solution

Step 1 Show that S_1 is true. Statement S_1 is

$$1 = \frac{1(1+1)}{2}.$$

Simplifying on the right, we obtain $1 = 1$. This true statement shows that S_1 is true.

Step 2 Show that if S_k is true, then S_{k+1} is true. Using S_k and S_{k+1} from Example 1a, show that the truth of S_k,

$$1 + 2 + 3 + \cdots + k = \frac{k(k+1)}{2}$$

implies the truth of S_{k+1},

$$1 + 2 + 3 + \cdots + (k+1) = \frac{(k+1)(k+2)}{2}.$$

Visualizing Summation Formulas

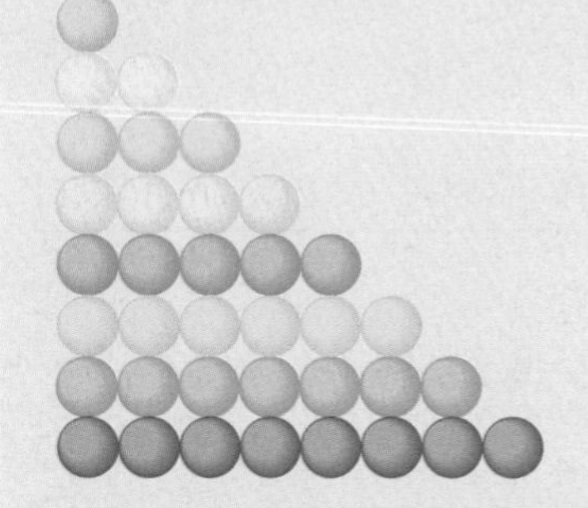

Finding the sum of consecutive positive integers leads to **triangular numbers** of the form $\frac{n(n+1)}{2}$.

$\frac{n(n+1)}{2}$	$\frac{n(n+1)}{2}$
$n = 1$: 1	$n = 2$: 3

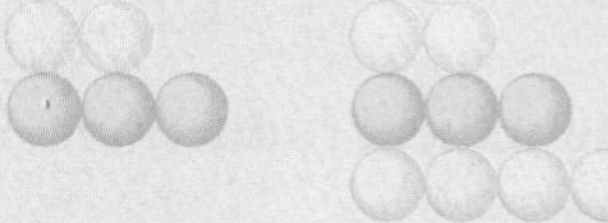

$\frac{n(n+1)}{2}$	$\frac{n(n+1)}{2}$
$n = 3$: 6	$n = 4$: 10

We will work with S_k. Because we assume that S_k is true, we add the next consecutive integer after k—namely, $k + 1$—to both sides.

$$1 + 2 + 3 + \cdots + k = \frac{k(k+1)}{2}$$

This is S_k, which we assume is true.

$$1 + 2 + 3 + \cdots + k + (k+1) = \frac{k(k+1)}{2} + (k+1)$$

Add $k + 1$ to both sides of the equation.

We do not have to write this k because k is understood to be the integer that precedes $k + 1$.

$$1 + 2 + 3 + \cdots + (k+1) = \frac{k(k+1)}{2} + \frac{2(k+1)}{2}$$

Write the right side with a common denominator of 2.

$$1 + 2 + 3 + \cdots + (k+1) = \frac{(k+1)}{2}(k+2)$$

Factor out the common factor $\frac{k+1}{2}$ on the right.

$$1 + 2 + 3 + \cdots + (k+1) = \frac{(k+1)(k+2)}{2}$$

This final result is the statement S_{k+1} at the bottom of page 637.

We have shown that if we assume that S_k is true, and we add $k + 1$ to both sides of S_k, then S_{k+1} is also true. By the principle of mathematical induction, the statement S_n, namely,

$$1 + 2 + 3 + \cdots + n = \frac{n(n+1)}{2}$$

is true for every positive integer n.

Check Point 2 Use mathematical induction to prove that

$$2 + 4 + 6 + \cdots + 2n = n(n+1).$$

EXAMPLE 3 Proving a Formula by Mathematical Induction

Use mathematical induction to prove that

$$1^2 + 2^2 + 3^2 + \cdots + n^2 = \frac{n(n+1)(2n+1)}{6}$$

for all positive integers n.

Solution

Step 1 Show that S_1 is true. Statement S_1 is

$$1^2 = \frac{1(1+1)(2 \cdot 1 + 1)}{6}.$$

Simplifying, we obtain $1 = \frac{1 \cdot 2 \cdot 3}{6}$. Further simplification on the right gives the statement $1 = 1$. This true statement shows that S_1 is true.

Step 2 Show that if S_k is true, then S_{k+1} is true. Using S_k and S_{k+1} from Example 1b, show that the truth of

$$S_k\colon 1^2 + 2^2 + 3^2 + \cdots + k^2 = \frac{k(k+1)(2k+1)}{6}$$

implies the truth of

$$S_{k+1}\colon 1^2 + 2^2 + 3^2 + \cdots + (k+1)^2 = \frac{(k+1)(k+2)(2k+3)}{6}.$$

We will work with S_k. Because we assume that S_k is true, we add the square of the next consecutive integer after k—namely, $(k+1)^2$—to both sides of the equation.

$$1^2 + 2^2 + 3^2 + \cdots + k^2 = \frac{k(k+1)(2k+1)}{6}$$

This is S_k, assumed to be true. We must work with this and show S_{k+1} is true.

$$1^2 + 2^2 + 3^2 + \cdots + k^2 + (k+1)^2 = \frac{k(k+1)(2k+1)}{6} + (k+1)^2$$

Add $(k+1)^2$ to both sides.

$$1^2 + 2^2 + 3^2 + \cdots + (k+1)^2 = \frac{k(k+1)(2k+1)}{6} + \frac{6(k+1)^2}{6}$$

It is not necessary to write k^2 on the left. Express the right side with the least common denominator, 6.

$$= \frac{(k+1)}{6}\left[k(2k+1) + 6(k+1)\right]$$

Factor out the common factor $\frac{k+1}{6}$.

$$= \frac{(k+1)}{6}(2k^2 + 7k + 6)$$

Multiply and combine like terms.

$$= \frac{(k+1)}{6}(k+2)(2k+3)$$

Factor $2k^2 + 7k + 6$.

$$= \frac{(k+1)(k+2)(2k+3)}{6}$$

This final statement is S_{k+1}.

We have shown that if we assume that S_k is true, and we add $(k+1)^2$ to both sides of S_k, then S_{k+1} is also true. By the principle of mathematical induction, the statement S_n, namely,

$$1^2 + 2^2 + 3^2 + \cdots + n^2 = \frac{n(n+1)(2n+1)}{6}$$

is true for every positive integer n.

Check Point 3 Use mathematical induction to prove that

$$1^3 + 2^3 + 3^3 + \cdots + n^3 = \frac{n^2(n+1)^2}{4}.$$

Example 4 illustrates how mathematical induction can be used to prove statements about positive integers that do not involve sums.

EXAMPLE 4 Using the Principle of Mathematical Induction

Prove that 2 is a factor of $n^2 + 5n$ for all positive integers n.

Solution

Step 1 **Show that S_1 is true.** Statement S_1 reads

$$2 \text{ is a factor of } 1^2 + 5 \cdot 1.$$

Simplifying the arithmetic, the statement reads

$$2 \text{ is a factor of } 6.$$

This statement is true: that is, $6 = 2 \cdot 3$. This shows that S_1 is true.

Step 2 **Show that if S_k is true, then S_{k+1} is true.** Let's write S_k and S_{k+1}:

S_k: 2 is a factor of $k^2 + 5k$.

S_{k+1}: 2 is a factor of $(k + 1)^2 + 5(k + 1)$.

We can rewrite statement S_{k+1} by simplifying the algebraic expression in the statement as follows:

$$(k + 1)^2 + 5(k + 1) = k^2 + 2k + 1 + 5k + 5 = k^2 + 7k + 6.$$

Use the formula $(A+B)^2 = A^2 + 2AB + B^2$.

Statement S_{k+1} now reads

$$2 \text{ is a factor of } k^2 + 7k + 6.$$

We wish to use statement S_k—that is, 2 is a factor of $k^2 + 5k$—to prove statement S_{k+1}. We do this as follows:

$$k^2 + 7k + 6 = (k^2 + 5k) + (2k + 6) = (k^2 + 5k) + 2(k + 3).$$

We know that 2 is a factor of $k^2 + 5k$ because we assume S_k is true.

Factoring the last two terms shows that 2 is a factor of $2k + 6$.

The voice balloons show that 2 is a factor of $k^2 + 5k$ and of $2(k + 3)$. Thus, 2 is a factor of the sum $(k^2 + 5k) + 2(k + 3)$, or of $k^2 + 7k + 6$. This is precisely statement S_{k+1}. We have shown that if we assume that S_k is true, then S_{k+1} is also true. By the principle of mathematical induction, the statement S_n, namely 2 is a factor of $n^2 + 5n$, is true for every positive integer n.

Check Point 4 Prove that 2 is a factor of $n^2 + n$ for all positive integers n.

EXERCISE SET 8.4

Practice Exercises

In Exercises 1–4, a statement S_n about the positive integers is given. Write statements S_1, S_2, and S_3, and show that each of these statements is true.

1. S_n: $1 + 3 + 5 + \cdots + (2n - 1) = n^2$

2. S_n: $3 + 4 + 5 + \cdots + (n + 2) = \frac{n(n + 5)}{2}$

3. S_n: 2 is a factor of $n^2 - n$.

4. S_n: 3 is a factor of $n^3 - n$.

In Exercises 5–10, a statement S_n about the positive integers is given. Write statements S_k and S_{k+1}, simplifying statement S_{k+1} completely.

5. S_n: $4 + 8 + 12 + \cdots + 4n = 2n(n + 1)$

6. S_n: $3 + 4 + 5 + \cdots + (n + 2) = \frac{n(n + 5)}{2}$

7. S_n: $3 + 7 + 11 + \cdots + (4n - 1) = n(2n + 1)$

8. S_n: $2 + 7 + 12 + \cdots + (5n - 3) = \frac{n(5n - 1)}{2}$

9. S_n: 2 is a factor of $n^2 - n + 2$.

10. S_n: 2 is a factor of $n^2 - n$.

In Exercises 11–30, use mathematical induction to prove that each statement is true for every positive integer n.

11. $4 + 8 + 12 + \cdots + 4n = 2n(n + 1)$

12. $3 + 4 + 5 + \cdots + (n + 2) = \frac{n(n + 5)}{2}$

13. $1 + 3 + 5 + \cdots + (2n - 1) = n^2$

14. $3 + 6 + 9 + \cdots + 3n = \frac{3n(n + 1)}{2}$

15. $3 + 7 + 11 + \cdots + (4n - 1) = n(2n + 1)$

16. $2 + 7 + 12 + \cdots + (5n - 3) = \frac{n(5n - 1)}{2}$

17. $1 + 2 + 2^2 + \cdots + 2^{n-1} = 2^n - 1$

18. $1 + 3 + 3^2 + \cdots + 3^{n-1} = \frac{3^n - 1}{2}$

19. $2 + 4 + 8 + \cdots + 2^n = 2^{n+1} - 2$

20. $\frac{1}{2} + \frac{1}{4} + \frac{1}{8} + \cdots + \frac{1}{2^n} = 1 - \frac{1}{2^n}$

21. $1 \cdot 2 + 2 \cdot 3 + 3 \cdot 4 + \cdots + n(n + 1) = \frac{n(n + 1)(n + 2)}{3}$

22. $1 \cdot 3 + 2 \cdot 4 + 3 \cdot 5 + \cdots + n(n + 2) = \frac{n(n + 1)(2n + 7)}{6}$

23. $\frac{1}{1 \cdot 2} + \frac{1}{2 \cdot 3} + \frac{1}{3 \cdot 4} + \cdots + \frac{1}{n(n + 1)} = \frac{n}{n + 1}$

24. $\frac{1}{2 \cdot 3} + \frac{1}{3 \cdot 4} + \frac{1}{4 \cdot 5} + \cdots + \frac{1}{(n + 1)(n + 2)} = \frac{n}{2n + 4}$

25. 2 is a factor of $n^2 - n$.

26. 2 is a factor of $n^2 + 3n$.

27. 6 is a factor of $n(n + 1)(n + 2)$.

28. 3 is a factor of $n(n + 1)(n - 1)$.

29. $(ab)^n = a^n b^n$

30. $\left(\frac{a}{b}\right)^n = \frac{a^n}{b^n}$

Writing in Mathematics

31. Explain how to use mathematical induction to prove that a statement is true for every positive integer n.

32. Consider the statement S_n given by

$$n^2 - n + 41 \text{ is prime.}$$

Although $S_1, S_2, \ldots, S_{40}$ are true, S_{41} is false. Describe how this is illustrated by the dominoes in the figure. What does this tell you about a pattern, or formula, that seems to work for several values of n?

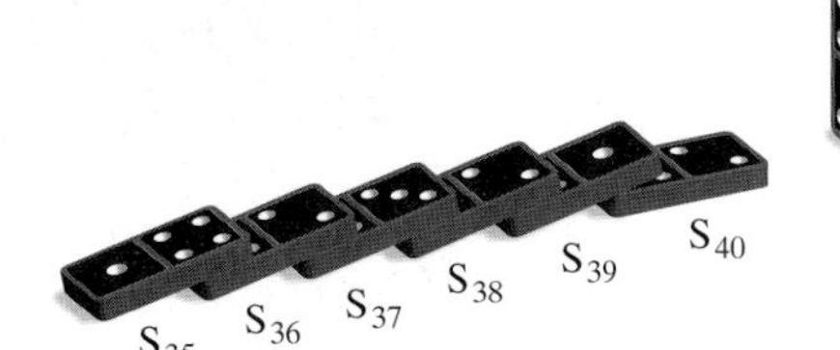

Critical Thinking Exercises

Some statements are false for the first few positive integers, but true for some positive integer on. In these instances, you can prove S_n for $n \geq k$ by showing that S_k is true and that S_k implies S_{k+1}. Use this extended principle of mathematical induction to prove that each statement in Exercises 33–34 is true.

33. Prove that $n^2 > 2n + 1$ for $n \geq 3$. Show that the formula is true for $n = 3$ and then use step 2 of mathematical induction.

34. Prove that $2^n > n^2$ for $n \geq 5$. Show that the formula is true for $n = 5$ and then use step 2 of mathematical induction.

In Exercises 35–36, find S_1 through S_5 and then use the pattern to make a conjecture about S_n. Prove the conjectured formula for S_n by mathematical induction.

35. S_n: $\frac{1}{4} + \frac{1}{12} + \frac{1}{24} + \cdots + \frac{1}{2n(n+1)}$

36. S_n: $\left(1 - \frac{1}{2}\right)\left(1 - \frac{1}{3}\right)\left(1 - \frac{1}{4}\right)\cdots\left(1 - \frac{1}{n+1}\right)$

Group Exercise

37. Fermat's most notorious theorem baffled the greatest minds for more than three centuries. In 1994, after ten years of work, Princeton University's Andrew Wiles proved Fermat's Last Theorem. *People* magazine put him on its list of "the 25 most intriguing people of the year," the Gap asked him to model jeans, and Barbara Walters chased him for an interview. "Who's Barbara Walters?" asked the bookish Wiles, who had somehow gone through life without a television.

Using the 1993 PBS documentary "Solving Fermat: Andrew Wiles" or information about Andrew Wiles on the Internet, research and present a group seminar on what Wiles did to prove Fermat's Last Theorem, problems along the way, and the role of mathematical induction in the proof.

SECTION 8.5 *The Binomial Theorem*

Objectives

1. Recognize patterns in binomial expansions.
2. Evaluate a binomial coefficient.
3. Expand a binomial raised to a power.
4. Find a particular term in a binomial expansion.

Galaxies are groupings of billions of stars bound together gravitationally. Some galaxies, such as the Centaurus galaxy shown here, are elliptical in shape.

Is mathematics discovered or invented? For example, planets revolve in elliptical orbits. Does that mean that the ellipse is out there, waiting for the mind to discover it? Or do people create the definition of an ellipse just as they compose a song? And is it possible for the same mathematics to be discovered/invented by independent researchers separated by time, place, and culture? This is precisely what occurred when mathematicians attempted to find efficient methods for raising binomials to higher and higher powers, such as

$$(x + 2)^3, (x + 2)^4, (x + 2)^5, (x + 2)^6,$$

and so on. In this section, we study higher powers of binomials and a method first discovered/invented by great minds in Eastern and Western culture working independently.

1 Recognize patterns in binomial expansions.

Patterns in Binomial Expansions

When we write out the *binomial expression* $(a + b)^n$, where n is a positive integer, a number of patterns begin to appear.

$$(a+b)^1 = a+b$$
$$(a+b)^2 = a^2+2ab+b^2$$
$$(a+b)^3 = a^3+3a^2b+3ab^2+b^3$$
$$(a+b)^4 = a^4+4a^3b+6a^2b^2+4ab^3+b^4$$
$$(a+b)^5 = a^5+5a^4b+10a^3b^2+10a^2b^3+5ab^4+b^5$$

Discovery

Each expanded form of the binomial expression is a polynomial. Study the five polynomials and answer the following questions.

1. For each polynomial, describe the pattern for the exponents on a. What is the largest exponent on a? What happens to the exponent on a from term to term?
2. Describe the pattern for the exponents on b. What is the exponent on b in the first term? What is the exponent on b in the second term? What happens to the exponent on b from term to term?
3. Find the sum of the exponents on the variables in each term for the polynomials in the five rows. Describe the pattern.
4. How many terms are there in the polynomials on the right in relation to the power of the binomial?

How many of the following patterns were you able to discover?

1. The first term is a^n. The exponent on a decreases by 1 in each successive term.
2. The exponents on b increase by 1 in each successive term. In the first term, the exponent on b is 0. (Because $b^0 = 1$, b is not shown in the first term.) The last term is b^n.
3. The sum of the exponents on the variables in any term is equal to n, the exponent on $(a+b)^n$.
4. There is one more term in the polynomial expansion than there is in the power of the binomial, n. There are $n+1$ terms in the expanded form of $(a+b)^n$.

Using these observations, the variable parts of the expansion of $(a+b)^6$ are

$$a^6,\quad a^5b,\quad a^4b^2,\quad a^3b^3,\quad a^2b^4,\quad ab^5,\quad b^6.$$

The first term is a^6, with the exponent on a decreasing by 1 in each successive term. The exponents on b increase from 0 to 6, with the last term being b^6. The sum of the exponents in each term is equal to 6.

We can generalize from these observations to obtain the variable parts of the expansion of $(a+b)^n$. They are

$$a^n, a^{n-1}b, a^{n-2}b^2, a^{n-3}b^3, \ldots, ab^{n-1}, b^n.$$

Exponents on a are decreasing by 1.
Exponents on b are increasing by 1.

Sum of exponents: $n-1+1=n$

Sum of exponents: $n-3+3=n$

Sum of exponents: $1+n-1=n$

Let's now establish a pattern for the coefficients of the terms in the binomial expansion. Notice that each row in the figure on page 644 begins and ends with 1. Any other number in the row can be obtained by adding the two numbers immediately above it.

Study Tip

We have not shown the number in the top row of Pascal's triangle on the right. The top row is *row zero* because it corresponds to $(a + b)^0 = 1$. With row zero, the triangle appears as

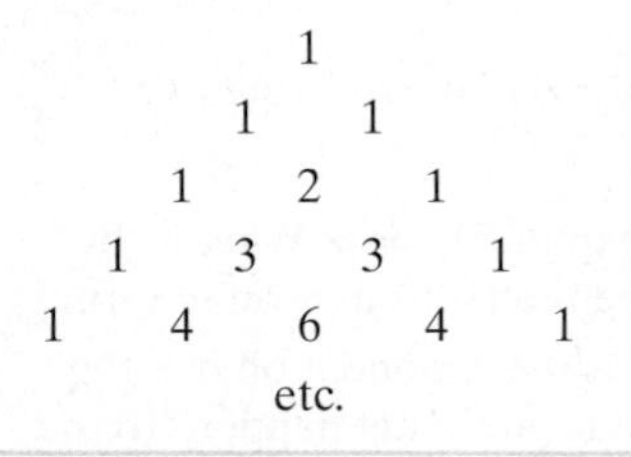

Coefficients for $(a+b)^1$.	1 1
Coefficients for $(a+b)^2$.	1 2 1
Coefficients for $(a+b)^3$.	1 3 3 1
Coefficients for $(a+b)^4$.	1 4 6 4 1
Coefficients for $(a+b)^5$.	1 5 10 10 5 1

The following triangular array of coefficients is called **Pascal's triangle.** If we continue with the sixth row, the first and last numbers are 1. Each of the other numbers is obtained by finding the sum of the two closest numbers above it in the fifth row.

1 1

1 2 1

1 3 3 1

1 4 6 4 1

1 5 10 10 5 1

1+5 5+10 10+10 10+5 5+1

1 6 15 20 15 6 1

We can use the numbers in the sixth row and the variable parts we found to write the expansion for $(a + b)^6$. It is

$$(a + b)^6 = a^6 + 6a^5b + 15a^4b^2 + 20a^3b^3 + 15a^2b^4 + 6ab^5 + b^6.$$

2 Evaluate a binomial coefficient.

Binomial Coefficients

Pascal's triangle becomes cumbersome when a binomial contains a relatively large power. Therefore, the coefficients in a binomial expansion are instead given in terms of factorials. The coefficients are written in a special notation, which we define next.

Definition of a Binomial Coefficient $\binom{n}{r}$

For nonnegative integers n and r, with $n \geq r$, the expression $\binom{n}{r}$ (read "n above r") is called a **binomial coefficient** and is defined by

$$\binom{n}{r} = \frac{n!}{r!(n-r)!}.$$

Technology

Graphing utilities can compute binomial coefficients. For example, to find $\binom{6}{2}$, many utilities require the sequence

6 [nCr] 2 [ENTER].

The graphing utility will display 15. Consult your manual and verify the other evaluations in Example 1.

The symbol ${}_nC_r$ is often used in place of $\binom{n}{r}$ to denote binomial coefficients.

EXAMPLE 1 Evaluating Binomial Coefficients

Evaluate: **a.** $\binom{6}{2}$ **b.** $\binom{3}{0}$ **c.** $\binom{9}{3}$ **d.** $\binom{4}{4}$.

Solution In each case, we apply the definition of the binomial coefficient.

a. $\binom{6}{2} = \frac{6!}{2!(6-2)!} = \frac{6!}{2!\,4!} = \frac{6 \cdot 5 \cdot \cancel{4!}}{2 \cdot 1 \cdot \cancel{4!}} = 15$

b. $\binom{3}{0} = \dfrac{3!}{0!(3-0)!} = \dfrac{\cancel{3!}}{0!\cancel{3!}} = \dfrac{1}{1} = 1$

Remember that $0! = 1$.

c. $\binom{9}{3} = \dfrac{9!}{3!(9-3)!} = \dfrac{9!}{3!\,6!} = \dfrac{9 \cdot 8 \cdot 7 \cdot \cancel{6!}}{3 \cdot 2 \cdot 1 \cdot \cancel{6!}} = 84$

d. $\binom{4}{4} = \dfrac{4!}{4!(4-4)!} = \dfrac{\cancel{4!}}{\cancel{4!}0!} = \dfrac{1}{1} = 1$

Check Point 1 Evaluate: **a.** $\binom{6}{3}$ **b.** $\binom{6}{0}$ **c.** $\binom{8}{2}$ **d.** $\binom{3}{3}$.

3 Expand a binomial raised to a power.

The Binomial Theorem

If we use binomial coefficients and the pattern for the variable part of each term, a formula called the **Binomial Theorem** can be written for any positive integral power of a binomial.

A Formula for Expanding Binomials: The Binomial Theorem

For any positive integer n,

$$(a+b)^n = \binom{n}{0}a^n + \binom{n}{1}a^{n-1}b + \binom{n}{2}a^{n-2}b^2 + \binom{n}{3}a^{n-3}b^3 + \cdots + \binom{n}{n}b^n.$$

The Universality of Mathematics

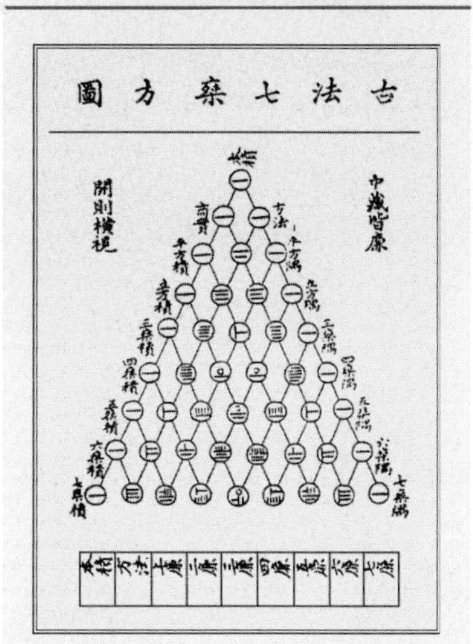

"Pascal's" triangle, credited to French mathematician Blaise Pascal (1623–1662), appeared in a Chinese document printed in 1303. The Binomial Theorem was known in Eastern cultures prior to its discovery in Europe. The same mathematics is often discovered/invented by independent researchers separated by time, place, and culture.

EXAMPLE 2 Using the Binomial Theorem

Expand: $(x+2)^4$.

Solution We use the Binomial Theorem

$$(a+b)^n = \binom{n}{0}a^n + \binom{n}{1}a^{n-1}b + \binom{n}{2}a^{n-2}b^2 + \binom{n}{3}a^{n-3}b^3 + \cdots + \binom{n}{n}b^n$$

to expand $(x+2)^4$. In $(x+2)^4$, $a = x$, $b = 2$, and $n = 4$.

$$(x+2)^4 = \binom{4}{0}x^4 + \binom{4}{1}x^3 \cdot 2 + \binom{4}{2}x^2 \cdot 2^2 + \binom{4}{3}x \cdot 2^3 + \binom{4}{4}2^4$$

These binomial coefficients are evaluated using $\binom{n}{r} = \dfrac{n!}{r!(n-r)!}$.

$$= \frac{4!}{0!\,4!}x^4 + \frac{4!}{1!\,3!}x^3 \cdot 2 + \frac{4!}{2!\,2!}x^2 \cdot 4 + \frac{4!}{3!\,1!}x \cdot 8 + \frac{4!}{4!\,0!} \cdot 16$$

$\dfrac{4!}{2!\,2!} = \dfrac{4 \cdot 3 \cdot \cancel{2!}}{\cancel{2!} \cdot 2 \cdot 1} = \dfrac{12}{2} = 6$
Take a few minutes to verify the other factorial evaluations.

$$= 1 \cdot x^4 + 4x^3 \cdot 2 + 6x^2 \cdot 4 + 4x \cdot 8 + 1 \cdot 16$$
$$= x^4 + 8x^3 + 24x^2 + 32x + 16$$

Check Point 2 Expand: $(x + 1)^4$.

EXAMPLE 3 Using the Binomial Theorem

Expand: $(2x - y)^5$.

Solution Because the Binomial Theorem involves the addition of two terms raised to a power, we rewrite $(2x - y)^5$ as $[2x + (-y)]^5$. We use the Binomial Theorem

$$(a + b)^n = \binom{n}{0}a^n + \binom{n}{1}a^{n-1}b + \binom{n}{2}a^{n-2}b^2 + \binom{n}{3}a^{n-3}b^3 + \cdots + \binom{n}{n}b^n$$

to expand $[2x + (-y)]^5$. In $[2x + (-y)]^5$, $a = 2x$, $b = -y$, and $n = 5$.

$$(2x - y)^5 = [2x + (-y)]^5$$

$$= \binom{5}{0}(2x)^5 + \binom{5}{1}(2x)^4(-y) + \binom{5}{2}(2x)^3(-y)^2 + \binom{5}{3}(2x)^2(-y)^3 + \binom{5}{4}(2x)(-y)^4 + \binom{5}{5}(-y)^5$$

Evaluate binomial coefficients using $\binom{n}{r} = \frac{n!}{r!(n-r)!}$.

$$= \frac{5!}{0!\,5!}(2x)^5 + \frac{5!}{1!\,4!}(2x)^4(-y) + \frac{5!}{2!\,3!}(2x)^3(-y)^2 + \frac{5!}{3!\,2!}(2x)^2(-y)^3 + \frac{5!}{4!\,1!}(2x)(-y)^4 + \frac{5!}{5!\,0!}(-y)^5$$

$\frac{5!}{2!\,3!} = \frac{5 \cdot 4 \cdot 3!}{2 \cdot 1 \cdot 3!} = 10$

Take a few minutes to verify the other factorial evaluations.

$$= 1(2x)^5 + 5(2x)^4(-y) + 10(2x)^3(-y)^2 + 10(2x)^2(-y)^3 + 5(2x)(-y)^4 + 1(-y)^5$$

Raise both factors in these parentheses to the indicated powers.

$$= 1(32x^5) + 5(16x^4)(-y) + 10(8x^3)(-y)^2 + 10(4x^2)(-y)^3 + 5(2x)(-y)^4 + 1(-y)^5$$

Now raise $-y$ to the indicated powers.

$$= 1(32x^5) + 5(16x^4)(-y) + 10(8x^3)y^2 + 10(4x^2)(-y^3) + 5(2x)y^4 + 1(-y^5)$$

Multiplying factors in each of the six terms gives us the desired expansion:

$$(2x - y)^5 = 32x^5 - 80x^4y + 80x^3y^2 - 40x^2y^3 + 10xy^4 - y^5.$$

Check Point 3 Expand: $(x - 2y)^5$.

4 Find a particular term in a binomial expansion.

Finding a Particular Term in a Binomial Expansion

The Binomial Theorem can be used to write any single term of a binomial expansion.

Finding a Particular Term in a Binomial Expansion

The rth term of the expansion of $(a + b)^n$ is

$$\binom{n}{r-1}a^{n-r+1}b^{r-1}.$$

EXAMPLE 4 Finding a Single Term of a Binomial Expansion

Find the fourth term in the expansion of $(3x + 2y)^7$.

Solution We will use the formula for the rth term of the expansion of $(a + b)^n$,

$$\binom{n}{r-1}a^{n-r+1}b^{r-1}$$

to find the fourth term of $(3x + 2y)^7$. For the fourth term of $(3x + 2y)^7$, $n = 7$, $r = 4$, $a = 3x$ and $b = 2y$. Thus, the fourth term is

$$\binom{7}{4-1}(3x)^{7-4+1}(2y)^{4-1} = \binom{7}{3}(3x)^4(2y)^3 = \frac{7!}{3!(7-3)!}(3x)^4(2y)^3$$

We use $\binom{n}{r} = \frac{n!}{r!(n-r)!}$ to evaluate $\binom{7}{3}$.

Now we need to evaluate the factorial expression and raise $3x$ and $2y$ to the indicated powers. We obtain

$$\frac{7!}{3!\,4!}(81x^4)(8y^3) = \frac{7 \cdot 6 \cdot 5 \cdot \cancel{4!}}{3 \cdot 2 \cdot 1 \cdot \cancel{4!}}(81x^4)(8y^3) = 35(81x^4)(8y^3) = 22{,}680x^4y^3.$$

The fourth term of $(3x + 2y)^7$ is $22{,}680x^4y^3$.

Check Point 4 Find the fifth term in the expansion of $(2x + y)^9$.

EXERCISE SET 8.5

Practice Exercises

In Exercises 1–8, evaluate the given binomial coefficient.

1. $\binom{8}{3}$
2. $\binom{7}{2}$
3. $\binom{12}{1}$
4. $\binom{11}{1}$
5. $\binom{6}{6}$
6. $\binom{15}{2}$
7. $\binom{100}{2}$
8. $\binom{100}{98}$

In Exercises 9–30, use the Binomial Theorem to expand each binomial and express the result in simplified form.

9. $(x + 2)^3$
10. $(x + 4)^3$
11. $(3x + y)^3$
12. $(x + 3y)^3$
13. $(5x - 1)^3$
14. $(4x - 1)^3$
15. $(2x + 1)^4$
16. $(3x + 1)^4$
17. $(x^2 + 2y)^4$
18. $(x^2 + y)^4$

19. $(y - 3)^4$
20. $(y - 4)^4$
21. $(2x^3 - 1)^4$
22. $(2x^5 - 1)^4$
23. $(c + 2)^5$
24. $(c + 3)^5$
25. $(x - 1)^5$
26. $(x - 2)^5$
27. $(x - 2y)^5$
28. $(x - 3y)^5$
29. $(2a + b)^6$
30. $(a + 2b)^6$

In Exercises 31–38, write the first three terms in each binomial expansion, expressing the result in simplified form.

31. $(x + 2)^8$
32. $(x + 3)^8$
33. $(x - 2y)^{10}$
34. $(x - 2y)^9$
35. $(x^2 + 1)^{16}$
36. $(x^2 + 1)^{17}$
37. $(y^3 - 1)^{20}$
38. $(y^3 - 1)^{21}$

In Exercises 39–46, find the term indicated in each expansion.

39. $(2x + y)^6$; third term
40. $(x + 2y)^6$; third term
41. $(x - 1)^9$; fifth term
42. $(x - 1)^{10}$; fifth term
43. $(x^2 + y^3)^8$; sixth term
44. $(x^3 + y^2)^8$; sixth term
45. $(x - \frac{1}{2})^9$; fourth term
46. $(x + \frac{1}{2})^8$; fourth term

Application Exercises

47. The percentage of people taking the SAT whose intended college major is engineering, $f(t)$, can be modeled by

$$f(t) = 0.002t^3 - 0.9t^2 + 1.27t + 6.76, \qquad 0 \le t \le 20,$$

where $t = 0$ represents 1975. How can we adjust this model so that $t = 0$ corresponds to 1985 rather than 1975? We shift the graph of f ten units to the left. We obtain $g(t) = f(t + 10)$. Use the Binomial Theorem to express $g(t)$ in descending powers of t.

48. The personal income per capita in the United States, $f(t)$, in constant 1992 dollars, can be modeled by

$$f(t) = 3.75t^3 - 115.23t^2 + 1229.81t + 16{,}025.65, \qquad 0 \le t \le 15,$$

where $t = 0$ represents 1979. How can we adjust this model so that $t = 0$ corresponds to 1989 rather than 1979? We shift the graph of f ten units to the left. We obtain $g(t) = f(t + 10)$. Use the Binomial Theorem to express $g(t)$ in descending powers of t.

Writing in Mathematics

49. Describe the pattern on the exponents on a in the expansion of $(a + b)^n$.

50. Describe the pattern on the exponents on b in the expansion of $(a + b)^n$.

51. What is true about the sum of the exponents on a and b in any term in the expansion of $(a + b)^n$?

52. How do you determine how many terms there are in a binomial expansion?

53. What is Pascal's triangle? How do you find the numbers in any row of the triangle?

54. Explain how to evaluate $\binom{n}{r}$. Provide an example with your explanation.

55. Explain how to use the Binomial Theorem to expand a binomial. Provide an example with your explanation.

56. Explain how to find a particular term in a binomial expansion without having to write out the entire expansion.

57. Are there situations in which it is easier to use Pascal's triangle than binomial coefficients? Describe these situations.

58. Describe how you would use mathematical induction to prove

$$(a + b)^n = \binom{n}{0}a^n + \binom{n}{1}a^{n-1}b + \binom{n}{2}a^{n-2}b^2 + \cdots + \binom{n}{n-1}ab^{n-1} + \binom{n}{n}b^n.$$

What happens when $n = 1$? Write the statement that we assume true. Write the statement that we must prove. What must be done to the left side of the assumed statement to make it look like the left side of the statement that must be proved? (More detail on the actual proof is found in Exercise 71.)

Technology Exercises

59. Use the nCr key on a graphing utility to verify your answers in Exercises 1–8.

In Exercises 60–61, graph each of the functions in the same viewing rectangle. Describe how the graphs illustrate the Binomial Theorem.

60. $f_1(x) = (x + 2)^3$
$f_2(x) = x^3$
$f_3(x) = x^3 + 6x^2$
$f_4(x) = x^3 + 6x^2 + 12x$
$f_5(x) = x^3 + 6x^2 + 12x + 8$
Use a $[-10, 10, 1]$ by $[-30, 30, 10]$ viewing rectangle.

61. $f_1(x) = (x + 1)^4$
$f_2(x) = x^4$
$f_3(x) = x^4 + 4x^3$
$f_4(x) = x^4 + 4x^3 + 6x^2$
$f_5(x) = x^4 + 4x^3 + 6x^2 + 4x$
$f_6(x) = x^4 + 4x^3 + 6x^2 + 4x + 1$
Use a $[-5, 5, 1]$ by $[-30, 30, 10]$ viewing rectangle.

In Exercises 62–64, use the Binomial Theorem to find a polynomial expansion for each function. Then use a graphing utility and an approach similar to the one in Exercises 60 and 61 to verify the expansion.

62. $f_1(x) = (x - 1)^3$

63. $f_1(x) = (x - 2)^4$

64. $f_1(x) = (x + 2)^6$

65. Graphing utilities capable of symbolic manipulation, such as the TI-92, will expand binomials. On the TI-92, to expand $(3a - 5b)^{12}$, input the following:

[EXPAND] ((3a [−] 5b) [^] 12) [ENTER].

Use a graphing utility with this capability to verify any five of the expansions you performed by hand in Exercises 9–30.

Critical Thinking Exercises

66. Which one of the following is true?

a. The binomial expansion for $(a + b)^n$ contains n terms.

b. The Binomial Theorem can be written in condensed form as $(a + b)^n = \sum_{r=0}^{n} \binom{n}{r} a^{n-r} b^r$.

c. The sum of the binomial coefficients in $(a + b)^n$ cannot be 2^n.

d. There are no values of a and b such that $(a + b)^4 = a^4 + b^4$.

67. Use the Binomial Theorem to expand and then simplify the result: $(x^2 + x + 1)^3$. [*Hint*: Write $x^2 + x + 1$ as $x^2 + (x + 1)$].

68. Find the term in the expansion of $(x^2 + y^2)^5$ containing x^4 as a factor.

69. Prove that

$$\binom{n}{r} = \binom{n}{n-r}.$$

70. Show that

$$\binom{n}{r} + \binom{n}{r+1} = \binom{n+1}{r+1}.$$

Hints:

$$(n - r)! = (n - r)(n - r - 1)!$$

$$(r + 1)! = (r + 1)r!$$

71. Follow the outline below to use mathematical induction to prove that

$$(a + b)^n = \binom{n}{0}a^n + \binom{n}{1}a^{n-1}b + \binom{n}{2}a^{n-2}b^2 + \cdots + \binom{n}{n-1}ab^{n-1} + \binom{n}{n}b^n.$$

a. Verify the formula for $n = 1$.

b. Replace n with k and write the statement that is assumed true. Replace n with $k + 1$ and write the statement that must be proved.

c. Multiply both sides of the statement assumed to be true by $a + b$. Add exponents on the left. On the right, distribute a and b, respectively.

d. Collect like terms on the right. At this point, you should have

$$(a + b)^{k+1} = \binom{k}{0}a^{k+1} + \left[\binom{k}{0} + \binom{k}{1}\right]a^k b + \left[\binom{k}{1} + \binom{k}{2}\right]a^{k-1}b^2 + \left[\binom{k}{2} + \binom{k}{3}\right]a^{k-2}b^3 + \cdots + \left[\binom{k}{k-1} + \binom{k}{k}\right]ab^k + \binom{k}{k}b^{k+1}.$$

e. Use the result of Exercise 70 to add the binomial sums in brackets. For example, because $\binom{n}{r} + \binom{n}{r+1} = \binom{n+1}{r+1}$ then $\binom{k}{0} + \binom{k}{1} = \binom{k+1}{1}$ and $\binom{k}{1} + \binom{k}{2} = \binom{k+1}{2}$.

f. Because $\binom{k}{0} = \binom{k+1}{0}$ (why?) and $\binom{k}{k} = \binom{k+1}{k+1}$ (why?), substitute these results and the results from part (e) into the equation in part (d). This should give the statement that we were required to prove in the second step of the mathematical induction process.

SECTION 8.6 Counting Principles, Permutations, and Combinations

Objectives

1. Use the Fundamental Counting Principle.
2. Use the permutations formula.
3. Distinguish between permutation problems and combination problems.
4. Use the combinations formula.

Have you ever imagined what your life would be like if you won the lottery? What changes would you make? Before you fantasize about becoming a person of leisure with a staff of obedient elves, think about this: The probability of winning top prize in the lottery is about the same as the probability of being struck by lightning. There are millions of possible number combinations in lottery games, and only one way of winning the grand prize. Determining the probability of winning involves calculating the chance of getting the winning combination from all possible outcomes. In this section, we begin preparing for the surprising world of probability by looking at methods for counting possible outcomes.

1 Use the Fundamental Counting Principle.

The Fundamental Counting Principle

It's early morning, you're groggy, and you have to select something to wear for your 8 A.M. class. (What *were* you thinking of when you signed up for a class at that hour?!) Fortunately, your "lecture wardrobe" is rather limited—just two pairs of jeans to choose from (one blue, one black), three T-shirts to choose from (one beige, one yellow, and one blue), and two pairs of sneakers to select from (one black, one red). Your possible outfits are shown in Figure 8.5.

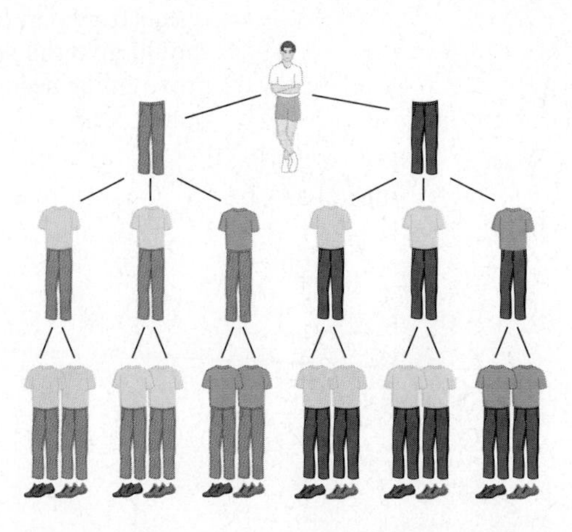

Figure 8.5 Selecting a wardrobe

The number of possible ways of playing the first four moves on each side in a game of chess is 318,979,564,000.

The **tree diagram**, so named because of its branches, shows that you can form 12 outfits from your two pairs of jeans, three T-shirts, and two pairs of sneakers. Notice that the number of outfits can be obtained by multiplying the number of choices for jeans, 2, the number of choices for T-shirts, 3, and the number of choices for sneakers, 2:

$$2 \cdot 3 \cdot 2 = 12.$$

We can generalize this idea to any two or more groups of items—not just jeans, T-shirts, and sneakers—with the **Fundamental Counting Principle**.

The Fundamental Counting Principle

The number of ways in which a series of successive things can occur is found by multiplying the number of ways in which each thing can occur.

For example, if you own 30 pairs of jeans, 20 T-shirts, and 12 pairs of sneakers, you have

$$30 \cdot 20 \cdot 12 = 7200$$

choices for your wardrobe!

EXAMPLE 1 Options in Planning a Course Schedule

Next semester you are planning to take three courses—math, English, and humanities. Based on time blocks and highly recommended professors, there are 8 sections of math, 5 of English, and 4 of humanities that you find suitable. Assuming no scheduling conflicts, how many different three-course schedules are possible?

Solution This situation involves making choices with three groups of items.

MATH	ENGLISH	HUMANITIES
8 choices	5 choices	4 choices

We use the Fundamental Counting Principle to find the number of three-course schedules. Multiply the number of choices for each of the three groups.

$$8 \cdot 5 \cdot 4 = 160$$

Thus, there are 160 different three-course schedules.

A pizza can be ordered with three choices of size (small, medium, or large), four choices of crust (thin, thick, crispy, or regular), and six choices of toppings (ground beef, sausage, pepperoni, bacon, mushrooms, or onions). How many different one-topping pizzas can be ordered?

EXAMPLE 2 A Multiple-Choice Test

You are taking a multiple-choice test that has ten questions. Each of the questions has four choices, with one correct choice per question. If you select one of these options per question and leave nothing blank, in how many ways can you answer the questions?

Solution We use the Fundamental Counting Principle to determine the number of ways you can answer the test. Multiply the number of choices, 4, for each of the ten questions.

$$4 \cdot 4 \cdot 4 \cdot 4 \cdot 4 \cdot 4 \cdot 4 \cdot 4 \cdot 4 \cdot 4 = 4^{10} = 1{,}048{,}576$$

Thus, you can answer the questions in 1,048,576 different ways.

Are you surprised that there are over one million ways of answering a ten-question multiple-choice test? Of course, there is only one way to answer the test and receive a perfect score. The probability of guessing your way into a perfect score involves calculating the chance of getting a perfect score, just one way, from all 1,048,576 possible outcomes. In short, prepare for the test and do not rely on guessing!

Check Point 2 You are taking a multiple-choice test that has six questions. Each of the questions has three choices, with one correct choice per question. If you select one of these options per question and leave nothing blank, in how many ways can you answer the questions?

Permutations and Rubik's Cube

First developed in Hungary in the 1970s by Erno Rubik, a Rubik's cube contains $3 \times 3 \times 3 = 27$ small cubes. The square faces of the cubes are colored in six different colors. The cubes can be twisted horizontally or vertically. When first purchased, the cube is arranged so that each face shows a single color. To do the puzzle, you first turn columns and rows in a random way until all of the six faces are multicolored. To solve the puzzle, you must return the cube to its original state—that is, a single color on each of the six faces. With 43,252,003,274,489,856,000 arrangements, this is no easy task! If it takes one-half second for each of these arrangements, it would require over 681,000,000,000 years to move the cube into all possible arrangements.

EXAMPLE 3 Telephone Numbers in the United States

Telephone numbers in the United States begin with three-digit area codes followed by seven-digit local telephone numbers. Area codes and local telephone numbers cannot begin with 0 or 1. How many different telephone numbers are possible?

Solution This situation involves making choices with ten groups of items.

Area Code ☐ ☐ ☐ Local Telephone Number ☐ ☐ ☐ ☐ ☐ ☐ ☐

You cannot use 0 or 1 in these groups. There are only 8 choices: 2, 3, 4, 5, 6, 7, 8, or 9.

You can use 0, 1, 2, 3, 4, 5, 6, 7, 8, or 9 in these groups. There are 10 choices per group.

We use the Fundamental Counting Principle to determine the number of different telephone numbers that are possible. The total number of telephone numbers possible is

$$8 \cdot 10 \cdot 10 \cdot 8 \cdot 10 \cdot 10 \cdot 10 \cdot 10 \cdot 10 \cdot 10 = 6{,}400{,}000{,}000.$$

There are six billion four hundred million different telephone numbers that are possible.

Check Point 3 License plates in a particular state display two letters followed by three numbers, such as AT-887 or BB-013. How many different license plates can be manufactured?

2 Use the permutations formula.

Permutations

You are the coach of a little league baseball team. There are 13 players on the team (and lots of parents hovering in the background, dreaming of stardom for their little "Mark McGwire"). You need to choose a batting order having 9 players. The order makes a difference, because, for instance, if bases are loaded and "Little Mark" is fourth or fifth at bat, his possible home run will drive in three additional runs. How many batting orders can you form?

You can choose any of 13 players for the first person at bat. Then you will have 12 players from which to choose the second batter, then 11 from which to choose the third batter, and so on. The situation can be shown as follows:

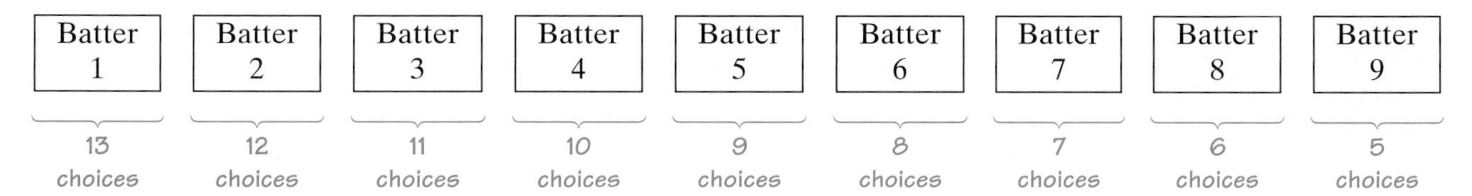

We use the Fundamental Counting Principle to find the number of batting orders. The total number of batting orders is

$$13 \cdot 12 \cdot 11 \cdot 10 \cdot 9 \cdot 8 \cdot 7 \cdot 6 \cdot 5 = 259{,}459{,}200.$$

Nearly 260 million batting orders are possible for your 13-player little league team. Each batting order is called a **permutation** of 13 players taken 9 at a time. The number of permutations of 13 players taken 9 at a time is 259,459,200. A permutation is an ordered arrangement of items that occurs when

- No item is used more than once. (Each of the 9 players in the batting order bats exactly once.)
- The order of arrangement makes a difference.

We can obtain a formula for finding the number of permutations by rewriting our computation:

$$13 \cdot 12 \cdot 11 \cdot 10 \cdot 9 \cdot 8 \cdot 7 \cdot 6 \cdot 5$$
$$= \frac{13 \cdot 12 \cdot 11 \cdot 10 \cdot 9 \cdot 8 \cdot 7 \cdot 6 \cdot 5 \cdot \boxed{4 \cdot 3 \cdot 2 \cdot 1}}{\boxed{4 \cdot 3 \cdot 2 \cdot 1}} = \frac{13!}{4!} = \frac{13!}{(13-9)!}.$$

Thus, the number of permutations of 13 things taken 9 at a time is $\frac{13!}{(13-9)!}$. The special notation ${}_{13}P_9$ is used to replace the phrase "the number of permutations of 13 things taken 9 at a time." Using this new notation, we can write

$${}_{13}P_9 = \frac{13!}{(13-9)!}.$$

The numerator of this expression is the number of items, 13 team members, expressed as a factorial: 13! The denominator is also a factorial. It is the factorial of the difference between the number of items, 13, and the number of items in each permutation, 9 batters: $(13 - 9)!$.

The notation ${}_nP_r$ means the **number of permutations of n things taken r at a time.** We can generalize from the situation in which 9 batters were taken from 13 players. By generalizing, we obtain the following formula for the number of permuations if r items are taken from n items.

Permutations of n Things Taken r at a Time

The number of possible permuations if r items are taken from n items is

$${}_nP_r = \frac{n!}{(n-r)!}.$$

Because all permutation problems are also Fundamental Counting problems, they can be solved using the formula for $_nP_r$, or using the Fundamental Counting Principle.

Technology

Graphing utilities have a key for calculating permuations, usually labeled $\boxed{_nP_r}$. For example, to find $_{20}P_3$, the keystrokes are:

20 $\boxed{_nP_r}$ 3 $\boxed{\text{ENTER}}$.

If you are using a scientific calculator, check your manual for the location of the key for calculating permutations and the required keystrokes.

EXAMPLE 4 Using the Formula for Permutations

You and 19 of your friends have decided to form an Internet marketing consulting firm. The group needs to choose three officers—a CEO, an operating manager, and a treasurer. In how many ways can those offices be filled?

Solution Your group is choosing $r = 3$ officers from a group of $n = 20$ people (you and 19 friends). The order in which the officers are chosen matters because the CEO, the operating manager, and the treasurer each have different responsibilities. Thus, we are looking for the number of permutations of 20 things taken 3 at a time. We use the formula

$$_nP_r = \frac{n!}{(n-r)!}$$

with $n = 20$ and $r = 3$.

$$_{20}P_3 = \frac{20!}{(20-3)!} = \frac{20!}{17!} = \frac{20 \cdot 19 \cdot 18 \cdot 17!}{17!} = \frac{20 \cdot 19 \cdot 18 \cdot \cancel{17!}}{\cancel{17!}} = 20 \cdot 19 \cdot 18 = 6840$$

Thus, there are 6840 different ways of filling the three offices.

Check Point 4 A corporation has seven members on its board of directors. In how many different ways can it elect a president, vice-president, secretary, and treasurer?

How to Pass the Time for $2\frac{1}{2}$ Million Years

If you were to arrange 15 different books on a shelf and it took you one minute for each permuation, the entire task would take 2,487,965 years.

Source: Isaac Asimov's Book of Facts.

EXAMPLE 5 Using the Formula for Permutations

You need to arrange seven of your favorite books along a small shelf. How many different ways can you arrange the books, assuming that the order of the books makes a difference to you?

Solution Because you are using all seven of your books in every possible arrangement, you are arranging $r = 7$ books from a group of $n = 7$ books. Thus, we are looking for the number of permutations of 7 things taken 7 at a time. We use the formula

$$_nP_r = \frac{n!}{(n-r)!}$$

with $n = 7$ and $r = 7$.

$$_7P_7 = \frac{7!}{(7-7)!} = \frac{7!}{0!} = \frac{7!}{1} = 5040$$

Thus, you can arrange the books in 5040 ways. There are 5040 different possible permuations.

Check Point 5 In how many ways can 6 books be lined up along a shelf?

3 Distinguish between permutation problems and combination problems.

Combinations

As the twentieth century drew to a close, *Time* magazine presented a series of special issues on the most influential people of the century. In their issue on heroes and icons (June 14, 1999), they discussed a number of people whose careers became more profitable after their tragic deaths, including Marilyn Monroe, James Dean, Jim Morrison, Kurt Cobain, and Selena.

Imagine that you ask your friends the following question: "Of these five people, which three would you select to be included in a documentary featuring the best of their work?" You are not asking your friends to rank their three favorite artists in any kind of order—they should merely select the three to be included in the documentary.

One friend answers, "Jim Morrison, Kurt Cobain, and Selena." Another responds, "Selena, Kurt Cobain, and Jim Morrison." These two people have the same artists in their group of selections, even if they are named in a different order. We are interested *in which artists are named, not the order in which they are named* for the documentary. Because the items are taken without regard to order, this is not a permutation problem. No ranking of any sort is involved.

Marilyn Monroe, actress (1927–1962)

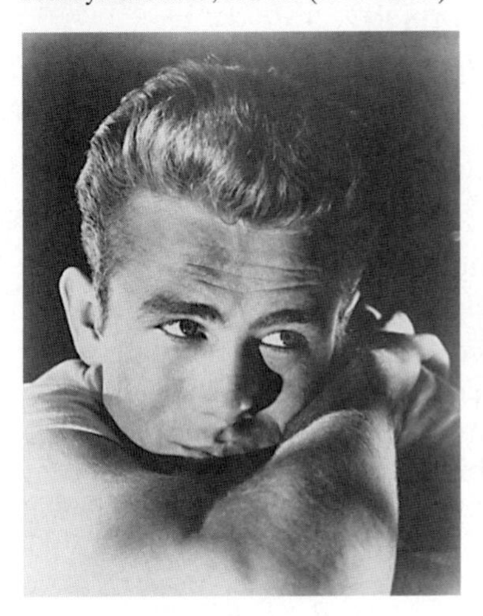

James Dean, actor (1931–1955)

Jim Morrison, musician and lead singer of The Doors (1943–1971)

Kurt Cobain, musician and front man for Nirvana (1967–1994)

Selena, musician of Tejano music (1971–1995)

Later on, you ask your roommate which three artists she would select for the documentary. She names Marilyn Monroe, James Dean, and Selena. Her selection is different from those of your two other friends because different entertainers are cited.

Mathematicians describe the group of artists given by your roommate as a *combination*. A **combination** of items occurs when

- The items are selected from the same group (the five stars who died young and tragically).
- No item is used more than once. (You may adore Selena, but your three selections cannot be Selena, Selena, and Selena).
- The order of items makes no difference. (Morrison, Cobain, Selena is the same group in the documentary as Selena, Cobain, Morrison.)

Do you see the difference between a permutation and a combination? A permutation is an ordered arrangement of a given group of items. A combination is a group of items taken without regard to their order. *Permutation* problems involve situations in which *order matters. Combination* problems involve situations in which the *order* of items *makes no difference.*

EXAMPLE 6 Distinguishing between Permutations and Combinations

For each of the following problems, explain whether the problem is one involving permutations or combinations. (It is not necessary to solve the problem.)

a. Six candidates are running for president, chief technology officer, and director of marketing of an Internet company. The candidate with the greatest number of votes becomes the president, the second biggest vote-getter becomes chief technology officer, and the candidate who gets the third largest number of votes will be director of marketing. How many different outcomes are possible for these three positions?

b. From the six candidates who desire to hold office in an Internet company, a three-person committee is formed to study ways of finding new investors. How many different committees could be formed?

Solution

a. Voters are choosing three officers from six candidates. The order in which the officers are chosen makes a difference because each of the offices (president, chief technology officer, and director of marketing) is different. Order matters. This is a problem involving permutations. (How many permutations are possible if three candidates are elected from six candidates?)

b. A three-person committee is to be formed from the six candidates. The order in which the three people are selected does not matter because they are not filling different roles on the committee. Because order makes no difference, this is a problem involving combinations. (How many different combinations of three people can be chosen from a group of six people?)

Check Point 6

For each of the following problems, explain if the problem is one involving permutations or combinations. (It is not necessary to solve the problem.)

a. How many ways can you select 6 free videos from a list of 200 videos?

b. In a race in which there are 50 runners and no ties, in how many ways can the first three finishers come in?

4 Use the combinations formula.

The notation ${}_nC_r$ means the **number of combinations of n things taken r at a time**. In general, there are $r!$ times as many permutations of n things taken r at a time as there are combinations of n things taken r at a time. Thus, we find the number of combinations of n things taken r at a time by dividing the number of permutations of n things taken r at a time by $r!$.

$${}_nC_r = \frac{{}_nP_r}{r!} = \frac{\frac{n!}{(n-r)!}}{r!} = \frac{n!}{(n-r)!\,r!}$$

Combinations of *n* Things Taken *r* at a Time

The number of possible combinations if r items are taken from n items is

$$_{n}C_{r} = \frac{n!}{(n-r)!\,r!}.$$

Notice that the formula for $_{n}C_{r}$ is the same as the formula for the binomial coefficient $\binom{n}{r}$.

We cannot find the number of combinations if r items are taken from n items using the Fundamental Counting Principle. We must use the formula shown in the box to do so.

EXAMPLE 7 Using the Formula for Combinations

A three-person committee is needed to study ways of improving public transportation. How many committees could be formed from the eight people on the board of supervisors?

Solution The order in which the three people are selected does not matter. This is a problem of selecting $r = 3$ people from a group of $n = 8$ people. We are looking for the number of combinations of eight things taken three at a time. We use the formula

$$_{n}C_{r} = \frac{n!}{(n-r)!\,r!}$$

with $n = 8$ and $r = 3$.

$$_{8}C_{3} = \frac{8!}{(8-3)!\,3!} = \frac{8!}{5!\,3!} = \frac{8 \cdot 7 \cdot 6 \cdot 5!}{5! \cdot 3 \cdot 2 \cdot 1} = \frac{8 \cdot 7 \cdot 6 \cdot \cancel{5!}}{\cancel{5!} \cdot 3 \cdot 2 \cdot 1} = 56$$

Thus, 56 committees of three people each can be formed from the eight people on the board of supervisors.

Check Point 7 From a group of 10 physicians, in how many ways can four people be selected to attend a conference on acupuncture?

EXAMPLE 8 Using the Formula for Combinations

In poker, a person is dealt 5 cards from a standard 52-card deck. The order in which you are dealt the 5 cards does not matter. How many different 5-card poker hands are possible?

Solution Because the order in which the 5 cards are dealt does not matter, this is a problem involving combinations. We are looking for the number of combinations of $n = 52$ cards drawn $r = 5$ at a time. We use the formula

$$_{n}C_{r} = \frac{n!}{(n-r)!\,r!}$$

with $n = 52$ and $r = 5$.

$$_{52}C_5 = \frac{52!}{(52-5)!\,5!} = \frac{52!}{47!\,5!} = \frac{52 \cdot 51 \cdot 50 \cdot 49 \cdot 48 \cdot \cancel{47!}}{\cancel{47!} \cdot 5 \cdot 4 \cdot 3 \cdot 2 \cdot 1} = 2{,}598{,}960$$

Thus, there are 2,598,960 different 5-card poker hands possible. It surprises many people that more than 2.5 million 5-card hands can be dealt from a mere 52 cards.

Figure 8.6 A royal flush

If you are a card player, it does not get any better than to be dealt the 5-card poker hand shown in Figure 8.6. This hand is called a *royal flush*. It consists of an ace, king, queen, jack, and 10, all of the same suit: all hearts, all diamonds, all clubs, or all spades. The probability of being dealt a royal flush involves calculating the number of ways of being dealt such a hand: just 4 of all 2,598,960 possible hands. In the next section, we move from counting possibilities to computing probabilities.

Check Point 8 How many different 4-card hands can be dealt from a deck that has 16 different cards?

EXERCISE SET 8.6

Practice Exercises

In Exercises 1–8, use the formula for $_nP_r$ to evaluate each expression.

1. $_9P_4$

2. $_7P_3$

3. $_8P_5$

4. $_{10}P_4$

5. $_6P_6$

6. $_9P_9$

7. $_8P_0$

8. $_6P_0$

In Exercises 9–16, use the formula for $_nC_r$ to evaluate each expression.

9. $_9C_5$

10. $_{10}C_6$

11. $_{11}C_4$

12. $_{12}C_5$

13. $_7C_7$

14. $_4C_4$

15. $_5C_0$

16. $_6C_0$

In Exercises 17–20, does the problem involve permutations or combinations? Explain your answer. (It is not necessary to solve the problem.)

17. A medical researcher needs 6 people to test the effectiveness of an experimental drug. If 13 people have volunteered for the test, in how many ways can 6 people be selected?

18. Fifty people purchase raffle tickets. Three winning tickets are selected at random. If first prize is $1000, second prize is $500, and third prize is $100, in how many different ways can the prizes be awarded?

19. How many different four-letter passwords can be formed from the letters A, B, C, D, E, F, and G if no repetition of letters is allowed?

20. Fifty people purchase raffle tickets. Three winning tickets are selected at random. If each prize is $500, in how many different ways can the prizes be awarded?

Application Exercises

Use the Fundamental Counting Principle to solve Exercises 21–32.

21. The model of the car you are thinking of buying is available in nine different colors and three different styles (hatchback, sedan, or station wagon). In how many ways can you order the car?

22. A popular brand of pen is available in three colors (red, green, or blue) and four writing tips (bold, medium, fine, or micro). How many different choices of pens do you have with this brand?

23. An ice cream store sells two drinks (sodas or milk shakes), in four sizes (small, medium, large, or jumbo), and five flavors (vanilla, strawberry, chocolate, coffee, or pistachio). In how many ways can a customer order a drink?

24. A restaurant offers the following lunch menu.

Main Course	Vegetables	Beverages	Desserts
Ham	Potatoes	Coffee	Cake
Chicken	Peas	Tea	Pie
Fish	Green beans	Milk	Ice cream
Beef		Soda	

If one item is selected from each of the four groups, in how many ways can a meal be ordered? Describe two such orders.

25. You are taking a multiple-choice test that has five questions. Each of the questions has three choices, with one correct choice per question. If you select one of these options per question and leave nothing blank, in how many ways can you answer the questions?

26. You are taking a multiple-choice test that has eight questions. Each of the questions has three choices, with one correct choice per question. If you select one of these options per question and leave nothing blank, in how many ways can you answer the questions?

27. In the original plan for area codes in 1945, the first digit could be any number from 2 through 9, the second digit was either 0 or 1, and the third digit could be any number except 0. With this plan, how many different area codes were possible?

28. How many different four-letter radio station call letters can be formed if the first letter must be W or K?

29. Six performers are to present their comedy acts on a weekend evening at a comedy club. One of the performers insists on being the last stand-up comic of the evening. If this performer's request is granted, how many different ways are there to schedule the appearances?

30. Five singers are to perform at a night club. One of the singers insists on being the last performer of the evening. If this singer's request is granted, how many different ways are there to schedule the appearances?

31. In the *Cambridge Encyclopedia of Language* (Cambridge University Press, 1987), author David Crystal presents five sentences that make a reasonable paragraph regardless of their order. The sentences are:

> Mark had told him about the foxes.
> John looked out the window.
> Could it be a fox?
> However, nobody had seen one for months.
> He thought he saw a shape in the bushes.

How many different five-sentence paragraphs can be formed if the paragraph begins with "He thought he saw a shape in the bushes" and ends with "John looked out of the window"?

32. A television programmer is arranging the order that five movies will be seen between the hours of 6 P.M. and 4 A.M. Two of the movies have a G rating, and they are to be shown in the first two time blocks. One of the movies is rated NC-17, and it is to be shown in the last of the time blocks, from 2 A.M. until 4 A.M. Given these restrictions, in how many ways can the five movies be arranged during the indicated time blocks?

Use the formula for $_nP_r$ to solve Exercises 33–40.

33. A club with ten members is to choose three officers—president, vice-president, and secretary-treasurer. If each office is to be held by one person and no person can hold more than one office, in how many ways can those offices be filled?

34. A corporation has ten members on its board of directors. In how many different ways can it elect a president, vice-president, secretary, and treasurer?

35. For a segment of a radio show, a disc jockey can play 7 records. If there are 13 records to select from, in how many ways can the program for this segment be arranged?

36. Suppose you are asked to list, in order of preference, the three best movies you have seen this year. If you saw 20 movies during the year, in how many ways can the three best be chosen and ranked?

37. In a race in which six automobiles are entered and there are no ties, in how many ways can the first three finishers come in?

38. In a production of *West Side Story*, eight actors are considered for the male roles of Tony, Riff, and Bernardo. In how many ways can the director cast the male roles?

39. Nine bands have volunteered to perform at a benefit concert, but there is only enough time for five of the bands to play. How many lineups are possible?

40. How many arrangements can be made using four of the letters of the word COMBINE if no letter is to be used more than once?

Use the formula for $_nC_r$ to solve Exercises 41–48.

41. An election ballot asks voters to select three city commissioners from a group of six candidates. In how many ways can this be done?

42. A four-person committee is to be elected from an organization's membership of 11 people. How many different committees are possible?

43. Of 12 possible books, you plan to take 4 with you on vacation. How many different collections of 4 books can you take?

44. There are 14 standbys who hope to get seats on a flight, but only 6 seats are available on the plane. How many different ways can the 6 people be selected?

45. You volunteer to help drive children at a charity event to the zoo, but you can fit only 8 of the 17 children present in your van. How many different groups of 8 children can you drive?

46. Of the 100 people in the U.S. Senate, 18 serve on the Foreign Relations Committee. How many ways are there to select Senate members for this committee (assuming party affiliation is not a factor in selection)?

47. To win at LOTTO in the state of Florida, one must correctly select 6 numbers from a collection of 49 numbers (1 through 49). The order in which the selection is made does not matter. How many different selections are possible?

48. To win in the New York State lottery, one must correctly select 6 numbers from 54 numbers. The order in which the selection is made does not matter. How many different selections are possible?

In Exercises 49–58, solve by the method of your choice.

49. In a race in which six automobiles are entered and there are no ties, in how many ways can the first four finishers come in?

50. A book club offers a choice of 8 books from a list of 40. In how many ways can a member make a selection?

51. A medical researcher needs 6 people to test the effectiveness of an experimental drug. If 13 people have volunteered for the test, in how many ways can 6 people be selected?

52. Fifty people purchase raffle tickets. Three winning tickets are selected at random. If first prize is $1000, second prize is $500, and third prize is $100, in how many different ways can the prizes be awarded?

53. From a club of 20 people, in how many ways can a group of three members be selected to attend a conference?

54. Fifty people purchase raffle tickets. Three winning tickets are selected at random. If each prize is $500, in how many different ways can the prizes be awarded?

55. How many different four-letter passwords can be formed from the letters A, B, C, D, E, F, and G if no repetition of letters is allowed?

56. Nine comedy acts will perform over two evenings. Five of the acts will perform on the first evening. How many ways can the schedule for the first evening be made?

57. Using 15 flavors of ice cream, how many cones with three different flavors can you create if it is important to you which flavor goes on the top, middle, and bottom?

58. Baskin-Robbins offers 31 different flavors of ice cream. One of their items is a bowl consisting of three scoops of ice cream, each a different flavor. How many such bowls are possible?

Writing in Mathematics

59. Explain the Fundamental Counting Principle.

60. Write an original problem that can be solved using the Fundamental Counting Principle. Then solve the problem.

61. What is a permutation?

62. Describe what ${}_nP_r$ represents.

63. Write a word problem that can be solved by evaluating ${}_7P_3$.

64. What is a combination?

65. Explain how to distinguish between permutation and combination problems.

66. Write a word problem that can be solved by evaluating ${}_7C_3$.

Technology Exercises

67. Use a graphing utility with an $\boxed{{}_nP_r}$ key to verify your answers in Exercises 1–8.

68. Use a graphing utility with an $\boxed{{}_nC_r}$ key to verify your answers in Exercises 9–16.

Critical Thinking Exercises

69. Which one of the following is true?

a. The number of ways to choose four questions out of ten questions on an essay test is ${}_{10}P_4$.

b. If $r > 1$, ${}_nP_r$ is less than ${}_nC_r$.

c. ${}_7P_3 = 3!{}_7C_3$

d. The number of ways to pick a winner and first runner-up in a piano recital with 20 contestants is ${}_{20}C_2$.

70. Five men and five women line up at a checkout counter in a store. In how many ways can they line up if the first person in line is a woman, and the people in line alternate woman, man, woman, man, and so on?

71. How many four-digit odd numbers less than 6000 can be formed using the digits 2, 4, 6, 7, 8, and 9?

72. If a collection of n objects has n_1 identical objects of the same type, n_2 identical objects of a second kind, n_3 of a third kind, and so on for a total of $n = n_1 + n_2 + \cdots + n_k$ objects, the number of distinguishable permutations of the n objects is given by

$$\frac{n!}{n_1!\,n_2!\,n_3!\cdots n_k!}.$$

Use this formula to find the number of different signals consisting of eight flags that can be made using three white flags, four red flags and one blue flag.

Group Exercise

73. The group should select real-world situations where the Fundamental Counting Principle can be applied. These could involve the number of possible student ID numbers on your campus, the number of possible phone numbers in your community, the number of meal options at a local restaurant, the number of ways a person in the group can select outfits for class, the number of ways a condominium can be purchased in a nearby community, and so on. Once situations have been selected, group members should determine in how many ways each part of the task can be done. Group members will need to obtain menus, find out about telephone-digit requirements in the community, count shirts, pants, shoes in closets, visit condominium sales offices, and so on. Once the group reassembles, apply the Fundamental Counting Principle to determine the number of available options in each situation. Because these numbers may be quite large, use a calculator.

SECTION 8.7 *Probability*

Objectives

1. Compute empirical probability.
2. Compute theoretical probability.
3. Find the probability that an event will not occur.
4. Find the probability of one event or a second event occurring.
5. Find the probability of one event and a second event occurring.

Table 8.3 Number of Americans and the Hours of Sleep They Get on a Typical Night

Hours of Sleep	Number of Americans, in millions
4 or less	11
5	24.75
6	68.75
7	82.5
8	74.25
9	8.25
10 or more	5.5
Total:	275

Source: Discovery Health Media

How many hours of sleep do you typically get each night? Table 8.3 indicates that 11 million out of 275 million Americans are getting four hours or less sleep on a typical night. The *probability* of an American getting four hours or less sleep on a typical night is $\frac{11}{275}$. This fraction can be reduced to $\frac{1}{25}$, or expressed as 0.04 or 4%. Thus, 4% of Americans get four hours or less sleep each night.

We find a probability by dividing one number by another. Probabilities are assigned to an *event*, such as getting four hours or less sleep on a typical night. Events that are certain to occur are assigned probabilities of 1, or 100%. For example, the probability that a given individual will eventually die is 1. Regrettably, taxes and death are always certain! By contrast, if an event cannot occur, its probability is 0. For example, the probability that Elvis will return from the dead and serenade us with one final reprise of "Heartbreak Hotel" is 0.

Probabilities of events are expressed as numbers ranging from 0 to 1, or 0% to 100%. The closer the probability of a given event is to 1, the more likely it is that the event will occur. The closer the probability of a given event is to 0, the less likely it is that the event will occur.

1 Compute empirical probability.

Empirical Probability

Empirical probability applies to situations in which we observe how frequently an event occurs. We use the following formula to compute the empirical probability of an event.

Computing Empirical Probability

The **empirical probability** of event E is

$$P(E) = \frac{\text{observed number of times } E \text{ occurs}}{\text{total number of observed occurrences}}.$$

EXAMPLE 1 Computing Empirical Probability

An American is randomly selected. Use Table 8.3 to find the probability of that person getting eight hours sleep on a typical night.

Solution The probability of getting eight hours sleep is the observed number of Americans who do this, 74.25 million, divided by the total number of Americans, 275 million.

$$P(\text{eight hours sleep}) = \frac{\text{number of Americans who sleep 8 hours}}{\text{total numbers of Americans}}$$
$$= \frac{74.25}{275} = \frac{297}{1100} = 0.27$$

The empirical probability of randomly selecting an American who gets eight hours sleep on a typical night is $\frac{297}{1100}$, or 0.27.

Check Point 1 Use Table 8.3 to find the probability of randomly selecting an American who gets seven hours sleep on a typical night.

2 Compute theoretical probability.

Theoretical Probability

You toss a coin. Although it is equally likely to land either heads up, denoted by H, or tails up, denoted by T, the actual outcome is uncertain. Any occurrence for which the outcome is uncertain is called an **experiment**. Thus, tossing a coin is an example of an experiment. The set of all possible equally likely outcomes of an experiment is the **sample space** of the experiment, denoted by S. The sample space for the coin-tossing experiment is

$$S = \{H, T\}.$$

lands heads up — lands tails up

We can define an event more formally using these concepts. An **event**, denoted by E, is any subcollection, or subset, of a sample space. For example, the subset $E = \{T\}$ is the event of landing tails up when a coin is tossed.

Theoretical probability applies to situations like this, in which the sample space of all equally likely outcomes is known. To calculate the theoretical probability of an event, we divide the number of outcomes in the event by the number of outcomes in the sample space.

Computing Theoretical Probability

If an event E has $n(E)$ equally likely outcomes and its sample space S has $n(S)$ equally likely outcomes, the theoretical probability of event E, denoted by $P(E)$, is

$$P(E) = \frac{\text{number of outcomes in event } E}{\text{number of outcomes in sample space } S} = \frac{n(E)}{n(S)}.$$

The sum of the theoretical probabilities of all possible outcomes in the sample space is 1.

How can we use this formula to compute the probability of a coin landing tails up? We use the following sets:

$$E = \{T\} \qquad S = \{H, T\}$$

This is the event of landing tails up.

This is the sample space with all equally possible outcomes.

Figure 8.7 Outcomes when a die is rolled

The probability of a coin landing tails up is

$$P(E) = \frac{n(E)}{n(S)} = \frac{1}{2}.$$

Theoretical probability applies to many games of chance, including dice rolling, lotteries, card games, and roulette. The next example deals with the experiment of rolling a die. Figure 8.7 illustrates that when a die is rolled, there are six equally likely outcomes. The sample space can be shown as

$$S = \{1, 2, 3, 4, 5, 6\}.$$

EXAMPLE 2 Computing Theoretical Probability

A die is rolled. Find the probability of getting a number less than 5.

Solution The sample space of equally likely outcomes is $S = \{1, 2, 3, 4, 5, 6\}$. There are six outcomes in the sample space, so $n(S) = 6$.

We are interested in the probability of getting a number less than 5. The event of getting a number less than 5 can be represented by

$$E = \{1, 2, 3, 4\}.$$

There are four outcomes in this event, so $n(E) = 4$.

The probability of rolling a number less than 5 is

$$P(E) = \frac{n(E)}{n(S)} = \frac{4}{6} = \frac{2}{3}.$$

Check Point 2 A die is rolled. Find the probability of getting a number greater than 4.

EXAMPLE 3 Computing Theoretical Probability

Two ordinary six-sided dice are rolled. What is the probability of getting a sum of 8?

Solution Each die has six equally likely outcomes. By the Fundamental Counting Principle, there are $6 \cdot 6$, or 36, equally likely outcomes in the sample space. That is, $n(S) = 36$. The 36 outcomes are shown here as ordered pairs. The five ways of rolling a sum of 8 appear in the highlighted diagonal as follows.

First Die \ Second Die	⚀	⚁	⚂	⚃	⚄	⚅
⚀	(1, 1)	(1, 2)	(1, 3)	(1, 4)	(1, 5)	(1, 6)
⚁	(2, 1)	(2, 2)	(2, 3)	(2, 4)	(2, 5)	(2, 6)
⚂	(3, 1)	(3, 2)	(3, 3)	(3, 4)	(3, 5)	(3, 6)
⚃	(4, 1)	(4, 2)	(4, 3)	(4, 4)	(4, 5)	(4, 6)
⚄	(5, 1)	(5, 2)	(5, 3)	(5, 4)	(5, 5)	(5, 6)
⚅	(6, 1)	(6, 2)	(6, 3)	(6, 4)	(6, 5)	(6, 6)

$$S = \{(1,1), (1,2), (1,3), (1,4), (1,5), (1,6), (2,1), (2,2), (2,3), (2,4), (2,5), (2,6), (3,1), (3,2), (3,3), (3,4), (3,5), (3,6), (4,1), (4,2), (4,3), (4,4), (4,5), (4,6), (5,1), (5,2), (5,3), (5,4), (5,5), (5,6), (6,1), (6,2), (6,3), (6,4), (6,5), (6,6)\}$$

The phrase "getting a sum of 8" describes the event

$$E = \{(6, 2), (5, 3), (4, 4), (3, 5), (2, 6)\}.$$

This event has 5 outcomes, so $n(E) = 5$. Thus, the probability of getting a sum of 8 is

$$P(E) = \frac{n(E)}{n(S)} = \frac{5}{36}.$$

Check Point 3 What is the probability of getting a sum of 5 when two six-sided dice are rolled?

Computing Theoretical Probability Without Listing an Event and the Sample Space

In some situations, we can compute theoretical probability without having to write out each event and each sample space. For example, suppose you are dealt one card from a standard 52-card deck, illustrated in Figure 8.8. The deck has four suits: Hearts and diamonds are red, and clubs and spades are black. Each suit has 13 different face values—A(ace), 2, 3, 4, 5, 6, 7, 8, 9, 10, J(jack), Q(queen), and K(king). Jacks, queens, and kings are called picture cards.

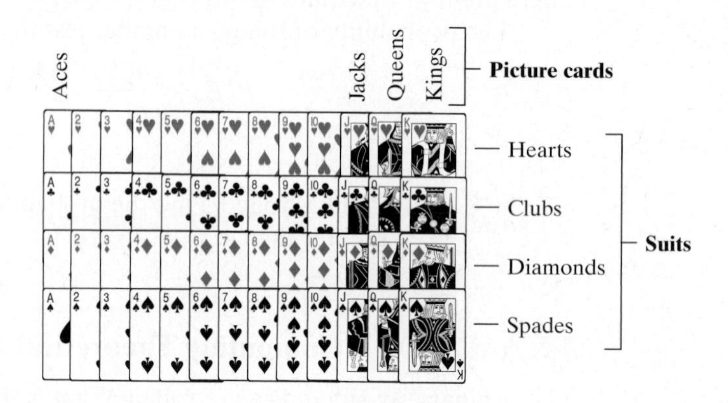

Figure 8.8 A standard 52-card bridge deck

EXAMPLE 4 Probability and a Deck of 52 Cards

You are dealt one card from a standard 52-card deck. Find the probability of being dealt a heart.

Solution Let E be the event of being dealt a heart. Because there are 13 hearts in the deck, the event of being dealt a heart can occur in 13 ways. The number of outcomes in event E is 13: $n(E) = 13$. With 52 cards in the deck, the total number of possible ways of being dealt a single card is 52. The number of outcomes in the sample space is 52: $n(S) = 52$. The probability of being dealt a heart is

$$P(E) = \frac{n(E)}{n(S)} = \frac{13}{52} = \frac{1}{4}.$$

Check Point 4 If you are dealt one card from a standard 52-card deck, find the probability of being dealt a king.

State lotteries keep 50 cents on the dollar, resulting in $10 billion a year for public funding.

If your state has a lottery drawing each week, the probability that someone will win the top prize is relatively high. If there is no winner this week, it is virtually certain that eventually someone will be graced with millions of dollars. So how come you are unlucky compared to this undisclosed someone? In Example 5, we provide an answer to this question, using the counting principles discussed in Section 8.6.

EXAMPLE 5 Probability and Combinations: Winning the Lottery

Florida's lottery game, LOTTO, is set up so that each player chooses six different numbers from 1 to 49. If the six numbers chosen match the six numbers drawn randomly each Saturday evening, the player wins (or shares) the top cash prize. (As of this writing, the top cash prize has ranged from $7 million to $106.5 million.) With one LOTTO ticket, what is the probability of winning this prize?

Solution Because the order of the six numbers does not matter, this is a situation involving combinations. Let E be the event of winning the lottery with one ticket. With one LOTTO ticket, there is only one way of winning. Thus, $n(E) = 1$. The sample space is the set of all possible six-number combinations. We can use the combinations formula

$$_nC_r = \frac{n!}{(n-r)!\,r!}$$

to find the number of outcomes in the sample space. We are selecting $r = 6$ numbers from a collection of $n = 49$ numbers.

$$_{49}C_6 = \frac{49!}{(49-6)!\,6!} = \frac{49!}{43!\,6!} = \frac{49 \cdot 48 \cdot 47 \cdot 46 \cdot 45 \cdot 44 \cdot \cancel{43!}}{\cancel{43!} \cdot 6 \cdot 5 \cdot 4 \cdot 3 \cdot 2 \cdot 1} = 13{,}983{,}816$$

There are nearly 14 million number combinations possible in LOTTO. If a person buys one LOTTO ticket, the probability of winning is

$$P(E) = \frac{n(E)}{n(S)} = \frac{1}{13{,}983{,}816} \approx 0.0000000715.$$

The probability of winning the top prize with one LOTTO ticket is $\frac{1}{13{,}983{,}816}$, or about 1 in 14 million.

In 1997, Americans spent nearly 17 billion dollars on lotteries set up by revenue-hungry states. If a pigeon, er, person, buys, say 5000 different tickets in Florida's LOTTO, that person has selected 5000 different combinations of the six numbers. The probability of winning is

$$\frac{5000}{13{,}983{,}816} \approx 0.000358.$$

The chances of winning top prize are about 358 in a million. At $1 per LOTTO ticket, it is highly probable that Mr. or Ms. Pigeon will be $5000 poorer.

Check Point 5 In a state lottery, a player chooses five different numbers from 1 to 30. If the five numbers chosen match the five numbers drawn each week, the player wins (or shares) the top cash prize. With one lottery ticket, what is the probability of winning this prize?

Surprising Probabilities

Imagine that one person is randomly selected from all 6 billion people on planet Earth. The following empirical probabilities, each rounded to two decimal places, might surprise you.

Probability of selecting
- a woman = 0.51
- a non-white = 0.7
- a non-Christian = 0.7
- a person who cannot read = 0.7
- a person suffering from malnutrition = 0.5
- a person with a college education = 0.01
- a person who is near death = 0.01

When viewing our world from the perspective of these probabilities, the need for both tolerance and understanding becomes apparent.

Source: United Nations

3 Find the probability that an event will not occur.

Probability of an Event Not Occurring

A survey (*source*: Penn, Schoen, and Berland, 1999) asked 500 Americans to rate their health. Of those surveyed, 270 rated their health as good/excellent. This means that 500 − 270, or 230, people surveyed did not rate their health as good/excellent. Notice that

$$P(\text{good/excellent}) + P(\text{not good/excellent}) = \frac{270}{500} + \frac{230}{500} = \frac{500}{500} = 1.$$

In general, because the sum of the probabilities of all possible outcomes in any situation is 1,

$$P(E) + P(\text{not } E) = 1.$$

We now solve this equation for $P(\text{not } E)$, the probability that event E will not occur, by subtracting $P(E)$ from both sides. The resulting formula is given in the following box.

> **The Probability of an Event Not Occurring**
>
> The probability that an event E will not occur is equal to one minus the probability that it will occur.
>
> $$P(\text{not } E) = 1 - P(E)$$

EXAMPLE 6 The Probability of Not Winning the Lottery

We have seen that the probability of winning Florida's LOTTO with one ticket is $\frac{1}{13{,}983{,}816}$. What is the probability of not winning?

Solution

$$P(\text{not winning}) = 1 - P(\text{winning})$$

$$= 1 - \frac{1}{13{,}983{,}816} = \frac{13{,}983{,}816}{13{,}983{,}816} - \frac{1}{13{,}983{,}816}$$

$$= \frac{13{,}983{,}815}{13{,}983{,}816} \approx 0.9999999$$

The probability of not winning is close to 1. It is almost certain that with one LOTTO ticket, a person will not win top prize.

Check Point 6 With one lottery ticket, what is the probability of not winning the lottery described in Checkpoint 5?

4 Find the probability of one event or a second event occurring.

Or Probabilities with Mutually Exclusive Events

Suppose that you randomly select one card from a deck of 52 cards. Let A be the event of selecting a king and B be the event of selecting a queen. Only one card is selected, so it is impossible to get both a king and a queen. The outcomes of selecting a king and a queen cannot occur simultaneously. They are called *mutually*

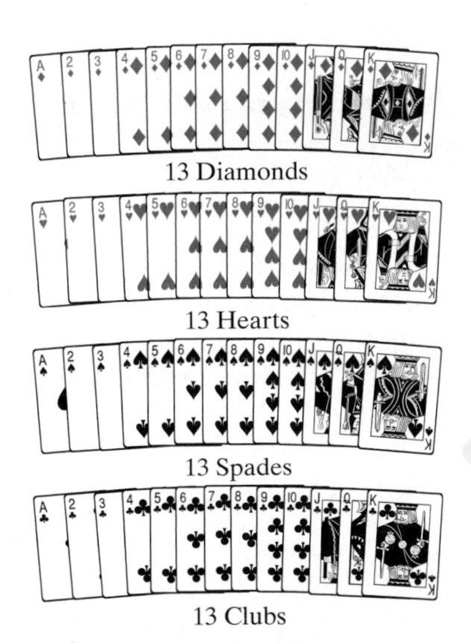

Figure 8.9 A deck of 52 cards

exclusive events. If it is impossible for any two events, A and B, to occur simultaneously, they are said to be **mutually exclusive**. If A and B are mutually exclusive events, the probability that either A or B will occur is determined by adding their individual probabilities.

_Or_ Probabilities with Mutually Exclusive Events

If A and B are mutually exclusive events, then

$$P(A \text{ or } B) = P(A) + P(B).$$

EXAMPLE 7 The Probability of Either of Two Mutually Exclusive Events Occurring

If one card is randomly selected from a deck of cards, what is the probability of selecting a king or a queen?

Solution We find the probability that either of these mutually exclusive events will occur by adding their individual probabilities.

$$P(\text{king or queen}) = P(\text{king}) + P(\text{queen}) = \frac{4}{52} + \frac{4}{52} = \frac{8}{52} = \frac{2}{13}$$

The probability of selecting a king or a queen is $\frac{2}{13}$.

Check Point 7 If you roll a single, six-sided die, what is the probability of getting either a 4 or a 5?

_Or_ Probabilities with Events That Are Not Mutually Exclusive

Figure 8.10 Three diamonds are picture cards

Consider the deck of 52 cards shown in Figure 8.9. Suppose that these cards are shuffled and you randomly select one card from the deck. What is the probability of selecting a diamond or a picture card (jack, queen, king)? Begin by adding their individual probabilities:

$$P(\text{diamond}) + P(\text{picture card}) = \frac{13}{52} + \frac{12}{52}.$$

There are 13 diamonds in the deck of 52 cards.

There are 12 picture cards in the deck of 52 cards.

However, this is not the probability of selecting a diamond or a picture card. The problem is that there are three cards that are simultaneously diamonds and picture cards, shown in Figure 8.10. The events of selecting a diamond and selecting a picture card are not mutually exclusive. It is possible to select a card that is both a diamond and a picture card.

The situation is illustrated in the diagram in Figure 8.11. Why can't we find the probability of selecting a diamond or a picture card by adding their individual probabilities? The diagram shows that three of the cards, the three diamonds that are picture cards, get counted twice when we add the individual probabilities. First the three cards get counted as diamonds, and then they get counted as picture cards. In order to avoid the error of counting the three cards twice, we need to subtract the probability of getting a diamond and a picture card, $\frac{3}{52}$, as follows:

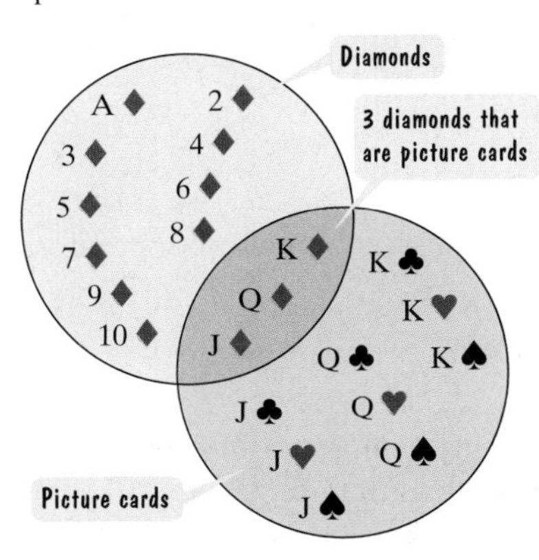

Figure 8.11

$$P(\text{diamond or picture card})$$
$$= P(\text{diamond}) + P(\text{picture card}) - P(\text{diamond and picture card})$$
$$= \frac{13}{52} + \frac{12}{52} - \frac{3}{52} = \frac{13 + 12 - 3}{52} = \frac{22}{52} = \frac{11}{26}.$$

Thus, the probability of selecting a diamond or a picture card is $\frac{11}{26}$.

In general, if A and B are events that are not mutually exclusive, the probability that A or B will occur is determined by adding their individual probabilities and then subtracting the probability that A and B occur simultaneously.

_Or_ Probabilities with Events That Are Not Mutually Exclusive

If A and B are not mutually exclusive events, then

$$P(A \text{ or } B) = P(A) + P(B) - P(A \text{ and } B).$$

EXAMPLE 8 An _Or_ Probability with Events That Are Not Mutually Exclusive

Figure 8.12 It is equally probable that the pointer will land on any one of the eight regions.

Figure 8.12 illustrates a spinner. It is equally probable that the pointer will land on any one of the eight regions, numbered 1 through 8. If the pointer lands on a borderline, spin again. Find the probability that the pointer will stop on an even number or a number greater than 5.

Solution It is possible for the pointer to land on a number that is even and greater than 5. Two of the numbers, 6 and 8, are even and greater than 5. These events are not mutually exclusive. The probability of landing on a number that is even and greater than 5 is

$$P\begin{pmatrix}\text{even or}\\ \text{greater than 5}\end{pmatrix} = P(\text{even}) + P(\text{greater than 5}) - P\begin{pmatrix}\text{even and}\\ \text{greater than 5}\end{pmatrix}$$
$$= \frac{4}{8} + \frac{3}{8} - \frac{2}{8}$$

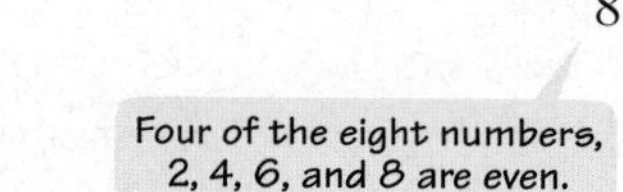

Three of the eight numbers, 6, 7, and 8 are greater than 5.

Two of the eight numbers, 6 and 8, are even and greater than 5.

$$= \frac{4 + 3 - 2}{8} = \frac{5}{8}.$$

The probability that the pointer will stop on an even number or a number greater than 5 is $\frac{5}{8}$.

Use Figure 8.12 to find the probability that the pointer will stop on an odd number or a number less than 5.

EXAMPLE 9 An *Or* Probability with Events That Are Not Mutually Exclusive

A group of people is comprised of 15 U.S. men, 20 U.S. women, 10 Canadian men, and 5 Canadian women. If a person is selected at random from the group, find the probability that the selected person is a man or a Canadian.

Solution The group is comprised of $15 + 20 + 10 + 5$, or 50 people. It is possible to select a man who is Canadian. We are given that there are 10 Canadian men, so these events are not mutually exclusive.

$$P(\text{man or Canadian}) = P(\text{man}) + P(\text{Canadian}) - P(\text{man and Canadian})$$

$$= \frac{25}{50} + \frac{15}{50} - \frac{10}{50}$$

Of the 50 people, 25 are men–15 U.S. men and 10 Canadian men.

Of the 50 people, 15 are Canadian–10 Canadian men and 5 Canadian women.

Of the 50 people, 10 are Canadian men.

$$= \frac{25 + 15 - 10}{50} = \frac{30}{50} = \frac{3}{5}$$

The probability of selecting a man or a Canadian is $\frac{3}{5}$.

Check Point 9 In a group of 25 baboons, 18 enjoy picking fleas off their neighbors, 16 enjoy screeching wildly, while 10 enjoy picking fleas off their neighbors and screeching wildly. If one baboon is selected at random from the group, find the probability that it enjoys picking fleas off its neighbors or screeching wildly.

5 Find the probability of one event and a second event occurring.

And Probabilities with Independent Events

Suppose that you toss a fair coin two times in succession. The outcome of the first toss, heads or tails, does not affect what happens when you toss the coin a second time. For example, the occurrence of tails on the first toss does not make tails more likely or less likely to occur on the second toss. The repeated toss of a coin produces **independent events** because the outcome of one toss does not affect the outcome of others. Two events are *independent* if the occurrence of either of them has no effect on the probability of the other.

If two events are independent, we can calculate the probability of the first occurring and the second occurring by multiplying their probabilities.

Figure 8.13 A U.S. roulette wheel

And Probabilities with Independent Events

If A and B are independent events, then

$$P(A \text{ and } B) = P(A) \cdot P(B).$$

EXAMPLE 10 Independent Events on a Roulette Wheel

Figure 8.13 shows a U.S. roulette wheel that has 38 numbered slots (1 through 36, 0, and 00). Of the 38 compartments, 18 are black, 18 are red, and 2 are green.

Each play consists of spinning the wheel and a small ball in opposite directions. As the ball slows to a stop, it can land with equal probability on any one of the 38 numbered slots. Find the probability of red occurring on two consecutive plays.

Solution The wheel has 38 equally likely outcomes and 18 are red. Thus, the probability of red occurring on a play is $\frac{18}{38}$, or $\frac{9}{19}$. The result that occurs on each play is independent of all previous results. Thus,

$$P(\text{red and red}) = P(\text{red}) \cdot P(\text{red}) = \frac{9}{19} \cdot \frac{9}{19} = \frac{81}{361} \approx 0.224.$$

The probability of red occurring on two consecutive plays is $\frac{81}{361}$.

Some roulette players incorrectly believe that if red occurs on two consecutive plays, then another color is "due." Because the events are independent, the outcomes of previous spins have no effect on any other spins.

Check Point 10 Find the probability of green occurring on two consecutive plays on a roulette wheel.

The *and* rule for independent events can be extended to cover three or more events. Thus, if A, B, and C are independent events, then

$$P(A \text{ and } B \text{ and } C) = P(A) \cdot P(B) \cdot P(C).$$

EXAMPLE 11 Independent Events in a Family

The picture in the margin shows a family that has had nine girls in a row. Find the probability of this occurrence.

Solution If two or more events are independent, we can find the probability of them all occurring by multiplying the probabilities. The probability of a baby girl is $\frac{1}{2}$, so the probability of nine girls in a row is $\frac{1}{2}$ used as a factor nine times.

$$P(\text{nine girls in a row}) = \frac{1}{2} \cdot \frac{1}{2} \cdot \frac{1}{2} \cdot \frac{1}{2} \cdot \frac{1}{2} \cdot \frac{1}{2} \cdot \frac{1}{2} \cdot \frac{1}{2} \cdot \frac{1}{2}$$

$$= \left(\frac{1}{2}\right)^9 = \frac{1}{512}$$

The probability of a run of nine girls in a row is $\frac{1}{512}$. (If another child is born into the family, this event is independent of the other nine, and the probability of a girl is still $\frac{1}{2}$.)

Check Point 11 Find the probability of a family having four boys in a row.

EXERCISE SET 8.7

Practice and Application Exercises

Exercises 1–4 involve empirical probability. Use the empirical probability formula to solve each exercise. Express answers as fractions. Then use a calculator to express probabilities as decimals, rounded to the nearest thousandth.

Use the table showing U.S. family size to solve Exercises 1–2.

U.S. Families (includes only a householder and his/her relatives) by Size, 1997

Total: 70,241,000 Families	
Size	**Number of Families**
2 people	29,780,000
3 people	16,239,000
4 people	14,602,000
5 people	6,326,000
6 people	2,108,000
7 people or more	1,186,000

Source: U.S. Bureau of the Census

Find the probability that a U.S. family has:

1. 2 people.

2. 3 people.

Use the table showing world population for selected regions to solve Exercises 3–4.

Populations of Selected Regions of the World

Total World Population: 5926 million	
Region	**Population in millions**
Africa	761
Near East	165
Asia	3363
Latin America	508
Europe	799
North America	301

Source: U.S. Bureau of the Census

If one person is randomly selected from all people on planet Earth, find the probability of selecting a person from:

3. Africa.

4. North America.

Exercises 5–20 involve theoretical probability. Use the theoretical probability formula to solve each exercise. Express each probability as a fraction reduced to lowest terms.

In Exercises 5–10, a die is rolled. The sample space of equally likely outcomes is $\{1, 2, 3, 4, 5, 6\}$. *Find the probability of getting:*

5. a 4.

6. a 5.

7. an odd number.

8. a number greater than 3.

9. a number greater than 4.

10. a number greater than 7.

In Exercises 11–14, you are dealt one card from a standard 52 card deck. Find the probability of being dealt:

11. a queen.

12. a diamond.

13. a picture card.

14. a card greater than 3 and less than 7.

In Exercises 15–16, a fair coin is tossed two times in succession. The sample space of equally likely outcomes is $\{HH, HT, TH, TT\}$. *Find the probability of getting:*

15. two heads.

16. the same outcome on each toss.

In Exercises 17–18, you select a family with three children. If M represents a male child and F a female child, the sample space of equally likely outcomes is $\{MMM, MMF, MFM, MFF, FMM, FMF, FFM, FFF\}$. *Find the probability of selecting a family with:*

17. at least one male child.

18. at least two female children.

In Exercises 19–20, a single die is rolled twice. The 36 equally likely outcomes are shown as follows:

First Roll \ *Second Roll*	1	2	3	4	5	6
1	(1, 1)	(1, 2)	(1, 3)	(1, 4)	(1, 5)	(1, 6)
2	(2, 1)	(2, 2)	(2, 3)	(2, 4)	(2, 5)	(2, 6)
3	(3, 1)	(3, 2)	(3, 3)	(3, 4)	(3, 5)	(3, 6)
4	(4, 1)	(4, 2)	(4, 3)	(4, 4)	(4, 5)	(4, 6)
5	(5, 1)	(5, 2)	(5, 3)	(5, 4)	(5, 5)	(5, 6)
6	(6, 1)	(6, 2)	(6, 3)	(6, 4)	(6, 5)	(6, 6)

Find the probability of getting:

19. two numbers whose sum is 4.

20. two numbers whose sum is 6.

21. To play the California lottery, a person has to correctly select 6 out of 51 numbers, paying $1 for each six-number selection. If you pick six numbers that are the same as the ones drawn by the lottery, you win mountains of money. What is the probability that a person with one combination of six numbers will win? What is the probability of winning if 100 different lottery tickets are purchased?

22. A state lottery is designed so that a player chooses six numbers from 1 to 30 on one lottery ticket. What is the probability that a player with one lottery ticket will win? What is the probability of winning if 100 different lottery tickets are purchased?

23. A poker hand consists of five cards.
   a. Find the total number of possible five-card poker hands that can be dealt from a deck of 52 cards.
   b. A diamond flush consists of a five-card hand containing all diamonds. Find the number of possible five-card diamond flushes.
   c. Find the probability of being dealt a diamond flush.

24. A committee of five people is to be formed from six lawyers and seven teachers. Find the probability that all are lawyers.

Use these figures for the U.S. population in 2000 to answer Exercises 25–30.

Total U.S. Population: 274,634 Thousand

Age	under 5	5–13	14–17	18–24	25–34	35–44	45–64	65–84	85 and older
Population (in thousands)	18,987	36,043	15,752	26,258	37,233	44,659	60,992	30,378	4332

Source: U.S. Bureau of the Census

If a U.S. citizen is chosen at random, find the probability that this person is not:

25. under 5.

26. in the 18–24 age group.

27. in the 25–34 age group.

28. 85 and older.

Exercises 29–32 involve or *probabilities with mutually exclusive events.*

29. If a U.S. citizen is chosen at random, find the probability that this person is in the 14–17 or 18–24 age group.

30. If a U.S. citizen is chosen at random, find the probability that this person is in the 25–34 or 35–44 age group.

If one card is randomly selected from a 52-card deck of cards, find the probability of selecting:

31. a 2 or a 3.

32. a red 7 or a black 8.

Exercises 33–40 involve or *probabilities with events that are not mutually exclusive.*

In Exercises 33–34, a single die is rolled. Find the probability of getting:

33. an even number or a number less than 5.

34. an odd number or a number less than 4.

In Exercises 35–36, you are dealt one card from a 52-card deck. Find the probability that you are dealt:

35. a 7 or a red card.

36. a 5 or a black card.

In Exercises 37–38, it is equally probable that the pointer on the spinner shown will land on any one of the eight regions, numbered 1 through 8. If the pointer lands on a borderline, spin again.

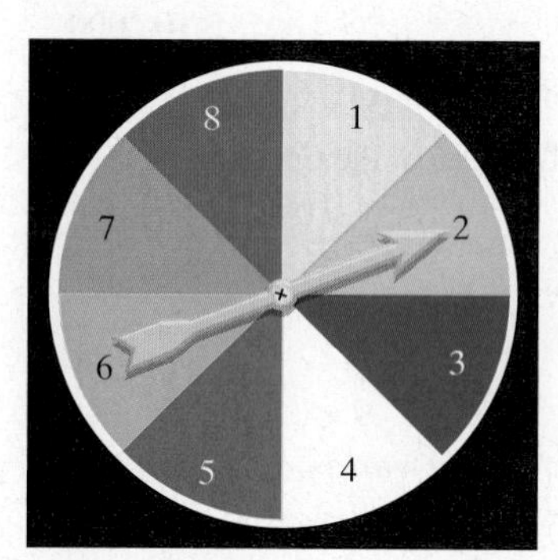

Find the probability that the pointer will stop on:

37. an odd number or a number less than 6.

38. an odd number or a number greater than 3.

Use this information to solve Exercises 39–40. The mathematics department of a college has 8 male professors, 11 female professors, 14 male teaching assistants, and 7 female teaching assistants. If a person is selected at random from the group, find the probability that the selected person is:

39. a professor or a male.

40. a professor or a female.

Exercises 41–46 involve and *probabilities with independent events.*

In Exercises 41–44, a single die is rolled twice. Find the probability of getting:

41. a 2 the first time and a 3 the second time.

42. a 5 the first time and a 1 the second time.

43. an even number the first time and a number greater than 2 the second time.

44. an odd number the first time and a number less than 3 the second time.

45. If you toss a fair coin six times, what is the probability of getting all heads?

46. If you toss a fair coin seven times, what is the probability of getting all tails?

47. The probability that South Florida will be hit by a major hurricane (category 4 or 5) in any single year is $\frac{1}{16}$. (*Source*: National Hurricane Center)

a. What is the probability that South Florida will be hit by a major hurricane two years in a row?

b. What is the probability that South Florida will be hit by a major hurricane in three consecutive years?

c. What is the probability that South Florida will not be hit by a major hurricane in the next ten years?

d. What is the probability that South Florida will be hit by a major hurricane at least once in the next ten years?

Writing in Mathematics

48. Describe the difference between theoretical probability and empirical probability.

49. Give an example of an event whose probability must be determined empirically rather than theoretically.

50. Write a probability word problem whose answer is one of the following fractions: $\frac{1}{6}$ or $\frac{1}{4}$ or $\frac{1}{3}$.

51. Explain how to find the probability of an event not occurring. Give an example.

52. What are mutually exclusive events? Give an example of two events that are mutually exclusive.

53. Explain how to find *or* probabilities with mutually exclusive events. Give an example.

54. Give an example of two events that are not mutually exclusive.

55. Explain how to find *or* probabilities with events that are not mutually exclusive. Give an example.

56. Explain how to find *and* probabilities with independent events. Give an example.

57. The president of a large company with 10,000 employees is considering mandatory cocaine testing for every employee. The test that would be used is 90% accurate, meaning that it will detect 90% of the cocaine users who are tested, and that 90% of the nonusers will test negative. This also means that the test gives 10% false positive. Suppose that 1% of the employees actually use cocaine. Find the probability that someone who tests positive for cocaine use is, indeed, a user. (See the hint at the top of the next column.)

Hint: Find the following probability fraction:

$$\frac{\text{the number of employees who test positive and are cocaine users}}{\text{the number of employees who test positive}}$$

$$= \frac{90\% \text{ of } 1\% \text{ of } 10{,}000}{\text{the number who test positive who actually use cocaine plus the number who test positive who do not use cocaine.}}$$

What does this probability indicate in terms of the percentage of employees who test positive who are not actually users? Discuss these numbers in terms of the issue of mandatory drug testing. Write a paper either in favor of or against mandatory drug testing, incorporating the actual percentage accuracy for such tests.

Critical Thinking Exercises

58. The target in the figure shown contains four squares. If a dart thrown at random hits the target, find the probability that it will land in a colored region.

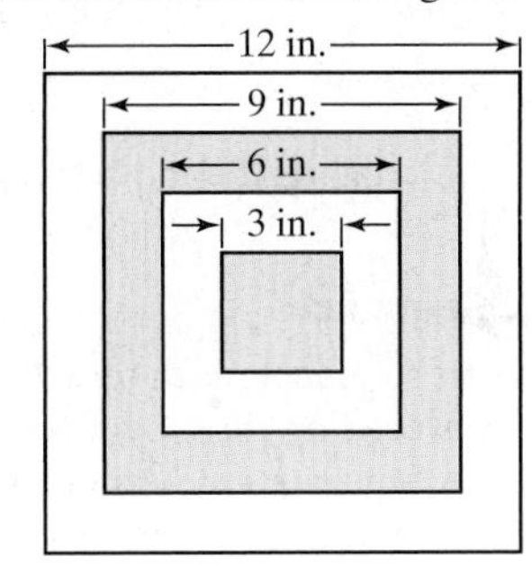

59. Suppose that it is a week in which the cash prize in Florida's LOTTO is promised to exceed $50 million. If a person purchases 13,983,816 tickets in LOTTO at $1 per ticket (all possible combinations), isn't this a guarantee of winning the lottery? Because the probability in this situation is 1, what's wrong with doing this?

60. a. If two people are selected at random, the probability that they do not have the same birthday (day and month) is $\frac{365}{365} \cdot \frac{364}{365}$. Explain why this is so. (Ignore leap years and assume 365 days in a year.)

b. If three people are selected at random, find the probability that they all have different birthdays.

c. If three people are selected at random, find the probability that at least two of them have the same birthday.

d. If 20 people are selected at random, find the probability that at least 2 of them have the same birthday.

e. How large a group is needed to give a 0.5 chance of at least two people having the same birthday?

Group Exercise

61. Research and present a group report on state lotteries. Include answers to some or all of the following questions: Which states do not have lotteries? Why not? How much is spent per capita on lotteries? What are some of the lottery games? What is the probability of winning top prize in these games? What income groups spend the greatest amount of money on lotteries? If your state has a lottery, what does it do with the money it makes? Is the way the money is spent what was promised when the lottery first began?

CHAPTER SUMMARY, REVIEW, AND TEST

Summary

8.1 Sequences and Summation Notation

a. An infinite sequence $\{a_n\}$ is a function whose domain is the set of positive integers. The function values, or terms, are represented by

$$a_1, a_2, a_3, a_4, \ldots, a_n, \ldots.$$

b. Sequences can be defined using recursion formulas that define the nth term as a function of the previous term.

c. Factorial Notation:

$$n! = n(n-1)(n-2)\cdots(3)(2)(1) \quad \text{and} \quad 0! = 1$$

d. Summation Notation:

$$\sum_{i=1}^{n} a_i = a_1 + a_2 + a_3 + a_4 + \cdots + a_n$$

8.2 Arithmetic Sequences

a. In an arithmetic sequence, each term after the first differs from the preceding term by a constant, the common difference. Subtract any term from the term that directly follows to find the common difference.

b. General term or nth term: $a_n = a_1 + (n-1)d$. The first term is a_1 and the common difference is d.

c. Sum of the first n terms: $S_n = \frac{n}{2}(a_1 + a_n)$

8.3 Geometric Sequences

a. In a geometric sequence, each term after the first is obtained by multiplying the preceding term by a nonzero constant, the common ratio. Divide any term after the first by the term that directly precedes it to find the common ratio.

b. General term or nth term: $a_n = a_1 r^{n-1}$. The first term is a_1 and the common ratio is r.

c. Sum of the first n terms: $S_n = \frac{a_1(1 - r^n)}{1 - r}, \quad r \neq 1$

d. An annuity is a sequence of equal payments made at equal time periods. The value of an annuity, A, is the sum of all deposits made plus all interest paid, given by

$$A = P\frac{\left(1 + \frac{r}{n}\right)^{nt} - 1}{\frac{r}{n}}.$$

The deposit made at the end of each period is P, the annual interest rate is r compounded n times per year, and t is the number of years deposits have been made.

e. Sum of the infinite geometric series $a_1 + a_1 r + a_1 r^2 + a_1 r^3 + \cdots$ is $S = \frac{a_1}{1 - r}$; $|r| < 1$. If $|r| \geq 1$, the infinite series does not have a sum.

8.4 Mathematical Induction

To prove that S_n is true for all positive integers n:

a. Show that S_1 is true.

b. Show that if S_k is assumed true, then S_{k+1} is also true, for every positive integer k.

8.5 The Binomial Theorem

a. Binomial coefficient: $\binom{n}{r} = \frac{n!}{r!\,(n-r)!}$

b. Binomial Theorem: $(a + b)^n = \binom{n}{0}a^n + \binom{n}{1}a^{n-1}b + \binom{n}{2}a^{n-2}b^2 + \cdots + \binom{n}{n}b^n$

c. The rth term in a binomial expansion:

$$\binom{n}{r-1}a^{n-r+1}b^{r-1}$$

8.6 Counting Principles, Permutations, and Combinations

a. The Fundamental Counting Principle: The number of ways in which a series of successive things can occur is found by multiplying the number of ways in which each thing can occur.

b. A permutation from a group of items occurs when no item is used more than once and the order of arrangement makes a difference.

c. Permutations Formula: The number of possible permutations if r items are taken from n items is

$${}_nP_r = \frac{n!}{(n-r)!}.$$

d. A combination from a group of items occurs when no item is used more than once and the order of items makes no difference.

e. Combinations Formula: The number of possible combinations if r items are taken from n items is

$${}_nC_r = \frac{n!}{(n-r)!r!}.$$

8.7 Probability

a. Empirical probability applies to situations in which we observe the frequency of occurrence of an event. The empirical probability of event E is

$$P(E) = \frac{\text{observed number of times } E \text{ occurs}}{\text{total number of observed occurrences}}.$$

b. Theoretical probability applies to situations in which the sample space of all equally likely outcomes is known. The theoretical probability of event E is

$$P(E) = \frac{\text{number of outcomes in event } E}{\text{number of outcomes in sample space } S} = \frac{n(E)}{n(S)}.$$

c. Probability of an event not occurring: $P(\text{not } E) = 1 - P(E)$.

d. If it is impossible for events A and B to occur simultaneously, the events are mutually exclusive.

e. If A and B are mutually exclusive events, then $P(A \text{ or } B) = P(A) + P(B)$.

f. If A and B are not mutually exclusive events, then $P(A \text{ or } B) = P(A) + P(B) - P(A \text{ and } B)$.

g. Two events are independent if the occurrence of either of them has no effect on the probability of the other.

h. If A and B are independent events, then

$$P(A \text{ and } B) = P(A) \cdot P(B).$$

i. The probability of a succession of independent events is the product of each of their probabilities.

Review Exercises

8.1

In Exercises 1–6, write the first four terms of each sequence whose general term is given.

1. $a_n = 7n - 4$

2. $a_n = (-1)^n \frac{n+2}{n+1}$

3. $a_n = \frac{1}{(n-1)!}$

4. $a_n = \frac{(-1)^{n+1}}{2^n}$

5. $a_1 = 9$ and $a_n = \frac{2}{3a_{n-1}}$

6. $a_1 = 4$ and $a_n = 2a_{n-1} + 3$

7. Evaluate: $\frac{40!}{4!38!}$.

In Exercises 8–9, find each indicated sum.

8. $\sum_{i=1}^{5} (2i^2 - 3)$

9. $\sum_{i=0}^{4} (-1)^{i+1} i!$

In Exercises 10–11, express each sum using summation notation. Use i for the index of summation.

10. $\frac{1}{3} + \frac{2}{4} + \frac{3}{5} + \cdots + \frac{15}{17}$

11. $4^3 + 5^3 + 6^3 + \cdots + 13^3$

8.2

In Exercises 12–15, write the first six terms of each arithmetic sequence.

12. $a_1 = 7, d = 4$

13. $a_1 = -4, d = -5$

14. $a_1 = \frac{3}{2}, d = -\frac{1}{2}$

15. $a_{n+1} = a_n + 5, a_1 = -2$

In Exercises 16–18, find the indicated term of the arithmetic sequence with first term, a_1, and common difference, d.

16. Find a_6 when $a_1 = 5, d = 3$.

17. Find a_{12} when $a_1 = -8, d = -2$.

18. Find a_{14} when $a_1 = 14, d = -4$.

In Exercises 19–21, write a formula for the general term (the nth term) of each arithmetic sequence. Do not use a recursion formula. Then use the formula for a_n to find a_{20}, the 20th term of the sequence.

19. $-7, -3, 1, 5, \ldots$

20. $a_1 = 200, d = -20$

21. $a_n = a_{n-1} - 5, a_1 = 3$

22. Find the sum of the first 22 terms of the arithmetic sequence: $5, 12, 19, 26, \ldots$.

23. Find the sum of the first 15 terms of the arithmetic sequence: $-6, -3, 0, 3, \ldots$.

24. Find $3 + 6 + 9 + \cdots + 300$, the sum of the first 100 positive multiples of 3.

In Exercises 25–27, use the formula for the sum of the first n terms of an arithmetic sequence to find the indicated sum.

25. $\sum_{i=1}^{16} (3i + 2)$

26. $\sum_{i=1}^{25} (-2i + 6)$

27. $\sum_{i=1}^{30} (-5i)$

28. In 1911, the world record for the men's mile run was 1043.04 seconds. The world record has decreased by approximately 0.4118 second each year since then.

a. Write the general term for the arithmetic sequence modeling record times for the men's mile run n years after 1910.

b. Use the model to predict the record time for the men's mile run for the year 2010.

29. A company offers a starting salary of $31,500 with raises of $2300 per year. Find the total salary over a ten-year period.

30. A theater has 25 seats in the first row and 35 rows in all. Each successive row contains one additional seat. How many seats are in the theater?

8.3

In Exercises 31–34, write the first five terms of each geometric sequence.

31. $a_1 = 3, r = 2$ **32.** $a_1 = \frac{1}{2}, r = \frac{1}{2}$

33. $a_1 = 16, r = -\frac{1}{2}$ **34.** $a_n = -5a_{n-1}, a_1 = -1$

In Exercises 35–37, find the indicated term of the geometric sequence with first term, a_1, and common ratio, r.

35. Find a_7 when $a_1 = 2, r = 3$.

36. Find a_6 when $a_1 = 16, r = \frac{1}{2}$.

37. Find a_5 when $a_1 = -3, r = 2$.

In Exercises 38–40, write a formula for the general term (the nth term) of each geometric sequence. Then use the formula for a_n to find a_8, the eighth term of the sequence.

38. $1, 2, 4, 8, \ldots$ **39.** $100, 10, 1, \frac{1}{10}, \ldots$

40. $12, -4, \frac{4}{3}, -\frac{4}{9}, \ldots$

41. Find the sum of the first 15 terms of the geometric sequence: $5, -15, 45, -135, \ldots$.

42. Find the sum of the first 7 terms of the geometric sequence: $\frac{1}{3}, \frac{1}{9}, \frac{1}{27}, \frac{1}{81}, \ldots$.

In Exercises 43–45, use the formula for the sum of the first n terms of a geometric sequence to find the indicated sum.

43. $\sum_{i=1}^{6} 5^i$ **44.** $\sum_{i=1}^{7} 3(-2)^i$

45. $\sum_{i=1}^{5} 2\left(\frac{1}{4}\right)^{i-1}$

In Exercises 46–49, find the sum of each infinite geometric series.

46. $9 + 3 + 1 + \frac{1}{3} + \cdots$ **47.** $2 - 1 + \frac{1}{2} - \frac{1}{4} + \cdots$

48. $-6 + 4 - \frac{8}{3} + \frac{16}{9} - \cdots$ **49.** $\sum_{i=1}^{\infty} 5(0.8)^i$

In Exercises 50–51, express each repeating decimal as a fraction in lowest terms.

50. $0.\overline{6}$ **51.** $0.\overline{47}$

52. A job pays $32,000 for the first year with an annual increase of 6% per year beginning in the second year. What is the salary in the sixth year? What is the total salary paid over this six-year period?

53. You decide to deposit $200 at the end of each month into an account paying 10% interest compounded monthly to save for your child's education. How much will you save over 18 years?

54. A factory in an isolated town has an annual payroll of $4 million. It is estimated that 70% of this money is spent within the town, that people in the town receiving this money will again spend 70% of what they receive in the town, and so on. What is the total of all this spending in the town each year?

8.4

In Exercises 55–59, use mathematical induction to prove that each statement is true for every positive integer n.

55. $5 + 10 + 15 + \cdots + 5n = \frac{5n(n+1)}{2}$

56. $1 + 4 + 4^2 + \cdots + 4^{n-1} = \frac{4^n - 1}{3}$

57. $2 + 6 + 10 + \cdots + (4n - 2) = 2n^2$

58. $1 \cdot 3 + 2 \cdot 4 + 3 \cdot 5 + \cdots + n(n+2) = \frac{n(n+1)(2n+7)}{6}$

59. 2 is a factor of $n^2 + 5n$.

8.5

In Exercises 60–61, evaluate the given binomial coefficient.

60. $\binom{11}{8}$ **61.** $\binom{90}{2}$

In Exercises 62–65, use the Binomial Theorem to expand each binomial and express the result in simplified form.

62. $(2x + 1)^3$ **63.** $(x^2 - 1)^4$

64. $(x + 2y)^5$ **65.** $(x - 2)^6$

In Exercises 66–67, write the first three terms in each binomial expansion, expressing the result in simplified form.

66. $(x^2 + 3)^8$ **67.** $(x - 3)^9$

In Exercises 68–69, find the term indicated in each expansion.

68. $(x + 2)^5$; fourth term **69.** $(2x - 3)^6$; fifth term

8.6

In Exercises 70–73, evaluate each expression.

70. ${}_8P_3$ **71.** ${}_9P_5$

72. ${}_8C_3$ **73.** ${}_{13}C_{11}$

In Exercises 74–80, solve by the method of your choice.

74. A popular brand of pen comes in red, green, blue, or black ink. The writing tip can be chosen from extra bold, bold, regular, fine, or micro. How many different choices of pens do you have with this brand?

75. A stock can go up, go down, or stay unchanged. How many possibilities are there if you own five stocks?

76. A club with 15 members is to choose four officers—president, vice-president, secretary, and treasurer. In how many ways can these offices be filled?

77. How many different ways can a director select 4 actors from a group of 20 actors to attend a workshop on performing in rock musicals?

78. From the 20 CDs that you've bought during the past year, you plan to take 3 with you on vacation. How many different sets of three CDs can you take?

79. How many different ways can a director select from 20 male actors and cast the roles of Mark, Roger, Angel, and Collins in the musical *Rent*?

80. In how many ways can five airplanes line up for departure on a runway?

8.7

Exercises 81–82 involve empirical probabilities. Express each probability as a fraction. Then use a calculator to express the probability in decimal form, rounded to the nearest thousandth. The table shows the two states with the largest Hispanic populations. Find the probability that:

81. a person randomly selected from California is Hispanic.

82. a person randomly selected from Texas is Hispanic.

Largest Hispanic Population, 1997

State	Total Population	Hispanic Population
California	31,878,234	9,630,188
Texas	19,128,261	5,503,372

Source: Bureau of the Census

In Exercises 83–84, a die is rolled. Find the probability of:

83. getting a number less than 5.

84. getting a number less than 3 or greater than 4.

In Exercises 85–86, you are dealt one card from a 52-card deck. Find the probability of:

85. getting an ace or a king.

86. getting a queen or a red card.

In Exercises 87–88, it is equally probable that the pointer on the spinner shown will land on any one of the six regions, numbered 1 through 6, and colored as shown. If the pointer lands on a borderline, spin again. Find the probability of:

87. not stopping on yellow.

88. stopping on red or a number greater than 3.

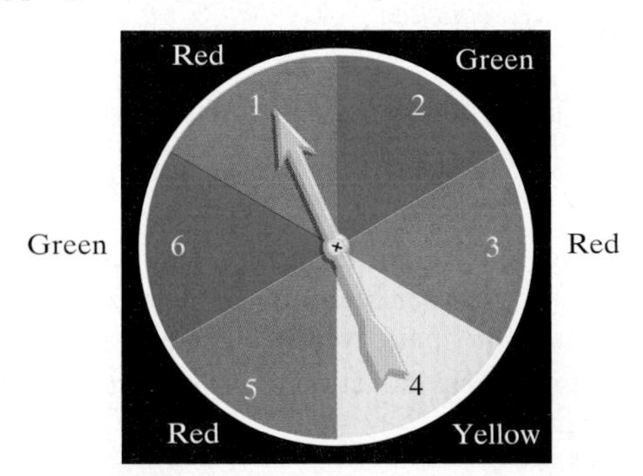

89. A lottery game is set up so that each player chooses five different numbers from 1 to 20. If the five numbers match the five numbers drawn in the lottery, the player wins (or shares) the top cash prize. What is the probability of winning the prize:

a. with one lottery ticket?

b. with 100 different lottery tickets?

Use this information to solve Exercises 90–91. At a workshop on police work and the black community, there are 50 black male police officers, 20 black female police officers, 90 white male police officers, and 40 white female police officers. If one police officer is selected at random from the people at the workshop, find the probability that the selected person is:

90. black or male. **91.** female or white.

92. What is the probability of a family having five boys born in a row?

93. The probability of a flood in any given year in a region prone to floods is 0.2.

a. What is the probability of a flood two years in a row?

b. What is the probability of a flood for three consecutive years?

c. What is the probability of no flooding for four consecutive years?

Chapter 8 Test

1. Write the first five terms of the sequence whose general term is $a_n = \frac{(-1)^{n+1}}{n^2}$.

In Exercises 2–4, find each indicated sum.

2. $\sum_{i=1}^{5} (i^2 + 10)$ **3.** $\sum_{i=1}^{20} (3i - 4)$ **4.** $\sum_{i=1}^{15} (-2)^i$

In Exercises 5–7, evaluate each expression.

5. $\binom{9}{2}$ **6.** ${}_{10}P_3$ **7.** ${}_{10}C_3$

8. Express the sum using summation notation. Use i for the index of summation.

$$\frac{2}{3} + \frac{3}{4} + \frac{4}{5} + \cdots + \frac{21}{22}$$

In Exercises 9–10, write a formula for the general term (the nth term) of each sequence. Do not use a recursion formula. Then use the formula to find the twelfth term of the sequence.

9. 4, 9, 14, 19, ...

10. 16, 4, 1, $\frac{1}{4}$, ...

In Exercises 11–12, use a formula to find the sum of the first ten terms of each sequence.

11. 7, −14, 28, −56, ...

12. −7, −14, −21, −28, ...

13. Find the sum of the infinite geometric series:

$$4 + \frac{4}{2} + \frac{4}{2^2} + \frac{4}{2^3} + \cdots.$$

14. A job pays $30,000 for the first year with an annual increase of 4% per year beginning in the second year. What is the total salary paid over an eight-year period?

15. Use mathematical induction to prove that for every positive integer n,

$$1 + 4 + 7 + \cdots + (3n - 2) = \frac{n(3n - 1)}{2}.$$

16. Use the Binomial Theorem to expand and simplify: $(x^2 - 1)^5$.

17. A human resource manager has 11 applicants to fill three different positions. Assuming that all applicants are equally qualified for any of the three positions, in how many ways can this be done?

18. From the ten books that you've recently bought but not read, you plan to take four with you on vacation. How many different sets of four books can you take?

19. How many seven-digit local telephone numbers can be formed if the first three digits are 279?

20. A lottery game is set up so that each player chooses six different numbers from 1 to 15. If the six numbers match the six numbers drawn in the lottery, the player wins (or shares) the top cash prize. What is the probability of winning the prize with 50 different lottery tickets?

21. One card is randomly selected from a deck of 52 cards. Find the probability of selecting a black card or a picture card.

22. A group of students consists of 10 male freshmen, 15 female freshmen, 20 male sophomores, and 5 female sophomores. If one person is randomly selected from the group, find the probability of selecting a freshman or a female.

23. A quiz consisting of four multiple-choice questions has four available options (a, b, c, or d) for each question. If a person guesses at every question, what is the probability of answering all questions correctly?

24. If the spinner shown is spun twice, find the probability that the pointer lands on red on the first spin and blue on the second spin.

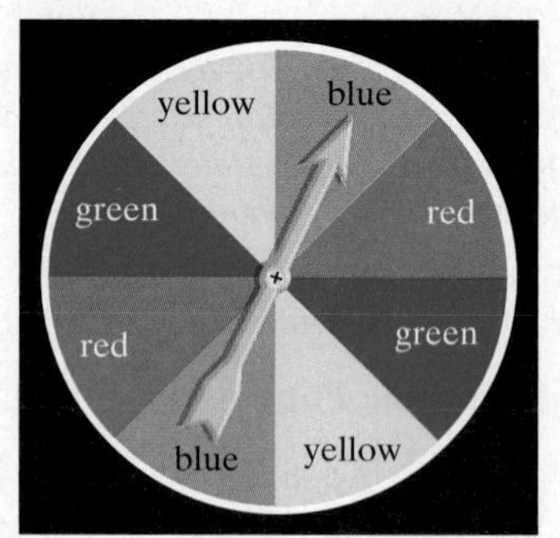

Cumulative Review Exercises (Chapters 1–8)

Solve each equation or inequality in Exercises 1–10.

1. $-2(x - 5) + 10 = 3(x + 2)$

2. $3x^2 - 6x + 2 = 0$

3. $\log_2 x + \log_2(2x - 3) = 1$

4. $x^{1/2} - 6x^{1/4} + 8 = 0$

5. $\sqrt{2x + 4} - \sqrt{x + 3} - 1 = 0$

6. $|2x + 1| \leq 1$

7. $6x^2 - 6 < 5x$

8. $\frac{x - 1}{x + 3} \leq 0$

9. $30e^{0.7x} = 240$

10. $2x^3 + 3x^2 - 8x + 3 = 0$

Solve each system in Exercises 11–13.

11. $4x^2 + 3y^2 = 48$
$3x^2 + 2y^2 = 35$

12. (Use matrices.)
$x - 2y + z = 16$
$2x - y - z = 14$
$3x + 5y - 4z = -10$

13. $x - y = 1$
$x^2 - x - y = 1$

In Exercises 14–19, graph each equation, function, or system in the rectangular coordinate system.

14. $100x^2 + y^2 = 25$

15. $4x^2 - 9y^2 - 16x + 54y - 29 = 0$

16. $f(x) = \frac{x^2 - 1}{x - 2}$

17. $2x - y \geq 4$
$x \leq 2$

18. $f(x) = x^2 - 4x - 5$

19. $y = \log_2 x$

20. Find $f^{-1}(x)$ if $f(x) = \sqrt[3]{x + 4}$.

21. If $A = \begin{bmatrix} 4 & 2 \\ 1 & -1 \\ 0 & 5 \end{bmatrix}$ and $B = \begin{bmatrix} 2 & 4 \\ 3 & 1 \end{bmatrix}$, find $AB - 4A$.

22. Find the partial fraction decomposition for

$$\frac{2x^2 - 10x + 2}{(x-2)(x^2+2x+2)}.$$

23. Expand and simplify: $(x^3 + 2y)^5$.

24. Use the formula for the sum of the first n terms of an arithmetic sequence to find $\sum_{i=1}^{50}(4i - 25)$.

25. Mailings in the United States increased by more than 40% from 1983 to 1993.

x **(Number of Years after 1983)**	y **(Number of Pieces of Mail, in Billions)**
0	119.4
10	171.1

a. Write the point-slope form of the line on which these measurements fall.
b. Use the point-slope form of the equation to write the slope-intercept form of the equation.
c. Use the slope-intercept model from part (b) to predict the number of pieces of mail, in billions, for the year 2000.

26. Most of the world's very tall buildings are in the United States, where the skyscraper was first conceived. The height of the World Trade Center in New York is 790 feet less than twice that of New York's Empire State Building. If the mean (average) height of the two buildings is 980 feet, determine the height of each building.

27. The perimeter of a soccer field is 300 yards. If the length is 50 yards longer than the width, what are the field's dimensions?

28. If 10 pens and 12 pads cost $42, and 5 of the same pens and 10 of the same pads cost $29, find the cost of a pen and a pad.

29. A ball is thrown vertically upward from the top of a 96-foot tall building with an initial velocity of 80 feet per second. The height of the ball above ground is modeled by the position function

$$s(t) = -16t^2 + 80t + 96.$$

a. After how many seconds will the ball strike the ground?
b. When does the ball reach its maximum height? What is the maximum height?

30. The current, I, in amperes, flowing in an electrical circuit varies inversely as the resistance, R, in ohms, in the circuit. When the resistance of an electric percolator is 22 ohms, it draws 5 amperes of current. How much current is needed when the resistance is 10 ohms?

Appendix
Where Did That Come From? Selected Proofs

SECTION 4.3 Properties of Logarithms

The Product Rule

Let b, M, and N be positive real numbers with $b \neq 1$.

$$\log_b(MN) = \log_b M + \log_b N$$

Proof

We begin by letting $\log_b M = R$ and $\log_b N = S$.

Now we write each logarithm in exponential form.

$$\log_b M = R \quad \text{means} \quad b^R = M.$$
$$\log_b N = S \quad \text{means} \quad b^S = N.$$

By substituting and using a property of exponents, we see that

$$MN = b^R b^S = b^{R+S}.$$

Now we change $MN = b^{R+S}$ to logarithmic form.

$$MN = b^{R+S} \quad \text{means} \quad \log_b(MN) = R + S.$$

Finally, substituting $\log_b M$ for R and $\log_b N$ for S gives us

$$\log_b(MN) = \log_b M + \log_b N,$$

the property that we wanted to prove.

The quotient and power rules for logarithms are proved using similar procedures.

The Change-of-Base Property

For any logarithmic bases a and b, and any positive number M,

$$\log_b M = \frac{\log_a M}{\log_a b}.$$

Proof

To prove the change-of-base property, we let x equal the logarithm on the left side:

$$\log_b M = x.$$

Now we rewrite this logarithm in exponential form.

$$\log_b M = x \quad \text{means} \quad b^x = M.$$

Because b^x and M are equal, the logarithms with base a for each of these expressions must be equal. This means that

$$\log_a b^x = \log_a M$$

$$x \log_a b = \log_a M \qquad \text{Apply the power rule for logarithms on the left side.}$$

$$x = \frac{\log_a M}{\log_a b} \qquad \text{Solve for } x \text{ by dividing both sides by } \log_a b.$$

In our first step we let x equal $\log_b M$. Replacing x on the left side by $\log_b M$ gives us

$$\log_b M = \frac{\log_a M}{\log_a b},$$

which is the change-of-base property.

SECTION 7.2 *The Hyperbola*

The Asymptotes of a Hyperbola Centered at the Origin

The hyperbola

$$\frac{x^2}{a^2} - \frac{y^2}{b^2} = 1$$

with a horizontal transverse axis has the two asymptotes

$$y = \frac{b}{a}x \quad \text{and} \quad y = -\frac{b}{a}x.$$

Proof

Begin by solving the hyperbola's equation for y.

$$\frac{x^2}{a^2} - \frac{y^2}{b^2} = 1$$ This is the standard form of the equation of a hyperbola.

$$\frac{y^2}{b^2} = \frac{x^2}{a^2} - 1$$ We isolate the term involving y^2 to solve for y.

$$y^2 = \frac{b^2x^2}{a^2} - b^2$$ Multiply both sides by b^2.

$$y^2 = \frac{b^2x^2}{a^2}\left(1 - \frac{a^2}{x^2}\right)$$ Factor out $\frac{b^2x^2}{a^2}$ on the right. Verify that this result is correct by multiplying using the distributive property and obtaining the previous step.

$$y = \pm\sqrt{\frac{b^2x^2}{a^2}\left(1 - \frac{a^2}{x^2}\right)}$$ Solve for y using the square root method: If $u^2 = d$, then $u = \pm\sqrt{d}$.

$$y = \pm\frac{b}{a}x\sqrt{1 - \frac{a^2}{x^2}}$$ Simplify.

As $|x| \to \infty$, the value of $\frac{a^2}{x^2}$ approaches 0. Consequently, the value of y can be approximated by

$$y = \pm \frac{b}{a} x.$$

This means that the lines whose equations are $y = \frac{b}{a} x$ and $y = -\frac{b}{a} x$ are asymptotes for the graph of the hyperbola.

Answers to Selected Exercises

CHAPTER P

Section P.1

Check Point Exercises

1. a. $\sqrt{2} - 1$ **b.** $\pi - 3$ **c.** 1 **2.** 9 **3.** 414.5; In 2080, the population of the United States will be 414.5 million. **4.** $38x - 19y$

Exercise Set P.1

1. a. $\sqrt{100}$ **b.** 0, $\sqrt{100}$ **c.** $-9, 0, \sqrt{100}$ **d.** $-9, -\frac{4}{5}, 0, 0.25, 9.2, \sqrt{100}$ **e.** $\sqrt{3}$ **3. a.** $\sqrt{64}$ **b.** 0, $\sqrt{64}$ **c.** $-11, 0, \sqrt{64}$ **d.** $-11, -\frac{5}{6}, 0, 0.75, \sqrt{64}$ **e.** $\sqrt{5}, \pi$ **5.** 0 **7.** Answers may vary. **9.** true **11.** true **13.** true **15.** 300 **17.** $12 - \pi$ **19.** $5 - \sqrt{2}$ **21.** -1 **23.** 15 **25.** 7 **27.** 15 **29.** 2.2 **31.** 27 **33.** -19 **35.** 25 **37.** 10 **39.** commutative property of addition **41.** associative property of addition **43.** commutative property of addition **45.** distributive property of multiplication over addition **47.** $15x + 16$ **49.** $27x - 10$ **51.** $29y - 29$ **53.** $14x$ **55.** $-2x + 3y + 6$ **57.** x **59.** yes **61.** Answers may vary. **63.** 25,401; In 1997, the average yearly earnings in the United States was $25,401. **65. a.** $132 - 0.6a$ **b.** 120 **73.** (c) is true. **75.** $<$ **77.** $>$

Section P.2

Check Point Exercises

1. -256 **2. a.** $\frac{1}{8}$ **b.** 36 **3. a.** 243 **b.** $\frac{1}{8}$ **c.** x^6 **4. a.** 729 **b.** y^{28} **c.** $\frac{1}{x^8}$ **5. a.** 9 **b.** $\frac{1}{x^7}$ **c.** y^9 **6.** $-64x^3$ **7. a.** $\frac{27}{64}$ **b.** $-\frac{32}{y^5}$ **8. a.** $16x^{12}y^{24}$ **b.** $-18x^3y^8$ **c.** $\frac{5y^6}{x^4}$ **d.** $\frac{y^8}{25x^2}$ **9. a.** 7,400,000,000 **b.** 0.000003017 **10. a.** 7.41×10^9 **b.** 9.2×10^{-8} **11.** 5.2×10^5 mi

Exercise Set P.2

1. 50 **3.** 64 **5.** -64 **7.** 1 **9.** -1 **11.** $\frac{1}{64}$ **13.** 32 **15.** 64 **17.** 16 **19.** $\frac{1}{9}$ **21.** $\frac{1}{16}$ **23.** $\frac{y}{x^2}$ **25.** y^5 **27.** x^{10} **29.** x^5 **31.** x^{21} **33.** x^{-15} **35.** x^7 **37.** x^{21} **39.** $64x^6$ **41.** $-\frac{64}{x^3}$ **43.** $9x^4y^{10}$ **45.** $6x^{11}$ **47.** $18x^9y^5$ **49.** $4x^{16}$ **51.** $-5a^{11}b$ **53.** $\frac{2}{b^7}$ **55.** $\frac{1}{16x^6}$ **57.** $\frac{3y^{14}}{4x^4}$ **59.** $\frac{y^2}{25x^6}$ **61.** 4700 **63.** 4,000,000 **65.** 0.000786 **67.** 0.00000318 **69.** 3.6×10^3 **71.** 2.2×10^8 **73.** 2.7×10^{-2} **75.** 7.63×10^{-4} **77.** 600,000 **79.** 0.123 **81.** 30,000 **83.** 0.021 **85.** 1.694×10^{12} **87.** 6.0×10^{10} **89.** 3.24×10^{10} **99.** $\frac{1}{4}$ **100.** A = C + D

Section P.3

Check Point Exercises

1. a. 3 **b.** $5|x|\sqrt{2}$ **2. a.** $\frac{5}{4}$ **b.** $5|x|\sqrt{3}$ **3. a.** $17\sqrt{13}$ **b.** $-19\sqrt{17x}$ **4. a.** $17\sqrt{3}$ **b.** $10\sqrt{2x}$ **5. a.** $\frac{5\sqrt{3}}{3}$ **b.** $\sqrt{3}$ **6.** $\frac{32 - 8\sqrt{5}}{11}$ **7. a.** $2\sqrt[3]{5}$ **b.** $2\sqrt[5]{2}$ **c.** $\frac{5}{3}$ **8.** $5\sqrt[3]{3}$ **9. a.** 9 **b.** 3 **c.** $\frac{1}{2}$ **10. a.** 8 **b.** $\frac{1}{4}$ **11. a.** $10x^4$ **b.** $4x^{5/2}$ **12.** $\sqrt{x}$

Exercise Set P.3

1. 6 **3.** not a real number **5.** 13 **7.** $5\sqrt{2}$ **9.** $3|x|\sqrt{5}$ **11.** $2|x|\sqrt{3}$ **13.** $|x|\sqrt{x}$ **15.** $2|x|\sqrt{3x}$ **17.** $\frac{1}{9}$ **19.** $\frac{7}{4}$ **21.** $4|x|$ **23.** $5|x|\sqrt{2x}$ **25.** $13\sqrt{3}$ **27.** $-2\sqrt{17x}$ **29.** $5\sqrt{2}$ **31.** $3\sqrt{2x}$ **33.** $34\sqrt{2}$ **35.** $\frac{\sqrt{7}}{7}$ **37.** $\frac{\sqrt{10}}{5}$

39. $\frac{13(3-\sqrt{11})}{-2}$ **41.** $7(\sqrt{5}+2)$ **43.** $3(\sqrt{5}-\sqrt{3})$ **45.** 5 **47.** −2 **49.** not a real number **51.** 3 **53.** −3 **55.** $2\sqrt[3]{4}$
57. $x\sqrt[3]{x}$ **59.** $3\sqrt[3]{2}$ **61.** $2x$ **63.** 6 **65.** 2 **67.** 25 **69.** $\frac{1}{16}$ **71.** $14x^{7/12}$ **73.** $4x^{1/4}$ **75.** x^2 **77.** $5x^2|y|^3$
79. $\sqrt{5}$ **81.** x^2 **83.** $|x|^{2/3}$ **85.** $20\sqrt{2}$ mph **87.** $\frac{\sqrt{5}+1}{2}\approx 1.62$ **89.** $\frac{7\sqrt{2\cdot 2\cdot 3}}{6}=\frac{7\sqrt{2^2\cdot 3}}{6}=\frac{7\sqrt{2^2}\sqrt{3}}{6}=\frac{7\cdot 2\sqrt{3}}{6}=\frac{7}{3}\sqrt{3}$
91. The duration of a storm whose diameter is 9 miles is 1.89 hours. **99.** during 1990 **101.** (d) is true.
103. Let □ = 25 and □ = 14. **105. a.** > **b.** >

Section P.4

Check Point Exercises

1. a. $-x^3+x^2-8x-20$ **b.** $20x^3-11x^2-2x-8$ **2.** $15x^3-31x^2+30x-8$ **3.** $28x^2-41x+15$
4. a. $49x^2-64$ **b.** $4y^6-25$ **5. a.** $x^2+20x+100$ **b.** $25x^2+40x+16$ **6. a.** $x^2-18x+81$ **b.** $49x^2-42x+9$
7. $2x^2y+5xy^2-2y^3$ **8. a.** $21x^2-25xy+6y^2$ **b.** $x^4+10x^2y+25y^2$

Exercise Set P.4

1. yes; $3x^2+2x-5$ **3.** no **5.** 2 **7.** 4 **9.** $11x^3+7x^2-12x-4$; 3 **11.** $12x^3+4x^2+12x-14$; 3 **13.** $6x^2-6x+2$; 2
15. x^3+1 **17.** $2x^3-9x^2+19x-15$ **19.** $x^2+10x+21$ **21.** $x^2-2x-15$ **23.** $6x^2+13x+5$ **25.** $10x^2-9x-9$
27. $15x^4-47x^2+28$ **29.** x^2-9 **31.** $9x^2-4$ **33.** $25-49x^2$ **35.** $16x^4-25x^2$ **37.** x^2+4x+4 **39.** $4x^2+12x+9$
41. x^2-6x+9 **43.** $16x^4-8x^2+1$ **45.** $4x^2-28x+49$ **47.** x^3+3x^2+3x+1 **49.** $8x^3+36x^2+54x+27$
51. $x^3-9x^2+27x-27$ **53.** $27x^3-108x^2+144x-64$ **55.** $7x^2y-4xy$ is of degree 3 **57.** $2x^2y+13xy+13$ is of degree 3
59. $-5x^3+8xy-9y^2$ is of degree 3 **61.** $x^4y^2+8x^3y+y-6x$ is of degree 6 **63.** $7x^2+38xy+15y^2$ **65.** $2x^2+xy-21y^2$
67. $15x^2y^2+xy-2$ **69.** $49x^2+70xy+25y^2$ **71.** $x^4y^4-6x^2y^2+9$ **73.** x^3-y^3 **75.** $9x^2-25y^2$
77. 7.567; A person earning $40,000 feels underpaid $7567. **79.** 54; 72; 54; Performance increases as enthusiasm goes from 1 to 50, then performance decreases as enthusiasm goes from 50 to 100. **81.** $4t-2t^2+\frac{2}{3}t^3$ **83.** $6x+22$ **93.** during 1992 and 1993
95. $49x^2+70x+25-16y^2$ **97.** x^4-y^4

Section P.5

Check Point Exercises

1. a. $2x^2(5x-2)$ **b.** $(x-7)(2x+3)$ **2.** $(x+5)(x^2-2)$ **3. a.** $(x+8)(x+5)$ **b.** $(x-7)(x+2)$ **4.** $(3x-1)(2x+7)$
5. a. $(x+9)(x-9)$ **b.** $(6x+5)(6x-5)$ **6.** $(9x^2+4)(3x+2)(3x-2)$ **7. a.** $(x+7)^2$ **b.** $(4x-7)^2$
8. a. $(x+1)(x^2-x+1)$ **b.** $(5x-2)(25x^2+10x+4)$ **9. a.** $2x(x-6)^2$ **b.** $(x-4)(x+3)(x-3)$

Exercise Set P.5

1. $9(2x+3)$ **3.** $3x(x+2)$ **5.** $9x^2(x^2-2x+3)$ **7.** $(x+5)(x+3)$ **9.** $(x-3)(x^2+12)$ **11.** $(x^2+5)(x-2)$
13. $(x-1)(x^2+2)$ **15.** $(3x-2)(x^2-2)$ **17.** $(x+2)(x+3)$ **19.** $(x-5)(x+3)$ **21.** $(x-5)(x-3)$
23. $(3x+2)(x-1)$ **25.** $(3x-28)(x+1)$ **27.** $(2x-1)(3x-4)$ **29.** $(2x+3)(2x+5)$ **31.** $(x+10)(x-10)$
33. $(6x+7)(6x-7)$ **35.** $(3x+5y)(3x-5y)$ **37.** $(x^2+4)(x+2)(x-2)$ **39.** $(4x^2+9)(2x+3)(2x-3)$
41. $(x+1)^2$ **43.** $(x-7)^2$ **45.** $(2x+1)^2$ **47.** $(3x-1)^2$ **49.** $(x+3)(x^2-3x+9)$ **51.** $(x-4)(x^2+4x+16)$
53. $(2x-1)(4x^2+2x+1)$ **55.** $(4x+3)(16x^2-12x+9)$ **57.** $3x(x+1)(x-1)$ **59.** $4(x+2)(x-3)$
61. $2(x^2+9)(x+3)(x-3)$ **63.** $(x-3)(x+3)(x+2)$ **65.** $2(x-8)(x+7)$ **67.** $x(x-2)(x+2)$ **69.** prime
71. $(x-2)(x+2)^2$ **73.** $y(y^2+9)(y+3)(y-3)$ **75.** $5y^2(2y+3)(2y-3)$ **77.** $-16(t-2)(t+1)$
79. $(3x+2)(3x-2)$ **89.** (d) is true. **91.** $-(x+5)(x-1)$ **93.** $(x-y)^3(x+y)$
95. $b=0,3,4,-c(c+4)$, where $c>0$ is an integer.

Section P.6

Check Point Exercises

1. a. −5 **b.** 6, −6 **2. a.** $x^2, x\neq -3$ **b.** $\frac{x-1}{x+1}, x\neq -1$ **3.** $\frac{x-3}{(x-2)(x+3)}, x\neq 2, x\neq -2, x\neq -3$
4. $\frac{3(x-1)}{x(x+2)}, x\neq 1, x\neq 0, x\neq -2$ **5.** $-2, x\neq -1$ **6.** $\frac{2(4x+1)}{(x+1)(x-1)}, x\neq 1, x\neq -1$ **7.** $(x-3)(x-3)(x+3)$
8. $\frac{-x^2+11x-20}{2(x-5)^2}, x\neq 5$ **9.** $\frac{2(2-3x)}{4+3x}, x\neq 0, x\neq -\frac{4}{3}$

Exercise Set P.6

1. 3 **3.** 5, −5 **5.** −1, −10 **7.** $\frac{3}{x-3}, x \neq 3$ **9.** $\frac{x-6}{4}, x \neq 6$ **11.** $\frac{y+9}{y-1}, y \neq 1, 2$ **13.** $\frac{x+6}{x-6}, x \neq 6, -6$ **15.** $\frac{1}{3}, x \neq 2, -3$ **17.** $\frac{(x-3)(x+3)}{x(x+4)}, x \neq 0, -4, 3$ **19.** $\frac{x-1}{x+2}, x \neq -2, -1, 2, 3$ **21.** $\frac{x^2+2x+4}{3x}, x \neq -2, 0, 2$ **23.** $\frac{7}{9}, x \neq -1$ **25.** $\frac{(x-2)^2}{x}, x \neq 0, -2$ **27.** $\frac{2(x+3)}{3}, x \neq 3, -3$ **29.** $\frac{x-5}{2}, x \neq 1, -5$ **31.** $2, x \neq -\frac{5}{6}$ **33.** $\frac{2x-1}{x+3}, x \neq 0, -3$ **35.** $3, x \neq 2$ **37.** $\frac{3}{x-3}, x \neq 3, -4$ **39.** $\frac{9x+39}{(x+4)(x+5)}, x \neq -4, -5$ **41.** $-\frac{3}{x(x+1)}, x \neq -1, 0$ **43.** $\frac{3x^2+4}{(x+2)(x-2)}, x \neq -2, 2$ **45.** $\frac{2x^2+50}{(x-5)(x+5)}, x \neq -5, 5$ **47.** $\frac{4x+16}{(x+3)^2}, x \neq -3$ **49.** $\frac{x^2-x}{(x+5)(x-2)(x+3)}, x \neq -5, 2, -3$ **51.** $\frac{1}{3}, x \neq 3$ **53.** $\frac{x+1}{3x-1}, x \neq 0, \frac{1}{3}$ **55.** $\frac{1}{xy}, x \neq 0, y \neq 0, x \neq -y$ **57.** $\frac{x}{x+3}, x \neq -2, -3$ **59.** $-\frac{x-14}{7}, x \neq -2, 2$ **61.** $\frac{540t^2+12{,}640t+107{,}100}{-0.14t^2+0.51t+31.6}$ **63. a.** 86.67, 520, 1170; It costs \$86,670,000 to inoculate 40% of the population against this strain of flu, and \$520,000,000 to inoculate 80% of the population, and \$1,170,000,000 to inoculate 90% of the population. **b.** $x = 100$ **c.** increases rapidly; impossible inoculate 100% of the population. **65.** $\frac{2r_1r_2}{r_1+r_2}$; 24 mph **79.** 1990 **81.** $-4x - 1$ **83.** yields the third power of x

Section P.7

Check Point Exercises

1. a. $8 + i$ **b.** $-10 + 10i$ **2. a.** $63 + 14i$ **b.** $58 - 11i$ **3.** $\frac{3}{5} + \frac{13}{10}i$ **4. a.** $7i\sqrt{3}$ **b.** $1 - 4i\sqrt{3}$ **c.** $-7 + i\sqrt{3}$

Exercise Set P.7

1. $8 - 2i$ **3.** $-2 + 9i$ **5.** $24 + 7i$ **7.** $-14 + 17i$ **9.** $21 + 15i$ **11.** $-43 - 23i$ **13.** $-29 - 11i$ **15.** 34 **17.** 34 **19.** $-5 + 12i$ **21.** $\frac{3}{5} + \frac{1}{5}i$ **23.** $1 + i$ **25.** $-\frac{24}{25} + \frac{32}{25}i$ **27.** $\frac{7}{5} + \frac{4}{5}i$ **29.** $3i$ **31.** $47i$ **33.** $-8i$ **35.** $2 + 6i\sqrt{7}$ **37.** $-\frac{1}{3} + \frac{\sqrt{2}}{6}i$ **39.** $-\frac{1}{8} - \frac{\sqrt{3}}{24}i$ **41.** $-2\sqrt{6} - 2i\sqrt{10}$ **43.** $24\sqrt{15}$ **53.** (d) is true. **55.** $\frac{14}{25} - \frac{2}{25}i$ **57.** 0

Section P.8

Check Point Exercises

1.

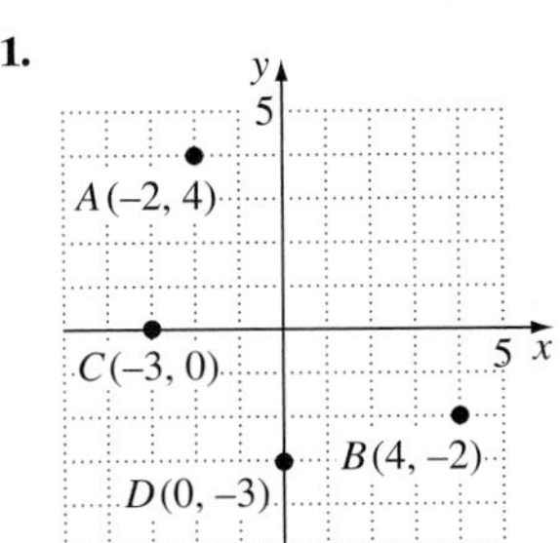

2.

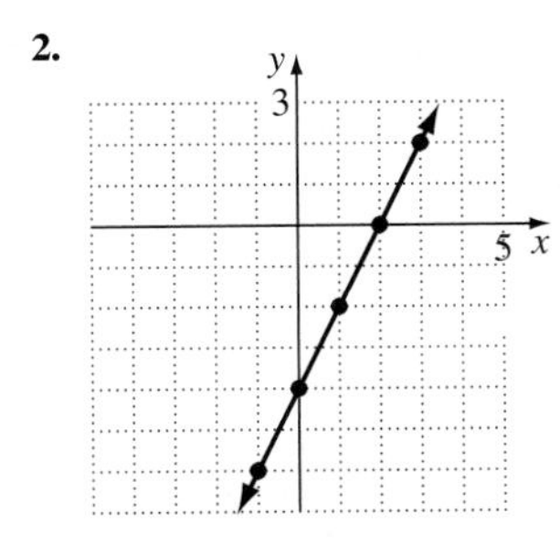

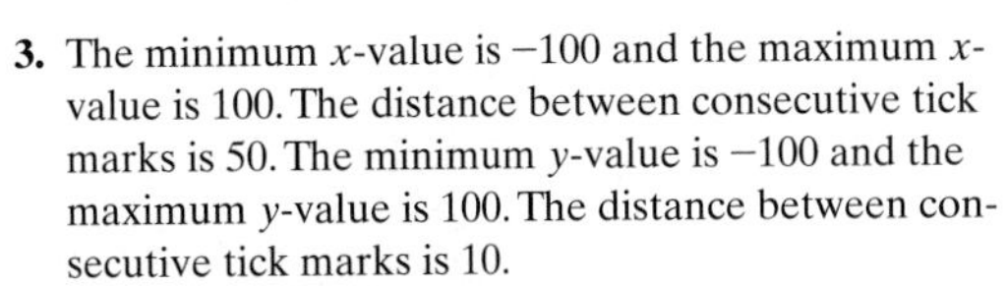
3. The minimum x-value is −100 and the maximum x-value is 100. The distance between consecutive tick marks is 50. The minimum y-value is −100 and the maximum y-value is 100. The distance between consecutive tick marks is 10.

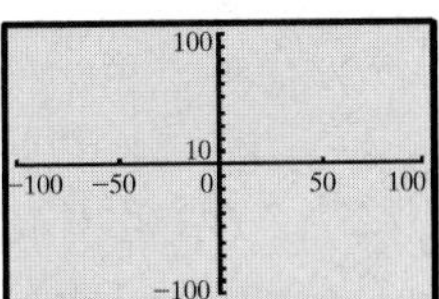

4. 5 **5.** $\left(4, -\frac{1}{2}\right)$ **6.** 1991; about \$800 million

Exercise Set P.8

1.

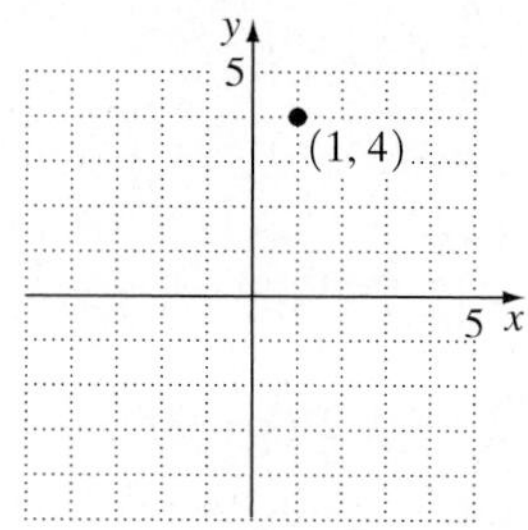

3.

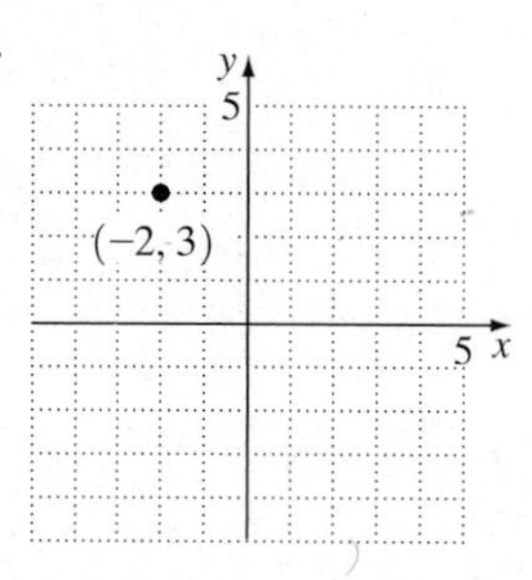

5.

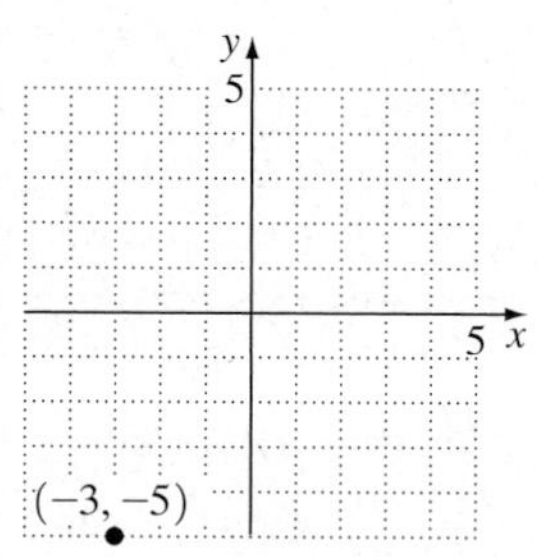

7.

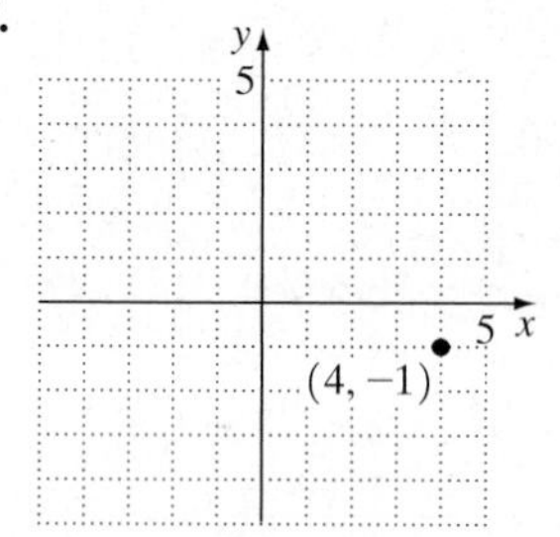

9.

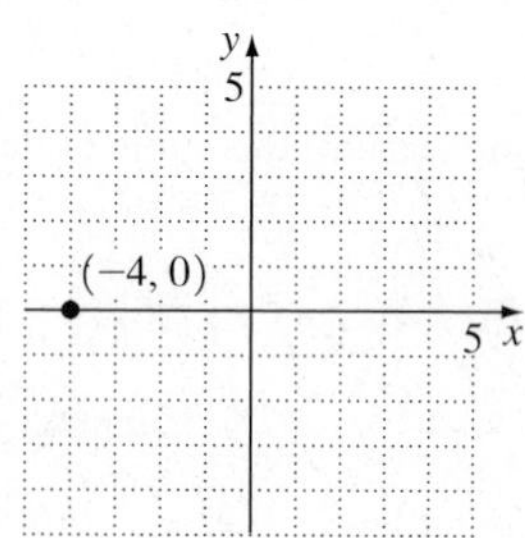

11.

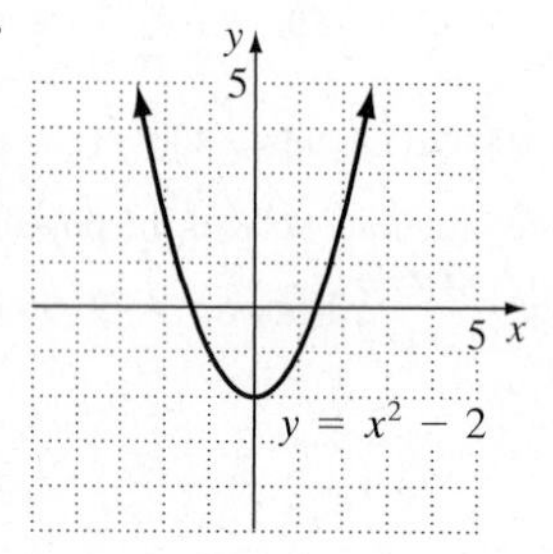

13.

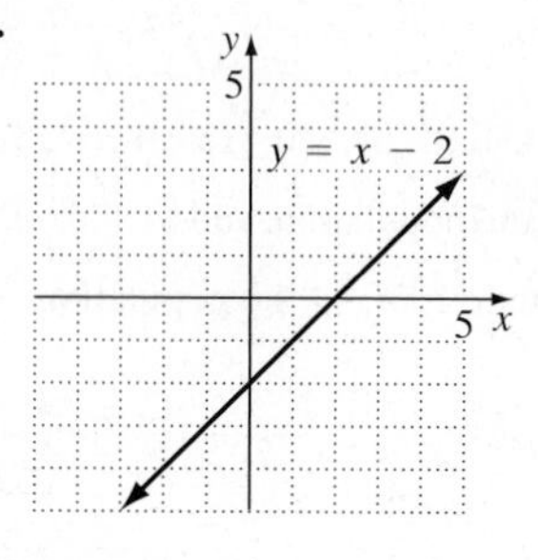

15.

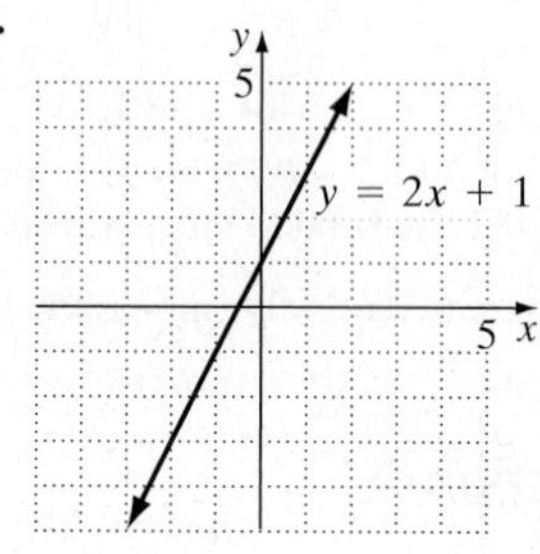

17.

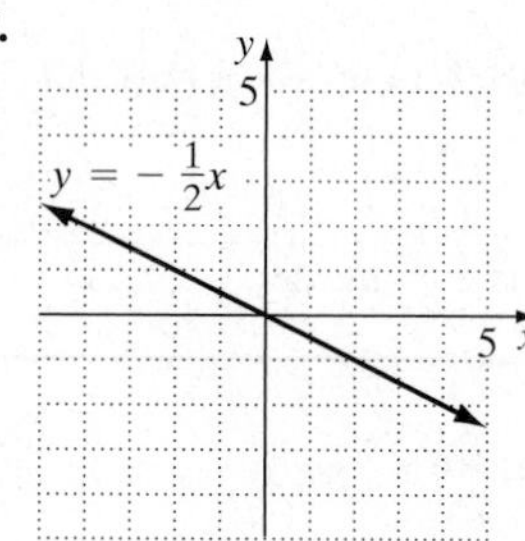

19.

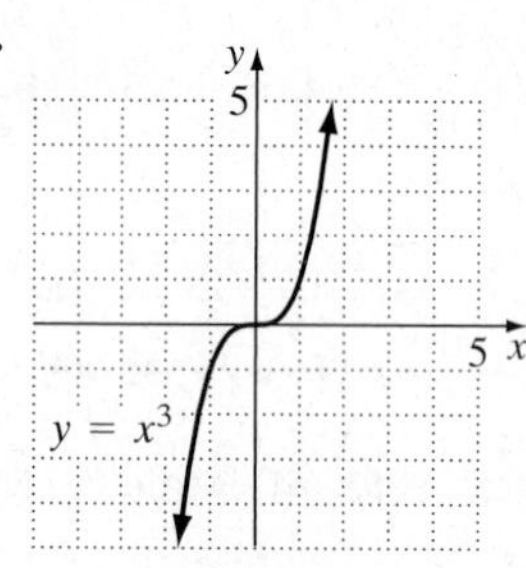

21.

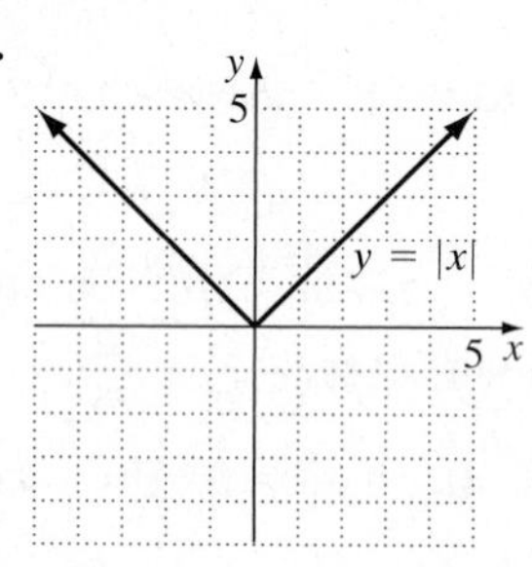

23. (c) **25.** (b) **27. a.** 2 **b.** −4 **29. a.** 1, −2 **b.** 2 **31. a.** −1 **b.** None

33. a.

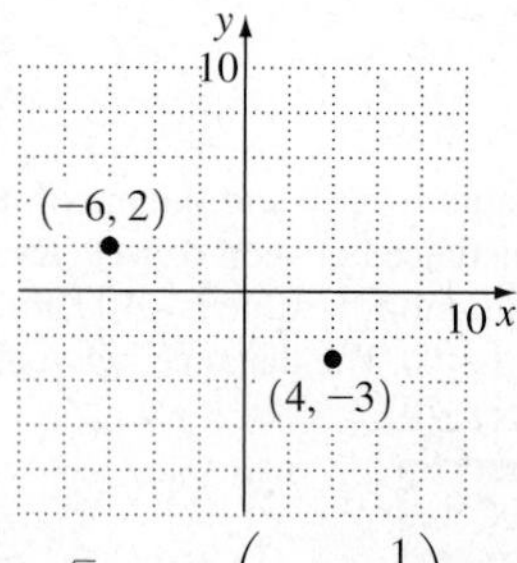

b. $5\sqrt{5}$ **c.** $\left(-1, -\frac{1}{2}\right)$

35. a.

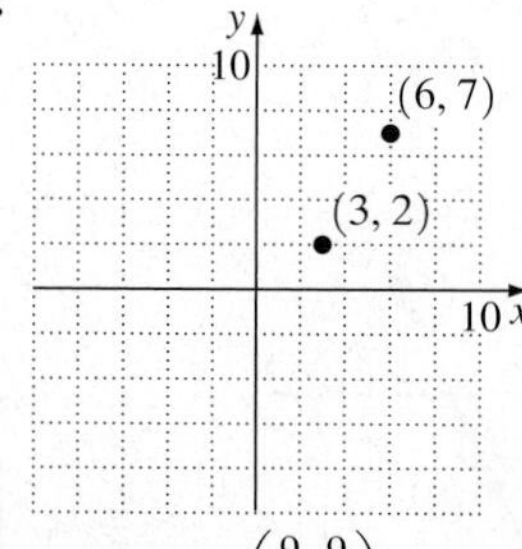

b. $\sqrt{34}$ **c.** $\left(\frac{9}{2}, \frac{9}{2}\right)$

37. a.

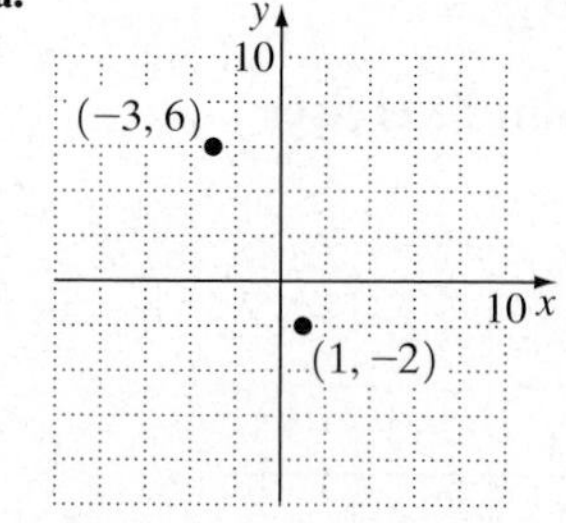

b. $4\sqrt{5}$ **c.** $(-1, 2)$

39. about 5% **41.** around 1982, about 9.7% **43.** (1970, 61); In 1970, the population of United States for people under 16 was about 61 million. **45.** (1990, 60); In 1990, the population of United States. for people under 16 was about 60 million. **47. a.** 1250 ft **b.** 8.8 sec **57.** (c) gives a complete graph. **59.** (b) gives a complete graph. **61.** (d) is true.

Chapter P Review Exercises

1. a. $\sqrt{81}$ **b.** $0, \sqrt{81}$ **c.** $-17, 0, \sqrt{81}$ **d.** $-17, -\frac{9}{13}, 0, 0.75, \sqrt{81}$ **e.** $\sqrt{2}, \pi$ **2.** 103 **3.** $\sqrt{2} - 1$ **4.** $\sqrt{17} - 3$ **5.** $|4 - (-17)|$; 21 **6.** 20 **7.** 4 **8.** commutative property of addition **9.** associative property of multiplication **10.** distributive property of multiplication over addition **11.** commutative property of multiplication **12.** commutative property of multiplication **13.** commutative property of addition **14.** $23x - 23y - 2$ **15.** $2x$ **16.** −108 **17.** $\frac{5}{16}$ **18.** $\frac{1}{25}$ **19.** $\frac{1}{27}$ **20.** $-8x^{12}y^9$ **21.** $\frac{10}{x^8}$ **22.** $\frac{1}{16x^{12}}$ **23.** $\frac{y^8}{4x^{10}}$ **24.** 37,400 **25.** 0.0000745 **26.** 3.59×10^6

27. 7.25×10^{-3} **28.** 3.9×10^5 **29.** 2.3×10^{-2} **30.** 10^3 or 1000 yr **31.** $\$4.05 \times 10^{10}$ **32.** $10\sqrt{3}$ **33.** $2|x|\sqrt{3}$ **34.** $2|x|\sqrt{5}$ **35.** $|r|\sqrt{r}$ **36.** $\frac{11}{2}$ **37.** $4|x|\sqrt{3}$ **38.** $20\sqrt{5}$ **39.** $16\sqrt{2}$ **40.** $24\sqrt{2} - 8\sqrt{3}$ **41.** $6\sqrt{5}$ **42.** $\frac{\sqrt{6}}{3}$ **43.** $\frac{5(6 - \sqrt{3})}{33}$ **44.** $7(\sqrt{7} + \sqrt{5})$ **45.** 5 **46.** -2 **47.** not a real number **48.** 5 **49.** $3\sqrt[3]{3}$ **50.** $y\sqrt[3]{y^2}$ **51.** $2\sqrt[4]{5}$ **52.** $13\sqrt[3]{2}$ **53.** $|x|\sqrt[4]{2}$ **54.** 4 **55.** $\frac{1}{5}$ **56.** 5 **57.** $\frac{1}{3}$ **58.** 16 **59.** $\frac{1}{81}$ **60.** $20x^{11/12}$ **61.** $3x^{1/4}$ **62.** $25x^4$ **63.** $y^{1/2}$ **64.** $8x^3 + 10x^2 - 20x - 4$; degree 3 **65.** $8x^4 - 5x^3 + 6$; degree 4 **66.** $12x^3 + x^2 - 21x + 10$ **67.** $6x^2 - 7x - 5$ **68.** $16x^2 - 25$ **69.** $4x^2 + 20x + 25$ **70.** $9x^2 - 24x + 16$ **71.** $8x^3 + 12x^2 + 6x + 1$ **72.** $125x^3 - 150x^2 + 60x - 8$ **73.** $-x^2 - 17xy - 3y^2$; degree 2 **74.** $24x^3y^2 + x^2y - 12x^2 + 4$; degree 5 **75.** $3x^2 + 16xy - 35y^2$ **76.** $9x^2 - 30xy + 25y^2$ **77.** $9x^4 + 12x^2y + 4y^2$ **78.** $49x^2 - 16y^2$ **79.** $a^3 - b^3$ **80.** $3x^2(5x + 1)$ **81.** $(x - 4)(x - 7)$ **82.** $(3x + 1)(5x - 2)$ **83.** $(8 - x)(8 + x)$ **84.** prime **85.** $3x^2(x - 5)(x + 2)$ **86.** $4x^3(5x^4 - 9)$ **87.** $(x + 3)(x - 3)^2$ **88.** $(4x - 5)^2$ **89.** $(x^2 + 4)(x + 2)(x - 2)$ **90.** $(y - 2)(y^2 + 2y + 4)$ **91.** $(x + 4)(x^2 - 4x + 16)$ **92.** $3x^2(x - 2)(x + 2)$ **93.** $(3x - 5)(9x^2 + 15x + 25)$ **94.** $x(x - 1)(x + 1)(x^2 + 1)$ **95.** $(x^2 - 2)(x + 5)$ **96.** $x^2, x \neq -2$ **97.** $\frac{x - 3}{x - 6}, x \neq -6, 6$ **98.** $\frac{x}{x + 2}, x \neq -2$ **99.** $\frac{(x + 3)^3}{(x - 2)^2(x + 2)}, x \neq 2, -2$ **100.** $\frac{2}{x(x + 1)}, x \neq 0, 1, -1, -\frac{1}{3}$ **101.** $\frac{x + 3}{x - 4}, x \neq -3, 4, 2, 8$ **102.** $\frac{1}{x - 3}, x \neq 3, -3$ **103.** $\frac{4x(x - 1)}{(x + 2)(x - 2)}, x \neq 2, -2$ **104.** $\frac{x(2x + 1)}{(x - 3)(x + 3)(x - 2)}, x \neq 3, -3, 2$ **105.** $\frac{11x^2 - x - 11}{(2x - 1)(x + 3)(3x + 2)}, x \neq \frac{1}{2}, -3, -\frac{2}{3}$ **106.** $\frac{3}{x}, x \neq 0, 2$ **107.** $\frac{3x}{x - 4}, x \neq 0, 4, -4$ **108.** $\frac{3x + 8}{3x + 10}, x \neq -3, -\frac{10}{3}$ **109.** $-9 + 4i$ **110.** $-12 - 8i$

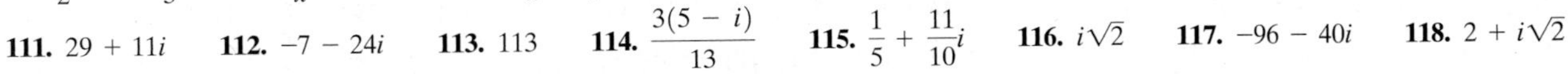

111. $29 + 11i$ **112.** $-7 - 24i$ **113.** 113 **114.** $\frac{3(5 - i)}{13}$ **115.** $\frac{1}{5} + \frac{11}{10}i$ **116.** $i\sqrt{2}$ **117.** $-96 - 40i$ **118.** $2 + i\sqrt{2}$

119.

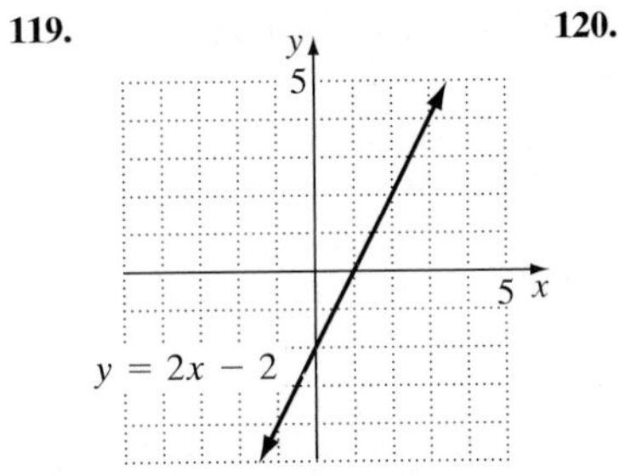

120.

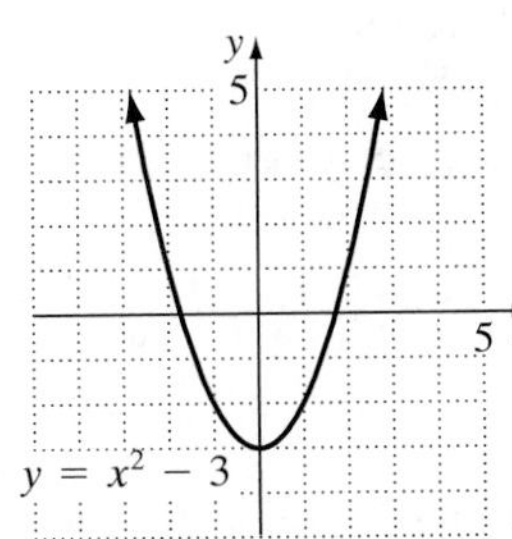

121.

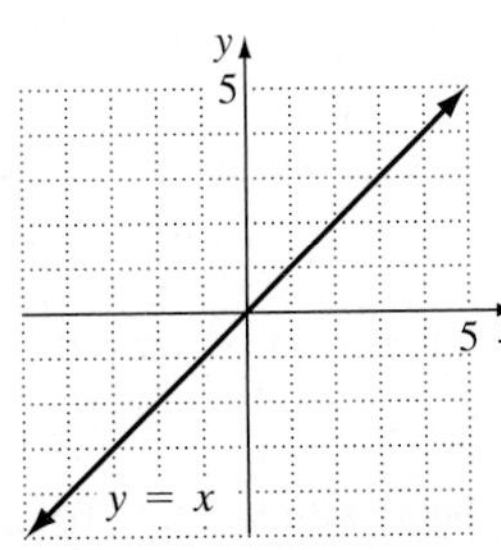

122.

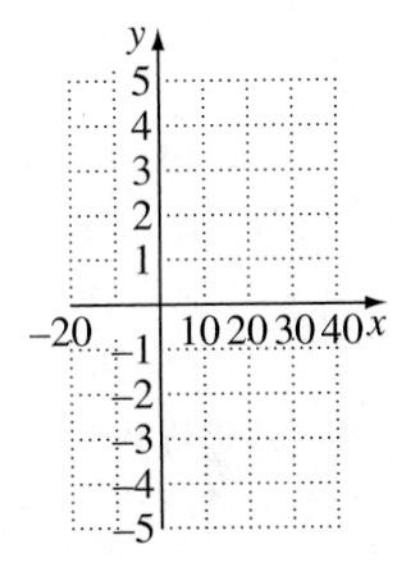

123. x-intercept: -2; y-intercept: 2 **124.** x-intercepts: 2, -2; y-intercept: -4 **125.** x-intercept: 5; y-intercept: none

126. a.

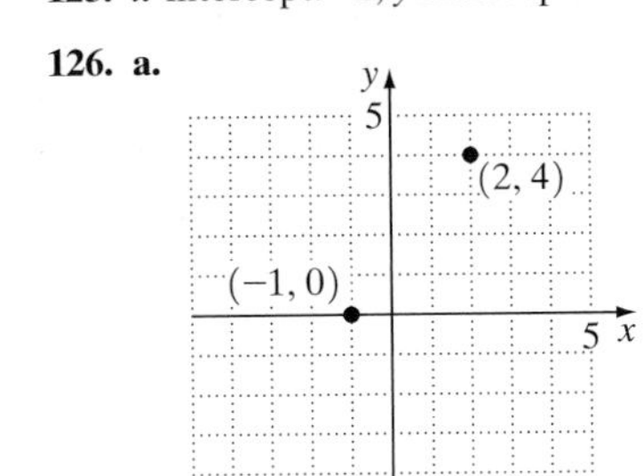

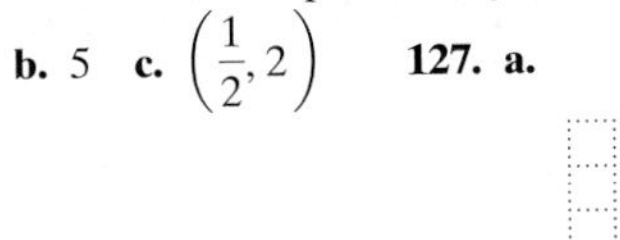

b. 5 **c.** $\left(\frac{1}{2}, 2\right)$

127. a.

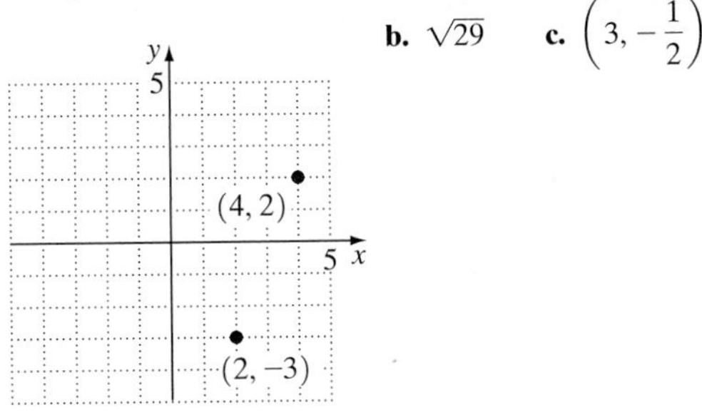

b. $\sqrt{29}$ **c.** $\left(3, -\frac{1}{2}\right)$

128. a. 1991; 100,000 **b.** 30,000 **c.** $A(1998, 70)$; In 1998, there were about 70,000 law school applicants.

Chapter P Test

1. $-7, -\frac{4}{5}, 0, 0.25, \sqrt{4}, \frac{22}{7}$ **2.** commutative property of addition **3.** distributive property of multiplication over addition **4.** 7.6×10^{-4} **5.** $85x + 2y - 15$ **6.** $\frac{5y^8}{x^6}$ **7.** $3|r|\sqrt{2}$ **8.** $11\sqrt{2}$ **9.** $\frac{3(5 - \sqrt{2})}{23}$ **10.** $2x\sqrt[3]{2x}$ **11.** $\frac{x + 3}{x - 2}, x \neq 2, 1$ **12.** $\frac{1}{243}$ **13.** $2x^3 - 13x^2 + 26x - 15$ **14.** $25x^2 + 30xy + 9y^2$ **15.** $(x - 3)(x - 6)$ **16.** $(x^2 + 3)(x + 2)$ **17.** $(5x - 3)(5x + 3)$ **18.** $(6x - 7)^2$ **19.** $(y - 5)(y^2 + 5y + 25)$ **20.** $\frac{2(x + 3)}{x + 1}, x \neq 3, -1, -4, -3$ **21.** $\frac{x^2 + 2x + 15}{(x + 3)(x - 3)}, x \neq 3, -3$ **22.** $\frac{11}{(x - 3)(x - 4)}, x \neq 3, 4$ **23.** $\frac{3 - x}{3}, x \neq 0$ **24.** $47 + 16i$ **25.** $2 + i$ **26.** $38i$

27.

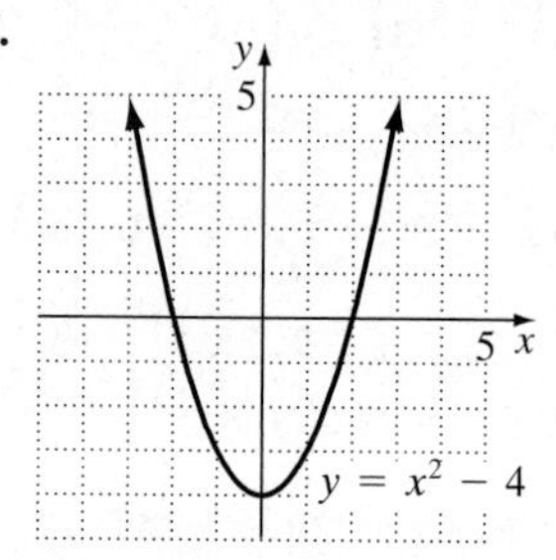

28. $2\sqrt{13}$

CHAPTER 1

Section 1.1

Check Point Exercises

1. {16} **2.** {5} **3.** {−2} **4.** {3} **5.** ∅ **6.** identity

Exercise Set 1.1

1. {16} **3.** {7} **5.** {13} **7.** {2} **9.** {9} **11.** {−5} **13.** {6} **15.** {−2} **17.** {12} **19.** {24} **21.** {−15} **23.** {5} **25.** $\left\{\frac{33}{2}\right\}$ **27.** {−12} **29.** $\left\{\frac{46}{5}\right\}$ **31. a.** 0 **b.** $\left\{\frac{1}{2}\right\}$ **33. a.** 0 **b.** {−2} **35. a.** 0 **b.** {2} **37. a.** 0 **b.** {4} **39. a.** 1 **b.** {3} **41. a.** −1 **b.** ∅ **43. a.** 1 **b.** {2} **45. a.** −2, 2 **b.** ∅ **47. a.** −1, 1 **b.** {−3} **49. a.** −2, 4 **b.** ∅ **51.** identity **53.** inconsistent equation **55.** conditional equation **57.** inconsistent equation **59.** {−7} **61.** ∅ **62.** not true for any real number **63.** {−4} **65.** {8} **67.** {−1} **69. a.** 205 mg/dl **b.** 375,000 annual deaths; 125,000 saved lives **71.** $409\frac{1}{5}$ ft **85.** inconsistent **87.** conditional; {−5} **89.** $x = \frac{c - b}{a}$ **91.** Answers may vary. **93.** 20

Section 1.2

Check Point Exercises

1. 100 g **2.** Bee Gees = 11 million albums; Morissette = 16 million albums **3.** about 57 hr **4.** $15,000 at 9%; $10,000 at 12% **5.** width = 40 ft; length = 120 ft **6.** $m = \frac{y - b}{x}$

Exercise Set 1.2

1. $x + 9$ **3.** $20 - x$ **5.** $8 - 5x$ **7.** $15 \div x$ **9.** $2x + 20$ **11.** $7x - 30$ **13.** $4(x + 12)$ **15.** $x + 40 = 450$; {410} **17.** $5x - 7 = 123$; {26} **19.** $9x = 3x + 30$; {5} **21.** 40 years old; Find 117 on the vertical axis and follow it over to the graph for female. **23.** 30 years after 1980; 2010 **25.** yes; The height (about 61.2 in.) is greater than 5 ft. **27.** Sosa = 66 home runs; McGwire = 70 home runs **29.** 19, 20 **31.** United States = 41%; Sweden = 13.5% **33.** 800 mi **35.** 6 months **37.** 7 oz **39. a.** total monthly cost with coupon book = $21 + 0.50x$; total monthly cost without coupon book = $1.25x$ **b.** 28 times **41.** $25,000 at 9%; $0 at 12% **43.** length = 78 ft; width = 36 ft **45.** length = 12 ft; height = 4 ft **47.** 11 hr **49.** $126 **51.** $740 **53.** $467.20 **55.** 5 ft 7 in. **57.** $w = \frac{A}{l}$; area of rectangle **59.** $b = \frac{2A}{h}$; area of triangle **61.** $p = \frac{I}{rt}$; interest **63.** $m = \frac{E}{c^2}$; energy **65.** $p = \frac{T - D}{m}$ **67.** $\frac{2A}{h} - b = a$; trapezoid area **69.** $\frac{S - P}{Pt} = r$; interest **71.** $S = \frac{F}{B} + V$

77. a. $y = 3.82 + 0.3x$

b.

8
0 12
0

c. The trace feature shows x to be about 8 when $y = 6.22$, so 1988.
d. When $y = 6.22$, $x = 88$.

79. 600 students from the school with 10% African American students; 400 students from the school with 90% African American students **81.** Coburn = 60 years old; woman = 20 years old **83.** $4000 = mother; $8000 = boy; $2000 = girl

Section 1.3

Check Point Exercises

1. a. $\{0, 3\}$ **b.** $\left\{-1, \frac{1}{2}\right\}$ **2. a.** $\{-\sqrt{7}, \sqrt{7}\}$ **b.** $\{-5 + \sqrt{11}, -5 - \sqrt{11}\}$ **3.** 49; $(x - 7)^2$ **4.** $\{1 + \sqrt{3}, 1 - \sqrt{3}\}$
5. $\left\{\frac{-1 + \sqrt{3}}{2}, \frac{-1 - \sqrt{3}}{2}\right\}$ **6.** $\{1 + i, 1 - i\}$ **7.** -56; two complex imaginary solutions **8.** 10 yr **9.** 12 in.

Exercise Set 1.3

1. $\{-2, 5\}$ **3.** $\{3, 5\}$ **5.** $\left\{-\frac{5}{2}, \frac{2}{3}\right\}$ **7.** $\left\{-\frac{4}{3}, 2\right\}$ **9.** $\{-4, 0\}$ **11.** $\left\{0, \frac{1}{3}\right\}$ **13.** $\{-3, 1\}$ **15.** $\{-3, 3\}$
17. $\{-\sqrt{10}, \sqrt{10}\}$ **19.** $\{-7, 3\}$ **21.** $\left\{-\frac{5}{3}, \frac{1}{3}\right\}$ **23.** $\left\{\frac{1 - \sqrt{7}}{5}, \frac{1 + \sqrt{7}}{5}\right\}$ **25.** $\left\{\frac{4 - 2\sqrt{2}}{3}, \frac{4 + 2\sqrt{2}}{3}\right\}$ **27.** 36; $(x + 6)^2$
29. 25; $(x - 5)^2$ **31.** $\frac{9}{4}$; $\left(x + \frac{3}{2}\right)^2$ **33.** $\frac{49}{4}$; $\left(x - \frac{7}{2}\right)^2$ **35.** $\frac{1}{9}$; $\left(x - \frac{1}{3}\right)^2$ **37.** $\frac{1}{36}$; $\left(x - \frac{1}{6}\right)^2$ **39.** $\{-7, 1\}$
41. $\{1 + \sqrt{3}, 1 - \sqrt{3}\}$ **43.** $\{3 + 2\sqrt{5}, 3 - 2\sqrt{5}\}$ **45.** $\{-2 + \sqrt{3}, -2 - \sqrt{3}\}$ **47.** $\left\{\frac{-3 + \sqrt{13}}{2}, \frac{-3 - \sqrt{13}}{2}\right\}$ **49.** $\left\{\frac{1}{2}, 3\right\}$
51. $\left\{\frac{1 + \sqrt{2}}{2}, \frac{1 - \sqrt{2}}{2}\right\}$ **53.** $\left\{\frac{1 + \sqrt{7}}{3}, \frac{1 - \sqrt{7}}{3}\right\}$ **55.** $\{-5, -3\}$ **56.** $\{-6, -2\}$ **57.** $\left\{\frac{-5 + \sqrt{13}}{2}, \frac{-5 - \sqrt{13}}{2}\right\}$
59. $\left\{\frac{3 + \sqrt{57}}{6}, \frac{3 - \sqrt{57}}{6}\right\}$ **61.** $\left\{\frac{1 + \sqrt{29}}{4}, \frac{1 - \sqrt{29}}{4}\right\}$ **63.** $\{3 + i, 3 - i\}$ **65.** 36; 2 unequal real solutions
67. 97; 2 unequal real solutions **69.** 0; 1 real solution **71.** 37; 2 unequal real solutions **73.** $\left\{-\frac{1}{2}, 1\right\}$ **75.** $\left\{\frac{1}{5}, 2\right\}$
77. $\{-2\sqrt{5}, 2\sqrt{5}\}$ **79.** $\{1 + \sqrt{2}, 1 - \sqrt{2}\}$ **81.** $\left\{\frac{-11 + \sqrt{33}}{4}, \frac{-11 - \sqrt{33}}{4}\right\}$ **83.** $\left\{0, \frac{8}{3}\right\}$ **85.** $\{2\}$ **87.** $\{-2, 2\}$
89. $\{3 + 2i, 3 - 2i\}$ **91.** $\{2 + i\sqrt{3}, 2 - i\sqrt{3}\}$ **93.** $\left\{0, \frac{7}{2}\right\}$ **95.** 2024 **97.** 1999; very well **99.** 1986
101. 1990; 739,980; very well **103.** 1995; 340,000; fairly well **105.** 127.28 ft **107.** 34 ft **109.** width = 15 ft; length = 20 ft
111. 10 in. **123.** (c) is true. **125.** $x^2 - 2x - 15 = 0$ **127.** 1144; It is possible, so the applicant should be hired.

Section 1.4

Check Point Exercises

1. $\{-\sqrt{3}, 0, \sqrt{3}\}$ **2.** $\left\{-2, -\frac{3}{2}, 2\right\}$ **3.** $\{-1, 3\}$ **4.** $\{4\}$ **5.** $\{\sqrt[3]{25}\}$ or $\{5^{2/3}\}$ **6.** $\{-\sqrt{3}, -\sqrt{2}, \sqrt{2}, \sqrt{3}\}$ **7.** $\left\{-\frac{1}{27}, 64\right\}$
8. $\{-2, 3\}$

Exercise Set 1.4

1. $\{-4, 0, 4\}$ **3.** $\{0, 2\}$ **5.** $\left\{-2, -\frac{2}{3}, 2\right\}$ **7.** $\left\{-\frac{1}{2}, \frac{1}{2}, \frac{3}{2}\right\}$ **9.** $\left\{-2, -\frac{1}{2}, \frac{1}{2}\right\}$ **11.** $\{6\}$ **13.** $\{6\}$ **15.** $\{-6\}$
17. $\{10\}$ **19.** $\{12\}$ **21.** $\{8\}$ **23.** $\varnothing$ **25.** $\varnothing$ **27.** $\left\{\frac{13 + \sqrt{105}}{6}\right\}$ **29.** $\{4\}$ **31.** $\{13\}$ **33.** $\{\sqrt[5]{4}\}$ **35.** $\{-4, 5\}$
37. $\{-2, -1, 1, 2\}$ **39.** $\left\{-\frac{4}{3}, -1, 1, \frac{4}{3}\right\}$ **41.** $\{-\sqrt[3]{5}, -\sqrt[3]{3}\}$ **43.** $\left\{-\sqrt[3]{2}, \sqrt[3]{\frac{9}{5}}\right\}$ **45.** $\{-8, 27\}$ **47.** $\{1\}$ **49.** $\left\{\frac{1}{4}, 1\right\}$
53. $\{-3, -1, 2, 4\}$ **55.** $\{-8, -2, 1, 4\}$ **57.** $\{-8, 8\}$ **59.** $\{-5, 9\}$ **61.** $\{-2, 3\}$ **63.** $\{1\}$ **65.** $\{0\}$ **67.** $\left\{\frac{5}{2}\right\}$
69. $\{-8, -6, 4, 6\}$ **71.** $\{-1, 1, 2\}$ **73.** 36 years old **75.** 1952 **77.** 161,081 mi/sec **79. a.** \$29 **b.** 402 CD sets
81. a. 8 **b.** The road should be positioned so that they meet on the expressway at a point 8 miles from the point on the expressway closest to A and 4 miles from the point closest to B. **91.** $\{0, 2\}$ **93.** $\{-2\}$

95.

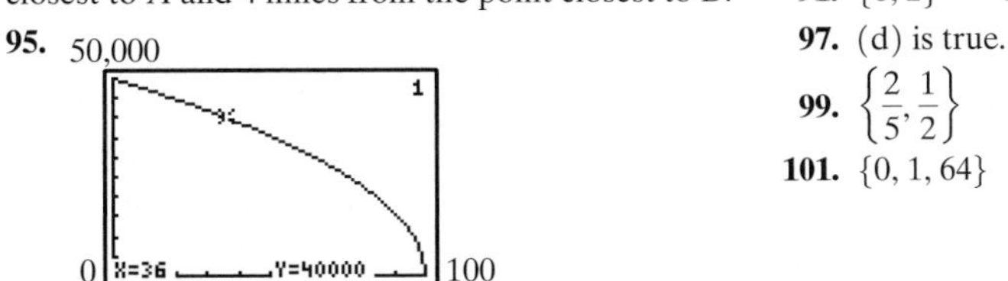

Tracing along the curve shows the point (36, 40,000) on the graph.

97. (d) is true.
99. $\left\{\frac{2}{5}, \frac{1}{2}\right\}$
101. $\{0, 1, 64\}$

Section 1.5

Check Point Exercises

1. a. **b.** **c.**

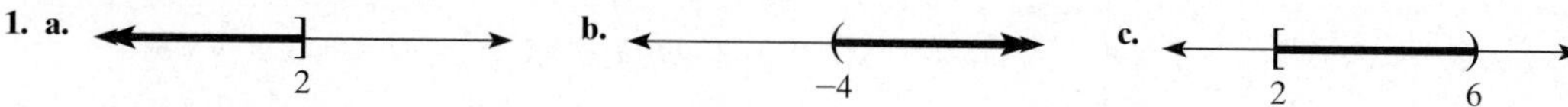

2. a. $\{x|-2 \le x < 5\}$ **b.** $\{x|1 \le x \le 3.5\}$ **c.** $\{x|x < -1\}$ **3.** $[-1, \infty)$ or $\{x|x \ge -1\}$

4. $[1, \infty)$ or $\{x|x \ge 1\}$ **5.** $[-1, 4)$ or $\{x|-1 \le x < 4\}$ **6.** $(-3, 7)$ or $\{x|-3 < x < 7\}$ **7.** $(-\infty, 1)$ or $[4, \infty)$ or $\{x|x \le 1 \text{ or } x \ge 4\}$

8. more than 720 mi

Exercise Set 1.5

1. **3.** **5.** **7.** **9.** **11.** **13.** $1 < x \le 6$ **15.** $-5 \le x < 2$ **17.** $-3 \le x \le 1$ **19.** $x > 2$

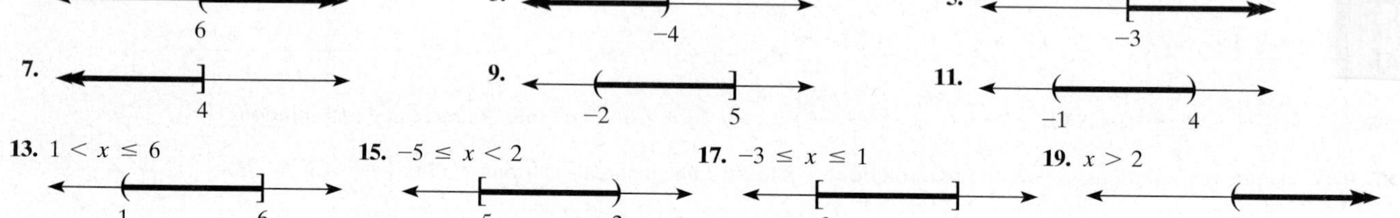

21. $x \ge -3$ **23.** $x < 3$ **25.** $x < 5.5$ **27.** $(-\infty, 3)$

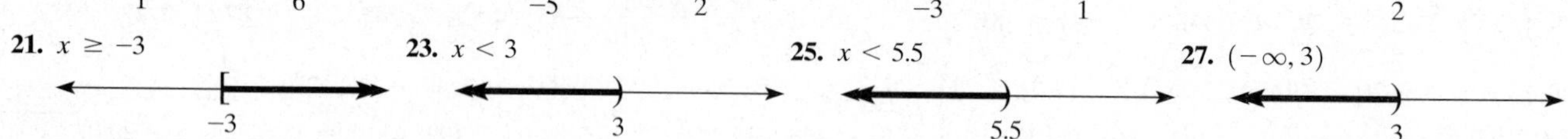

29. $\left[\frac{20}{3}, \infty\right)$ **31.** $(-\infty, -4]$ **33.** $\left(-\infty, -\frac{2}{5}\right]$ **35.** $[0, \infty)$

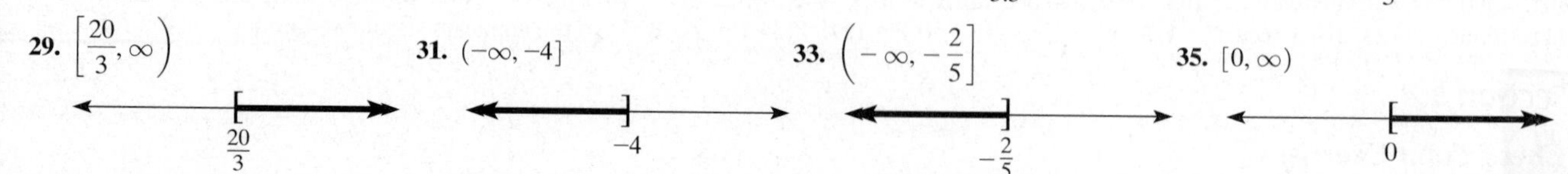

37. $(-\infty, 1)$ **39.** $[6, \infty)$ **41.** $\left[-\frac{32}{5}, \infty\right)$ **43.** $(-\infty, -6)$

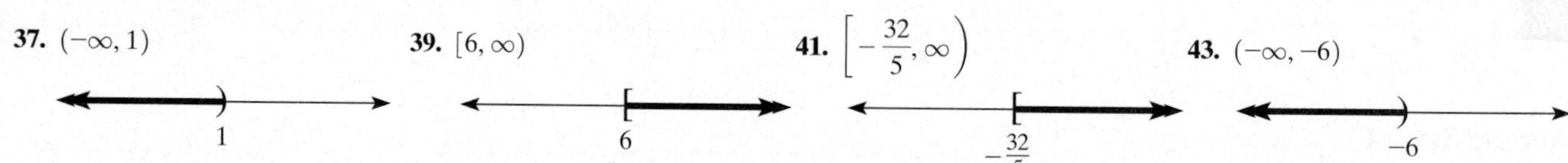

45. $[13, \infty)$ **47.** $(-\infty, \infty)$ **49.** $(3, 5)$ **51.** $[-1, 3)$

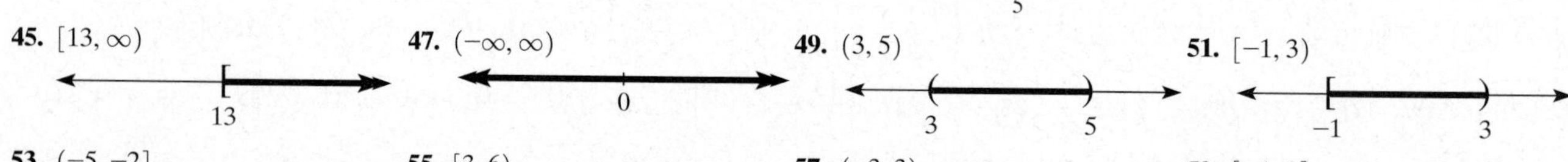

53. $(-5, -2]$ **55.** $[3, 6)$ **57.** $(-3, 3)$ **59.** $[-1, 3]$

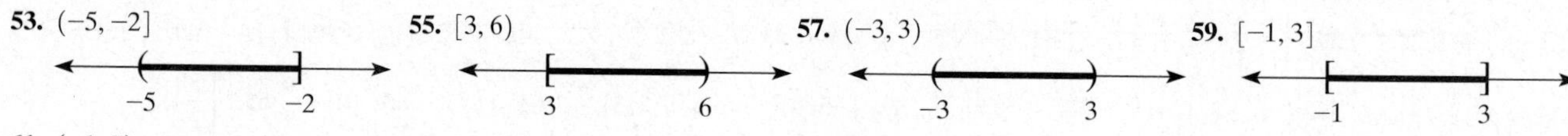

61. $(-1, 7)$ **63.** $[-5, 3]$ **65.** $(-6, 0)$ **67.** $(-\infty, -3)$ or $(3, \infty)$

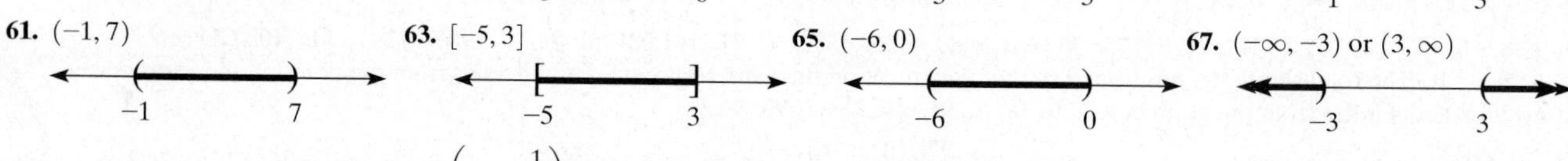

69. $(-\infty, -1]$ or $[3, \infty)$ **71.** $\left(-\infty, \frac{1}{3}\right)$ or $(5, \infty)$ **73.** $(-\infty, -5]$ or $[3, \infty)$ **75.** $(-\infty, -3)$ or $(12, \infty)$

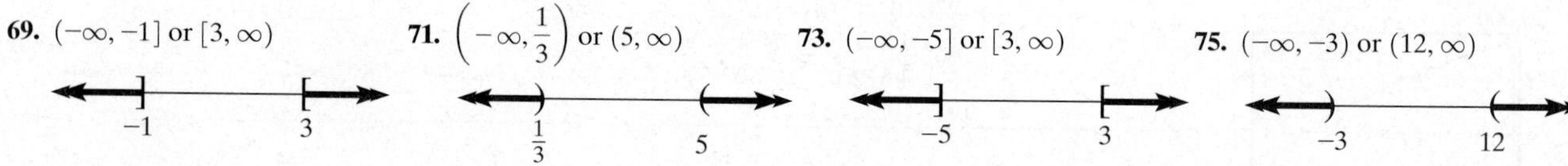

77. $(-\infty, -1]$ or $[3, \infty)$ **79.** $(-\infty, -1)$ or $(2, \infty)$ **81.** $\left(-\infty, -\frac{75}{14}\right)$ or $\left(\frac{87}{14}, \infty\right)$ **83.** $(-\infty, -6]$ or $[24, \infty)$

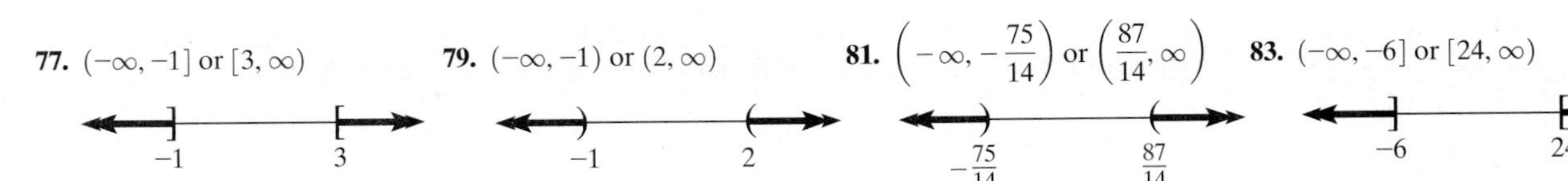

85. Raleigh, NC, Seattle, San Francisco, Austin, TX **87.** San Diego **89.** Austin, TX, Washington, DC, Lexington-Fayette, KY, Minneapolis, Boston, Arlington, TX **91.** severe cognitive impairment, substance abuse disorders, and depressive: manic, major depressive **93.** 2013 **95.** between 2004 and 2009 **97.** 199 checks or less **99. a.** at least a 96 **b.** a grade less than 66 **101.** 1001 or more pairs

113.

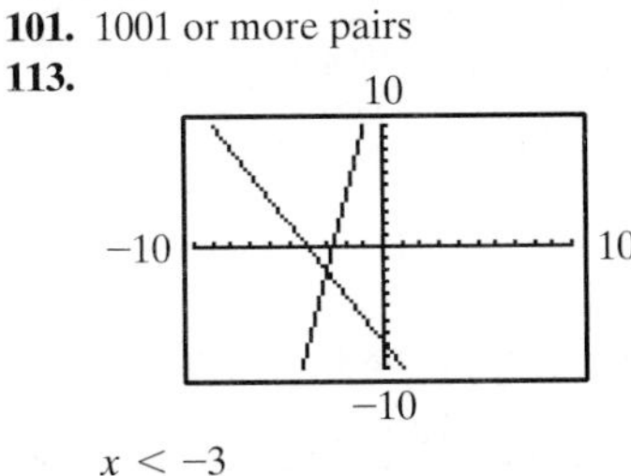

$x < -3$

115.

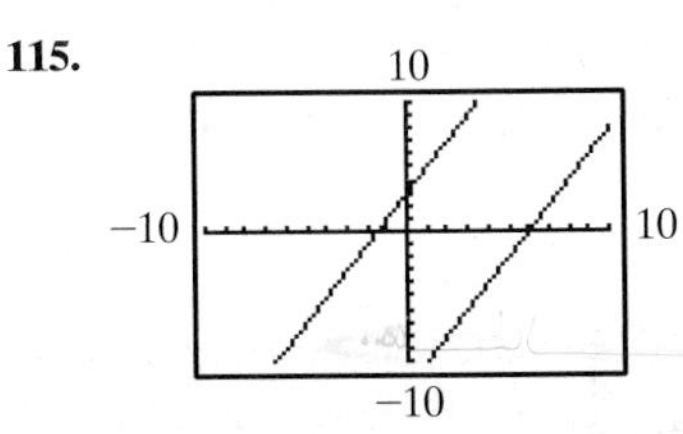

The graph of the left side of the inequality is always above the graph of the right side, therefore all values of x are included in the solution; You get a statement that is always true.

117. (c) is true. **119.** $(-10, -8)$

Section 1.6

Check Point Exercises

1. $(-3, 1)$ **2.** $(-\infty, -4]$ or $[5, \infty)$ **3.** $(-\infty, -2)$ or $(5, \infty)$ **4.** $(-1, 1]$

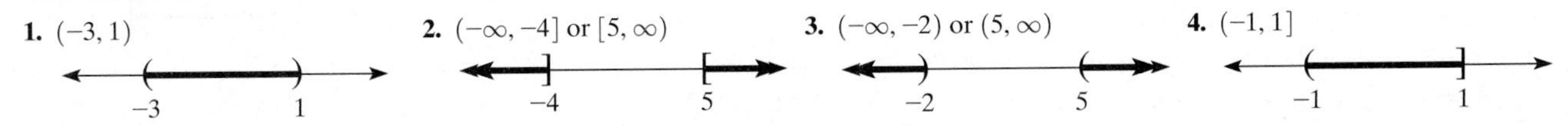

5. between 1 and 4 sec

Exercise Set 1.6

1. $(-\infty, -2)$ or $(4, \infty)$ **3.** $[-3, 7]$ **5.** $(-\infty, 1)$ or $(4, \infty)$ **7.** $(-\infty, -4)$ or $(-1, \infty)$ **9.** $\varnothing$ **11.** $[2, 4]$ **13.** $\left[-4, \frac{2}{3}\right]$ **15.** $\left(-3, \frac{5}{2}\right)$

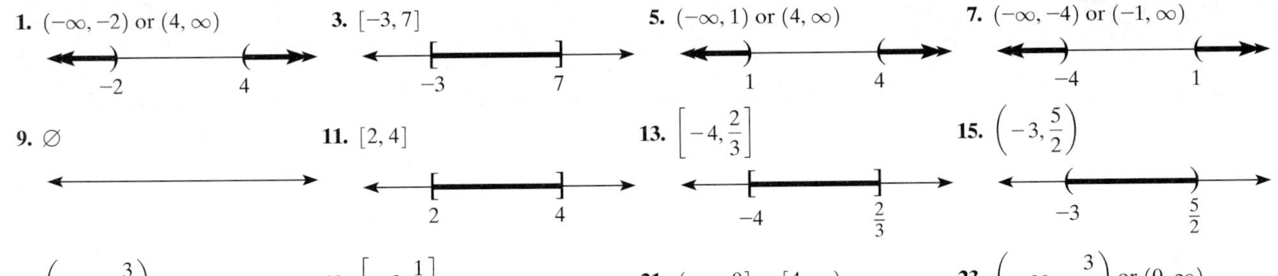

17. $\left(-1, -\frac{3}{4}\right)$ **19.** $\left[-2, \frac{1}{3}\right]$ **21.** $(-\infty, 0]$ or $[4, \infty)$ **23.** $\left(-\infty, -\frac{3}{2}\right)$ or $(0, \infty)$ **25.** $[0, 1]$ **27.** $(-\infty, -3)$ or $(4, \infty)$ **29.** $(-4, -3)$ **31.** $[2, 4)$

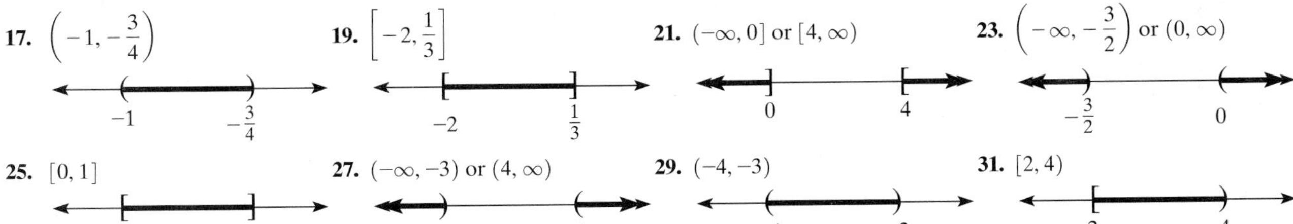

33. $\left(-\infty, -\frac{4}{3}\right)$ or $[2, \infty)$ **35.** $(-\infty, 0)$ or $(3, \infty)$ **37.** $(-\infty, -5)$ or $(-3, \infty)$ **39.** $\left(-\infty, \frac{1}{2}\right)$ or $\left[\frac{7}{5}, \infty\right)$

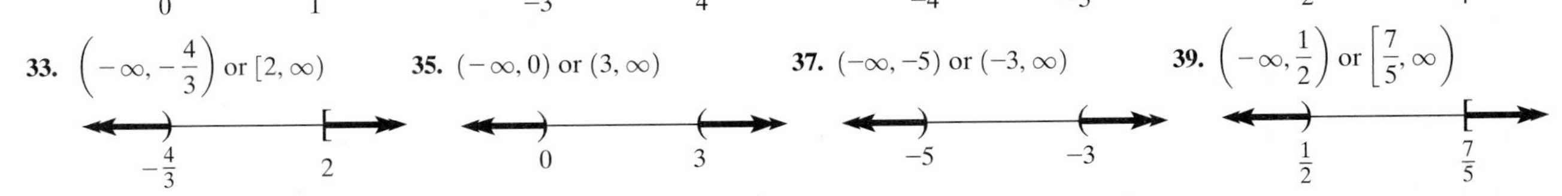

41. $(-\infty, -6]$ or $(-2, \infty)$

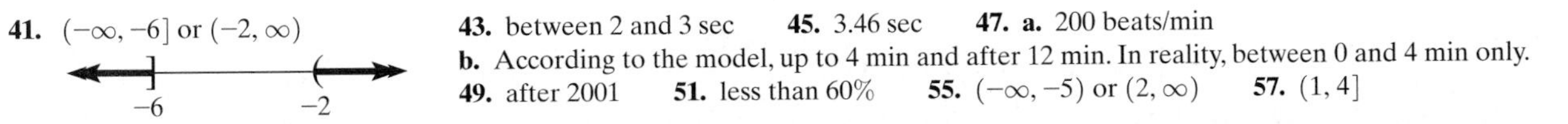

43. between 2 and 3 sec **45.** 3.46 sec **47. a.** 200 beats/min **b.** According to the model, up to 4 min and after 12 min. In reality, between 0 and 4 min only. **49.** after 2001 **51.** less than 60% **55.** $(-\infty, -5)$ or $(2, \infty)$ **57.** $(1, 4]$

59. $(-4, -1)$ or $[2, \infty)$ **61.** from 1.7 mm to 3.5 mm **63.** Answers may vary. **65.** Because the square of any number other than zero is positive, the solution includes all real numbers except 2. **67.** Because the square of any number is positive, the solution is $\varnothing$.
69. a. $(-\infty, \infty)$ **b.** $\varnothing$

Chapter 1 Review Exercises

1. $\{6\}$ **2.** $\{-10\}$ **3.** $\{5\}$ **4.** $\{-13\}$ **5.** $\{-3\}$ **6.** $\{-1\}$ **7.** $\{2\}$ **8.** $\{2\}$ **9.** $\left\{\frac{72}{11}\right\}$ **10.** $\{-12\}$ **11.** $\left\{\frac{77}{15}\right\}$
12. a. 0 **b.** $\{2\}$ **13. a.** 5 **b.** $\varnothing$ **14. a.** $-1, 1$ **b.** all real numbers except ± 1 **15. a.** $-2, 4$ **b.** $\{7\}$
16. inconsistent equation **17.** identity **18.** conditional equation **19.** 1997 **20.** 2000 **21.** 20 times
22. LA = 159; NY = 26 **23.** \$6250 at 8%; \$3750 at 12% **24.** width = 53 m; length = 120 m **25.** \$10,000 **26.** \$450
27. 95 concerts **28.** $h = \frac{3V}{B}$ **29.** $M = \frac{f - F}{f}$ **30.** $\left\{-8, \frac{1}{2}\right\}$ **31.** $\{-4, 0\}$ **32.** $\{-8, 8\}$ **33.** $\left\{\frac{4 + 3\sqrt{2}}{3}, \frac{4 - 3\sqrt{2}}{3}\right\}$
34. $100; (x + 10)^2$ **35.** $\frac{9}{4}; \left(x - \frac{3}{2}\right)^2$ **36.** $\{3, 9\}$ **37.** $\left\{2 + \frac{\sqrt{3}}{3}, 2 - \frac{\sqrt{3}}{3}\right\}$ **38.** $\{1 + \sqrt{5}, 1 - \sqrt{5}\}$
39. $\{1 + 3i\sqrt{2}, 1 - 3i\sqrt{2}\}$ **40.** $\left\{\frac{-2 + \sqrt{10}}{2}, \frac{-2 - \sqrt{10}}{2}\right\}$ **41.** -36; 2 complex imaginary solutions **42.** 81; 2 unequal real solutions
43. $\left\{\frac{1}{2}, 5\right\}$ **44.** $\left\{-2, \frac{10}{3}\right\}$ **45.** $\left\{\frac{7 + \sqrt{37}}{6}, \frac{7 - \sqrt{37}}{6}\right\}$ **46.** $\{-3, 3\}$ **47.** $\{-2, 8\}$ **48.** $\left\{\frac{1 + i\sqrt{23}}{6}, \frac{1 - i\sqrt{23}}{6}\right\}$
49. 20 weeks **50.** 1989 **51.** 12 ft by 27 ft **52.** approximately 134.16 m **53.** $\{-5, 0, 5\}$ **54.** $\left\{-3, \frac{1}{2}, 3\right\}$ **55.** $\{2\}$
56. $\{8\}$ **57.** $\{16\}$ **58.** $\{32\}$ **59.** $\{-2, -1, 1, 2\}$ **60.** $\{16\}$ **61.** $\{-4, 3\}$ **62.** $\left\{-\frac{11}{2}, \frac{23}{2}\right\}$ **63.** $\left\{-1, -\frac{2\sqrt{6}}{9}, \frac{2\sqrt{6}}{9}, 1\right\}$
64. $\{2\}$ **65.** $\{1, 4\}$ **66.** $\{-3, -2, 3\}$ **67.** 1250 ft
68. **69.** **70.** **71.** $-2 < x \le 3$

72. $-1.5 \le x \le 2$ **73.** $x > -1$ **74.** $[-2, \infty)$ **75.** $\left[\frac{3}{5}, \infty\right)$

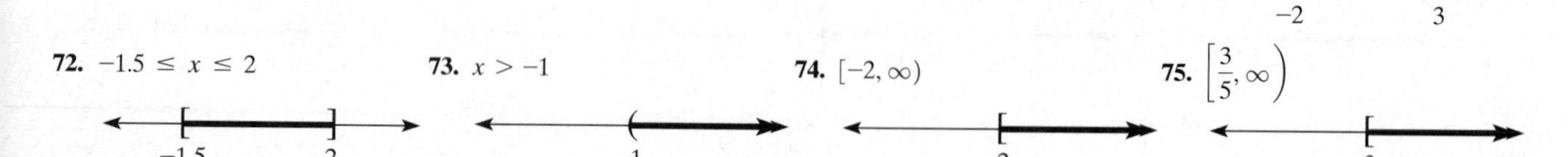

76. $\left(-\infty, -\frac{21}{2}\right)$ **77.** $(-3, \infty)$ **78.** $(-\infty, -2]$ **79.** $(2, 3]$

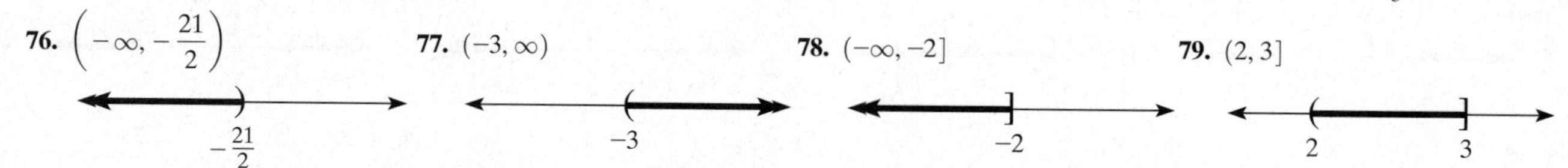

80. $[-9, 6]$ **81.** $(-\infty, -6)$ or $(0, \infty)$ **82.** $(-\infty, -3]$ or $[-2, \infty)$

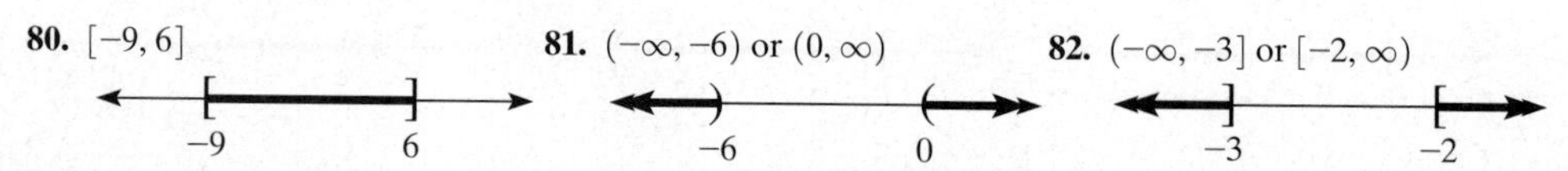

83. Canada, Former Soviet Union **84.** Australia, Canada, United States **85.** Most people sleep between 5.5 and 7.5 hours.
86. between 59° and 95° inclusively **87.** more than 50 checks **88.** 1986
89. $\left[-4, \frac{1}{2}\right]$ **90.** $\left(-\infty, \frac{3 - \sqrt{3}}{2}\right)$ or $\left(\frac{3 + \sqrt{3}}{2}, \infty\right)$ **91.** $(-\infty, -2)$ or $(6, \infty)$
92. $(-\infty, 4)$ or $\left[\frac{23}{4}, \infty\right)$ **93.** from 1 to 2 sec

Chapter 1 Test

1. $\{-1\}$ **2.** $\{-6\}$ **3.** $\{5\}$ **4.** $\left\{-\frac{1}{2}, 2\right\}$ **5.** $\left\{\frac{1 - 5\sqrt{3}}{3}, \frac{1 + 5\sqrt{3}}{3}\right\}$ **6.** $\{1 - \sqrt{5}, 1 + \sqrt{5}\}$ **7.** $\left\{\frac{2 - i}{2}, \frac{2 + i}{2}\right\}$
8. $\{-1, 1, 4\}$ **9.** $\{7\}$ **10.** $\{5\}$ **11.** $\{\sqrt[3]{4}\}$ **12.** $\{1, 512\}$ **13.** $\{6, 12\}$

14. $(-\infty, 12]$ **15.** $\left[\frac{21}{8}, \infty\right)$ **16.** $\left[-7, \frac{13}{2}\right)$ **17.** $\left(-\infty, -\frac{5}{3}\right]$ or $\left[\frac{1}{3}, \infty\right)$

18. $(-3, 4)$ **19.** $(3, 10)$ **20. a.** $n = 125B - 50w$ **b.** 24,000 years! **21.** 1986 **22.** 1997 **23.** \$42,000 **24.** \$4000 at 9%; \$2000 at 6% **25.** 10 feet up the pole **26.** at least 92 **27.** less than 100 hr

CHAPTER 2

Section 2.1

Check Point Exercises

1. a. 6 **b.** $-\frac{7}{5}$ **2.** $y + 5 = 6(x - 2)$; $y = 6x - 17$ **3.** $y + 1 = -5(x + 2)$; $y = -5x - 11$

4.

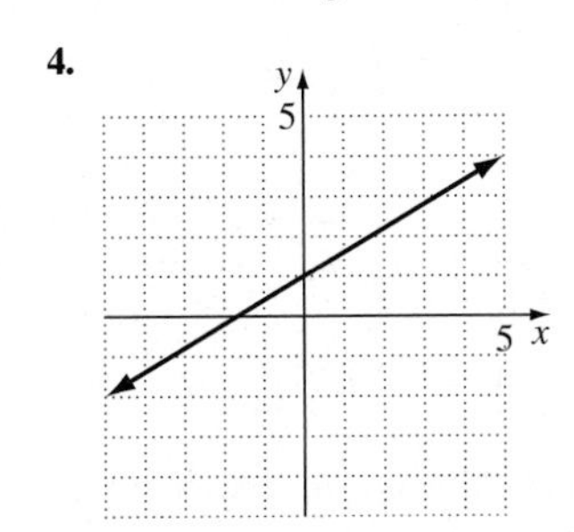

5.

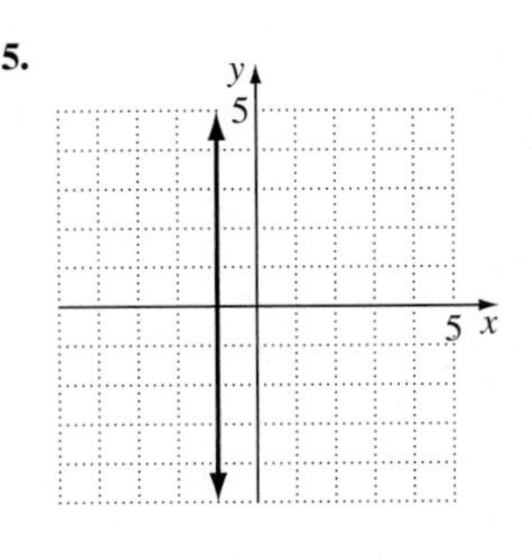

6. slope: $-\frac{1}{2}$; y-intercept: 2

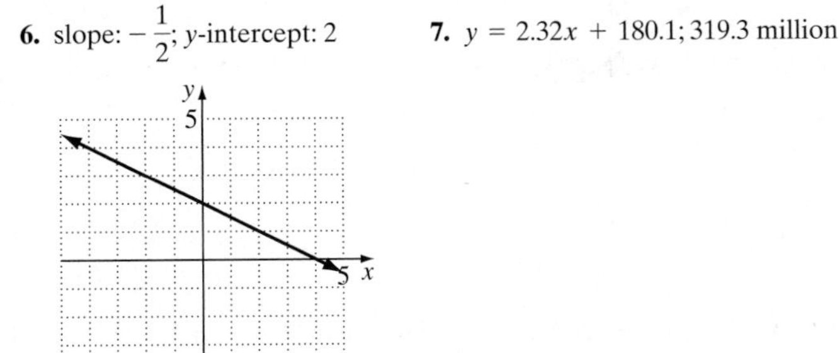

7. $y = 2.32x + 180.1$; 319.3 million

Exercise Set 2.1

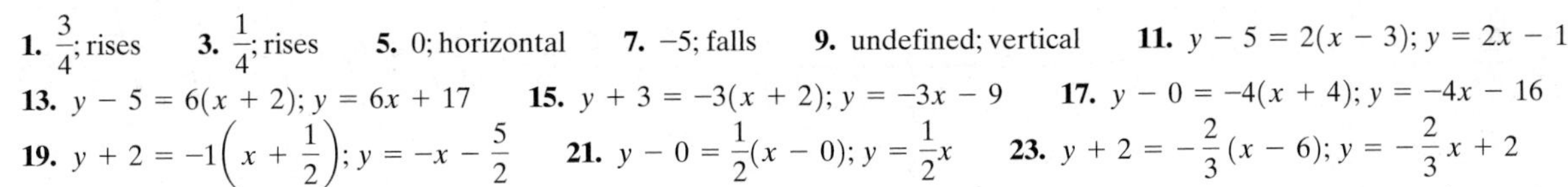

1. $\frac{3}{4}$; rises **3.** $\frac{1}{4}$; rises **5.** 0; horizontal **7.** -5; falls **9.** undefined; vertical **11.** $y - 5 = 2(x - 3)$; $y = 2x - 1$ **13.** $y - 5 = 6(x + 2)$; $y = 6x + 17$ **15.** $y + 3 = -3(x + 2)$; $y = -3x - 9$ **17.** $y - 0 = -4(x + 4)$; $y = -4x - 16$ **19.** $y + 2 = -1\left(x + \frac{1}{2}\right)$; $y = -x - \frac{5}{2}$ **21.** $y - 0 = \frac{1}{2}(x - 0)$; $y = \frac{1}{2}x$ **23.** $y + 2 = -\frac{2}{3}(x - 6)$; $y = -\frac{2}{3}x + 2$ **25.** using $(1, 2)$, $y - 2 = 2(x - 1)$; $y = 2x$ **27.** using $(-3, 0)$, $y - 0 = 1(x + 3)$; $y = x + 3$ **29.** using $(-3, -1)$, $y + 1 = 1(x + 3)$; $y = x + 2$ **31.** using $(-3, -2)$, $y + 2 = \frac{4}{3}(x + 3)$; $y = \frac{4}{3}x + 2$ **33.** using $(-3, -1)$, $y + 1 = 0(x + 3)$; $y = -1$ **35.** using $(2, 4)$, $y - 4 = 1(x - 2)$; $y = x + 2$ **37.** using $(0, 4)$, $y - 4 = 8(x - 0)$; $y = 8x + 4$

39. $m = 2$; $b = 1$

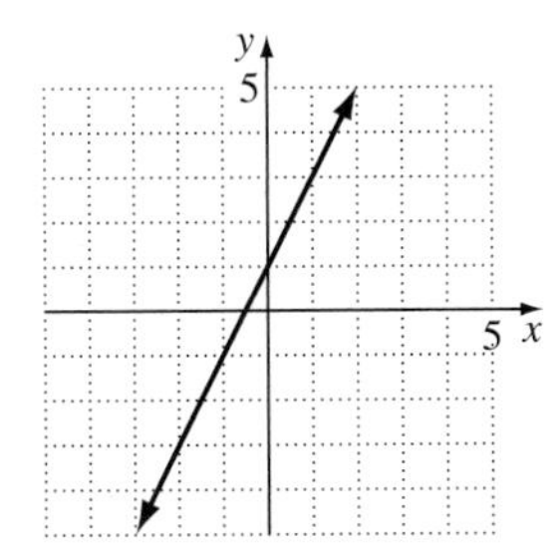

41. $m = -2$; $b = 1$

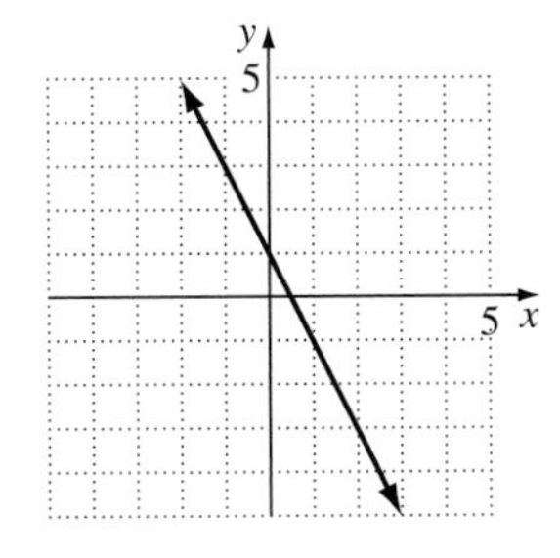

43. $m = \frac{3}{4}$; $b = -2$

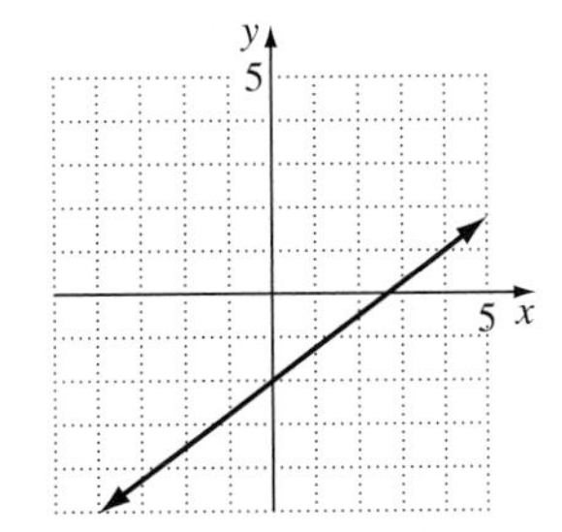

45. $m = -\frac{3}{5}$; $b = 7$

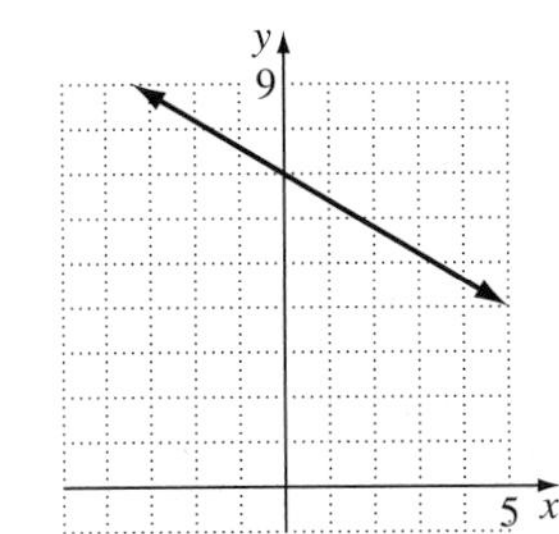

47.

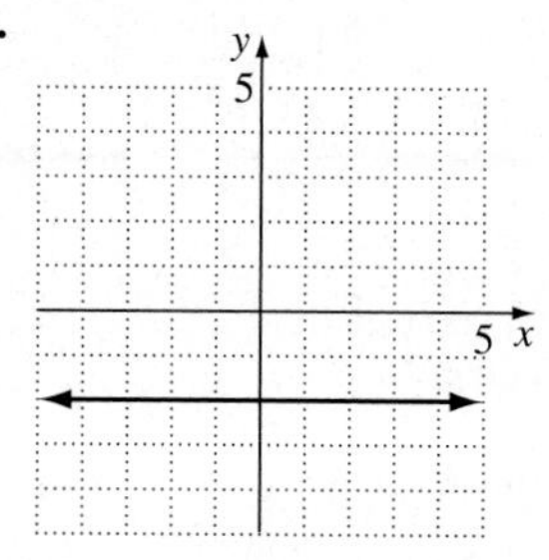

49.

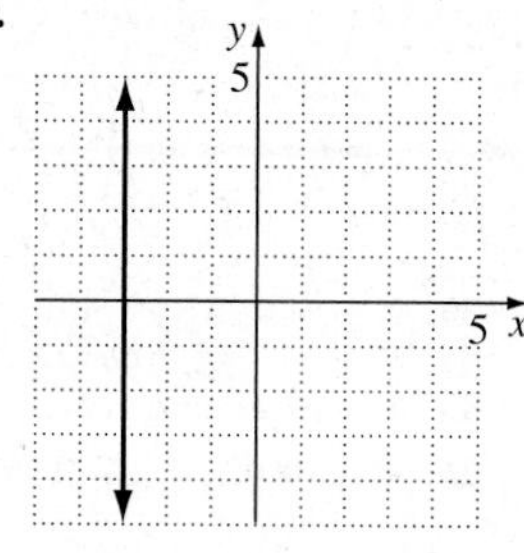

51.

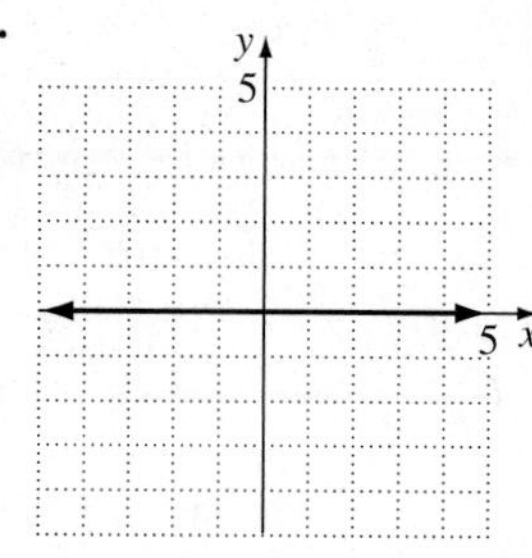

53. a. $y = -3x + 5$
b. $m = -3; b = 5$
c.

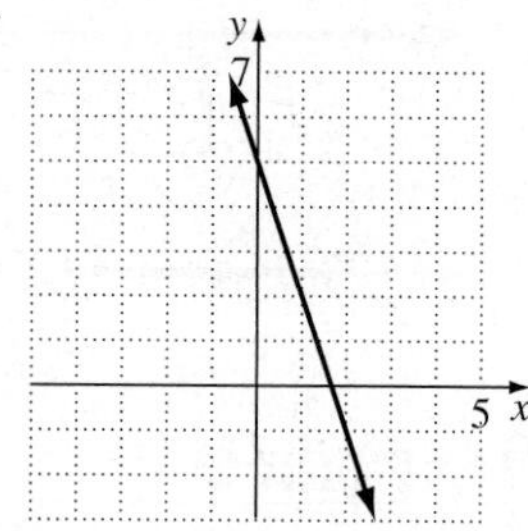

55. a. $y = -\frac{2}{3}x + 6$
b. $m = -\frac{2}{3}; b = 6$
c.

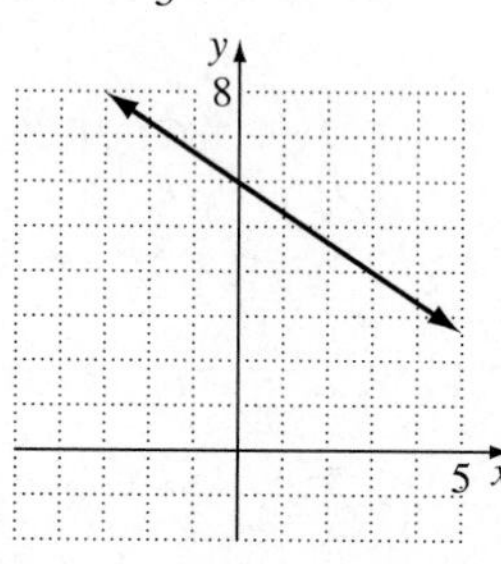

57. a. $y = 2x - 3$
b. $m = 2; b = -3$
c.

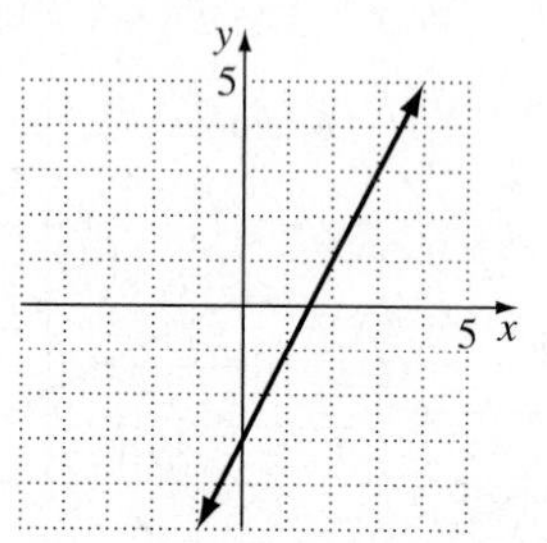

59. a. $x = 3$
b. m is undefined; no y-intercept
c.

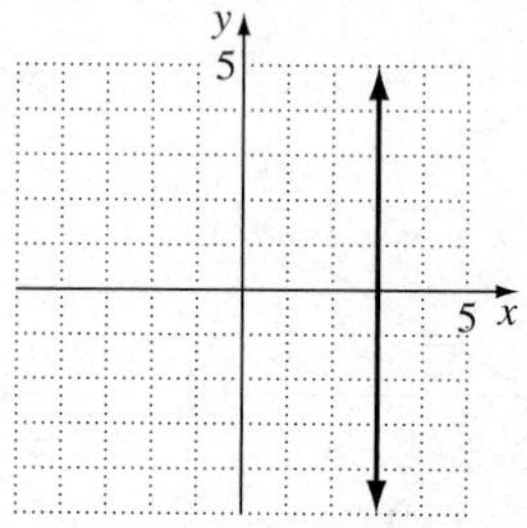

61. $y = 80$ **63.** In 2030, $y = 344$ million; In 2040; $y = 367.5$ million; In 2050, $y = 391$ million; The equation models the projections well, but is slightly lower than the projections. **65.** $y - 4459.2 = 258.625(x - 3)$; $y = 258.625x + 3683.325$; \$11,442.075 billion
67. $y = -0.002x + 59$; 4500 shirts **69.** Point-slope form: $y - 3 = -0.65(x - 12)$; Slope-intercept form: $y = -0.65x + 10.8$; about 6.25

79. $m = -3$

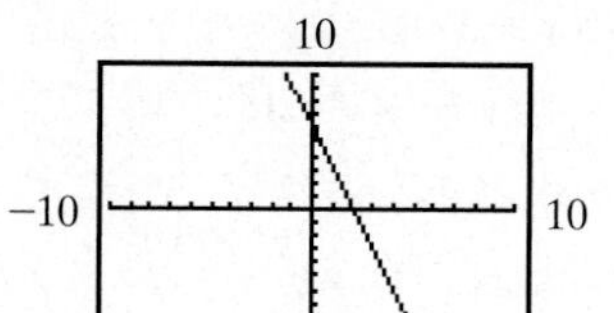

81. $m = \frac{3}{4}$

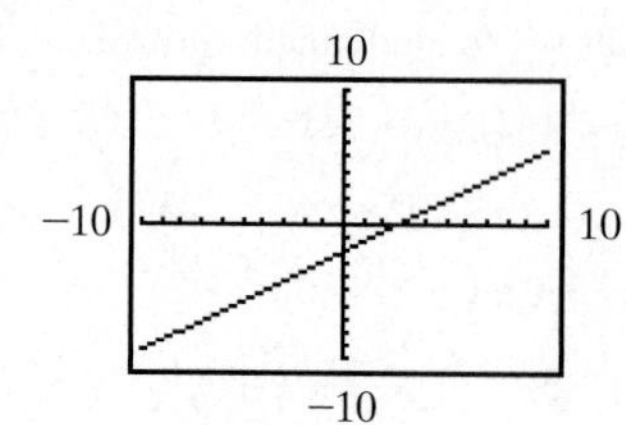

83. (c) is true.

85. a. m_1, m_3, m_2, m_4
b. b_2, b_1, b_4, b_3

Section 2.2

Check Point Exercises

1. $y - 5 = 3(x + 2)$; $y = 3x + 11$ **2.** 3 **3.** $x^2 + y^2 = 16$ **4.** $(x - 5)^2 + (y + 6)^2 = 100$
5. center: $(-3, 1)$; radius: 2 **6.** $(x + 2)^2 + (y - 2)^2 = 9$

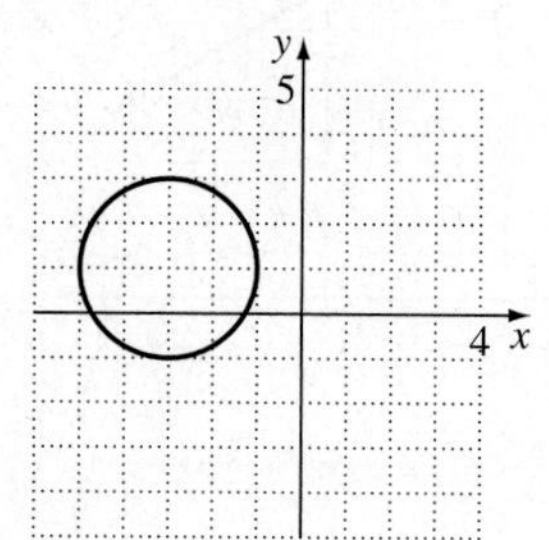

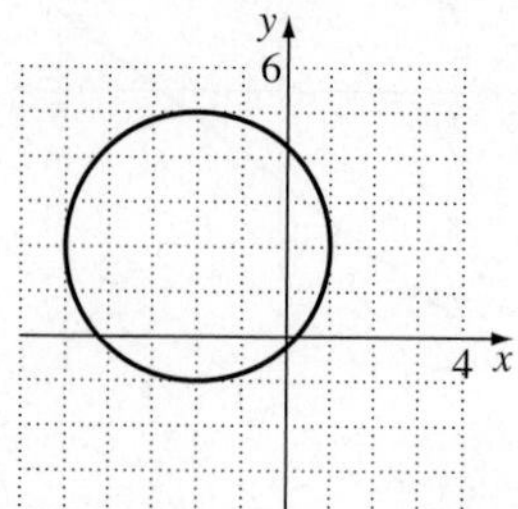

Exercise Set 2.2

1. a. 5 **b.** $-\frac{1}{5}$ **3. a.** -7 **b.** $\frac{1}{7}$ **5. a.** $\frac{1}{2}$ **b.** -2 **7. a.** $-\frac{2}{5}$ **b.** $\frac{5}{2}$ **9. a.** -4 **b.** $\frac{1}{4}$ **11. a.** $-\frac{1}{2}$ **b.** 2 **13. a.** $\frac{2}{3}$ **b.** $-\frac{3}{2}$ **15. a.** undefined **b.** 0 **17.** $y - 2 = 2(x - 4); y = 2x - 6$ **19.** $y - 4 = -\frac{1}{2}(x - 2); y = -\frac{1}{2}x + 5$ **21.** $y + 10 = -4(x + 8); y = -4x - 42$ **23.** $y + 3 = -5(x - 2); y = -5x + 7$ **25.** $y - 2 = \frac{2}{3}(x + 2); y = \frac{2}{3}x + \frac{10}{3}$ **27.** $y + 7 = -2(x - 4); y = -2x + 1$ **29.** $x^2 + y^2 = 49$ **31.** $(x - 3)^2 + (y - 2)^2 = 25$ **33.** $(x + 1)^2 + (y - 4)^2 = 4$ **35.** $(x + 3)^2 + (y + 1)^2 = 3$ **37.** $(x + 4)^2 + (y - 0)^2 = 100$

39. center: (0, 0)
radius: 4

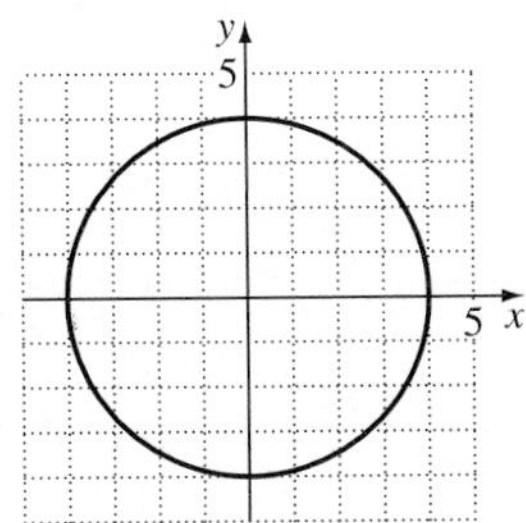

41. center: (3, 1)
radius: 6

43. center: (−3, 2)
radius: 2

45. center: (−2, −2)
radius: 2

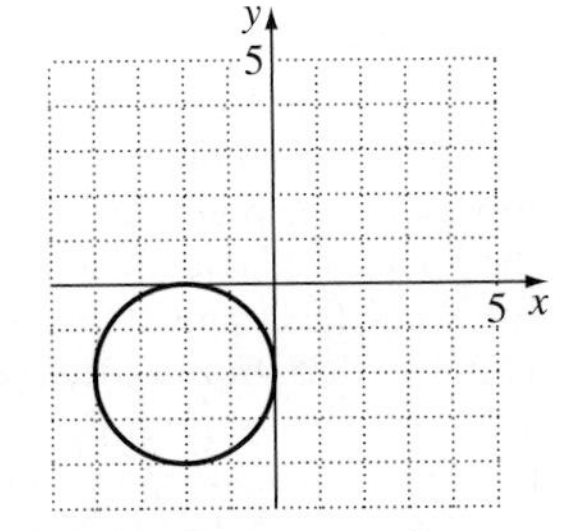

47. $(x + 3)^2 + (y + 1)^2 = 4$
center: (−3, −1)
radius: 2

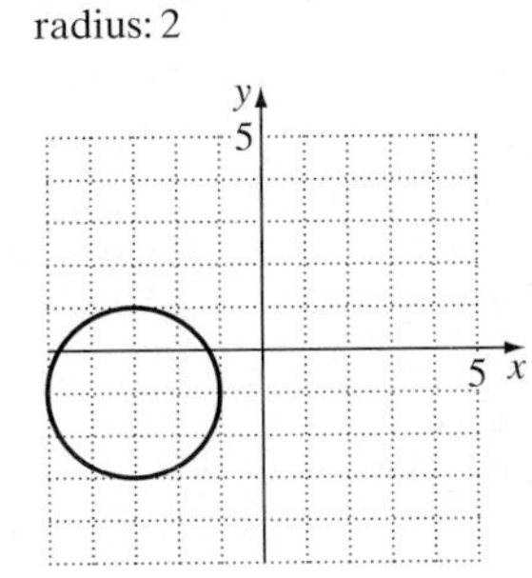

49. $(x - 5)^2 + (y - 3)^2 = 64$
center: (5, 3)
radius: 8

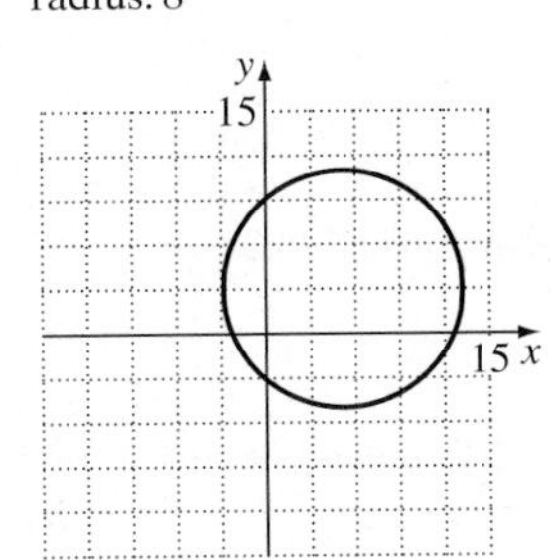

51. $(x + 4)^2 + (y - 1)^2 = 25$
center: (−4, 1)
radius: 5

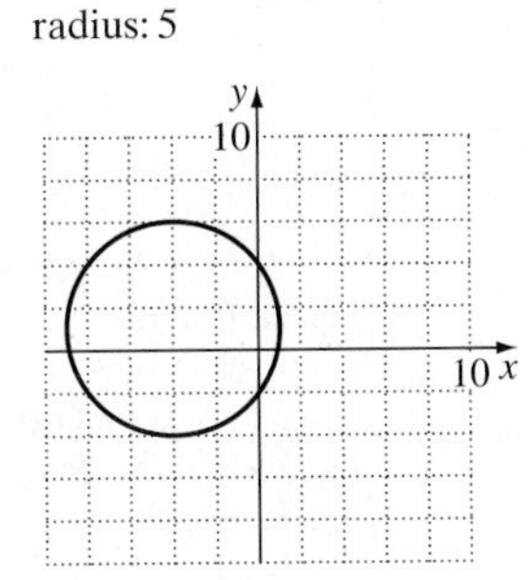

53. $(x - 1)^2 + (y - 0)^2 = 16$
center: (1, 0)
rradius: 4

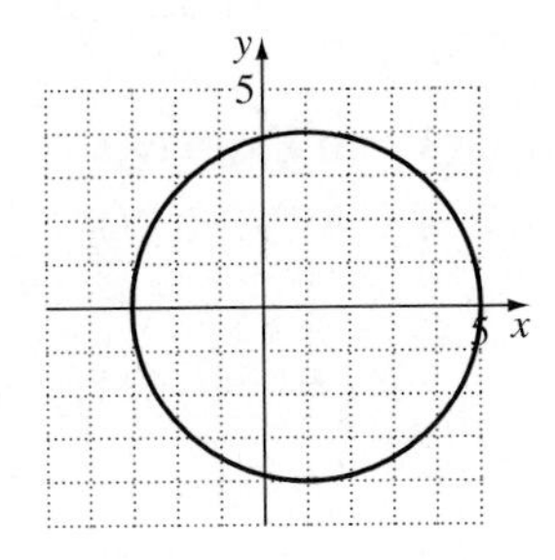

55. a. $\frac{1}{5}$ **b.** $\frac{2}{15}$ **c.** no; The yearly increase for women is greater than the yearly increase for men.
57. a. $x^2 + y^2 = 1444$ **b.** $x^2 + y^2 = 2704$

69.

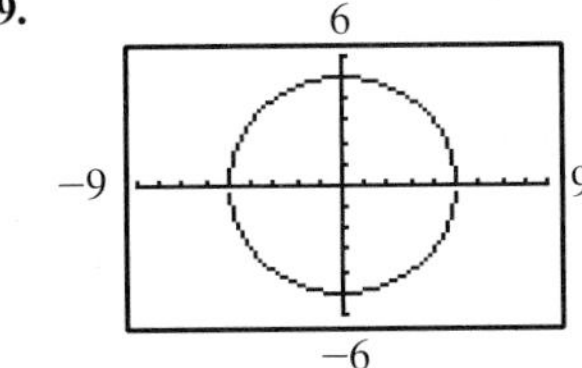

71.

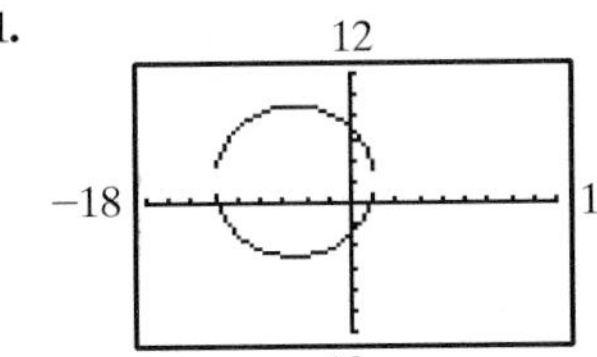

73. $y - 0 = 3(x + 3)$ or $y = 3x + 9$
75. $(x - 3)^2 + (y + 5)^2 = 61;$
$x^2 + y^2 - 6x + 10y - 27 = 0$
77. 11π

Section 2.3

Check Point Exercises

1. domain: {20, 30, 100, 200}; range: {157.4, 231.8, 752.6, 1496.6} **2. a.** not a function **b.** function **3. a.** $y = 6 - 2x$; function **b.** $y = \pm\sqrt{1 - x^2}$; not a function **4. a.** 42 **b.** $x^2 + 6x + 15$ **c.** $x^2 + 2x + 7$ **5. a.** 28 **b.** 33 **6. a.** $(-\infty, \infty)$ **b.** $\{x|x \neq -7, x \neq 7\}$ **c.** $[3, \infty)$ **7.** $f(50) = 20.19$; In 1990, United States automobiles averaged 20.19 miles per gallon.

Exercise Set 2.3

1. function; {1, 3, 5}; {2, 4, 5} **3.** not a function; {3, 4}; {4, 5} **5.** function; {−3, −2, −1, 0}; {−3, −2, −1, 0} **7.** not a function; {1}; {4, 5, 6} **9.** y is a function of x. **11.** y is a function of x. **13.** y is not a function of x. **15.** y is not a function of x. **17.** y is a function of x. **19.** y is a function of x. **21. a.** 29 **b.** $4x + 9$ **c.** $-4x + 5$ **23. a.** 2 **b.** $x^2 + 12x + 38$

c. $x^2 - 2x + 3$ **25. a.** 13 **b.** 1 **c.** $x^4 - x^2 + 1$ **d.** $81a^4 - 9a^2 + 1$ **27. a.** 3 **b.** 7 **c.** $\sqrt{x} + 3$ **29. a.** $\frac{15}{4}$
b. $\frac{15}{4}$ **c.** $\frac{4x^2 - 1}{x^2}$ **31. a.** 1 **b.** −1 **c.** 1 **33. a.** $4a$ **b.** $4a + 4h$ **c.** $4, h \neq 0$ **d.** $4a + 4h$ **35. a.** $3a + 7$
b. $3a + 3h + 7$ **c.** $3, h \neq 0$ **d.** $3a + 3h + 14$ **37. a.** $-5a - 3$ **b.** $-5a - 5h - 3$ **c.** $-5, h \neq 0$ **d.** $-5a - 5h - 6$
39. a. a^2 **b.** $a^2 + 2ah + h^2$ **c.** $2a + h, h \neq 0$ **d.** $a^2 + h^2$ **41. a.** 6 **b.** 6 **c.** $0, h \neq 0$ **d.** 12
43. a. $\frac{1}{a}$ **b.** $\frac{1}{a + h}$ **c.** $-\frac{1}{a(a + h)}, h \neq 0$ **d.** $\frac{h + a}{ah}$ **45. a.** −1 **b.** 7 **c.** 19 **47. a.** 3 **b.** 3 **c.** 0
49. a. 8 **b.** 3 **c.** 6 **51.** $(-\infty, \infty)$ **53.** $(-\infty, 4)$ or $(4, \infty)$ **55.** $(-\infty, -4)$ or $(-4, 4)$ or $(4, \infty)$
57. $(-\infty, -3)$ or $(-3, 7)$ or $(7, \infty)$ **59.** $(-\infty, -8)$ or $(-8, -3)$ or $(-3, \infty)$ **61.** $(-\infty, \infty)$ **63.** $[3, \infty)$ **65.** $(3, \infty)$
67. $[-7, \infty)$ **69.** $(-\infty, 12]$ **71.** $(-\infty, -2]$ or $[7, \infty)$ **73.** Answers may vary.
75. $f(16) = 5.22$; There were 5.22 million women enrolled in United States colleges in the year 2000.
77. $f(20) - g(20) = 1.4$; There will be 1.4 million more women than men enrolled in United States colleges in the year 2004.
79. $f(0) = 200$; There were 200 thousand lawyers in the United States in 1951.
81. $f(50) = 1058$; There will be 1058 thousand or 1,058,000 lawyers in the United States in the year 2001.
83. $f(0) = 7$ represents the point (0, 7) on the graph. It means that an average infant girl weighs 7 pounds at birth.; $f(2) = 10$ represents the point (2, 10) on the graph. It means that an average infant girl weighs 10 pounds at age 2 months.; $f(4) = 13$ represents the point (4, 13) on the graph. It means that an average infant girl weighs 13 pounds at age 4 months.; $f(6) = 16$ represents the point (6, 16) on the graph. It means that an average infant girl weighs 16 pounds at age 6 months.
85. $f(10) = 3375.97$; In 1984, 3375.97 calories per person were consumed each day in the United States.
87. $f(15) - f(10) = 101.6$; Between 1984 and 1989, the number of calories per person consumed each day in the United States increased by 101.6.
89. a. $f(15) = 9.3$; In 1955, an average of 9.3 thousand miles were driven per car in the United States each year.
b. $f(50) = 10.7$; In 1990, an average of 10.7 thousand miles were driven per car in the United States each year.
91. $V = 17{,}900 - 2100x$; $V(4) = 9500$; The value of the car after 4 years is \$9500. **99.** $[1, \infty)$ **101.** $(-\infty, 5]$
103. Answers may vary. **105.** Answers may vary.

Section 2.4

Check Point Exercises

1. $(-3, 7), (-2, 2), (-1, -1), (0, -2), (1, -1), (2, 2), (3, 7)$

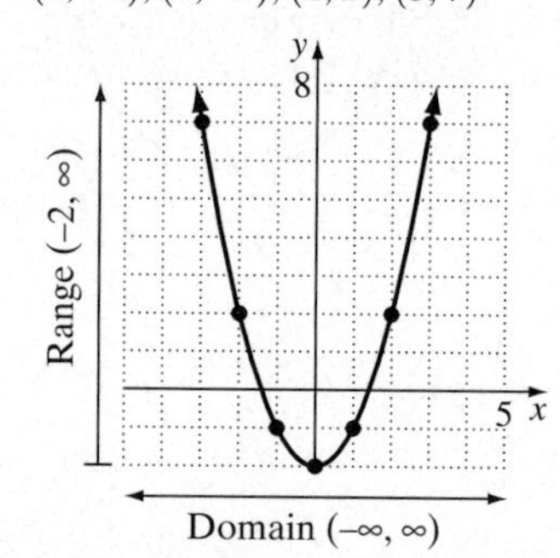

2. $f(4) = 1$; domain: $[0, 6)$; range: $(-2, 2]$
3. a. function
b. function
c. not a function
4. increasing on $(-\infty, -1)$, decreasing on $(-1, 1)$, increasing on $(1, \infty)$
5. a. even
b. odd
c. neither
6. a. (0, 3); Drug concentration increases during the first three hours after an injection.
b. (3, 13); Drug concentration decreases between the third and thirteenth hours after an injection.
c. 0.05 milligrams per 100 milliliters at 3 hours
d. By the end of the 13 hours, there is no more of the drug in the body.

Exercise Set 2.4

1. $(-3, 11), (-2, 6), (-1, 3), (0, 2), (1, 3), (2, 6), (3, 11)$

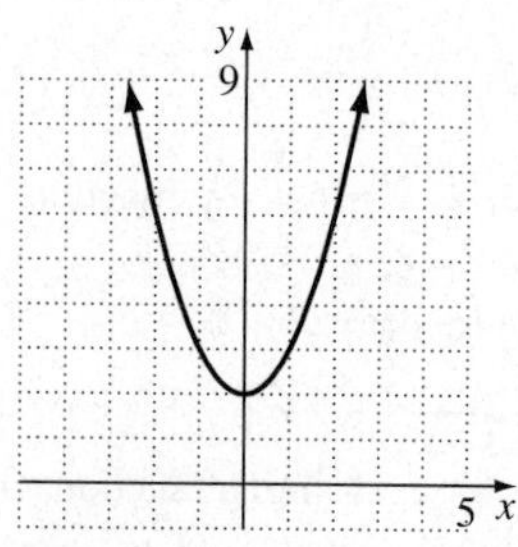

Domain: $(-\infty, \infty)$
Range: $[2, \infty)$

3. $(0, -1), (1, 0), (4, 1), (9, 2)$

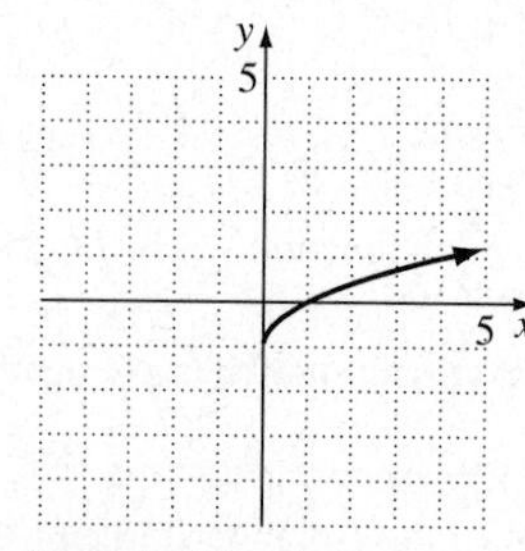

Domain: $[0, \infty)$
Range: $[-1, \infty)$

5. $(1, 0), (2, 1), (5, 2), (10, 3)$

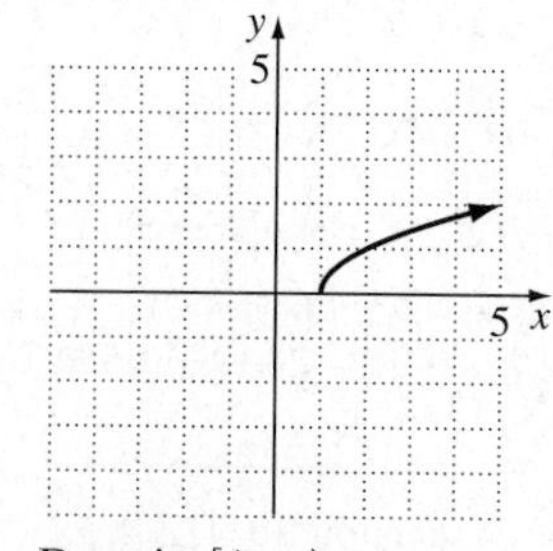

Domain: $[1, \infty)$
Range: $[0, \infty)$

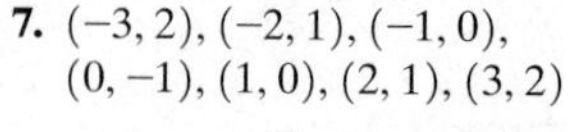
7. $(-3, 2), (-2, 1), (-1, 0), (0, -1), (1, 0), (2, 1), (3, 2)$

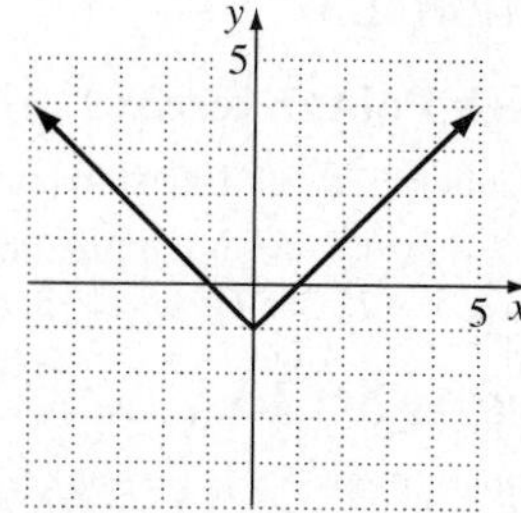

Domain: $(-\infty, \infty)$
Range: $[-1, \infty)$

9. $(-3, 4), (-2, 3), (-1, 2), (0, 1), (1, 0), (2, 1), (3, 2)$

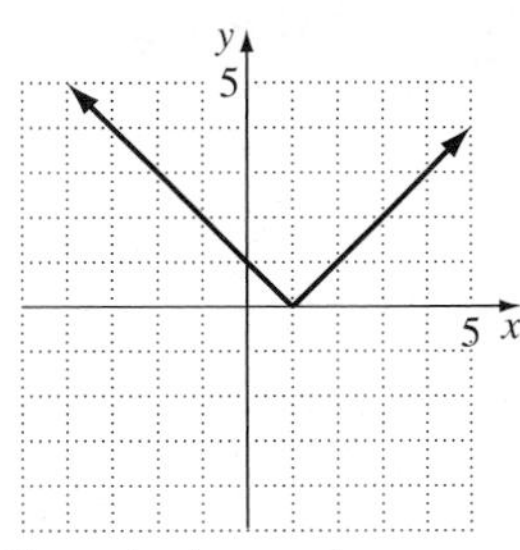

Domain: $(-\infty, \infty)$
Range: $[0, \infty)$

11. $(-3, 5), (-2, 5), (-1, 5), (0, 5), (1, 5), (2, 5), (3, 5)$

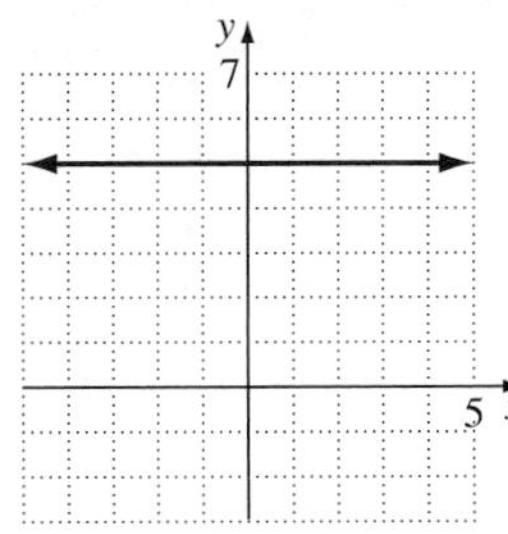

Domain: $(-\infty, \infty)$
Range: $\{5\}$

13. $(-2, -10), (-1, -3), (0, -2), (1, -1), (2, 6)$

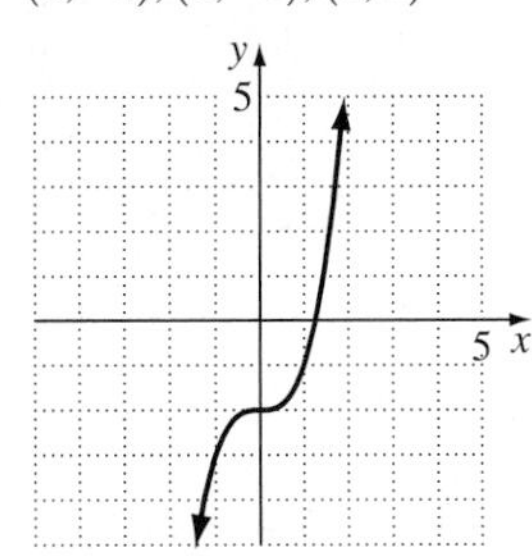

Domain: $(-\infty, \infty)$
Range: $(-\infty, \infty)$

15. a. $(-\infty, \infty)$ **b.** $[-4, \infty)$ **c.** -3 and 1 **d.** -3
17. a. $(-\infty, \infty)$ **b.** $[1, \infty)$ **c.** none **d.** 1 **e.** $f(-1) = 2$ and $f(3) = 4$
19. a. $[0, 5)$ **b.** $[-1, 5)$ **c.** 2 **d.** -1 **e.** $f(3) = 1$

21. a. $[0, \infty)$ **b.** $[1, \infty)$ **c.** none **d.** 1 **e.** $f(4) = 3$ **23. a.** $[-2, 6]$ **b.** $[-2, 6]$ **c.** 4 **d.** 4 **e.** $f(-1) = 5$ **25. a.** $(-\infty, \infty)$ **b.** $(-\infty, -2]$ **c.** none **d.** -2 **e.** $f(-4) = -5$ and $f(4) = -2$ **27. a.** $(-\infty, \infty)$ **b.** $(0, \infty)$ **c.** none **d.** 1 **29. a.** $\{-5, -2, 0, 1, 3\}$ **b.** $\{2\}$ **c.** none **d.** 2 **31.** function **33.** function **35.** not a function **37.** function **39. a.** increasing: $(-1, \infty)$ **b.** decreasing: $(-\infty, -1)$ **c.** constant: none **41. a.** increasing: $(0, \infty)$ **b.** decreasing: none **c.** constant: none **43. a.** increasing: none **b.** decreasing: $(-2, 6)$ **c.** constant: none **45. a.** increasing: $(-\infty, -1)$ **b.** decreasing: none **c.** constant: $(-1, \infty)$ **47. a.** increasing: $(-\infty, 0)$ or $(1.5, 3)$ **b.** decreasing: $(0, 1.5)$ or $(3, \infty)$ **c.** constant: none **49. a.** increasing: $(-2, 4)$ **b.** decreasing: none **c.** constant: $(-\infty, -2)$ or $(4, \infty)$ **51.** odd **53.** neither **55.** even **57.** even **59.** even **61.** neither **63.** even **65.** odd **67.** identity function **69.** square root function **71.** standard cubic function **73.** $f(1.06) = 1$ **75.** $f\left(\frac{1}{3}\right) = 0$ **77.** $f(-2.3) = -3$ **79.** $f(1989) \approx 294$ billion dollars. This is the maximum function value. **81.** Defense spending is increasing from 1988 to 1989, from 1991 to 1992, and from 1996 to 1997.
83. a. increasing: $(45, 74)$; decreasing: $(16, 45)$; The number of accidents occurring per 50,000 miles driven increases with age starting at age 45, while it decreases with age starting at age 16.
b. $x = 45$ and $f(45) = 190$; The fewest number of accidents per 50 million miles driven occurs at age 45.
c. $[190, 526.4]$; Between the ages of 16 and 74, the number of accidents per 50 million miles driven is between 190 and 526.4.

85.

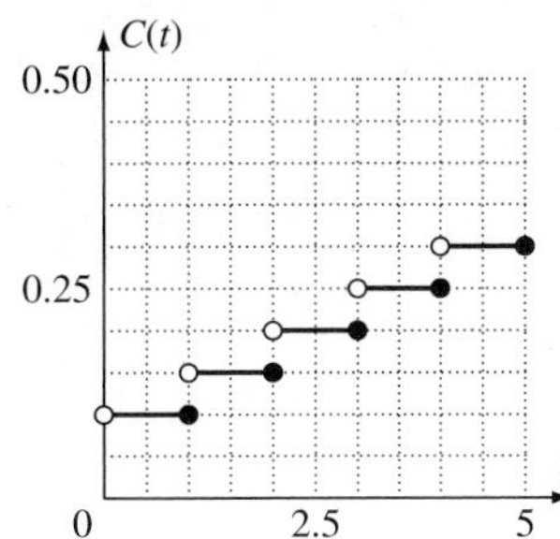

87. Answers may vary.

97. a.

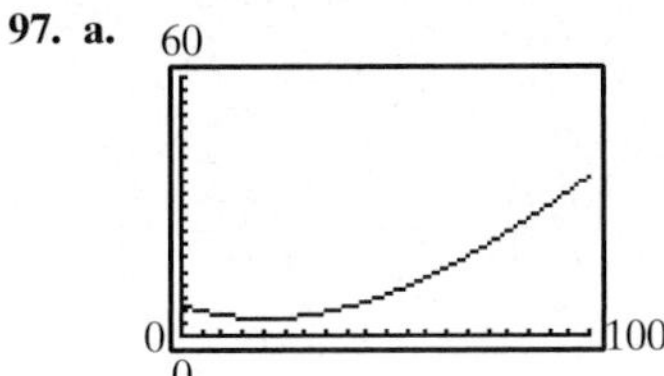

b. The number of doctor visits decreases during childhood and then increases as you get older.
c. The minimum is (20.29, 3.99), which means that the minimum number of doctor visits, about 4, occurs at around age 20.

99.

Increasing: $(-\infty, 1)$ or $(3, \infty)$
Decreasing: $(1, 3)$

101.

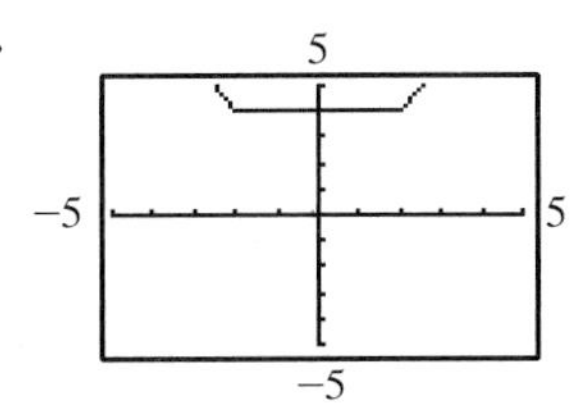

Increasing: $(2, \infty)$
Decreasing: $(-\infty, -2)$
Constant: $(-2, 2)$

103.

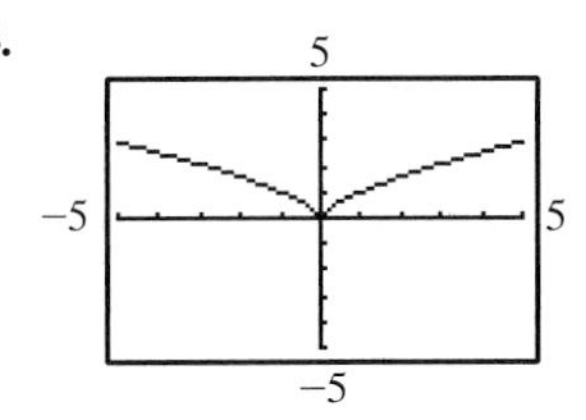

Increasing: $(0, \infty)$
Decreasing: $(-\infty, 0)$

105. a.

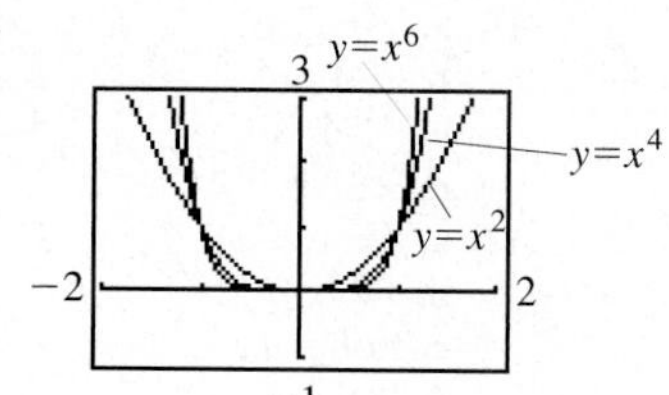

b.

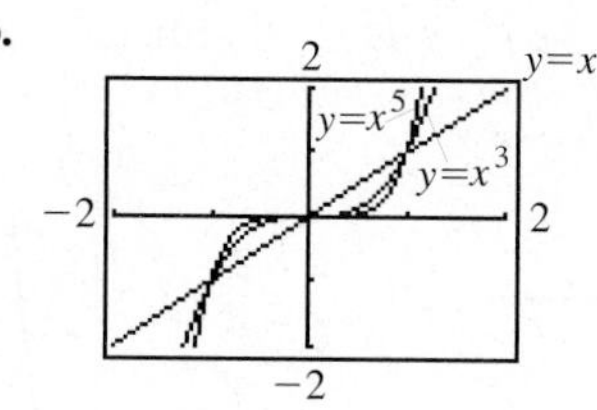

c. Increasing: $(0, \infty)$; Decreasing: $(-\infty, 0)$
d. $f(x) = x^n$ is increasing from $(-\infty, \infty)$ when n is odd.
e.

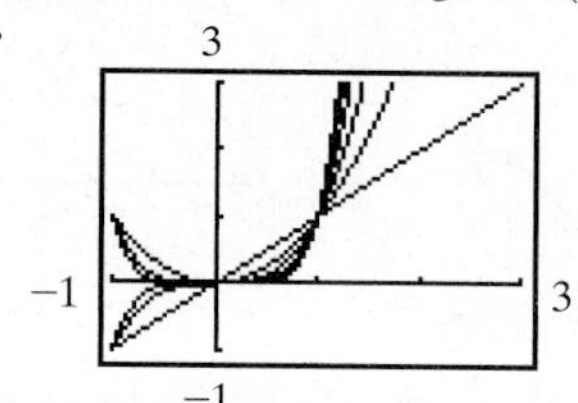

As n increases the steepness increases.

111.

Weight at least	Cost
0 oz.	\$0.33
1	0.55
2	0.77
3	0.99
4	1.21

107. Answers may vary. **109.** Answers may vary.

Section 2.5

Check Point Exercises

1.

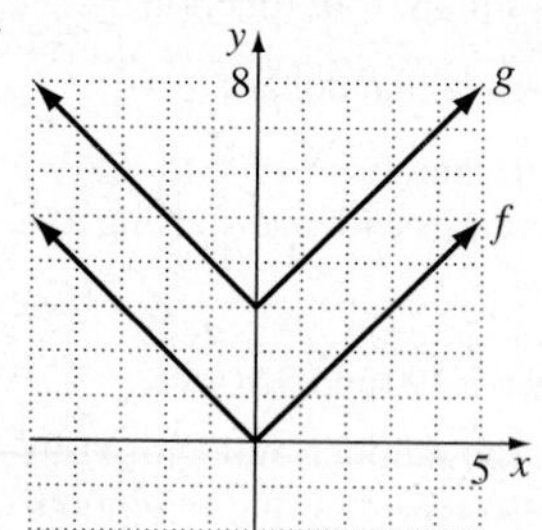

2.

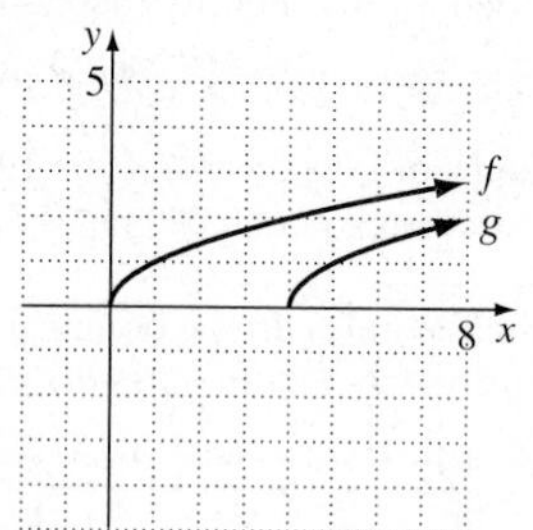

3.

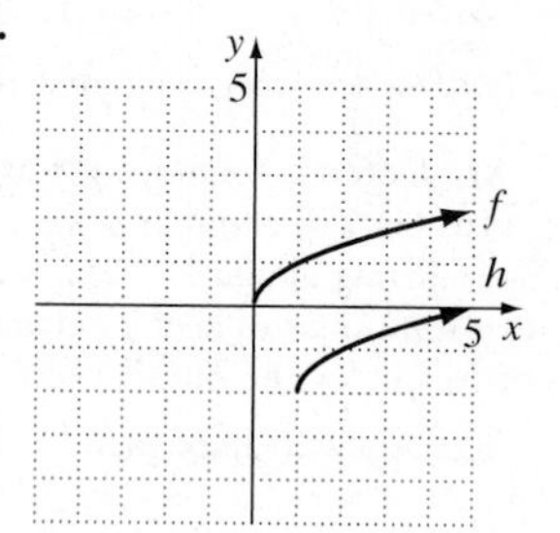

4.

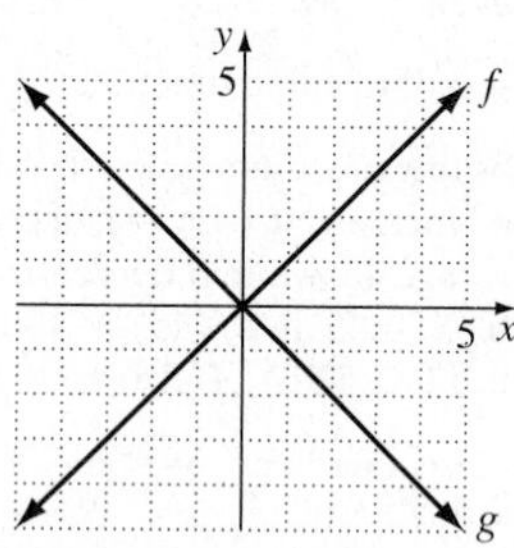

5.

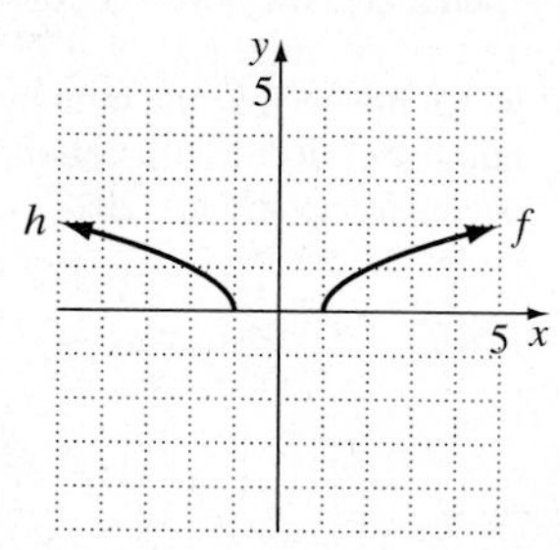

6.

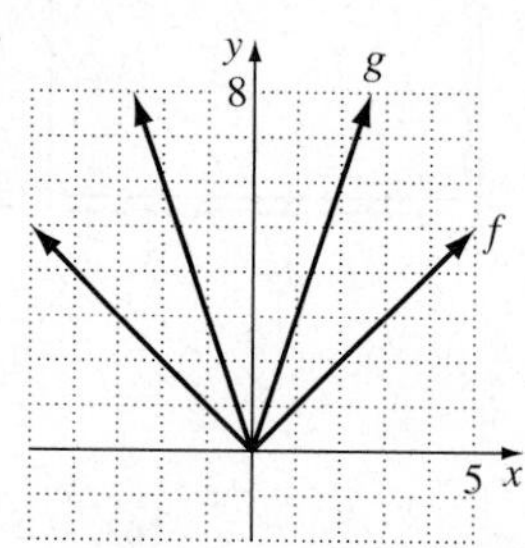

7.

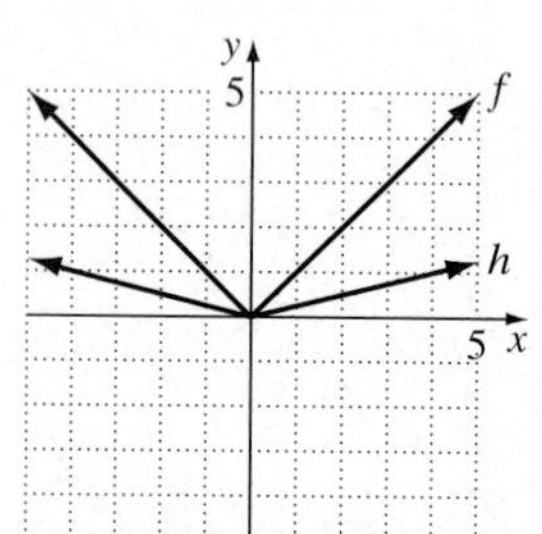

8.

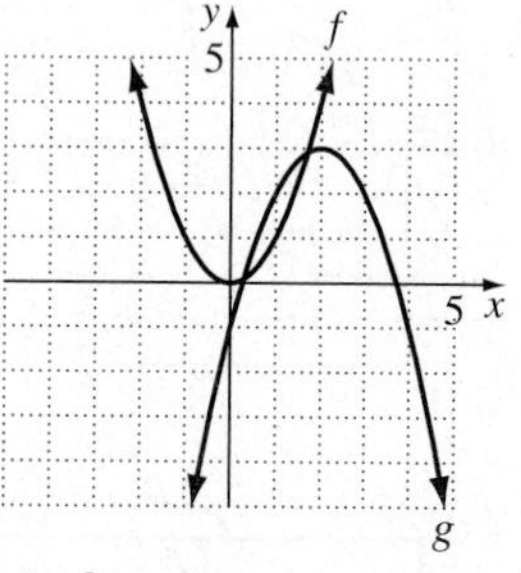

9. a. $(f + g)(x) = 3x^2 + 6x + 6$ **b.** $(f + g)(4) = 78$ **10. a.** $(f + g)(x) = \sqrt{x - 3} + \sqrt{x + 1}$ **b.** $[3, \infty)$

11. a. $(f - g)(x) = -x^2 + x - 4$ **b.** $(fg)(x) = x^3 - 5x^2 - x + 5$ **c.** $\left(\frac{f}{g}\right)(x) = \frac{x - 5}{x^2 - 1}, x \neq \pm 1$

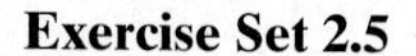

Exercise Set 2.5

1.

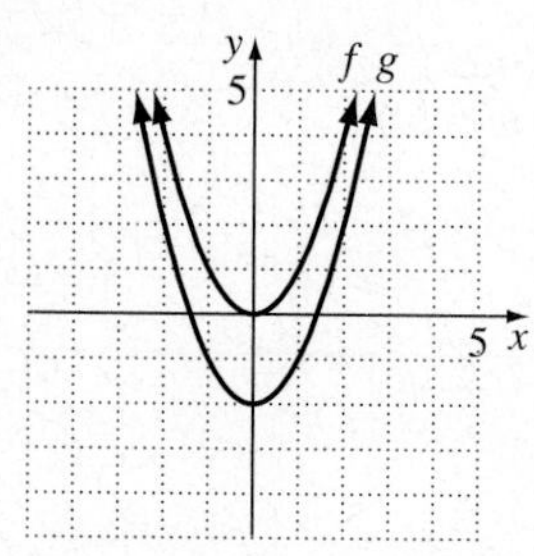

3.

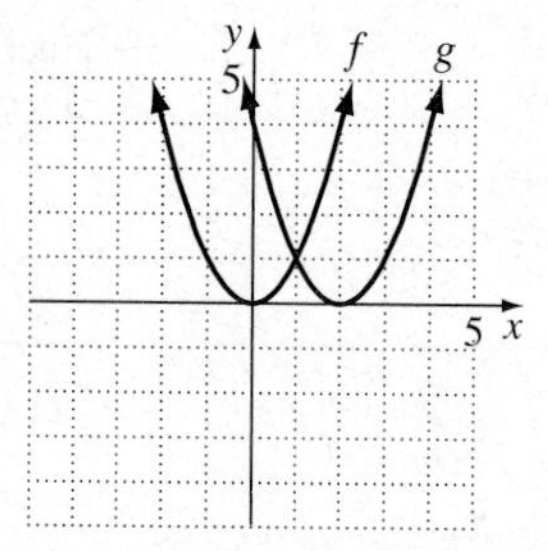

5.

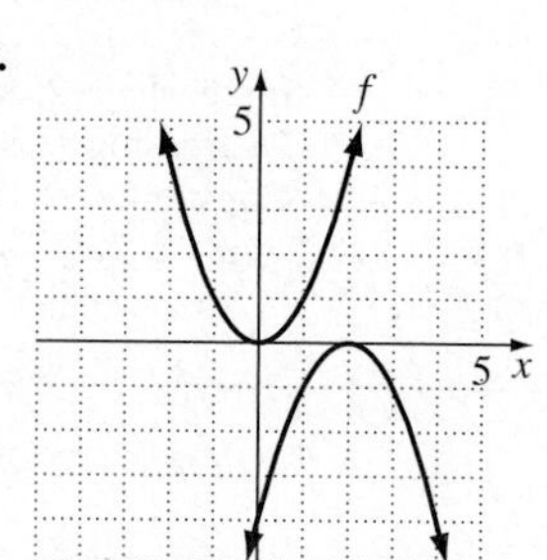

7.

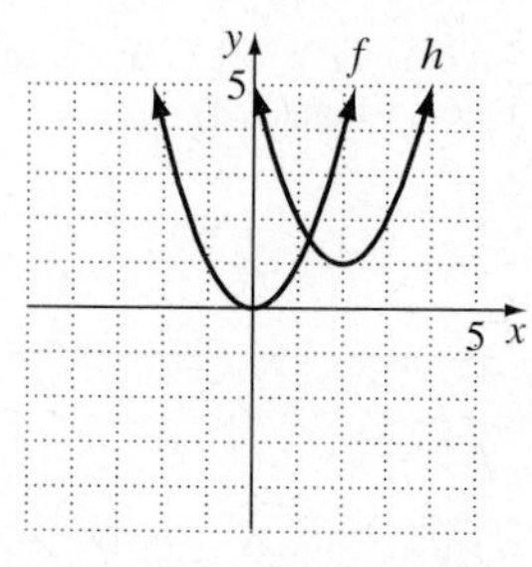

9.

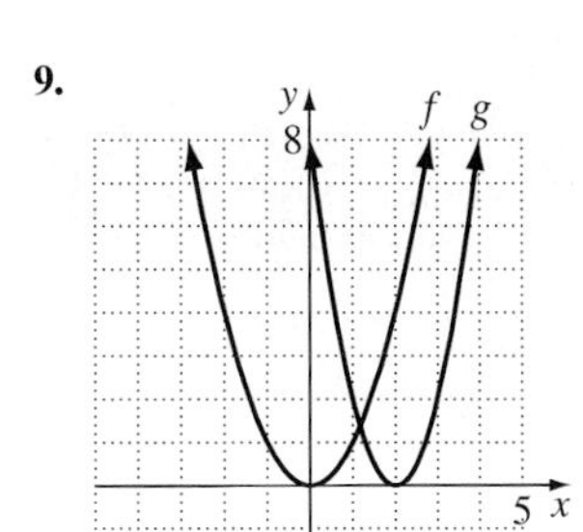

11.

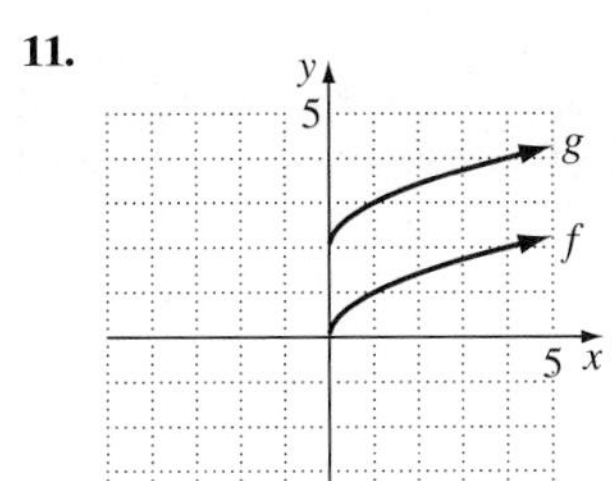

13.

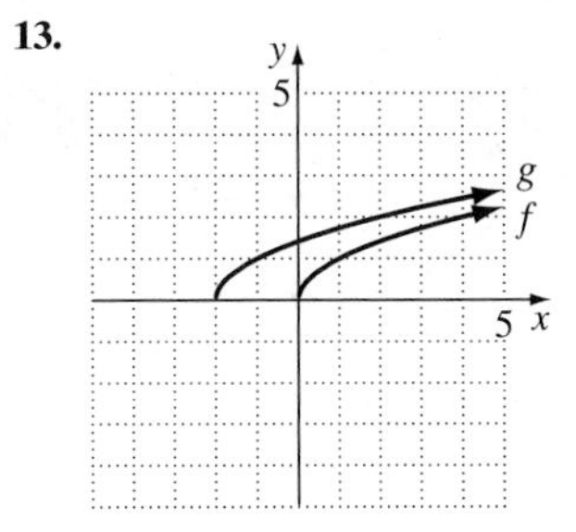

15.

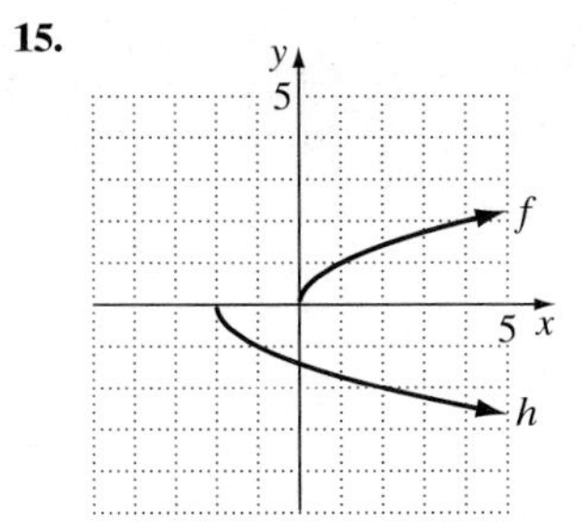

17.

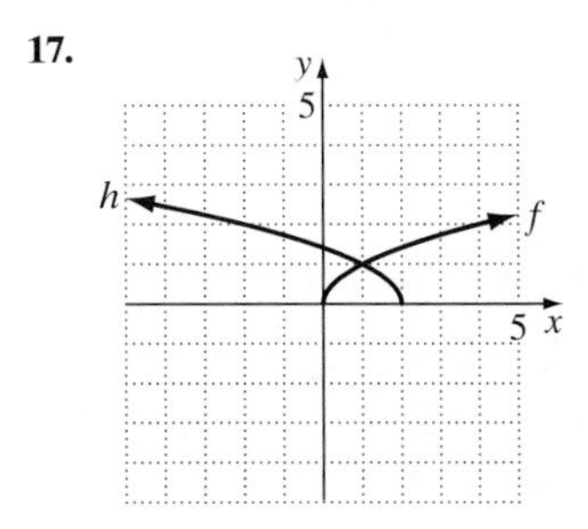

19.

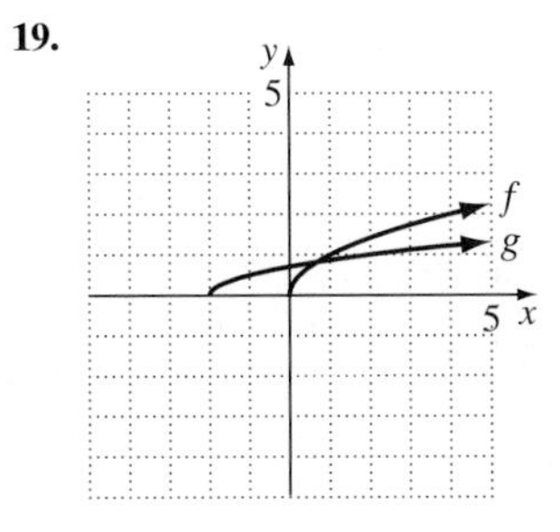

21.

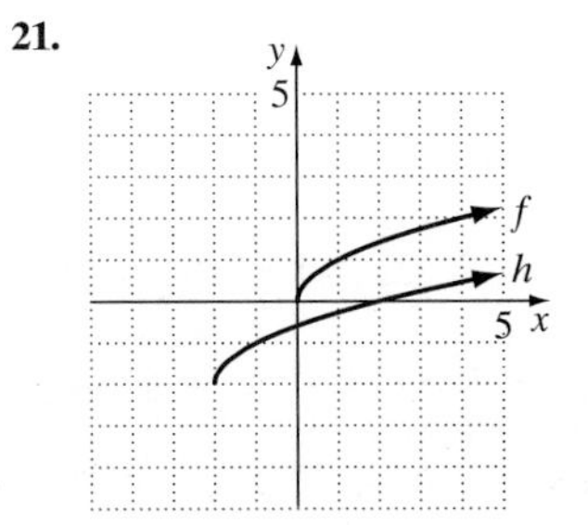

23.

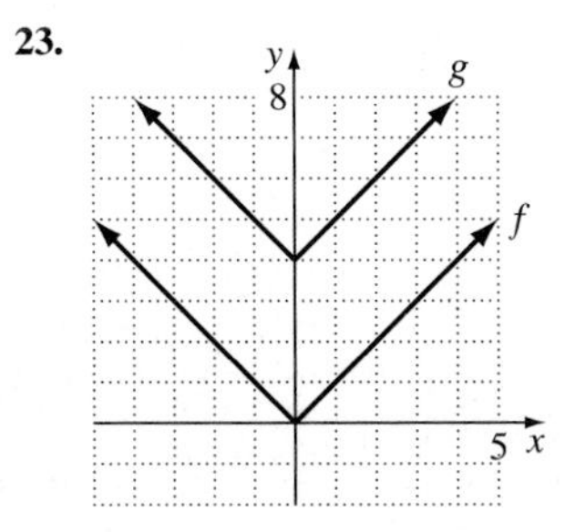

25.

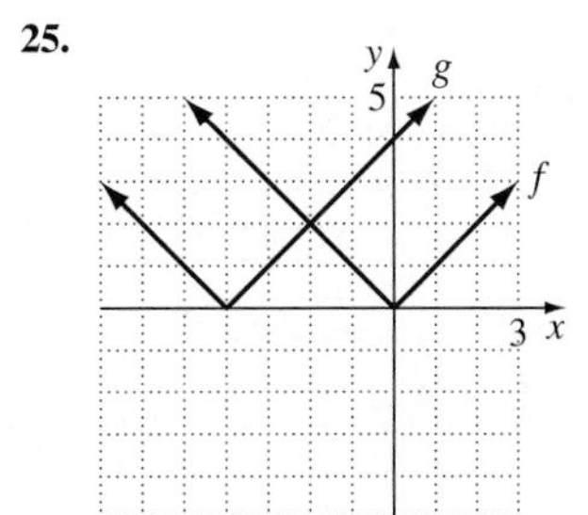

27.

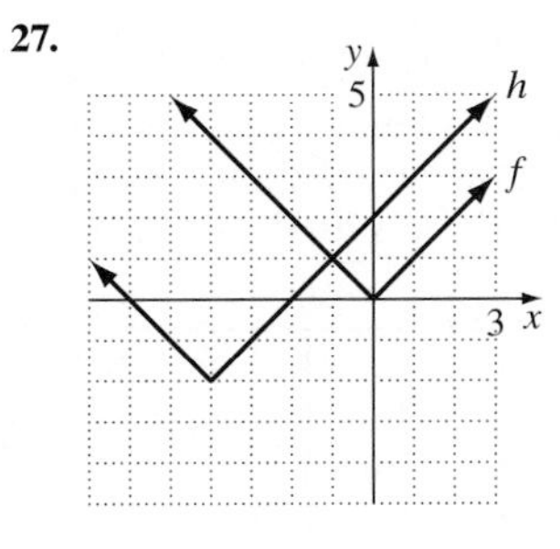

29.

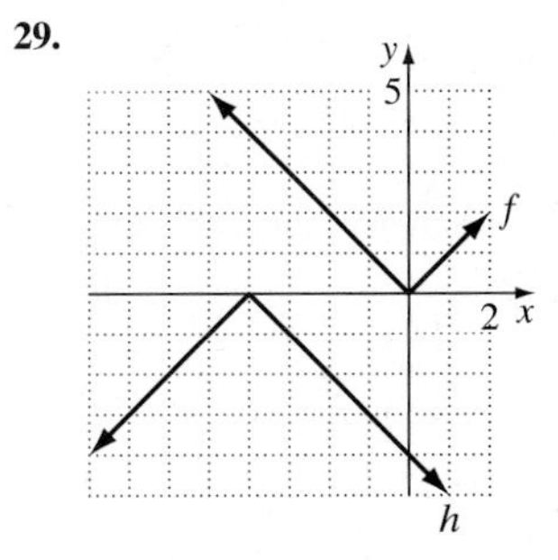

31.

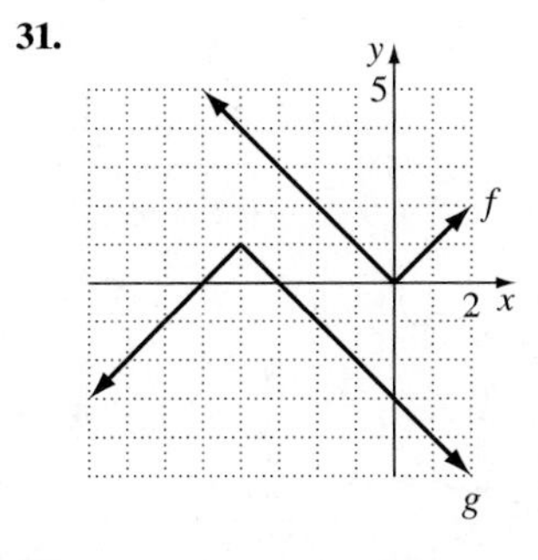

33.

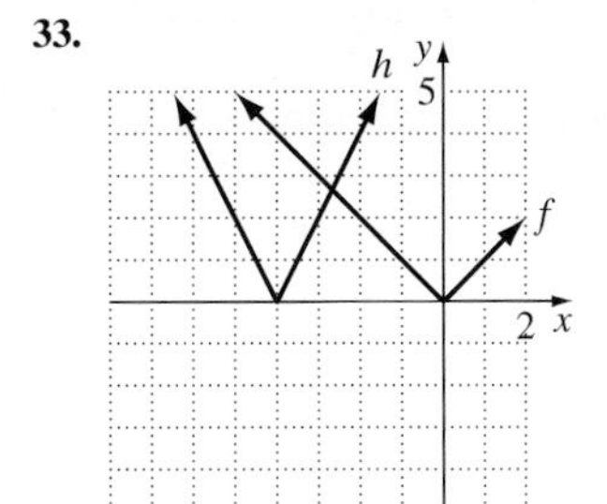

35.

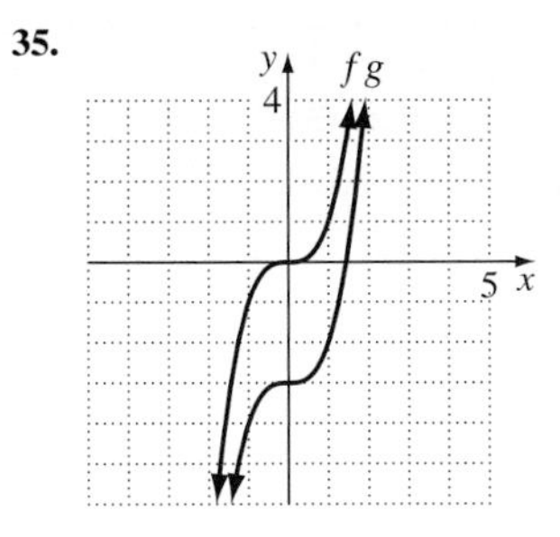

37.

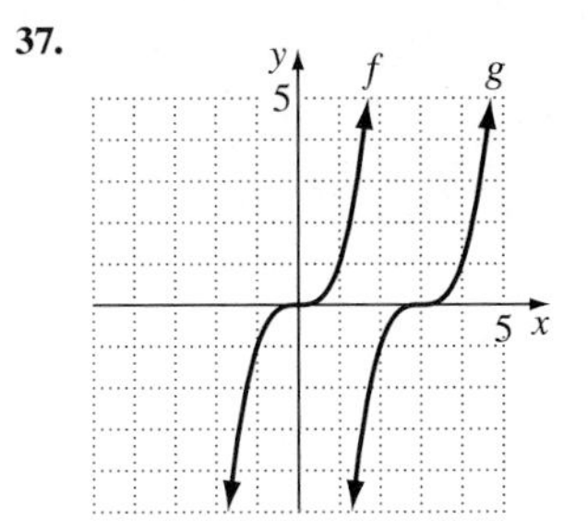

39.

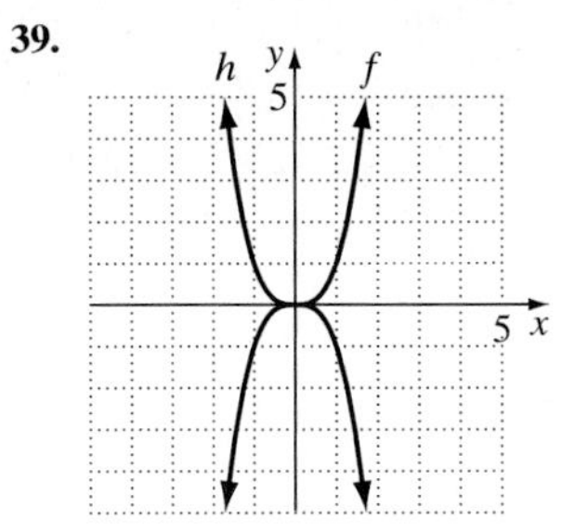

41.

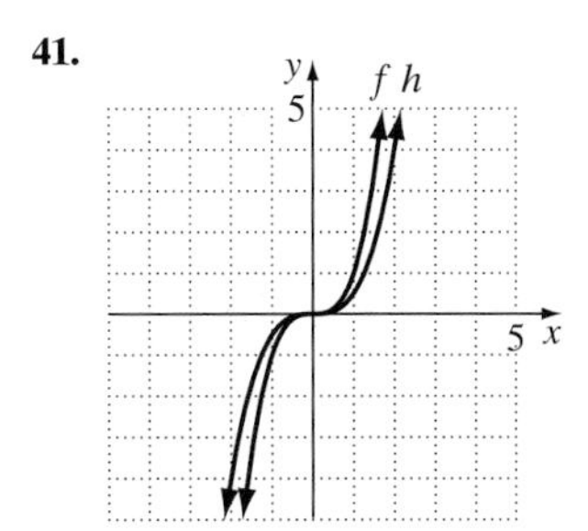

43.

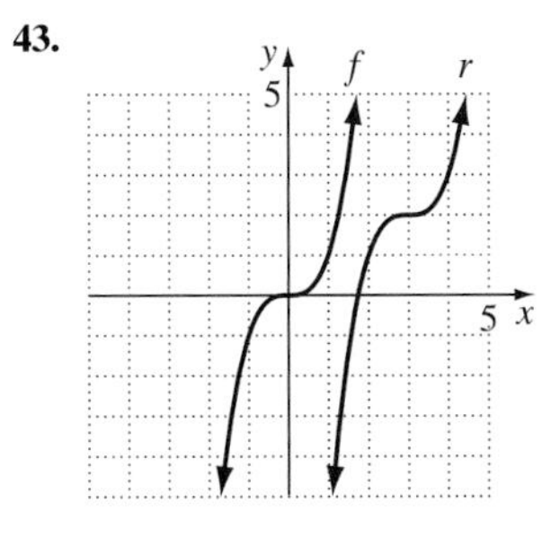

45.

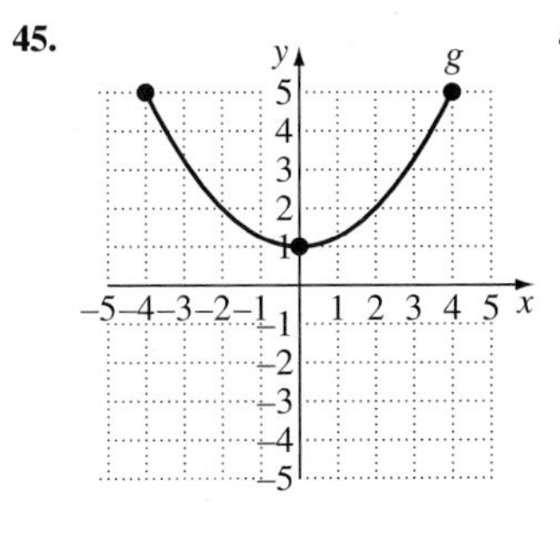

47.

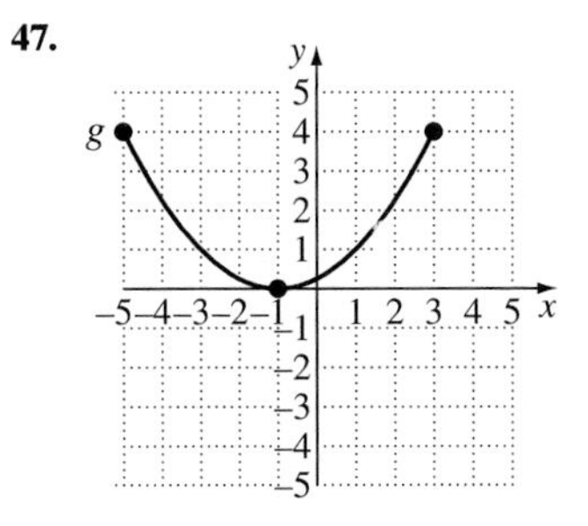

49.

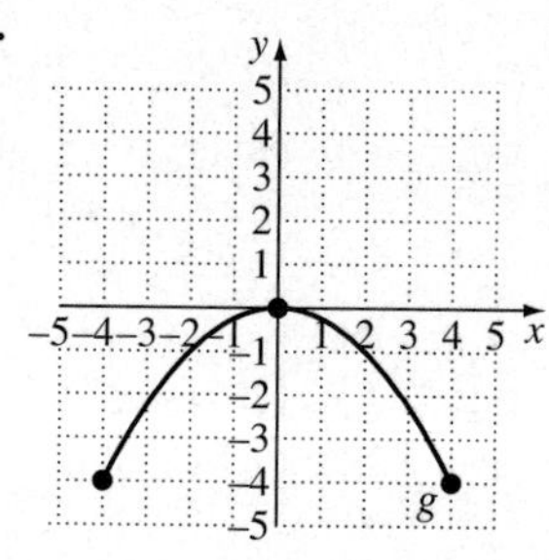

51.

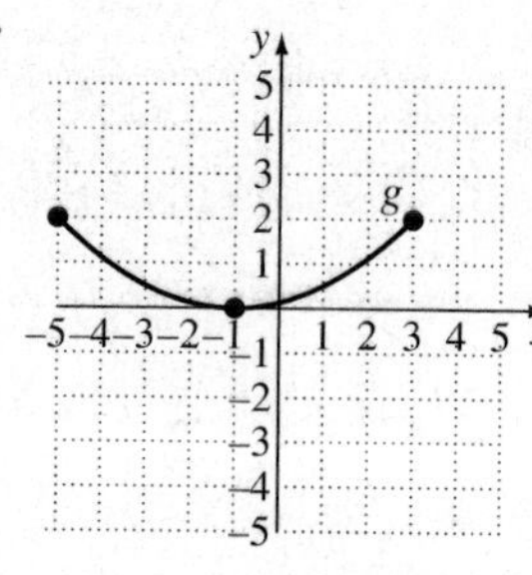

53. a. $(f + g)(x) = 2x^2 + 3x + 2$ **b.** $(f + g)(4) = 46$
55. a. $(f + g)(x) = \sqrt{x - 6} + \sqrt{x + 2}$ **b.** Domain: $[6, \infty)$
57. $(f + g)(x) = 3x + 2$; Domain: $(-\infty, \infty)$; $(f - g)(x) = x + 4$; Domain: $(-\infty, \infty)$; $(fg)(x) = 2x^2 + x - 3$; Domain: $(-\infty, \infty)$; $\left(\frac{f}{g}\right)(x) = \frac{2x + 3}{x - 1}$; Domain: $(-\infty, 1)$ or $(1, \infty)$
59. $(f + g)(x) = 3x^2 + x - 5$; Domain: $(-\infty, \infty)$; $(f - g)(x) = -3x^2 + x - 5$; Domain: $(-\infty, \infty)$; $(fg)(x) = 3x^3 - 15x^2$; Domain: $(-\infty, \infty)$; $\left(\frac{f}{g}\right)(x) = \frac{x - 5}{3x^2}$; Domain: $(-\infty, 0)$ or $(0, \infty)$

61. $(f + g)(x) = 2x^2 - 2$; Domain: $(-\infty, \infty)$; $(f - g)(x) = 2x^2 - 2x - 4$; Domain: $(-\infty, \infty)$; $(fg)(x) = 2x^3 + x^2 - 4x - 3$; Domain: $(-\infty, \infty)$; $\left(\frac{f}{g}\right)(x) = 2x - 3$; Domain: $(-\infty, -1)$ or $(-1, \infty)$ **63.** $(f + g)(x) = \sqrt{x} + x - 4$; Domain: $[0, \infty)$; $(f - g)(x) = \sqrt{x} - x + 4$; Domain: $[0, \infty)$; $(fg)(x) = \sqrt{x}(x - 4)$; Domain: $[0, \infty)$; $\left(\frac{f}{g}\right)(x) = \frac{\sqrt{x}}{x - 4}$; Domain: $[0, 4)$ or $(4, \infty)$
65. $(f + g)(x) = \frac{2x + 2}{x}$; Domain: $(-\infty, 0)$ or $(0, \infty)$; $(f - g)(x) = 2$; Domain: $(-\infty, 0)$ or $(0, \infty)$; $(fg)(x) = \frac{2x + 1}{x^2}$; Domain: $(-\infty, 0)$ or $(0, \infty)$; $\left(\frac{f}{g}\right)(x) = 2x + 1$; Domain: $(-\infty, 0)$ or $(0, \infty)$ **67.** $(f + g)(x) = \sqrt{x + 4} + \sqrt{x - 1}$; Domain: $[1, \infty)$; $(f - g)(x) = \sqrt{x + 4} - \sqrt{x - 1}$; Domain: $[1, \infty)$; $(fg)(x) = \sqrt{x^2 + 3x - 4}$; Domain: $[1, \infty)$; $\left(\frac{f}{g}\right)(x) = \frac{\sqrt{x + 4}}{\sqrt{x - 1}}$; Domain: $(1, \infty)$
69. $f + g$ represents the total world population in year x. **71.** $f(2000) \approx 1.5$ billion people; $g(2000) \approx 6$ billion people; $(f + g)(2000) \approx 7.5$ billion people **73.** $(R - C)(20{,}000) = -200{,}000$; The company lost \$200,000 since costs exceeded revenues; $(R - C)(30{,}000) = 0$; The company broke even since revenues equaled cost; $(R - C)(40{,}000) = 200{,}000$; The company made a profit of \$200,000.

75.

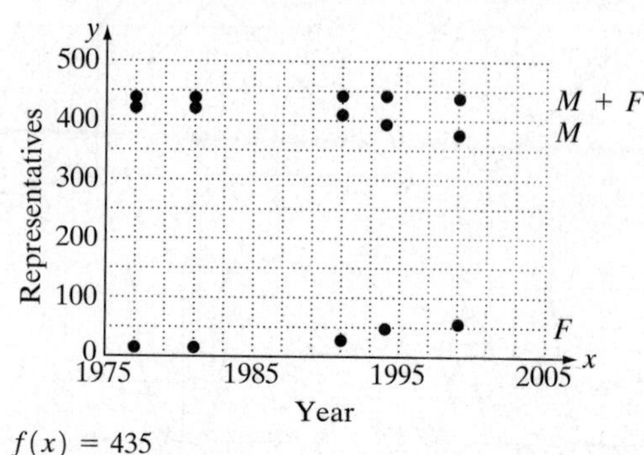

$f(x) = 435$

85. a.

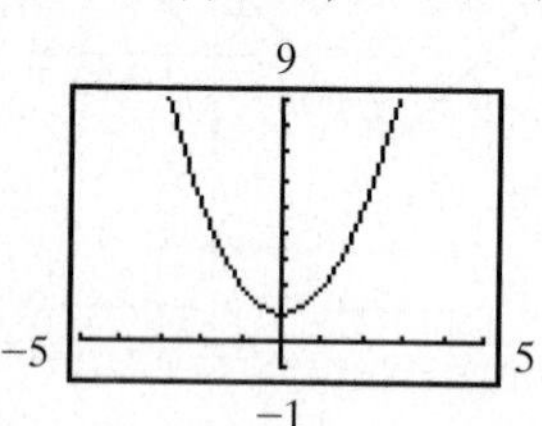

b.

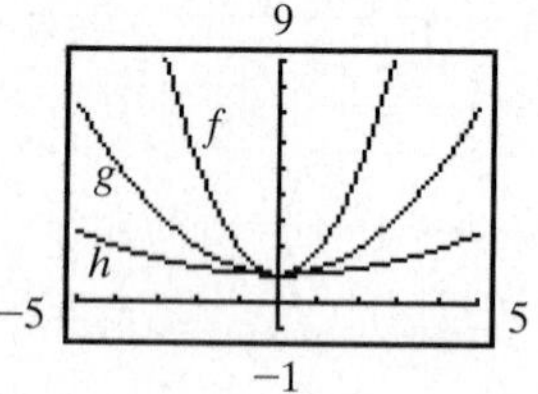

c. Answers may vary. **d.** Answers may vary. **e.** Answers may vary.

87. $g(x) = -(x + 4)^2$ **89.** $g(x) = -\sqrt{x - 2} + 2$ **91.** $(-a, b)$
93. $(a + 3, b)$

Section 2.6

Check Point Exercises

1. a. $(f \circ g)(x) = 5x^2 + 1$ **b.** $(g \circ f)(x) = 25x^2 + 60x + 35$ **2.** $f(g(x)) = x$; $g(f(x)) = x$; f and g are inverses.
3. $f(g(x)) = x$; $g(f(x)) = x$; f and g are inverses. **4.** $f^{-1}(x) = \frac{x - 7}{2}$ **5.** $f^{-1}(x) = \sqrt[3]{\frac{x + 1}{4}}$
6. (b) and (c) have inverse functions. **7.**

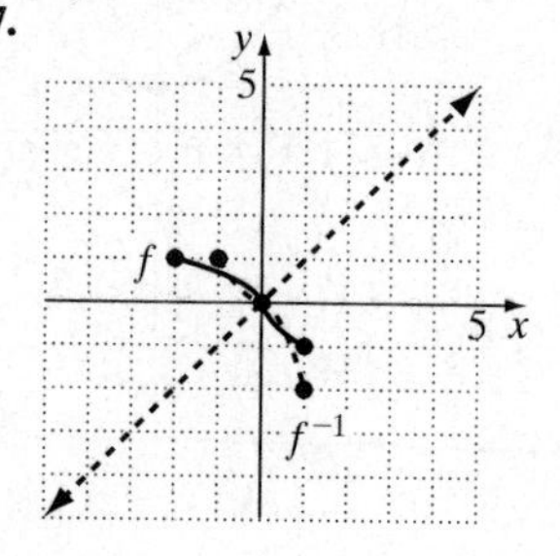

Exercise Set 2.6

1. a. $(f \circ g)(x) = 2x + 14$ **b.** $(g \circ f)(x) = 2x + 7$ **c.** $(f \circ g)(2) = 18$ **3. a.** $(f \circ g)(x) = 2x + 5$ **b.** $(g \circ f)(x) = 2x + 9$ **c.** $(f \circ g)(2) = 9$ **5. a.** $(f \circ g)(x) = 20x^2 - 11$ **b.** $(g \circ f)(x) = 80x^2 - 120x + 43$ **c.** $(f \circ g)(2) = 69$ **7. a.** $(f \circ g)(x) = x^4 - 4x^2 + 6$ **b.** $(g \circ f)(x) = x^4 + 4x^2 + 2$ **c.** $(f \circ g)(2) = 6$ **9. a.** $(f \circ g)(x) = \sqrt{x - 1}$ **b.** $(g \circ f)(x) = \sqrt{x} - 1$ **c.** $(f \circ g)(2) = 1$ **11. a.** $(f \circ g)(x) = x$ **b.** $(g \circ f)(x) = x$ **c.** $(f \circ g)(2) = 2$ **13. a.** $(f \circ g)(x) = x$ **b.** $(g \circ f)(x) = x$ **c.** $(f \circ g)(2) = 2$ **15.** $f(g(x)) = x$; $g(f(x)) = x$; f and g are inverses. **17.** $f(g(x)) = x$; $g(f(x)) = x$; f and g are inverses. **19.** $f(g(x)) = \frac{5x - 56}{9}$; $g(f(x)) = \frac{5x - 4}{9}$; f and g are not inverses. **21.** $f(g(x)) = x$; $g(f(x)) = x$; f and g are inverses. **23.** $f(g(x)) = x$; $g(f(x)) = x$; f and g are inverses. **25.** $f^{-1}(x) = x - 3$ **27.** $f^{-1}(x) = \frac{x}{2}$ **29.** $f^{-1}(x) = \frac{x - 3}{2}$ **31.** $f^{-1}(x) = \sqrt[3]{x - 2}$ **33.** $f^{-1}(x) = \sqrt[3]{x} - 2$ **35.** $f^{-1}(x) = \frac{1}{x}$ **37.** $f^{-1}(x) = x^2, x \geq 0$ **39.** $f^{-1}(x) = \sqrt{x - 1}$ **41.** $f^{-1}(x) = \frac{3x + 1}{x - 2}$ **43.** $f^{-1}(x) = (x - 3)^3 + 4$ **45.** The function is not one-to-one, so it does not have an inverse function. **47.** The function is not one-to-one, so it does not have an inverse function. **49.** The function is one-to-one, so it does have an inverse function.

51.

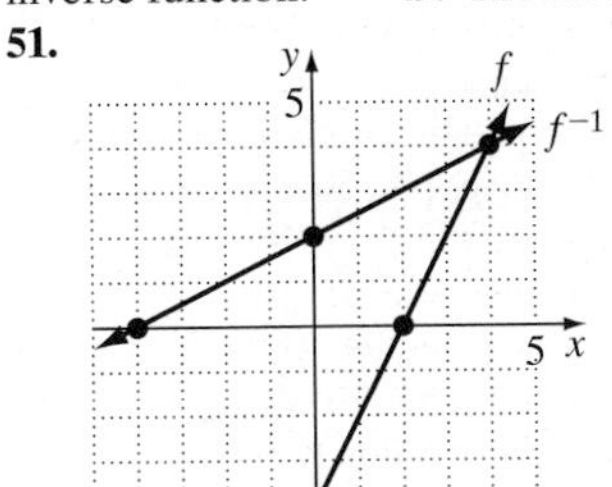

53.

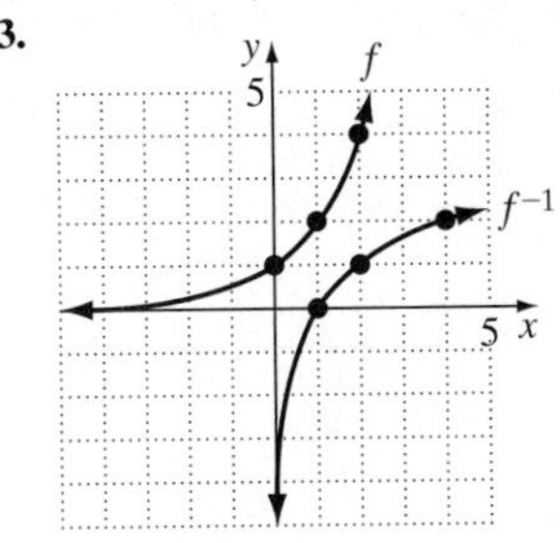

55. a. f gives the price of the computer after a \$400 discount. g gives the price of the computer after a 25% discount. **b.** $(f \circ g)(x) = 0.75x - 400$. This models the price of a computer after first a 25% discount and then a \$400 discount. **c.** $(g \circ f)(x) = 0.75(x - 400)$. This models the price of a computer after first a \$400 discount and then a 25% discount. **d.** The function $f \circ g$ models the greater discount, since the 25% discount is taken on the regular price first. **e.** $f^{-1}(x) = x + 400$; If x is the discount price of the computer, then $f^{-1}(x)$ is the regular price.

57. a. f is a one-to-one function. **b.** $f^{-1}(0.25)$ is the number of people in a room for a 25% probability of two people sharing a birthday. $f^{-1}(0.5)$ is the number of people in a room for a 50% probability of two people sharing a birthday. $f(0.7)$ is the number of people in a room for a 70% probability of two people sharing a birthday. **59.** No. The graph does not pass the horizontal line test, so it is not one-to-one. This means that the average age at which United States women marry has been the same during more than one year.

69.

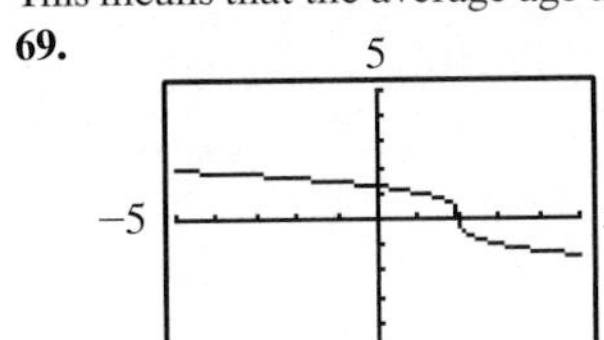

one-to-one

71.

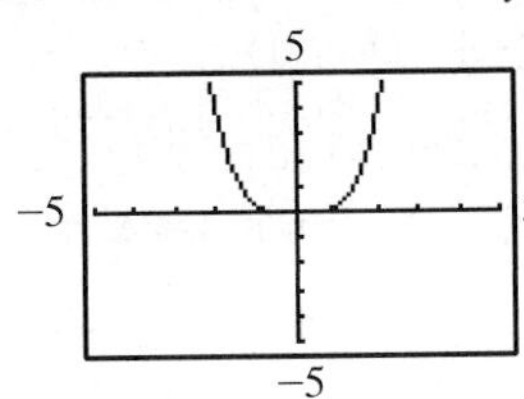

not one-to-one

73.

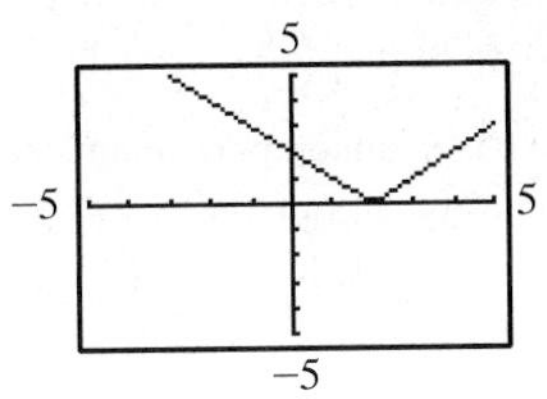

not one-to-one

75.

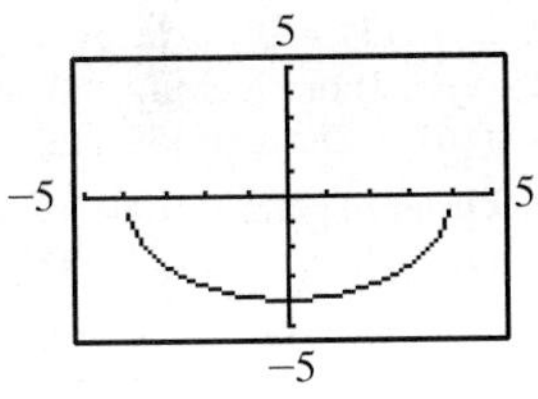

not one-to-one

77.

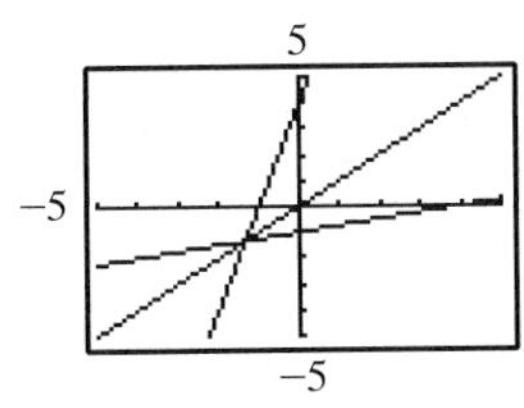

f and g are inverses.

79. 5 −5 5 −5

f and g are inverses.

81. Answers may vary. **83.** $(f \circ f)(x) = x$, so f is its own inverse.

Chapter 2 Review Exercises

1. $m = -\frac{1}{2}$; falls **2.** $m = 1$; rises **3.** $m = 0$; horizontal **4.** $m =$ undefined; vertical **5.** $y - 2 = -6(x + 3)$; $y = -6x - 16$ **6.** using $(1, 6)$, $y - 6 = 2(x - 1)$; $y = 2x + 4$

7. Slope: $\frac{2}{5}$; y-intercept: -1

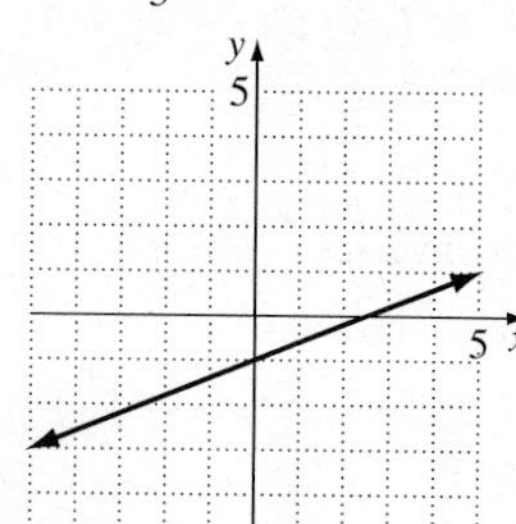

8. Slope: -4; y-intercept: 5

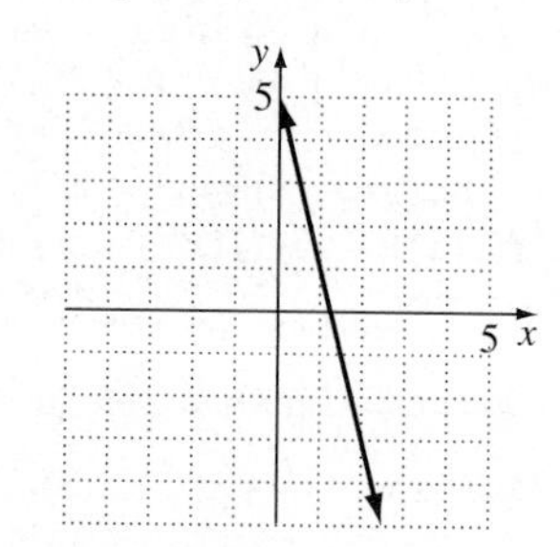

9. Slope: $-\frac{2}{3}$; y-intercept: -2

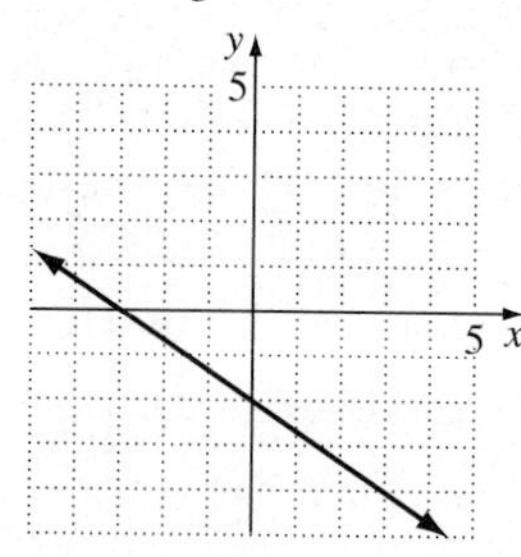

10. Slope: 0; y-intercept: 4

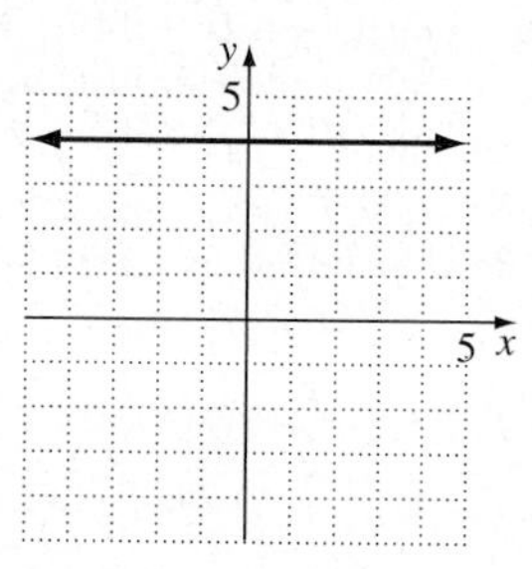

11. a. using $(0, 16)$, $y - 16 = -0.13(x - 0)$ **b.** $y = -0.13x + 16$ **c.** 6.9 ft; 5.6 ft **d.** In 2000, the equation predicts the surfboard length to be 3 feet, which is not reasonable. **12. a.** Answers may vary. **b.** Answers may vary. **c.** Answers may vary. **13.** $y + 7 = -3(x - 4)$; $y = -3x + 5$ **14.** $y - 6 = -3(x + 3)$; $y = -3x - 3$ **15.** $x^2 + y^2 = 9$ **16.** $(x + 2)^2 + (y - 4)^2 = 36$

17. Center: $(0, 0)$; radius: 1

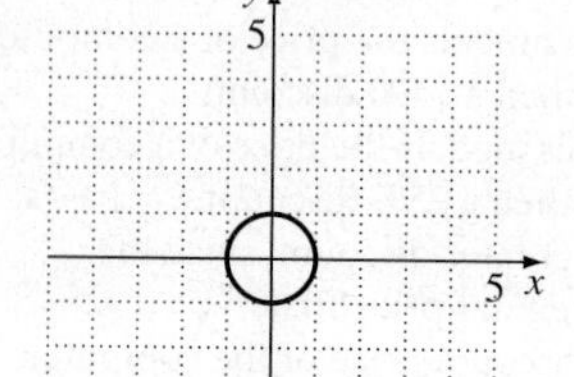

18. Center: $(-2, 3)$; radius: 3

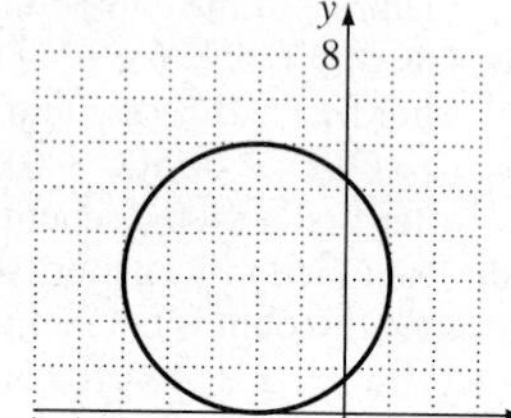

19. Center: $(2, -1)$; radius: 3

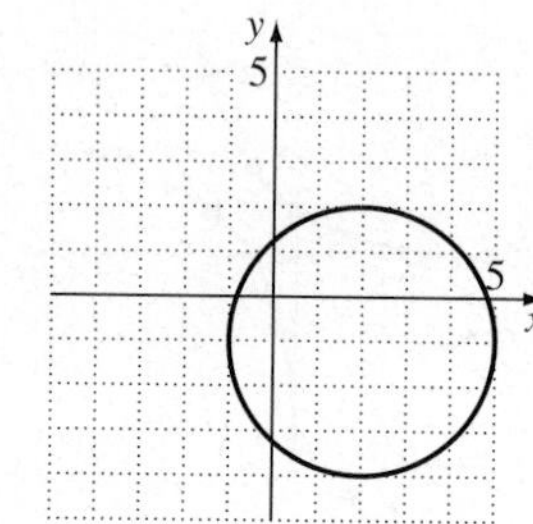

20. Function; Domain: $\{2, 3, 5\}$; Range: $\{7\}$
21. Function; Domain: $\{1, 2, 13\}$; Range: $\{10, 500, \pi\}$
22. Not a function; Domain: $\{12, 14\}$; Range: $\{13, 15, 19\}$
23. y is a function of x.
24. y is a function of x.
25. y is not a function of x.
26. a. $f(4) = -23$
b. $f(x + 3) = -7x - 16$
c. $f(-x) = 5 + 7x$

27. a. $g(0) = 2$ **b.** $g(-2) = 24$ **c.** $g(x - 1) = 3x^2 - 11x + 10$ **d.** $g(-x) = 3x^2 + 5x + 2$
28. a. $f(a) = 4a - 3$ **b.** $f(a + h) = 4a + 4h - 3$ **c.** $\frac{f(a + h) - f(a)}{h} = 4$ **d.** $f(a) + f(h) = 4a + 4h - 6$
29. a. $g(13) = 3$ **b.** $g(0) = 4$ **c.** $g(-3) = 7$ **30. a.** $f(-2) = -1$ **b.** $f(1) = 12$ **c.** $f(2) = 3$ **31.** $(-\infty, \infty)$
32. $(-\infty, 7)$ or $(7, \infty)$ **33.** $(-\infty, 4]$ **34.** $(-\infty, -1)$ or $(-1, 1)$ or $(1, \infty)$ **35.** $[2, 5)$ or $(5, \infty)$
36. $f(6) = 25.48$; In 1996, there were 25.48 million participants in the Federal Food Stamp Program.

37. Ordered pairs: $(-1, 9)$, $(0, 4)$, $(1, 1)$, $(2, 0)$, $(3, 1)$, $(4, 4)$.

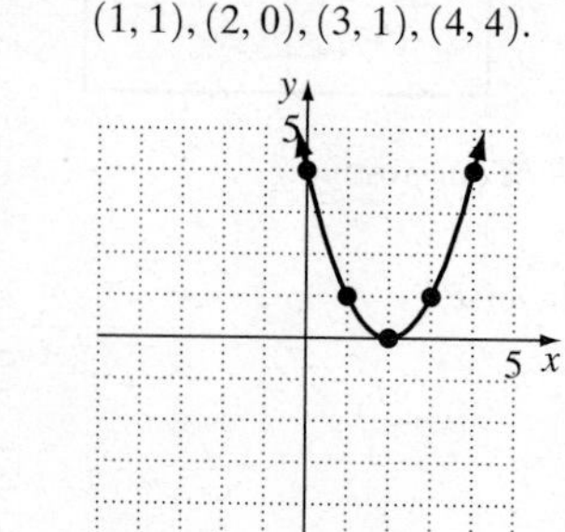

Domain: $(-\infty, \infty)$
Range: $[0, \infty)$

38. Ordered pairs: $(-1, 3)$, $(0, 2)$, $(1, 1)$, $(2, 0)$, $(3, 1)$, $(4, 2)$.

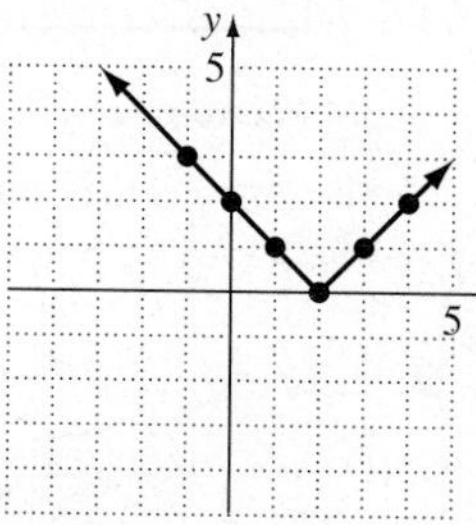

Domain: $(-\infty, \infty)$
Range: $[0, \infty)$

39. a. Domain: $[-3, 5)$
b. Range: $[-5, 0]$
c. x-intercept: -3
d. y-intercept: -2
e. increasing: $(-2, 0)$ or $(3, 5)$
decreasing: $(-3, -2)$ or $(0, 3)$
f. $f(-2) = -3$ and $f(3) = -5$

40. a. Domain: $(-\infty, \infty)$
b. Range: $(-\infty, \infty)$
c. x-intercepts: -2 and 3
d. y-intercept: 3
e. increasing: $(-5, 0)$; decreasing: $(-\infty, -5)$ or $(0, \infty)$
f. $f(-2) = 0$ and $f(6) = -3$

41. a. Domain: $(-\infty, \infty)$ **b.** Range: $[-2, 2]$ **c.** x-intercept: 0 **d.** y-intercept: 0 **e.** increasing: $(-2, 2)$; constant: $(-\infty, -2)$ or $(2, \infty)$ **f.** $f(-9) = -2$ and $f(14) = 2$ **42.** not a function **43.** function **44.** function **45.** not a function
46. odd; symmetric with respect to the origin **47.** even; symmetric with respect to the y-axis **48.** odd; symmetric with respect to the origin
49. a. yes; The graph passes the vertical line test. **b.** Decreasing: $(3, 12)$; The vulture descended.
c. Constant: $(0, 3)$ and $(12, 17)$; The vulture's height held steady during the first 3 seconds and the vulture was on the ground for 5 seconds.
d. Increasing: $(17, 30)$; The vulture was ascending.

CHAPTER 3

Section 3.1

Check Point Exercises

1.

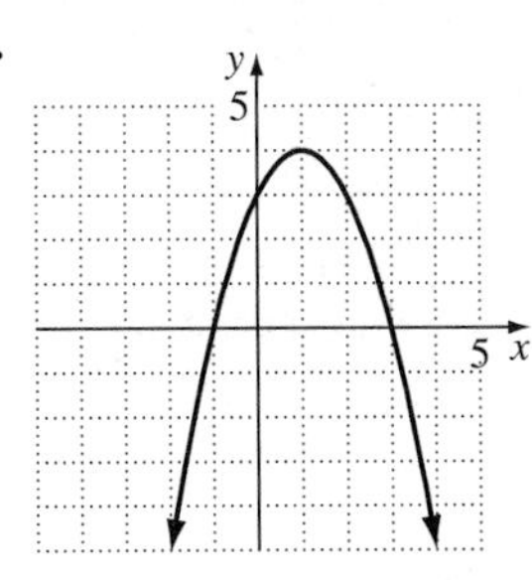

2.

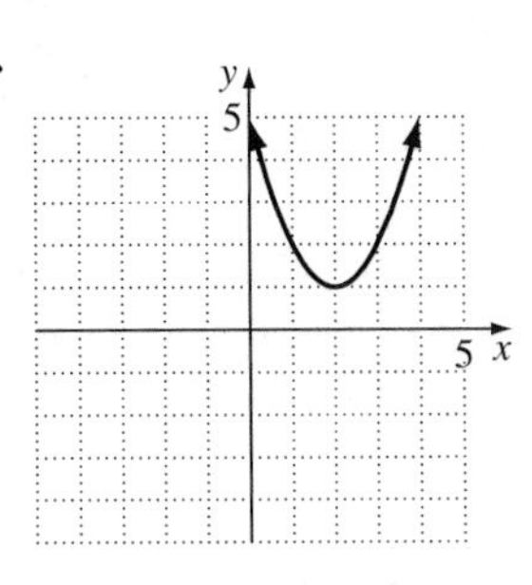

3.

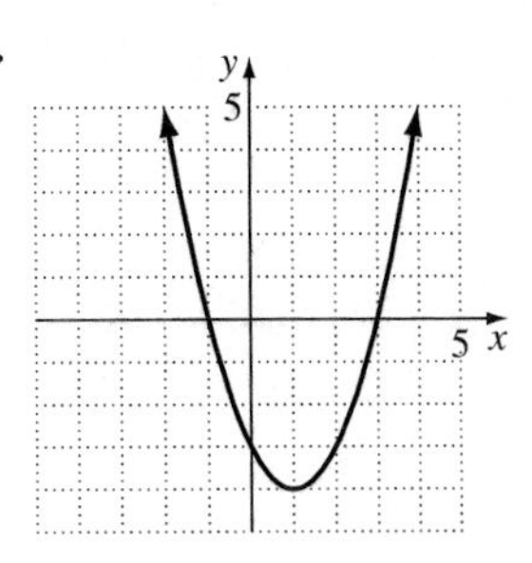

4. 45; 190

Exercise Set 3.1

1. $h(x) = (x - 1)^2 + 1$ **3.** $j(x) = (x - 1)^2 - 1$ **5.** $h(x) = x^2 - 1$ **7.** $g(x) = x^2 - 2x + 1$ **9.** $(3, 1)$ **11.** $(-1, 5)$ **13.** $(2, -5)$ **15.** $(-1, 9)$

17. axis of symmetry: $x = 4$

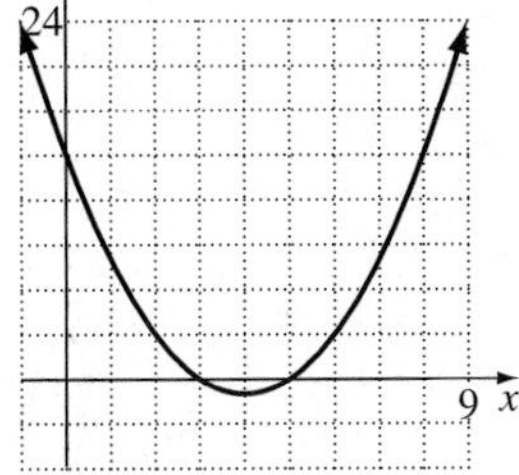

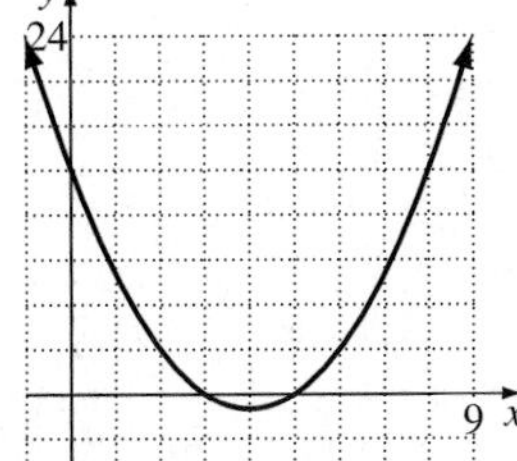

19. axis of symmetry: $x = 1$

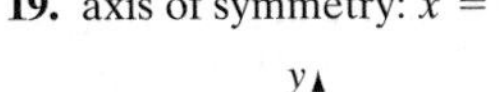

21. axis of symmetry: $x = 3$

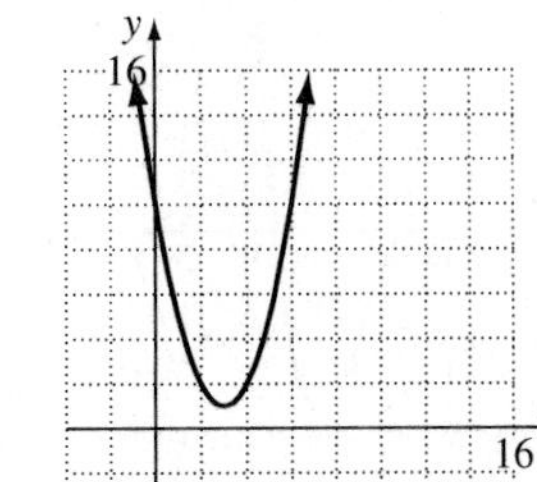

23. axis of symmetry: $x = -2$

25. axis of symmetry: $x = 1$

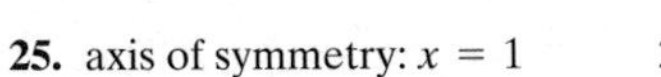

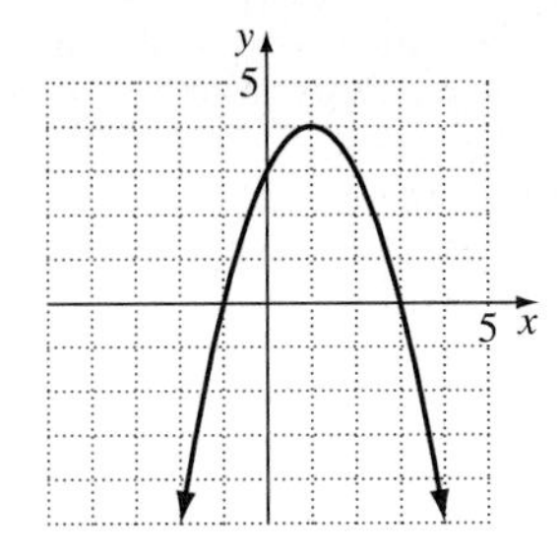

27. axis of symmetry: $x = 1$

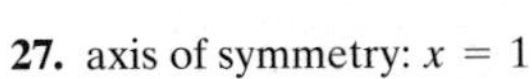

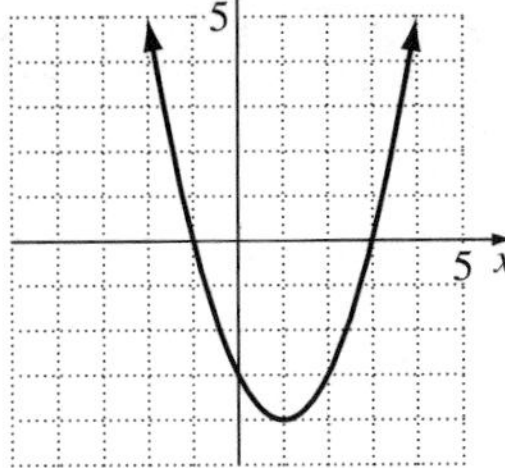

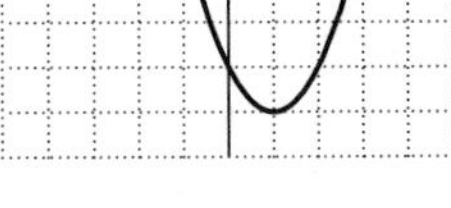

29. axis of symmetry: $x = -\frac{3}{2}$

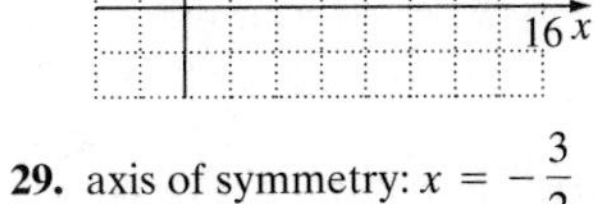

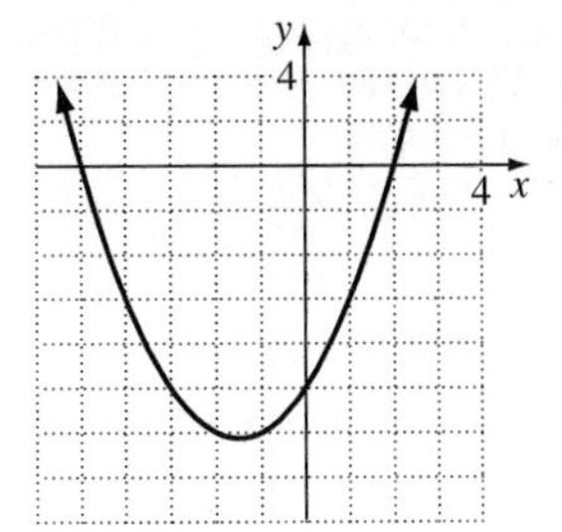

31. axis of symmetry: $x = 1$

33. axis of symmetry: $x = 1$

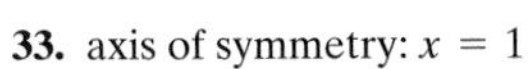

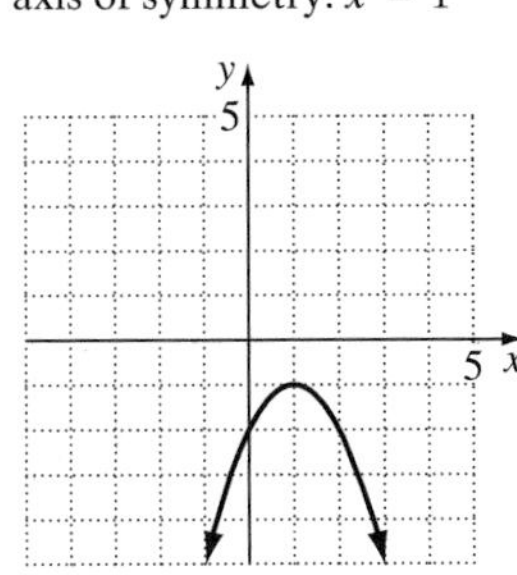

35. minimum; $(2, -13)$

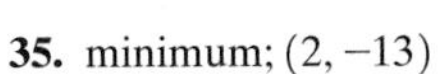

37. maximum; $(1, 1)$ **39.** minimum; $\left(\frac{1}{2}, -\frac{5}{4}\right)$

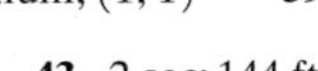

41. 1968; 4238 cigarettes per person **43.** 2 sec; 144 ft **45.** $(9, 2100)$ **47.** 30 ft; 60 ft; 1800 ft^2

57. a.

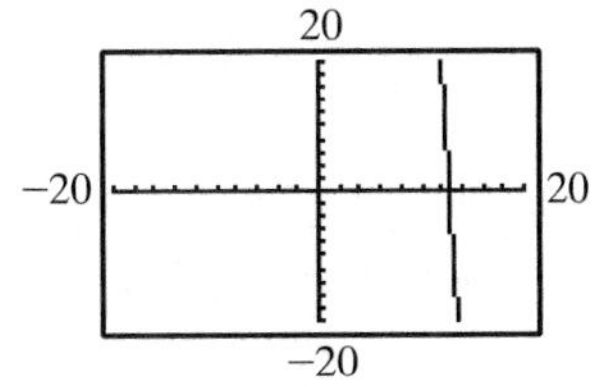

You can only see a little of the parabola.

b. $(20.5, -120.5)$

c. Ymax = 750

d. You can choose Xmin and Xmax so the x-value of the vertex is in the center of the graph. Choose Ymin to include the y-value of the vertex.

59. (2.5, 185)

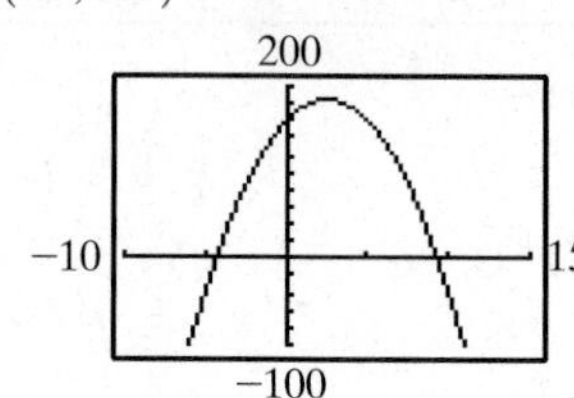

61. (−30, 91)

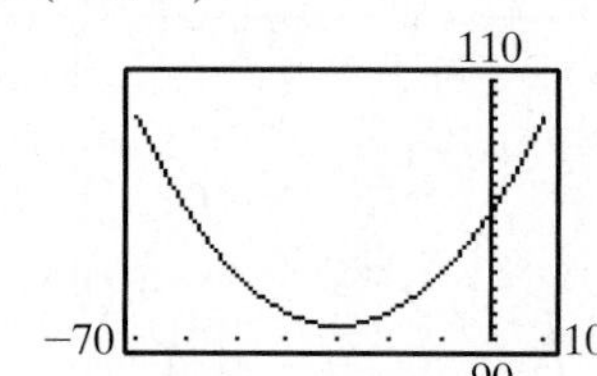

63.

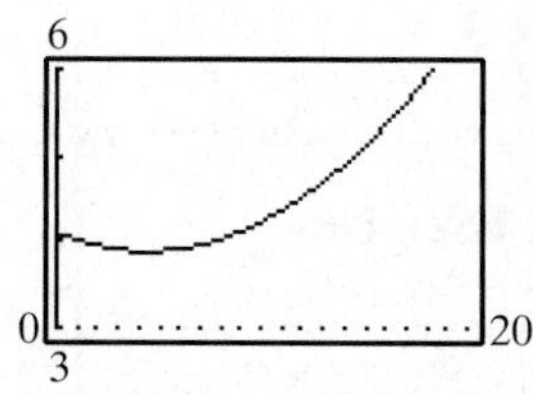

Vertex: about (4.41, 3.89);
The minimum number of people in the United States holding more than one job was 3.89 million in 1974.

65. (a) is true. **67.** $x = -2$; $(-3, -2)$ **69.** $x = 166\frac{2}{3}$ ft, $y = 125$ ft; approximately 20,833 ft^2

Section 3.2

Check Point Exercises

1. The graph rises to the left and to the right. **2.** Since n is odd and the leading coefficient is negative, the function falls to the right. Since the ratio cannot be negative, the model won't be appropriate. **3.** No; the graph should fall to the left, but doesn't appear to. **4.** $\{-2, 2\}$ **5.** $\{-2, 0, 2\}$ **6.**

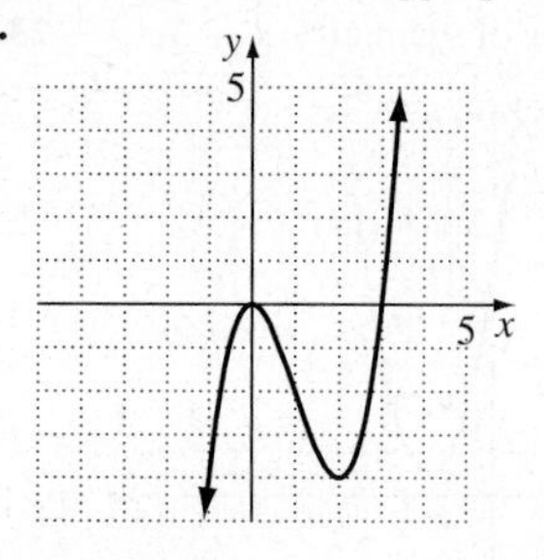

Exercise Set 3.2

1. polynomial function; degree: 3 **3.** polynomial function; degree: 5 **5.** not a polynomial function **7.** not a polynomial function **9.** not a polynomial function **11.** polynomial function **13.** not a polynomial function **15.** (c) **17.** (b) **19.** (a) **21.** falls to the left and rises to the right **23.** rises to the left and to the right **25.** falls to the left and to the right **27.** $x = 5$ has multiplicity 1; The graph crosses the x-axis; $x = -4$ has multiplicity 2; The graph touches the x-axis and turns around. **29.** $x = 3$ has multiplicity 1; The graph crosses the x-axis; $x = -6$ has multiplicity 3; The graph crosses the x-axis. **31.** $x = 0$ has multiplicity 1; The graph crosses the x-axis; $x = 1$ has multiplicity 2; The graph touches the x-axis and turns around. **33.** $x = 2, x = -2$ and $x = -7$ have multiplicity 1; The graph crosses the x-axis.

35. a. $f(x)$ rises to the right and falls to the left.
b. $x = -2, x = 1, x = -1$;
$f(x)$ crosses the x-axis at each.
c. The y-intercept is −2.
d. neither
e.

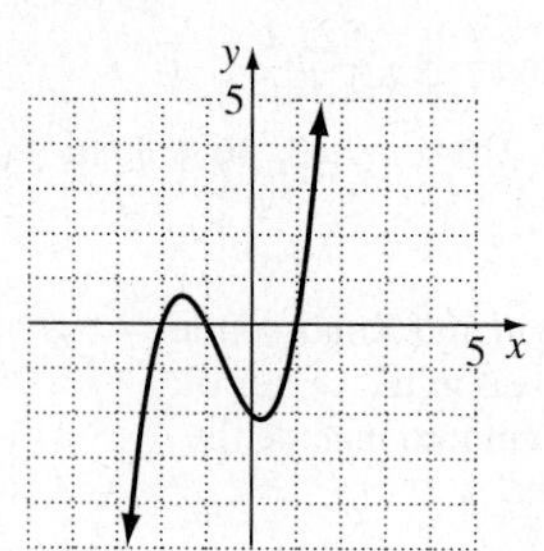

37. a. $f(x)$ rises to the left and the right.
b. $x = 0, x = 3, x = -3$;
$f(x)$ crosses the x-axis at −3 and 3;
$f(x)$ touches the x-axis at 0.
c. The y-intercept is 0.
d. y-axis symmetry
e.

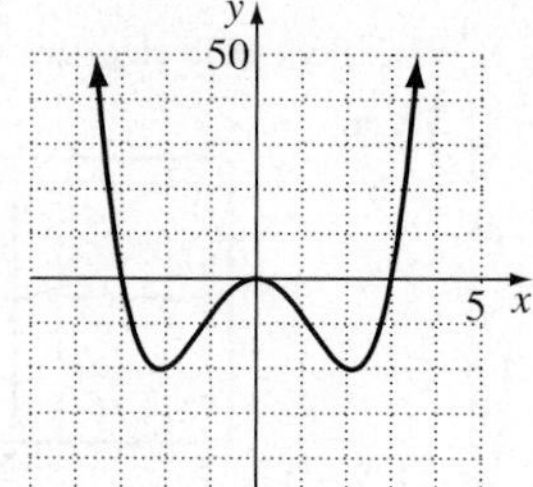

39. a. $f(x)$ falls to the left and the right.
b. $x = 0, x = 4, x = -4$;
$f(x)$ crosses the x-axis at -4 and 4;
$f(x)$ touches the x-axis at 0.
c. The y-intercept is 0.
d. y-axis is symmetry
e.

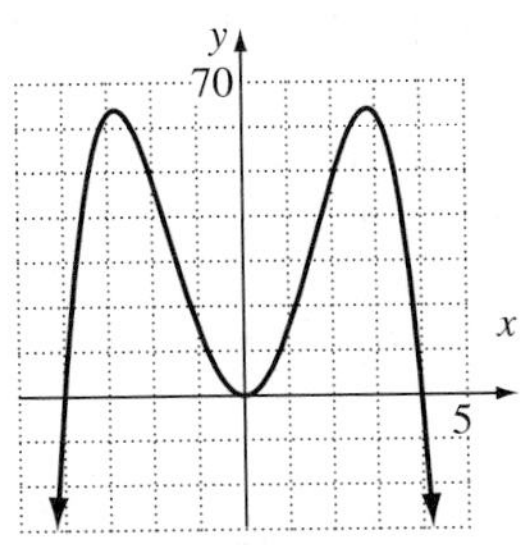

41. a. $f(x)$ rises to the left and the right.
b. $x = 0, x = 1$;
$f(x)$ touches the x-axis at 0 and 1.
c. The y-intercept is 0.
d. neither
e.

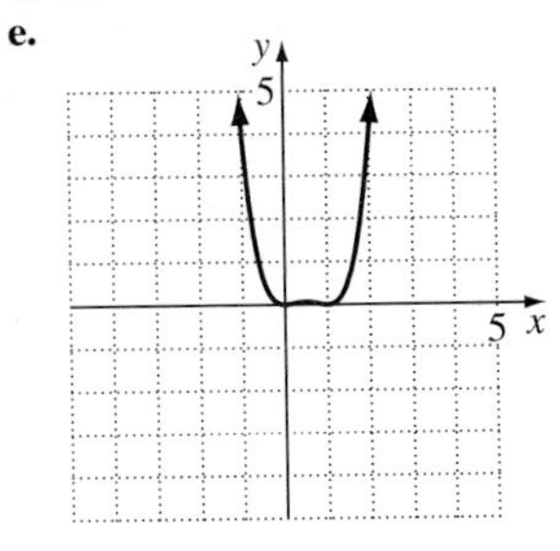

43. a. $f(x)$ falls to the left and the right.
b. $x = 0, x = 2$;
$f(x)$ crosses the x-axis at 0 and 2.
c. The y-intercept is 0.
d. neither
e.

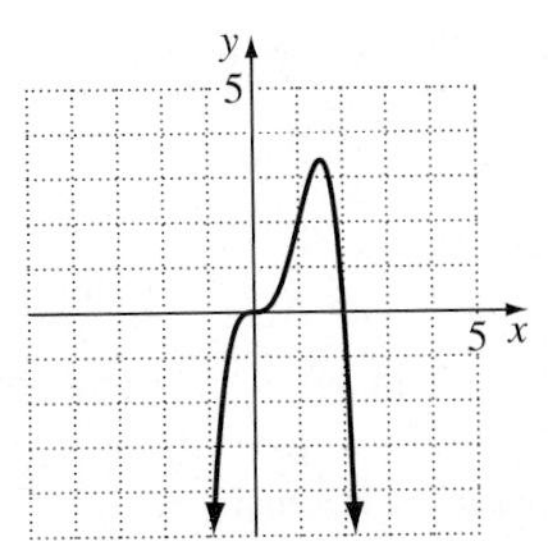

45. a. $f(x)$ rises to the left and falls to the right.
b. $x = 0, x = \pm\sqrt{3}$;
$f(x)$ crosses the x-axis at $(0, 0)$;
$f(x)$ touches the x-axis at $\sqrt{3}$ and $-\sqrt{3}$.
c. The y-intercept is 0.
d. origin symmetry
e.

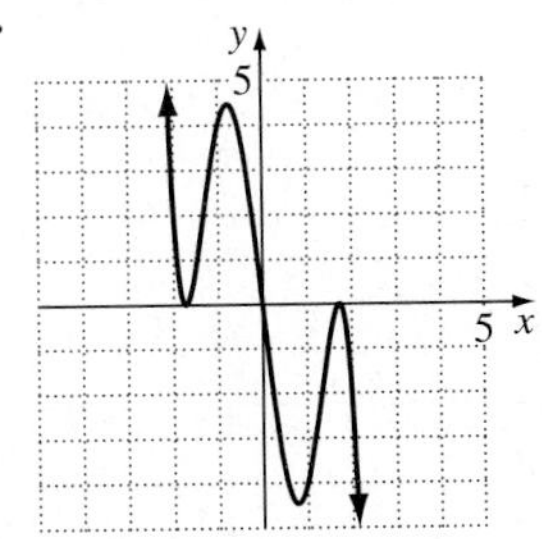

47. a. $f(x)$ rises to the left and falls to the right.
b. $x = 0, x = 3$;
$f(x)$ crosses the x-axis at 3;
$f(x)$ touches the axis at $(0, 0)$.
c. The y-intercept is 0.
d. neither
e.

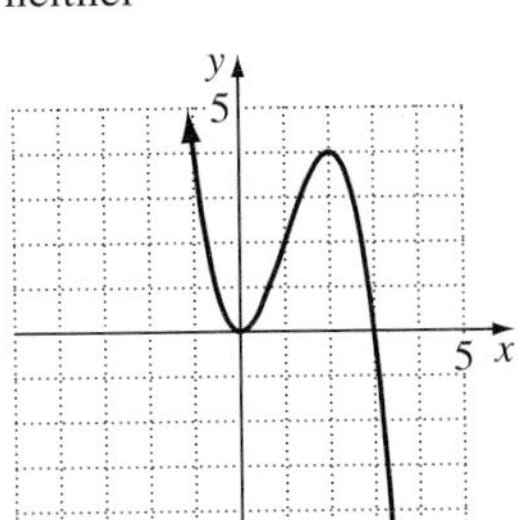

49. a. $f(x)$ falls to the left and the right.
b. $x = 1, x = -2, x = 2$;
$f(x)$ crosses the x-axis at -2 and 2;
$f(x)$ touches the x-axis at $(1, 0)$.
c. The y-intercept is 12.
d. neither
e.

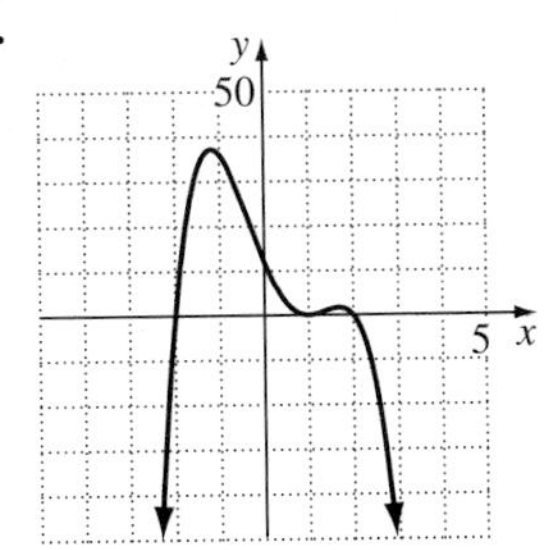

51. a. Leading coefficient test suggests the elk population will decline and eventually will die off.

b.

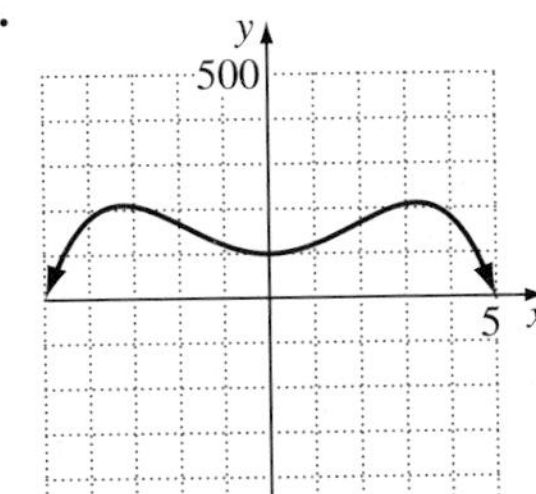

c.

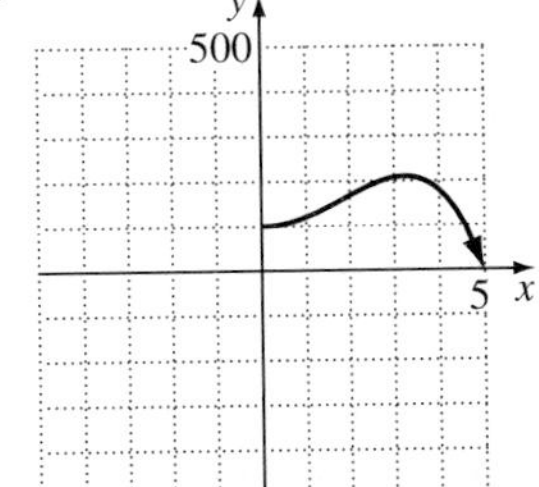

The population reaches extinction at the end of 5 years.

53. a. about 4194 larceny thefts **b.** No; eventually the function would predict a negative number of larceny thefts, which is impossible.
55. a. 55.178 yr **b.** No; since $a_n < 0$ and n is even, $H(x)$ eventually starts decreasing as x increases, which is impossible.
69. Answers may vary. **71.** Answers may vary.

73.
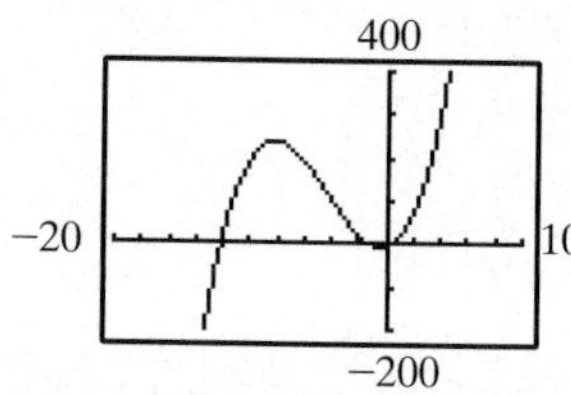

75.
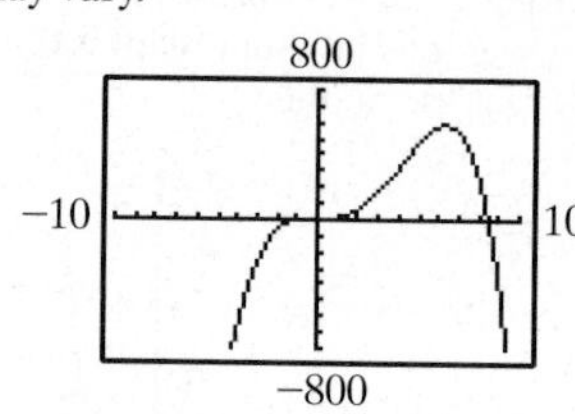

77.
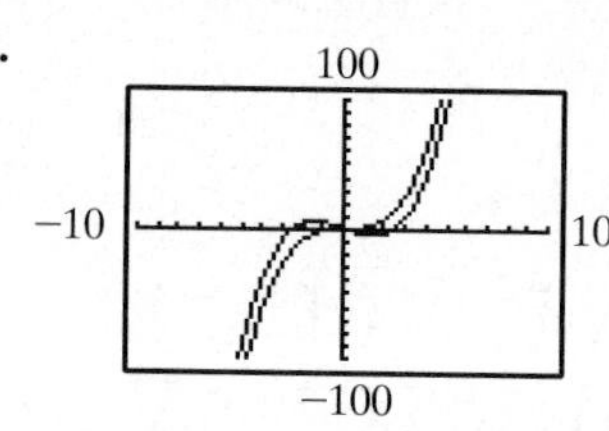

79. (c) is true. **81.** $f(x) = x^3 - 2x$

Section 3.3

Check Point Exercises

1. $x + 5$ **2.** $2x^2 + 3x - 2 + \frac{1}{x - 3}$ **3.** $2x^2 + 7x + 14 + \frac{21x - 10}{x^2 - 2x}$ **4.** $x^2 - 2x - 3$ **5.** -105 **6.** $\left\{-1, -\frac{1}{3}, \frac{2}{5}\right\}$

Exercise Set 3.3

1. $x + 3$ **3.** $x^2 + 3x + 1$ **5.** $2x^2 + 3x + 5$ **7.** $4x + 3 + \frac{2}{3x - 2}$ **9.** $2x^2 + x + 6 - \frac{38}{x + 3}$
11. $4x^3 + 16x^2 + 60x + 246 + \frac{984}{x - 4}$ **13.** $2x + 5$ **15.** $6x^2 + 3x - 1 - \frac{3x - 1}{3x^2 + 1}$ **17.** $2x + 5$ **19.** $3x - 8 + \frac{20}{x + 5}$
21. $4x^2 + x + 4 + \frac{3}{x - 1}$ **23.** $6x^4 + 12x^3 + 22x^2 + 48x + 93 + \frac{187}{x - 2}$ **25.** $x^3 - 10x^2 + 51x - 260 + \frac{1300}{x + 5}$
27. $x^4 + x^3 + 2x^2 + 2x + 2$ **29.** $x^3 + 4x^2 + 16x + 64$ **31.** $2x^4 - 7x^3 + 15x^2 - 31x + 64 - \frac{129}{x + 2}$
33. -25 **35.** 4729 **37.** $x^2 - 5x + 6; x = -1, x = 2, x = 3$ **39.** $\left\{-\frac{1}{2}, 1, 2\right\}$ **41.** $\left\{-\frac{3}{2}, -\frac{1}{3}, \frac{1}{2}\right\}$ **43.** $x^3 + 5x^2 - 9x - 45$

45. a. $1 + \frac{5}{x + 20}$ **b.**

x	0	5	10	25	50	75
$\frac{x + 25}{x + 20}$	1.25	1.2	≈ 1.17	≈ 1.11	≈ 1.07	≈ 1.05

c. The ratio decreases as x increases.

55.
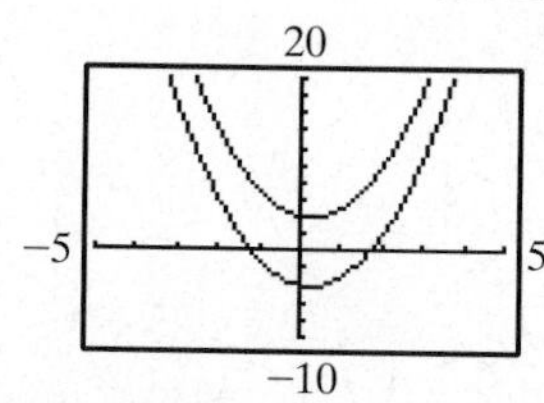

The division is not correct.

57.
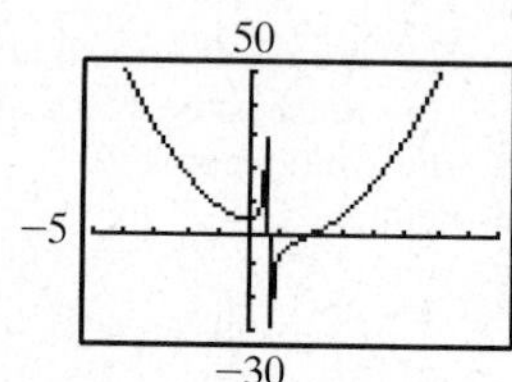

The division is correct.

59. $k = -12$ **61.** $x^{2n} - x^n + 1$

Section 3.4

Check Point Exercises

1. $\pm 1, \pm 2, \pm 3, \pm 6$ **2.** $\pm 1, \pm 3, \pm\frac{1}{2}, \pm\frac{1}{4}, \pm\frac{3}{2}, \pm\frac{3}{4}$ **3.** $\{-5, -4, 1\}$ **4.** $\{1, 2 - 3i, 2 + 3i\}$
5. 4, 2, or 0 positive zeros, no possible negative zeros

Exercise Set 3.4

1. $\pm 1, \pm 2, \pm 4$ **3.** $\pm 1, \pm 2, \pm 3, \pm 6, \pm\frac{1}{3}, \pm\frac{2}{3}$ **5.** $\pm 1, \pm 2, \pm 3, \pm 6, \pm\frac{1}{2}, \pm\frac{1}{4}, \pm\frac{3}{2}, \pm\frac{3}{4}$ **7.** $\pm 1, \pm 2, \pm 3, \pm 4, \pm 6, \pm 12$ **9. a.** $\pm 1, \pm 2, \pm 4$
b. 2 is a zero **c.** $\{2, -2, -1\}$ **11. a.** $\pm 1, \pm 2, \pm 3, \pm 6, \pm\frac{1}{2}, \pm\frac{3}{2}$ **b.** 3 is a zero **c.** $\left\{3, \frac{1}{2}, -2\right\}$
13. a. $\pm 1, \pm 2, \pm 4, \pm 8, \pm\frac{1}{3}, \pm\frac{2}{3}, \pm\frac{4}{3}, \pm\frac{8}{3}$ **b.** 2 is a zero **c.** $\left\{2, -\frac{1}{3}, -4\right\}$ **15. a.** $\pm 1, \pm 2, \pm 3, \pm 4, \pm 6, \pm 12$ **b.** 4 is a root
c. $\{-3, 1, 4\}$ **17. a.** $\pm 1, \pm 2, \pm 3, \pm 4, \pm 6, \pm 12$ **b.** -2 is a root **c.** $\{-2, 1 + \sqrt{7}, 1 - \sqrt{7}\}$

19. a. $\pm1, \pm5, \pm\frac{1}{2}, \pm\frac{5}{2}, \pm\frac{1}{3}, \pm\frac{5}{3}, \pm\frac{1}{6}, \pm\frac{5}{6}$ **b.** -5 is a root **c.** $\left\{-5, \frac{1}{2}, \frac{1}{3}\right\}$ **21. a.** $\pm1, \pm2, \pm4$ **b.** 2 is a root **c.** $\{-2, 2, 1+\sqrt{2}, 1-\sqrt{2}\}$ **23.** no positive real roots; 3 or 1 negative real roots **25.** 3 or 1 positive real roots; no negative real roots **27.** 2 or 0 positive real roots; 2 or 0 negative real roots **29.** $x=-2, x=5, x=1$ **31.** $\left\{-\frac{1}{2}, \frac{1+\sqrt{17}}{2}, \frac{1-\sqrt{17}}{2}\right\}$ **33.** $\{-1, -2, 3+\sqrt{13}, 3-\sqrt{13}\}$ **35.** $x=-1, x=2, x=-\frac{1}{3}, x=3$ **37.** $\left\{1, -\frac{3}{4}, i\sqrt{2}, -i\sqrt{2}\right\}$ **39.** $\left\{-2, \frac{1}{2}, \sqrt{2}, -\sqrt{2}\right\}$ **41. a.** $x=3, x\approx4.2$ **b.** degree: 4; leading coefficient: negative **43.** $W=3$ mm **45.** 2 in. by 9 in. by 4 in. **55.** $\frac{1}{2}, \frac{2}{3}, 2$ **57.** $\pm\frac{1}{2}$ **59.** 5, 3, or 1 positive real roots exist **61.** (d) is true. **63.** 3 in.

Section 3.5

Check Point Exercises

1.

2	2	11	−7	−6
		4	30	46
	2	15	23	40

All the numbers are nonnegative.

−7	2	11	−7	−6
		−14	21	−98
	2	−3	14	−104

The signs alternate.

2. $f(-3)=-42; f(-2)=5$ **3.** $\{-3, 7, 2+i, 2-i\}$ **4. a.** $(x^2-5)(x^2+1)$ **b.** $(x+\sqrt{5})(x-\sqrt{5})(x^2+1)$ **c.** $(x+\sqrt{5})(x-\sqrt{5})(x+i)(x-i)$ **5.** $f(x)=x^3+3x^2+x+3$

Exercise Set 3.5

1.

−4	1	−5	11	33	−18
		−4	36	−188	620
	1	−9	47	−155	602

Since signs alternate, −4 is a lower bound.

7	1	−5	11	33	−18
		7	14	175	1456
	1	2	25	208	1438

Since no sign is negative, 7 is an upper bound.

3.

−4	2	5	−8	7
		−8	12	−16
	2	−3	4	−9

Since signs alternate, −4 is a lower bound.

2	2	5	−8	7
		4	18	20
	2	9	10	27

Since no sign is negative, 2 is an upper bound.

5. a. $\pm1, \pm2, \pm3, \pm4, \pm6, \pm12$ **b.** 1 is not a root. 1 is an upper bound. **c.** Eliminate all positive possible rational roots. **d.** -3 is not a root. -3 is a lower bound. **e.** Eliminate $-3, -4, -6$ and -12. **7.** $f(1)=-1; f(2)=5; 1.3$ **9.** $f(-1)=-1; f(0)=1; -0.5$ **11.** $f(-3)=-11; f(-2)=1; -2.1$ **13.** $f(-3)=-42; f(-2)=5; -2.2$ **15.** $\{-2i, 2i, 2\}$ **17.** $\left\{1-i, 1+i, \frac{1}{3}\right\}$ **19.** $\{2-i, 2+i, -2+i, -2-i\}$ **21.** $\{2-i, 2+i, -3, 7\}$ **23. a.** $(x^2-5)(x^2+4)$ **b.** $(x+\sqrt{5})(x-\sqrt{5})(x^2+4)$ **c.** $(x+\sqrt{5})(x-\sqrt{5})(x+2i)(x-2i)$ **25. a.** $(x^2-2)(x^2+3)$ **b.** $(x+\sqrt{2})(x-\sqrt{2})(x^2+3)$ **c.** $(x+\sqrt{2})(x-\sqrt{2})(x+i\sqrt{3})(x-i\sqrt{3})$ **27. a.** $(x-3)(x+1)(x^2+4)$ **b.** $(x-3)(x+1)(x^2+4)$ **c.** $(x-3)(x+1)(x+2i)(x-2i)$ **29.** $f(x)=2x^3-2x^2+50x-50$ **31.** $f(x)=x^3-3x^2-15x+125$ **33.** $f(x)=x^4+10x^2+9$ **35.** $f(x)=x^4-9x^3+21x^2+21x-130$ **37.** $x=1; x=\pm5i; f(x)=(x-1)(x-5i)(x+5i)$ **39.** $x=2; x=3\pm2i; f(x)=(x-2)(x-3+2i)(x-3-2i)$ **41.** $x=\pm6i; x=\pm i; f(x)=(x-6i)(x+6i)(x-i)(x+i)$ **43.** $x=-2; x=\frac{3}{4}; x=-\frac{1}{2}\pm i;$ $f(x)=(x+2)(4x-3)(2x+1-2i)(2x+1+2i)$ **45.** about 2.5 yr **47.** Answers may vary.
49. According to the linear model, spending will reach \$458 billion, 10.93 or almost 11 years after 1995; According to the quadratic model, spending will reach \$458 billion 10.12 years after 1995; There is a solution between 10.1 and 10.2; According to the third-degree polynomial model, spending will reach \$458 billion just over 10 years after 1995; The third degree polynomial is best.

55.

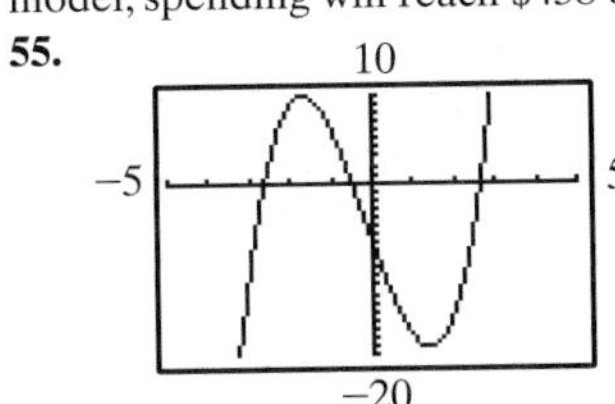

−3 is a lower bound;
3 is an upper bound

57. a. As x, a person's age, increases, y, the number of visits, increases. **b.** 60 **c.**

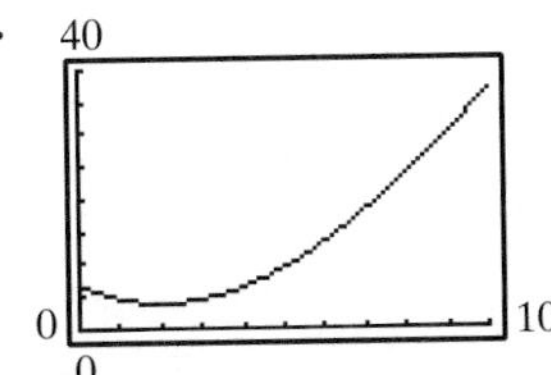

59.

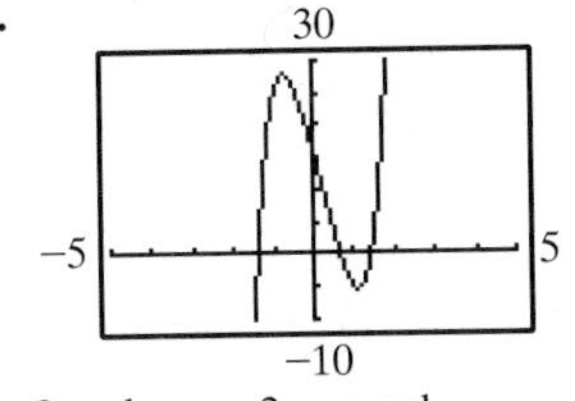

3 real zeros, 2 nonreal complex zeros

61.

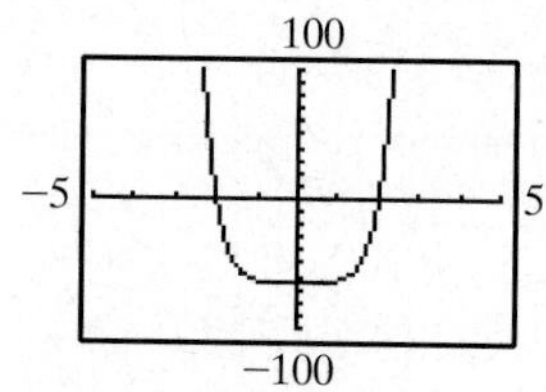

2 real zeros, 4 nonreal complex zeros

63. 3
65. Answers may vary.
67. Answers may vary.

Section 3.6

Check Point Exercises

1. a. $\{x|x \neq 5\}$ **b.** $\{x \mid x \neq -5, x \neq 5\}$ **c.** all real numbers **2. a.** $x = 1, x = -1$ **b.** $x = -1$ **c.** none
3. a. $y = 3$ **b.** $y = 0$ **c.** none

4.

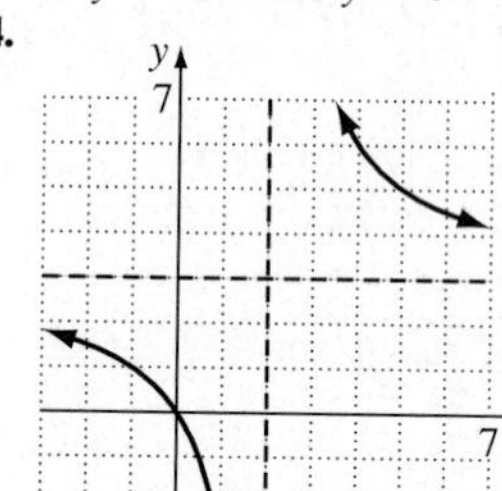

5.

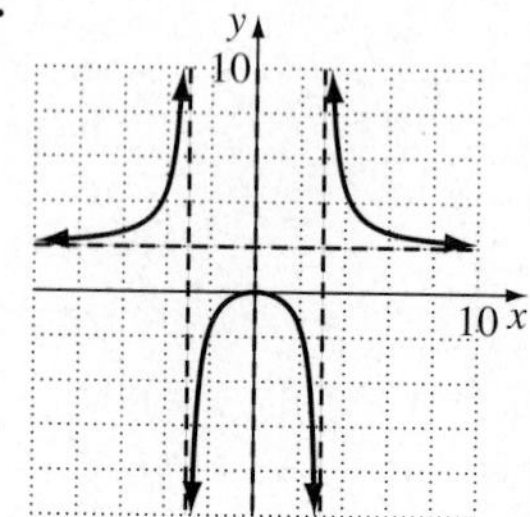

6.

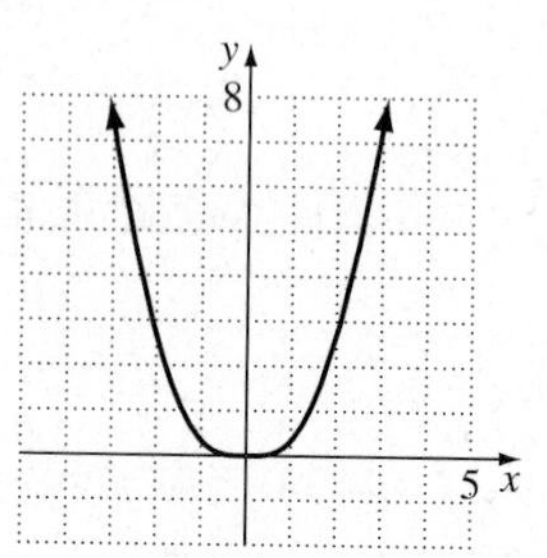

7. $y = 2x - 1$

8. a. $\overline{C}(1000) = 330$, when 1000 pairs of shoes are produced, it costs \$330 to produce each pair; $\overline{C}(10{,}000) = 60$, when 10,000 pairs of shoes are produced, it costs \$60 to produce each pair; $\overline{C}(100{,}000) = 33$, when 100,000 pairs of shoes are produced, it costs \$33 to produce each pair.
b. $y = 30$; The cost per pair of shoes approaches \$30 as more shoes are produced.

Exercise Set 3.6

1. $\{x|x \neq 4\}$ **3.** $\{x|x \neq 5, x \neq -4\}$ **5.** $\{x|x \neq 7, x \neq -7\}$ **7.** All real numbers **9.** $-\infty$ **11.** $-\infty$ **13.** 0 **15.** $+\infty$
17. $-\infty$ **19.** 1 **21.** $x = -4$ **23.** $x = 0, x = -4$ **25.** $x = -4$ **27.** no vertical asymptotes **29.** $y = 0$ **31.** $y = 4$
33. no horizontal asymptote **35.** $y = -\frac{2}{3}$

37.

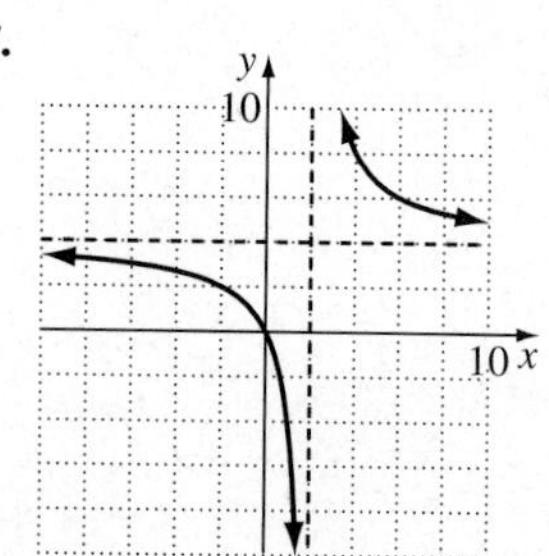

39.

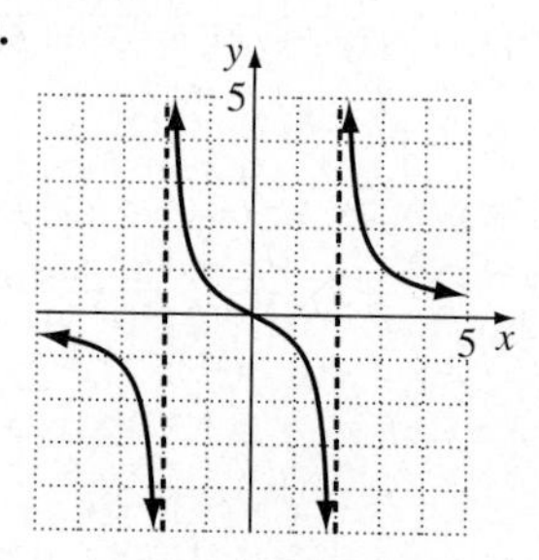

41.

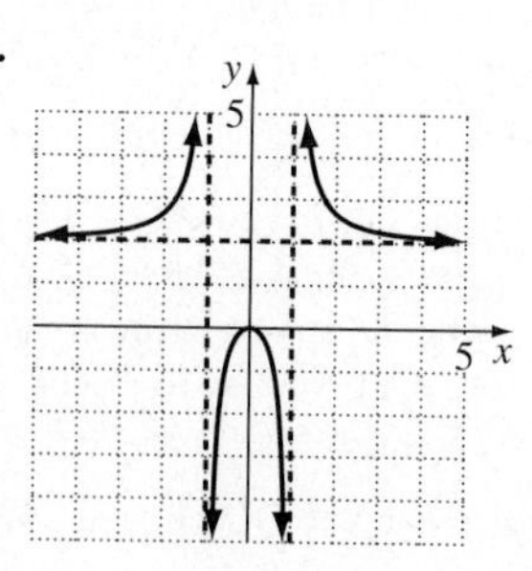

43.

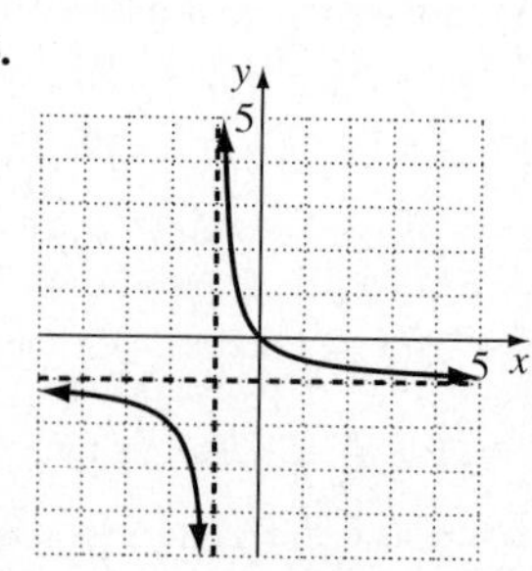

45.

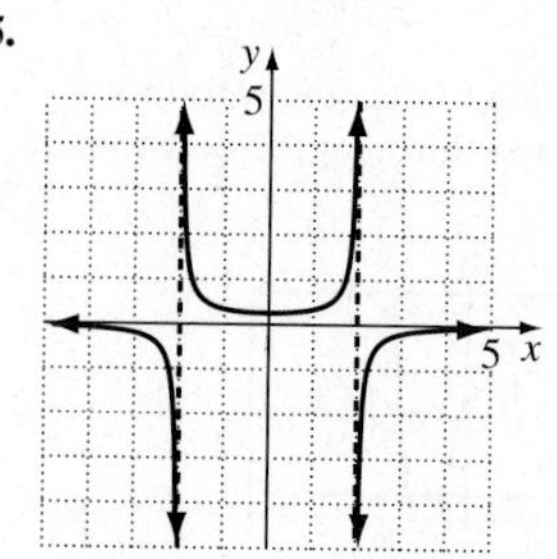

47.

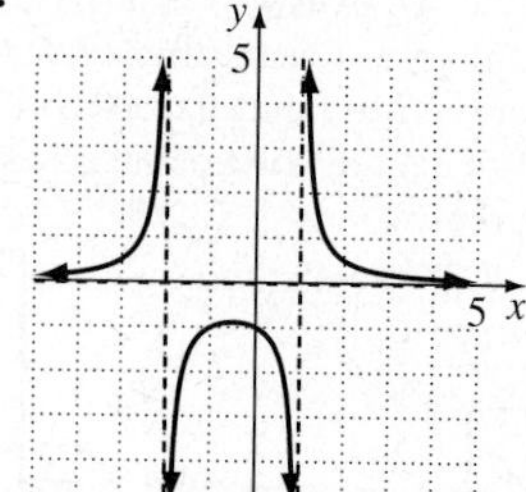

49.

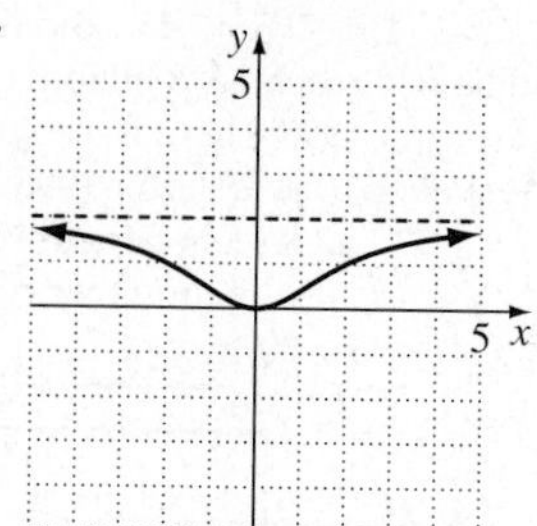

51.

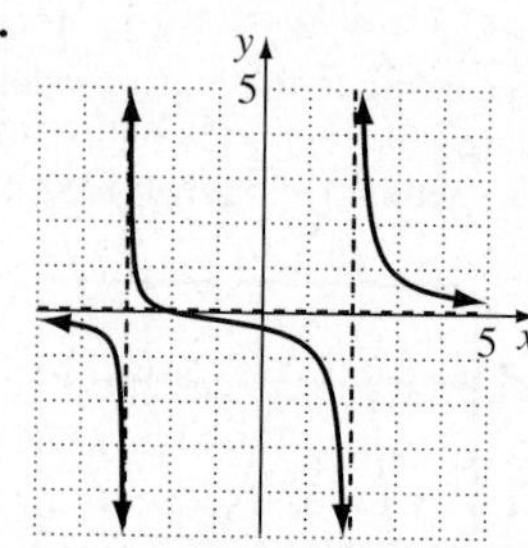

53.

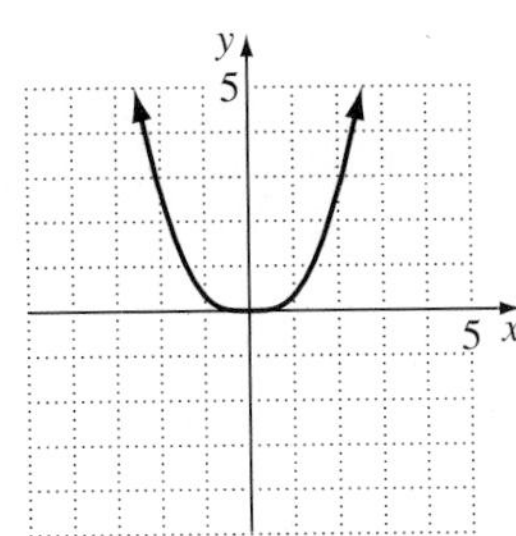

55.

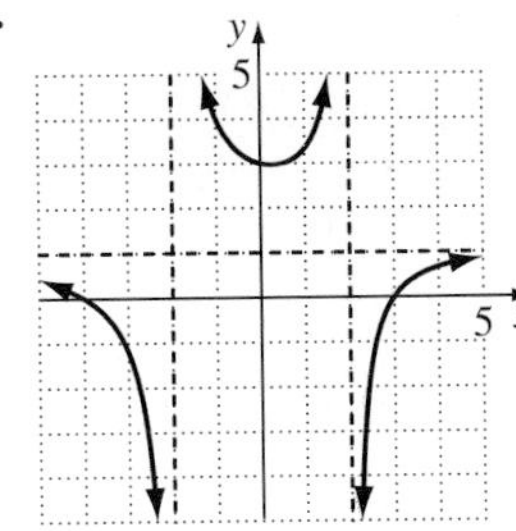

57.

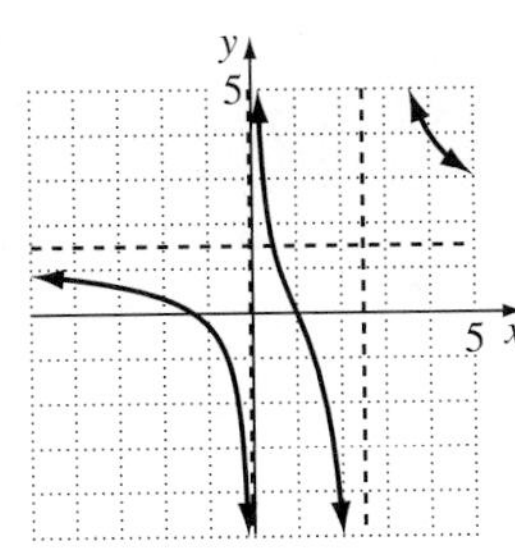

59. a. Slant asymptote: $y = x$

b.

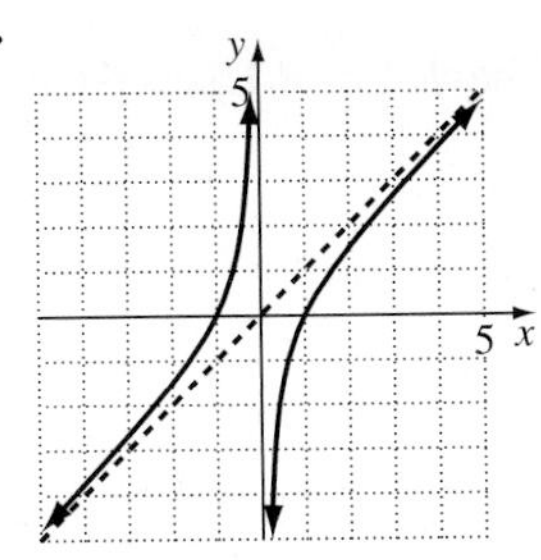

61. a. Slant asymptote: $y = x$

b.

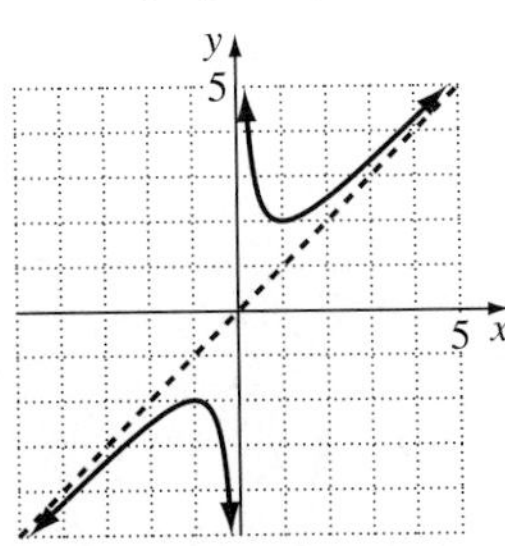

63. a. Slant asymptote: $y = x + 4$

b.

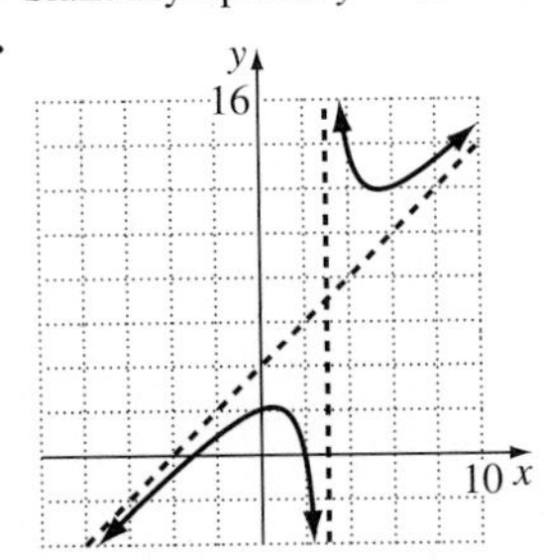

65. a. Slant asymptote: $y = x - 2$

b.

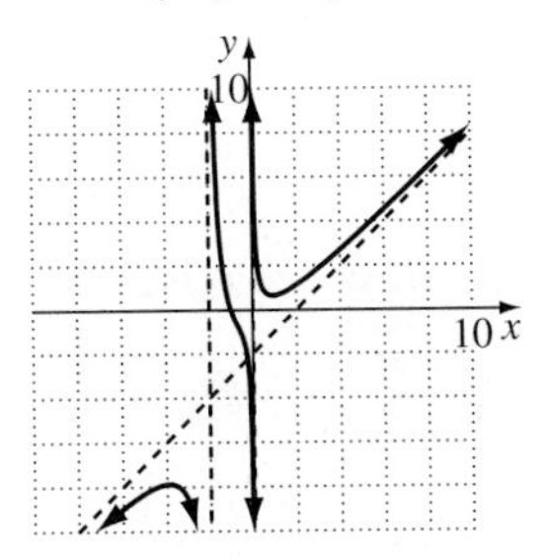

67. a. $\overline{C}(100) = \$220; \overline{C}(1000) = \$40; \overline{C}(10{,}000) = \$22; \overline{C}(100{,}000) = \$20.2$
b. $y = 20$; \$20 is the minimum average cost of producing a canoe. As more canoes are manufactured, the average cost approaches \$20.

69. a. \$100,000
b. No; the model indicates that no amount of money can remove 100% of the pollutants since $C(p)$ increases without bound as p approaches 100.

71. a. $F(0) = 80$; When the dessert is placed in the icebox, its temperature is 80°F.
b. $F(1) = 13.3$°F; $F(2) = 6.2$°F; $F(3) \approx 3.6$°F; $F(4) \approx 2.4$°F; $F(5) \approx 1.7$°F
c. $y = 0$; The temperature will approach but not reach 0°F.
d.

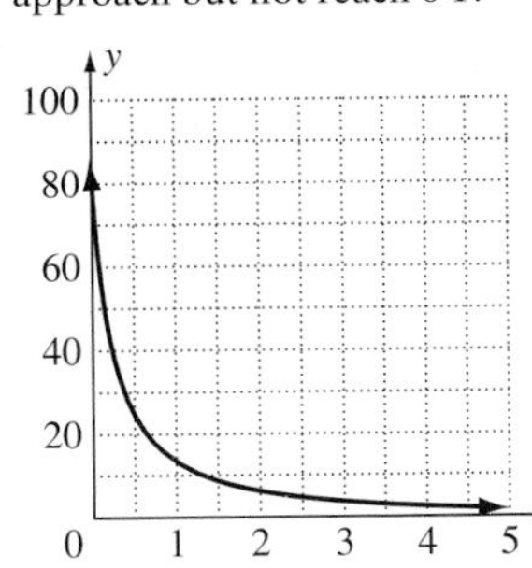

73. a. After 1 day: 35 words; after 5 days: about 12 words; after 15 days: about 7 words
b. $N(1) = 35$ words; This is the same as the estimate from the graph.; $N(5) = 11$ words; This is a little less than the estimate from the graph.; $N(15) = 7$ words; This is the same as the estimate from the graph.
c. The graph indicates the students will remember 5 words over a long period of time.
d. $y = 5$; The horizontal asymptote indicates the students will remember 5 words over a long period of time.

85.

100 | 0 | 5 | 0   40 | 0 | 40 | 0

87.

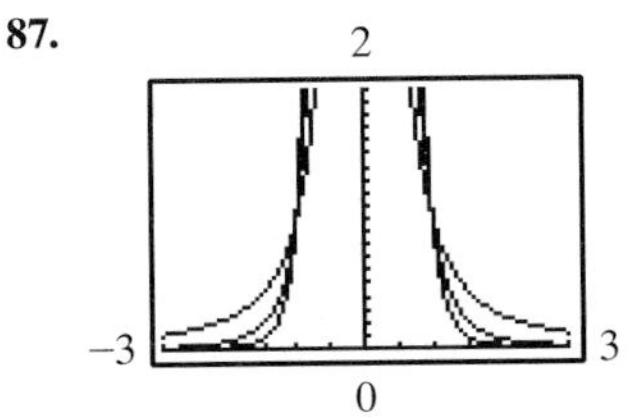

The graph approaches the horizontal asymptote faster and the vertical asymptote slower as n increases.

89.

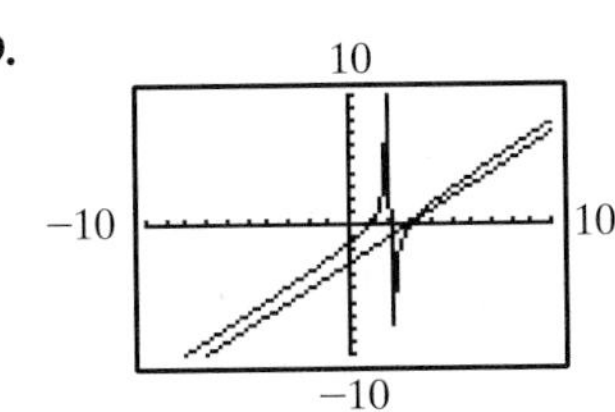

$g(x)$ is the graph of a line whereas $f(x)$ is the graph of a rational function with a slant asymptote; In $g(x)$, $x - 2$ is a factor of $x^2 - 5x + 6$.

91. (d) is true.
93. Answers may vary.
95. Answers may vary.

Section 3.7

Check Point Exercises

1. a. $L = kN$ **b.** $L = 4N$ **c.** 68 in. **2. a.** $W = kL$ **b.** $k = \frac{75}{6}$ **c.** $W = \frac{75L}{6}$ **d.** 200 lb **3.** 137.5 lb/in^2
4. about 556 ft **5.** about 4.36 lb/in^2 **6.** 24 min **7.** 96π ft^3

Exercise Set 3.7

1. $g = kh$ **3.** $a = kb^2$ **5.** $r = \frac{k}{t}$ **7.** $a = \frac{k}{b^3}$ **9.** $r = \frac{ks}{v}$ **11.** $s = kgt^2$ **13.** $k = 25$ **15.** $k = 5$ **17.** $k = 5000$
19. $k = 3$ **21.** $k = 2$ **23.** 84 **25.** 25 **27.** $\frac{5}{6}$ **29.** 240 **31. a.** $L = kW$ **b.** $L = 0.02W$ **c.** 1.04 in. **33.** \$60
35. 2442 mph **37.** 607 lb **39.** 0.5 hr **41.** 6.4 lb **43.** 31.78; index: about 32; not in the desirable range **45.** 11.11 foot-candles
47. 72 erg **59.** The destructive power is four times as much. **61.** Reduce the resistance by a factor of $\frac{1}{3}$.

Chapter 3 Review Exercises

1.

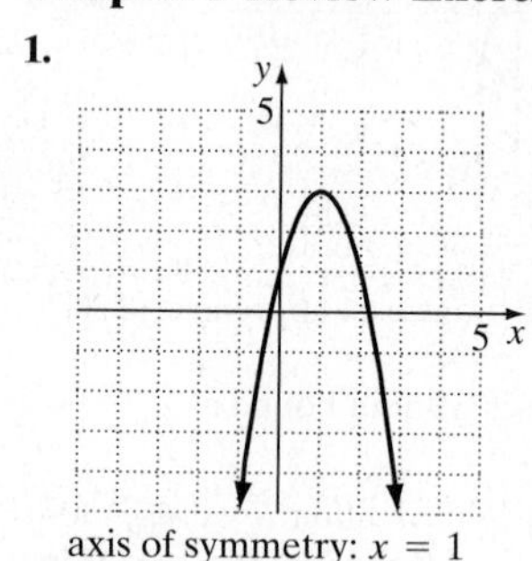

axis of symmetry: $x = 1$

2.

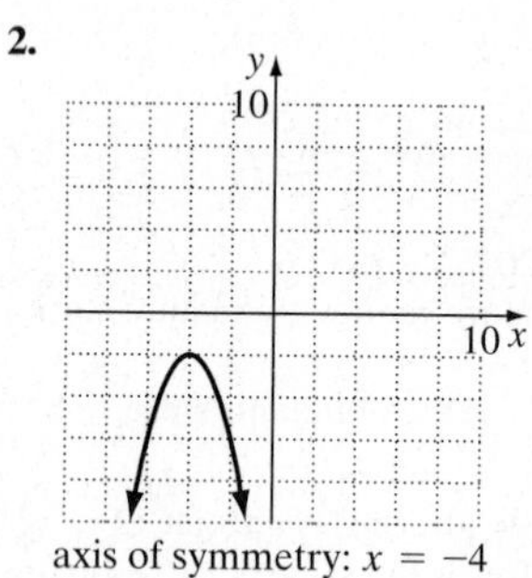

axis of symmetry: $x = -4$

3.

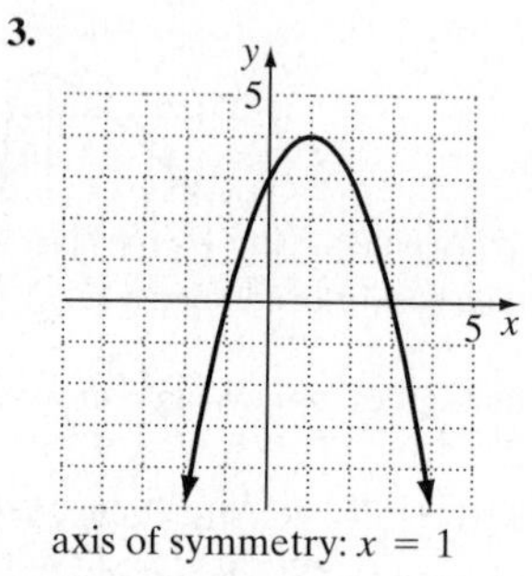

axis of symmetry: $x = 1$

4.

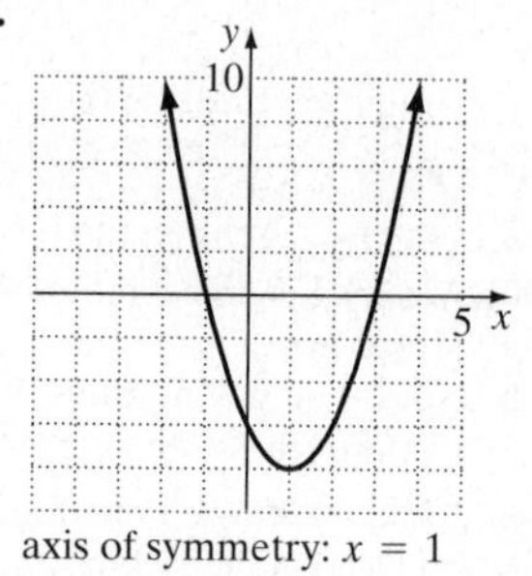

axis of symmetry: $x = 1$

5. 7.2 hr; 622 deaths **6.** 2 sec; 144 ft **7. c.** **8. b.** **9. a.** **10. d.** **11.** Because the degree is odd and the leading coefficient is negative, the graph falls to the right. Therefore, the model indicates that the percentage of families below the poverty level will eventually be negative, which is impossible. **12.** Since the degree is even and the leading coefficient is negative, the graph falls to the right. Therefore, the model indicates a patient will eventually have a negative number of viral bodies, which is impossible.
13. $x = 1$, multiplicity 1, crosses; $x = -2$, multiplicity 2, touches; $x = -5$, multiplicity 3, crosses
14. $x = -5$, multiplicity 1, crosses; $x = 5$, multiplicity 3, crosses

15. a. The graph falls to the left and rises to the right.
b. no symmetry
c.

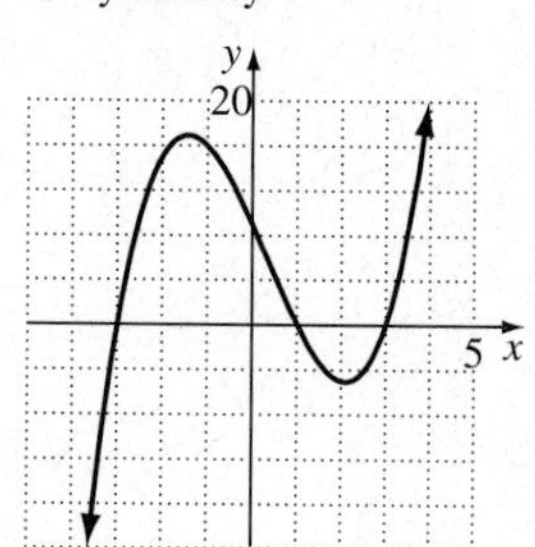

16. a. The graph rises to the left and falls to the right.
b. origin symmetry
c.

17. a. The graph falls to the left and rises to the right.
b. no symmetry
c.

18. a. The graph falls to the left and rises to the right. **b.** y-axis symmetry **c.**

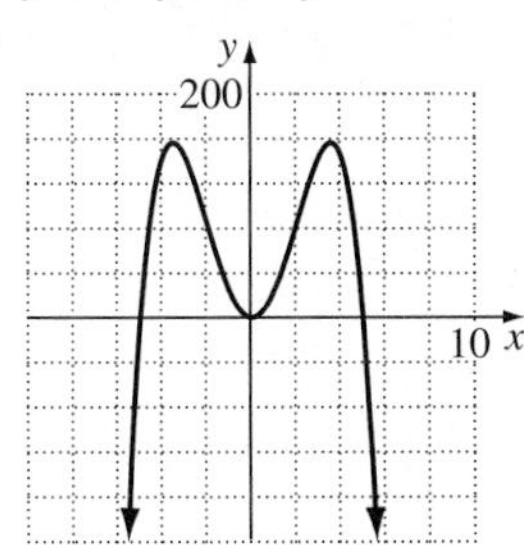

19. a. The graph falls to the left and to the right. **b.** no symmetry **c.**

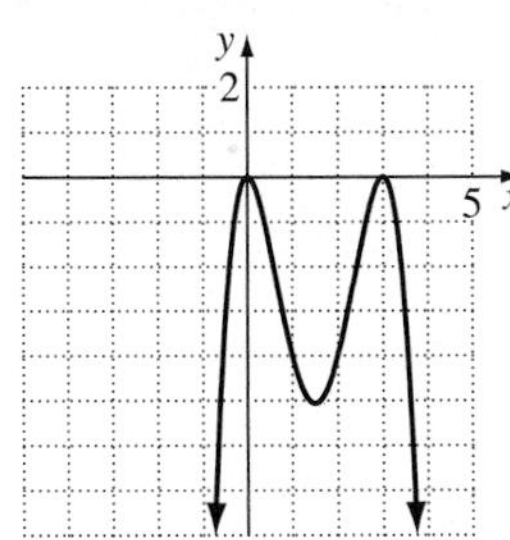

20. a. The graph rises to the left and to the right. **b.** no symmetry **c.**

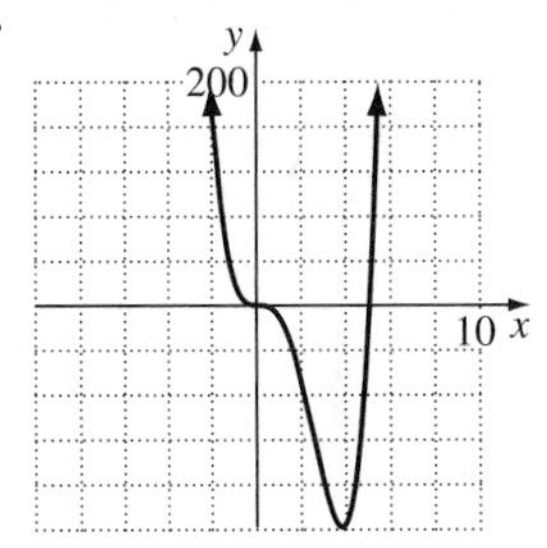

21. $4x^2 - 7x + 5 - \frac{4}{x+1}$ **22.** $2x^2 - 4x + 1 - \frac{10}{5x-3}$ **23.** $2x^2 + 3x - 1$ **24.** $3x^3 - 4x^2 + 7$

25. $3x^3 + 6x^2 + 10x + 10 + \frac{20}{x-2}$ **26.** -5697 **27.** $2, \frac{1}{2}, -3$ **28.** $\{4, -2 \pm \sqrt{5}\}$ **29.** $\pm 1, \pm 5$

30. $\pm 1, \pm 2, \pm 4, \pm 8, \pm \frac{8}{3}, \pm \frac{4}{3}, \pm \frac{2}{3}, \pm \frac{1}{3}$ **31.** 2 or 0 positive solutions; no negative solutions

32. 3 or 1 positive real roots; 2 or 0 negative solutions **33.** No sign variations exist for either $f(x)$ or $f(-x)$, so no real roots exist.

34. a. $\pm 1, \pm 2, \pm 4$ **b.** 1 positive real zero; 2 or no negative real zeros **c.** 1 is a zero **d.** $\{1, -2\}$

35. a. $\pm 1, \pm \frac{1}{2}, \pm \frac{1}{3}, \pm \frac{1}{6}$ **b.** 2 or 0 positive real zeros; 1 negative real zero **c.** -1 is a zero **d.** $\left\{-1, \frac{1}{3}, \frac{1}{2}\right\}$

36. a. $\pm 1, \pm 3, \pm 5, \pm 15, \pm \frac{1}{2}, \pm \frac{1}{4}, \pm \frac{1}{8}, \pm \frac{3}{2}, \pm \frac{3}{4}, \pm \frac{3}{8}, \pm \frac{5}{2}, \pm \frac{5}{4}, \pm \frac{5}{8}, \pm \frac{15}{2}, \pm \frac{15}{4}, \pm \frac{15}{8}$

b. 3 or 1 positive real solutions; no negative real solutions **c.** $\frac{1}{2}$ is a zero **d.** $\left\{\frac{1}{2}, \frac{3}{2}, \frac{5}{2}\right\}$

37. a. $\pm 1, \pm 2, \pm 3, \pm 6$ **b.** 2 or zero positive real solutions; 2 or zero negative real solutions **c.** -2 is a zero **d.** $\{-2, -1, 1, 3\}$

38. a. $\pm 1, \pm 2, \pm \frac{1}{2}, \pm \frac{1}{4}$ **b.** 1 positive real root; 1 negative real root **c.** $\frac{1}{2}$ is a zero **d.** $\left\{\frac{1}{2}, -\frac{1}{2}, i\sqrt{2}, -i\sqrt{2}\right\}$

39. a. $\pm 1, \pm 2, \pm 4, \pm \frac{1}{2}$ **b.** 2 or no positive zeros; 2 or no negative zeros **c.** $x = 2$ is a zero **d.** $\left\{2, -2, \frac{1}{2}, -1\right\}$

40.

−2	2	−7	−5	28	−12
		−4	22	−34	12
	2	−11	17	−6	0

-2 is a root and a lower bound.

6	2	−7	−5	28	−12
		12	30	150	1068
	2	5	25	178	1056

6 is an upper bound, but not a zero.

$\pm 1, \pm 2, 3, 4, \pm \frac{1}{2}, \pm \frac{3}{2}$

41. a. $\pm 1, \pm 2, \pm 3, \pm 4, \pm 6, \pm 12, \pm \frac{1}{2}, \pm \frac{3}{2}$ **b.** 2 is not a root but is an upper bound. **c.** -2 is not a root but is a lower bound.

d. Possible roots are $\pm 1, \pm \frac{1}{2}$, and $\pm \frac{3}{2}$. **42.** $f(1) = -2; f(2) = 3; x \approx 1.6$ **43.** $f(-3) = -32; f(-2) = 7; x \approx -2.3$

44. $\left\{-\frac{1}{4}, 6 \pm 5i\right\}$ **45.** $\{1 \pm 3i, 1 \pm i\}$ **46.** $\left\{-\frac{1}{2}, 1, 4 \pm 7i\right\}$ **47.** $f(x) = x^3 - 6x^2 + 21x - 26$

48. $f(x) = 2x^4 + 12x^3 + 20x^2 + 12x + 18$ **49.** $f(x) = x^4 - 3x^3 + 6x^2 + 2x - 60$

50. $-2, \frac{1}{2}, \pm i; f(x) = (x - i)(x + i)(x + 2)\left(x - \frac{1}{2}\right)$ **51.** $-1, 4; g(x) = (x + 1)^2(x - 4)^2$

52. 4 real zeros, one with multiplicity two **53.** 3 real zeros; 2 nonreal complex zeros

54. 2 real zeros, one with multiplicity two; 2 nonreal complex zeros **55.** 1 real zero; 4 nonreal complex zeros

56. Vertical asymptote: $x = 3$ and $x = -3$
horizontal asymptote: $y = 0$

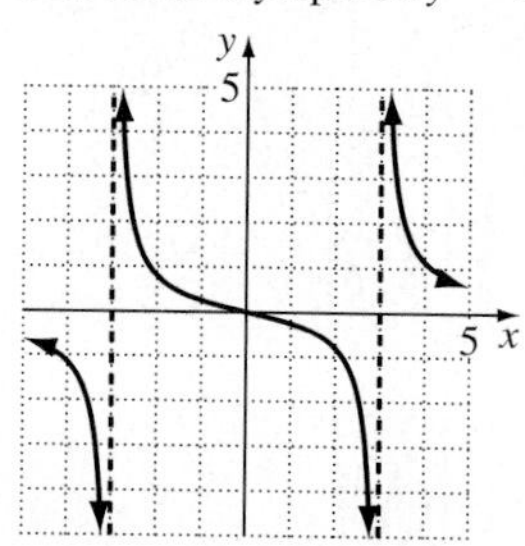

57. Vertical asymptote: $x = -3$
horizontal asymptote: $y = 2$

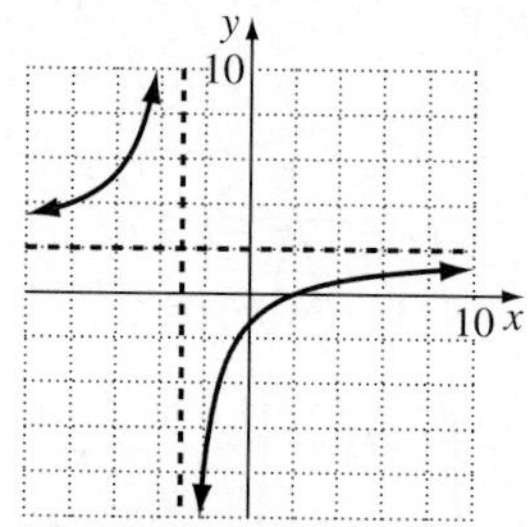

58. Vertical asymptotes: $x = 3, -2$
horizontal asymptote: $y = 1$

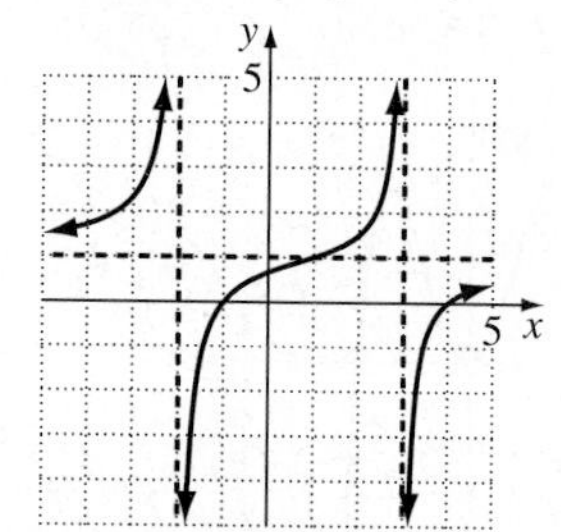

59. Vertical asymptote: $x = -2$
horizontal asymptote: $y = 1$

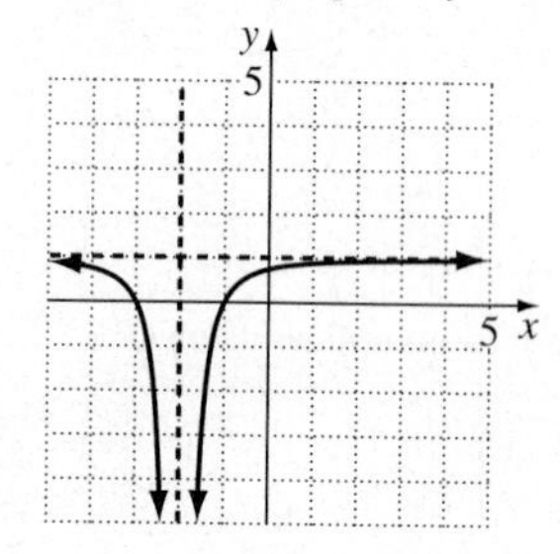

60. Vertical asymptote: $x = -1$
no horizontal asymptote
slant asymptote: $y = x - 1$

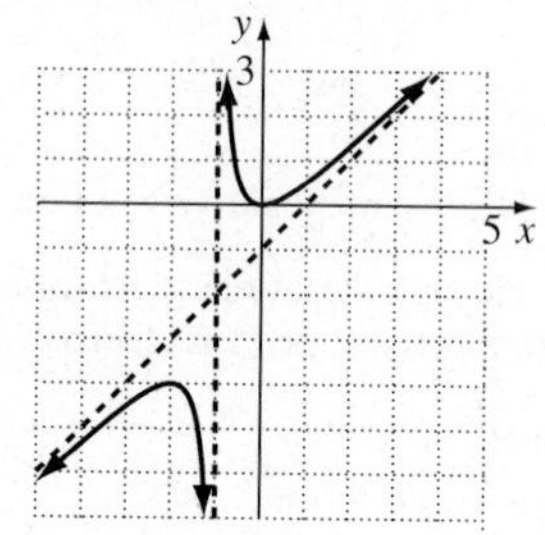

61. Vertical asymptote: $x = 3$
no horizontal asymptote
slant asymptote: $y = x + 5$

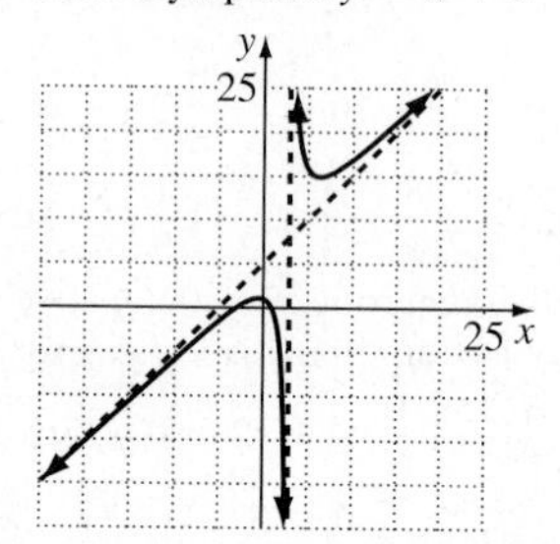

62. No vertical asymptote

no horizontal asymptote
slant asymptote: $y = -2x$

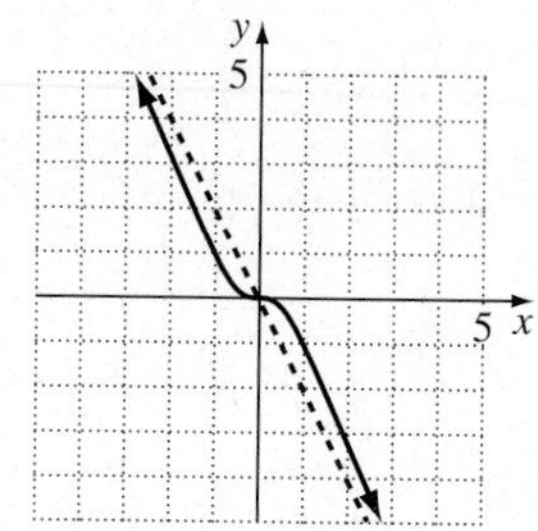

63. Vertical asymptote: $x = \frac{3}{2}$

no horizontal asymptote
slant asymptote: $y = 2x - 5$

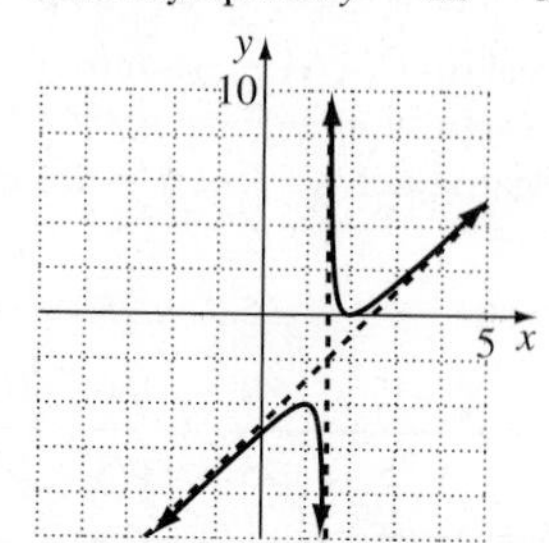

64. a. $\overline{C}(50) = 1025$, when 50 calculators are manufactured, it costs \$1025 to manufacture each; $\overline{C}(100) = 525$, when 100 calculators are manufactured, it costs \$525 to manufacture each; $\overline{C}(1000) = 75$, when 1000 calculators are manufactured, it costs \$75 to manufacture each; $\overline{C}(100{,}000) = 25.5$, when 100,000 calculators are manufactured, it costs \$25.50 to manufacture each.
b. $y = 25$; Minimum costs will approach \$25.

65. a. 1600; The difference in cost of removing 90% versus 50% of the contaminants is 16 million dollars.
b. $x = 100$; No amount of money can remove 100% of the contaminants, since $C(x)$ increases without bound as x approaches 100.
66. $y = 3000$; The number of fish in the pond approaches 3,000,000.
67. $y = 0$; As the number of years of education increases the percentage rate of unemployment approaches zero.
68. Since $C(p)$ increases without bound as p approaches 100, the politician will not be able to keep his promise.
69. \$154 **70.** 1600 ft **71.** 5 hr **72.** 112 decibels **73.** 16 hr **74.** 800 ft^3

Chapter 3 Test

1. axis of symmetry: $x = -1$

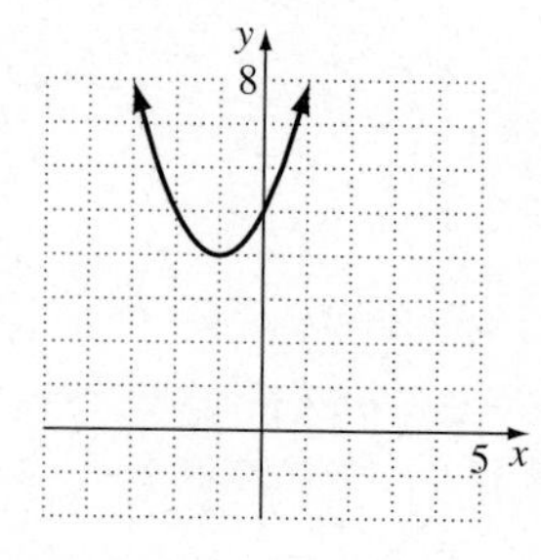

2. axis of symmetry: $x = 1$

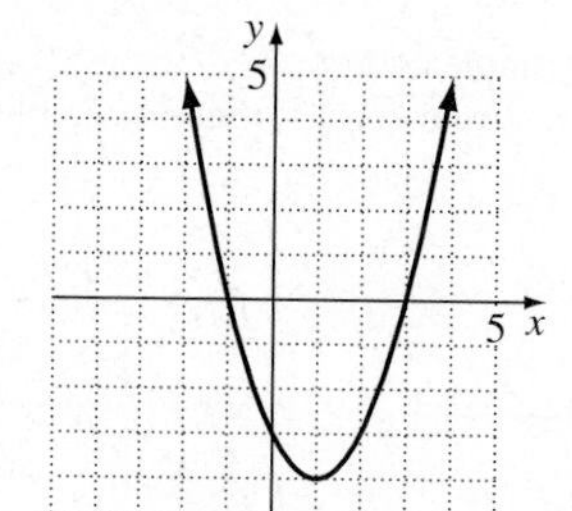

3. maximum; (3, 2)
4. 23 VCRs; maximum daily profit = \$16,900
5. a. 5, 2, −2
b.

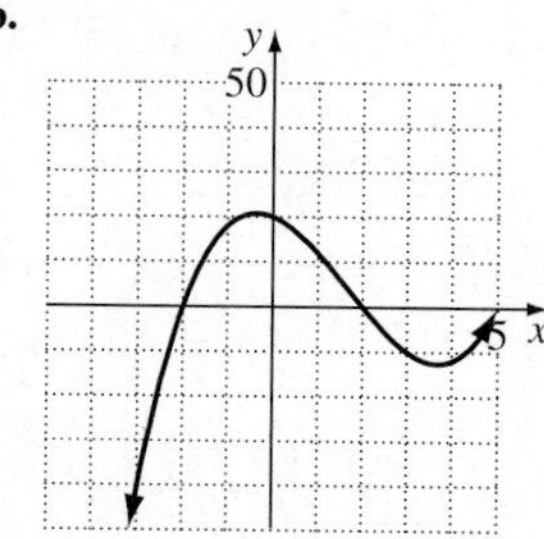

6. Since the degree of the polynomial is odd and the leading coefficient is positive, the graph of f should fall to the left and rise to the right. The x-intercepts should be −1 and 1.

7. a. 2 **b.** $\frac{1}{2}, \frac{2}{3}$ **8.** $\pm 1, \pm 2, \pm 3, \pm 6, \pm\frac{1}{2}, \pm\frac{3}{2}$ **9.** 3 or 1 positive real zeros; no negative real zeros. **10.** $\{-5, -3, 2\}$

11. a. $\pm 1, \pm 3, \pm 5, \pm 15, \pm\frac{1}{2}, \pm\frac{3}{2}, \pm\frac{5}{2}, \pm\frac{15}{2}$ **b.** $\left\{-1, \frac{3}{2}, \pm\sqrt{5}\right\}$

12.

−3	3	4	−7	−2	−3
		−9	15	−24	78
	3	−5	8	−26	75

−3 is a lower bound.

2	3	4	−7	−2	−3
		6	20	26	48
	3	10	13	24	45

2 is an upper bound.

13. $\{2, 3, 1 + i\}$ **14.** $(x - 1)(x + 2)^2$

15. domain: $\{x | x \neq 4, x \neq -4\}$

16. domain: $\{x | x \neq 2\}$

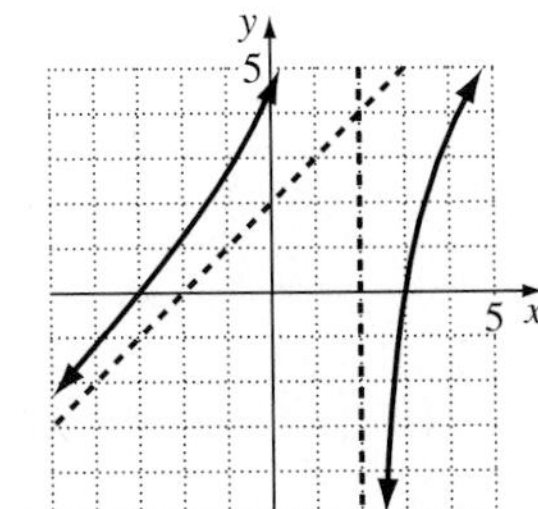

17. domain: $\{x | x \neq -3, x \neq 1\}$

18. domain: all real numbers

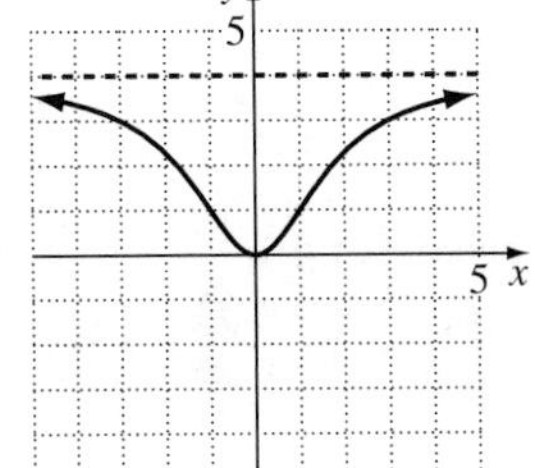

19. a. 50 **b.** 150 **c.** $y = 225$; The deer population will approach a maximum of 225 deer. **20.** 45 foot-candles

Cumulative Review Exercises (Chapters P–3)

1. $2 + \sqrt{3}$ **2.** $-3x^2 - 11x + 11$ **3.** $15\sqrt{2}$ **4.** $x^5(x - 1)(x + 1)$ **5.** $\{2, -1\}$ **6.** $\left\{\frac{5 + \sqrt{13}}{6}, \frac{5 - \sqrt{13}}{6}\right\}$ **7.** $\left\{\frac{1}{3}, -\frac{2}{3}\right\}$

8. $\{-3, -1, 2\}$ **9.** $\{-\infty, 1)$ or $(4, \infty)$ **10.** $(-\infty, -1)$ or $\left(\frac{5}{3}, \infty\right)$

11. Center: $(1, -2)$; radius: 3

12. $t = 1 - \frac{V}{C}$

13. $(-\infty, 5]$

14. $x^2 - 2x - 4$

15. $16x^2 - 6$

16. −9

17. a. $\{-1, 1, 4\}$

b.

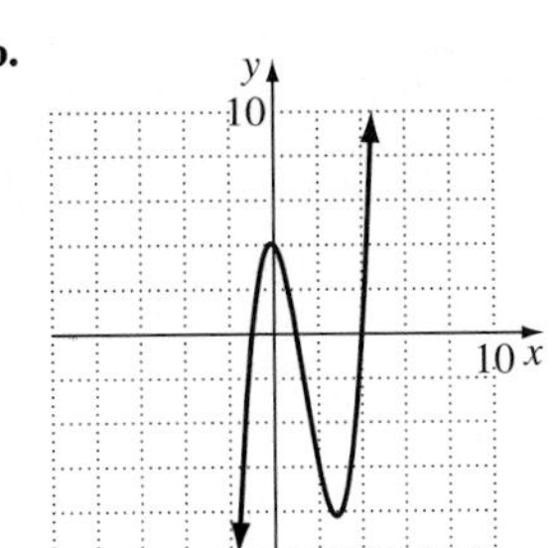

18.

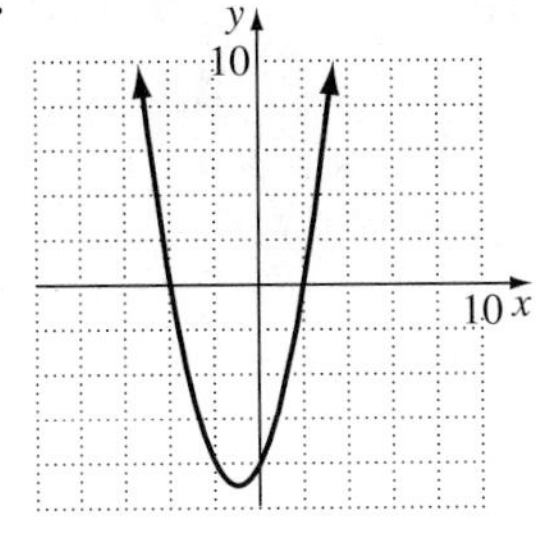

19.

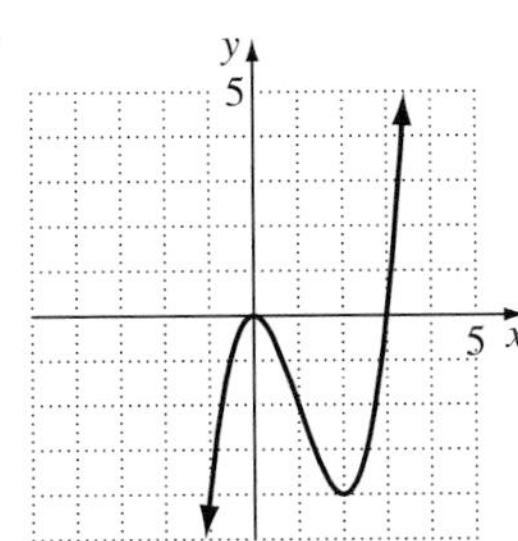

20.

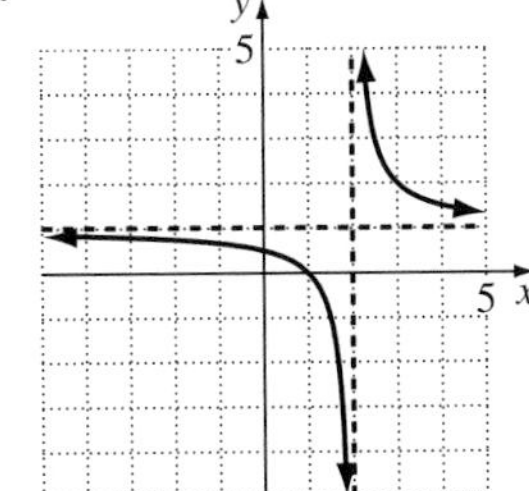

CHAPTER 4

Section 4.1

Check Point Exercises

1. 1 O-ring

2.

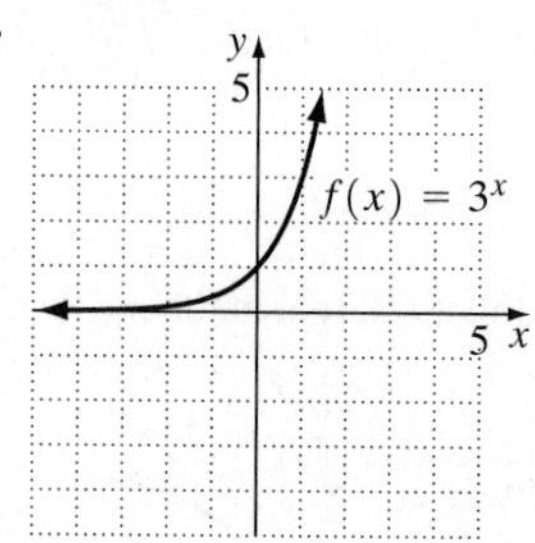

3.

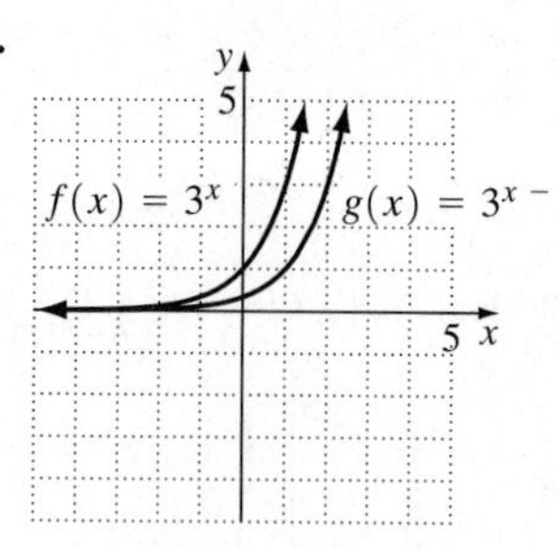

4.

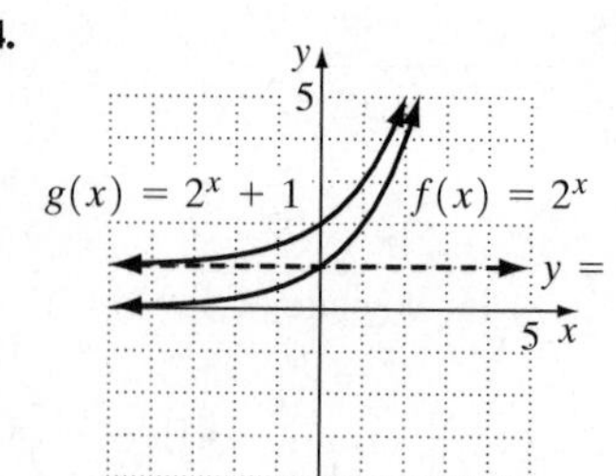

5. 11.64 billion
6. a. $14,859.47
b. $14,918.25

Exercise Set 4.1

1. 10.556 **3.** 11.665 **5.** 0.125 **7.** 9.974 **9.** 0.387

11. **13.**

15.

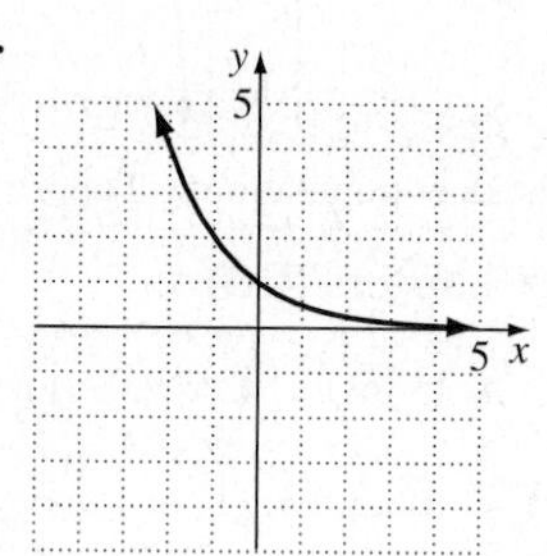

17.

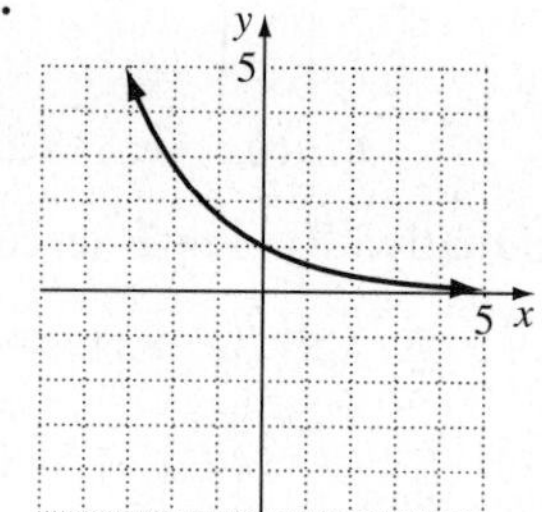

19. $H(x) = -3^{-x}$ **21.** $F(x) = -3^x$ **23.** $h(x) = 3^x - 1$

25.

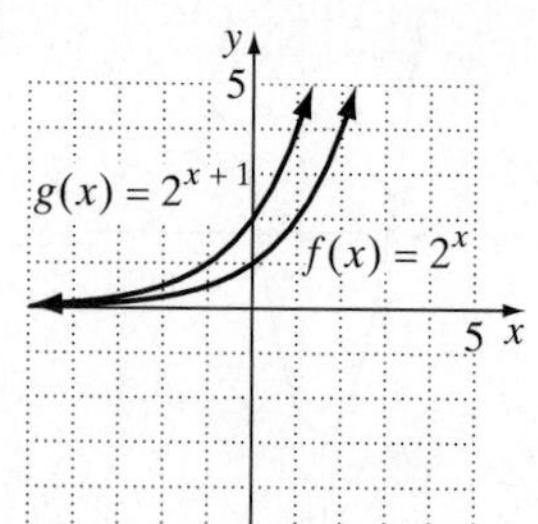

27.

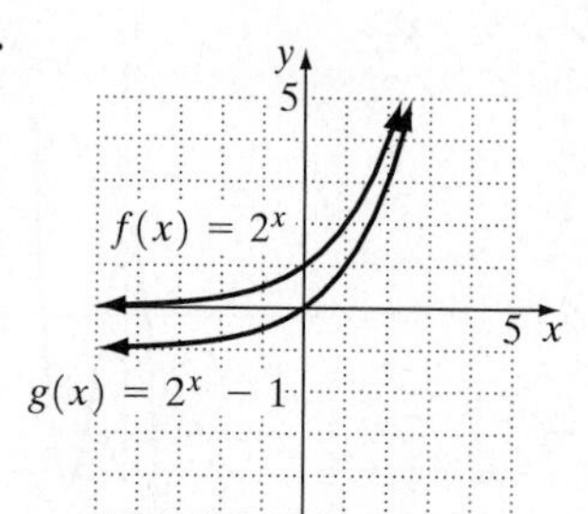

29.

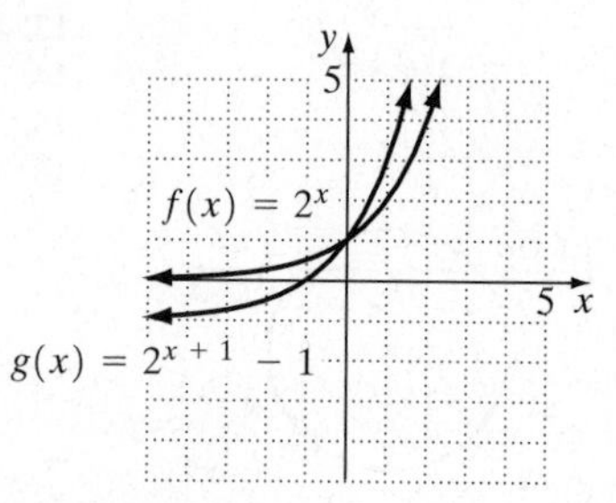

31.

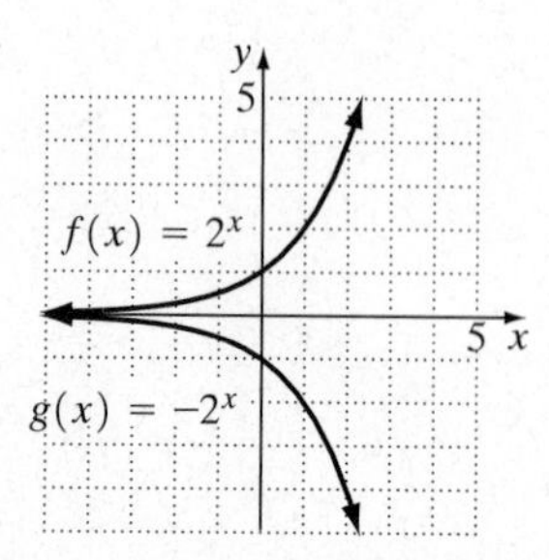

33.

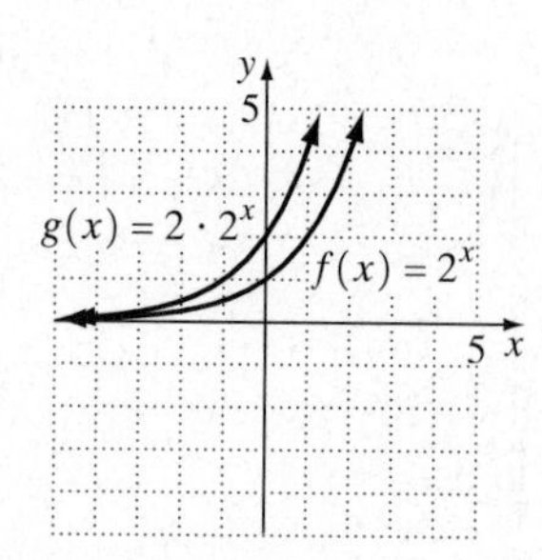

35. a. $13,116.51
b. $13,157.04
c. $13,165.31
37. 7% compounded monthly
39. a. 67.38 million
b. about 134.74 million
c. about 269.46 million
d. 538.85 million
e. appears to double every 27 yr

41. $f(25) \approx 0.0653$; About 6.5% of 25-year-olds have some coronary heart disease. **43.** $116,405.10 **45.** 3.249009585; 3.317278183; 3.321880096; 3.321995226; 3.321997068; $2^{\sqrt{3}} \approx 3.321997085$; The closer the exponent is to $\sqrt{3}$, the closer the value is to $2^{\sqrt{3}}$.
47. $f(11) \approx 241{,}786.19$; The number of AIDS cases among IV drug users in 2000 will be about 241,786. **49. a.** 100% **b.** 68.5% **c.** 30.8% **d.** 20% **51. a.** 1429 **b.** 24,546 **c.** Growth is limited by the population; The entire population will eventually become ill.

59.

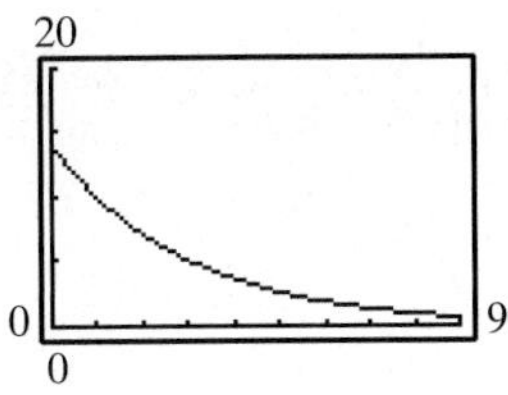

no; Nearly 4 O-rings are expected to fail.

61. a. $A = 10{,}000\left(1 + \frac{0.05}{4}\right)^{4t}$; $A = 10{,}000\left(1 + \frac{0.045}{12}\right)^{12t}$

b.

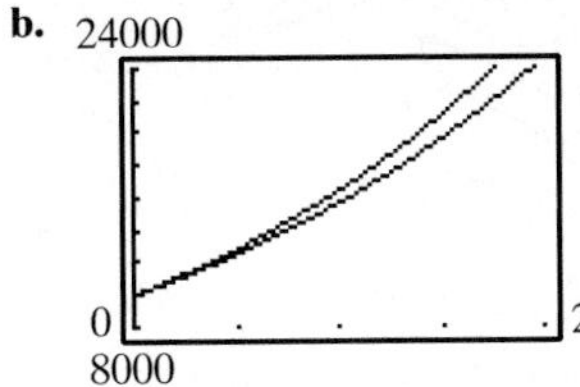

5% interest compounded quarterly

63. (d) is true.

65.

$$\left(\frac{e^x + e^{-x}}{2}\right)^2 - \left(\frac{e^x - e^{-x}}{2}\right)^2 \stackrel{?}{=} 1$$

$$\frac{e^{2x} + 2 + e^{-2x}}{4} - \frac{e^{2x} - 2 + e^{-2x}}{4} \stackrel{?}{=} 1$$

$$\frac{e^{2x} + 2 + e^{-2x} - e^{2x} + 2 - e^{-2x}}{4} \stackrel{?}{=} 1$$

$$\frac{4}{4} \stackrel{?}{=} 1$$

$$1 = 1$$

Section 4.2

Check Point Exercises

1. a. $7^3 = x$ **b.** $b^2 = 25$ **c.** $4^y = 26$ **2. a.** $5 = \log_2 x$ **b.** $3 = \log_b 27$ **c.** $y = \log_e 33$ **3. a.** 2 **b.** 1 **c.** $\frac{1}{2}$

4. a. 1 **b.** 0 **5. a.** 8 **b.** 17 **6.**

$f(x) = 3^x$; $g(x) = \log_3 x$

7. $(5, \infty)$ **8.** 80% **9.** 4.0
10. a. $(-\infty, 4)$
b. $(-\infty, 0)$ or $(0, \infty)$
11. a. $25x$
b. $\sqrt{x}$
12. 4.6 ft per sec

Exercise Set 4.2

1. $2^4 = 16$ **3.** $3^2 = x$ **5.** $b^5 = 32$ **7.** $6^y = 216$ **9.** $\log_2 8 = 3$ **11.** $\log_2 \frac{1}{16} = -4$ **13.** $\log_8 2 = \frac{1}{3}$ **15.** $\log_{13} x = 2$

17. $\log_b 1000 = 3$ **19.** $\log_7 200 = y$ **21.** 2 **23.** 6 **25.** $\frac{1}{2}$ **27.** −3 **29.** $\frac{1}{2}$ **31.** 1 **33.** 0 **35.** 7 **37.** 19

39.

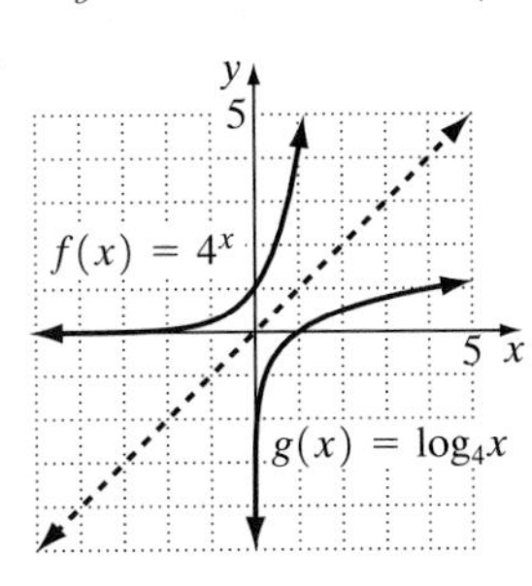

41.

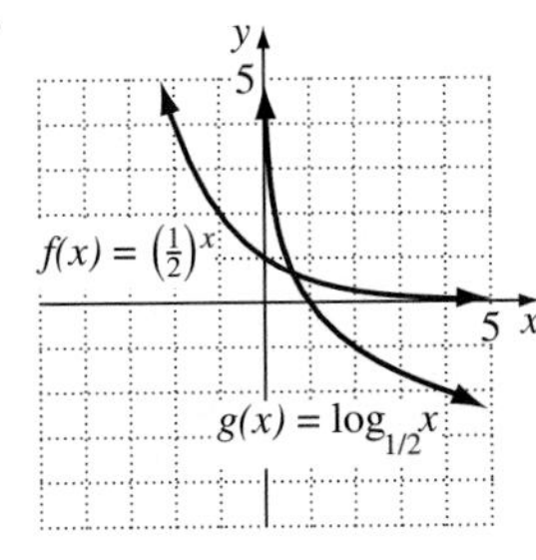

43. $H(x) = 1 - \log_3 x$
45. $h(x) = \log_3 x - 1$
47. $g(x) = \log_3(x - 1)$

49.

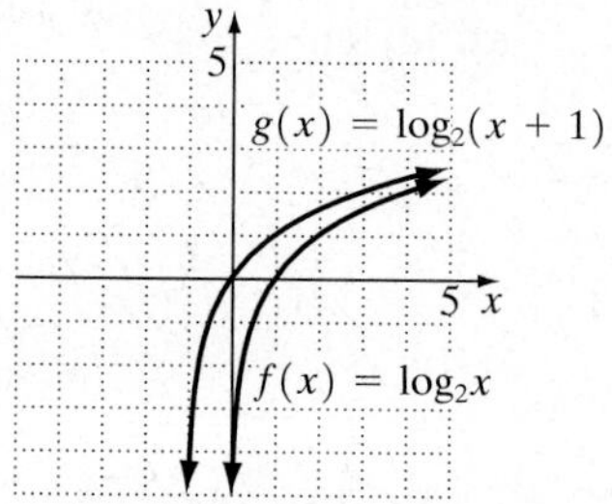

x-intercept: (0, 0)
vertical asymptote: $x = -1$

51.

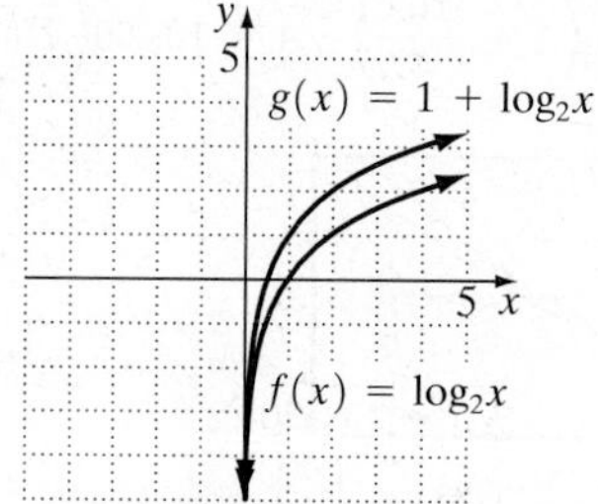

x-intercept: (0.5, 0)
vertical asymptote: $x = 0$

53.

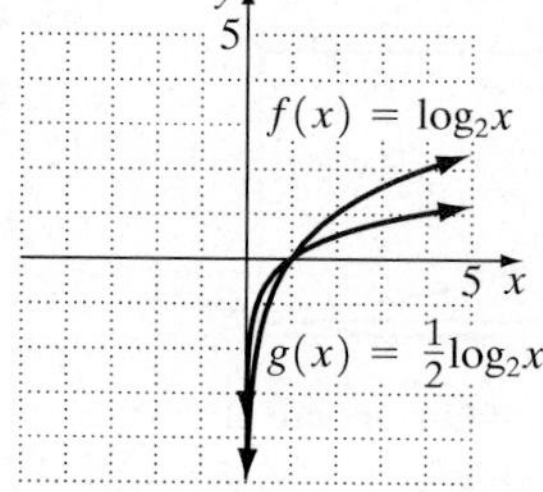

x-intercept: (1, 0)
vertical asymptote: $x = 0$

55. $(-4, \infty)$ **57.** $(-\infty, 2)$ **59.** $(-\infty, 2)$ or $(2, \infty)$ **61.** 2 **63.** 7 **65.** 33 **67.** 0 **69.** 6 **71.** −6 **73.** 125 **75.** $9x$ **77.** $5x^2$ **79.** $\sqrt{x}$ **81.** 95.4% **83.** $5.65 billion **85.** $\approx$ 188 db; yes

87. a. 88
b. 71.5; 63.9; 58.8; 55; 52; 49.5
c.

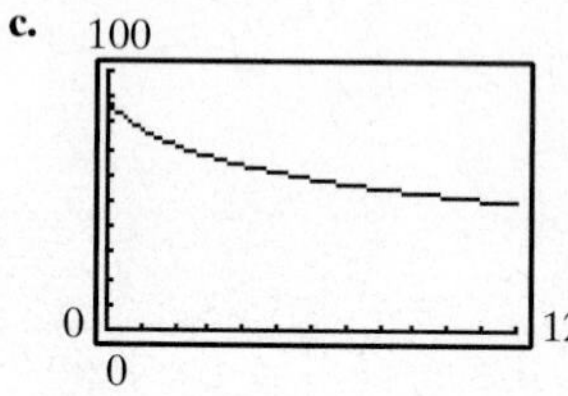

Material retention decreases as time passes.

97.

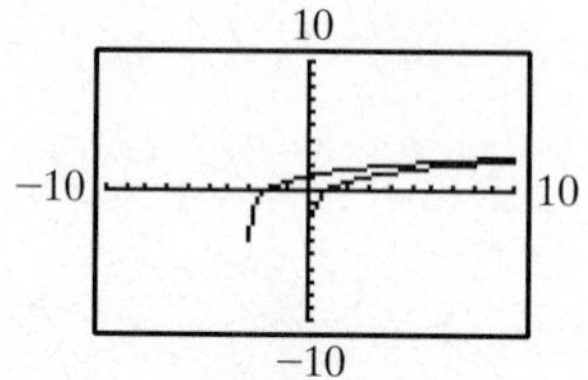

$g(x)$ is $f(x)$ shifted 3 units left.

99.

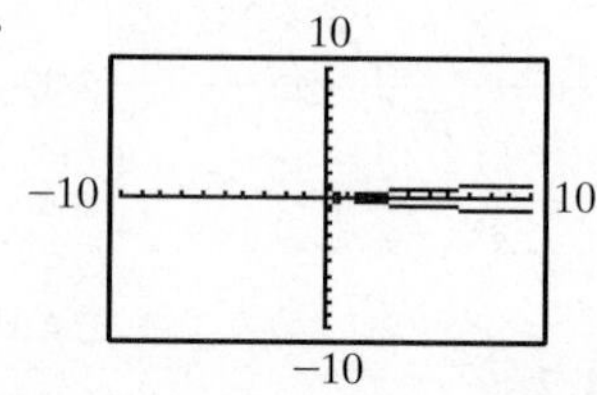

$g(x)$ is $f(x)$ reflected about the x-axis.

101.

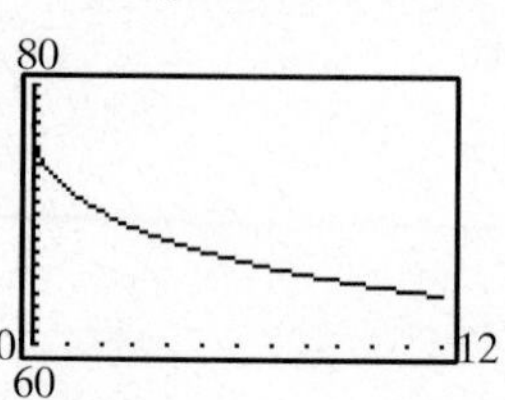

The score falls below 65 after 9 months.

103. $y = \ln x$, $y = \sqrt{x}$, $y = x$, $y = x^2$, $y = e^x$, $y = x^x$

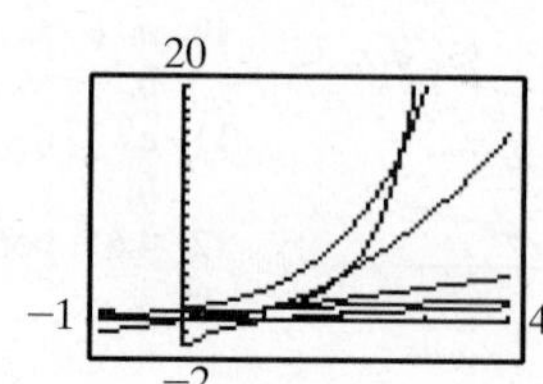

105. $\frac{4}{5}$

107. $\log_3 40 > \log_4 60$

Section 4.3

Check Point Exercises

1. a. $\log_6 10 + \log_6 9$ **b.** $2 + \log x$ **2. a.** $\log_8 23 - \log_8 x$ **b.** $5 - \ln 11$ **3. a.** $9 \log_6 8$ **b.** $\frac{1}{3} \ln x$ **4. a.** $4 \log_b x + \frac{1}{3} \log_b y$ **b.** $\frac{1}{2} \log_5 x - 2 - 3 \log_5 y$ **5. a.** 2 **b.** $\log \frac{7x + 6}{x}$ **6. a.** $\ln x^2 \sqrt[3]{x + 5}$ **b.** $\log \frac{(x - 3)^2}{x}$ **7.** 4.02 **8.** 4.02 **9.**

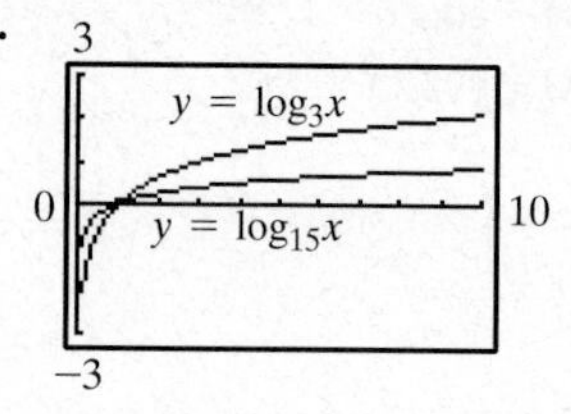

Exercise Set 4.3

1. $\log_5 12 + \log_5 3$ **3.** $1 + \log_7 x$ **5.** $3 + \log x$ **7.** $1 - \log_7 x$ **9.** $\log x - 2$ **11.** $3 - \log_4 y$ **13.** $2 - \ln 5$ **15.** $3 \log_b x$ **17.** $-b \log N$ **19.** $\frac{1}{5} \ln x$ **21.** $2 \log_b x + \log_b y$ **23.** $\frac{1}{2} \log_4 x - 3$ **25.** $2 - \frac{1}{2} \log_6(x + 1)$ **27.** $2 \log_b x + \log_b y - 2 \log_b z$ **29.** $1 + \frac{1}{2} \log x$ **31.** $\frac{1}{3} \log x - \frac{1}{3} \log y$ **33.** 1 **35.** $\ln(7x)$ **37.** 5 **39.** $\log\left(\frac{2x + 5}{x}\right)$

41. $\log(xy^3)$ **43.** $\ln(x^{1/2}y)$ or $\ln(y\sqrt{x})$ **45.** $\log_b(x^2y^3)$ **47.** $\ln\left(\frac{x^5}{y^2}\right)$ **49.** $\ln\left(\frac{x^3}{y^{1/3}}\right)$ or $\ln\left(\frac{x^3}{\sqrt[3]{y}}\right)$ **51.** $\ln\frac{(x+6)^4}{x^3}$

53. 1.5937 **55.** 1.6944 **57.** −1.2304 **59.** 3.6193 **61. a.** $D = 10\log\frac{I}{I_0}$ **b.** 20 decibels louder

71. a.

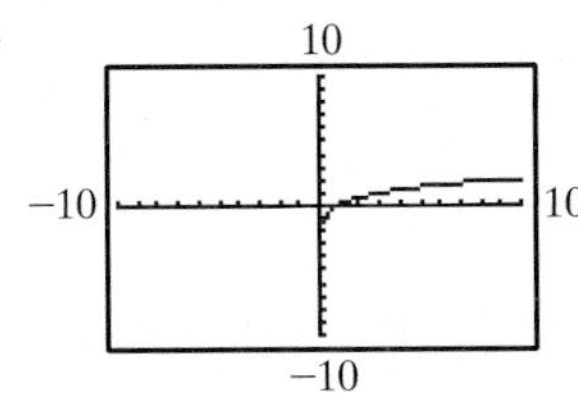

b.

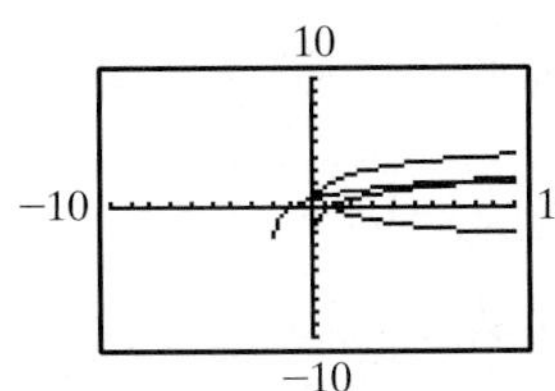

$y = 2 + \log_3 x$ shifts the graph of $y = \log_3 x$ two units upward; $y = \log_3(x + 2)$ shifts the graph of $y = \log_3 x$ two units left; $y = -\log_3 x$ reflects the graph of $y = \log_3 x$ about the x-axis.

73.

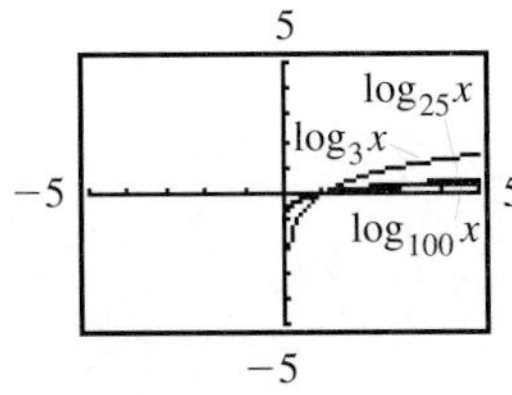

a. top graph: $y = \log_{100} x$; bottom graph: $y = \log_3 x$
b. top graph: $y = \log_3 x$; bottom graph: $y = 100\log_{100} x$
c. The graph of the equation with the largest b will be on the top in the interval $(0, 1)$ and on the bottom in the interval $(1, \infty)$.

79. (d) is true. **81.** $\frac{2A}{B}$

Section 4.4

Check Point Exercises

1. $\left\{\frac{\ln 134}{\ln 5}\right\}$; ≈ 3.04 **2.** $\left\{\frac{\ln 9}{2}\right\}$; ≈ 1.10 **3.** $\left\{\frac{\ln 2088 + 4\ln 6}{3\ln 6}\right\}$; ≈ 2.76 **4.** $\{0, \ln 7\}$; $\ln 7 \approx 1.95$ **5.** $\{12\}$ **6.** $\{5\}$
7. $\left\{\frac{e^2}{3}\right\}$ **8.** 0.01 **9.** 16.2 yr **10.** 2149

Exercise Set 4.4

1. $\left\{\frac{\ln 3.91}{\ln 10}\right\}$; ≈ 0.59 **3.** $\{\ln 5.7\}$; ≈ 1.74 **5.** $\left\{\frac{\ln 17}{\ln 5}\right\}$; ≈ 1.76 **7.** $\left\{\ln\frac{23}{5}\right\}$; ≈ 1.53 **9.** $\left\{\frac{\ln 659}{5}\right\}$; ≈ 1.30
11. $\left\{\frac{\ln 793 - 1}{-5}\right\}$; ≈ -1.14 **13.** $\left\{\frac{\ln 10{,}478 + 3}{5}\right\}$; ≈ 2.45 **15.** $\left\{\frac{\ln 410}{\ln 7} - 2\right\}$; ≈ 1.09 **17.** $\left\{\frac{\ln 813}{0.3\ln 7}\right\}$; ≈ 11.48
19. $\{0, \ln 2\}$; $\ln 2 \approx 0.69$ **21.** $\left\{\frac{\ln 3}{2}\right\}$; ≈ 0.55 **23.** $\{81\}$ **25.** $\{59\}$ **27.** $\left\{\frac{109}{27}\right\}$ **29.** $\left\{\frac{62}{3}\right\}$ **31.** $\left\{\frac{5}{4}\right\}$ **33.** $\{6\}$
35. $\{6\}$ **37.** $\{e^2\}$; ≈ 7.39 **39.** $\left\{\frac{e^4}{2}\right\}$; ≈ 27.30 **41.** $\{e^{-1/2}\}$; ≈ 0.61 **43.** $\{e^2 - 3\}$; ≈ 4.39 **45.** about 0.22
47. a. 18.2 million **b.** 2010 **49.** 8 yr **51.** 16.8% **53.** 9 yr **55.** 15.7% **57.** 1995 **59.** 2.8 days; Yes, the point (2.8, 50) appears to lie on the graph of P. **61.** $10^{-2.4}$; 0.004 moles per liter **67.** $\{2\}$ **69.** $\{4\}$ **71.** $\{2\}$ **73.** $\{-1.391606, 1.6855579\}$

75.

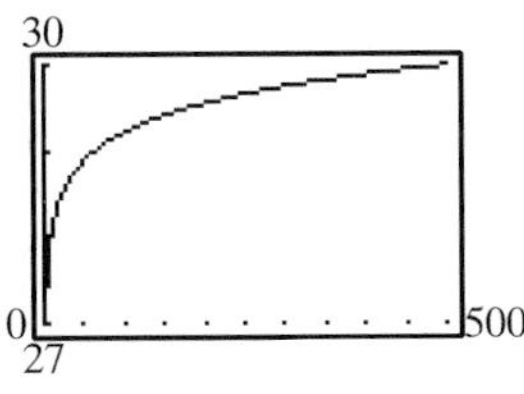

As distance from eye increases, barometric air pressure increases, leveling off at about 30 inches of mercury.

77.

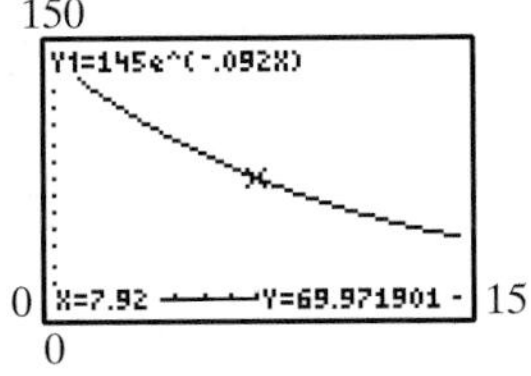

about 7.92 min

79. (c) is true.
81. $\{1, e^2\}$, $e^2 \approx 7.389$
83. $\{e\}$, $e \approx 2.718$

Section 4.5

Check Point Exercises

1. a. $A = 491e^{0.027t}$ **b.** 2006 **2. a.** $A = A_0e^{-0.0248t}$ **b.** about 72 yr **3. a.** 0.4 correct responses **b.** 0.7 correct responses **c.** 0.8 correct responses **4.** $y = 4e^{(\ln 7.8)x}$; $y = 4e^{2.054x}$

Exercise Set 4.5

1. 208 million **3.** 2016 **5.** 2.6% **7.** 2014 **9.** $140,000 **11.** 2005 **13.** 0.175; $A = 200e^{0.175t}$; 17.5% **15.** 8.01 g

17. 8 g; 4 g; 2 g; 1 g; 0.5 g **19.** 15,679 years old **21. a.** $\frac{A_0}{2} = A_0e^{k(1.31)}; \frac{1}{2} = e^{1.31k}; \ln\frac{1}{2} = \ln e^{1.31k}; \ln\frac{1}{2} = 1.31k; k = \frac{\ln\frac{1}{2}}{1.31} \approx -0.52912$

b. 107 million years **23.** $2A_0 = A_0e^{kt}; 2 = e^{kt}; \ln 2 = \ln e^{kt}; \ln 2 = kt; t = \frac{\ln 2}{k}$ **25.** 63 yr **27. a.** about 20 people

b. about 1080 people **c.** 100,000 people **29.** about 3.7% **31.** about 48 years old **33.** $y = 100e^{(\ln 4.6)x}$; $y = 100e^{1.526x}$

35. $y = 2.5e^{(\ln 0.7)x}$; $y = 2.5e^{-0.357x}$ **47.** $y = 51.75985638 + 109.7788574 \ln x$; $r = 0.8974781617$, a good fit

49. $y = 98.06189365x^{0.4398361087}$; $r = 0.9546621296$, a good fit

51.

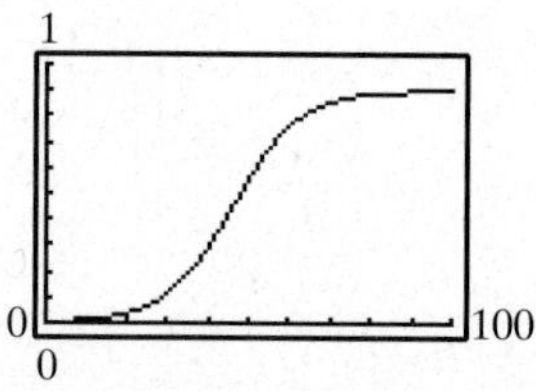

The probability of coronary heart disease starts increasing at a more rapid rate at about age 20. At about age 60, the rate of increase starts to slow down.

53. The linear model, $y = 74.52833333x + 214.7694444$, best fits the data. Answers for prediction may vary.

55. about 126°F

Chapter 4 Review Exercises

1. $g(x) = 4^{-x}$ **2.** $h(x) = -4^{-x}$ **3.** $r(x) = -4^{-x} + 3$ **4.** $f(x) = 4^{x}$

5. $f(x) = 2^x$; $g(x) = 2^{x-1}$

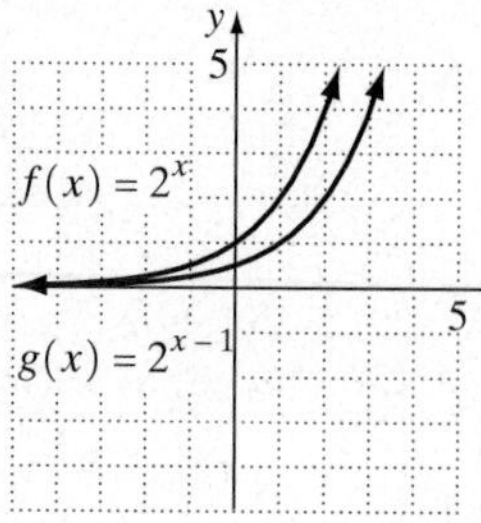

6.

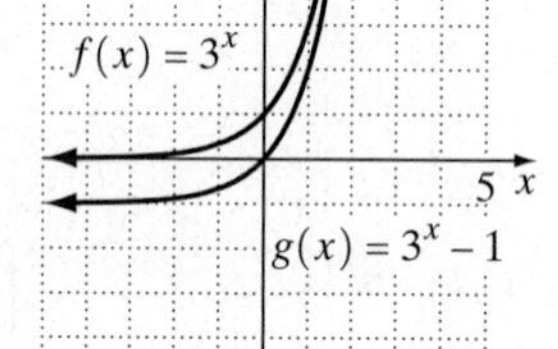

7.

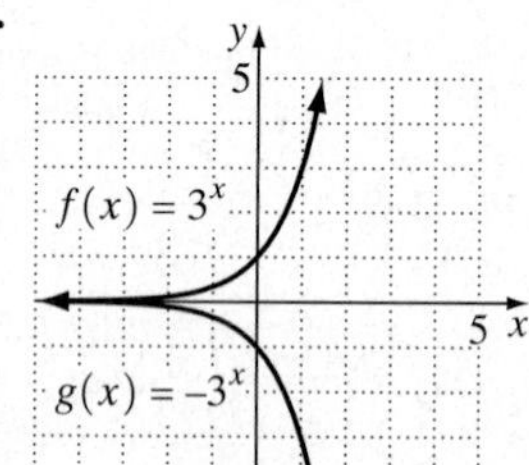

8.

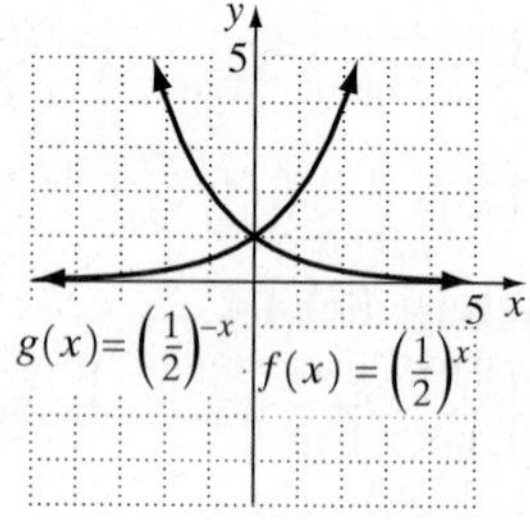

9. 5.5% compounded semiannually **10.** 7% compounded monthly **11. a.** 200° **b.** 120°; 119°

c. 70°; The temperature in the room is 70°. **12.** $49^{1/2} = 7$ **13.** $4^3 = x$ **14.** $3^y = 81$ **15.** $\log_6 216 = 3$ **16.** $\log_b 625 = 4$

17. $\log_{13} 874 = y$ **18.** 3 **19.** -2 **20.** $\varnothing$; $\log_b x$ is defined only for $x > 0$. **21.** $\frac{1}{2}$ **22.** 1 **23.** 8

24. 5 **25.** 0 **26.** $f(x) = 2^x$; $g(x) = \log_2 x$

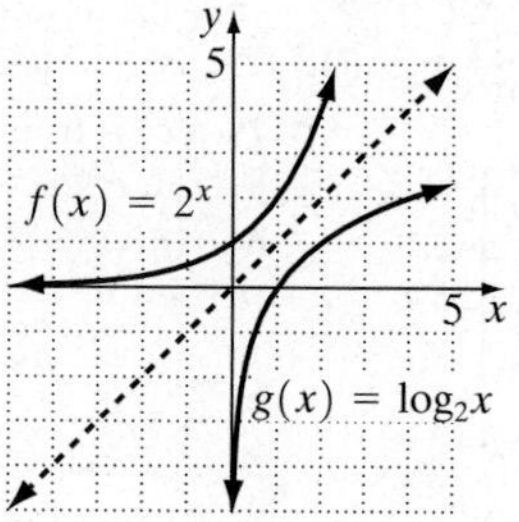

27.

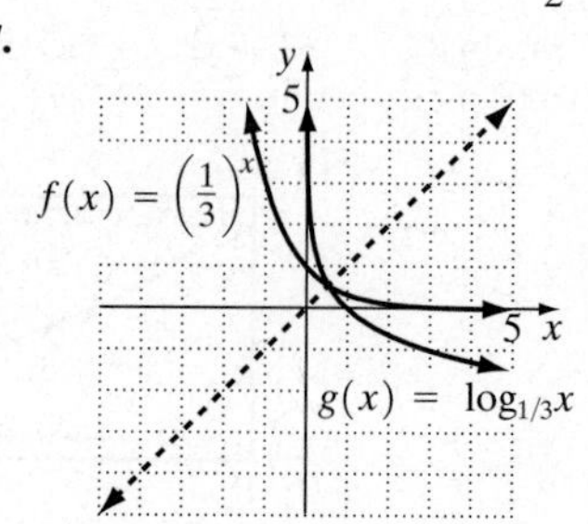

28. $g(x) = \log(-x)$
29. $r(x) = 1 + \log(2 - x)$
30. $h(x) = \log(2 - x)$
31. $f(x) = \log x$

32.

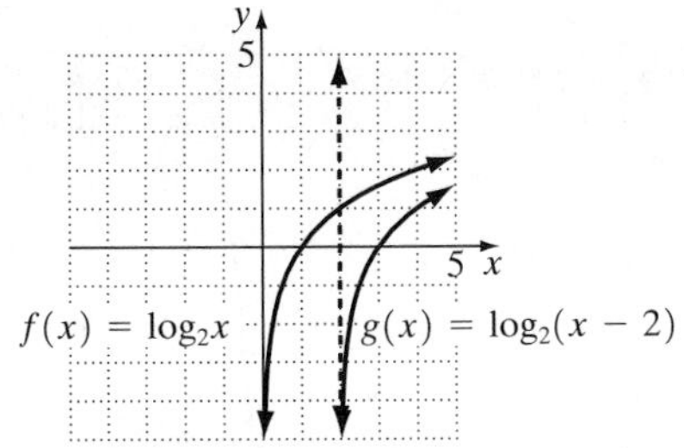

x-intercept: (3, 0)
vertical asymptote: $x = 2$

33.

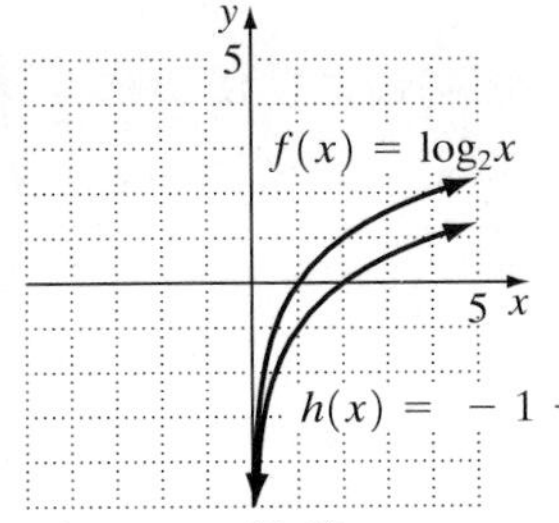

x-intercept: (2, 0)
vertical asymptote: $x = 0$

34.

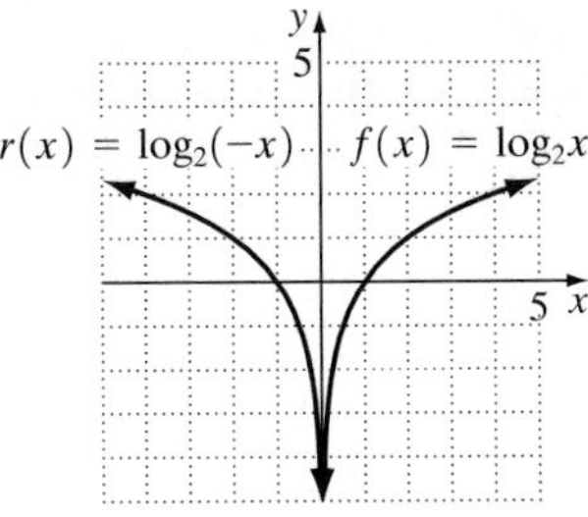

x-intercept: (−1, 0)
vertical asymptote: $x = 0$

35. $(-5, \infty)$ **36.** $(-\infty, 3)$ **37.** $(-\infty, 1)\cup(1, \infty)$ **38.** $6x$ **39.** $\sqrt{x}$ **40.** $4x^2$ **41.** 3.0

42. a. 76
b. ≈ 67, ≈ 63, ≈ 61, ≈ 59, ≈ 56
c.

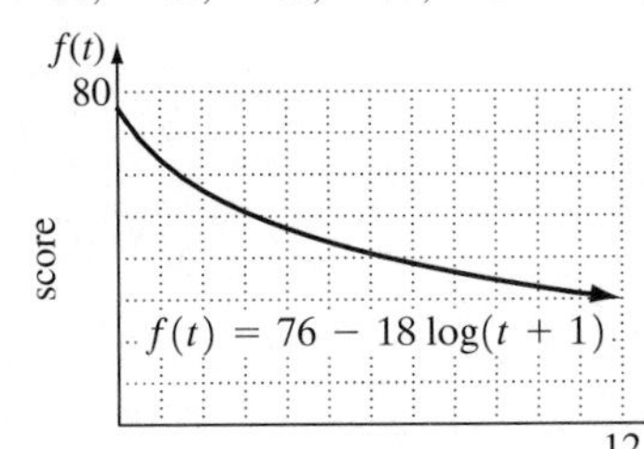

Retention decreases as time passes.

43. about 9 weeks **44.** $2 + 3\log_6 x$ **45.** $\frac{1}{2}\log_4 x - 3$ **46.** $\log_2 x + 2\log_2 y - 6$ **47.** $\frac{1}{3}\ln x - \frac{1}{3}$ **48.** $\log_b 21$ **49.** $\log\frac{3}{x^3}$ **50.** $\ln(x^3y^4)$ **51.** $\ln\frac{\sqrt{x}}{y}$ **52.** 6.2448 **53.** −0.1063 **54.** $\left\{\frac{\ln 12{,}143}{\ln 8}\right\}$; ≈ 4.523 **55.** $\left\{\frac{1}{5}\ln 141\right\}$; ≈ 0.990 **56.** $\left\{\frac{12 - \ln 130}{5}\right\}$; ≈ 1.426 **57.** $\left\{\frac{\ln 37{,}500 - 2\ln 5}{4\ln 5}\right\}$; ≈ 1.136 **58.** $\{\ln 3\}$; ≈ 1.099 **59.** $\{23\}$ **60.** $\{5\}$ **61.** $\varnothing$ **62.** $\left\{\frac{1}{e}\right\}$ or $\{0.368\}$ **63.** $\left\{\frac{e^3}{2}\right\}$ or $\{10.043\}$ **64.** 2042 **65.** 2086 **66.** 2005 **67.** 7.3 yr

68. 14.6 yr **69.** about 21.97% **70. a.** 0.041 **b.** 40.7 million **c.** 2010 **71.** about 15,679 years old **72. a.** about 9 people **b.** about 104 people **c.** 171 people; yes; The limiting size is 171; however, 178 people died. **73.** $y = 73e^{(\ln 2.6)x}$; $y = 73e^{0.956x}$ **74.** $y = 6.5e^{(\ln 0.43)x}$; $y = 6.5e^{-0.844x}$ **75.** high: exponential; medium: linear; low: quadratic; Explanations will vary; negative; The parabola opens downward. **76.** The exponential model, $y = (3.38051786)(1.0235357)^x$, is the best fit; 113.4 million

Chapter 4 Test

1.

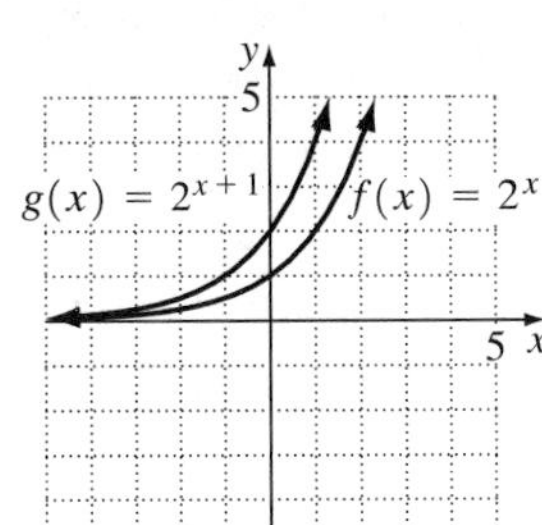

2.

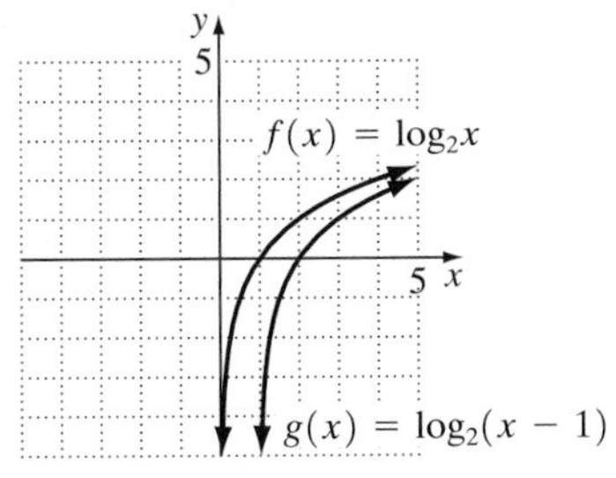

3. $5^3 = 125$ **4.** $\log_{36}6 = \frac{1}{2}$ **5.** $(-\infty, 3)$ **6.** $3 + 5\log_4 x$ **7.** $\frac{1}{3}\log_3 x - 4$ **8.** $\log(x^6y^2)$ **9.** $\ln\frac{7}{x^3}$ **10.** 1.5741 **11.** $\left\{\frac{\ln 1.4}{\ln 5}\right\}$ or $\{0.2091\}$ **12.** $\left\{\frac{\ln 4}{0.005}\right\}$ or $\{277.2589\}$ **13.** $\{0, \ln 5\}$ or $\{0, 1.6094\}$ **14.** $\{54.25\}$ **15.** $\{5\}$ **16.** $\left\{\frac{e^4}{3}\right\}$ or $\{18.1993\}$

17. 6.5% compounded semiannually; \$221.15 more **18.** 120 db **19. a.** about 89% **b.** decreasing; $k = -0.004 < 0$ **c.** 1995 **20.** $A = 484e^{0.005t}$ **21.** about 24,758 years ago **22. a.** 14 elk **b.** about 51 elk **c.** 140 elk

Cumulative Review Exercises (Chapters 1–4)

1. $\left\{\frac{2}{3}, 2\right\}$ **2.** $\{3, 7\}$ **3.** $\{-2, -1, 1\}$ **4.** $\{0.9704\}$ **5.** $\{3\}$ **6.** $(-\infty, 4]$ **7.** $[1, 3]$

8. using $(1, 3)$, $y - 3 = -3(x - 1)$; $y = -3x + 6$ **9.** $(f \circ g)(x) = (x + 2)^2$; $(g \circ f)(x) = x^2 + 2$ **10.** $f^{-1}(x) = \frac{1}{2}x + \frac{7}{2}$

11. $x^2 + 3x - 3 + \frac{-4}{x + 2}$ **12.** $\pm 1, \pm\frac{1}{2}, \pm\frac{1}{4}, \pm 3, \pm\frac{3}{2}, \pm\frac{3}{4}$ **13.** 300 **14.** $\{1 + i, 1 - i, 2\}$

15.

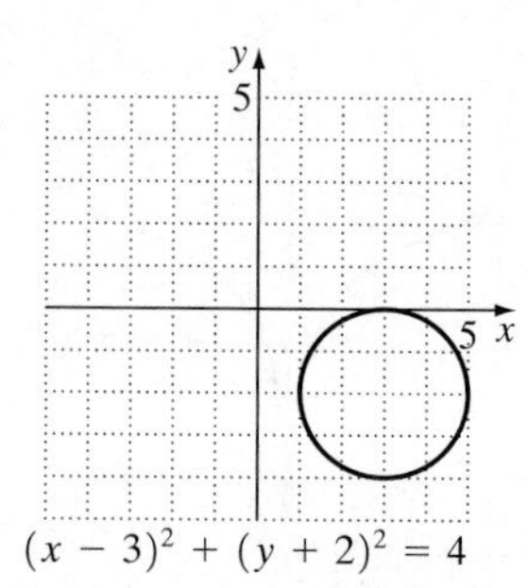

$(x-3)^2 + (y+2)^2 = 4$

16.

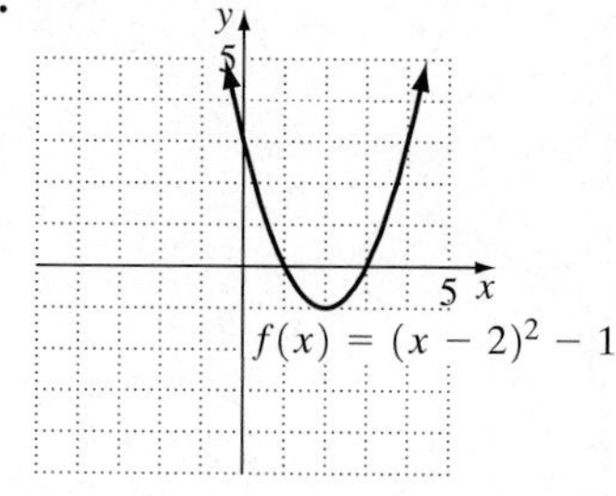

$f(x) = (x-2)^2 - 1$

17.

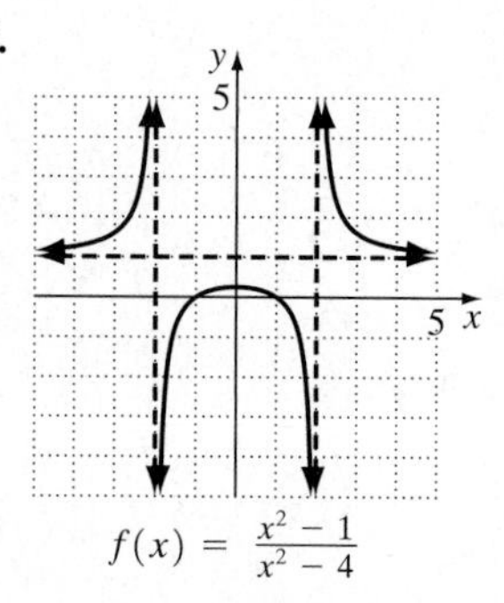

$f(x) = \frac{x^2-1}{x^2-4}$

18.

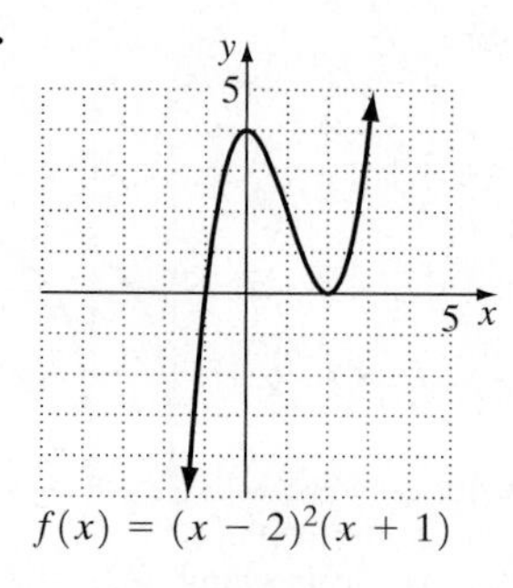

$f(x) = (x-2)^2(x+1)$

19. \$12 per hr **20.** $\frac{0.5}{\ln 4} \approx 0.361$; about $\frac{3}{10}$ of the people

CHAPTER 5

Section 5.1

Check Point Exercises

1. solution **2.** $\{(3, 2)\}$ **3.** $\{(1, -2)\}$ **4.** $\{(2, -1)\}$ **5.** $\left\{\left(\frac{23}{16}, \frac{3}{8}\right)\right\}$ **6.** $\varnothing$ **7.** $\{(x, y) \mid y = 4x - 4\}$ **8.** \$30; 400 units

Exercise Set 5.1

1. solution **3.** not a solution **5.** $\{(1, 3)\}$ **7.** $\{(5, 1)\}$ **9.** $\{(2, 1)\}$ **11.** $\{(-1, 3)\}$ **13.** $\{(4, 5)\}$ **15.** $\left\{\left(-4, \frac{5}{4}\right)\right\}$ **17.** $\{(2, -1)\}$ **19.** $\{(3, 0)\}$ **21.** $\{(-4, 3)\}$ **23.** $\{(3, 1)\}$ **25.** $\{(1, -2)\}$ **27.** $\left\{\left(\frac{7}{25}, -\frac{1}{25}\right)\right\}$ **29.** $\varnothing$ **31.** $\{(x, y) \mid y = 3x - 5\}$ **33.** $\{(1, 4)\}$ **35.** $\{(x, y) \mid x + 3y = 2\}$ **37.** $x + y = 7$; $x - y = -1$; 3 and 4 **39.** $3x - y = 1$; $x + 2y = 12$; 2 and 5 **41. a.** 6500 tickets can be sold. 6200 tickets can be supplied. **b.** \$50; 6250 tickets **43.** 2700 gal **45.** Quarter Pounder: 77 mg; Whopper with cheese: 122 mg **47.** 5.8 million pounds of potato chips; 4.6 million pounds of tortilla chips **49.** 2032; less than 1 death in 1000 live births **63.** $y = \frac{a_1c_2 - a_2c_1}{a_1b_2 - a_2b_1}$; $x = \frac{b_2c_1 - b_1c_2}{a_1b_2 - a_2b_1}$ **65.** Yes; 8 hexagons and 4 squares

Section 5.2

Check Point Exercises

2. $\{(1, 4, -3)\}$ **3.** $\{(4, 5, 3)\}$ **4.** $y = 3x^2 - 12x + 13$

Exercise Set 5.2

1. not a solution **3.** solution **5.** $\{(2, 3, 3)\}$ **7.** $\{(2, -1, 1)\}$ **9.** $\left\{\left(\frac{1}{3}, -\frac{2}{5}, \frac{1}{2}\right)\right\}$ **11.** $\{(3, 1, 5)\}$ **13.** $\{(1, 0, -3)\}$ **15.** $\{(1, -5, -6)\}$ **17.** $\left\{\left(\frac{1}{2}, \frac{1}{3}, -1\right)\right\}$ **19.** 7, 4 and 5 **21.** $y = 2x^2 - x + 3$ **23.** $y = 2x^2 + x - 5$ **25.** chemical engineer: \$42,758; mechanical engineer: \$39,852; electrical engineer: \$38, 811 **27.** $A = -8$; $B = 50$; $C = 0$; $y = 156$ when $x = 6$; When a car is in motion for 6 seconds after the brakes are applied, it travels 156 feet. **29.** 200 \$8 tickets; 150 \$10 tickets; 50 \$12 tickets **31.** \$1200 at 8%, \$2000 at 10%, and \$3500 at 12% **41.** 13 triangles, 21 rectangles, and 6 pentagons

Section 5.3

Check Point Exercises

1. $\frac{2}{x-3} + \frac{3}{x+4}$ **2.** $\frac{2}{x} - \frac{2}{x-1} + \frac{3}{(x-1)^2}$ **3.** $\frac{2}{x+3} + \frac{6x-8}{x^2+x+2}$ **4.** $\frac{2x}{x^2+1} + \frac{-x+3}{(x^2+1)^2}$

Exercise Set 5.3

1. $\frac{A}{x-2} + \frac{B}{x+1}$ **3.** $\frac{A}{x+2} + \frac{B}{x-3} + \frac{C}{(x-3)^2}$ **5.** $\frac{A}{x-1} + \frac{Bx+C}{x^2+1}$ **7.** $\frac{Ax+B}{x^2+4} + \frac{Cx+D}{(x^2+4)^2}$ **9.** $\frac{3}{x-3} - \frac{2}{x-2}$ **11.** $\frac{7}{x-9} - \frac{4}{x+2}$ **13.** $\frac{24}{7(x-4)} + \frac{25}{7(x+3)}$ **15.** $\frac{3}{x} + \frac{2}{x-1} - \frac{1}{x+3}$ **17.** $\frac{3}{x} + \frac{4}{x+1} - \frac{3}{x-1}$ **19.** $\frac{6}{x-1} - \frac{5}{(x-1)^2}$

21. $\frac{1}{x-2} - \frac{2}{(x-2)^2} - \frac{5}{(x-2)^3}$ **23.** $\frac{7}{x} - \frac{6}{x-1} + \frac{10}{(x-1)^2}$ **25.** $-\frac{2}{x+1} + \frac{3}{(x+1)^2} + \frac{7}{x-3}$ **27.** $\frac{3}{x-1} + \frac{2x-4}{x^2+1}$
29. $\frac{2}{x+1} + \frac{3x-1}{x^2+2x+2}$ **31.** $\frac{4}{x+1} + \frac{2x-3}{x^2+1}$ **33.** $\frac{x+1}{x^2+2} - \frac{2x}{(x^2+2)^2}$ **35.** $\frac{x-2}{x^2-2x+3} + \frac{2x+1}{(x^2-2x+3)^2}$
37. $\frac{3}{x-2} + \frac{x-1}{x^2+2x+4}$ **39.** $\frac{1}{x} - \frac{1}{x+1}; \frac{99}{100}$ **49.** $\frac{a}{x-c} + \frac{b+ac}{(x-c)^2}$

Section 5.4

Check Point Exercises

1. $\{(0, 1), (4, 17)\}$ **2.** $\left\{\left(-\frac{6}{5}, \frac{3}{5}\right), (2, -1)\right\}$ **3.** $\{(3, 2), (3, -2), (-3, 2), (-3, -2)\}$ **4.** $\{(0, 5)\}$
5. length: 7 ft; width: 3 ft or length: 3 ft; width: 7 ft

Exercise Set 5.4

1. $\{(-3, 5), (2, 0)\}$ **3.** $\left\{\left(\frac{-1+\sqrt{17}}{2}, \frac{1+\sqrt{17}}{2}\right), \left(\frac{-1-\sqrt{17}}{2}, \frac{1-\sqrt{17}}{2}\right)\right\}$ **5.** $\{(4, -10), (-3, 11)\}$ **7.** $\{(4, 3), (-3, -4)\}$
9. $\left\{\left(-\frac{3}{2}, -4\right), (2, 3)\right\}$ **11.** $\{(-5, -4), (3, 0)\}$ **13.** $\{(3, 1), (-3, -1), (1, 3), (-1, -3)\}$ **15.** $\{(4, -3), (-1, 2)\}$
17. $\{(0, 1), (4, -3)\}$ **19.** $\{(3, 2), (3, -2), (-3, 2), (-3, -2)\}$ **21.** $\{(3, 2), (3, -2), (-3, 2)\ (-3, -2)\}$
23. $\{(2, 1), (2, -1), (-2, 1), (-2, -1)\}$ **25.** $\{(3, 4), (3, -4)\}$ **27.** $\{(0, 2), (0, -2), (-1, \sqrt{3}), (-1, -\sqrt{3})\}$
29. $\{(2, 1), (2, -1), (-2, 1), (-2, -1)\}$ **31.** $\{(-2\sqrt{2}, -\sqrt{2}), (-1, -4), (1, 4), (2\sqrt{2}, \sqrt{2})\}$ **33.** $\{(2, 2), (4, 1)\}$ **35.** $\{(0, 0), (-1, 1)\}$
37. $\{(0, 0), (-2, 2), (2, 2)\}$ **39.** $\left\{(-4, 1), \left(-\frac{5}{2}, \frac{1}{4}\right)\right\}$ **41.** $\left\{\left(\frac{12}{5}, -\frac{29}{5}\right), (-2, 3)\right\}$ **43.** 4 and 6
45. 2 and 1, 2 and −1, −2 and 1, or −2 and −1 **47.** $(0, -4), (-2, 0), (2, 0)$ **49.** 11 ft and 7 ft **51.** width: 6 in.; length: 8 in.
53. $x = 5$ m, $y = 2$ m **61.** (b) is true. **63.** $b = 6, a = 8$ **65.** $\{(10{,}000, 5)\}$

Section 5.5

Check Point Exercises

1.

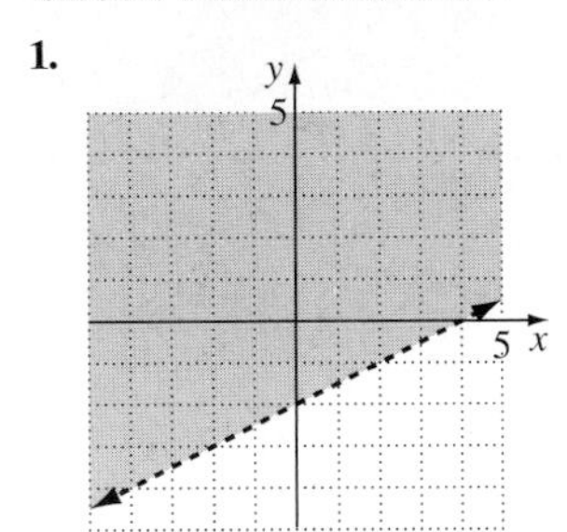

2.

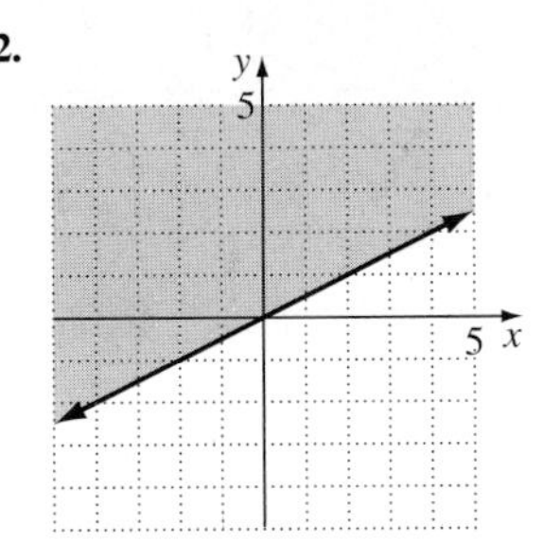

3.

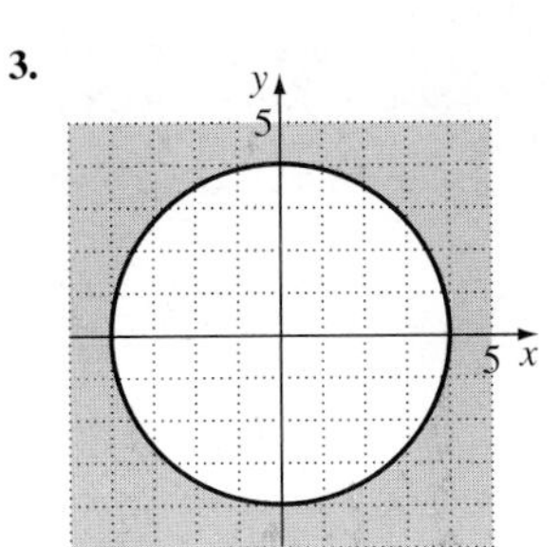

4.

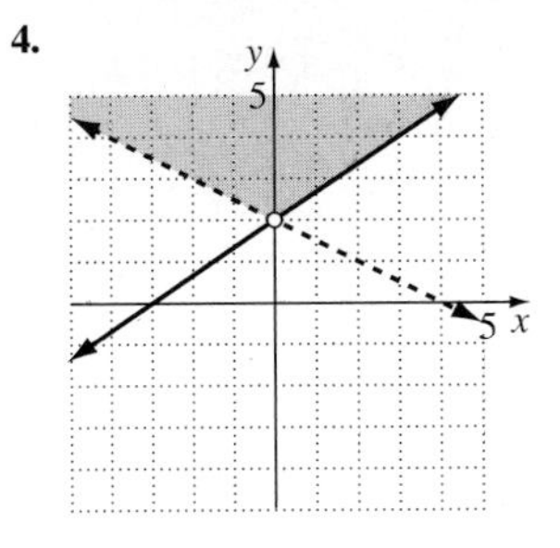

5.

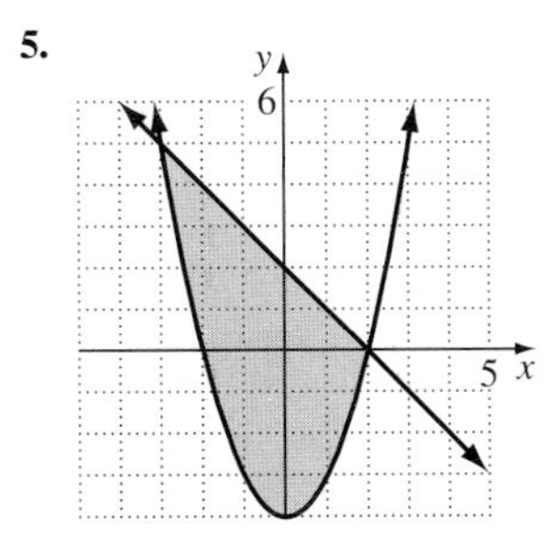

6.

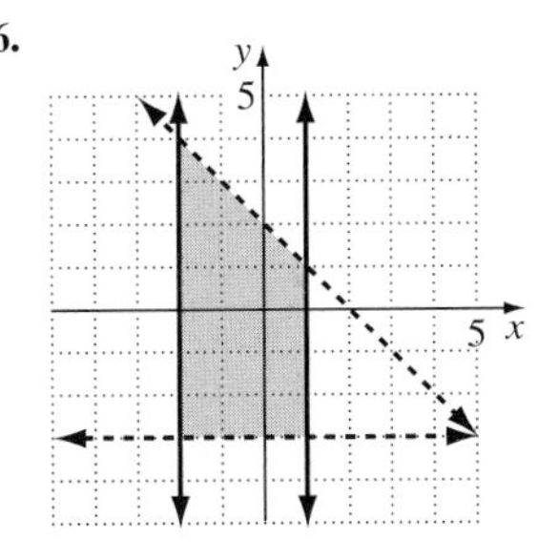

Exercise Set 5.5

1.

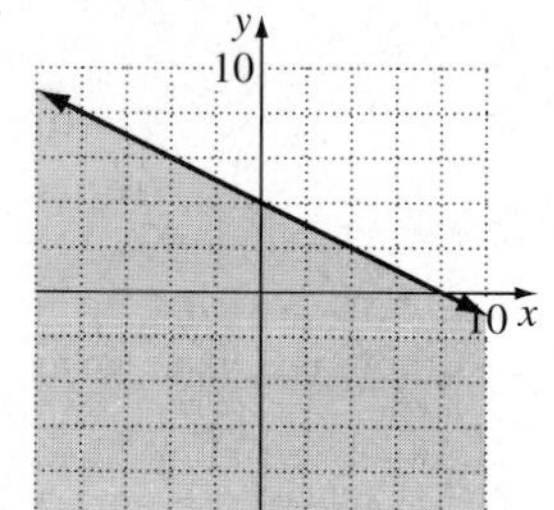

3.

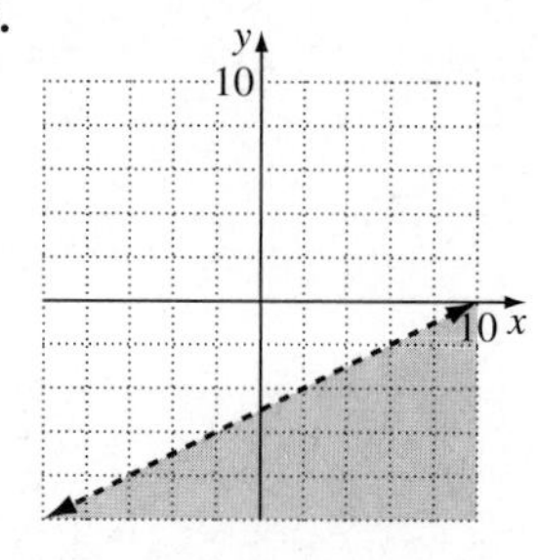

5.

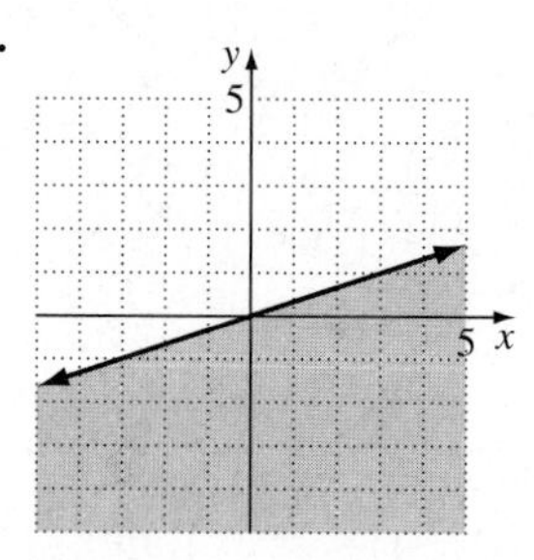

7.

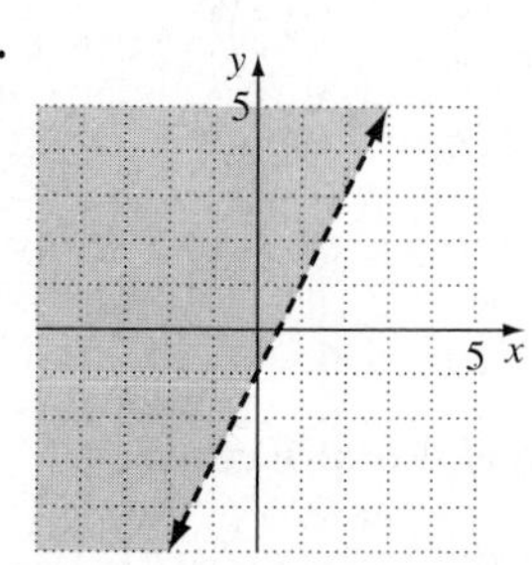

9.

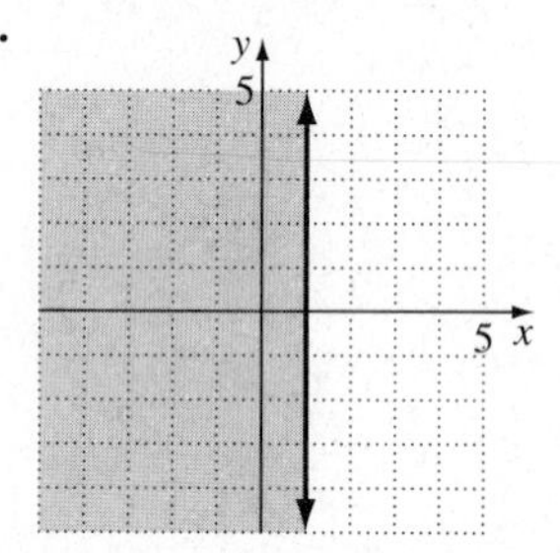

11.

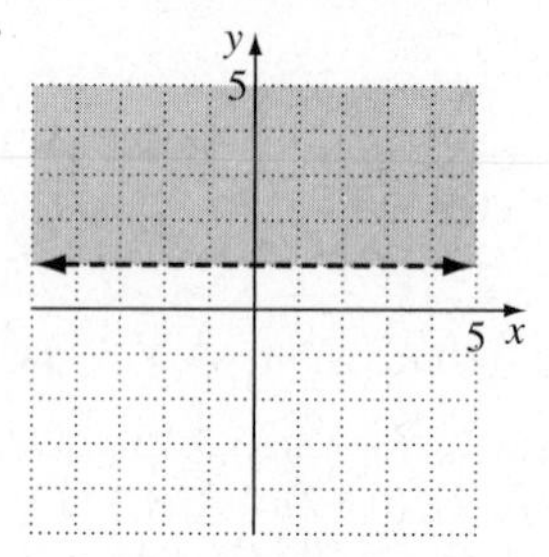

13.

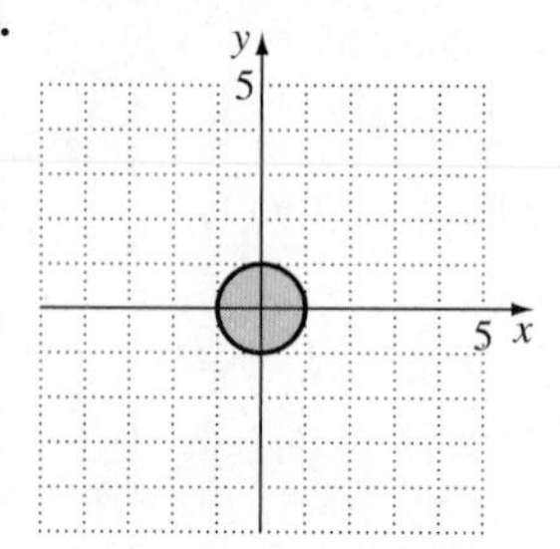

15.

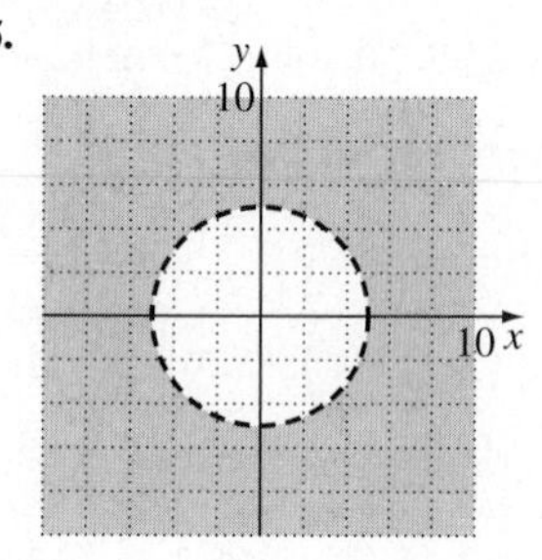

17.

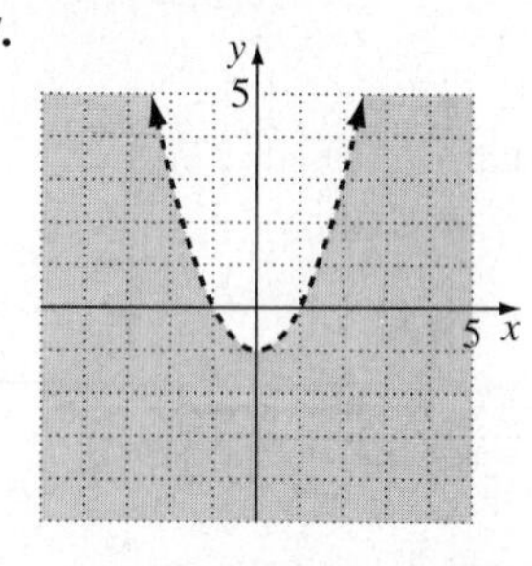

19.

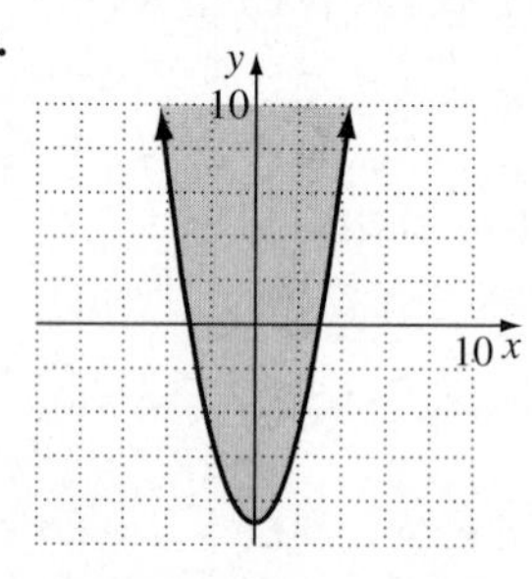

21.

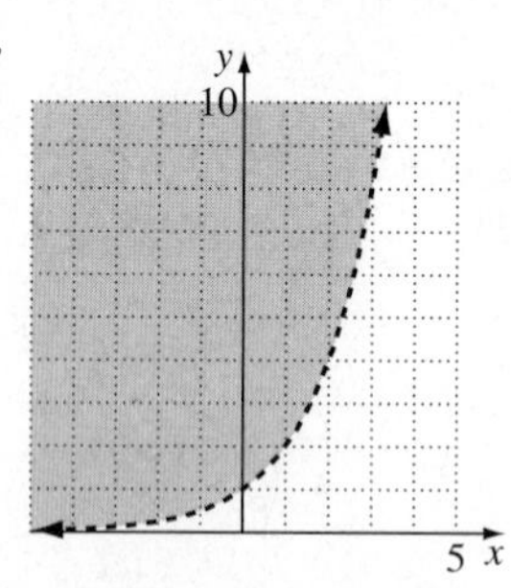

23.

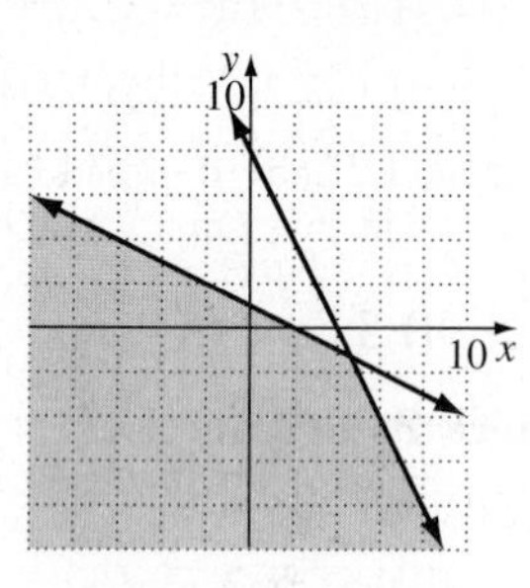

25.

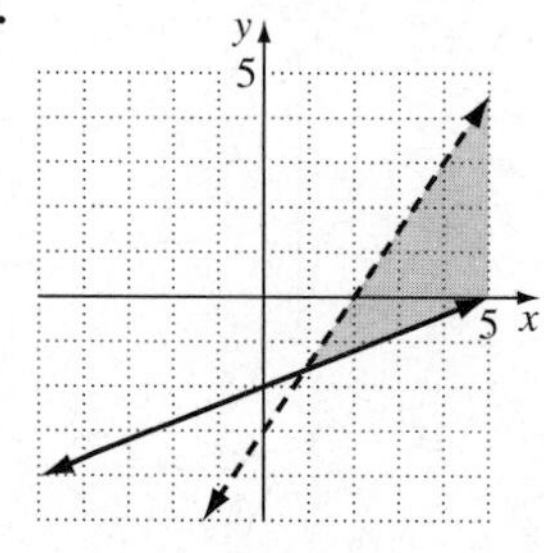

27.

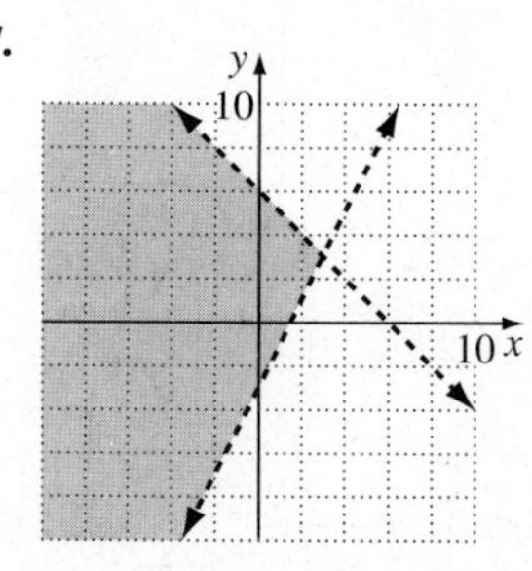

29.

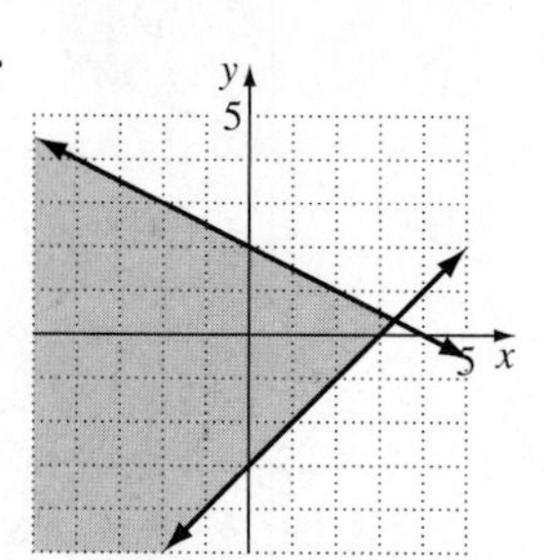

31.

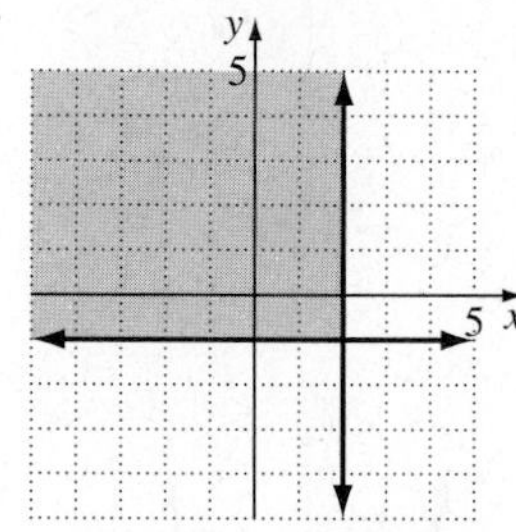

33.

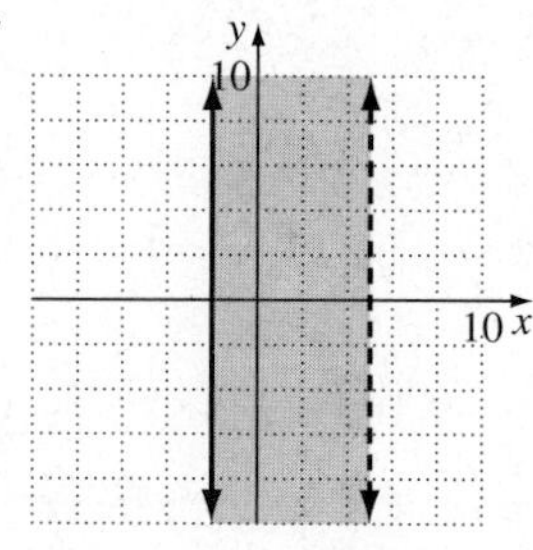

35.

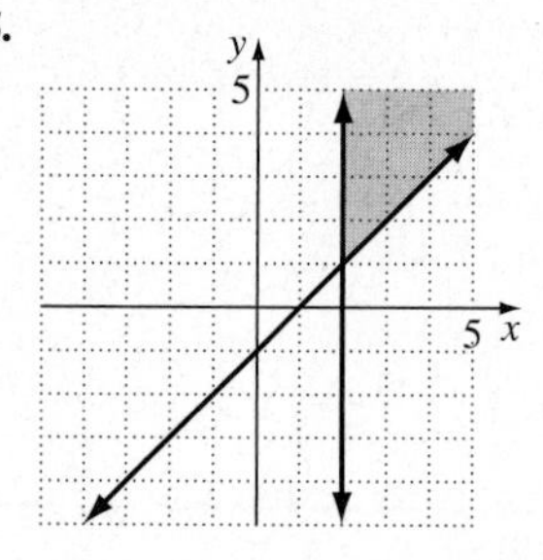

37. no solution

39.

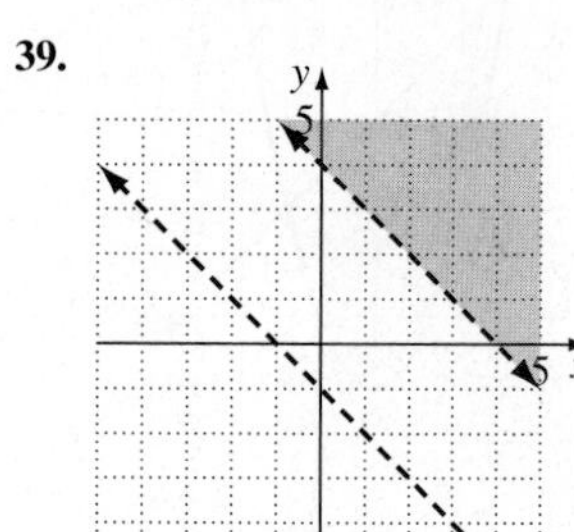

41.

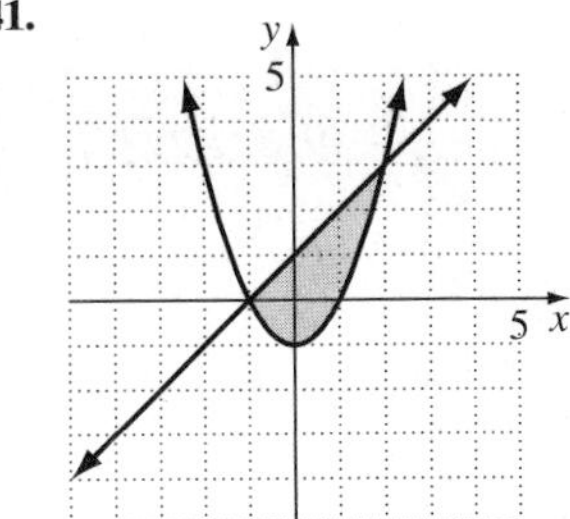

43.

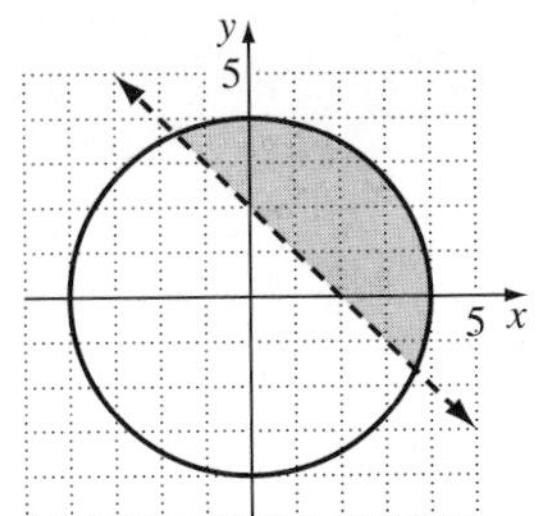

45.

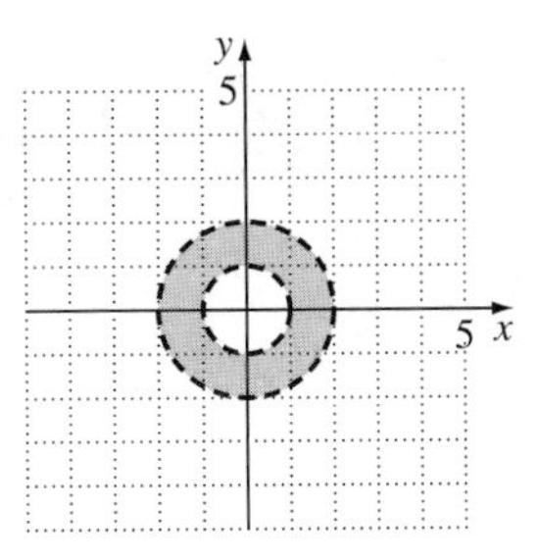

47.

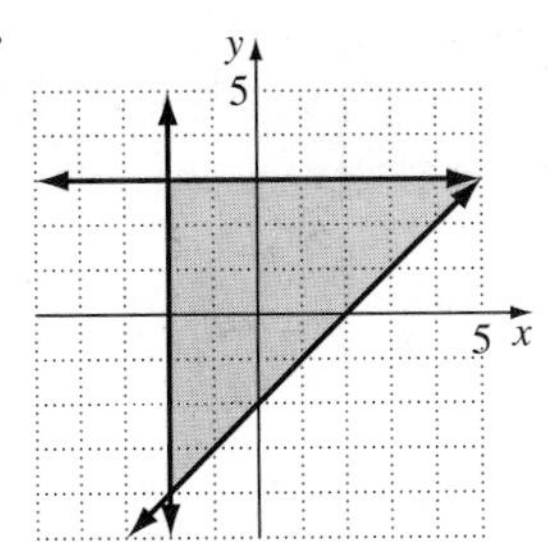

49.

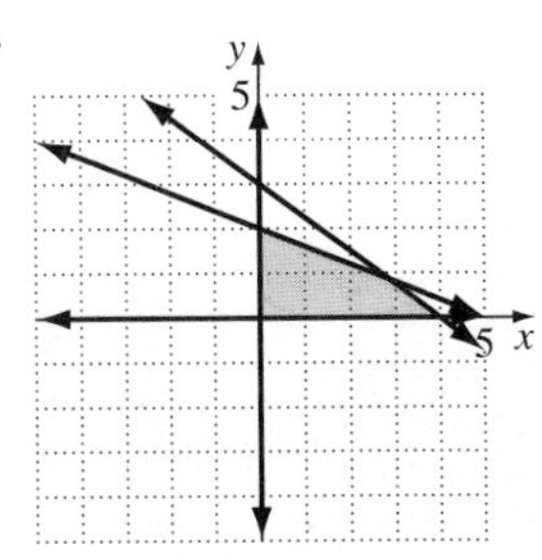

51.

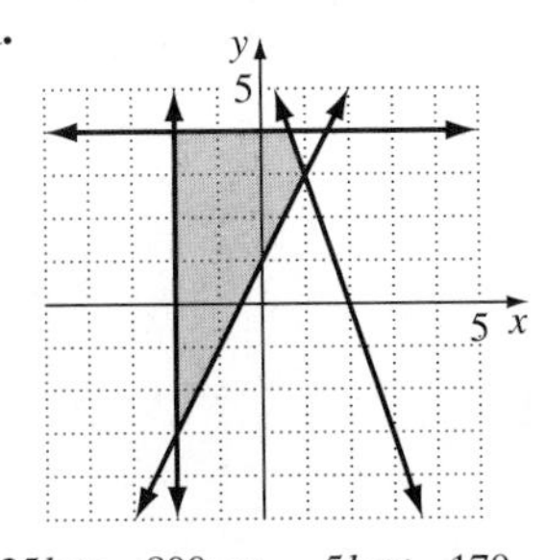

52.

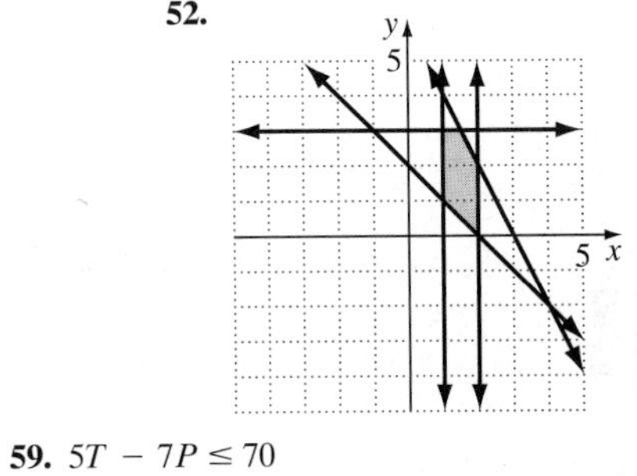

53. Answers may vary. **55.** no **57.** $7w - 25h \geq -800;\ w - 5h \leq -170$ **59.** $5T - 7P \leq 70$

61. $50x + 150y > 2000$

y
50
50 x

63. $x + y \leq 15{,}000$
$x \geq 2000$
$y \geq 3x$
$x \geq 0$
$y \geq 0$

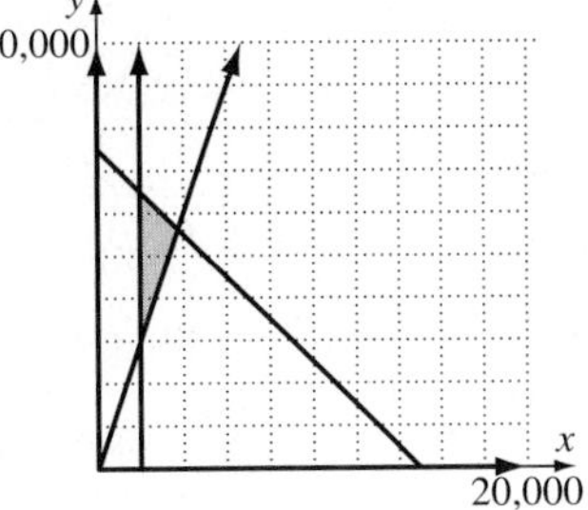

71.

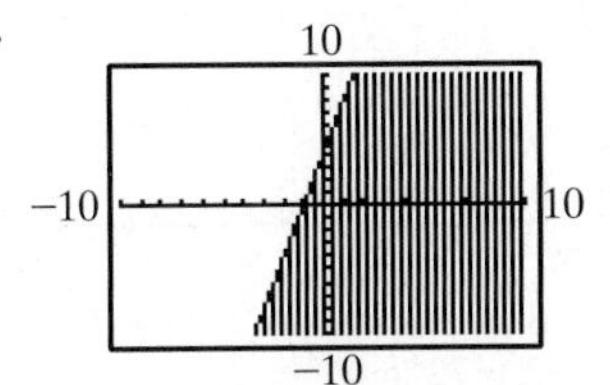

73.

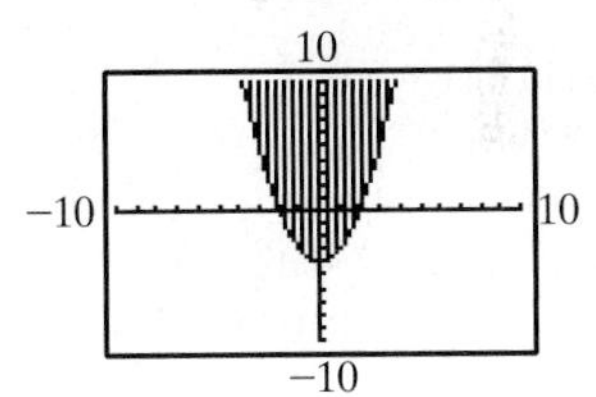

75. Answers may vary. **79.** $x^2 + y^2 \leq 9;\ y < x^2$ **81.**

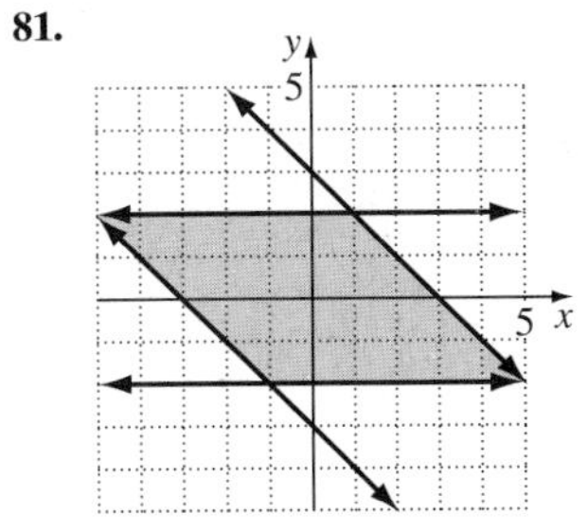

Section 5.6

Check Point Exercises

1. $z = 25x + 55y$ **2.** $x + y \leq 80$ **3.** $30 \leq x \leq 80; 10 \leq y \leq 30$; objective function: $z = 25x + 55y$; constraints: $x + y \leq 80$; $30 \leq x \leq 80; 10 \leq y \leq 30$ **4.** 50 bookshelves and 30 desks; $2900 **5.** 30

Exercise Set 5.6

1. (1, 2): 17; (2, 10): 70; (7, 5): 65; (8, 3): 58; maximum: $z = 70$; minimum: $z = 17$
3. (0, 0): 0; (0, 8): 400; (4, 9): 610; (8, 0): 320; maximum: $z = 610$; minimum: $z = 0$

5. a.

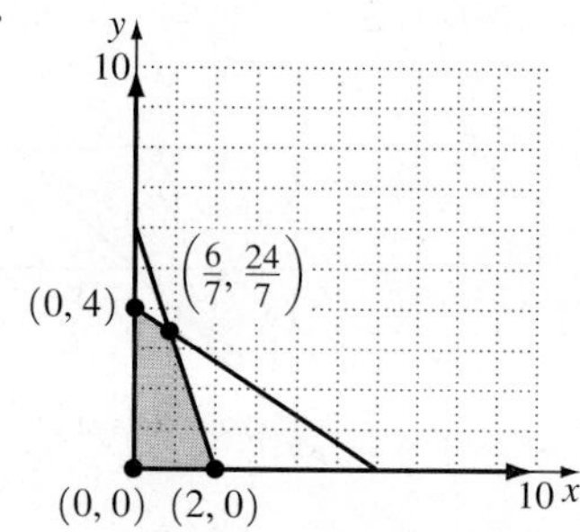

b. (0, 0): 0; (0, 4): 12; $\left(\frac{6}{7}, \frac{24}{7}\right)$: 12; (2, 0): 4

c. maximum value: 12 at $x = 0$ and $y = 4$ and at $x = \frac{6}{7}$ and $y = \frac{24}{7}$

7. a.

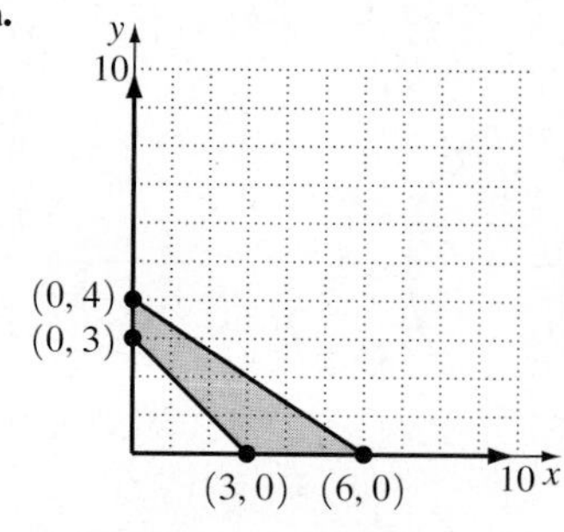

b. (0, 4): 4; (0, 3): 3; (3, 0): 12; (6, 0): 24

c. maximum value: 24 at $x = 6$ and $y = 0$

9. a.

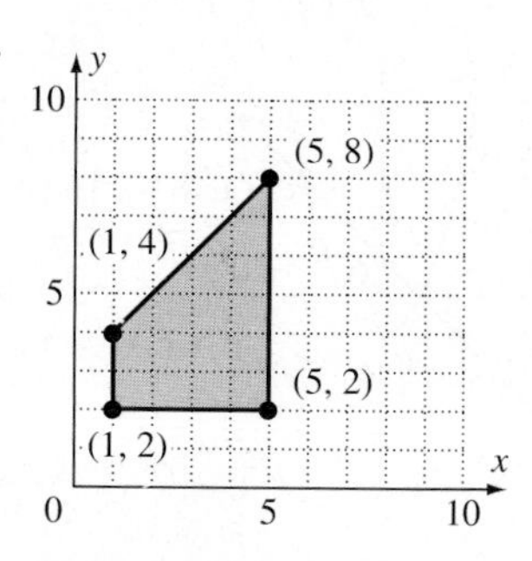

b. (1, 2): −1; (1, 4): −5; (5, 8): −1; (5, 2): 11

c. maximum value: 11 at $x = 5$ and $y = 2$

11. a.

y
10
(0, 4)
$\left(\frac{12}{5}, \frac{12}{5}\right)$
(0, 2)
(2, 0) (4, 0)
10 x

b. (0, 4): 8; (0, 2): 4; (2, 0): 8; (4, 0): 16; $\left(\frac{12}{5}, \frac{12}{5}\right)$: $\frac{72}{5}$

c. maximum value: 16 at $x = 4$ and $y = 0$

13. a.

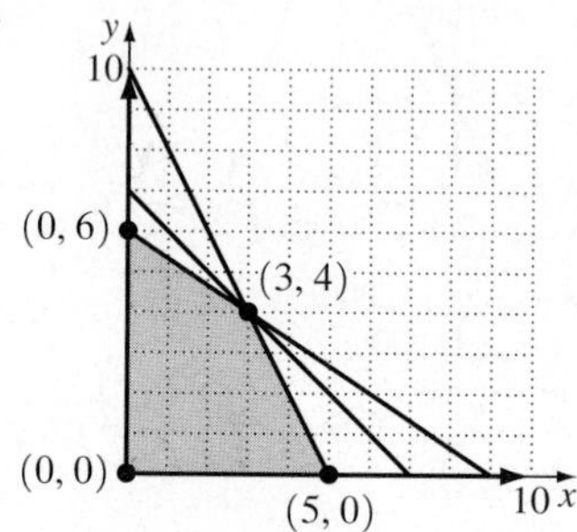

b. (0, 6): 72, (0, 0): 0; (5, 0): 50; (3, 4): 78

c. maximum value: 78 at $x = 3$ and $y = 4$

15. a. $z = 125x + 200y$

b. $x \leq 450$; $y \leq 200$; $600x + 900y \leq 360{,}000$

c.

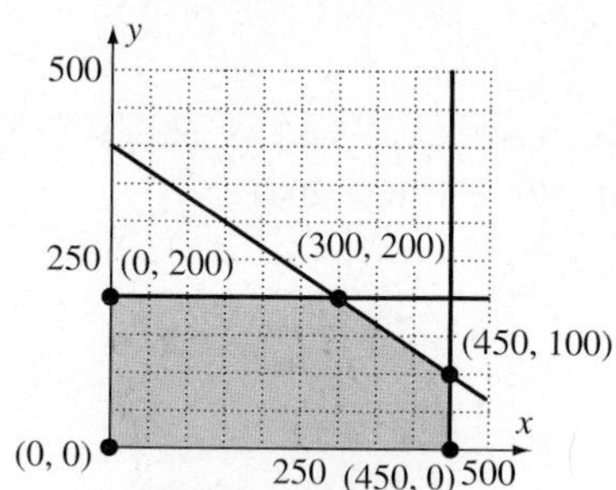

d. (0, 0): 0; (0, 200): 40,000; (300, 200): 77,500; (450, 100): 76,250; (450, 0): 56,250

e. 300; 200; $77,500

17. 40 model A bicycles and no model B bicycles **19.** No cartons of food and 600 cartons of clothing; 3600 people

21. 50 students and 100 parents **23.** 10 Boeing 727s and 42 Falcon 20s

29.

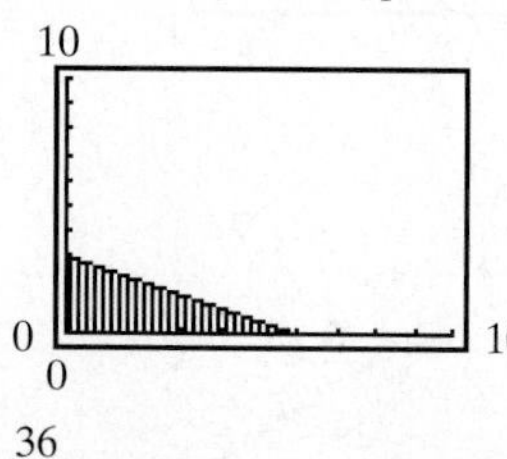

31.

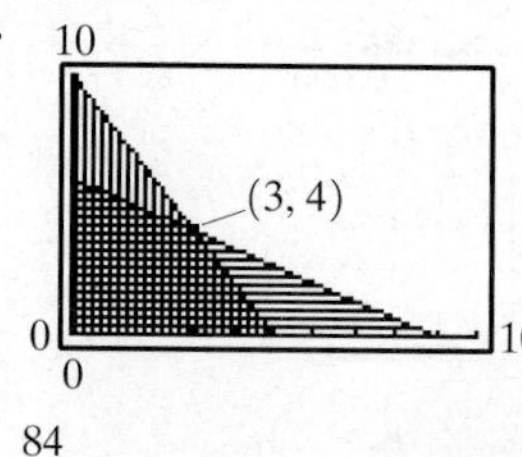

33. $5000 in stocks and $5000 in bonds

34. (0, 0): 0; (0, 3): 3B; (3, 1): 3B; (2, 0): $\frac{4}{3}B$; maximum value: 3B at (0, 3) and (3, 1)

Chapter 5 Review Exercises

1. $\{(1, 5)\}$ **2.** $\{(2, 3)\}$ **3.** $\{(2, -3)\}$ **4.** $\varnothing$ **5.** $\{(x, y) | 3x - 6y = 12\}$ **6.** No, $(-1, 3)$ is not a solution to $2x + y = -5$. **7.** Shrimp: 42 mg of cholesterol per oz; Scallops: 15 mg of cholesterol per oz **8.** 3 apples and 2 avocados **9.** 250 copies can be supplied and sold for $12.50 each. **10.** $\{(0, 1, 2)\}$ **11.** $\{(2, 1, -1)\}$ **12.** $y = 3x^2 - 4x + 5$ **13.** Japan: 16%; Germany: 15%; France: 14% **14.** Substitute the ordered pairs into $x = ax^2 + bx + c$, solve the resulting system for a, b, and c, then substitute these values into the equation to form a quadratic model. **15.** $\frac{3}{5(x-3)} + \frac{2}{5(x+2)}$ **16.** $\frac{6}{x-4} + \frac{5}{x+3}$ **17.** $\frac{2}{x} + \frac{3}{x+2} - \frac{1}{x-1}$ **18.** $\frac{2}{x-2} + \frac{5}{(x-2)^2}$ **19.** $-\frac{4}{x-1} + \frac{4}{x-2} - \frac{2}{(x-2)^2}$ **20.** $\frac{6}{5(x-2)} + \frac{-6x+3}{5(x^2+1)}$ **21.** $\frac{5}{x-3} + \frac{2x-1}{x^2+4}$ **22.** $\frac{x}{x^2+4} - \frac{4x}{(x^2+4)^2}$ **23.** $\frac{4x+1}{x^2+x+1} + \frac{2x-2}{(x^2+x+1)^2}$ **24.** $\{(4, 3), (1, 0)\}$ **25.** $\{(0, 1), (-3, 4)\}$ **26.** $\{(1, -1), (-1, 1)\}$ **27.** $\{(3, \sqrt{6}), (3, -\sqrt{6}), (-3, \sqrt{6}), (-3, -\sqrt{6})\}$ **28.** $\{(2, 2), (-2, -2)\}$ **29.** $\{(9, 6), (1, 2)\}$

30. $\{(-3, -1), (1, 3)\}$ **31.** $\left\{\left(\frac{1}{2}, 2\right), (-1, -1)\right\}$ **32.** $\left\{\left(\frac{5}{2}, -\frac{7}{2}\right), (0, -1)\right\}$ **33.** $\{(2, -3), (-2, -3), (3, 2), (-3, 2)\}$

34. $\{(3, 1), (3, -1), (-3, 1), (-3, -1)\}$ **35.** 8 m and 5 m **36.** $(1, 6), (3, 2)$ **37.** $x = 46$ and $y = 28$ or $x = 50$ and $y = 20$

38.

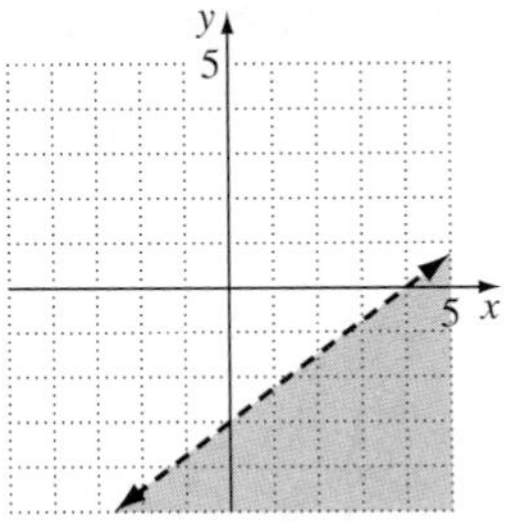

39.

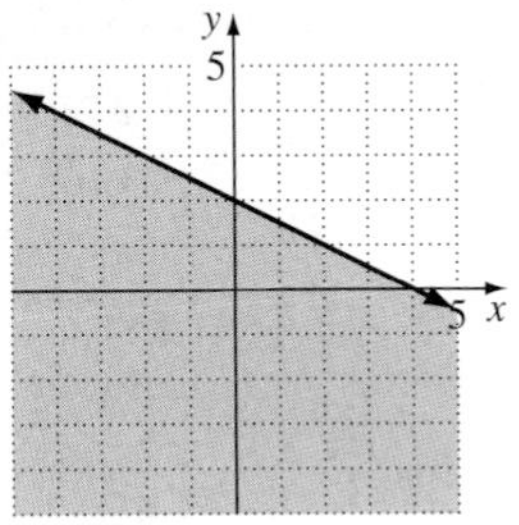

40.

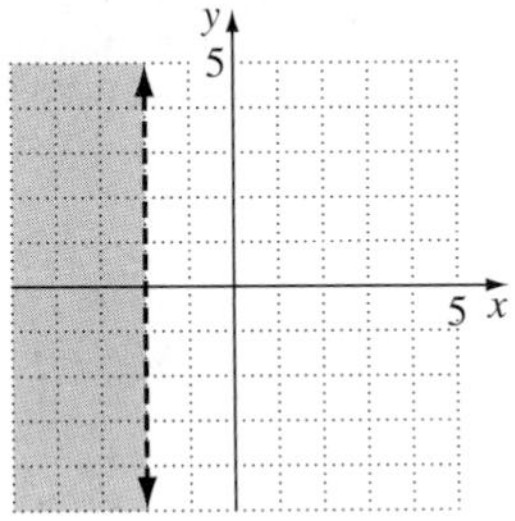

41.

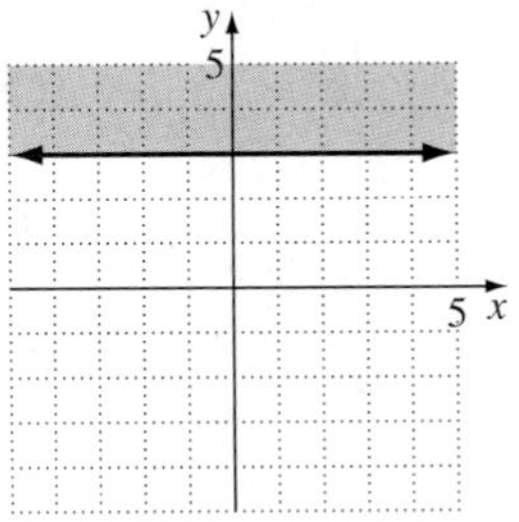

42.

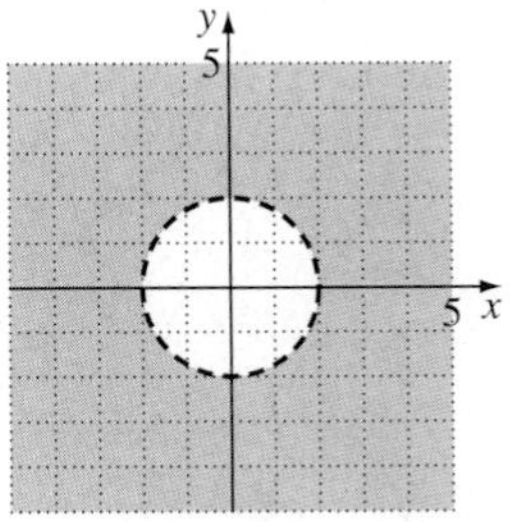

43.

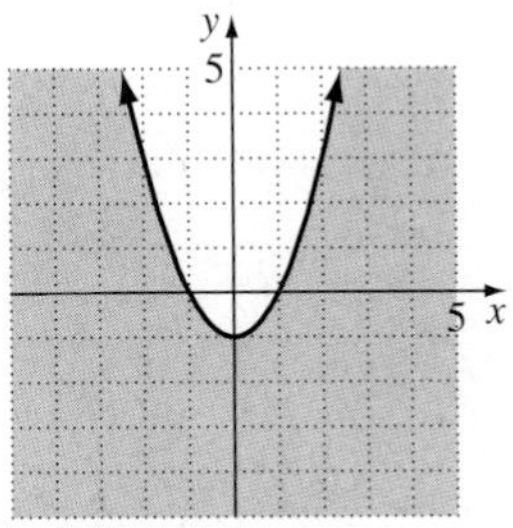

44.

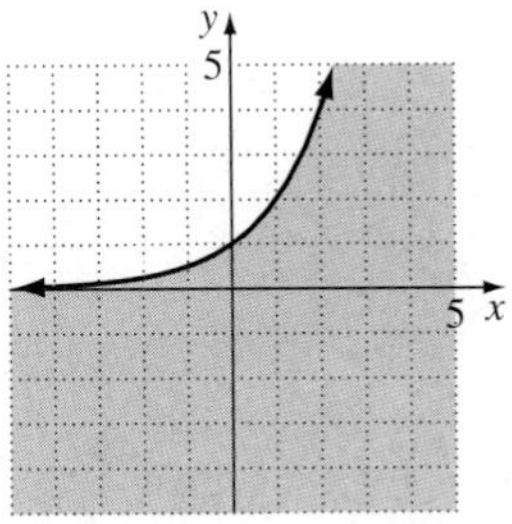

45.

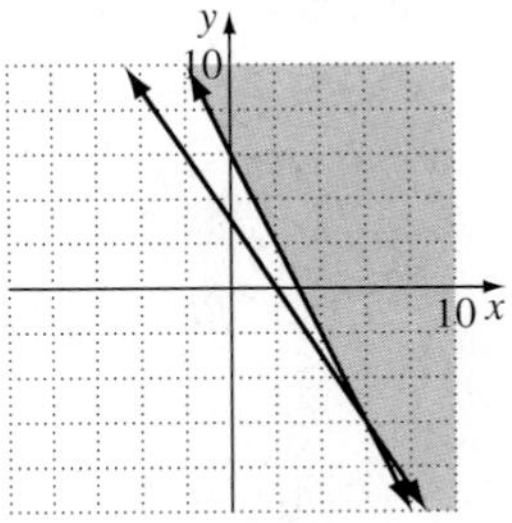

46.

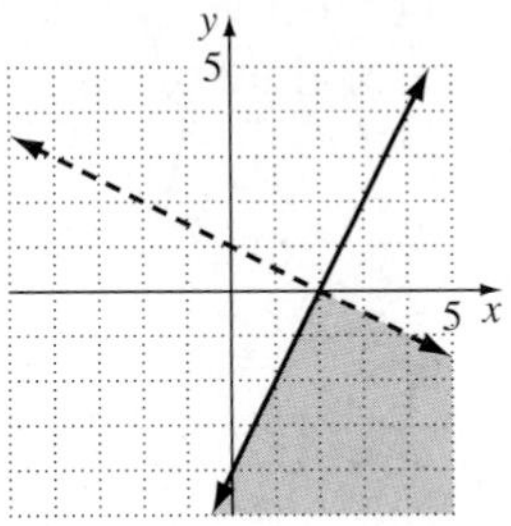

47.

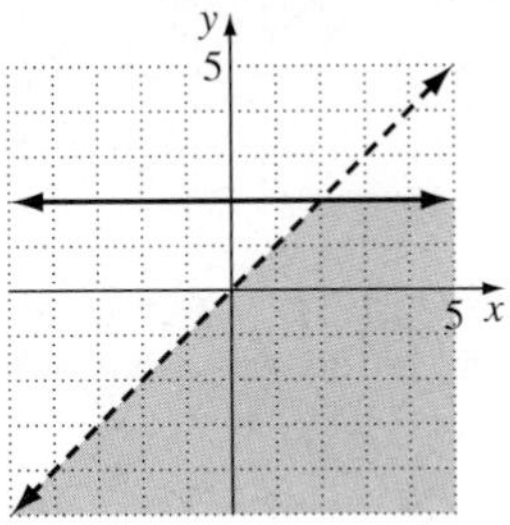

48.

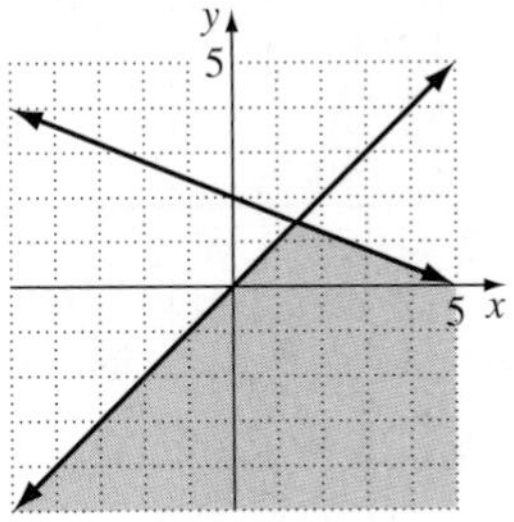

49.

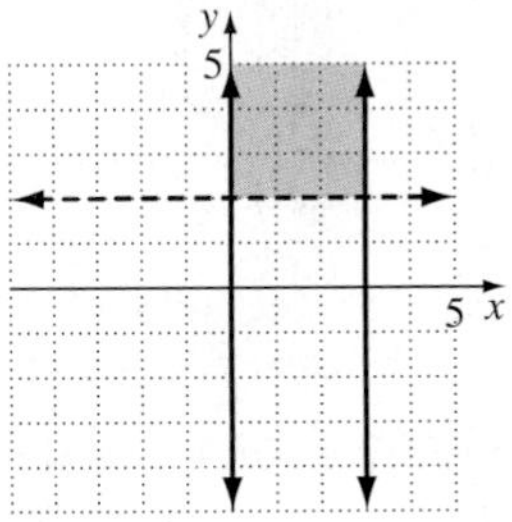

50. no solution

51.

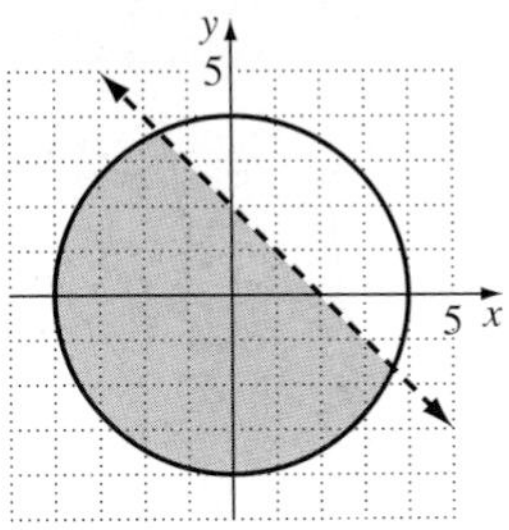

52.

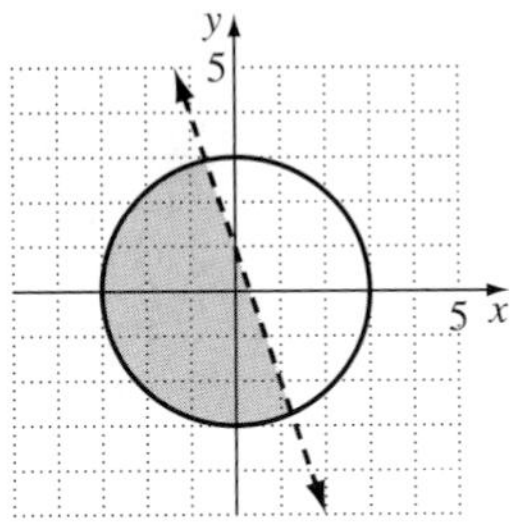

53.

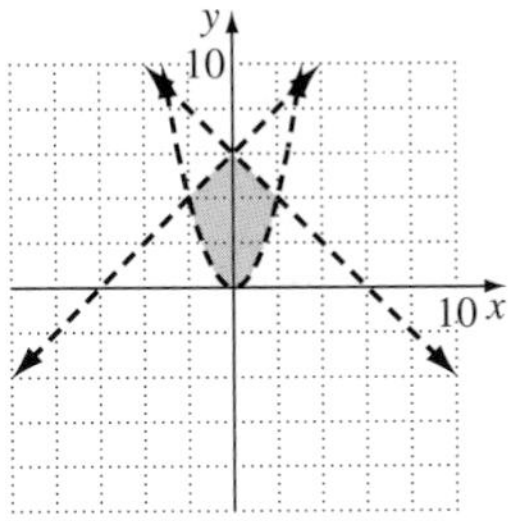

54.

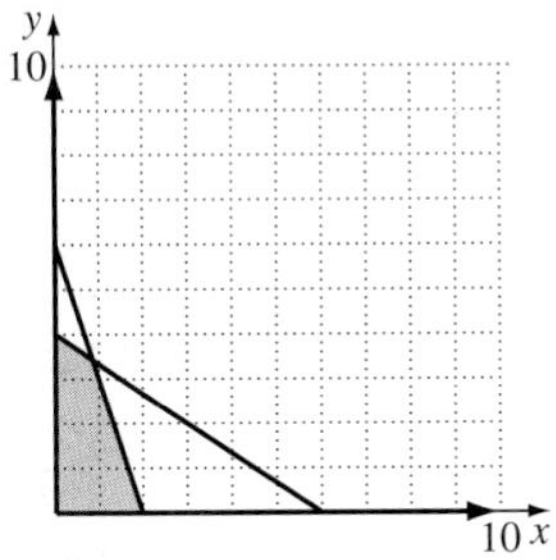

55. $(2, 2)$: 10; $(4, 0)$: 8; $\left(\frac{1}{2}, \frac{1}{2}\right)$: $\frac{5}{2}$; $(1, 0)$: 2; maximum value: 10; minimum value: 2

56.

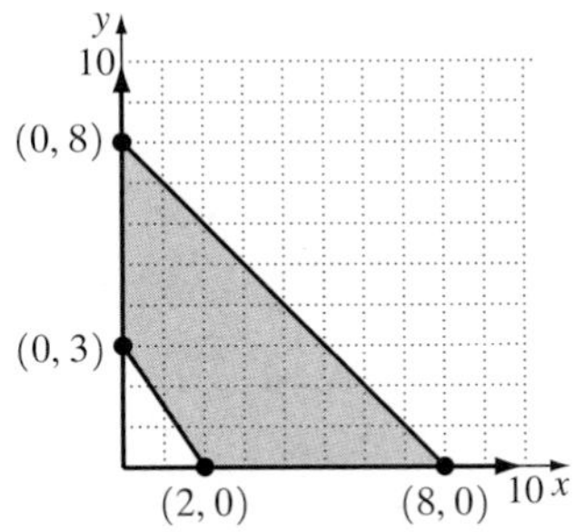

(0, 8) (0, 3) (2, 0) (8, 0)

24

57.

(0, 7) (5, 7) (0, 3) (3, 0) (5, 0)

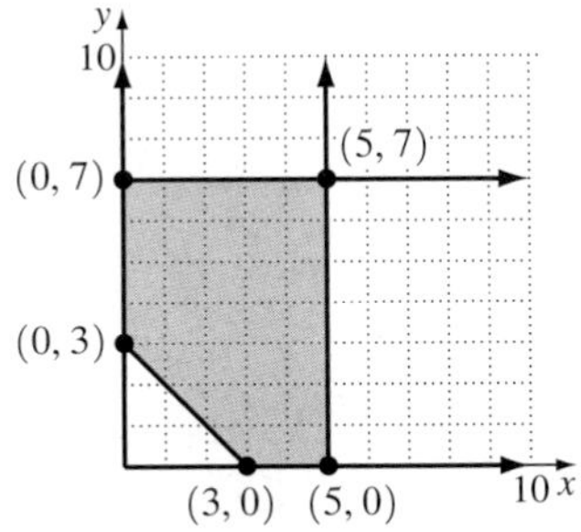

33

58.

(4, 4) $\left(\frac{6}{5}, \frac{6}{5}\right)$ (3, 0) (6, 0)

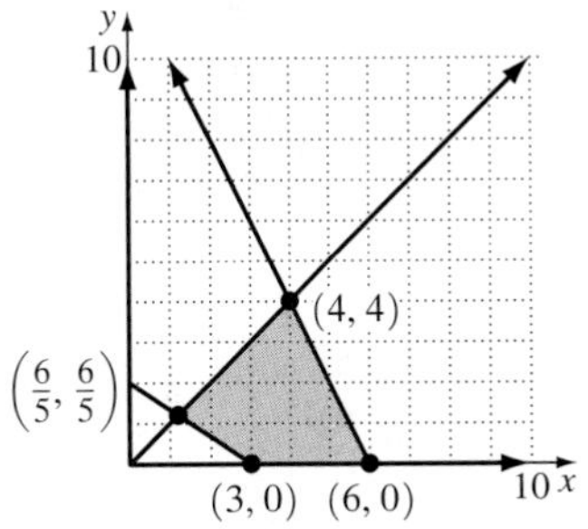

44

59. a. $z = 500x + 350y$
b. $x + y \leq 200$; $x \geq 10$; $y \geq 80$
c.

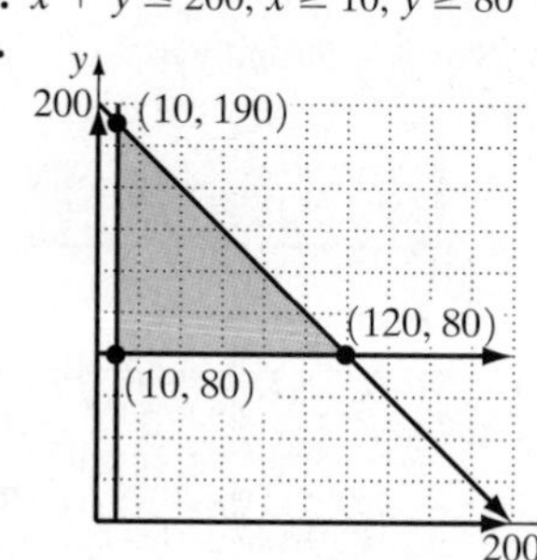

d. (10, 80): 33,000; (10, 190): 71,500; (120, 80): 88,000
e. 120 units of writing paper and 80 units of newsprint; $88,000

60. 480 of model A and 240 of model B

Chapter 5 Test

1. $\{(1, -3)\}$ **2.** $\{(4, -2)\}$ **3.** $\{(1, 3, 2)\}$ **4.** $\{(4, -3), (-3, 4)\}$ **5.** $\{(3, 2), (3, -2), (-3, 2), (-3, -2)\}$
6. $\frac{-1}{10(x + 1)} + \frac{x + 9}{10(x^2 + 9)}$
7.

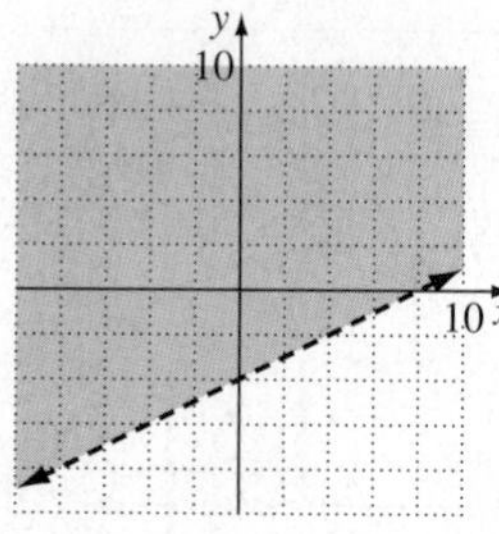

8.

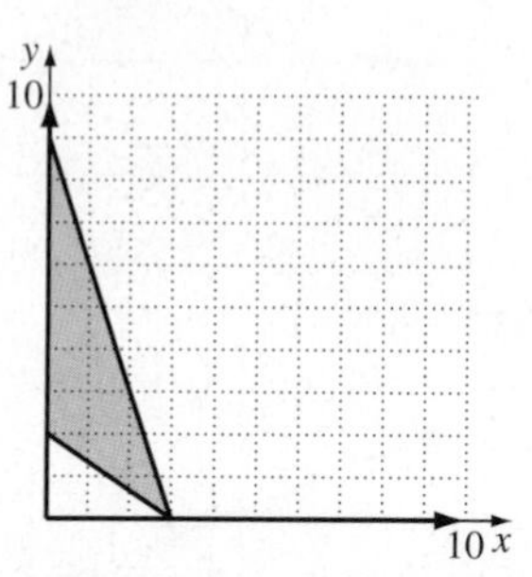

9.

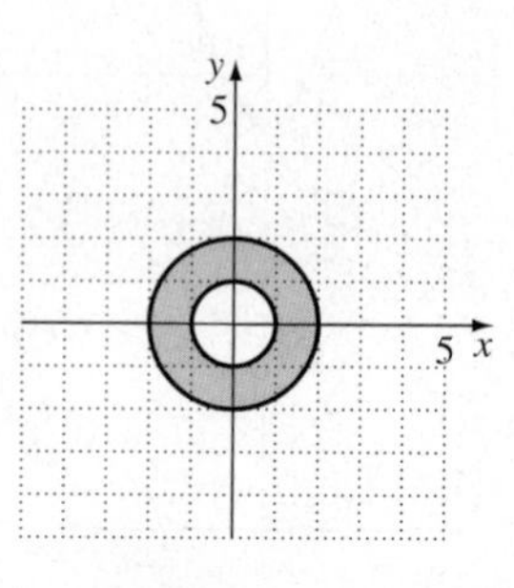

10.

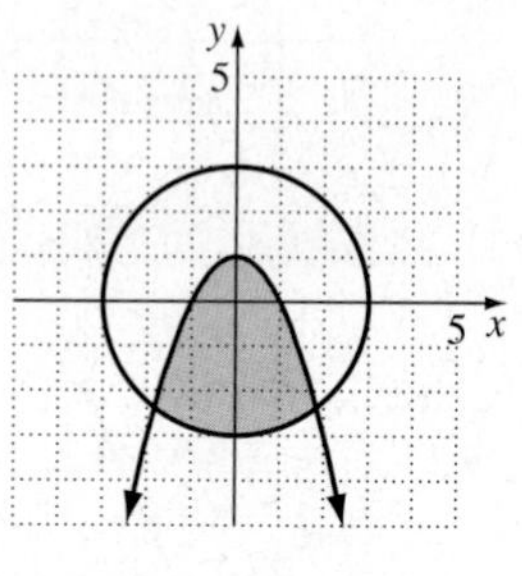

11. 26 **12.** orchestra ticket: $23; mezzanine ticket: $14 **13.** 400 units; $30 each **14.** $y = x^2 - 3$
15. $x = 7.5$ ft and $y = 24$ ft or $x = 12$ ft and $y = 15$ ft **16.** 50 regular and 100 deluxe jet skis; $35,000

Cumulative Review Exercises (Chapters 1–5)

1. $\{3, 4\}$ **2.** $\left\{\frac{2 + i\sqrt{3}}{2}, \frac{2 - i\sqrt{3}}{2}\right\}$ **3.** $(-18, 6)$ **4.** $(1, 7)$ **5.** $\left\{-3, \frac{1}{2}, 2\right\}$ **6.** $\{-2\}$ **7.** $\{2\}$
8. $\{-2 + \log_3 11\}$, or approximately 0.18
9.

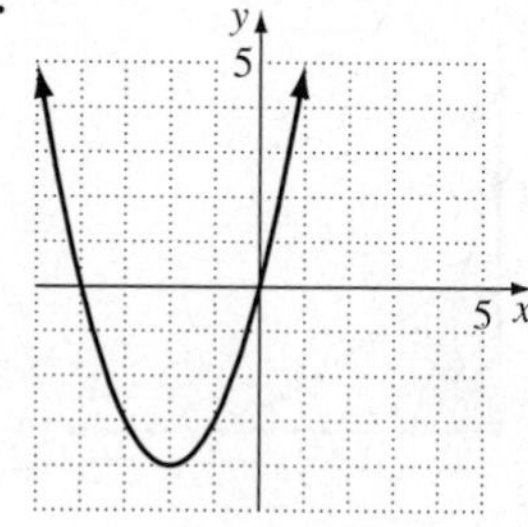

10.

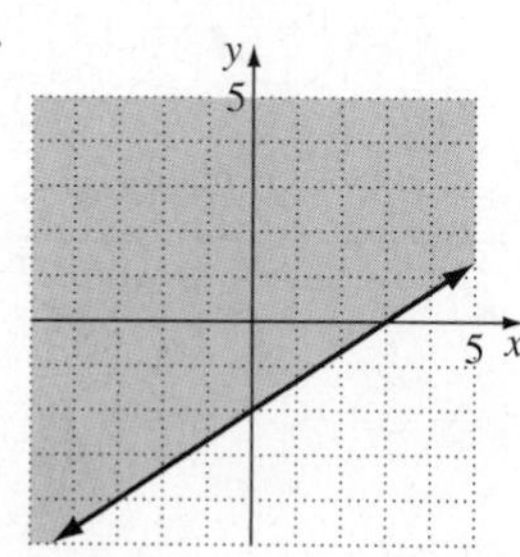

11.

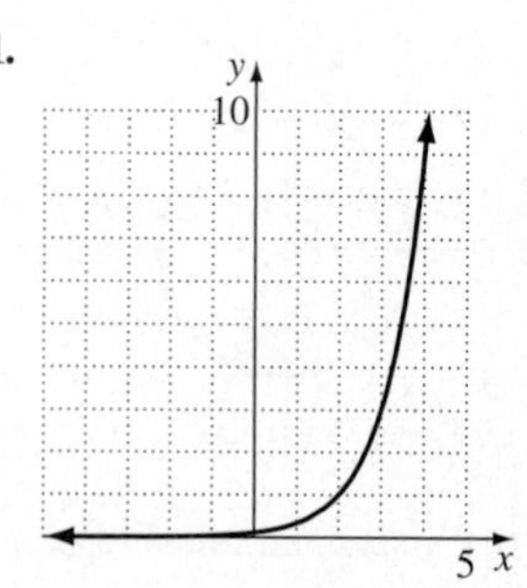

12.

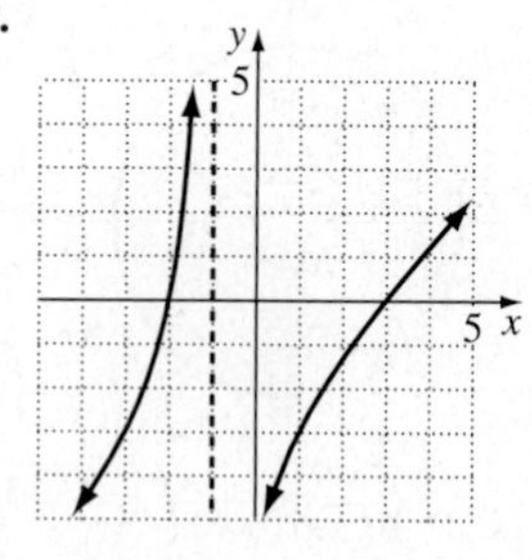

13. $3 + 5 \log_2 x$ **14.** 10.99% **15.** $f^{-1}(x) = \frac{1}{7}x + \frac{3}{7}$ **16.** $g(f(x)) = 21x - 16$ **17.** Answers may vary. **18.** $\left\{\left(-\frac{1}{2}, \frac{1}{2}\right), (2, 8)\right\}$

19. 4 m by 9 m **20.** A plane with an initial landing speed of 90 ft per second needs 562 ft to land. There is a problem since 550 ft is not enough.

CHAPTER 6

Section 6.1

Check Point Exercises

1. $\{(4,-3,1)\}$ **2. a.** $\left[\begin{array}{ccc|c} 1 & 6 & -3 & 7 \\ 4 & 12 & -20 & 8 \\ -3 & -2 & 1 & -9 \end{array}\right]$ **b.** $\left[\begin{array}{ccc|c} 1 & 3 & -5 & 2 \\ 1 & 6 & -3 & 7 \\ -3 & -2 & 1 & -9 \end{array}\right]$ **c.** $\left[\begin{array}{ccc|c} 4 & 12 & -20 & 8 \\ 1 & 6 & -3 & 7 \\ 0 & 16 & -8 & 12 \end{array}\right]$

3. $\{(5,2,3)\}$ **4.** $\{(1,-1,2,-3)\}$ **5.** $\left\{\left(\frac{5}{2},\frac{8}{7},\frac{11}{14}\right)\right\}$

Exercise Set 6.1

1. $\left[\begin{array}{ccc|c} 2 & 1 & 2 & 2 \\ 3 & -5 & -1 & 4 \\ 1 & -2 & -3 & -6 \end{array}\right]$ **3.** $\left[\begin{array}{ccc|c} 1 & -1 & 1 & 8 \\ 0 & 1 & -12 & -15 \\ 0 & 0 & 1 & 1 \end{array}\right]$ **5.** $\left[\begin{array}{ccc|c} 5 & -2 & -3 & 0 \\ 1 & 1 & 0 & 5 \\ 2 & 0 & -3 & 4 \end{array}\right]$ **7.** $\left[\begin{array}{cccc|c} 2 & 5 & -3 & 1 & 2 \\ 0 & 3 & 1 & 0 & 4 \\ 1 & -1 & 5 & 0 & 9 \\ 5 & -5 & -2 & 0 & 1 \end{array}\right]$ **9.** $\begin{aligned} 5x + 3z &= -11 \\ y - 4z &= 12 \\ 7x + 2y &= 3 \end{aligned}$

11. $\begin{aligned} x + y + 4z + w &= 3 \\ -x + y - z &= 7 \\ 2x + 5w &= 11 \\ 12z + 4w &= 5 \end{aligned}$ **13.** $\begin{aligned} x - 4z &= 5 \\ y - 12z &= 13 \\ z &= -\frac{1}{2} \end{aligned}$; $\left\{\left(3,7,-\frac{1}{2}\right)\right\}$ **15.** $\begin{aligned} x + \frac{1}{2}y + z &= \frac{11}{2} \\ y + \frac{3}{2}z &= 7 \\ z &= 4 \end{aligned}$; $(\{1,1,4)\}$

17. $\begin{aligned} x - y + z + w &= 3 \\ y - 2z - w &= 0 \\ z + 6w &= 17 \\ 1w &= 3 \end{aligned}$; $\{(2,1,-1,3)\}$ **19.** $\left[\begin{array}{ccc|c} 1 & -3 & 2 & 5 \\ 1 & 5 & -5 & 0 \\ 3 & 0 & 4 & 7 \end{array}\right]$ **21.** $\left[\begin{array}{ccc|c} 1 & -3 & 2 & 0 \\ 0 & 10 & -7 & 7 \\ 2 & -2 & 1 & 3 \end{array}\right]$ **23.** $\left[\begin{array}{cccc|c} 1 & -1 & 1 & 1 & 3 \\ 0 & 1 & -2 & -1 & 0 \\ 0 & 2 & 1 & 2 & 5 \\ 0 & 6 & -3 & -1 & -9 \end{array}\right]$

25. R_2: $-3, -18$; R_3: $-12, -15$; R_2: $-\frac{3}{5}, -\frac{18}{5}$; R_3: $-12, -15$ **27.** $\{(1,-1,2)\}$ **29.** $\{(3,-1,-1)\}$ **31.** $\{(1,2,-1)\}$
33. $\{(1,2,3,-2)\}$ **35.** $\{(0,-3,0,-3)\}$ **37.** $\{(-1,2,3,4)\}$ **39.** $\{(1,-1,2,-2,0)\}$ **41. a.** $a = 2; b = 22; c = 320$
b. 2780 **c.** Answers may vary. **43.** $\begin{aligned} x + y + z &= 100 \\ x - y + z &= 2 \\ -x + 2z &= -3 \end{aligned}$; under 30: 35%; 30–49: 49%; 50 and over: 16%

45. $\begin{aligned} 40x + 200y + 400z &= 660 \\ 5x + 2y + 4z &= 25 \\ 30x + 10y + 300z &= 425 \end{aligned}$; 4 oz of Food A: $\frac{1}{2}$ oz of Food B; 1 oz of Food C **57.** $y = x^3 + 2x^2 + 5x - 3$

Section 6.2

Check Point Exercises

1. $\varnothing$ **2.** $\{(11t + 13, 5t + 4, t)\}$ **3.** $\{(t + 50, -2t + 10, t)\}$
4. a. $\begin{aligned} x + w &= 15 \\ x + y &= 30 \\ y + z &= 45 \\ z + w &= 30 \end{aligned}$ **b.** $\{(-t + 15, t + 15, -t + 30, t)\}$ **c.** $x = 5; y = 25; z = 20$

Exercise Set 6.2

1. $\varnothing$ **3.** $\left\{\left(-2t + 2, 2t + \frac{1}{2}, t\right)\right\}$ **5.** $\{(-3,4,-2)\}$ **7.** $\{(5 - 2t, -2 + t, t)\}$ **9.** $\{(-1,2,1,1)\}$ **11.** $\{(1,3,2,1)\}$

13. $\{(1,-2,1,1)\}$ **15.** $\left\{\left(1 + \frac{1}{3}t, \frac{1}{3}t, t\right)\right\}$ **17.** $\{(-13t + 5, 5t, t)\}$ **19.** $\left\{\left(2t - \frac{5}{4}, \frac{13}{4}, t\right)\right\}$ **21.** $\{(1, -t - 1, 2, t)\}$

23. $\left\{\left(-\frac{2}{11}t + \frac{81}{11}, \frac{1}{22}t + \frac{10}{11}, \frac{4}{11}t - \frac{8}{11}, t\right)\right\}$ **25.** $z + 12 = x + 6$ **27.** $\{(t + 6, t + 2, t)\}$

29. a. $x + w = 380$
$x + y = 600$
$z - w = 50$
$y - z = 170$
b. $\{(380 - t, 220 + t, 50 + t, t)\}$
c. $x = 330, y = 270, z = 100, w = 50$

31. a. The system has no solution, so there is no way to satisfy these dietary requirements with no Food 1 available.
b. 4 oz of Food 1, 0 oz of Food 2, 10 oz of Food 3; 2 oz of Food 1, 5 oz of Food 2, 9 oz of Food 3 (other answers are possible).
37. $a = 1$ or $a = 3$

Section 6.3

Check Point Exercises

1. a. 3×2 **b.** $a_{12} = -2; a_{31} = 1$ **2. a.** $\begin{bmatrix} 2 & 0 \\ 9 & -10 \end{bmatrix}$ **b.** $\begin{bmatrix} 9 & -4 \\ -9 & 7 \\ 5 & -2 \end{bmatrix}$ **3. a.** $\begin{bmatrix} 6 & 12 \\ -48 & -30 \end{bmatrix}$ **b.** $\begin{bmatrix} -14 & -1 \\ 25 & 10 \end{bmatrix}$ **4.** $\begin{bmatrix} 7 & 6 \\ 13 & 12 \end{bmatrix}$

5. $[30]$; $\begin{bmatrix} 2 & 0 & 4 \\ 6 & 0 & 12 \\ 14 & 0 & 28 \end{bmatrix}$ **6. a.** $\begin{bmatrix} 2 & 18 & 11 & 9 \\ 0 & 10 & 8 & 2 \end{bmatrix}$ **b.** The product is undefined. **7.** $\begin{bmatrix} 2 & 2 & 2 \\ 1 & 2 & 1 \\ 1 & 2 & 1 \end{bmatrix} + \begin{bmatrix} -1 & -1 & -1 \\ 2 & -1 & 2 \\ 2 & -1 & 2 \end{bmatrix} = \begin{bmatrix} 1 & 1 & 1 \\ 3 & 1 & 3 \\ 3 & 1 & 3 \end{bmatrix}$

8. \$2548

Exercise Set 6.3

1. a. 2×3 **b.** a_{32} does not exist; $a_{23} = -1$ **3. a.** 3×4 **b.** $a_{32} = \frac{1}{2}; a_{23} = -6$ **5.** $x = 6; y = 4$ **7.** $x = 4; y = 6; z = 3$

9. a. $\begin{bmatrix} 9 & 10 \\ 3 & 9 \end{bmatrix}$ **b.** $\begin{bmatrix} -1 & -8 \\ 3 & -5 \end{bmatrix}$ **c.** $\begin{bmatrix} -16 & -4 \\ -12 & -8 \end{bmatrix}$ **d.** $\begin{bmatrix} 22 & 21 \\ 9 & 20 \end{bmatrix}$ **11. a.** $\begin{bmatrix} 3 & 2 \\ 6 & 2 \\ 5 & 7 \end{bmatrix}$ **b.** $\begin{bmatrix} -1 & 4 \\ 0 & 6 \\ 5 & 5 \end{bmatrix}$ **c.** $\begin{bmatrix} -4 & -12 \\ -12 & -16 \\ -20 & -24 \end{bmatrix}$

d. $\begin{bmatrix} 7 & 7 \\ 15 & 8 \\ 15 & 20 \end{bmatrix}$ **13. a.** $\begin{bmatrix} -3 \\ -1 \\ 0 \end{bmatrix}$ **b.** $\begin{bmatrix} 7 \\ -7 \\ 2 \end{bmatrix}$ **c.** $\begin{bmatrix} -8 \\ 16 \\ -4 \end{bmatrix}$ **d.** $\begin{bmatrix} -4 \\ -6 \\ 1 \end{bmatrix}$ **15. a.** $\begin{bmatrix} 8 & 0 & -4 \\ 14 & 0 & 6 \\ -1 & 0 & 0 \end{bmatrix}$ **b.** $\begin{bmatrix} -4 & -20 & 0 \\ 14 & 24 & 14 \\ 9 & -4 & 4 \end{bmatrix}$

c. $\begin{bmatrix} -8 & 40 & 8 \\ -56 & -48 & -40 \\ -16 & 8 & -8 \end{bmatrix}$ **d.** $\begin{bmatrix} 18 & -10 & -10 \\ 42 & 12 & 22 \\ 2 & -2 & 2 \end{bmatrix}$ **17. a.** $\begin{bmatrix} 0 & 16 \\ 12 & 8 \end{bmatrix}$ **b.** $\begin{bmatrix} -7 & 3 \\ 29 & 15 \end{bmatrix}$ **19. a.** $[30]$ **b.** $\begin{bmatrix} 1 & 2 & 3 & 4 \\ 2 & 4 & 6 & 8 \\ 3 & 6 & 9 & 12 \\ 4 & 8 & 12 & 16 \end{bmatrix}$

21. a. $\begin{bmatrix} 4 & -5 & 8 \\ 6 & -1 & 5 \\ 0 & 4 & -6 \end{bmatrix}$ **b.** $\begin{bmatrix} 5 & -2 & 7 \\ 17 & -3 & 2 \\ 3 & 0 & -5 \end{bmatrix}$ **23. a.** $\begin{bmatrix} 6 & 8 & 16 \\ 11 & 16 & 24 \\ 1 & -1 & 12 \end{bmatrix}$ **b.** $\begin{bmatrix} 38 & 27 \\ -16 & -4 \end{bmatrix}$ **25. a.** $\begin{bmatrix} 0 & 0 \\ 0 & 0 \end{bmatrix}$ **b.** $\begin{bmatrix} 4 & -1 & -3 & 1 \\ -1 & 4 & -3 & 2 \\ 14 & -11 & -3 & -1 \\ 25 & -25 & 0 & -5 \end{bmatrix}$

27. $\begin{bmatrix} 17 & 7 \\ -5 & -11 \end{bmatrix}$ **29.** $\begin{bmatrix} 11 & -1 \\ -7 & -3 \end{bmatrix}$ **31.** $A - C$ is not defined because A is 3×2 and C is 2×2. **33.** $\begin{bmatrix} 16 & -16 \\ -12 & 12 \\ 0 & 0 \end{bmatrix}$

35. $\begin{bmatrix} 1 & 3 & 1 \\ 3 & 3 & 3 \\ 1 & 3 & 1 \end{bmatrix} + \begin{bmatrix} -1 & -1 & -1 \\ -1 & -1 & -1 \\ -1 & -1 & -1 \end{bmatrix} = \begin{bmatrix} 0 & 2 & 0 \\ 2 & 2 & 2 \\ 0 & 2 & 0 \end{bmatrix}$ **37.** $\begin{bmatrix} 1 & 3 & 1 \\ 3 & 3 & 3 \\ 1 & 3 & 1 \end{bmatrix} + \begin{bmatrix} 1 & -2 & 1 \\ -2 & -2 & -2 \\ 1 & -2 & 1 \end{bmatrix} = \begin{bmatrix} 2 & 1 & 2 \\ 1 & 1 & 1 \\ 2 & 1 & 2 \end{bmatrix}$ **39. a.** $\begin{bmatrix} 0 & 3 & 0 \\ 0 & 3 & 0 \\ 0 & 3 & 0 \end{bmatrix}$ **b.** $\begin{bmatrix} 1 & 0 & 1 \\ 1 & 0 & 1 \\ 1 & 0 & 1 \end{bmatrix}$

41. a. $\begin{bmatrix} 589.5 & 586 \\ 1556 & 1521 \\ 1234.5 & 1183 \end{bmatrix}$ **b.** 1521 **c.** 1235

43. a. System 1: The midterm and final both count for 50% of the course grade.
System 2: The midterm counts for 30% of the

b. $\begin{bmatrix} 84 & 87.2 \\ 79 & 81 \\ 90 & 88.4 \\ 73 & 68.6 \\ 69 & 73.4 \end{bmatrix}$ System 1 grades are listed first (if different).
Student 1: B; Student 2: C or B; Student 3: A or B;
Student 4: C or D; Student 5: D or C

57. Answers may vary. **59.** $AB = -BA$ so they are anticommutative.

Section 6.4

Check Point Exercises

1. $AB = I_2; BA = I_2$ **2.** $\begin{bmatrix} 3 & -7 \\ -2 & 5 \end{bmatrix}$ **3.** $\begin{bmatrix} 1 & 2 \\ 1 & 3 \end{bmatrix}$ **4.** $\begin{bmatrix} 3 & -2 & -4 \\ 3 & -2 & -5 \\ -1 & 1 & 2 \end{bmatrix}$ **5.** $\{(4, -2, 1)\}$

6. The encoded message is −7, 10, −53, 77. **7.** The decoded message is 2, 1, 19, 5 or BASE.

Exercise Set 6.4

1. $AB = I_2; BA = I_2; B = A^{-1}$ **3.** $AB = \begin{bmatrix} 8 & -16 \\ -2 & 7 \end{bmatrix}; BA = \begin{bmatrix} 12 & 12 \\ 1 & 3 \end{bmatrix}; B \neq A^{-1}$ **5.** $AB = I_2; BA = I_2; B = A^{-1}$

7. $AB = I_3; BA = I_3; B = A^{-1}$ **9.** $AB = I_3; BA = I_3; B = A^{-1}$ **11.** $AB = I_4; BA = I_4; B = A^{-1}$

13. $\begin{bmatrix} \frac{2}{7} & -\frac{3}{7} \\ \frac{1}{7} & \frac{2}{7} \end{bmatrix}$ **15.** $\begin{bmatrix} 1 & \frac{1}{2} \\ 2 & \frac{3}{2} \end{bmatrix}$ **17.** A does not have an inverse. **19.** $\begin{bmatrix} 1 & 0 & 1 \\ 1 & 1 & 2 \\ 3 & 2 & 6 \end{bmatrix}$ **21.** $\begin{bmatrix} -3 & 2 & -4 \\ -1 & 1 & -1 \\ 8 & -5 & 10 \end{bmatrix}$ **23.** $\begin{bmatrix} 1 & 0 & 0 & 0 \\ 0 & -1 & 0 & 0 \\ 0 & 0 & \frac{1}{3} & 0 \\ -1 & 0 & 0 & 1 \end{bmatrix}$

25. $\begin{bmatrix} 6 & 5 \\ 5 & 4 \end{bmatrix}\begin{bmatrix} x \\ y \end{bmatrix} = \begin{bmatrix} 13 \\ 10 \end{bmatrix}$ **27.** $\begin{bmatrix} 1 & 3 & 4 \\ 1 & 2 & 3 \\ 1 & 4 & 3 \end{bmatrix}\begin{bmatrix} x \\ y \\ z \end{bmatrix} = \begin{bmatrix} -3 \\ -2 \\ -6 \end{bmatrix}$ **29.** $4x - 7y = -3$; $2x - 3y = 1$ **31.** $2x - z = 6$; $3y = 9$; $x + y = 5$ **33. a.** $\begin{bmatrix} 2 & 6 & 6 \\ 2 & 7 & 6 \\ 2 & 7 & 7 \end{bmatrix}\begin{bmatrix} x \\ y \\ z \end{bmatrix} = \begin{bmatrix} 8 \\ 10 \\ 9 \end{bmatrix}$

b. $\{(1, 2, -1)\}$ **35. a.** $\begin{bmatrix} 1 & -1 & 1 \\ 0 & 2 & -1 \\ 2 & 3 & 0 \end{bmatrix}\begin{bmatrix} x \\ y \\ z \end{bmatrix} = \begin{bmatrix} 8 \\ -7 \\ 1 \end{bmatrix}$ **b.** $\{(2, -1, 5)\}$ **37. a.** $\begin{bmatrix} 1 & -1 & 2 & 0 \\ 0 & 1 & -1 & 1 \\ -1 & 1 & -1 & 2 \\ 0 & -1 & 1 & -2 \end{bmatrix}\begin{bmatrix} x \\ y \\ z \\ w \end{bmatrix} = \begin{bmatrix} -3 \\ 4 \\ 2 \\ -4 \end{bmatrix}$

b. $\{(2, 3, -1, 0)\}$ **39.** The encoded message is 27, −19, 32, −20.; The decoded message is 8, 5, 12, 16 or HELP.

41. The encoded message is 14, 85, −33, 4, 18, −7, −18, 19, −9. **53.** $\begin{bmatrix} 1 & 1 \\ 2 & 3 \end{bmatrix}$ **55.** $\begin{bmatrix} 1 & 0 & 1 \\ 2 & 1 & 3 \\ -1 & 1 & 1 \end{bmatrix}$ **57.** $\begin{bmatrix} 0 & -1 & 0 & 1 \\ -1 & -5 & 0 & 3 \\ -2 & -4 & 1 & -2 \\ -1 & -4 & 0 & 1 \end{bmatrix}$

59. $\{(2, 3, -5)\}$ **61.** $\{(1, 2, -1)\}$ **63.** $\{(2, 1, 3, -2, 4)\}$ **65.** Answers may vary. **67.** (c) is true. **69.** Answers may vary.
71. $a = 3$ or $a = -2$

Section 6.5

Check Point Exercises

1. a. −4 **b.** −17 **2.** $\{(4, -2)\}$ **3.** 80 **4.** −24 **5.** $\{(2, -3, 4)\}$ **6.** −250

Exercise Set 6.5

1. 1 **3.** −29 **5.** 0 **7.** 33 **9.** $-\frac{7}{16}$ **11.** $\{(5, 2)\}$ **13.** $\{(2, -3)\}$ **15.** $\{(3, -1)\}$ **17.** The system is dependent.

19. $\{(4, 2)\}$ **21.** $\{(7, 4)\}$ **23.** The system is inconsistent. **25.** $\{(0, 4)\}$ **27.** 72 **29.** −75 **31.** 0 **33.** $\{(-5, -2, 7)\}$
35. $\{(2, -3, 4)\}$ **37.** $\{(3, -1, 2)\}$ **39.** $\{(2, 3, 1)\}$ **41.** −200 **43.** 195

45. a. 28 sq units

b.

47. yes

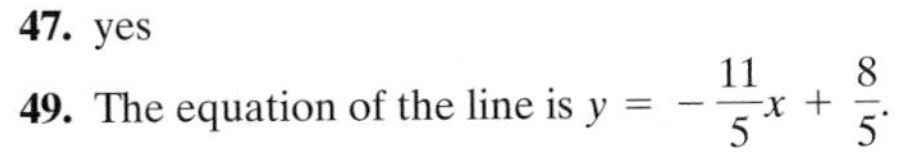

49. The equation of the line is $y = -\frac{11}{5}x + \frac{8}{5}$.

61. 13,200

63. a. a^2 **b.** a^3 **c.** a^4

d. Each determinant has zeros below the main diagonal and a's everywhere else.

e. Each determinant equals a raised to the power equal to the order of the determinant.

65. The sign of the value is changed when 2 columns are interchanged in a 2nd order determinant.

Chapter 6 Review Exercises

1. $\begin{aligned} x + y + 3z &= 12 \\ y - 2z &= -4 \\ z &= 3 \end{aligned}$; $\{(1, 2, 3)\}$ **2.** $\begin{aligned} x - 2z + 2w &= 1 \\ y + z - w &= 0 \\ z - \frac{7}{3}w &= -\frac{1}{3} \\ w &= 1 \end{aligned}$; $\{(3, -1, 2, 1)\}$ **3.** $\left[\begin{array}{ccc|c} 1 & 2 & 2 & 2 \\ 0 & 1 & -1 & 2 \\ 0 & 0 & 9 & -9 \end{array}\right]$ **4.** $\left[\begin{array}{ccc|c} 1 & -1 & \frac{1}{2} & -\frac{1}{2} \\ 1 & 2 & -1 & 2 \\ 6 & 4 & 3 & 5 \end{array}\right]$

5. $\{(1, 3, -4)\}$ **6.** $\{(-2, -1, 0)\}$ **7.** $\{(2, -2, 3, 4)\}$ **8. a.** $a = -2; b = 32; c = 42$ **b.** 2:00 P.M.; 170 parts per million **9.** $\varnothing$ **10.** $\{(2t + 4, t + 1, t)\}$ **11.** $\{(-37t + 2, 16t, -7t + 1, t)\}$ **12.** $\{(7t + 18, -3t - 7, t)\}$

13. a. $\begin{aligned} x + z &= 750 \\ y - z &= -250 \\ x + y &= 500 \end{aligned}$ **b.** $\{(-t + 750, t - 250, t)\}$ **c.** $x = 350; y = 150$ **14.** $x = -5; y = 6; z = 6$ **15.** $\begin{bmatrix} 0 & 2 & 3 \\ 8 & 1 & 3 \end{bmatrix}$ **16.** $\begin{bmatrix} 0 & -4 \\ 6 & 4 \\ 2 & -10 \end{bmatrix}$

17. $\begin{bmatrix} -4 & 4 & -1 \\ -2 & -5 & 5 \end{bmatrix}$ **18.** Not possible since B is 3×2 and C is 3×3. **19.** $\begin{bmatrix} 2 & 3 & 8 \\ 21 & 5 & 5 \end{bmatrix}$ **20.** $\begin{bmatrix} -12 & 14 & 0 \\ 2 & -14 & 18 \end{bmatrix}$

21. $\begin{bmatrix} 0 & -10 & -15 \\ -40 & -5 & -15 \end{bmatrix}$ **22.** $\begin{bmatrix} -1 & -16 \\ 8 & 1 \end{bmatrix}$ **23.** $\begin{bmatrix} -10 & -6 & 2 \\ 16 & 3 & 4 \\ -23 & -16 & 7 \end{bmatrix}$ **24.** $\begin{bmatrix} -6 & 4 & -8 \\ 0 & 5 & 11 \\ -17 & 13 & -19 \end{bmatrix}$ **25.** $\begin{bmatrix} 10 & 5 \\ -2 & -30 \end{bmatrix}$

26. Not possible since AB is 2×2 and BA is 3×3. **27.** $\begin{bmatrix} 7 & 6 & 5 \\ 2 & -1 & 11 \end{bmatrix}$ **28.** $\begin{bmatrix} -6 & -22 & -40 \\ 9 & 43 & 58 \\ -14 & -48 & -94 \end{bmatrix}$ **29.** $\begin{bmatrix} 2 & 1 & 1 \\ 2 & 1 & 1 \\ 2 & 2 & 2 \end{bmatrix}$ **30.** $\begin{bmatrix} 1 & -1 & -1 \\ 1 & -1 & -1 \\ 1 & 1 & 1 \end{bmatrix}$

31. a. $\begin{bmatrix} 360{,}000 & 444{,}000 \\ 556{,}000 & 685{,}000 \\ 178{,}000 & 218{,}500 \end{bmatrix}$ **b.** The rows of AB correspond to the outlets, the columns represent the wholesale and retail prices. The entries tell how much value in wholesale or retail is at each outlet. **c.** \$360,000 **d.** \$685,000 **e.** \$40,500

32. $AB = \begin{bmatrix} 1 & 7 \\ 0 & 5 \end{bmatrix}; BA = \begin{bmatrix} 1 & 0 \\ 1 & 5 \end{bmatrix}; B \neq A^{-1}$ **33.** $AB = I_3; BA = I_3; B = A^{-1}$ **34.** $\begin{bmatrix} 3 & 1 \\ 2 & 1 \end{bmatrix}$ **35.** $\begin{bmatrix} -\frac{3}{5} & \frac{1}{5} \\ 1 & 0 \end{bmatrix}$ **36.** $\begin{bmatrix} 3 & 0 & -2 \\ -6 & 1 & 4 \\ 1 & 0 & -1 \end{bmatrix}$

37. $\begin{bmatrix} 8 & -8 & 5 \\ -3 & 2 & -1 \\ -1 & -1 & 1 \end{bmatrix}$ **38. a.** $\begin{bmatrix} 1 & 1 & 2 \\ 0 & 1 & 3 \\ 3 & 0 & -2 \end{bmatrix}\begin{bmatrix} x \\ y \\ z \end{bmatrix} = \begin{bmatrix} 7 \\ -2 \\ 0 \end{bmatrix}$ **b.** $\{(-18, 79, -27)\}$ **39. a.** $\begin{bmatrix} 1 & -1 & 2 \\ 0 & 1 & -1 \\ 1 & 0 & 2 \end{bmatrix}\begin{bmatrix} x \\ y \\ z \end{bmatrix} = \begin{bmatrix} 12 \\ -5 \\ 10 \end{bmatrix}$ **b.** $\{(4, -2, 3)\}$

40. The encoded message is 96, 135, 46, 63; The decoded message is 18, 21, 12, 5 or RULE. **41.** 17 **42.** 4 **43.** −86 **44.** −236 **45.** 4 **46.** 16 **47.** $\left\{\left(\frac{7}{4}, -\frac{25}{8}\right)\right\}$ **48.** $\{(2, -7)\}$ **49.** $\{(23, -12, 3)\}$ **50.** $\{(-3, 2, 1)\}$ **51.** $a = \frac{5}{8}; b = -50; c = 1150;$ 30- and 50-year-olds are involved in an average of 212.5 automobile accidents per day.

Chapter 6 Test

1. $\left\{\left(-3, \frac{1}{2}, 1\right)\right\}$ **2.** $\{(t, t - 1, t)\}$ **3.** $\begin{bmatrix} 5 & 4 \\ 1 & 11 \end{bmatrix}$ **4.** $\begin{bmatrix} 5 & -2 \\ 1 & -1 \\ 4 & -1 \end{bmatrix}$ **5.** $\begin{bmatrix} \frac{3}{5} & -\frac{2}{5} \\ \frac{1}{5} & \frac{1}{5} \end{bmatrix}$ **6.** $\begin{bmatrix} -1 & 2 \\ -5 & 4 \end{bmatrix}$

7. $AB = I_3; BA = I_3$ **8. a.** $\begin{bmatrix} 3 & 5 \\ 2 & -3 \end{bmatrix}\begin{bmatrix} x \\ y \end{bmatrix} = \begin{bmatrix} 9 \\ -13 \end{bmatrix}$ **b.** $\begin{bmatrix} \frac{3}{19} & \frac{5}{19} \\ \frac{2}{19} & -\frac{3}{19} \end{bmatrix}$ **c.** $\{(-2, 3)\}$ **9.** 18 **10.** $x = 2$

Cumulative Review Exercises (Chapters 1–6)

1. $\left\{\frac{-1 + \sqrt{33}}{4}, \frac{-1 - \sqrt{33}}{4}\right\}$ **2.** $\left[\frac{1}{2}, \infty\right)$ **3.** $\{6\}$ **4.** $\left\{-4, \frac{1}{3}, 1\right\}$ **5.** $\{\ln 5, \ln 9\}$ **6.** $\{1\}$ **7.** $\{(7, -4, 6)\}$

8. $y = -1$ **9.** $f^{-1}(x) = \frac{x^2 + 7}{4}\ (x \geq 0)$

10.

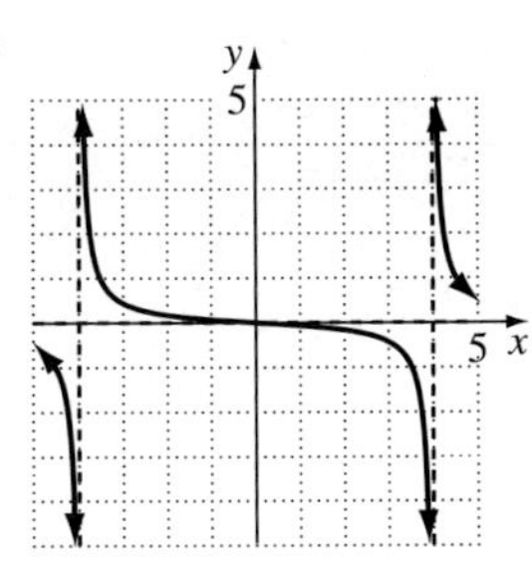

11. $f(x) = (x + 2)(x - 3)(2x + 1)(2x - 1)$

12.

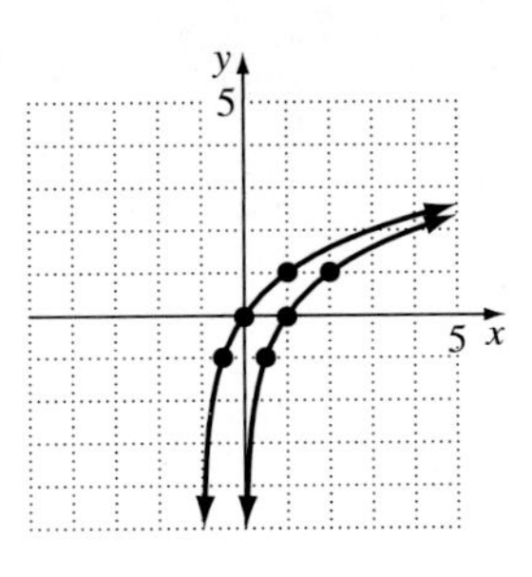

13. a. $A = A_0e^{-0.017t}$ **b.** 759.30 g

14. $\begin{bmatrix} 2 & -1 \\ 13 & 1 \end{bmatrix}$ **15.** $\frac{8}{x-3} + \frac{-2}{x-2} + \frac{-3}{x+2}$

16.

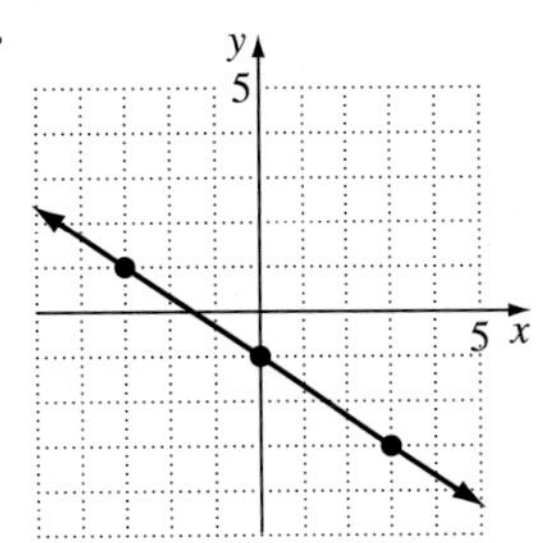

17.

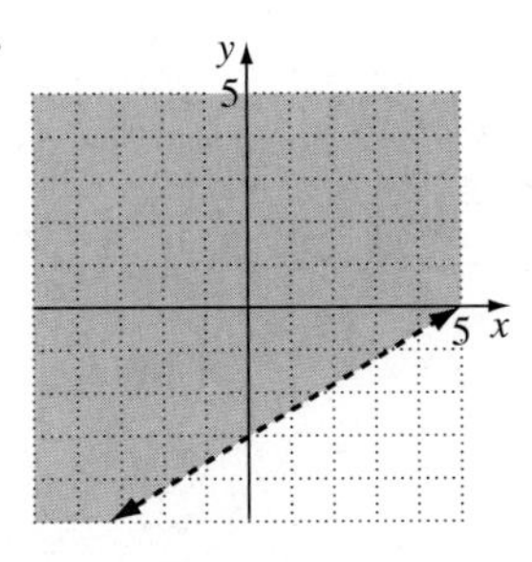

18.

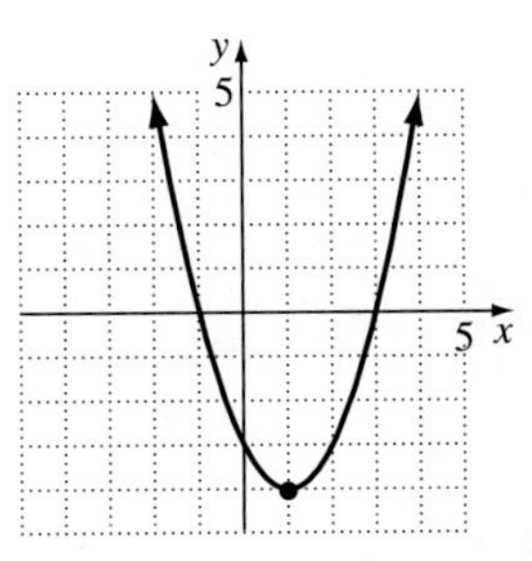

19.

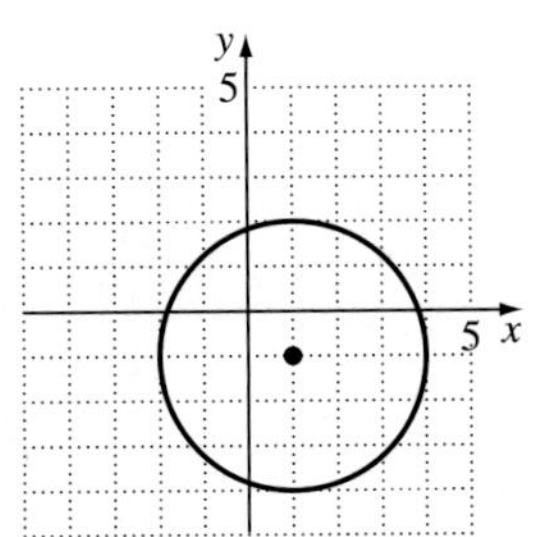

20. $= x^2 + 2x - 2$

CHAPTER 7

Section 7.1

Check Point Exercises

1. foci at $(-3\sqrt{3}, 0)$ and $(3\sqrt{3}, 0)$

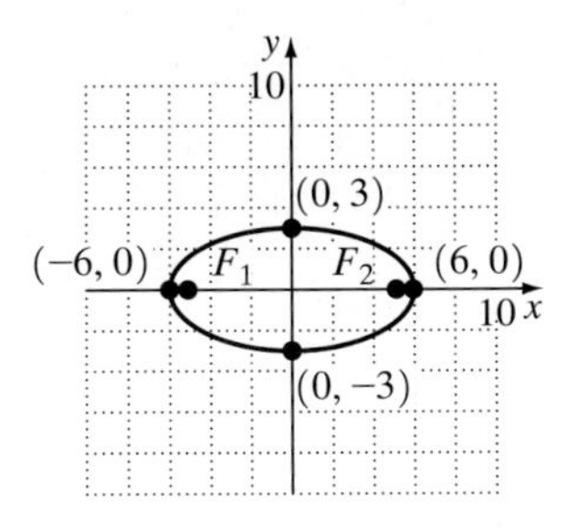

2. foci at $(0, -\sqrt{7})$ and $(0, \sqrt{7})$

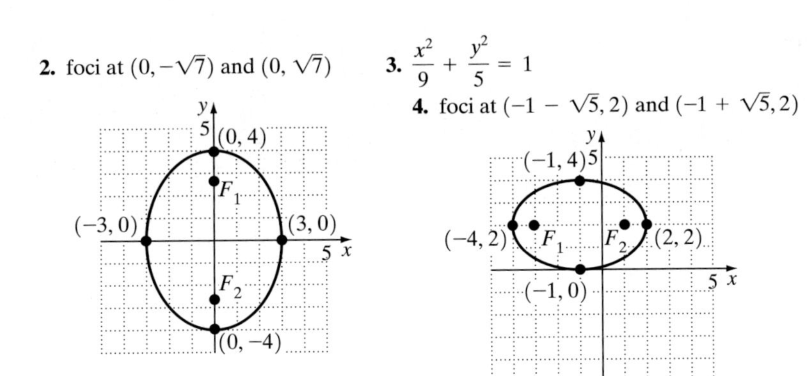

3. $\frac{x^2}{9} + \frac{y^2}{5} = 1$

4. foci at $(-1 - \sqrt{5}, 2)$ and $(-1 + \sqrt{5}, 2)$

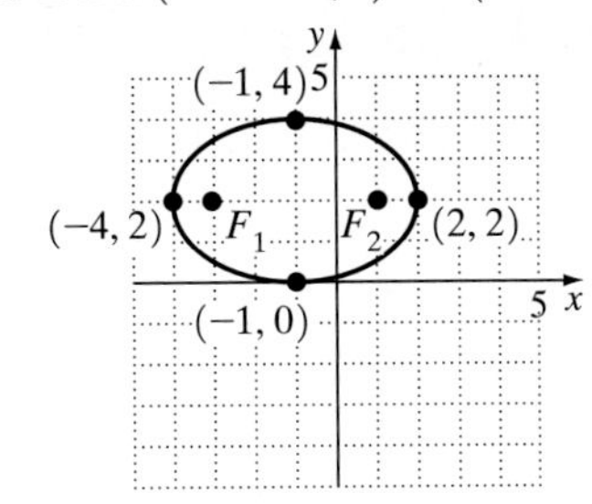

5. Yes

Exercise Set 7.1

1. foci at $(-2\sqrt{3}, 0)$ and $(2\sqrt{3}, 0)$

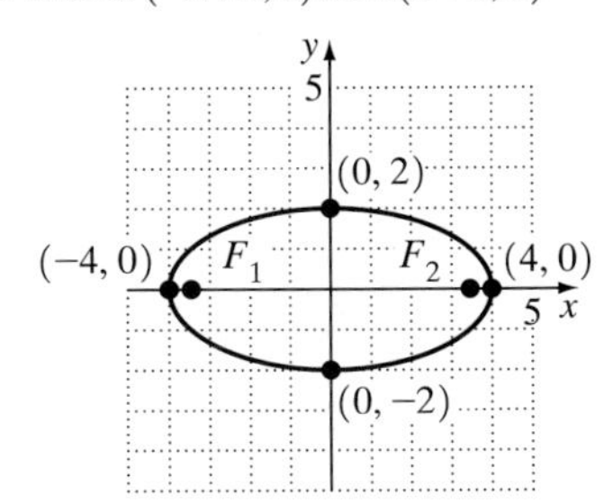

3. foci at $(0, -3\sqrt{3})$ and $(0, 3\sqrt{3})$

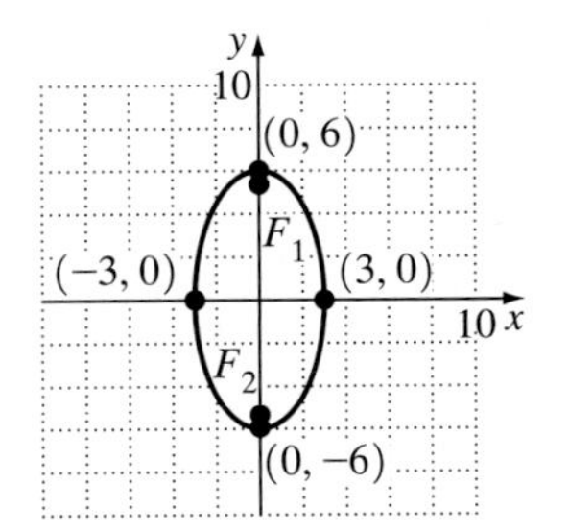

5. foci at $(0, -\sqrt{39})$ and $(0, \sqrt{39})$

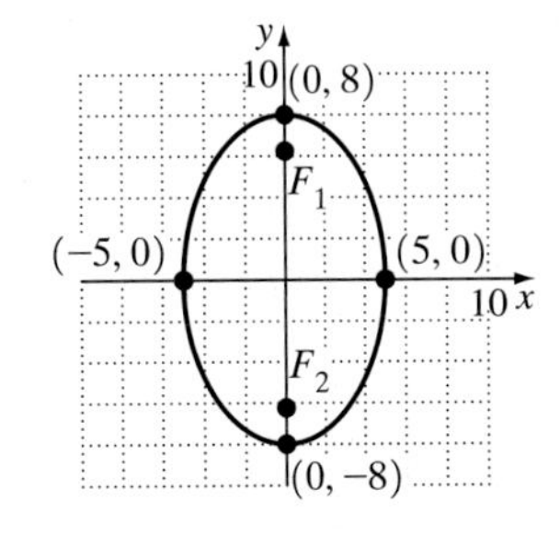

7. foci at $(0, -4\sqrt{2})$ and $(0, 4\sqrt{2})$

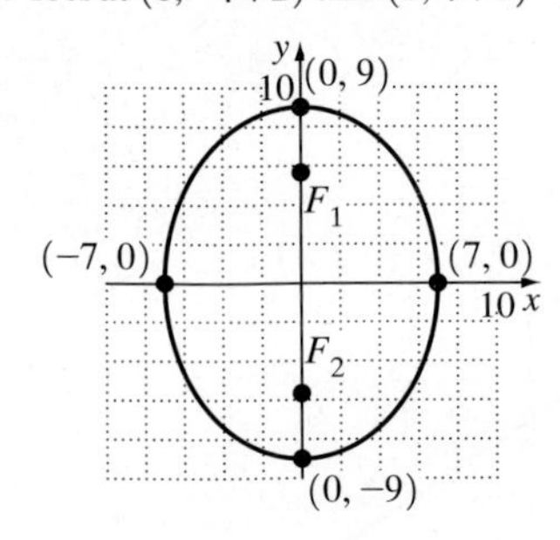

9. foci at $(0, -\sqrt{21})$ and $(0, \sqrt{21})$

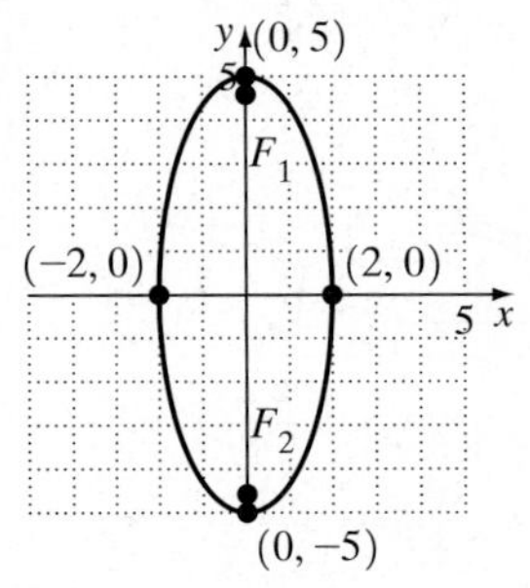

11. foci at $(-2\sqrt{3}, 0)$ and $(2\sqrt{3}, 0)$

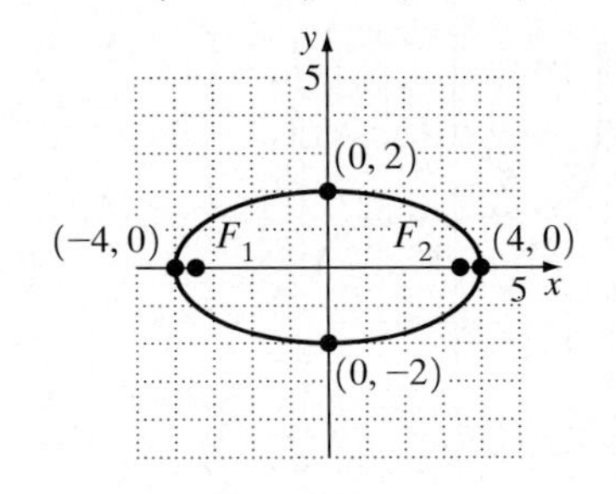

13. foci at $(0, 4)$ and $(0, -4)$

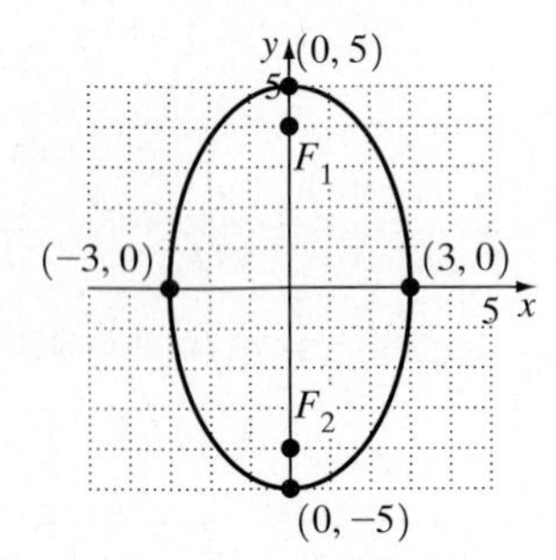

15. foci at $(2, 0)$ and $(-2, 0)$

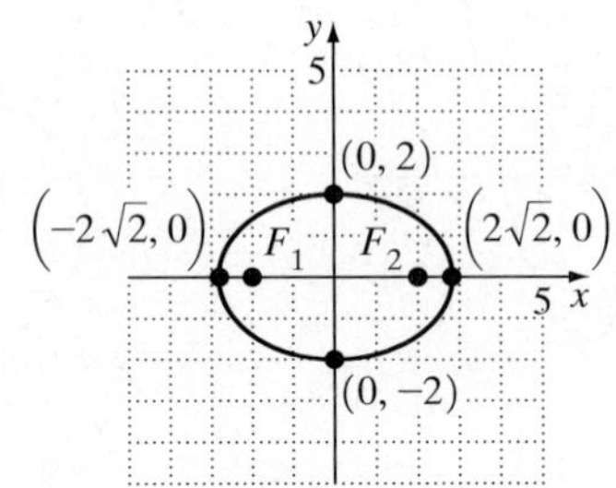

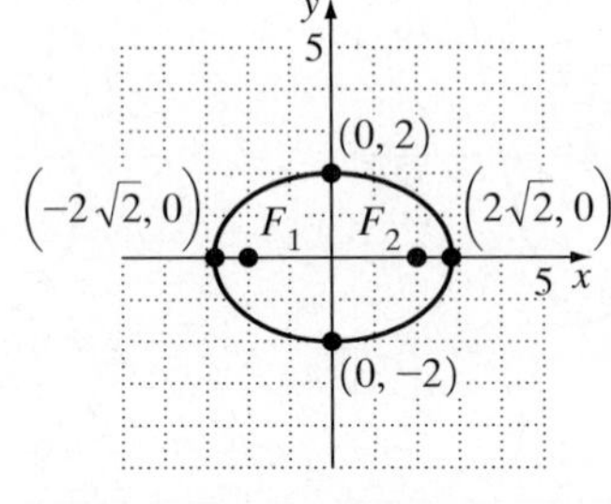

17. $\frac{x^2}{4} + \frac{y^2}{1} = 1$; foci at $(-\sqrt{3}, 0)$ and $(\sqrt{3}, 0)$

19. $\frac{x^2}{1} + \frac{y^2}{4} = 1$; foci at $(0, \sqrt{3})$ and $(0, -\sqrt{3})$

21. $\frac{x^2}{64} + \frac{y^2}{39} = 1$

23. $\frac{x^2}{33} + \frac{y^2}{49} = 1$

25. $\frac{x^2}{13} + \frac{y^2}{9} = 1$

27. $\frac{x^2}{16} + \frac{y^2}{4} = 1$

29. $\frac{x^2}{4} + \frac{y^2}{25} = 1$

31. foci at $(2 - \sqrt{5}, 1)$ and $(2 + \sqrt{5}, 1)$

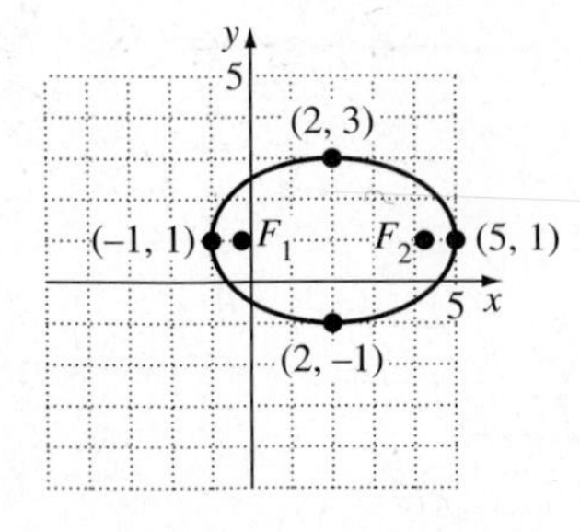

33. foci at $(-3 - 2\sqrt{3}, 2)$ and $(-3 + 2\sqrt{3}, 2)$

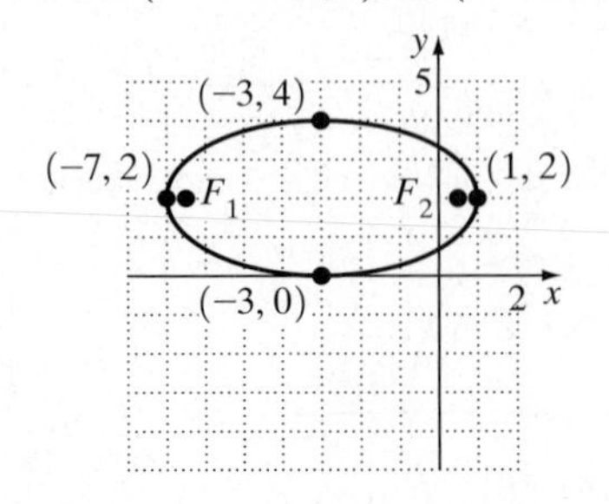

35. foci at $(4, 2)$ and $(4, -6)$

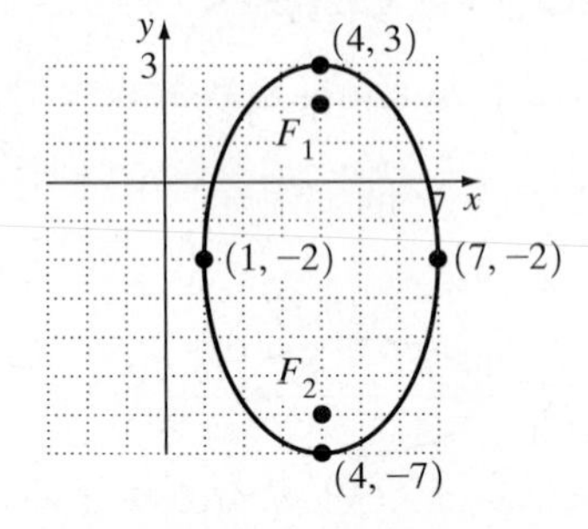

37. foci at $(0, 2 + \sqrt{11})$, $(0, 2 - \sqrt{11})$

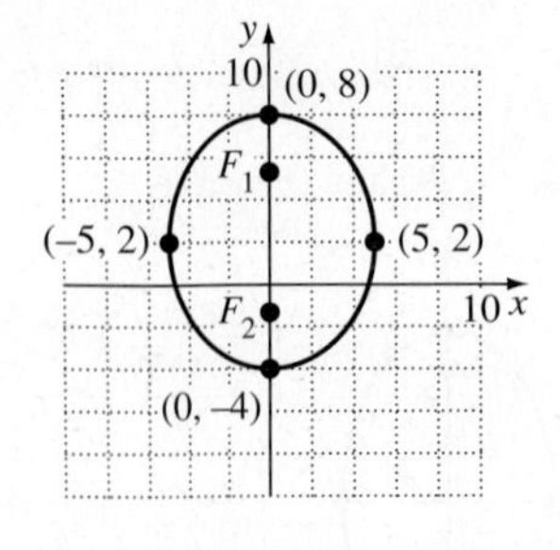

39. foci at $(-3 - 2\sqrt{2}, 2)$ and $(-3 + 2\sqrt{2}, 2)$

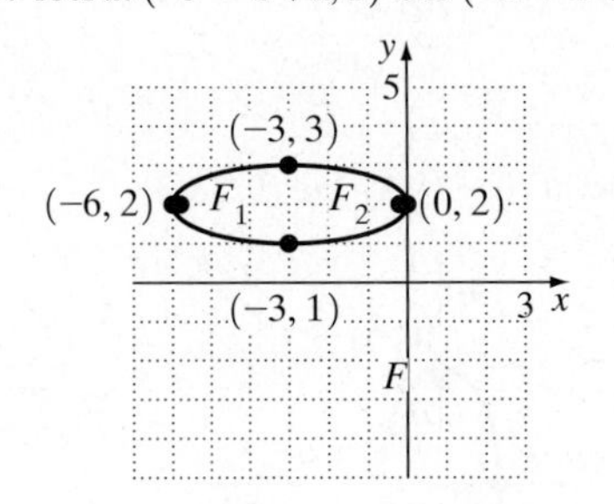

41. foci at $(1, -3 + \sqrt{5})$ and $(1, -3 - \sqrt{5})$

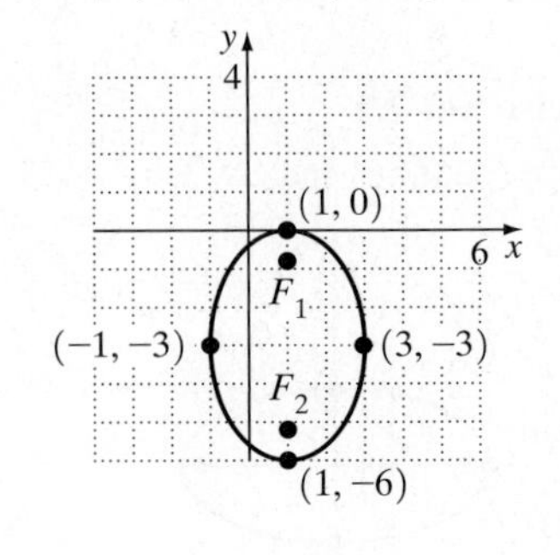

43. $\dfrac{(x-2)^2}{25}+\dfrac{(y+1)^2}{9}=1$
foci at $(-2,-1)$ and $(6,-1)$

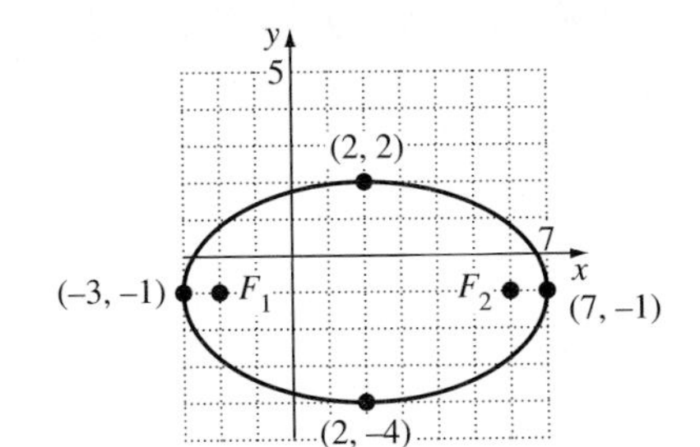

45. $\dfrac{(x-1)^2}{16}+\dfrac{(y+2)^2}{9}=1$
foci at $(1-\sqrt{7},-2)$ and $(1+\sqrt{7},-2)$

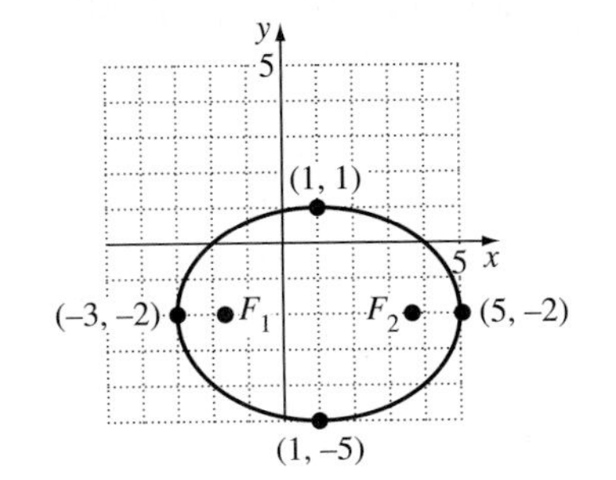

47. $\dfrac{(x+2)^2}{16}+\dfrac{(y-3)^2}{64}=1$
foci at $(-2, 3+4\sqrt{3})$ and $(-2, 3-4\sqrt{3})$

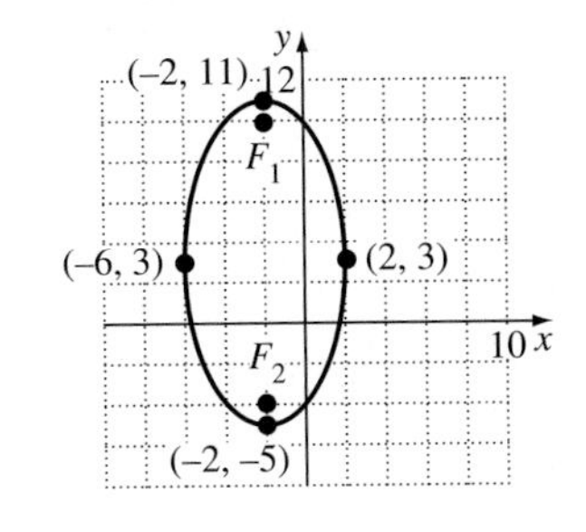

49. Yes **51. a.** $\dfrac{x^2}{2304}+\dfrac{y^2}{529}=1$ **b.** about 42 feet **63.** $\dfrac{x^2}{\frac{36}{5}}+\dfrac{y^2}{36}=1$ **65.** The large circle has radius 5 with center $(0, 0)$. Its equation is $x^2+y^2=25$. The small circle has radius 3 with center $(0, 0)$. Its equation is $x^2+y^2=9$.

Section 7.2

Check Point Exercises

1. a. vertices at $(5, 0)$ and $(-5, 0)$; foci at $(\sqrt{41}, 0)$ and $(-\sqrt{41}, 0)$ **b.** vertices at $(0, 5)$ and $(0, -5)$; foci at $(0, \sqrt{41})$ and $(0, -\sqrt{41})$

2. $\dfrac{y^2}{9}-\dfrac{x^2}{16}=1$ **3.** foci at $(-3\sqrt{5}, 0)$ and $(3\sqrt{5}, 0)$ **4.** foci at $(0, \sqrt{5})$ and $(0, -\sqrt{5})$ **5.** foci at $(3-\sqrt{5}, 1)$ and $(3+\sqrt{5}, 1)$

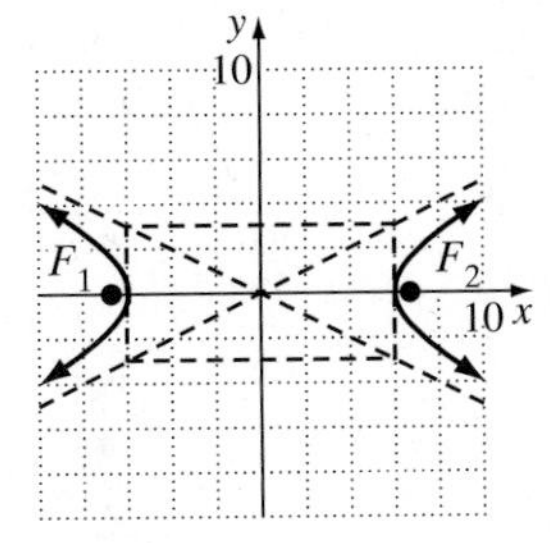

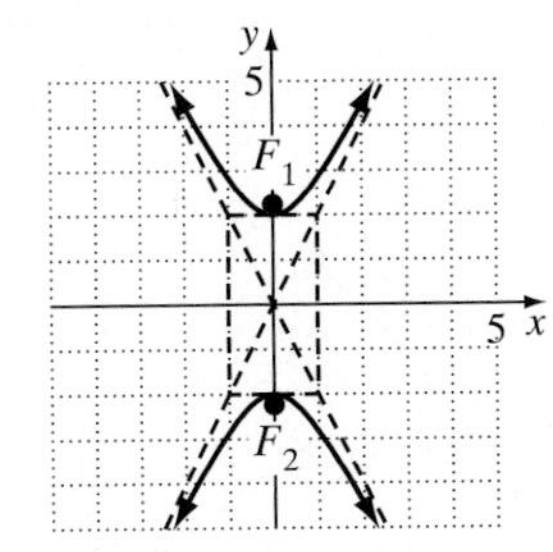

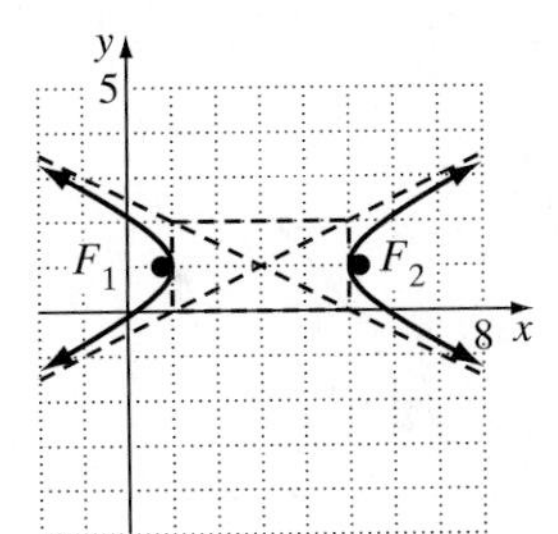

6. $\dfrac{x^2}{2{,}722{,}500}-\dfrac{y^2}{25{,}155{,}900}=1$

Exercise Set 7.2

1. vertices at $(2, 0)$ and $(-2, 0)$; foci at $(\sqrt{5}, 0)$ and $(-\sqrt{5}, 0)$; graph (b)

3. vertices at $(0, 2)$ and $(0, -2)$; foci at $(0, \sqrt{5})$ and $(0, -\sqrt{5})$; graph (a) **5.** $y^2-\dfrac{x^2}{8}=1$ **7.** $\dfrac{x^2}{9}-\dfrac{y^2}{7}=1$

9. foci: $(\pm\sqrt{34}, 0)$ **11.** foci: $(\pm 2\sqrt{41}, 0)$ **13.** foci: $(0, \pm 2\sqrt{13})$ **15.** foci: $(0, \pm\sqrt{61})$

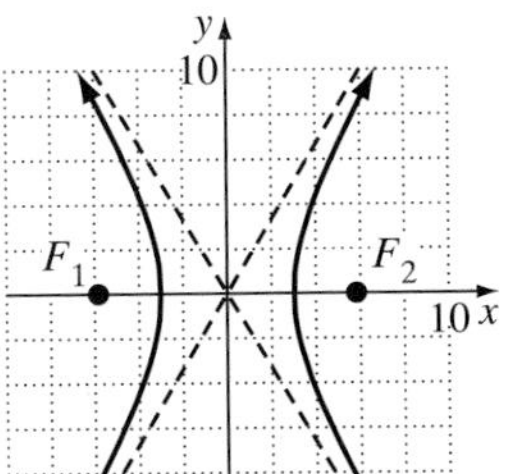

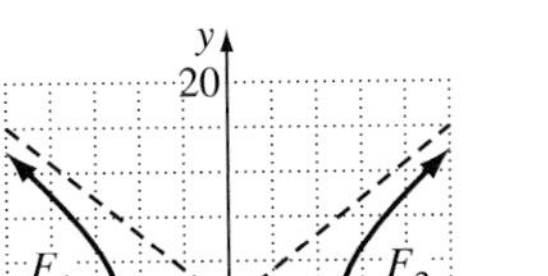

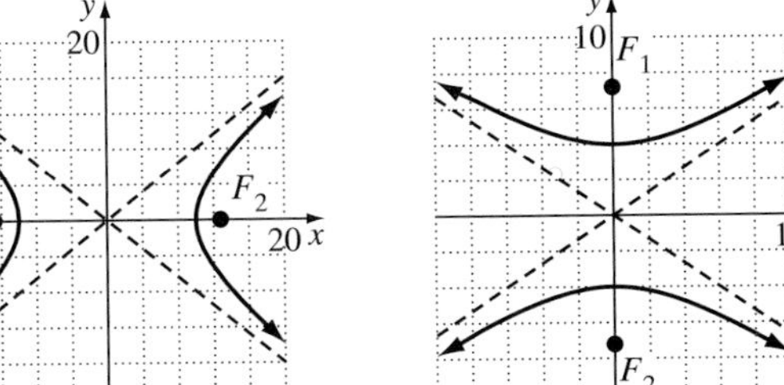

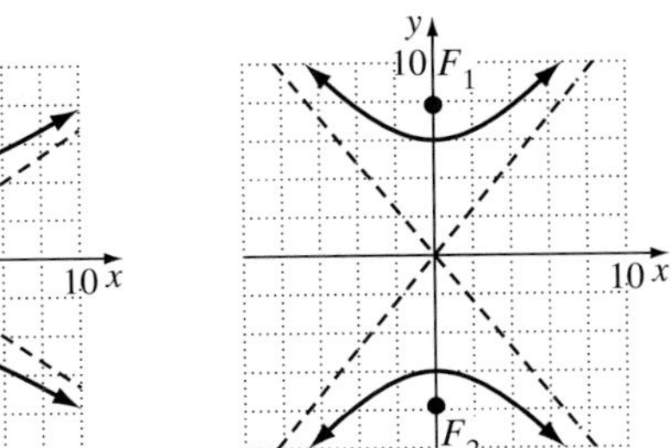

17. foci: $(\pm\sqrt{13}, 0)$

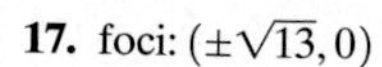

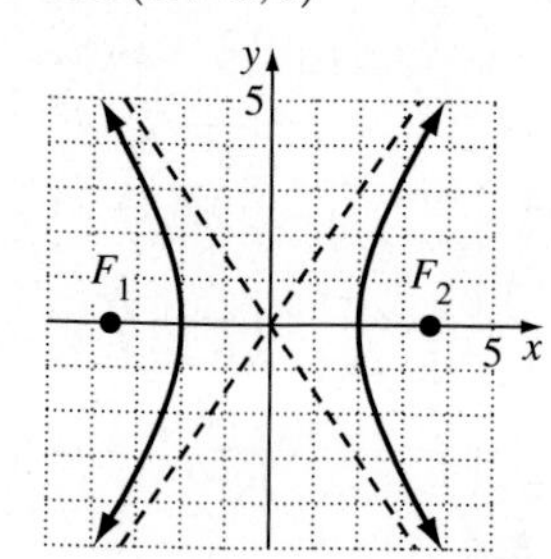

19. foci: $(0, \pm\sqrt{34})$

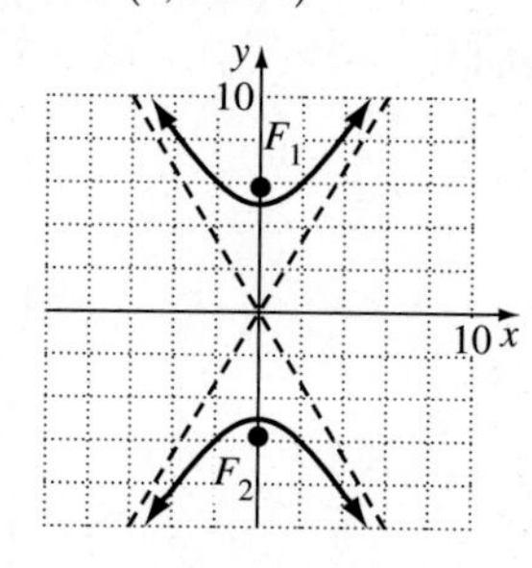

21. foci: $(\pm\sqrt{5}, 0)$

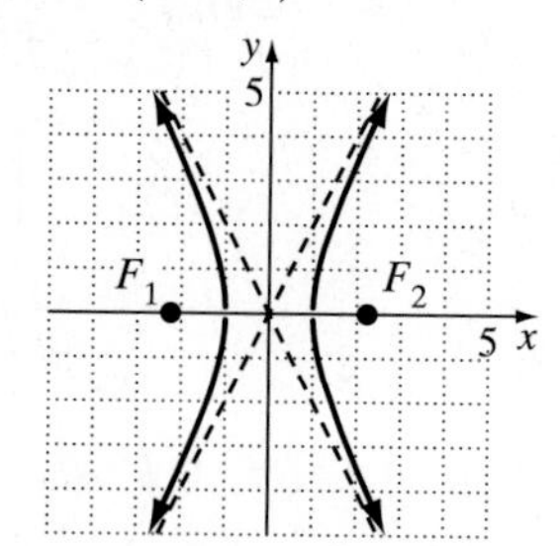

23. $\frac{x^2}{9} - \frac{y^2}{25} = 1$

25. $\frac{y^2}{4} - \frac{x^2}{9} = 1$

27. foci: $(-9, -3), (1, -3)$

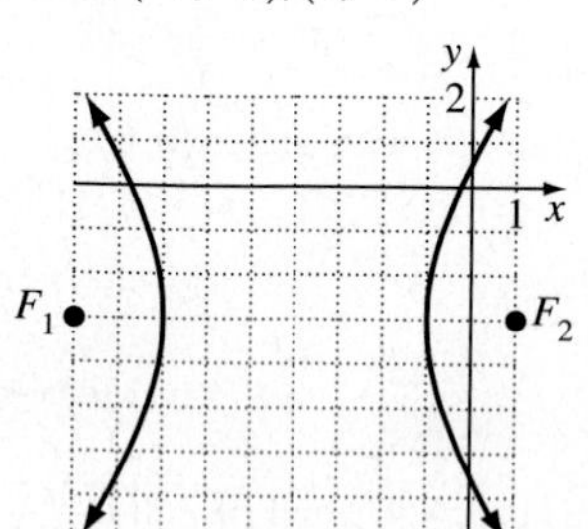

29. foci: $(-3 \pm \sqrt{41}, 0)$

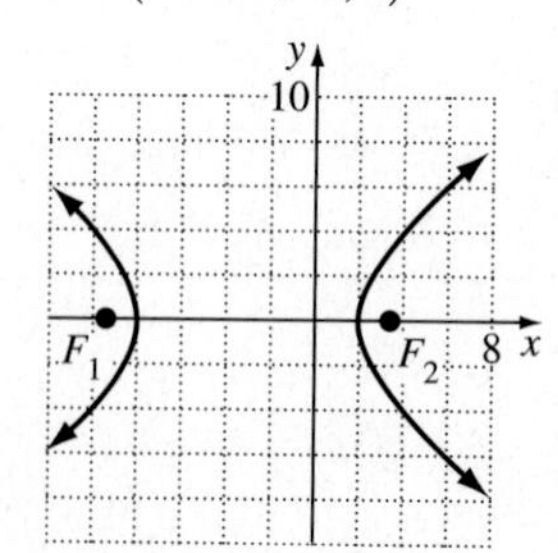

31. foci: $(1, -2 \pm 2\sqrt{5})$

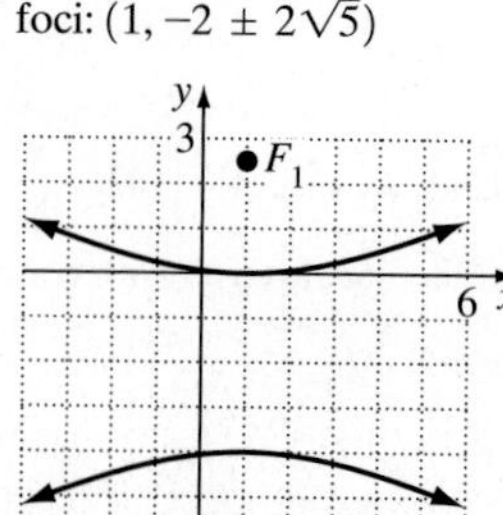

33. foci: $(3 \pm \sqrt{5}, -3)$

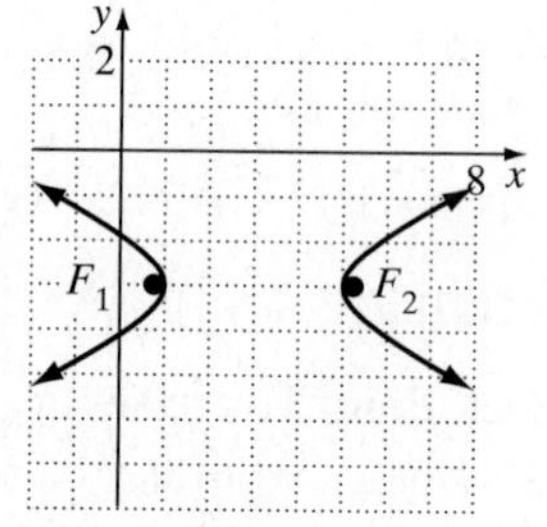

35. foci: $(1 \pm 2\sqrt{2}, 2)$

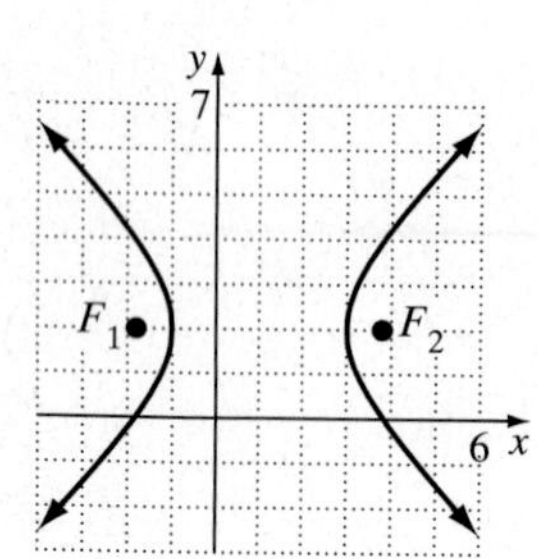

37. $(x - 1)^2 - (y + 2)^2 = 1$

foci: $(1 \pm \sqrt{2}, -2)$

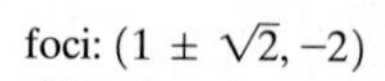

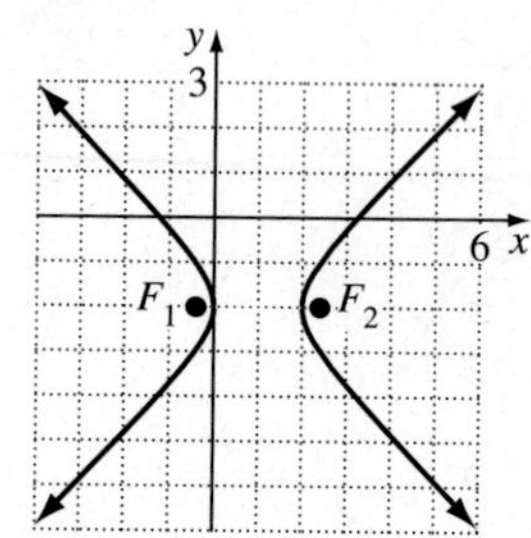

39. $\frac{(y + 1)^2}{4} - \frac{(x + 2)^2}{0.25} = 1$

foci: $(-2, -1 \pm \sqrt{4.25})$

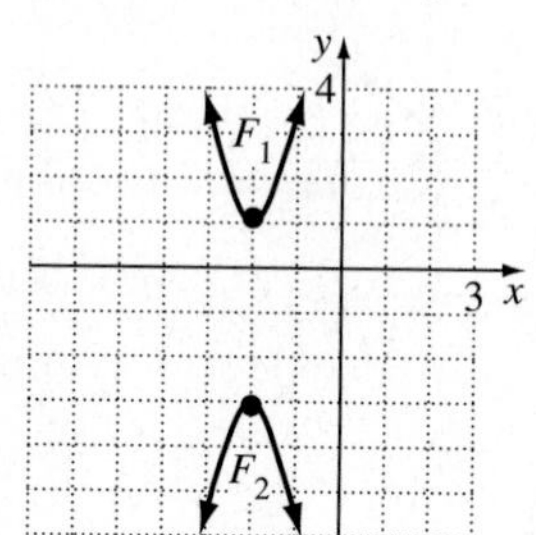

41. $\frac{(x - 2)^2}{9} - \frac{(y - 3)^2}{4} = 1$

foci: $(2 \pm \sqrt{13}, 3)$

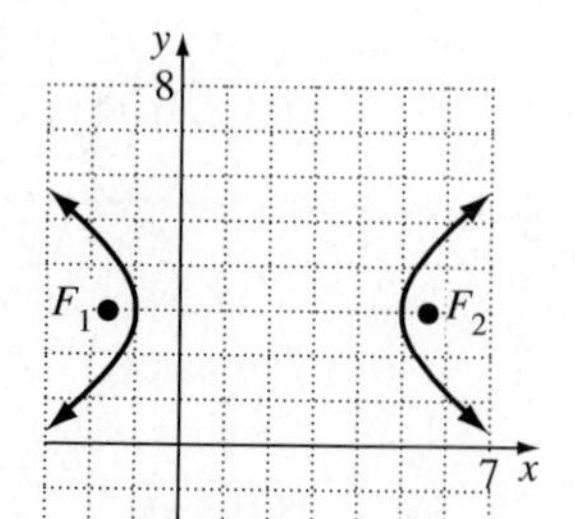

43. $\frac{y^2}{4} - \frac{(x - 4)^2}{25} = 1$

foci: $(4, \pm\sqrt{29})$

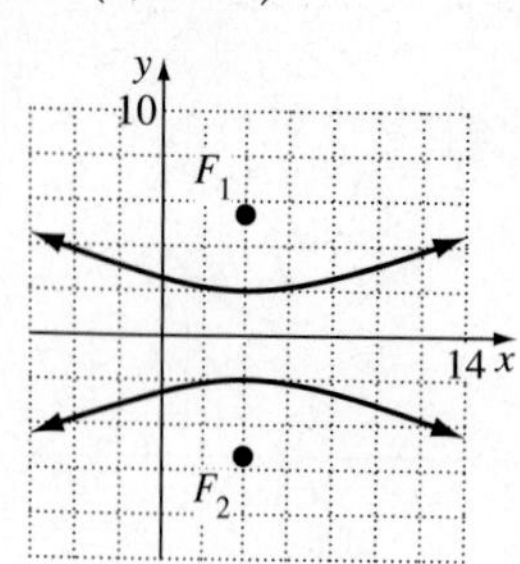

45. If M_1 is located 2640 feet to the right of the origin on the x-axis, the explosion is located on the right branch of the hyperbola given by the equation $\frac{x^2}{1{,}210{,}000} - \frac{y^2}{5{,}759{,}600} = 1$.

47. 40 yd **59.**

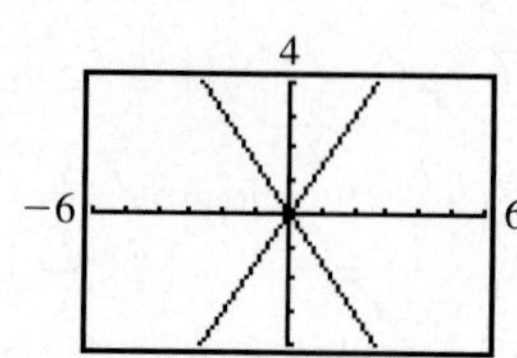

; No. Two intersecting lines.

61. $2y^2 + (10 - 6x)y + (4x^2 - 3x - 6) = 0$

$$y = \frac{3x - 5 \pm \sqrt{x^2 - 24x + 37}}{2}$$

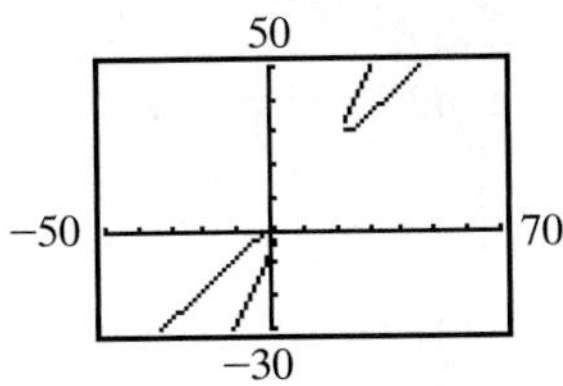

The xy-term rotates the hyperbola.

63. (c) is true.

65. $\frac{y^2}{36} - \frac{(x-5)^2}{20} = 1$

Section 7.3

Check Point Exercises

1. focus: $(2, 0)$
directrix: $x = -2$

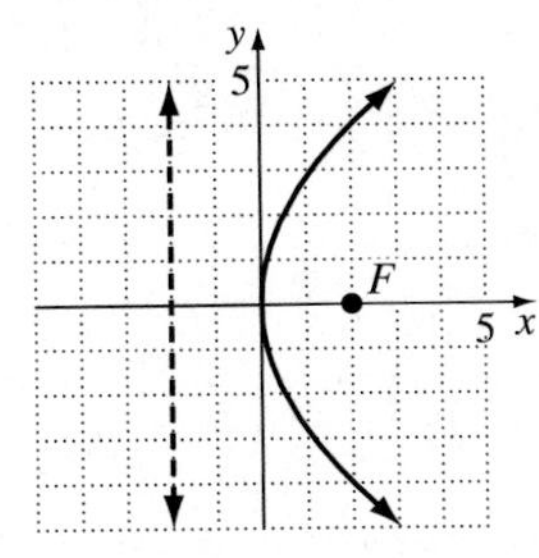

2. focus: $(0, -3)$
directrix: $y = 3$

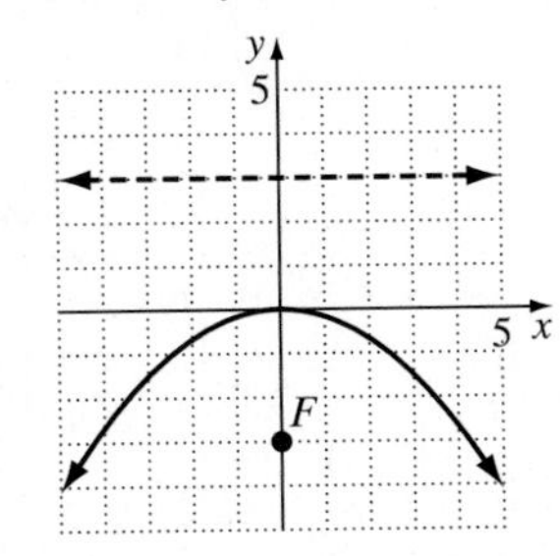

3. $y^2 = 32x$

4. vertex: $(2, -1)$; focus: $(2, 0)$;
directrix: $y = -2$

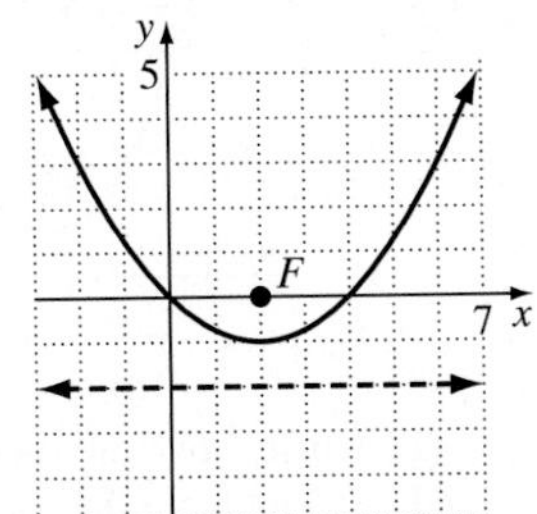

5. vertex: $(2, -1)$; focus: $(1, -1)$;
directrix: $x = 3$

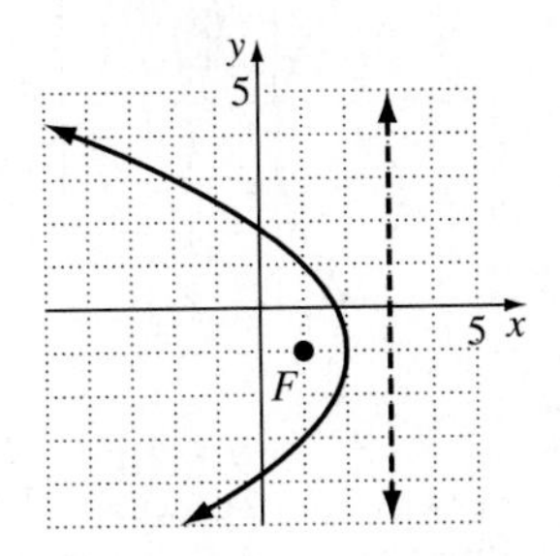

6. $x^2 = \frac{9}{4}y$; The light should be placed at $\left(0, \frac{9}{16}\right)$, or $\frac{9}{16}$ inch above the vertex.

Exercise Set 7.3

1. focus: $(1, 0)$; directrix: $x = -1$; graph (c)

3. focus: $(0, -1)$, directrix: $y = 1$; graph (b)

5. focus: $(4, 0)$; directrix: $x = -4$

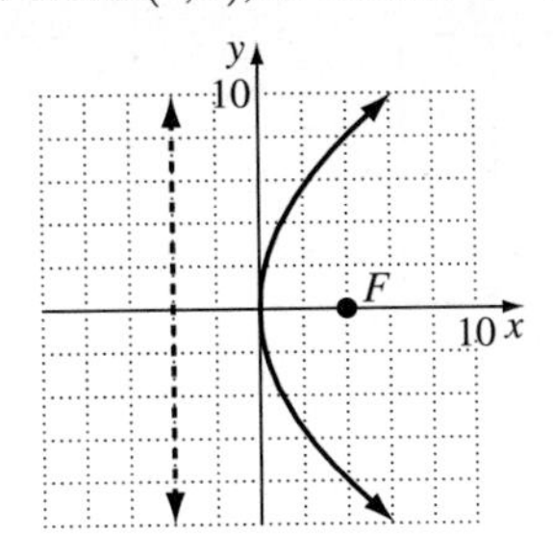

7. focus: $(-2, 0)$; directrix: $x = 2$

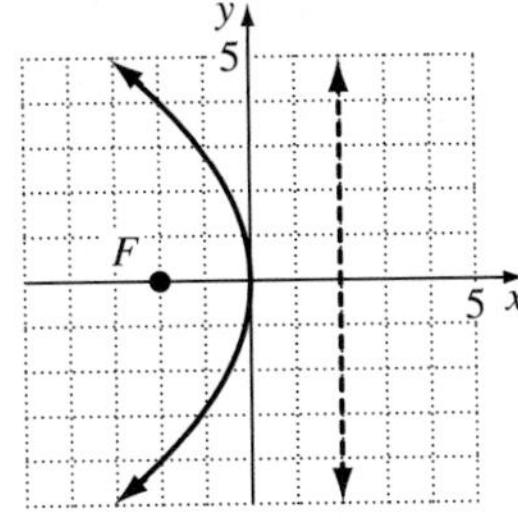

9. focus: $(0, 3)$; directrix: $y = -3$

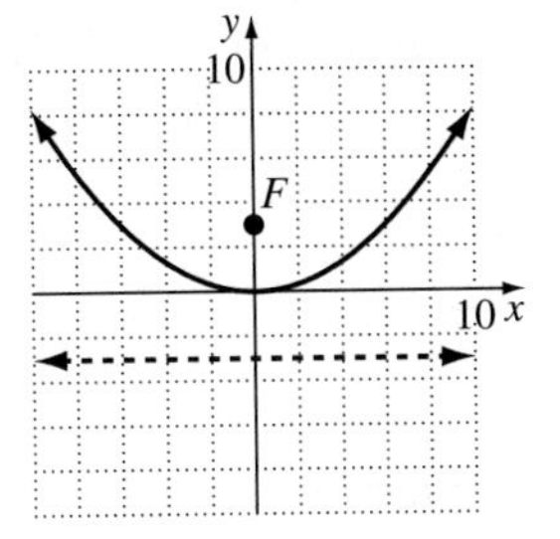

11. focus: $(0, -4)$; directrix: $y = 4$

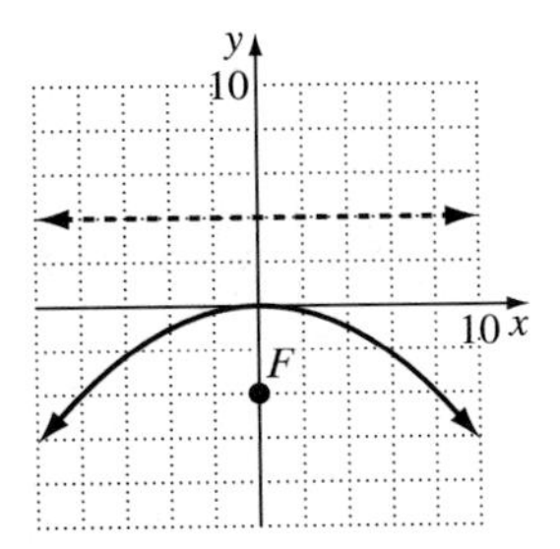

13. focus: $\left(\frac{3}{2}, 0\right)$; directrix: $x = -\frac{3}{2}$

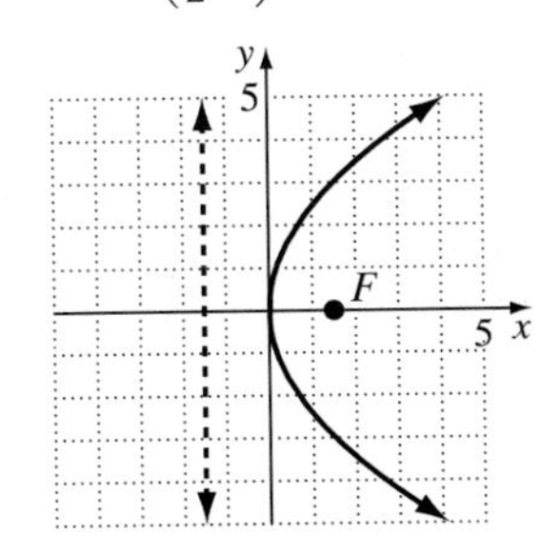

15. $y^2 = 28x$

17. $y^2 = -20x$

19. $x^2 = 60y$

21. $x^2 = -100y$

23. vertex: $(1, 1)$; focus: $(2, 1)$;
directrix: $x = 0$; graph (c)

25. vertex: $(-1, -1)$; focus: $(-1, -2)$;
directrix: $y = 0$; graph (d)

27. vertex: $(2, 1)$; focus: $(2, 3)$; directrix: $y = -1$

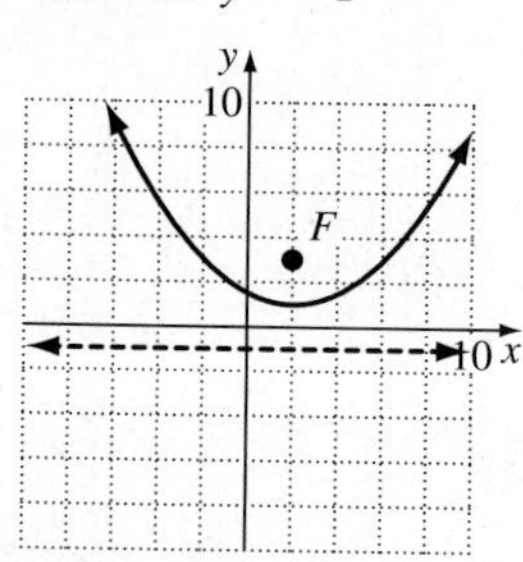

29. vertex: $(-1, -1)$; focus: $(-1, -3)$; directrix: $y = 1$

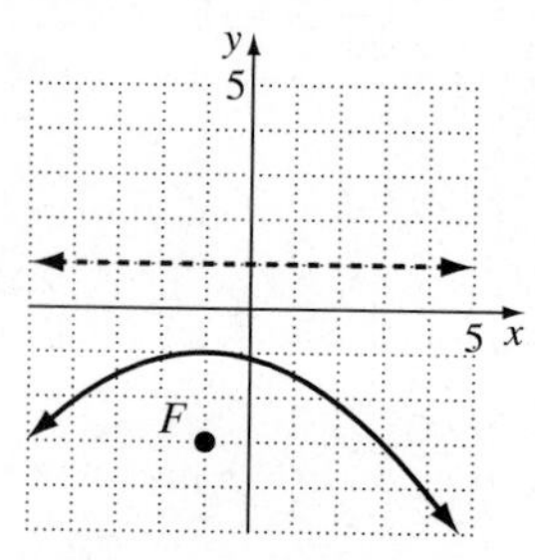

31. vertex: $(-1, -3)$; focus: $(2, -3)$; directrix: $x = -4$

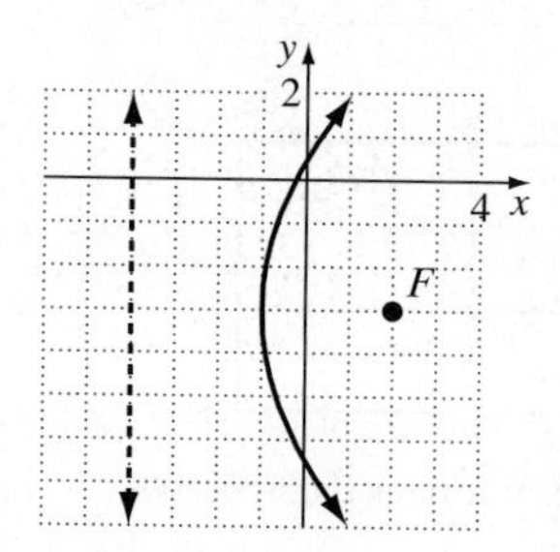

33. vertex: $(0, -1)$; focus: $(-2, -1)$; directrix: $x = 2$

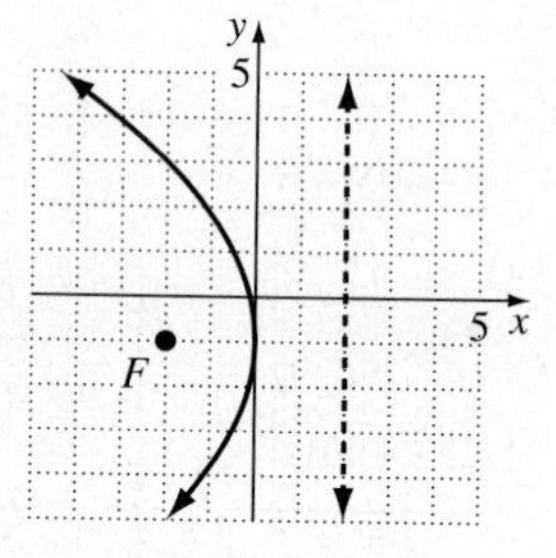

35. $(x - 1)^2 = 4(y - 2)$; vertex: $(1, 2)$; focus: $(1, 3)$; directrix: $y = 1$

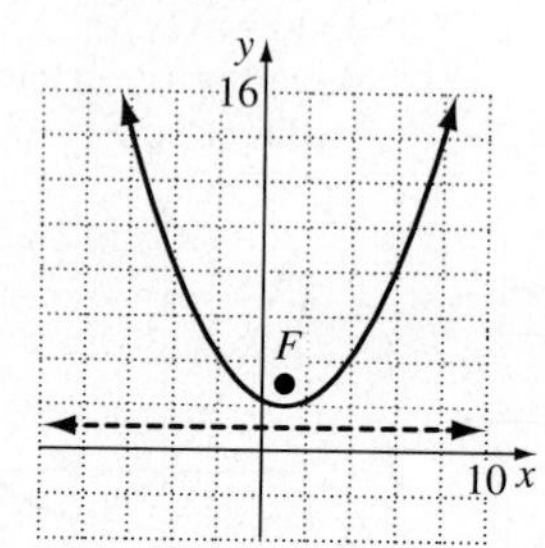

37. $(y - 1)^2 = -12(x - 3)$; vertex: $(3, 1)$; focus: $(0, 1)$; directrix: $x = 6$

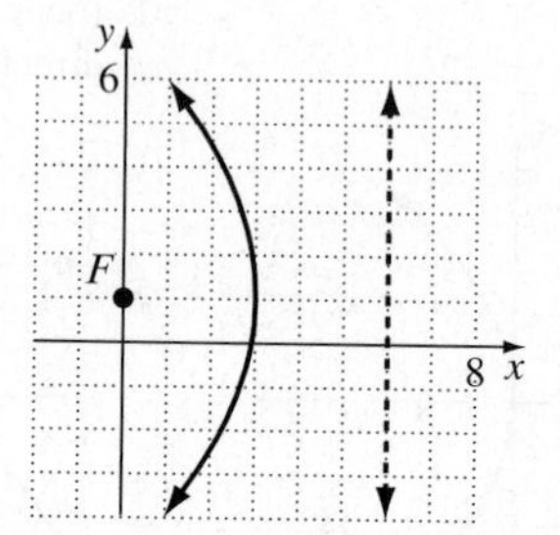

39. $(x + 3)^2 = 4(y + 2)$; vertex: $(-3, -2)$; focus: $(-3, -1)$; directrix: $y = -3$

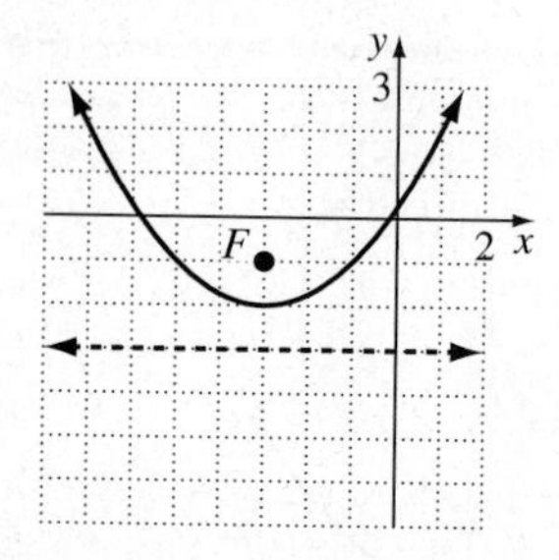

41. 1 inch above the vertex.

43. 4.5 feet from the base of the dish.

45. 75.625 m

47. yes

57. $y = -5 \pm \sqrt{x}$

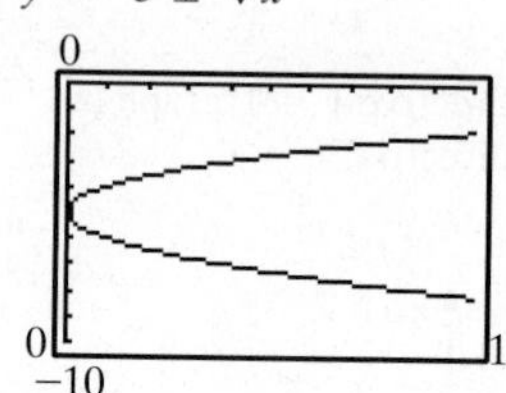
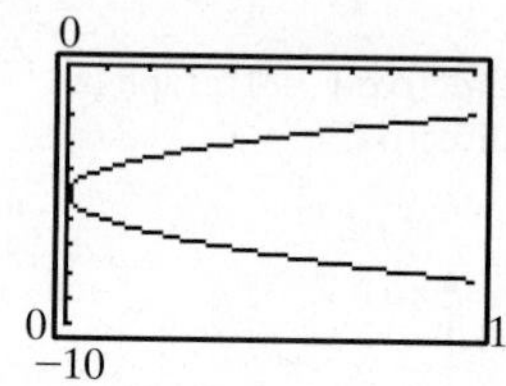

59. $3y^2 + (2\sqrt{3}x - 8)y + x^2 + 8\sqrt{3}x + 32 = 0$

$$y = \frac{-\sqrt{3}x + 4 \pm 4\sqrt{-2\sqrt{3}x - 5}}{3}$$

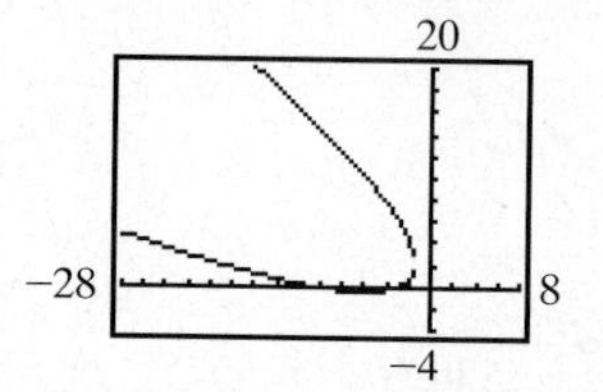

61. $\frac{25}{6}$ ft

Chapter 7 Review Exercises

1. foci: $(\pm\sqrt{11}, 0)$

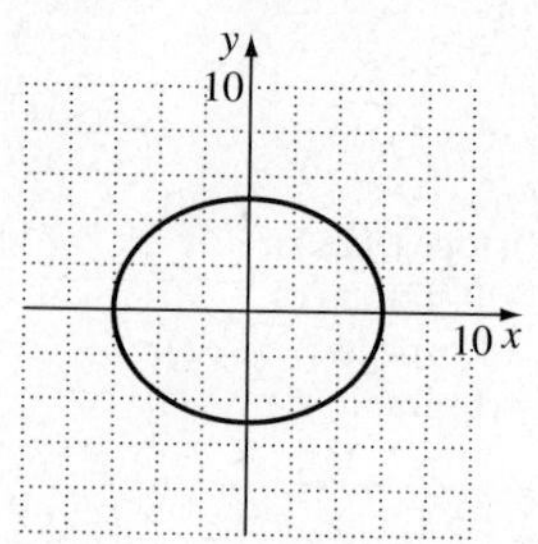

2. foci: $(0, \pm 3)$

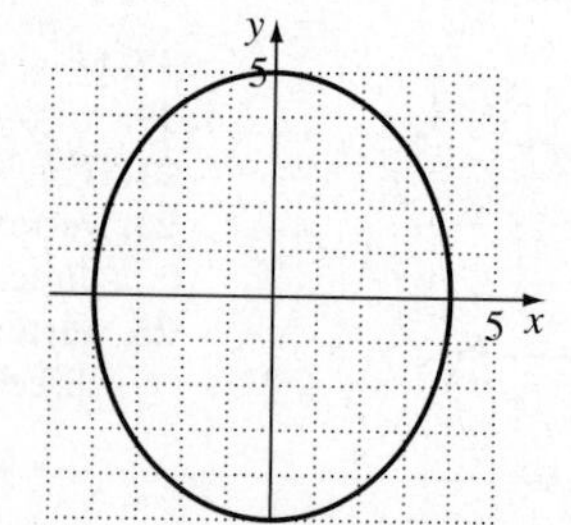

3. foci: $(0, \pm 2\sqrt{3})$

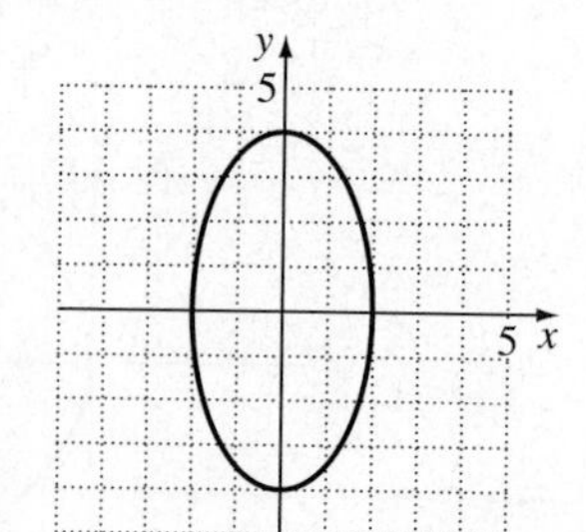

4. foci: $(\pm\sqrt{5}, 0)$

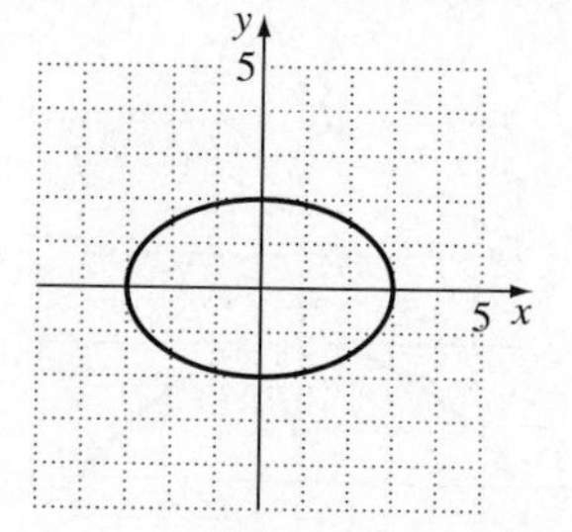

5. foci: $(1 \pm \sqrt{7}, -2)$

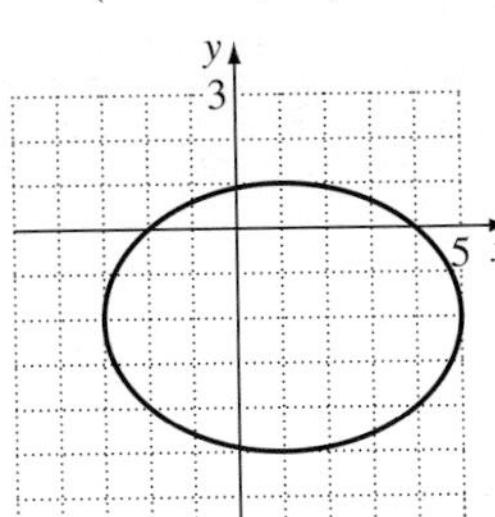

6. foci: $(-1, 2 \pm \sqrt{7})$

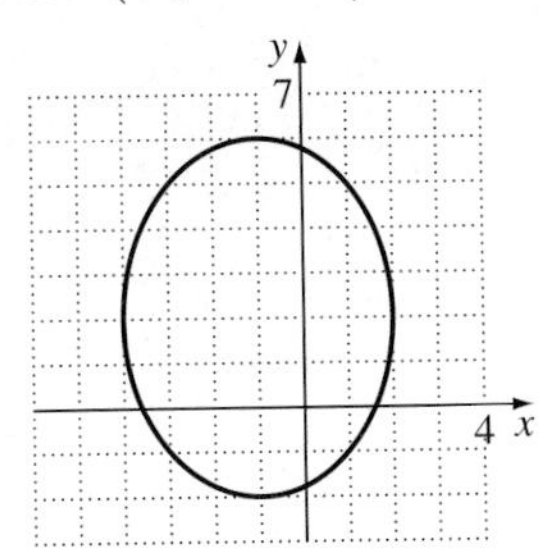

7. foci: $(-3 \pm \sqrt{5}, 2)$

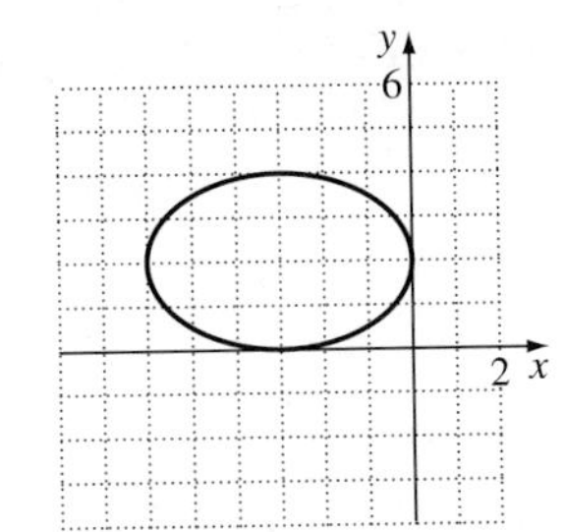

8. foci: $(1, -1 \pm \sqrt{5})$

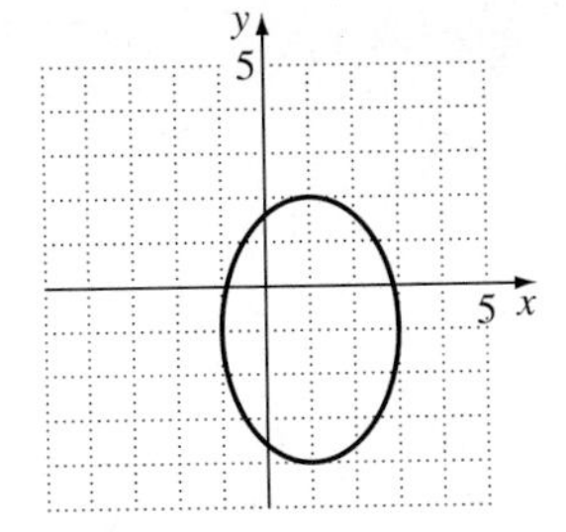

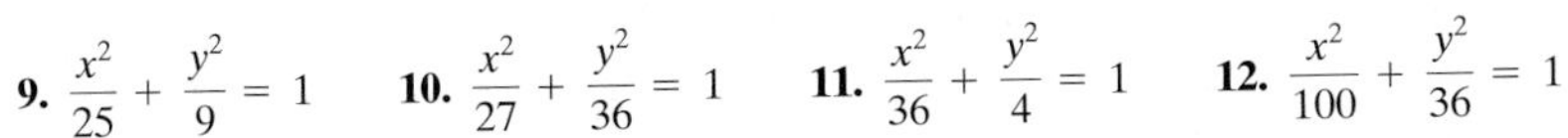

9. $\frac{x^2}{25} + \frac{y^2}{9} = 1$ **10.** $\frac{x^2}{27} + \frac{y^2}{36} = 1$ **11.** $\frac{x^2}{36} + \frac{y^2}{4} = 1$ **12.** $\frac{x^2}{100} + \frac{y^2}{36} = 1$ **13.** yes

14. The hit ball will collide with the other ball.

15. foci: $(\pm\sqrt{17}, 0)$

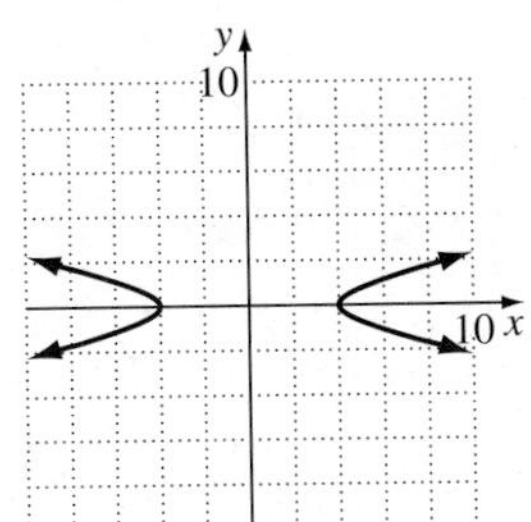

16. foci: $(0, \pm\sqrt{17})$

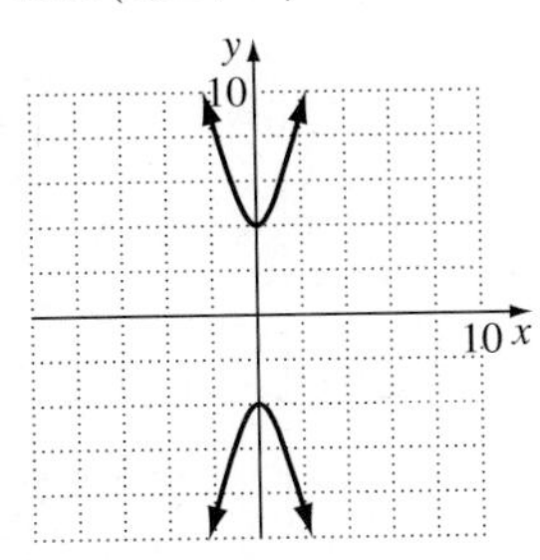

17. foci: $(\pm 5, 0)$

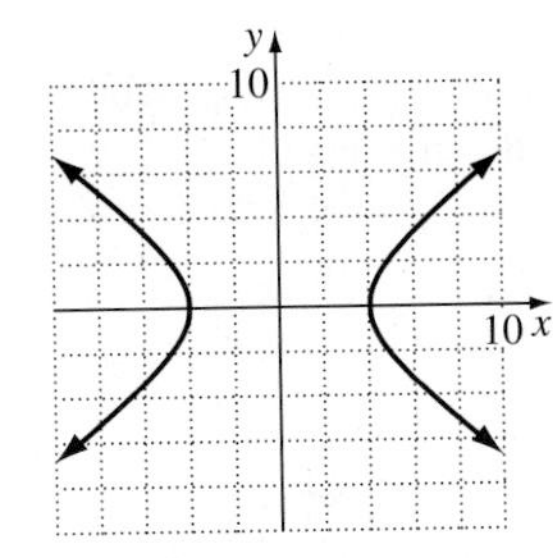

18. foci: $(0, \pm 2\sqrt{5})$

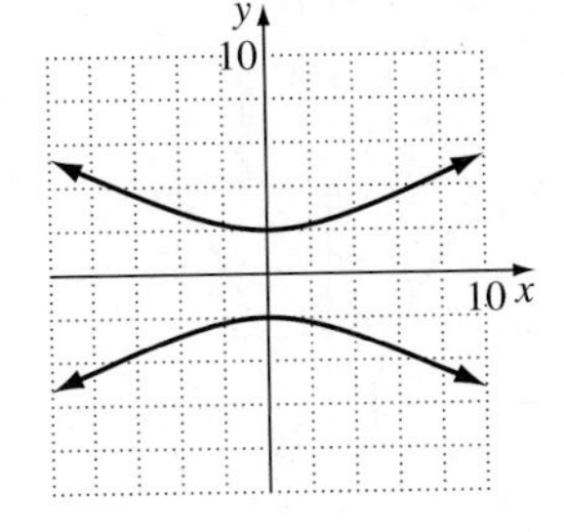

19. foci: $(2 \pm \sqrt{41}, -3)$

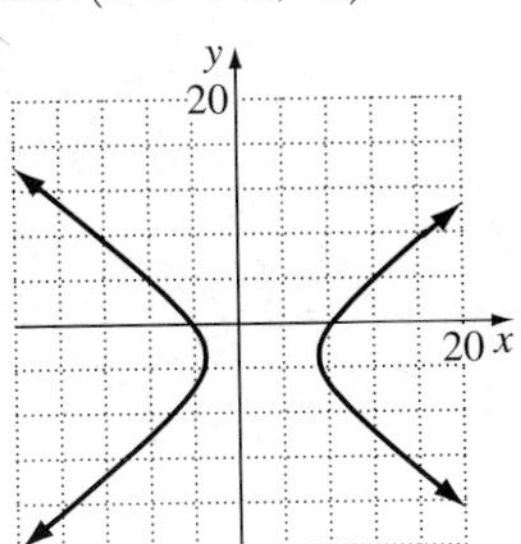

20. foci: $(3, -2 \pm \sqrt{41})$

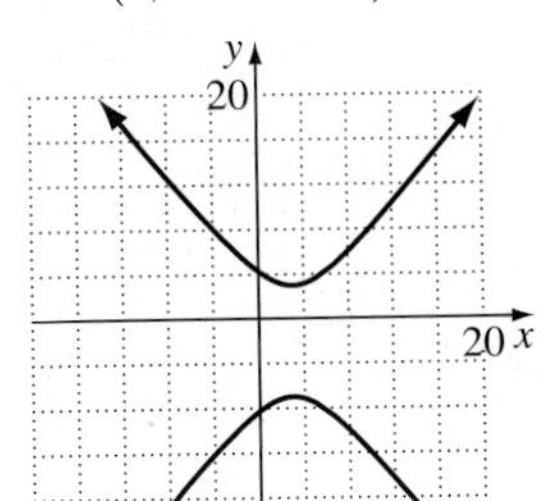

21. foci: $(1, 2 \pm \sqrt{5})$

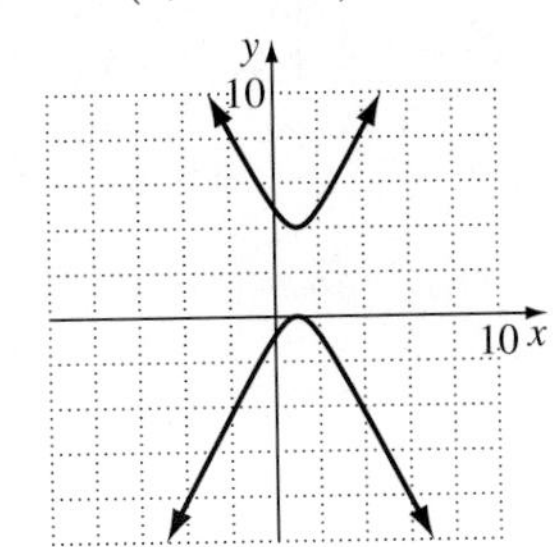

22. foci: $(1 \pm \sqrt{2}, -1)$

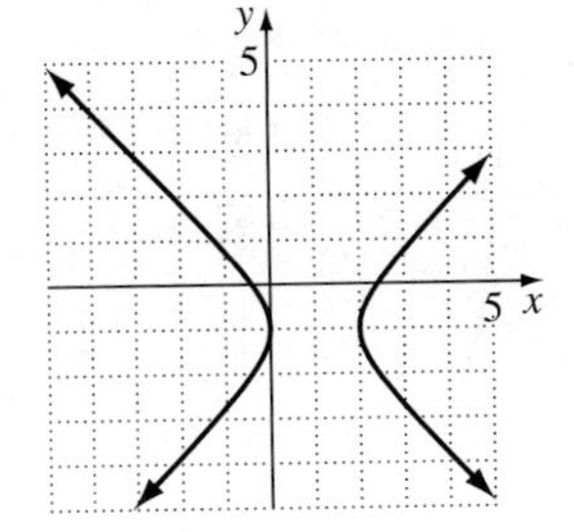

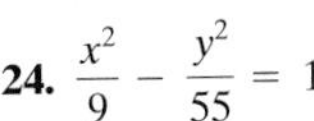

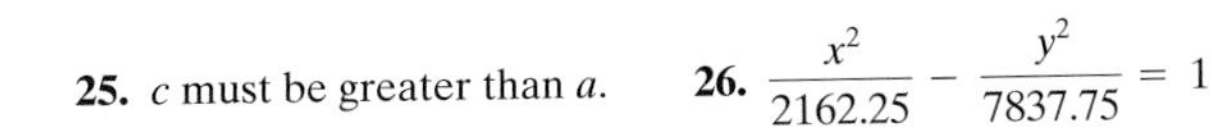

23. $\frac{y^2}{4} - \frac{x^2}{12} = 1$ **24.** $\frac{x^2}{9} - \frac{y^2}{55} = 1$ **25.** c must be greater than a. **26.** $\frac{x^2}{2162.25} - \frac{y^2}{7837.75} = 1$

27. vertex: $(0, 0)$; focus: $(2, 0)$; directrix: $x = -2$

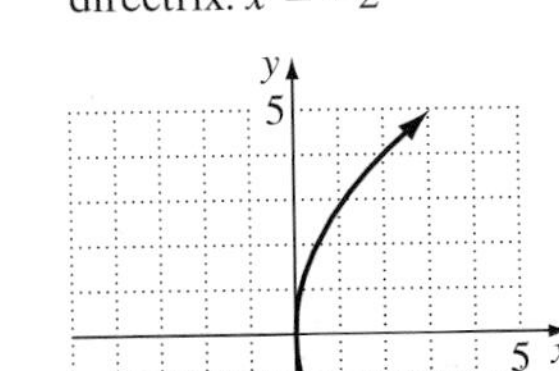

28. vertex: $(0, 0)$; focus: $(0, -4)$; directrix: $y = 4$

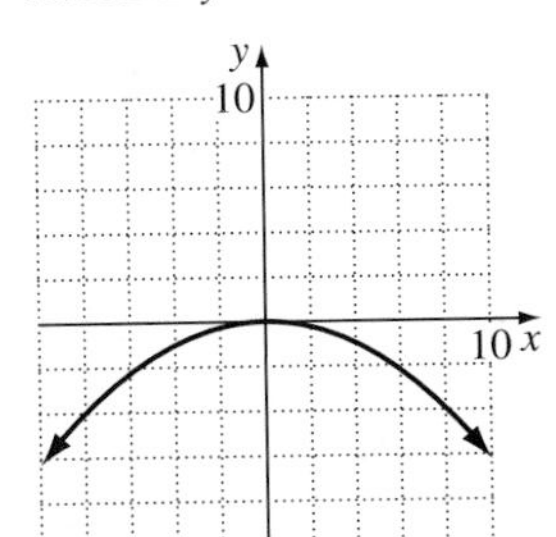

29. vertex: $(0, 2)$; focus: $(-4, 2)$; directrix: $x = 4$

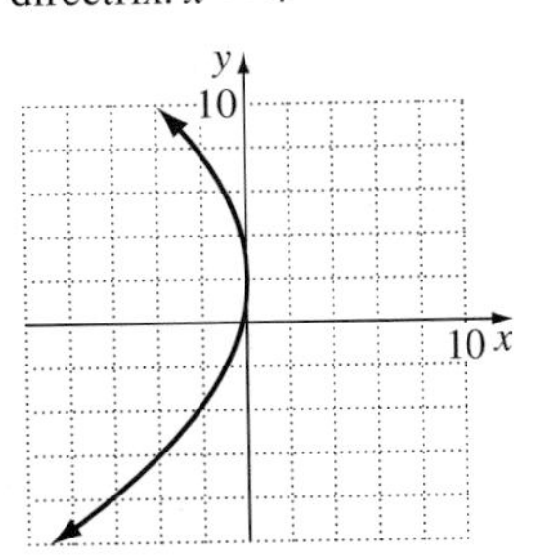

30. vertex: $(4, -1)$; focus: $(4, 0)$; directrix: $y = -2$

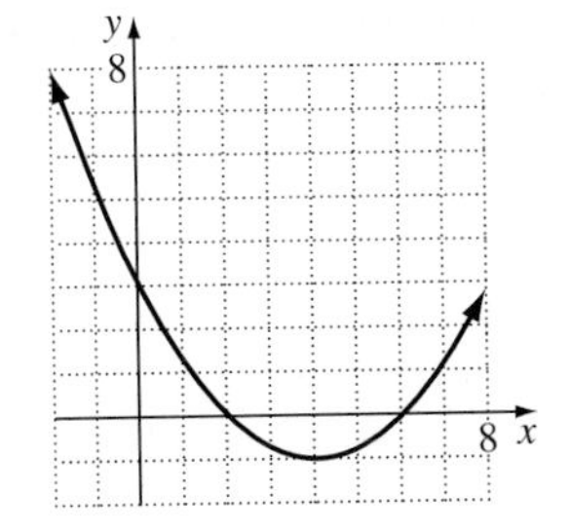

31. vertex: $(0, 1)$; focus: $(0, 0)$; directrix: $y = 2$

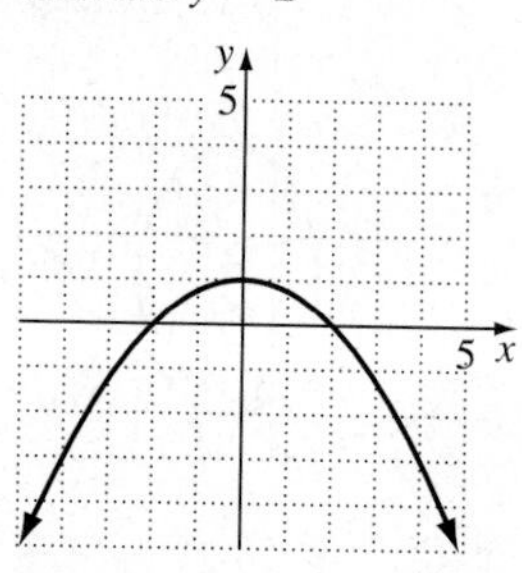

32. vertex: $(-1, 5)$; focus: $(0, 5)$; directrix: $x = -2$

33. vertex: $(2, -2)$; focus: $\left(2, -\frac{3}{2}\right)$; directrix: $y = -\frac{5}{2}$

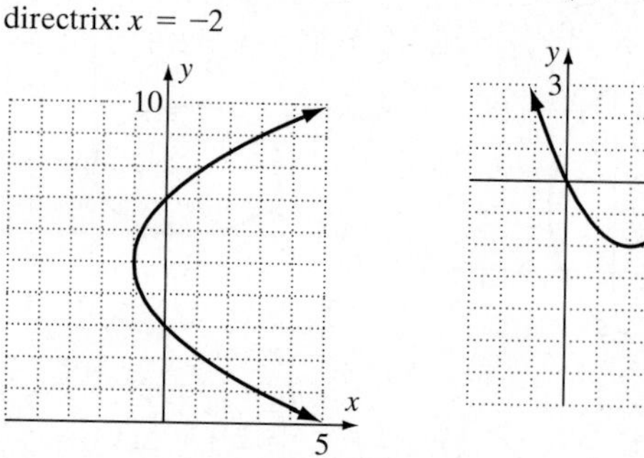

34. $y^2 = 48x$ **35.** $x^2 = -44y$ **36.** $x^2 = 12y$; Place the light 3 inches from the vertex at $(0, 3)$. **37.** approximately 58 ft **38.** approximately 128 ft

Chapter 7 Test

1. foci: $(\pm\sqrt{13}, 0)$

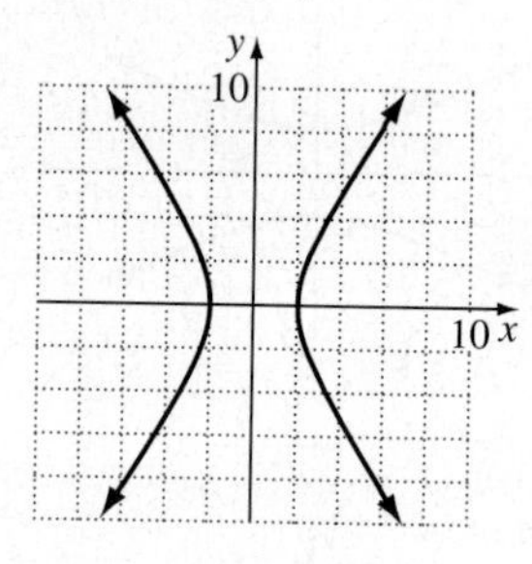

2. vertex: $(0, 0)$; focus: $(0, -2)$; directrix: $y = 2$

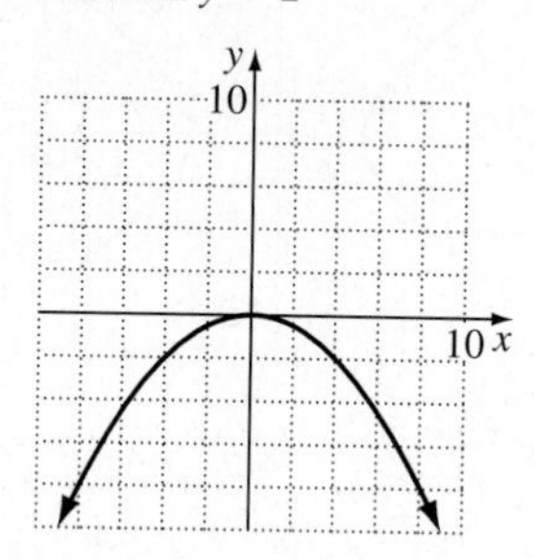

3. foci: $(-6, 5)$, $(2, 5)$

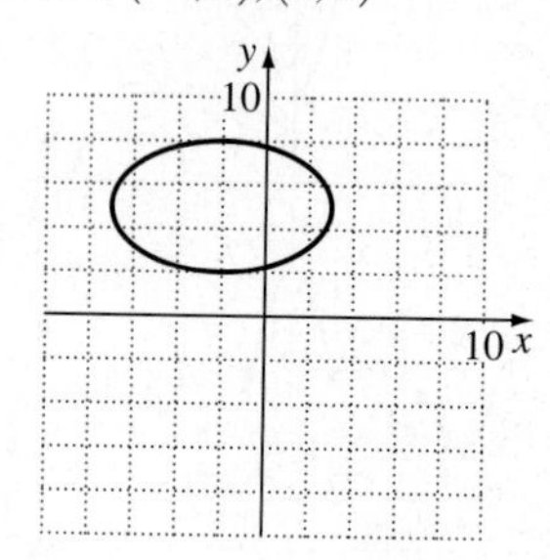

4. foci: $(-1, 1 \pm \sqrt{5})$

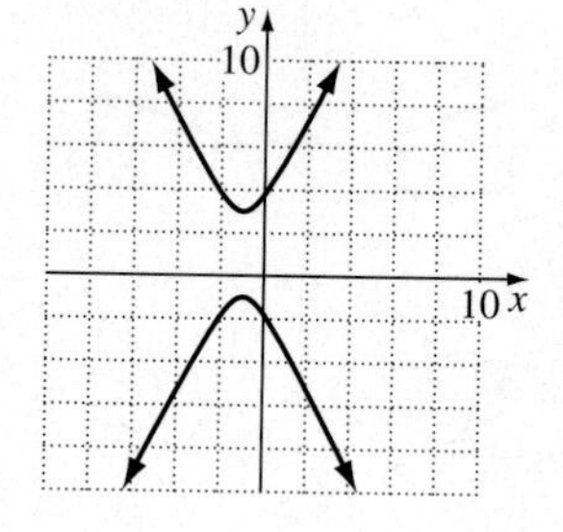

5. vertex: $(-5, 1)$; focus: $(-5, 3)$; directrix: $y = -1$

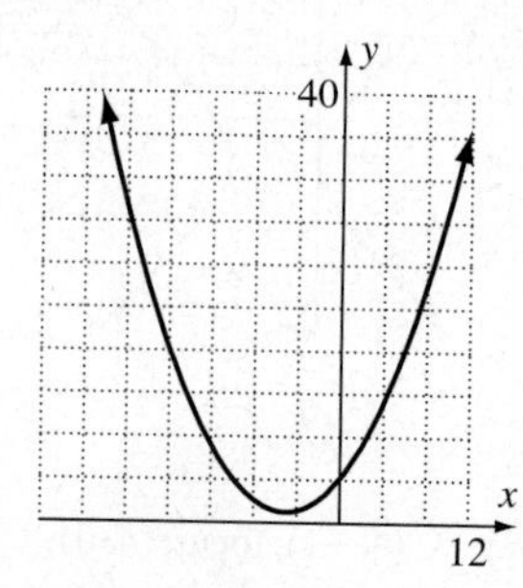

6. $\frac{x^2}{100} + \frac{y^2}{51} = 1$

7. $\frac{y^2}{49} - \frac{x^2}{51} = 1$

8. $y^2 = 200x$

9. 32 ft

10. a. $x^2 = 3y$

b. Light is placed $\frac{3}{4}$ inch above the vertex.

Cumulative Review Exercises (Chapters 1–7)

1. $x = 2$ **2.** $x < 2$ **3.** $x = 9$ **4.** $x = 2 \pm 2\sqrt{5}$ **5.** $x \geq 4$ or $x \leq -3$ **6.** $\left\{\frac{2}{3}, -1 \pm \sqrt{2}\right\}$

7. $\{3\}$ **8.** $(2, -1)$

9. $(2, -4)$ and $(-14, -20)$ **10.** $(7, -4, 6)$

11.

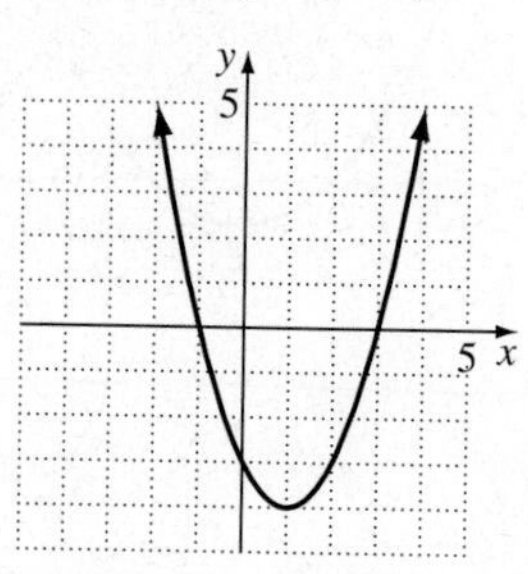

12.

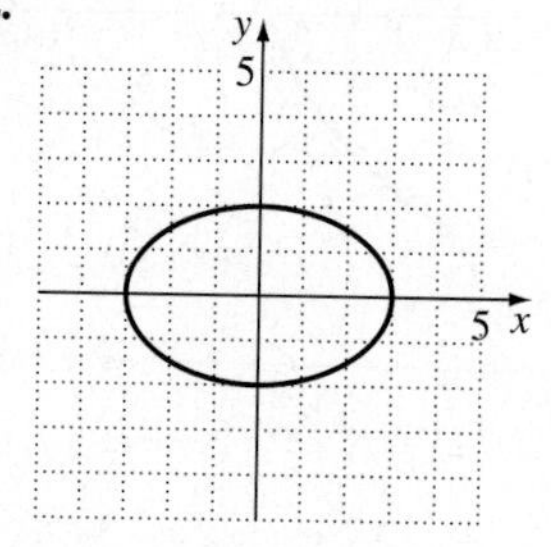

13.

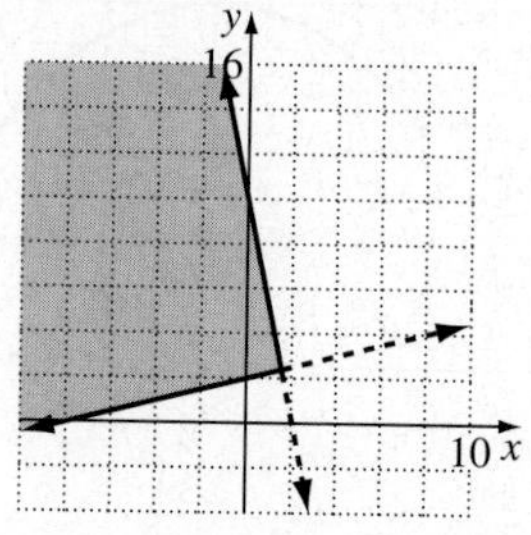

14. a. $\pm1, \pm3, \pm\frac{1}{2}, \pm\frac{3}{2}, \pm\frac{1}{4}, \pm\frac{3}{4}, \pm\frac{1}{8}, \pm\frac{3}{8}, \pm\frac{1}{16}, \pm\frac{3}{16}, \pm\frac{1}{32}, \pm\frac{3}{32}$ **b.** $\left\{-\frac{1}{8}, \frac{3}{4}, 1\right\}$
15. a. 1980–1991 **b.** 1991–2025 **c.** 1950–1980 **d.** $f(x) = 98$ **e.** The scale is not uniformly spaced.
16. $(g \circ f)(x) = x^2 - 2$ **17.** $3 \log_5 x + \frac{1}{2} \log_5 y - 3$ **18.** $y = -2x - 2$
19. The costs will be the same when the number of miles driven is 175 miles. The cost will be \$67.
20. Richard Nixon lived 81 years, Thomas Jefferson lived 83 years, and James Madison lived 85 years.

CHAPTER 8

Section 8.1

Check Point Exercises

1. a. 7, 9, 11, 13 **b.** $-\frac{1}{3}, \frac{1}{5}, -\frac{1}{9}, \frac{1}{17}$ **2.** 3, 11, 27, 59 **3.** $10, \frac{10}{3}, \frac{5}{6}, \frac{1}{6}$ **4. a.** 91 **b.** n **5. a.** 182 **b.** 47 **c.** 20
6. a. $\sum_{i=1}^{9} i^2$ **b.** $\sum_{i=1}^{n} \frac{1}{2^{i-1}}$

Exercise Set 8.1

1. 5, 8, 11, 14 **3.** 3, 9, 27, 81 **5.** −3, 9, −27, 81 **7.** −4, 5, −6, 7 **9.** $\frac{2}{5}, \frac{2}{3}, \frac{6}{7}, 1$ **11.** $1, -\frac{1}{3}, \frac{1}{7}, -\frac{1}{15}$ **13.** 7, 12, 17, 22
15. 3, 12, 48, 192 **17.** 4, 11, 25, 53 **19.** $1, 2, \frac{3}{2}, \frac{2}{3}$ **21.** 4, 12, 48, 240 **23.** 272 **25.** 120 **27.** $(n + 2)(n + 1)$ **29.** 105
31. 60 **33.** 115 **35.** $-\frac{5}{16}$ **37.** 55 **39.** $\frac{3}{8}$ **41.** 15 **43.** $\sum_{i=1}^{15} i^2$ **45.** $\sum_{i=1}^{11} 2^i$ **47.** $\sum_{i=1}^{30} i$ **49.** $\sum_{i=1}^{14} \frac{i}{i+1}$ **51.** $\sum_{i=1}^{n} \frac{4^i}{i}$
53. $\sum_{i=1}^{n} (2i - 1)$ **55.** $\sum_{k=1}^{14} (2k + 3)$ **57.** $\sum_{k=0}^{12} ar^k$ **59.** $\sum_{k=0}^{n} (a + kd)$ **61. a.** 4613; the total number of children, in thousands, home-educated in the years 1992–1997 **b.** 769; the average number of children, in thousands, home-educated each year
63. 30.15; Americans spent \$30.15 billion on recreational boating from 1991 through 1995. **65.** \$8081.13 **75.** 39,800
77. 8,109,673,360,588,800 **79.** 24,804

83.

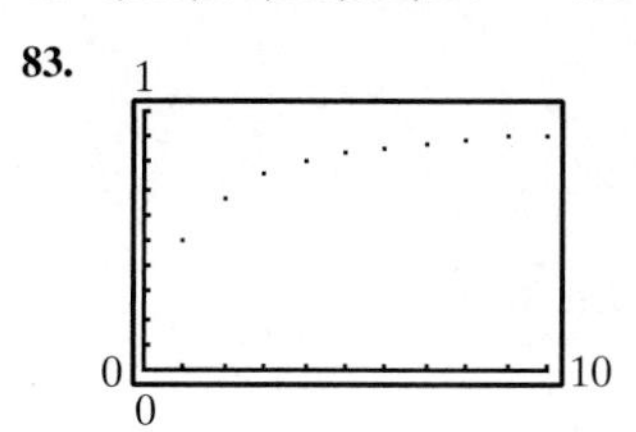

As n gets larger, a_n approaches 1.

85.

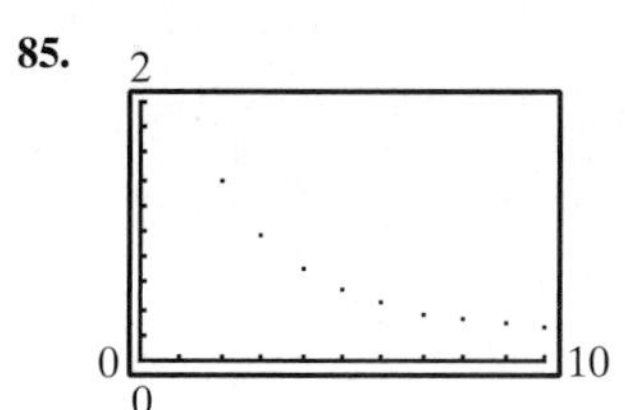

As n gets larger a_n approaches 0.

87. (b) is true.

Section 8.2

Check Point Exercises

1. 100, 70, 40, 10, −20 **2.** −34 **3. a.** $a_n = 2350n + 10{,}458$ **b.** \$73,908 million **4.** 360 **5.** 2460 **6.** \$596,300

Exercise Set 8.2

1. 200, 220, 240, 260, 280, 300 **3.** −7, −3, 1, 5, 9, 13 **5.** 300, 210, 120, 30, −60, −150 **7.** $\frac{5}{2}, 2, \frac{3}{2}, 1, \frac{1}{2}, 0$ **9.** −9, −3, 3, 9, 15, 21
11. 30, 20, 10, 0, −10, −20 **13.** 1.6, 1.2, 0.8, 0.4, 0, −0.4 **15.** 33 **17.** 252 **19.** 955 **21.** −142 **23.** $a_n = 4n - 3; a_{20} = 77$
25. $a_n = 11 - 4n; a_{20} = -69$ **27.** $a_n = 7 + 2n; a_{20} = 47$ **29.** $a_n = -16 - 4n; a_{20} = -96$ **31.** $a_n = 1 + 3n; a_{20} = 61$
33. $a_n = 40 - 10n; a_{20} = -160$ **35.** 1220 **37.** 4400 **39.** 5050 **41.** 3660 **43.** 396 **45.** $8 + 13 + 18 + \cdots + 88; 816$
47. $2 - 1 - 4 - \cdots - 85; -1245$ **49.** $4 + 8 + 12 + \cdots + 400; 20{,}200$ **51. a.** $a_n = 125{,}159 + 1265n$ **b.** 145,399 thousand
53. Company A will pay \$1400 more. **55. a.** $a_n = 3.204 + 0.576n$ **b.** 627.3 million tons **57.** \$442,500 **59.** 1430 seats
69. the 200th term **71.** $S_n = \frac{n}{2}(1 + 2n - 1) = \frac{n}{2}(2n) = n^2$

Section 8.3

Check Point Exercises

1. $12, 6, 3, \frac{3}{2}, \frac{3}{4}, \frac{3}{8}$ **2.** 3645 **3.** $a_n = 3(2)^{n-1}$; 384 **4.** 9842 **5.** 19,680 **6.** \$2,371,746 **7.** \$1,327,778 **8.** 9 **9.** 1 **10.** \$4000

Exercise Set 8.3

1. 5, 15, 45, 135, 405 **3.** $20, 10, 5, \frac{5}{2}, \frac{5}{4}$ **5.** 10, −40, 160, −640, 2560 **7.** −6, 30, −150, 750, −3750 **9.** $a_8 = 768$ **11.** $a_{12} = -10{,}240$ **13.** $a_{40} \approx -0.000000002$ **15.** $a_8 = 0.1$ **17.** $a_n = 3(4)^{n-1}$; $a_7 = 12{,}288$ **19.** $a_n = 18\left(\frac{1}{3}\right)^{n-1}$; $a_7 = \frac{2}{81}$ **21.** $a_n = 1.5(-2)^{n-1}$; $a_7 = 96$ **23.** $a_n = 0.0004(-10)^{n-1}$; $a_7 = 400$ **25.** 531,440 **27.** 2049 **29.** $\frac{16{,}383}{2}$ **31.** 9840 **33.** 10,230 **35.** $\frac{63}{128}$ **37.** $\frac{3}{2}$ **39.** 4 **41.** $\frac{2}{3}$ **43.** $S_\infty \approx 6.15385$ **45.** $\frac{5}{9}$ **47.** $\frac{47}{99}$ **49.** $\frac{257}{999}$ **51.** arithmetic, $d = 1$ **53.** geometric, $r = 2$ **55.** neither **57.** \$16,384 **59.** \$3,795,957 **61. a.** 1.04, 1.04, 1.04; The population is increasing geometrically with $r = 1.04$. **b.** $a_n = 20.6(1.04)^{n-1}$ **c.** 30.49 million people **63.** \$32,767 **65.** \$793,582.90 **67.** 130.26 in. **69.** \$844,706.11 **71.** \$94,834.21 **73.** \$9 million **75.** $\frac{1}{3}$

87.

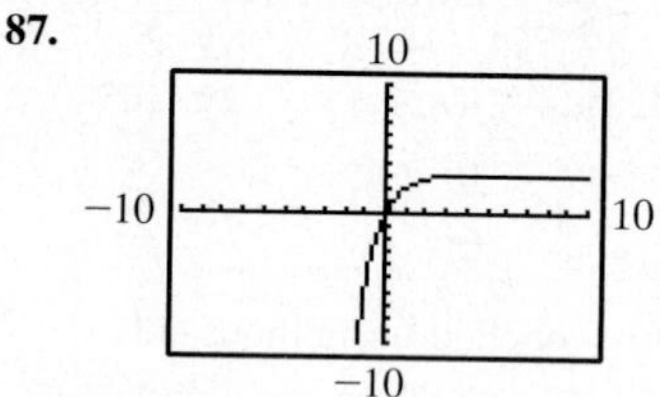

Horizontal asymptote at $y = 3$

$\sum_{n=0}^{\infty} 2\left(\frac{1}{3}\right)^n = 3$

89. (d) is true. **91.** \$442.38

Section 8.4

Check Point Exercises

1. a. $S_1: 2 = 1(1 + 1)$; $S_k: 2 + 4 + 6 + \cdots + 2k = k(k + 1)$; $S_{k+1}: 2 + 4 + 6 + \cdots + 2(k + 1) = (k + 1)(k + 2)$
b. $S_1: 1^3 = \frac{1^2(1 + 1)^2}{4}$; $S_k: 1^3 + 2^3 + 3^3 + \cdots + k^3 = \frac{k^2(k + 1)^2}{4}$; $S_{k+1}: 1^3 + 2^3 + 3^3 + \cdots + (k + 1)^3 = \frac{(k + 1)^2(k + 2)^2}{4}$
2. $S_1: 2 = 1(1 + 1)$; $S_k: 2 + 4 + 6 + \cdots + 2k = k(k + 1)$; $S_{k+1}: 2 + 4 + 6 + \cdots + 2k + 2(k + 1) = (k + 1)(k + 2)$; S_{k+1} can be obtained by adding $2k + 2$ to both sides of S_k.
3. $S_1: 1^3 = \frac{1^2(1 + 1)^2}{4}$; $S_k: 1^3 + 2^3 + 3^3 + \cdots + k^3 = \frac{k^2(k + 1)^2}{4}$; $S_{k+1}: 1^3 + 2^3 + 3^3 + \cdots + k^3 + (k + 1)^3 = \frac{(k + 1)^2(k + 2)^2}{4}$; S_{k+1} can be obtained by adding $k^3 + 3k^2 + 3k + 1$ to both sides of S_k.
4. S_1: 2 is a factor of $1^2 + 1$; S_k: 2 is a factor of $k^2 + k$; S_{k+1}: 2 is a factor of $(k + 1)^2 + (k + 1) = k^2 + 3k + 2$; S_{k+1} can be obtained from S_k by writing $k^2 + 3k + 2$ as $(k^2 + k) + 2(k + 1)$.

Exercise Set 8.4

1. $S_1: 1 = 1^2$; $S_2: 1 + 3 = 2^2$; $S_3: 1 + 3 + 5 = 3^2$ **3.** S_1: 2 is a factor of $1 - 1 = 0$; S_2: 2 is a factor of $2^2 - 2 = 2$; S_3: 2 is a factor of $3^2 - 3 = 6$ **5.** $S_k: 4 + 8 + 12 + \cdots + 4k = 2k(k + 1)$; $S_{k+1}: 4 + 8 + 12 + \cdots + (4k + 4) = 2(k + 1)(k + 2)$
7. $S_k: 3 + 7 + 11 + \cdots + (4k - 1) = k(2k + 1)$; $S_{k+1}: 3 + 7 + 11 + \cdots + (4k + 3) = (k + 1)(2k + 3)$
9. S_k: 2 is a factor of $k^2 - k + 2$; S_{k+1}: 2 is a factor of $k^2 + k + 2$
11. $S_1: 4 = 2(1)(1 + 1)$; $S_k: 4 + 8 + 12 + \cdots + 4k = 2k(k + 1)$; $S_{k+1}: 4 + 8 + 12 + \cdots + 4(k + 1) = 2(k + 1)(k + 2)$; S_{k+1} can be obtained by adding $4k + 4$ to both sides of S_k.
13. $S_1: 1 = 1^2$; $S_k: 1 + 3 + 5 + \cdots + (2k - 1) = k^2$; $S_{k+1}: 1 + 3 + 5 + \cdots + (2k + 1) = (k + 1)^2$; S_{k+1} can be obtained by adding $2k + 1$ to both sides of S_k.
15. $S_1: 3 = 1[2(1) + 1]$; $S_k: 3 + 7 + 11 + \cdots + (4k - 1) = k(2k + 1)$; $S_{k+1}: 3 + 7 + 11 + \cdots + (4k + 3) = (k + 1)(2k + 3)$; S_{k+1} can be obtained by adding $4k + 3$ to both sides of S_k.

17. $S_1: 1 = 2^1 - 1; S_k: 1 + 2 + 2^2 + \cdots + 2^{k-1} = 2^k - 1; S_{k+1}: 1 + 2 + 2^2 + \cdots + 2^k = 2^{k+1} - 1;$ S_{k+1} can be obtained by adding 2^k to both sides of S_k.
19. $S_1: 2 = 2^{1+1} - 2; S_k: 2 + 4 + 8 + \cdots + 2^k = 2^{k+1} - 2; S_{k+1}: 2 + 4 + 8 + \cdots + 2^{k+1} = 2^{k+2} - 2;$ S_{k+1} can be obtained by adding 2^{k+1} to both sides of S_k.
21. $S_1: 1 \cdot 2 = \dfrac{1(1+1)(1+2)}{3}; S_k: 1 \cdot 2 + 2 \cdot 3 + 3 \cdot 4 + \cdots + k(k+1) = \dfrac{k(k+1)(k+2)}{3};$ $S_{k+1}: 1 \cdot 2 + 2 \cdot 3 + 3 \cdot 4 + \cdots + (k+1)(k+2) = \dfrac{(k+1)(k+2)(k+3)}{3};$ S_{k+1} can be obtained by adding $(k+1)(k+2)$ to both sides of S_k.
23. $S_1: \dfrac{1}{1 \cdot 2} = \dfrac{1}{1+1}; S_k: \dfrac{1}{1 \cdot 2} + \dfrac{1}{2 \cdot 3} + \dfrac{1}{3 \cdot 4} + \cdots + \dfrac{1}{k(k+1)} = \dfrac{k}{k+1}; S_{k+1}: \dfrac{1}{1 \cdot 2} + \dfrac{1}{2 \cdot 3} + \dfrac{1}{3 \cdot 4} + \cdots + \dfrac{1}{(k+1)(k+2)} = \dfrac{k+1}{k+2};$ S_{k+1} can be obtained by adding $\dfrac{1}{(k+1)(k+2)}$ to both sides of S_k.
25. S_1: 2 is a factor of 0; S_k: 2 is a factor of $k^2 - k$; S_{k+1}: 2 is a factor of $k^2 + k$; S_{k+1} can be obtained from S_k by rewriting $k^2 + k$ as $(k^2 - k) + 2k$.
27. S_1: 6 is a factor of 6; S_k: 6 is a factor of $k(k+1)(k+2)$; S_{k+1}: 6 is a factor of $(k+1)(k+2)(k+3)$; S_{k+1} can be obtained from S_k by rewriting $(k+1)(k+2)(k+3)$ as $k(k+1)(k+2) + 3(k+1)(k+2)$ and noting that either $k+1$ or $k+2$ is even, so 6 is a factor of $3(k+1)(k+2)$.
29. $S_1: (ab)^1 = a^1b^1; S_k: (ab)^k = a^kb^k; S_{k+1}: (ab)^{k+1} = a^{k+1}b^{k+1};$ S_{k+1} can be obtained by multiplying both sides of S_k by (ab).
33. $S_3: 3^2 > 2(3) + 1; S_k: k^2 > 2k + 1$ for $k \geq 3$; $S_{k+1}: (k+1)^2 > 2(k+1) + 1$ or $k^2 + 2k + 1 > 2k + 3$; S_{k+1} can be obtained from S_k by noting that S_{k+1} is the same as $k^2 > 2$ which is true for $k \geq 3$.
35. $S_1: \frac{1}{4}; S_2: \frac{1}{3}; S_3: \frac{3}{8}; S_4: \frac{2}{5}; S_5: \frac{5}{12}; S_n: \frac{n}{2n+2};$ Use S_k to obtain the conjectured formula.

Section 8.5

Check Point Exercises

1. a. 20 **b.** 1 **c.** 28 **d.** 1 **2.** $x^4 + 4x^3 + 6x^2 + 4x + 1$ **3.** $x^5 - 10x^4y + 40x^3y^2 - 80x^2y^3 + 80xy^4 - 32y^5$
4. $4032x^5y^4$

Exercise Set 8.5

1. 56 **3.** 12 **5.** 1 **7.** 4950 **9.** $x^3 + 6x^2 + 12x + 8$ **11.** $27x^3 + 27x^2y + 9xy^2 + y^3$ **13.** $125x^3 - 75x^2 + 15x - 1$
15. $16x^4 + 32x^3 + 24x^2 + 8x + 1$ **17.** $x^8 + 8x^6y + 24x^4y^2 + 32x^2y^3 + 16y^4$ **19.** $y^4 - 12y^3 + 54y^2 - 108y + 81$
21. $16x^{12} - 32x^9 + 24x^6 - 8x^3 + 1$ **23.** $c^5 + 10c^4 + 40c^3 + 80c^2 + 80c + 32$ **25.** $x^5 - 5x^4 + 10x^3 - 10x^2 + 5x - 1$
27. $x^5 - 10x^4y + 40x^3y^2 - 80x^2y^3 + 80xy^4 - 32y^5$ **29.** $64a^6 + 192a^5b + 240a^4b^2 + 160a^3b^3 + 60a^2b^4 + 12ab^5 + b^6$
31. $x^8 + 16x^7 + 112x^6 + \cdots$ **33.** $x^{10} - 20x^9y + 180x^8y^2 - \cdots$
35. $x^{32} + 16x^{30} + 120x^{28} + \cdots$ **37.** $y^{60} - 20y^{57} + 190y^{54} - \cdots$ **39.** $240x^4y^2$ **41.** $126x^5$ **43.** $56x^6y^{15}$ **45.** $-\frac{21}{2}x^6$
47. $g(t) = 0.002t^3 - 0.84t^2 - 16.13t - 68.54$

61.

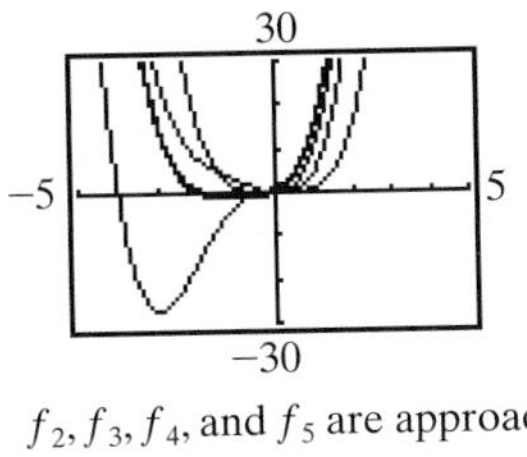

30

−5 5

−30

f_2, f_3, f_4, and f_5 are approaching $f_1 = f_6$.

62. $f_1(x) = x^3 - 3x^2 + 3x - 1$
63. $f_1(x) = x^4 - 8x^3 + 24x^2 - 32x + 16$
67. $x^6 + 3x^5 + 6x^4 + 7x^3 + 6x^2 + 3x + 1$
69. $\binom{n}{r} = \dfrac{n!}{r!\,n-r!}; \binom{n}{r-1} = \dfrac{n!}{(n-r)![n-(n-r)]!} = \dfrac{n!}{(n-r)!r!} = \binom{n}{r}$

Section 8.6

Check Point Exercises

1. 72 **2.** 729 **3.** 676,000 **4.** 840 **5.** 720 **6. a.** combinations **b.** permutations **7.** 210 **8.** 1820

Exercise Set 8.6

1. 3024 **3.** 6720 **5.** 720 **7.** 1 **9.** 126 **11.** 330 **13.** 1 **15.** 1 **17.** combinations **19.** permutations
21. 27 ways **23.** 40 ways **25.** 243 ways **27.** 144 area codes **29.** 120 ways **31.** 6 paragraphs **33.** 720 ways
35. 8,648,640 ways **37.** 120 ways **39.** 15,120 lineups **41.** 20 ways **43.** 495 collections **45.** 24,310 groups
47. 13,983,816 selections **49.** 360 ways **51.** 1716 ways **53.** 1140 ways **55.** 840 passwords **57.** 2730 cones
69. (c) is true. **71.** 144 numbers

Section 8.7

Check Point Exercises

1. $\frac{82.5}{275} = 0.3$ **2.** $\frac{1}{3}$ **3.** $\frac{1}{9}$ **4.** $\frac{1}{13}$ **5.** $\frac{1}{142{,}506}$ **6.** $\frac{142{,}505}{142{,}506} \approx 0.999993$ **7.** $\frac{1}{3}$ **8.** $\frac{3}{4}$ **9.** $\frac{24}{25}$ **10.** $\frac{1}{361} \approx 0.003$ **11.** $\frac{1}{16}$

Exercise Set 8.7

1. 0.42 **3.** $\frac{761}{5926} \approx 0.13$ **5.** $\frac{1}{6}$ **7.** $\frac{1}{2}$ **9.** $\frac{1}{3}$ **11.** $\frac{1}{13}$ **13.** $\frac{3}{13}$ **15.** $\frac{1}{4}$ **17.** $\frac{7}{8}$ **19.** $\frac{1}{12}$ **21.** $\frac{1}{18{,}009{,}460}; \frac{5}{900{,}473}$
23. a. 2,598,960 **b.** 1287 **c.** $\frac{1287}{2{,}598{,}960} \approx 0.0005$ **25.** $\frac{255{,}647}{274{,}634} \approx 0.93$ **27.** $\frac{237{,}401}{274{,}634} \approx 0.86$ **29.** $\frac{42{,}010}{274{,}634} \approx 0.15$ **31.** $\frac{2}{13}$
33. $\frac{5}{6}$ **35.** $\frac{7}{13}$ **37.** $\frac{3}{4}$ **39.** $\frac{33}{40}$ **41.** $\frac{1}{36}$ **43.** $\frac{1}{3}$ **45.** $\frac{1}{64}$ **47. a.** $\frac{1}{256}$ **b.** $\frac{1}{4096}$ **c.** $\left(\frac{15}{16}\right)^{10}$ **d.** $1 - \left(\frac{15}{16}\right)^{10}$
59. Answers may vary.

Chapter 8 Review Exercises

1. $a_1 = 3; a_2 = 10; a_3 = 17; a_4 = 24$ **2.** $a_1 = -\frac{3}{2}; a_2 = \frac{4}{3}; a_3 = -\frac{5}{4}; a_4 = \frac{6}{5}$ **3.** $a_1 = 1; a_2 = 1; a_3 = \frac{1}{2}; a_4: \frac{1}{6}$
4. $a_1 = \frac{1}{2}; a_2 = -\frac{1}{4}; a_3 = \frac{1}{8}; a_4 = -\frac{1}{16}$ **5.** $a_1 = 9; a_2 = \frac{2}{27}; a_3 = 9; a_4 = \frac{2}{27}$ **6.** $a_1 = 4; a_2 = 11; a_3 = 25; a_4 = 53$ **7.** 65
8. 95 **9.** −20 **10.** $\sum_{i=1}^{15} \frac{i}{i+2}$ **11.** $\sum_{i=1}^{10} (i+3)^3$ **12.** 7, 11, 15, 19, 23, 27 **13.** −4, −9, −14, −19, −24, −29
14. $\frac{3}{2}, 1, \frac{1}{2}, 0, -\frac{1}{2}, -1$ **15.** −2, 3, 8, 13, 18, 23 **16.** $a_6 = 20$ **17.** $a_{12} = -30$ **18.** $a_{14} = -38$ **19.** $a_n = 4n - 11; a_{20} = 69$
20. $a_n = 220 - 20n; a_{20} = -180$ **21.** $a_n = 8 - 5n; a_{20} = -92$ **22.** 1727 **23.** 225 **24.** 15,150 **25.** 440 **26.** −500
27. −2325 **28. a.** $a_n = 1043.4518 - 0.4118n$ **b.** 1002.2718 sec **29.** \$418,500 **30.** 1470 seats **31.** 3, 6, 12, 24, 48
32. $\frac{1}{2}, \frac{1}{4}, \frac{1}{8}, \frac{1}{16}, \frac{1}{32}$ **33.** 16, −8, 4, −2, 1 **34.** −1, 5, −25, 125, −625 **35.** $a_7 = 1458$ **36.** $a_6 = \frac{1}{2}$ **37.** $a_5 = -48$
38. $a_n = 2^{n-1}; a_8 = 128$ **39.** $a_n = 100\left(\frac{1}{10}\right)^{n-1}; a_8 = \frac{1}{100{,}000}$ **40.** $a_n = 12\left(-\frac{1}{3}\right)^{n-1}; a_8 = -\frac{4}{729}$ **41.** 17,936,135 **42.** $\frac{1093}{2187}$
43. 19,530 **44.** −258 **45.** $\frac{341}{128}$ **46.** $\frac{27}{2}$ **47.** $\frac{4}{3}$ **48.** $-\frac{18}{5}$ **49.** 20 **50.** $\frac{2}{3}$ **51.** $\frac{47}{99}$ **52.** \$42,823.22; \$223,210.19
53. \$120,112.64 **54.** $\$9\frac{1}{3}$ million

55. $S_1: 5 = \frac{5(1)(1+1)}{2}; S_k: 5 + 10 + 15 + \cdots + 5k = \frac{5k(k+1)}{2}; S_{k+1}: 5 + 10 + 15 + \cdots + 5(k+1) = \frac{5(k+1)(k+2)}{2};$
S_{k+1} can be obtained by adding $5(k+1)$ to both sides of S_k.

56. $S_1: 1 = \frac{4^1 - 1}{3}; S_k: 1 + 4 + 4^2 + \cdots + 4^{k-1} = \frac{4^k - 1}{3}; S_{k+1}: 1 + 4 + 4^2 + \cdots + 4^k = \frac{4^{k+1} - 1}{3};$
S_{k+1} can be obtained by adding 4^k to both sides of S_k.

57. $S_1: 2 = 2(1)^2; S_k: 2 + 6 + 10 + \cdots + (4k-2) = 2k^2; S_{k+1}: 2 + 6 + 10 + \cdots + (4k+2) = 2k^2 + 4k + 2;$
S_{k+1} can be obtained by adding $4k + 2$ to both sides of S_k.

58. $S_1: 1 \cdot 3 = \frac{1(1+1)[2(1)+7]}{6}; S_k: 1 \cdot 3 + 2 \cdot 4 + 3 \cdot 5 + \cdots + k(k+2) = \frac{k(k+1)(2k+7)}{6};$
$S_{k+1}: 1 \cdot 3 + 2 \cdot 4 + 3 \cdot 5 + \cdots + (k+1)(k+3) = \frac{(k+1)(k+2)(2k+9)}{6};$
S_{k+1} can be obtained by adding $(k+1)(k+3)$ to both sides of S_k.

59. S_1: 2 is a factor of 6; S_k: 2 is a factor of $k^2 + 5k$; S_{k+1}: 2 is a factor of $k^2 + 7k + 6$; S_{k+1} can be obtained from S_k by rewriting $k^2 + 7k + 6$ as $(k^2 + 5k) + 2(k + 3)$.

60. 165 **61.** 4005 **62.** $8x^3 + 12x^2 + 6x + 1$ **63.** $x^8 - 4x^6 + 6x^4 - 4x^2 + 1$
64. $x^5 + 10x^4y + 40x^3y^2 + 80x^2y^3 + 80xy^4 + 32y^5$ **65.** $x^6 - 12x^5 + 60x^4 - 160x^3 + 240x^2 - 192x + 64$
66. $x^{16} + 24x^{14} + 252x^{12} + \cdots$ **67.** $x^9 - 27x^8 + 324x^7 - \cdots$ **68.** $80x^2$ **69.** $4860x^2$ **70.** 336 **71.** 15,120 **72.** 56
73. 78 **74.** 20 choices **75.** 243 possibilities **76.** 32,760 ways **77.** 4845 ways **78.** 1140 sets **79.** 116,280 ways
80. 120 ways **81.** $\frac{9{,}630{,}188}{31{,}878{,}234} \approx 0.302$ **82.** $\frac{5{,}503{,}372}{19{,}128{,}261} \approx 0.288$ **83.** $\frac{2}{3}$ **84.** $\frac{2}{3}$ **85.** $\frac{2}{13}$ **86.** $\frac{7}{13}$ **87.** $\frac{5}{6}$ **88.** $\frac{5}{6}$
89. a. $\frac{1}{15{,}504}$ **b.** $\frac{25}{3876}$ **90.** $\frac{4}{5}$ **91.** $\frac{3}{4}$ **92.** $\frac{1}{32}$ **93. a.** 0.04 **b.** 0.008 **c.** 0.4096

Chapter 8 Test

1. $a_1 = 1; a_2 = -\frac{1}{4}; a_3 = \frac{1}{9}; a_4 = -\frac{1}{16}; a_5 = \frac{1}{25}$ **2.** 105 **3.** 550 **4.** −21,846 **5.** 36 **6.** 720 **7.** 120 **8.** $\sum_{i=1}^{20} \frac{i+1}{i+2}$ **9.** $a_n = 5n - 1; a_{12} = 59$ **10.** $a_n = 16\left(\frac{1}{4}\right)^{n-1}; a_{12} = \frac{1}{262{,}144}$ **11.** −2387 **12.** −385 **13.** 8 **14.** \$276,426.79 **15.** $S_1: 1 = \frac{1[3(1) - 1]}{2}; S_k: 1 + 4 + 7 + \cdots + (3k - 2) = \frac{k(3k - 1)}{2}; S_{k+1}: 1 + 4 + 7 + \cdots + (3k + 1) = \frac{(k + 1)(3k + 2)}{2};$ S_{k+1} can be obtained by adding $3k + 1$ to both sides of S_k. **16.** $x^{10} - 5x^8 + 10x^6 - 10x^4 + 5x^2 - 1$ **17.** 990 ways **18.** 210 sets **19.** $10^4 = 10{,}000$ **20.** $\frac{10}{1001}$ **21.** $\frac{8}{13}$ **22.** $\frac{3}{5}$ **23.** $\frac{1}{256}$ **24.** $\frac{1}{16}$

Cumulative Review Exercises (Chapters 1–8)

1. $x = \frac{14}{5}$ **2.** $\left\{\frac{3 + \sqrt{3}}{3}, \frac{3 - \sqrt{3}}{3}\right\}$ **3.** {2} **4.** {16, 256} **5.** {6} **6.** $-1 \le x \le 0$ or $[-1, 0]$ **7.** $\left(-\frac{2}{3}, \frac{3}{2}\right)$ **8.** $(-3, 1]$ **9.** {2.9706} **10.** $\left\{1, \frac{1}{2}, -3\right\}$ **11.** {(3, 2), (3, −2), (−3, 2), (−3, −2)} **12.** {(6, −4, 2)} **13.** {(0, −1), (2, 1)}

14.

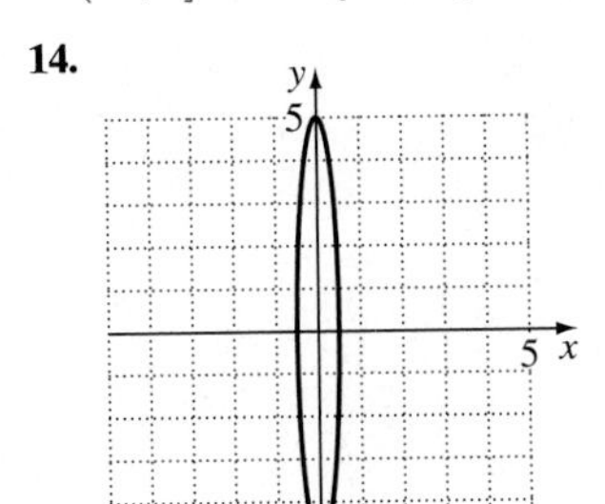

15.

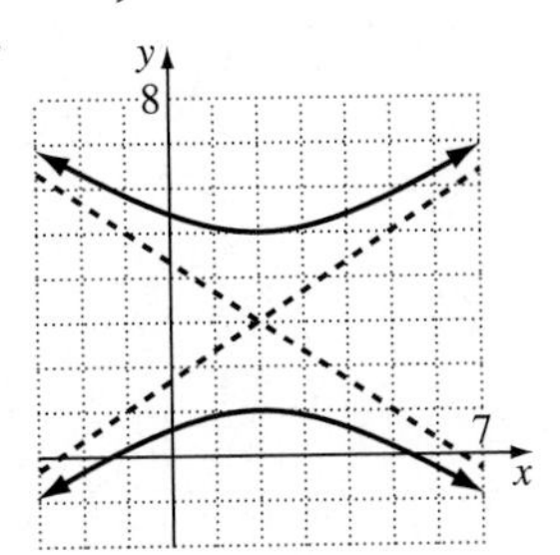

16.

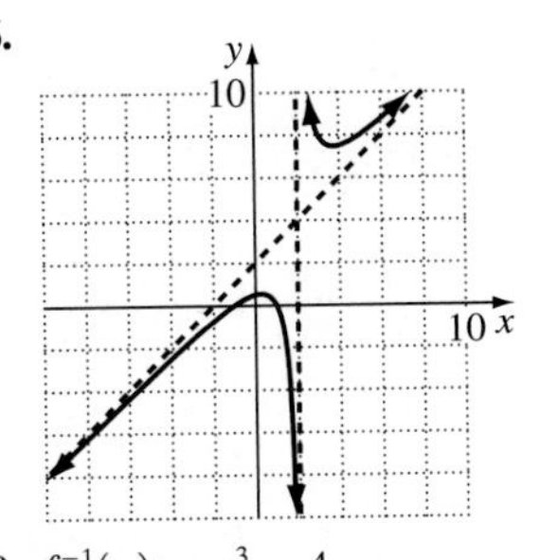

17.

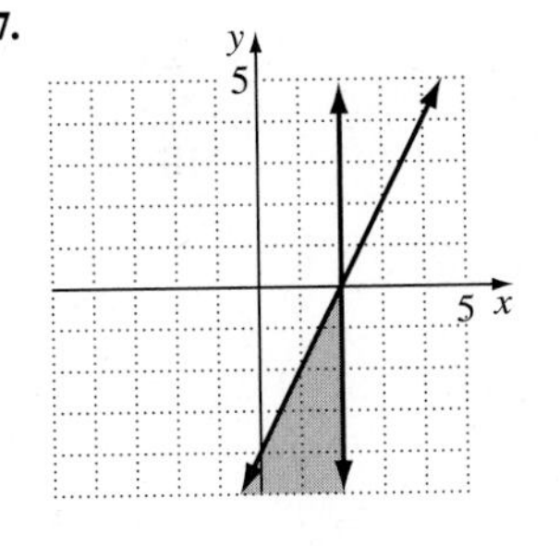

18.

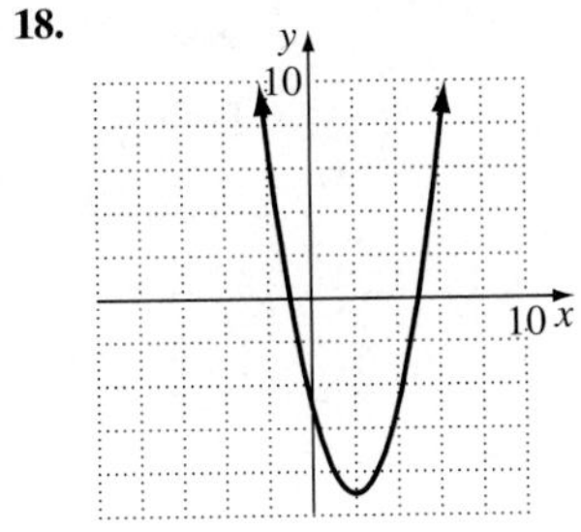

19.

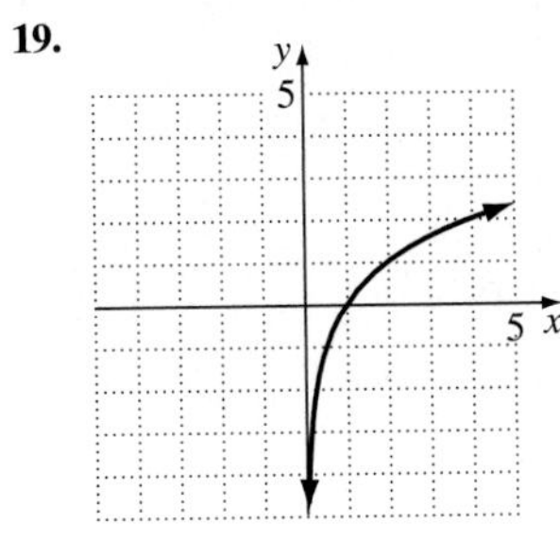

20. $f^{-1}(x) = x^3 - 4$

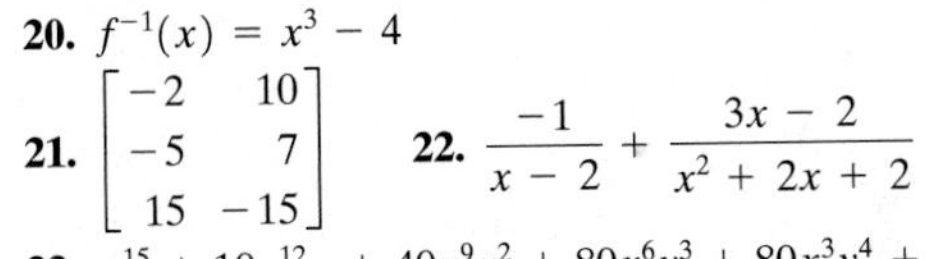

21. $\begin{bmatrix} -2 & 10 \\ -5 & 7 \\ 15 & -15 \end{bmatrix}$ **22.** $\frac{-1}{x - 2} + \frac{3x - 2}{x^2 + 2x + 2}$

23. $x^{15} + 10x^{12}y + 40x^9y^2 + 80x^6y^3 + 80x^3y^4 + 32y^5$

24. 3850

25. a. $y - 119.4 = 5.17x$ **b.** $y = 5.17x + 119.4$ **c.** 207.29 billion pieces of mail

26. The Empire State Building is about 916.7 feet tall and the World Trade Center is about 1043.4 feet tall.

27. length: 100 yd; width: 50 yd **28.** pen: \$1.80; pad: \$2 **29. a.** 6 sec **b.** 2.5 sec; 196 ft **30.** 11 amps

Subject Index

D

P

Q

T

U

V

W

X

Y

Z

Photo Credits

CHAPTER P **CO** cmcd/PhotoDisc, Inc **p. 2** (United Nations) SuperStock, Inc. **p. 13** Tom Stewart/The Stock Market **p. 23** William Sallaz/Duomo Photography Incorporated **p. 34** (PEANUTS reprinted by permission of United Feature Syndicate, Inc.) United Media/United Feature Syndicate, Inc. **p. 35** Bruce Ayres/Stone **p. 46** Gabe Palmer/Mug Shots/The Stock Market **p. 54** Jon Ortner/Stone **p. 65** (C) 1999 Roz Chast from Cartoonbank.com. All Rights Reserved/The Cartoon Bank **p. 68** Gary J. Shulfer/C. Marvin Lang/Stamp from the private collection of Professor C. M. Lang, photography by Gary J. Shulfer, University of Wisconsin, Stevens Point. *Germany: #5*; Scott Standard Postage Stamp Catalogue, Scott Pub. Co., Sidney, Ohio **p. 71** S.S. Archives/Shooting Star International Photo Agency. All rights reserved.

CHAPTER 1 **CO** Nita Winter Photography **p. 86** A. Ramu/Stock Boston **p. 88** (Squeak Carnwath "Equations" 1981, oil on canvass) Squeak Carnwath **p. 97** (Man barefoot water-skiing) Stone **p. 106** (PEANUTS reprinted by permission of United Features Syndicate, Inc.) United Media/United Feature Syndicate, Inc. **p. 109** Joel Sartore/Sharpshooters **p. 117** (Pencil sketch of Evariste Galois) Corbis **p. 121** (left) Steve Smith/FPG International LLC; (right) Peter Cade/Stone **p. 127** Schafer & Hill/Stone **p. 132** Paul Katz/The Image Bank **p. 139** (The Beverly Hillibillies) Corbis **p. 151** Tom Sanders/Adventure Photo & Film

CHAPTER 2 **CO** Chris Salvo/FPG International LLC **p. 168** (Online shopping) SuperStock, Inc. **p. 169** Carol Simowitz/San Francisco Convention and Visitors Bureau **p. 180** (left) Paul Avis/FPG International LLC; (right) Bob Schatz/Stone **p. 183** Skip Moody/Dembinsky Photo Associates **p. 190** (Couple on cell phones) Bob Daemmrich Photography, Inc. **p. 193** Paul & Lindamarie Ambrose/FPG International LLC **p. 203** Hans Neleman/The Image Bank **p. 211** (left) Joyce Photographics/Photo Researchers, Inc.; (right) Gott/Rapho/Photo Researchers, Inc. **p. 221** Douglas Kirkland/Corbis Sygma **p. 225** (Burnside Bridge, Antietam National Battlefield, Maryland) SuperStock, Inc. **p. 238** Brad Hitz/Stone **p. 252** (Surfboards) Bishop Museum

CHAPTER 3 **CO** Tim Davis/Stone **p. 259** Keith Brofsky/PhotoDisc, Inc. **p. 260** (left) Simon Bruty/Allsport Photography (USA), Inc.; (right) Illustration of the word Mirror/© 1981 Scott Kim, scottkim.com. All rights reserved. **p. 272** NIBSC/Science Photo Library/Photo Researchers, Inc. **p. 283** (left) Golden Retriever puppy/Townsend P. Dickinson/The Image Works; (right) Golden Retriever panting/Townsend P. Dickinson/The Image Works **p. 285** Barbara Penoyar/PhotoDisc, Inc **p. 294** Paul Silverman/Fundamental Photographs **p. 299** (drawing of Rene Descartes) Library of Congress **p. 304** (drawing of Girolamo Cardano) Smithsonian Institute **p. 314** Bob Daemmrich/Stock Boston **p. 332** Stephen Simpson/FPG International LLC **p. 341** (top) David Madison/Duomo Photography Incorporated; (bottom) David Madison/Duomo Photography Incorporated **p. 342** UPI/Corbis

CHAPTER 4 **CO** Tony Neste/Anthony Neste **p. 351** Bruce Weaver/AP/Wide World Photos **p. 362** David Weintraub/Photo Researchers, Inc. **p. 375** Ron Chapple/FPG International LLC **p. 384** Shoneman, Stanley R/Omni-Photo Communications, Inc. **p. 395** Bullit Marquez/AP/World Wide Photos **p. 398** Jean-Marie Chauvet/Corbis Sygma

CHAPTER 5 **CO** Travelpix/FPG International LLC **p. 416** Dave Martin/AP/Wide World Photos **p. 430** David W. Hamilton/The Image Bank **p. 439** Pekka Parviainen/Science Photo Library/Photo Researchers, Inc. **p. 448** (Illustration of dipiodocus) Index Stock Photography **p. 457** Simon Bruty/Stone **p. 467** (U.S airlift) AP/World Wide Photos **p. 472** Robert Patterson/Donna Cox/NCSA Media Technology Resources

CHAPTER 6 **CO** (Microsoft Chairman Bill Gates) AP/World Wide Photos **p. 482** John Dominis/Index Stock Photography **p. 496** Walter Geiersperger/The Stock Market **p. 505** Greg Pease/Stone **p. 513** (Arthur Cayley engraving) The Granger Collection **p. 515** (False color image of Saturn) Genesis Space Photo Library **p. 519** (Copy of an RCA Radiogram) RCA Communications, Inc. **p. 534** David Parker/Science Museum/Science Photo Library/Photo Researchers, Inc.

CHAPTER 7 **CO** (Doctor and patient) SuperStock, Inc. **p. 553** Kevin Fleming/Corbis **p. 563** David Austen/FPG International LLC **p. 566** Andrea Pistolesi/The Image Bank **p. 582** J. Hester/P. Scowen (Arizona State)/NASA Headquaters **p. 589** © Berenice Abbott/Commerce Graphics Ltd., Inc. **p. 590** (Hubble Telescope deployment) Space Telescope Science Institute

CHAPTER 8 **CO** Reuters/Barbara L. Johnston/Archive Photos **p. 600** (Sunflower) Dick Morton **p. 611** David Young-Wolff/PhotoEdit **p. 619** (Business people shaking hands) SuperStock, Inc. **p. 622** Richard Lord/The Image Works **p. 625** (bills) U.S. Bureau of Engraving and Printing **p. 633** Charles Rex Arbogast/AP/Wide World Photos **p. 642** Dr. Rudolph Schild/Science Photo Library/PhotoResearchers, Inc. **p. 645** Reprinted with permission from Science and Civilization in China, Vol. 3, by Joseph Needham, 1959, Cambridge University Press **p. 650** L. Schwartzwald/Corbis Sygma **p. 651** (Chessplayer) SuperStock, Inc. **p. 652** Sue Klemen/Stock Boston **p. 655** (Marilyn Monroe) Kobal Collection (James Dean) Imapress/Globe Photos, Inc. (Jim Morrison) Michael Ochs Archives (Kurt Cobain) Sin/Corbis (Selena) AP/World Wide Photos **p. 661** (Man sleeping in hammock) SuperStock, Inc. **p. 665** (Lotto tickets) Karen Furth **p. 670** UPI/Corbis

2. Quadratic Function: $f(x) = ax^2 + bx + c, a \neq 0$

Graph is a parabola with vertex at $x = -\dfrac{b}{2a}$.

Quadratic Function: $f(x) = a(x - h)^2 + k$

In this form, the parabola's vertex is (h, k).

3. nth-Degree Polynomial Function: $f(x) = a_n x^n + a_{n-1}x^{n-1} + a_{n-2}x^{n-2} + \cdots + a_1 x + a_0, a_n \neq 0$

For n odd and $a_n > 0$, graph falls to the left and rises to the right.

For n odd and $a_n < 0$, graph rises to the left and falls to the right.

For n even and $a_n > 0$, graph rises to the left and to the right.

For n even and $a_n < 0$, graph falls to the left and to the right.

4. Rational Function: $f(x) = \dfrac{p(x)}{q(x)}$, $p(x)$ and $q(x)$ are polynomials, $q(x) \neq 0$

5. Exponential Function: $f(x) = b^x, b > 0, b \neq 1$

Graphs:

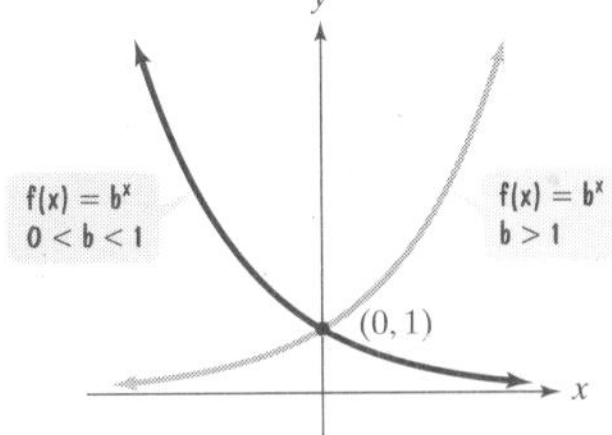

6. Logarithmic Function: $f(x) = \log_b x, b > 0, b \neq 1$

$y = \log_b x$ is equivalent to $x = b^y$.

Graph:

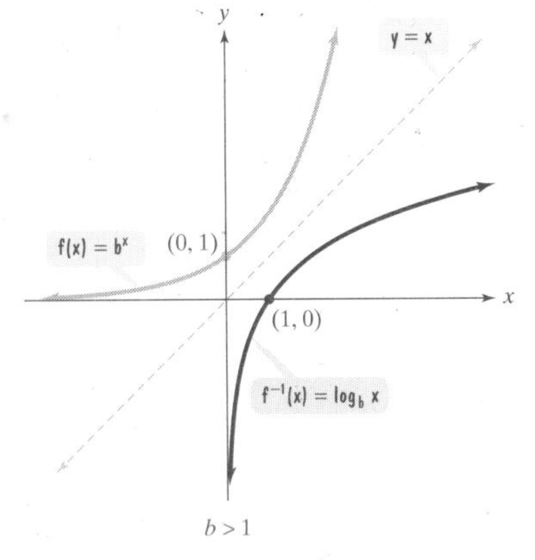

PROPERTIES OF LOGARITHMS

1. $\log_b(MN) = \log_b M + \log_b N$
2. $\log_b\left(\dfrac{M}{N}\right) = \log_b M - \log_b N$
3. $\log_b M^p = p \log_b M$
4. $\log_b M = \dfrac{\log_a M}{\log_a b} = \dfrac{\ln M}{\ln b} = \dfrac{\log M}{\log b}$
5. $\log_b b^x = x; \quad \ln e^x = x$
6. $b^{\log_b x} = x; \quad e^{\ln x} = x$

INVERSE OF A 2 × 2 MATRIX

If $A = \begin{bmatrix} a & b \\ c & d \end{bmatrix}$, then $A^{-1} = \dfrac{1}{ad - bc}\begin{bmatrix} d & -b \\ -c & a \end{bmatrix}$, where $ad - bc \neq 0$.

CRAMER'S RULE

If

$$\begin{aligned} a_{11}x_1 + a_{12}x_2 + a_{13}x_3 + \cdots + a_{1n}x_n &= b_1 \\ a_{21}x_1 + a_{22}x_2 + a_{23}x_3 + \cdots + a_{2n}x_n &= b_2 \\ a_{31}x_1 + a_{32}x_2 + a_{33}x_3 + \cdots + a_{3n}x_n &= b_3 \\ &\vdots \\ a_{n1}x_1 + a_{n2}x_2 + a_{n3}x_3 + \cdots + a_{nn}x_n &= b_n \end{aligned}$$

then $x_i = \dfrac{D_i}{D}, D \neq 0$.

D: determinant of the system's coefficients

D_i: determinant in which coefficients of x_i are replaced by $b_1, b_2, b_3, \ldots, b_n$.

CONIC SECTIONS

Circle

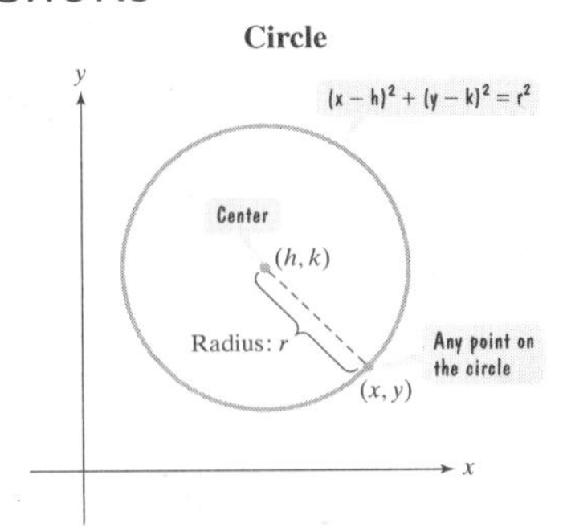

Ellipse

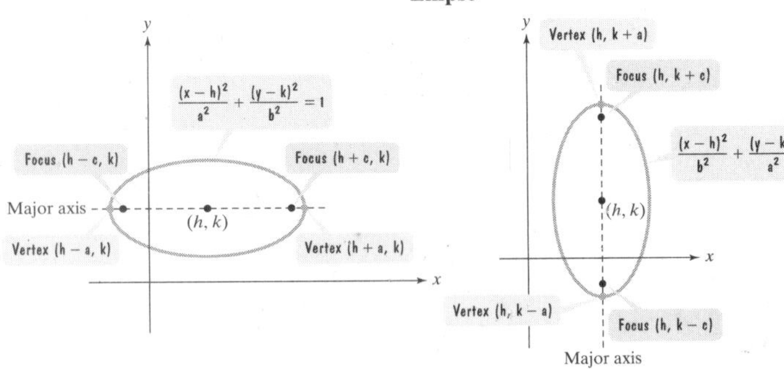

Hyperbola

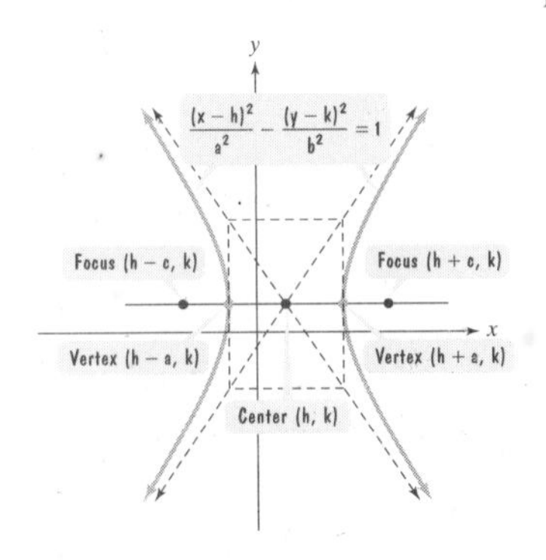

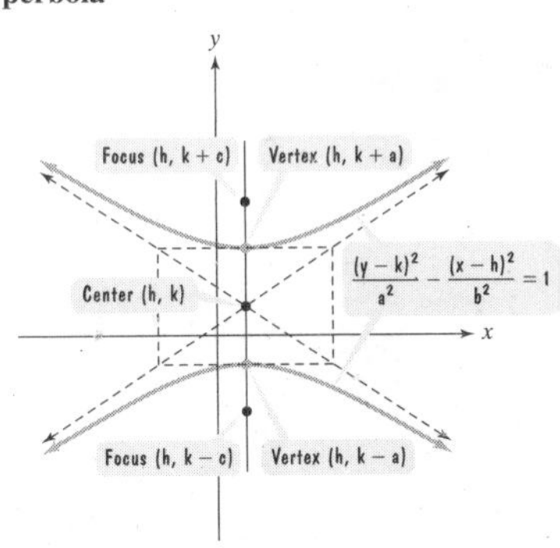